ADLER'S

Physiology of the eye

ADLER'S

Physiology of the eye

CLINICAL APPLICATION

Edited by

ROBERT A. MOSES, M.D.

Professor, Department of Ophthalmology,
Washington University School of Medicine,
St. Louis, Missouri

SEVENTH EDITION

With 844 illustrations

The C. V. Mosby Company

ST. LOUIS • TORONTO • LONDON 1981

SEVENTH EDITION

Printed in the United States of America

The C. V. Mosby Company
11830 Westline Industrial Drive, St. Louis, Missouri 63141

Library of Congress Cataloging in Publication Data

Adler, Francis Heed, 1895-
Adler's Physiology of the eye.

Includes bibliographies and index.
1. Eye. I. Moses, Robert A., 1917-
II. Title. III. Title: Physiology of the eye.
[DNLM: 1. Eye—Physiology. WW103 A237p]
RE67.A3 1980 612'.84 80-16862
ISBN 0-8016-3541-1

GW/CB/B 9 8 7 6 01/B/018

Contributors

ELIOT L. BERSON, M.D.

Director, Berman-Gund Laboratory for the Study of Retinal Degenerations and Associate Professor of Ophthalmology, Harvard University Medical School; Director, Electroretinography Service, Massachusetts Eye and Ear Infirmary, Boston, Massachusetts

ANDERS BILL, M.D.

Professor in Physiology, Biomedical Centre, Uppsala University, Uppsala, Sweden

PETER O. BISHOP, M.B., B.S., D.Sc. (Sydney), F.A.A., F.R.S.

Professor of Physiology, The John Curtin School of Medical Research, The Australian National University, Canberra, Australia

RONALD M. BURDE, M.D.

Professor of Ophthalmology and Neurology, Washington University School of Medicine, St. Louis, Missouri

HAROLD BURTON, Ph.D.

Associate Professor of Anatomy and Neurobiology, Washington University School of Medicine, St. Louis, Missouri

ADOLPH I. COHEN, Ph.D.

Professor of Anatomy-Neurobiology in Ophthalmology, Professor of Anatomy-Neurobiology, Washington University School of Medicine, St. Louis, Missouri

EDWARD COTLIER, M.D.

Professor of Ophthalmology and Visual Science, Yale University School of Medicine, New Haven, Connecticut

NIGEL W. DAW, Ph.D.

Professor, Department of Physiology and Biophysics, Professor of Physiology in Ophthalmology, Washington University School of Medicine, St. Louis, Missouri

†SIDNEY FUTTERMAN, Ph.D.

Professor, Department of Ophthalmology, University of Washington School of Medicine, Seattle, Washington

BALDER P. GLOOR, M.D.

Professor and Head of Department of Ophthalmology, University of Basel, Basel, Switzerland

BENJAMIN MILDER, M.D.

Associate Professor, Department of Ophthalmology, Washington University School of Medicine, St. Louis, Missouri

ROBERT A. MOSES, M.D.

Professor, Department of Ophthalmology, Washington University School of Medicine, St. Louis, Missouri

ALAN L. PEARLMAN, M.D.

Professor, Department of Neurology, Associate Professor, Department of Physiology and Biophysics, Washington University School of Medicine, St. Louis, Missouri

GILL ROPER-HALL, D.B.O. (T).

Clinical Orthoptist and Research Assistant in Neuro-Ophthalmology, Department of Ophthalmology, Washington University School of Medicine, St. Louis, Missouri

---

†Deceased, May 19, 1979.

†W. A. H. RUSHTON, Sc.D., F.R.S.

Emeritus Professor of Visual Physiology, University of Cambridge, Cambridge, England; Distinguished Research Professor, The Florida State University, Tallahassee, Florida

JOHN C. SAARI, Ph.D.

Associate Professor of Ophthalmology, Department of Ophthalmology, University of Washington School of Medicine, Seattle, Washington

MARVIN L. SEARS, M.D.

Professor and Chairman, Department of Ophthalmology and Visual Science, Yale University School of Medicine, New Haven, Connecticut

†Deceased June 21, 1980.

GERARD M. SHICKMAN, Ph.D.

Formerly Research Instructor, Department of Ophthalmology, Washington University Medical School; Wide Range Electronics Corporation, St. Louis, Missouri

H. STANLEY THOMPSON, M.D.

Professor of Ophthalmology, Director of Neuro-Ophthalmology Unit, Department of Ophthalmology, University of Iowa College of Medicine, Iowa City, Iowa

STEPHEN R. WALTMAN, M.D.

Professor of Ophthalmology, Washington University School of Medicine, St. Louis, Missouri

GERALD WESTHEIMER, Ph.D.

Department of Physiology-Anatomy, University of California, Berkeley, California

To

Fred, Ed, Joel, Jonas,

Tom, and Bruce

Preface

Several years ago when Dr. Adler suggested that I bring his *Physiology of the Eye* up to date I experienced acute panic. Dr. Adler had written the first four editions entirely by himself, and he recommended that I continue the pattern. It rapidly became clear to me that I needed help, so despite expert advice to keep the book a single-author work I asked several of my friends to write chapters. The resulting sixth edition was well received, and this has encouraged me to seek more help in preparing the present edition. Quite a few persons have kindly contributed chapters.

One of the problems of a multiauthored work is that of redundancy. It is not surprising that some overlap of material occurs. The editor must then decide either to delete the redundant material from one or the other chapter or to let the repetition stand. In several instances I have chosen the second alternative, believing that it is more convenient to the reader to reintroduce material pertinent to a topic than to refer to another section of the book.

I wish to thank the contributors. They have worked hard to make this book useful to clinicians.

It gives me pleasure to acknowledge the meticulous help of my secretary, Mrs. Leona Aberle.

Robert A. Moses

Contents

ADLER'S

Physiology of the eye

Chapter 1

The eyelids

ROBERT A. MOSES, M.D.

The eyelids form one of the most important elements in the protective system of the eye. This protective function is mediated by (1) the screening and sensing action of the cilia, (2) the secretions of the glands of the eyelids, and (3) the movements of the lids.

The skin of the eyelids is thinner than that of any other part of the body. This thin skin folds easily and permits rapid opening and closing of the palpebral fissure. The lids are covered with fine lanugo hairs, which may escape notice on casual inspection. These fine hairs may be responsible for corneal irritation if the skin of the lids comes into contact with the eyeball.

In each lid there is a wide fibrous tarsal plate that follows the curvature of the eyeball and is firmly attached to the medial and lateral palpebral ligaments and to the orbital septum. The orbicularis muscle and the skin lie anterior to the tarsal plate. The tarsus is covered by conjunctiva posteriorly (Fig. 1-1).

THE CILIA AND EYEBROWS

The first line of defense in the protective function of the lids is made up of the cilia and the eyebrows (Fig. 1-2).

The cilia, or eyelashes, are hairs situated on the margin of the lids. These are disposed in two rows, totaling about 100 to 150 cilia in the upper lid and half that number in the lower lid. In the upper lid the cilia are curved outward and upward and in the lower lid, outward and downward. Each cilium is a short, stout, cylindric hair growing from a typical hair follicle. Each follicle is surrounded by a nerve plexus with a very low threshold of excitation. Touching a cilium is sufficient to excite the nerve plexus of the follicle and produce a reflex blink.

The base of each cilium is surrounded by sebaceous glands (the glands of Zeis), which open into the hair follicle by short, wide ducts. Infection of these glands results in the common stye, or hordeolum. Excessive and altered secretion of the glands produces marginal blepharitis. It is not surprising that the lid margins and the cilia can be involved in many of the same disease processes that affect the scalp and its hair—for example, seborrhea.

The pigmentation of the cilia is often deeper than that of the scalp hair throughout adult life. Occasionally cilia turn gray or white with advancing age, and in some disease conditions the newly formed lashes fail to become pigmented and stay white. This is known as poliosis. The average life of each cilium is from 3 to 5 months, after which it falls out and a new one grows in to take its place. If a cilium is pulled out, the new one replacing it reaches full size in about 2 months. If the cilia are cut short, as is often done preceding operations on the eye, growth is so rapid that the lashes appear nearly normal in a few weeks.

Not infrequently a loose cilium will find its way into the upper or lower lacrimal punctum, and the protruding end scratches the cornea, causing considerable pain. When a patient complains of a foreign body in the eye and none is found, it is always wise to look for such cilia—they may be easily missed. Occasionally a cilium becomes twisted and turned back toward the skin of the lid, which it punctures, and then

□ This work was supported in part by Research Grant EY 00256 from the National Eye Institute, National Institutes of Health.

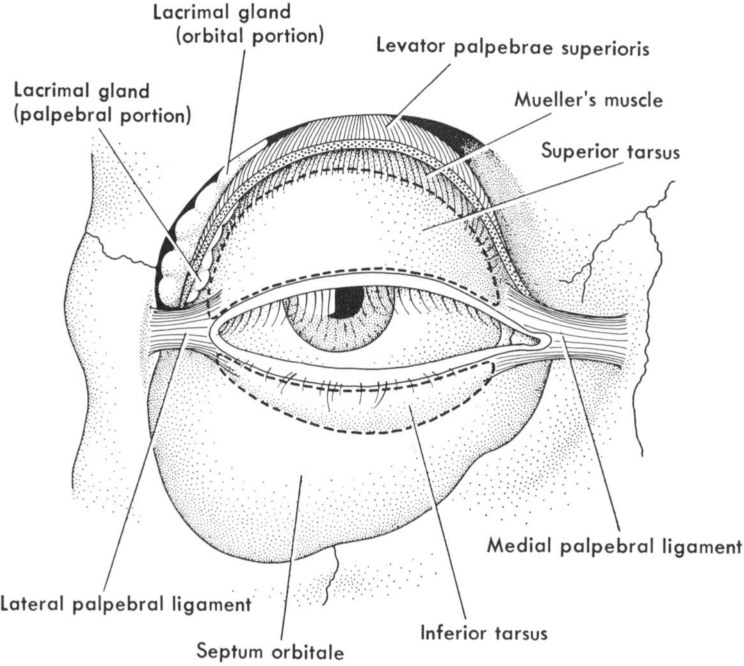

Fig. 1-1. Dissection of orbit from front. Upper portion of orbital septum has been removed. Broken line, extent of superior and inferior tarsi. Note that entire front of orbit is closed in by firm, resistant tissues except for palpebral fissure, and this can be completely closed by lid and accessory muscles. Palpebral ligaments reinforce diaphragm. (Modified from Wolff.)

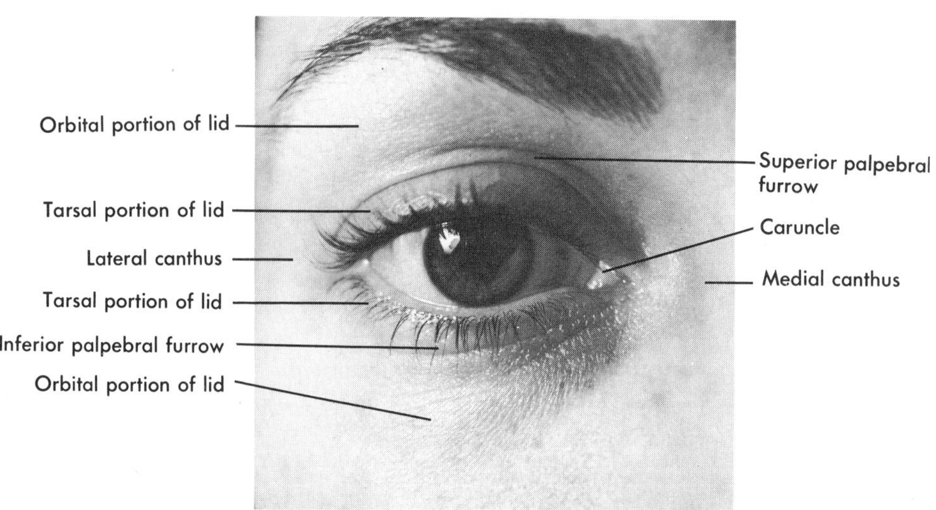

Fig. 1-2. Outer surface of eyelids. (From Newell, F. W., and Ernest, J. T.: Ophthalmology, principles and concepts, ed. 4, St. Louis, 1978, The C. V. Mosby Co.)

it grows into the skin. Cilia have often been reported in the anterior chamber following penetrating injuries.

The eyebrows are elevated by the frontalis muscle, depressed by the orbicularis muscle in forced lid closure, and drawn together in the act of frowning by the corrugator supercilii; they are not moved in the ordinary act of blinking. The eyebrows can be voluntarily elevated without the globes turning upward, but extreme upward gaze carries with it an elevation of the brows. Extreme upward gaze therefore involves three synergistic actions: upward rotation of the globes, elevation of the lids, and elevation of the eyebrows.

In nuclear or infranuclear seventh (facial) nerve paralysis the eyebrow on the paralyzed side is lower than that on the unaffected side; in the presence of unilateral ptosis the eyebrow on the involved side is frequently elevated as the frontalis muscle contracts in an effort to keep the upper lid raised.

SECRETIONS OF THE EYELIDS

The secretions of the glands of the eyelids form the second complex of their protective system. Chief among these is the oily layer secreted by the meibomian glands, which are large sebaceous glands located in the tarsal plates; there are approximately 30 of these glands in each tarsus. They are oriented perpendicular to the lid margins, with the ostia of the glands lying in a single row on the lid margin, posterior to the two rows of cilia (Fig. 1-3). When the eyelid is everted, these glands may appear as light parallel stripes beneath the tarsal conjunctiva. Mild pressure over the glands causes sebum to exude from the openings on the lid margins. The oily layer secreted by these glands forms the most superficial element of the precorneal tear film and functions to prevent evaporation of tear fluid as well as to prevent spillage of tears at the lid margins.

The eyelids also contain accessory lacrimal gland tissue, the glands of Krause and Wolfring (discussed in Chapter 2).

NORMAL MOVEMENTS OF THE EYELIDS

The third and most important element in the protective action of the eyelids is the motor system, which depends on the levator palpebrae

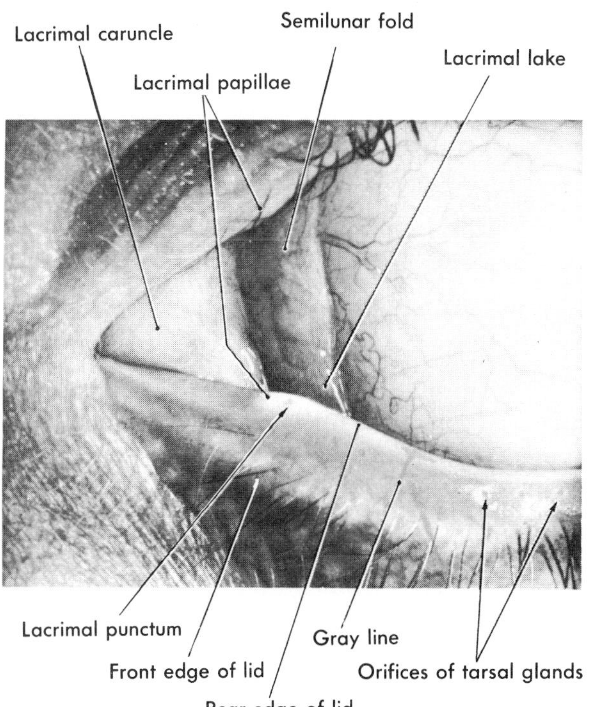

Fig. 1-3. Lacrimal portion of lid margin. (From Gibson, H. L.: Med. Radiogr. Photogr. **28:**126, 1952.)

superioris, the orbicularis oculi, and the smooth muscles of Müller in the upper and lower eyelids. The movements of the eyelids, opening and closing, may be voluntary or reflex.

Elevation

When the eyelids are opened, the upper lid is raised approximately 10 mm against gravity and it is drawn back under the orbital rim at the fold of the lid. This action takes place by contraction of the levator palpebrae superioris, which is innervated by the third cranial nerve. The levator tendon inserts into a broad area of the skin, from the free edge of the lid to the top of the tarsal plate (Fig. 1-4). A few fibers insert into the anterior surface of the tarsal plate. In their passage to the skin, tendinous fibers of the levator interdigitate with bundles of fibers of the orbicularis muscle.

The fold of the upper eyelid is formed by retraction of the skin when the lid is raised. In Oriental people this fold is absent because the terminal fibers of the levator palpebrae do not run forward to the skin, but rather insert only on the anterior surface of the tarsal plate.[13]

Elevation of the upper lid is assisted by smooth muscle fibers that arise from the undersurface of the levator and insert in the upper margin of the tarsus. These fibers, supplied by nerves of the sympathetic system, are called Müller's muscle of the eyelid. The lower lid has an inferior palpebral muscle of Müller, which arises from a fascial expansion of the inferior rectus and breaks into two parts, one inserting in the conjunctiva of the lower fornix and the other entering the lower lid to insert in the tarsal plate. Contraction of these fibers exerts a direct pull between the tarsus and the undersurface of the globe.

The levator is the chief muscle producing elevation of the upper lid; its activity is usually associated with contraction of the superior rectus.[3] In extreme upward gaze the frontalis muscle aids the levator in further elevating the lid.

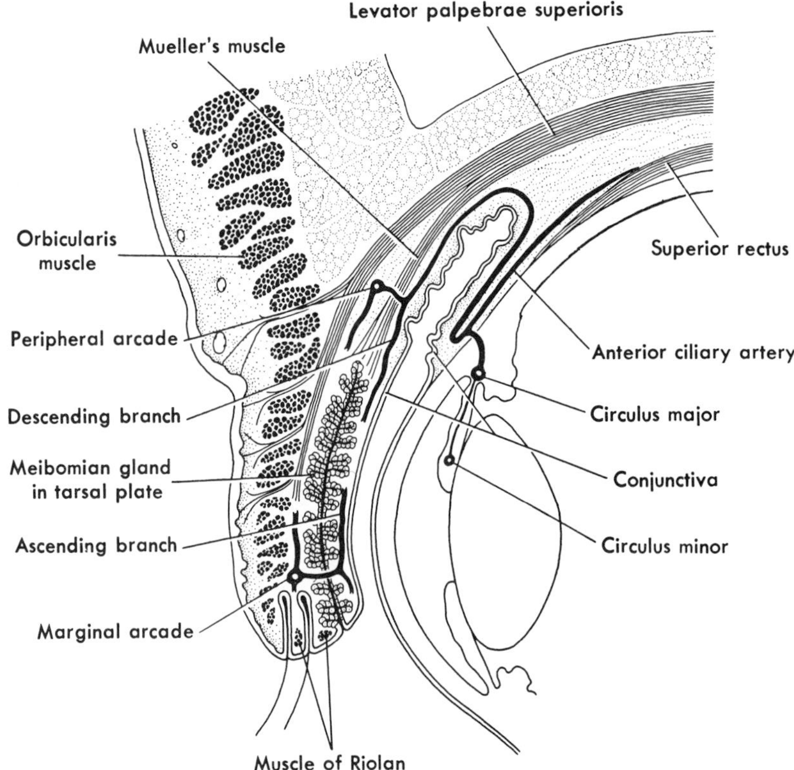

Fig. 1-4. Section of upper lid and anterior portion of eye, showing fibers of levator tendon passing between bundles of orbicularis muscle to insert in skin of lids. (Modified from Wolff.)

When the gaze is directed upward, the lid follows the upward movement of the globe. The levator and superior rectus muscles are innervated by branches of the same nerve and are connected by a common fascial sheath. These relations account for the frequent association of congenital ptosis and paralysis of the superior rectus, a fact especially important to remember when surgery for ptosis is contemplated.

There is excellent evidence derived from clinical observation that the levator muscles of the two upper lids behave as yoke muscles in that they act as a team, or pair, and that Hering's law of equal innervation of yoke muscles applies. When the levator on one side is weak, as in unilateral myasthenia gravis or unilateral congenital ptosis, the lid on the unaffected side may be retracted in an unconscious attempt to elevate the ptotic lid. If the eye with ptosis is covered, relieving the necessity of effort to elevate the lid, retraction on the normal side disappears, showing that there has been ab-

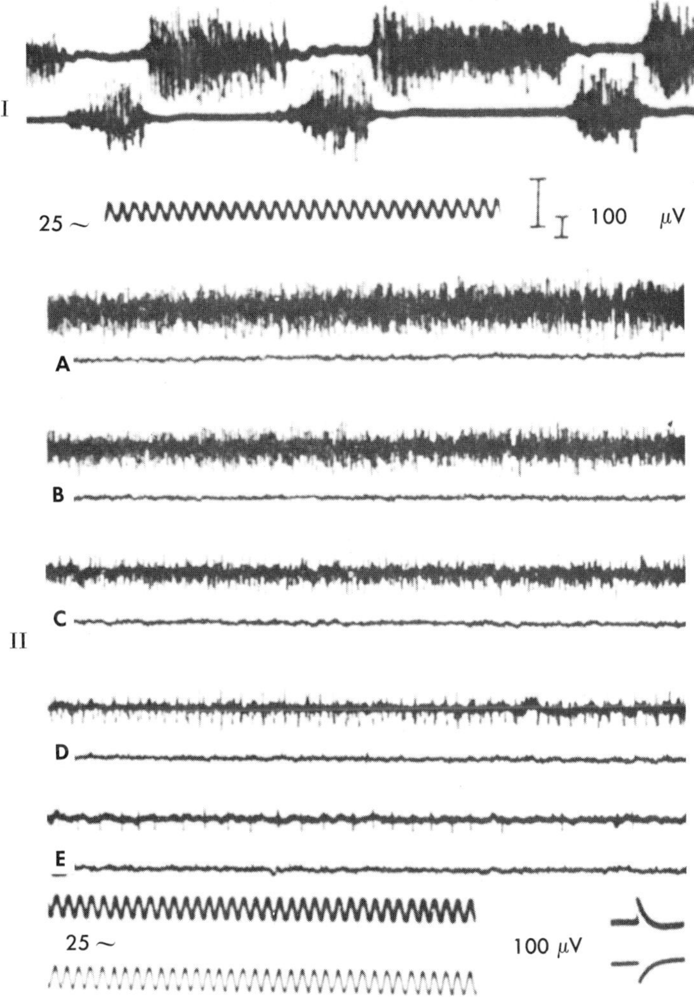

Fig. 1-5. *I*, Simultaneous electromyographic recording from levator muscle (upper trace) and orbiculars (lower trace) during repeated voluntary blinking. Note reciprocal activity in muscles, with absence of activity in levator muscle during active contraction of orbicularis. Compare with *II*, which shows activity during gradual lid descent. *II*, Simultaneous electromyographic recordings of same muscles as in *I*. **A,** Intense levator activity with eyes looking straight ahead. Other strips show decreasing activity in levator as lids are gradually lowered in 10-degree steps. **E,** Only one motor unit still active, with lids dropped 40 degrees from horizontal. Note lack of activity in orbicularis as upper lid is lowered. (From Björk, A.: Br. J. Ophthalmol. **38:**605, 1954.)

normally great innervation stimulus to *both* levators. On the other hand, lid retraction of local origin may result in pseudoptosis of the sound eye; when the eye with retraction is covered, relieving the unconscious attempt to lower the affected lid, the pseudoptosis of the sound eye disappears.[4]

Although the upper lid follows the globe on voluntary upward gaze, in the reflex act of blinking the globe and lid move in opposite directions. The globe turns upward as the eyelids close and downward as they open again. The upward movement of the eyeball during the act of blinking rotates the anteroposterior axis approximately 15 degrees above the horizontal.

In operations on the superior rectus the fascial connections of this muscle with the levator must be kept in mind. Thus if the insertion of the superior rectus is recessed on the globe, the levator will be carried with it, so the upper lid will be raised and the fissure will be widened. On the other hand, if the superior rectus is advanced or resected, the levator will be pulled forward and the eyelid will droop, creating ptosis. The synergism between the superior rectus, frontalis, and levator is sometimes employed in operations for the relief of ptosis.

It might be supposed that the eyelid would be kept from drooping by the action of the sympathetically innervated smooth muscle when the levator is paralyzed. This does not occur, however, because the fibers of Müller's muscle have their origin from the undersurface of the levator. When the latter is paralyzed, the smooth muscle fibers have no firm origin from which to pull and their contraction becomes ineffective; the ptosis is therefore usually complete. If the sympathetic nerve supply is paralyzed, only a slight droop of the lid results, as seen in Horner's syndrome. Degeneration or interruption of the sympathetic nerves results in hypersensitivity to adrenergic agents, so the lid elevates when 1:1000 epinephrine is instilled into the conjunctival sac. This dosage does not affect normally innervated smooth muscle. Pharmacologic paralysis of Müller's muscle by topical application of guanethidine is used in the treatment of lid retraction.[5]

Not only does the upper lid follow upward rotation of the eyeball, but it also follows downward movements by relaxation of tonus in the levator muscle. The studies of Björk[2] very clearly demonstrate this point (Fig. 1-5). Downward motion of the lid is not caused solely by gravity; it also occurs when a person is placed on a tilted table with his head lower than his feet. The mechanism is not clear.

Closure

The eyelids are closed by the action of the orbicularis oculi, supplied by the seventh cranial (facial) nerve (Fig. 1-6). The function of the orbicularis is reviewed in detail by Hoyt and Loeffler.[8] Anatomically there are two main portions of the orbicularis muscle: the palpebral and the orbital. The palpebral portion of the muscle, overlying the tarsus and the orbital septum, is used in the acts of blinking and voluntary winking. When the eyelids are forced shut, such as occurs in blepharospasm, the orbital portion of the muscle is brought into play along with the muscles of the eyebrow. The two portions of the orbicularis are differentiated physiologically by their chronaxie.* The chronaxie of the palpebral portion is about half that of the orbital portion; this agrees well with the general rule that muscles designed for rapid movement have a lower chronaxie than those of slower but more forcible action.

Three distinct types of closure of the eyelid are effected by different combinations of fiber bundles of the orbicularis working together with the muscles that control the eyebrows. These are blinking, voluntary winking, and blepharospasm.

Blinking

There are two types of blinking: blinking of reflex origin and spontaneous blinking, probably of central origin.

Reflex blinking. Many different stimuli cause reflex blinking. Strong lights; the sudden approach of an object toward the eyes; loud noises; touching of the cornea, conjunctiva, or lashes; and other stimuli call forth rapid reflex closure of both eyes. It is possible, therefore, to divide reflex blinking into at least three dif-

---

*Chronaxie is the duration of time that a current of twice the rheobasic (galvanic threshold) intensity must flow to excite a tissue being tested. Chronaxie is related to irritability and is used as a measure of changes in irritability of a nerve or muscle.

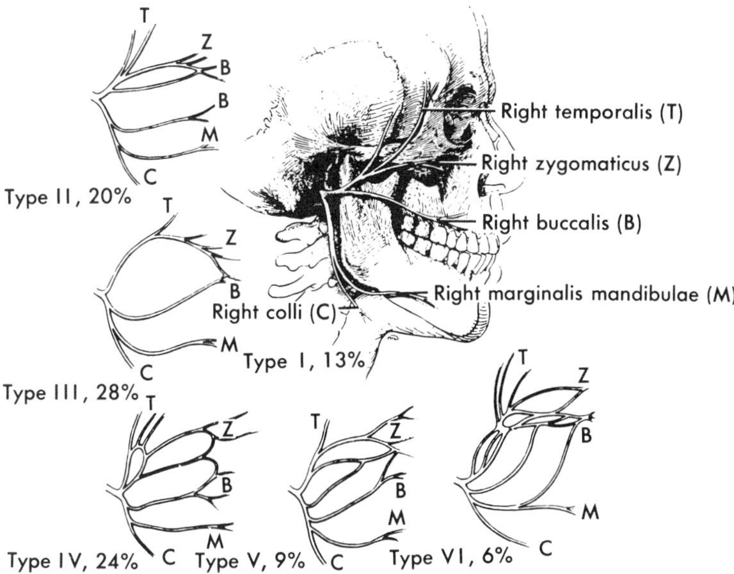

Fig. 1-6. Six basic patterns of branching of facial nerve. Note relation of temporal and zygomatic branches to zygoma and extreme degree of anastomosis that is possible. (From Davis, R. A., et al.: Surg. Gynecol. Obstet. **102:**385, 1956.)

ferent subtypes—tactile, optic, and auditory—depending on the nature of the stimulus. The corneal reflex is an example of a tactile stimulus. Any object touching the unanesthetized cornea will produce a reflex blink. The dazzle reflex is produced by shining a bright light into the eye. (In some persons this elicits a fit of sneezing as an additional reflex. It is not uncommon for a patient to complain that he sneezes when he first goes outdoors into bright sunlight, and occasionally a patient sneezes when an ophthalmoscope light is turned on his eye.) The so-called menace reflex is produced by an unexpected or threatening object coming suddenly into the near field of vision.

In the tactile corneal reflex the afferent pathway is the fifth cranial nerve and the efferent pathway the seventh cranial nerve. The reflex persists in the thalamic animal. In humans connections with the cortex are shown by the pain felt on touching the cornea and the strong spasm of the eyelids that occurs when such pain is produced. The reflex may be lost if there is a cortical lesion in the rolandic area. The corneal reflex is characteristically lost or impaired on the side of a tumor of the cerebellopontile angle. This is a diagnostic sign of great importance because the corneal reflex is lost before impairment of the other branches of the fifth nerve is clinically detectable. When testing for corneal sensitivity or blink reflex, it is important to remember that the cornea is easily traumatized. A wisp of cotton may be introduced from the side so close to the eye that the patient is hardly aware of its presence and does not blink in response to the menace reflex.

Contact lens wearers frequently experience abolition or diminution of the tactile corneal reflex. If it were not for this decreased corneal response, failures in adapting to contact lenses would be much more frequent than they are.

The afferent pathway of menace and dazzle reflex blinking is the optic nerve. The dazzle reflex appears to be subcortical. Association fibers to the facial nucleus form the efferent path. The dazzle reflex may be lost in certain mesencephalic lesions that give no other external signs. The menace reflex is cortical and requires not only the presence of the occipital lobe, but also of its connections with the rolandic area. Because of the cortical involvement, this reflex may be lost when the corneal tactile and dazzle reflexes persist.

A fourth type of blink is in response to stretching or striking structures near the orbit.[8] The orbicularis reflex blink has two components. A fast component occurs as a response to proprioceptor afferents in the fifth and seventh nerves, and a slow, nociceptive component is afferent in the fifth nerve.

Spontaneous blinking. Spontaneous blinking is the common form of blinking that occurs in a normal human being at frequent intervals during the waking hours and without any obvious stimuli. Each person seems to have his own individual rate of blinking, which is maintained as long as the external environment is not changed. The rate may be altered by changes in the surroundings or by the mental state of the subject; if there is any cause for excitement, the blink rate is usually increased considerably.

Spontaneous blinking does not occur or is very infrequent during the first few months of life; yet the delicate infant cornea does not suffer from dryness. Blinking occurs in all vertebrates possessing eyelids and living in air. Its rate varies considerably in the animal kingdom. The lion blinks at a rate of less than 1 blink/min, whereas some species of monkeys have a rate of 45 blinks/min. The blink rate continues in persons who are blind and thus does not depend on retinal stimulation. It has been suggested that blinking provides rest for the ocular muscles in that blinking allows a momentary upturning of the eyes, which is regarded as a position of rest. This is analogous to the position of the eyes during sleep. Blinking has also been regarded as a conditioned reflex because it is a frequently repeated response to indifferent sensory stimuli.[11]

Motion pictures of blinking show that the lower lid remains almost stationary during the act. The upper lid begins to close as a shade would drop. The movement is completed by a narrowing of the palpebral fissures in a zipper-like action from the lateral canthus toward the medial canthus. This aids the displacement of the tear film toward the lacrimal puncta.[1] Most persons blink about 15 times/min. The duration of a full blink is approximately 0.3 to 0.4 second. The average period between blinks is about 2.8 seconds in men and just under 4 seconds in women. During the period of blinking, vision must be in abeyance; therefore if blinking is occurring frequently enough, it might have some practical bearing on occupations in which constant perception is a necessity, such as piloting high-speed aircraft.

Spontaneous blinking does not produce a discontinuity of visual sensation, despite the fact that vision is obviously interrupted during the blink. In experimental situations, obscuration of vision of 0.003 second or 1/100 of a blink duration is barely detectable by the subject at daylight levels of illumination; a darkening of the entire visual field for 0.03 second by means other than spontaneous blinking is easily noticed. The continuity of visual sensation during the spontaneous blink is similar to the continuity of visual sensation during saccadic (rapid) eye movements, when vision is also suspended to a large degree.

By means of electromyographic studies, it has been shown that there are three main functional groups of motor units in the orbicularis oculi: (1) those responding in blinking and in the corneal reflex, (2) those responding in blinking and in sustained activity, and (3) those responding only in sustained activity.[7] There is considerable anatomic overlap among the distributions of these three groups of motor units. Those in the first group lie mainly in the pretarsal region, and those in the second group lie mainly in the preseptal region, whereas those in the third group extend from the preseptal region to the orbital region. During blinking the units in the first group discharge brief bursts of impulses at very high impulse frequency—up to 182/sec in a single unit. This is of the order of the highest frequencies of discharge observed by Reid[12] in the extraocular muscle motoneurons of the cat under reflex drive. The extraocular muscles have a very brief contraction time (p. 114); this may also be true of the white fibers of the orbicularis, which are mainly in the pretarsal portion. This extremely rapid movement is capable of intermittence at very high frequencies, since voluntary blinking in humans has been recorded at a maximum frequency of 390/min. The maintenance of the narrowed palpebral fissure after a blink depends on the activity of units in other parts of the muscle and on reciprocal innervation (relaxation of the levator). The first change to occur in a blink is relaxation of the levator rather than contraction of the pretarsal part of the orbicularis (Fig. 1-1). The preliminary relaxation of the antagonist is not a recognized feature of muscular contractions generally, but it may have significance in this unusual movement in that it allows the orbicularis to contract from the start against reduced resistance; this tends to shorten the time during which vision is disturbed.

Voluntary winking

A wink is a forced closure of one eye. Voluntary winking is produced by simultaneous contraction of the palpebral and orbital portions of the orbicularis muscle. Many persons cannot wink until they are taught, and some never learn to close one eye at a time. Occasionally a subject may learn to wink with one eye but not with the other. It has been noted that more people have difficulty in voluntary winking with the right eye than with the left. This has been attributed to the predominance of right-handedness, the theory being that there is greater development of the uncrossed fibers in the path of the facial nerve from the cortical centers to the subcortical centers on the left side in right-handed persons.[14] The act can be repeated at frequent intervals, but even at a maximum rate there must be a minimum period between winks of about 0.3 second.

Blepharospasm

Squeezing the eyelids together entails contraction of both portions of the orbicularis as well as of the muscles of the brow. Since this invariably raises the intraocular pressure, it is dangerous to have a patient squeeze the eyelids shut during any surgical procedure in which the globe is opened, such as in a cataract extraction. Even the speculum best designed for keeping the lids apart during eye surgery will not prevent the pressure of squeezing the eyelids from being transmitted to the globe, thereby making it possible to forcibly expel the vitreous. For this reason the branches of the seventh cranial nerve supplying the orbicularis muscle are usually temporarily paralyzed by injections of anesthetic solution prior to opening of the globe. The branches of the nerve may be injected as they course into the eyelids, or the nerve trunk may be injected instead. The trunk of the temporal and zygomatic branches is found by following the zygoma to a point just in front of and slightly below the external auditory meatus, where the anesthetic may be injected.

Blepharospasm is a frequent accompaniment of inflammatory diseases of the anterior segment of the globe and makes examination of the eye difficult. Many patients find it extremely hard to keep both eyes open when one is being examined, particularly if that eye is light sensitive, as most inflamed eyes are. They tend to keep the uninvolved eye closed, with the result that both eyes turn upward (Bell's phenomenon). The patient should be told to keep both eyes open at all times and, if possible, to fix his attention on some object in the room. If a patient is afforded a definite fixation point, he will find it easier to keep the unexamined eye open than if he is allowed to look off into space.

Fibrillary twitching of the eyelids is a common complaint of some persons. It may be related to fatigue, thyrotoxicosis, or mental tension, but refractive error is frequently the cause. When marked, it is called myokymia of the eyelids. The focus of irritation is thought to be in the nerve fibers of the orbicularis muscle.[6]

Weakness of the orbicularis is seen in myotonic dystrophy and other diseases that affect facial musculature. In myasthenia gravis the orbicularis may fatigue during the day, so ectropion and epiphora appear by afternoon.

PATHWAYS FOR MOVEMENTS OF THE EYELIDS

In the frontal cortex close to the oculogyric centers there is an area where stimulation produces raising of one eyelid or both. This eyelid elevation is generally greater on the side opposite the stimulated side. Stimulation of the region of the motor cortex close to the representation for the thumb results in closure of the eyelid, usually bilaterally, but greater on the side opposite the stimulation (Figs. 1-7 and 1-8). In monkeys lesions of this general region result in inability to close the eyes. The fact that closure of the eyes is not obtained from the same portion of the cortex that elicits movements of the eyeballs, but rather from the portion that causes movements of the face, is un-

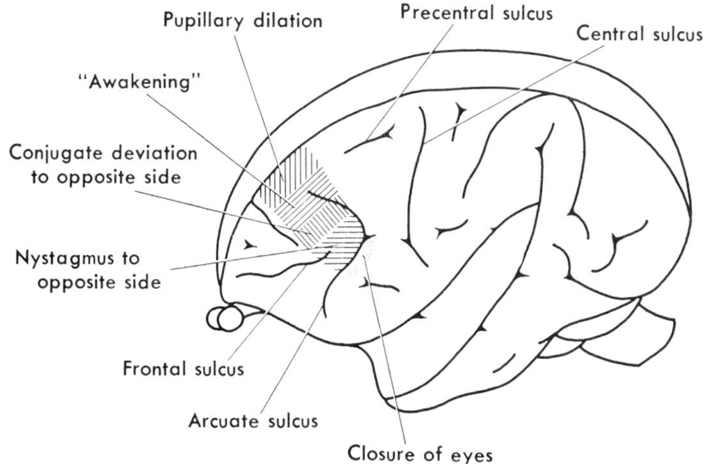

Fig. 1-7. Indicated subdivisions of frontal eye field and area yielding closure of eyes in monkey *(Macaca mulatta)*. (According to W. K. Smith.)

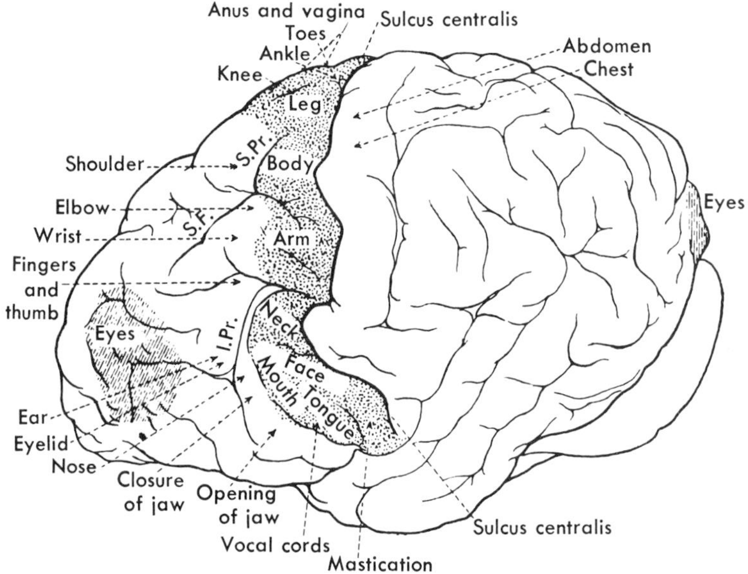

Fig. 1-8. Outer aspect of brain of chimpanzee, showing position of motor centers. Electric stimulation at parts indicated causes coordinate movements of corresponding muscle groups. (After Sherrington.)

derstandable when it is recalled that the extrinsic eye muscles are supplied by the third, fourth, and sixth cranial nerves; the orbicularis muscle is supplied by the seventh cranial nerve, which also supplies the muscles of the face.

The pathway from the frontal cortex to the brain stem nuclei that control movements of the eyelids is not known. Lesions in the region of the posterior commissure frequently produce an association of paralysis of upward gaze and changes in motility of the eyelids—either ptosis or retraction. On the other hand, the paths for conjugate lateral or vertical movements of the eyes and those for the eyelids must be separate, since lesions in the pons or tectum may separately abolish movements of either the eyelids

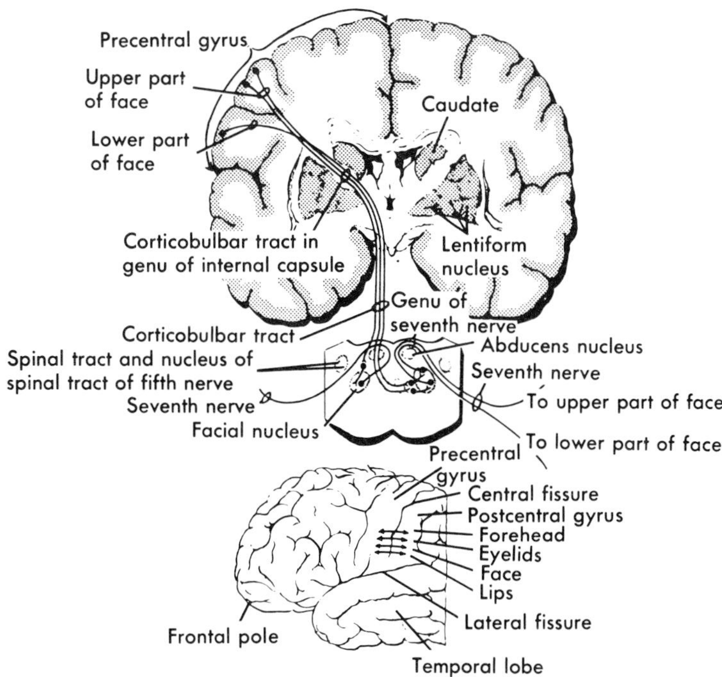

Fig. 1-9. Diagram of corticobulbar fibers serving facial nuclei. Note direct path through internal capsule and bilateral input for upper facial muscles. (From Crosby, E. C., and DeJonge, B. R.: Ann. Otol. Rhinol. Laryngol. **72:**735, 1963.)

or the eyes (Fig. 1-9). In lower animals the centers for lid movement may continue to function in response to visual stimuli after removal of the occipital lobes. This has been demonstrated in dogs and in monkeys.[9,10]

ASSOCIATED MOVEMENTS OF THE EYELIDS

Associated movements of the upper lid and the superior rectus muscle have been mentioned previously. Whenever the gaze is directed upward from the horizontal, the upper lid follows the movements of the globe. There is also an association between the movements of the eyelids and the globes when the eyes are closed. In the majority of persons the eyes are turned sharply upward when the eyelids are forcibly closed. This is a protective action that brings the cornea up under the covering eyelid and away from impending danger. It is called Bell's phenomenon. Bell's phenomenon is not present in 10% of otherwise healthy persons, and therefore its absence is not necessarily a sign of disease. Although the pathway for Bell's phenomenon is not known, it is known that the pathway is different from that for voluntary upward gaze, since Bell's phenomenon may be present in supranuclear paralysis of upward gaze.

Other movements of the eyelids and the simultaneous contraction of the ocular or facial muscles form associated movements under pathologic conditions. An interesting phenomenon of the eyelid known as the pseudo-Graefe phenomenon occasionally occurs following recovery from paralysis of the third cranial nerve. In straight-ahead gaze there is slight ptosis on the side of the previous palsy of the cranial nerve. The ptosis persists when this eye is abducted, but disappears and is even replaced by excessive widening of the fissure when the eye is adducted. The cause of the pseudo-Graefe phenomenon is misdirection of regenerating nerve fibers intended for the medial rectus muscle into the third nerve bundle going to the levator muscle. The ptosis in the primary and abducted positions represents a residual weakness of the levator. In adduction, on the other hand, the medial rectus is innervated and the mis-

Table 1-1. Distribution of average length of palpebral fissure according to age groups*

	Length of fissure (mm)															
	18	19	20	21	22	23	24	25	26	27	28	29	30	31	32	33
Under 1 year	2	3	2	3												
Ages 1 to 10		5	2	9	19	31	20	24	47	14	4	2				
Ages 11 to 60						4	5	29	80	173	337	206	259	117	22	9
Over age 60							5	17	47	43	75	55	28	25	9	
TOTAL	2	8	4	12	19	35	30	70	174	230	416	263	287	142	31	9

*From Fox, S. A.: Am. J. Ophthalmol. **62**:73, 1966.

directed fibers that have now reached the levator succeed in elevating the lid normally or even excessively.

Another abnormal phenomenon of the eyelid is that known as jaw winking, or the Marcus Gunn phenomenon. It consists of opening and shutting of the eye on one side during the act of chewing. The spontaneous movements of the eyelid are best produced by requesting the subject to move the lower jaw horizontally across the upper jaw. Apparently the pterygoid muscle, which is involved in this motion, is abnormally linked neurologically in some fashion with the levator muscle on that side. Thus when the pterygoids are innervated, the surplus innervation flows to the levator and causes the lid retraction. Although it has frequently been accepted that this association movement is caused by the misdirection of fibers intended for the pterygoid into the levator, it is difficult to see how such misdirection could occur. It may be significant that the cortical locus for elevation of the eyelid lies very close to the locus for jaw movements, and the fact that the condition is congenital may indicate that the defect is in the cortex and not further down in the pathways for innervation of these two areas.

THE PALPEBRAL FISSURE

The normal adult palpebral fissure is usually 27 to 30 mm long and, when the eyes are in primary position, 8 to 11 mm wide.[3] The maximum width is medial to the center of the palpebral fissure, making the fissure slightly almond shaped. The maximum excursion of the upper lid without frontalis action widens the fissure to about 15 mm. Frontalis action adds another 2 to 3 mm of upward movement. The lower lid moves about 5 mm from downward to upward gaze. No difference in these measurements was found between men and women or between blacks and whites. Although Orientals were not included in the series, they show a characteristic almond- or slit-shaped palpebral fissure.

Table 1-2. Width of palpebral fissure in 1732 normal individuals*

Width (mm)	Distribution (%)
7	3.3
8	22.0
9	29.2
10	32.5
11	10.0
12	2.1
13	0.9

*From Fox, S. A.: Am. J. Ophthalmol. **62**:73, 1966.

In children the fissures are not so long, but are relatively wider, and in infants the fissures may be nearly circular (Tables 1-1 to 1-5).

Normally the edge of the upper lid covers the upper limbus from 10 to 2 o'clock. The edge of the lower lid is more variable in position, but usually lies less than 1 mm below the lower limbus. If the upper limbus is completely exposed, the fissure is wider than normal. If this exposure is bilaterally symmetric, it may have little significance, the eyeballs being equally prominent. Unequal palpebral fissures generally have pathologic significance. Since the position of the eyeball in the orbit determines the width of the fissure to a large degree, the more prominent globe will produce a wider fissure. Hence, anything that causes one eyeball to protrude will lead to a wider fissure on that side. Similar-

Table 1-3. Average width of palpebral fissure at various lengths in various age groups*

Number of patients	Length of fissure (mm)	Width of fissure (mm)			
		Under 1 year	Ages 1 to 10	Ages 11 to 60	Over age 60
2	18	8			
8	19	8	8.7		
4	20	8.2	9		
12	21	8.5	9		
20	22		8.7		
34	23		8.5	8	
30	24		8.5	9	7.3
70	25		8.8	8.6	8.1
174	26		8.8	8.6	8.6
230	27		9	8.9	8.8
416	28		8.8	9	9
262	29		9	9.3	9.1
288	30			9.9	9.1
142	31			10.2	10.0
31	32			10.3	10.0
9	33			11.2	
TOTAL 1732					

*From Fox, S. A.: Am. J. Ophthalmol. **62**:73, 1966.

Table 1-4. Upward excursion of upper lid in 1638 normal individuals*

Upward movement (mm)	Distribution without frontalis action (%)	Distribution with frontalis action (%)
12	9.5	0.3
13	20.8	1.3
14	27.5	2.8
15	16.7	9.5
16	13.5	20.0
17	9.3	26.7
18	2.7	16.0
19		13.1
20		9.3
21		1.0

*From Fox, S. A.: Am. J. Ophthalmol. **62**:73, 1966.

ly, if one eye is abnormally recessed in the orbit or is smaller in size, the fissure on that side will be narrower.

The width of the fissure may reflect the psychologic state of the individual; for example, the wide-open eyes of surprise and fright are often seen in chronic anxiety states. The intensity of light incident on the eyes may account for narrowing of the palpebral fissures.

Table 1-5. Upward excursion of lower lid in 1638 normal individuals*

Upward movement (mm)	Distribution (%)
3	1.7
4	21.1
5	36.8
6	26.0
7	13.0
8	1.4

*From Fox, S. A.: Am. J. Ophthalmol. **62**:73, 1966.

The width of the fissures also depends on the combined tonus of the muscles—that is, the levator palpebrae superioris and the sympathetic-innervated Müller's muscle, which raise the upper eyelids, and the orbicularis muscle, which closes the eyelids. When a person is fatigued, the levator muscle loses its tonus, the fissure becomes narrower, and the lids feel heavy because the levator muscles have to be activated by a willed effort. One thus feels the annoying sensation of the upper lids drooping as one tries to stay awake. People remark that they are so tired that they can "hardly keep their eyes open." (See p. 15 for a discussion of the eyelids during sleep.)

Widening of the palpebral fissures caused by retraction of the upper eyelids is responsible for the greater part of the apparent exophthalmos seen in thyrotoxicosis. The distribution curves of exophthalmometer measurements in normal persons and in a group of patients of the same age with diffuse toxic goiter are not dissimilar. Normal persons measure from 12 to 21 mm, with a mean of 16 mm. Those with toxic goiter measure from 12 to 24 mm, with a mean of 18 mm. However, only 5% of normal persons have measurements greater than 19 mm, whereas in 32% of patients with toxic goiter the measurements are greater than this. Nevertheless, the principal element in the prominent appearance of the eye of patients with diffuse toxic goiter is the lid retraction. The amount of exophthalmos should be determined only by exophthalmometer measurements and should not be estimated by the width of the palpebral fissure.

Widening of the palpebral fissure, known as Dalrymple's sign, may be caused by either excessive tonus of the sympathetically innervated Müller's muscle or excessive tonus of the levator palpebrae. Most of the evidence supports the former, although there are no accompanying pupillary signs of increased sympathetic tonus. The fact that both the upper and the lower lids are retracted in thyrotoxicosis supports the opinion that the sympathetically innervated Müller's muscles are primarily concerned. Additional evidence is found in the fact that the lid retraction in endocrine exophthalmos can be reduced by instillation of topical sympatholytic agents such as guanethidine.[5] Von Graefe's sign, or "lid-lag"—inability of the upper lid to follow the movements of the globe in downward gaze—is also caused by retraction of the upper eyelid, resulting from the sympathicotonia. When marked signs of thyrotoxicosis are present, the widening of the palpebral fissure and slowness of the eyelid to close cause a characteristic appearance of apprehension or fright. In the type of disease in which there may be little evidence of active thyrotoxicosis, edema and lymphocytic infiltration of the ocular muscles and the orbital tissues create a true protrusion of the globe. The exophthalmometer measurements in this type of exophthalmos are typically greater than in toxic thyroid disease.

Retraction of the eyelid (tucked lid, Collier's sign), if thyroid disease has been excluded, may be caused by an increased tonus of the levator resulting from a lesion of the posterior commissure. A lesion lower in the brain stem results in ptosis rather than lid retraction.

The eyelids owe their form chiefly to the tarsal plates, which are held against the globe by the overlying portion of the orbicularis oculi, a thin, striated muscle. The palpebral portion of the orbicularis is strongly attached to the medial palpebral ligament, which moors it to the nasal orbital margin. The domed form of the normal tarsus depends on pressure of the lid against the globe. A prosthesis must be fitted after enucleation of the eye; otherwise the lids bridging the empty eyesocket will slowly contract.

Changes in position of the margins of the eyelids in respect to their apposition to the globe are important. The margins of the eyelids normally remain closely applied to the globe as the eyes move in various directions or as the lids open and close. Numerous pathologic states cause the eyelid margins to lose normal contact with the globe and to turn either in toward the globe *(entropion)* or away from the globe *(ectropion)*. Ectropion of the lower lid is a common consequence of seventh nerve paralysis (for example, Bell's palsy), in which, in the absence of orbicularis support, even the firm tissues of young people are insufficient to keep the lower lid against the globe. In senile ectropion the lower lid may sag outward as the orbicularis loses tone. Tears spilling over the lid margin macerate the thin skin, and the ensuing inflammation enhances the ectropion. One surgical procedure for senile ectropion is tightening of the band of orbicularis fibers overlying the tarsus.

In some persons the pretarsal fascia holding the orbicularis over the tarsus becomes loose with age. The orbicularis can then slide toward the lid margin and leave the orbital edge of the tarsus unsupported. Concomitantly, diminution of orbital fat, which is common in older people, allows the globe to recede into the orbit, and the laxity of the lids is compounded. Lid closure then presses the palpebral margin against the globe, while allowing the unsupported tarsal edge to tilt forward (entropion). Digital traction on the skin may reposit the lid, but it inverts again at the next blink. The lower lid, in which the tarsal plate is less developed, is more prone to senile entropion than is the upper lid. Entropion may be initiated by blepharospasm secondary to inflammation or photophobia. The irritation of inturned lashes against the globe causes further orbicularis spasm and perpetuates the condition.

The orbicularis muscle at the very margin of the lid consists of fine striated muscle fibers. This anatomically distinct portion of the orbicularis is known as the muscle of Riolan. Because of its position at the lid margin, the muscle of Riolan has been indicted by some authors as a particular culprit in the development of spastic entropion.

A surgical procedure for entropion utilizes the principle of shortening the orbicularis fibers at the orbital margin of the tarsus and securing them to the tarsus to prevent sliding toward the lid margin. Spastic entropion seldom occurs in the young despite long-continued blepharospasm.

Trauma and disease often lead to malposition of the lids. Contracting scars can evert or invert the lid to produce cicatricial ectropion or entropion. Entropion and resultant trichiasis are characteristic of such diseases as late trachoma and ocular pemphigus.

THE EYELIDS DURING SLEEP

During sleep, closure of the eyelids is not a simple relaxation of the muscles that keep the eyes open; rather, it is a tonic stimulation of the orbicularis together with an inhibition of the levator—the converse of what occurs during waking hours. With closure of the eyelids at the onset of sleep, the receptors for the dazzle and menace blinking reflexes cease to act, but the corneal reflex is not eliminated until the deeper stages of sleep have been reached.

Some otherwise normal persons do not completely close their eyes during sleep, and in these persons the lower cornea may become eroded from exposure. Closure of the eyelids that accompanies the onset of sleep is an inherited automatic act that takes place when cortical inhibition reaches a certain degree. It serves the purposes of protecting the eye from injury and of allowing the ocular muscles to rest. Voluntary closure of the eyelids in preparation for sleep has been called the ritual of going to sleep. When sleep is desired under difficult circumstances, shutting the eyes is one of the few available acts of this ritual. It is not always successful in its purpose, but at least it does reduce visual stimuli that would otherwise prevent or at least delay the onset of unconsciousness.

REFERENCES

1. Anantanarayana, A.: Note on the mechanism of eyelid closure in blinking, Proc. All India Ophthalmol. Soc. **10:**154, 1949.
2. Björk, A.: Electromyographic studies on coordination of antagonistic muscles in cases of abducens and facial palsy, Br. J. Ophthalmol. **38:**605, 1954.
3. Fox, S. A.: The palpebral fissure, Am. J. Ophthalmol. **62:**73, 1966.
4. Gay, A. J., Salmon, M. L., and Windsor, C. E.: Hering's law, the levators and their relationship in disease states, Arch. Ophthalmol. **77:**157, 1967.
5. Gay, A. J., Salmon, M. L., and Wolkstein, M. A.: Topical sympatholytic therapy for pathological lid retraction, Arch. Ophthalmol. **77:**341, 1967.
6. Givner, I., and Jaffe, N.: Myokymia of the eyelids, Am. J. Ophthalmol. **32:**51, 1949.
7. Gordon, G.: Observations upon the movements of the eyelids, Br. J. Ophthalmol. **35:**339, 1951.
8. Hoyt, W. F., and Loeffler, J. D.: Neurology of the orbicularis oculi. In Smith, J. L., editor: Neuro-ophthalmology: Symposium of the University of Miami and the Bascom Palmer Eye Institute, vol. 2, St. Louis, 1965, The C. V. Mosby Co.
9. Marquis, D., and Hilgard, E.: Conditioned lid responses to light in dogs after removal of the visual cortex, J. Comp. Psychol. **22:**157, 1936.
10. Marquis, D., and Hilgard, E.: Conditioned responses to light in monkeys after removal of the occipital lobes, Brain **60:**1, 1937.
11. Martino, G.: The conditioned reflex of blinking, J. Neurophysiol. **2:**173, 1939.
12. Reid, G.: Rate of discharge of extraocular motorneurons, J. Physiol. (Lond) **110:**217, 1949.
13. Sayoc, B.: Absence of superior palpebral fold in slit eyes, Am. J. Ophthalmol. **42:**298, 1956.
14. Suda, K., and Kitani, K.: Rinsho Ganka **9:**222, 1955; quoted from Ophthalmic Literature **9:**670, 1955.

Chapter 2

The lacrimal apparatus

BENJAMIN MILDER, M.D.

The optical integrity and normal function of the eye depend on an adequate supply of fluid covering its surface. This moist layer serves (1) an optical function by maintaining an optically uniform corneal surface, (2) a mechanical function by flushing cellular debris and foreign matter from the cornea and conjunctival sac and by lubricating the surface, (3) a corneal nutritional function, and (4) an antibacterial function.

The maintenance of such a moist layer depends on a secretory mechanism and an excretory, or elimination, mechanism. The secretory component includes the lacrimal gland and accessory lacrimal gland tissue, as well as the sebaceous glands of the eyelids and the goblet cells and other mucin-secreting elements of the conjunctiva. The elimination of the lacrimal secretions depends on a movement of tears across the eye, aided by the act of blinking, and a drainage system consisting of the lacrimal puncta, canaliculi, sac, and nasolacrimal duct.

The integrity of the cornea as the most important refractive element of the eye depends on the normal functioning of each of these two mechanisms, as well as on a proper balance between them. Thus a disturbing excess of tear fluid may result from obstruction in the drainage system or, alternatively, from an excessive secretion of tears despite a normal drainage mechanism. Conversely, normal comfort and normal hydration of the cornea may exist in the face of some relative impairment of the outflow mechanism, if this is accompanied by a diminution of lacrimal secretion. Such a comfortable balance of these two components is seen in older patients who have relative insufficiency of the lacrimal puncta caused by mild eversion of the lacrimal papillae, but who at the same time exhibit a decrease in lacrimal secretory activity, which is normal with advancing years.

The preservation of this balance between tear formation and tear outflow must be borne in mind in therapeutic programs. The importance of such a balance is illustrated by a 65-year-old woman who presented typical findings of keratoconjunctivitis sicca—a dry eye—associated with other evidences of Sjögren's syndrome. In addition, she had a chronic dacryocystitis. Although the treatment of dacryocystitis by removal of the tear sac has long been superseded by surgical procedures directed at restoration of the lacrimal drainage, in this instance the decision to remove the infected tear sac was predicated on the need to conserve, rather than eliminate, the already inadequate supply of tears.

THE TEAR FILM

The tear film (Fig. 2-1) that covers the surface of the eye is composed of three layers.

The *superficial oily layer* is derived from the meibomian glands and the accessory sebaceous glands of Zeis (Fig. 2-2). This oily layer reduces the rate of evaporation of the underlying tear layer and forms a barrier along the lid margins that retains the lid margin tear strip and prevents its overflow onto the skin. The oily layer is so thin that there are no interference color patterns such as one normally sees on oily surfaces; however, if one squints, the oily layer thickens and distinct interference colors may be seen.

The middle layer, *tear fluid,* is secreted by the lacrimal gland and the accessory glands of Krause and Wolfring. This aqueous layer is 6.5

to 7.5 μm thick, many times thicker than the fine superficial oily layer.[61] It contains inorganic salts, glucose, and urea, as well as biopolymers, proteins, and glycoproteins. Uptake of oxygen through the tear film is essential to normal corneal metabolism.

The *mucoid layer,* elaborated by goblet cells of the conjunctiva, is the deepest stratum of the precorneal tear film.

The tear film is not visible grossly on the surface of the eye, but at the upper and lower lid margins a 1 mm strip of tear fluid with concave outer surface can be seen. It is here that the oily surface prevents spillage of the tear fluid over the lid margin. The tears forming the upper tear strip are conducted nasally from the upper temporal fornix. At the lateral canthus the tears fall by gravity to form the lower strip. Spreading nasally, the upper and lower strips reach the plica and caruncle, where they join. The tear fluid does not flow over the eye by gravity, but rather a thin film is spread over the cornea by blinking and eye movements.

It has been suggested by Brown and Dervichian[6,7] that the oily component of the precorneal tear film is actually a multimolecular layer, which in itself cannot spread over an aqueous solution and prevent evaporation in that manner. In vitro experiments have led them to believe that blinking produces a biphasic movement of tears across the cornea. As the upper eyelid elevates, there is first an upward movement of the aqueous component; this is followed by a slower spreading of the oily layer, which drags with it additional tears, thus thickening and stabilizing the entire precorneal tear film.

Lemp and associates[52,53] have stressed the importance of the mucin layer in the normal wetting of the cornea. Emphasizing the hydrophobic nature of the corneal epithelium, they have demonstrated that the conjunctival mucin layer is spread over and adsorbs on the corneal epithelium by the act of blinking, thus providing a hydrophilic surface over which the tear fluid spreads evenly. As the lids open in the

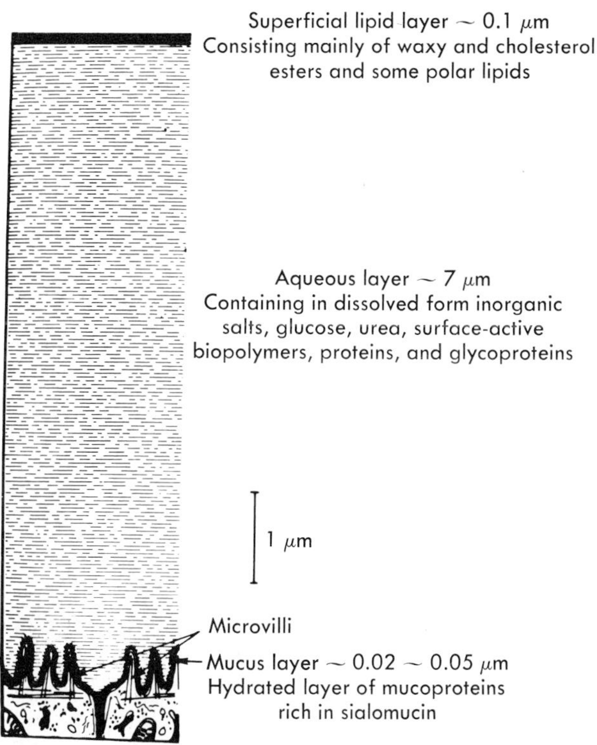

Fig. 2-1. Structure and composition of the tear film. (From Holly, F. J., and Lemp, M. A.: Surv. Ophthalmol. **22:**69, 1977.)

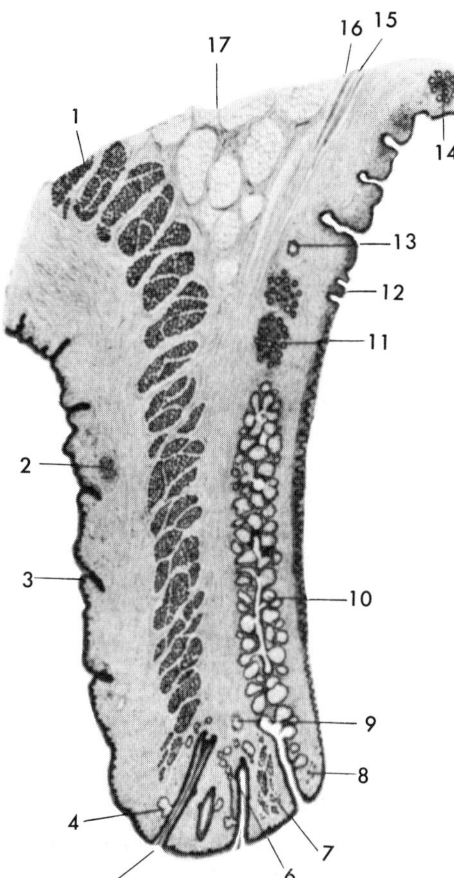

1. Orbicularis muscle
2. Sweat gland
3. Hair follicle
4. Gland of Zeis
5. Cilium
6. Gland of Moll
7. Pars marginalis of orbicularis muscle
8. Pars subtarsalis of orbicularis muscle
9. Inferior arterial arcade
10. Meibomian gland
11. Gland of Wolfring
12. Conjunctival crypts
13. Superior arterial arcade
14. Gland of Krause
15. Müller's muscle
16. Levator palpebrae superioris
17. Fat

Fig. 2-2. Sagittal section through upper eyelid, showing sebaceous and various accessory lacrimal glands. (From Duke-Elder, W. S.: Text-book of ophthalmology, vol. 5, St. Louis, 1952, The C. V. Mosby Co.)

relaxation phase of the blink, the high surface tension at the air–aqueous layer interface facilitates a rapid spread of the lipid layer.

As the tear film is decreased by evaporation between blinks, the oily superficial layer becomes admixed with the mucinous film, recreating the hydrophobic state of the corneal epithelium and favoring the breakup of the tear film. Thus each blink aids in maintaining normal wetting of the cornea by spreading the mucin layer. Normally the tear film breakup time (BUT) is longer than the interval between blinks, and no corneal drying occurs. A deficiency in conjunctival secretions could produce "dry eye" symptoms, even in the presence of an adequate aqueous tear component.

Generally there is a balance between the secretion and excretion of tears, and the rate of tear drainage increases with increased tear volume. However, in tear-deficient states retention of tears can be facilitated by the use of certain synthetic polymers. The surface-spreading characteristics of these polymers, by virtue of their surface-wetting properties or their viscosity, stabilize the tear film, thus dragging a thicker layer of tears as they spread with each blink.[5]

Evaporation of tears

In the normal tear film the evaporation rate is low because of the protective oily surface. Between 10% and 25% of the total tears secreted are lost by evaporation. In the absence of the protective oily layer, the rate of evaporation is increased 10 to 20 times.[37,61]

The tonicity of human tears is subject to a

dynamic change because of the evaporation process and the rate of tear flow. When evaporation is prevented, the osmotic pressure of tears is equivalent to 0.9% sodium chloride solution. As tear flow increases, the effect of evaporation lessens. Mastman and co-workers[56] state that when tear flow decreases, as in Sjögren's syndrome, the tears become markedly hypertonic (0.97% sodium chloride solution or more) and corneal dehydration results.

When the eyes are closed, there is no evaporation of tears, and the precorneal tear film is in osmotic equilibrium with the cornea. With the eyes open, evaporation takes place, increasing the tonicity of the tear film and producing an osmotic gradient from the aqueous through the cornea to the tear film. This direction of flow will continue as long as evaporation maintains the hypertonicity of the tear film.

COMPOSITION OF TEARS

Data comparing physical properties and chemical composition of tears with those of blood plasma are summarized in Table 2-1.[2,69]

Average glucose in tears, 2.5 to 4.1 mg/100 ml, represents 40% of the total reducing substance, whereas it makes up 80% of the total reducing substance in blood.[25,27] Changes in the concentrations of glucose in tears and blood show a close correlation. It is possible to make a semiquantitative determination of hyperglycemia with various commercially available reagent test strips moistened with tears.[16] Such a method is useful for a comatose patient or one from whom, for some reason, a urine sample cannot be obtained.

Table 2-1 shows that the concentrations of potassium and chloride are greater in tears than in plasma, which suggests that these are secredted by the lacrimal gland. The total amount of urea in tears is small compared to plasma, but it increases with tear flow, although the concentration drops as tear secretion increases.[4,78]

Alterations in the physical-chemical properties of tears in contact lens wearers provide some insights into the ability of the eye to tolerate these foreign bodies. When polymethyl methacrylate (PMMA, ''hard'') contact lenses are first worn, chloride ions in tears decrease, but return to normal in 5 or 6 days, correlating well with the visual adaptation period for these lenses.

Protein content of tears

The protein content of tears differs in several respects from that of blood plasma, lending additional evidence that tear fluid is a true secretion. In human tears the protein fractions, determined by various electrophoretic techniques, are albumin, globulin fractions, and lysozyme. Albumin represents about 60% of the total protein in tears, as it does in plasma. The remainder is divided equally between globulin and lysozyme fractions.

Immunoglobulins in tears

In the globulin component of blood plasma and various body secretions, including tears, a number of different immunoglobulin fractions have been identified (Table 2-2).* These immunoglobulins constitute important elements of the body's defense against certain diseases, and their role varies with the site of antigenic stimulation. Thus IgG is the most prominent circulating (serum) immunoglobulin, present in concentration five times that of IgA. However, in tears and saliva IgA is dominant and is found in concentrations approximately equal to IgG.

The IgA of external secretions differs from circulating IgA, having an additional ''secretory piece'' molecule. Plasma cells that secrete IgA have been found in the lacrimal gland and ''secretory piece'' in the tips of the acinar cells. Judging from the presence of ''secretory piece'' and the differing IgG/IgA ratios in tears and plasma, it is probable that immunoglobulins in normal tears are not a transudate from serum, but rather are produced locally. While the circulating immunoglobulins function in blocking systemic infection, locally produced antibody provides a more effective defense against externally invasive viral or bacterial antigen.

Enzymes in tears

Lysozyme is a long-chain, high–molecular weight proteolytic enzyme produced by lyso-

*See references 1, 5, 13, 22, 38, 57, and 74.

Table 2-1. Physical properties and chemical composition of human tears and plasma

	Tears	Plasma
Physical properties		
Osmotic pressure	0.9% NaCl (300 mOsm/L)	6.62 atm
pH	7.4 (7.3 to 7.7)	7.39
Refractive index	1.357	1.35
Volume	0.50 to 0.67 g/16 hr (waking)	
General chemical components		
Ash	1.05 g/100 ml	0.6 to 1.0 g/100 ml
Solids, total	1.8 g/100 ml	8.6 g/100 ml
Water	98.2 g/100 ml	94 g/100 ml
Electrolytes		
Bicarbonate	26 mEq/L	24.3 mEq/L
Chloride	120 to 135 mEq/L	102 mEq/L
Potassium	15 to 29 mEq/L	5 mEq/L
Sodium	142 mEq/L	137 to 142 mEq/L
Calcium	2.29 mg/100 ml[81]	
Nitrogenous substances		
Total protein	0.669 to 0.800 g/100 ml[57]	6.7 g/100 ml
Albumin	0.394 g/100 ml	4.0 to 4.8 g/100 ml
Globulin	0.275 g/100 ml	2.3 g/100 ml
Ammonia	0.005 g/100 ml	0.047 mg/100 ml
Uric acid		3 to 5 mg/100 ml
Urea	0.04 mg/100 ml	26.8 mg/100 ml
Nitrogen		
Total nitrogen	158 mg/100 ml	1140 mg/100 ml
Nonprotein nitrogen	51 mg/100 ml	27 (15 to 42) mg/100 ml
Carbohydrates		
Glucose	2.5 (0 to 5.0) mg/100 ml	80 to 90 mg/100 ml
Sterols		
Cholesterol and cholesterol esters	8 to 32 mg/100 ml[36]	200 to 300 mg/100 ml
Miscellaneous organic acids, vitamins, enzymes		
Citric acid	0.6 mg/100 ml	2.2 to 2.8 mg/100 ml
Ascorbic acid	0.14 mg/100 ml	0.1 to 0.7 mg/100 ml
Lysozyme	1438 (viscosimetric)[78] (800 to 2500) units/ml	

Table 2-2. Levels of immunoglobulins and other proteins in tears and plasma

Class	Tears	Plasma
Total protein	800 mg/100 ml	6500 mg/100 ml
IgA	14 to 24 mg/100 ml	170 to 200 mg/100 ml
IgG	17 mg/100 ml	1000 mg/100 ml
IgM	<1 mg/100 ml	—
IgD	—	—
IgE	250 ng/ml	2000 ng/ml

somes, a known cellular ultrastructure. It has the ability to dissolve bacterial walls by enzymatic digestion of tissue mucopolysaccharides. Although lysozyme has long been considered to be the antibacterial substance in tears, it has been shown that when tears are heated they lose their antistaphylococcal action, although the lysozyme content is unaffected.[79] This heat-sensitive component has been found to be a low–molecular weight protein not related to the immunoglobulins, but apparently possessing a much greater antibacterial activity than lysozyme.[24]

The nonlysozymal bactericidal protein, β lysin, is found in tears and aqueous humor.[23] It is reported to be derived chiefly from platelets, but exists in higher concentration in tears than in blood plasma. The lysozyme and β lysin protein fractions can be separated by filtering the tears. The antibacterial activity of the filtrate is due to lysozyme, but in whole tears β lysin is responsible for three-fourths of the bactericidal effect. β Lysin acts primarily on the cellular membrane, whereas lysozyme dissolves bacterial cell walls.

Twelve enzymes, including significant amounts of lactate dehydrogenase, pyruvate kinase, malate dehydrogenase, and amylase, have been identified in tears. These are present in concentrations similar to those in the lacrimal gland but unlike serum concentrations.[82]

The protein content of tears is a species-specific characteristic, and the lysozyme fraction varies widely in different species. In human tears, lysozyme makes up 21% to 25% of the total protein.[60] Although lysozyme is present in most animal tissues and secretions, it is present in sufficiently high concentrations to be bactericidal only in white blood cells, nasal secretions, and tears. In the presence of epiphora, lysozyme is decreased in concentration. It is decreased or absent in hypolacrimia found in keratitis sicca (Sjögren's syndrome). Tear lysozyme is not significantly reduced in those corneal diseases not associated with keratoconjunctivitis sicca, nor is it reduced in autoimmune diseases without eye findings. The lowered tear lysozyme level in Sjögren's syndrome may precede the classic eye signs, forming a useful diagnostic test.[54]

In suppurative eye diseases the lytic principle increases, presumably because of the lysozyme content of the leukocytes.[15] The amount of lysozyme in tears also increases after drinking water as well as after administration of systemic epinephrine or histamine. It is decreased by systemic corticosteroids, glucose, urea, and insulin.[71] Lysozyme levels decrease with age, although not necessarily in proportion to the decrease in tear volume.[65]

The electrophoretic activity of tear lysozyme is identical with that of egg white lysozyme. Because of this similarity, beaten, refrigerated egg white has been used successfully in the form of eyedrops as a treatment for keratoconjunctivitis sicca.[66] Practically, however, there are other types of commercially available preparations that are equally efficacious and less troublesome.

The action of lysozyme depends on pH. The optimum pH for lysis varies with the solubility of the bacterial proteins, but in general lies between 6.0 and 7.4. Low salt concentrations favor lysis by increasing solubility. Certain chemical antiseptics and lysozyme are mutually antagonistic and reduce the antibacterial properties of each.[62]

Cotlier[14] reported the biochemical assay of tears for the enzyme hexosaminidase A as a means of diagnosing Tay-Sachs disease and identifying carriers of the gene for this inherited, generally fatal illness. The absence of hexosaminidase A is associated with abnormal fat deposition in the retinas and brain of affected infants. Tears collected on a small strip of absorbent paper can be analyzed for this enzyme, yielding information as accurately as blood specimens. This use of tears in diagnostic biochemistry for the study of diseases other than those generally associated with lacrimal secretory function may find further application in enzyme disorders.[11]

pH of tears

The pH of tears, 7.4, approximates that of blood plasma. Although wide variations have been found in normal individuals (between 5.2 and 8.35), the usual range is from 7.3 to 7.7.[2] Tear pH is lowest on awakening due to acid by-products associated with the relatively anaerobic conditions in prolonged lid closure and increases due to loss of CO_2 as the eyes are open. Tear pH is characteristic for each individual, and the normal buffering mechanism maintains the pH at a relatively constant level during waking hours.[12,35,68] When solutions having a pH level below 6.6 or above 7.8 are instilled into the conjunctival sac, subjective discomfort occurs. Corneal injuries tend to produce an alkaline reaction in the tears, and an acid-buffered solution adds to relief of symptoms. The permeability of the corneal epithelium does not seem to be affected by wide variations in the pH of tear fluid or instilled solutions.

Osmotic concentration of tears

The osmotic concentration of tears is of prime importance, since optical integrity of the cornea is significantly influenced by the tonicity of the tears. The normal osmotic pressure of tears is equivalent to 0.9% to 0.95% sodium chloride solution.[48,56,73] The total osmotic pressure is proportional to the dissolved crystalloids, whereas the tear proteins, of high molecular weight, contribute little to the osmotic pressure of the lacrimal fluid. Variations in osmotic pressure between 0.6% and 1.3% sodium chloride equivalent appear to be well tolerated by the eye, but beyond these limits discomfort is experienced. Contact lens solutions seem to be tolerated best at an osmotic pressure equal to 0.9% sodium chloride solution. Corneal healing is not affected by variation in osmotic pressures from 0.8% to 1.3% sodium chloride (NaCl).[73] Although these findings provide guidelines in the preparation of ophthalmic solutions, it has been found that when solutions as hypertonic as 2.5% sodium chloride are instilled into the conjunctival sac, they are diluted to the concentration of tears within 1½ to 2 minutes—most likely by elimination and replacement of the hypertonic agent.[56] The tonicity of tears is quite similar to that of blood plasma. Ridley[69] found that the osmotic pressure of tears is slightly below that of blood and concluded, with regard to diffusible components, that tears are a dialysate from blood. He found no difference in osmotic pressure between tears and aqueous humor.[73]

Osmotic pressure is sensitive to changes in tear flow.[28] Thus reflex stimulation of tears in early adaptation to contact lenses results in a decrease in electrolytes and in total protein. Decrease in tear protein has been found in afternoon tear samples of the first week of contact lens wear. The resulting relative hypotonicity may account for the corneal edema often seen in early stages of contact lenses adaptation.[9]

SECRETION OF TEARS

Lacrimal fluid is made up of the secretions of the lacrimal gland and accessory lacrimal gland tissue, together with the secretions of the meibomian glands and the mucous glands of the conjunctiva. The lacrimal gland is situated in the superior temporal angle of the orbit. It lies under the orbital rim in the shallow lacrimal fossa, between the globe and the lateral process of the frontal bone. The lateral aponeurosis of the levator tendon produces an indentation in the gland so deep that the gland appears to be composed of two separate lobes—an orbital portion lying above the aponeurosis and a palpebral portion below (Fig. 2-3).

The larger, orbital lobe empties by a variable number of ducts (two to eight) into the conjunctival sac at the upper temporal fornix, and one or two of these ducts may open inferior to the lateral canthus. The ducts from the orbital lobe pass through the smaller, palpebral lobe (which has six to ten ducts); thus surgical removal of the palpebral lobe will abolish secretion from the entire lacrimal gland.

Tears are also secreted by the accessory lacrimal glands of Krause. The upper conjunctival fornix has some 20 of these glands, and the lower fornix has six to eight. Some lacrimal secretion is produced by the glands of Wolfring, located in the supratarsal conjunctiva of the upper lid, with an occasional one in the infratarsal conjunctiva of the lower lid.

The oily layer of the precorneal tear film is secreted chiefly at the lid margin by the meibomian glands and also by the lash follicle glands of Zeis. The deep, mucous layer of the tear film is elaborated by the goblet cells of the conjunctiva and the crypts of Henle in the fornices.

Jones[45] delegates the function of reflex secretion of tears to the main lacrimal gland and says that the accessory glands and the mucin- and oil-producing glands are the basic secretors responsible for the tear film. He believes that it is possible to maintain the tear film solely through the activity of the basic secretors. However, enthusiasm for extirpating the main gland itself to control hypersecretion may be dampened by reported cases in which the basic secretors did not provide adequate tear formation following such surgery and in which keratitis sicca resulted.[85]

Newborn babies secrete tears in the first 24 hours of life. In one study 82% of full-term in-

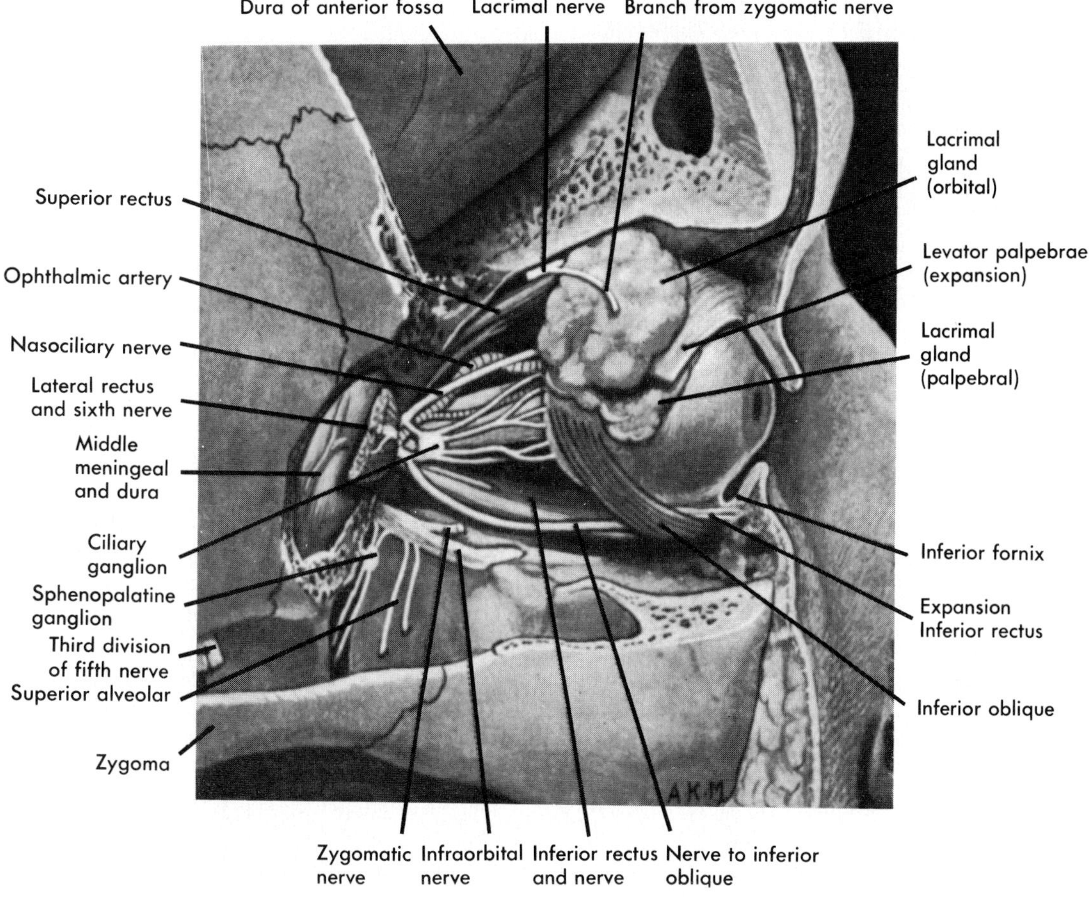

Fig. 2-3. Lacrimal gland in relation to globe. (From Last, R. J.: Eugene Wolff's anatomy of the eye and orbit, ed. 6, Philadelphia, 1968, W. B. Saunders Co., p. 304.)

fants were found to secrete tears in the first day, and this number increased to 96% if the infants were stimulated to cry. Premature infants may fail to secrete tears at birth, depending on the degree of prematurity.[3,64]

A quantitative evaluation of tear flow is difficult, but after allowing for different investigative methods and for such variables as the rate of evaporation, estimates of tear production range from 0.5 to 1.25 g over a 16-hour waking period. Up to 25% of tears are said to be lost by evaporation. A negligible amount of tears is formed during sleep.[19,47,77,88]

A useful clinical test for the rate of tear formation was devised by Schirmer in 1903. This test utilizes a 5 by 30 mm strip of no. 41 Whatman filter paper. A 5 mm tab is folded over at one end and inserted into the lower conjunctival sac at the junction of the middle and temporal third of the lower lid (Fig. 2-4).

The moistening of the exposed portion of the paper strip (excluding the folded over tab) over a 5-minute period is a measure of the rate of tear secretion. If the test is performed after instilling a topical anesthetic, only the basic secretion is measured, since a principal stimulus to reflex secretion has been eliminated. Should inadequate production of tears be indicated from the initial test, a Schirmer II test may be performed by repeating the process while stimulating the nasal mucosa with fumes of aromatic spirit of ammonia, a wisp of cotton, or other mechanical irritant. Prepackaged Schirmer strips are available, and blue litmus paper has been used for the same purpose.

Although the Schirmer test leaves much to be

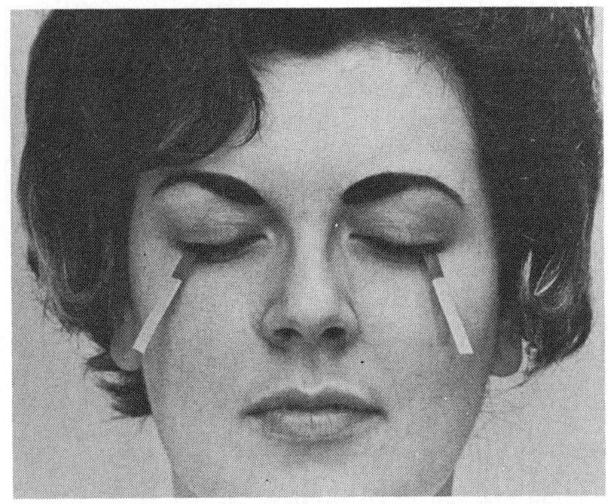

Fig. 2-4. The Schirmer test.

desired as a quantitative test, it is rapid, simple, and adequately repeatable in a given patient. The normal range of Schirmer findings is from 10 to 25 mm, decreasing with advancing age. However, the Schirmer results must be interpreted in terms of the patient's symptoms and objective findings. Results of less than 5 mm are commonly, but not always, associated with symptoms such as burning, scratchiness, and dry feeling and with accumulation of mucus in the conjunctival sac. Statistical surveys utilizing this method have revealed that the two eyes usually secrete equally, although one study indicated that in suspected unilateral secretory deficiency a difference between the two eyes of less than 27% was not clinically significant.[31] There is no significant diurnal rhythm in lacrimal secretion, and in patients under the age of 30 the rate of lacrimal secretion is higher in females. In those over age 30 there is little difference in rate of tear formation between the sexes, but the rate decreases with each decade, somewhat more rapidly in women.[32] A recent study examined Schirmer tests in relation to corneal sensitivity; when the test was performed after use of topical anesthetic, no decrease in Schirmer values was found with increasing age. The authors suggest that a progressive decrease in Schirmer test values in the nonanesthetized eye was actually due to diminishing corneal sensitivity with advancing age.[49]

Although the Schirmer test is a useful device for evaluating tear flow, it actually measures the aqueous component of the tear film. Deficiencies of the mucous layer can be detected by an indirect method, the tear film BUT. If, as has been suggested, uniform normal wetting of the corneal surface depends on even spreading of the adsorbed mucin layer, the rapidity of appearance of dry spots on the cornea between blinks becomes an index of the adequacy of the mucin layer.[50,83]

The BUT is measured by instilling fluorescein solution into the conjunctival sac and scanning the cornea with cobalt blue illumination at the slit-lamp microscope for the first sign of dry (fluorescein-free) areas. The normal tear film BUT is 15 to 35 seconds, and readings of less than 10 seconds suggest mucin deficiency. The test is reproducible for a given eye.

In the study of known dry-eye syndromes the BUT is an aid in the appropriate selection of therapeutic agents. Useful mucomimetic wetting compounds are large molecule polymers that adsorb onto the epithelial surface and resist removal by the shearing force of the lids in the act of blinking.[51]

The innervational control of tear secretion is derived from three sources: the trigeminal nerve, the facial nerve, and cervical sympathetic nerve fibers (Fig. 2-5).

The lacrimal nerve, a division of the ophthalmic branch of the fifth nerve, provides the principal afferent pathway in the reflex arc. The efferent pathway of the reflex lacrimal secretory mechanism is made up of fibers of the seventh nerve that originate in the pons above the superior salivary nucleus. These fibers join the sensory root of the facial nerve and pass through the facial nucleus and the geniculate ganglion. Exiting in the nervus intermedius of the facial nerve, they join the greater superficial petrosal nerve, and together with the deep petrosal nerve (sympathetic fibers from the internal carotid plexus) they become the vidian nerve (nerve of the pterygoid canal). The parasympathetic fibers synapse at the sphenopalatine ganglion. From this ganglion, lacrimal postganglionic fibers enter the zygomatic nerve, a branch of the maxillary division of the fifth nerve, to reach the lacrimal nerve and terminate in the lacrimal gland.

The lacrimal gland is also supplied by sympathetic fibers. These fibers originate in the hypothalamus and, by way of the superior cervical ganglion and the carotid plexus, reach the lacrimal gland by several routes. Some

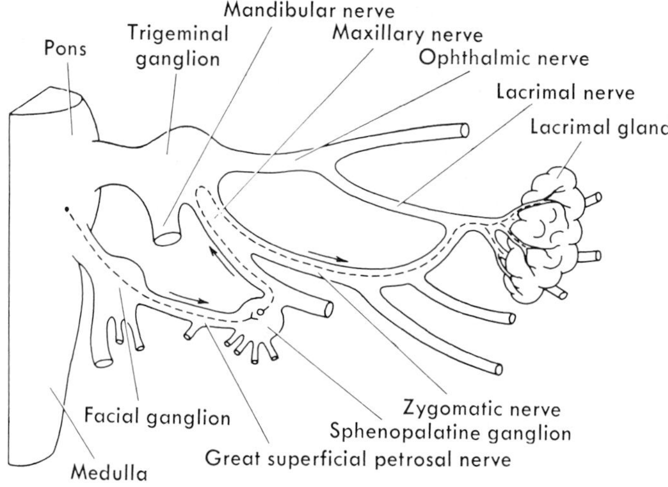

Fig. 2-5. Innervation of lacrimal gland. (After Mutch.)

pass along the lacrimal artery, and some join the parasympathetics in the vidian nerve by way of the deep petrosal nerve and reach the gland via the lacrimal nerve.[55]

The complex neurogenic control of the secretion of tears may be best understood in terms of Jones'[45] concept of "basic" and "reflex" secretions. The basic secretors are the accessory lacrimal glands of Krause and Wolfring, together with the mucous and sebaceous glands; these structures provide all three layers of the tear film. The reflex production of tears is the function of the lacrimal gland, which secretes lacrimal fluid only. Reflex secretion may be of *peripheral sensory origin* through fifth nerve stimulation (cornea, conjunctiva, skin, nose) or of *central sensory origin*. In the latter the stimulation may be retinal, varying with intensity of light, or psychogenic, as in weeping caused by emotional disturbances or by various central nervous system diseases. The production of tears on an emotional basis is unique to humans among all vertebrates.

If the stimulus to the fifth nerve on one side is slight, the reflex tearing will be unilateral, but stronger stimuli will involve both eyes; thus a patient with corneal irritation in one eye may have profuse bilateral tearing. Such tearing can be inhibited by surface anesthesia or by paralysis of the ophthalmic division of the fifth nerve. Psychogenic reflex tearing, on the other hand, will always be bilateral and cannot be abolished by surface anesthesia or fifth nerve paralysis. Psychogenic and peripheral sensory lacrimation can be abolished by blocking the sphenopalatine ganglion (efferent pathway). Psychogenic tearing, however, remains normal if the sensory root of the fifth nerve is cut.[63]

The anatomy of the nerve supply of the lacrimal gland is better understood than is the function of each of the innervation components. Peripheral sensory reflex tearing is certainly mediated through the afferent fibers of the ophthalmic division of the trigeminal nerve. However, central sensory reflex tearing due to retinal stimulation is not as fully understood. The pathways for psychogenic, or emotional, weeping are not clear, although the frontal lobes, the basal ganglia, and the hypothalamus play a role. Whatever the stimulus to lacrimation, the efferent pathway appears to be parasympathetic via the seventh nerve fibers and the sphenopalatine ganglion.

Blocking the sphenopalatine ganglion will reduce tear flow.[30] After lesions of the facial ganglion, regenerating salivary gland nerve fibers may be misdirected to the lacrimal gland, so mastication produces tearing as well as salivation—the phenomenon of "crocodile tears."*[29,43]

While reflex secretion of the lacrimal gland is controlled by the parasympathetic supply,

*The crocodile is said to weep as it eats its victims.

the sympathetic fibers may control the basic secretors in the lids and conjunctiva.[21] The role of the sympathetics may be the regulation of blood flow through the gland, indirectly influencing secretion. Section of the cervical sympathetic chain was found to sensitize the lacrimal gland to sympathomimetic *and* parasympathomimetic drugs.[55] This apparently paradoxical pharmacologic reaction of the lacrimal gland may be caused by increased permeability of the cells of the gland following interruption of the sympathetic supply.

Supportive evidence for the efferent role of the parasympathetics is found in the pharmacologic responses of the lacrimal gland. Directly acting parasympathomimetic drugs such as pilocarpine produce an increase in lacrimal flow, whereas inhibitors such as atropine and scopolamine reduce tear flow. DeHaas[17] found that after parasympathetic denervation, the lacrimal gland is hypersensitive to direct parasympathomimetics such as pilocarpine and methacholine but not to indirect stimulators such as neostigmine. However, when the fifth nerve is destroyed and the facial nerve is intact, the lacrimal gland is not sensitized to pilocarpine.

Hypersecretion and hyposecretion

In practice, when a patient complains of a wet eye, hypersecretion must be differentiated from failure of the tear evacuation system. If the wet eye is caused by hypersecretion, the Schirmer test results will be increased and the Jones fluorescein dye test will reveal normal outflow function.

The most frequent causes of hypersecretion are irritative stimuli affecting the ophthalmic division of the trigeminal nerve. These include trauma and foreign body or disease of cornea, conjunctiva, eyelids, and nasal mucosa. Gaynor[26] has described widely flared nares, which alter the air currents in the nose, as a cause of lacrimal hypersecretion. Both hypothyroid and hyperthyroid disease have been implicated as causes of lacrimation. The tearing of hypothyroidism has been checked by administration of thyroid extract.[76] It has been suggested that lacrimal hypersecretion may occur in association with anovulatory drugs used as oral contraceptives.[84]

Lacrimal hyposecretion may be congenital, although this is quite rare.[20,75] Acquired lacrimal hyposecretion may be the result of local inflammatory disease of the conjunctiva, usually conjunctival scarring secondary to bacterial or viral infection, or it may be the result of chronic inflammatory disease of the salivary and lacrimal glands (Mikulicz's syndrome). Other diseases, such as sarcoidosis, may involve the lacrimal gland. Hyposecretion may also occur as a result of senile atrophy of the lacrimal gland.

The syndrome of Sjögren includes lacrimal hyposecretion as a result of mononuclear infiltration of the lacrimal gland with B (bone marrow) lymphocytes. A cytotoxic immune mechanism leads to eventual atrophy and fibrosis of the gland. The salivary glands undergo the same pathologic process. There may also be achlorhydria and associated rheumatoid arthritis. This syndrome predominates in women (85%) in the premenopausal and menopausal years. More than 90% of these patients have a positive rheumatoid factor, and the evidence indicates that this is an autoimmune disease. In its milder forms Sjögren's syndrome represents the most frequently encountered cause for the symptom of a "dry eye."

ELIMINATION OF TEARS

It has been pointed out that some 25% of tears secreted is lost by evaporation. The remainder leaves the conjunctival sac through the lacrimonasal excretory system. The tears secreted into the upper temporal fornix are conducted to the lacrimal puncta in three ways. At the lateral canthus the tears fall by *gravity* to form the lower tear strip. The lower canaliculus is said to collect four times as much of the tear flow as the upper canaliculus. However, studies suggest that as many as 45% of patients have greater outflow through the upper canaliculus.[39] *Capillary attraction* plays a role in conducting the tears into the punctum and the vertical limb of the canaliculus. *Lid movements* form an important mechanism in the transport of tears to the puncta by the act of blinking. Blinking not only spreads the tear strips over the eye as a film but also moves the tears toward the puncta with each blink. The nasally directed movement of the tears results from the fact that the orbicularis muscle is more firmly fixed at its nasal attachment, thus moving the temporal part of the orbicularis ring in a nasal direction during the act of blinking; also, the temporal end of the palpebral aperture closes more rapidly in blinking.

Since the frequency of blinking varies widely, from 29.3/min during conversation to 3.57 for men when reading (less in women), it follows that maintaining a normal tear film and a normal rate of elimination of the tears does not depend entirely on the blink mechanism. One does note, however, an increased rate of blinking in response to excessive tear formation.

As the tears enter the lacrimal puncta they are propelled through the canaliculi into the tear sac by the same blinking movements. Each canaliculus has a short vertical and a longer horizontal segment. At the junction of the two segments the canaliculus widens into an ampulla. Orbicularis fibers are intimately disposed around the punctum and the canaliculus, so when this muscle contracts in blinking, the punctum is drawn nasally, the ampulla is compressed, and the horizontal limb of the canaliculus is shortened, thus driving tears into the lacrimal sac[44] (Fig. 2-6).

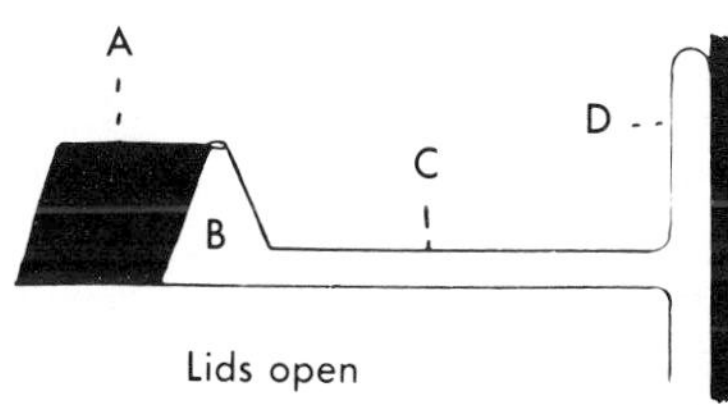

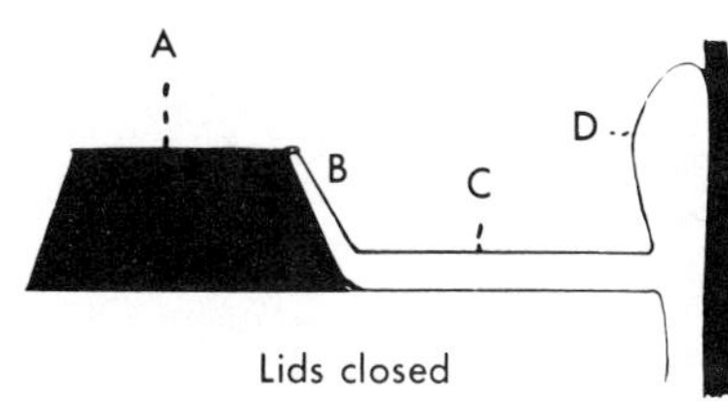

Fig. 2-6. Schematic drawing of lacrimal pump. (From Jones, L. T.: Trans. Am. Acad. Ophthalmol. Otolaryngol. **66:**506, 1962.)

Careful dissections of the nasal orbicularis attachments by Jones have demonstrated that fibers of the upper preseptal portion of the muscle insert into the fascia overlying the fundus of the lacrimal sac, so in blinking the contraction of the orbicularis draws the lateral wall of the sac laterally, thus creating a negative pressure and aspirating into the sac the tears, which are forced along the canaliculus by the same orbicularis contraction. When the orbicularis relaxes, the sac collapses and drives the accumulated tears into the lacrimonasal duct.

Other investigators have relegated the lacrimal sac to a lesser role in tear outflow dynamics.[33,34] They believe that the rise in pressure within the conjunctival sac on lid closure is the sole propelling force driving tear fluid into the lacrimal sac. Reflux is prevented by the valve action at the internal punctum. Their studies revealed a higher positive pressure in the canaliculus than in the lacrimal sac on lid closure, and the pressure increased as the manometric device was moved away from the sac toward the punctum.

The same authors were able to correlate their intracanalicular pressure measurements with orbicularis muscle tone as a means of evaluating outflow function. Pressures within the system were found to decrease with age and were improved by lateral canthoplasty.[34]

The membranous lacrimonasal duct plays little or no role in the active transport of tears, but the variable folds, or valves, in the duct form a baffle that prevents air currents within the nose from being drawn up into the drainage system.

It is convenient to think of an upper segment of the excretory system consisting of the lid margins, puncta, and canaliculi, and a lower segment consisting of the lacrimal sac and lacrimonasal duct (Fig. 2-7). In a patent drainage system, dysfunction may be present because of failure of either segment. When one has eliminated hypersecretion of tears as the cause of a watering eye, attention is turned to the elimination mechanism, only a limited part of which can be evaluated by inspection and palpation. The integrity of the punctum must be noted with regard to patency and to apposition to the globe. Tonus of the orbicularis muscle can be

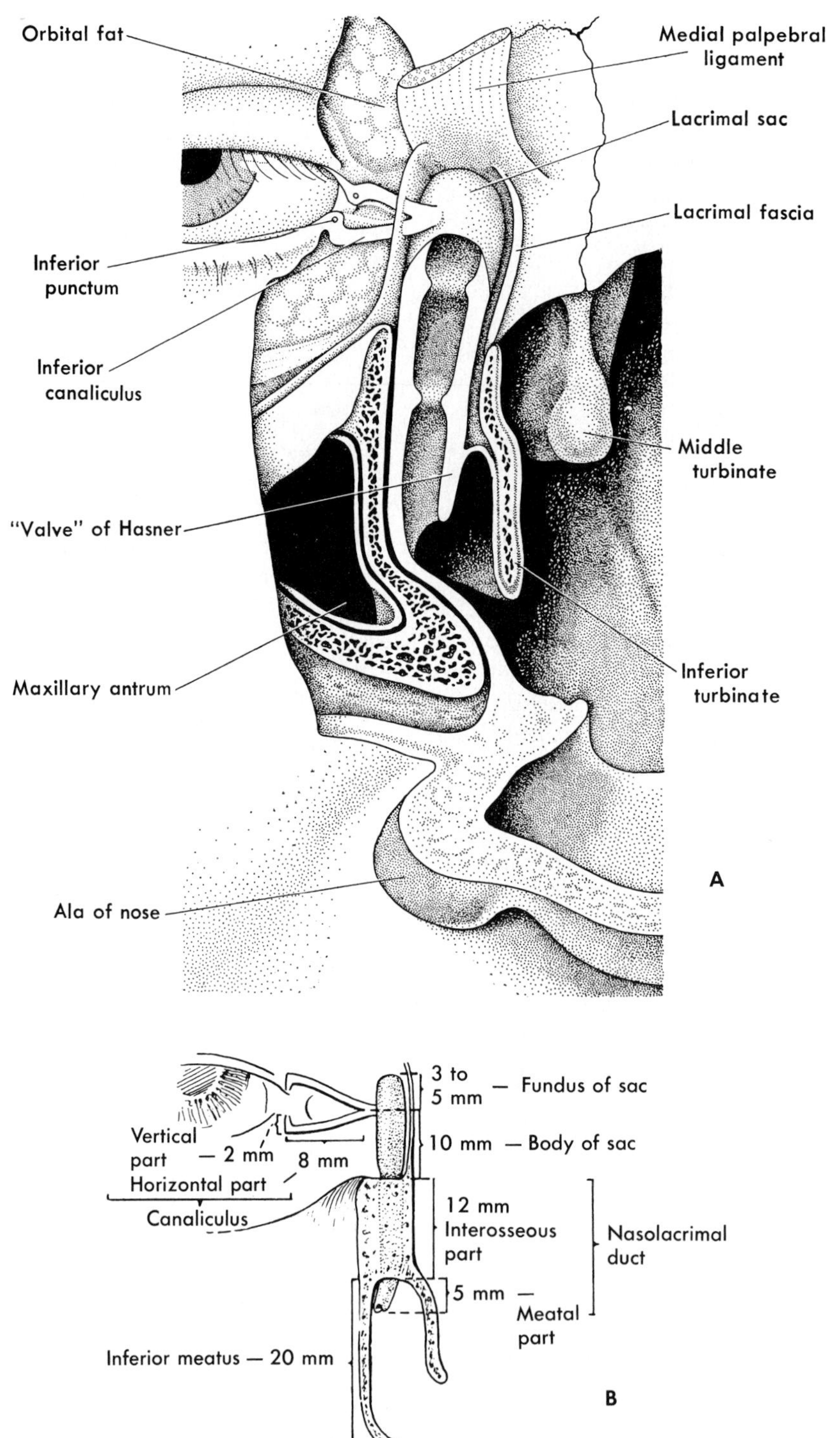

Fig. 2-7. A, Lacrimal excretory system. **B,** Dimensions of lacrimal excretory system. (**A** modified from Wolff; **B** from Jones, L. T.: Trans. Am. Acad. Ophthalmol. Otolaryngol. **66:**506, 1962.)

evaluated by pulling the lower lid down while the patient rotates the eyes upward without blinking. The lag time in regaining normal apposition of the lower lid margin to the globe is an index of the muscle tonus. If both of these and the orbicularis function are grossly normal, the clinician has reached the diagnostic limits by visual means. Two methods of further evaluating the function of this system are then available: (1) instillation of a dye into the conjunctival sac and notation of its appearance in the nose and (2) radiographic appraisal of the emptying time for this lacrimal drainage system by instillation of a contrast medium of known viscosity into the lacrimal sac.

The Jones tests

In the fluorescein dye test of Jones, 1 drop of 1% fluorescein solution is instilled into the conjunctival sac.[44] A cotton-tipped applicator moistened with 1:1000 epinephrine and 5% cocaine is inserted under the inferior turbinate after 1 minute and at subsequent intervals up to 5 minutes. Aqueous solutions will pass from the conjunctival sac to the nose in 1 minute if the elimination system is functioning normally,[46] and the dye is recovered on the applicator. However, failure to identify the dye in the nose occurs in 22% of normal patients.[86] Thus a positive test indicates normal function, but a negative test does not exclude it. Similar tests using other colored substances have been described.[80]

If the Jones I test is negative, the Jones II test is performed. Residual fluorescein is flushed from the conjunctival sac after 5 minutes, and clear saline solution is instilled into the lacrimal sac. If the irrigant enters the nose heavily stained with fluorescein, the upper segment of the system is normal; that is, orbicularis, puncta, and canaliculi are functioning sufficiently to transport the fluorescein into the lacrimal sac. If the irrigant enters the nose clear, then none of the fluorescein has entered the excretory

CLINICAL INTERPRETATION OF TESTS OF LACRIMAL EXCRETORY FUNCTION IN A PATENT SYSTEM*

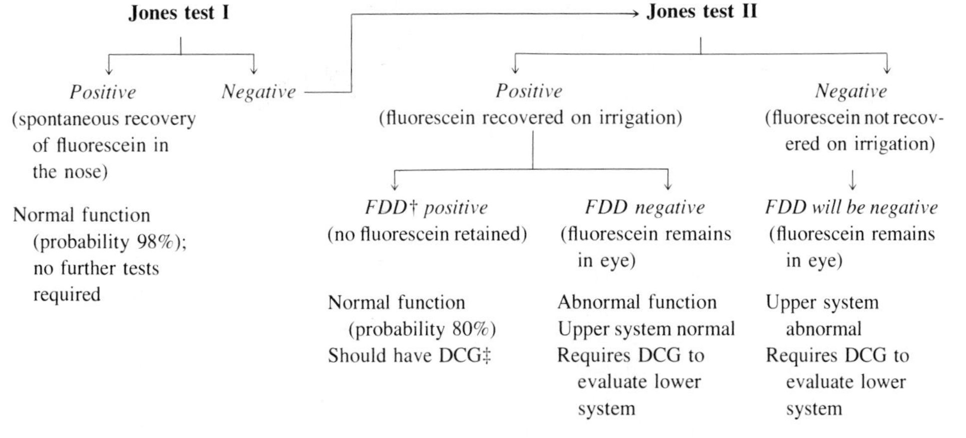

*Where the excretory system cannot be irrigated successfully, the Jones I, Jones II, and FDD will all be negative, and definitive study will rest with probing and DCG.

†Fluorescein dye disappearance. After instillation of one drop of 2% fluorescein into the conjunctival sac, the fluorescein remaining after 5 minutes is graded as 0 to 4+ retained. A residual of 2+ to 4+, indicating failure of the dye to flow into the lacrimal outflow system, is abnormal. Retention of 0 to 1+ fluorescein is normal.

‡Dacryocystography.

system and the upper system is faulty. If no fluid is recovered from the nose on irrigation, there is complete obstruction.

Fluorescein dye disappearance test

An accurate appraisal of the excretory capability can be obtained by observing the behavior of a single drop of 2% fluorescein solution instilled into the inferior conjunctival cul-de-sac.[87] The color intensity remaining after 5 minutes can be graded on a scale of 0 to 4+. In normal excretion the retained fluorescein will be 0 to 1+. Any greater residual is indicative of impaired outflow. This test does not enable one to distinguish between impairment of the upper and lower segments of the system, but may complement the Jones test and dacryocystography in the diagnosis as shown on p. 29.

Dacryocystography

Radiographic visualization of the lacrimonasal excretory system utilizes a low-viscosity contrast medium.[18,59] The test is performed with the patient seated at an x-ray head unit. After application of the usual topical anesthetic as for lacrimal probing, 1 ml of the radiopaque dye is instilled into the lacrimal sac (Fig. 2-8). Posterior-anterior exposures are made at the time of instillation and again after 30 minutes. An initial lateral view is also made for one side; for the second eye a 30-degree oblique view will prevent superimposition of the radiographic shadows of the two sides.

Various modifications of this method have been employed for radiographic visualization of the lacrimal excretory system. Aqueous contrast agents have the advantage of simulating tear outflow in a more physiologic fashion than oily media; however, difficulties are encountered in interpretation of films, since the aqueous media are diluted in varying degrees, depending on the rate of tear secretion. In macrodacryocystography, the patient is placed at a greater distance from the x-ray cassette to provide greater magnification of the area under investigation. Subtraction techniques eliminate confusing bone shadows.

With adequate knowledge of the normal outflow system, dacryocystography becomes the most effective means of studying both the pathologic anatomy and the functional state of the lacrimal sac and nasolacrimal duct.

A

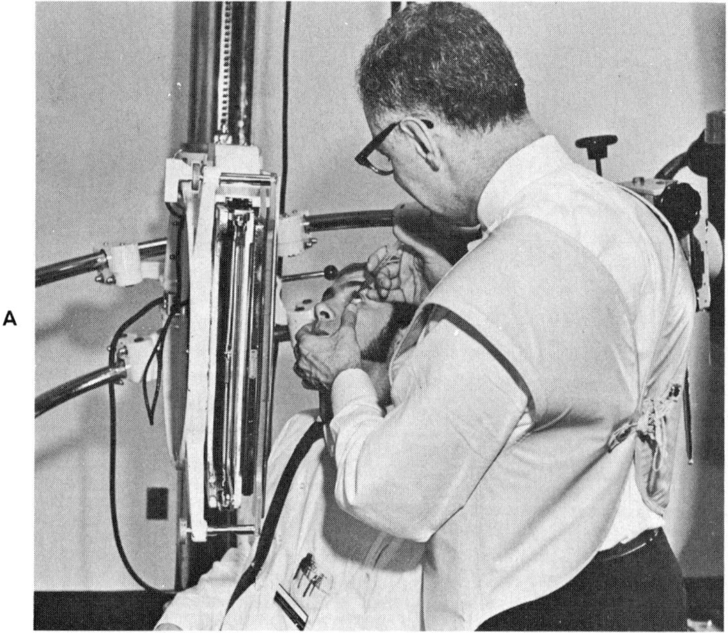

Fig. 2-8. Dacryocystography. **A,** Instilling contrast medium. **B,** Caldwell exposure. **C,** Lateral exposure.

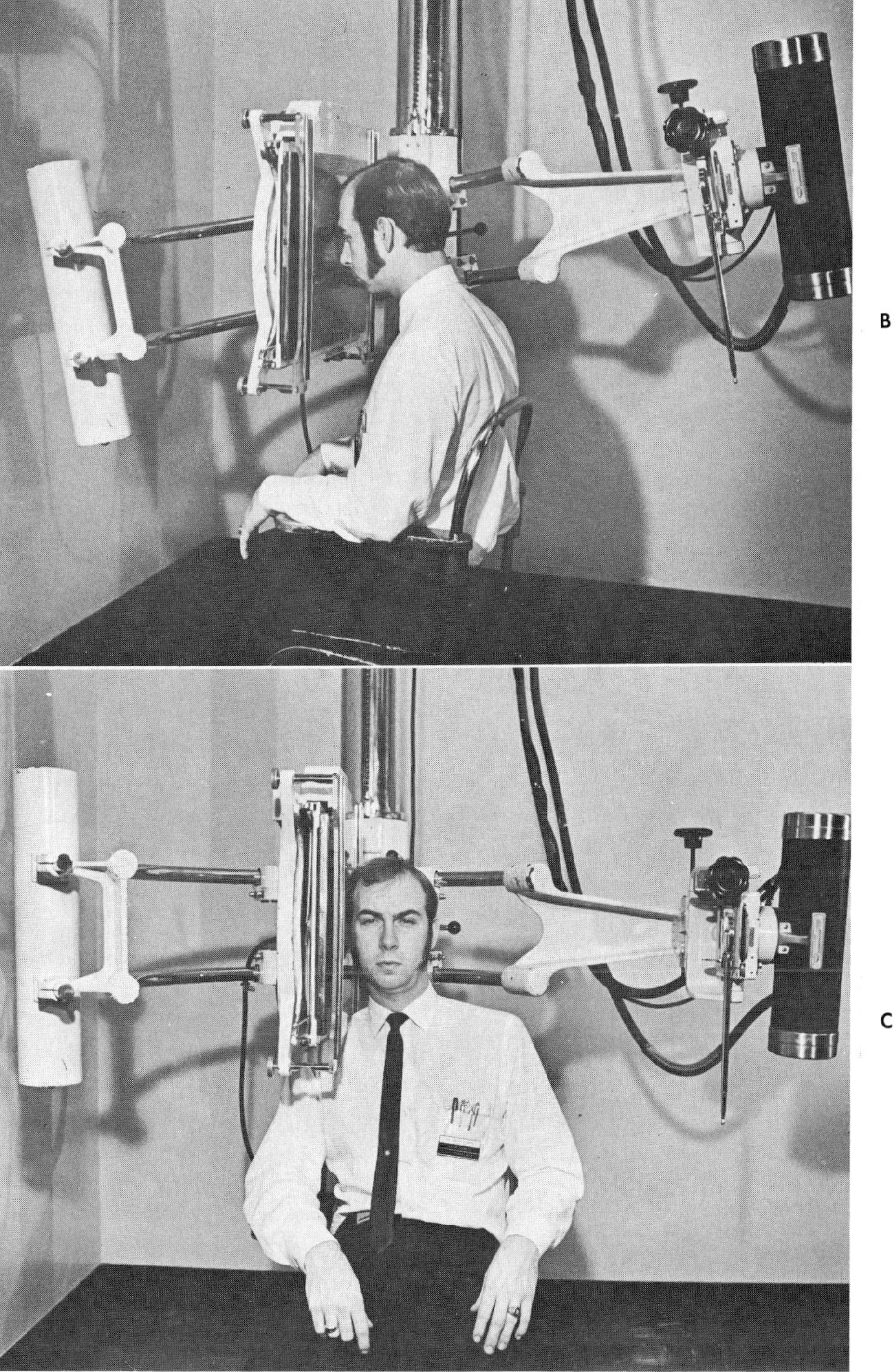

Fig. 2-8, cont'd. For legend see opposite page.

The obvious benefits of dacryocystography are visual; one can identify abnormalities in outline, size, or position of the lacrimal sac. Complete or partial obstructions, diverticula, and dacryoliths can be localized (Fig. 2-9). Most important, one can visualize and record the functional state of the lacrimal apparatus.

Functional block

When ethyl iodophenylundecylate (Pantopaque) is used, the normal lacrimal sac will be emptied of contrast medium in less than 15 minutes, and the normal lacrimonasal duct will be emptied within 30 minutes (Fig. 2-10). Therefore any substantial dye residual in the sac after 30 minutes indicates malfunction or nonfunction of the lacrimal pump, even in a patent system.[58] Such impaired function in a patent system has been termed functional block and may be indicative of failure of the lacrimal pump (as in paralysis of the orbicularis oculi) or of partial obstruction in the lacrimal sac or the duct (Fig. 2-11).

Epiphora caused by functional block is characterized by a positive Jones II test and a dacryocystogram that shows abnormal contrast medium retention after 30 minutes. Of those patients who show epiphora without demonstrable abnormality of the upper lacrimal apparatus, 25% will show functional block radiographically.

Callahan and associates,[8] from analysis of a plastic model of the lacrimal apparatus, have suggested that differentiation between obstructions in the canalicular system and those in the sac or duct can be made by use of a perfusion system, with strain-gauge measurements of resistance to flow through the system.

Numerous variations in dacryocystography techniques have been employed. These include magnification roentgenography, distention of the lacrimal sac, and the use of aqueous contrast media.[10,42,72] Various head positions have been employed.[67]

In lacrimal scintillography (dacryoscintigraphy) the lacrimal drainage system has been

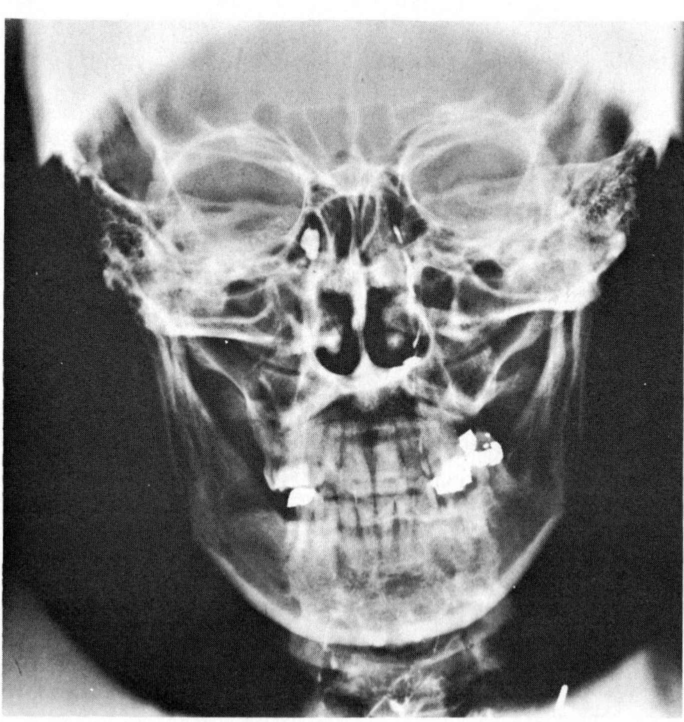

Fig. 2-9. Normal lacrimal sac and duct at right of illustration; dilated sac with obstruction at junction of sac and duct at left of illustration.

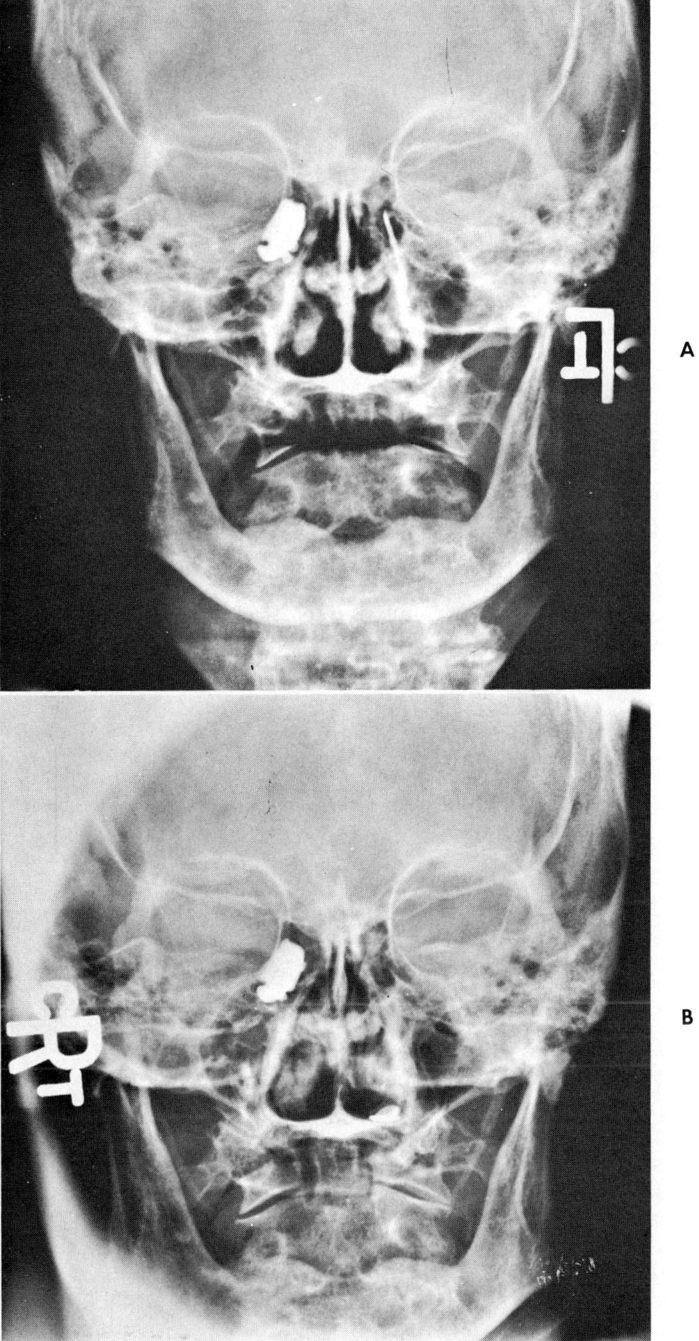

Fig. 2-10. Elimination of contrast medium (Pantopaque). Normal lacrimal system at right of illustration in **A** and **B**; at left of illustration, sac is dilated and there is a block at sac-duct junction. **A,** Immediately after instillation of dye. **B,** After 30 minutes; note dye in inferior meatus on right.

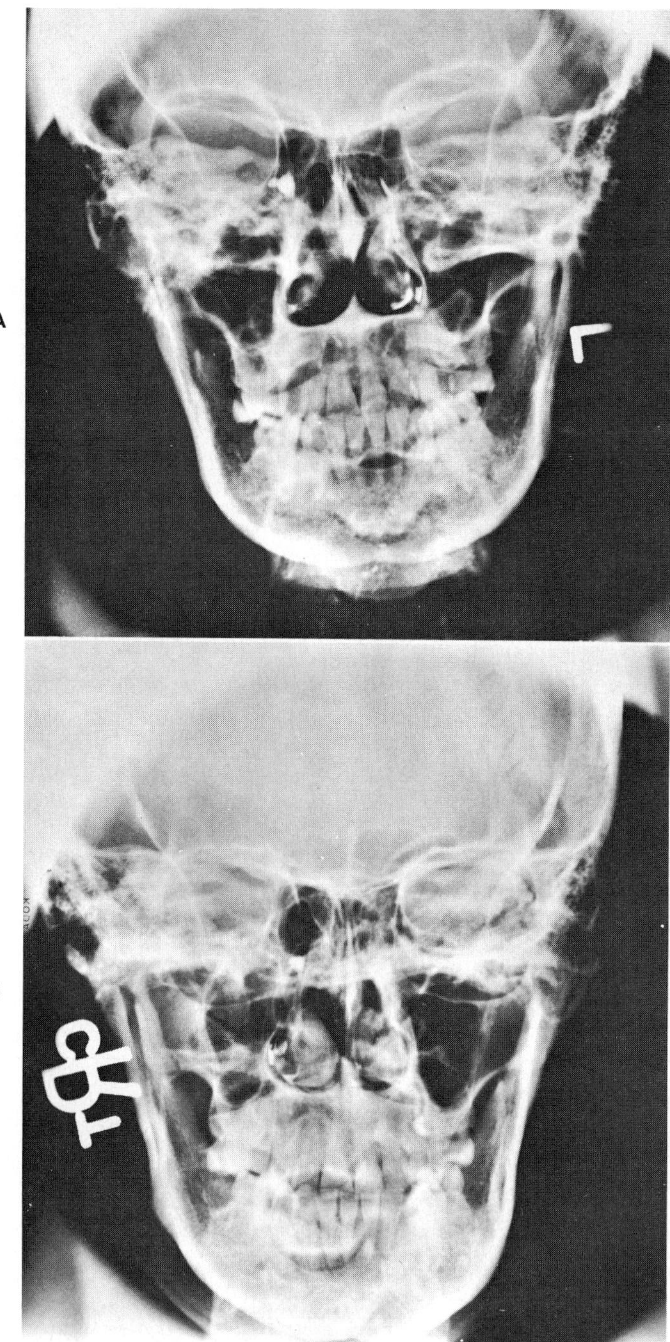

Fig. 2-11. Functional block of lacrimal excretory system at left of illustration in **A** and **B**; right side normal. **A,** Immediately after instillation of Pantopaque. Sac at left of illustration is dilated. **B,** After 30 minutes there is partial retention of contrast medium in sac at left, demonstrating delay in elimination, but contrast medium is seen below inferior turbinate, demonstrating anatomic patency. Sac and duct at right of illustration are clear.

visualized using a gamma camera to photograph the behavior of a single drop of a radioactive tracer ([^{99m}Tc] sodium pertechnetate) as it moved into and through the excretory system. This method gives promise of better understanding of excretory physiology,[70] although it is not a satisfactory device for visualizing the pathologic anatomy of the excretory system. Lacrimal scanning seems to confirm that there is a "canalicular pump" action associated with blinking. This finding is useful in explaining the mechanism of successful tear outflow after dacryocystorhinostomy.[39,40,41]

Fundamentally, the usefulness of dacryocystography depends on a knowledge of the radiographic appearance of the normal system for the specific contrast agent used.[59]

The Jones dye tests and dacryocystography provide complementary information, and by combining these with the fluorescein dye disappearance test, one can arrive at an explanation for epiphora in more than 95% of cases. The use of these functional methods of evaluating a patient with epiphora in a patent system is summarized on p. 29.

It is evident that the appropriate treatment for epiphora must be based on proper assessment of the primary etiologic mechanism involved. With such assessment, patients who demonstrate patency of the drainage system by irrigation, but functional block by dacryocystography, are relieved of their symptoms by dacryocystorhinostomy,[21] with the same success that is achieved in those patients who exhibit a complete block in the lacrimal sac or duct.

REFERENCES

1. Allansmith, M.: Immunology of the tears. In Holly, F. J., and Lemp, M. A., editors: The preocular tear film and dry eye syndromes, Int. Ophthalmol. Clin. **13:**47, 1973.
2. Altman, P. L.: Blood and other body fluids, Washington, D.C., 1961, Federation of American Society for Experimental Biology, pp. 488-490.
3. Apt, L., and Cullen, B. F.: Newborns do secrete tears, J.A.M.A. **189:**951, 1964.
4. Balik, J.: Ueber die Ausscheidung des Harnstoffes in die Tränen, Acta Ophthalmol. **37:**103, 1959.
5. Benedetto, D. A., Shah, D. O., and Kaufman, H. E.: The instilled fluid dynamics and surface chemistry of polymers in the preocular tear film, Invest. Ophthalmol. **14:**887, 1975.
6. Brown, S. I., and Dervichian, D. G.: The oils of the meibomian glands, Arch. Ophthalmol. **82:**537, 1969.
7. Brown, S. I., and Dervichian, D. G.: Hydrodynamics of blinking, Arch. Ophthalmol. **82:**541, 1969.
8. Callahan, W. P., Forbath, P., and Besser, W. D. S.: A method of determining the patency of the nasolacrimal apparatus, Am. J. Ophthalmol. **60:**475, 1965.
9. Callender, M., and Morrison, P. E.: A quantitative study of human tear proteins before and after adaptation to non-flexible contact lenses, Am. J. Optom. Physiol. Opt. **51:**939, 1974.
10. Campbell, W.: The radiology of the lacrimal system, Br. J. Radiol. **37:**1, 1964.
11. Carmody, P. J., Rattazzi, M. C., and Davidson, R. G.: Tay Sachs disease: the use of tears for the detection of heterozygotes, N. Engl. J. Med. **289:**1072, 1973.
12. Carney, L. G., and Hill, R. M.: Human tear pH, Arch. Ophthalmol. **94:**821, 1976.
13. Carney, L. G., and Hill, R. M.: Human tear buffering capacity, Arch. Ophthalmol. **97:**951, 1979.
14. Cotlier, E.: Tears for diagnosis of Tay-Sachs' and other genetic diseases. The Fourth National Science Writers Seminar in Ophthalmology, New York, 1973, Research to Prevent Blindness, Inc., p. 16.
15. Cutinelli, C., and Bonavolonta, A.: Lysozyme in suppurative diseases of the eye, G. Ital. Oftalmol. **2:**19, 1949.
16. Daubs, J. G.: A comparison of enzyme reagent strip tests for lacrimal glucose, Am. J. Optom. Physiol. Opt. **53:**232, 1976.
17. de Haas, E.: Lacrimal gland response to parasympathicomimetics after parasympathetic denervation, Arch. Ophthalmol. **64:**34, 1960.
18. Demorest, B. H., and Milder, B.: Dacryocystography. II, The pathologic lacrimal apparatus, Arch. Ophthalmol. **54:**410, 1955.
19. de Roetth, A., Sr.: Lacrimation in normal eyes, Arch. Ophthalmol. **49:**185, 1953.
20. Deutsch, A. K.: Congenital alacrimia, Am. J. Opthalmol. **59:**325, 1965.
21. Duke-Elder, S.: Text-book of ophthalmology, vol. 5, St. Louis, 1952, The C. V. Mosby Co.
22. Editorial: The immunoglobulins, N. Engl. J. Med. **272:**376, 1965.
23. Ford, L. C., DeLange, R. J., and Petty, R. W.: Identification of a nonlysozymal bactericidal factor (beta lysin) in human tears and aqueous humor, Am. J. Ophthalmol. **81:**30, 1976.
24. Friedland, B. R., Anderson, D. R., and Forster, R. K.: Non-lysozyme antibacterial factor in human tears, Am. J. Ophthalmol. **74:**52, 1972.
25. Gasset, A. R., et al.: Tear glucose detection in hyperglycemia, Am. J. Ophthalmol. **65:**414, 1968.
26. Gaynor, I. E.: Lacrimal insufficiency, keratoconjunctivitis sicca and malfunction of the inferior turbinate, Am. J. Ophthalmol. **53:**614, 1962.
27. Giardini, A., and Roberts, J. R. E.: Concentration of glucose and total chloride in tears, Br. J. Ophthalmol. **34:**737, 1950.
28. Gilbard, J. P., Farris, R. L., and Santamaria, J., II: Osmolarity of tear microvolumes in keratoconjunctivitis sicca, Arch. Ophthalmol. **96:**677, 1978.

29. Golding-Wood, P. H.: Crocodile tears, Br. Med. J. **1:**1518, 1963.
30. Gottesfeld, B. H., and Leavitt, F. H.: "Crocodile tears" treated by injection into the spheno-palatine ganglion, Arch. Neurol. **47:**314, 1942.
31. Hanson, J., Fikentscher, R., and Roseburg, B.: Schirmer test of lacrimal secretions, its clinical importance, Arch. Otolaryngol. **101:**293, 1975.
32. Henderson, J. W., and Prough, W. A.: Influence of age and sex on flow of tears, Arch. Ophthalmol. **43:** 224, 1950.
33. Hill, J. C., Bethell, W., and Smirinaul, H. J.: Lacrimal drainage—a dynamic evaluation. Part I, Mechanics of tear transport, Can. J. Ophthalmol. **9:**411, 1974.
34. Hill, J. C., Bethell, W., and Smirinaul, H. J.: Lacrimal drainage—a dynamic evaluation. Part II, Clinical aspects, Can. J. Ophthalmol. **9:**417, 1974.
35. Hill, R. M., and Carney, L. G.: Tear pH, how predictable? J. Am. Optom. Soc. **49:**269, 1970.
36. Hill, R. M., and Terry, J. E.: Human tear cholesterol levels, Arch. Ophtalmol. (Paris) **36:**155, 1976.
37. Holly, F. J., and Lemp, M. A.: Tear physiology and dry eyes, Surv. Ophthalmol. **22:**69, 1977.
38. Horwitz, B. L., Christensen, G. R., and Ritzmann, S. R.: Diurnal profiles of tear lysozyme and gamma A globulin, Ann. Ophthalmol. **10:**75, 1978.
39. Hurwitz, J. J., and Welham, R. A. N.: Radiography in functional lacrimal testing, Br. J. Ophthalmol. **59:**323, 1975.
40. Hurwitz, J. J., Maisey, M. N., and Welham, R. A. N.: Quantitative lacrimal scintillography. I, Method and physiological application, Br. J. Ophthalmol. **59:** 308, 1975.
41. Hurwitz, J. J., Maisey, M. N., and Welham, R. A. N.: Quantitative lacrimal scintillography. II, Lacrimal pathology, Br. J. Ophthalmol. **59:**313, 1975.
42. Iba, G. B., and Hanafee, W. N.: Distention dacryocystography, Radiology **90:**1020, 1968.
43. Jacklin, H. N.: The gusto-lacrimal reflex, Am. J. Ophthalmol. **61:**1521, 1966.
44. Jones. L. T.: The cure of epiphora due to canalicular disorders, trauma and surgical failures on the lacrimal passages, Trans. Am. Acad. Ophthalmol. Otolaryngol. **66:**506, 1962.
45. Jones, L. T.: The lacrimal secretory system and its treatment, Am. J. Ophthalmol. **62:**47, 1966.
46. Jones, L. T., and Marquis, M. M.: Lacrimal function, Am. J. Ophthalmol. **73:**658, 1972.
47. Kirchner, C.: Untersuchungen über das Ausmass der Tränensekretion beim Menschen, Klin. Monatsbl. Augenheilkd. **144:**412, 1964.
48. Krogh, A., Lund, A. C., and Pedersen-Bjergaard, K.: The osmotic concentration of human lacrymal fluid, Acta Physiol. Scand. **10:**88, 1945.
49. Lamberts, D. W., Foster, C. S., and Perry, H. D.: Schirmer test after topical anesthesia and the tear meniscus height in normal eyes, Arch. Ophthalmol. **97:**1082, 1979.
50. Lemp, M. A., and Hamill, J. R.: Factors affecting the tear film breakup in normal eyes, Arch. Ophthalmol. **89:**103, 1973.
51. Lemp, M. A., and Szymanski, E. S.: Polymer adsorption at the ocular surface, Arch. Ophthalmol. **93:** 134, 1975.
52. Lemp, M. A., et al.: The precorneal tear film: factors in spreading and maintaining a continuous tear film over the corneal surface. Arch. Ophthalmol. **83:**89, 1970.
53. Lemp, M. A., et al.: Dry eye secondary to mucus deficiency, Trans. Am. Acad. Ophthalmol. Otolaryngol. **75:**1223, 1971.
54. Mackie, I. A., and Seal, D. V.: Beta-blockers, eye complaints, and tear secretion (letter), Lancet **2:**1027, 1977.
55. Maes, J.: The effect of removal of superior cervical ganglion on lachrymal secretion, Am. J. Physiol. **123:**359, 1938.
56. Mastman, G. J., Baldes, E. J., and Henderson, J. W.: The total osmotic pressure of tears in normal and various pathologic conditions, Arch. Ophthalmol. **65:**509, 1961.
57. McClellan, B. H., et al.: Immunoglobulins in tears, Am. J. Ophthalmol. **76:**89, 1973.
58. Milder, B.: Functional block in the lacrimal drainage system. Eighteenth Councilium Ophthalmologicum, Brussels, 1958, Acta Medica Belgica, p. 1111.
59. Milder, B., and Demorest, B. H.: Dacryocystography. I, The normal lacrimal apparatus, Arch. Ophthalmol. **51:**180, 1954.
60. Minton, L. R.: Para-limbal ring keratitis and absence of lysozyme in lupus erythematosus, Am. J. Ophthalmol. **60:**532, 1965.
61. Mishima, S.: Some physiological aspects of the precorneal tear film, Arch. Ophthalmol. **73:**233, 1965.
62. Modell, W.: Pharmacologic action of some ophthalmic drugs, Arch. Ophthalmol. **37:**160, 1947.
63. Mutch, J. R.: The lacrimation reflex, Br. J. Ophthalmol. **28:**318, 1944.
64. Patrick, R. K.: Lacrimal secretion in full-term and premature babies, Trans. Ophthalmol. Soc. U.K. **94:**283, 1974. (Cited in Am. J. Ophthalmol. **79:**713, 1975.)
65. Pietsch, R. L., and Pearlman, M. E.: Human tear lysozyme variables, Arch. Ophthalmol. **90:**94, 1973.
66. Pistocchi, P., and Apponi, G.: The variation in activity of lacrimal lysozyme in subjects given the water provocative test, Arch. Otolaryngol. **66:**519, 1962.
67. Radnót, M., and Gáll, J.: Die Rontgendiagnostik der Tränenableitenden Wege, Budapest, 1966, Akadémiai Kiadó.
68. Rengstorff, R. H.: The precorneal tear film: breakup time and location in normal subjects, Am. J. Optom. Physiol. Opt. **51:**765, 1974.
69. Ridley, F., and Sorsby, A., editors: Modern trends in ophthalmology, New York, 1940, Paul B. Hoeber, Inc.
70. Rossomondo, R. M., et al.: A new method of evaluating lacrimal drainage, Arch. Ophthalmol. **88:**523, 1972.

71. Rubin, M. R.: Egg white in the treatment of keratoconjunctivitis sicca, Eye Ear Nose Throat Mon. **34:**50, 1955.
72. Sargent, E. N., and Ebersole, C.: Dacryocystography. The use of sinografin for visualization of nasolacrimal passages, Am. J. Roentgenol. Radium Ther. Nucl. Med. **102:**831, 1968.
73. Schaeffer, A. J.: Osmotic pressure of the extraocular and intraocular fluids, Arch. Ophthalmol. **43:**1026, 1950.
74. Sen, D. K., et al.: Biological variation of immunoglobulin concentrations in normal human tear related to age and sex, Acta Ophthalmol. **56:**439, 1978.
75. Sjögren, H.: Congenital alacrimia and keratoconjunctivitis sicca in children, N. Z. Med. J. **51:**32, 1952.
76. Smith, W. H., Howsman, K. G., and Billings, J. J.: Epiphora, a symptom of early hypothyroidism, Br. J. Ophthalmol. **43:**622, 1959.
77. Szmyt, J.: The lacrimation tests, Klin. Oczna **28:**195, 1958.
78. Thaysen, J. H., and Thorn, N. A.: Excretion of urea, sodium, potassium and chloride in human tears, Am. J. Physiol. **178:**160, 1954.
79. Thompson, R., and Gallardo, E.: The antibacterial action of tears on staphylococci, Am. J. Ophthalmol. **24:**635, 1941.
80. Toth, Z.: Functional test for the pre-lacrimal sac segment of the lacrimal apparatus, Klin. Monatsbl. Augenheilkd. **113:**158, 1948.
81. Uotila, M. H., Soble, R. E., and Savory, J.: Measurement of tear calcium levels, Invest. Ophthalmol. **11:** 258, 1972.
82. van Haeringen, N. J., and Glasius, E.: Enzymatic studies in lacrimal secretion, Exp. Eye Res. **19:**135, 1974.
83. Vanley, G. T., Leopold, I. H., and Gregg, T. H.: Interpretation of tear film breakup, Arch. Ophthalmol. **95:**445, 1977.
84. Werb, A.: Unusual causes of epiphora, Br. J. Ophthalmol. **55:**559, 1971.
85. Whitwell, J.: Denervation of the lacrimal gland, Br. J. Ophthalmol. **42:**518, 1958.
86. Zappia, R., and Milder, B.: Lacrimal drainage function. I, The Jones fluorescein test, Am. J. Ophthalmol. **74:**154, 1972.
87. Zappia, R. J., and Milder, B.: Lacrimal drainage function. II, The fluorescein dye disappearance test, Am. J. Ophthalmol. **74:**160, 1972.
88. Zintz, R., and Schilling, T.: Ein Kolorimetrisches Verfahren Zur Messung des Flüssigkeitsvolumens in Bindeshautsack, Klin. Monatsbl. Augenheilkd. **144:** 393, 1964.

Chapter 3

The cornea

STEPHEN R. WALTMAN, M.D.

ANATOMIC AND PHYSICAL PROPERTIES OF THE CORNEA

The cornea is the first and most powerful refracting surface of the optical system of the eye. Production of a sharp image at the retinal receptors requires that the cornea be transparent and of appropriate refractive power.

The refractive power of the cornea depends on its curvature and the difference in refractive indices between it and the air. The radius of curvature of the anterior surface of the average cornea in the central region is 7.8 mm. A rather wide range of variation, from 7.0 to 8.5 mm, is compatible with good visual function, and in pathologic conditions the range is broadened still further. Short radii, resulting in high refractive power and myopia, may be the first indications of keratoconus. The curvature of the central cornea is often not the same in all meridians, the shape of the surface then resembling the side of an egg. The graduated focusing power of the cornea in such cases produces corneal astigmatism. (Stigma means point; astigmatism is a condition in which a point-object is not imaged as a point.)

The refractive index of the cornea is 1.376, which gives the average anterior central region a refractive power of 48.8 D.

The concave posterior surface of the cornea faces the aqueous, which has a lower refractive index (1.336), so the refractive power of this surface is −5.8 D, giving the entire cornea a refractive power of 43.0 D, or about 70% of the total refractive power of the eye.

The anterior surface of the cornea is steepest centrally and somewhat flattened peripherally, giving the cornea a hyperboloidal shape. The central area of 3 to 4 mm of the cornea is nearly spherical.

The corneal curvature changes somewhat with aging. It is more spherical in infancy and changes to astigmatism with the rule* during childhood and adolescence; it again becomes more spherical in middle age and astigmatic against the rule in senility.[53] The thickness of the cornea also increases somewhat with age. The average central thickness of a normal cornea is 0.56 mm in persons under 25 years of age; this thickness slowly increases with age to become 0.57 mm in persons over 65 years of age. The cornea is somewhat thicker in the periphery. At 20 degrees from the center, where a 6 mm trephine would cut for a corneal transplant, the cornea is almost 0.7 mm thick. The thickness of the cornea is largely determined by the degree of corneal hydration, which will be discussed more fully later. In diseases of the anterior segment, however, there may be a slight increase in corneal thickness without any clinical evidence of corneal edema or corneal scarring.[72]

The thickness of the cornea is greatest after the eyes have been closed for some time, as after sleeping. When the eyes are opened and exposed to the dehydrating effect of the air, the thickness decreases slightly. If the eye is kept open for several minutes, dry spots appear on

□ This work was supported in part by grant EY-02198 from the National Eye Institute, National Institutes of Health.

*The rule is that the cornea has a shorter radius of curvature, and in consequence is more highly refracting, in the vertical meridian than in the horizontal. The rule holds in about 90% of cases.

the epithelium. The temperature of the cornea is relatively low. In the rabbit, for example, the difference between the temperature of the cornea and that of the iris amounts to 5° C.

ANATOMY OF THE CORNEA

Until recently the cornea was thought to be composed of only five layers: epithelium, Bowman's layer or membrane, substantia propria or stroma, Descemet's membrane, and endothelium. It is now recognized that a thin basement membrane lies beneath the epithelium. Bowman's membrane, distinct in light microscopy, loses its identity in the electron micrograph.

Epithelium

The epithelium consists of five or six layers of cells. The most superficial cells are flat overlapping squamous cells, similar to the most superficial epithelial cells of the skin. Unlike these cells, however, normal corneal epithelium is not keratinized. The middle layers consist of cells that become more columnar as the deeper layers are approached. The innermost (basal) layer is made up of columnar cells packed closely together. All the cells are held together by a cement substance. Also, the surfaces of the cells form processes that are fitted into corresponding indentations of adjacent cells and connected in places by attachment bodies (desmosomes).[39] The basal cells are connected to the basement membrane by hemidesmosomes.

Between the columnar epithelial cells and Bowman's membrane is a basement membrane from 60 to 65 nm thick. The basement membrane has been examined histochemically and has been found to be similar to other basement membranes in various tissues.[51]

The epithelial cells form a layer of uniform thickness and great regularity. The surface epithelial cells are uniform and smooth, and when covered with tears, these cells act as a convex mirror that can be observed by flashlight or slit lamp illumination. By electron microscopy the outer cell membranes of epithelial cells show fingerlike projections known as microvilli.[52] These microvilli projecting into the tear film may trap tear fluid and thus prevent drying of the epithelial cells.

Bowman's layer or membrane

Bowman's layer is a sheet of transparent tissue about 12 μm thick, without structure as seen by light microscopy. Under electron microscopy it appears to be made up of uniform fibrils, probably of collagenous material, running parallel to the surface. Bowman's layer is acellular; it is a modified superficial stromal layer found only in primates. Its absence in lower animals is associated with a plasticity of the corneal stroma, so calibration of eye pressure–measuring instruments (particularly applanation tonometers) on these eyes is different from that on primate eyes.

Substantia propria, or stroma

The substantia propria is composed of layers of lamellae, each of which runs the full length of the cornea; although the bundles interlace with one another, they are nearly parallel to the surface. The lamellae are only loosely adherent to each other; the layers of an excised cornea gripped between thumb and forefinger may be slid over each other for some distance without meeting appreciable resistance. The layered structure of the stroma makes corneal splitting, as in superficial keratectomy, technically easy. The cell bodies, called corneal corpuscles or keratocytes, are flattened, so they too lie parallel to the surface, and their cell processes interlace with one another. This arrangement of the fibers gives optical uniformity to the cornea. The stroma comprises about 90% of the whole cornea.

The lamellae are made up of bundles of collagen fibrils (64 nm banding)* separated from each other by a ground substance.

Descemet's membrane

Descemet's membrane, which separates the deep stroma from the endothelium, is about 10 μm thick. It is considered to be the product of secretion of the endothelial cells.

Endothelium

The endothelium is a single layer of cells lining Descemet's membrane. Its inner surface

*The exact spacing depends on fixation and varies from 64 to 66 nm.

Text continued on p. 44.

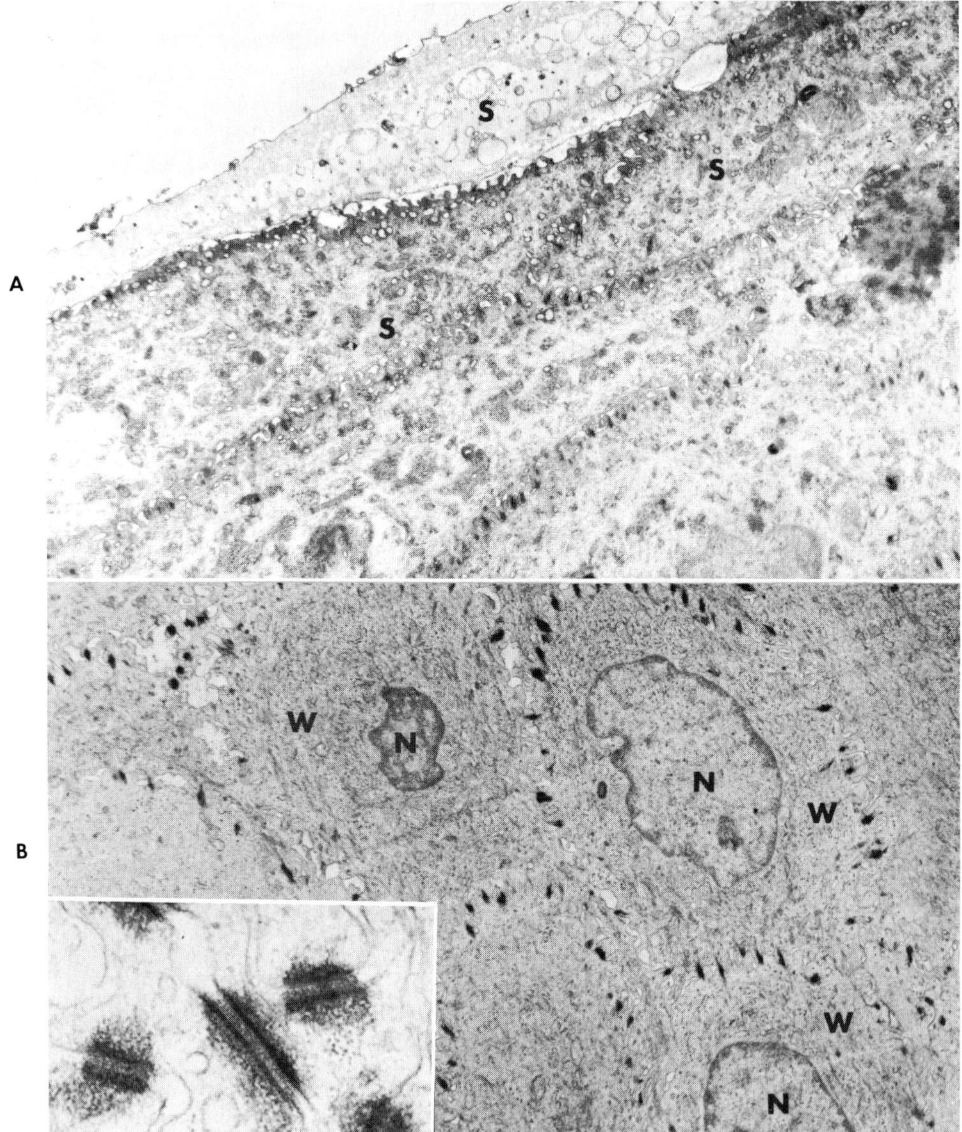

Fig. 3-1. A, Corneal epithelium, outer surface, shows surface cells, *S,* without nuclei. (Magnification ×3350.) **B,** Next layer deeper, wing cells, *W,* shows much rounder appearance of cells in which nuclei, *N,* which can easily be seen. Note marked interdigitation of cell borders. (Magnification ×3840.) Inset shows mutilayered demosomes joining cell borders. (Magnification ×44,400.) (Courtesy Dr. Jack Kayes.)

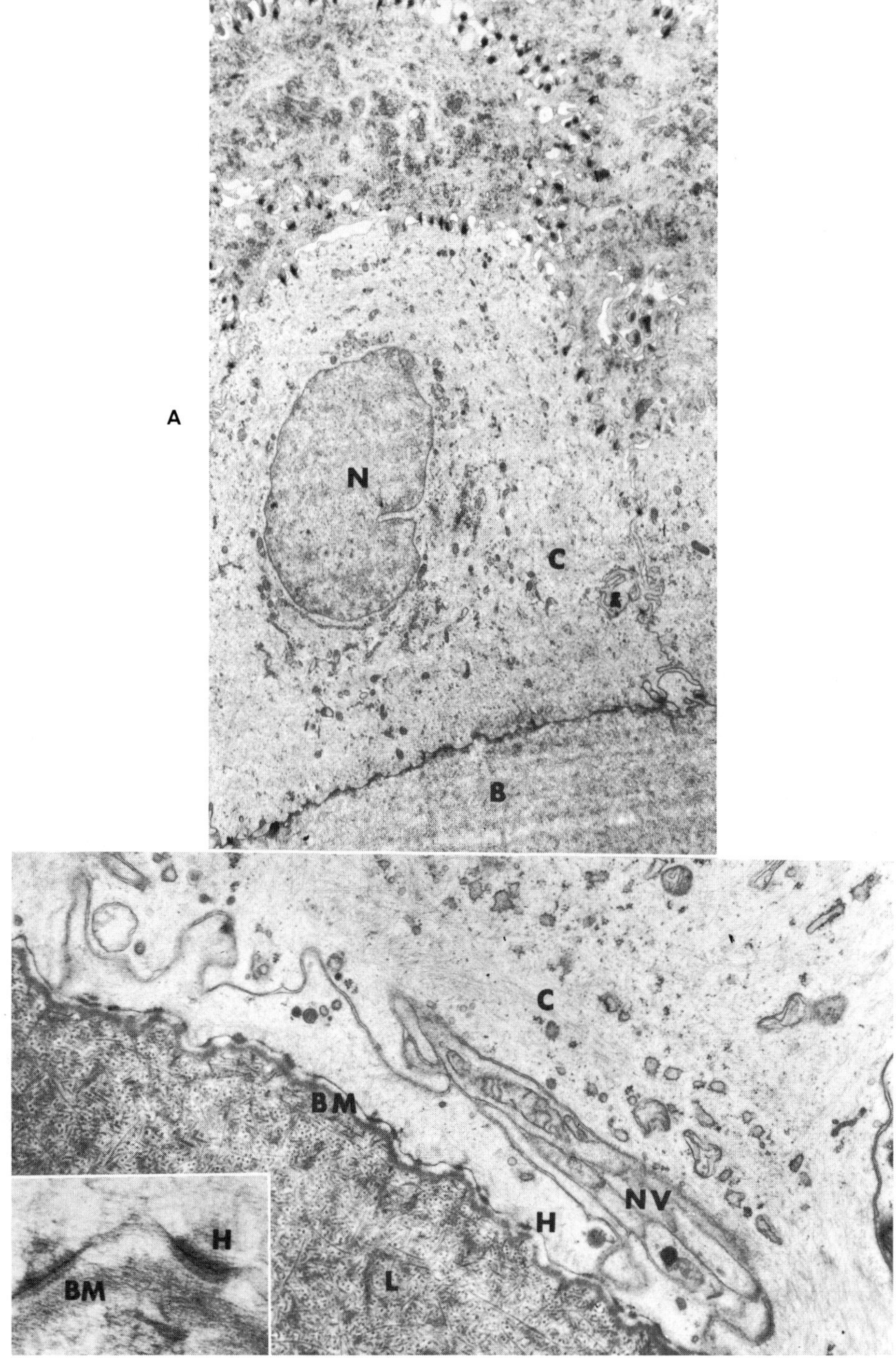

Fig. 3-2. A, Basal cell of corneal epithelium, *C,* shows nucleus, *N,* and Bowman's layer, *B.* (Magnification ×5950.) **B,** Higher magnification of basal cell, *C,* shows nerve, *NV,* invaginating into cell. Basement membrane, *BM,* is now visible, as well as hemidesmosomes, *H.* Fine filaments within cell should be contrasted with thicker fibers forming feltwork of Bowman's layer, *L.* Inset shows more detail of hemidesmosomes and basement membrane. (Magnification ×54,000.) (Courtesy Dr. Jack Kayes.)

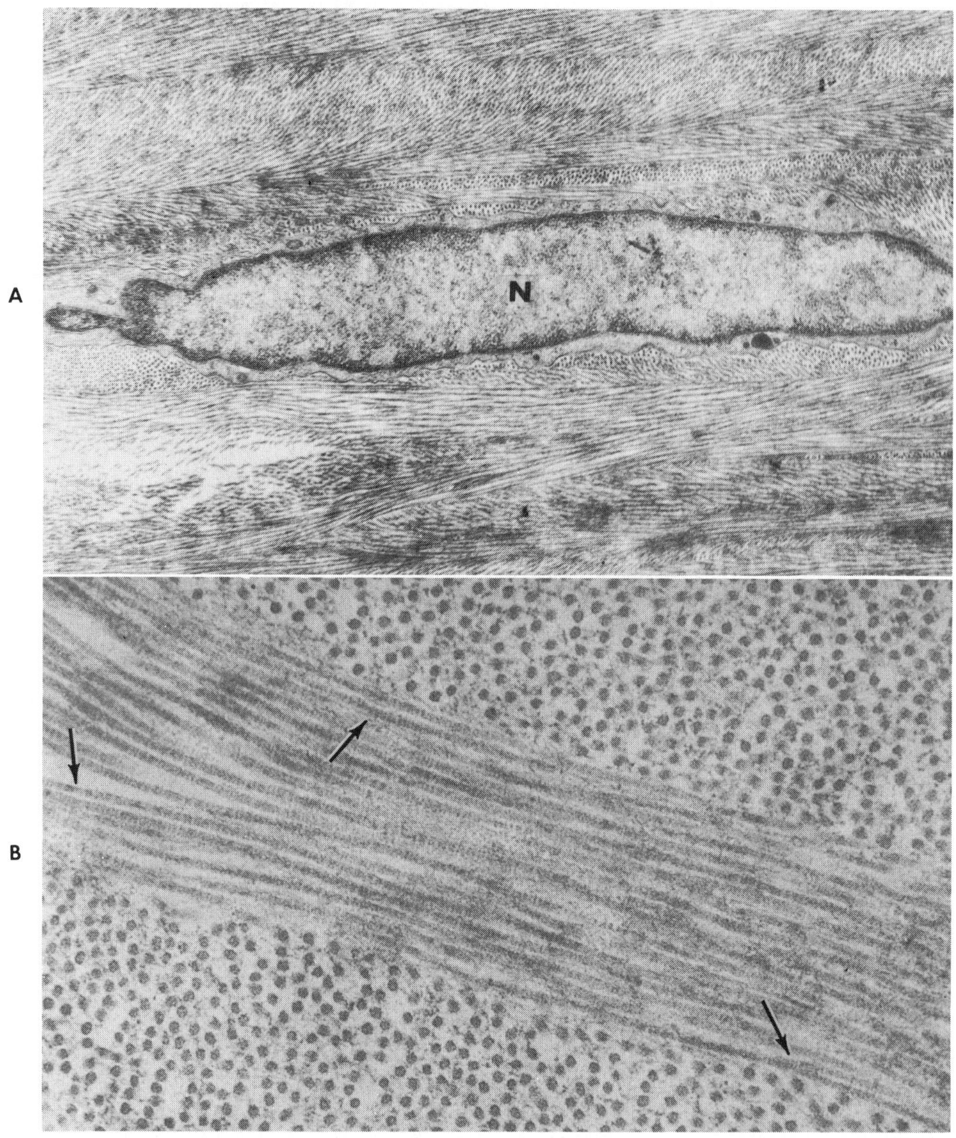

Fig. 3-3. A, Nucleus, *N*, of corneal stromal cell; layers of corneal stromal collagen are seen from various angles. (Magnification ×9000.) **B,** Higher magnification of stromal collagen, showing collagen fibrils cut on end and from side with their 640 Å banding visible (arrows). (Magnification ×72,000.) (Courtesy Dr. Jack Kayes.)

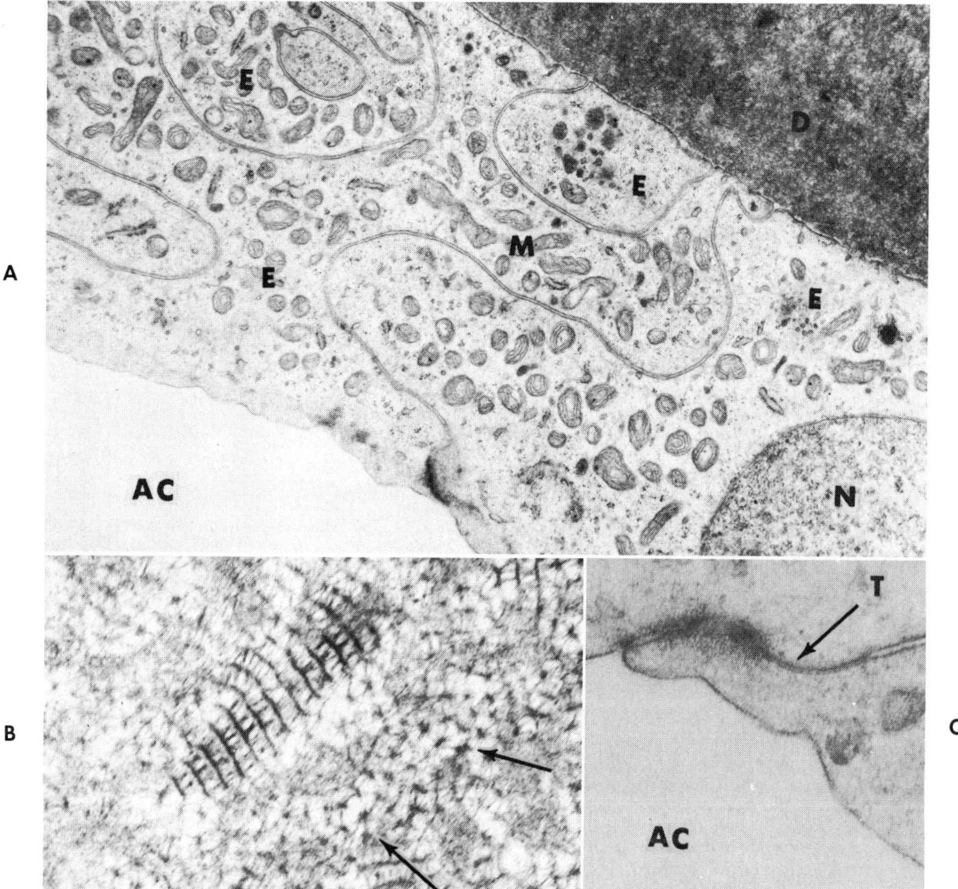

Fig. 3-4. A, Corneal endothelial cells, *E,* with tortuous borders, nucleus, *N,* mitochondria *M,* Descemet's membrane, *D,* and anterior chamber, *AC.* (Magnification ×14,000.) **B,** Descemet's membrane, tangentially cut, shows hexagonal pattern of fibers (arrows). (Magnification ×34,000.) **C,** Tight junction, *T,* of endothelial cells in detail, anterior chamber, *AC.* (Magnification ×30,000.) (Courtesy Dr. Jack Kayes.)

is bathed by the aqueous humor. The endothelium can be viewed and photographed in vivo with the clinical specular microscope. In humans this cell layer has limited, if any, reproductive capacity. Aging causes cell loss, and the remaining cells enlarge and spread so that Descemet's membrane remains completely covered.[8] Therefore endothelial cell density, expressed as cells per unit area, decreases with age. Similarly, cell loss from trauma, inflammation, or surgery is compensated for by increased cell size and decreased cell density.

In the rabbit cornea, terminal bars are found in the endothelium.[42,43] In the frog cornea these terminal bars are missing, and in addition there is a complex interweaving of the lateral margins of the endothelial cells.[41] In these respects the human cornea more nearly resembles that of the frog.[39] The significance of these fine details of structure of the endothelium will be considered further on pp. 48 to 49. The fine structure of the cornea is illustrated in Figs. 3-1 to 3-4.

NERVES OF THE CORNEA

The cornea is richly supplied with sensory nerves. These are derived from the ciliary nerves, which are the end-branches of the ophthalmic division of the fifth cranial nerve. Nerves enter the cornea in the middle and anterior stromal layers and run forward in a radial fashion toward the center of the cornea. The nerve filaments generally do not lose their myelin sheaths until they have traversed a millimeter or so of the cornea; thus in the periphery of the cornea they can often be seen as fairly thick fibrils. From this point on they are more difficult to follow without high magnification. As they run into the cornea, the nerves divide dichotomously. Emerging from the deeper parts of the cornea, nerve fibrils perforate Bowman's membrane to form a plexus just underneath the epithelium.

In their further course, the free nerve endings run between the epithelial cells. The cornea is one of the most sensitive tissues of the body, and this sensitivity serves to protect it. It is exquisitely painful when the nerve endings are exposed in corneal ulcers or abrasions. The corneal endothelium of rabbits has been found to contain nerve fibers ending within endothelial cells, but this has not been confirmed in human beings.[77]

THE CORNEA IN DISEASE

The reactions of the cornea are quite important in disease processes. It is avascular and therefore reacts differently from those tissues that have a blood supply. Bowman's layer possesses little resistance to any pathologic process; therefore it is easily destroyed and never regenerates. Descemet's membrane, on the other hand, is highly resistant and elastic and may remain in the form of a bulging balloon-like structure, called a *descemetocele,* after all the other layers of the cornea are destroyed. Even after it has been destroyed, a new membrane can be secreted by the endothelium. Ruptures are rare in Bowman's layer, but they are common in Descemet's membrane.

Many diseases of the cornea result in its vascularization, and the type of vessels and their location in the cornea are often clues to the nature of the disease process. The diseases that commonly affect epithelial structures are accompanied by superficial vascularizations. Those diseases affecting parenchymatous tissues, on the other hand, are characterized by deep vessels. Once the cornea has become vascularized, the vessels remain throughout life. They may become empty of blood ("ghost vessels"), but their presence can always be detected with suitable magnification and illumination and is testimony to the fact that the cornea has been the seat of a previous inflammation. Corneal sensation may be decreased in diabetes mellitus or after repeated infections with herpes simplex virus.

BIOCHEMISTRY OF THE CORNEA

The chemical composition of the cornea is the sum of its parts—namely, the stroma, the epithelium, endothelium, and Descement's membrane. The number of cells and morphology varies among these areas, and this is reflected in chemical dissimilarities; chemically, the cornea is a heterogeneous tissue.

Corneal stroma

The corneal stroma forms the bulk of the cornea. It is a differentiated connective tissue

Table 3-1. Chemical composition of glycosaminoglycans (GAG)

GAG	Molecular weight	Sulfate	Carbohydrate	Aminosugar	Uronic acid
Keratan sulfate	10,000 to 19,999	1	Galactose	Glucosamine	
Chondroitin	40,000	0		Galactosamine	Glucuronic acid
Chondroitin sulfate A		1		Galactosamine	Glucuronic acid

containing 75% to 80% water on a wet weight basis. Of the remaining solids (20% to 25%), collagen, other proteins, and glycosaminoglycans or mucopolysaccharides constitute the major part. Corneal fibrils are neatly organized and present the typical 64 to 66 nm periodicity of collagen.[39] These collagen fibrils form the skeleton of the corneal stroma. The physicochemical properties of corneal collagen do not differ from those of tendon and skin collagen. Like collagen from these other sources, corneal collagen has a high nitrogen, glycine, proline, and hydroxyproline content. In boiling water or acid, corneal collagen is converted to gelatin. Collagen is dissolved by proteolytic enzymes such as collagenase, pepsin, or papain.

Glycosaminoglycans (GAG, mucopolysaccharides) represent 4% to 4.5% of the dry weight of the cornea. GAG are localized in the interfibrillar or interstitial space, probably attached to the collagen fibrils or to soluble proteins of the cornea. The GAG in the interstitial space play a role in corneal hydration through interactions with the electrolytes and water. GAG act as anions and bind cations and water. Three major GAG fractions are found in the corneal stroma: keratan sulfate (50%), chondroitin (25%), and chondroitin sulfate A (25%).[22,57] The chemical composition of these GAG is summarized in Table 3-1. Corneal keratan sulfate is unique and differs from the keratan sulfate found in cartilage by the length of its polymer chain, the degree of branching, and the predominant amino acids. Specific enzymes degrade corneal GAG. These glycosidases cleve terminal carbohydrates, aminosugars, and uronic acid.

Chrondroitin is found exclusively in the cornea. Incorporation of ^{35}S into GAG of corneal stroma is used as a biochemical or autoradiographic technique for measuring the rate of formation, or turnover of the two sulfated GAG (keratan sulfate and chondroitin sulfate A) in experimental animals.[24] GAG stain metachromatically with such stains as periodic acid–Schiff (neutral GAG) or alcian blue (acid GAG) in histologic corneal sections. A new nomenclature identifies GAG containing galactosamine as galactosaminoglycans (for example, chondroitin and chrondroitin sulfate A). GAG accumulate in the cornea and cause corneal haziness in patients affected by the inborn errors of GAG metabolism known as mucopolysaccharidoses. Corneal cloudiness is found in the following subtypes of mucopolysaccharidoses: Hurler (type I), Morquio (type IV), Scheie (type V), and Maroteaux-Lamy (type VI).[44] It is now well established that a deficiency of α-L-iduronidase, a glycosidase degrading GAG, is present in both the Hurler and Scheie syndromes.

Chemical comparisons between the cornea and the sclera show that GAG are practically absent from the sclera. It is natural, then, that GAG have been implicated in the maintenance of the corneal hydration level and transparency, since the sclera is not clear and reflects and scatters light under normal conditions.

The number of cells in the ox corneal stroma (keratocytes) has been estimated as 200 million to 1.5 billion, but there are relatively few cells per unit volume, or per gram, of tissue. Although these cells contain glycolytic and Krebs cycle enzymes, when the enzymatic activities are expressed on a weight basis, the stroma is low in enzymatic activities. The adenosine triphosphate (ATP) content of the stroma is low (10 to 15 μmol/kg wet tissue).[34,35]

Descemet's membrane

Descemet's membrane is made of type IV collagen with a high content of glycine, hydroxyglycine, and hydroxyproline, a chemical characteristic of basement membranes in other tissues (such as kidney glomeruli, blood vessels, and the lens capsule). Glycoproteins are tightly bound to the protein moiety. Carbohydrates and aminosugars of the glycoproteins can be obtained by acid hydrolysis of Descemet's membrane. In contrast to the corneal stroma, there are no significant amounts of sulfated GAG. In addition, collagen in Descemet's membrane is insoluble except in strong alkali or acid and is more resistant to collagenase than corneal stroma collagen. Jakus[39] has observed with the electron microscope that this membrane has collagen-like structure of great regularity. However, such morphology has not been found in other tissues. Descemet's membrane is highly elastic and represents a barrier to perforation in deep corneal ulcers. In histologic sections the glycoproteins of Descemet's membrane stain intensely with the periodic acid–Schiff stain.

Corneal epithelium

The epithelium represents 10% of the total wet weight of the cornea. The chemical composition of epithelial cells could be compared to that of liver parenchyma cells. Water in the epithelium represents 70% of the wet weight. The solids are (1) nucleic acids (deoxyribonucleic acid [DNA] and ribonucleic acid [RNA]), (2) lipids (phospholipids and cholesterol) in the cell membranes, and (3) proteins. The epithelium contains high activities of enzymes of glycolysis, Krebs cycle, and Na^+, K^+-activated ATPase. The epithelium contains ATP, 2000 μmol/kg wet weight; glycogen, 10 mg/g; glutathione, 75 to 180 mg/g; and ascorbic acid, 47 to 94 mg/100 g.[34,35] High concentrations of acetylcholine (ACh) and cholinesterases are also found in the epithelium. Although the functions of ACh and cholinesterases are uncertain, they may have a role in cation transport or in trophic nerve function.

CORNEAL METABOLISM

The term metabolism embraces a series of chemical processes by which energy is obtained and utilized to provide for the normal functions of a tissue. In the cornea, energy is needed for maintenance of its transparency and dehydration. Energy in the form of ATP is generated by the breakdown of glucose into lactic acid (glycolysis) and into CO_2 and H_2O (Krebs cycle) (Fig. 3-5). The cornea obtains glucose mainly from the aqueous humor.[69] The tears and limbal capillaries appear to contribute minimal amounts of glucose or O_2 for corneal metabolism. With the breakdown of 1 mol glucose into pyruvic acid and with its conversion into 2 mol lactic acid, 2 mol ATP are generated. The utilization of pyruvic acid throughout the Krebs cycle, with utilization of O_2, results in the formation of 36 mol ATP/mol glucose.

Most of the O_2 consumed by the cornea is taken in by the epithelium and the endothelium. The O_2 consumption of the epithelium and endothelium, expressed as QO_2, is 5 to 6, whereas that of the stroma alone is 0.23.[23,50] The corneal endothelium gets most of its required O_2 from the aqueous humor, while the corneal epithelium gets much of its O_2 from either the capillaries at the limbus or from the O_2 dissolved in the precorneal film. This is also borne out by experimental and clinical experience: when large, tight contact lenses are worn for prolonged periods of time, the cornea becomes edematous.

The corneal stroma and endothelium in living human eyes will remain clear for months if the epithelium is replaced by a plastic contact lens glued to the stroma.[40] If access of O_2 to the epithelium is abolished by tight contact lenses or by replacement of the air in goggles with N, the cornea swells and becomes cloudy. Under such anaerobic conditions the corneal epithelium produces lactic acid. The success of epikeratoprosthesis suggests that the epithelium may be responsible for the stromal changes, perhaps by producing excess lactic acid.

Glucose can be stored in the epithelium as glycogen to be broken down in certain emergency situations when the tissue needs additional energy, such as during reparative process following traumatic or surgical wounds. In the epithelium, glucose can also be metabolized through the hexose monophosphate shunt (pentose shunt),[45] but without a net gain in ATP. A characteristic of the metabolism through the

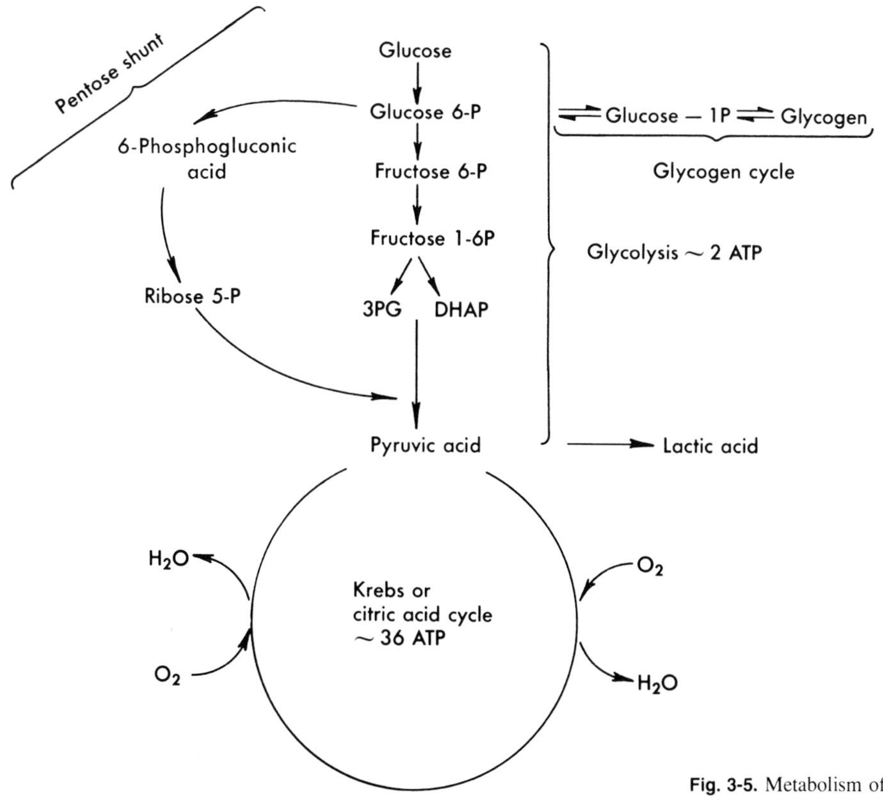

Fig. 3-5. Metabolism of glucose in cornea.

pentose shunt is the formation of ribose-5-phosphate (ribose-P) and reduced triphosphopyridine nucleotide (TPNH). The synthesis of lipids by the corneal epithelium needs TPNH formed through this pathway, while the ribose-P formed may build into the nucleic acids (DNA or RNA).

The epithelium contains much higher levels of ATP, glycogen, and oxidative enzymes than the stroma does, and the QO_2 of both epithelium and endothelium is 25 to 30 times higher than that of the stroma.[48] Comparisons of epithelial-endothelial areas with the stroma on a wet weight basis may not be altogether adequte, since the epithelial and endothelial cells are tightly packed while the keratocytes are distributed in a much larger volume. To establish meaningful comparisons, a correction factor must be introduced. Since the epithelium and endothelium contain 15 to 20 times the number of cells in stroma of equal weight, the values obtained for stromal metabolism should be multiplied by this factor. When expressed as cells per unit weight, the metabolic activities of the epithelium do not differ markedly from those of the stroma.

ELECTROLYTE COMPOSITION OF CORNEAL STROMA AND EPITHELIUM

A summary of the Na^+, K^+, and Cl^- composition of the rabbit cornea and its epithelium and stroma as compared to aqueous humor and tears is shown in Table 3-2. The corneal stroma has a high concentration of Na^+ ions, while the epithelium is rich in K^+ ions. The three fluids surrounding the cornea (plasma, aqueous, and tears) are also rich in Na^+ ions. In the corneal stroma it is evident that the sum of cations Na^+ and K^+ is in excess of that of the anion Cl^- by 85 mEq/kg H_2O. Part of the osmotic forces may be provided by bicarbonate ions (25 to 35 mEq/kg H_2O). The remaining anionic forces, 50 to 60 mEq/kg H_2O, may be provided by acidic GAG acting as anions (GAG^-). It is known that each molecule of keratan sulfate and chondroitin sulfate has between 27 and 160 binding sites for univalent cations. In addition

Table 3-2. Electrolyte composition of rabbit cornea, plasma, and aqueous humor (mEq/LH_2O)

	Na^+	K^+	Cl^-
Cornea			
Whole cornea	156	28	97
Stroma	172	21	108
Epithelium	75	142	30
Aqueous humor	143.5	5.2	108
Plasma	151	5.2	109
Tears (human)	149	17	131

to these GAG, the collagen fibrils and free protein may act as anions to provide for electrochemical neutrality in the cornea.

Corneal dehydration

The normal cornea maintains a fairly constant thickness during life. It keeps its water content at a steady level of about 75% to 80% of its weight. However, excised pieces of corneal tissue are markedly hydrophilic and swell in vitro, even when immersed in isotonic solutions. The corneal swelling results from the marked osmotic force of the stromal GAG. Minimal alterations in corneal swelling can be quantitated by direct measurements of corneal thickness with instruments such as the corneal pachometer attachment to the Haag-Streit slit lamp. Because it is fixed at the limbus, the cornea can only swell in the anterior-posterior direction. Therefore increases in corneal thickness can be linearly correlated with increases in hydration.

The water equilibrium existing in the living cornea has raised questions regarding the following factors that preserve the relative dehydration of the tissue: anatomic integrity of endothelium and epithelium, electrolyte and osmotic equilibrium, metabolism, evaporation of water through the anterior surface, and intraocular pressure. These mechanisms will be dealt with as separate entities in the following paragraphs; however, equilibrium among them is needed for dehydration of the cornea.

Anatomic integrity of endothelium and epithelium

When either the corneal epithelium or endothelium is damaged by chemical or physical means (including ionizing radiation), swelling of the stroma follows. Abrasion of the cornea or any condition leading to loss of the epithelium is likely to produce localized areas of corneal swelling and cloudiness. Fortunately, the corneal epithelium regenerates rapidly, and ordinarily the excessive hydration of the cornea is slight and transient. Damage to the endothelium is far more serious. Extensive destruction of endothelial cells will cause marked and sometimes permanent swelling and loss of transparency. This can result as a consequence of mechanical damage to the endothelium during ocular surgery. The transient and reversible corneal edema following cataract surgery (striate keratopathy) is attributed to endothelial damage.

In the rabbit removal of the epithelium produced an average increase in the corneal thickness of 200% in 24 hours, whereas removal of the endothelium produced an increase of 500%.[56] The average rate of increase is initially 0.5% of the original thickness/min when the epithelium is removed, and 2%/min when the endothelium is removed. The fact that endothelial damage results in much more corneal swelling and more rapid swelling than epithelial damage supports the premise that the endothelium is of great importance in maintaining dehydration. It has been pointed out, however, that this might be the result of the additive effect of the intraocular pressure on the endothelial side of the cornea.

Harris[32] suggested that the major cause of hydration of the cornea following endothelial damage was influx of water from the aqueous humor rather than decreased transport of water out of the cornea. According to his view, the endothelium provides a barrier to the influx of water from the anterior chamber, and Descemet's membrane is the barrier across which outflow occurs.

The barrier function can be measured in vivo using fluorescein as a tracer.[73] The role of the endothelium as a barrier is further supported in recent studies by Refojo and co-workers[64]; a silicone rubber membrane was fixed behind the corneal endothelium of human corneal grafts. The physical barrier provided by the silicone membrane was sufficient to maintain the cornea in a state of dehydration. Because of the impor-

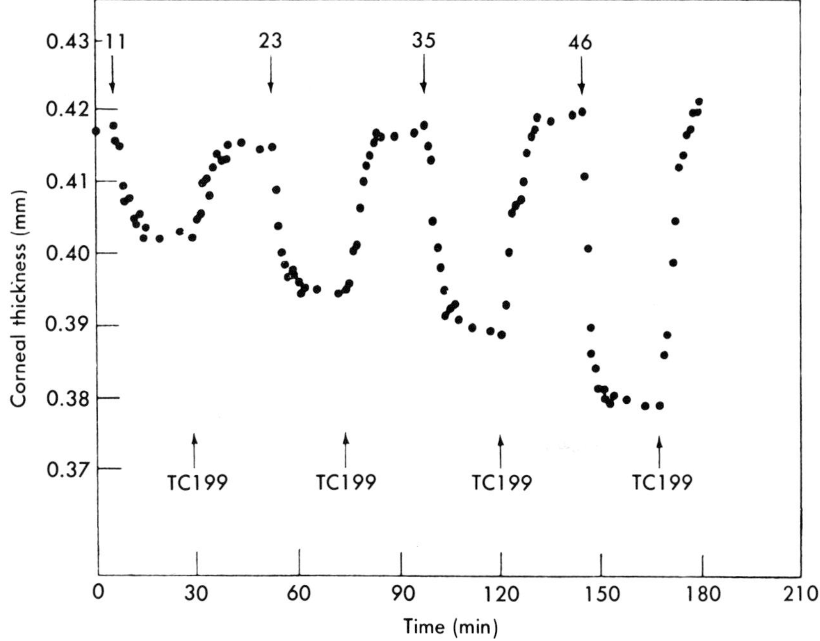

Fig. 3-6. Changes in corneal thickness resulting from hypertonicity of solutions used in anterior chamber perfusion. Hypertonicities were produced by addition of glucose to TC199. Numerals indicate hypertonicity in milliosmoles per liter; arrows indicate when solution was changed. (From Mishima, S., and Hedbys, B.: Exp. Eye Res. **6:**10, 1967.)

tance of the corneal endothelium in maintaining corneal dehydration, if keratoplasty is indicated, emphasis must be placed on obtaining a donor graft with viable, healthy endothelium.

Electrolyte and osmotic balances

The corneal stroma is markedly hydrophilic due to the osmotic force of the stromal GAG, of which keratan sulfate may be the most important. The maintenance of the deturgescent state is necessary for normal vision because the edematous cornea quickly loses its transparency. The corneal epithelium and endothelium are the limiting membranes of the stroma and play critical roles in maintaining its dehydration. The endothelium has a barrier function, and an active metabolic pump in the endothelium removes fluid that leaks into the stroma. The corneal endothelium has been thought to pump water, sodium, and possibly other electrolytes from the stoma. However, recent evidence indicates that, at least in certain species, principally bicarbonate ions are pumped from the stroma into the aqueous humor.[37]

In addition to the endothelial pumping function a Cl^- pump stimulated by epinephrine and cyclic AMP has been found in the epithelium of the rabbit and frog[78] and may play a role in corneal dehydration.

The osmotic pressure of the stromal fluid and electrolytes may also be important in regulating water movement. The relationship between the tonicity of stromal fluid and that of plasma, aqueous humor, and tears has not been resolved. If aqueous humor and tears are slightly hypertonic to the stroma, they may play an active role in corneal dehydration by constant removal of water from the anterior and posterior surfaces. This theory was first proposed by Cogan and Kinsey,[13-19,38] but its validity has recently been questioned.

Topical application of glycerine reduces corneal swelling by increasing the osmolarity of the precorneal film, a simple clinical means of temporarily clearing an edematous cornea. Similarly, if the osmolarity of fluid perfusing the anterior chamber is increased, the cornea will dehydrate, the amount of water lost being proportional to the hyperosmolarity of the perfusing fluid (Fig. 3-6). The in vitro swelling of the cornea in a solution isotonic to plasma (300 mOsm) can be prevented by increasing the osmolarity of the bathing fluid. Dohlman and coworkers[25] were able to maintain the normal corneal water content by incubating corneas in media 250% hyperosmotic to plasma. All these studies tend to support the concept that osmotic forces may operate on the living cornea to maintain the tissue in a relative state of dehydration.

Metabolism

Metabolic activities of the cornea contribute to the maintenance of the dehydration of the tissue. When iodoacetate, a metabolic poison that inhibits glycolysis and most of the oxidative metabolism of the cornea, is injected into the anterior chamber of rabbits, an 80% increase in corneal hydration will result. A similar effect, but at higher concentrations of iodoacetate, was found by Harris[33] in corneas in which the poison was applied to the epithelium for 5 minutes. Probably the inhibition of metabolism of the endothelium and epithelium due to lack of ATP, needed for the functioning of the electrolyte "pump," is largely responsible for such corneal overhydration.

It is currently accepted that the active extrusion of Na^+ ions from most tissues is mediated by the enzyme Na^+, K^+-activated ATPase. This enzyme, by breaking down ATP, provides sufficient energy for cation extrusion.

Na^+, K^+-ATPase is present in relatively high concentrations in the corneal epithelium and endothelium, but is practically absent in the corneal stroma. Ouabain, a cardiac glycoside, is a specific inhibitor of Na^+, K^+-ATPase; when injected into the anterior chamber or applied topically to the eye, it results in inhibition of corneal Na^+, K^+-ATPase and corneal overhydration.

The cornea becomes hydrated when the ambient temperature is lowered as the metabolic activities are decreased. However, if enough glucose is present, when the cornea is brought back to room temperature, it becomes dehydrated again.[32] Thus corneas from autopsy eyes stored in eye banks at 4° C are slightly hydrated, and grafted corneas may be slightly swollen. The thickness of corneal grafts decreases during the days following transplantation, indicating the cornea can adapt quickly and restore its ability to dehydrate.

$$ATP + Na^+ + K^+ \underset{\text{(Ouabain inhibition)}}{\overset{Na^+,\,K^+\text{-ATPase}}{\rightleftarrows}} ADP + P_i + Na^+ + K^+$$

Evaporation of water through the anterior surface

Von Bahr,[72] in a series of elegant experiments in rabbits, first called attention to the evaporation of water from the precorneal tear film and anterior corneal surface. He calculated the rate of evaporation for the rabbit at 60 μl/hr. Mishima and Maurice[60] found it to be 6 μl/hr for the undisturbed rabbit eye and 100 μl/hr when the anterior surface of the cornea had been washed. The evaporation of water from the precorneal tear film would concentrate this fluid and increase its osmolarity relative to the cornea. The hypertonicity of the tear film could draw water from the cornea and maintain the relative state of corneal dehydration.

Intraocular pressure

The fact that increased intraocular pressure results in corneal edema is well known to clinicians as a useful sign in the diagnosis of glaucoma. It is accepted that intraocular pressures in the human eye of 50 mm Hg or higher often result in corneal edema detectable with the slit lamp. Corneal swelling caused by increased intraocular pressure is reversible when the pressure is lowered to normal values by surgical or medical procedures if there has not been permanent endothelial damage. In infants increased intraocular pressure can result in enlargement of the cornea and breaks in Descemet's membrane in addition to the edema; this occurs in congenital glaucoma.

PERMEABILITY AND TRANSPORT OF DRUGS ACROSS THE CORNEA

The permeability of the cornea is significant from several viewpoints. First, the nutrition of the cornea depends on the diffusion of oxygen and glucose and other substances from the surrounding fluids. Second, the transport of drugs across the cornea is determined by the permeability of the corneal layers. This is clinically significant because many of the topically applied preparations are intended to reach the anterior chamber of the eye or to affect pathologic processes taking place in the various areas of the cornea. Finally, the exquisite nerve supply of the cornea determines the degree of tolerance of medications applied topically for diseases of the eye or as wetting agents for contact lenses.

Many factors affect the drug penetration through the cornea. These can be divided into those related to the drug solution itself (solvent and solute), and those determined by the corneal layers.

Drug solution

Chemical structure of the solute and the solvent

Barriers rich in lipid are traversed by nonpolar compounds, while structures rich in water are easily crossed by polar compounds. The term polar refers to its water-lipid solubility. A nonpolar compound is more soluble in organic solvents (lipids) than in water; a more polar compound is more soluble in water than in lipids.

The epithelial and endothelial cell membranes, being barriers rich in lipid, are more rapidly crossed by lipid-soluble compounds, while the stroma is more readily crossed by water-soluble compounds. Since drugs have to go across barriers with different structural specificities, substances combining both solubilities will penetrate more freely. The studies of Swan and White,[67] who used procaine and tetracaine as model substances, support this concept. Although these anesthetic agents have the same basic structure, changes in the relationship between polar and nonpolar groups change their penetration through the cornea significantly. Tetracaine, which has a less polar group than procaine at one end of the molecule (Fig. 3-7) and a more polar group than procaine at the other end, penetrated into the cornea more rapidly than procaine. The rate of penetration of procaine through the cornea was only 14.8% of that of tetracaine. The difference in penetration is in accord with the clinical fact that tetracaine is an effective corneal anesthetic at 0.5% concentration, while procaine is ineffective at 2% concentration.

Recent studies have determined that the permeability of corticosteroids across excised rabbit corneas is determined by their relative lipid-water solubility.[65] Modifying dexamethasone or prednisolone by the addition of an acetate group significantly changed the octanol-water partition coefficients and increased the penetration of the applied corticosteroids across the excised corneas eight times. This was also confirmed in vivo using rabbits and dexamethasone. Correlation of in vitro and in vivo studies should allow other drugs to be modified to increase intraocular penetration.

Molecular weight and concentration of transported molecules

As in other biologic membranes, the molecular weight and concentration of the drug af-

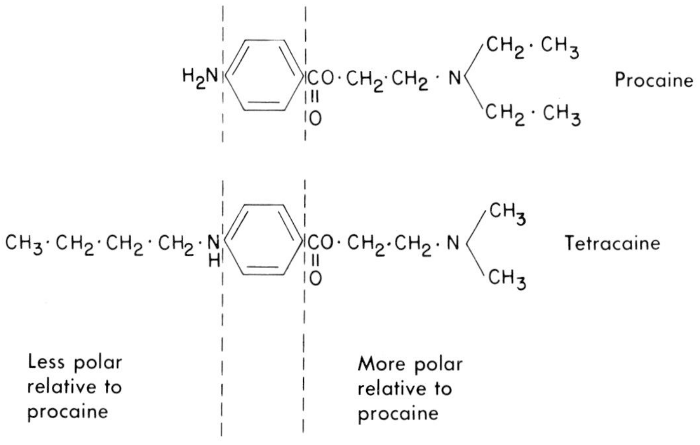

$H_2N-C_6H_4-CO\cdot CH_2\cdot CH_2\cdot N(CH_2\cdot CH_3)_2$ Procaine

$CH_3\cdot CH_2\cdot CH_2\cdot CH_2\cdot NH-C_6H_4-CO\cdot CH_2\cdot CH_2\cdot N(CH_3)_2$ Tetracaine

Less polar relative to procaine

More polar relative to procaine

Fig. 3-7. Penetration of anesthetic related to polarity.

fects its penetration through the cornea. In the experiments by Swan and White, procaine reached the maximum rate of penetration at concentrations of 1 g/100 ml. The rate of penetration through the cornea and the effects on the anterior segment of the eye of such drugs as pilocarpine, atropine, homatropine, epinephrine, echothiophate iodide, and steroids depend on their concentration in the solution. The preparations of such drugs used for topical applications are very concentrated, since only a small percentage reaches the anterior chamber through the cornea.

pH and osmolarity of solutions

The pH and osmolarity of solutions have a definite effect on the penetration of the solutes through the cornea. pH affects the electrical charges and stability of solutions. Most commonly used ophthalmic medications are buffered with combinations of weak acids with their salts. Such buffers as citric, acetic, boric, or phosphoric acid–salt combinations are common.

Kinsey[46] has evolved the following concept of how a weak base, such as the alkaloid homatropine, may penetrate the intact cornea (Fig. 3-8). As it is used clinically, a solution of homatropine hydrobromide consists of homatropine ions bearing a positive charge, $R_3N:H^+$, bromide ions bearing a negative charge, Br^-, and a small amount of homatropine (free base), R_3N, and hydrogen ion, H^+. At a pH of 6.0, which is the approximate pH of the usual solution of homatropine, the ratio of homatropine ions to free base is approximately 1000 : 1. When a drop of solution is placed on the cornea, the amount of dissociation will not be the same as that in the bottle because the pH of the tears is about 7.4 rather than 6.0, but the homatropine will exist in both the charged and free base forms. Free homatropine base penetrates the epithelium because of its lipid solubility, as previously explained. The homatropine ion penetrates the epithelium poorly because it is relatively insoluble in lipid. At the stroma the drug reaches a second water phase; homatropine ion readily dissolves in the stroma water, and equilibrium again shifts toward the charged particle. The $R_3N:H^+$ diffuses through the stroma until it reaches the endothelium. At the endothelium the same process that took place at the surface of the epithelium must occur again. Finally, charged homatropine leaves the endothelium for the aqueous humor.

In each of these steps the substance is able to penetrate because of the difference in solubility of the two forms in which it exists. The pro-

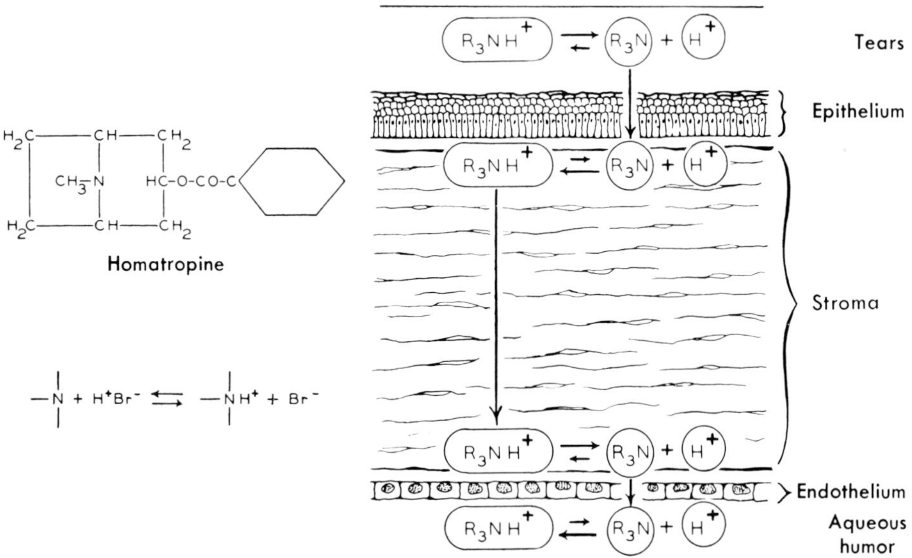

Fig. 3-8. Transfer of homatropine through cornea. (According to Kinsey.)

cess is not limited by exhaustion of the amount of homatropine present in one form because equilibrium between the forms is reestablished as soon as a particle of either form (charged or free base) leaves the layer.

The permeability of the epithelium is increased considerably if the solution of the test substances is hypotonic, that is, below 0.9%. Potts[62] found the tonicity that does least damage to the corneal epithelium to be 1.35% sodium chloride, which he infers to be the tonicity of the tears. The pH of the solution may be varied from 4.0 to 10.0 without affecting the permeability of the epithelium, but solutions outside this range increase the permeability.

Surface tension–reducing agents

Agents that reduce the surface tension increase the permeability of membranes. O'Brien and Swan[61] showed that the use of benzalkonium chloride (Zephiran), a surface tension–reducing agent, increases the absorption of carbachol (Doryl).[61] Use of benzalkonium chloride results in lysis of cell membranes and, if in high concentrations, may cause corneal edema.[70] Substances that reduce surface tension are called wetting agents. The relative impermeability of the corneal epithelium to carbachol is due to the fact that this compound has a low lipid solubility, that is, it is highly polar.

Layers of the cornea

The epithelium and endothelium represent the lipid barriers of the cornea, and the removal of these layers will increase the penetration of most drugs into the stroma. As shown in Table 3-3, 0.1% dexamethasone sodium phosphate, 50 μl to the cornea, penetrates into the stroma and the aqueous humor only when the epithelium is removed. However, if the eye is inflamed, dexamethasone will enter the corneal stroma and aqueous even with an intact epithelium (Table 3-3).

Table 3-3. Effect of the epithelium and inflammation on corneal and aqueous humor penetration of dexamethasone*

	Inflammation	Corneal stroma† (μg/g)	Aqueous† (μg/ml)
Corneal epithelium intact	No	0	0
Corneal epithelium removed	No	42	4.8
Corneal epithelium intact	Yes	14	1.5

*From Cox, W. V., Kupferman, A., and Leibowitz, H. M.: Arch. Ophthalmol. **88:**308, 1972.

†Concentrations at 60 minutes mean.

The harmful penetration of acid through the cornea is retarded by the epithelium. The immediate precipitation of protein acts as a barrier to the further penetration of the acid. When the eye is exposed to strong alkaline solutions, however, the first effect is swelling and then desquamation of the epithelium. Hence the preliminary removal of the epithelium has less effect on the intensity of experimental alkali burns, and even the intact epithelium forms only a soft jellylike barrier to further passage of the alkali.[28]

CORNEAL TRANSPARENCY

The normal cornea is transparent, and any change in this property seriously interferes with the clarity of the retinal image (Fig. 3-9). The anatomic peculiarities of the corneal structure, such as the uniformity and regularity in the arrangement of the epithelial cells, the closely packed corneal lamellae of uniform size running almost parallel to each other, and the absence of blood vessels, all contribute to the efficiency of the eye as an optical instrument. The transparency of the cornea depends on its physical makeup and, as we have seen from the discussion in the previous section, on the mechanisms that keep it from swelling.

Physical factors

An explanation of corneal transparency based on the physical arrangement of the corneal micelles has been proposed by Maurice.[54,55] He finds that the corneal collagen fibrils form a lattice structure so arranged that scattering of light is eliminated by mutual interference from individual fibrils.

The basic structure of the corneal stroma is collagen, saturated with a solution of GAG. The index of refraction of the corneal collagen is 1.380 in cattle and 1.373 in pigs.[4]

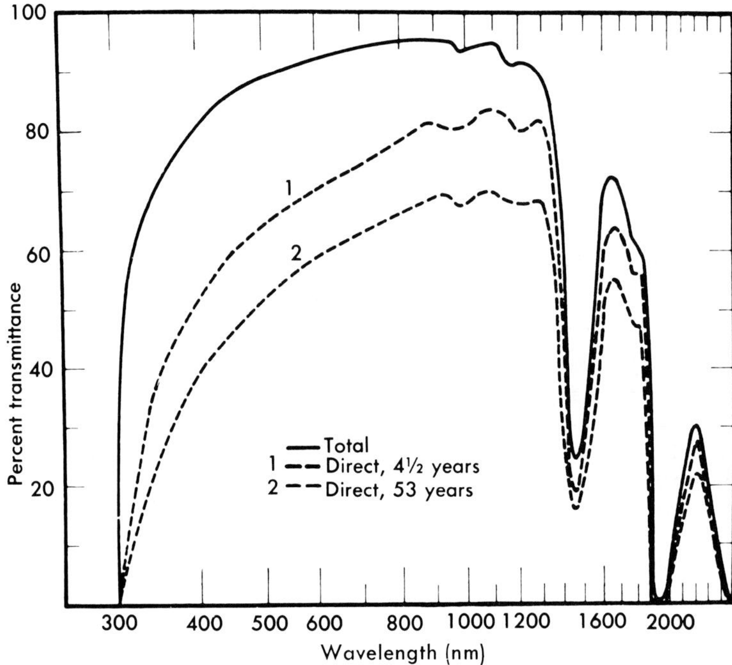

Fig. 3-9. Transmittance of cornea. Difference between total and direct transmittance is scattered light. (From Boettner, E. A., and Wolter, J. R.: Invest. Ophthalmol. **1**:776, 1962.)

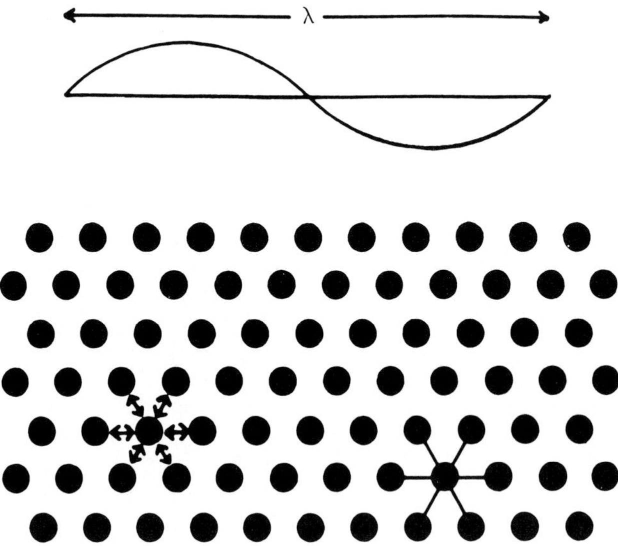

Fig. 3-10. Cross-sectional view of fibrils arranged in lattice. Size of wavelength shown above for comparison. Forces of repulsion and rigid links between fibrils shown schematically. (From Maurice, D.: The physics of corneal transparency. In Duke-Elder, S., editor: Transparency of the cornea, Oxford, 1960, Blackwell Scientific Publication, Ltd.)

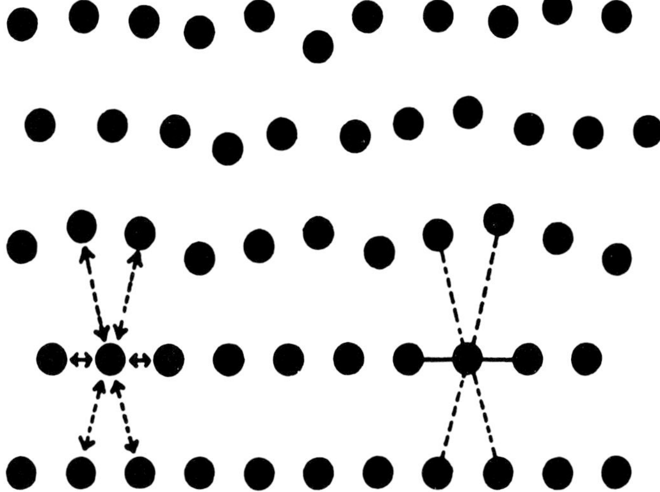

Fig. 3-11. Diagram of swelling of cornea and disorder of rows of fibrils consequent on weakening of forces of alignment from neighboring rows. (From Maurice, D.: The physics of corneal transparency. In Duke-Elder, S., editor: Transparency of the cornea, Oxford, 1960, Blackwell Scientific Publications, Ltd.)

The cornea remains clear when immersed in fluids of different refractive indexes up to 1.564. Therefore the cornea can maintain its transparency even when the index of refraction of its collagen differs considerably from that of the surrounding medium. This property of remaining clear is believed to be due to the fibrils' being tightly and regularly packed together; it vanishes if the fibrils are broken up by artificial means. As long as the fibrils are regularly arranged in a lattice and separated by less than a wavelength of light, the cornea will remain transparent. Fig. 3-10 shows the arrangement of the fibrils in a regular lattice, each separated from its neighbors by a regular spacing. The arrangements of the fibers when the cornea is swollen or distorted is shown in Fig. 3-11. The distance between fibrils, that is, the lattice space, is greater; destructive interference no longer occurs, light is scattered, and the cornea appears hazy. Another theory of corneal transparency has been proposed by Goldman and Benedek.[30] They believe that the cornea is transparent because the fibers are small with reference to the wavelength of light, and consequently the incident light *is not scattered* as much as if larger particles were in suspension. Indeed, the shark cornea with regions of disorganized fibers and random distribution of interfibrillar distances is quite transparent.

The transparency of the cornea becomes temporarily impaired when abnormal pressures are applied to it. During the rise of intraocular pressure in acute glaucoma, the cornea becomes cloudy, and it is quite evident that this loss of transparency is not caused entirely by imbibition of fluid, which does occur, but also by physical changes in the stroma and epithelial edema, since the cloudiness disappears immediately when the pressure is lowered. This can be demonstrated easily in an enucleated eye by applying pressure to the globe, whereupon the cornea immediately loses its transparency. As soon as the pressure is relieved, the transparency returns.

Effect of swelling on corneal transparency

The cornea is a lyophilic colloid system inasmuch as its behavior is similar to that of gelatin and fibrin. Various physiocochemical factors affect its transparency, acting through changes in the refractive index either of its intercellular fluid, of the corneal micelles, or of both. The regulation of the water content of the cornea has already been discussed (p. 48). The absorbing power of the connective tissue must be due to its two chief components, collagen and GAG. The collagen, split up into loose bundles of fine fibers, lies enmeshed in the GAG. The GAG, as a gel, not only fills all spaces between the fiber bundles, but also penetrates the bundles and takes part in the formation of the micelles of the intermediate substance itself. These two components must be considered separately with regard to their contribution to the total absorptive power of the connective tissue. When dried corneas are exposed to water vapor of increasing concentration up to a water content of 30%, the water is taken up in the micelles of collagen, causing a demonstrable expansion of the crystalline lattice. Further absorption of water is ascribed mostly to the interfibrillar GAG. Imbibition of collagen is intensified by the addition of acid and imbibition of GAG by the addition of alkali. From the observations just mentioned, it should be expected that the GAG of the cornea is also important to its transparency. This is confirmed by the fact that the increase in transparency of the corneas of chick embryos coincides with the appearance of the metachromatic GAG staining. Van Walbeek and Neumann[71] investigated the transparency of the cornea after it was swollen in water. Maximal transparency occurred when the amount of water was 200% to 400% of the weight of the dry matter. This corresponds to the physiologic water content of fresh corneas. Their experiments led them to believe that there is a complex relationship between collagen and GAG and that these substances are not physicochemically independent.

To maintain its transparency, the cornea must be bathed with a fluid having an osmotic pressure as high as interstitial fluid. If the cornea is bathed with a hypotonic solution, it becomes cloudy due to the loss of the osmotic forces acting at the corneal epithelium. When the cornea is cloudy because it is swollen, it can be temporarily cleared by bathing the eye with a hypertonic solution, such as 10% salt or glycerine solution. Solutions of drugs to be used in ocular therapeutics should be made equal to a 1% to 1.5% salt solution (p. 49).

In several pathologic conditions, such as glaucoma or endothelial dystrophy, fluid collects in the epithelial cells to such an extent that bullae are formed. This occurs because the epithelium is much more resistant to water movement than the stroma. Aqueous easily passes through the corneal stroma, but its escape is retarded by the epithelium, which becomes edematous. Normally the tear film over the surface of the cornea is concentrated by evaporation. When evaporation at the surface does not occur, the edema becomes considerably worse. Cogan was able to prove this in bilateral endothelial dystrophy by having the patient wear airtight goggles, one side of which contained calcium chloride to absorb the moisture. The eye on that side showed improvement in corneal edema after the goggles had been worn for an hour. If evaporation was prevented, the edema became worse.

HEALING OF CORNEAL WOUNDS

Accidental or surgical injury to one or all layers of the cornea results in increased hydration and loss of transparency. Thus rapid and proper healing of wounds is needed to avoid formation of corneal scars or opacities (leukomas). The loss of the corneal epithelium by trauma is generally healed in 24 to 48 hours in a normal eye. However, ulcers produced by bacterial, viral, or fungal infections or involving the deep layers of the stroma may take weeks to heal. Similarly, although a superficial corneal laceration heals rapidly, a deep corneal laceration or surgical incision takes several weeks to heal completely. Sutures are needed to maintain apposition of the wound edges in large, deep lacerations. The following factors determine the speed and type of wound healing:

1. Anatomic location: limbal vs. central corneal wound
2. Size of wound
3. Layer of cornea involved: epithelium, stroma, endothelium
4. Presence or absence and type of sutures: silk (6-0, 7-0, 8-0, 10-0), catgut (6-0, 7-0, 8-0), or nylon (10-0)
5. Associated bacterial, viral, or fungal infection
6. Topical drugs administered for therapeutic purposes (steroids, iododeoxyuridine, etc.)

Table 3-4. Corneal wound healing

Stage	Cytologic and tissue response	Biochemical and histochemical response
I	Stromal edema, epithelial cell mitosis	Swelling
II	Polymorphonuclear and mononuclear stromal invasion	Increased hydrolases activities; degradation of GAG and collagen
III	New fibroblasts	Same as II, plus new precollagen and collagen formation; appearance of new GAG (dermatan sulfate)
IV	Collagen	Collagen formation
V	Healed stroma	Normal hydrolases activity; normal concentrations of GAG and collagen

We will discuss here the normal process of healing of sharp cuts into the central area of the rabbit cornea. By and large the findings are also applicable to limbal incisions such as those made during cataract surgery. The cytologic and tissue response to a central corneal injury involves the following[74]:

1. Stromal edema
2. Epithelial cell mitosis and sliding
3. Polymorphonuclear and monocytic migration
4. Appearance of new fibroblasts and mitosis
5. Formation of precollagen
6. Formation of collagen

Occasionally, and for unknown reasons, blood vessels may grow into healed corneal tissue. Chemical changes associated with the cytologic response involve (1) initial hydration, (2) increased activities of lysosomal hydrolases (glycosidases, collagenases, 5-nucleotidases, oxidases),[75,76] and (3) decrease in the levels of mucopolysaccharide (keratan sulfate and chondroitin sulfate).[63] After about 4 weeks a new mucopolysaccharide not normally found in the rabbit cornea, dermatan sulfate, appears[1] (Table 3-4). Corticosteroids administered in high dosages have a retarding effect on corneal wound healing. Gasset and co-workers[29] found that dexamethasone 0.1% administered topically to the cornea every hour decreased by half the tensile strength of 11-day-old corneal wounds. Previously, Ashton and Cook[2] found that cortisone in dosages of 2.5 to 15 mg administered subconjunctivally inhibited the cellular infiltration and fibroblastic repair, thus retarding corneal wound healing. However, topical dexamethasone in lower dosages, 0.01% every hour or 0.1% four times a day, does not affect tensile strength.[29]

Although cocaine has been singled out as the anesthetic agent most damaging to the cornea, most synthetic anesthetic agents are as capable as cocaine of producing corneal erosions.[5] In addition to the toxic effect of repeated anesthetic agents on the cornea, Behrendt[5] has found that further instillations of most anesthetic agents after the anesthesia has worn away show a progressive diminishing anesthetic effect until finally the effect is completely lost. It is not known just how this comes about. It does point up the fact that prescribing anesthetic drugs for patients to be used at home is dangerous and probably accounts for the fact that, although patients are directed not to use the anesthetic more often than at 1- or 2-hour intervals, they frequently ignore these directions and begin to use the drops at more frequent intervals to obtain relief from pain.

An explanation of the basic mechanism involved must undoubtedly take into consideration the action of drugs on the intracellular processes, as shown by the studies of Michaelis and Quastel[58] and of Herrmann, Moses, and Friedenwald.[36] Another factor that must be considered in the delay in healing of corneal wounds caused by anesthetics is the fact that some of the anesthetics used are good detergents, that is, wetting agents. This property may be partly responsible for the inhibition of epithelization. Strong detergents act in this manner and, if drugs are incorporated in ointments in which their detergent action is reduced, they fail to delay the migration of cells significantly. The least harmful anesthetics, as far as epithelial repair is concerned, seem to be dibucaine and phenacaine ointments. These hardly impair mitosis in normal epithelium or retard division and migration of cells in regenerating corneal epithelium if used in moderation.

Although the studies of Smelser and Ozanics[66] indicate that ointments rather than solutions of anesthetics are the most satisfactory in permitting epithelial repair, ointments have generally been found to slightly delay wound healing.

Since cocaine and ephedrine were found to inhibit the mitotic activity of the corneal epithelium, and since both of these substances sensitize various effector organs to the action of epinephrine, Friedenwald and Buschke[27] investigated the effect of epinephrine itself on the mitotic activity of epithelium in the rat. Ether inhibits mitotic activity and is known to stimulate the secretion of epinephrine by the adrenal glands during the stage of excitation. They found that excitement itself, particularly by painful stimuli, diminished the mitotic rate in the normal corneal epithelium, and that this decrease is an adrenergic response that could be stimulated by either local or systemic administration of epinephrine. If ergotamine or nicotine was given prior to the period of excitement, the inhibition of mitosis was eliminated or at least diminished. The effect of epinephrine is not due to a decrease in temperature of tissue or to local circulatory disturbances. Removal of the superior cervical ganglion led to a decreased mitotic rate in the epithelium after an interval of about 20 hours.

Ultraviolet light has an appreciable effect on the rate of corneal mitosis.[9] Low dosages of an unfiltered quartz mercury arc stimulate mitotic activity in the corneal epithelium. With large dosages this activity is inhibited. Severe exposures lead to nuclear fragmentation, mainly in the superficial layers of the corneal epithelium, and to loss of cohesion between the epithelium and the stroma. This occurs after unprotected exposure to a sunlamp or welder's arc. Since both these phenomena are suppressed under anaerobic incubation, the pathologic processes involved are in part, at least, oxidative. The two phenomena appear to be independent of one another, since they have different temperature coefficients. At proper incubation temperatures the loss of cohesion between the epithelium and the stroma can be demonstrated in the absence of nuclear fragmentation, which may be a form of pathologic mitosis. Healing of very small wounds was not inhibited after exposure to ultraviolet rays in the dosage used. This is in keeping with the general finding that posttraumatic cell movements are much more resistant to interference by various toxic agents than is mitotic activity.

CORNEAL VASCULARIZATION

Capillaries extend into the cornea 1 or 2 mm beyond the limbus; otherwise, the normal cornea is entirely devoid of blood vessels. If at any time blood vessels are present beyond the limbal zone, the cornea is, or has been, the seat of some pathologic process. The condition that produced the vascularization may have been either a local process or some more generalized pathologic condition such as a nutritional disease. Certain diseases produce early and intense vascularization, and the situation and character of the newly formed blood vessels are sometimes guides to their diagnosis. For example, the straight branching vessels that enter the cornea in the deeper layers at the limbus are characteristic of interstitial keratitis, suggesting congenital syphilis, whereas the superficial tortuous vessels that can be traced onto the cornea from the conjunctival surface across the limbus are characteristic of various forms of keratoconjunctivitis. Vessels entering the cornea from the upper limbus associated with granulation tissue in the conjunctiva of the upper lid are highly suggestive of trachoma.

In many nutritional diseases vascularization of the cornea occurs early and is typical. Vascularization has been observed in rats on diets deficient in riboflavin,[7,68] vitamin A, tryptophan, lycine, zinc, sodium,[21] or methionine[6] and in diets devoid of protein.[26,31] Vascularization is considered a defensive mechanism, by which means the cornea tries to eliminate toxic influences.

Many different theories have been proposed to account for the absence of blood vessels in the normal cornea and in their ingrowth in diseased states. Among these are the following:

1. A substance is elaborated in the cornea as a result of the corneal pathologic process that stimulates the vessels to grow into the cornea to the site where this substance is manufactured. The substance is supposed to possess the power to initiate neovasculogenesis from already existing blood vessels and to have positive chemotaxis. While some experiments suggest that such a substance might exist, it has not been isolated.[3]

The accumulation of acid metabolites, histamine, a deficiency in ascorbic acid, and local anoxia have also been suggested as factors that induce new vascularization. The relation of anoxia to the development of blood vessels in other ocular tissues is especially well documented by the work of Michaelson and coworkers,[59] who showed that a capillary-free zone exists around the normal retinal arterioles, whereas no such capillary-free zone exists around the normal retinal venules. Campbell[10] found that litters of rats exposed to a low-oxygen environment showed an encroachment of capillaries on the retinal arterioles, indicating that oxygen tension is a factor that determines the growth of these normal vessels. In many disease conditions anoxia has been held responsible for the characteristic development of new vessels, as in Eales' disease, diabetic retinopathy, and central vein occlusion. It goes without saying that anoxia may be effective in this respect by releasing some growth-promoting substance from the anoxic tissue, and carbon dioxide is not only a vasodilating agent, but also a strong stimulus to vasculogenesis.

2. The failure of blood vessels to grow into the normal cornea has been attributed to the properties of the cornea itself—either that it contains some substance which repels the growth of vessels or that it is so compact vessels are unable to penetrate it. In disease states associated with neovascularization either the inhibiting substance is lost to the cornea or the swelling of the tissue allows the blood vessels to grow in. GAG have been investigated as vessel-inhibiting substances, since the cartilage, umbilical cord, and cornea (avascular tissues) are rich in GAG, but no growth-inhibiting substance has yet been isolated from the cornea.

Cogan[11,12] was the first to suggest that the compactness of normal corneal tissue was the reason for its avascularity, and he pointed out that cartilage and the nails cannot be vascularized because they have no invadable intercellular substance. Reasoning along these lines, he suggested that the factor responsible for vasculogenesis in diseased states is a release of corneal tissue pressure. When the cornea becomes swollen and therefore less compact, spaces are opened up through which vessels can grow. He reasoned that as a result of swelling of the cornea at the limbus the vessels normally present there become engorged and form saccular aneurysms that ultimately burst. This leads to the development of new vessels in these areas, with the spread of the newly formed vessels into the now swollen corneal stroma. He also suggested that swelling of the peripheral parts of the cornea sets up a self-corrective cycle through neovasculogenesis. The swelling of the cornea initiates the formation of new vessels, but this in turn corrects the swelling. He has observed that edematous corneas tend to experience deturgescence as new blood vessels grow into them. Those corneas, on the other hand, that do not become vascularized remain edematous and develop epithelial bullae. Luetic interstitial keratitis is an example of an edematous cornea accompanied by bullae formation that undergoes deturgescence as newly formed vessels grow into it. On the other hand, the edematous corneas in endothelial dystrophy remain swollen and afflicted with bullae and never become vascularized. In endothelial dystrophy the cornea becomes swollen with subsequent turgescence as a result of the impairment of endothelial function; the swelling is usually localized to the central regions of the cornea. Even when swelling extends to the limbus, however, vascularization usually does not develop.

The normal cornea is relatively acellular, and an increase in cells usually precedes vascularization. Various processes, such as silver nitrate burns, alkali burns, and injection of intracorneal antigens, all result in local leukocyte accumulation prior to vascular ingrowth. Leukocytes also accumulate prior to vascularization in certain vitamin deficiencies. In most models vascular ingrowth occurs at the level of previous leukocyte infiltration. Total body x-radiation, which suppresses leukocyte infiltration, prevents corneal vascularization in most models. Polymorphonuclear leukocytes rather than lymphocytes appear to be the cells involved.[47]

Further experiments of Ashton and Cook[3] have shown that vascularization is usually produced if the corneal compactness is reduced by any means, provided that the zone of corneal swelling extends to the limbus, the site of the normal blood supply. The swelling of the stroma must persist for some time; for example, the injection of saline solution into the stroma must be repeated a number of times to produce neovascularization, whereas viscous substances that remain in the cornea spreading the lamellae need to be injected only once.

We may accept the conclusion that blood vessels are kept out of the cornea largely by the compactness of this tissue, and if this is so modified that the stroma becomes looser, vessels can gain access. This does not prove, however, that neovasculogenesis from the limbal vessels is initiated by changing the compactness of the stroma, and it seems very unlikely that this factor alone would stimulate the limbal vessels to send out new shoots. New vessels do not grow from previously existing ones in other parts of the eye where the tissue is loose, for example, in the retina. It seems likely that some agent must be present in the cornea that reaches the limbal blood vessels when the stroma is loosened and acts on the vessels. Evidence pointing in this direction has been offered by Langham.[49] He was able to induce swelling of the cornea by injecting alloxan into the anterior chamber of rabbits. By proper selection of dosage this swelling could be reversed. In some cases he was able to show that after the vessels had grown into the stroma from the periphery a distance of 1 to 2 mm, they ceased to grow, although the cornea underwent extreme turgescence; in some cases vascularization ceased at a time when the cornea was still twice its normal thickness.

The problem is not solved, and it still seems likely that, along with the change in the physical state of the stroma, changes in the chemistry of the cornea occur, including cellular infiltration, which is necessary for vascularization.

REFERENCES

1. Anseth, A., and Fransson, L.: Studies on corneal polysaccharides. VI, Isolation of dermatan sulfate from corneal scars, Exp. Eye Res. **8**:302, 1969.
2. Ashton, N., and Cook, C.: Effect of cortisone on healing of corneal wounds, Br. J. Ophthalmol. **35**:708, 1951.
3. Ashton, N., and Cook, C.: Mechanism of corneal vascularization, Br. J. Ophthalmol. **37**:193, 1953.
4. Aurell, G., and Holmgren, H.: Metachromatic substance in cornea, with special reference to question of transparency, Nord. Med. **30**:1277, 1946.
5. Behrendt, T.: Corneal lesions after topical anesthesia, Am. J. Ophthalmol. **41**:99, 1956.
6. Berg, J., et al.: Formation of capillaries and other tissue changes in the cornea of methionine-deficient rats, J. Nutr. **33**:271, 1947.
7. Bessey, O., and Wolbach, S.: Vascularization of the cornea of the rat in riboflavin deficiency, with note on corneal vascularization in vitamin A deficiency, J. Exp. Med. **69**:1, 1939.
8. Bourne, W. M., and Kaufman, H. E.: Specular microscopy of human corneal endothelium in vivo, Am. J. Ophthalmol. **81**:319, 1976.
9. Buschke, W., Friedenwald, J., and Moses, S.: Effects of ultraviolet irradiation on corneal epithelium, J. Cell. Comp. Physiol. **26**:147, 1945.
10. Campbell, F.: The influence of a low atmospheric pressure on the development of the retinal vessels in the rat, Trans. Ophthalmol. Soc. U. K. **71**:287, 1951.
11. Cogan, D.: Vascularization of the cornea, Trans. Am. Ophthalmol. Soc. **46**:457, 1948.
12. Cogan, D.: Studies on the clinical physiology of the cornea, Am. J. Ophthalmol. **32**:625, 1949.
13. Cogan, D., and Hirsch, E.: The cornea; permeability to weak electrolytes, Arch. Ophthalmol. **32**:276, 1944.
14. Cogan, D., and Kinsey, V.: Transfer of water and sodium chloride by osmosis and diffusion through the excised cornea, Arch. Ophthalmol. **27**:466, 1942.
15. Cogan, D., and Kinsey, V.: Transfer of water and sodium chloride by hydrostatic pressure through the excised cornea, Arch. Ophthalmol. **27**:696, 1942.
16. Cogan, D., and Kinsey, V.: Hydration properties of excised corneal pieces, Arch. Ophthalmol. **28**:272, 1942.
17. Cogan, D., and Kinsey, V.: Hydration properties of excised corneal pieces, Arch. Ophthalmol. **28**:449, 1942.
18. Cogan, D., and Kinsey, V.: Cornea: physiological aspects. Arch. Ophthalmol. **28**:661, 1942.
19. Cogan, D., Hirsch, E., and Kinsey, V.: Permeability characteristics of the excised cornea, Arch. Ophthalmol. **31**:408, 1944.
20. Cox. W. V., Kupferman, A., and Leibowitz, H. M.: Topically applied steroids in corneal disease, Arch. Ophthalmol. **88**:308, 1972.

21. Dann, W., and Darby, W.: Appraisal of nutritional status (nutriture) of humans, with especial reference to vitamin deficiency diseases, Physiol. Rev. **25:**326, 1945.
22. Davidson, E. A., and Meyer, K.: Chondroitin, a new mucopolysaccharide, J. Biol. Chem. **211:**605, 1954.
23. deRoetth, A.: Glycolytic activity of the cornea, Arch. Ophthalmol. **45:**1239, 1951.
24. Dohlman, C. H.: Incorporation of radioactive sulfate into the rabbit eye, Acta Ophthalmol. **35:**115, 1957.
25. Dohlman, C. H., Hedbys, B. O., and Mishima, S.: The swelling pressure of the corneal stroma, Invest. Ophthalmol. **1:**158, 1962.
26. Ferraro, A., and Roizin, L.: Ocular changes in rats on amino acid (valine)—deficient diet, Am. J. Ophthalmol. **30:**330, 1947.
27. Friedenwald, J., and Buschke, W.: Effects of excitement, of epinephrine, and of sympathectomy on the mitotic activity of corneal epithelium in the rat, Am. J. Physiol. **141:**689, 1944.
28. Friedenwald, J., Hughes, W., and Herrmann, H.: Acid base tolerance of the cornea, Arch. Ophthalmol. **31:** 279, 1944.
29. Gassett, A. R., et al.: Quantitative corticosteroid effect on corneal wound healing, Arch. Ophthalmol. **81:** 589, 1969.
30. Goldman, J. N., and Benedek, G. B.: The relationship between morphology and transparency in the nonswelling corneal stroma of the shark, Invest. Ophthalmol. **6:**574, 1968.
31. Hall, W., Sydenstricker, V., Hock, C., and Bowles, L.: Protein deprivation as cause of vascularization of cornea in rat, J. Nutr. **32:**509, 1946.
32. Harris, J.: Transport of fluid from the cornea. In Duke-Elder, S., editor: Transparency of the cornea, Springfield, Ill., 1960, Charles C Thomas, Publisher, p. 73.
33. Harris, J. E.: Factors influencing corneal hydration, Invest. Ophthalmol. **1:**151, 1962.
34. Herrmann, H., and Hickman, F.: Exploratory studies on corneal metabolism, Bull. Johns Hopkins Hosp. **82:** 225, 1948.
35. Herrmann, H., and Hickman, F.: The utilization of ribose and other pentoses by the cornea, Bull. Johns Hopkins Hosp. **82:**287, 1948.
36. Herrmann, H., Moses, S., and Friedenwald, J.: Influence of pontocaine hydrochloride and chlorobutanol on respiration and glycolysis of cornea, Arch. Ophthalmol. **28:**652, 1942.
37. Hodson, S.: The endothelial pump of the cornea, Invest. Ophthalmol. **16:**589, 1977.
38. Holt, M., and Cogan, D.: Permeability of the excised cornea to ions as determined by measurements of impedance, Arch. Ophthalmol. **35:**292, 1946.
39. Jakus, M.: The fine structure of the human cornea. In Smelser, G., editor: The structure of the eye, New York, 1961, Academic Press Inc., p. 344.
40. Kaufman, H. E., and Gasset, A. R.: Clinical experience with the epikeratoprosthesis, Am. J. Ophthalmol. **67:**38, 1969.
41. Kaye, G.: Studies on the cornea, III, J. Cell. Biol. **15:**241, 1962.
42. Kaye, G., and Pappas, G.: Studies on the cornea, I, J. Cell. Biol. **12:**457, 1962.
43. Kaye, G., and Pappas, G.: Studies on the cornea, II, J. Cell. Biol. **12:**481, 1962.
44. Kenyon, K. R., et al.: The systemic mucopolysaccharides, Am. J. Ophthalmol. **73:**811, 1972.
45. Kinoshita, J., and Masurat, T.: Aerobic pathways of glucose metabolism in bovine corneal epithelium, Am. J. Ophthalmol. **48**(2):47, 1959.
46. Kinsey, V. E.: Personal communication.
47. Klintworth, G. K.: The cornea—structure and macromolecules in health and disease, Am. J. Pathol. **89:** 719, 1977.
48. Langham, M.: Utilization of oxygen by the component layers of the living cornea, J. Physiol. **117:**461, 1952.
49. Langham, M.: Observations on the growth of blood vessels into the cornea. Application of a new experimental technique, Br. J. Ophthalmol. **37:**210, 1953.
50. Langham, M.: Glycolysis in cornea of rabbit, J. Physiol. **126:**396, 1954.
51. La Tessa, A., Teng, C., and Katzin, H.: The histochemistry of the basement membrane of the cornea, Am. J. Ophthalmol. **58:**171, 1954.
52. Levenson, J. E.: The effect of short term drying on the surface ultrastructure of the rabbit cornea: a scanning electron microscopic study, Ann. Ophthalmol. **5:**865, 1973.
53. Marin-Amat, M.: The physiological variations of the corneal curvature during life; their significance in ocular refraction, Bull. Soc. Belge Ophthalmol. **113:** 251, 1956.
54. Maurice, D.: The structure and transparency of the cornea, J. Physiol. **136:**263, 1957.
55. Maurice, D.: The physics of corneal transparency. In Duke-Elder, S., editor: Transparency of the cornea, Springfield, Ill, 1960, Charles C Thomas, Publisher.
56. Maurice, D., and Giardini, A.: Swelling of the cornea in vivo after the destruction of its limiting layers, Br. J. Ophthalmol. **35:**791, 1951.
57. Meyer, K., et al.: The mucopolysaccharides of bovine cornea, J. Biol. Chem. **205:**611, 1953.
58. Michaelis, M., and Quastel, J.: Site of action of narcotics in respiratory processes, Biochem. J. **35:**518, 1941.
59. Michaelson, I. C., and Campbell, A. C. P.: Anatomy of finer retinal vessels, and some observations on their significance in certain retinal diseases, Trans. Ophthalmol. Soc. U. K. **60:**71, 1940.
60. Mishima, S., and Maurice, D.: The effect of normal evaporation on the eye, Exp. Eye Res. **1:**46, 1961.
61. O'Brien, C., and Swan, K.: Carbaminoylcholine chloride in the treatment of glaucoma simplex, Arch. Ophthalmol. **27:**253, 1942.
62. Potts, A. M.: The nutritional supply of corneal regions in experimental animals, II, Am. J. Ophthalmol. **36:** 127, 1953.
63. Praus, R., and Dohlman, C. H.: Changes in the biosynthesis of corneal glycosaminoglycans during wound healing, Exp. Eye Res. **8:**69, 1969.

64. Refojo, M. F., et al.: Evaluation of adhesives for corneal surgery, Arch. Ophthalmol. **80:**645-656, 1968.
65. Schoenwald, R., and Ward, R.: Relationship between steroid permeability across excised rabbit cornea and octanol-water partition coefficients, J. Pharm. Sci. **67:** 786, 1978.
66. Smelser, G., and Ozanics, V.: Effect of local anesthetics on cell division and migration following thermal burns of cornea, Arch. Ophthalmol. **34:**271, 1945.
67. Swan, K., and White, N.: Corneal permeability: factors affecting penetration of drugs into the cornea, Am. J. Ophthalmol. **25:**1043, 1942.
68. Sydenstricker, V., et al.: Ocular manifestations of ariboflavinosis, J.A.M.A. **114:**2437, 1940.
69. Thoft, R. A., Friend, J., and Dohlman, C. H.: Corneal glucose flux, Arch. Ophthalmol. **86:**685, 1971.
70. Tonjum, A.: Permeability of rabbit corneal epithelium to horseradish peroxidase after the influences of benzalkonium chloride, Acta Ophthalmol. **53:**335, 1975.
71. vanWalbeek, K., and Neumann, H.: Studies of corneal transparency under various experimental conditions, Arch. Ophthalmol. **46:**482, 1951.
72. Von Bahr, G.: Corneal thickness; its measurement and changes, Am. J. Ophthalmol. **42:**251, 1956.
73. Waltman, S. R., and Kaufman, H. E.: In vivo studies of human corneal endothelial permeability, Am. J. Ophthalmol. **70:**45, 1970.
74. Weimar, V. L.: The sources of fibroblasts in corneal wound repair, Arch. Ophthalmol. **60:**93, 1958.
75. Weimar, V. L., and Haraguchi, K. H.: The development of enzyme activities in corneal connective tissue cells during the lag phase of wound repair. I, 5-nucleotidase and succinic dehydrogenase, Invest. Ophthalmol. **4:**853, 1965.
76. Weimar, V. L., and Haraguchi, K. H.: The development of enzyme activities in corneal connective tissue cells during the lag phase of wound repair. II, Formalin-resistant oxidase-like reaction Invest. Ophthalmol. **5:**14, 1966.
77. Wolter, J.: Innervation of the corneal endothelium of the eye of the rabbit, Arch. Ophthalmol. **58:**246, 1957.
78. Zaduniasky, J. A., et al.: Ion pumps in the cornea and their stimulation by epinephrine and cyclic AMP, Exp. Eye Res. **15:**577, 1973.

Chapter 4

Somatosensory sensations from the eye

HAROLD BURTON, Ph.D.

The body is richly innervated with receptors in the skin and deeper somatic structures that collectively mediate sensations of touch, light pressure, cold and warmth, pain, position, and movement. All of these sensations are also mediated by the somatic afferent receptors located in the eye and its adnexa; therefore a consideration of the general somatic afferent system is applicable to the special somatic information from the eye. However, a separate discussion is needed for the somatic innervation of the cornea because the density, distribution, and types of receptors in this structure differ to such an extent that only limited somatosensory sensations arise from the cornea. The somatic afferent pathways from the eye, which are associated with the trigeminal system, also present features that warrant a separate analysis from the general somatosensory pathways relaying through the spinal cord. Before considering the somatosensory receptors, the following section reviews some general concepts of receptor physiology.

THE GENERATOR POTENTIAL

In all receptors the critical response to an effective stimulus is a movement of electrical charge across the receptor's membrane. This flow of current produces a potential change that is localized to the membrane of the receptor. Such local potentials have been called generator potentials. This potential change is a *nonpropagated event*. This means that the amplitude of the generator potential is maximum at the receptor terminal and diminishes rapidly with distance from this point because the potential does not regenerate itself. In contrast, the action potential of a nerve fiber is cyclically reactivated as it propagates along the nerve fiber, and the nerve impulse is thereby conducted without decrement. In addition to being a *stationary* event, the generator potential differs further from the all-or-none action potential because its amplitude is *graded* as a direct function of the stimulus intensity. This characteristic of the generator potential allows receptor responses to vary faithfully with changes in the stimulating events; thus the output from sense organs is a reasonable analog of the stimulus (Fig. 4-1).

A stimulus generally causes positive current to flow across the membrane of the receptor axon terminal or specialized receptor cell from the outside to the inside of the receptor. This direction of current flow *reduces* the transmembrane potential in the area of the membrane surrounding the activated point, that is, the receptor is *depolarized*. The generator potential in rods and cones, however, arises from positive current flowing out of the receptor. In this situation the transmembrane potential increases or is *hyperpolarized*. Depolarizing generator potentials appear to be due to the same ionic mechanisms governing the depolarizing excitatory postsynaptic potentials (EPSP) of postsynaptic membranes. Thus the stimulus initiates a *nonselective* increase in the receptive membrane permeability to the monovalent ions Na^+, K^+, and Cl^-. This contrasts with the specific increased conductance for Na^+ ions

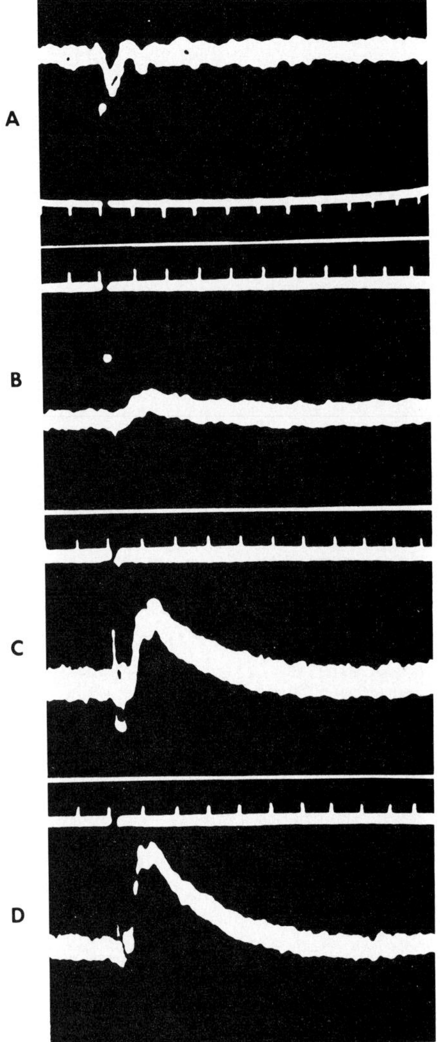

Fig. 4-1. Change in generator potentials in a pacinian corpuscle with increasing strength of mechanical compression. Stimulus level increases from **A** to **D** and occurrence of stimulus is indicated by break in 1 msec time marker trace. (From Gray, J. A. B., and Sato, M.: J. Physiol. [Lond] **122:**610, 1953.)

that triggers the action potential in nerve fibers. The transducer process, although not definitely understood, may be visualized as an opening of small holes in the receptor membrane (permeability change) by the application of an appropriate stimulus. There may be intermediate steps, such as the biochemical changes in visual pigments, that precede the permeability changes. Once the membrane's conductance is

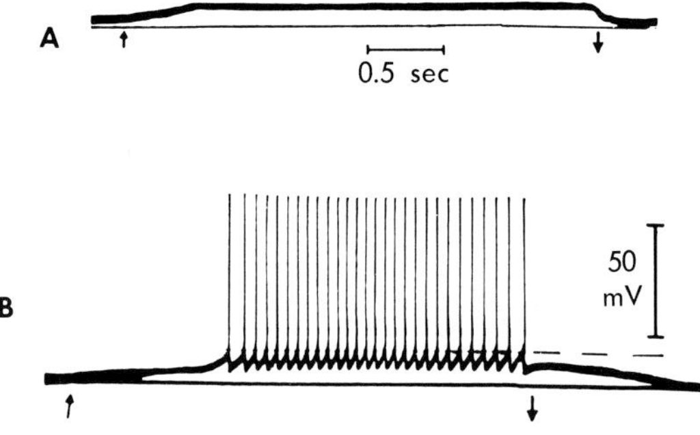

Fig. 4-2. Membrane potential changes recorded from crayfish stretch receptor. **A,** Weak stretch, onset of which is indicated by first arrow, sets up depolarizing generator potential that reflects intensity and duration of extension. **B,** When stronger stretch is imposed, generator potential reaches critical level of depolarization and impulses are initiated. (From Kuffler, S. W.: Excitation and inhibition in single nerve cells. In The Harvey lectures, series 54, New York, 1960, Academic Press, Inc, copyright © Academic Press, Inc.)

shifted, the ions will move through the membrane and down their respective electrochemical concentration gradients. If a more intense stimulus is applied, the conductance increases and more charged particles flow across the membrane. The degree of depolarization or hyperpolarization thus varies directly with the membrane's permeability and the latter varies with stimulus intensity. The process is limited by the total transmembrane potential. Once zero potential is reached, however, further increases in the membrane's permeability will be ineffective because there will be no net flow of current across the membrane. The conditions controlling the adaptation of the receptor will also limit the size of the generator potential.

The importance of the generator potential is that it is essential for initiating the afferent nerve discharge (Fig. 4-2). In many situations involving somatosensory receptors the terminal part of the sensory nerve fiber is the receptor element that is specialized to form generator potentials. Conducted nerve impulses then originate in the nerve fiber; in myelinated fibers the impulses usually start at the first node of Ranvier proximal to the unmyelinated receptor portion of the axon. Most mechanoreceptors belong to this class of receptors. In the retina, however, the receptor cell is specialized to form only a generator potential. Under these circumstances the electronic changes in the receptor cells are transmitted across synaptic junctions to cells capable of forming action potentials.

CODING STIMULUS INTENSITY

Increased stimulus intensity is expressed directly within some receptors as a larger generator potential; this is signaled to the central nervous system by the number of nerve impulses activated per unit of time. In these receptors stimulus intensity is coded by frequency modulation of the firing rates. A second method of coding intensity is to vary the number of sensory units that are active as a direct function of stimulus intensity. This is relevant to coding intensity because all afferents in a given area of a sensory organ are not equally sensitive to the application of a stimulus. This may be because the receptors in the area have real differences in threshold or because of the spatial distribution of receptive fields. The latter is significant, since sensory units are most sensitive only in the centers of their receptive fields. Consequently, the receptors activated from one locus may appear to have different thresholds because only a few of the sensory units will have the centers of their receptive fields activated and only when the stimulus intensity rises will additional receptors be engaged. Perceived intensity correlates best with the number of sensory receptors responding multiplied by their impulse frequency. Thus the nervous system uses both the mechanisms of increased rate of discharge and increased number of active peripheral receptors to encode stimulus intensity. Both of these mechanisms of coding intensity are not used by all receptors because either the rate of firing or the number of active sensory units is used to signal some other aspect of the stimulus. For example, vibration-sensitive neurons utilize firing rate to signal stimulus frequency; therefore intensity of vibration can be signaled only by the number of active units.

ADAPTATION IN SENSORY RECEPTORS

When a suprathreshold stimulus is applied to a sensory receptor, the response is a burst of one or more impulses. However, if the stimulus is maintained at a constant level, the rate of firing is not maintained, but after the initial response it either ceases immediately or grows progressively slower until it reaches a minimum rate, which is held constant (Fig. 4-3). This property, called *adaptation,* is present in all sensory receptors. The mechanisms responsible for adaptation differ in various receptors, but generally changes in the amplitude of the generator potential are responsible for changes in the frequency of firing. Thus adaptation may result from a failure to maintain the magnitude of the generator potential during prolonged stimulation. In rapidly adapting receptors the amplitude of the generator potential quickly falls below the threshold level for triggering impulses; in slowly adapting receptors the generator potential declines but is maintained above the threshold for impulse initiation.

The degree of adaptation has an important influence on the nature of the message the brain receives from the sense organs. Rapidly adapting receptors react only to the onset and cessation of stimulation, whereas slowly adapting receptors additionally respond to the duration of the stimulus. The adaptive properties of mechanoreceptors are quite broad, and classifications based only on the duration of a response to a physical displacement of the end-organ do not adequately represent the diversity of information provided by these receptors.

In general, the various receptors that are necessary for the maintenance of homeostasis adapt very little. Thus the pressure receptors in the carotid sinus and aortic arch, tension receptors in the lungs, and muscle spindles are all slowly adapting (Fig. 4-4). Many of the receptors for pain in the cornea have persistent responsiveness to a maintained stimulus. Many tactile responses, like those found in the conjunctiva and eyelids, adapt very quickly, so after a contact occurs, the end-organ is immediately ready to transmit a new signal.

The precise mechanism of adaptation is unknown, although it is likely that in many instances the physical or chemical properties of the structure surrounding receptor terminals are somehow responsible. For example, it has been found that the lamellated connective tissue cap-

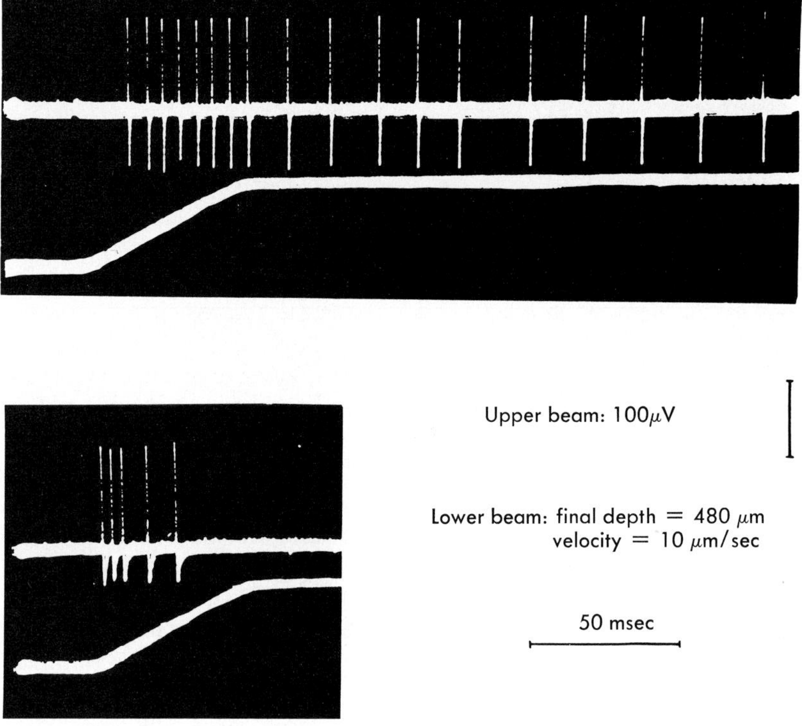

Fig. 4-3. Examples of slowly (upper) and rapidly (lower) adapting responses to 100 and 480 μm indentations, respectively. Responses are from single median nerve fibers innervating glabrous forepaw skin of raccoon. (From Munger, B. L., and Pubols, L. M.: Brain Behav. Evol. **5**:367, 1972.)

sule, which surrounds the peripheral nerve terminal in pacinian corpuscles, only transmits velocity-dependent forces to the terminal axon within the center of the capsule; static pressures are dissipated uniformly around the layers of the capsule such that no net static pressures are available to excite the nerve terminal.[27] In addition to the contribution of nonneural structures to adaptation, alteration in the membrane properties of the receptor terminals and the receptor segments of sensory nerve fibers must also be considered. For example, a sustained generator potential may be artificially imposed on the nerve terminals of pacinian corpuscles that have had their capsules removed. However, even under these conditions, only a rapidly adapting response is excited in the nerve fiber. The nerve fibers "accommodate" to the sustained depolarization, and this property may be partly responsible for the behavior of these receptors. The precise mechanism for accommodation in nerve fibers is obscure, but may be related to interruption of Na^+ conductance inactivation or delayed onset of increased K^+ conductance.[27]

CUTANEOUS RECEPTORS

In comparison with the photoreceptors, somatosensory receptors are structurally simple because the majority are not associated with specialized receptor cells like rods and cones. However, a diverse population of mechanoreceptors, thermoreceptors, nociceptors, and proprioceptors have long been recognized, and these have been classified in a variety of ways that reflect morphologic and functional distinctions. The significance of many of these classifications can be traced to Johannes Müller's "law of specific irritability," which held that perceived sensory quality is bound to the activity of a particular sensory organ because each type of receptor is specifically evolved to re-

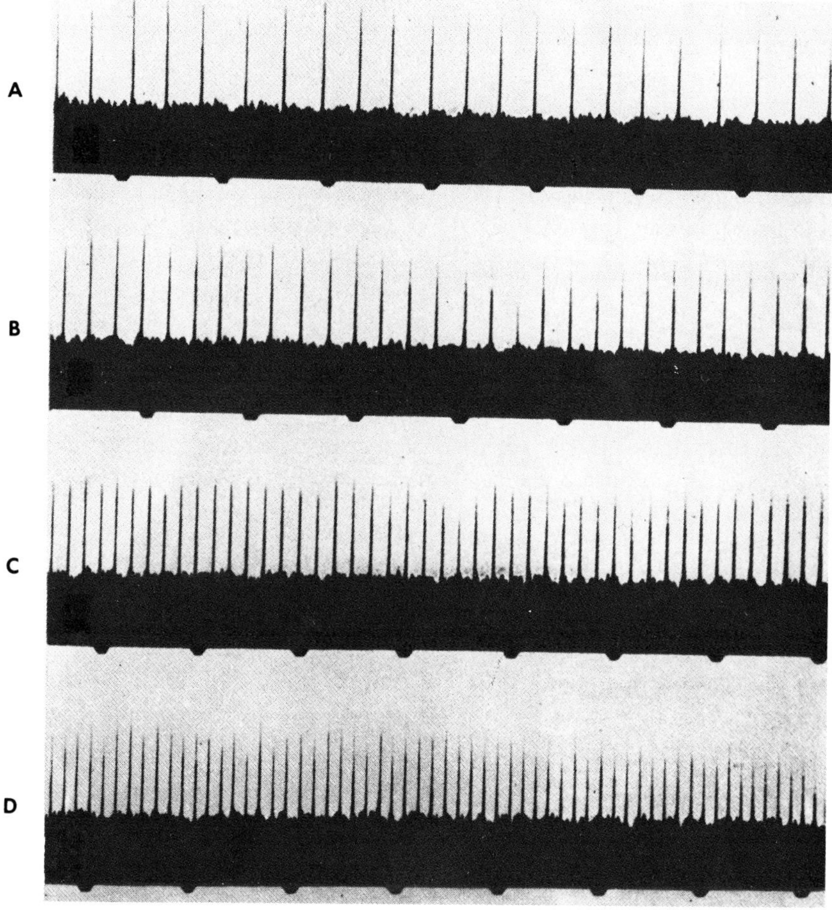

Fig. 4-4. Afferent impulses from single end organ in carotid sinus stimulated by constant pressures within sinus. **A,** 40 mm Hg. **B,** 80 mm Hg. **C,** 140 mm Hg. **D,** 200 mm Hg. Time marked in fifths of seconds. (From Bronk, D., and Stella, G.: Am. J. Physiol. **110:**708, 1935; The Rockefeller University Archives.)

spond to a definite kind of stimulus. The stimulus that is most readily detected by a receptor is called the *adequate stimulus*. Usually this stimulus excites the receptor at low intensity levels, while the same receptor has a high threshold to inadequate stimuli. For example, retinal receptors have very low thresholds to light, but are also excited by mechanical deformation. In both instances the identical sensation of light is perceived and no perceptual distinction exists between the adequate and inadequate stimulus. The result is due to the existence of unique and specific central connections for each of the major classes of peripheral receptors. Thus Müller's doctrine regarding specialized sensory fibers includes the notion that different sensations occur by virtue of the central connections of different sets of sensory nerve fibers. This arrangement has been called the principle of the "labeled line."[32]

A major interest in the study of cutaneous sensation has been the demonstration of correlations between the structural types of receptors in all kinds of skin and the types of sensations perceived from stimulation of these tissues.* A difficulty has been identifying the common morphologic elements that are responsible for equivalent sensations in such diverse structures as the conjunctiva of the eye and the hairy skin of the body or glabrous skin of the

*See references 1, 3, 8, 18, 21, and 22.

hands and feet. One of the problems with earlier attempts to obtain these correlations was excessive emphasis on minor variations in end-organ morphology and failure to recognize that the distribution of cutaneous receptor endings varies with age, occupation, and region of skin. In addition, purely technical circumstances added to the confusion.[50]

Current classifications are based on the morphology of cutaneous nerves and the structure of somatosensory receptors. The nerves vary in the degree of myelinization and in diameter from a maximum of 20 μm to a minimum of 0.2 μm. The morphology of receptors may be classified by subdividing the end-organs into those associated with free nerve endings and encapsulated endings.[1,8,18,33] The free nerve endings are sometimes associated with specialized epithelial cells or structures, whereas the encapsulated endings are enclosed within a covering of connective tissue.

CLASSIFICATION ACCORDING TO FIBER DIAMETER

Gasser[15] recognized that fibers could be grouped according to their diameters and that these separate groups, because of their different average conduction velocities, formed separate components of the compound action potential recorded from a peripheral nerve bundle. Each group's existence was inferred from the fact that only a synchronously conducted discharge in one group of fibers could summate algebraically and thereby form a specific hump in the waveform of the compound action potential. In the original scheme proposed by Gasser[15] A fibers, with conduction velocities of 5 to 120 m/sec and 2 to 20 μm diameters, cover the range of myelinated fibers that are both motor and sensory. B fibers refer to autonomic motor fibers (although these were subsequently deleted in considerations of mixed cutaneous nerves). The C fibers are unmyelinated, include the most slowly conducting (0.4 to 2.2 m/sec), and are the smallest fibers (0.2 to 1.5 μm). The A wave is presently divided into α, β, and δ components. A second classification of axonal diameters, which was originally based on muscle afferent nerves of the cat,[26] has occasionally been designated for cutaneous nerves. According to this schema, four groups exist: I (72 to 120 m/sec, 12 to 20 μm), II (24 to 66 m/sec, 5 to 11 μm), III (6 to 20 m/sec, 2 to 4 μm), and IV (0.4 to 2.0 m/sec, 0.2 to 1.5 μm). The correspondences between the two classifications are not exact, but with minor boundary adjustments, groups I to IV are representative of, respectively, A-α, A-β, A-δ, and C fibers.

Many subsequent experiments on single cutaneous nerve fibers generally confirmed the conduction velocity–diameter classification scheme and further correlated these with the functional specificity of the single fibers. The exclusiveness of the modality–fiber diameter correlations is directly related to fiber diameter. For example, the largest fibers (A-α) serve only for afferent fibers from muscles; the next largest myelinated fibers (A-β) provide input for a number of mechanoreceptive sensations such as vibration, joint position, and light touch. However, contact sensibility and some muscle afferent input are also provided by the smaller myelinated fibers. In addition, touch, temperature, and pain are signaled by separate fibers within the A-δ and C fiber groups. These classifications were originally designated from analyses of segmental nerves innervating the limbs, but they are also appropriate for the nerve branches of the trigeminal root, which innervates the eye.

INNERVATION OF THE CORNEA—FREE NERVE ENDINGS

In humans approximately 70 to 80 bundles of nerve filaments supply over 1000 small (0.5 to 5 μm) axons to the cornea. The majority of these fibers pass to the cornea from ciliary branches of the ophthalmic nerve. A third of the ocular fibers in primates may also come from an orbitocillary branch of the maxillary nerve.[41] The fibers pass between the sclera and choroid in the stratum perichorioideum or directly in the sclera together with fibers destined for the ciliary body, iris, and sclera. The fibers destined for the cornea diverge from those passing to the other structures just proximal to the limbus. On entering the cornea directly from the sclera, episclera, and conjunctiva, the bundles of axons are arranged in a radial manner[52] (Fig. 4-5). Only C fibers and

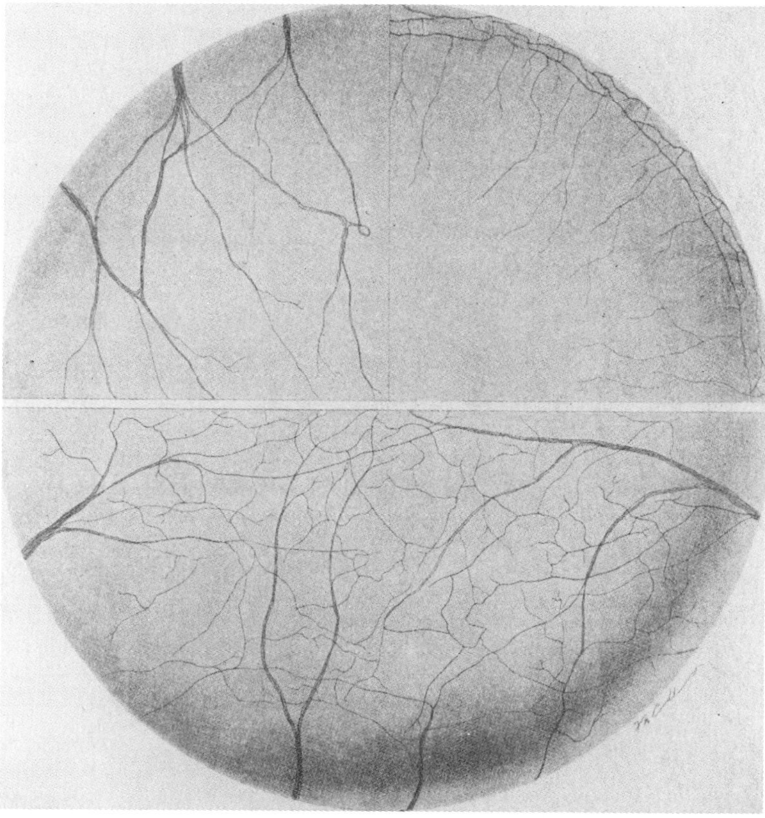

Fig. 4-5. Arrangement and behavior of the various nerve bundles that enter the periphery of the cornea in different planes. Composite drawing based on observations from a large number of methylene blue–stained rabbit corneas. Upper left quadrant, form and behavior of nerve bundles entering the cornea from the scleral position. Upper right quadrant, form and behavior of nerve bundles entering the cornea from the episcleral and subconjunctival positions. Lower half of drawing, manner in which the plexiform pattern of nerve fibers arises from the nerve bundles; passage of nerve bundles across the center of the cornea from limbus to limbus. Note that not all the nerve bundles in the cornea are radially disposed. (From Zander, E., and Weddel, G.: J. Anat. **85:**68, 1951, Cambridge University Press.)

A-δ fibers innervate the cornea. All the myelinated nerve fibers to the cornea lose their multiple myelin wrappings within 1 mm of their entrance from the limbus; thus the entire plexus within the cornea is formed by fine unmyelinated fibers. The ramification of these fibers forms a plexus with twigs, loops, or brushes whose branches are distributed throughout the substantia propria from Descemet's to Bowman's membrane. A second plexus forms before individual axon terminals enter the epithelial layer.

A single sensory axon distributes over 50 to 200 mm^2 of the corneal surface; its terminal field may lie up to three fourths of the distance across the cornea from the axon's entry point at the limbus.[52] Recordings from single nerve fibers have confirmed this innervation pattern by finding that a fiber can be activated from receptive fields covering a fourth to a third of the corneal surface, although some fibers have very small receptive fields of less than 1mm^2.[29] The innervation density of the cornea is greatest across a central region approximately 5 mm in diameter and falls off gradually towards the limbus. The ability to localize stimuli on the cornea is directly related to this innervation pattern.

Except for the extreme density of endings, the fine structure of the free nerve end-organs in the cornea is comparable to descriptions of these endings in other parts of the body.[8] The axons in the stroma are still encased in a Schwann cell covering. When a collateral branch enters the epithelium, it loses this remaining sheath entirely. However, the sensory endings are never directly exposed to intercellular spaces because they course within invaginations of the epidermal cells. A smaller number of the axons also terminate within the stroma. The final end-organ may be swollen and filled with mitochondria and occasional vesicles.[48] Some investigators[20] have suggested that there are two types of sensory endings in the cornea. One is associated with thicker axons and has numerous mitochondria, a dense whirl of neurofilaments, many neurotubules, and only an occasional small, clear, round vesicle. These are found on intraepithelial and stromal axons, but are much less numerous in the center of the cornea. A second type contains few neurotubules, some mitochondria, a reticular network of neurofilaments, and usually two types of vesicles consisting of one to three large (100 nm), dense-cored vesicles, and three to four small (60 nm) clear vesicles. These endings are associated with thin, beaded axons that are especially prevalent in the intraepithelial plexus.

The endings containing agranular vesicles resemble cholinergic endings.[30] This is consistent with histochemical demonstrations of acetylcholinesterase activity in the axolemma of nerves in the stroma, epithelial cell membranes, and intraepithelial endings in the cornea of the rat. These endings are not likely to come from an autonomic innervation of the cornea because the cholinesterase reaction goes away only after ophthalmic nerve transections and not following cervical sympathectomy and ciliary ganglionectomy.[45] Corneal sensitivity can also be lowered following application of hemicholinium.[14] However, these results do not positively demonstrate that stimuli are transduced in the free nerve endings of the cornea through a cholinergic step. No agranular vesicles have been seen in free nerve endings in the skin.[8]

The cornea also receives endings with small, dense-cored vesicles from the superior cervical and ciliary ganglion.[46-48] Most of these sympathetic endings are distributed in the stroma. They reach the cornea through the posterior ciliary nerve. The epithelial layer contains an adrenergic innervation in primates, but only in embryos.[11] However, the persistence of intraepithelial adrenergic fluorescence has been demonstrated in adult rats when the endings have been preloaded with norepinephrine and nialamide and protected with propranolol prior to fixation.[46,47]

SENSATIONS FROM THE CORNEA

Pain

The correlation between the presence of free nerve endings and sensations of pain reaches its highest level in the cornea because this structure has one of the greatest concentrations of free nerve endings and one of the highest sensitivities to painful stimuli. Sensitivity to pain increases from the periphery to the center of the cornea. At first, sensitivity increases quickly until a fairly high value is reached, and then more slowly. This pattern corresponds to the distribution of the corneal nerves, the center of the cornea being richly supplied with free nerve endings. The horizontal meridian is more richly supplied than the vertical meridian, and the temporal half is more sensitive than the nasal half. Measurements of the sensitivity of the eye with Cochet and Bonnet's esthiometer indicated that the cornea has the greatest sensitivity and in successively declining order were the eyelids, caruncle, and conjunctiva (e.g., 12, 53 to 58, 63, and 83 to 96 mg/0.0113 mm^2.[35]) The sensitivity of the cornea decreases with age. Reversible declines in sensitivity are also seen in normal contact lens wearers.

Pain from the cornea and from all the diseases of the outer ocular coats that produce irritation of the conjunctiva and cornea is quite distinctive and localized. It differs entirely from the deep-seated neuralgic pain of iritis and iridocyclitis. The latter pain is deep and throbbing and is always worse at night or early in the morning. It is frequently widespread in its reference, and patients may complain of earache, pain in the teeth, or pain over one of the sinuses.

There is extensive evidence that the skin is endowed with two types of receptors for responding to noxious stimuli. One receptor type gives rise to the quality of pricking, which reaches consciousness rapidly, and the other gives rise to the quality of burning, which reaches consciousness less rapidly, but which persists longer. Similar experiences accompany stimulation of the cornea. Numerous experiments in humans and animals have shown that a class of A-δ fibers produce the sensation of pricking pain and some of the C fibers are responsible for burning pain. For example, high-threshold A-δ fibers respond to an intense mechanical stimulus like a pinprick with a brief high-frequency burst of impulses that does not persist (Fig. 4-6). Repeated stimulation of the same receptive spot results in inactivation of an A-δ nociceptor.[38] The more slowly conducting C fibers may also respond only to intense mechanical stimuli, although some of these fibers are polymodal and will be activated by any damaging stimuli. The response from C fibers does not adapt, and the discharge may persist after the noxious stimulus is withdrawn. In addition, hyperalgesic responses may be noted following the initial activation of the C fiber's receptive field (Fig. 4-7). The free nerve endings in the cornea also arise from the same classes of fibers. The major difference between stimulation of the cornea and skin is that even low mechanical forces, which would be perceived as innocuous tactile stimuli on the skin, frequently elicit sensations of irritation and pain from the cornea.

Sense of touch and pressure

The question of a true sense of touch in the cornea has been the subject of lively dispute. It was maintained by earlier workers that there is no true sensation of touch or pressure in the cornea. Mechanical stimulation of the cornea, according to these observers, results only in the sensation of pain. They showed that a vibratory stimulus applied to the cornea results in a steady persistent sensation and not one of true vibration. It was claimed that the failure to react to vibratory stimuli proved the absence of touch as a mode of corneal sensation. Further proof of the absence of a pressure sense was

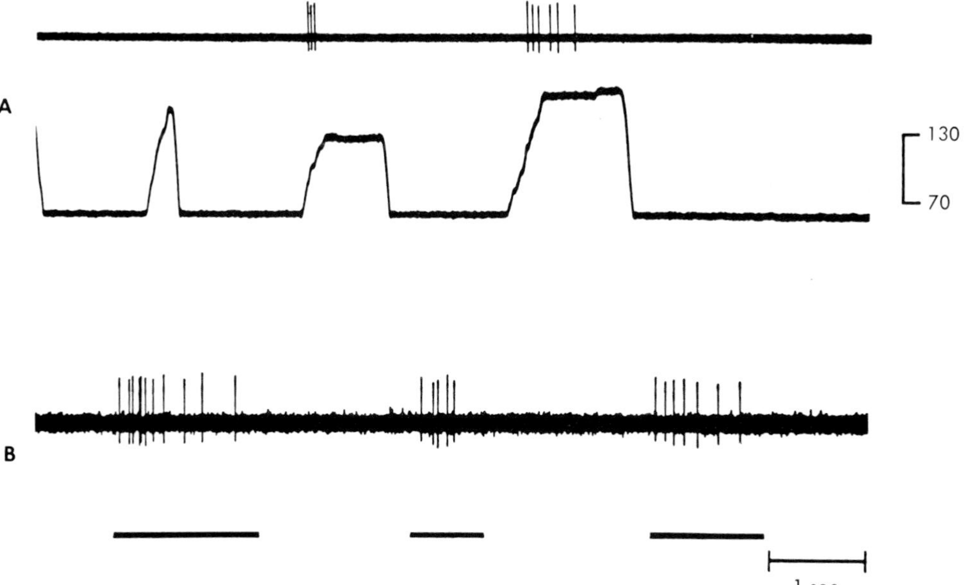

Fig. 4-6. A, Response of A-δ fiber to high-intensity mechanical stimulation. Receptive field of this nociceptive fiber was in cat hairy skin. Stimulus consisted of needle mounted in conjunction with strain gauge, output of which is noted in grams in lower record. Note that response occurs only at high force levels and does not persist for duration of stimulus. **B,** Response during pinching of skin in center of receptive field by serrated forceps. Bars indicate duration of pinching. (From Burgess, P. R., and Perl, E. R.: J. Physiol. [Lond] **190:**541, 1967.)

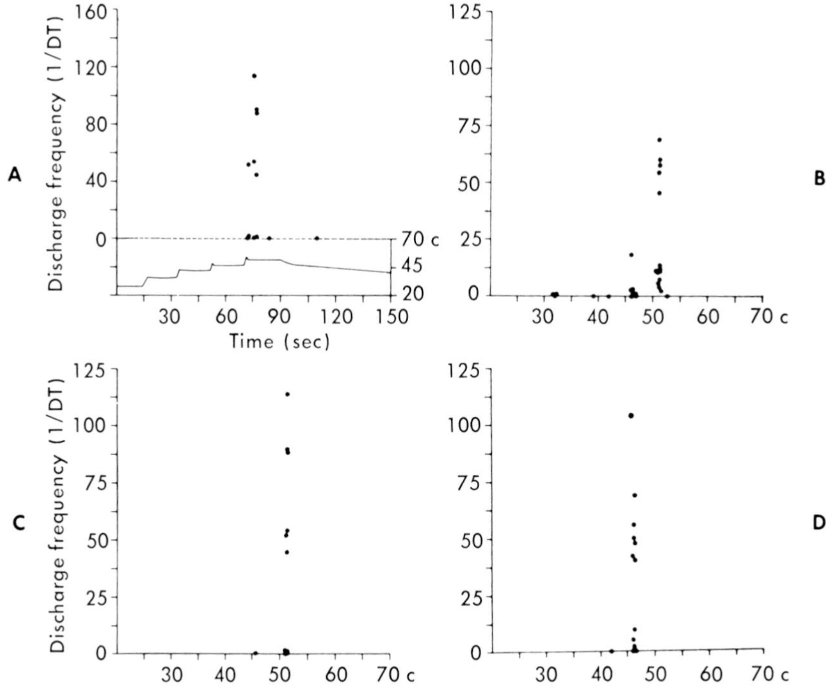

Fig. 4-7. Response of C fiber, polymodal nociceptor during repeated heating of receptive field with thermode fixed in contact with one area of skin. **A,** Discharge frequency is plotted on ordinate with respect to time on abscissa, and 5° C step changes in thermode temperature are noted below dashed lines. **B** to **D,** Graphs of discharge rate versus thermode temperature for three sequential heating cycles. **B** was taken from data in **A.** Note that discharges during third cycle began at 46° C in **D** in contrast to threshold of 51° C in **B** during initial cycle. (From Bessou, P., and Perl, E. R.: J. Neurophysiol. **32:**1025, 1969.)

suggested by poor stereognostic sensation in the cornea. It has also been found that the rate of disappearance of sensation in the cornea after the administration of cocaine follows a simple arithmetic curve. It has been argued that if, in addition to the pain nerves, there were others carrying a pressure sense, the curve of the fall in sensation would be complex rather than the simple one obtained. The latter assumes that tactile sensations must travel only in large diameter myelinated fibers; however, tactile inputs may also be conveyed by A-δ fibers.

Quantitative studies of the tactile sensibilities of the cornea showed that only one threshold could be determined for the corneal eyelid closure reflex in various laboratory animals.[43] This threshold was found to be 0.5 to 1 g/mm^2 in rhesus monkeys and was usually higher by a factor of less than 10 in all other animals. These results were comparable to threshold values in humans of 0.2 to 0.3 g/mm^2. These thresholds were based on determinations over the more sensitive center of the cornea; the margins of the cornea were found to be about four to five times less sensitive. The force levels that caused the corneal reflex in humans also corresponded to the threshold for pain. Consequently, because of the similarities in thresholds for the corneal reflex in all instances, and because only a single threshold level was observed, the results were interpreted as indicating that only pain sensations could be elicited from the cornea. In contrast, separate force levels for touch and pain sensations could be determined when the skin of the eyelids was tested.

Not all observers are in accord with this point of view. Nafe and Wagoner[34] believe that the methods used in the earlier studies were faulty and account for their erroneous conclusions. The stimulus used by the earlier investigators was a hair, capable of arousing touch or pressure. On the cornea it aroused only painful sensations; hence, it was concluded that the cornea was sensitive only to pain. Nafe and Wagoner pointed out that it is possible to stimulate the cornea either painfully or painlessly at will when other means of mechanical stimulation are used, and they concluded that some free nerve endings, which are the only sensory endings present, may transmit impulses associated with touch. The result depends on one factor only, the intensity of stimulation. Recent single-fiber recordings from the isolated cornea have clearly shown that different sensory units in the cornea are responsive to touch or pain even though receptors for both sensations exist as free nerve endings.[29] It has been suggested[20] that the free nerve endings associated with larger diameter axons and with many mitochondria are responsible for touch sensitivity in the cornea because the ultrastructure of these terminals resembles the axonal endings within the encapsulated endings of mechanoreceptors like pacinian corpuscles.

The reality of the sense of touch or pressure has been corroborated by the employment of Sjöqvist's operation of medullary tractotomy.[13,17,42,51] In this operation the descending tract of the trigeminal nerve is cut in the medulla for the relief of trigeminal neuralgia. Rowbotham[40] examined two patients after this operation and established the fact that although the cornea was insensitive to pain, the contact of a wisp of cotton or of a camel's hair brush was appreciated as touch. This has been confirmed by Grant, Groff, and Lewey.[17] Lele and Weddell[25] also found that if carefully graded stimuli are employed, one can readily distinguish both touch and pain in the cornea.

It is now generally agreed that the cornea does possess a true sense of touch, and that contact with the cornea by a foreign body produces either a sensation of touch or a painful sensation, depending on the intensity of stimulation.

Cold and warmth

In regular skin, innocuous thermal sensations have been correlated with the responses of two classes of thermoreceptors that are primarily associated with A-δ fibers, especially in skin innervated by the trigeminal nerve.[18,21] For example, cold and warm receptors discharge at different levels, respectively, to cold and warm temperatures over the range of 17° to 39° C or 27° to 50° C. In addition, cold and warm receptors respond phasically with a high-frequency burst of activity to step changes of temperature, respectively, in the cold and warm directions; the rate of discharge is directly correlated with the magnitude of the temperature change. The minimal thermal change needed to activate these receptors is the same as the stimulus levels that provide a threshold thermal sensation. For example, lowering skin temperature by 0.15° C causes a cold receptor to change its response rate and also elicits a sensation of cold.

It has generally been assumed that because of the presence of free nerve endings, some spots on the cornea would be specifically sensitive to innocuous temperatures. Part of this assumption is based on the common observation that cold sensations resist the action of local anesthetics more than the sensation of pain. For example, when local anesthetics such as phenacaine (Holocaine) are instilled in the eye, a disagreeable sensation of cold increases at first with each drop instilled. The cornea may be totally insensitive to pain after the first drop or two of a 1% solution of phenacaine; yet the instillation of another drop still produces a decidedly unpleasant cold sensation. Eventually this sensation of cold disappears, and further drops fail to produce it. It has been concluded from this that the cold receptors must lie at a deeper level in the cornea than the free nerve endings that yield the sensation of pain. However, Kenshalo[23] has argued that no thermoreceptors are present on the cornea that are comparable to those in the skin because the thresholds for detecting changes in temperature are much greater than ±0.15° C.

The presence of a sensation of heat or warmth from the cornea has also been controversial. One group[34] showed only a sense of pain when the cornea was stimulated by placing an applicator at 51.5° C 2 mm in front of the limbus; another group reported a sensation of heat with infrared heat or a fine jetstream of warmed air.[25] Subsequent experiments that involved the use of greater control over the restriction of thermal stimuli have clearly shown only irritation or pain can be elicited directly from the cornea even with very minor shifts away from 33° C.[2] Previous reports of thermal sensations apparently were due to stimulation of the large number of very sensitive temperature spots on the eyelids and conjunctiva[2] or possibly infrared transmission through the cornea to heat-sensitive spots on the iris.[10]

SENSATIONS ASSOCIATED WITH FREE NERVE ENDINGS IN OTHER PARTS OF THE EYE

The sensibilities of the other parts of the eye are important clinically, especially pain, and may be considered here briefly. The extraocular muscles produce very marked, painful sensations when they are stretched. Both the retina and the optic nerve are said to be devoid of pain; in this respect they are like brain tissue.

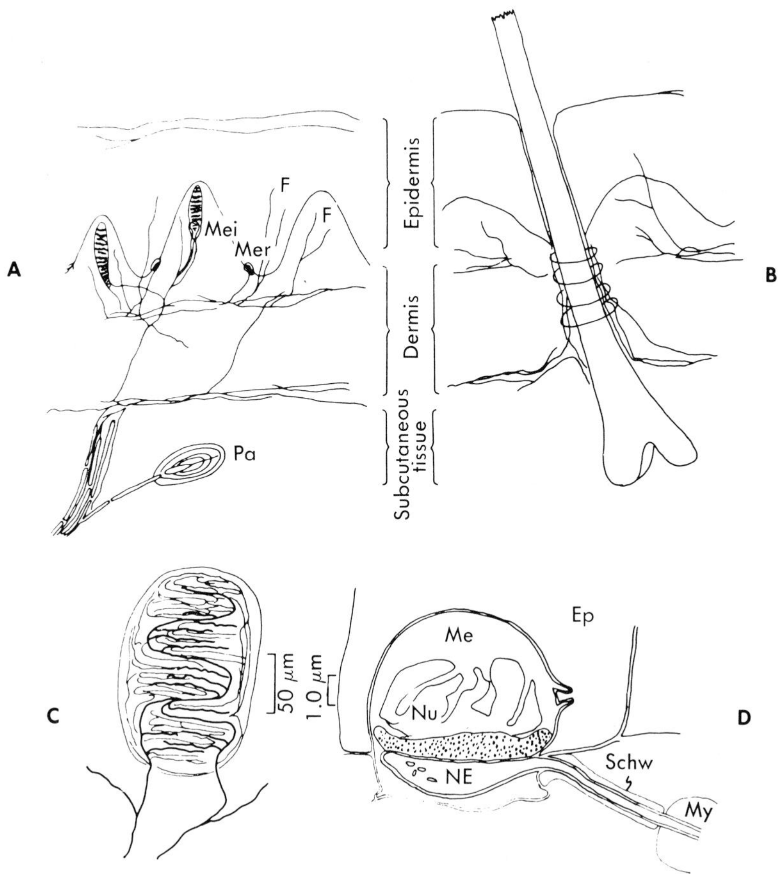

Fig. 4-8. Schematic drawings of location and morphology of some somesthetic receptor organs. **A,** Free nerve endings, *F,* are found within and below the epidermal layer; Meissner's corpuscles, *Mei,* are frequently found in the dermal ridges; Merkel cell–neurite complexes, *Mer,* are generally located in the basal layer of the epidermis; pacinian corpuscles, *Pa,* which are located in nonvisceral regions, occur in subcutaneous tissue, around the periosteum, and in or near joints. **B,** Hair follicles have several nerve nets from different sensory units entwined around the root of the hair, especially above the level of the hair bulb. Many of these terminals lie parallel to the root; some endings wrap around the air. **C,** Enlarged view of a Meissner corpuscle shows the sandwiching of sensory nerve terminals between connective tissue leaflets. **D,** Drawing taken from an electron micrograph shows the Merkel cell, *Me,* and its associated nerve terminal, *Ne.* The Merkel cell is a modified epithelial cell characterized by its large, lobulated nucleus, *Nu.* The sensory terminal arises from a myelinated fiber, *My;* the Schwann cell covering, *Schw,* extends up to the region of contact between the nerve terminal and the Merkel cell. (From Somjen, G.: Sensory coding in the mammalian nervous system, New York, 1972, Appleton-Century-Crofts.)

The sheaths of the optic nerve, that is, the meninges, must be supplied with pain fibers, since patients with retrobulbar neuritis characteristically complain of pain behind the eyeball on moving the eye. Pain can be referred to the ophthalmic nerve distribution in cases of severe supraorbital pain associated with acute glaucoma or with frontal and ethmoidal sinusitis.

In experimental animals Tower[49] tested the sensitivity of the sclera, iris, and lens and the spontaneous discharge in the long ciliary nerves that took place when intraocular pressure was artificially elevated. The sclera was found to be poorly supplied with nerve fibers, gauged by the response to equivalent stimulation applied to the cornea, but the form of the response to pressure, pinching, and pulling was similar to that in the cornea. The iris, in contrast to the sclera, was found to be exceedingly sensitive. This is borne out well in clinical experience, since the pain of acute iritis is very severe. The conjunctiva and eyelids are nearly as sensitive to painful stimuli as the cornea. The lens was not found to be demonstrably sensitive to pricking or to touch, but pushing on its anterior curvature by means of a blunt glass rod provoked a small burst of nerve impulses. The lens appeared to be without surface sensibility but seemed to possess what might be characterized as a position sensibility in modest proportions. The receptors in question are probably located in the ciliary body, since no nerve fibers reach the lens itself. Raising the intraocular pressure produced an increase in the spontaneous activity of these nerve fibers, suggesting the production of painful impulses.

SOMATIC SENSATIONS ASSOCIATED WITH SPECIALIZED CUTANEOUS RECEPTORS

Many specialized mechanoreceptors are found in the skin, some of which are also common to the innervation of the eyelids, conjunctiva, and eye muscles. A complete discussion of these receptors is beyond the scope of this chapter, but several recent excellent reviews are available.[1,3,21] The following will consider the morphology and responses of some of the best known mechanoreceptors (Fig. 4-8).

Traditionally, mechanoreceptors have been classified by the rate of adaptation of their responses to constant physical distortion of the tissue. Differences in the duration of activity led to identification of "rapidly" and "slowly" adapting receptors. Advances in the control of mechanical stimulation procedures have provided the basis for a newer and simpler classification system that covers more aspects of the continuum of responses to tactile stimuli. It has become apparent that different mechanoreceptors respond to various aspects of the time course of a physical displacement. Some respond only to the position or displacement of the sense organ; others respond to the first and second time derivatives of displacement (e.g., the velocity or the acceleration of the displacement). Consequently, mechanoreceptors, as a population in any skin, can be viewed as providing a continuum of sensitivity that includes position detectors, slow to fast velocity detectors, and very fast transient or acceleration detectors.[3]

MORPHOLOGY OF SOME MECHANORECEPTORS

Expanded tip endings—specialized free nerve endings

The expanded tip terminals are from myelinated fibers. These specialized endings consist of a close contact between some epidermal cell or structure, like hairs, and an enlarged nerve terminal that surrounds or is itself partially encased within the nonneural element. The most prevalent example of these endings are hair follicle–neurite systems (Fig. 4-8). A single hair follicle may have 10 to 15 individual nerve fibers entwined like a palisade around its shaft above the level of the hair bulb. A single myelinated fiber may send terminals to either one or several hundred hair follicles. Differences exist in the caliber of hairs innervated and the extent of neural ramification around the hairs.[33] The majority of the receptors directly associated with hairs are either velocity or acceleration detectors whose excitation is triggered by hair bending.

Another important member of the expanded tip terminals is Merkel's disc. The nonneural element here consists of modified epithelial cells that are found among the cells of the stratum germinativum at the base of the rete pegs of the epidermis (Fig. 4-8). In the electron microscope they are distinguished by accumulations of dense secretory vesicles close to the surface contacted by the nerve terminal.[33] In many mammals Merkel's cell–neurite complexes are found clustered together within elevated, translucent domes in the hairy skin. These receptors are sensitive to static displacement and velocity; this suggests that Merkel's cell–neurite complexes signal sensations of pressure and touch.[22]

Encapsulated endings

The encapsulated nerve endings consist of a nerve terminal completely enveloped by several layers of connective tissue. The pacinian corpuscle is the most extensively studied encapsulated receptor. The Krause end-bulb is a related structure that has long been recognized in the conjunctiva of the eye.[52] Encapsulated structures are located in the deep connective tissue and periosteum, within the mesenteries and along the sides of blood vessels. The pacinian receptor is a whitish, somewhat translucent oval that may reach 1 mm in length; it has the appearance of an onion because it consists of many successive concentric fibroblastic lamellae (Fig. 4-8). The outer lamellae are separated by a viscous fluid, while the innermost core consists of several layers of flattened sheets of lamellar cell cytoplasm with collagen fibers interspersed between the layers. The central neural terminal is swollen and filled, especially along its perimeter, with numerous mitochondria. The terminal has no Schwann cell covering, although the first node of Ranvier is also contained within the corpuscle. The parent fiber is large and responds to single mechanical displacements of the corpuscle with one or two impulses. It is thought that the fluid between the cellular lamellae, which surround the nerve terminal, functions as a high-pass filter such that only high-velocity transient components of a stimulus are transmitted to the terminal. Consequently, the frequency of an oscillating tuning fork would be transferred, but the static pressure of the fork on the skin would not be conveyed by this receptor.

Another example of an encapsulated receptor is Meissner's corpuscle, which is usually located at the apex of a dermal papilla in glabrous skin (Fig. 4-8). A single Meissner corpuscle receives the terminals of two to five myelinated fibers, and one nerve fiber may supply Meissner's corpuscles over a field of several papillary ridges. These terminals are especially prevalent in the fingers. They may also be present on the inner portion of the eyelid. The corpuscle consists of a vertical cylinder of interlaced stacks of nerve terminals and flat but turgid connective tissue cells. It is likely that the Meissner corpuscle is sensitive only to transient stimulation.

CORRELATION OF RECEPTOR RESPONSES WITH SOMESTHETIC SENSIBILITIES

Sense of touch and pressure

Some of the mechanoreceptors responsible for the sensation of "light" touch or pressure have already been mentioned, such as the hair follicle receptors, the Merkel cell–neurite complex, the Meissner corpuscles, and many of the free nerve endings from the A-δ fibers. However, the sensation of light touch is not just the consequence of exciting one or all of these receptor groups, but involves discrimination of movement, degrees of steady pressure, and the duration of the contact. Quantitative analyses of some of these receptors have shown that several aspects of "light" touch may be due to the stimulus-response functions of the peripheral receptors. For example, in Fig. 4-9 the response to a slight skin indentation was a high-frequency burst of impulses that was followed by a maintained but lower frequency discharge. The latter was present throughout the indentation period. The rate of discharge in the initial or dynamic phase is related to the velocity plus the instantaneous intensity of the indentation; the rate of activity during the second or static phase is related to the indentation amplitude, and the persistence of the response is coupled to the duration of the stimulus. Within physiologic ranges, the level of activity during these two phases of this receptor's response increases as the intensity and velocity of indentation increases. These receptors thus

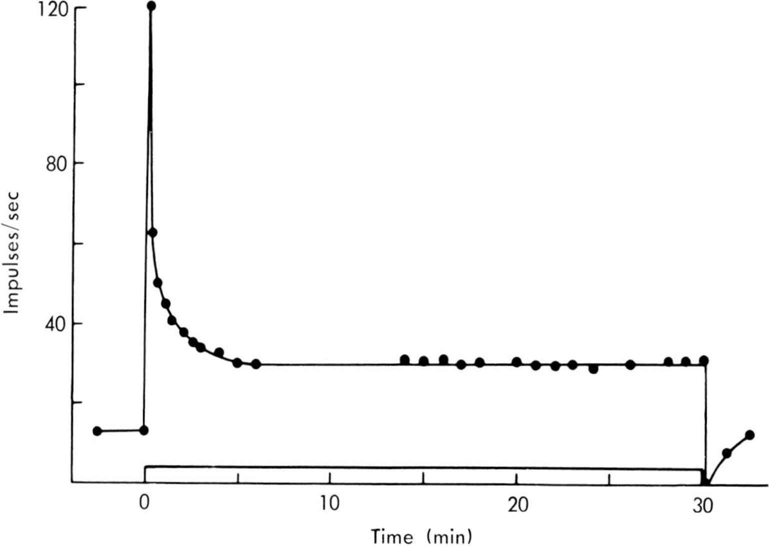

Fig. 4-9. Response of slowly adapting cutaneous mechanoreceptor to persistent and constant pressure to its receptive field in hairy skin of monkey. Note initial high-frequency burst of activity when stimulus is applied and subsequent decline to constant discharge rate, which is maintained for duration of stimulus. (From Iggo, A.: Acta Neuroveg. **24:**225, 1963.)

may serve to signal the degree and duration of pressure on the skin. They may also provide some information regarding stimulus movement because the initial phasic response is tied to the speed of stimulus application.

Sense of vibration

The sensation of vibration has also been carefully examined at the receptor level. For example, pacinian corpuscles are capable of responding to stimuli that oscillate up to 500 times/sec. Once the intensity of the oscillation is sufficient to trigger a response, these receptors will fire an impulse during each cycle of the stimulus. The requisite stimulus intensities vary for different vibration frequencies before a receptor's response will be entrained for each stimulus cycle. Pacinian corpuscles generally require the least stimulus intensity at 200 Hz and higher intensities for frequencies above or below this point. A second group of receptors is most sensitive to low-frequency vibrations, that is, below 100 Hz. Minimal human detection of oscillatory mechanical stimulation corresponds to the minimal stimulus levels necessary to entrain these receptors.[44] This suggests that the sensation of vibration depends on the one to one responses to mechanical oscillation shown by these receptors.

TRIGEMINAL NERVE AND ITS NUCLEI

Peripheral nerves

Almost all the nerve terminals originating from somatosensory receptors throughout the face, including all the inputs from the eye, are gathered together into the sensory root of the trigeminal nerve. This root arises from the central processes of the *semilunar* or *gasserian ganglion*. The trigeminal root may be regarded as a compressed group of afferent dorsal roots, and the semilunar ganglion contains the cell bodies of the three principal peripheral divisions of the trigeminal nerve: *ophthalmic, maxillary,* and *mandibular* nerves.

The first, or ophthalmic, division of the fifth cranial nerve supplies the somatic afferent innervation to the eyeball and surrounding structures via three main branches: the lacrimal, frontal, and nasociliary nerves. The *lacrimal nerve* supplies sensory branches to the lacrimal gland, the conjunctiva, and the skin of the lateral part of the upper eyelid. Section of the lacrimal nerve may stop secretion of tears due to the interruption of secretomotor fibers that are conveyed to the lacrimal gland via a communicating parasympathetic branch, which passes from the maxillary nerve to the lacrimal nerve. The *frontal nerve* supplies the forehead, scalp, upper eyelids, and frontal sinus via the supraorbital and supratrochlear nerves. The *nasociliary nerve* provides innervation for the mucous lining of the nasal cavity, the skin in and around the nose, the lower eyelids, and the cornea. The long ciliary nerves, which lie in continuity with the nasociliary nerve, convey sympathetic innervation to the dilator pupillae muscle in the iris along with the afferent fibers to the iris and ciliary body. Vasodilator fibers may also be present in this nerve and may account for the reflex ciliary flush caused by an imbedded corneal foreign body.

Trigeminal nuclei

The trigeminal root fibers enter the brain at the pons and distribute projections to various divisions of the sensory nuclei of the trigeminal nerve (Fig. 4-10). Since many clinical syndromes are associated with the trigeminal nerve and nuclei and these frequently include somatic motor and sensory disturbances of the ocular tissue, it is necessary to review some of the connections and physiology of the trigeminal nuclei.

There are three components to the sensory nuclei of the trigeminal nerve situated laterally in the brain stem in an extensive caudal to rostral column of cells that reaches from the second cervical segment to the mesencephalon (Fig. 4-11). In the discussion that follows, only the caudal *nucleus of the spinal tract* and the *main or principal sensory nucleus* in the pons will be considered; the *mesencephalic sensory nucleus* will not be discussed. Many of the sensory fibers that enter at the pons divide into a short ascending and long descending branch, and thus collaterals of many sensory fibers are sent to both divisions of the trigeminal nuclei.

On the basis of clinical studies, it has gener-

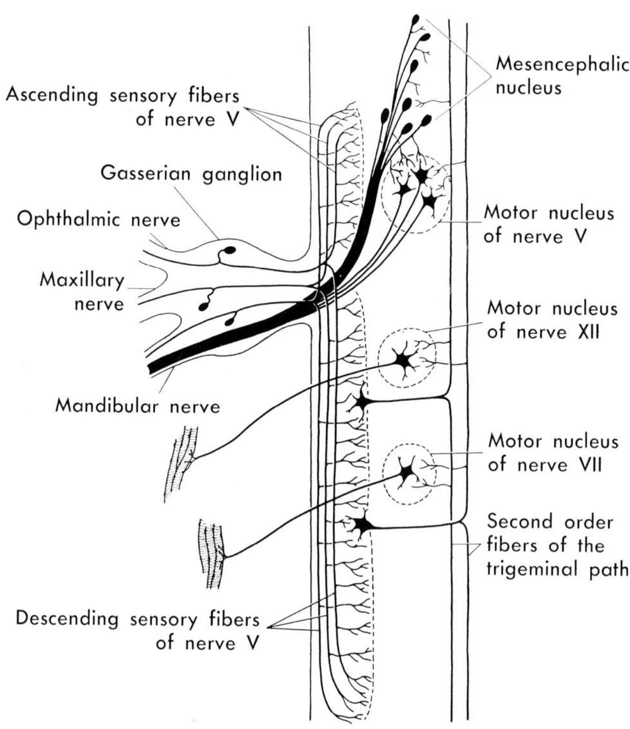

Fig. 4-10. Diagram of nuclei and central connections of trigeminal nerve. (Modified from Ranson.)

ally been proposed that the main sensory nucleus provides the brain stem relay for tactile sensations from the head and that facial pain and temperature sensibilities are transmitted by all of the nucleus of the spinal tract. Recent findings, however, indicate that this is an oversimplification of the complex connections and physiology of the trigeminal nuclei.

The nucleus of the spinal tract is divisible into three distinct portions called the nucleus caudalis, interpolaris, and oralis.[36] Each of these nuclei receive input from all the trigeminal roots via the descending spinal trigeminal tract of the fifth nerve. The fibers from each of the three branches of the nerve are somatotopically arranged in this tract with those from the ophthalmic found most ventrally and the mandibular most dorsally. It has also been found in humans, but not in lower animals, that the ophthalmic division extends furthest caudally in the tract beyond the spinomedullary junction. Afferent projections to all trigeminal nuclei are also somatotopically ordered dorsoventrally with the ophthalmic representation most ventral.

The organization of the nucleus caudalis is of special interest because it appears to be a major terminus for specific pain and temperature afferents from the head.[31,37,39] Structurally, this nucleus resembles the dorsal horn of the spinal cord in having a marginal zone, a subjacent substantia gelatinosa, and a deep proprius known as the magnocellular zone. In contrast to the rest of the trigeminal nuclei, a medial to lateral order has been found within the marginal zone with the ophthalmic representation furthest lateral.[37] However, the specific location for the cornea within the nucleus caudalis has not yet been determined.

The clinical demonstration that pain from the head was localized within the extreme caudal aspects of the spinal trigeminal tract[13,51] was probably based on selective deafferentation of projections to the nucleus caudalis and its associated marginal zone. However, examination of patients whose trigeminal neuralgia has

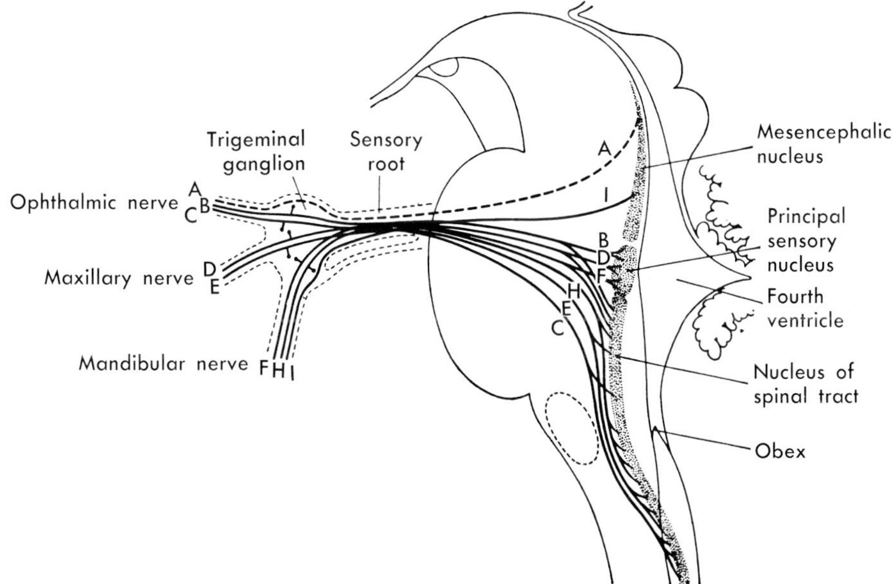

Fig. 4-11. Nuclei receiving the primary afferent fibers of the trigeminal nerve. *A,* Proprioceptive fibers from ocular muscles. *B,* Tactile and pressure fibers from ophthalmic area. *C,* Pain and temperature fibers from ophthalmic area. *D,* Tactile and pressure fibers from maxillary area. *E,* Pain and temperature fibers from maxillary area. *F,* Tactile and pressure fibers from mandibular area. *H,* Pain and temperature fibers from mandibular area. *I,* Proprioceptive fibers from muscles of mastication. Proprioceptive fibers are believed to occur in all three divisions of the trigeminal nerve. (From Williams, P. L., and Warwick, R.: Functional neuroanatomy of man being the neurology section from Gray's antatomy, British ed. 35, Philadelphia, 1975, W. B. Saunders Co.)

been treated with transections of the medulla at these caudal levels showed that some tactile sensitivity was also eliminated. This is not surprising in view of the large number of mechanoreceptor projections to all of the trigeminal nuclei. (See Darian-Smith[9] for review.) Reappearance of neuralgia in the trigeminal region in some of these patients also emphasizes that the functional organization of the trigeminal nuclei is complex and that pain transmission from the face may involve more than just a relay in nucleus caudalis. For example, it is known that the caudal part of this nucleus sends a significant intranuclear projection to the main sensory nucleus and to rostral parts of the nucleus of the spinal tract.[6,7,16] Finally, painful stimuli on the face may elicit responses from cells in the adjoining caudal medullary reticular formation[4] and in or near the main sensory nucleus.[12] These results suggest that it is not plausible to regard the main sensory nucleus as only a center for tactile sensations. It is important, however, to recognize that the caudalmost extent of the nucleus of the spinal tract is vitally concerned with facial pain.

Studies of the central connections from the trigeminal nuclei have shown that these projections mainly join into the appropriate somatotopic portion of the somatosensory pathways from the spinal cord and medulla to the thalamus. All parts of the trigeminal nuclei contribute to these projections, but to a variable degree, with the major outflow from the main sensory nucleus.[5] There is no evidence that the projections from the eye occupy a separate and distinct portion within these connections. The face is found within the medial portion of the ventroposterior nuclei of the thalamus and along the lateral aspect of the postcentral gyrus in the cortex. Within the thalamus the representation from the eye is located dorsally, and within the cortical map it is medial and posterior to the mandibular representation. These maps are entirely based on the projection from mechanoreceptors. Recent studies of the pain and temperature projections from the nucleus caudalis have indicated that these modalities may have additional separate relationships within the thalamus.[6] Again, however, knowledge of the corneal representation is still absent.

The connections responsible for the corneal blink reflex provide a further example of the complexity of the central neural mechanisms associated with the somatosensory inputs from the eye. The blink reflex in humans and lower animals is based on the fusion of a short and long latency excitation of the orbicularis oculi muscle. The short latency response probably is mediated by direct connections from the nucleus caudalis to the facial nucleus.[6,19] The bases for the long latency responses are not known, but may involve additional connections from the trigeminal nuclei to the reticular formation and from the latter to the facial nucleus. Relays through the reticular formation from visual or auditory relay nuclei or other central structures could also be the bases for the initiation of blinking.

Clinical symptomatology

A number of diseases of the eye are caused by lesions of the fifth cranial nerve. These may be divided into those in which the peripheral nerve terminals and peripheral fibers are affected and those in which the primary lesion is central—in the brain stem nuclei or in the gasserian ganglion. Recurrent erosion of the corneal epithelium, which sometimes follows abrasions of the cornea, belongs in the first group. The epithelium heals over an abrasion in the usual fashion and the eye seems cured, but a few days (occasionally weeks or months) later the patient experiences severe pain in the eye, usually on awakening in the morning, and the epithelium is lifted up in a bleb that strips off easily. This process of healing and recurrent formation of blebs may occur many times before the condition finally yields to appropriate treatment. It is possible that the superficial plexus of nerve fibrils is damaged as the result of the original injury and forms small neuromas instead of allowing the fibers to make their way among the new epithelial cells. As a result of this, the epithelium is without its normal nerve supply. Because of the absence of some trophic influence or other unknown factor, these cells never become firmly united to the underlying stroma.

Another example of a peripheral neuropathy is superficial punctate keratitis, which is caused by a lesion (probably virus induced) of the fifth cranial nerve endings in the corneal epithelium. The punctate character and position of the infiltrates suggest that they represent the distribution of nerve fibrils passing through Bowman's membrane from the deep plexus to the superficial plexus. Dendritic keratitis, which is also an inflammatory condition resulting from a herpetic virus infection, owes its characteristic features to the pattern of the nerve trunks as they course through the stroma from the limbus toward the center of the cornea. Corneal sensitivity decreases in these conditions.

Herpes zoster ophthalmicus and neuroparalytic keratitis are examples of conditions caused by disease or injury to the gasserian ganglion, sensory root, or first division. The former is due to infection of the ganglion with a filtrable virus. The latter condition occurs after section of the ophthalmic division or of the sensory root of the fifth cranial nerve in the operation for major trigeminal neuralgia. In the condition of neuroparalytic keratitis the cornea becomes edematous, blebs are formed, and iridocyclitis is observed. In addition, protective reactions to the presence of foreign bodies on the cornea are no longer available because the cornea is insentient. Various theories have been proposed to account for neuroparalytic keratitis. One hypothesis is that due to the sectioning of the ophthalmic division, the trophic influence of the nerve is reduced and the corneal nutrition subsequently suffers in much the same way as the skin becomes atrophic in the distribution of a peripheral neuropathy. It has also been suggested that histamine is released into the tissues after the nerve fibers are destroyed and the trophic effects of this substance may then be responsible for the keratitis. However, antihistaminics do not adequately treat this condition and some other substance (bradykinin or 5-hydroxytryptamine) may be released locally.

It has also been suggested that the primary cause of neuroparalytic keratitis is interference with the formation of tears due to damage of the greater superficial petrosal nerve during operation. In support of this hypothesis it has been observed that the incidence of neuroparalytic keratitis was greatest in those patients in whom operative trauma was considerable, especially when the seventh cranial nerve was damaged along with the fifth cranial nerve. To avoid this serious complication in the operation for trigeminal neuralgia, surgeons may merely section the posterior two thirds of the sensory root.

Measurement of corneal sensitivity can often be used to distinguish between a benign episcleritis that may require local therapy and scleritis. The latter is a severe granulomatous condition of the sclera in which the collagen fibers are severely disrupted. Corneal sensitivity is rarely affected in episcleritis because the condition is a superficial inflammation of the sclera, whereas 75% of the patients with scleritis show decreased sensitivity.[28] The latter probably occurs because the nerve fibers running within the sclera are compromised by the disruption of the sclera.

Various complex reflex phenomena are associated with the ophthalmic inputs to the trigeminal nuclei; most of these cannot be explained adequately at the present time. Some examples of these clinical phenomena follow.

Reflex pain in ocular lesions, such as acute glaucoma, usually follows the distribution of the fifth cranial nerve, but when it is severe, it may be localized in areas supplied by other nerves. In these situations, the pain is often accompanied by sweating, vomiting, or diffuse headaches.

Pressure on the eyeball slows the heart rate (so-called oculocardiac reflex). This may be of such severity as to be critical—cardiac arrest has occurred during ocular surgery. To better evaluate the factors producing a serious type of oculocardiac reflex, Kirsch and associates[24] investigated the electrocardiographic changes induced during various types of operations performed on the eye. In 30% of the patients significant changes occurred in the electrocardiogram, and in two patients (4%) temporary cardiac arrest occurred. Digital pressure on the globe, manipulation of the ocular muscles, and direct pressure on the orbital contents following enucleation were implicated. The reflex could be eliminated entirely by retrobulbar anesthesia; therefore these authors recommend that retrobulbar anesthesia be used in all operations performed for strabismus, retinal detachment, and enucleation, whether or not general anesthesia is employed.

Other reflex phenomena include reflex weeping, sneezing, and disturbances in gastric motility with incorrect refractions and headaches plus photophobia when the retina is stimulated with excessive light. The term photophobia is sometimes erroneously used to denote dazzling, which is not true photophobia. The term should be only used to describe pain arising from exposure to excessive light. It has frequently been noted that in those corneal lesions accompanied by extreme photophobia, the lesion is situated in the superficial layers of the cornea, with the degree of pathology related to the superficiality of the lesion. For example, extreme photophobia is always present with imbedded foreign bodies. This may be accounted for by the superficial arrangement of the corneal nerves, the plexuses being in the anterior stroma and subepithelium, with the fibrils from the latter almost reaching the surface. Photophobia may occur in other than local ocular diseases, such as epidemic meningitis or trigeminal neuralgia. Photophobia may be due largely to the vasodilation that occurs locally as an antidromic reflex. Most of this vasodilation, which takes place in the ciliary muscle, accounts for the fact that the photophobia generally disappears when cycloplegia is induced.

REFERENCES

1. Andres, K. H., and von Düring, M.: Morphology of cutaneous receptors. In Iggo, A., editor: Handbook of sensory physiology. Vol. II, The somatosensory system, Berlin, 1973, Springer-Verlag, p. 1.
2. Beuerman, R. W., Maurice, D. M., and Tanelian, D. L.: Thermal stimulation of the cornea. In Anderson, D. J., and Matthews, B., editors: Pain in the trigeminal region, Amsterdam, 1977, Elsevier, pp. 413-422.
3. Burgess, P. R., and Perl, E. R.: Cutaneous mechanoreceptors and nociceptors. In Iggo, A., editor: Handbook of sensory physiology. Vol. II, The somatosensory system, Berlin, 1973, Springer-Verlag, p. 29.
4. Burton, H.: Somatic sensory properties of caudal bulbar reticular neurons in the cat, Brain Res. **11:**357, 1968.
5. Burton, H., and Craig, A. D., Jr.: Distribution of trigeminothalamic projection cells in cat and monkey, Brain Res. **161:**515, 1979.
6. Burton, H., et al.: Efferent projections from temperature sensitive recording loci within the marginal zone of the nucleus caudalis of the spinal trigeminal complex in the cat, J. Comp. Neurol. **183:**753, 1979.
7. Carpenter, M. B., and Hanna, G. R.: Fiber projections from the spinal trigeminal nucleus in the cat, J. Comp. Neurol. **117:**117, 1961.
8. Cauna, N.: The fine structure of the sensory receptor organs in the auricle of the rat, J. Comp. Neurol. **136:** 81, 1969.
9. Darian-Smith, I.: The trigeminal system. In Iggo, A., editor: Handbook of sensory physiology. Vol. II, The somatosensory system, Berlin, 1973, Springer-Verlag, pp. 271-314.
10. Dawson, W. W.: The thermal excitation of afferent neurones in the mammalian cornea and iris. In Hardy, J. D., editor: Temperature. Its measurement and control in science and industry, New York, 1963, Reinhold.
11. Ehinger, G., and Sjöberg, N.-O.: Development of the adrenergic nerve supply in man and guinea pig, S. Zellforsch. **118:**579, 1971.
12. Eisenman, J., Landgrem, S., and Novin, D.: Functional organization in the main sensory trigeminal nucleus and in the rostral subdivision of the nucleus of the spinal trigeminal tract in the cat, Acta Physiol. Scand. **59:**5, 1963.
13. Flaconer, M. A.: Intramedullary trigeminal tractotomy and its place in the treatment of facial pain, J. Neurol. Neurosurg. Psychiatry **12:**297, 1949.
14. Fitzgerald, G. G., and Cooper, J. R.: Acetylcholine as a possible sensory mediator in rabbit corneal epithelium, Biochem. Pharmacol. **20:**2741, 1971.
15. Gasser, H. S.: Conduction in nerves in relation to fiber types, A. Res. Nerv. Ment. Dis. Proc. **15:**35, 1935.
16. Gobel, S., and Purvis, M. B.: Anatomical studies of the organization of the spinal V nucleus: the deep bundles and the spinal V tract, Brain Res. **48:**27, 1972.
17. Grant, F., Groff, R., and Lewey, F.: Section of decending spinal root of the fifth cranial nerve, Arch. Neurol. **43:**489, 1940.
18. Hensel, H. H.: Cutaneous thermoreceptors. In Iggo, A., editor: Handbook of sensory physiology. Vol. II, The somatosensory system, Berlin, 1973, Springer-Verlag, p. 79.
19. Hiraoka, M., and Shimamura, M.: Neural mechanisms of the corneal blinking reflex in cats, Brain Res. **125:** 265, 1977.
20. Hoyes, A. D., and Barber, P.: Ultrastructure of the corneal nerves in the rat, Cell Tissue Res. **172:**133, 1976.
21. Iggo, A.: Cutaneous receptors. In Hubbard, J. I., editor: The peripheral nervous system, New York, 1974, Plenum Publishing Corp., pp. 347-404.
22. Iggo, A., and Muir, A. R.: The structure and function of a slowly adapting touch corpuscle in hairy skin, J. Physiol. (Lond) **200:**763, 1969.
23. Kenshalo, D. R.: Comparison of thermal sensitivity of the forehead, lip, conjunctiva, and cornea, J. Appl. Physiol. **15:**987, 1960.
24. Kirsch, R., et al.: Electrocardiographic changes during ocular surgery and their prevention by retrobulbar injection, Arch. Ophthalmol. **58:**348, 1957.
25. Lele, P., and Weddell, G.: The relationship between neurohistology and corneal sensibility, Brain, **79:**119, 1956.
26. Lloyd, D. P. C.: Neuron patterns controlling transmission of ipsilateral hind limb reflexes in cat, J. Neurophysiol. **6:**293, 1943.
27. Loewenstein, W. R.: Mechano-electric transduction in the Pacinian corpuscle. Initiation of sensory im-

pulses in mechanoreceptors. In Loewenstein, W. R., editor: Handbook of sensory physiology. Vol. I, Principles of receptor physiology, Berlin, 1971, Springer-Verlag.

28. Lyne, A. J.: Corneal sensation in scleritis and episcleritis, Br. J. Ophthalmol. **61:**650, 1977.
29. Mark, D., and Maurice, D.: Sensory recording from the isolated cornea, Invest. Ophthalmol. **16:**541, 1977.
30. Matsuda, H.: Electron microscopic study of the corneal nerve with special reference to the nerve endings, Acta Soc. Ophthalmol. Jpn. **72:**880, 1968.
31. Mosso, J. A., and Kruger, L.: Spinal trigeminal neurons excited by noxious and thermal stimuli, Brain Res. **38:**206, 1972.
32. Mountcastle, V. B.: Sensory receptors and neural encoding: introduction to sensory processes. In Mountcastle, V. B., editor: Medical physiology, ed. 14, St. Louis, 1980, The C. V. Mosby Co.
33. Munger, B. L.: Pattern of organization of peripheral sensory receptors. In Loewenstein, W. R., editor: Handbook of sensory physiology. Vol. I, Principles of receptor physiology, Berlin, 1971, Springer-Verlag.
34. Nafe, J., and Wagoner, K.: Insensitivity of cornea to heat and pain derived from high temperatures, Am. J. Psychol. **49:**631, 1937.
35. Norn, M. S.: Conjunctival sensitivity in normal eyes, Acta Ophthalmol. **51:**58, 1973.
36. Olszewski, J.: On the anatomical and functional organization of the spinal trigeminal nucleus, J. Comp. Neurol. **92:**401, 1950.
37. Poulos, D. A., et al.: Localization of specific thermoreceptors in spinal trigeminal nucleus of the cat, Brain Res. **165:**144, 1979.
38. Perl, E. R.: Is pain a specific sensation? J. Psychiatr. Res. **8:**273, 1971.
39. Price, D. D., Dubner, R., and Hu, J. W.: Trigeminothalamic neurons in nucleus caudalis responsive to tactile, thermal, and nociceptive stimulation of the monkey's face, J. Neurophysiol. **39:**936, 1976.
40. Rowbotham, G.: Observations on effects of trigeminal denervation, Brain **62:**364, 1939.
41. Ruskell, G. L.: Ocular fibers of the maxillary nerves in monkeys, J. Anat. **118:**195, 1974.
42. Sjöqvist, O.: Studies on the pain conduction in the trigeminal nerve, Acta Psychiatr. Scand. **17**(suppl.):1, 1938.
43. Strughold, H.: Mechanical threshold of cornea-reflex of the usual laboratory animals, Am. J. Physiol. **94:** 235, 1930.
44. Talbot, W. H., et al.: The sense of flutter-vibration: comparison of the human capacity with response patterns of mechanoreceptive afferents from the monkey hand, J. Neurophysiol. **31:**301, 1968.
45. Tervo, T.: Histochemical demonstration of cholinesterase activity in the cornea of the rat and the effect of various denervations on the corneal nerves, Histochemistry **47:**133, 1976.
46. Tervo, T., and Palkama, A.: Sympathetic nerves to the rat cornea, Acta Ophthalmol. **54:**75, 1976.
47. Tervo, T., and Palkama, A.: Adrenergic innervation of the rat corneal epithelium, Invest. Ophthalmol. **15:** 147, 1976.
48. Tervo, T., et al.: Fine structure of sensory nerves in the rat cornea: an experimental nerve degeneration study, Pain **6:**57, 1979.
49. Tower, S.: Unit for sensory reception in the cornea, J. Neurophysiol. **3:**486, 1940.
50. Weddell, G., Palmer, E., and Pallie, W.: Nerve endings in the mammalian skin, Biol. Rev. **30:**159, 1955.
51. Weinberger, L. N., and Grant, F. C.: Experiences with intramedullary tractotomy. III, Studies in sensation, Arch. Neurol. Psychiatry **48:**355, 1942.
52. Zander, E., and Weddell, G.: Observations on the innervation of the cornea, J. Anat. **85:**68, 1951.

Chapter 5

The extraocular muscles

Part one

Anatomy, physiology, and pharmacology

RONALD M. BURDE, M.D.

Any naturally occurring condition under which humans do not coordinate and control the movements of both eyes to produce a unified image is considered abnormal. This section will explore the components involved in the initiation, coordination, and completion of eye movements.

ARTICULATION OF THE EYEBALL IN THE SOCKET—ORBIT AND FASCIAE

Each eyeball is suspended in its bony orbital cavity by a complex matrix of fascia and muscle (Fig. 5-1). The two bony orbits are, roughly, quadrilateral pyramids whose medial walls are parallel to each other. The lateral walls produce approximately a 90 degree angle (Fig. 5-2). The widest part of each pyramid is about 1.5 cm posterior to the orbital margin, and the walls taper to the orbital opening giving the cavity a somewhat pear-shaped appearance.

The fasciae of the orbit extend from the orbital apex to the orbital rim and by dissection may be divided into check ligaments, septa, muscle sheaths, and suspensory ligaments. In reality the fascia is one structure that, along with the viscoelastic properties of the extraocular muscles, resists the initiation of ocular motion and acts as a fluid brake to assist the smooth cessation of motion once it has been started.

The orbital fascia is continuous with the dural sheaths surrounding the optic nerve as it enters the orbit through the optic foramen. The dura mater splits at this point. The external leaf becomes the periosteum of the bones forming the orbital cavity (periorbita), which in turn is continuous with the periosteum of the facial bones at the orbital rim. From the periosteum at the orbital rim a fascial sheath, known as the orbital septum, extends frontally into the lids (Fig. 5-3). This septum acts as a barrier to the escape of the orbital fat and as a volume-limiting tambour when the orbital contents are traumatized with subsequent hemorrhage or as a "fire wall" in the presence of infection.

At the apex of the orbit the inner layer of dura continues forward as the dural sheath of the optic nerve. At the globe the dura participates in the formation of, or blends into, the sclera.

Tenon's capsule is a fascial sheath enclosing the sclera, which becomes distinctly evident anterior to the equator. This sheath continues to become more identifiable, especially around the insertions of the rectus muscles and anterior to their insertions, until it fuses with the conjunctiva at the corneoscleral limbus. In addition, the dura forms a fascial sheath surrounding each of the muscles (muscle sheaths) arising at the apex of the orbit. The muscle sheaths blend with Tenon's capsule anteriorly. A ring-shaped thickening of the periorbita (anulus of Zinn) at the apex of the orbit serves as the origin for all four rectus muscles.

The fascia surrounding each muscle is continuous with the fasciae of the other muscles, being quite tenuous posteriorly but dense anteriorly near the tendons of the rectus muscles. The fasciae between the muscles are called the intermuscular septa. The intermuscular septa, in conjunction with the muscle "footplates," restrain the various rectus muscles from slipping over the globe during ocular movements.[65] Jampel[33] has obtained experimental evidence

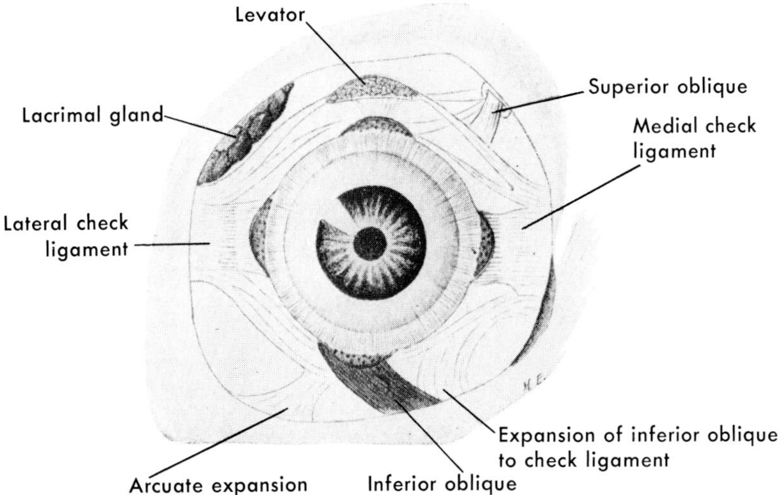

Fig. 5-1. Orbital expansions of muscles of eye, slightly schematized. The recti sectioned with their sheaths embrace the globe and are joined with capsule of Tenon, orbital extensions of levator, and superior rectus. Present are expansion of superior rectus to levator and arcuate expansion of inferior oblique to floor of orbit. Check ligaments from medial and lateral recti are strengthened above and below by superior and inferior expansions. (From Last, R. L., editor: Eugene Wolff's anatomy of the eye and orbit, ed. 6, Philadelphia, 1968, W. B. Saunders Co.)

that these so-called check ligaments play very little role in the normal physiology of eye movement mechanics. The extraocular muscles and their interconnecting septa form the muscle cone. The space within this cone (Fig. 5-4) is known as the central space. Blood vessels and sensory and motor nerves to the extraocular muscles and globe enter the central space (Fig. 5-5). The remainder of the central space is filled with orbital fat. The ciliary ganglion (site of parasympathetic synapse) is located lateral to the optic nerve.

Medial and lateral fascial expansions of the muscle sheaths that extend to the nearby bone and that are assumed to limit the muscle action have been called check ligaments. The check ligament of the lateral rectus muscle extends to the region of the orbital tubercle of the zygomatic bone, and the check ligament of the medial rectus attaches to the lacrimal bone. It is implied that these check ligaments are surgically important, failure to sever a ligament impairing the efficacy of a recession procedure (Fig. 5-6). Ligamentous expansions of the inferior rectus muscle extend into the lower lid, coordinating the lid position with the level of gaze. A fascial expansion between the inferior

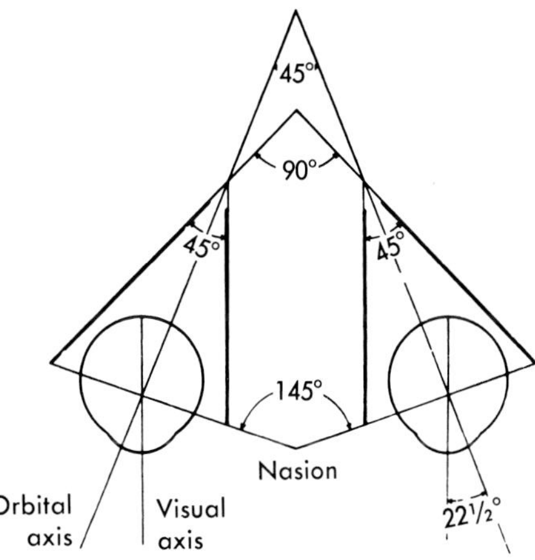

Fig. 5-2. Angles formed by walls of orbit. (From Duke-Elder, S., and Wybar, K. C.: The anatomy of the visual system. In Duke-Elder, S., editor: System of ophthalmology, St. Louis, 1961, The C. V. Mosby Co., vol. 2.)

rectus and the inferior oblique muscles serves as an inferior sling for the globe in the orbit (Lockwood's ligament). Lockwood's ligament acts as the effective insertion of the inferior oblique muscle.[65] The fascia of the superior rectus muscle is attached to the fascia of the levator palpe-

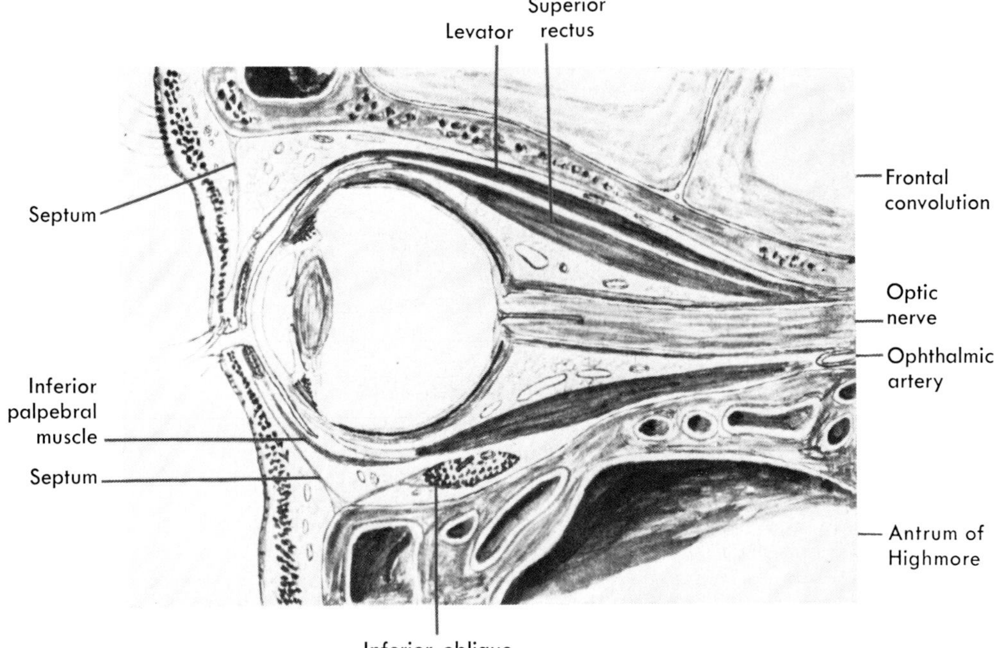

Fig. 5-3. Vertical anteroposterior section of orbit. (From Last, R. L., editor: Eugene Wolff's anatomy of the eye and orbit, ed. 6, Philadelphia, 1968, W. B. Saunders Co.)

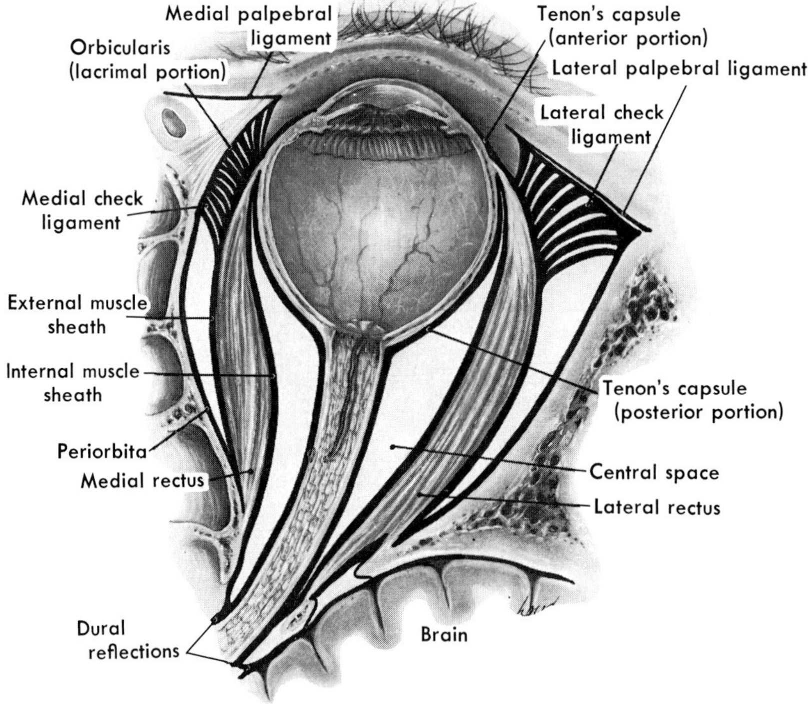

Fig. 5-4. Horizontal section through orbit, showing fascial sheaths and check ligaments.

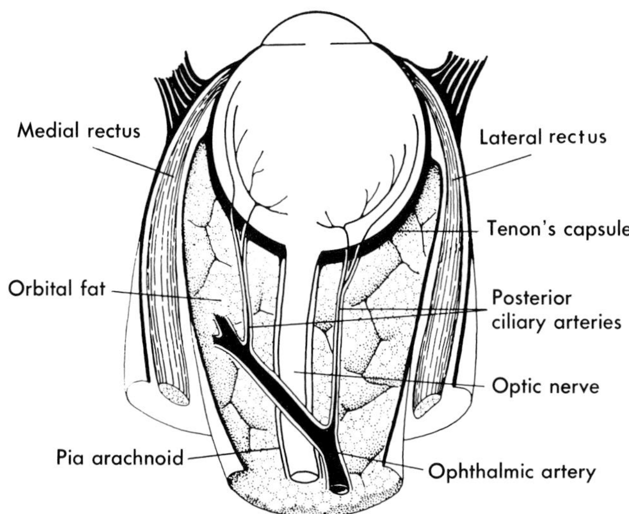

Fig. 5-5. Horizontal section through orbit, showing orbital contents.

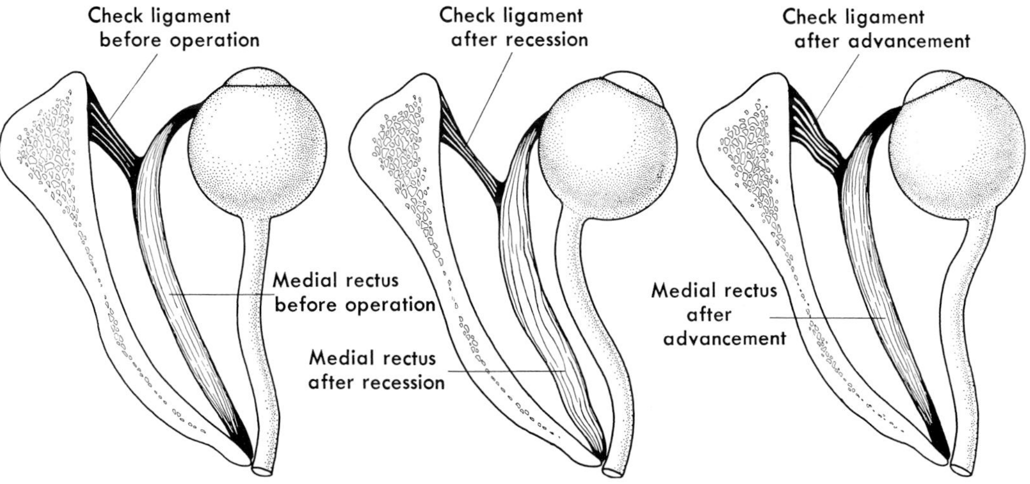

Fig. 5-6. Effect of recession and advancement of medial rectus on check ligament and on position of globe.

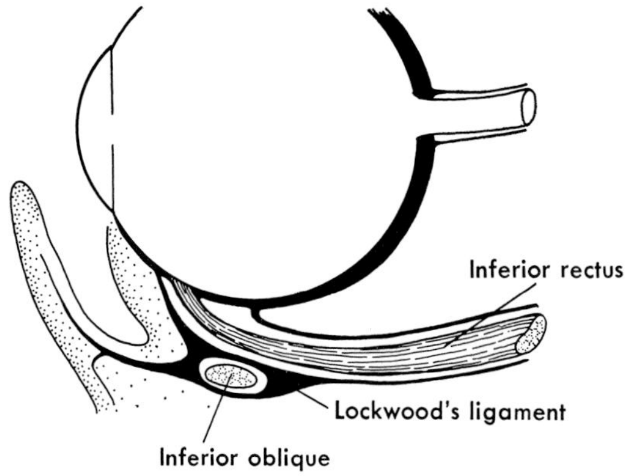

Fig. 5-7. Vertical longitudinal section through lower anterior portion of orbit, showing Lockwood's ligament. (Modified from Maddox.)

brae superioris by multiple slips. The movement of these two muscles with respect to each other is therefore limited.[11] Surgery of the vertical rectus muscles may produce undesirable alterations of lid position (ptosis or retraction) if these fascial structures are ignored (Fig. 5-7).

During operations in which the inferior oblique is picked up on a muscle hook, care must be exercised that the inferior rectus is not included with it. If this should occur, the inferior rectus might be cut, either along with or in place of the inferior oblique when a myectomy of the inferior oblique muscle is attempted.

Active clinical debate continues as to whether the most effective weakening procedure for overaction of the inferior oblique muscle is recession or myectomy, but further discussion of this topic is beyond the scope of this work.

MECHANICS OF THE MOVEMENT OF EACH EYE IN ITS ORBIT

There is no movement of the eyes that is produced by a single extraocular muscle acting alone. All the extraocular muscles have a given tone for every position of gaze. To discuss eye movements a frame of reference against which the movement may be quantitated is necessary. The primary position of the eyes is that position from which all other ocular movements are initiated or that position of the eyeball in its socket against which all torsional, rotational, or translatory movements are measured. The primary position has been defined by Scobee[64] as

> . . . that position of the eyes in binocular vision when, with the head erect, the object of regard is at infinity and lies at the intersection of the sagittal plane of the head and a horizontal plane passing through the centers of rotation of the two eyeballs.

Center of rotation

Early investigators assumed that all ocular motions occurred around a fixed point in the orbit called the center of rotation. It has been shown, however, that translatory movements do take place to some extent (i.e., laterally, vertically, or in or out), and there is no fixed center of rotation that does not move when the globe rotates.[58,72] Backward displacements of the eye in humans are much more limited than those possible for lower animals because of the absence of a retractor bulbi muscle. In humans a demonstrable movement of the globe backward may occur when an object such as a tonometer is placed on the cornea. Although the eye is anesthetized, some patients attempt to close the lids in a natural defensive reaction when the instrument is brought close to the cornea. At the same time, the external ocular muscles may be cocontracted, so the globe is pulled back into the orbit approximately 1 to 2 mm. Cocontraction refers to the simultaneous contraction of muscles that are normally antagonists.

Cocontraction of the extraocular muscles may also be seen clinically in Duane's syndrome and in certain pathologic states involving the midbrain. In addition, the eye may be displaced backward or forward in the orbit under certain abnormal conditions. Exophthalmos (caused by an increase in the volume of the orbital contents) and enophthalmos (caused by herniation of the orbital contents into the maxillary antrum following orbital fracture) are often encountered. In paralysis of the extraocular muscles, particularly if all four recti are paralyzed, the globe will be exophthalmic 2 to 3 mm because of loss of normal tone of these muscles.

Horizontal movements rotate the center of the globe in a semicircle in the plane of eye rotation. This locus has been called the space centroid (Fig. 5-8), but for all practical purposes the globe can be considered to rotate around a fixed

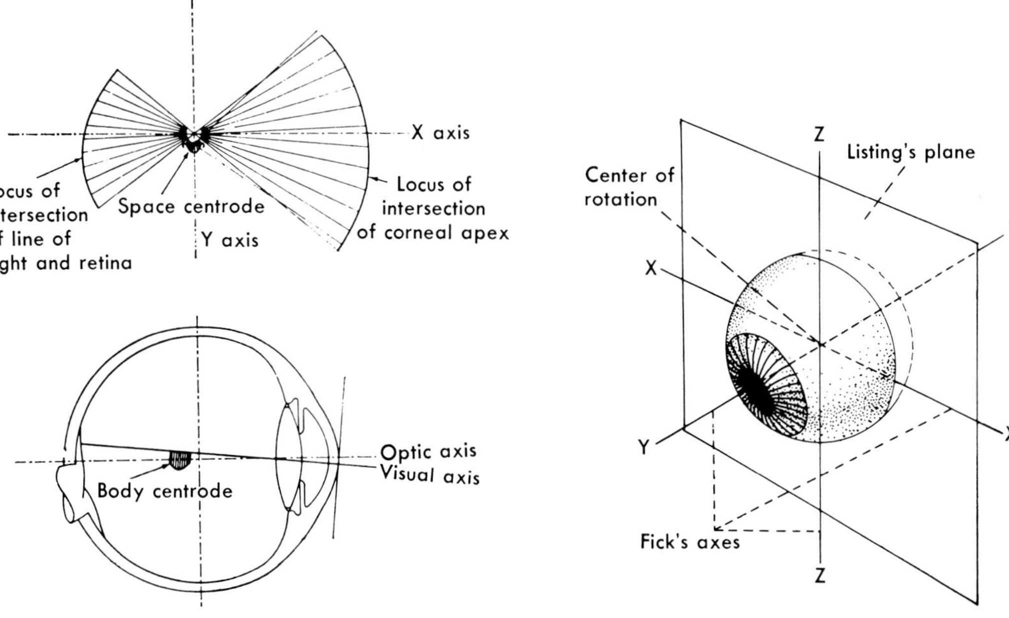

Fig. 5-8. Position of space centrode as computed by Park and Park.[58]

Fig. 5-9. Fick's axes and Listing's plane.

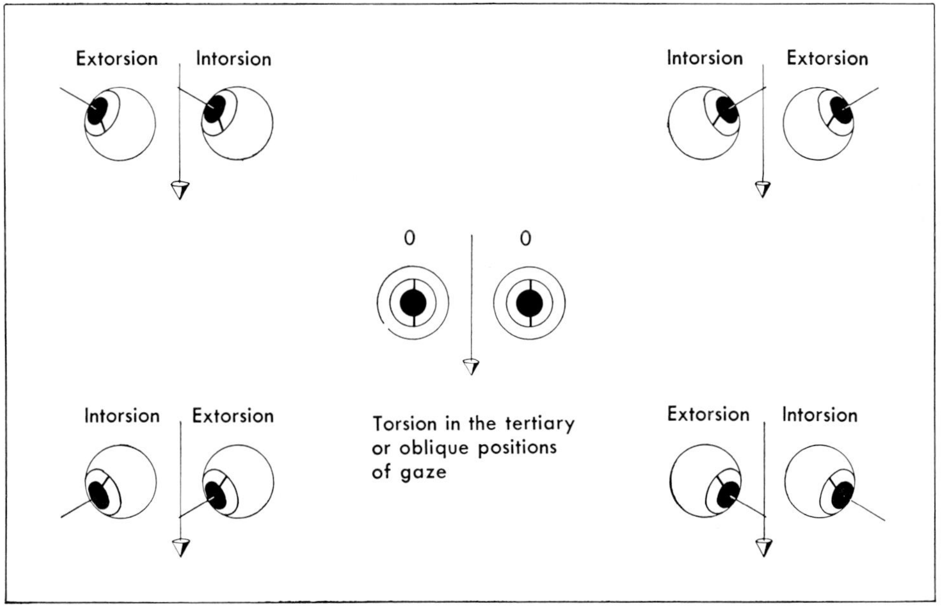

Fig. 5-10. Globe in primary and tertiary positions of gaze to demonstrate false torsion (in terms of a planar coordinate system).

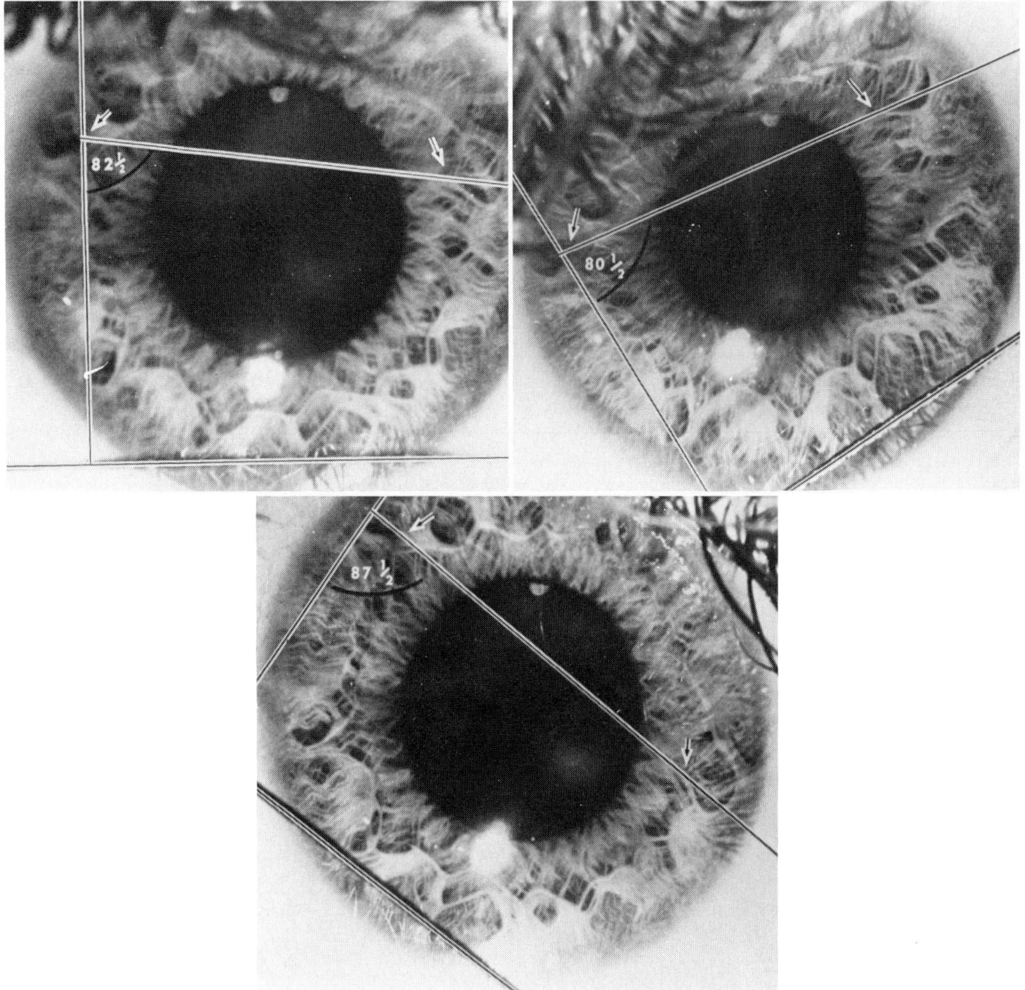

Fig. 5-11. Right eye. Lower line represents string stretched across eye. Line is drawn perpendicular to this. Upper line connects two landmarks (arrows) on iris and forms an angle with perpendicular line. By measuring angle, it was found that head tilt to right (top right) caused 2 degrees of incycloduction, and head tilt to left (bottom) caused 5 degrees of excycloduction. (From Linwong, M., and Herman, S. J.: Arch. Ophthalmol. **85:**570, 1971.)

point that lies 13.5 mm in back of the corneal apex and 1.6 mm to the nasal side of the geometric center of the globe.

Fick's axes

All movements of the globe around the hypothetical center of rotation can be analyzed in terms of a coordinate system with three axes perpendicular to each other and intersecting at the center of rotation. These three axes were described by Fick[20] and bear his name as the X, Y, Z axes of Fick (Fig. 5-9). These axes are stable in relation to a frontal plane fixed in the skull that corresponds roughly with the equatorial plane of the eye when it is directed straight ahead (Listing's plane).

Rotations of either eye alone without attention to the movements of its mate are called ductions. Horizontal rotation (rotation around the Z axis) is termed a*d*duction if the anterior pole of the eye is rotated nasally (medially, in) or a*b*duction if the anterior pole of the eye is rotated temporally (laterally, out). Vertical rotation (rotation round the X axis) is called elevation (sursumduction) if the anterior pole moves up or depression (deorsumduction) if the anterior pole moves down.

Rotation purely around either the horizontal or vertical axis places the eye in a secondary position of gaze. In achieving secondary positions of gaze there is no rotation of the globe around the Y axis. Therefore secondary positions are not associated with torsion.

The oblique positions of gaze are called tertiary positions. Tertiary positions are achieved, with reference to Fick's axes, by a simultaneous rotation of the eyeball around the horizontal and vertical axes, a movement that can be considered to occur around an oblique axis lying in Listing's plane.[47] As the eye rotates obliquely out of the primary position, the vertical axis of the globe is seen to tilt with respect to the X and Z axes of the fixed planar coordinate system (Fig. 5-10). This tilt is termed false torsion, for there is no real rotation around the Y axis, only an apparent movement with respect to the planar coordinate system. The amount of false torsion associated with any particular oblique position is constant regardless of how the eye reaches the position (Donder's law). Tertiary positions of gaze are thus positions of gaze associated with false torsion.

Movements around the X and Z axes to secondary and tertiary positions of gaze are under voluntary control. True torsion (or wheel rotation) of the eye around the Y axis is not under voluntary control and does not occur except for compensatory torsional movements of the eye in response to labyrinthine and tonic neck reflex arcs. Relative movement of the sagittal axis of the eye with respect to the sagittal plane of Fick's axis, although not a true wheel rotation strictly around the Y axis but around a point at the lateral limbus,[29-31,34] is termed torsion (Fig. 5-11). These "torsional" movements work within a small range to keep the sensory vertical raphe of the retinae perpendicular to the horizon. If the upper end of the vertical meridian of the eye tilts toward the nose, this is termed incycloduction, and if it tilts toward the temple, excycloduction. With a 30-degree head tilt there is a mean incycloduction of the ipsilateral eye of 7.00 ± 3.10 degrees and an excycloduction of 8.36 ± 2.50 degrees of the contralateral eye.[46] Thus torsional movements of the eye compensate for small degrees of lateral head tilt from the erect position and tend to keep the vertical meridians of the retinae perpendicular to the horizon. Head tilts of greater magnitude cannot be compensated by eye torsion, and central adjustment mechanisms must be postulated. In addition, torsional movements occur in response to torsional disparities. This cyclofusional response is binocularly coordinated.[12,39,69]

SUMMARY OF ANATOMY OF EXTRAOCULAR MUSCLES

Rotation of each globe is controlled by six extraocular muscles: four rectus muscles and two oblique muscles. These muscles, with the exception of the inferior oblique, take origin at the apex of the orbit in an oval around the optic foramen and part of the superior orbital fissure. Through this oval, the superior and inferior wings of which are made up of the superior and inferior ligaments (the so-called circle of Zinn), pass the optic nerve, ophthalmic artery, the two divisions of the third nerve, the sixth nerve, and the nasociliary branch of the fifth cranial nerve.

The recti

The lateral (external) and medial (internal) rectus muscles follow the corresponding walls of the orbit forward until they pass the equator of the globe. The inferior rectus follows the floor of the orbit only in its posterior half. The superior rectus is separated from the roof of the orbit by the levator palpebrae superioris muscle. Anterior to the equator the muscles follow the curve of the globe and insert by tendinous expansions on the sclera. The curve formed by connecting the insertions of the medial, inferior, lateral, and superior rectus muscles each increasing in distance from the limbus is called the spiral of Tillaux (Fig. 5-12).

The obliques

The superior oblique muscle runs forward for a short distance from its origin at the orbital apex before forming a long tendon that passes through the trochlea, a fibrous cartilaginous tube anchored in the trochlear fossa of the

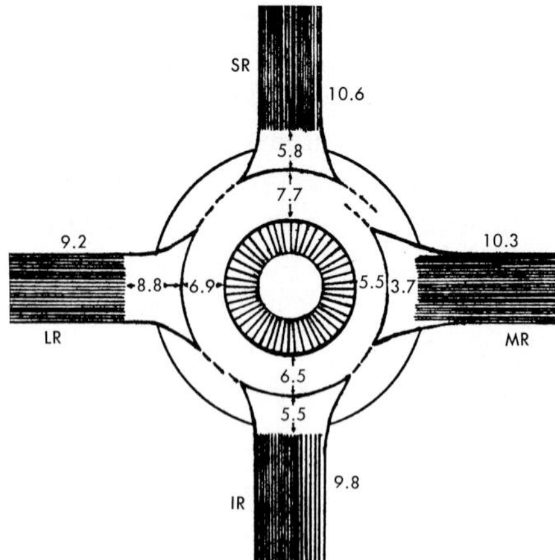

Fig. 5-12. Diagrammatic view of front of right eye drawn to scale, showing insertions of rectus muscles. All figures are in millimeters. Spiral made by insertions of oculorotary muscle tendons is indicated by dotted line; this is called spiral of Tillaux. (From Scobee, R. G.: The oculorotary muscles, St. Louis, 1952, The C. V. Mosby Co.)

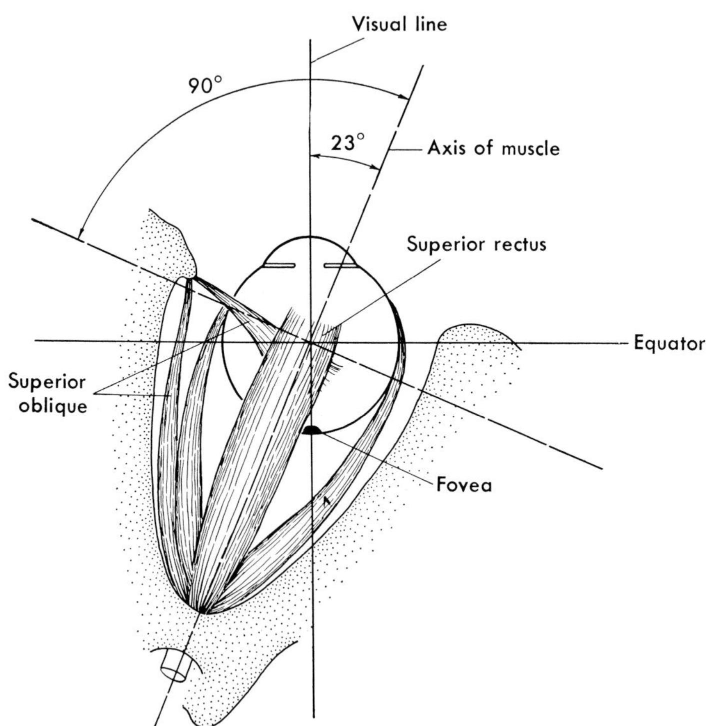

Fig. 5-13. Muscles seen from above, showing origin and insertion of superior rectus.

frontal bone just inside the superomedial orbital rim. This tendon, after passing through the trochlea, turns obliquely backward and outward (under the superior rectus muscle) to attach to the upper sclera behind the equator. The inferior oblique muscle arises from the inferonasal aspect of the orbit just inside its rim and passes obliquely backward and outward (beneath the inferior rectus muscle, where it is intimately bound to it in a fascial condensation) to insert on the lower portion of the globe behind the equator (Fig. 5-13). As mentioned previously, the effective insertion of the inferior oblique muscle is Lockwood's ligament.

The muscle plane

The ocular muscles are paired, each pair having a common muscle plane. These pairs are

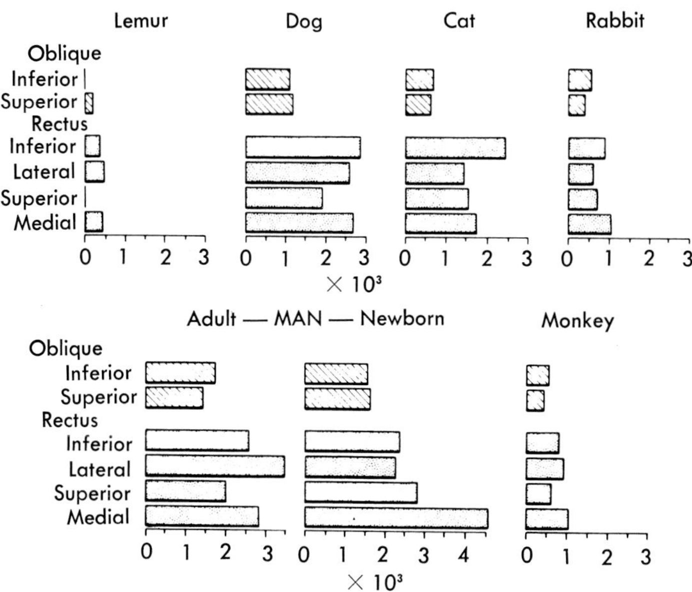

Fig. 5-14. Number of muscle fibers in extraocular muscles of various species, according to histologic studies of Kato.[35] No values are given for two of the muscles of the lemur. (From Peachey, L.: In Bach-y-Rita, P., Collins, C. C., and Hyde, J. E.: The control of eye movements, New York, 1971, Academic Press, Inc.)

the superior and inferior recti, the medial and lateral recti, and the superior oblique tendon from the trochlea to the globe and the inferior oblique. The muscle plane contains the midpoint of the origin of each muscle of the pair and the midpoints of their tendinous insertions. Thus each plane runs through the long axes of its muscles. The angle formed by a muscle plane with the visual direction or line of fixation depends on the position of the globe. When the line of fixation lies in or parallel to the muscle plane, the angle is zero. The muscle plane of the horizontal recti coincides with the horizontal plane of the globe; the superior and inferior recti define an approximately vertical plane, and this muscle plane makes an angle of 23 to 25 degrees with the line of fixation when the eye is directed straight ahead (in the primary position). The superior oblique tendon and inferior oblique muscle lie only approximately in a common plane, and their muscle plane makes an angle of approximately 51 to 53 degrees with the line of fixation when the eye is in primary position.

The muscle structure

The individual extraocular muscles are composed of a variable number of fibers, the largest muscle mass being in the medial rectus muscle and the smallest muscle mass being in the inferior oblique (Fig. 5-14).[59] Although the individual muscles vary in overall structure with regard to such features as length, mass, and tendon size, certain features are common to the group. Each muscle mass may be roughly divided into two layers: an orbital layer consisting of small fibers and a central mass consisting of larger fibers. In addition, the perimesium and septa dividing the muscles into fascicles contain an unusually high number of elastic fibers. The high number of elastic fibers contributes to the viscoelastic properties of the muscles.

INDIVIDUAL MUSCLES

Medial rectus

The medial rectus muscle arises from the anulus of Zinn and runs forward to insert on the sclera about 5.5 mm from the limbus (Table 5-1). It is the thickest and most powerful of the ocular muscles and has a very short tendon, slightly under 4 mm in length. The tendon is about 10 mm broad where it inserts on the globe. The muscle is supplied by a branch of the inferior division of the third cranial nerve. This muscle has the best developed check ligament.

Table 5-1. Distance of rectus muscle insertions from limbus as measured in 55 cadaver eyes (in millimeters)*

Muscle	31 emmetropic eyes			20 myopic eyes			4 hyperopic eyes (mean)
	Maximum	Minimum	Mean	Maximum	Minimum	Mean	
Medial rectus	6.7	4.3	5.5	6.2	4.8	5.5	5.2
Inferior rectus	8.2	5.3	6.5	8.5	6.3	6.9	6.0
Lateral rectus	8.2	5.3	6.9	8.2	5.8	6.9	6.4
Superior rectus	9.0	6.8	7.7	9.7	6.3	7.7	7.1

*From Fuchs, E.: Graefes Arch. Ophthalmol. **30:**1, 1884.

Lateral rectus

The lateral rectus muscle arises from the anulus of Zinn and runs forward on the temporal side of the globe to insert approximately 7 mm from the limbus. It has a very long and relatively thin tendon, approximately 9 mm in length and the same in breadth. It is supplied by the sixth cranial nerve.

Superior rectus

The superior rectus muscle arises from the anulus of Zinn and runs forward above the globe beneath the levator palpebrae superioris muscle, with which it is intimately bound by fascial slips, to insert on the globe 7.7 mm in back of the limbus. As it passes over the equator of the globe, it lies above the superior oblique tendon. The superior rectus is supplied by a branch of the superior division of the third cranial nerve. There is no well-defined check ligament, but the intermuscular septa connect the superior rectus muscle to the horizontal recti, thus forming a superior envelope.[11] The superior rectus muscle and superior oblique tendon are connected by a common fascial sheath, which is sometimes well developed.

Inferior rectus

The inferior rectus muscle also arises from the anulus of Zinn and runs forward beneath the globe to attach 6.5 mm from the limbus. Its tendon is 5.5 mm long and about 10 mm wide. The sheath of the inferior rectus muscle splits anteriorly, the inferior layer of this sheath inserting into the posterior border of the sheath of the inferior oblique muscle and contributing to the suspensory ligament of Lockwood. At this point the inferior rectus and inferior oblique muscles are firmly bound together. It is believed that this union acts as part of a physiologic insertion of the inferior oblique muscle and assumes primary importance when weakening procedures are performed on the inferior oblique muscle lateral to the temporal border of the inferior rectus. The muscle plane containing the superior and inferior rectus muscles makes an angle of 23 degrees with the visual line when the eye is in primary position.

Superior oblique

The superior oblique muscle arises from the lesser wing of the sphenoid bone above the anulus of Zinn and runs forward superomedial to the eyeball to the trochlear fossa in the frontal bone. A plate of cartilage attached here forms a U-shaped pulley (the trochlea) through which the tendon of the muscle passes. The tendon then turns sharply backward and outward over the globe to insert into the sclera just behind the equator, posterior to the center of rotation of the globe. In its course backward it passes underneath the superior rectus. The superior oblique muscle is supplied by the fourth cranial nerve. The trochlea must be considered to be its physiologic origin, and its "muscle plane" is the plane through the tendon of the superior oblique between the trochlea and the globe, and the inferior oblique muscle. This plane forms an angle of 51 to 55 degrees with the visual line when the eye is in primary position (Fig. 5-15). The superior oblique tendon is surrounded by a sheath, which almost doubles its diameter. The sheath is connected to the tendon by many fine areolar slips, which prevent the tendon from sliding freely back into the orbit if the tendon

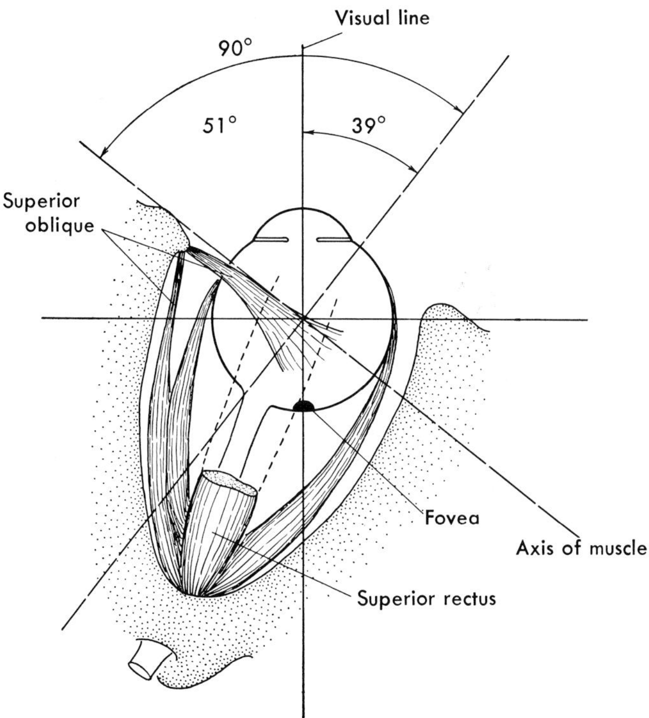

Fig. 5-15. Origin and insertion of muscles from above, with superior rectus cut away. Physiologic origin of superior oblique at trochlea.

and not the sheath is severed. Graded surgery by intrasheath tenotomy is thus possible.

Inferior oblique

The inferior oblique muscle, unlike the other extraocular muscles, arises from a shallow depression in the anterior portion of the floor of the orbit near the lacrimal fossa. It passes backward and temporally beneath the inferior rectus to insert on the globe just below the lateral rectus. Care must be taken not to pick up the inferior oblique on the muscle hook when isolating the lateral rectus muscle. This clinical error can for the most part be avoided by hooking the lateral rectus muscle from its superior border. The inferior oblique is supplied by a branch of the third cranial nerve.

The anatomic relations of the inferior oblique to the other ocular muscles and to other structures of the globe are important from a surgical standpoint. In particular, the insertion of the inferior oblique is very near other structures, notably the macula, the inferotemporal vortex vein, and the ciliary vessels and nerves. The vortex vein is usually 10 to 12 mm below the posterior end of the attachment of the inferior oblique and immediately under the posterior border of the muscle. The nerve to the inferior oblique enters the muscle just lateral to the point at which this muscle crosses the inferior rectus. The ciliary ganglion is fused to the nerve to the inferior oblique muscle at the point that the inferior division subdivides. Care must be taken when isolating this muscle not to be overly aggressive, since injury to the motor root of the ganglion can cause parasympathetic paralysis.

PHYSIOLOGY OF THE MOTOR UNIT

Until recently the concepts of individual extraocular muscle function were for the most part based on the writings of Duane.[13-15] Duane proposed that opposed vertical muscles act as synergists in certain gaze positions and that their functions vary with change in direction of gaze (e.g., with increasing abduction the oblique muscles are stated to cause increasing torsion and abduction). Continued development[41,42,64,71] of these concepts was based mainly on mathematic analysis of static anatomic relationships and the construction of ophthalmotropes (models of the eye in the orbit in which straps and springs represented extraocular muscles). Clinical observations in normal people were compatible with these models when movements in proximity to the primary position were studied.

In 1961, Boeder[7] challenged these concepts in an incisive thesis, and his ideas have been supported by the experimental[29,30,34] and clinical[31] works of Jampel. More recent clinical evidence by Stager and Parks[68] adds credence to these newer concepts as best fitting the data in clinical situations. Since the old concepts are widely held, the student of extraocular function will be expected to be familiar with them. In truth, using the accepted concepts the clinician will be able to isolate the involved muscle in patients with extraocular motor palsies, for the actual physiology and Duane's theories are not entirely mutually exclusive. First Boeder's and Jampel's concepts will be presented as those that will gain acceptance in the future, and then the more widely held concepts will be presented summarily at the end of this section.

Boeder's analysis of the function of the extraocular muscles was based on the observations of Volkmann.[22,73] Volkmann determined the average coordinates of the effective origins and insertions of the six extraocular muscles. These measurements were taken in relationship to Fick's axes. Volkmann's coordinates of the muscle insertions refer to the primary position of the eye and should seemingly assume different values with every change in eye position. But because of the anatomic construction of the extraocular muscle insertion, the effective insertions resist being rotated out of their primary muscle planes. The ocular insertions are broad and arclike, and with ocular motion the effective insertions shift from the medial to the temporal extended fibers of the tendon.[30,31] This shift restricts the displacement of the effective insertions and also the displacement of the muscle planes with respect to ocular rotation. Similarly, the intermuscular septa function to stabilize the extraocular muscles with respect to the globe during ocular rotation.

To discuss the independent action of any individual extraocular muscle or any pair of extraocular muscles is strictly a hypothetical convenience. For in any actual rotation of the eye all six extraocular muscles are involved. To quote Boeder, referring to antagonist pairs, "the pairs in turn have to surrender their theoretical independence and merge into a single muscle unit with a single axis of rotation at any given moment."[7]

The complete muscle unit can produce an infinite variety of rotations consistent with Listing's law,[47] which states that when the line of fixation passes from the primary to any other position, the angle of false torsion in this second position is the same as if the eye had arrived at this position by turning around a fixed axis perpendicular to the initial and final positions of the line of fixation. Analysis of antagonist pairs may be more reasonable because we may assume the tone of the other extraocular muscles to be held constant to prevent the eye from wobbling during movement.

Boeder assumes, and rightly so, that the contraction of one member of an antagonist pair is generally associated with an extension or lengthening of the opposite muscle in the pair. From a mechanical viewpoint this lengthening must be regarded as a muscle action quite on par with contraction. When a muscle pair functions, the eye is rotated around an axis that lies midway between the two muscles. (For more detailed discussion, see Boeder.[7]) Boeder, in his analysis, uses a spherical coordinate system with ϕ representing horizontal and δ representing vertical deviations.

Analysis of the action of individual antagonist pairs reveals the following (Fig. 5-16):

I. Vertical recti
   A. Primary action
      1. Superior rectus muscle: elevation
      2. Inferior rectus muscle: depression
   B. Secondary actions
      1. Adduction
      2. Torsion

Adduction increases with increasing medial movement. Note that there is relative abduction

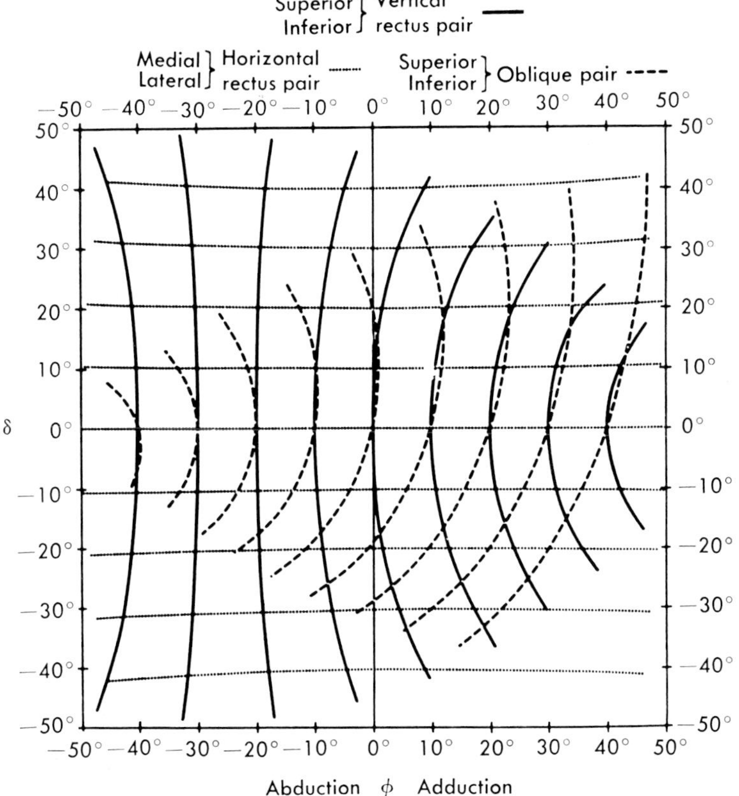

Fig. 5-16. Traces of line of fixation when the three muscle pairs act individually. (From Boeder, P.: Am. J. Ophthalmol. **51:**469, 1961.)

(lateral movement) in moving from inferior or superior gaze to the primary position. Theoretically, using Volkmann's figures for cross-sectional areas and calculating torque (force times radius), it is possible for the vertical muscle pair, by cocontracting, to exert greater than 50% of the force required for adduction once the globe has been adducted slightly less than 30 degrees. However, there is no physiologic evidence of increased firing of these muscles with horizontal rotation. Even at a fixed firing rate the mechanical advantage for adduction by the vertical recti increases with increasing adduction. In abduction both muscles must lengthen (relax). When the eye is externally rotated 25 to 30 degrees, these muscles have minimal abducting function. With increasing adduction the vertical recti must take up the slack (physically).

The superior rectus causes intorsion and is estimated to contribute about 30% of the force required in primary position to keep the horizontal raphe parallel to the horizon. The inferior rectus causes extorsion and contributes approximately 30% of the required force in primary position to keep the horizontal raphe parallel to the horizon. The torsional effect of the vertical recti is almost nil in lateral gaze and continues to increase with increasing adduction.

II. Oblique muscle pair
  A. Primary action
    1. Superior oblique: intorsion
    2. Inferior oblique: extorsion
  B. Secondary actions

The fundamental finding[30-32] on which the newer concepts of oblique muscle action are founded is that the axis about which the oblique muscles rotate the eye in the orbit remains constant when the eye rotates horizontally from any position of gaze. This constancy is due to

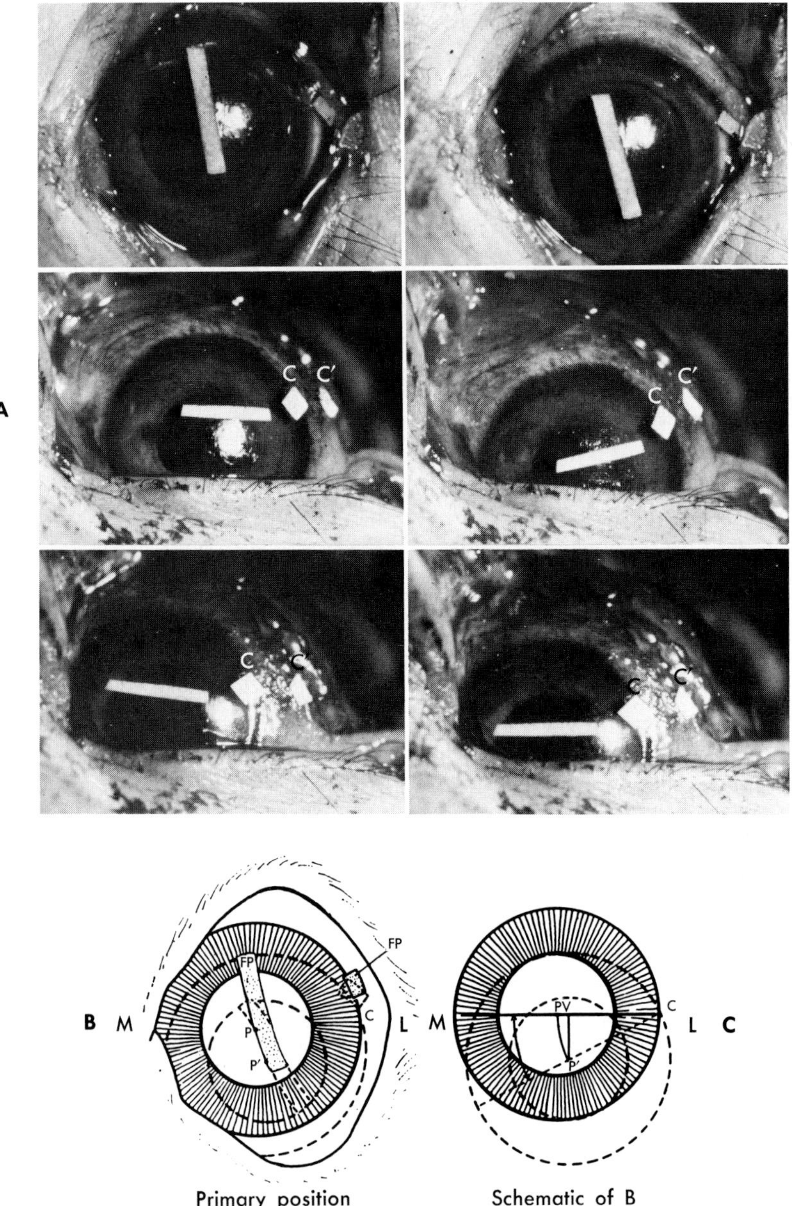

Fig. 5-17. Left eye of monkey. **A,** Prints taken from motion picture frames illustrating technique for demonstrating that anterior axis pole of rotational axis is fixed in space during horizontal eye movements. Top left and right, before and after contraction of superior oblique with eye in primary position. Center left and right, same situation as in top left and right except that lateral and superior walls of orbit had been removed along with lid and levator muscle. Note that marker *C* moved very little in relation to adnexal structures because it was placed near pole of superior oblique rotational axis. Bottom left and right, before and after contraction of superior oblique with eye in adduction. Note that marker *C'* now moved very little because it was near pole of rotational axis. **B,** Tracing taken from successive motion picture frames (upper left and right in **A**), showing eye movement produced by superior oblique from primary position. Anterior pole of rotational axis is at *c* near lateral limbus. Fixation line moves from *P* to *P'*, transcribing an arc. **C,** Schematic drawing of **B,** showing components of movement. *M,* Medial; *L,* lateral; *FP,* filter paper markers; *PV,* outward component; *VP'*, downward component. (From Jampel, R. S.: Arch. Ophthalmol. **75:**535, 1966.)

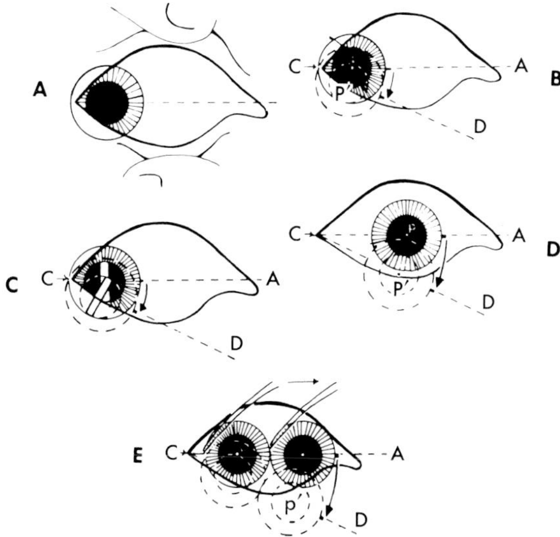

Fig. 5-18. Action of the superior oblique (SO) muscle in humans is demonstrated in a patient with oculomotor nerve paralysis. **A,** In oculomotor nerve paralysis the eye deviates outward and downward (as has been taught). **B,** On command to look down, the eye rotates around point *C* from *P* to *P'*, transcribing an arc of latitude. **C,** Cornea is marked with a strip of eggshell membrane to make the type of movement evident. **D,** Eye from the straight-ahead position also rotates around point *C*, transcribing an arc of latitude. **E,** Eye can be passively moved into different positions in the horizontal plane by means of a corneoscleral limbal suture. Eye movements produced by the SO have the same characteristics from all horizontal gaze positions and an estimate of the magnitude of these movements is possible. (From Jampel, R. S.: Am. J. Ophthalmol. **79:**292, 1975.)

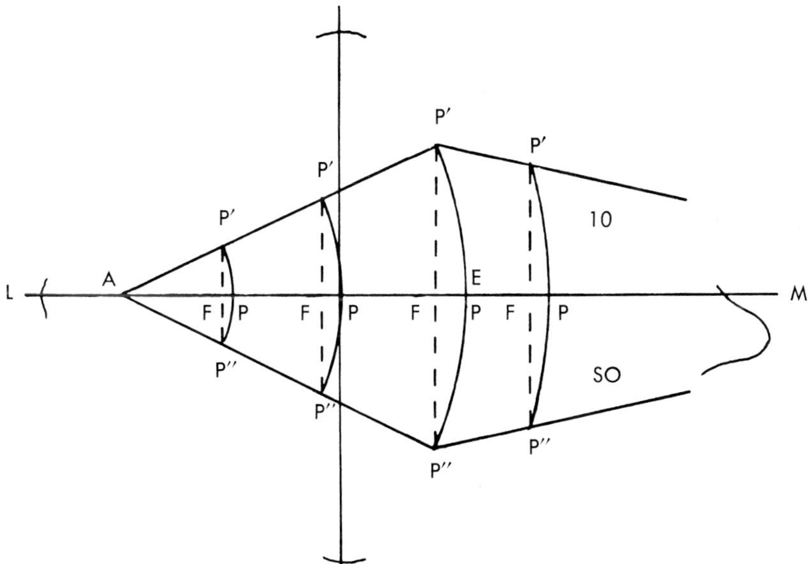

Fig. 5-19. Eye movements produced in humans by the inferior oblique *(IO)* and superior oblique *(SO)* muscles projected onto a frontal plane. When the eye (*P*, the center of the pupil) rotates around the axis pole *(A)* from the horizontal plane, and arc of latitude *(P* to *P'* or *P* or *P")* is produced. The arcs increase in magnitude until the eye moves horizontally to *E*, which is 90 degrees from *A*, or 30 degrees from the straight-ahead position. The arcs decreased in magnitude with further medial rotation from *E*. The horizontal displacement *(PF)* also increases until point *E* is reached and then decreases. *L,* Lateral; *M,* medial. (From Jampel, R. S.: Am. J. Ophthalmol. **79:**292, 1975.)

the structure of the oblique muscle tendons, which insert obliquely onto the globe.

The inner fibers of the fanlike tendon are elongated in adduction while the outer fibers are shortened, and conversely in abduction. Hence the contractile force of the muscle tends to remain concentrated at the same site on the line of insertion regardless of the horizontal rotation of the globe.[31]

The torsional movement is not a wheel movement around the visual, or Y, axis but rather is a motion relative to a fixed spot on the lateral limbus (Figs. 5-17 and 5-18). For a given contraction of the oblique muscles the torsional component is the same throughout the range of eye movements.

Because of the stability of the rotational axis in the orbit, a given contraction of one of the obliques (with relaxation of its antagonist) produces a constant rotation around its axis independent of the fixation line in the horizontal plane. With increasing adduction the obliques cause the fixation line to transcribe increasing arcs of latitude that, when projected onto the frontal plane by vector analysis, can be broken down into a horizontal and vertical displacement (Figs. 5-19 and 5-20). The magnitude of the horizontal and vertical displacement increases as the eye adducts until a point 90 degrees (30 degrees of adduction) from the oblique axis—60 degrees lateral to the "Y" axis of Fick—is reached.[32] At this point, further medial movement is accompanied by a decrease in magnitude of the arc movement (Figs. 5-19 and 5-20). Whether the horizontal displacement is abduction or adduction depends on whether the fixation line is above or below the horizontal plane. From these analyses it is clear that the vector component for vertical displacement increases with increasing adduction (to 30 degrees). The vertical recti remain, in all positions of gaze, the main elevators and depressors; this is consistent with Boeder's calculations (Figs. 5-21 and 5-22).

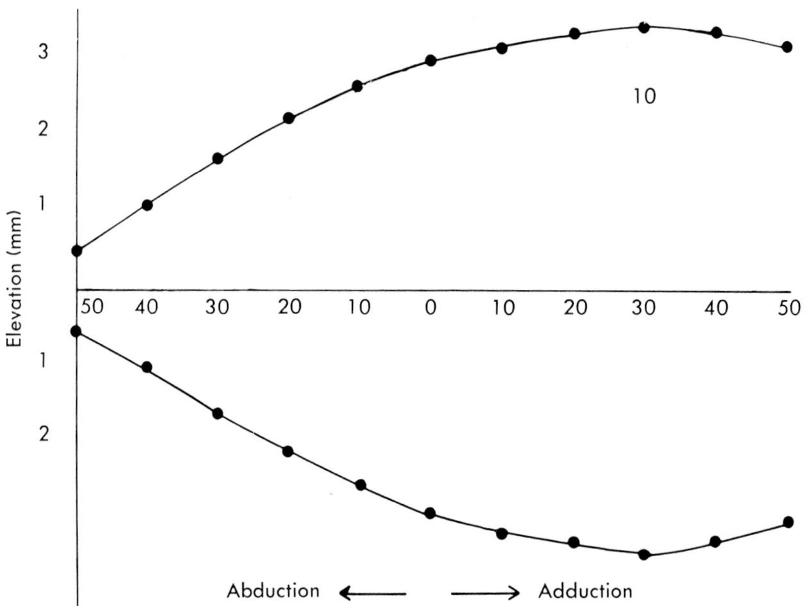

Fig. 5-20. Approximate amount of depression produced by the superior oblique (SO) and elevation produced by the inferior oblique (IO) muscles in humans projected onto a frontal plane with the head in the normal erect position and the eye in different positions in the horizontal plane. Maximum rotation of the eye (θ) is about 17 degrees around the axis of the oblique muscles. Note that the maximum elevation or depression occurs when the eye is adducted about 30 degrees. (SO measurements were obtained from the experiment in Fig. 5-18, *E*.) (From Jampel, R. S.: Am. J. Ophthalmol. **79**:292, 1975.)

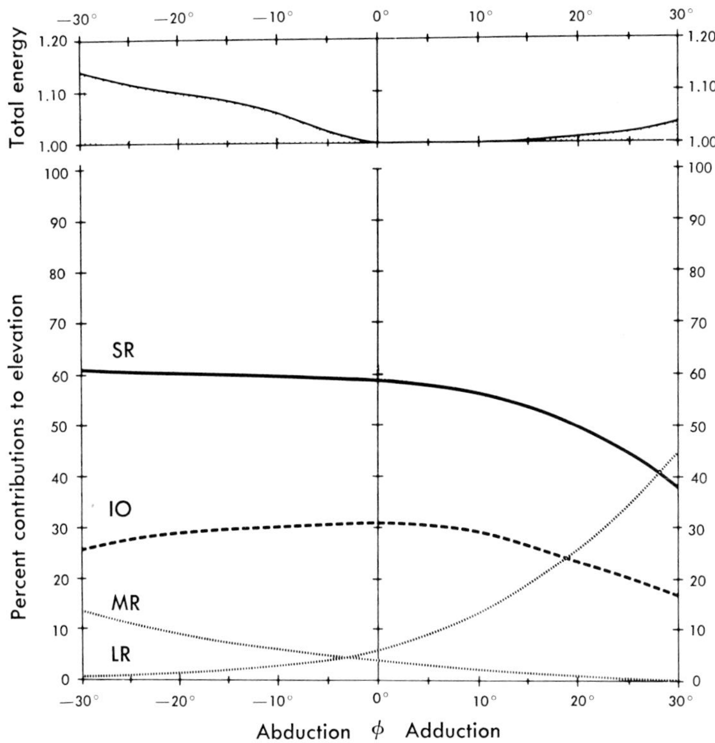

Fig. 5-21. Contributions in percent to 30-degree elevations. Upper graph, relative total muscular energy spent in these elevations. (From Boeder, P.: Am. J. Ophthalmol. **51:**469, 1961.)

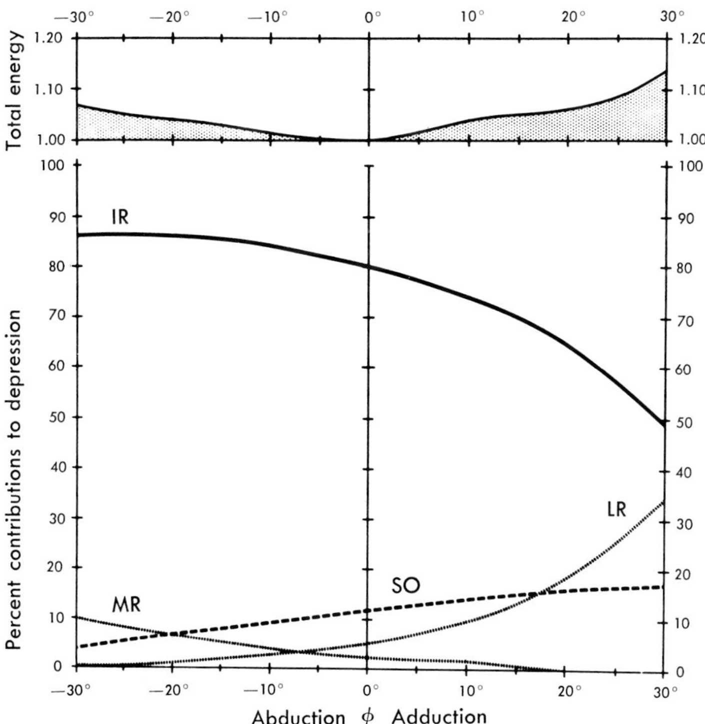

Fig. 5-22. Contributions in percent to 30-degree depressions. Upper graph, relative total energy spent in these depressions. (From Boeder, P.: Am. J. Ophthalmol. **51:**469, 1961.)

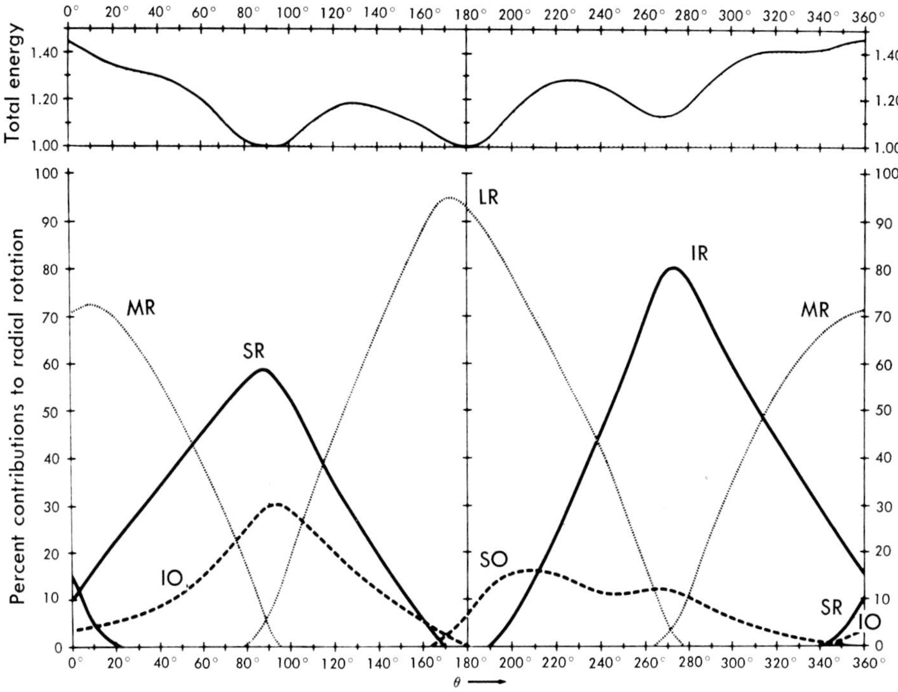

Fig. 5-23. Contributions in percent to radial rotations of 30 degrees from primary position. Upper graph, relative total muscular energy spent in these rotations. (From Boeder, P.: Am. J. Ophthalmol. **51:**469, 1961.)

III. Horizontal recti
  A. Primary action
    1. Medial rectus: adduction
    2. Lateral rectus: abduction

The secondary actions of the horizontal recti are minimal (Fig. 5-16). For initial positions of elevation there is a component of excycloduction in adduction and incycloduction in abduction. The opposite is true in depression. In addition, a change in mechanical advantage of these muscles due to upward or downward displacement of the insertions with vertical rotation can aid elevation in upgaze and depression in downgaze. Alteration of the insertion of the horizontal rectus muscles with respect to their vertical placement can be used to treat vertical deviations as well as A and V pattern syndromes.

As stated at the outset, the question of what contribution an individual muscle or muscle pair makes to eye movement is hypothetical, and one must attempt to analyze what changes take place in the contractile state of the individual muscles and what the individual muscles contribute toward a given excursion of the eye. Boeder has calculated the changes in muscle length for each eye position. From this he has deduced each muscle's relative contribution to a given excursion (Figs. 5-21 to 5-23).

The older concepts of muscle function may be summarized as follows.

Medial rectus

The muscle plane of the medial rectus is exactly in the horizontal plane of the globe; hence, when the globe is in the primary position, contraction of the medial rectus results in internal rotation only, that is, adduction (medial rotation). When the visual line is directed above the horizon, contraction of the medial rectus will aid in further elevation, and when the visual line is directed below the horizon, its contraction will further depress the globe (Fig. 5-24).

Lateral rectus

The plane of the lateral rectus is the same as that of the medial rectus, and when the eye is

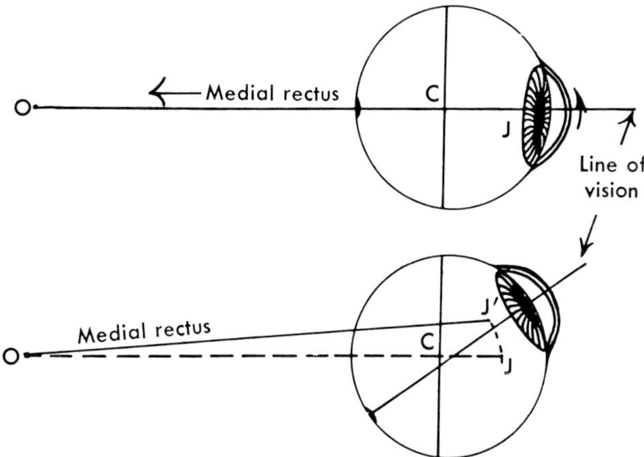

Fig. 5-24. Upper drawing, *J*, insertion of medial rectus muscle when eye is in primary position. Lower drawing, when eye turns up, *J* moves to *J'*.

in the primary position, contraction of the lateral rectus results in abduction alone. Like the medial rectus, when the eyeball is elevated or depressed, contraction of the lateral rectus will further elevate or depress the globe.

Superior rectus

The superior rectus muscle makes an angle of 23 to 25 degrees with the visual line when the eye is in primary position. Its primary action is elevation. Additional, or secondary, actions due to its angle of insertion are adduction and intorsion. Contraction of this muscle will therefore result in a movement of the globe around several axes.

The relative strength of the individual vector forces (i.e., motions around Fick's axes) vary according to the direction of gaze. When the visual line is directed horizontally outward to 23 degrees from the primary position, contraction of the superior rectus will produce elevation only. If the eye is rotated 67 degrees nasally from the primary position, so the visual line is at a right angle to the muscle plane, contraction of the muscle will produce adduction and intorsion only.[41]

Inferior rectus

The inferior rectus muscle has the same muscle plane as the superior rectus muscle (i.e., makes an angle of 23 degrees with the visual line in primary position). Accordingly, when the eye is in the primary position, contraction of the inferior rectus, like that of the superior rectus, results in a compound rotation of the globe consisting primarily of depression and secondarily of adduction and excyclotorsion. When the visual line is turned 23 degrees temporally from the primary position, contraction of the muscle produces depression of the globe alone. When the globe is turned 67 degrees nasally from the primary position, contraction of the muscle produces adduction and extorsion alone.[41]

Superior oblique

In primary position contraction of the superior oblique muscle produces a compound rotation of the globe consisting of three components: primary action—intorsion, secondary actions—depression and abduction. When the eye is turned 51 degrees nasally from the primary position, so the visual line is parallel to the muscle pull, contraction of the muscle will produce depression of the globe alone. If the eye is turned temporally 39 degrees from the primary position, so the visual line is at right angles to the muscle pull, contraction will produce almost entirely intorsion.[41]

Inferior oblique

The muscle plane makes an angle of 51 degrees with the visual line when the eye is in the primary position; hence, when the eye is in the primary position, contraction of the inferior oblique will produce a compound movement of the globe consisting of three components: primary action—extorsion, secondary actions—abduction and elevation. When the eye is turned 39 degrees temporally from the primary position, the pull of the inferior oblique produces extorsion and some abduction, whereas when the eye is turned 51 degrees nasally from the primary position, the pull of the muscle produces elevation alone.

BINOCULAR MOVEMENTS

The movements of each single eye from the primary position into the secondary and tertiary positions of gaze are termed ductions. When we speak of the movements of the two eyes together, the term versions (conjugate movements) is used if the movement of the two eyes is in the same direction, and the term vergence (disjugate movements) if the movement of the two eyes is in opposite directions.

Versions

The following terms are applied when the two eyes move from the primary position to a secondary position of gaze:

Looking to the right: dextroversion, right gaze
Looking to the left: levoversion, left gaze
Looking up: sursumversion, elevation
Looking down: deorsumversion, depression

When one speaks of movement of the eyes from the primary position into a tertiary position of gaze, simple descriptive terms are employed, such as up and to the right or down and to the left.

Hering's law

When both eyes are directed into a field of gaze (e.g., into right gaze), it becomes obvious that different, noncorresponding, muscles of the muscle unit of each eye are activated (the exceptions being straight up and down). When an individual looks voluntarily to the right, both eyes must turn an equal amount to the right to maintain binocular vision (assuming fixation is at infinity). This is the basis of Hering's law,[23,24] which states that in all voluntary conjugate movements of the eyes equal and

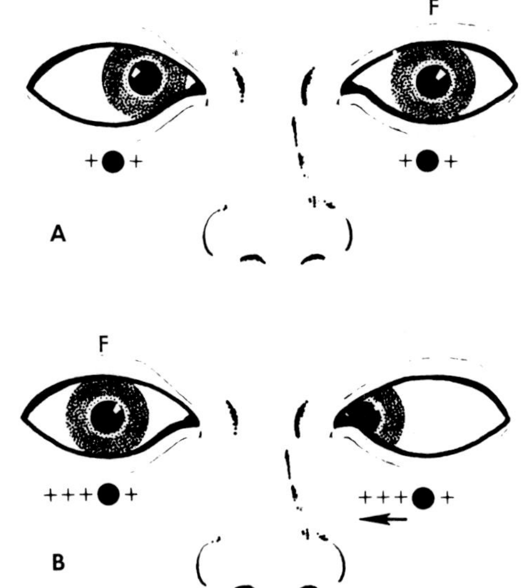

Fig. 5-25. A, Paretic right lateral rectus muscle, primary deviation. Left eye fixing, *F,* in straight-ahead gaze. Equal innervation to all four horizontal rectus muscles. Since right lateral rectus muscle cannot respond normally to input, designated by + signs, right eye will be deviated medially by a given amount, *X.* **B,** Paretic right lateral rectus muscle, secondary deviation. Right eye fixing, *F,* in straight-ahead position. Since it will require a greater input to right lateral rectus muscle to maintain right eye in straight-ahead position, a greater innervation will be sent to the yoke or left medial rectus muscle and result in a greater absolute deviation. (From Burde, R. M.: In Eliasson, S. G., Prensky, A., and Hardin, W.: Neurological pathophysiology, New York, 1974, Oxford University Press.)

simultaneous innervation flows from the oculogyric centers to the muscles concerned in establishing the direction of gaze. It is important to remember that the innervation to both eyes is equal; that is, under normal conditions the amount of nervous energy required to turn the right eye to the right is also sent to the muscles that turn the left eye to the right. The law does not state that each muscle in the two eyes receives the same amount of innervation. However, the law does imply that, under normal conditions, that amount of nervous energy is sent to the muscles of the two eyes which causes the eyes to turn equally in a particular direction. Furthermore, any abnormal condition in one eye that necessitates greater or lesser than normal innervation to turn that eye in one direc-

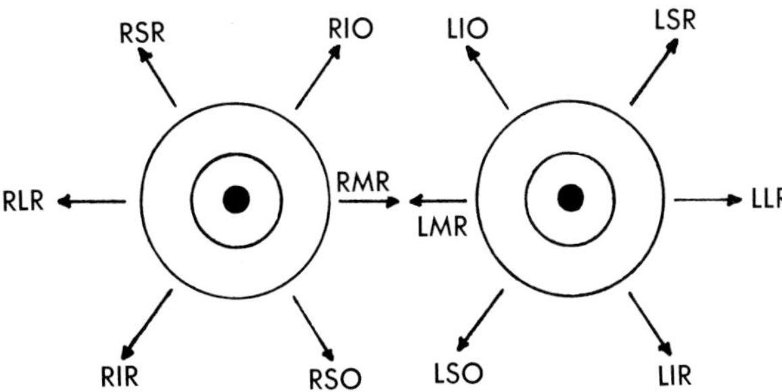

Fig. 5-26. Diagrammatic representation of direction of action of oculorotary muscles of each eye. View is from the front. Arrows indicate direction of maximum action of each muscle. Parallel arrows running in same direction represent yoke muscles. (From Scobee, R. G.: The oculorotary muscles, St. Louis, 1952, The C. V. Mosby Co.)

tion will be reflected in a correspondingly greater or lesser innervation sent to the muscles of the fellow eye performing the same function.

Since the muscle units of each eye function according to Hering's law, a paresis of one of the extraocular muscles will produce a differential deviation that will depend on which eye is used for fixation. That is, there will be a difference in the deviation measured if the uninvolved eye is used for fixation ("primary deviation"), or if the involved eye is used for fixation ("secondary deviation") (Fig. 5-25). The secondary deviation measured with the eye with the paretic muscle fixing will always be greater than the primary deviation. With the passing of time after the acute lesion, this differential in deviation becomes less prominent ("spread of comitance"). Maximum deviation, as well as primary and secondary deviations, can be demonstrated for all the extraocular muscles in the gaze fields that allow relative isolation of their function (Fig. 5-26). Patients with comitant strabismus such as primary esotropia or exotropia do not demonstrate such a differential deviation.

Yoke muscles

If one considers the muscles involved in a right conjugate gaze movement, it is easy to see that primarily there is activation of the right lateral rectus and left medial rectus muscles with concomitant inhibition of the right medial rectus muscle and the left lateral rectus muscle. The remainder of the extraocular muscles receive the proper innervation to aid in this movement and to keep the eye from wobbling. The right lateral rectus muscle and the left medial rectus muscle act as if they are yoked together as a pair of oxen and are therefore called yoke muscles. Because of the previously accepted but inaccurate interpretation of isolated extraocular muscle functions, the following yoke muscle pairs were postulated (Table 5-2 and Fig. 5-26). These groupings did not represent the action of the individual extraocular muscles near the primary position but in their fields of maximum function; that is, it is a grouping of muscles that could most successfully maintain the eyes in any of the six cardinal directions of gaze. Depending on the reference source, four, six, or eight cardinal positions of gaze were listed. Many authors believed that since there are six extraocular muscles for each eye, with respect to both eyes they could be grouped into six pairs of muscles. Each of these pairs would work maximally in one position of gaze, so there must be six cardinal positions of gaze (Fig. 5-26). Six cardinal positions of gaze are necessary to clinically isolate extraocular muscle dysfunction.

Table 5-2. Yoke muscles of the eyes

Right eye	Left eye
Medial rectus	Lateral rectus
Lateral rectus	Medial rectus
Superior rectus	Inferior oblique
Inferior rectus	Superior oblique
Superior oblique	Inferior rectus
Inferior oblique	Superior rectus

In light of Boeder's and Jampel's mathematic and physiologic analyses, the concept of yoke muscles must be modified from a physiologic to a clinical concept. This is especially true with regard to oblique function. If an individual moves the eyes down from right gaze, the major, or only, depressor of the right eye is the right inferior rectus. In the adducted left eye the inferior rectus muscle is the major depressor, but the left superior oblique muscle makes a substantial contribution. Similarly, in movement up from right gaze the major elevator of the right eye is the right superior rectus muscle. While the left superior rectus supplies most of the elevating force, the left inferior oblique muscle makes a contribution of about 20% of the force needed to elevate the adducted left eye. The function of the superior and inferior recti can be isolated in lateral gaze (abduction) as the major elevators or depressors. The obliques, on the other hand, cannot be truly isolated clinically with respect to an easily measured action. However, since the obliques play a significant role in elevation and depression in adduction, an underaction or overaction of one of these muscles can be measured in terms of a vertical deviation. Therefore a working definition of yoke muscles is that pair of muscles in a particular cardinal position of gaze in which dysfunction is most easily recognized clinically. Although conceptually this changes the original definition of yoke muscles, it does not change the classic pairings.

Vergences

When the movement of the two eyes is in opposite directions, the term vergence is applied to the movement. Convergence is a rotation of the eyes toward one another, whereas divergence is a rotation of the eyes away from one another. Vertical divergence is a term used for a rare condition in which the eyes move vertically in opposite directions. Convergence and divergence will be considered in more detail later.

Relation of versions and vergences

Although the same muscles may be concerned in carrying out a version movement at one time and a vergence movement at another, for example, the right medial rectus producing levoversion (adduction of the right eye) and convergence, the neuromuscular mechanisms underlying these movements must be quite separate. The characteristics of the two movements are different, and under pathologic conditions one may be lost or impaired while the other remains intact. For example, in internuclear ophthalmoplegia the right eye may fail to follow the left on left gaze, which would suggest paralysis of the right medial rectus, but when the patient is asked to converge, both eyes readily take up fixation on a near object. This proves that the right medial rectus muscle is not paralyzed, but that there is an interruption of the impulses to it from the left gaze pathways. The impulses for convergence, on the other hand, are not blocked and arrive at both medial recti. The lesion that causes this characteristic syndrome lies in the medial longitudinal fasciculus between the nuclei of the third and sixth cranial nerves. The condition is called internuclear ophthalmoplegia (INO) and is associated with nystagmus of the abducting eye.

Binocular torsions

If both corneal vertical meridians tilt to the right, that is, the right eye showing "extorsion" and the left eye "intorsion," the eyes are said to have undergone dextrocycloversion. If, however, the right eye and also the left eye undergo "intorsion," the eyes are said to have undergone incyclovergence. If both corneal meridians tilt out away from each other, that is, "extort," the condition is called excyclovergence.

CHARACTERISTICS OF MUSCLES

The extraocular muscles consist of a number of different types of fiber that vary from mammalian-type striated muscle (voluntary or fast fibers) to a type of striated muscle cell also found in fish, amphibians, and reptiles; such fibers are called slow fibers. The fast fibers at the one end of the spectrum have so-called en plaque motor end-plates and conduct propagated impulses, each of which is followed by a twitch movement. The slow fibers, on the other hand, have multiple nerve endings, so-called en grappe, and produce graded contractions in the absence of muscle action potentials. Before the specific anatomy, physiology, and pharmacology of the extraocular muscles themselves are discussed, a brief review of the structure and physiology of striated muscles is given.

In 1965 Huxley[28] reviewed the mechanisms of muscle contraction. Under the microscope, voluntary muscle fibers appear regularly striated at right angles to their length, whereas muscles responsible for involuntary movements, such as those of the gut, appear smooth (i.e., without striation). Under the higher resolution and magnification afforded by the electron microscope, striated muscle is seen to consist of a number of individual fibers (Fig. 5-27). An individual fiber consists of parallel elements called myofibrils, each about 1 μm in diameter. In turn, each myofibril is made of parallel thin actin (50 to 70 Å) and thick myosin (160 Å) filaments. The arrays of thick and thin filaments produce the characteristic banding pattern of striate muscle.

This pattern is constructed with a succession of dense bands called A bands and light bands called I bands. The A bands consist of the thicker myosin filaments overlapped at each end by actin filaments. The H zone of the A band appears relatively less dense because the myo-

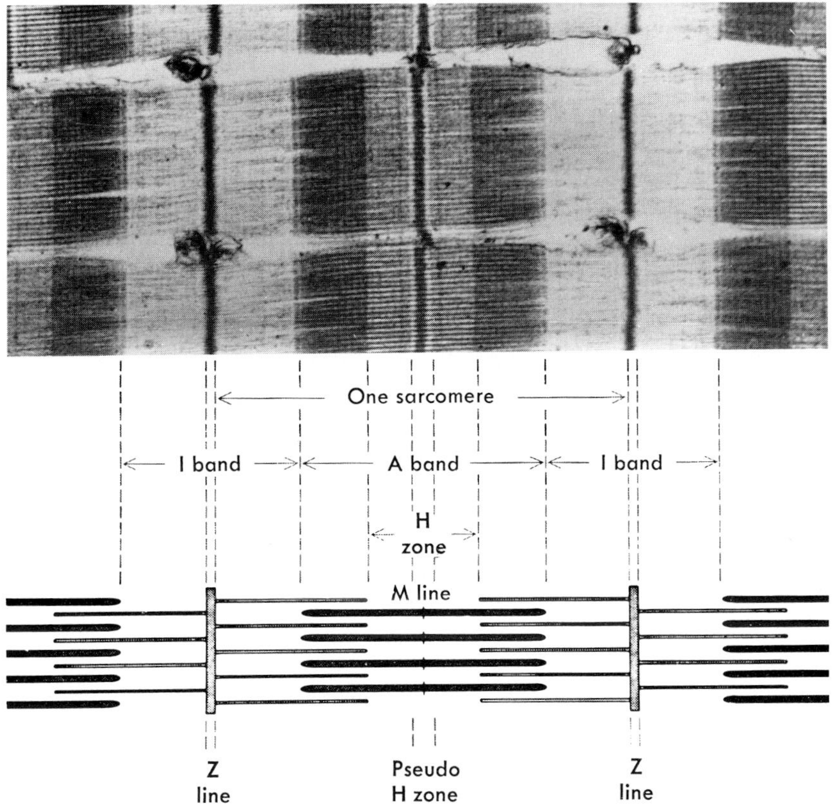

Fig. 5-27. Striated muscle from leg of frog is shown in longitudinal section in electron micrograph (top), and overlap of filaments that gives rise to its band pattern is illustrated schematically (bottom). Part of two myofibrils (long parallel strands organized into muscle fiber) are enlarged some 23,000 diameters in micrograph. Myofibrils are separated by a gap running horizontally across micrograph. Major features of the sarcomere (a functional unit enclosed by two membranes, the Z lines) are labeled. I band is light because it consists of thin filaments. A band is dense (and thus dark) where it consists of overlapping thick and thin filaments; it is lighter in H zone, where it consists solely of thick filaments. M line is caused by a bulge in center of each thick filament, and pseudo H zone by a bare region immediately surrounding the bulge. (From The mechanism of muscular contraction by H. E. Huxley. Copryright © 1965 by Scientific American, Inc. All rights reserved.)

sin filaments are not overlapped by the actin filaments. A dense central line in the A band, called the M line, is caused by a slight increase in the diameter of each myosin filament. The low density of the I band is due to the fact that only thin actin filaments are present. Two sets of actin filaments are attached, one on each side, to a dense transverse structure called the Z line, which divides the I band. The structure between two Z lines is the functional unit of skeletal muscle, the sarcomere.

With active contraction of the muscle fiber, a distinct change is noted in the banding pattern. First the H zone closes as the actin filaments move over the myosin filaments, and then a new dense zone develops in the center of the A band when approaching actin filaments overlap (Fig. 5-28). The individual filaments do not change

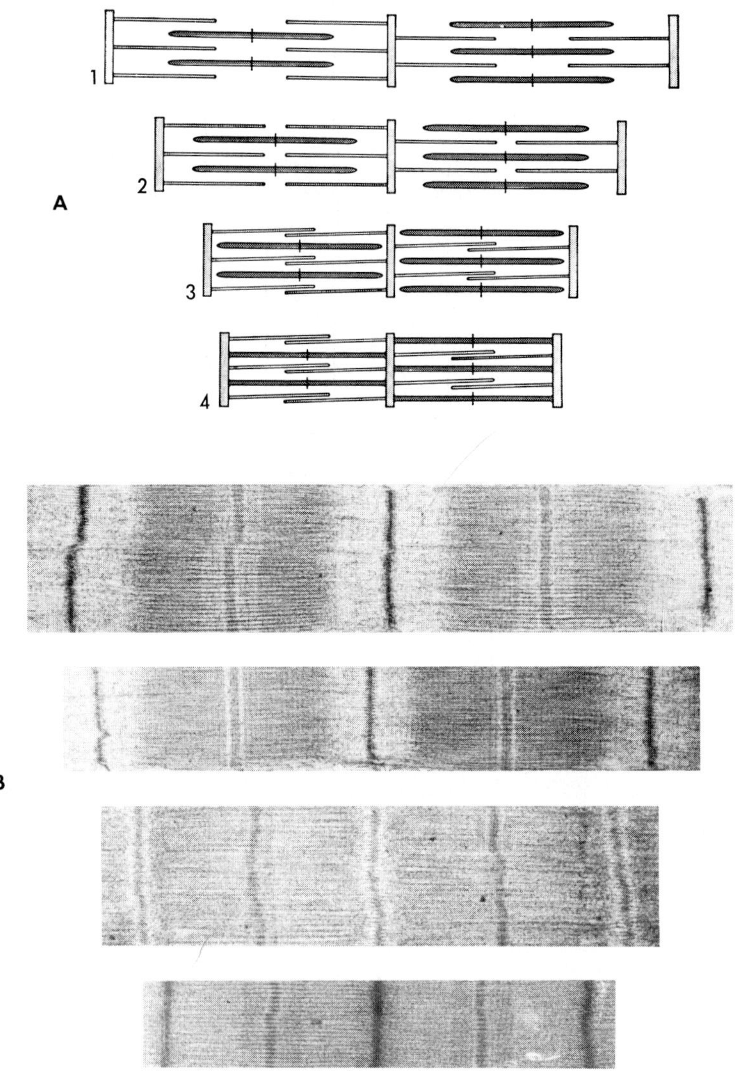

Fig. 5-28. Contraction of muscle entails change in relative position of thick and thin filaments that comprise myofibril (top left and right). Effect of contraction on band pattern of muscle is indicated by four electron micrographs and accompanying schematic illustrations of muscle in longitudinal section, fixed at consecutive stages of contraction. First the H zone closes, *2;* then a new dense zone develops in the center of the A band, *3* and *4,* as thin filaments from each end of the sarcomere overlap. (From The mechanism of muscular contraction by H. E. Huxley. Copyright © 1965 by Scientific American, Inc. All rights reserved.)

in length but slide past each other in a ''ratchet-like'' fashion.

The muscle fibers (cells) are surrounded by the surface membrane, the sarcolemma, from which invaginations in the form of tubules permeate the cell structure. Abundant sarcoplasmic reticulum (SR) may be seen surrounding individual myofibrils and associated with the surface tubules at special structures called triads[55] (Fig. 5-29). Active contraction and relaxation of the muscle units depend on the release and binding of calcium by the sarcoplasmic reticulum.

As mentioned previously, muscle contraction takes place when myosin and actin filaments are actively moved past each other. The origin of the sheer force[19] probably lies in the myosin heads that project in helical arrays from the surface of the filament. The myosin heads lie close to the thick strands in the presence of ATP and the absence of calcium. In the absence of ATP the myosin heads move out and attach to actin. Myosin can be considered to act like an ATPase. In relaxed muscle the relative concentration of ATP is maintained by the relaxing system composed of troponin and tropomyosin, proteins associated with the thin filaments.[9] This relaxing system is inhibited in the presence of calcium, which binds to troponin.

The proteins tropomyosin and troponin lie along the grooves of the two-stranded actin helix of the thin filaments. In resting muscle the troponin-tropomyosin strands appear to lie close to the actin helix, but in contraction when the myosin heads are attached to actin, this complex is displaced toward the center of the groove. It is postulated that troponin-tropomyosin combination acts to block actin and myosin interaction in the absence of calcium. When calcium is released from the SR, a configurational alteration occurs in the structure of troponin that causes the troponin-tropomyosin to move, unblocking the access of myosin heads to the actin.

The extraocular muscle fibers were originally subdivided anatomically into large- and small-

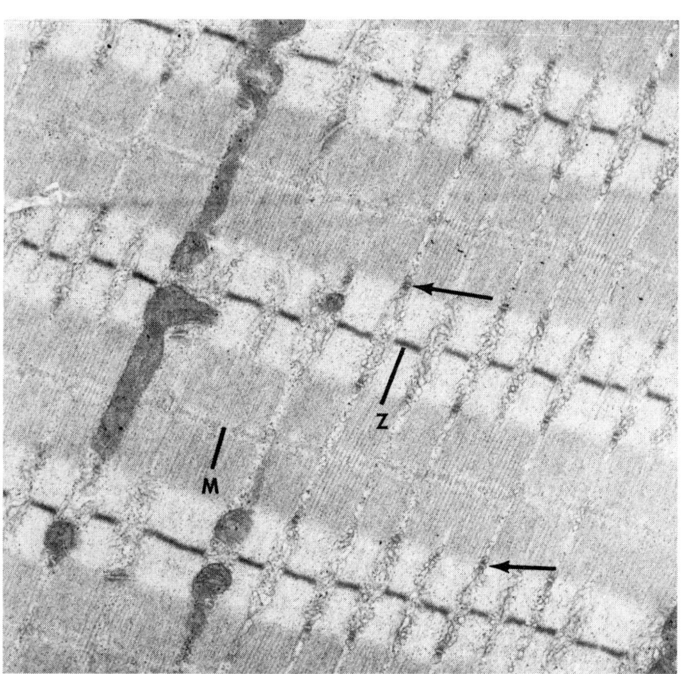

Fig. 5-29. Cat twitch muscle fiber in superior oblique muscle. Separation of fibrils by sarcoplasmic reticulum, straight Z line, *Z*, regularly occurring T system (arrows), and M line, *M*, are seen. Compare with Fig. 5-31. (Glutaraldehyde fixation; longitudinal section; magnification × 16,000.) (From Hess, A.: Invest. Ophthalmol. **6:**217, 1967.)

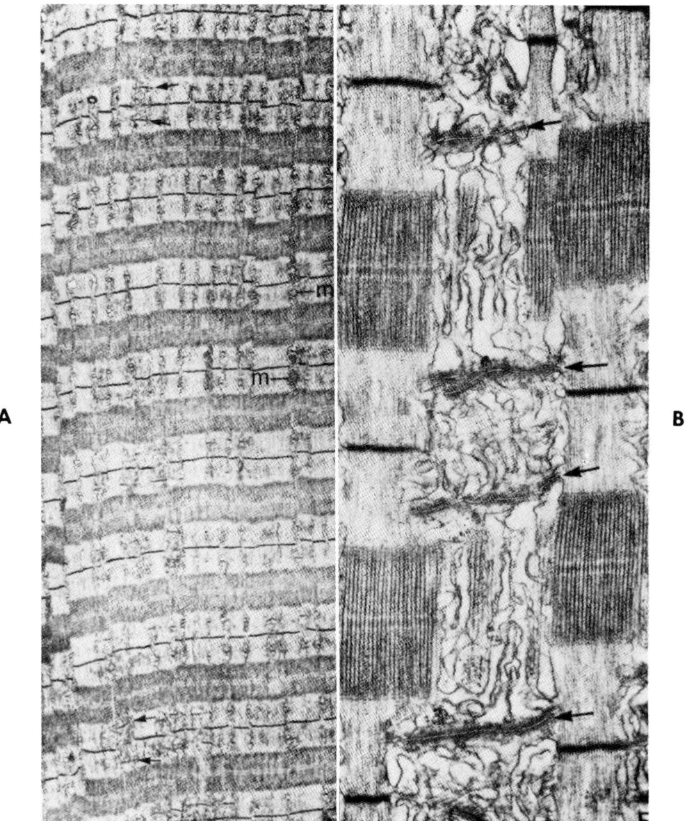

Fig. 5-30. Type I muscle fiber. **A,** Myofibrils are well aligned, clearly separated from each other by an extensive transverse tubular system (arrows) and sarcoplasmic reticulum (SR); the sarcomere has thin Z lines and a prominent M line; the mitochondria *(M)* are few and very small. (Magnification ×13,200.) **B,** Detail of the sarcomere showing the well-developed SR, especially along the I-band region. Note the complex interconnections among the SR tubules as well as the triads (arrows) formed with the transverse tubules. (Magnification ×27,000.) (From Alvarado, J. A., and van Horn, C.: Muscle cell types of the cat inferior oblique. In Lennerstrand, G., and Bach-y-Rita, P., editors: Basic mechanisms of ocular motility and their clinical implications. Proceedings of International Symposium, Wenner-Gren Center, Stockholm, June 4 to 6, 1974, Oxford, 1975, Pergamon Press Ltd., p. 15.)

diameter fibers by Kato.[35] In 1961, Hess,[25] using the electron microscope, established the presence of two distinct morphologic fiber types, one of which resembled the so-called slow fiber of certain vertebrates. In 1963, Hess and Pilar[26] correlated these anatomic findings of two distinct morphologic fiber types with the physiologic evidence of two forms of contractile activity in extraocular muscle: twitch or fast activity and slow or tonic activity. This finding offered for the first time a basis for the observation that eye movements can vary in a controlled fashion from very slow (10 to 20 degrees of arc/sec) to extremely rapid (400 to 500 degrees of arc/sec).

The twitch fibers of extraocular muscle resemble the twitch fibers of skeletal muscle. A twitch fiber is innervated by one large motoneuron, which may have several endplates, and the fiber conducts propagated action potentials. The typical twitch fiber has a prominent SR and a well-developed tubule system (Fig. 5-30). The neuromuscular junction consists of a typical motor end-plate or so-called en plaque ending with many postjunctional folds.

The slow fiber is not found in mammalian skeletal muscle but is found in the muscles of the middle ear. The slow fiber is supplied along its length by small-diameter motor nerves with

Table 5-3. Fiber types in extraocular muscles*

	Type 1	Type 2	Type 3	Type 4	Type 5
Location	Surface	Surface	Interior	Interior	Interior
Diameter	5 to 20 μm (small)	5 to 10 μm (small)	10 to 50 μm (large)	25 to 50 μm (large)	18 μm (medium)
Mitochondria/lipid/SD	High	Low	High	Low	Low (mitochondria elongated transversely)
Fibrils	Larger, indistinct	Smaller, indistinct	Small, distinct	Small, distinct	Large, indistinct
Z line	Thin, straight	Irregular	Description varies	Thin, straight	Broad
M line	Absent	Absent	Absent	Present	Weak or absent
SR	Moderate amount, mostly I, triads A/I	Moderate amount, mostly I, triads A/I	Much, triads A/I	Much, triads A/I	Little, few triads
GPD	Weak	Weak	Low	Moderately high	Low to moderate
Innervation	Single, folds	Multiple, short folds	Single, folds	Single, folds	Multiple, no folds

*From Peachey, L.: In Bach-y-Rita, P., Collins, C. C., and Hyde, J. E.: The control of eye movements, New York, 1971, Academic Press, Inc.

''en grappe'' endings. These endings lie in proximity to the sarcolemma without development of a refined postjunctional apparatus. Unfortunately, the extraocular muscle fibers cannot be so easily characterized into two morphologic groups as these early studies suggested. Neither can function be easily correlated with anatomy.

In 1971, Peachey[59] subdivided the extraocular muscle fiber types into five groups (Table 5-3). Similar classifications have been reported by Mayr[50-52] and Alvarado and van Horn.[2] This classification depends on variations in banding patterns, amount of SR, number of mitochondria, cell size, and so on. At one end of the spectrum is the typical twitch fiber (Fig. 5-30) and at the other the typical slow fiber (Fig. 5-31), with atypical fibers having intermediate characteristics completing the spectrum. Miller[56] has demonstrated in both monkeys and humans that the extraocular muscles undergo sequential changes with increasing age. The muscle fiber that atrophied most was the predominate cell within the central core having fewer mitochondria and an M line in the center of the A band (Peachey's Type 4). Miller astutely recommends that correlates from experiments involving eye movements or extraocular muscle fibers be limited to given age groups.

As mentioned previously, the extraocular muscles are composed of two layers, an outer orbital layer (OL) and a global central layer (GL). The orbital layer consists of two fiber types, both small in diameter, one with multiple and one with focal innervation. The GL contains three fiber types, all larger than fibers in the OL, two of which are singly and one of which is multiply innervated.

The works of Mayr[50] and Alvarado and van Horn[2] demonstrate that many individual muscle fibers do not extend the entire length of the muscle. In the OL the fibers are generally shorter than those in the GL, and a staggered arrangement of muscle tendon junctions is generally found. Large motor end-plates are confined to a large band in the middle third of the muscle, and myomyous junctions are very rare, existing only at either end of the muscle.

In the GL large motor end-plates are found along the entire length of the muscle. There are many myomyous junctions dispersed throughout both ends of the fiber bundles, since only a few of the singly innervated fibers run the entire length of the muscle, most being approximately one-third the length of the bundle. The myomyous junctions can be one to one or multiply connected. Myomyous junctions exhibit cholinesterase activity[53,70] and are the sites of mechanical linkage between different muscle fibers.

Multi-innervated muscle fibers run the entire length of the muscle and extend about 0.5 mm further into the distal tendon than the

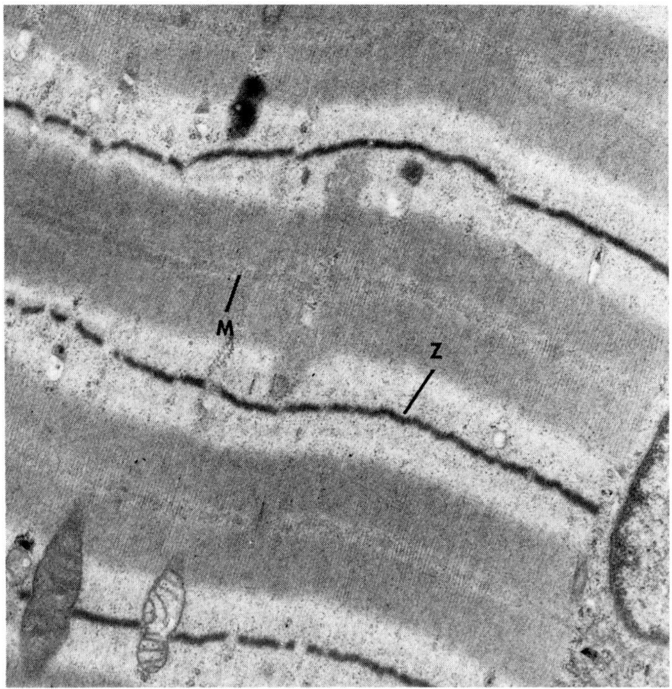

Fig. 5-31. Cat slow muscle fiber in superior oblique muscle. Poor separation of fibrils by sparse amounts of sarcoplasmic reticulum, zig-zag Z line, *Z,* virtual absence of T system elements (only aberrant T system elements occur), and M line, *M,* are illustrated. Compare with Fig. 5-29. (Glutaraldehyde fixation; longitudinal section; magnification ×16,000.) (From Hess, A.: Invest. Ophthalmol. **6:**217, 1967.)

focally innervated fibers. Mayr[52] has emphasized that the internal structure of the extraocular muscles (i.e., the presence of serially linked muscle fibers) must have functional importance.

Miller,[55] in 1967, showed that the orbital surface layer of extraocular muscle consisted of two small fiber types. By histochemical means he determined that this surface layer corresponded with the classic red muscle, high in mitochondrial content (ergo, oxidative enzymes). The central mass of muscle consisted of three fiber types corresponding to so-called white muscle, which is relatively low in mitochondrial content and high in glycolytic enzymes. Miller supports the theory that there is a close correlation between histochemistry and function. Muscles that have the characteristics of red fibers are utilized for prolonged contractions (i.e., slow movements), whereas muscles that have the characteristics of white fibers perform phasic or twitch movements. Peachey[59] warns against the easy acceptance of classifications according to muscle color (i.e., histochemistry or ultrastructure) in implying physiologic function. More recently, Kugelberg[43] has demonstrated that the histochemical reaction for myosin ATPase is associated with contraction speed and mitochondrial oxidative activity with resistance to fatigue. The independent variability of both enzyme types enables a wide range of specifications with regard to speed and endurance. According to Kugelberg, careful histochemical analysis should make it possible to correlate with some accuracy the functional competence of muscle fibers and motor units.

Afferent control mechanisms

The question of proprioceptive mechanisms in the control of eye movements has been the subject of long and intensive debate. To add to the confusion, there is wide intraspecies variation in the number and type of receptors,

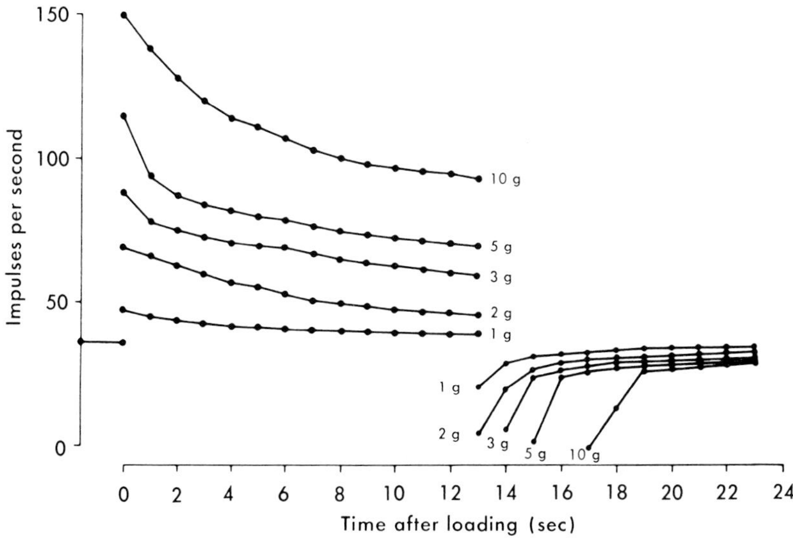

Fig. 5-32. Frequency of response of spontaneously discharging receptor after various loads were hung on and removed from inferior oblique muscle. (From Bach-y-Rita, P., and Ito, F.: J. Physiol. [Lond] **186:**663, 1966.)

as well as their central connections.[4,6,48,49] It now appears that the muscle spindles found in human extraocular muscle give no conscious perception of eye position, but are involved with reflex or subconscious control of muscular contractions.[54,67]

The extraocular muscle spindle organs differ from skeletal muscle spindles in that they have small, thin capsules, are distributed in the proximal and distal thirds of the muscle, do not have nuclear bag regions, and have intrafusal fibers similar in size to the extrafusal fibers. Physiologically, these muscle spindles appear to represent a single type of stretch receptor.[3] The minimum load that produces a change in frequency of firing of the stretch receptor is 0.4 g. The change in frequency of firing in response to various loads is seen in Fig. 5-32.

It appears that the first-order neurons arise in the semilunar ganglion and the second-order neurons in the descending trigeminal nucleus. These neurons project to the ventrobasilar complex of the thalamus.[50] Baker and associates[6] have confirmed and extended the work of Kornhuber,[40] demonstrating mossy and climbing fiber projections to the cerebellum, while Abrahams and co-workers[1] have found that afferents from extraocular muscles and neck muscles constitute the richest projection to the superior colliculus.

The concept of stretch receptors in extraocular muscle is still far from being universally accepted.[38] The functional role of the muscle spindles is probably related to a cerebellum-mediated proprioceptive loop providing information to the cerebellum as to the magnitude of saccades.[6,21] Furthermore, Reinecke and Simons[61] have postulated that the presence of phorias is the result of a discrepancy at the vergence control center between the ocular alignment signal arising from visual retinal input and the alignment signal arising from extraocular muscle afferent inputs. Phoria is the difference between what binocular visual input specifies as normal alignment and what extraocular muscle afferences indicate as balanced tonus—remove visual cues and the eyes assume a nonaligned configuration.

In addition to muscle spindles, other specialized organs assumed to deal with proprioception have been identified in humans (e.g., Golgi tendon organs and certain spiral endings). Sas and Appletauer[63] have described an additional receptor adjacent to arterioles in human extraocular muscle that may be temperature sensitive. Certain reflexes present in humans and other animals must have their basis in receptors in extraocular muscles. These reflexes include:

1. The oculocardiac reflex with which the ophthalmic surgeon is all too familiar (i.e., extreme bradycardia in association with tension on the extraocular muscles); this reflex can only be blocked by intravenous atropine immediately preceding muscle manipulation
2. An oculorespiratory reflex leading to apnea, well established in cats but not yet in humans.

The receptors for these reflexes are unknown. The peripheral afferent pathways are still the subject of controversy, but the most convincing evidence seems to indicate that the afferents leave the orbit via the fifth cranial nerve and travel to the mesencephalic nucleus of the fifth cranial nerve.

Efferent control mechanisms

The nerves to the extraocular muscles contain a large number of small gamma-range fibers as well as large motor fibers. The small motoneurons supply the multiple, or en grappe, endings of the slow fibers. The large motoneurons innervate the motor end-plates on the singly innervated twitch fibers. Some of the muscle fibers receive polyneuronal innervation from both large and small motor fibers. The conduction velocities for the small motor fibers are in the range of 6 to 40 m/sec versus 40 to 80 m/sec for the large fibers.

Bach-y-Rita[3] warns that at this time it is not possible to specifically determine the ratios of motor nerves to muscle fibers for two reasons: (1) The number of motor fibers per nerve is not known because the nerves are mixed nerves carrying sensory and autonomic fibers as well as motor fibers and (2) the question of the number of muscle fibers has not yet been definitely settled. This is complicated by the fact that muscle fibers do not extend the entire length of muscle.[2] Adler estimated the nerve to muscle ratio in the extraocular muscles of humans to be 1:10, basing his calculations on the assumptions that all sensory fibers were 4 μm or less in diameter and that the muscle fibers extended the entire length of the muscle. He believes that this high ratio of motor nerves to muscle fibers accounts in part for the fine gradations of ocular motility.

PHYSIOLOGY AND PHARMACOLOGY OF EXTRAOCULAR MUSCLE FIBERS

Although Miller[55] and subsequently Peachey[59] and Alvarado and van Horn[2] identified at least five morphologic types of muscle fibers, so far only three[44,45] types of fiber have been identified using sophisticated intracellular and extracellular recording methods. As mentioned previously, Hess and Pilar[26] indicated that there were two types of extraocular muscle fiber in mammals that could be differentiated by their response to nerve stimulation.

The large, singly innervated fiber has one motor nerve ending on one motor end-plate, has a high resting membrane potential (65 mV), and responds to nerve stimulation with a propagated action potential of 80 mV. The rapid depolarization is associated with a Na^+/K^+ shift and a rapid release of Ca^{2+} by the SR as the depolarization wave spreads into the muscle via the tubule system. This type of muscle fiber responds with a rapid contraction, having a rise time of 5 to 8 milliseconds with a decay time of 7 milliseconds (Fig. 5-33). These fibers have a fusion frequency to tetanic contraction between 175 to 350 pps.

The small multiply innervated fibers are now divided into two groups,[45] those that do not propagate action potentials (Fig. 5-34), or nonconducting (MINC) units, and those that propagate action potentials, or conducting (MIC) units. These findings obviate the long-standing argument as to whether or not multiply innervated fibers propagate spike impulses.[5,57,60] MINC units demonstrate only local responses that summate at a rate of stimulation greater than 50 pps (Fig. 5-35). Excitation-coupling in these fibers depends on Ca^{2+} release from the surface membrane rather than from the SR. These muscle fibers are innervated by small diameter nerve fibers. MIC fibers conduct action potentials at a significantly lower velocity than singly innervated fibers (1.32 vs. 2.93 m/sec). In addition, the membrane potential is lower (Fig. 5-34), approximately 40 mV. These findings indicate that the fibers in the MIC units should be thinner, and this is compatible with

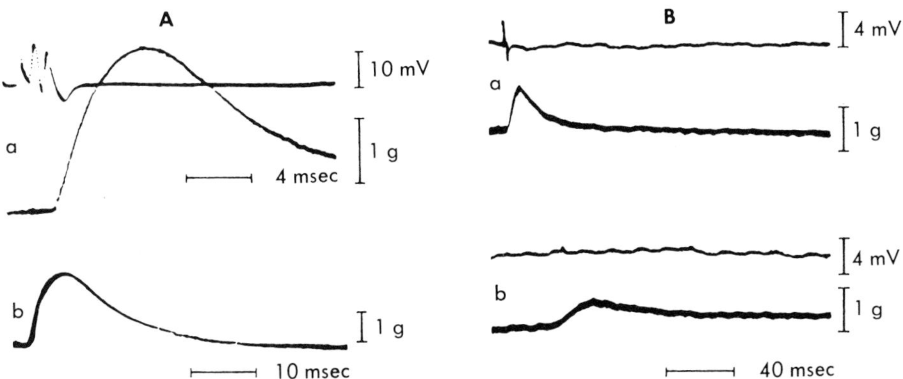

Fig. 5-33. Isometric twitches in response to nerve stimulation of inferior oblique muscle. **A,** Time courses of maximal twitch responses to cathode excitation of a nerve in which none of the three branches had been cut; initial tension 2.6 g. *a,* Single twitch (lower trace) and simultaneously recorded surface potential indicating both fast and slow fibers were activated (upper trace); *b,* total contraction time course of single twitch. **B,** Single twitch response when proximal and central branches of innervating nerve had been cut; initial tension 5.5 g. *a,* Single twitch of fast fibers selectively stimulated by threshold cathode excitation to the nerve (lower trace) and simultaneously recorded extracellular action potential showing only fast fiber activation (upper trace); *b,* single twitch of slow fibers selectively stimulated by anode block excitation (lower trace) and simultaneously recorded extracellular action potential showing only slow fiber activation (upper trace). (From Bach-y-Rita, P., and Ito, F.: J. Gen. Physiol. **49:**1177, 1966.)

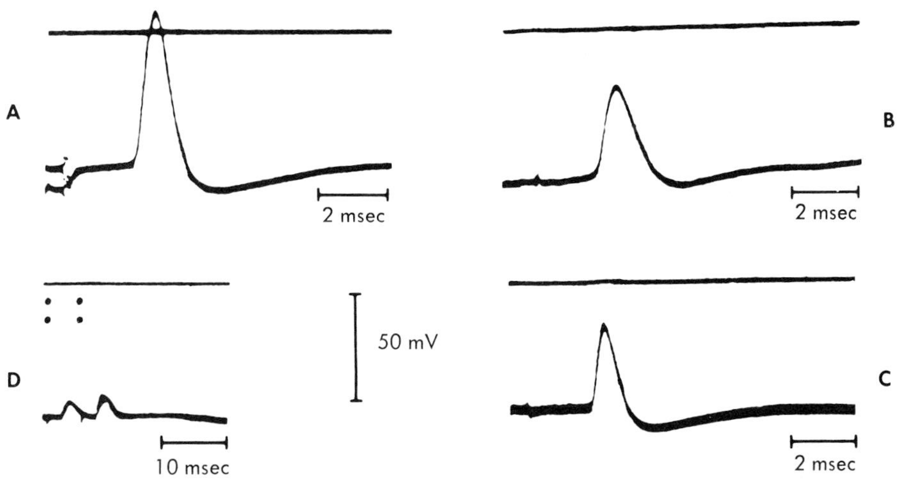

Fig. 5-34. Intracellular responses to single shocks in a SI unit, **A,** and in two different MIC units, **B** and **C,** and to double shocks in a MINC unit, **D.** Top line is reference (zero) potential. Action potentials are seen in **A, B,** and **C.** In the MINC unit, **D,** no action potentials were fired even at increased stimulus strength (marked by stimulus dots); two superimposed traces coincide. (From Lennerstrand, G.: Acta Physiol. Scand. **91:**458, 1974.)

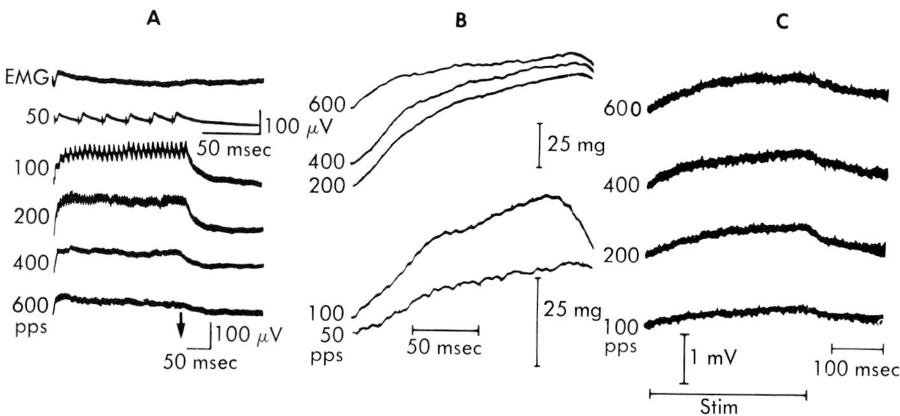

Fig. 5-35. Electrical, **A** and **C,** and mechanical, **B,** responses of MINC units to stimulation at the rates marked. In **A,** monopolar, dc-EMG is recorded from one unit. Local responses to single stimulus (top trace) summate at repetitive stimulation of 60 pps and above. Arrow in lowest trace signifies stimulus removal. **B,** and **C,** From another unit. The tetanus fuses at approximately 60 pps **B.** The electrical activity presented in **C** is recorded with a micropipette close to the unit. Because of the high noise level of this dc-recording, individual local responses cannot be separated. (From Lennerstrand, G.: Acta Physiol. Scand. **91:**458, 1974.)

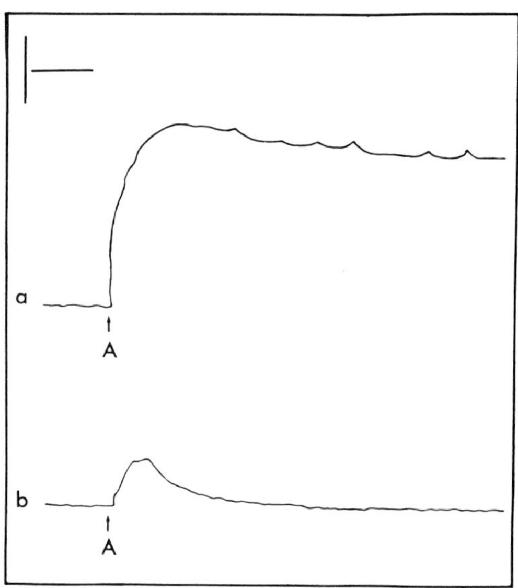

Fig. 5-36. "Contractures" induced by 0.05 μg acetylcholine/ml bath solution. Calibration, 20 mg tension, 1 minute. *a,* From dissection comprising Felderstruktur fibers exclusively; weight of muscle piece, 1.1 mg. *b,* Same muscle as *a,* from dissection comprising Fibrillenstruktur fibers with a few Felderstruktur fibers; weight of muscle piece, 5.2 mg. (From Kern, R.: Invest. Ophthalmol. **4:** 901, 1965.)

the morphologic findings. MIC units have a slower contraction time (5 to 18 milliseconds) and lower fusion frequencies (120 to 225 pps) than singly innervated units.

Tetanic tension is noted at stimulation frequencies of 400 to 600 Hz, a frequency range in which individual twitches blend into a sustained, smooth contraction. Tetany can be divided into two parts: a rapid force rise and a slowly rising gradient of force noted about 40 milliseconds after initiation of stimulation. The low gradient does not reach a maximum until 125 to 150 milliseconds (Fig. 5-36). This long, late rise in tension is due to contracture of the slow fibers; that is, the rate of rise to tetanic stimulation was higher in fast units with high fusion frequencies than in slow units (Fig. 5-37).

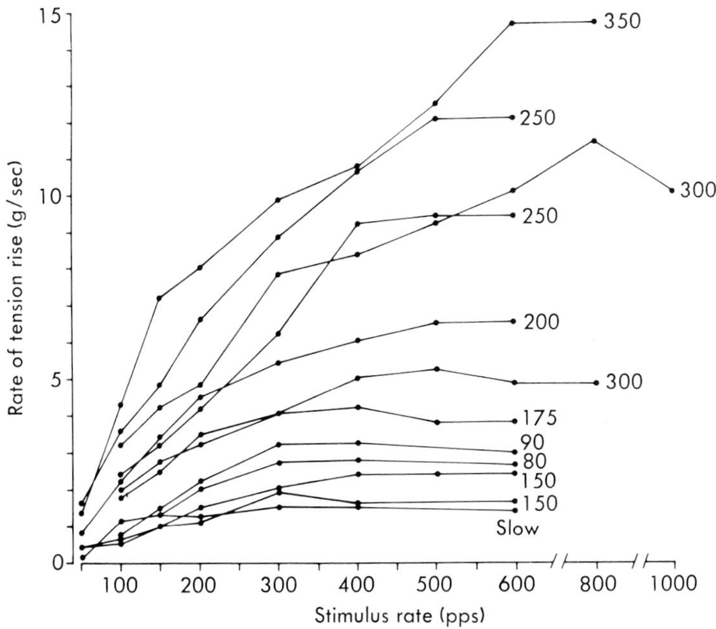

Fig. 5-37. Rate of tension rise in tetanic contraction plotted against stimulus rate in a representative sample of inferior oblique units, identified by their fusion frequencies (marked at the end of each curve; the unit marked "slow" was a MINC unit in which no fusion frequency could be determined). The fastest units show the highest rates of tension rise and their responses also increase over a wider range of stimulus rates (up to 800 pps) than those of the slower units (up to 400 pps). (From Lennerstrand, G.: Acta Physiol. Scand. **91**:458, 1974.)

In the absence of eye movements the extraocular muscles have a resting tone seen as electrical potentials when studied by electromyography. It has been stated that only slow fibers demonstrate spontaneous activity, and Breinin[8] inferred from this statement that the constant tonic activity characteristic of electromyography of extraocular muscle reflected potentials of the slow system only. This posed a paradox. The motor unit activity of tonic contraction as recorded by electromyography consists of propagated spikes, implying contraction of twitch fibers, and cannot arise from nonpropagated junction potentials of slow fibers. It seemed apparent that the view that the fast twitch fibers are responsible for saccadic movements and that slow fibers are responsible for resting tone, vergences, and pursuit is too simplistic. Neither twitch nor slow fibers can be isolated to one function, but both probably have roles in all ocular movements. Collins and Scott,[10,66] by recording simultaneously from fibers in the different layers of human eye muscles (Fig. 5-38), have demonstrated that all fibers are active at all times and are recruited differentially during different ocular movements. At this time it still appears that multi-innervated mechanically slow units are responsible for the majority of tonic activity in all gaze positions, including the primary position. Singly innervated units generate most of the initial force necessary for saccadic eye movements, but also have a supportive role during fixation and pursuit conditions. The physiologic correlation of fiber type with movement characteristics remains to be elucidated in greater detail.

Analysis of available pharmacologic data also leads to the postulation of a dual neuromuscular system in extraocular muscle consisting of a twitch, or fast, system and a tonic, or slow, system. Earlier reports of the apparent hypersensitivity of these muscles to acetylcholine (ACh), choline, and nicotine were due to the failure to recognize the duality of the system. ACh, choline, and nicotine cause a slow

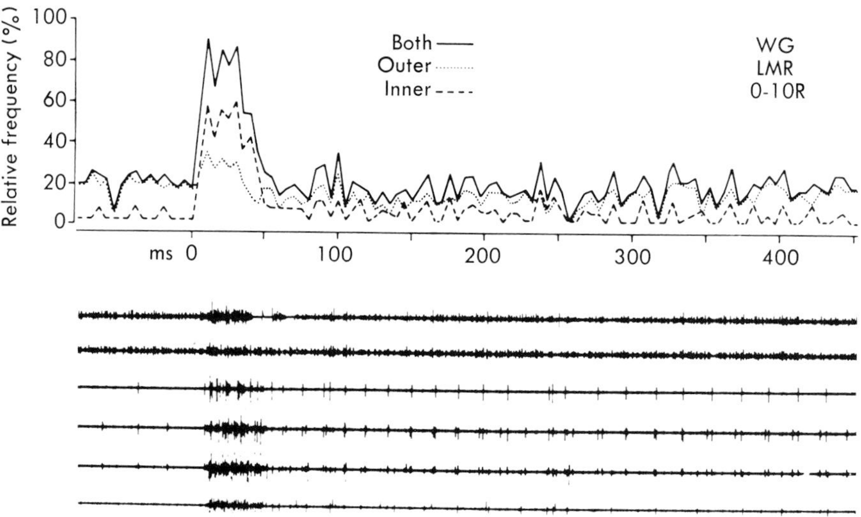

Fig. 5-38. Bottom, record from six channels of the multiple electrode array. The top two channels are recording from the superficial, small, orbital fibers and the bottom four channels are recording from the predominantly large fibers deeper in the central part of the muscle. Top, instantaneous frequencies of activity of both the large and small fibers and the total muscle activity corresponding to the bottom EMG single-unit records. These represent the eye movement control signal for a 10-degree "on" saccade. (From Collins, C. C.: The human oculomotor control system. In Lennerstrand, G., and Bach-y-Rita, P., editors: Basic mechanisms of ocular motility and their clinical implications. Proceedings of International Symposium, Wenner-Gren Center, Stockholm, June 4 to 6, 1974, Oxford, 1975, Pergamon Press Ltd., p. 145.)

chronic contraction of extraocular muscle because of the presence of the tonic, or slow, system. At the time it was originally observed,[16,17] this type of response had only been seen in denervated voluntary mammalian muscle, and it was interpreted as an unusual sensitivity of the extraocular muscle to these cholinergic agents. ACh applied to isolated fast fibers produces a characteristic twitch, whereas when ACh is applied to isolated slow fibers, an immediate maintained contraction is produced (Fig. 5-36).[18] Other pharmacologic findings are reported by Sanghvi and Smith.[62] Bach-y-Rita,[4] using elegant intracellular labeling techniques, has demonstrated that only multi-innervated muscle fibers in the orbital layer, as well as scattered fibers in other layers, respond to succinylcholine with a depolarization-repolarization response.

The effect of neuromuscular blocking agents on the extraocular muscles is of interest, since these agents are frequently used in conjunction with general anesthesia. Blocking agents can be classified into two groups based on their effect on the postjunctional membrane.

1. Depolarizing agents combine with the receptors of the postjunctional membrane and cause a sustained depolarization. Thus ACh in large quantities can act as a depolarizing agent. Clinically useful depolarizing agents include succinylcholine (SCh) and decamethonium. In 1953 it was demonstrated that SCh increased intraocular pressure in humans.[27] This pressure-raising effect, the result of cocontraction of the extraocular muscles, is associated with divergence and immobility of the eyes and is dose dependent (Fig. 5-39). The action of SCh is not blocked by atropine.

Katz and Eakins[36,37] contrasted the effects of various neuromuscular agents on the twitch and tonic systems of extraocular muscle. Small dosages of SCh increased baseline muscle tension but did not depress the twitch response. Larger dosages of SCh continued to increase the baseline tension while depressing the twitch system. The dosage of SCh required to depress the twitch response in extraocular muscle is greater than that required to depress the twitch response of skeletal muscle. The raised baseline muscle tension with consequent increased intra-

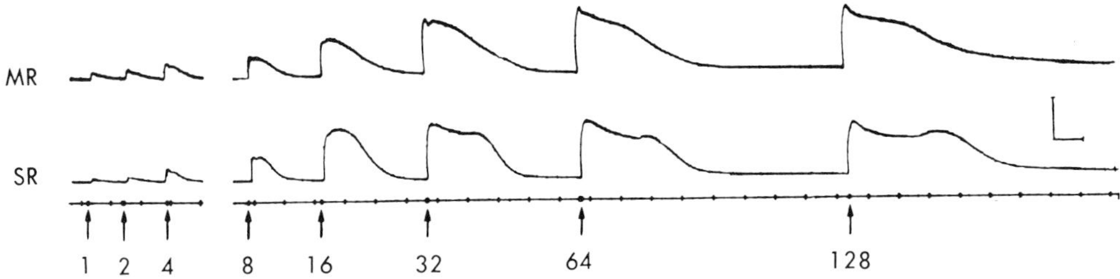

Fig. 5-39. Cat, 3.7 kg, pentobarbital anesthetic. Effect of increasing intravenous dosage (in μg/kg) of succinylcholine on tension of medical rectus muscle, *MR*, and superior rectus muscle, *SR*. Calibrations, 10 g. tension, 1 minute. (From Eakins, K. E., and Katz, R. L.: Br. J. Pharmacol. **26:**205, 1966.)

ocular pressure should be considered a contraindication to clinical use of SCh as an adjunct to general anesthesia when the globe is perforated.

2. Nondepolarizing agents combine with the postjunctional receptor sites to produce a block but do not cause depolarization. This is a competitive block, whereas depolarizing agents produce a noncompetitive block. Clinically, the most commonly used nondepolarizing agent is *d*-tubocurarine. It was widely held that mammalian extraocular muscles are more sensitive to curare than are other skeletal muscles, but this concept has been proved erroneous.[36,62] Depolarizing and nondepolarizing agents may be differentiated as follows:

Depolarizing agent	Nondepolarizing agent
Initial fasciculations	None
Tetanus is well sustained	Tetanus is poorly sustained
Posttetanic facilitation is not seen	Posttetanic facilitation is present
Cholinesterase inhibitors enhance the block	The block is antagonized by cholinesterase inhibitors

The statement that the extraocular muscles are unusually sensitive to neuromuscular blocking agents can no longer be considered valid.

Cholinesterase inhibitors stimulate both the twitch and tonic systems. The twitch response is facilitated, and the baseline muscle tension is increased.

Administration of sympathetic amines produces a modest increase in the baseline tension of the extraocular muscles, as does stimulation of the superior cervical ganglion. These responses are independent of changes in intraorbital smooth muscle and cardiovascular effects. Sympathetic amines produce only one-tenth the effect of cholinergics. The sympathetic response is blocked by the α-adrenergic blocking agents phentolamine and phenoxybenzamine but not by the β blocking agent propranolol. The physiologic implications of this response, although probably bearing on the tonic system, do not seem clear at this time.

REFERENCES

1. Abrahams, V. C., Rancier, F., and Rose, P. K.: Neck muscle and extraocular receptors and their relationship to the tectospinal tract. In Stein, R. B., et al., editors: Control of posture and locomotion, New York, 1973, Plenum Publishing Corp.
2. Alvarado, J., and van Horn, C.: Muscle cell types of the cat inferior oblique. In Lennerstrand, G., and Bach-y-Rita, P., editors: Basic mechanisms of ocular motility and their clinical implications. Proceedings of International Symposium, Wenner-Gren Center, Stockholm, June 4 to 6, 1974, Oxford, 1975, Pergamon Press Ltd., pp. 15-43.
3. Bach-y-Rita, P.: Neurophysiology of eye movements. In Bach-y-Rita, P., Collins, C. C., and Hyde, J. E.: The control of eye movements, New York, 1971, Academic Press, Inc., p. 7.
4. Bach-y-Rita, P.: Structural-functional correlations in eye muscle fibers. Eye muscle proprioception. In Lennerstrand, G., and Bach-y-Rita, P., editors: Basic mechanisms of ocular motility and their clinical implications. Proceedings of International Symposium, Wenner-Gren Center, Stockholm, June 4 to 6, 1974, Oxford, 1975, Pergamon Press Ltd., pp. 91-109.
5. Bach-y-Rita, P., and Ito, F.: In vivo studies on fast and slow muscle fibers in cat extraocular muscles, J. Gen. Physiol. **49:**1177, 1966.
6. Baker, R., Precht, W., and Llinas, R.: Mossy and climbing fiber projections of extraocular muscle afferents to the cerebellum, Brain Res. **38:**440, 1972.
7. Boeder, P.: The cooperation of extraocular muscles, Am. J. Ophthalmol. **51:**469, 1961.
8. Breinin, G. M.: The structure and function of extra-

ocular muscle. An appraisal of the duality concept, Am. J. Ophthalmol. **72:**1, 1971.
9. Bremel, R., and Weber, A.: Cooperation within actin filament in vertebrate skeletal muscle, Nature (New Biol.) **238:**97, 1972.
10. Collins, C. C.: The human oculomotor control system. In Lennerstrand, G., and Bach-y-Rita, P., editors: Basic mechanisms of ocular motility and their clinical implications. Proceedings of International Symposium, Wenner-Gren Center, Stockholm, June 4 to 6, 1974, Oxford, 1975, Pergamon Press Ltd., pp. 145-180.
11. Crone, R. A.: Diplopia, Amsterdam, 1973, Excerpta Medica BV.
12. Crone, R. A., and Everhard-Halin, Y.: Optically induced eye torsion, Albrecht Von Graefes Arch. Klin. Exp. Ophthalmol. **195:**231, 1975.
13. Duane, A.: A new classification of the motor anomalies of the eye, Ann. Ophthalmol. Otolaryngol. **5:**969, 1896.
14. Duane, A.: The basic principles of diagnosis in motor anomalies of the eye, Arch. Ophthalmol. **48:**2, 1919.
15. Duane, A.: Anomalies of the ocular muscles, Arch. Ophthalmol. **11:**394, 1934.
16. Duke-Elder, S.: New observations on the physiology of the extraocular muscles, Trans. Ophthalmol. Soc. U. K. **50:**181, 1930.
17. Duke-Elder, S., and Duke-Elder, P.: Contraction of the extrinsic muscles of the eye by choline and nicotine, Proc. R. Soc. Lond. **107B:**332, 1930.
18. Eakins, K. E., and Katz, R.: The pharmacology of extraocular muscle. In Bach-y-Rita, P., Collins, C. C., and Hyde, J. E.: The control of eye movements, New York, 1971, Academic Press, Inc., p. 237.
19. Editorial: Towards a unified theory of muscle contraction, Nature **238:**187, 1972.
20. Fick, A.: Die Bewegungen des menschlichen Augapfels, Henle und Pfeufer, Zeitschrift **IV:**101, 1854.
21. Fuchs, A. F., and Kornhuber, H. H.: Extraocular muscle afferents to the cerebellum of the cat, J. Physiol. **200:**713,1969.
22. Helmholtz, H. von: Handbuch der Physiologischen Optik, ed. 3, Hamburg and Leipzig, 1910, Verlag von Leopold Voss, vol. 3, p. 47.
23. Hering, E.: Die Lehre v. binokularen Sehen, Leipzig, 1868, Engelman.
24. Hering, E.: Dr. Raumsinn und die Bewegungen der Auges. In Hermann, Ludimar: Handbuch der Physiologie, vol. 3 (pt. 1), 1879; English translation by Radde, C. A., Baltimore, 1942, American Academy of Optometry.
25. Hess, A.: The structure of vertebrate slow and fast extrafusal muscle fibers in the extraocular muscles and their nerve endings in guinea pigs, J. Cell. Physiol. **58:**63, 1961.
26. Hess, A., and Pilar, G.: Slow fibers in the extraocular muscles of the cat, J. Physiol. **169:**780, 1963.
27. Hoffman, H., and Holzer, H.: Die Wirkung von Muskelrelaxantien auf den intraokularen Druk, Klin. Monatsbl. Augenheilkd. **123:**1, 1953.
28. Huxley, H. E.: The mechanism of muscular contraction, Sci. Am. **213:**18, Dec. 1965.
29. Jampel, R. S.: Extraocular muscle action from brain stimulation in the macaque, Invest. Ophthalmol. **1:**565, 1962.
30. Jampel, R. S.: The action of the superior oblique muscle, Arch. Ophthalmol. **75:**535, 1966.
31. Jampel, R. S.: The fundamental principle of the action of the oblique ocular muscles, Am. J. Ophthalmol. **69:**623, 1970.
32. Jampel, R. S.: Ocular torsion and the function of the vertical extraocular muscles, Am. J. Ophthalmol. **79:** 292, 1975.
33. Jampel, R. S.: Personal communication, 1978.
34. Jampel, R. S., and Bloomgarden, C.: Individual extraocular muscle function from faradic stimulation of the oculomotor and trochlear nerves in the macaque, Invest. Ophthalmol. **2:**265, 1963.
35. Kato, T.: Ueber histologische Untersuchungen der Augenmuskeln von Menschen und Säugetieren, Okajimas Folia Anat. Jpn. **16:**131, 1938.
36. Katz, R. L., and Eakins, K. E.: A comparison of the effects of neuromuscular blocking agents and cholinesterase inhibitors on the tibialis anterior and superior rectus muscle of the cat, J. Pharmacol. Exp. Ther. **152:**304, 1966.
37. Katz, R. L., and Eakins, K. E.: The effects of succinylcholine, decamethonium, hexacarbacholine, gallamine, and dimethyl tubocurarine on the twitch and tonic neuromuscular systems of the cat, J. Pharmacol. Exp. Ther. **154:**303, 1966.
38. Keller, E. L., and Robinson, D. A.: Absence of a stretch reflex in extraocular muscles of the monkey, J. Neurophysiol. **34:**908, 1971.
39. Kertesz, A. E., and Sullivan, M. J.: The effect of stimulus size on human cyclofusional response, Vision Res. **18:**567, 1978.
40. Kornhuber, H. H., editor: Handbook of sensory physiology. Vestibular system, Berlin, 1974, Springer-Verlag, vol. 6, parts 1-2.
41. Krewson, W. E.: Action of the extraocular muscles, Trans. Am. Ophthalmol. Soc. **48:**443, 1950.
42. Krewson, W. E.: Comparison of the oblique extraocular muscles, Arch. Ophthalmol. **32:**204, 1964.
43. Kugelberg, F.: The motor unit: histochemical and functional correlations. In Lennerstrand, G., and Bach-y-Rita, P., editors: Basic mechanisms of ocular motility and their clinical implications. Proceedings of International Symposium, Wenner-Gren Center, Stockholm, June 4 to 6, 1974, Oxford, 1975, Pergamon Press Ltd., pp. 85-89.
44. Lennerstrand, G.: Electrical activity and isometric tension in motor units of the cat's inferior oblique muscles, Acta Physiol. Scand. **91:**458, 1974.
45. Lennerstrand, G.: Motor units in eye muscles. In Lennerstrand, G., and Bach-y-Rita, P., editors: Basic mechanisms of ocular motility and their clinical implications. Proceedings of International Symposium, Wenner-Gren Center, Stockholm, June 4 to 6, 1974, Oxford, 1975, Pergamon Press Ltd., pp. 119-143.
46. Linwong, M., and Herman, S. J.: Cycloduction of the eyes with head tilt, Arch. Ophthalmol. **85:**570, 1971.

47. Listing, J. B.: Moleschott's Untersuch, vol. 193, 1854.
48. Manni, E.: Discussion remarks on Bach-y-Rita, P.: Structural-functional correlations in eye muscle fibers. Eye muscle proprioception. In Lennerstrand, G., and Bach-y-Rita, P., editors: Basic mechanisms of ocular motility and their clinical implications, Proceedings of International Symposium, Wenner-Gren Center, Stockholm, June 4 to 6, 1974, Oxford, 1975, Pergamon Press Ltd., pp. 109-110.
49. Manni, E., Bortolami, R., and Desole, C.: Peripheral pathway of eye muscle proprioception, Exp. Neurol. **22:**1, 1968.
50. Mayr, R.: Structure and distribution of fiber types in the external eye muscles of the rat, Tissue Cell **3:**433, 1971.
51. Mayr, R.: Morphometrie von Ratten-Augemuskelfasern, Verh. Anat. Ges. **67:**353, 1973.
52. Mayr, R.: Discussion remarks on two aspects of eye muscle morphology. In Lennerstrand, G., and Bach-y-Rita, P., editors: Basic mechanisms of ocular motility and their clinical implications. Proceedings of International Symposium, Wenner-Gren Center, Stockholm, June 4 to 6, 1974, Oxford, 1975, Pergamon Press Ltd., pp. 44-45.
53. Mayr, R., Zenker, W., and Gruber, H.: Zwischensehnenfreie Skeletmuskelfaser-Vergindugen, Z. Zellforsch. **79:**319, 1967.
54. Merton, P. A.: Absence of conscious position in the human eyes. In Bender, M. B., editor: The oculomotor system, New York, 1964, Harper & Row, Publishers.
55. Miller, J. E.: Cellular organization of rhesus extraocular muscle, Invest. Ophthalmol. **6:**18, 1967.
56. Miller, J. E.: Aging changes in extraocular muscle. In Lennerstrand, G., and Bach-y-Rita, P., editors: Basic mechanisms of ocular motility and their clinical implications. Proceedings of International Symposium, Wenner-Gren Center, Stockholm, June 4 to 6, 1974, Oxford, 1975, Pergamon Press Ltd., pp. 47-61.
57. Ozawa, T., et al.: Correlation of potential and fiber type in extraocular muscle, Doc. Ophthalmol. **26:**192, 1969.
58. Park, R., and Park, G.: The center of ocular rotation in the horizontal plane, Am. J. Physiol. **104:**545, 1933.
59. Peachey, L.: The structure of the extraocular muscle fibers of mammals. In Bach-y-Rita, P., Collins, C. C., and Hyde, J. E.: The control of eye movements, New York, 1971, Academic Press, Inc., p. 47.
60. Pilar, G.: Further studies of the electrical and mechanical responses of slow fibers in cat extraocular muscle, J. Gen. Physiol. **50:**2289, 1967.
61. Reinecke, R. D., and Simons, K.: Phoria and EOM afference: preliminary support for a new theory. In Lennerstrand, G., and Bach-y-Rita, P., editors: Basic mechanisms of ocular motility and their clinical implications. Proceedings of International Symposium, Wenner-Gren Center, Stockholm, June 4 to 6, 1974, Oxford, 1975, Pergamon Press Ltd., pp. 113-117.
62. Sanghvi, I. S., and Smith, C. M.: Characterization of stimulation of extraocular muscles by cholinomimetics, J. Pharmacol. Exp. Ther. **167:**351, 1969.
63. Sas, J., and Appletauer, C.: Atypical muscle spindles in the extrinsic eye muscles of man, Acta Anat. **55:**311, 1963.
64. Scobee, R. G.: The oculorotary muscles, ed. 2, St. Louis, 1952, The C. V. Mosby Co.
65. Scott, A. B.: Strabismus—muscle forces and innervations. In Lennerstrand, G., and Bach-y-Rita, P., editors: Basic mechanisms of ocular motility and their clinical implications. Proceedings of International Symposium, Wenner-Gren Center, Stockholm, June 4 to 6, 1974, Oxford, 1975, Pergamon Press Ltd., pp. 181-191.
66. Scott, A. B., and Collins, C. C.: Division of labor in the human extraocular muscles, Arch. Ophthalmol. **90:**319, 1973.
67. Skavinski, A. A.: Inflow as a source of extra retinal eye position information, Vision Res. **12:**221, 1972.
68. Stager, D. R., and Parks, M. M.: Inferior oblique weakening procedures. Effect on primary position horizontal alignment, Arch. Ophthalmol. **90:**15, 1973.
69. Sullivan, M. J., and Kertesz, A. E.: Binocular coordination of torsional eye movements in cyclofusional response, Vision Res. **18:**943, 1978.
70. Teräväinen, H.: Localization of acetylcholinesterase activity in myotendinous and myomyous junctions of the striated skeletal muscles of the rat, Experientia (Basel) **25:**524, 1969.
71. Van der Hoeve, J.: Ocular movements, Trans. Ophthalmol. Soc. U. K. **52:**1, 1932.
72. Verrijp, C.: Movements of the eyes. In Behrens, C., editor: The eye and its diseases, Philadelphia, 1949, W. B. Saunders Co.
73. Volkmann, A. W.: Zur Mechanik der Augenmuskeln, Trans. Leipzig Soc. Sci. **21:**28, 1869.

Part two

Control of eye movements

RONALD M. BURDE, M.D.

Conceptually, the control mechanisms for eye movements fit into five functionally distinct groups. These are as follows:

1. The saccadic system
2. The smooth pursuit system
3. The vestibular system
4. The vergence system
5. The position maintenance fixation system

These systems are defined by the functional stimuli that evoke the activation of these systems and the response characteristics of the ocular motor movements initiated (Table 5-4). These control mechanisms are represented in supranuclear centers that specify the function of coordinated muscle groups and not of individual muscles. Destruction of supranuclear centers or pathways results in a loss of gaze function, not in diplopia.

The output of these systems is translated into action in one of two modes: rapid eye movements or slow eye movements.[94,125] Rapid eye movements are due to a pulsatile output whose amplitude is determined by burst activity (Fig. 5-40). Slow eye movements, on the other hand, are mediated by a continuous graded response. The slow system can be analyzed in terms of closed loop continuous feedback and monitoring system that can function smoothly at speeds up to 40 degrees of arc/sec. At speeds exceeding this the target will fall on nonfoveal points, and a corrective saccade will ensue (foveation) (Fig. 5-41).

Classically, since the works of Young and Stark[131] in 1962, the saccadic system has been considered a sampled data model. Fuchs,[44] in a recent review, challenges this concept as being a great oversimplification that does not fit all the available data. This system can best be described as *discontinuous*. The system must account for the ability of an individual to change saccades in midflight, have predictive characteristics, as well as be generated by a pulse step–type response.[93,101] Fuchs further states, "the real future for the control systems engineer involves plunging *into* the oculomotor system

Table 5-4. Control mechanism

	Position maintenance	Pursuit	Saccadic	Vergence	Nonoptic reflex (vestibular)
Function	Maintain eye position vis-à-vis target	Maintain object of regard near fovea; matches eye and target	Place object of interest on fovea rapidly	Align visual axes to maintain bifoveal fixation	Maintain eye position with respect to changes in head and body posture
Stimulus	Visual interest and attention(?)	Moving object near fovea	Object of interest in peripheral field	Retinal disparity	Stimulation of semicircular canals, utricle, and saccule
Latency (from stimulus to onset of eye movement)		125 milliseconds	200 milliseconds	160 milliseconds	Very short
Velocity	Both rapid (flicks, microsaccades) and slow (drifts)	To 100 degrees/sec, accurately to 30 degrees/sec	To 400 degrees/sec	Around 20 degrees/sec	To 300 degrees/sec*
Feedback		Continuous	Sampled data		
Substrate	Occipitoparietal junction	Occipitoparietal junction	Frontal lobe; occipitoparietal junction; superior colliculus(?)	Unknown	Vestibular apparatus; muscle receptors in neck → cerebellum(?)

*Slow phase only. The fast phase, although initiated in the pontine reticular formation, is discharged via the saccadic mechanism.

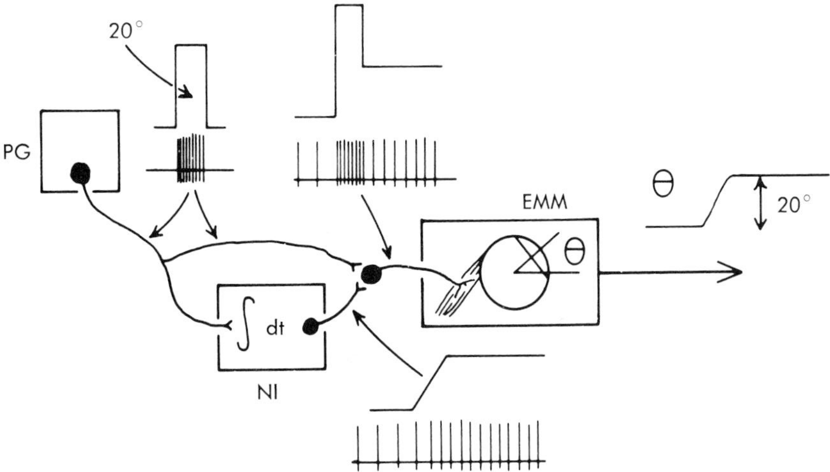

Fig. 5-40. Scheme proposed for making saccades and quick phases. A neural pulse generator, *PG*, produces a pulse of intense activity characterized by the high discharge rates of burst neurons. This pulse is a signal proportional to desired eye velocity. (From Robinson, D. A.: Oculomotor control signals. In Lennerstrand, G., and Bach-y-Rita, P., editors: Basic mechanisms of ocular motility and their clinical implications. Proceedings of International Symposium, Wenner-Gren Center, Stockholm, June 4 to 6, 1974, Oxford, 1975, Pergamon Press Ltd., p. 337.)

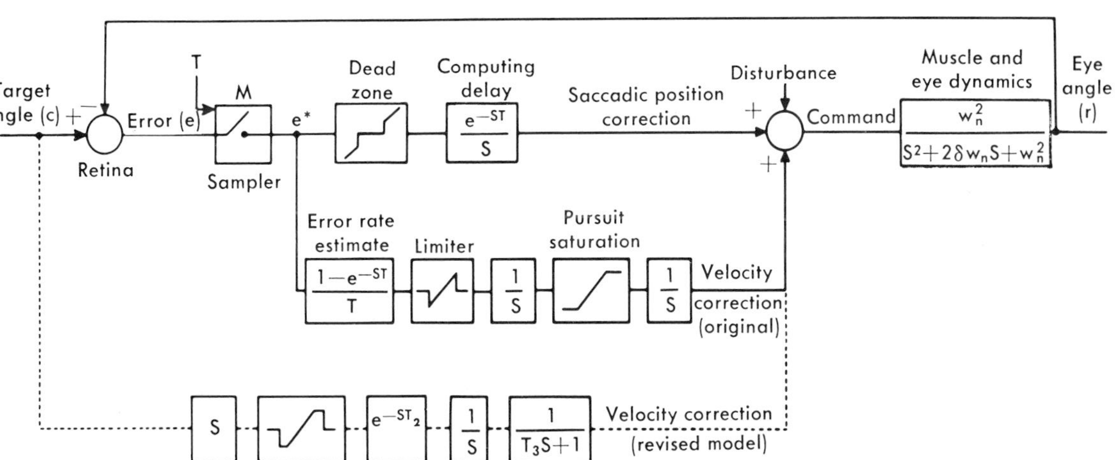

Fig. 5-41. Sampled data model for voluntary eye tracking. The latest (revised) model replaces the original sampled smooth pursuit path by a continuous path with no feedback. Both models have a sampled path for saccadic eye movements. (From Fuchs, A. F.: In Bach-y-Rita, P., Collins, C. C., and Hyde, J. E.: The control of eye movements, New York, 1971, Academic Press, Inc.)

to try to locate and quantify some of the black boxes."[44]

EXTRAOCULAR MUSCLES

The final common pathway for eye movements involves the nuclei of cranial nerves III, IV, and VI and the extraocular muscles subservient to these nuclei.

Contraction of the extraocular muscles as described previously is associated with a change in electrical charge, propagated or not, that can be measured with appropriate apparatus. The recording of this change in electrical activity is called an electromyogram. The amplitude and type[30] of the discharge depends on many factors, chief of which is the distance of the recording electrode from the motor unit. The closer the electrode to the motor unit, the greater the amplitude of the discharge.

The electrical activity in the ocular muscles differs in some respects from that found in skeletal muscle. The ocular muscle motor unit has a much lower amplitude, the duration of the discharge is much shorter, and the rate of firing or frequency is very much higher. The individual motor units can also recruit faster than the skeletal motor units. Furthermore, as long as the ocular muscles are attached to the globe in the unanesthetized subject, the motor units are constantly firing, whereas skeletal motor units in the so-called positions of rest (not actively contracting) are electrically silent (Table 5-5).

Almost all the current classic laboratory and clinical works utilizing electromyography are based on recordings from single monopolar electrodes, which sample only those few fibers with which the electrode is in contact. Anatomically, histologically, and physiologically, we have shown that the extraocular muscles

Table 5-5. Electrical activity of ocular muscle compared with that of skeletal muscle

Characteristics	Ocular muscle	Skeletal muscle
Amplitude at rest	20 to 150 μV	0
Duration at rest	1 to 2 milliseconds	0
Frequency at rest	Up to 150/sec	0
Amplitude during activity	400 to 600 μV	100 to 3000 μV
Duration during activity	1 to 2 milliseconds	5 to 10 milleseconds
Frequency during activity	Several hundred/sec	5 to 30/sec
Minimal rate of stimulation to produce maximal tetanus	250/sec	50/sec
Minimal number of shocks to produce maximal tetanus	20 to 30	10

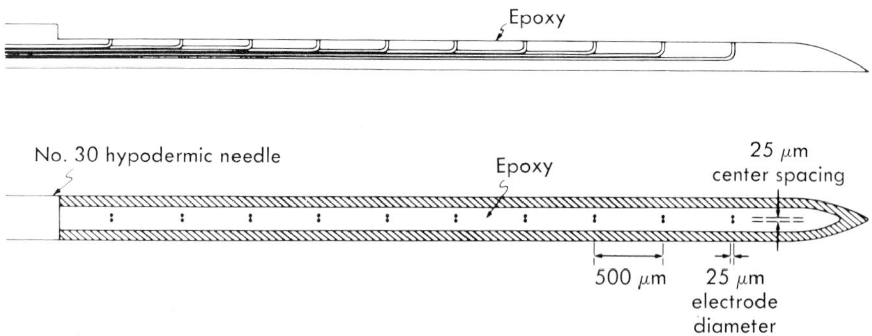

Fig. 5-42. Miniature multielectrode needle array. Drawing of the 10-pair multiple electrode array assembled in a no. 30 hypodermic needle. A very flexible 20-conductor miniature cable is attached to the needle electrode to isolate external forces from the point of insertion in the muscle. (From Collins, C. D.: The human oculomotor control system. In Lennerstrand, G., and Bach-y-Rita, P., editors: Basic mechanisms of ocular motility and their clinical implications. Proceedings of International Symposium, Wenner-Gren Center, Stockholm, June 4 to 6, 1974, Oxford, 1975, Pergamon Press Ltd., p. 145.)

are very complex, and therefore, using localized recording techniques, it is difficult to draw conclusions about the differential activity of various muscle fiber types under physiologic or pathophysiologic conditions. Collins and Scott[30,31] devised an elegant miniature multielectrode array (Fig. 5-42), which has allowed them to sample up to 50 single units from an overall cross section of muscle simultaneously.

The major problem in synthesizing a rational overview is that this type of recording has remained limited to one laboratory group. As mentioned previously, all the classic work and most of the continuing research utilize less elegant techniques that cannot tell us what the muscle as a whole is doing; that is, no interpretation of total muscle activity can be made. A paretic muscle might have more than half its fibers inoperative, yet the record appears normal because the electrode lies adjacent to some unaffected motor units. The possible errors of recording and interpretation have been pointed out by Jampolsky and co-workers.[67,69]

ELECTRICAL ACTIVITY OF MUSCLES

Fixation

Collins,[30] by recording the electrical activity of both large and small fibers at different eccentricities of fixation from the primary position (Fig. 5-38), has determined the relative contribution of each fiber type at gaze positions (Fig. 5-43). It is evident from his records that the large fibers are inactive outside the muscle's field of action and assume an increasing role as the muscle approaches its field of maximum action. The smaller fibers are active in all fields of gaze, increasing their activity nonlinearly as the eye moves into the field of action of the muscle. Thus in fixation the central nervous system seems to use an eye movement control strategy involving a variable ratio of large to small fiber activity (from 0:1 out of the field of action to 1:1 in the field of action). The small fiber activity outside the field of action of the individual muscle seems to serve the function of keeping the muscles on stretch and thereby preventing slack. It is assumed that motor units

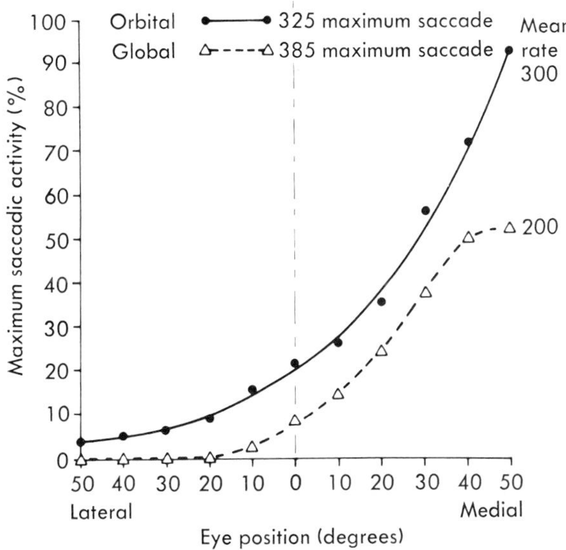

Fig. 5-43. LMR innervation—global versus orbital fibers fixation activity. Quantitative representation of the relative contributions of the large and small oculorotary muscle fibers to fixation and slow tracking movements. It can be seen that the small fibers contribute the greater activity during fixation, progressively increasing their activity across the entire gamut of eye positions. The large fibers appear to saturate at extreme gaze fixation, which suggests that they may be fatiguing. (From Collins, C. C.: The human oculomotor control system. In Lennerstrand, G., and Bach-y-Rita, P., editors: Basic mechanisms of ocular motility and their clinical implications. Proceedings of International Symposium, Wenner-Gren Center, Stockholm, June 4 to 6, 1974, Oxford, 1975, Pergamon Press Ltd., p. 145.)

must discharge out of phase, so one group of motor units is contracting while another group is relaxing; thus muscular fatigue is obviated.

Saccadic movements

When a rapid eye movement is made to the right or left (voluntary and fast phase of nystagmus), there is an abrupt increase in firing of the agonist lateral and medial rectus muscle pair (Fig. 5-38). Both large and small fibers are turned on maximally during the first portion of the saccade and decay logarithmically to their new equilibrium state, depending on the gaze position. The rise time of the saccadic innervation is about 10 milliseconds (the initial movement portion of the saccade being attributed to immediate relaxation of the antagonist[74]). The amplitude and duration of saccadic innervation to the central fibers increases with the magnitude of the saccade. The differential activity of the orbital and central fibers for a series of 10-degree saccades is demonstrated in Fig. 5-44. The large central fibers contribute primarily to saccadic innervation (pulse) and very little toward fixation, whereas the smaller orbital fibers are active even far out of the field of action of the muscle during fixation (step), but contribute less to the saccade itself. The proportion of active orbital and global fibers depends on the final resting position of the eye (Fig. 5-45). The innervational requirements to attain a given orbital eccentricity depend on the direction from which it is approached, the innervation compensating for the internal mechanical properties of the muscle (Fig. 5-46). Action of the antagonist is accompanied by an immediate relative cessation of activity in the antagonists (i.e., right medial and left lateral rectus muscles) (Fig. 5-47). This drop in activity in antagonist muscles follows Sherrington's law of reciprocal innervation. When a decerebrate or spinal preparation executes a muscular movement, augmentation of the contraction never proceeds concurrently in antagonistic muscles, and similarly diminution of the contraction in antagonists does not occur concomitantly. The law merely states that "increase of contraction (or the converse) does not proceed simultaneously in opposed muscles." As contraction occurs in an extensor muscle, for example, contractile activity diminishes simultaneously but in a reciprocal relation in the flexor. This relationship allows steadiness of movement with antagonism.

During horizontal versions the electrical activity of the superior and inferior recti and the superior and inferior obliques shows no change from the resting potential.[16] There is no evidence, therefore, for the belief that active cocontraction of these pairs of muscles takes

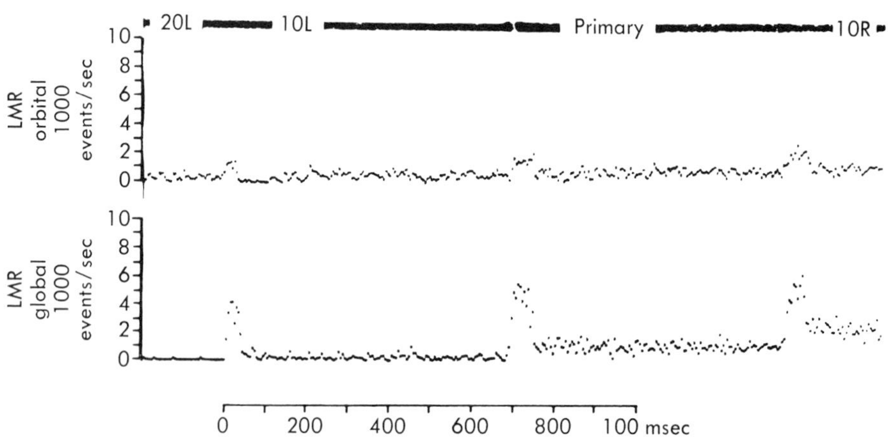

Fig. 5-44. Recording of the motor unit activity of the small, orbital fibers and predominantly large, central, global fibers of an agonist oculorotary muscle during a series of 10-degree "on" saccades. (From Collins, C. C.: The human oculomotor control system. In Lennerstrand, G., and Bach-y-Rita, P., editors: Basic mechanisms of ocular motility and their clinical implications. Proceedings of International Symposium, Wenner-Gren Center, Stockholm, June 4 to 6, 1974, Oxford, 1975, Pergamon Press Ltd., p. 145.)

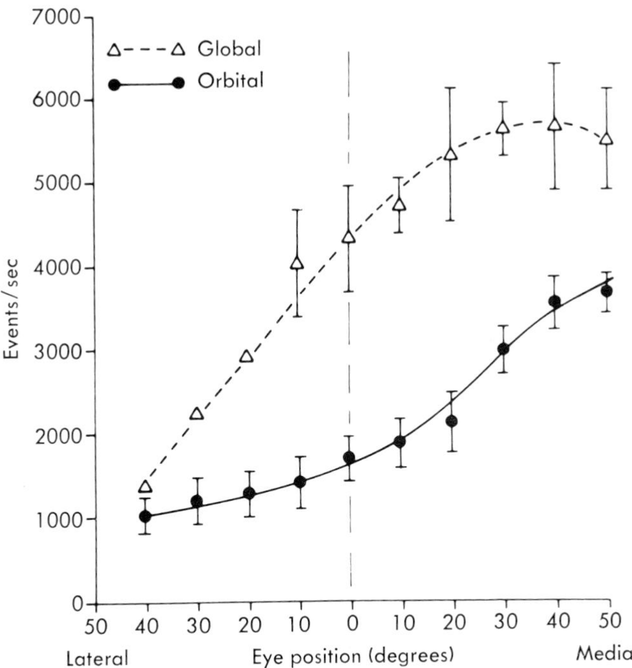

Fig. 5-45. LMR innervation—global versus orbital fibers saccadic activity. Comparison of the peak saccadic activity level of the orbital and central (or global) oculorotary muscle fibers as a function of the final eye position for a complete series of 10-degree "on" saccades. Mean values and standard deviations of the measurements of several saccades at each eye position are shown. (From Collins, C. C.: The human oculomotor control system. In Lennerstrand, G., and Bach-y-Rita, P., editors: Basic mechanisms of ocular motility and their clinical implications. Proceedings of International Symposium, Wenner-Gren Center, Stockholm, June 4 to 6, 1974, Oxford, 1975, Pergamon Press Ltd., p. 145.)

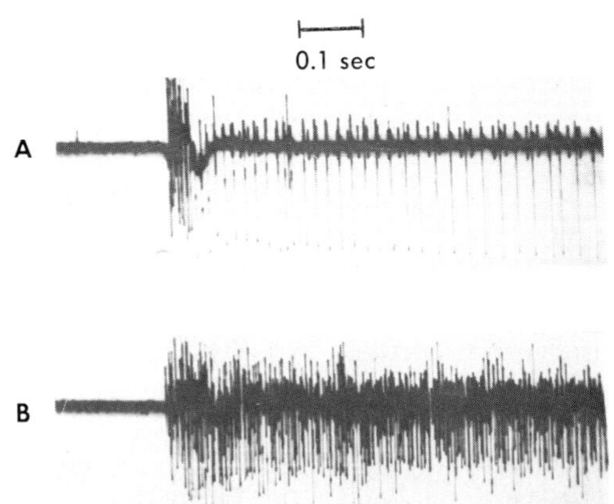

Fig. 5-46. Electromyogram of medial rectus. **A,** Twenty-five–degree saccade. Note immediate burst of high-frequency activity followed by steady but lower frequency of firing. **B,** Twenty-five–degree convergence. (From Miller, J. E.: Arch. Ophthalmol. **62:**790, 1959.)

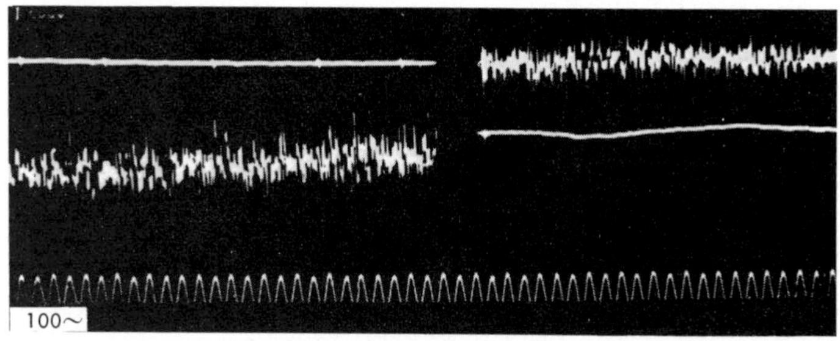

Fig. 5-47. Reciprocal relationship, lateral recti of two eyes. Upper tracing, left eye; lower tracing, right eye. On extreme gaze right there is marked activity of lateral rectus of right eye, with almost complete inhibition of left lateral rectus. On extreme gaze left, left lateral rectus fires maximally, whereas there is complete inhibition of right lateral rectus. (From Breinin, G., and Moldaver, J.: Arch. Ophthalmol. **54**:204, 1955.)

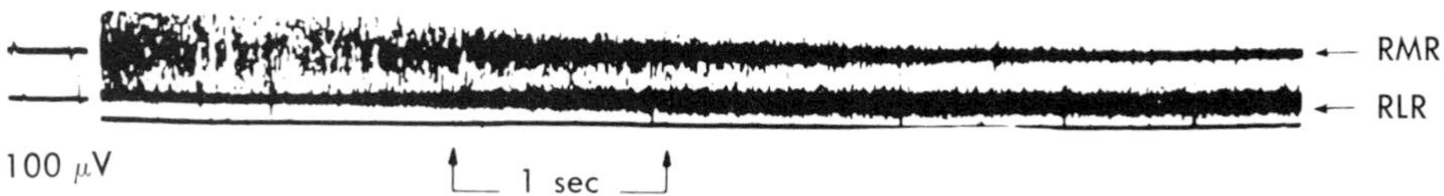

Fig. 5-48. Abducting right eye pursuit movement to the right. (From Jampolsky, A.: Invest. Ophthalmol. **9**:570, 1970.)

place during horizontal versions to aid in either adduction or abduction. It is true, however, as stated in the discussion of muscles concerned in ductions, that these muscles do maintain their normal resting potentials, and insofar as this tenses each muscle, the opposing pairs aid in horizontal movement by preventing the eye from either wobbling up and down or tilting. Furthermore, by their tonic contraction, they may modify horizontal movement on a mechanical basis.

Following movements

During following movements there is neither an abrupt increase nor decrease in firing of agonist (Fig. 5-48) and antagonist muscles. Rather, there is a slow increase in the frequency of firing in agonists and decrease in antagonists, forming a characteristic "summated" envelope of activity (Fig. 5-48).

ELECTRICAL ACTIVITY OF EYE MUSCLES DURING DISJUNCTIVE MOVEMENTS

Convergence

After a given stimulus for convergence, there is a short latent period, followed by a gradual increase in the activity of the medial rectus muscle, which reaches a maximum and then gradually declines. The lateral rectus shows reciprocity throughout the movement.[84] A priori reasoning would suggest that the innervation force patterns seen during fixation and following movements would be similarly recorded during convergence movements in each eye individually.

Asymmetric convergence

Asymmetric convergence is characterized by a saccadic burst in "yoke muscles" followed by a convergence pattern in the medial recti (Fig. 5-49). If the subject is required to follow a slowly approaching target and the activity in the stationary eye is recorded, increasing lateral rectus activity accompanying increasing medial rectus activity is seen as the object approaches the near point of convergence.

Divergence

Electromyographic activity suggests two types of divergence. The first type consists of coactivity of the medial and lateral recti with a saccadic burst seen concomitantly in both muscles followed by reciprocal excitation and in-

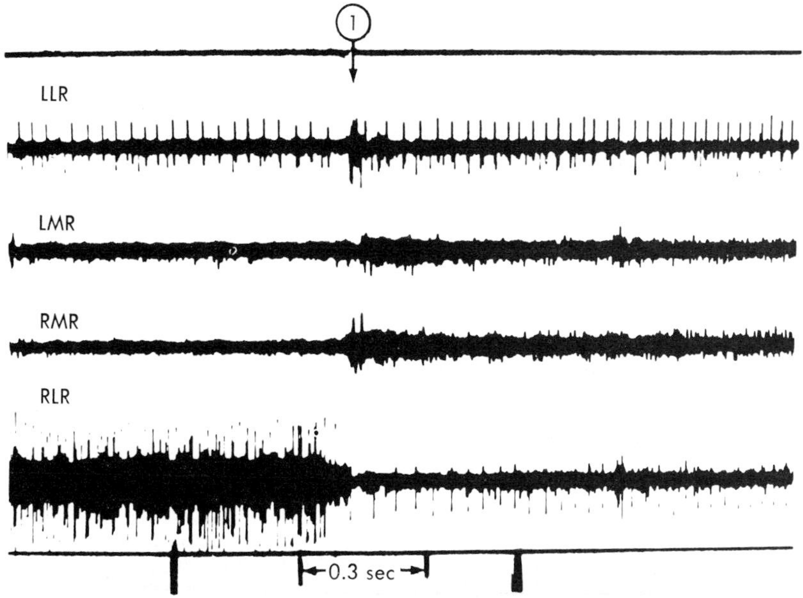

Fig. 5-49. Arrow *1* shows uncover of the exotropic right eye, followed by a convergence refusion of 40Δ of intermittent exotropia. It clearly shows both the reciprocity of the moving eye and the maintained cocontraction of the fixing eye. Note the fixation disengagement of the fixing left eye in this brief but "powerful" refusion vergence. This electromyogram clearly shows all the details of refusion in intermittent exotropia. (From Jampolsky, A.: Coordination of extraocular muscles; clinical aspects. In Lennerstrand, G., and Bach-y-Rita, P., editors: Basic mechanisms of ocular motility and their clinical implications. Proceedings of International Symposium, Wenner-Gren Center, Stockholm, June 4 to 6, 1974, Oxford, 1975, Pergamon Press Ltd., p. 209.)

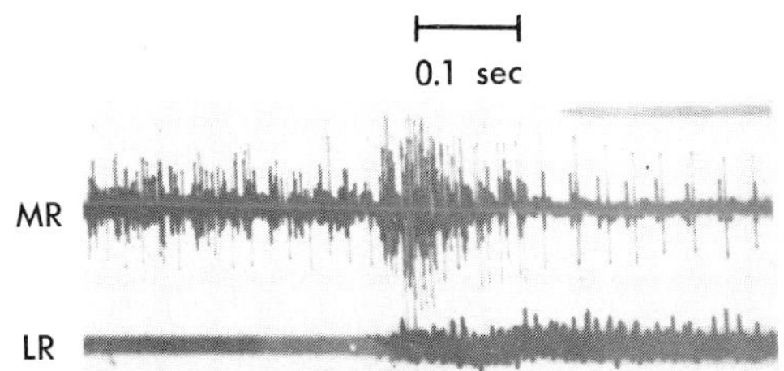

Fig. 5-50. Divergence from 20 degrees. (From Miller, J. E.: Arch. Ophthalmol. **62:**790, 1959.)

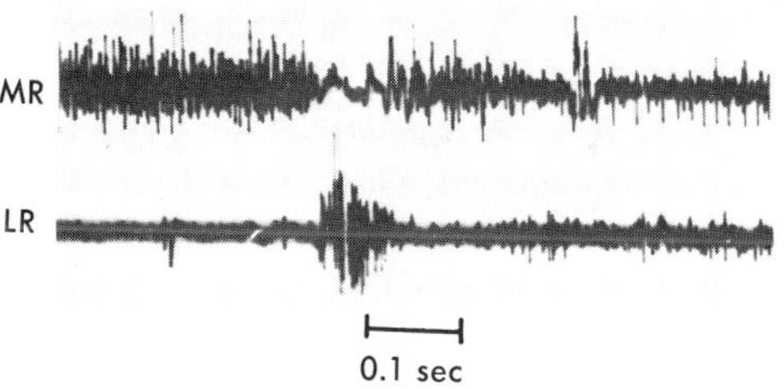

Fig. 5-51. Twenty-five–degree divergence. (From Miller, J. E.: Arch. Ophthalmol. **62:**790, 1959.)

hibition of agonist and antagonist (Fig. 5-50). The second type of pattern seen consists of an immediate saccadic burst of activity in the lateral rectus with concomitant inhibition of the medial rectus followed by irregular activity in both muscles (Fig. 5-51).

Jampolsky[68] has demonstrated a slightly different pattern for divergence associated with exophoria (Fig. 5-52). Occlusion of one eye is associated with a slow increase in activity of the lateral rectus muscle and decrease in activity of the medial rectus muscle of the covered (deviating) eye. Interestingly, there is a concomitant and parallel decrease in activity of the medial and lateral rectus muscles of the fixating eye.

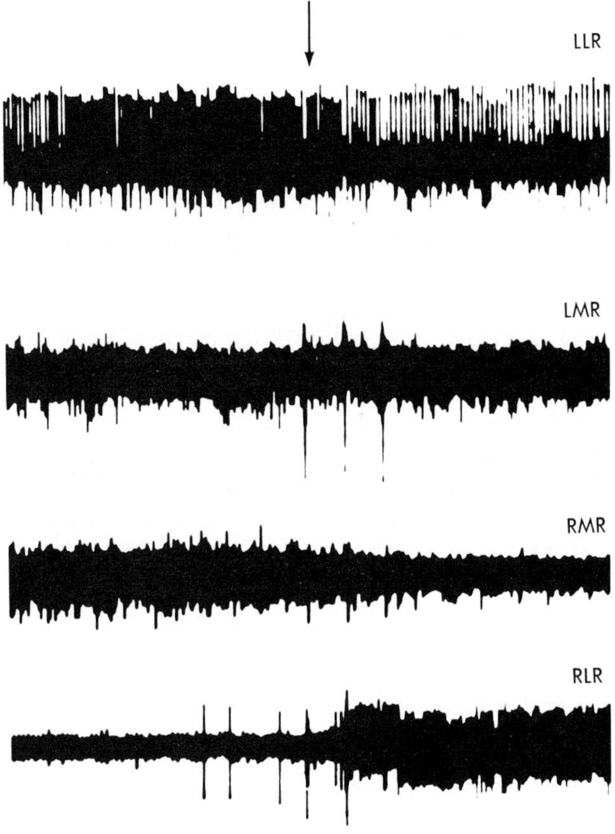

Fig. 5-52. Arrow shows covering the right eye in a patient with 25Δ of intermittent exotropia. Note the smooth reciprocity of the abducting right eye following the break of fusion. When this movement is free of a blink artifact, as it is here, one clearly sees the usual expected physiologic reciprocity in the moving eye during the break from fusion. Note especially the very clear *decrease* in the lateral rectus of the fixing left eye (at arrow). (From Jampolsky, A.: Coordination of extraocular muscles; clinical aspects. In Lennerstrand, G., and Bach-y-Rita, P., editors: Basic mechanisms of ocular motility and their clinical implications. Proceedings of International Symposium, Wenner-Gren Center, Stockholm, June 4 to 6, 1974, Oxford, 1975, Pergamon Press Ltd., p. 209.)

CRANIAL NERVES III, IV, AND VI AND THEIR NUCLEAR ORGANIZATION

Anatomy

The third cranial nerve supplies the superior, inferior, and medial recti as well as the inferior oblique and levator palpebrae superioris muscles. It also supplies the presynaptic parasympathetic outflow to the internal muscles of the eye (the ciliary muscle and the sphincter muscle of the iris). The oculomotor nuclear complex is a collection of cell columns and discrete nuclei (Fig. 5-53). In 1953 Warwick,[122] studying retrograde chromatolysis in the third nerve complex after extirpation of each of the ocular muscles in monkeys, proposed that the cell columns in this nucleus had both a rostrocaudal and dorsoventral arrangement. The levator has only a single midline nucleus that supplies both sides. The medial rectus, inferior rectus, and inferior oblique are innervated directly, whereas the superior rectus is innervated contralaterally (similar to the superior oblique). The visceral nuclei consist of two distinct parts that are in continuity rostrally. The Edinger-Westphal nucleus consists of two slender columns dorsal to the rostral part of the oculomotor complex. Caudally each of these columns divides into two smaller columns that gradually fade away.

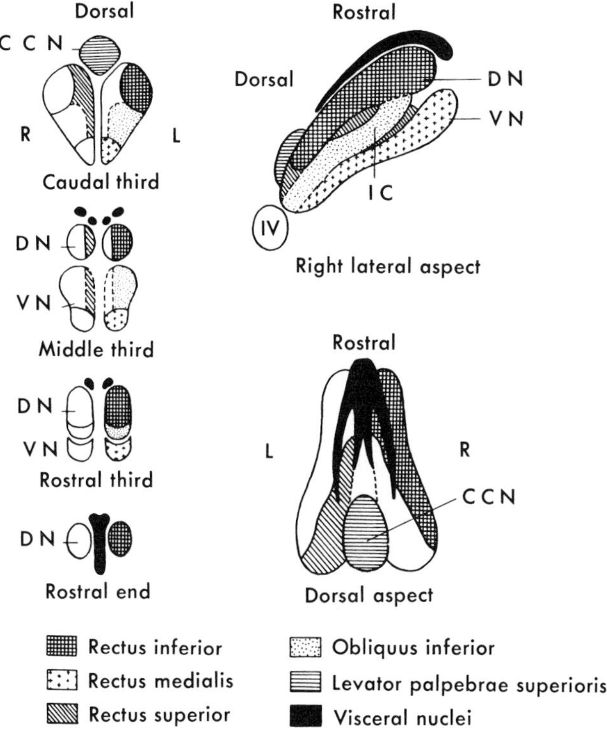

Fig. 5-53. Representation of right extraocular muscles in oculomotor nucleus of monkey. Transverse sections at levels as indicated in the complex. *DN*, Dorsal nucleus; *VN*, ventral nucleus; *CCN*, caudal central nucleus; *IC*, intermediate column; *IV*, trochlear nucleus. (From Warwick, R.: J. Comp. Nuerol. **98:**449, 1953.)

Rostrally the two columns merge into the anterior median nucleus. Both of these *subnuclei* supply the presynaptic parasympathetic outflow.

The oculomotor complex lies ventral to the central gray of the midbrain (Fig. 5-54) at the level of the superior colliculus in a V-shaped trough formed by the medial longitudinal fasciculus (Fig. 5-55). Fibers arising from the oculomotor complex course ventrally through the medial longitudinal fasciculus, the red nucleus, the substantia nigra, and the basis pedunculi to emerge as the third cranial nerve in the interpeduncular fossa. The nerve travels anteriorly in the subarachnoid space being suspended under and over the anterior superior cerebellar and posterior cerebral arteries, to then enter the lateral wall of the cavernous sinus. Pressure on the nerve suspended between these two arteries leads to the pupillary signs in threatened uncal herniation. The pupillary fibers travel superficially in the superior portion of the nerve. The nerve enters the orbit through the superior orbital fissure and divides into a superior branch (superior rectus and levator palpebrae superioris muscles) and an inferior branch carrying the presynaptic parasympathetic fibers as well as motor fibers to the inferior rectus, medial rectus, and inferior oblique muscles.

In 1970 Robinson[100] reported his findings on single-unit behavior of motor nuclei in the oculomotor nucleus of the alert rhesus monkey. Advancement of the recording electrode in a coronal plane identified cell groups with fairly sharp boundaries whose activity correlated with movements in a given direction. Robinson's cell groups in a consecutive order were associated with down, up, left, and then right movements, and he believed these to be motoneurons of the inferior rectus, inferior oblique, and superior rectus in adjacent groups and most ventrally the medial recti, supporting the concept

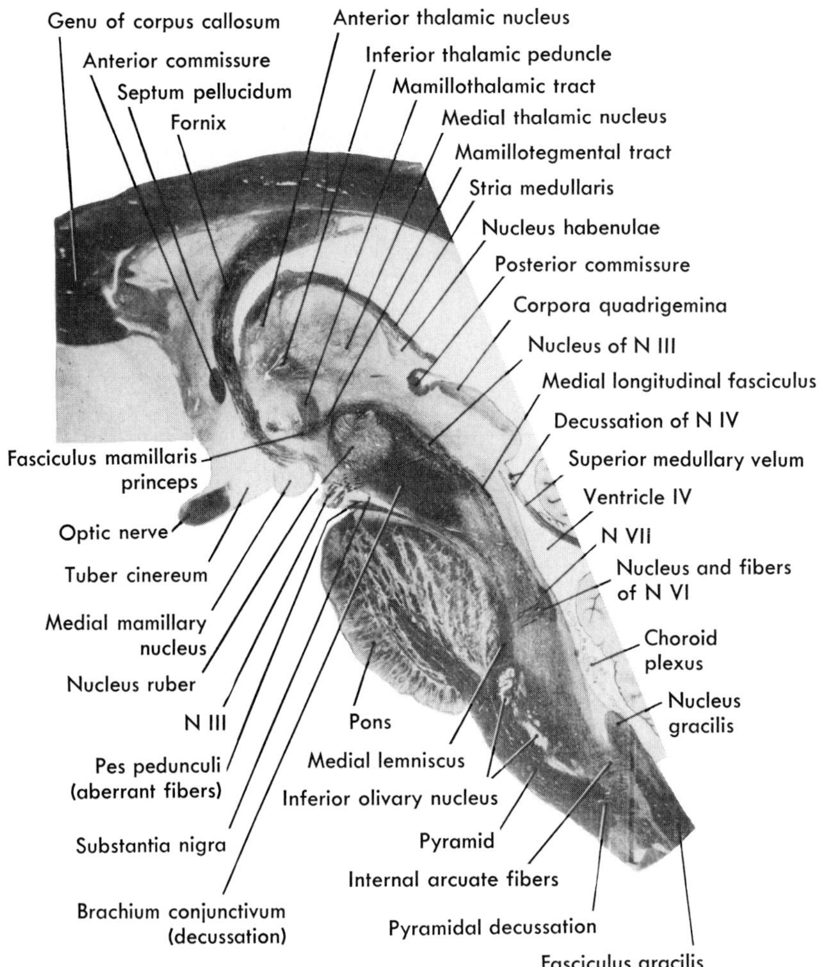

Fig. 5-54. Sagittal section of brain stem through pillar of fornix and root of third nerve. (Weigert's myelin stain; photograph.) (From Strong, O. S., and Ellwyn, A.: Human neuroanatomy, ed. 3, Baltimore, 1953, The Williams & Wilkins Co.)

of rostrocaudal columns defined by Warwick.[122]

The trochlear nerve (cranial nerve IV) arises from a pair of small compact cell groups ventral to the periaqueductal gray at the level of the inferior tectal plate (Fig. 5-56). The cell groups are almost continuous with the rostral part of the third nerve nucleus and indent the dorsal surface of the medial longitudinal fasciculus. The fibers course dorsolaterally and caudally to turn medially and cross completely in the superior medullary velum. The fourth cranial nerve is the only cranial nerve to exit dorsally as well as decussate completely. It travels anteriorly and ventrally in the subarachnoid space to pierce the dura and enter the lateral wall of the cavernous sinus just caudal to the posterior clinoid process. It enters the orbit through the superior orbital fissure to innervate the superior oblique muscle.

The sixth cranial nerves arise from paired groups of multipolar motoneurons situated beneath the floor of the fourth ventricle (Fig. 5-57). These nuclei lie within the complex loops of the internal genu of the facial nerves and produce eminences called the facial colliculi on the floor of the fourth ventricle. The sixth nerve fibers course ventrally through the pons to emerge at its base at the medullopontine junc-

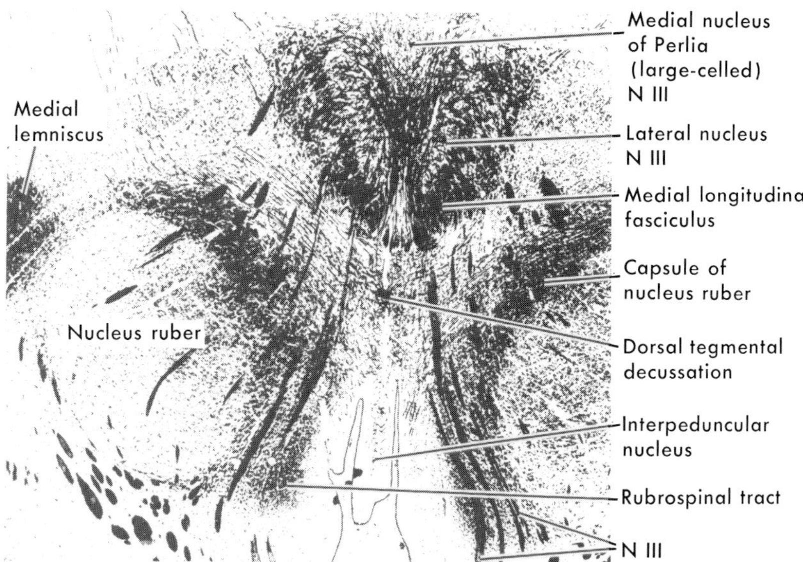

Fig. 5-55. Portion of midbrain tegmentum. One-month-old infant. (Weigert's myelin stain; photograph.) (From Strong, O. S., and Ellwyn, A.: Human neuroanatomy, ed. 3, Baltimore, 1953, The Williams & Wilkins Co.)

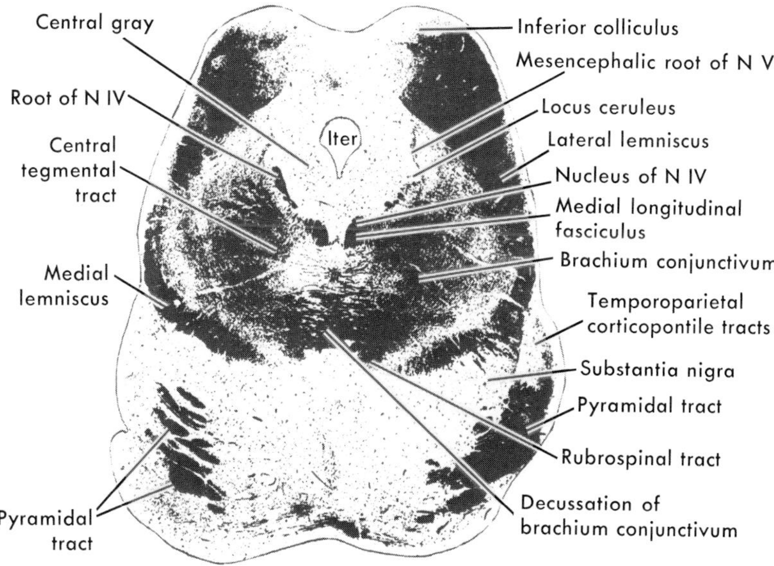

Fig. 5-56. Section of midbrain through inferior colliculi. One-month-old infant. (Weigert's myelin stain; photograph; magnification ×7.) (From Strong, O. S., and Ellwyn, A.: Human neuroanatomy, ed. 3, Baltimore, 1953, The Williams & Wilkins Co.)

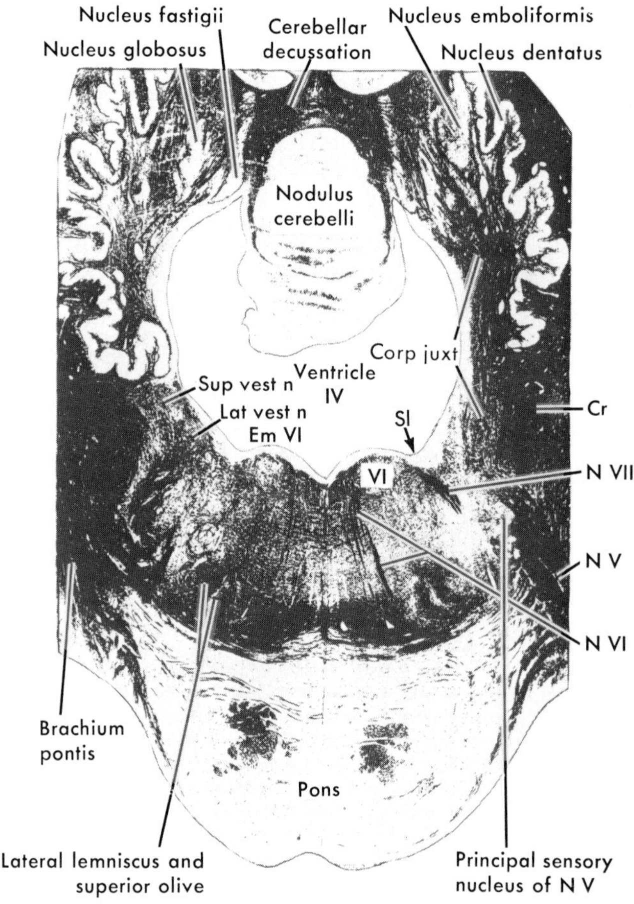

Fig. 5-57. Section through pons, pontine tegmentum, and part of cerebellum, just below entrance of trigeminal nerve. One-month-old infant. (Weigert's myelin stain; photograph.) *Cr,* Restiform body; *Corp juxt,* juxtarestiform body; *Em VI,* eminentia abducentis; *Sl,* sulcus limitans; *VI,* nucleus of sixth nerve. (From Strong, O. S., and Ellwyn, A.: Human neuroanatomy, ed. 3, Baltimore, 1953, The Williams & Wilkins Co.)

tion. The abducens nerve turns rostrally and laterally in the subarachnoid space, crossing over the tip of the petrous portion of the temporal bone. It is anchored there by a reflection of dura, the petrosphenoidal ligament (Gruber's ligament). The reflection of dura at the tip of the petrous pyramid resembles a canal and bears the eponym Dorello's canal. Inflammation of the temporal bone, usually from mastoiditis, can cause an ipsilateral sixth nerve paresis associated with severe pain (Gradenigo's syndrome).

The abducens nerve pierces the dura at the level of the dorsum sellae to enter the cavernous sinus. It travels within the sinus near the carotid artery. The nerve enters the orbit through the superior orbital fissure and supplies the lateral rectus muscle. It has the longest intracranial course of any nerve and is often subject to embarrassment with raised intracranial pressure. Because of its long intracranial course, an isolated lesion of the abducens nerve has no anatomically localizing significance.

Activity of motoneurons

Robinson,[100,102] Schiller,[108] Fuchs and Luschei,[46] Keller,[70] and Keller and Robinson[72] have concluded that there is only one type of motoneuron that subserves all different types of ocular movement in the third, fourth, and sixth cranial nerve nuclei. Their analyses of individual units have revealed the following:

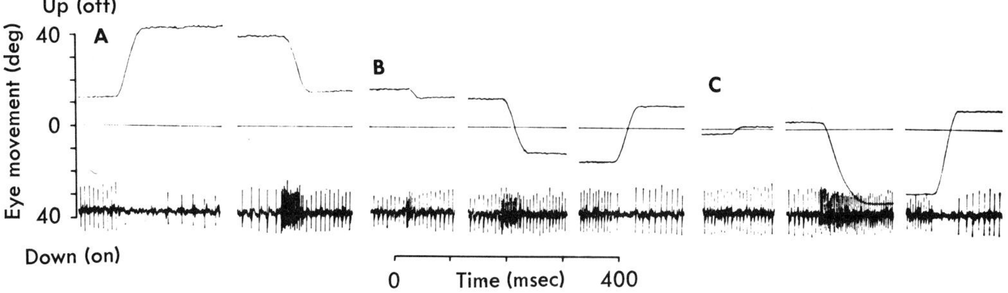

Fig. 5-58. Behavior typical of discharge activity of single units (lower trace) in oculomotor nucleus of intact, alert monkey during fixation and saccadic eye movements (upper trace). Sections were selected from a continuous recording of activity of unit associated with downward movements and thought to be a motoneuron of inferior rectus muscle. (From Robinson, D. A.: J. Neurophysiol. **33:**393, 1970.)

1. No difference other than direction was found in the way activity was related to eye movement.
2. For each position of gaze the discharge rate was fairly constant.
3. The frequency of firing paralleled the functional direction of movement, that is, increased firing in the medial rectus subnucleus with increasing adduction and vice versa.
4. The discharge rate increased in a burst or was inhibited during appropriate saccadic movement (Fig. 5-58).
5. An appropriate change in unit activity always preceded the initiation or cessation of a saccadic movement.
6. All units participated in following movements by increasing or decreasing their discharge rates smoothly.
7. Firing rates were unaffected by saccades perpendicular to the field of action of a particular subnucleus.
8. The same motoneurons were active in vergence movements.

In contradistinction to these authors, Henn and Cohen[54-56] have demonstrated that motoneurons can be classified according to their behavior during saccades and fixation. Their analysis revealed the following characteristics:

1. All motoneurons had clear on-off directions (in direction of agonist activity).
2. On-off directions for medial rectus and abducens units lay on the horizontal plane, whereas on-off directions for the other muscles were tilted 10 to 20 degrees off the vertical plane.
3. Most units were active during periods of fixation and saccades (Fig. 5-59).

These units could be subdivided into tonic units (Fig. 5-59, *A* and *B*), burst units (Fig. 5-59, *F*), and burst-tonic units. Motoneurons form a continuous spectrum of graded response, with burst units at one end and tonic units at the other.

Tonic units are active during periods of fixation and increase their firing patterns in a steplike fashion during saccades. These units tend to have a lower threshold and a lower maximum frequency than phasic units. The tonic units are often not totally inhibited during movement in the off direction. Burst units are active over the entire range of eye movements, but fire a train of spikes only at the onset of ocular movement in the on direction. Similar conclusions have been reached by Barmack.[7] Thus it appears that while specific eye movements are not produced by one motoneuron, there exists a physiologic hierarchy, the more tonic motoneurons (as a broad class) being responsible for holding the eyes steady or moving them slowly and the more phasic or burstlike motoneurons responsible for moving the eyes rapidly.

Furthermore, using antidromic stimuli,[56] it has been shown that the burstlike motoneurons have faster conduction velocities than those with more tonic activity. This suggests that the more phasic neurons have larger axons and are

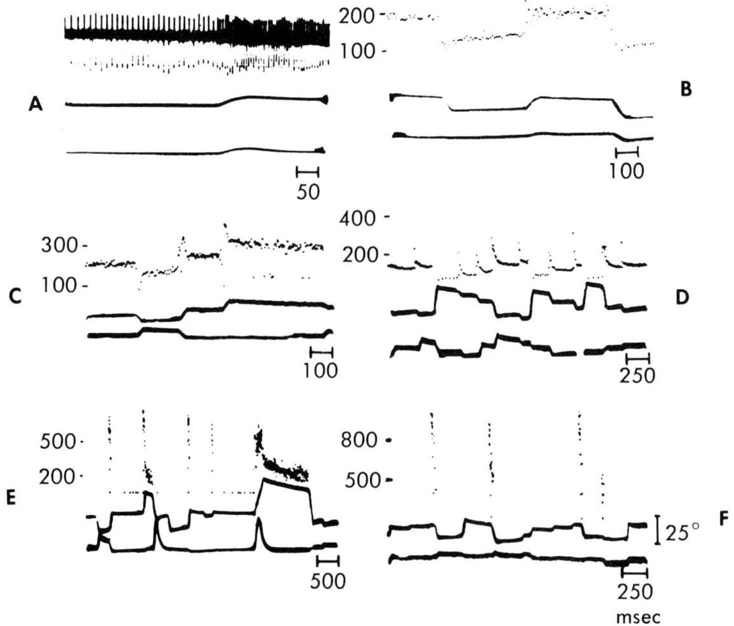

Fig. 5-59. Medial rectus units with different characteristics. **A,** Top trace is spike activity; **B** to **F,** dot displays represent instantaneous frequency. Height of each dot is the reciprocal of the last spike interval. Numbers on the left refer to frequencies in hertz. Middle and bottom traces are the horizontal and vertical EOG respectively. Up corresponds to right or upward movements. **A** and **B** are samples of the same tonic unit; **C** and **D** are two predominantly tonic units; **E** is predominantly phasic unit; **F** is a phasic unit. Note that from **C** to **F** the phasic part of the unit activity becomes more prominent until in **F** the unit fires only during quick eye movements in the on direction. (From Henn, V., and Cohen, B.: Activity in eye muscle motoneurons and brainstem units during eye movements. In Lennerstrand, G., and Bach-y-Rita, P., editors: Basic mechanisms of ocular motility and their clinical implications. Proceedings of International Symposium, Wenner-Gren Center, Stockholm, June 4 to 6, 1974, Oxford, 1975, Pergamon Press Ltd., p. 303.)

therefore the same neuron that singly innervates the larger eye muscle twitch fibers.

Clinically, the hallmark of disturbances of the efferent pathways, that is, nuclear or infranuclear mechanisms to the extraocular muscles, is diplopia. A few supranuclear disorders such as skew deviation and internuclear ophthalmoplegia (transient) can also produce this symptom.

Summary

Anatomically at least five different types of muscle fibers have been identified (maybe a continuum exists). These fiber types can be more simply categorized according to their innervational pattern, that is, true motor end-plate versus en grappe endings. Physiologically (studying the muscles themselves) and pharmacologically there appear to be three types of fiber: so-called twitch or phasic fibers, which can be correlated anatomically with fibers having a true motor end-plate, and two types of tonic fibers—a multiply innervated, nonconducting unit (en grappe?) and multiply innervated conducting fibers(?). These findings have reasonably led to the hypothesis that the saccadic system, or fast-movement system, is associated with twitch fibers, whereas following and vergence movements are associated with the tonic fibers. Jampel[66] succinctly states:

> In man either saccadic eye movements and tonic eye movements are different modes of action of the same neurologic and peripheral muscular apparatus or there are separate neuronal pathways and peripheral motor units responsible for these functions.

The latter concept would imply two separate parallel pathways, one for saccadic movements and one for tonic eye movements, each having its own separate supranuclear components, subnuclei within the ocular motoneuron complex, lower motoneuron fibers, and muscle fibers (motor units).

SUPRANUCLEAR CONTROL MECHANISMS

The concept of centers implying an anatomically isolated volume of neurons acting as a repository of preprogramed ocular movements in the cortex and brain stem is, at best, simplistic. This concept is merely a hypothetical convenience with pathophysiologic correlates that allows a rational discussion of supranuclear control mechanisms.

Saccadic system

Examples of different types of saccadic, or rapid, eye movements have been tabulated by Daroff[33]; they are shown below:

1. Command; random
2. Voluntary refixation
3. Sensory evoked (visual, auditory)
4. Nystagmus fast phase (a) pathological (b) induced: optokinetic nystagmus (OKN), vestibular, OKAN
5. Rapid eye movement (REM) (sleep)
6. Rapid pursuit (above 45 degrees/sec)
7. Microsaccades

Command, random movements, and involuntary movements toward visual and auditory stimuli are always saccades. The fast phases of all types of jerk nystagmus and the REM associated with dreaming are also saccadic movements. Saccadic movements are necessary to track objects moving at velocities exceeding 40 degrees/sec. Microsaccades are small, high-velocity flicks (4.5 minutes of arc) associated with fixation.

All saccades, except for foveation movements, probably originate in the contralateral frontal eye fields. Recent evidence suggests that each hemisphere is capable of producing ipsilateral saccades as well.[106,118] Electrical stimulation of the frontal cortex in area 8 of Brodmann (Fig. 5-60) in the caudal part of the middle frontal gyrus and extending into the contiguous part of the inferior frontal gyrus will cause contralateral conjugate movement of the eyes; that is, right frontal eye field yields left conjugate gaze. This movement is an "all or nothing" phenomenon producing a given movement independent of stimulus as long as the threshold is exceeded. The direction of the movement is determined by the site of the stimulation.[106] Stimulation of various regions within this area will produce lateral movement up, lateral movement, and lateral movement down. Stimulation of the frontal eye fields is associated with similar movements of the head and neck. Strictly vertical eye movements are produced only by bilateral stimulation of the frontal eye fields.[92,95] The final trajectory in simultaneous stimulation is the result of the individual vector forces of the vector direction of movement invoked by either stimulus alone.

Recently Kurtzberg and Vaughan[76] have used computer averaging techniques to extract ongoing electroencephalographic activity related to eye movements in humans. These investigators demonstrated eye movement potentials that begin 200 milleseconds before the initiation of eye movement. These eye movement potentials have a maximum amplitude over Brodmann areas 8 and 19. A subsequent study by Melamed and Larsen,[80] which measured regional blood flow, noted focal increases over the middle precentral and premotor regions as well as in the temporoparietal visual association area with voluntary eye movements.

Bizzi,[10] Bizzi and Schiller,[12] and Mohler and associates,[85] recording from single motoneurons in the frontal eye fields in monkeys, could not demonstrate neurons that fired prior to voluntary saccades. Bizzi and Schiller[10,12] found two types of cells in the frontal eye fields. Type I fired during voluntary saccadic movement in a given direction and also during the fast phase of optokinetic and vestibular nystagmus, and type II cells fired during smooth pursuit movements and the slow phase of nystagmus. These two cell types maintained their firing patterns irrespective of head position. The failure of these researchers to identify cells that fire preceding eye movements has been used as an argument against a saccadic gaze center in the frontal lobe. The problem with their work is that visual stimuli rather than voluntary responses to commands were used to in-

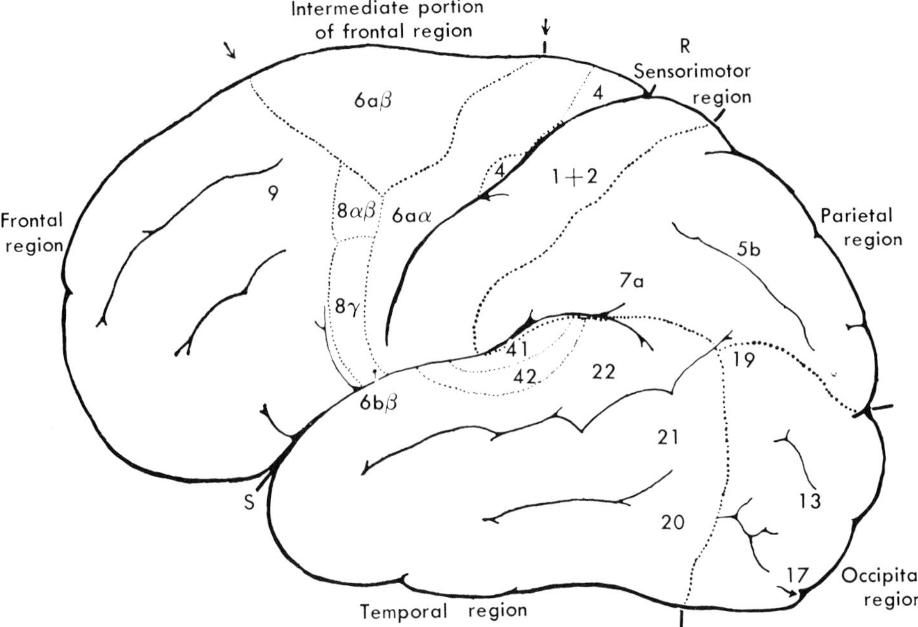

Fig. 5-60. Surface view of left motor cortex, slightly tilted to show Vogt's areas 8 alpha, beta, and gamma. These are areas of voluntary ocular movements to right and possibly upward. No cortical areas have been discovered where stimulation produces either downward movements or movements of convergence. (From Penfield, W., and Erickson, T.: Epilepsy and cerebral localization, Springfield, Ill., 1941, Charles C Thomas, Publisher.)

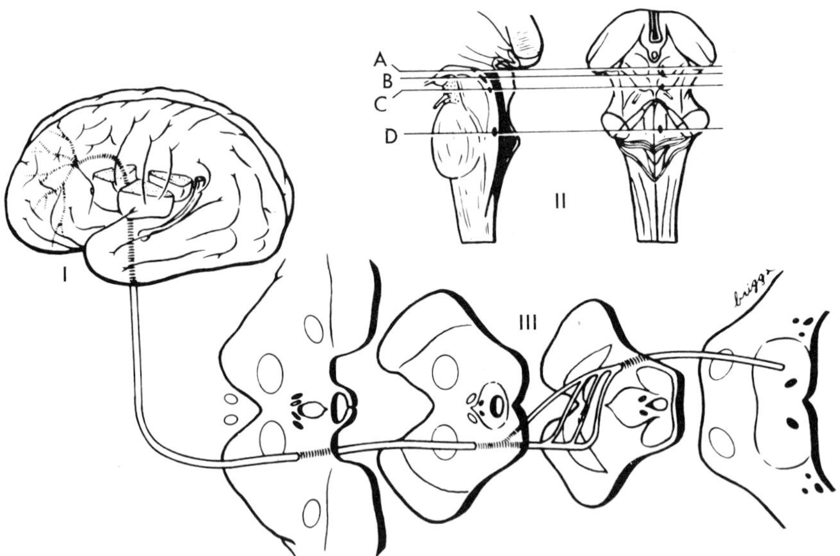

Fig. 5-61. Frontomesencephalic pathways for saccadic mechanism (saccades, fast eye movements, voluntary eye movements). Path for horizontal movements. *A*, Pretectal area; *B*, oculomotor nucleus; *C*, trochlear nucleus; *D*, abducens nucleus. (From Gay, A. J., and Newman, N. M.: In Critchley, M., O'Leary, J. L., and Jennett, B., editors: Scientific foundations of neurology, London, 1972, William Heinemann Medical Books, Ltd.)

duce eye movements. Therefore what would at first appear to be contradictory information really represents the current inability of investigators to elicit voluntary nonvisually evoked eye movements in the primate. Clinically, saccadic dysfunction is noted in patients with frontocortical disease[34] and therefore is indicative of some role for the frontal motor area in saccadic movement control in humans.

A separate group of cells was found to be exclusively related to head turning. The firing of the eye movement cells only occurred after the initiation of eye movements, whereas in the latter group firing preceded head movement. In the primate the appearance of a target in the visual field is always followed by an orderly sequence of eye and head movements. Initially there is a saccadic movement of the eyes followed by a head movement in the same direction with a latency of 20 to 40 milliseconds. This coupled movement would suggest teleologically that the programed generation of these movements would be produced by substrates in close contiguity. Furthermore, once the foveas fixate on the target, the eyes maintain fixation by performing a rotational movement counter to that of the head.[13] The adjustment of the amplitude of the saccade necessitated by the head movement is reflexly achieved by vestibular afferents.[11,35,86]

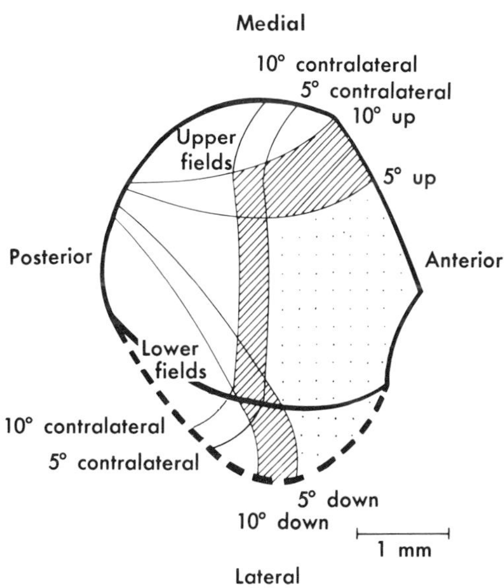

Fig. 5-62. Representation of visual field on surface of right colliculus (as viewed from above in Horsley-Clarke plane). Strippled area represents parts of contralateral visual field within 5 degrees of fovea. Striped and stippled areas combined represent parts of contralateral visual field within 10 degrees of fovea. (From Cynader, M., and Berman, N.: J. Neurophysiol. **35:**187, 1972.)

The precise pathway of the corticopontine fibers subserving conjugate gaze remains speculative (Fig. 5-61). The best evidence would suggest that the fibers for conjugate gaze are grouped in the anterior limb of the internal capsule and adjacent globus pallidus as part of the corticobulbar projection. Thereafter, the fibers separate into two bundles.

1. The major pathway courses caudomedially, passing along the ventrolateral surface of the thalamus amid the zona incerta and the fields of Forel. In the rostral midbrain these fibers are situated primarily in the ipsilateral reticular formation and then decussate in the lower midbrain and upper pons to terminate in a special area of the paramedian pontine reticular formation (PPRF) of the pontine tegmentum termed the pontine center for conjugate horizontal gaze.

2. The secondary bundle descends through the internal capsule and cerebral peduncle to the basis pontis before it turns dorsally to enter the pontine tegmentum. This pathway constitutes Déjèrine's aberrant pyramidal pathway. The significance of this pathway, either physiologically or pathologically, is unknown.

Recent evidence reported by Schiller[109] demonstrates that saccadic movements may be produced by stimulation of the occipital cortex. The corticofugal fibers from this area pass to the superior colliculus.

Although there is disagreement on the role of the superior colliculus in pattern discrimination, there is general agreement that it plays an important role in orienting responses. Direct retinotectal and visual cortical pathways have been identified. Cynader and Berman[32] have constructed a sensory map reflecting spatial receptive fields in the superficial layers of the superior colliculus (Fig. 5-62). The representation of the visual field is not uniform; the central 10 degrees occupy about 30% of the surface area with little if any foveal representation. A similar nonuniform representation of the contralateral (major) and ipsilateral (minor) eye is also evident.[63] Using microelectrode stimulation, Robinson[102] constructed a motor map in the deeper layers of the colliculus that produced the ocular movement of foveation compatible with the eccentricity of the superficial sensory map. This finding implies that one of the functions of the tectal plate is to act as a sensory motor way station in the act of turning the eyes to look at a visual stimulus of potential interest. The temporal series of events in the act of foveation[110] is initiated by a visual stimulus falling on the retinas on some nonfoveal point in the visual field. This stimulus induces neuronal activity via the retinotectal projection to a localized region of the superior colliculus, whose orientation is relative to the fovea. The impulses descend into the deeper layers of the superior colliculus via a columnar organization and from there to the pontine gaze center. The tectofugal activity creates a programed saccade of a specific direction and amplitude, bringing the original visual stimulus onto the fovea. It is obvious that if the motor-sensory map is oriented with relation to eccentricity from the fovea, these collicular eye movements are independent of initial eye position.

Studies of evoked saccades in the superior colliculus allowed construction of a motor map (Fig. 5-63) on the tectal surface. Saccades had an up component medially, a down component laterally, and became larger caudally and smaller rostrally. The major vector is contralaterally horizontal for the most part. These movements could vary from 42 seconds to 51 degrees of arc. Pure vertical movements can be produced only by bilateral stimulation.

Within the colliculus the amplitude and direction of an evoked saccade did not vary markedly with depth. This is compatible with the idea that there is a columnar organization of the retinal locations in the colliculus. Analyzing the data in Fig. 5-63, Robinson constructed a motor map that roughly predicts by interpolation the amplitude and direction of a saccade evoked by stimulation of any point in the superior colliculus. Schiller and Koerner[110] found that deep collicular units apparently had both sensory and motor activity. These cells have visual receptive fields and also fire before saccades, most vigorously for saccades that bring the stimulus onto the fovea. Robinson believes that exact registration of a sensory map in the upper layers with a motor map in the lower layers is unlikely, but the point to point association between the two maps is well demonstrated by the duality cells of Schiller and Koerner mentioned earlier. A similar view of motor function is held by Sparks and Pollack.[115] It must be said at this point that these concepts are extremely new and must bear the weight of continued investigation.

An interesting feature of species evolution is the increasing percentage of space taken up by corticotectal projections.[115] Wurtz and associates[129,130] believe that the superior colliculus is not a precise guidance system or foveator. They postulate that the superior colliculus of primates contributes to a shift in visual attention and a facilitation of eye movements toward areas of attention. This might explain the increasing presence of corticotectal and tectothalamic projections in primates.

Investigators have long recognized that there is a variability in the accuracy of saccadic eye movements. In 1971 Weber and Daroff[123]

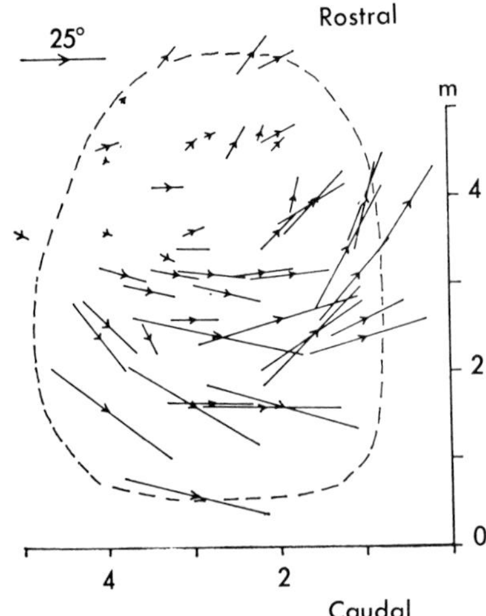

Fig. 5-63. Map of saccade direction and amplitude on dorsal view of left superior colliculus. Scale is in millimeters; *m*, midline. A composite of the saccades (arrows) evoked from 50 electrode tracks in two monkeys. Dotted line is reconstructed boundary of superior colliculus. (From Robinson, D. A.: Vision Res. **12:**1795, 1972.)

studied the metric characteristics of saccadic movements in 25 individuals. They found that the major influence on saccadic precision was the amplitude of refixation (Figs. 5-64 and 5-65). With increasing amplitudes the incidence of dysmetric conjugate undershoots or overshoots increases. The incidence of undershoots increases, and the incidence of overshoots decreases, while the size of the error increases. The incidence of disconjugacy, usually overshooting of the abducting eye or undershooting of the adducting eye, increases. For example, among 10-degree saccades, nearly 70% are normometric and require no corrective movement; 12.8% are conjugate undershoots, 9.3% conjugate overshoots, and 9.5% brief disconjugacy requiring corrective movements. For 30-degree saccades only 20% of the movements were normometric, 50% were dysmetric undershoots, 3% were dysmetric overshoots, and 27% were terminated disconjugately.

All dysmetric and disconjugate saccades are followed by appropriate small correctional movements (CM). Correctional movements in normal individuals are divided into two types[124] One, a designated saccadic CM, was fast (> 400 degrees/sec), had a definite latency (125 milliseconds), and always followed conjugate errors (Fig. 5-66). The size of the saccadic CM increased monotonically with the interfixation distance (Fig. 5-64). The second type of corrective movement was termed a glissade, or glissadic CM. This type of movement was of low velocity (~ 20 degrees/sec) and inseparable from the terminal portion of the saccade (no latency) (Figs. 5-67 and 5-68). The glissadic error also increased monotonically with the interfixational distance (Fig. 5-65). The mechanisms involved in these corrective movements are unknown, but Weber and Daroff postulate a prenuclear feedback system for glissades.

Recently Troost and co-workers[118] reported finding two additional CM in the eye movements of a patient who had a total hemispherectomy: (1) mixed glissadic-saccadic and (2) a short latency (45 milliseconds) saccadic correction of dysmetric undershoot. These types of corrective movements have never been seen in normal individuals.

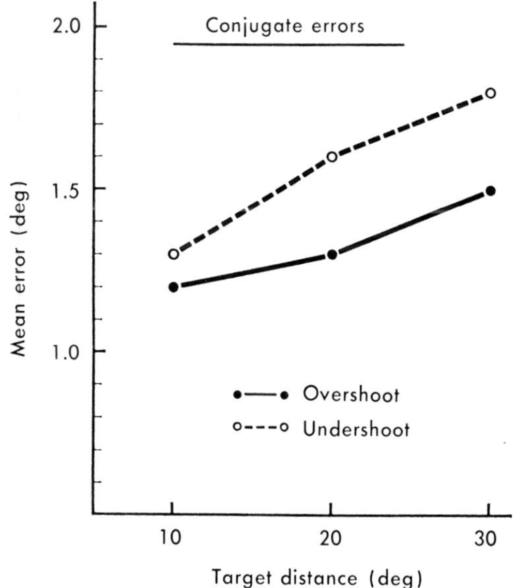

Fig. 5-64. Relationship of interfixational distance to mean size of binocular (conjugate) error. (From Weber, R. B., and Daroff, R. B.: Vision Res. **12:**467, 1972.)

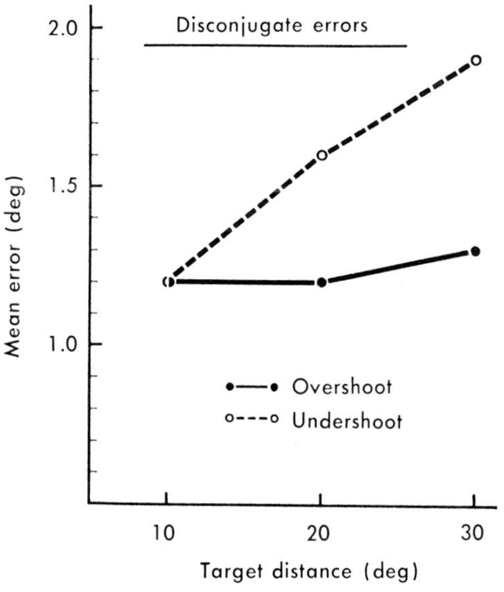

Fig. 5-65. Relationship of interfixational distance to mean size of disconjugate uniocular error. (From Weber, R. B., and Daroff, R. B.: Vision Res. **12:**467, 1972.)

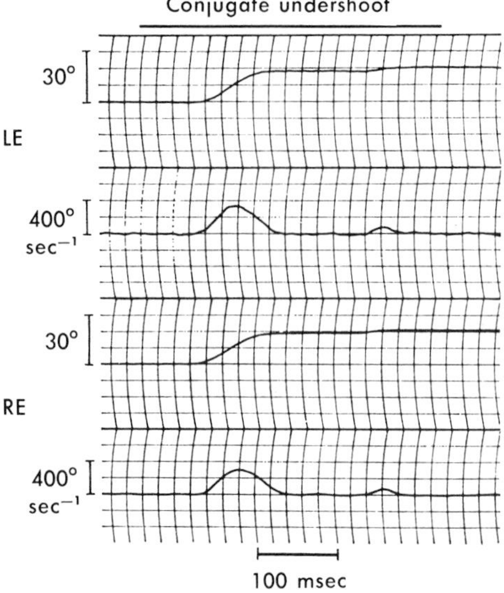

Fig. 5-66. Refixational saccade from primary position to target 20 degrees to the right. First and third channels represent position of left eye, *LE*, and right eye, *RE*, respectively. Second and fourth channels demonstrate differentiated peak velocities. Both eyes undershoot by approximately 2 degrees, and after 125 milliseconds binocular conjugate saccadic corrective movements align eyes with target. (From Weber, R. B., and Daroff, R. B.: Vision Res. **12**:467, 1972.)

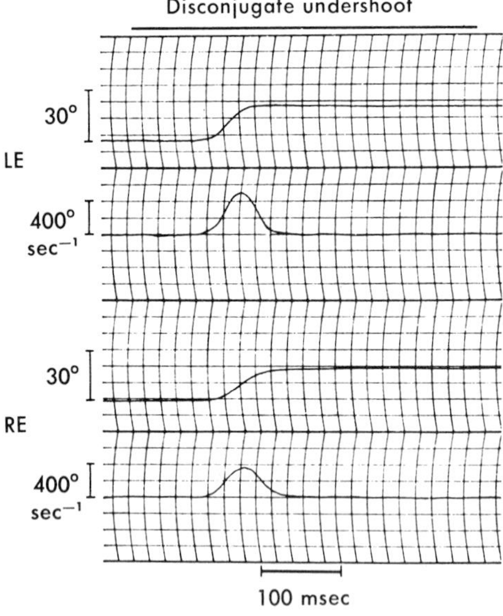

Fig. 5-67. Saccade from 20 degrees left to primary position. Right eye, *RE*, undershoots target by approximately 2 degrees, followed by a glissadic corrective movement. Left eye overshoots by less than 1 degree. (From Weber, R. B., and Daroff, R. B.: Vision Res. **12**:467, 1972.)

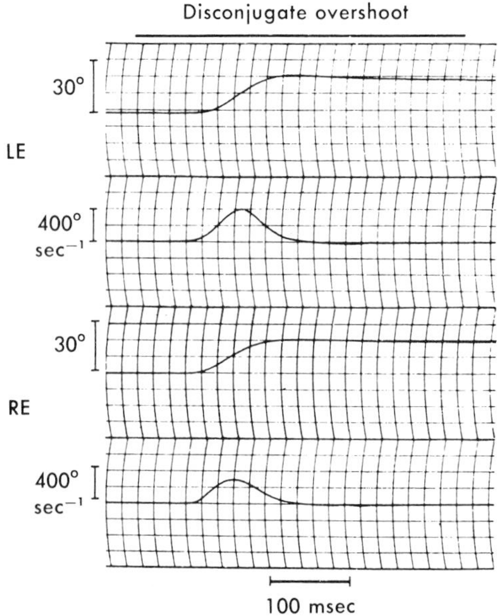

Fig. 5-68. Uniocular overshoot of left eye followed by glissadic corrective movement. (From Weber, R. B., and Daroff, R. B.: Vision Res. **12**:467, 1972.)

Pursuit system

It is fairly well agreed that the center for smooth tracking or pursuit resides in the parieto-occipital junction (Brodmann's area 19). The classic concept of contralateral control based on the original work of Foerster[43] has recently been challenged by Daroff and Hoyt[34] and Troost and co-workers[119] based on clinical findings. Smooth pursuit movements maintain foveal fixation on moving targets with speeds of 30 to 50 degrees/sec. A target motion is necessary to elicit smooth pursuit, for without such a stimulus subjects attempting to move their eyes smoothly actually generate a series of small saccades. It has been shown recently that the eyes can make sustained smooth movement indistinguishable from normal pursuit in the absence of a moving stimulus if an afterimage is tracked.[58] It is postulated that this phenomenon depends on the suppression of the saccadic system.

The corticofugal pathways dealing with horizontal pursuit are for the most part conjectural. Fibers from Brodmann's areas 18 and 19 of the visual cortex have been shown to travel in the internal sagittal striatum medial to the visual radiation, pass through the posterior end of the internal capsule, and terminate in the thalamus, pons, and superior colliculi. Classic theory would hold that these fibers decussate in the cerebral peduncles and terminate in the contralateral PPRF. Daroff and Hoyt[34] postulate the following pathway:

1. Occipitocollicular projections lie medial to the visual radiations (internal sagittal striatum) and pass through the retrolenticular portion of the internal capsule to the ipsilateral superior colliculus and synapse.
2. Fibers presumably leave the superior colliculus with the mediotectospinal tract and cross to the opposite side of the midbrain tegmentum in the dorsal fountain decussation of Meynert.

These fibers then accompany the decussation from the mesencephalic saccadic tract and terminate in the PPRF. This requires a double decussation. Vertical pursuit movements are probably bilaterally represented similar to the saccadic system.[8]

Smooth movements have not been elicited in stimulation studies of the primate superior colliculus. Collewijn[29] has produced smooth eye movements in the superior colliculus in the rabbit. He found that in this animal removal of the tectal plate abolished optokinetic nystagmus. Robinson[102] believes that the superior colliculus undoubtedly mediates the detection of slip velocity in executing smooth movements, but that this would logically require a subset of directionally selective neurons scattered throughout the colliculus. Therefore among primates in whom the saccadic system has become predominant punctate stimulation is unlikely to demonstrate these units.

The argument of Daroff and Hoyt[34] that the oculomotor pursuit centers are ipsilateral is based on certain clinical observations. In 1964 Cogan[23] reported an ocular deviation of oblique or horizontal nature opposite the side of the lesion during forcible lid closure in patients with midcerebral lesions.[113] This is accompanied by loss of smooth pursuit and optokinetic nystagmus toward the side of the lesion. Cogan termed this "spasticity of conjugate gaze" and postulated an innervational hypertonus of the occipitoparietal center. Daroff and Hoyt argued that this could be more logically explained by assuming ipsilateral pursuit control; that is, when the ipsilateral center is damaged, by breaking fixation with the eyes closed the normal tonus of the contralateral center will deviate the eyes to the side contralateral to the damaged center in the field of the functioning center. Similarly, pursuit movements would be normal contralateral to the lesion and abnormal (microsaccades) in the field of the lesion (that is, ipsilaterally). A 1972 paper by Troost and co-workers[119] would seem to add credence to this argument. Hoyt and Frisén[62] believe (a priori) that both occipital lobes participate to produce smooth pursuit in all directions.

INTERCONNECTIONS BETWEEN CEREBRAL EYE FIELDS

Extensive callosal fiber systems connect the occipital, parietal, and temporal lobes. Similar callosal connections exist for the frontal fields. The occipital eye fields and frontal eye fields are connected ipsilaterally via fiber pathways in the superior longitudinal bundle. The frontal eye fields have a suppressor effect over the activity of the preoccipital fields both ipsilaterally and contralaterally. These tracts are important in understanding optokinetic nystagmus. Optokinetic nystagmus (OKN) is a type of induced nystagmus easily elicited in all persons with a normal visual mechanism although not necessarily a high degree of visual acuity. It is a manifestation of the fixation and following mechanisms with a fast phase that may either represent foveation or a corrective movement. It is usually induced clinically by a rotating drum with stripes or by a moving tape with stripes. The eyes involuntarily follow a stripe of interest until it passes out of the visual field or the central mechanism attempts to foveate a more distinct image, and a fast movement is made in the opposite direction. This results in a jerk nystagmus whose slow-phase velocity depends on the speed of the stimulus and whose fast phase depends on the innate saccadic velocity of the individual. OKN is usually produced in the horizontal plane and is less well sustained in the vertical plane. The well-known railroad car nystagmus is a manifestation of OKN.

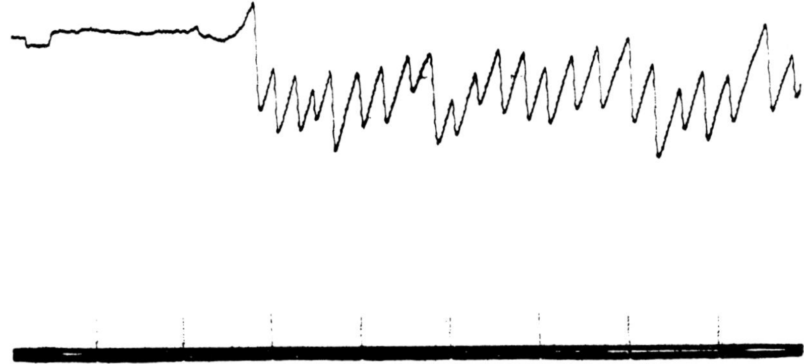

Fig. 5-69. Recording showing initiation of optokinetic nystagmus (normal subject) on illumination of revolving optokinetic drum. Note initial slow component. (From Dix, M. R., and Hood, J. D.: Acta Otolaryngol. **71:**217, 1971.)

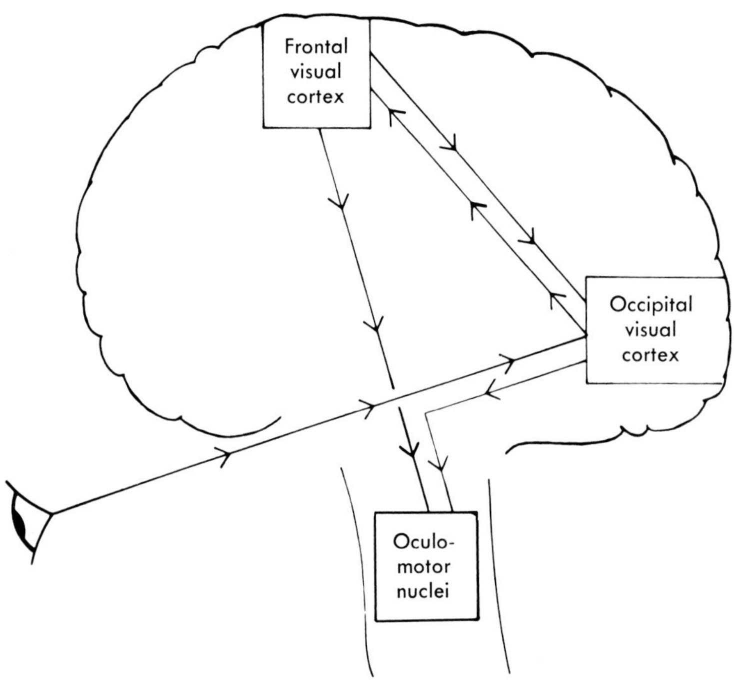

Fig. 5-70. Schematic illustration of suggested nervous pathways subserving optokinetic nystagmus. (From Dix, M. R., and Hood, J. D.: Acta Otolaryngol. **71:**217, 1971.)

Many factors determine a normal individual's response to an OKN stimulus. Angular subtense of the stimulus is somewhat related to visual acuity, and if the visual angle is too small for a given visual apparatus, no response will be elicited. If the speed of the stimulus is too fast or if there is observer inattention, a negative response may be seen also. OKN is an involuntary reflex, and the eye movements cannot be voluntarily inhibited for prolonged periods. For example, if a malingerer or hysteric with simulated blindness is confronted with an OKN stimulus that completely envelopes his visual field, an appropriate response may be elicited. A good ploy is to use the stock page of a newspaper at the bedside.

The normal response to an OKN stimulus is an initial slow component that deviates the eyes in the direction of the movement of the stripe. This slow component lasts some 200 milliseconds, and its velocity is considerably less than the subsequent slow components. This is followed by a fast phase that deviates the eyes in the opposite direction of gaze, and it is in this field of gaze that all subsequent nystagmus takes place (Fig. 5-69).

The slow and fast phases are distinct and separable. The slow phase is obviously mediated through the afferent visual tracks to area 19, where the pursuit mechanisms are activated. This is involuntary in that the eyes cannot be kept stationary if gaze is fixated on the drum. The pathway for the fast phase remains conjectural at best. Ling and Gay[77] and Dix and Hood[39] believe that the fast phase is mediated through the ipsilateral frontal lobe center. These authors postulated an occipitoparietal frontal association tract (Fig. 5-70). This pathway has been tentatively identified.[14] In face of the recent work on the function of the superior colliculus described in the previous two discussions, this foveator may play a role in the fast phase of OKN. In any case OKN abnormalities can be found in both supratentorial and infratentorial defects involving pathways of pursuit or saccadic mechanisms.

Vergence mechanism

Vergences, or disjunctive eye movements, enable the subject to fixate points at various distances in visual space. This calls for either active convergence or divergence. Vergence movements have a maximum velocity of approximately 21 degrees of arc/sec,[128] very much slower than the speed of pursuit or saccadic movements.

The stimuli eliciting vergence movements are two in number: (1) the change in blur patterns of images on the retinas stimulates accommodative convergence, and (2) a displacement of the image away from the fovea, that is, to noncorresponding retinal points by a change in the real or apparent distance of the object of regard from the observer stimulates fusional vergence. "There are clearly very few real world situations in which these two stimulus characteristics are separated."[133]

The substrates underlying vergence mechanisms, for the most part, are not yet even speculative. Jampel[65] has demonstrated asymmetric convergence movements variably associated with other parts of the near reflex. Jampel states, and Cogan concurs, that convergence movements are synthesized bilaterally. The seat of this movement is probably represented somewhere in the peristriate area. Convergence impulses are postulated to be projected independdently to both sides of the pretectum and tectum of the midbrain, where they are modified at some unknown location. At this time, specification of an association of convergence with a specific subcortical center would be more than presumptive. Since convergence is known to be retained after transection of the posterior end of the midbrain and clinically retained despite extensive lesions of the pons, it is reasonable to infer that the corticofugal pathways for convergence reach the oculomotor nuclei by way of the superior brachia. Convergence results from activation of the medial recti with inhibition of the lateral recti.

Under normal conditions the act of convergence is associated with miosis and accommodation. This triad forms the near reflex, but where in the brain these three components become associated is unknown. The basic tonus of the convergence mechanisms has been termed tonic convergence; this is the tone that keeps the eyes aligned in the primary position.

AC/A ratio

As mentioned at the outset, one of the stimuli for convergence is the change in the blur circle on the retinas stimulating accommodation. In the normal individual any change in accommodation will effect a change in the position of the eyes if they are dissociated sensorially. The relationship of convergence associated with the *stimulus* to accommodate is commonly measured clinically and is recorded in the form of a ratio: accommodative convergence/accommodation ratio (AC/A ratio) expressed as prism diopters of convergence per diopter of accommodation. The stimulus may be a target brought physically closer to the subject or varied in optical distance by lenses, as recommended by Sloan[112] and by Sears.[111] The stimulus AC/A ratio, then, is the change in phoria with change in stimulus to accommodate.

The actual accommodative response of the eye in dioptric adjustment may not match the stimulus demand, and so the response AC/A may differ markedly from the stimulus AC/A. Indeed, this has been shown to be so for pupil diameters of 1 mm or less.[99] In other words, a change in accommodation will invariably change the phoria. Since an increase in accommodation always produces an increased tendency for the eyes to converge, the effect on the two visual axes is called the accommodative convergence. The magnitude of the change in convergence in prism diopters caused by a given increase in the accommodation measured in diopters, as mentioned previously, is called the response AC/A ratio. It is obviously also true that as the accommodation is diminished, the relative convergence is decreased.

The simplest way to determine the relationship between accommodative convergence and accommodation is to determine the phoria at a given fixation distance and then add plus or minus lenses and again measure the phoria. The amount of change in the phoria brought about per diopter change in accommodation demanded gives us the stimulus AC/A ratio.

The AC/A ratio has been found to be the same for the same individual at all viewing distances and for any change in stimulus to accommodation, provided the retinal image is not too blurred. The relationship between accommodative convergence and accommodation is linear throughout the entire range of response to changes in the accommodative stimulus in most individuals; that is, 1 diopter change in the stimulus to accommodation always produces the same amount of change in the phoria within the entire range of the accommodative response.

Approximately 10% of normal persons have a nonlinear AC/A ration, according to Martens and Ogle,[79] and a significant nonlinearity was found in a considerable proportion of normal persons by Flom.[41] Martens and Ogle believe that the change in the stimulus to accommodation rather than the actual change in accommodation that has taken place gives rise to the change in accommodative convergence because of the following:

1. The same ratio is found, even when the powers of the lenses used are such that retinal images are blurred to the point of obscuring the discrimination of the test targets
2. A linear AC/A ratio is found in absolute presbyopes within their limited response range.

According to these authors it is the innervation to the ciliary muscle, rather than any response of the shape of the lens to that innervation, which gives rise to the AC/A ratio. Alpern[3] believes that the accommodative vergence follows the changes in accommodative response rather than the stimulus to accommodation, since he found that the accommodative vergence could be changed even when the accommodative stimulus remained fixed.

In the normal individual the ratio is fairly stable, the mean being 3.5Δ/D. There may, however, be day to day variations around this mean. Cycloplegic drugs increase the ratio dramatically.[22] On the other hand, miotic drugs have little influence.[89] There is little evidence that the AC/A ratio can be altered by orthoptic training.

The second stimulus for vergence movements is disparate retinal images that depend on an innate reflex to keep the image of regard placed on both foveas or corresponding retinal points.

The limits of the fusional reflex are not well known, and it will be necessary to modify somewhat the preceding statement that images must be focused on corresponding receptive fields to be seen singly. This is not quite true. In fact, when images are focused on slightly disparate retinal receptive fields, the result is stereopsis or depth perception by parallax (Chapter 21). Thus, although there is a limit to the amplitude of motor fusion and although the images of an object are beyond the limits of identical retinal receptors, sensory fusion continues for a time.

Under experimental conditions it is possible to determine the effectiveness of peripheral fusional stimuli that cause the eyes to break away from bifoveal fixation. The subject is presented with similar targets, which are fixated with both foveas. Targets that create fusional impulses are then introduced into the peripheral field of each eye. For these targets to be fused, however, the eyes must break from bifoveal fixation. Thus an estimate may be obtained of the relative effectiveness of peripheral compared to foveal fusional impulses. Under these experimental conditions, which are rarely duplicated in nature,[20] it has been found that paramacular targets present a greater stimulus to fusion than do foveal targets, and that it is the retinal locus rather than the area stimulated that determines the effectiveness of the peripheral fusional impulses.[20,88]

It is of considerable importance to ophthalmologists to know whether the amplitude of fusional movements can or cannot be improved by various training methods. Many orthoptists are of the opinion that it is possible to do so. The evidence for this is not convincing. Sensory fusion is quite a different thing, and there is no question that individuals can be trained to acquire better sensory fusion in stereoscopic devices. Almost everyone has had the experience of not being able to fuse test targets that have very small disparity if they are looked at first in a stereoscope. If, however, targets with a large disparity are tried first, the knack of fusing difficult targets is soon learned. This does not mean, however, that one can build up motor fusional abilities, and the evidence to the present time suggests that the amplitude of fusion is fixed in any one individual and cannot be increased beyond that point. Further controlled experiments are needed to decide this.

Paralysis of convergence is an inability to converge with subsequent crossed diplopia at near. Paralysis of convergence is associated with midbrain pathology and can be associated with vertical gaze abnormalities as well as with pupillary abnormalities.

Divergence is an active process. The cortical centers are assumed to be associated with those cerebral centers involved with convergence. The corticofugal tracts are completely unknown, but divergence paralysis, a rare occurrence, cannot be explained easily without assuming a divergence mechanism. This clinical entity, which is often erroneously diagnosed as bilateral paralysis of the lateral rectus muscles, has such pathognomonic features that the diagnosis should never be missed if these are understood. At some middistance from the eyes the visual lines can fixate an object, but as the object is withdrawn into the distance in the midline, the visual axes cannot diverge to maintain fixation; therefore increasing esotropia develops as the object is removed, with consequent uncrossed diplopia. If at any one distance the separation of the double images is measured, and then at this distance the object being fixated is moved laterally in the arc of a circle with the patient at the center, the separation of the double images remains the same. This is the essential point that differentiates this condition from bilateral paralysis of the lateral rectus muscles. Unless this feature is present, the diagnosis of divergence paralysis cannot be made. The great difficulty in making the diagnosis is that late in the course of the disease many patients do develop true palsy of one rectus muscle, in addition to the typical divergence paralysis. If the patient is not seen early, before this develops, the diagnosis can only be provisional. The only easy explanation of how divergence paralysis can be produced is to assume that there is a divergence mechanism, separate from abduction for each lateral rectus, which actively diverges the eyes from the convergent position to the primary position. If such a mechanism can be postulated for convergence separate from adduction, there is no insurmountable reason why the same should not hold for divergence. The center for divergence must be near the midline in the neighborhood of each nucleus of the sixth cranial nerve, and this could explain why late in the disease, associated paralysis of one lateral rectus develops as the lesion spreads from the midline toward one side or the other. Several cases of typical divergence paralysis, in which the lesion was found in the midline at autopsy have been reported. From the point of view of pathogenesis, paralysis of divergence could be caused by a lesion anywhere in the supranuclear pathway for divergence from the cortex to the hypothetical subcortical center.

Fixation maintenance mechanism

The idea that a mechanism separate from the pursuit mechanism is involved in fixation has received recent support from Robinson's[100] studies of single motor neurons in the oculomotor nerve. The fixation mechanism works to maintain the object on the fovea once it is placed there by the foveation movement.

Fixation is a relative term. The eyes are never absolutely still; even under strict laboratory conditions with immobilization of the head, an image cannot be steadily fixed on the fovea. Even when the subject is strenuously fixating a stationary cross hair in an optical instrument, which has been found by experiment to be the best test object for securing steady fixation, the eye is not at rest. When it is steadiest, small trembling movements may be seen on the records obtained photographically[2] (Fig. 5-71). They probably represent the frequency of vibration of the extraocular muscles. The average movement of a point on the retina, as a result of these fine vibratory movements, is 0.0018 mm (38 seconds of arc). The diameter of a foveal cone is about 25 seconds of arc. Hence a point image must move back and forth over several cones. At no time will it come to lie still on a fixed cone pattern. In addition to these finer movements, the eyes cannot remain fixed for very long on any object, but invariably break from fixation in the form of drifts.

During a study of monocular fixation of a stationary object, Ratliff and Riggs[95] found four main types of movement:

1. Small rapid motions with a median extent of about a 17.5-second angle of rotation (peak to trough) and frequencies ranging from 30 to 70 Hz (vibratory movements)
2. Slow motions of irregular frequency and extent
3. Slow drifts in one direction or another on which the motions just mentioned are superimposed
4. Rapid jerks with an average extent of about 5.6 minutes of arc (micronystagmus), occurring at irregular intervals and at times apparently compensating for the drifts.

The total movement, because of the combined effects of these motions over a period of 3 to 4 seconds, is usually less than 10 minutes of arc. On the basis of these results they conclude that the retinal image of an object being fixated generally moves over 25 to 50 receptors. Slow motions, drifts, and jerks may carry the retinal image across about a dozen receptors. The vibratory motion moves the retinal image across two or three receptors at most. According to these observers the median extent of these motions is so small that the movement of the retinal image under optimal conditions of fixation is not likely to exceed the width of one receptor.

Higgins and Stultz[59] recorded the motions of an eye during steady fixation by photographing a blood vessel. They found very rapid motions with a frequency of about 50/sec and an average amplitude of 1.2 minutes of arc (vibratory movement). Riggs[96] modified Lord and Wright's method and found the ocular tremor to be in the neighborhood of 30 to 90 vibrations/sec with an amplitude of 15 to 20 seconds of arc. Nachmias[87] has been able to record eye movements in both the vertical and horizontal meridians simultaneously.

The fine ocular tremor probably represents a continuous variation in the total activity of

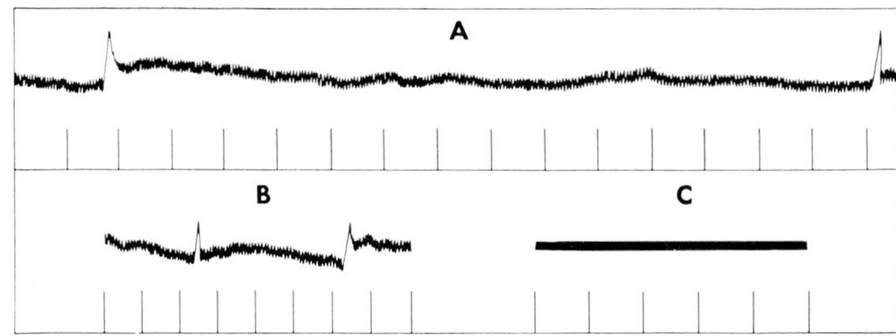

Fig. 5-71. A, Steady fixation of cross hair. **B,** Another section of record in same experiment. **C,** Control record (of artificial eye) taken before experiment. Time intervals shown are fifths of seconds.

the acting individual motor units of the muscles. The normal tonus of these muscles effectively damps potential pulse movement. Micronystagmus is probably mediated by efferent neural stimulation of the eye muscles. Saccades originate, at least in part, in the binocular motor areas, whereas the fine tremor is entirely peripheral to binocular control.

A recent symposium on extraocular muscle function[103] suggested that the microsaccades reflected the ability of the observer to maintain concentrated attention on the object of regard. Observers could be trained to obviate these movements. The underlying mechanism of the slow motion drift can also presumably be attributed to total involvement of the conscious mechanism with respect to attention.

Although views have been held from one extreme, that the ocular tremor is a useless and undesirable by-product of the ocular motility apparatus, to the other extreme, that it is helpful to vision,[91] recent research supports the latter concept. Ocular tremor helps to prevent the fading away of detailed vision that occurs when a retinal image is optically immobilized.*

In summary, it appears that the involuntary movements of the eye during steady fixation of a target are three in number:

1. A high-frequency tremor of amplitude less than 0.5 minute of arc and frequencies up to 150 Hz.
2. Rapid flicks of up to 50 minutes of arc that occur at irregular intervals
3. Slow motion drift at the rate of about 1 minute of arc/sec in the interflick periods. At the fovea, 1 minute of arc corresponds to a distance of about 5 μm.

Patients with pathologic nystagmus often have poor vision. It might be argued that the nystagmus is the result of the poor vision, which in some cases is undoubtedly true. There is, however, a type of nystagmus known as latent nystagmus, which occurs in some children with strabismus in whom there is no nystagmus when both eyes are open. When either eye is covered and the remaining eye is made to fixate, nystagmus appears. If the visual acuity is recorded when both eyes are open and then when one eye is closed, the acuity is always better with both eyes open. Hence in these patients the visual acuity of either eye shows a definite deterioration when nystagmus is present as compared to when it is quiet. In this condition the presence of the nystagmus is a deterrent to good form perception, whereas the small movements of physiologic nystagmus or ocular tremor help to maintain retinal function.

The fixation system may be isolated by having the subject fixate on a test object. The patient's head is slowly rotated either passively or actively. When the head is rotated around the vertical axis to the right, the eyes will be carried into a position of left lateral conjugate gaze and vice versa. When the movement of the head is performed slowly, the response of the eyes is due almost entirely to the optic fixation reflex.

NONOPTIC REFLEX MECHANISMS: VESTIBULAR APPARATUS

The vestibular apparatus plays a major role in coordinating the tone of body musculature, including the extraocular muscles. Maintained postural deviations are controlled by the otoliths in the utricles and saccules, while postural movements resulting from acceleration are controlled by impulses from the hair cells in the semicircular canals; that is, the sensory apparatus codes head position and angular and linear head movements.

ANATOMY: THE PERIPHERAL RECEPTORS

The labyrinth is composed of a series of interconnecting membranous structures known as the membranous labyrinth. The membranous labyrinth consists of three semicircular canals—the horizontal, the anterior vertical, and the posterior vertical, each lying in a plane approximately 90 degrees to the others—and adjacent saclike structures, the utricle and the saccule (Fig. 5-72). The anterior end of the horizontal canal is actually deviated 30 degrees superiorly to the horizontal (Fig. 5-73), and both left and right canals sit parallel to each other. The vertical canals form angles of approximately 45 degrees with the anteroposterior axis. The posterior vertical canal lies in a plane parallel to

*See references 36-38, 48, 97, and 98.

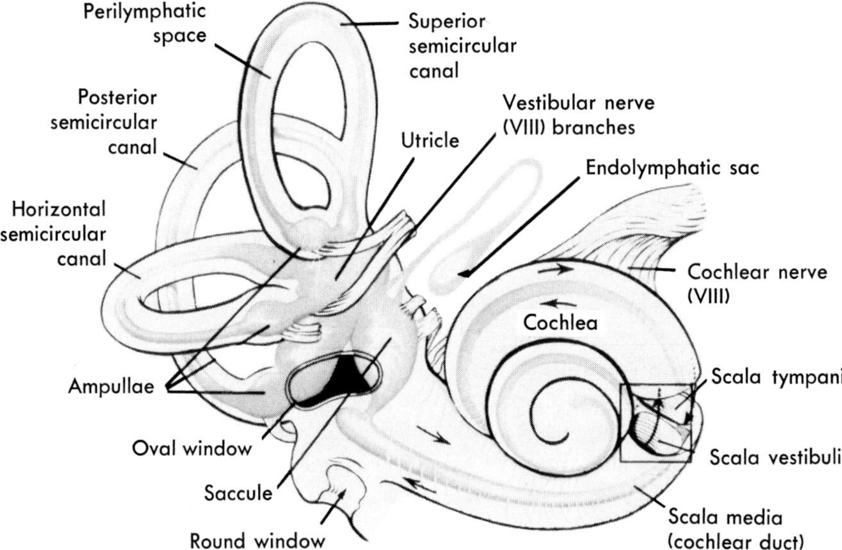

Fig. 5-72. Right auricular end-organs and their nerves. Labyrinth and cochlea are viewed from horizontal aspect. Note the three ampullary nerves that conduct impulses centrally from each semicircular canal; each of these nerves is associated with specific sets of ocular muscles by means of tegmental brain stem connections between vestibular nuclei and ocular motor nuclei. Sensory input from these nerves and the nerves from the utricle and saccule plays an essential role in visual spatial perception as well as in stabilization of ocular fixation during movements of the head. (Drawing by Ernest W. Beck; courtesy Beltone Electronics Corporation, Chicago, Ill.)

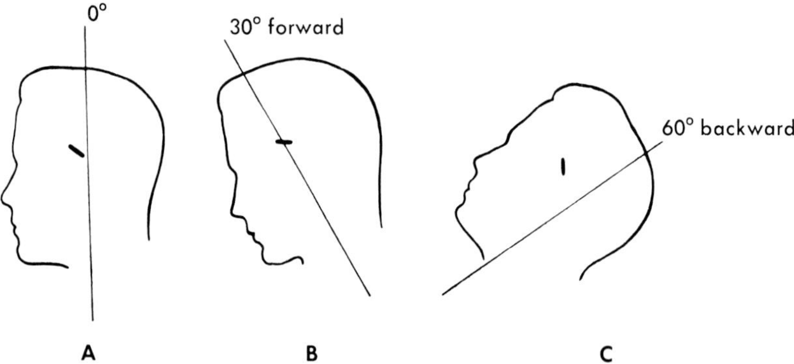

Fig. 5-73. Plane of the horizontal semicircular canal with different positions of the head. In erect position of the head, **A,** anterior end of horizontal canal is tilted upward approximately 30 degrees but may be made horizontal by tilting the head forward 30 degrees, **B,** or be made vertical by tilting the head backward 60 degrees, **C.** (From Cogan, D. G.: Neurology of the ocular muscles, ed. 2, Springfield, Ill., 1966, Charles C Thomas, Publisher.)

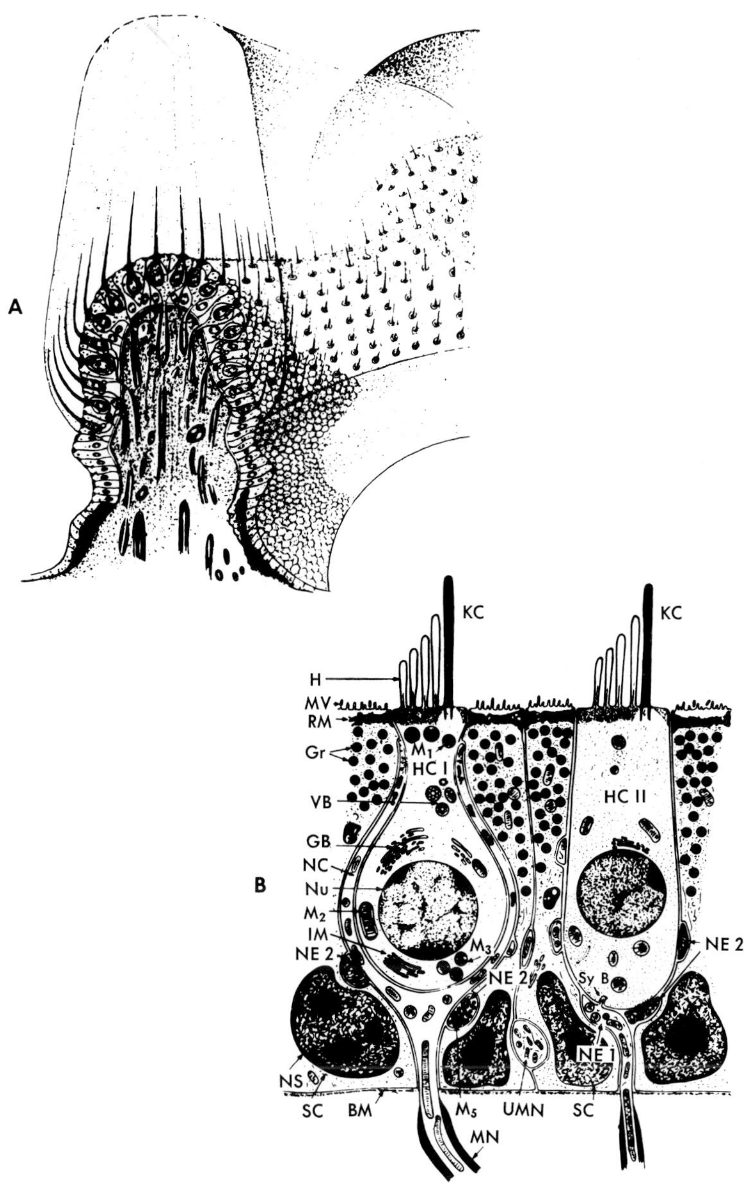

Fig. 5-74. A, Schematic drawing of crista ampullaris, the sense organ of the semicircular canal. Note ciliated vestibular sensory cells. **B,** Schematic details of two types of vestibular sensory cells. Flask-shaped type I cell, *HC I,* is surrounded by a nerve calyx, *NC,* which makes contact on its outer surface with granulated (presumably efferent) nerve endings, *NE 2.* Unmyelinated nerve fibers, *UMN,* are extensions of myelinated fibers, *MN,* which lose their myelin sheath as they pass through basement membrane. Type II sensory cell, *HC II,* is roughly cylindrical and is supplied by two types of nerve endings, *NE 1* and *NE 2,* which can be seen at its basal end. Several groups of mitochondria, M_1, M_2, M_3, M_5, are found in sensory cells and neural elements. Two types of hair project from surfaces of sensory cells, stereocilia, *H,* and kinocilium, *KC,* single kinocilium always being the longest on each cell. Supporting cells are easily distinguished from sensory cells by virtue of their numerous population of rather uniformly distributed granules, *Gr.* (**A** From Wersall, W.: Acta Otolaryngol. **126**(suppl.):1, 1956; **B** from Engström, H., Ades, H. W., and Hawkins, J. E., Jr.: J. Acoust. Soc. Am. **34:**1356, 1962.)

the opposite anterior vertical canal. The medial arms of the vertical canals merge to form a common canal and with the posterior end of the horizontal canal connect with the utricle. One end of each semicircular canal has a bullous expansion, the ampulla, containing a ridge of neuroepithelium. The ridge contains ciliated somatosensory cells innervated by vestibular nerve fascicles, and supporting cells. The cilia are embedded in a gelatinous dome-shaped structure, the cupola (Fig. 5-74). The membranous labyrinth is filled with endolymph, the formation of which is associated with a membranous structure of the auditory receiving apparatus, the cochlea. The fluid apparently flows to the membranous labyrinth through the saccule to the utricle.

The utricle is a sac connected to the semicircular canals superiorly and the saccule inferiorly. The receptor organs in these sacs are plaques of hair cells (maculae) embedded in a gelatinous mass containing crystals of calcium carbonate (otoliths). The hair cells in the utricle are positioned parallel to the horizon, and those in the saccule are positioned perpendicular to the horizon. The entire membranous labyrinth is enveloped in the petrous portion of the temporal bone, the bony labyrinth. The bony labyrinth contains perilymph and communicates with the subarachnoid space.

The cell bodies of the vestibular nerve lie in the internal auditory meatus, making up the vestibular (Scarpa's) ganglion. The ganglion contains bipolar cells whose afferent limb passes to the sensory organs of the membranous labyrinth and whose central efferents enter the upper medulla as the vestibular part of the eighth cranial nerve. After entering the brain stem, the fibers divide into two tracts: ascending and descending. The ascending tract

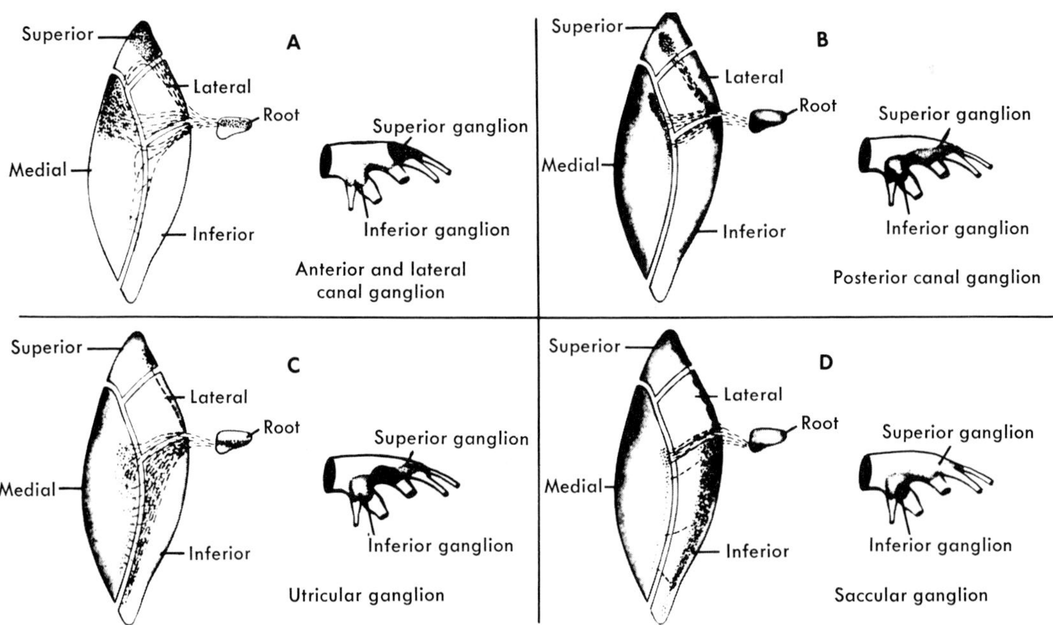

Fig. 5-75. Diagrammatic representation of central distribution of degeneration in vestibular nuclei resulting from lesions of various parts of vestibular ganglion. In **A,** lesion destroyed cells that innervate anterior and lateral canals; in **B,** posterior canal; in **C,** the utricle; in **D,** the saccule. Note that in **A** and **B,** the most profuse central degeneration is present in superior vestibular nucleus and in rostral parts of medial vestibular nucleus, whereas in **C** and **D,** central degeneration is most profuse in inferior vestibular nucleus and in caudal portions of medial vestibular nucleus. (From Stein, B. M., and Carpenter, M. B.: Am. J. Anat. **120:**281, 1967; modified by Cohen, B.: In Bach-y-Rita, P., Collins, C. C., and Hyde, J. E.: The control of eye movements, New York, 1971, Academic Press, Inc.)

is concerned with oculomotor control. Certain of these fibers ascend in the vestibulocerebellar fascicle to the vermis and roof nuclei, but most of the fibers terminate in the vestibular nuclei. Stein and Carpenter[116] have shown that neurons from the semicircular canals synapse primarily in the rostral medial and superior vestibular nuclei and those from the utricle and saccule go to the inferior and caudal medial vestibular nuclei (Fig. 5-75).

The connections of the vestibular nuclei are unbelievably complex. The second-order fibers from the vestibular nuclei project to the cerebellum, to the ipsilateral vestibulospinal tract, to the contralateral PPRF, to other brain stem nuclei dealing with horizontal, vertical, and torsional ocular movements. Zee[132] has recently shed some light on the organizational pathways subserving the tilt reaction and vertical canals in humans (Fig. 5-76).

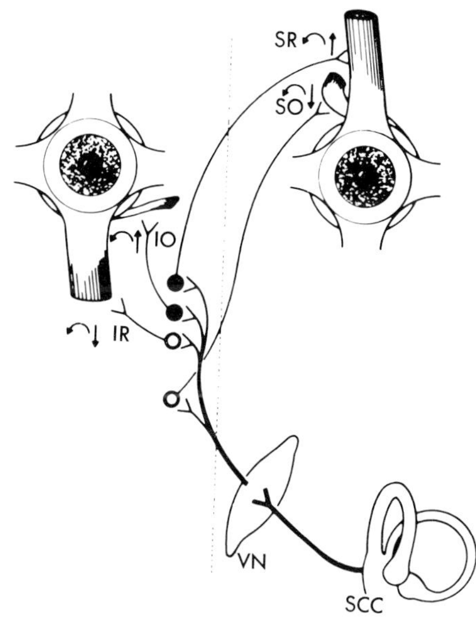

Fig. 5-76. Main *excitatory* vestibuloocular connections from the vertical semicircular canals. Dashed line indicates midline of brain stem. Arrows indicate pulling directions of ocular muscles in the frontal plane. Filled circles receive anterior canal projection; open circles receive posterior canal projection. *SR*, Superior rectus; *SO*, superior oblique; *IO*, inferior oblique; *IR*, inferior rectus; *VN*, vestibular nuclei; *SCC*, semicircular canals. (From Zee, D. S.: Ann. Neurol. **4**:384, 1978, copyright © Little, Brown & Co.)

PHYSIOLOGY

The semicircular canals are stimulated in everyday living by rotational acceleration or deceleration in their respective planes. Movement alone is insufficient if there is no change in rate. A given semicircular canal is stimulated when there is a relative flow of endolymph displacing the hair cells in the ampulla. The horizontal canal is stimulated when the relative flow of endolymph is toward the ampulla, whereas the two vertical canals are stimulated when the relative flow is away from the ampulla; for example, if a subject is stabilized in a rotating chair with his chin depressed 30 degrees and the chair rotated to his right, there will be a relative flow of endolymph toward the left; that is, the fluid suspended within the membranous labyrinth will have a certain inertia, resisting motion, causing a relative displacement of endolymph anteriorly in the right canal (excitatory) and posteriorly in the left horizontal canal (inhibitory) (Fig. 5-77). The opposite effect is brought about by deceleration.

Impulses arising in the ampullae of the semicircular canals pass centrally to the vestibular nuclei; that is, during angular acceleration, first-order neurons in the ampullae increase their firing. In the vestibular nucleus on the ipsilateral

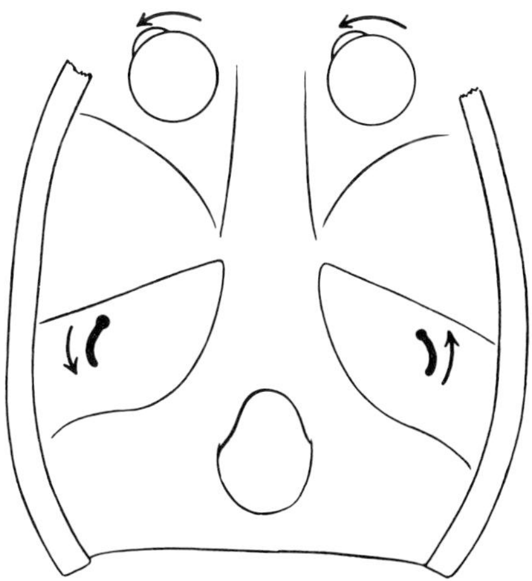

Fig. 5-77. Deviation of eyes and displacement of endolymph with rotation of head to the right. (From Cogan, D. G.: Neurology of the ocular muscles, ed. 2, Springfield, Ill., 1966, Charles C Thomas, Publisher.)

side cells are both excited and inhibited. Of the cells excited there are two types:

1. Tonic cells, which have continuous spontaneous activity that increases or decreases in response to ipsilateral or contralateral rotation respectively
2. Kinetic cells. which are silent in the resting state and only increase their firing with strong ipsilateral rotation

Inhibitory postsynaptic potentials are induced in both types of cells with stimulation of the opposite vestibular nerve via commissural fibers. The exact role of these cells in induced movements is not clear.

Nystagmus with a jerk pattern, that is, a slow movement in one direction and a fast movement in the other direction, is the ocular response to physiologic or pathologic vestibular stimulation. Flourens,[42] in a series of experiments, established the following rule: each semicircular canal gives rise to nystagmus in the plane of that canal (Flourens' law). The slow phase of the induced nystagmus originates from the vestibular apparatus. The horizontal canal probably projects via at least two synapses to the contralateral parapontine reticular formation (pontine gaze center). The fast phase of nystagmus is a corrective movement generated in the saccadic center in the frontal lobe.[15]

The deviation of the eyes corresponds to the slow phase of vestibular nystagmus and is always in the direction opposite the change in rate of motion. Thus with acceleration the eyes appear to lag, whereas with deceleration they anticipate the movement. This serves to stabilize the visual image. Simplistically, each semicircular canal can be thought of in terms of influencing a pair of eye muscles, one for each eye, that rotate each globe in approximately the same plane as that in which the semicircular canal is located.[117] For the vertical canals, the anterior canal excites the ipsilateral superior rectus and contralateral inferior oblique muscles; the posterior canal excites the ipsilateral superior oblique and contralateral inferior rectus muscles (Fig. 5-76); the horizontal canal excites the ipsilateral medial rectus and contralateral lateral rectus muscles. Stimulation of one anterior canal produces upward rotation of both eyes with ipsilateral incyclotorsional and contralateral excyclotorsional movement. Stimulation of one posterior canal elicits a mixture of downward rotation with incyclotorsion of the ipsilateral and excyclotorsion of the contralateral globe.[132] The relative degree of vertical or torsional movement depends on the position of the eye in the orbit.[25]

Thus the semicircular canals project through the vestibular nuclei to the contralateral inferior rectus and inferior oblique and to the ipsilateral superior oblique and superior rectus. All these muscles have their brain stem nuclei contralateral to the stimulated vestibular nuclei; that is, the superior oblique and superior rectus are contralaterally innervated (Fig. 5-76). This scheme also accounts for the counterrolling of the eyes seen with head tilt.

Semicircular canal function in humans is most often studied with caloric or rotatory stimuli. In normal individuals these stimuli do not excite individual canals. When two or more canals are simultaneously stimulated, the induced eye movements summate. Inducing caloric stimulation produces convection currents in the endolymph, causing relative displacement of the receptor cells. In this method the external auditory canal is irrigated with either warm (40° C) or cool (33° C) water. Ice water is often used clinically. A caloric stimulus in one ear simultaneously activates all the canals, but with appropriate positioning, for example, if the head is held back at 60 degrees, the horizontal canals will predominate. If hot water is used, the relative movement will be toward the elevated ampulla (Fig. 5-73), and the eyes will deviate toward the opposite side with a fast phase toward the ipsilateral side. The opposite is true for cold water stimuli; there is relative inhibition, with tonic deviation ipsilaterally with the fast phase to the contralateral side. Clinically, the mnemonic device COWS (Cold, Opposite, Warm, Same) should be remembered for the fast phase of induced nystagmus. Bilateral stimulation will induce vertical nystagmus.

Vestibular reactions are strictly reflexes and are not under voluntary control. The vestibular system is inhibited by the fixation maintenance mechanism. Optic fixation tends to inhibit vestibular nystagmus, and optokinetic stimuli can completely dominate this system. Vestibular nystagmus is exaggerated in the dark, when the eyes are closed, and by the use of high plus lenses.

The utricle and saccule respond to rectilinear acceleration and govern postural tone in relation to head position. These organs compensate for static head tilt by counterrolling the eyes. The amount of static compensatory eye rolling is small, but dynamic compensatory eye rolling can be large; for example, 20 degrees of ocular torsion can be induced in pilots rolling jet airplanes rapidly.[24] The utricular reflex of counterrolling forms the basis of Bielschowsky's head tilt test used in isolating paretic vertical muscles (Fig. 5-78). Rotation of the head up and down with contraversive rolling of the eyes,

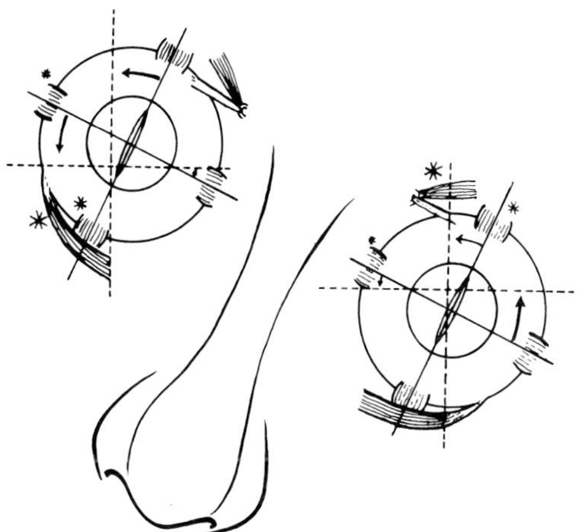

Fig. 5-78. Otostatic reflex compensating for left head tilt. Solid line is normal vertical axis. Dotted line is assumed vertical axis after head tilt to left shoulder. Superior oblique and superior rectus muscles are activated in the eye ipsilateral to the head tilt, with the inferior oblique and inferior rectus activated in the contralateral eye. If one were attempting to differentiate between a left superior oblique palsy and a right superior rectus palsy, both leading to a left hypertropia, the patient would be instructed to tilt his head first to one side then the other, and the manifest deviation would be measured in both positions. When the head is tilted to the left, a larger vertical deviation would be measured if the left superior oblique were involved because the left superior rectus would act unopposed.

that is, eyes up with chin down and vice versa, is under the control of the utricle and saccule.

Vestibuloocular illusions

Subjective phenomena resulting from vestibular stimuli are generally described as vertigo. At present, these subjective phenomena have been characterized as the following[121]:

autokinetic illusion Perception of movement of a fixated object when there are inadequate visual cues. This illusion represents the failure of the vestibular apparatus in a nonaccelerated subject to govern spatial perception without additional spatial cues.

oculogyral illusion Perception of movement of the environment following stimulation of the semicircular canals. This illusion represents an overwhelming of voluntary control mechanisms due to angular acceleration.

Corioles effect Disintegration of visual reference in space and simultaneous loss of the spatial relationship of the body to itself. Here vestibular stimulation with acceleration in one set of canals and deceleration in another set of canals produces virtual disintegration of sensory perception and motor performance.

oculogravic illusion Sensation of being tilted backward with rectilinear stimulation. This is thought to be due to stimulation of the saccule. Visual references are altered to coincide with the illusion.

oculoagravic illusion Sensation of the raising and lowering of observed objects as positive or negative acceleration is applied through the long axis of the body. The mechanism is not understoood.

Tonic neck reflexes

The importance of tonic neck reflexes in humans is questionable. Changes in the position of the body relative to the head produce simultaneous reflex changes in the eyes. It has been demonstrated that quantitated ocular counterrolling in patients with bilateral labyrinthine disease is markedly abnormal, that is, reduced, and implies that vestibulospinal reflexes play a subordinate role in primates.

Neural pathways originate in proprioceptors in neck muscles and tendons and pass through the dorsal spinal roots of the cervical cord to enter the spinocerebellar tracts. There is some question as to whether some of these fibers reach the inferior vestibular nucleus as well as the cerebellum. The central pathways of this reflex activity are at this time largely unknown.

OTHER SUPRANUCLEAR CENTERS

Basal ganglia

The evidence for inferring that certain ocular control mechanisms originate in the basal ganglia region is based on the clinical presence of ocular motor dysfunction in disease states affecting this region. Involvement of descending tracts through this region is the most likely explanation for the known oculomotor dysfunction.

Cerebellum

The role of the cerebellar system in the control of ocular movements is not well understood. The pathways leading to and away from the cerebellum are diffuse, and those dealing with eye movements are almost impossible to sort out. Recent experiments[1,50,51,104] have demonstrated that the vestibuloocular reflex (VOR) can be extensively modified by the use of reversing prisms. Indeed, within days of initiating the use of prisms either in humans or animals, the VOR is rewired backward, so the eyes move backward (to normal), reducing retinal slip and thus clearing vision during head movements. Similar, but less drastic, adaptive changes occur when people get a new prescription for glasses with a change in power of several diopters. They complain of dizziness and distortion, which usually disappears within a few days. The cause of this disorientation can be understood by using the concrete example given by Robinson.[105]

> Suppose the new glasses magnified everything by 20%. If the patient turns his head to the left at 100 degrees/sec (a modest velocity) the visual world turns to the right (relative to the head) at 120 degrees/sec with his new glasses on. But the vestibuloocular reflex is only moving the eyes to the right, just like it used to before the new glasses. Thus the images of the world appear to slip to the right at 20 degrees/sec.

This visual illusion is called oscillopsia. It occurs in motion sickness and in other acquired rhythmic ocular motor disorders. In those instances in which the oscillopsia disappears, something has affected the gain (eye velocity/head velocity) of the VOR. Experimentally, it has been shown that the substrate effectively demonstrating the ability to cause changes in the VOR is the vestibulocerebellum, specifically the flocculus and nodulus. Furthermore, Miles and Fuller[81-83] have demonstrated that the flocculus contributes little to the overall gain of the VOR in the normal monkey, but when faced with abnormal paradigms, that is, telescopic lenses or reversing prisms, the flocculus plays an important role in the plasticity of the system. Thus it appears that a repair mechanism exists to deal with the wear and tear changes associated with aging, allowing adjustment of the VOR as necessary, and that this mechanism exists within the vestibulocerebellum. The exact pathways remain to be defined.

Similar plasticity has been demonstrated in the saccadic system in humans[1,75] and animals[90] in response to partial paresis of the third or sixth cranial nerves. These adaptive changes are also believed to be under control of the vestibulocerebellum.

Furthermore, clinical observations have established that the cerebellum is an important part of the ocular motor system, but they are difficult to evaluate because the exact location, extent, and nature of the lesions are almost never certain. Ocular dyskinesia has been observed in several patients with unequivocal evidence of cerebellar disease clinically localized to the vermis.[40] These patients seemed to have lost control of their saccadic ocular motor system; that is, they were unable to produce saccades of appropriate amplitude for an intended voluntary refixation. Instead, their eyes overshot the target (ocular dysmetria), made wide, rapid, sinusoidal oscillations from a point of fixation (flutterlike oscillations), and at the height of their disease their eyes made continuous uncontollable chaotic, conjugate, saccadic movements in every direction (opsoclonus). Disorders of the smooth pursuit system have also been described in patients with familial cerebellar degeneration whose following movements were replaced by saccades.[120] Conjugate horizontal deviation of the eyes may occur after cerebellar hemorrhage or injury to the cerebellum.

Earlier experimental studies yielded conflicting results. Many investigators ablated all or part of the cerebellum and merely observed the eyes without the aid of any recording technique. Usually no abnormalities were noted, probably because the eyes remained in the resting position. The animals had not been trained to fix, follow, or gaze voluntarily; therefore effects of loss of cerebellar function, important in the control of movement, were not detected.

Recent experiments by Aschoff and Cohen[4] revealed that the neural organizations in the cortex of the vermis and paravermis participated in producing gaze shifts of positions of fixation to the ipsilateral side. When the cortex of the vermis and paravermis of the posterior lobe of the cerebellum were damaged on one side, the eyes would move more often into the contralateral hemifield. Furthermore, there was a difference in the amplitude and relative frequency of ipsilateral and contralateral saccades. The monkeys would make a normal saccade to the contralateral field and hypometric saccades to the midline. This implies a cerebellar role in the generation of saccades.

Westheimer and Blair[126,127] performed essentially total cerebellectomies, including the roof nuclei, on four monkeys. Their animals demonstrated normal saccades, but the following deficits: (1) failure to maintain eccentric gaze, that is, gaze-holding nystagmus and (2) absence of smooth pursuit movements to visual stimuli. In addition, Burde and associates,[18] in hemicerebellectomized animals, have demonstrated a failure of inhibition of vestibularly induced nystagmus. This further supports the contention that the flocculus may be serving as an inhibitory side loop of the vestibuloocular reflex. Since smooth movements could be induced by faradic stimulation of the pontine reticular formation and by caloric stimulation (i.e., nystagmus), Westheimer and Blair postulated that the motor apparatus is capable of executing smooth movements but that pursuit failure is due to failure of modulation of the motor mechanism.

Experimental studies based on cerebellar stimulation have been critically reviewed by Ron.[107] Stimulus strength was rarely controlled and often was too far above threshold to produce localized effects or slow following movements. Frequency and duration were selected arbitrarily. Anesthesia usually eliminated the possibility of evoking rapid movements. Recording methods, if used, were usually not precise and did not measure latency (important in determining the number of synapses in the pathways between cerebellum and ocular motor nuclei) and did not allow distinction between smooth pursuit and following movements. Because of the lack of a systematic approach, many regions of the cerebellum were not explored.

As a result of clinical observations, it has been suggested that the cerebellum is a spatiotemporal translator; that is, it translates the amount of displacement from the fovea of a point of potential interest into the proper duration of the burst of neural activity sent to the oculorotatory muscles from the pulse generator in the pons, so the resulting saccade is appropriate in amplitude. Breakdown of this function results in ocular dyskinesia. Ron, in his study of the effects of stimulation of the cerebellum of the monkey in which he found that most evoked saccades were goal directed, also confirmed the previous hypothesis.[107]

The cerebellum thus probably acts as a sampler of the displacement between the initial position of the eyes and the next point of potential interest. To do this it requires information. Such input may come from the visual system; visual evoked potentials have been recorded from the cerebellar cortex.[114] It may also come to the cerebellum in the form of an "efferent copy"; that is, the motor system informs the cerebellum where it has moved the eyes in relation to the head. It may come from the fusimotor afferents; the extraocular muscles have a higher concentration of muscle spindles than any other muscles in the body, except for the small muscles of the hands.

Some indication of the role of the gamma motor system in control of trunk and limb muscles has come from the work of Granit,[53] who regards this system as "governor of muscular performance both in tone and locomotion." Gilman[47] has recorded transiently decreased fusimotor efferent activity in decerebellate cats and postulates that this may be part of the mechanism of cerebellar movement disorders. Fuchs and Kornhuber[45] recorded short latency evoked potentials from the cerebellar vermis after stretching the extraocular muscles of the cat. The presence of a large number of intrafusal fibers in the extraocular muscles and present knowledge of the mechanism of cerebellar control suggest that the fusimotor system may be important in mediating cerebellar control of normal ocular movement.

In summary, there is evidence for the following assumptions. Midline structures, vermis and paravermis, are associated with saccadic control and the vestibuloocular reflexes. The cerebellar hemispheres are related to smooth pursuit and eccentric fixation. The basic motor mechanisms and integrators exist elsewhere, but in the normal individual their actions are modulated by cerebellar output.

INTEGRATION OF SUPRANUCLEAR GAZE INPUTS IN THE BRAIN STEM

The exact anatomy of the neuronal networks responsible for the processing of supranuclear commands is unknown. All the anatomy until recently has been inferred from clinical pathology and ablation experiments in mammals.

Anatomy

The brain stem reticular formation is a rostral continuation of the reticular formation of the spinal cord. It is composed of small and large neurons, which, in certain areas, are gathered in discrete groups considered functional nuclei. For the most part they make up the cement of the tegmentum, filling in between ocular nuclear structures.

The medial longitudinal fasciculus is a bi-

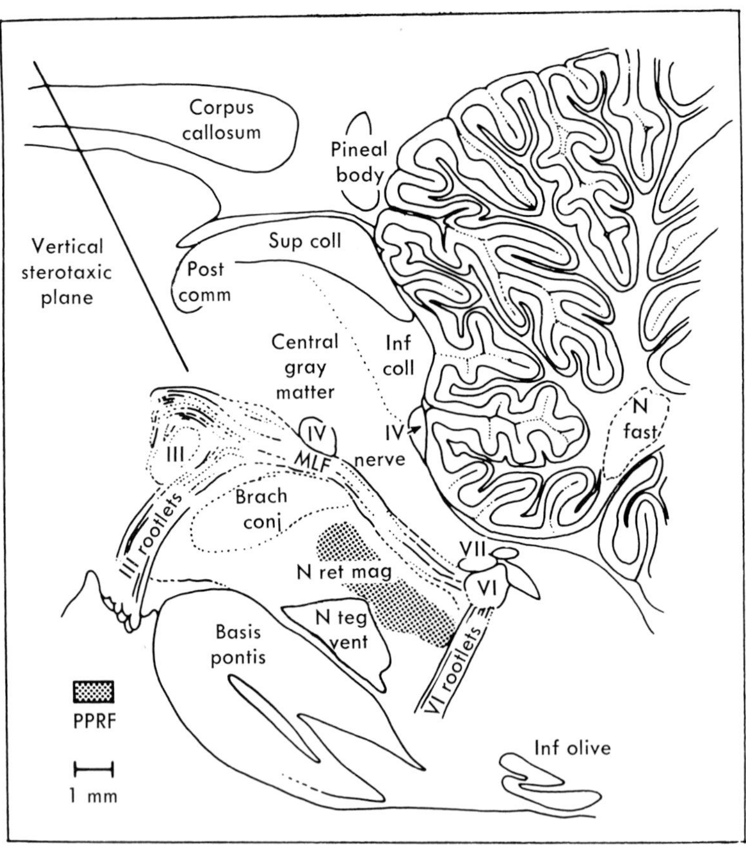

Fig. 5-79. Sagittal section of brain stem of *Macaca mulatta*, 0.75 mm from midline. Solid line shows a vertical stereotaxic plane anterior to AP-O. Dotted region shows extent of paramedian zone of pontine reticular formation. It lies in the nucleus reticularis magnocellularis, *N ret mag*, between level of trochlear and abducens nucleus. Destruction of this region causes paralysis of horizontal conjugate gaze. (From Goebel, H. H., et al.: Arch. Neurol. **24**:431, 1971.)

lateral fiber tract that runs from the thalamus to the anterior horn cells of the spinal column. It extends dorsally on either side of the midline, grossly forming a V-shaped pattern, which is especially well developed between the region of the vestibular (medulla) and oculomotor nuclei (midbrain). It is composed of many ascending and descending tracts, but its role of major importance in ocular motor function is carrying the fibers from the contralateral pontine gaze center to the ipsilateral oculomotor nucleus (ventral subnucleus of the medial rectus).

The paramedian tegmental reticular formation and the medial longitudinal fasciculus are the anatomic substrates underlying supranuclear gaze mechanisms. A recent study by Goebel and co-workers[49] demonstrated that lesions in the medial portions of the nucleus reticularis magnocellularis (NRM) between the trochlear and abducens nuclei produced paralysis of ipsilateral conjugate gaze, that is, paralysis of the ipsilateral lateral rectus and contralateral medial rectus (Fig. 5-79). This region is called the paramedian zone of the pontine reticular formation (PPRF) and can now be considered the site of the pontine center for horizontal conjugate gaze. For the eyes to move horizontally, there must be a pulse step to the agonists, ipsilateral lateral rectus, and contralateral medial rectus muscles and a negative pulse step to the antagonists, ipsilateral medial rectus, and contralateral lateral rectus muscles. In the primate there are direct pathways from the PPRF to the ipsilateral abducens motoneurons and to the neurons in the periabducens

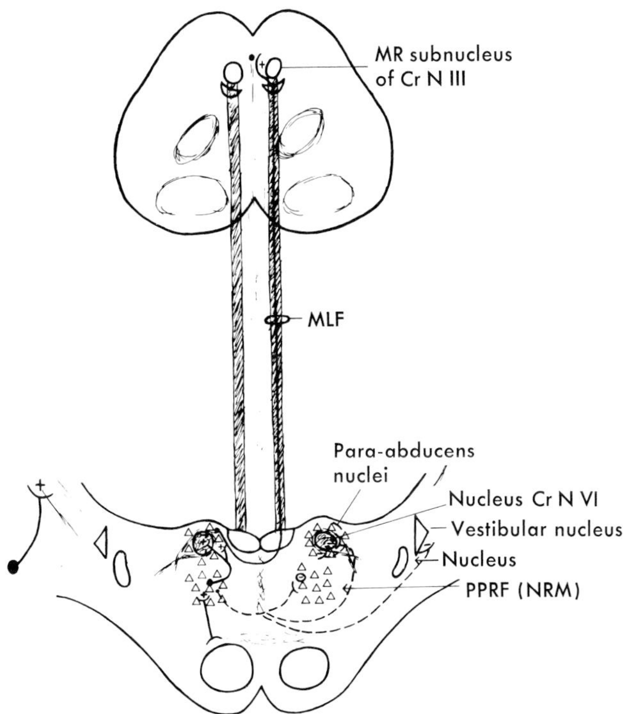

Fig. 5-80. Fiber is seen emanating from the area of the frontobulbar tract synapsing in the left PPRF. From the left PPRF, excitatory fibers travel to left sixth cranial nerve *(Cr N VI)* and to the left para-abducens nucleus, wherein a synapse occurs. Excitatory fibers arise from the left para-abducens nucleus, cross the midline dorsally, and travel by way of the right MLF to the medial rectus *(MR)* subnucleus of the right third cranial nerve *(Cr N III)*. Excitation of the ipsilateral sixth cranial nerve *(Cr N VI)* and contralateral third cranial nerve *(Cr N III)* MR produce a saccade to the ipsilateral side (left). Inhibitory fibers also arise from the PPRF, cross the midline ventrally, and synapse in the contralateral PPRF, para-abducens nucleus, abducens nucleus, and vestibular nucleus, thus abetting the ipsilateral saccade.

region.[17] Graybiel and Hartwieg[52] demonstrated neurons lying in the periabducens region project to the region of the oculomotor nucleus. Baker and Highstein[5] have extended this work by showing that these neurons produce excitation monosynaptically in contralateral medial rectus motoneurons. These internuclear neurons are activated directly the PPRF.[61] Their axons cross the brain stem at the level of the abducens nuclei to ascend in the medial longitudinal fasciculus (MLF)[19,60] (Fig. 5-80). Destruction of the axons of these interneurons is responsible for the paresis of adduction seen in so-called "internuclear ophthalmoplegia."

The vestibular system activates medial rectus motoneurons via the lateral ascending tract of Deiter.[6] This is compatible with the findings that there are no MLF units associated with horizontal head velocity. Since the vestibular system feeds directly to the vertical recti as well (Fig. 5-76), an attempt to overcome the weak adducting movement seen in internuclear ophthalmoplegia (plastic repair by the vestibulocerebellum?) may lead to an overflow into the vertical system producing the vertical nystagmus,[73] which sometimes accompanies lesions of the MLF. The inhibitory pathways are less well delineated but are assumed to be self-contained within the stem.

Slow eye movements can be induced by either vestibular or visual stimuli. Stimulus velocity is the primary signal driving the ocular motor system.[93] Rostral PPRF neurons do not carry information related to pursuit movements or slow-phase movements of vestibularly induced nystagmus. However, neurons with activity related to eye position and eye velocity are widely distributed in the periabducens region of the caudal PPRF, the rostral medulla, the vestibular nuclei, the prepositus nucleus, and the flocculi. The integrator that provides the step-ramp for driving motoneurons during slow movements appears to be located in the periabducens region. The immediate precursor receiving input from the descending pathways for organization of smooth pursuit has not yet been precisely identified but is probably a discrete region of the caudal PPRF. Lesions in the PPRF cause a paralysis of all ipsilateral movements, both fast and slow; that is, the eyes do not move past the midline. Lesions in this region produce the following:

1. Loss of crossing fibers
2. Demyelination and rostral gliosis in the region lateral to the MLF
3. A loss of cells in the PPRF lateral to the lesions.

The oculomotor and abducens nuclei appear normal. In addition, the stimulation experiments by Bender and Shanzer[9] and Carpenter and McMasters[21] confirm that the pathways

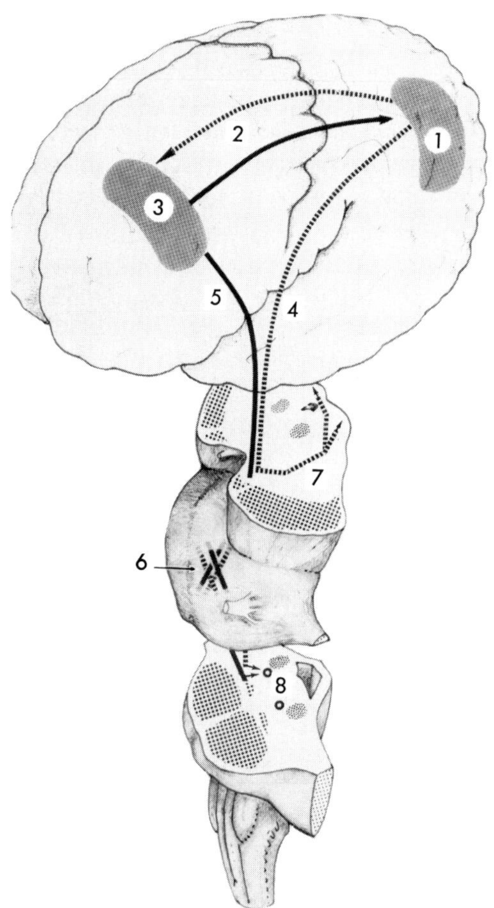

Fig. 5-81. Theoretical gaze centers (o) and associated pathways. Occipitoparietal junction—(?) Brodmann's area 19 parastriate area. *1*, Following movement center; *2*, association tracts—(?) internal sagittal striatum; *3*, saccadic frontal gaze center—Brodmann's areas 8 alpha and gamma; *4*, occipitobulbar tract; *5*, frontobulbar tract; *6*, decussation of corticobulbar tracts dealing with eye movements; *7*, occipitotectal projection; *8*, pontine gaze center. (From Burde, R. M.: In Eliasson, S. G., Prensky, A., and Hardin, W.: Neurological pathophysiology, New York, 1974, Oxford University Press.)

for horizontal gaze are concentrated in the PPRF. Stimulation experiments have further demonstrated that there is a decussation of the right and left supranuclear pathways between the third and fourth cranial nerves (Fig. 5-81). The medial longitudinal fasciculus conducts impulses for adduction as part of conjugate deviation but does not mediate control of supranuclear impulses.

The substrates for vertical movements are assumed to lie in the brain stem reticular formation, but the exact location is unknown. Clinicopathologic correlations have demonstrated that midline lesions affecting the dorsal midbrain under the superior collicular plate more often than not produce paresis of upgaze, either saccadic and following or saccadic movements alone. This does not necessarily mean that this is the location of vertical upgaze integration but more than likely suggests a confluence of supranuclear tracts for vertical upgaze. Lesions rostral to the red nuclei extending toward the thalamus in the ventral midbrain affect downgaze movements preferentially.[64] Similar results have been obtained by stimulation experiments. There is reasonable evidence that the mesencephalic center for downgaze is driven in part by the PPRF, most likely in its rostral extent. The center for upgaze may be no more than a contiguous region of the caudal PPRF.

Physiology

A study by Luschei and Fuchs[78] reported the result of single-cell recordings in the brain stem of monkeys from the posterior end of the trochlear nucleus to the most anterior end of the hypoglossal nucleus. Units whose activity related to eye movements or position were located, for the most part, dorsomedially in the area of the PPRF. The patterns of activity of cells outside the ocular motor nuclei fell into four main categories.

''burst-tonic'' units Units that exhibited high-frequency bursts related to saccadic eye movements in a particular direction and displayed tonic firing, the frequency of which was related to steady eye position. Most of these units were associated with ipsilateral horizontal gaze movements, but some were related to downward movement and position.

''burst'' units Units that exhibited bursts of activity to saccades in one or more directions. These units could be subdivided into three groups, depending on the time interval between the first spike and the initiation of a saccade. The activity of this group was associated with either ipsilateral saccades, ipsilateral and contralateral saccades, or burst activity preceding saccades in any direction. The cells could fire before or after the initiation of saccadic movement.

tonic units Units whose activity was related to eye position but did not burst during saccadic movement.

pausing units Units that fired at a relatively constant rate during fixation and stopped firing during saccadic eye movements. The vast majority of these units were associated with conjugate horizontal movements rather than with vertical movements.

Firing patterns of units found in rostral portions of the PPRF are different from those units found in the caudal PPRF around or caudal to the abducens nuclei. In the rostral PPRF unit activity is predominantly phasic (burst) and is associated with rapid eye movements.[26,27,56,57,78] These neurons have little tonic activity associated with eye position. The activity in the rostral PPRF is organized in a polar coordinate system; that is, individual units code direction, duration or amplitude, and the component of movement in the plane of the individual eye muscles.[56,57]

Units in the periabducens region have both burst and tonic units associated with saccades and eye position. Raphan and Cohen[93] believe that since lesions in the periabducens area produce paresis of the conjugate gaze, the periabducens area serves as the brain stem integrator of the pulse from the rostral PPRF. The necessity for an integrator in the ocular motor pathways has been inferred from stimulation studies[28,71] and VOR studies.

It is obvious that at each level of the so-called final common pathway, that is, the brain stem motoneurons and muscle fibers, there exists a continuum of motoneuron and fiber types that can transform supranuclear input into a spectacular array of movements—fast, medium, and slow; small, medium, and large; smooth or pulsatile; conjugate or disconjugate. This functional diversity, including the maintenance of both foveas on the object of regard, is certainly suggestive of Divine Providence.

REFERENCES

1. Abel, I. A, et al.: Saccadic system plasticity in humans, Ann. Neurol. **4:**313, 1978.
2. Adler, F., and Fliegelman, M.: Influence of fixation on the visual acuity, Arch. Ophthalmol. **12:**475, 1934.
3. Alpern, M.: Vergence and accommodation. II, Is accommodative vergence related merely to the accommodative stimulus? Arch. Ophthalmol. **60:**358, 1958.
4. Aschoff, J. C., and Cohen, B.: Changes in saccadic eye movements produced by cerebellar cortical lesions, Exp. Neurol. **32:**123, 1971.
5. Baker, R., and Highstein, S. M.: Physiological identification of interneurons and motoneurons in the abducens nucleus, Brain Res. **91:**292, 1975.
6. Baker, R., and Highstein, S. M.: Vestibular projections to medial rectus motoneurons in the cat, Neurosci. Soc. Abstr. **6:**226, 1976.
7. Barmack, N. H.: Saccadic discharges evoked by intracellular stimulation of extraocular motoneurons., J. Neurophysiol. **37:**395, 1974.
8. Bender, M. B.: Comments on the physiology and pathology of eye movements in the vertical plane, J. Nerv. Ment. Dis. **130:**456, 1960.
9. Bender, M. B., and Shanzer, S.: Effects of brainstem lesions on vertical gaze in monkeys, Fed. Proc. **19:**28, 1960.
10. Bizzi, E.: Discharge of frontal eye field neurons during saccadic and following eye movements in unanesthetized monkeys, Exp. Brain Res. **6:**69, 1968.
11. Bizzi, E.: Central control of eye and head movements in monkeys. In Lennerstrand, G., and Bach-y-Rita, P., editors: Basic mechanisms of ocular motility and their clinical implications. Proceedings of International Symposium, Wenner-Gren Center, Stockholm, June 4 to 6, 1974, Oxford, 1975, Pergamon Press Ltd., pp. 469-471.
12. Bizzi, E., and Schiller, P. H.: Single unit activity in the frontal eye fields of unanesthetized monkeys during eye and head movement, Exp. Brain Res. **10:**151, 1970.
13. Bizzi, E., Kalil, R. E., and Tagliasco, V.: Eye-head coordination in monkeys: evidence for centrally patterned organization, Science **173:**452, 1971.
14. Blackwood, W., Dix, M. R., and Rudge, P.: The cerebral pathways of optokinetic nystagmus: a neuroanatomical study, Brain **98:**297, 1975.
15. Blumenthal, H., et al.: A clinical physiologic correlate of voluntary gaze palsy, optokinetic nystagmus, and vestibular nystagmus, Am. J. Ophthalmol. **69:** 437, 1970.
16. Breinin, G. M.: Quantitation of extraocular muscle innervation, Arch. Ophthalmol. **57:**644, 1957.
17. Buettner-Ennever, J. A., and Henn, V.: An autoradiographic study of the pathways from the pontine reticular formation involved in horizontal eye movements, Brain Res. **108:**155, 1977.
18. Burde, R. M., et al.: Ocular motor dysfunction in total and hemicerebellectomized monkeys. Br. J. Ophthalmol. **59:**560, 1975.
19. Burde, R. M., et al.: Experimental internuclear ophthalmoplegia, Br. J. Ophthalmol. **61:**233, 1977.
20. Burgess, D., Roper-Hall, G., and Burde, R. M.: Binocular diplopia associated with subretinal neovascular membranes, Arch. Ophthalmol. **98:**311, 1980.
21. Carpenter, M. B., and McMasters, R. E.: Disturbances of conjugate horizontal eye movements in the monkey. II, Physiological effects and anatomical degeneration resulting from lesions in the medial longitudinal fasciculus, Arch. Neurol. **8:**347, 1963.
22. Christoferson, K., and Ogle, K.: The effect of homatropine on the accommodation-convergence association, Arch. Ophthalmol. **55:**779, 1956.
23. Cogan, D. G.: Brain lesions and eye movements in man. In Bender, M. B., editor: The oculomotor system, New York, 1964, Harper & Row, Publishers, p. 417.
24. Cohen, B.: Vestibulo-ocular relations. In Bach-y-Rita, P., Collins, C. C., and Hyde, J. E., editors: The control of eye movements, New York, 1971. Academic Press, Inc., p. 105.
25. Cohen, B.: The vestibulo-ocular reflex arc. In Kornhuber, H. H., editor: Handbook of sensory physiology. Vestibular system, Berlin, 1974, Springer-Verlag, vol. 6, part 1, pp. 477-540.
26. Cohen, B., and Henn, V.: The origin of quick phases of nystagmus in the horizontal plane, Bibl. Ophthalmol. **82:**36, 1972.
27. Cohen, B., and Henn, V.: Unit activity in the pontine reticular formation associated with eye movements, Brain Res. **46:**403, 1972.
28. Cohen, B., and Komatsuzaki, A.: Eye movements induced by stimulation of the pontine reticular formation: evidence for integration in oculomotor pathways, Exp. Neurol. **36:**101, 1972.
29. Collewijn, H.: The role of the superior colliculi in smooth eye movements in the rabbit. Presented at the Meeting of The Association for Research in Vision and Ophthalmology, Sarasota, Fla., April 24 to 28, 1972.
30. Collins, C. C.: The human oculomotor control system. In Lennerstrand, G., and Bach-y-Rita, P., editors: Basic mechanisms of ocular motility and their clinical implications. Proceedings of International Symposium, Wenner-Gren Center, Stockholm, June 4 to 6, 1974, Oxford, 1975, Pergamon Press Ltd., pp. 145-180.
31. Collins, C. C., and Scott, A. B.: The eye movement control signal. Proceedings of Second Bioengineering Conference, Ophthalmology Section, Milan, Italy, Nov. 1973.
32. Cynader, M., and Berman, N.: Receptive-field organization of monkey superior colliculus, J. Neurophysiol. **35:**187, 1972.

33. Daroff, R. B.: Physiologic, anatomic and pathophysiologic considerations of eye movements, Trans. Ophthalmol. Soc. U.K. **90:**409, 1970.
34. Daroff, R. B., and Hoyt, W. F.: Supranuclear disorders of ocular control systems in man: clinical, anatomical, and physiological correlations. In Bach-y-Rita, P., Collins, C. C., and Hyde, J. E.: The control of eye movements, New York, 1971, Academic Press, Inc., p. 175.
35. Dichgans, J., et al.: The role of vestibular and neck afferents during eye-head coordination in the monkey, Brain Res. **71:**225, 1974.
36. Ditchburn, R.: Eye movements in relation to retinal action, Opt. Acta **1:**171, 1955.
37. Ditchburn, R., and Ginsborg, B.: Vision with stabilized retinal image, Nature **170:**36, 1952.
38. Ditchburn, R., Fender, D., and Mayne, S.: Vision with controlled movements of the retinal image, J. Physiol. **145:**98, 1959.
39. Dix, M. R., and Hood, J. D.: Further observations upon the neurological mechanism of optokinetic nystagmus, Acta Otolaryngol. **71:**217, 1971.
40. Ellenberger, C., Jr., Keltner, J. L., and Stroud, M. H.: Ocular dyskinesia in cerebellar disease: evidence for the similarity of opsoclonus, ocular dysmetria, and flutter-like oscillations, Brain **95:**685, 1972.
41. Flom, M.: On the relationship between accommodation and accommodative convergence, Am. J. Optom. Physiol. Opt. **37:**517, 1960.
42. Flourens, P.: Recherches experimentales sur les propriétés et les fonctions due système nerveux dans les animaux vertébrés, ed. 2, Paris, 1842, J. B. Baillière, p. 516.
43. Foerster, O.: Symptomatologie der Erkrankungen des Grosshirns. Motorische Felder und Bahnen. In Handbuch der Neurologie, vol. 6, Berlin, 1936, Springer-Verlag, pp. 1-448.
44. Fuchs, A. F.: The saccadic system. In Bach-y-Rita, P., Collins, C. C., and Hyde, J. E.: The control of eye movements, New York, 1971, Academic Press, Inc., p. 343.
45. Fuchs, A. F., and Kornhuber, H. H.: Extraocular muscle afferents to the cerebellum of the cat, J. Physiol. **200:**713, 1969.
46. Fuchs, A. F., and Luschei, E. S.: The activity of single trochlear nerve fibers during eye movements in the alert monkey, Exp. Brain Res. **13:**78, 1971.
47. Gilman, S.: The nature of cerebellar dyssynergia. In In Williams, D.: Modern trends in neurology, New York, 1970, Appleton-Century-Crofts, vol. 5, p. 60.
48. Ginsborg, B.: Small involuntary movements of the eye, Br. J. Ophthalmol. **37:**746, 1953.
49. Goebel, H. H., et al.: Lesions of the pontine tegmentum and conjugate gaze paralysis, Arch. Neurol. **24:**431, 1971.
50. Gonshor, A., and Melvill Jones, G.: Plasticity in the adult human vestibulo-ocular reflex arc, Proc. Can. Fed. Biol. Sci. **14:**11, 1971.
51. Gonshor, A., and Melvill Jones, G.: Extreme vestibulo-ocular adaptation induced by prolonged optical reversal of vision, J. Physiol. (Lond.) **256:**381, 1976.
52. Graybiel, A. M., and Hartwieg, E. A.: Some afferent connections of the oculomotor complex in the cat: an experimental study with tracer techniques, Brain Res. **81:**543, 1974.
53. Granit, R.: The basis of motor control. London, 1970, Academic Press, Inc., p. 248.
54. Henn. V., and Cohen, B.: Eye muscle motor neurons with different functional characteristics, Brain Res. **45:**561, 1972.
55. Henn, V., and Cohen, B.: Quantitative analysis of activity in eye muscle motoneurons during saccadic eye movements and positions of fixation, J. Neurophysiol. **36:**115, 1973.
56. Henn, V., and Cohen, B.: Activity in eye muscle motoneurons and brainstem units during eye movements. In Lennerstrand, G., and Bach-y-Rita, P., editors: Basic mechanisms of ocular motility and their clinical implications. Proceedings of International Symposium, Wenner-Gren Center, Stockholm, June 4 to 6, 1974, Oxford, 1975, Pergamon Press Ltd., pp. 302-324.
57. Henn, V., and Cohen, B.: Coding of information about rapid eye movements in the pontine reticular formation of alert monkeys, Brain Res. **108:**307, 1976.
58. Heywood, S.: Voluntary control of smooth eye movements and their velocity, Nature **238:**408, 1972.
59. Higgins, J., and Stultz, K.: Frequency and amplitude of ocular tremor, J. Opt. Soc. Am. **52:**872, 1952.
60. Highstein, S. M., and Backer, R.: Termination of internuclear neurons of the abducens nuclei on medial rectus motoneurons, Neurosci. Soc. Abstr. **2:**278, 1976.
61. Highstein, S. M., et al.: Synaptic input from the pontine reticular nuclei to abducens motoneurons and internuclear neurons, Brain Res. **112:**162, 1976.
62. Hoyt, W. F., and Frisén, L.: Supranuclear ocular motor control. Some clinical considerations—1974. In Lennerstrand, G., and Bach-y-Rita, P., editors: Basic mechanisms of ocular motility and their clinical implications. Proceedings of International Symposim, Wenner-Gren Center, Stockholm, June 4 to 6, 1974, Oxford, 1975, Pergamon Press Ltd., pp. 379-392.
63. Hubel, D. H., LeVay, S., and Wiesel, T. N.: Mode of termination of retinotectal fibers in macaque monkey. An autoradiographic study, Brain Res. **96:**25, 1975.
64. Jacobs, L., Anderson, P. J., and Bender, M. B.: The lesions producing paralysis of downward but not upward gaze, Arch. Neurol. **28:**319, 1973.
65. Jampel, R. S.: Representation of the near-response on the cerebral cortex of the macaque, Am. J. Ophthalmol. **48:**573, 1959.
66. Jampel, R. S.: Multiple motor systems in the extraocular muscles of man. Invest. Ophthalmol. **6:**288, 1967.

67. Jampolsky, A.: What can electromyography do for the ophthalmologist? Invest. Ophthalmol. **9:**570, 1970.
68. Jampolsky, A.: Coordination of extraocular muscles: clinical aspects. In Lennerstrand, G., and Bach-y-Rita, P., editors: Basic mechanisms of ocular motility and their clinical implications. Proceedings of International Symposium, Wenner-Gren Center, Stockholm, June 4 to 6, 1974, Oxford, 1975, Pergamon Press Ltd., pp. 209-223.
69. Jampolsky, A., Tamler, E., and Marg, E.: Artifacts and normal variations in human ocular electromyography, Arch. Ophthalmol. **61:**402, 1959.
70. Keller, E. L.: Accommodative vergence in the alert monkey. Monkey unit analysis, Vision Res. **13:**1565, 1973.
71. Keller, E. L.: Participation of medial pontine reticular formation in eye movement generation in monkey, J. Neurophysiol. **37:**316, 1974.
72. Keller, E. L., and Robinson, D. A.: Abducens unit behavior in the monkey during vergence movements, Vision Res. **12:**369, 1972.
73. Kirkham, T. H., and Katsarkas, A.: An electrooculographic study of internuclear ophthalmoplegia, Ann. Neurol. **2:**385, 1977.
74. Kommerell, G.: Clinical cues for the organization of horizontal quick eye movements and subsequent periods of fixation. In Lennerstrand, G., and Bach-y-Rita, P., editors: Basic mechanisms of ocular motility and their clinical implications. Proceedings of International Symposium, Wenner-Gren Center, Stockholm, June 4 to 6, 1974, Oxford, 1975, Pergamon Press Ltd., pp. 325-335.
75. Kommerell, G., Olivier, D., and Theopold, M.: Adaptive programming of phasic and tonic components in saccadic eye movements. Investigations in patients with abducens palsy, Invest. Ophthalmol. **15:**657, 1976.
76. Kurtzberg, D., and Vaughan, H. G., Jr.: Electrocortical potentials associated with eye movement. In Zikmund, V., editor: The oculomotor system and brain functions, London, 1973, Butterworth & Co. Ltd., pp. 137-145.
77. Ling. W., and Gay, A. J.: Optokinetic nystagmus: a proposed pathway and its clinical application. In Smith, J. L., editor: Neuro-ophthalmology: Symposium of the University of Miami and The Bascom Palmer Eye Institute, St. Louis, 1968, The C. V. Mosby Co., vol. IV, p. 117.
78. Luschei, E. S., and Fuchs, A. F.: Activity of brain stem neurons during eye movements of alert monkeys, J. Neurophysiol. **35:**445, 1972.
79. Martens, T., and Ogle, K.: Observations on accommodative convergence, Am. J. Ophthalmol. **47:**455, 1959.
80. Melamed, E., and Larsen, B.: Cortical activation pattern during saccadic eye movements in humans: localization by focal cerebral blood flow increases, Ann. Neurol. **5:**79, 1979.
81. Miles, F. A.: The primate flocculus and eye-head coordination. In Brooks, B. A., and Bajandas, F, J., editors: Eye movements. ARVO Symposium, 1976, New York, 1977, Plenum Publishing Corp., pp. 75-92.
82. Miles, F. A., and Fuller, J. H.: Adaptive plasticity in the vestibulo-ocular responses of the rhesus monkey, Brain Res. **80:**512, 1974.
83. Miles, F. A., and Fuller, J. H.: Visual tracking and the primate flocculus, Science **189:**1000, 1975.
84. Miller, J. E.: The electromyography of vergence movement, Arch. Ophthalmol. **62:**790, 1959.
85. Mohler, C. W., Goldberg, M. E., and Wurtz, R. H.: Visual receptive fields of frontal eye field neurons, Brain **61:**385, 1973.
86. Morasso, P., Bizzi, E., and Dichgans, J.: Adjustment of saccade characteristics during head movements, Exp. Brain Res. **16:**492, 1973.
87. Nachmias, J.: Two dimensional motion of the retinal image during monocular fixation, J. Opt. Soc. Am. **49:**901, 1959.
88. Nauheim, J.: A preliminary investigation of retinal locus as a factor in fusion, Arch. Ophthalmol. **58:** 122, 1957.
89. Ogle, K., and Sabin, F.: The accommodation-convergence association: effects of Neo-Synephrine, pilocarpine and physostigmine; quoted in Christoferson, K., and Ogle, K.: The effect of homatropine on the accommodation convergence association, Arch. Ophthalmol. **55:**779, 1956.
90. Optican, L. M., and Robinson, D. A.: Plastic adaptations of saccadic dysmetria (abstracts), Soc. Neurosci. **3:**157, 1977.
91. Park, R., and Park, G.: The center of ocular rotation in the horizontal plane, Am. J. Physiol. **104:**545, 1933.
92. Pasik, P., and Pasik, T.: Oculomotor functions in monkeys with lesions of the cerebrum and superior colliculi. In Symposium on the Oculomotor System, Arch. Ophthalmol. **66:**430, 1961.
93. Raphan, T., and Cohen, B.: Brainstem mechanisms for rapid and slow eye movements, Ann. Rev. Physiol. **40:**527, 1978.
94. Rashbass, C.: The relationship between saccadic and smooth tracking eye movements, J. Physiol. (Lond.) **159:**326, 1961.
95. Ratliff, F., and Riggs, L. A.: Involuntary motions of the eye during monocular fixation, Nature **162:**25, 1948.
96. Riggs, L. A.: Measurement of normal ocular tremor by corneal reflection, J. Opt. Soc. Am. **42:**287, 1952.
97. Riggs, L. A., and Ratliff, F.: The effects of counteracting the normal movements of the eye, J. Opt. Soc. Am. **42:**872, 1952.
98. Riggs, L. A., et al.: The disappearance of steadily fixed visual test objects, J. Opt. Soc. Am. **43:**495, 1953.
99. Ripps, H., et al.: The effect of pupil size on accommodation, convergence, and the AC/A ratio, Invest. Ophthalmol. **1:**127, 1962.
100. Robinson, D. A.: Oculomotor unit behavior in the monkey, J. Neurophysiol. **33:**393, 1970.

101. Robinson, D. A.: Models of oculomotor organization. In Bach-y-Rita, P., Collins, C. C., and Hyde, J. E., editors: The control of eye movements, New York, 1971, Academic Press, Inc., pp. 519-538.
102. Robinson, D. A.: Eye movements evoked by collicular stimulation in the alert monkey, Vision Res. **12:**1795, 1972.
103. Robinson, D. A., chairman: Symposium on Oculomotor Physiology and Disorders. Presented at the Spring Meeting of the Association for Research in Vision and Ophthalmology, Sarasota. Fla., April 24 to 28, 1972.
104. Robinson, D. A.: Adaptive gain control of vestibulo-ocular reflex by cerebellum, J. Neurophysiol. **39:**954, 1976.
105. Robinson, D. A.: Is the cerebellum too old to learn? In Brooks, B. A., and Bajandas, F. J., editors: Eye movements. ARVO Symposium, 1976, New York, 1977, Plenum Publishing Corp., pp. 65-73.
106. Robinson, D. A., and Fuchs, A. F.: Eye movements evoked by stimulation of the frontal eye fields, J. Neurophysiol. **32:**637, 1969.
107. Ron, S.: A quantitative study of eye movements evoked by cerebellar stimulation in the alert monkey, Ph.D. dissertation, Baltimore, 1971, The Johns Hopkins University.
108. Schiller, P. H.: The discharge characteristics of single units in the oculomotor and abducens nuclei of the unanesthetized monkey, Exp. Brain Res. **10:**347, 1970.
109. Schiller, P. H.: The role of the monkey superior colliculus in eye movement and vision, Invest. Ophthalmol. **11:**451, 1972.
110. Schiller, P. H., and Koerner, K.: Discharge characteristics of single units in superior colliculus of the alert rhesus monkey, J. Neurophysiol. **34:**920, 1971.
111. Sears, M., and Guber, D.: The change in the stimulus AC/A ratio after surgery, Am. J. Ophthalmol. **64:** 872, 1967.
112. Sloan, L., Sears, M. L., and Jablonsky, M. D.: Convergence-accommodation relationship: description of a simple clinical test and its application to the evaluation of isoflurophate (DFP) therapy, Arch. Ophthalmol. **63:**283, 1960.
113. Smith, J. L., Gay, A. J., and Cogan, D. G.: The spasticity of conjugate gaze phenomenon, Arch. Ophthalmol. **62:**694, 1959.
114. Snider, R. S., and Stowell, A.: Receiving areas of the tactile, auditory and visual systems in the cerebellum, J. Neurophysiol. **7:**331, 1944.
115. Sparks, D. L., and Pollack, J. G.: The neural control of saccadic eye movements: the role of the superior colliculus. In Brooks, B. A., and Bajandas, F. J., editors: Eye movements. ARVO Symposium, 1976, New York, 1977, Plenum Publishing Corp., pp. 179-219.
116. Stein, B. M., and Carpenter, M. B.: Central projections of portions of the vestibular ganglia innervating specific parts of the labyrinth in the rhesus monkey, Am. J. Anat. **120:**281, 1967.
117. Szentágothai, J.: Pathways and synaptic articulation patterns connecting vestibular receptors and oculomotor nuclei. In Bender, M. B., editor: The oculomotor system, New York, 1964, Harper & Row, Publishers, pp. 205-223.
118. Troost, B. T., Weber, R. B., and Daroff, R. B.: Hemispheric control of eye movements. I, Quantitative analysis of refixation saccades in a hemispherectomy patient, Arch. Neurol. **27:**441, 1972.
119. Troost, B. T., et al.: Hemispheric control of eye movements. II, Quantitative analysis of smooth pursuit in a hemispherectomy patient, Arch. Neurol. **27:**449, 1972.
120. Wadia, N. H., and Swami, R. K.: A new form of heredofamilial spinocerebellar degeneration with slow eye movements (nine families), Brain **94:**359, 1971.
121. Walsh, F. B., and Hoyt, W. F.: Clinical neuroophthalmology, ed. 3, Baltimore, 1969, The Williams & Wilkins Co.
122. Warwick, R.: Representation of the extraocular muscles in the oculomotor nuclei of the monkey, J. Comp. Neurol. **98:**449, 1953.
123. Weber, R. B., and Daroff, R. B.: The metrics of horizontal saccadic eye movements in normal humans, Vision Res. **11:**921, 1971.
124. Weber, R. B., and Daroff, R. B.: Corrective movements following refixation saccades: type and control system analysis, Vision Res. **12:**467, 1972.
125. Westheimer, G.: Mechanisms of saccadic eye movement. Arch. Ophthalmol. **52:**710, 1954.
126. Westheimer, G., and Blair, S. M.: Oculomotor defects in cerebellectomized monkeys, Invest. Ophthalmol. **12:**618, 1973.
127. Westheimer, G., and Blair, S. M.: Functional organization of primate oculomotor system revealed by cerebellectomy, Exp. Brain Res. **21:**463, 1974.
128. Westheimer, G., and Mitchell, A.: Eye movement responses to convergence stimuli, Arch. Ophthalmol. **55:**848, 1956.
129. Wurtz, R. H., and Goldberg, M. E.: The primate superior colliculus and the shift of visual attention, Invest. Ophthalmol. **11:**441, 1972.
130. Wurtz, R. H., and Mohler, C. W.: Organization of monkey superior colliculus. Enhanced visual response of superficial layer cells, J. Neurophysiol. **89:**748, 1976.
131. Young, L. and Stark, L.: A sampled-data model for eye-tracking movements, Q. Progr. Rep. Res. Lab. Electr. M.I.T. **66:**370, 1962.
132. Zee, D. S.: The organization of the brainstem ocular motor subnuclei, Ann. Neurol. **4:**384, 1978.
133. Zuber, B. L.: Control of vergence eye movements. In Bach-y-Rita, P., Collins, C. C., and Hyde, J. E. The control of eye movements, New York, 1971, Academic Press, Inc., p. 447.

Part three

Clinical analysis of extraocular muscle paralysis

GILL ROPER-HALL, D.B.O.(T).
RONALD M. BURDE, M.D.

The onset of ocular motor paralysis is for the most part accompanied by the sudden awareness of diplopia, an annoying subjective experience secondary to the stimulation of noncorresponding retinal points causing maculoparamacular projection of the retinal images. There is a simultaneous abnormality of all movements of the involved eye when it moves into the field of action of the involved muscle. This resulting movement disorder is broadly referred to as incomitant strabismus.

Incomitant strabismus has two characteristics: (1) a difference in the measured deviation depending on the eye used for fixation and (2) a variation of the angle of deviation in different directions of gaze. The difference in the angle of deviation with respect to the fixing eye is the clinical manifestation of Hering's law of ocular rotations. If the normal (uninvolved) eye is used for fixation (the usual case), the measured deviation is termed the primary deviation. If the eye with the involved paralytic muscle is used for fixation, the deviation is termed the secondary deviation, and the secondary deviation is always greater in the acute stage than the primary deviation.

Example: In a right lateral rectus palsy there will be a moderate amount of esotropia present with the left (nonparetic) eye fixing in the primary position. If the right eye is used for fixation, additional output will be required to the right lateral rectus to obtain fixation in the midline. The output will go simultaneously to the left medial rectus (its contralateral synergist), which being normal will overact, producing a greater esotropia than was present with the left (nonparetic) eye fixing. The farther into the field of action of the paretic muscle either eye attempts to fixate, the greater the deviation, and it will be greatest when the paretic eye fixates.

The differences in the angle of deviation are obviously maximal in the field of action of the paretic muscle and are conversely minimal away from the field of action of the involved muscle. Therefore objective or subjective measurement of the deviation in a minimum of five of the nine positions of gaze (Fig. 5-82), with either eye fixating in turn, will isolate which eye is involved and which muscles are paretic.

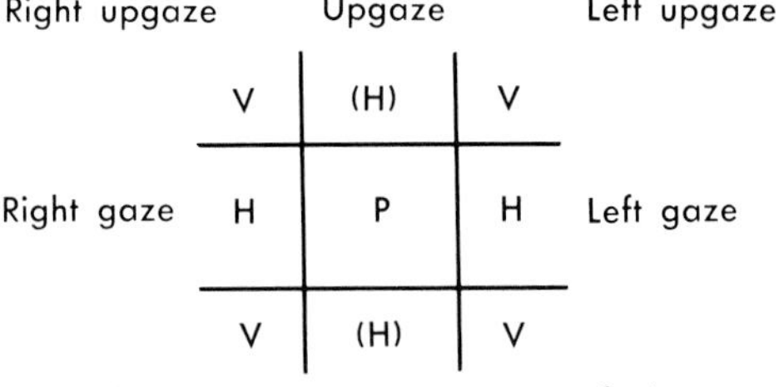

Fig. 5-82. The nine positions of gaze. Measurement of the deviation in primary gaze, *P*, and in those positions marked *V* are essential in vertical deviations. Measurements in primary gaze and the positions marked *H* are essential in horizontal deviations. Often only seven positions—all those shown excluding *(H)*—are measured.

DEVELOPMENT OF CONCOMITANCE

In an acute paresis the contralateral synergist of a recently paralyzed extraocular muscle will overact when the paretic eye is fixing (Hering's law). For example, in a paralysis of the right superior rectus muscle relative overaction of the left inferior oblique is found initially.

Over an indeterminate period of time the following sequence can develop: (1) an overaction of the ipsilateral antagonist (right inferior rectus) and (2) a secondary underaction of the contralateral antagonist of the nonparetic eye (left superior oblique). The underaction of the contralateral antagonist is not usually seen for several months after the onset of paralysis, but it may be seen as early as 2 weeks after the onset.

This development of a secondary underaction was termed by Chavasse[4] secondary inhibitional palsy of the contralateral antagonist. With some improvement in function of the paretic muscle and the development of a secondary underaction of the contralateral antagonist, the manifest deviation may appear to be almost concomitant. This change in the deviation is termed spread of concomitance. It may be impossible at this time to determine with assurance which muscle was primarily paretic.

OCULAR MOTILITY

Ocular movement is assessed by testing first versions and then ductions. In this manner the function (i.e., underaction and overaction) of individual muscles can be evaluated. Except in complete paralysis, it is not sufficient merely to observe ocular movements in diagnosing a muscle limitation. The examiner must be aware that such factors as facial asymmetry, unequal palpebral fissures, or proptosis can mislead one into thinking that a heterotropia or a muscle weakness is present when it is not. Version testing should be combined with a cover test, which precisely determines the type and direction of deviation present. This reveals any subtle ocular muscle underaction or overaction. It is particularly useful in those positions of gaze in which the patient claims single vision. It ensures that each eye can see the fixation object; neither has been occluded inadvertently by the nose or brow. It also reveals any muscle underaction, which produces a small enough ocular deviation that the patient can control it by fusing the images.

Objective measurements of the ocular deviation must be made and recorded in at least five, but usually seven or nine, positions of gaze (Fig. 5-82).

COVER TEST

The cover test is an objective dissociation test in which the patient fixates an accommodative target at a given distance while the examiner dissociates the eyes by placing an opaque occluder before each eye in turn. The test can be used to differentiate the type of deviation present (heterotropia, heterophoria, or orthophoria), its direction (convergent, divergent, vertical, or cyclodeviational), and to determine the approximate size of the deviation (minimal, moderate, or gross). It can also be used to evaluate the ability to fixate, particularly if one eye is amblyopic, the rate of recovery to binocular single vision in heterophoria, and to diagnose the presence of such conditions as incomitance, nystagmus, or an abnormal angle kappa.

Further details on the direct or indirect factors that can be determined from the cover test are described by Marshall[22] and Maurer.[23]

The three basic parts to the cover test are as follows:

1. The cover or single cover test used to detect a heterotropia
2. The cover-uncover test used to detect a heterophoria
3. The cross cover test (alternate cover test*) used to exaggerate the deviation by fully dissociating the visual axes

(Single) cover test

The patient fixates an accommodative target at 33 cm or 6 m (or any other position specified by the examiner). An opaque occluder is placed before one eye; the opposite eye is observed to see if it moves to take up fixation of the target. If the deviation is large enough, it will be obvious which eye the patient is using for fixation, and this eye should be occluded first. If movement is observed, its direction and speed should be noted and how well fixation is maintained.

Example: When the left eye is covered, the right eye is observed to move outward or temporally to take up fixation. This is a right esotropia. If, when the left eye is covered, the right eye moves inward (nasally) to take up fixation, a right exotropia would be present. If the right eye moves inward and downward to fixate, a right exotropia with right hypertropia would be present. If no movement of the right eye is observed when the left eye is occluded, the test should be repeated covering the right eye and observing the behavior of the left eye.

Three additional factors are occasionally present and can be detected during testing. These are alternation, eccentric fixation, and incomitance.

Alternation

After the cover is removed, the same eye should be observed to see whether it returns to its original position. If, however, that eye holds fixation and the originally fixing eye now uncovered is deviated, alternation of the deviation

---

*This term is sometimes used in British terminology for single cover test.

has occurred. The length of time this eye can hold fixation should be noted.

Eccentric fixation

Occasionally an obviously strabismic eye will either not move or will make a slow partial attempt to fixate in the midline when the fixing eye is covered. This may represent loss of central fixation, or eccentric fixation, a monocular condition in which a retinal point other than the fovea is used for fixation. The exact nature of this fixation should be further determined by visuoscopy.

Incomitance

A deviation that differs in type and degree according to the fixating eye represents incomitance. This is usually due to the influence of a muscle paresis, but in small amounts of horizontal deviation it can be related to a difference in the accommodative requirement of each eye produced by such factors as anisometropia.

• • •

If no movement can be found on repeated testing with either eye fixating, consideration having been given to any obvious deviation, the visual acuity, and state of fixation, a heterotropia can be said to be absent, and the following test should be performed.

Cover-uncover test

The patient fixates on an accommodative target. One eye is occluded, and the behavior of that eye is observed as the cover is removed. The direction of any deviation and the speed or rate of recovery to binocular fixation is noted. This indicates the desire for binocular single vision, and if, when the test is repeated covering the other eye, the rate of recovery is symmetrical, the visual acuity is probably equal.

Example: No heterotropia has been found using the cover test. The right eye is occluded; as the cover is removed, the right eye is seen to move rapidly outward or temporally to regain binocular fixation in the primary position. This is an esophoria. One does not refer to a right or left esophoria because in this example, if the left eye is covered, as the cover is removed the left eye will also be seen to move a symmetrical amount outward to take up fixation.

If no movement is found on repeated testing, a heterophoria is probably absent, but the cross-cover test should be completed before this assurance can be given.

Cross-cover test (alternate cover test)

A rapid movement of the occluder from one eye to the other is made, preventing the patient from using both eyes together during the test. The cover should remain in front of each eye long enough to allow the patient to take up fixation with the opposite eye. This test dissociates the visual axes, preventing fusion, and uncovers the full amount of the deviation. A small movement of the fixation target or the patient's head made by the examiner while one eye is covered ensures that the patient is fixating and disrupts the patient's kinesthetic sense of the straight-ahead position of the visual axes.

The cross-cover test can be used in four situations:

1. When no deviation has been revealed using either the single cover or cover-uncover test (the diagnosis of orthophoria or orthotropia is only accurately made if still no movement can be elicited by the full dissociation of the cross cover test)
2. In combination with prisms as part of the prism cover test to measure the full amount of the deviation
3. To dissociate a patient's binocularity deliberately to bring out his maximum deviation and observe his responses when both eyes are permitted to view the fixation object simultaneously (i.e., is there a fusion response or has the deviation now decompensated to a heterotropia?)
4. In the testing of ocular versions to exaggerate small defects in ocular movement

COMPENSATORY HEAD POSTURE

An objective appraisal of any unusual head position should always be made. The main aim of adopting a compensatory head posture is to obtain fusion and, if possible, place the area of binocular single vision in the straight-ahead position. It is suggested that the term compensatory head posture should be reserved for cases in which binocular single vision is achieved and should not be used for those abnormal head

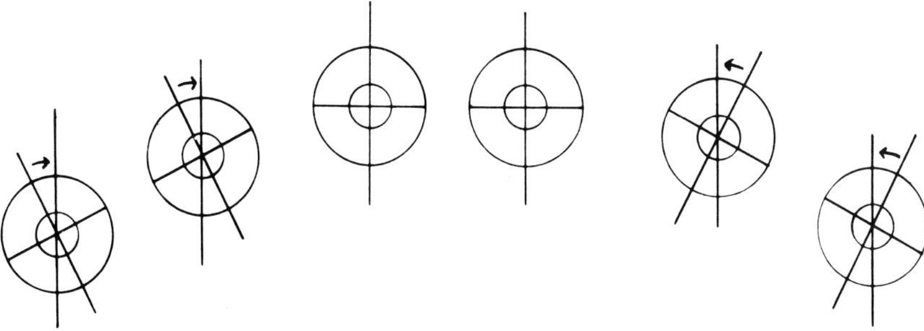

Fig. 5-83. Tilting head to right produces intorsion of right eye and extorsion of left eye.

postures adopted for a variety of other visual reasons such as nasal occlusion to avoid diplopia, separation of diplopic images, or placing the monocular field of vision straight ahead when the other eye is blind. In addition, the term should not be used to describe abnormal head postures adopted for nonvisual reasons such as deafness and nonocular torticollis. It can be used in cases in which a null point of nystagmus is found that benefits the visual acuity and often the binocular function.

If a horizontal deviation or a horizontal component of a combined deviation is present, the patient will attempt to compensate for the motor deficit by turning the face toward the defective area. This avoids the necessity of moving the eyes into the paretic field of action. For example, in a right lateral rectus palsy the patient, being unable to abduct his right eye fully, will appreciate diplopia on right gaze; this can be avoided by turning the face to the right, thus moving the eyes to the left. Similarly, a vertical deviation is diminished by elevating or depressing the chin. In paralysis of one of the elevators, such as the superior rectus, diplopia will be maximal on upgaze. Elevating the chin lowers the eyes away from the area of diplopia.

The area of minimal horizontal and vertical deviation can often be found by combining a face turn with a chin elevation or depression; this area is usually diagonally opposite the field of main action of the paralyzed muscle.

A torsional deviation (e.g., in superior oblique palsy), however, is more difficult to eliminate. It is accomplished by tilting the head toward one shoulder. If a normal individual tilts his head to one side, two physiologic movements result, forming the compensatory fixation reflex.

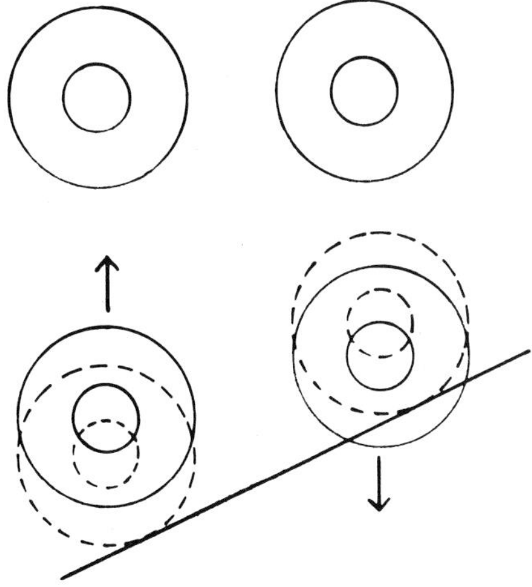

Fig. 5-84. Tilting head to right produces slight elevation of right eye and depression of left eye.

1. There is a compensatory movement of the true vertical meridian of each eye to maintain its normal position with respect to the horizon (perpendicular). There is intorsion of the lower eye (the side toward which the head is tilted) and extorsion of the upper eye (away from the head tilt) (Fig. 5-83). In most situations the amount of torsion involved is approximately one fifth to one sixth of the number of degrees through which the head is tilted.[20]
2. Simultaneously there is a minimal elevation of the lower eye and depression of the upper eye (Fig. 5-84). The head tilt is adopted to compensate predominantly for the torsional component of the deviation and not the vertical deviation.

A

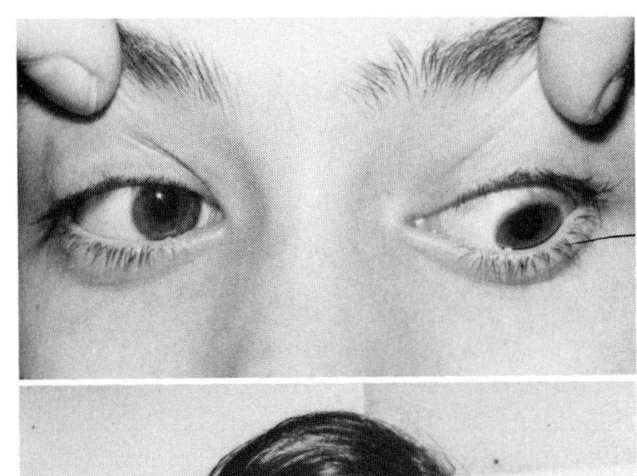

B

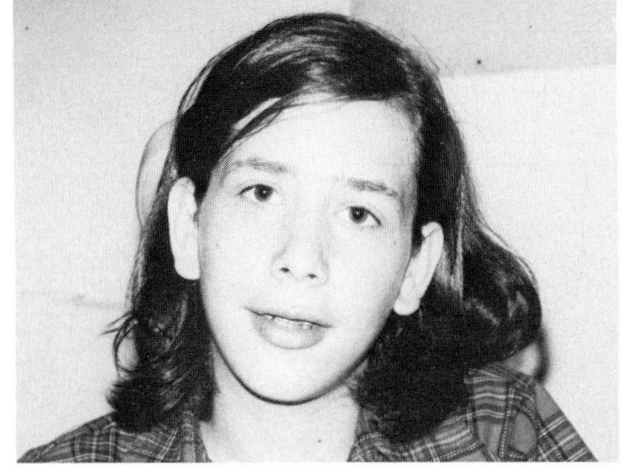

Fig. 5-85. A, Subject with eyes in left downgaze demonstrating right superior oblique paresis. **B,** Compensatory head posture in a right superior oblique paresis consisting of chin depression, face turn to the left and head tilt to the left, placing the eyes in right upgaze.

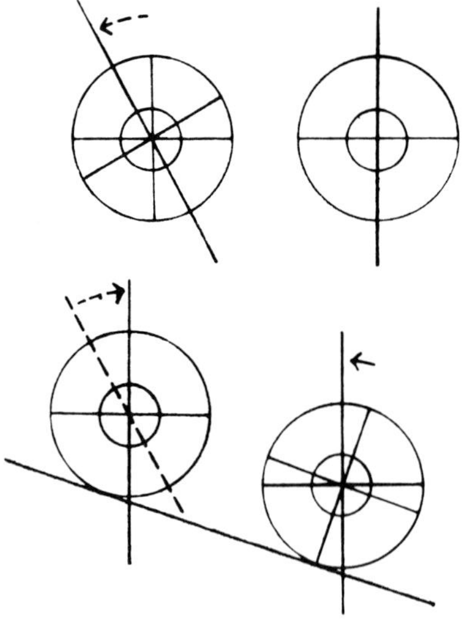

Fig. 5-86. Right excyclotropia produced by right superior oblique paresis. Tilting head to left places extorted image of right eye upright; left eye intorts to bring left vertical meridian parallel to the right, and perpendicular to the horizon.

Example: In a paralysis of the right superior oblique muscle the greatest deviation will typically be on left downgaze. The patient will adopt a head posture involving a chin depression (to elevate the eyes) and a face turn to the left (to move the eyes into right gaze). Since the main action of the superior oblique is intorsion, paralysis of this muscle will produce an extorsion of the right eye or a right excyclotropia. To compensate for this torsional deviation, the patient will tilt his head to the left, which will place the extorted image of the right eye upright (Fig. 5-85). The left eye will intort by the compensatory fixation reflex, and both vertical meridians will be parallel (Fig. 5-86).

Compensatory head postures may consist of a chin elevation or depression, a face turn, a head tilt, or a combination of these. Since the oblique muscles are the primary torters and the vertical recti have very little torsional effect, a patient with a head posture including a tilt characteristically has paralysis of an oblique muscle, usually the superior oblique.

BIELSCHOWSKY'S HEAD-TILT TEST

Based on the compensatory fixation reflex, Bielschowsky[2] described how a differential diagnosis could be made between a suspected superior oblique palsy and a contralateral superior rectus palsy in certain long-standing cases in which the spread of concomitance has produced an apparently concomitant deviation.

The patient is instructed to maintain fixation on a distant target and then tilt his head to the side of the suspected superior oblique palsy while the behavior of the eye on that side is observed. If the eye elevates, this is considered to be a positive response. The superior oblique muscle is then thought to have been the primarily paretic muscle. If there is no obvious movement of the eye, the test is considered negative, and no conclusion can be drawn as to which muscle was the originally paretic muscle.

The positive elevation of the involved eye in this test is explained as follows: When the head is being tilted to the side of the suspected superior oblique palsy, intorsion of that eye should occur to keep the vertical meridian perpendicular to the horizon. This intorsion is brought about by the synergic movement of the ipsilateral superior oblique and superior rectus muscles, which are the intorters. However, if the superior oblique is paretic, its secondary actions, one of which is depression, are also defective, and the main action of the superior rectus muscle, elevation, remains unopposed.

Bielschowsky's original description of this test relied on subjective rather than objective evaluation of the vertical deviation, but in recent years quantitative measurement of this vertical deviation has proved to be of value.

In practice today, objective measurements using prisms are usually made. The vertical deviation is measured in the primary position, with the patient maintaining fixation on a distance target. The measurements are repeated with the head tilted first to one side and then the other. An increased vertical deviation when the head is tilted to the side of the suspected superior oblique palsy indicates that the superior oblique is the primarily paretic muscle. This test was further modified by Moore[24] to diagnose paresis of any vertically acting muscle in all cases of concomitant vertical muscle paralysis. Use of the Bielschowsky head-tilt test is made in the effective three-step diagnostic test described first by Haagedorn[6] and popularized in this country by Parks[27] and later by Hardesty.[7] These tests are used in isolating the involved paretic muscle in concomitant or incomitant vertical deviations. They consist of determining the following:

1. Whether a right or left hypertropia is present in primary gaze; this eliminates all but four muscles (e.g., in a right hypertropia there could be paresis of the right inferior rectus, right superior oblique, left inferior oblique, or left superior rectus)
2. Whether this hypertropia increases in right or in left gaze, which eliminates another two muscles (e.g., if the right hypertropia increases on left gaze, either the right superior oblique or left superior rectus is paretic)
3. The differential diagnosis between these two muscles, one in each eye, is made with the head tilt test as described above (e.g., if the right hypertropia increases on right head tilt, then the right superior oblique is paretic)

Helveston[8] describes a similar method using a two-step procedure. Urist[31] uses a three-step procedure that includes evaluation of the A or V pattern as the third step.

DIPLOPIA

Diplopia is a very distressing symptom. Patients usually attempt to eliminate diplopia by one of the following mechanisms:

1. Fusing the images by adopting a compensatory head posture
2. Eliminating one image by suppression, by closing one eye, or by nasal occlusion
3. Increasing the separation of the images so that the troublesome image is projected peripherally where it can be ignored

The development of suppression and whether this occurs soon after the onset, eventually, or not at all depends partly on the age at which the paralysis occurs. The younger the patient, the more likely he is to develop suppression of one image.

Three types of diplopia can result from paralysis of an extraocular muscle—horizontal, vertical, and torsional. They can occur separately or in combination with each other. Of these, torsional diplopia is the most troublesome.

Diplopia, or maculoparamacular projection, results from the stimulation of noncorresponding points of the retina. The fovea projects images straight ahead; the nasal retina projects them temporally (i.e., it interprets images as having come from a temporal direction); the temporal retina projects nasally (Fig. 5-87).

In normal retinal correspondence an area of the nasal retina corresponds, or has a common visual direction with, a temporal area in the other eye (Fig. 5-88). The imaginary line in space joining the projected corresponding points is called the horopter. The imaginary area on either side of the horopter, within which slightly disparate points can be fused, is called Panum's fusional area. In the presence of a heterotropia images from an object will fall on noncorresponding areas of the retina and are seen double. For example, in a right esotropia images from an object will fall on the fovea of the left eye and a point on the nasal retina in the right eye. The left eye will project its image straight ahead; the image falling on the nasal retina in the right eye will be projected temporally,

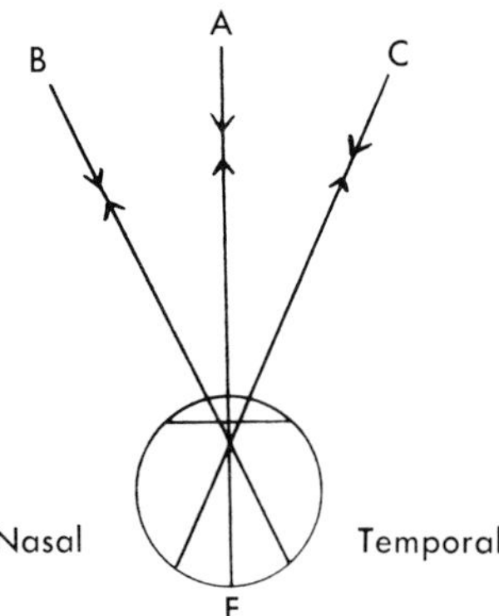

Fig. 5-87. *A*, Fixation object; *B*, object in nasal field; *C*, object in temporal field; *F*, fovea. Image from object *A* falling on fovea is interpreted as coming from straight ahead. Image from object *B* in the nasal field falls on temporal retina and is projected nasally. Image from object *C* in the temporal field falls on nasal retina and is projected temporally.

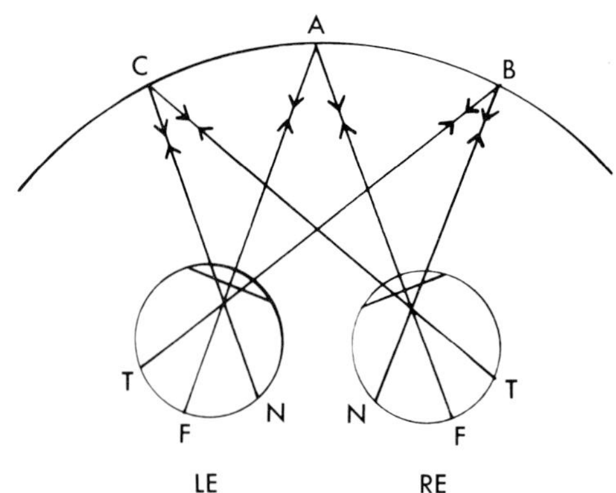

Fig. 5-88. *F*, Fovea; *A*, object in primary gaze; *B*, object in right field; *C*, object in left field; *T*, temporal retinal point; *N*, nasal retinal point. Image from object *A* falls on fovea of each eye and is seen straight ahead. Image from object *B* falls on nasal retina in right eye and temporal retina in left eye, which are corresponding retinal points, and is projected back to *B* in right gaze field. Image from object *C* acts similarly vice versa.

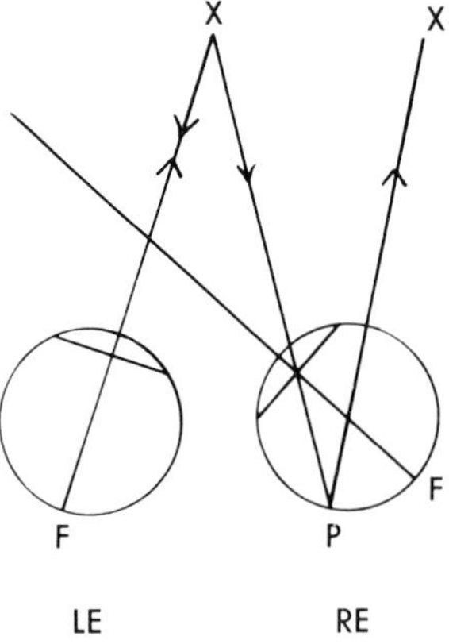

Fig. 5-89. *F*, Fovea; *X*, object; X_1, projected image of *X; P*, peripheral retinal point. Homonymous or uncrossed pathologic diplopia in a right esotropia. Image of *X* falling on fovea of left eye is projected straight ahead. Image of *X* falling on nasal retina of right eye is projected temporally, resulting in a diplopic mental image, X_1, of object *X*.

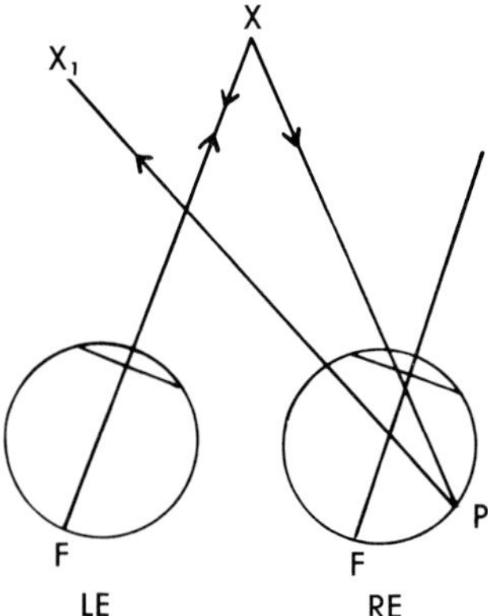

Fig. 5-90. *F*, Fovea; *X*, object; X_1, projected image of *X; P*, peripheral retinal point. Heteronymous or crossed pathologic diplopia in a right exotropia. Image of *X* falling on left fovea is projected straight ahead. Image of *X* falling on temporal retina of right eye is projected nasally, resulting in X_1.

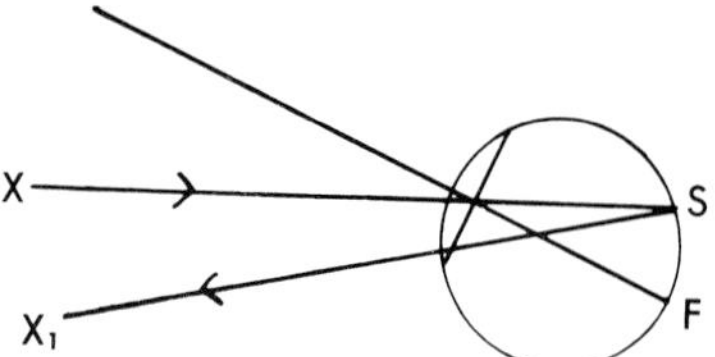

Fig. 5-91. *F*, Fovea; *X*, object in primary gaze; X_1, projected image of *X*, appears below *X; S*, superior retinal point. Vertical diplopia in a right hypertropia. Image from object *X* falling on superior retina in right eye will be projected inferiorly below *X*, the position where left eye (not shown) perceives image of object.

resulting in uncrossed or homonymous diplopia (Fig. 5-89). In a right exotropia images from an object will fall on the fovea of the left eye and a temporal point in the right eye. The image seen by the left eye will be projected straight ahead, and the image of the right eye will be projected nasally, producing crossed or heteronymous diplopia (Fig. 5-90). Vertical deviations respond in a similar manner. Normally, images received on the superior or upper retina are projected downward and are interpreted as having come from below; the lower or inferior retina projects upward. For example, in a right hypertropia the image falls on the superior retina in the right eye and is projected downward below the image seen in the straight-ahead position by the left eye (Fig. 5-91).

Because each extraocular muscle has certain primary and secondary actions that do not vary significantly among individuals, it is possible to calculate a "typical" diplopia field for each type of extraocular muscle paralysis. The "typical" diplopia fields currently available are based on the classic teachings of Duane. A set of theoretic diplopia fields based on the more recent concepts of extraocular muscle function as described by Boeder[3] and Jampel[12] is presented as follows (Fig. 5-92). Note that in the oblique paralyses the vertical and horizontal image separation varies but the torsional angle

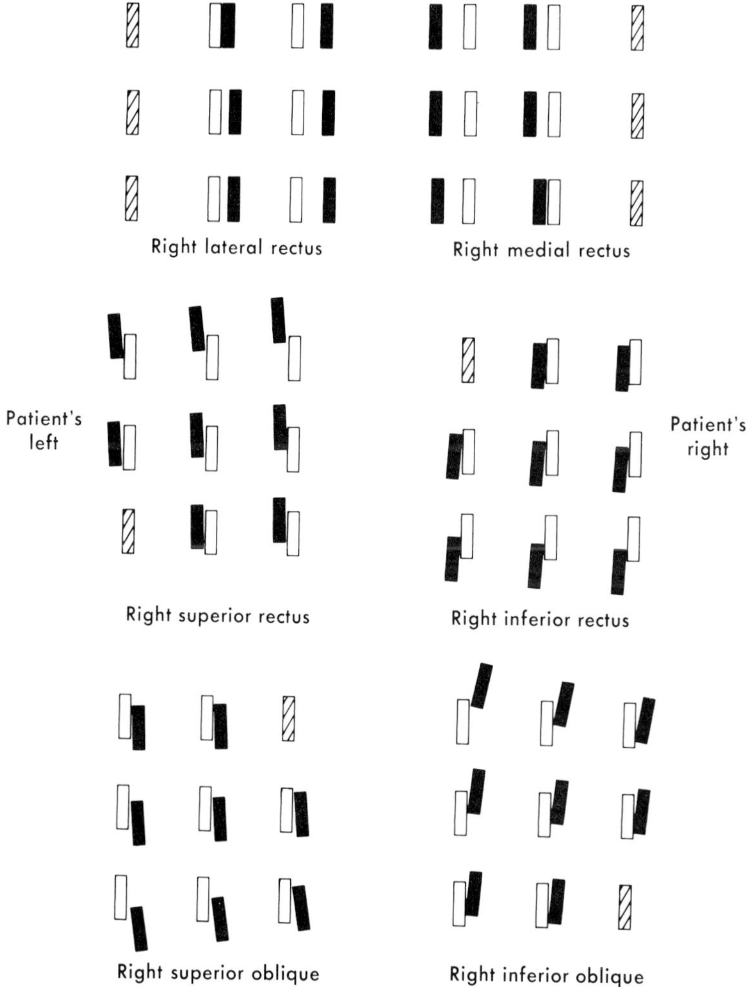

Fig. 5-92. Set of theoretic diplopia fields for each extraocular muscle paralysis in right eye. Fields for paralysis of left extraocular muscles would be reversed. Fields are presented as seen by patient. Solid bar, right eye; open bar, left eye; hatched bar, fused image.

of the image does not change across the field as in the classic charts. Practically, however, diplopia fields are usually more complex than the examples, for preexisting deviations and multiple muscle pareses are often present.

Practical points in plotting a diplopia field

1. Diplopia testing should always be preceded by a cross-cover test in all the positions of gaze where diplopia is to be evaluated. This gives the observer information on the type of diplopia that will be found and its angle of separation based on the size of the deviation.

2. Colored filters should be used to emphasize the images, dissociate the deviation, and differentiate between the images so that both the patient and the observer can relate to them. A red filter alone before one eye will suffice in most cases; the addition of a green one before the second eye is better in children, patients with a tendency to suppress (it is hard to differentiate between fusion or suppression with only one color), or in patients with a slight paresis and good fusion ability who can overcome their deviation. (They are *less* able to do this with the greater dissociation produced by the complementary colors, since there is no part of the spectrum common to both eyes.)

3. For reasons of standardization, it is preferable to place the red filter over the right eye; all questions may then be related to the position of the red image with regard to the white or green. A series of specific questions requiring simple answers prevents the patients from giving excessive or irrelevant answers. This enables the examiner to obtain just the necessary information. For example, the patient should first be asked whether one or two lights are seen. If the patient sees two lights, he is then asked what color they are and, after a satisfactory answer is given, on which side the red one is located and whether the red one is above or below the green one. The information thus received will be that the patient sees two lights (one red and one green), they are crossed or uncrossed, and the image relating to the right eye is higher or lower. If the patient claims single vision anywhere, verify that (1) both eyes can see the object and (2) the patient is not suppressing. This should be repeated in the nine positions of gaze.

A combination of a vertical and horizontal diplopia produces a compound, or *oblique*, diplopia. This is *not* torsion. Torsional diplopia cannot be detected by using a flashlight; a linear form of light such as the Armstrong Bar-lite[21] should be used for the diplopia test in cases in which the investigation of torsion is of primary interest. Quantitative methods of assessing torsion will be described further.

4. Having ascertained the type of diplopia and having noted whether it is consistent with the ocular deviation as seen on the preceding tests, one should then determine the area of *maximum vertical* (or horizontal) *separation*. This is done by asking the patient to compare, for example, the separation of images in right downgaze with those in right upgaze or right downgaze with left downgaze. When the position of greatest separation is found, the distal image is noted. This will be the upper of the two images in upgaze, the lower in downgaze, on the right in right gaze, and so on. This image relates to the defective muscle. Therefore if the maximum separation were on right downgaze in the direction of action of the right inferior rectus and left superior oblique and the distal (lower) image related to the left eye, this would suggest a left superior oblique paralysis.

5. It has been suggested that in obvious vertical paralyses the horizontal component can be ignored in performing diplopia fields, and in all probability this is true. The horizontal deviation may be of theoretic value as it reflects the secondary horizontal actions of the vertical muscles, which may be underacting or overacting, but clinically it is of lesser importance. The terms A and V pattern describe clinical conditions in which there is a relative convergence or divergence of the visual axes when the eyes move above or below the midline. There are four basic types: an exotropia that relatively diverges on elevation is called a V exotropia; an A exotropia is one in which the divergence increases on depression; an esotropia that relatively converges on depression is called a V esotropia; and A esotropia is one in which the convergence increases on elevation.

The presence of an A to V pattern in paralytic

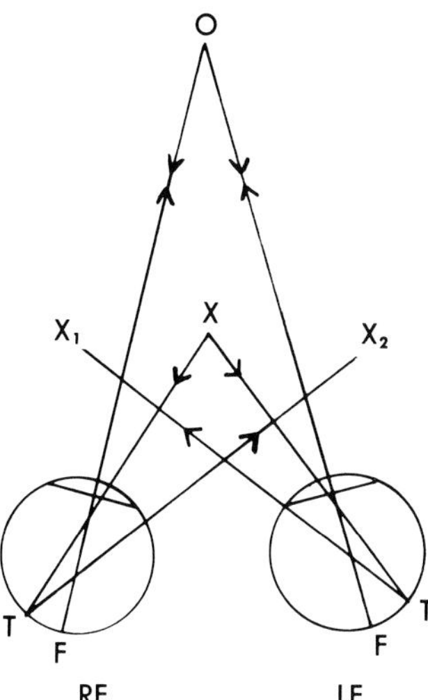

Fig. 5-93. *F*, Fovea; *O*, fixation object; *T*, temporal retinal point; *X*, object proximal to *O*; X_1 and X_2, projected images of object *X*. Heteronymous or crossed physiologic diplopia. Images of proximal object *X* fall on temporal retina and are projected nasally, resulting in diplopia at X_1 and X_2.

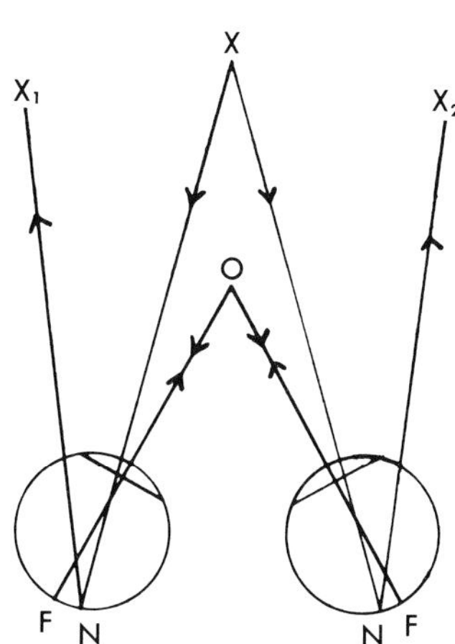

Fig. 5-94. *F*, Fovea; *O*, fixation object; *N*, nasal retinal point; *X*, object distal to *O*; X_1 and X_2, projections of object *X*. Homonymous or uncrossed physiologic diplopia. Images of distal object *X* fall on nasal retina and are projected temporally, resulting in diplopia at X_1 and X_2.

strabismus can help to identify the paretic muscle. A superior oblique paresis, even unilateral, tends to produce a relative convergence on downgaze and a superior rectus paresis a relative divergence on upgaze. Urist[31] uses this as the third step of his diagnostic test for identifying a paretic muscle.

6. While the patient is being tested subjectively, the area of *minimal separation* is also noted, particularly if there are any areas of single vision. These single-vision areas may be quantitated further by measuring a field of binocular single vision on a perimeter. If a reasonable area of single vision exists, a patient may be able to place this area straight ahead by the use of a compensatory head posture. Most patients find this for themselves, but some patients may need to be taught. In those patients whose deviations are not too incomitant, prisms can be utilized to facilitate fusion. Use of Fresnel prisms[13] is an excellent means of accomplishing this on a temporary basis because they can be changed or removed as necessary.

7. Patients with an ocular deviation occasionally complain of blurred vision instead of diplopia. This is eliminated when either eye is covered or when an appropriate prism is used to neutralize the deviation. This is binocular blurring and represents a type of unrecognized subjective diplopia. The diplopia test can be performed in the usual manner because the patient will give appropriate responses when the images are dissociated using red and green filters.

8. Patients occasionally complain of physiologic diplopia, a binocular condition in which two images are seen if an object lies proximal or distal to the object of regard. Images of a nearer object are crossed or heteronymous (Fig. 5-93); images of a more distant object are uncrossed or homonymous (Fig. 5-94). This condition is always horizontal and thus cannot be confused with a vertical muscle paralysis. It is also concomitant, not varying in different directions of gaze, but it may change with the fixation distance. The presence of pathologic diplopia may be easily ruled out by objective testing. These patients are usually somewhat introspective.

Atypical diplopia fields

Conditions other than paralysis of a muscle may cause limitation of ocular movement. For example, contracture of an inferior rectus muscle, as in restrictive thyroid ocular disease, will simulate paralysis of its antagonist muscle, the superior rectus. Similarly, if this muscle is restricted in its movement mechanically (e.g., by entrapment such as may occur in a fracture of the orbital floor), it may appear that the eye will not elevate. In the Brown superior oblique tendon sheath syndrome the eye will not elevate in adduction, clinically simulating an inferior oblique paresis. The limitation of elevation is caused by inability of the superior oblique tendon to move freely through the trochlea and not from any defect in inferior oblique action. This differentiation is made easily by forced (passive) duction testing with forceps. Therefore one cannot say from evaluation of a diplopia field alone that a specific muscle is paretic, but only that the diplopia field in this particular patient is the type of field one would expect in paralysis of a particular muscle.

A type of diplopia field commonly plotted in the paralysis of a vertically acting muscle is one in which an elevator in one eye and a depressor in the opposite eye appear to be paralyzed. This type of field usually demonstrates weakness of the superior oblique on one side and the contralateral superior rectus muscle. One should recognize this pattern as representing a primary paresis of one muscle with a secondary inhibitional palsy of its contralateral antagonist. Spread of concomitance leading to inhibitional palsy is less likely to occur (1) in a spontaneously recovering paralysis or (2) when the patient is maintaining binocular single vision by means of a compensatory head posture, thus keeping the eyes out of the area where the innervation is defective.

Diplopia fields due to paralysis of more than one muscle

The most common diplopia field depicting true paralysis of more than one ocular muscle is that caused by a third nerve palsy involving all the muscles except the lateral rectus and superior oblique. The next most common field is that resulting from paralysis of upward movement of one eye. In many cases this is due to actual paralysis of both the superior rectus and the inferior oblique muscles (double elevator palsy), but, again, differentiation should be made from a restrictive disorder involving the inferior musculature. Paralysis of downward movement in one eye may also occur, presumably because of involvement of the inferior rectus and superior oblique.

Although these diplopia fields strongly suggest a unilateral paralysis of both elevators or depressors in either upward or downward gaze, it cannot be assumed that these muscles are actually paralyzed. There does not exist an anatomic location where a single lesion might be placed to involve the nerves supplying solely these specific muscles.

A more complex type of diplopia field occurs when a double elevator palsy is associated with an inhibitional palsy of the contralateral depressors. The diplopia field looks as though four muscles were paralyzed.

There are certain types of acquired paralyses that can cause a concomitant deviation or fairly equal separation of diplopic images. These include skew deviation, convergence and divergence paralyses that are supranuclear in origin, bilateral paralyses such as both superior recti or both lateral recti, or decompensation of a longstanding heterophoria, either vertical or horizontal. These conditions should be carefully differentiated from concomitant nonparalytic strabismus, since their diagnostic significance and clinical management vary considerably.

Diplopia therefore varies in palsies of different muscles and is more or less annoying, depending on the muscle affected. In ordinary life the eyes are used in a limited field of binocular fixation. This seldom extends more than 20 degrees above the horizon, about 30 degrees to either side of the primary position, and downward about 60 degrees below the horizon. When the object of regard lies outside these limits the head is turned also, rather than the eyes alone. If the paralyzed muscle normally functions in an important area of the field of fixation (for example, the superior oblique, which is used especially in close work), the ensuing diplopia will be particularly distressing. A paralyzed inferior oblique or superior rectus, on the other hand, will not cause as much discomfort since their fields of action are well above the horizon and away from the important field of fixation.

QUANTIFICATION OF EXTRAOCULAR MUSCLE LIMITATION

Isolation of a specific muscle paralysis may be made with qualitative diagnostic methods such as diplopia fields. To document the condition for accurate follow-up and surgical planning in the management of the patient's ocular deviation, quantitative measurements are necessary.

Measurement of the deviation

The deviation should be measured fixing with each eye in turn in each of up to nine relevant gaze positions. Loose prisms, a prism bar, or a rotary prism may be used. Accuracy and repeatability in prism measurements may be facilitated by using a deviometer. This is a device providing a near fixation target at a fixed distance from the eyes in six or more different positions of gaze (Fig. 5-82). In 1903, Worth[33] designed a deviometer that combined the principles of the Maddox tangent scale and the Priestley Smith tape. The instrument incorporated a linear electric light bulb, but relied on the assessment of the corneal reflections by the examiner. It was not until later that quantitative measurements using prisms with the deviometer became standard.[17,26,32] The results may be recorded numerically or plotted graphically to resemble a Hess chart, as described by Lavat.[17] Alternatively, a haploscopic device such as the synoptophore may be used for repeatable measurements in any gaze position.

Measurement of torsion

The presence of torsion may be ascertained by the subjective description of slanting of the diplopic images. This is not usually volunteered by the patient and may not be appreciated by the patient on questioning. Although a cyclodeviation may be detected on the cover test, it is often too subtle to be noted. Therefore the presence of a torsional component should always be expected in all cases of oblique muscle dysfunction. The presence of this component can be confirmed by using an Armstrong Bar-lite or by the use of the double Maddox prism test. Torsion may also be noted during the plotting of each point on a Hess or Lees screen, but its quantitative value is doubtful. Quantitative methods of measuring torsion include the double Maddox rod test, in which the patient observes a flashlight through two Maddox rods of different colors placed in a trial frame. Burian's cyclophorometer incorporates this apparatus onto a fixed stand.[32] The rods produce two lines of light that can be adjusted until they appear parallel to the patient. The degree of rotation is then read by the examiner from the scale on the trial frame. A similar test utilizes two Bagolini striated glasses and a red filter. The synoptophore provides the possibility of measuring torsion in both eyes simultaneously while fixating in any gaze position.

Measurement of field of binocular single vision

One of the earliest descriptions of measurements of the field of binocular single vision was by Landolt[16] in 1881, in which an arc perimeter and candle were used. However, any perimetric instrument used routinely for the testing of monocular visual fields may be used to measure a field of binocular single vision.[5] An area of single vision found on diplopia field testing indicates that a quantitative assessment on the perimeter may be performed. The patient performs the test with both eyes open, fixating on a central target. The target is then moved out radially from the center to the point where diplopia is appreciated.

The target size used on the Goldmann perimeter is the III_4e. It is important to begin within an area the patient can see singly so that the fusion range is utilized. If diplopia is appreciated in primary gaze, any other position where single vision is present may be used as a starting point.

Measurement of fields of muscle action (Hess, Lancaster, and Lees screen tests)

Methods of documenting the action of ocular muscles in different gaze fields include an early description by Hirschberg.[11] A coordinate tangent scale was marked on a wall of the examining room. Measurements did not rely on corneal reflections, but on the separation of subjective diplopic images. The position of the images could be described by the patient and joined

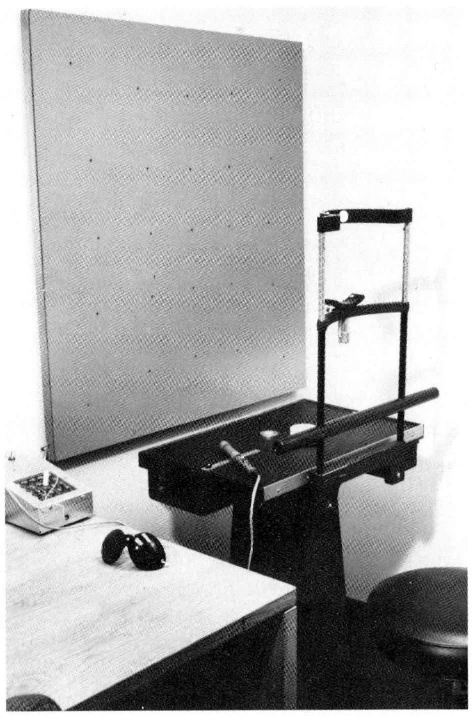

Fig. 5-95. Electric Hess screen. Patient sits facing the screen. Chin rest and head bar place the patient 0.5 m from the screen. Electric console on the left illuminates the red dots on the screen. The patient wears the Armstrong red/green goggles (foreground) and holds the green flashlight (lying on black table).

with an appropriate prism. This technique, however, did not permit full dissociation of the deviation.

Ohm[25] designed a black cloth screen with blue strings spaced to form the coordinate lines, based on Hirschberg's tangent scale. He used red/blue goggles and a red arrow for color dissociation. Krusius[14] designed a glass screen through which the examiner could observe the eye movements or corneal reflections of the patient. Ohm subsequently constructed his screen from wire mesh so that the patient could be observed through the screen. Hess designed his first screen from black cloth with red embroidered dots and green string attached to a pointer. The eyes were dissociated by red/green complementary colored filters.[9,10] Sattler[29] used a black screen and red/green color dissociation also, but with a coordinate system of green lines. Further details of these instruments, including others that are now obsolete, are described elsewhere.[30]

Several modifications were subsequently made to the Hess screen. In general, the test uses a gray or black screen marked with a tangent scale on which red targets are projected or positioned where the tangent lines cross. A green target or light is superimposed subjectively on the red fixation target. Complementary

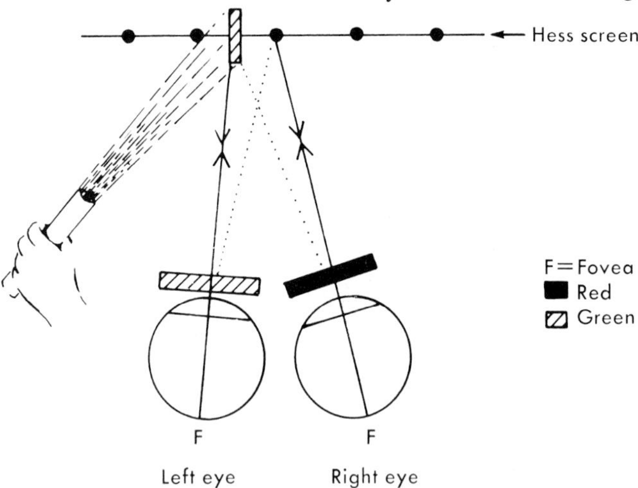

Fig. 5-96. Patient fixates the red dot with his right eye through the red filter (solid). He holds the flashlight, which projects a green line onto the screen. Left eye sees the green line through the green filter (hatched). Dotted lines indicate that the red dot cannot be seen through the complementary green filter and vice versa. The patient guides the flashlight into the position where the green line is superimposed subjectively on the red dot. This is maculomacular projection. However, due to an exodeviation brought out by color dissociation, the green line is seen displaced to the left of the red dot by the examiner.

red and green filters are worn to permit binocular dissociation, thus revealing the ocular deviation in each fixation position. The modern screens are gray with red fixation lights that can be illuminated consecutively. A modified flashlight (the Foster torch) is used, which produces a green linear beam of light. This green line is projected onto the red dot on the screen by the patient from a distance of 0.5 m. The patient wears Armstrong diplopia goggles (which consist of a red filter over one eye and a complementary green one over the other) to dissociate the deviation (Fig. 5-95).

The Lancaster red/green test,[15] also in popular use, incorporates the same principles as the Hess screen but uses a two-dimensional grid rather than a tangent screen and is performed at a distance of 1 or 2 m. Anderson[1] modified the Lancaster screen for use in a projector and describes the use of both red/green color dissociation and polarized glasses.

The Lees screen produces a chart identical to that obtained by plotting points on a Hess screen, but it does not use red/green color dissociation. Instead, the test uses two tangent screens at right angles and a doublesided mirror bisecting the screens to dissociate binocular fixation.[18,19]

Principle of Hess screen test

The Hess screen test utilizes red/green color dissociation as in the diplopia test, but differs

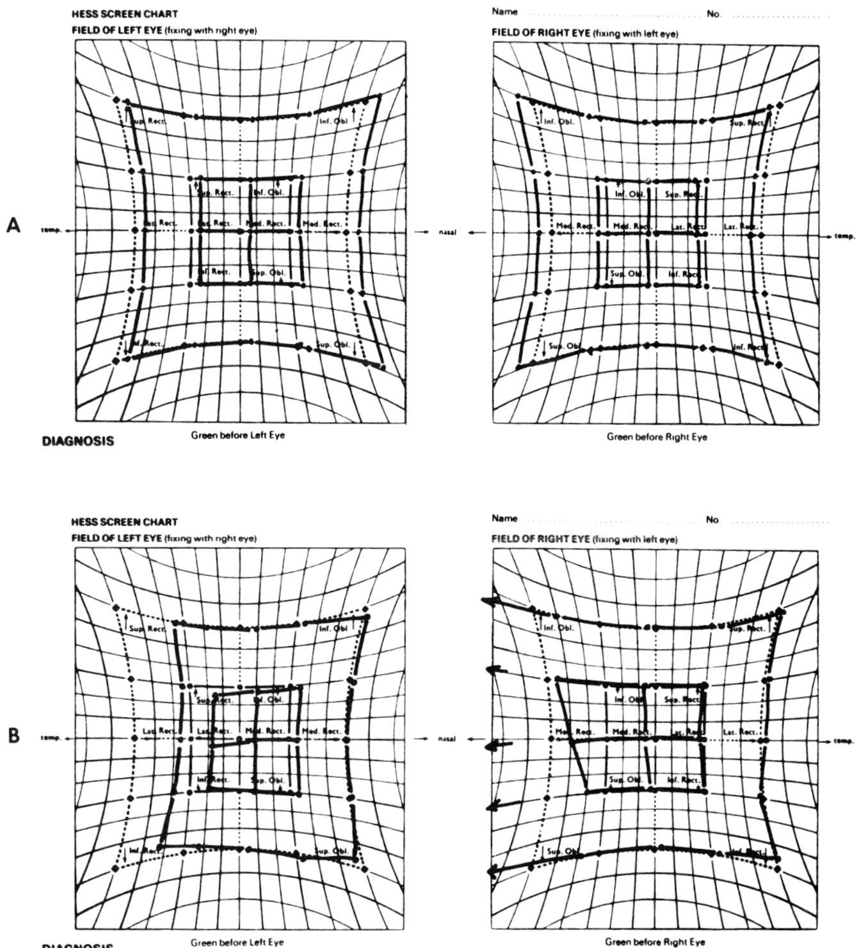

Fig. 5-97. A, Hess chart showing a concomitant esodeviation. **B,** Hess chart showing an incomitant esodeviation caused by a left sixth nerve paralysis.

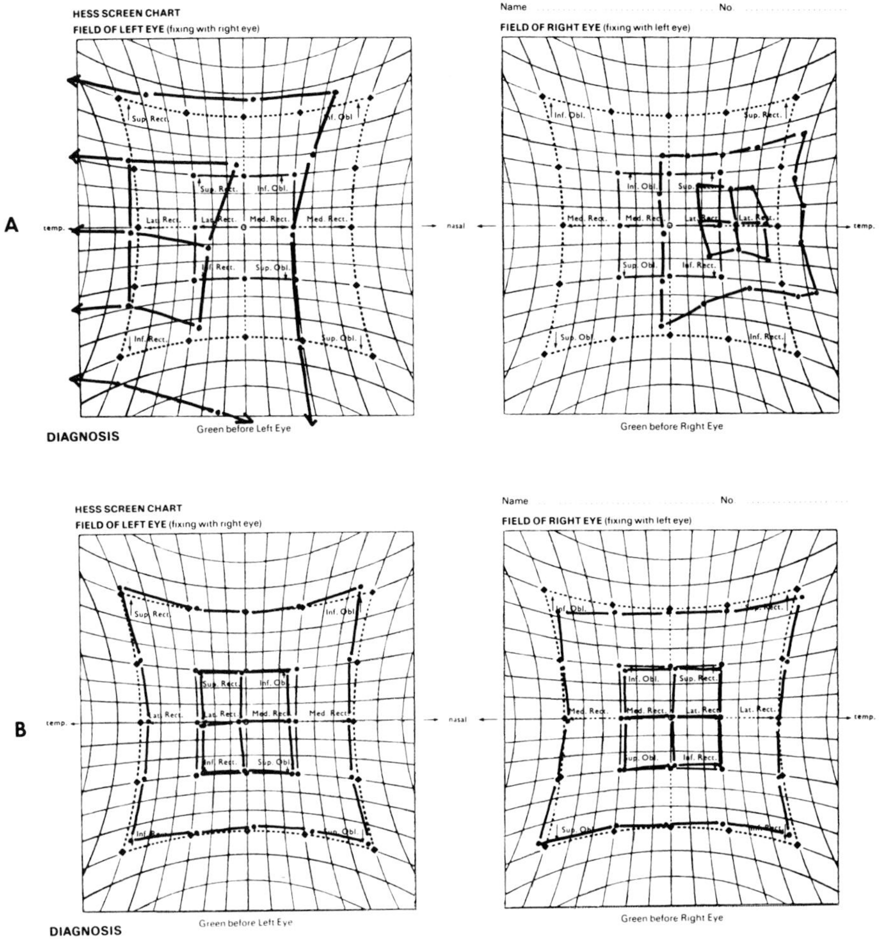

Fig. 5-98. A, Hess chart showing an incomitant exodeviation caused by a total right third nerve paralysis in a patient with diabetes mellitus. **B,** Hess chart of the same patient 8 weeks later after a spontaneous recovery, showing a concomitant exodeviation. Some vertical limitation is still present.

in principle. In the diplopia test one white fixation light is used, which can be seen by each eye through the red and green filters. In the Hess or Lancaster screen test, instead of a white light, two different color test objects are used. These colors are red and green and can only be perceived by the fovea of the eye behind the filter of the corresponding color. Each fovea perceives the image from its respective target as being located straight ahead. This is maculomacular projection (confusion). The binocular dissociation produced in this manner is sufficient to reveal even the most well-controlled heterophoria. The patient, wearing the red filter over the right eye, is asked to place the green line through the red dot as each one is illuminated consecutively. The nonfixing left eye is deviated behind the green filter, but the patient merely guides the flashlight so that the green line is positioned where the deviated fovea is pointing. At this point the patient sees the red and green lights superimposed; the examiner, however, can see the deviation of the green line from the red dot and plots the position of the green line onto a Hess chart (Fig. 5-96). The test is repeated with the Armstrong goggles reversed, so the left eye now fixates the red dot and the field of the deviating right eye may be plotted.

In the presence of a concomitant deviation

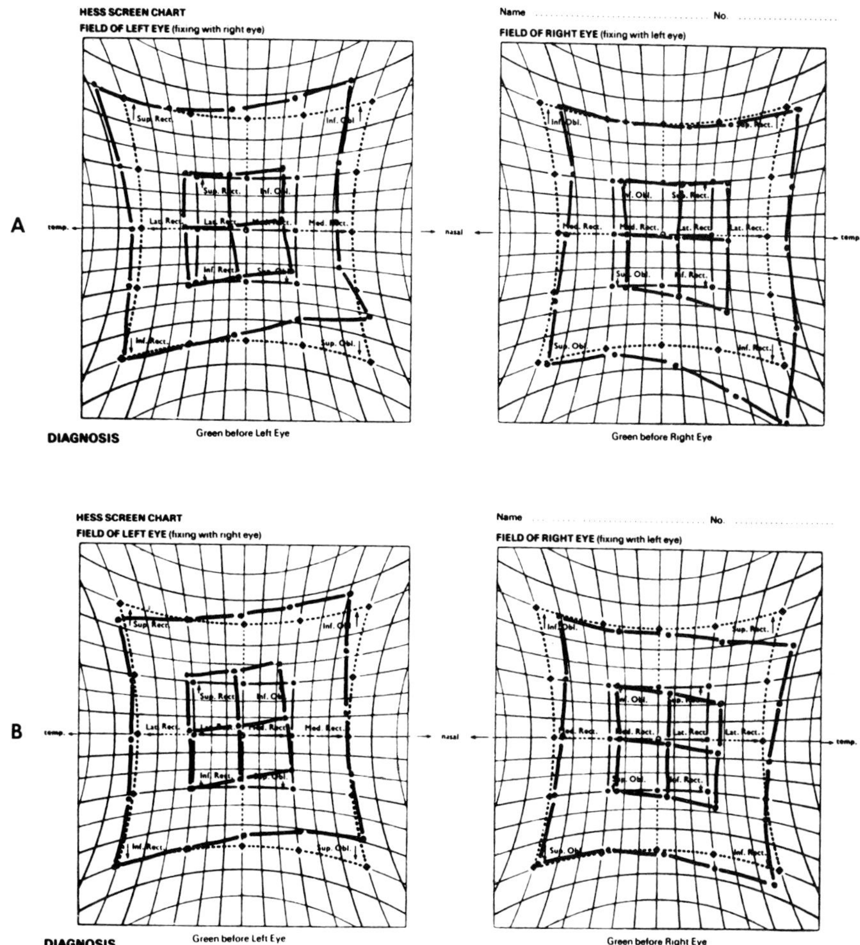

Fig. 5-99. A, Hess chart of an incomitant left fourth nerve palsy. It demonstrates left superior oblique underaction, with overaction of its contralateral synergist, the right inferior rectus, and of its ipsilateral antagonist, the left inferior oblique muscle. Note that the right superior rectus action is almost full. **B,** Hess chart showing concomitant fields, demonstrating both a left superior oblique underaction and a right superior rectus underaction of a similar degree. One of these underactions represents a primary paresis. The other, its contralateral antagonist, is secondarily inhibited. This represents a spread of concomitance.

(for example, an esophoria) the deviation seen as a separation of the red dot and green line by the examiner will be the same in all gaze positions; it will also measure the same amount in the opposite field when the other eye is fixating.

After each field has been plotted, the points are connected to form an inner and outer field. In concomitant deviations the fields will be of equal size and symmetric with the field obtained with the opposite eye fixing (Fig. 5-97, *A*).

In an incomitant deviation a difference will be noted in the relative sizes of the fields for each eye as well as differences in measurements in different gaze directions. This is basically a way of looking at a pictorial representation of a primary and secondary deviation (Figs. 5-97, *B*, 5-98, *A*, and 5-99, *A*).

When the normally innervated eye is fixating, the plotted field of the eye with the paralysis will be smaller. It can be thought of as the passive or primary deviation. The limitation of movement will be greatest in the direction of main action of the paralyzed muscle. When the involved eye is fixating, particularly when it is attempting to look into its field of main action, extra innervation causes the synergic muscles in the deviating eye to overact, producing a secondary deviation, or the larger field seen on the Hess chart (Figs. 5-97, *B*, 5-98, *A*, and 5-99, *A*).

Interpretation and utilization

1. The Hess chart should be examined to determine whether the fields are symmetric. If both fields are similar in shape and size, a concomitant deviation is present. The direction of deviation may be read directly from the chart. In an esodeviation both fields will be displaced inward (Fig. 5-97, *A*), exodeviation, outward, etc. The amount of the deviation may also be read from the chart because each small square represents 10 prism diopters. However, it should be remembered that the test is performed at a distance of 0.5 m, and measurements cannot be compared accurately with any other test except another Hess chart.

2. If the fields are not symmetric, an incomitant deviation is present. The smaller field will identify the eye containing the paretic or restricted muscle. The biggest limitation in that field represents the field of action of the paretic muscle or of the ipsilateral antagonist to a restricted muscle.

3. The limited or underacting muscle should show a corresponding overaction of its contralateral synergist. This may be identified in the opposite, larger field. If several muscles are paralyzed, as in an oculomotor nerve palsy, corresponding overactions may be seen in the contralateral field (Fig. 5-98, *A*).

4. A Hess chart can sometimes be used to differentiate a restrictive or mechanical palsy from an innervational palsy. For example, there are ''typical'' Hess charts for a blowout fracture, Duane's or Brown's syndrome, or a thyroid restrictive disorder. However, its diagnostic value here should only be to indicate that forced ductions or other investigative tests be done and to corroborate other clinical findings.

5. A Hess chart is particularly useful over a period of time in the management of patients with extraocular muscle limitations. A series of Hess charts provides an excellent quantitative and pictorial record from which to prognosticate recovery, plan surgery, etc. (Figs. 5-98, *A* and *B*).

6. If a paralysis does not recover spontaneously, a spread of concomitance may occur, resulting in a secondary inhibitional palsy of the contralateral antagonist to the originally paretic muscle (Fig. 5-99, *B*).

SUMMARY

Diplopia fields alone cannot be used to isolate a specific muscle paralysis. Other tests such as prism or synoptophore measurements in the relevant positions of gaze, fixing with each eye in turn, the plotting of a Hess chart, and a field of binocular single vision[28] should be completed. These results, considered with the ocular movements (both ductions and versions), the cover test, estimates of torsion, and the head-tilt test, should be sufficient to diagnose the paretic muscle and also form an excellent record of the recovery or progression of the paralysis on separate visits. Special tests such as forced ductions, saccadic velocities, or electromyography should be performed when necessary to obtain an accurate diagnosis.

ACKNOWLEDGEMENT

We would like to thank Mrs. Kathleen Vogt for her advice and assistance in the preparation of this chapter.

Ronald M. Burde
Gill Roper-Hall

REFERENCES

1. Anderson, J. R.: Ocular vertical deviations, Br. J. Ophthalmol. **12**(suppl.):80, 1947.
2. Bielschowsky, A.: Lectures on motor anomalies, Hanover, N.H., 1943, Dartmouth College Publications.
3. Boeder, P.: The co-operation of extraocular muscles, Am. J. Ophthalmol. **51**:469, 1961.
4. Chavasse, B.: In Lyle, T. K., editor: Worth and Chavasse's squint, ed. 8, Philadelphia, 1950, The Blakiston Co.
5. Feibel, R. M., and Roper-Hall, G.: Evaluation of the field of binocular single vision in incomitant strabismus, Am. J. Ophthalmol. **78**:800, 1974.
6. Haagedorn, A.: A new diagnostic motility scheme, Am. J. Ophthalmol. **25**:726, 1942.
7. Hardesty, H. H.: Diagnosis of paretic vertical rotators, Am. J. Ophthalmol. **56**:811, 1963.
8. Helveston, E. M.: A two-step test for diagnosing paresis of a single vertically acting extraocular muscle, Am. J. Ophthalmol. **64**:914, 1967.

9. Hess, W. R.: Eine neue Untersuchungsmethode bei Doppelbildern, Arch. Augenheilkd. **62:**233, 1909.
10. Hess, W. R.: Ein einfaches messendes Verfahren zur Motilitätsprüfung der Augen, Z. Augenheilkd. **35:**201, 1916.
11. Hirschberg, J.: Über Blickfeld messung, Arch. Augen Ohrenheilkd. **4:**273, 1874.
12. Jampel, R. S.: The fundamental principle of the action of the oblique ocular muscles, Am. J. Ophthalmol. **69:**623, 1970.
13. Jampolsky, A., Flom, M., and Thorson, J. C.: Membrane Fresnel prisms: a new therapeutic device. In Fells, P., editor: The First Congress of the International Strabismological Association, St. Louis, 1971, The C. V. Mosby Co., p. 183.
14. Krusius, F.: Über ein Scheiben-Deviometer, Arch. Augenheilkd. **59:**26, 1908.
15. Lancaster, W. B.: Detecting, measuring, plotting and interpreting ocular deviations, Arch. Ophthalmol. **22:**867, 1939.
16. Landolt, E.: Etude sur les mouvements des yeux. A l'état normal et a l'état pathologique, Arch. Ophthalmol. **1:**586, 1881.
17. Lavat, J.: Détails techniques dans le traitement orthoptique et chirurgical du strabisme, Bull. Mem. Soc. Fr. Ophtalmol. **70:**25, 1957.
18. Lees, V. T.: The Hess screen with mirror dissociation, Br. Orthopt. J. **6:**50, 1949.
19. Lees, V. T.: A new method of applying the screen test for interocular muscle balance, Br. J. Ophthalmol. **33:**54, 1949.
20. Linwong, M., and Herman, S. J.: Cycloduction of the eyes with head tilt, Arch. Ophthalmol. **85:**570, 1971.
21. Lyle, T. K., and Wybar, K. C.: Lyle and Jackson's practical orthoptics in the treatment of squint, ed. 5, London, 1967, H. K. Lewis & Co. Ltd., p. 473.
22. Marshall, E. B.: The purpose and significance of the cover test. In: Transactions of the First International Congress of Orthoptics, 1967, London, 1968, Henry Kimpton, Publishers.
23. Maurer, Y. M.: The cover test—a guide to its diagnostic and prognostic potential, Br. Orthopt. J. **26:**91, 1969.
24. Moore, S.: The head tilt test in comitant strabismus, Am. Orthopt. J. **17:**117, 1967.
25. Ohm, J.: Neue Instrumente, Medikamente usw, ein Apparat zur Untersuchung des Doppeltsehens, Centralbl. Prakt. Augenheilkd. **31:**201, 1907.
26. Owens, W. C.: A fixation light for the cardinal directions of gaze, Am. J. Ophthalmol. **30:**611, 1947.
27. Parks, M.: Isolated cyclovertical muscle palsy, Arch. Ophthalmol. **60:**1027, 1958.
28. Roper-Hall, G., and Feibel, R. M.: Measurement of the field of binocular single vision in the evaluation of incomitant paralytic strabismus, Am. Orthopt. J. **24:** 77, 1974.
29. Sattler, C. H.: Über die genaue Messung and Darstellung von Bewegungsstörungen der Augen, part 1, Klin. Monatsbl. Augenheilkd. **78:**161, 1927.
30. Sloane, A. E.: Analysis of methods for measuring diplopia fields, Arch. Ophthalmol. **46:**277, 1951.
31. Urist, M. J.: Head tilt in vertical muscle paresis, Am. J. Ophthalmol. **69:**440, 1970.
32. von Noorden, G. K.: Burian-von Noorden's binocular vision and ocular motility: theory and management of strabismus, ed. 2, St. Louis, 1980, The C. V. Mosby Co., pp. 189, 334-335.
33. Worth, C. A.: Squint: its causes, pathology and treatment, London, 1903, John Bale, Sons & Danielsson Ltd., p. 91.

GENERAL REFERENCES

Adler, F. H.: Physiologic factors in differential diagnosis of paralysis of superior rectus and superior oblique muscles, Arch. Ophthalmol. **36:**661, 1946.

Adler, F. H.: Some confusing factors in the diagnosis of paralysis of the vertically acting muscles, Am. J. Ophthalmol. **31:**387, 1948.

Crone, R. A.: Diplopia, Amsterdam, 1973, Excerpta Medica BV.

Duke-Elder, S.: Text-book of ophthalmology. The neurology of vision, motor and optical anomalies, St. Louis, 1949, The C. V. Mosby Co., vol. 4, p. 4051.

Hardesty, H. H.: Diagnosis and surgical treatment of paretic vertical muscles, Arch. Ophthalmol. **77:**147, 1967.

von Noorden, G. K.: von Noorden-Maumenee's atlas of strabismus, ed. 3, St. Louis, 1977, The C. V. Mosby Co., p. 136.

Chapter 6

Ocular circulation

ANDERS BILL, M.D.

BLOOD CIRCULATION

In humans and in many experimental animals, for example, rats, rabbits, cats, and monkeys, the eye has two separate systems of blood vessels, which anatomically and physiologically are very different: the retinal vessels, which supply part of the retina, and the uveal, or ciliary, blood vessels, which supply the rest of the eye.

Ocular arteries and veins

In humans both the retinal and uveal arteries branch from one ophthalmic artery, which is a branch of the internal carotid artery.[64,66,76] In lower animals, for example, rabbits, the eye is supplied by branches from both the internal and the external carotid arteries.

Fig. 6-1 shows schematically the blood vessels of the human eye. The central retinal artery, which is usually a direct branch from the ophthalmic artery, pierces the dural and arachnoid sheaths about 10 mm behind the eyeball and then runs in the subarachoid space for a few millimeters before it enters the optic nerve. The artery usually divides into two branches on the optic disc, and these divide irregularly into arterioles located in the innermost layers of the retina. Branches divide into capillaries distributed within the inner parts of the retina, including the inner nuclear layer (Fig. 6-2). The diameter of the retinal capillaries is about 5 to 6 μm.[76]

Michaelson's studies with ink infusion have contributed much to the knowledge of the arrangement of the retinal vessels.[80] They have shown that the arterioles are surrounded by a capillary-free zone with a maximum breadth of about 120 μm and that the outer 130 μm of the retina has no blood vessels at all (Fig. 6-3). In the central part of the fovea the retina is so thin that the avascular layer includes the whole thickness of the retina (Fig. 6-4). The small arterioles and veins interdigitate in a characteristic way.

Toussaint, Kuwabara, and Cogan[100] have visualized the three-dimensional arrangement of the retinal vessels by digesting the retinal tissue with trypsin and then embedding the blood vessels in plastic. This technique revealed a laminated arrangement of the retinal capillaries around the optic disc and in the peripheral part of the retina. Studies of the development of the retinal vessels have shown that capillaries originally present around the arterioles disappear during maturation, a phenomenon that is probably caused by a locally high oxygen tension.[12,68]

The arteries of the retina are end-arteries; that is, they have practically no anastomoses to each other except at the capillary level. This has been very well demonstrated in pigs. After occlusion of an arteriole with glass microspheres, blood passes via capillary anastomoses from the intact area into the distal part of the embolized vessel. Blood from the capillary anastomoses flows counter to the normal direction of flow in the occluded arteriole for a short distance and out into smaller arterioles and capillaries.[38] The collateral flow is usually not sufficient to prevent retinal damage.

A branch from the ciliary circulation, a cilio-

□ This work was supported in part by Research Grant PHS EY 00475 from the National Eye Institute, National Institutes of Health, and Grant B 79-14X-147 from the Swedish Medical Research Council.

retinal artery, sometimes contributes to the nutrition of the inner layers of the retina.

The blood is drained from the retinal capillaries via retinal venules and veins into the central retinal vein and from there into the cavernous sinus directly or via the orbital veins.

The uvea is supplied with blood from two long posterior ciliary arteries, several arterior ciliary arteries, and a variable number of short posterior ciliary arteries. In the choroid the short posterior ciliary arteries end as a large and very tight capillary net, the choriocapillaris. The transition from arteriole to capillaries is unusually abrupt (Fig. 6-5). In the ciliary body

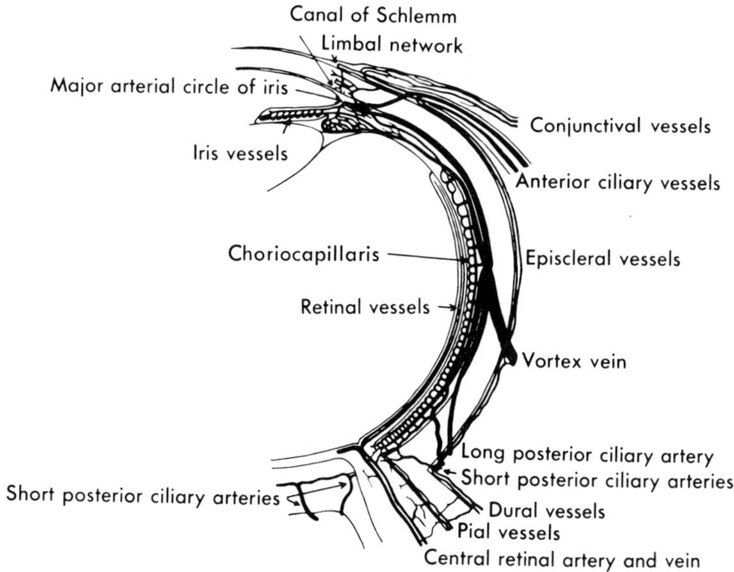

Fig. 6-1. Blood vessels of human eye. (Modified from Leber.)

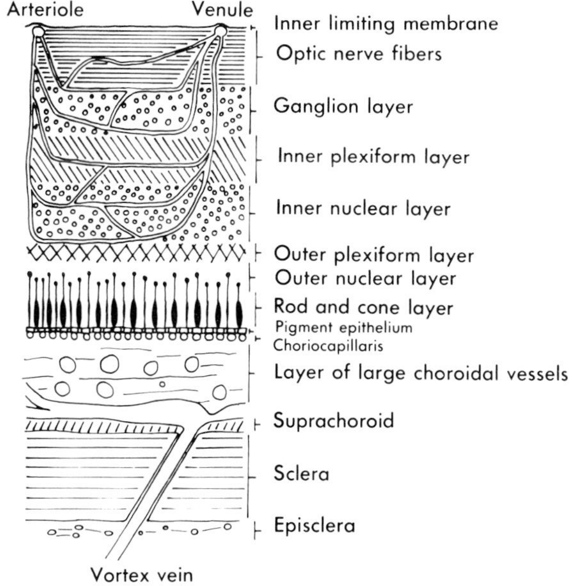

Fig. 6-2. Retinal capillaries are distributed within inner layers of retina. Outer layer, 130 μm thick, has no blood vessels. It is nourished mainly from choroidal capillaries.

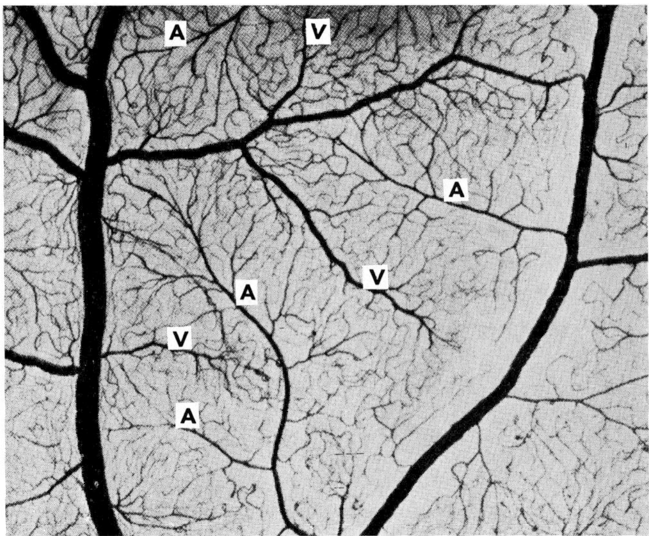

Fig. 6-3. Field from equatorial zone of retina. Left, vein; right, artery. Note interdigitation of venae efferentes, *V*, with arteriae afferentes, *A*. Note also capillary-free zone around artery. (From Michaelson, J., and Campbell, A. C. P.: Trans. Ophthalmol. Soc. U.K. **60:**71, 1940.)

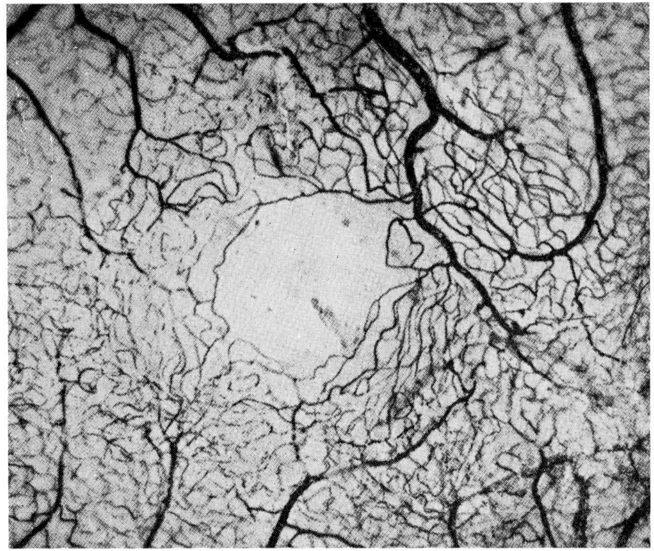

Fig. 6-4. Fovea with its capillary-free area. (From Michaelson, I., and Campbell, A. C. P.: Trans. Ophthalmol. Soc. U.K. **60:**71, 1940.)

at the base of the iris the anterior ciliary arteries and the long posterior arteries form the major arterial circle of the iris, from which the iris arteries and the arteries of the ciliary processes originate. Recurrent arteries from the circle supply the anterior part of the choriocapillaris. Occlusion of one of the ciliary arteries usually does not give dramatic effects because of the arterial ring and the manifold arterial supply to the choriocapillaris.[76]

In the sclera surrounding the optic nerve the short posterior ciliary arteries give rise to the

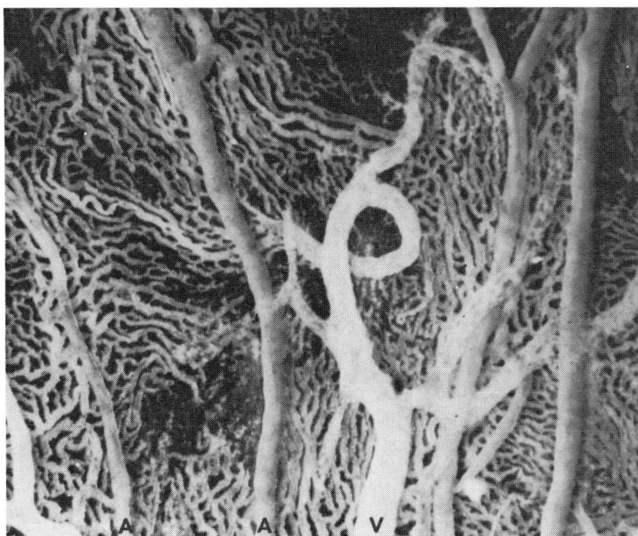

Fig. 6-5. In choroid, transition from arteriole to capillaries is unusually abrupt. *A*, Arteriole; *V*, vein. (Photograph by N. Ashton.)

intrascleral circle of Zinn. This circle is often incomplete. Branches from the circle and intrascleral arterioles contribute to the blood supply of the optic nerve head. There are contributions also from retinal and choroidal arterioles.[67]

Small anastomoses between the vessels of the optic nerve head of uveal origin and the central retinal artery connect the retinal and uveal circulations. Unfortunately the capacity of these anastomoses is too small to maintain significant retinal circulation after occlusion of the central retinal artery. In the nerve head there are venous anastomoses between the retinal and uveal circulations.

The capillaries of the uvea are unusually wide. In the choriocapillaris the diameters are often 20 to 30 μm. It is surprising then that microspheres with diameters of only 8 to 10 μm are retained in the choriocapillaris almost as efficiently as 15 and 25 μm spheres.[96] One explanation for this phenomenon might be that the capillaries are flat. Serial sections of the choroid indicate, however, that the spheres are caught in parts of the capillaries that are narrow in all directions.[28] In the anterior uvea the capillaries are so wide that more than 50% of 8 to 10 μm spheres can pass through the vascular bed.[96]

The blood from the choroid and part of the blood from the anterior uvea is drained via the vortex veins into the orbital veins. Small posterior ciliary veins drain the posterior sclera but may also contribute to the drainage of the choroid.

Small anterior ciliary veins that drain part of the blood from the anterior uvea pierce the sclera near the limbus, enter the eye muscles, and run within them accompanied by the anterior ciliary arteries (Figs. 6-1 and 6-6). Within the anterior sclera the veins anastomose with an intrascleral venous plexus, which collects blood from the scleral vessels proper; there are anastomoses with an episcleral venous plexus. The intrascleral and episcleral venous plexuses communicate with subconjunctival and conjunctival veins. The venous plexuses in the sclera, episclera, and conjunctiva receive aqueous humor drained by collector channels from the canal of Schlemm. Poor mixing of aqueous humor with the blood may result in visible laminated or aqueous-filled veins in episclera and conjunctiva. These are the aqueous veins of Ascher.[11]

Sometimes aqueous veins and small arteries share a common intrascleral pathway, but there are no direct communications between the arteries and the canal of Schlemm.

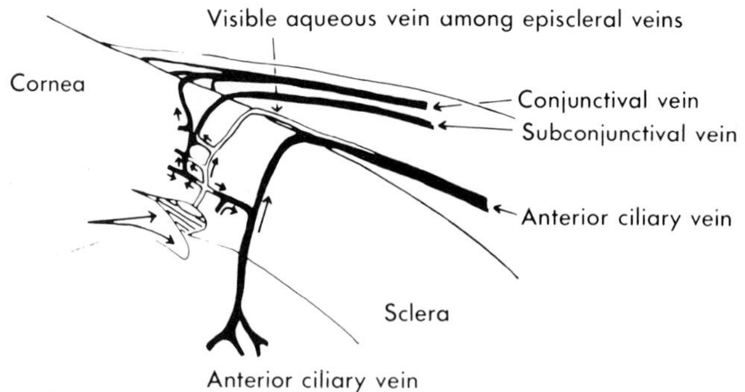

Fig. 6-6. Intrascleral venous plexus and its communications with episcleral and conjunctival plexuses. Part of aqueous humor drained via canal of Schlemm is mixed with blood within sclera; part leaves sclera unmixed to be drained into extraocular venous plexuses.

The arrangement of the ocular blood vessels in lower experimental animals is rather different from that in humans. In rabbits, for example, there are no anterior ciliary arteries,[88] and the retinal blood vessels are distributed only within two small wing-shaped areas containing myelinated nerve fibers. These vessels may therefore be regarded as an intraocular extension of the blood vessels of the optic nerve.[87]

The dependence of the retina on both the retinal and the choroidal vascular systems is shown very clearly by the lesions that sometimes occur spontaneously or that may be induced experimentally. It is well known clinically that occlusion of the central retinal artery gives loss of vision within seconds. In pigs the ganglion cells start to swell and become cloudy within 4 minutes after occlusion of the central retinal artery, but the changes are reversible until the death of the cells, which comes only after several hours.[38] It has been possible to experimentally produce massive occlusion of the choroidal capillaries, which produces necroses of the outer layers of the retina in both cats[36] and dogs.[60]

If the intraocular pressure is suddenly raised to a value that stops all blood flow through the eye, visual blackout comes within a few seconds,[10] but the loss of vision is reversible for a surprisingly long time. Thus, at least in cats[32] and monkeys,[58] vision is not permanently lost even after 1 hour of interrupted blood flow through the eye.

Pressure in the arteries

The arteries in the eye are small and have a pressure lower than that of the large arteries of the body. If the blood flow through the eye is reduced by raising the intraocular pressure, the pressure in the arteries rises. This makes it difficult to measure the pressure that prevails under normal conditions. Thus in ophthalmodynamometric measurements, when the intraocular pressure is raised to the level that stops diastolic flow and then to that which stops systolic flow, the two pressures measured are not representative for any normal pressures.[106] The pressure at the point of origin of the ciliary arteries in rabbits has been estimated to rise about 14 mm Hg when the blood flow through the eye is stopped.[24] It seems likely that the pressure rise produced in humans by ophthalmodynamometry is somewhat less since the ophthalmic artery in humans is a wider vessel than each of the two ophthalmic arteries in rabbits. At a mean arterial pressure of 100 mm Hg in the brachial artery, the mean ophthalmic pressure determined with ophthalmodynamometry is about 77 mm Hg.[106] The normal pressure prevailing at the origin of the retinal and ciliary arteries can thus be expected to be about 65 to 70 mm Hg. This is the situation in the sitting position. Lying down causes little change in mean arterial blood pressure at the level of the heart. At the level of the eye, however, there will be a rise in local mean arterial pressure corresponding roughly to the change in height of the eyes over the heart.

In normal eyes expansile pulsations of the retinal arteries can be seen only at a high magnification; in states with a large pulse pressure (systolic pressure minus diastolic pressure) or very high intraocular pressure, however, pulsation can be seen with an ordinary ophthalmoscope. At sites where the arteries are bent, the arterial pulse may produce small movements of the whole artery. Such a locomotion pulse is seen in healthy young children and in persons with a high pulse pressure.[51]

The blood volume within the eye varies with the pulse, and as a consequence the intraocular pressure also varies. The variation in intraocular pressure is about 1 to 2 mm Hg, which corresponds to volume changes of 1 to 4 μl.

Pressure in the veins

The pressure in the intraocular veins at the point where they enter the sclera is practically equal to the intraocular pressure under normal conditions. This has been determined in cats by direct measurements of the pressure in large choroidal veins.[23] In the retina the pulsations of the vein at its exit, which can be seen ophthalmoscopically, indicate a pressure just above the intraocular pressure. In normal eyes without venous pulsations, slight pressure on the globe produces the phenomenon.[41] The intrascleral part of the veins draining the uvea has a pressure that normally is lower than the intraocular pressure. In both cats and rabbits direct measurements have given values of about 8 mm Hg.[22,27]

The transmural pressure difference, that is, the pressure inside the vessel minus the pressure outside the vessel, in the intraocular veins and the venous part of the capillaries can be presumed to be very small, near zero or at most a few mm Hg. As a consequence, a local rise in external venous pressure great enough to be transmitted into the eye or a rise in the resistance to flow through the wall of the eye may raise the transmural pressure to several times its normal value, even if the pressure rise in the intraocular vessels is slight (Fig. 6-7). The increased transmural pressure will distend the vessels and tend to cause breaks in walls normally weak or weakened because of a disease.

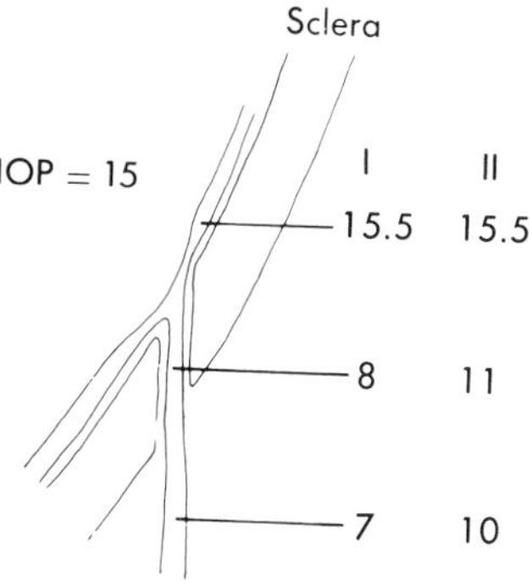

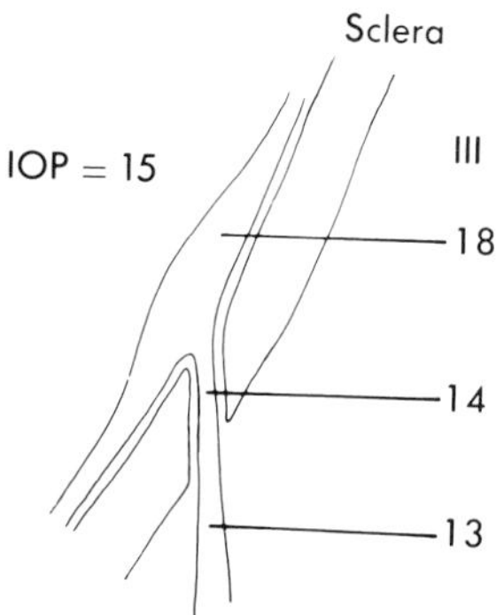

Fig. 6-7. At intraocular pressure of 15 mm Hg, vortex vein is partially collapsed at place where it enters sclera. Venous pressure at this point is practically equal to intraocular pressure, *I*. Small rise in extraocular venous pressure does not change uveal venous pressure unless there is change in intraocular pressure, *II*. Large rise in pressure in extraocular part of vortex vein raises pressure in choroid and produces intraocular venous congestion, *III*.

In eyes with high pressure within the sheath of the optic nerve there is obstruction of axoplasmic flow, swelling of the axons, and congestion of the blood vessels with capillary leakage and hemorrhages.[66]

It is possible that the development of microaneurysms in diabetic retinopathy may be due in part to a venous stasis; an early change in retinopathy is engorgement of the retinal vessels, which indicates that the transmural pressure in the capillaries is increased. Interestingly, clinical reports have suggested that in cases with impaired blood flow to one eye diabetic retinopathy in that eye is less pronounced than in the fellow eye.[59] In this situation it seems likely that a reduction in the transmural pressure in the veins and capillaries contributes to protection of the walls of the vessels.

Pressure in the capillaries

The capillary pressure in the intraocular blood vessels is not known, but it has to be higher than in most tissues, since the tissue pressure in the eye is 15 to 20 mm Hg higher than in most other tissues. In both the choroid and the ciliary processes the arterioles break up into capillaries rather suddenly, an arrangement that promotes a high capillary pressure. This is not the case in the retina; it is likely that the capillary pressure is lower there than in the choroid.

Rate of blood flow and oxygen supply

Many methods have been employed in studying the uveal and retinal circulations, but only a few will be mentioned here. Inspection of the retinal blood vessels is clinically the most important method to get information about vascular events, but detailed information about the blood flow cannot be obtained. During the last few years high-speed photography of the fundus during the passage of a slug of fluorescein (Fig. 6-8) has yielded much information,[38,39,69] but the method is not ideal for determinations of flow. One cannot tell if delayed passage from the retinal arteries into the retinal veins is caused by a reduction in flow rate or by an unchanged flow through dilated blood vessels. A combination of the technique just mentioned with measurements of vascular dimensions makes it possible to make approximative flow determinations, flow being expressed in arbitrary units.[34]

Direct observations of the large choroidal blood vessels through a scleral window[107] have given information about caliber changes in these vessels, and developments of this method have even made it possible to study the choroidal[54] and retinal capillary circulation.[57]

Changes in blood volume within the eye have been studied by means of radioactive erythrocytes[15] and direct monitoring over the sclera. Such monitoring during the passage of slugs of radioactive albumin has made it possible to determine transit times[55]; monitoring of the disappearance of radioactive krypton has permitted calculations of flow values.[56]

Calorimetric methods with continuous determination of the cooling effect of the blood flow through the uvea or parts of the uvea on a small heated body have been used by several investigators.[16,82,86] Ascorbic acid clearance has been used in studies of the blood flow through the ciliary processes.[78] Cannulation of the veins draining the uvea has made it possible to obtain quantitative data for uveal blood flow in rabbits[17,78] and cats[21]; with such direct

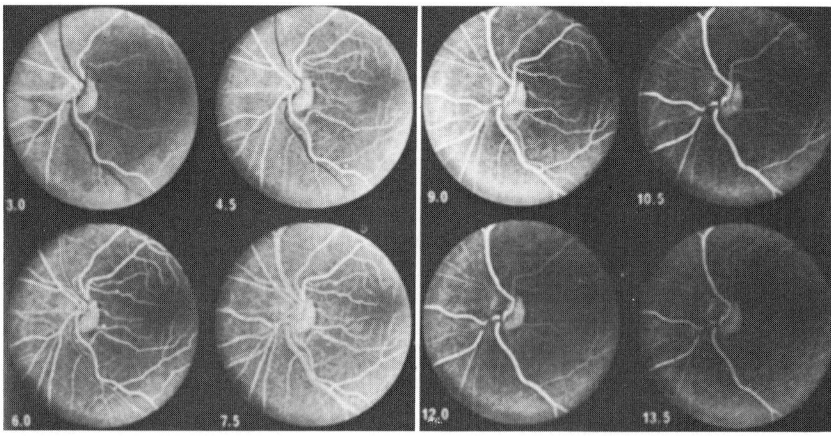

Fig. 6-8. Photographs taken at 1.5-second intervals during passage of fluorescein through retinal vessels. Numbers indicate time in seconds after first appearance of fluorescence in retinal arteries. Laminar streaming is prominent in veins. (From Hickam, J. B., and Frayser, R.: Invest. Ophthalmol. **4**:876, 1965.)

methods it is possible to determine the composition of mixed venous blood and blood from different parts of the uvea.[3] At present, fundus reflectometry[61] and infrared absorption angiography[71] seem to be the only methods that can be applied in studies of the choroidal circulation in humans. In experimental animals the labeled microsphere method offers unique possibilities to determine flow in different parts of the uvea, the retina, and the optic nerve.[5,8,83] In monkeys with spontaneous intraocular pressure Alm, Bill, and Young[8] found values for mean blood flow through the retina, the iris, the ciliary body, and the choroid of 34, 8, 81, and 677 mg/min, respectively. Blood flow through the ciliary processes was 227 g/min/100 g tissue, and through the ciliary muscle, 163 g/min/100 g tissue. In cats there are no marked regional differences in flow within the major part of the choroid and retina. In monkeys regional differences are marked in both tissues, blood flow near the fovea and around the papilla being much higher than that in peripheral parts[7] (Fig. 6-9).

As a consequence of the very high rate of blood flow through the uvea, the oxygen extraction from each milliliter of blood is very low. In cats, blood collected from the choroidal veins has an oxygen content about 3% lower than that in arterial blood. In the anterior uvea the venous oxygen saturation is somewhat lower, 4% to 5% less than that in arterial blood.[3] Similar data have been reported from experiments in dogs and pigs; mixed venous blood from the choroid and the anterior uvea had an oxygen content 3% to 4% lower than that in arterial blood.[35,44,98] In contrast, the oxygen extraction from each milliliter of blood passing through the retina is quite high; the oxygen content of retinal venous blood in humans is about 38% lower than that in arterial blood.[70] In pigs retinal venous blood is about 55% saturated with oxygen.[98]

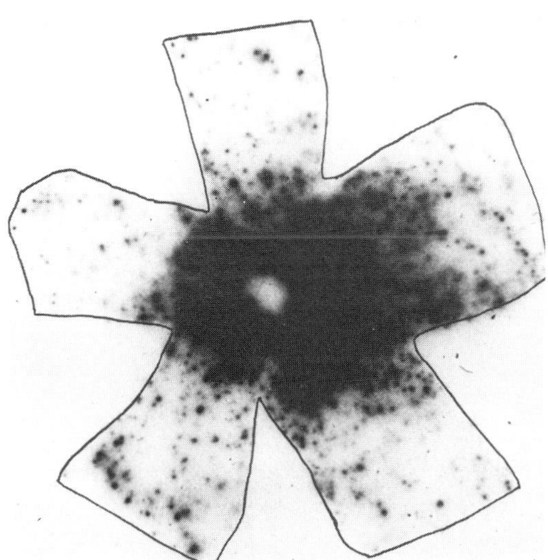

Fig. 6-9. Autoradiograph of flat mount of monkey choroid after injection of labeled microspheres into left heart ventricle. Flow is proportional to number of microspheres per unit area. High flow in central choroid corresponds to fovea and region around optic nerve. Five incisions were made from periphery into choroid. Optic nerve gave central hole. (Technique reported in Alm and Bill.[7])

In cats about 80% of the oxygen consumed by the retina is delivered by choroidal blood vessels; the rest comes from the retinal vessels.[5] Monkeys have a relatively higher rate of retinal blood flow, and reasonable values are 65% from the choroid and 35% from the retinal vessels.[7] The avascular part of the fovea is nourished mainly by the choroid. In pigs the oxygen delivery to the retina from the retinal vessels is about 140 nmol/min, 190 nmol coming from the choroid. About 30 nmol of glucose comes from the retinal vessels each minute, about 120 nmol from the choroidal vessels.[98]

The mean retinal circulation time in humans has been determined to be 4.7 ± 1.1 seconds,[69] and a mean transit time of 3 to 4 seconds was reported by Friedman and Smith[55] for the combined retinal and choroidal circulation in cats. The high rate of blood flow through the uvea is no doubt very important. Not only does it give a high Po_2 in the uvea, which facilitates the diffusion of oxygen into the retina, but it also helps to protect the eye from thermal damage even under rather extreme conditions—arctic snowstorm, Finnish sauna bath, and observation of very bright objects.

Control of circulation

As in other tissues, the factors that influence the blood flow through the eye are the pressure head promoting flow, the resistance of the blood vessels, and the viscosity of the blood.

In many tissues only some of the capillaries are open at a time, the number of simultaneously perfused capillaries depending mainly on the metabolic needs of the tissue. The structure causing opening and closing of the capillaries in most tissues is the precapillary sphincter. In the choroid there are no precapillary sphincters, and as a consequence the flow is steady and not alternating.[54] In the retina, too, Friedman, Smith, and Kuwabara[57] could find no precapillary sphincters; the blood flow appeared to be uninterrupted in all capillaries observed. Lemmingson,[77] investigating cats of different ages, found spontaneous contractions and dilations in the small arterioles and precapillary sphincters in young animals. Much less vasomotion was found in the retinal vessels of adult cats.

Perfusion pressure

In most tissues the pressure head for flow through the tissue can be defined as the difference in pressure between the arteries entering the tissue and the veins leaving it. In the eye a more distinct definition has to be used because of the high tissue pressure. The pressure in the veins at the points where they are about to leave the influence of the intraocular pressure is the pertinent venous pressure, not that in the sclera or just outside the sclera. As mentioned, the pertinent venous pressure normally is very near the intraocular pressure but may be higher if the intraocular pressure is very low or if the extraocular venous pressure is raised above the normal level.

A rise in intraocular pressure above the normal level gives a corresponding rise in intraocular venous pressure, and as a consequence the pressure head for flow through the eye is reduced. In the uvea the rate of flow decreases[20]; this tends to raise the blood pressure in the arteries just outside the eye. Thus the pressure head for flow through the retina and uvea decreases a little less than the rise in intraocular pressure. Fig. 6-10 shows the changes in pressure and total uveal blood flow that occur when the intraocular pressure is changed above and below the normal level.

A fall in arterial blood pressure produces a corresponding fall in perfusion pressure unless there is a fall in intraocular venous pressure secondary to a fall in intraocular pressure. From the point of view of perfusion pressure, a rise in intraocular pressure thus has very much the same effect as a fall in arterial blood pressure.

The blackout that unprotected airplane pilots

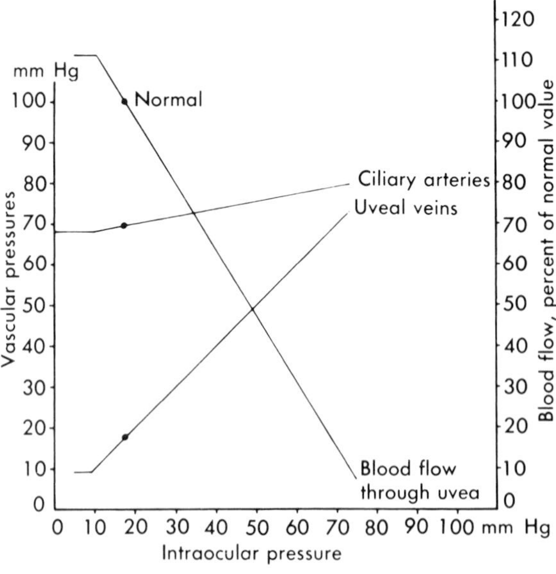

Fig. 6-10. Changes in vascular pressures and uveal blood flow produced by changes in intraocular pressure above and below normal level.

tend to experience in sharp turns and pullout from dives is a phenomenon that seems to be caused mainly by intraocular ischemia; at 3 to 4 *g* or more the pressure in the left ventricle is not great enough to overcome the effect of the *g* force on the blood column from the heart to the eye plus the intraocular pressure.[40]

Since a rise in intraocular pressure reduces the blood flow through the uvea, the venous pressure in the veins just outside the eye tends to fall, and a peculiar situation arises; the pressure head for flow through the scleral wall increases while the blood flow is reduced. This produces a passive adjustment of the resistance in the veins draining the eye; by partially collapsing, they can offer the increase in resistance that has to accompany the flow and pressure changes.[17]

In the kidney and in the brain a fall in perfusion pressure gives rise to vasodilation, and the flow is maintained at a steady level despite moderate changes in perfusion pressure (autoregulation of flow). In the choroid there is practically no such compensating adjustment; the flow falls very much in proportion to the fall in perfusion pressure. In the ciliary muscle, the ciliary processes, and the iris, there is vasodilation and little change in flow for moderate reductions in perfusion pressure.[5,7,8] Since there is autoregulation of the blood flow in the brain, one would expect that the blood flow through the retinal vessels is autoregulated, at least to some extent. Studies with labeled microspheres in cats[5] and monkeys[7] have shown that this is indeed the case. In cats a moderate rise in eye pressure increases retinal blood flow to compensate reduced choroidal blood flow (Fig. 6-11).

Autoregulation of blood flow may be accomplished in two ways: by a metabolic mechanism or by a myogenic mechanism.[47] In tissues with metabolic autoregulation the arterioles adjust their resistance to maintain the concentrations in the tissue of some critical metabolites at a constant level. In tissues with myogenic autoregulation, pacemaker cells in the arterioles sense the transmural pressure difference and cause adjustment of arteriolar tone by cell to cell propagation of signals through the smooth muscle cells. In the ocular tissues both types of mechanisms seem to operate.

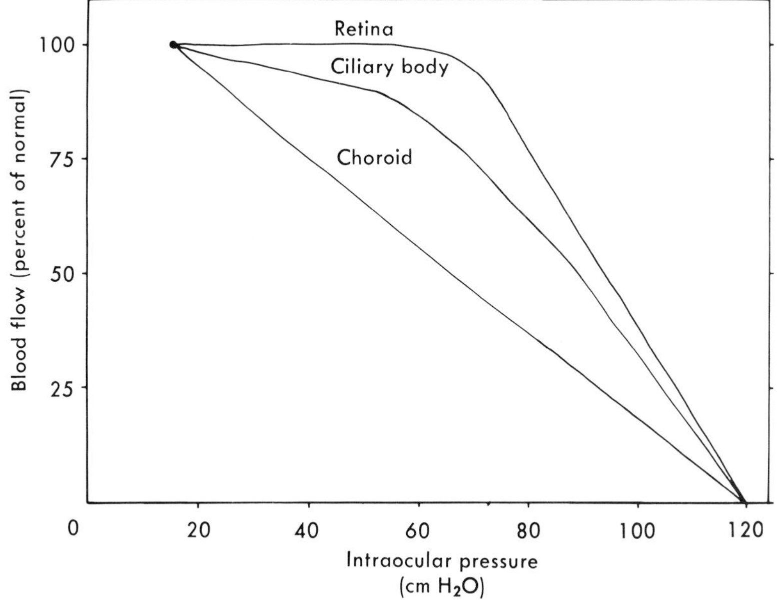

Fig. 6-11. When intraocular pressure is increased, there is no decrease in retinal blood flow up to a certain level. Change in flow in ciliary body is also small. In the choroid even moderate increments in eye pressure reduce blood flow. At high intraocular pressures further increments in pressure reduce flow in all intraocular tissues.

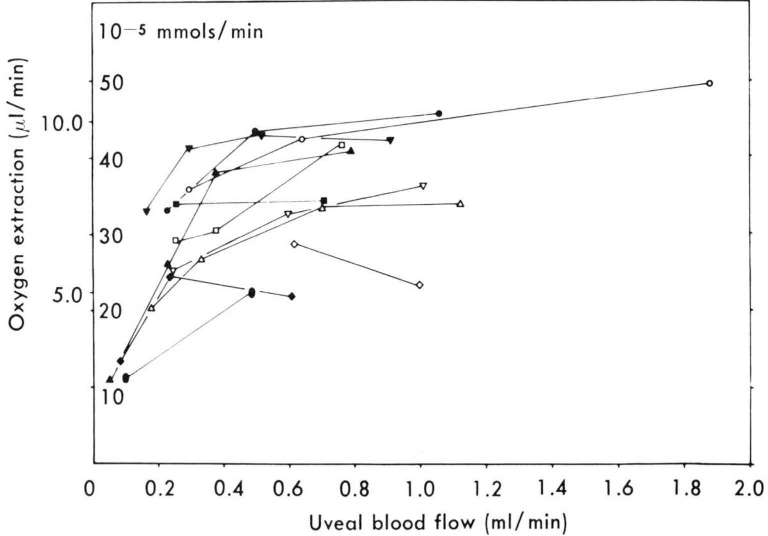

Fig. 6-12. Changes in oxygen extraction from blood passing through uvea when blood flow is reduced by increase in intraocular pressure. (From Alm, A., and Bill, A.: Acta Physiol. Scand. **80:**19, 1970.)

In cats, when the blood flow through the eye is reduced, the oxygen extraction from each milliliter of the blood perfusing the uvea is increased. As a consequence the net extraction of oxygen from the uveal blood may be maintained at about the same level despite large changes in blood flow[3] (Fig. 6-12). In the inner parts of the retina the oxygen tension is maintained at a constant level even during large variations in intraocular pressure[4]; this, of course, is an effect of autoregulation of retinal blood flow.

Nervous control of ocular blood flow

The ciliary arteries as well as the extraocular part of the central retinal artery are under sympathetic nervous control. In the uvea there is a rich sympathetic supply in the iris, ciliary processes, and choroid, while that in the ciliary muscle seems sparse. Only alpha receptors[1] seem to be present in the uvea.[6,18,52] In the retina there are no sympathetic nerves.[43,75] The physiologic role of the sympathetic nerves in ocular blood flow control has not been very clear. Recent experiments[30] indicate, however, that one important role—possibly the only one—is to help maintain the blood flow at a suitable level under conditions with sudden increments in blood pressure (Fig. 6-13). Such increments, occuring in everyday life in acute stress situations and during work, are the results of a general increase in sympathetic activity. They tend to cause overperfusion of the eye, resulting in breakdown of the blood-aqueous and blood-retinal barriers. However, with simultaneous activity in the sympathetic nerves to the eye, these effects are prevented. Sympathetic activity thus assists autoregulatory mechanisms in maintaining the intraocular blood flow and volume constant.

The role of the parasympathetic nerves in the eye is much less clear than that of the sympathetics. Uusitalo and Palkama[101] have described nerves in the ciliary processes that seem to be cholinergic, and Ruskell[89] has reported evidence for vasodilating nerves to the choroid passing through the sphenopalatine ganglion. Experiments with electrical stimulation of the oculomotor nerve in rabbits have revealed vasoconstrictive effects in the anterior uvea.[96] In monkeys and cats iridial blood flow was reduced but the flow through the ciliary body increased.[95]

Electrical stimulation of the intermediofacial nerve in rabbits resulted in increased flow in the whole uvea, most marked in the choroid.[94]

In monkeys topical neostigmine caused marked vasodilation in the ciliary muscle, the

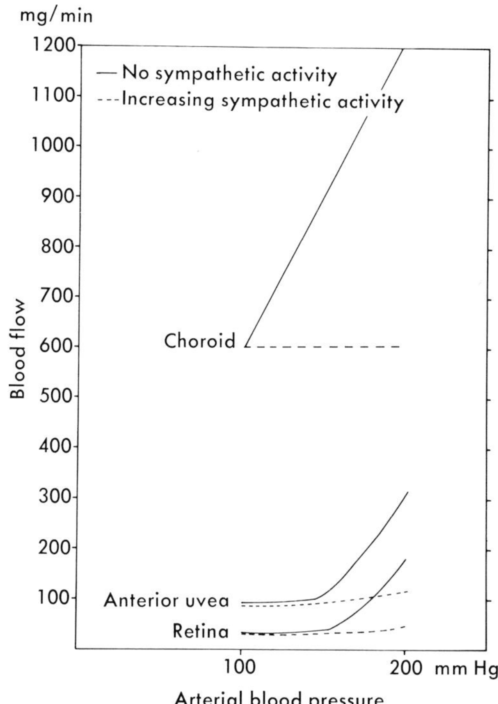

Fig. 6-13. Studies on the effect of sympathetic stimulation at different arterial blood pressures indicate that with increasing pressure due to sympathetic activation the blood flow in the different parts of the eye changes very little.[30] If acute blood pressure increments are not paralleled by increasing sympathetic tone to the eye, the blood flow in the choroid increases in proportion to the blood pressure. In the anterior uvea and the retina autoregulatory mechanisms prevent overperfusion at moderate pressure increments. However, they fail at very high pressures, and as a result the tissues are overperfused and there is small-vessel extravasation of plasma, resulting in a tendency toward breakdown of blood-aqueous and blood-retinal barriers.[72,79]

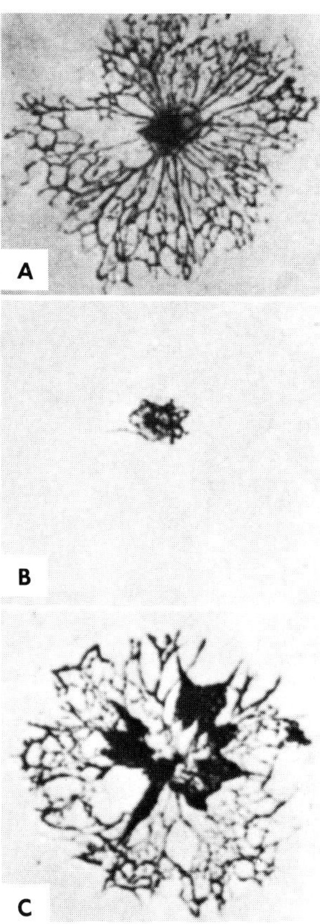

Fig. 6-14. A, Normal mouse retina at 1 day old. **B,** Total vasoobliteration after 5 days in 98% to 100% oxygen. **C,** Five days after return to air. New vessels have grown into retina and into vitreous (dense central vessel). (India ink injection.) (From Ashton, N.: Br. J. Ophthalmol. **52:**505, 1968.)

iris, and the ciliary processes, effects that probably are caused by acetylcholine release from parasympathetic nerves.[8]

A peculiar type of nervous influence on the blood vessels is particularly seen in rabbits. Stroking the iris in this animal gives rise to a local dilation of the arterioles that spreads all over the iris. The vascular permeability is increased, and there is prolonged miosis and an intraocular pressure rise.[91] Ambache, Kavanagh, and Withing[9] have shown that a substance with vasodilating and smooth muscle contracting properties is released when the iris has been stroked; they suggested that this substance, irin, is somehow involved in the reaction. Irin seems to be a prostaglandin or a mixture of several prostaglandins.[42] Interestingly, aspirin, which inhibits the synthesis of prostaglandins, modifies the effects of paracentesis in the rabbit eye.[81] Mechanical as well as electrical stimulation of the sensory nerve to the eye produces a reaction similar to that of stroking the iris, but surprisingly, nerve stimulation seems to cause little release of prostaglandins.[42] Recent experiments in rabbits indicate that the agent released may be substance P or a related peptide.[31] Mechanical irritation of the conjunctiva or the cornea gives widespread conjunctival vasodilation. This dilation can also be prevented to a large extent by local anesthesia.

Chemical control of ocular blood flow

In the brain, high concentrations of oxygen in the blood produce vasoconstriction, and low concentrations produce vasodilation; high concentrations of carbon dioxide in the blood produce vasodilation, and low concentrations produce vasoconstriction. In the uveas of adult cats and rabbits the effects of moderate oxygen excess or lack seem to be too small to be de-

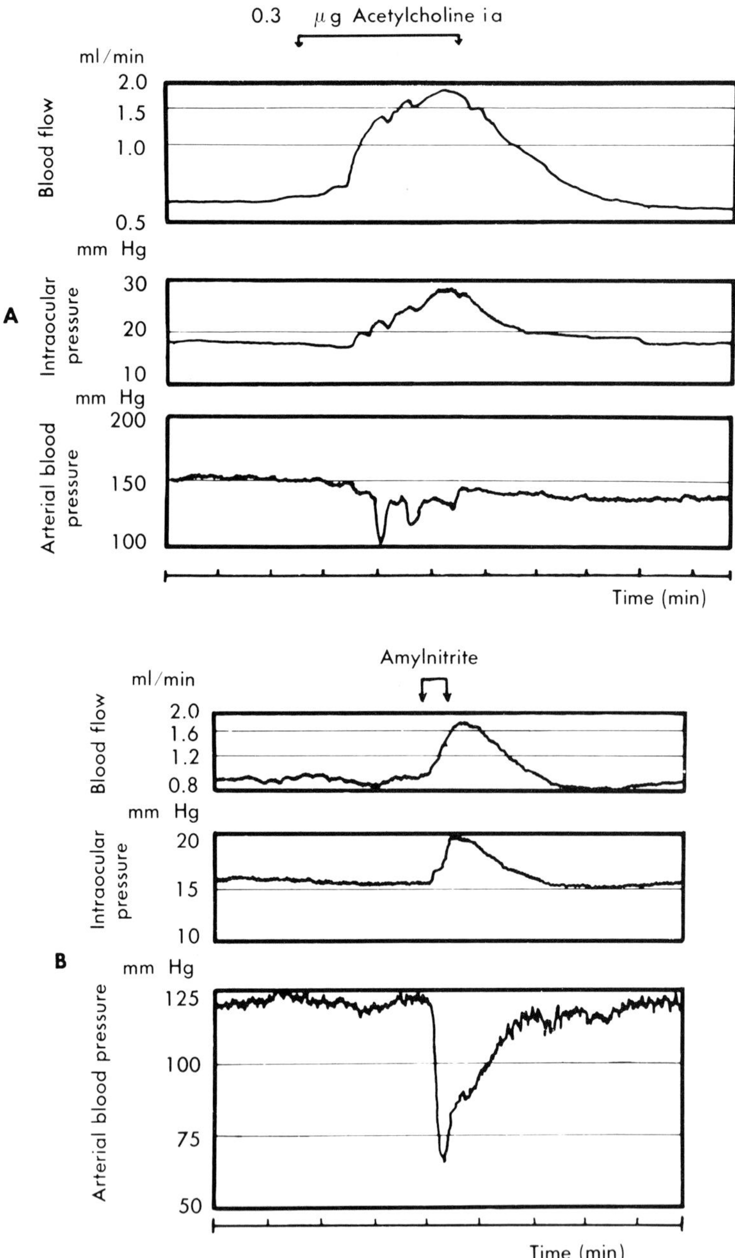

Fig. 6-15. A, Effects of intra-arterial acetylcholine on uveal blood flow, intraocular pressure, and arterial blood pressure. **B,** Effects of inhalation of amyl nitrite on uveal blood flow, intraocular blood press[illegible]l blood pressure. (**A** from Bill, A.: Acta Physiol. Scand. **56:**70, 1962; **B** from Bill, A.: Acta Soc. Med. Ups. **67:**122, 1962.)

tected.[19] Excessive carbon dioxide, however, gives marked vasodilation in all parts of the uvea.[5,53]

The effect on the retinal vessels of inhaling 100% oxygen is slight vasoconstriction.[49,69] Usually the vasoconstriction is not marked enough to prevent a rise in oxygen tension in the retina.[4]

In immature eyes a high oxygen concentration in the inhaled gas has dramatic effects in both humans and experimental animals[13]; vasoconstriction, inhibition of vascular development, and, after some time, obliteration of the vessels result (Fig. 6-14). After withdrawal of the extra oxygen, there may be vasoproliferation, retinal edema, and hemorrhages. These changes produce a scarlike traction on the retina, which may detach the retina and lead to blindness, with the clinical picture of retrolental fibroplasia. Therefore, if oxygen is given to premature infants, frequent ophthalmoscopic examination is necessary.[84] However, it has to be kept in mind that it is not the oxygen required by the infant to normalize arterial oxygen tension that is dangerous. It is the overdose, raising the oxygen tension of the arterial blood above the normal level, that has harmful effects.

Inhalation of 7% carbon dioxide in 21% oxygen gives moderate dilation of visible retinal vessels.[49,69]

In cats, at an arterial carbon dioxide tension of 80 mm Hg and normal oxygen tension, retinal blood flow is increased by 300% to 400%.[5] With combinations of carbon dioxide and oxygen the oxygen tension in the inner parts of the retina can be increased by more than 300%.[5]

Effects of drugs on the blood flow through the eye

Vasodilating drugs affect the eye, but their therapeutic value is doubtful. Thus acetylcholine,[18,56,82] papaverine, aminophylline,[19] theophylline,[82] and amyl nitrite[19,82,86] all dilate the uveal blood vessels (Fig. 6-15), but unfortunately general administration of the dilators reduces uveal vascular resistance only at dosages that also tend to produce a fall in arterial blood pressure. The reduction in perfusion pressure may lead to a reduced uveal blood flow. Intra-arterial or retrobulbar administration of vasodilating drugs may produce marked uveal vasodilation accompanied by a large increment in blood flow.[19] Vasoconstriction in the uvea is produced by dihydroergotamine.[19] Nicotinic acid has no immediate effect on uveal blood vessels.[19,82] Studies on the effect of drugs on retinal blood flow are complicated by the inaccessibility of the retinal blood vessels. Nitroglycerin has been reported to cause vasodilation, but the accompanying fall in blood pressure seems to reduce the rate of blood flow; the oxygen content of retinal venous blood falls. Aminophylline and papaverine in dosages that affect the blood flow through the brain did not affect retinal vascular diameters and retinal venous oxygen tension.[50] Studies by Alm[2] on the effects of various drugs on the oxygen tension of the vitreous body close to the retina indicate that none of the vasoconstricting drugs, norepinephrine, angiotensin, or dihydroergotamine has any effect on retinal vascular resistance, and this is also true for the vasodilating drugs isoproterenol, nicotinic acid, and xanthinol nicotinate. The drugs were given intra-arterially. Papaverine, on the other hand, invariably increased oxygen tension in the retina, indicating retinal vasodilation. Experimental spasm of the retinal arterioles in cats is relieved by papaverine, tolazoline, and a mixture of caffeine and sodium benzoate, as well as by inhalation of 10% carbon dioxide.[45]

TISSUE FLUID

Permeability of the intraocular blood vessels

The anatomy of the capillary wall differs from one tissue to another, as does the permeability of the blood vessels. Lipid-soluble substances such as oxygen and carbon dioxide penetrate the capillary wall in all tissues at very high rates, since they can pass right through the endothelial cells. This passage is so rapid that despite the oxygen consumption of the tissues the difference in oxygen tension between venous blood and the nearest tissue is slight, 1 to 3 mm Hg. Water can also pass through the endothelial cells.[108] Passage of substances soluble only in water is more or less restricted, since they have to pass through water-filled pores or spaces in the capillary wall.[74] In addition to movement through pores, there may be

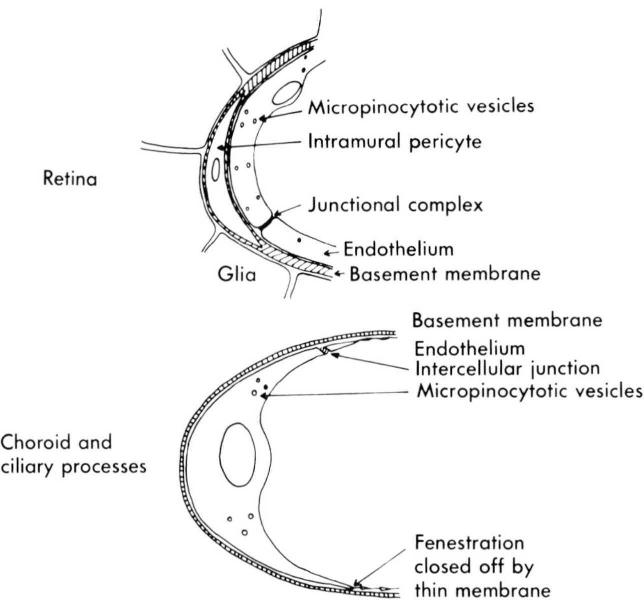

Fig. 6-16. Schematic representation of capillary wall in retina and in ciliary processes and choroid.

some transendothelial transport by means of small vesicles (pinocytosis or micropinocytosis). Physiologically, the capillary walls in most tissues seem to contain two populations of pores, one population with a diameter of about 9 nm and another with a diameter of 24 to 70 nm.[63,74] The smaller pores are practically impermeable to albumin and larger proteins, and the larger pores have a high permeability to proteins. The functional small pores are represented anatomically by slits between endothelial cells and small defects in the tiny membranes covering most fenestrae in capillaries with fenestrated walls. Open fenestrations, transendothelial pinocytosis, and occasional wide slits between the endothelial cells are the anatomic counterparts of the functional large-pore system. In the brain the pores that permit movement of water-soluble substances across the blood-brain barrier have been estimated to have a diameter as small as about 0.8 nm.[46]

The capillaries of the retina have a thick endothelium that lacks fenestrations; the endothelial cells seem to be held together by very tight junctional complexes (Fig. 6-16). Outside the endothelium is a thick basement membrane, and within this membrane are scattered cells, the mural cells of Kuwabara and Cogan,[73] intramural pericytes according to the terminology of Ashton and de Oliveira.[14] The basement membrane is in direct contact with the glial tissue. The role of the intramural pericytes is not known, but, interestingly, they tend to disappear in diabetic retinopathy and in states with endothelial proliferation and neovascularization.[73] They are also reduced in number under other unfavorable conditions such as cyanosis and polycythemia.[37] The permeability of the blood vessels is similar to that of the vessels in the central nervous system. The capillary wall is practically impermeable to proteins and fluorescein.[27,62] Even sodium ions pass very poorly through the capillary wall of the retinal vessels.[99] The cells of the retina need large amounts of glucose, which has to pass through the retinal capillaries using carrier mechanisms.[98]

In primates the capillaries of the iris have a continuous endothelial lining, and there are tight junctions between the cells and a thick basement membrane.[105] Water diffuses rapidly through the capillary wall, using paths both through and between the endothelial cells. The passage of sodium and larger water-soluble ions and molecules seems to be restricted to paths between cells unless there are mechanisms for transendothelial transport.[27] Proteins[25] and horseradish peroxidase[105] pass poorly if at all through the capillary wall in the iris.

The capillaries in the ciliary processes have very thin walls with large fenestrations bridged over by a thin membrane.[103] The junctions between the endothelial cells are wide and short. A continuous but very thin basement membrane surrounds the capillary. The capillary wall in the choroid is similar to that in the ciliary processes. Both in the ciliary processes and in the choroid the blood vessels have a high permeability, even to substances as large as myoglobin (molecular weight 17,000). The still larger albumin and gamma globulin molecules (molecular weight 67,000 and 156,000 respectively) also pass out at high rates when compared with the conditions in many other tissues.[25]

Pinocytotic vesicles are found in both the uveal and retinal capillaries; their role for transendothelial transport is unclear.

Blood-retinal barrier

The capillary wall in the retinal vessels seems to be a tight barrier, but this would not prevent the entrance of undesirable substances into the retina if there were no barrier between the uvea and the retina. The pigment epithelium of the retina is part of this additional barrier. The epithelial cells are attached to each other by tight junctions,[93] which appear to prevent diffusion of horseradish peroxidase[93] and fluorescein as well.[62] Water-soluble substances needed by the retina, such as glucose, have to be transported through the pigment epithelium by carriers.[98] Another route by which undesirable substances might enter the retina is by way of diffusion from the ciliary processes via the posterior chamber. Again, a competent barrier prevents such movement. The nonpigmented cells of the ciliary processes are attached to each other by tight junctions, and there is very little leakage of medium and high molecular weight substances from the stroma of the ciliary processes through this barrier into the posterior chamber.[92,93]

Grayson and Laties,[62] using fluorescein as a marker, have found one weak spot in the eye where fluorescein could enter nervous tissue; the dye was found to penetrate from the choroid into the prelaminar part of the optic nerve. The consequences of this defect in the barrier are not clear.

Formation of tissue fluid in the eye

A modified concept of Starling's hypothesis states that in the capillaries there is ultrafiltration and diffusion through the 9 nm pores, the net flow through each pore depending on the difference in hydrostatic pressure between the inside and outside of the capillary and also on the colloid osmotic pressure difference.

$$F = C \cdot [(P_{hc} - P_{ht}) + (P_{collt} - P_{collc})]$$

where

C = Constant
P_{hc} = Mean hydrostatic osmotic pressure in capillary
P_{collc} = Mean colloid osmotic pressure in capillary
P_{ht} = Hydrostatic pressure in tissue
P_{collt} = Colloid osmotic pressure in tissue

The small size of these pores makes net movement of fluid out of or into the capillaries very sluggish, despite a large total pore area (about 7000 cm^2/100 g muscle).[74] Diffusion of low molecular weight substances is hardly influenced by the smallness of the pores; the total area available is the important parameter. The net flow through the large pores also depends on the transmural difference in hydrostatic pressure, but the effect of differences in colloid osmotic pressure is most probably negligible (Fig. 6-17).

In rabbits the concentration of albumin and gammaglobulin in the tissue fluid of the ciliary processes and of the choroid is so high that the colloid osmotic pressure of the tissue fluid can be estimated to be about 50% to 60% of that in plasma.[26] This means that the mean transmural hydrostatic pressure difference in the capillaries has to be a little more than 40% to 50% of the colloid osmotic pressure of plasma to give a net outflow through the small pores. The large pores require only 1 to 2 mm Hg to give net outflow.

The protein concentration in the tissue fluid of the iris is very low. As a consequence the mean hydrostatic pressure in the capillaries has to be about the same as the colloid osmotic pressure of plasma to prevent net absorption of fluid

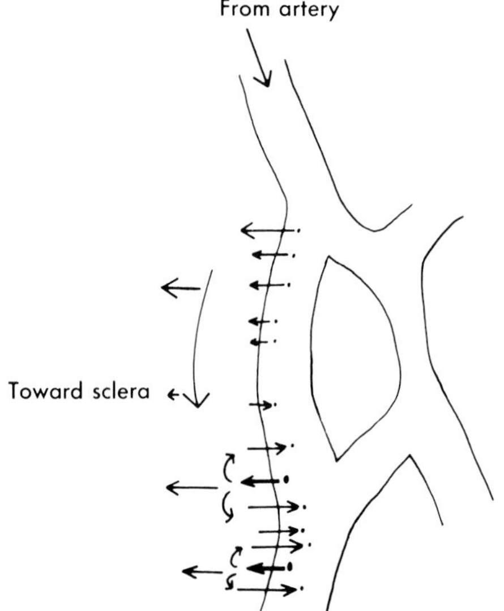

Fig. 6-17. Outward flow through small pores in capillary wall in arterial part of capillary; inward flow through such pores in venous part. Outward flow through large protein-leaking pores in all parts of capillary. Tissue fluid from choroid passes through suprachoroid and sclera into connective tissue outside eyeball.

from the anterior chamber. It is not known at present if diffusion and ultrafiltration through the iris capillaries give net inflow or outflow. Whatever the direction of flow may be, the net flow is negligible compared to the formation of aqueous humor.[27]

In the ciliary processes the net outflow from the capillaries has to be at least the same as the rate of aqueous secretion, otherwise the ciliary epithelium will rapidly pump out all fluid from the interstitial tissue of the processes. The protein concentration in the net ultrafiltrate is about 3% of that in plasma. However, the protein passes very poorly through the barrier constituted by the ciliary epithelium, and as a consequence the protein concentration in the tissue fluid in rabbits is as high as 70% of that in plasma.[26] It is not clear at present if the rate of aqueous secretion adjusts itself to the rate of net ultrafiltration from the blood vessels or if the rate of net ultrafiltration is adjusted to a level that corresponds to the rate of secretion; a likely alternative is a hybrid model,[25] in which the extravascular protein concentration plays a central role.

The high protein concentration of the tissue fluid in the choroid and the absence of protein in the tissue fluid in the retina has an interesting consequence. The colloid osmotic pressure of the choroidal tissue fluid will tend to cause water movement through the pigment epithelium from the retina into the choroid. This may be a mechanism of importance for the attachment of the rest of the retina to the pigment epithelium.

Drainage of intraocular tissue fluids

Proteins that have passed out of the capillaries return into the capillaries very poorly. In many other tissues the major part is drained with lymph vessels; the eye has no system of lymph vessels. The aqueous humor has been regarded as ocular lymph, and the fact that small amounts of proteins appear in the aqueous shows that this view is not altogether incorrect; small amounts of the tissue fluid from the ocular tissues may pass into the anterior and posterior chambers of the eye and be drained with the aqueous humor. However, at least in rabbits, the major part of the protein that passes out of the uveal blood vessels leaves the eye by passing through the scleral substance and through perivascular spaces.[25] Outside the eye the proteins enter the lymph vessels. Such vessels are present in conjunctiva and have been reported to be present also in the retrobulbar connective tissue.[90] The tissue pressure outside the eye can be presumed to be near zero or slightly negative; that in the suprachoroid is slightly lower than the intraocular pressure.[102] Hence the pressure that promotes the flow of tissue fluid through the sclera is nearly equal to the intraocular pressure. The sclera is so permeable to proteins that differences in protein concentration between the inside and the outside cannot be expected to have much influence on the flow through the sclera.

The fact that the tissue pressure in the suprachoroid is lower than the pressure in the anterior chamber results in the drainage of tissue fluid from the ciliary body into the suprachoroid and drainage of aqueous humor by such routes. In monkeys such flow is pronounced[27]; in adult humans it seems to be less than 20% of the total drainage.[29]

At very low intraocular pressures, for example, after intraocular operations, there is often an accumulation of fluid in the suprachoroid.[33] Most probably this is caused by increased filtration from the uveal blood vessels and by a reduction in the rate of elimination of tissue fluid through the sclera; the pressure gradient for elimination is obviously reduced.

In the rabbit there seems to be some movement of fluid from the posterior chamber into the vitreous humor. From the vitreous body the fluid is eliminated via the perivascular spaces of the retinal blood vessels and probably also by passing through the retina into the choroid.[48] In monkeys and humans there seems to be no drainage through the perivascular spaces of the retina[65]; in fact, it is not known if net movement of fluid is from the posterior chamber into the vitreous or in the reverse direction.

REFERENCES

1. Ahlqvist, R. P.: The adrenergic receptor, J. Pharm. Sci. **55:**359, 1966.
2. Alm, A.: Effects of norepinephrine, angiotensin, dihydroergotamine, papaverine, isoproterenol, histamine, nicotinic acid and xantinol nicotinate on retinal oxygen tension in cats, Acta Ophthalmol. **50:**707, 1972.
3. Alm, A., and Bill, A.: Blood flow and oxygen extraction in the cat uvea at normal and high intraocular pressure, Acta Physiol. Scand. **80:**19, 1970.
4. Alm, A., and Bill, A.: The oxygen supply to the retina. I, Effects of changes in intraocular and arterial blood pressures and in arterial Po_2 and Pco_2 on the oxygen tension in the vitreous body of the cat, Acta Physiol. Scand. **84:**261, 1972.
5. Alm, A., and Bill, A.: The oxygen supply to the retina. II, Effects of high intraocular pressure and of increased arterial carbon dioxide tension on uveal and retinal blood flow in cats. A study with labelled microspheres including flow determinations in brain and some other tissues, Acta Physiol. Scand. **84:**306, 1972.
6. Alm, A., and Bill, A.: The effect of stimulation of the sympathetic chain on retinal oxygen tension and uveal, retinal and cerebral blood flow in cats, Acta Physiol. Scand. **88:**84, 1973.
7. Alm, A., and Bill, A.: Ocular and optic nerve blood flow at normal and increased intraocular pressures in monkeys *(Macaca irus):* a study with radioactively labelled microspheres including flow determinations in brain and some other tissues, Exp. Eye Res. **15:**15, 1973.
8. Alm, A., Bill, A., and Young. F. A.: The effects of pilocarpine and neostigmine on the blood flow through the anterior uvea in monkeys. A study with radioactively labelled microspheres, Exp. Eye Res. **15:**31, 1973.
9. Ambache, N., Kavanagh, L., and Withing, J.: Effect of mechanical stimulation on rabbit eyes: release of active substance in anterior chamber perfusates, J. Physiol. **176:**378, 1965.
10. Anderson, B., Jr., and Saltzman, H. A.: Retinal oxygen utilization measured by hyperbaric blackout, Arch. Ophthalmol. **72:**792, 1964.
11. Ascher, K. W.: The aqueous veins, Springfield, Ill., 1961, Charles C Thomas, Publisher.
12. Ashton, N.: Lecture, William Mackenzie Centenary Symposium, Glasgow, Scotland, 1968.
13. Ashton, N.: Some aspects of the comparative pathology of oxygen toxicity in the retina, Br. J. Ophthalmol. **52:**505, 1968.
14. Ashton, N., and de Oliveira, F.: Nomenclature of pericytes, Br. J. Ophthalmol. **50:**119, 1966.
15. Bettman, J., and Fellows, V. A.: Technique for the determination of blood volume changes in the choroid and retina by the use of radioactive phosphorus, Am. J. Ophthalmol. **38:**513, 1954.
16. Bill, A.: Calorimetric procedures for the study of the blood flow through the ciliary region and the choroid in rabbits, Acta Ophthalmol. **40:**1, 1962.
17. Bill, A.: Quantitative determination of uveal blood flow in rabbits, Acta Physiol. Scand. **55:**101, 1962.
18. Bill, A.: Autonomic nervous control of uveal blood flow, Acta Physiol. Scand. **56:**70, 1962.
19. Bill, A.: Aspects of physiological and pharmacological regulation of uveal blood flow, Acta Soc. Med. Ups. **67:**122, 1962.
20. Bill, A.: Intraocular pressure and blood flow through the uvea, Arch. Ophthalmol. **67:**90, 1962.
21. Bill, A.: A method for quantitative determination of the blood flow through the cat uvea, Arch. Ophthalmol. **67:**156, 1962.
22. Bill, A.: Aspects of the regulation of the uveal venous pressure in rabbits, Exp. Eye Res. **1:**193, 1962.
23. Bill, A.: The uveal venous pressure, Arch. Ophthalmol. **69:**780, 1963.
24. Bill, A.: Blood pressure in the ciliary arteries of rabbits, Exp. Eye Res. **2:**20, 1963.
25. Bill, A.: Capillary permeability to and extravascular dynamics of myoglobin, albumin and gammaglobulin in the uvea, Acta Physiol. Scand. **73:**204, 1968.
26. Bill, A.: A method to determine osmotically effective albumin and gammaglobulin concentrations in tissue fluids, its application to the uvea and a note on the effects of capillary "leaks" on tissue fluid dynamics, Acta Physiol. Scand. **73:**511, 1968.
27. Bill, A.: Blood circulation and fluid dynamics in the eye, Physiol. Rev. **55:**383, 1975.
28. Bill, A.: Unpublished results.
29. Bill, A., and Phillips, C. I.: Uveoscleral drainage of aqueous humor in human eyes, Exp. Eye Res. **12:**275, 1971.
30. Bill, A., Linder, M., and Linder, J.: The protective role of ocular sympathetic vasomotor nerves in acute arterial hypertension. In Proceedings of the Ninth European Conference on Microcirculation, Antwerp, Belgium, 1976, Bibl. Anat. **16:**30, 1977.
31. Bill, A., et al.: Substance P—release on trigeminal nerve stimulation: effects on pupil, eye pressure and capillary permeability, Acta Physiol. Scand. **106:**371, 1979.

32. Böck, J., Bornschein, H., and Hommer, K.: Die Erholungslatenz der Helligkeitsempfindung und des Elektroretinogramms nach retinaler Ischämie, v. Graefes Arch. Ophthalmol. **167**:276, 1964.
33. Brav, S. S.: Serous choroidal detachment, Surv. Ophthalmol. **6**:395, 1961.
34. Bulpitt, C., and Dollery, C. T.: Estimation of retinal blood flow by measurement of the mean circulation time, Cardiovasc. Res. **5**:406, 1971.
35. Cohan, B. E., and Cohan, S.: Flow and oxygen saturation of blood in the anterior ciliary vein of the dog eye, Am. J. Physiol. **205**:60, 1963.
36. Collier, R. H.: Experimental embolic ischemia of the choroid, Arch. Ophthalmol. **77**:683, 1967.
37. de Oliveira, F.: Pericytes in diabetes retinopathy, Br. J. Ophthalmol. **50**:134, 1966.
38. Dollery, C. T., et al.: Focal retinal ischaemia. I, Ophthalmoscopic and circulatory changes in focal retinal ischaemia, Br. J. Ophthalmol. **50**:283, 1966.
39. Dollery, C. T., et al.: Effect of raised intraocular pressure on the retinal and choroidal circulation, Invest. Ophthalmol. **7**:191, 1968.
40. Duane, T.: Effect of anoxia: observations on the fundus oculi during blackout, Arch. Ophthalmol. **51**:343, 1954.
41. Duke-Elder, S.: The ocular circulation, its normal pressure relationships and their physiological significance, Br. J. Ophthalmol. **10**:513, 1926.
42. Eakins, K. E.: Prostaglandin and nonprostaglandin mediated breakdown of the blood-aqueous barrier, Exp. Eye Res. **25**(suppl.): 483, 1977.
43. Ehinger, B.: Adrenergic nerves to the eye and to related structures in man and the cynomolgus monkey, Invest. Ophthalmol. **5**:42, 1966.
44. Elgin, S. S.: Arteriovenous oxygen difference across the uveal tract of the dog eye, Invest. Ophthalmol. **3**:417, 1964.
45. Ellis, P. P., and Lende, R. A.: Induced spasm in the retinal arterioles of cats, Arch. Ophthalmol. **71**:760, 1964.
46. Fenstermacher, J. D., and Johnson, J. A.: Filtration and reflection coefficients of the rabbit blood-brain barrier, Am. J. Physiol. **211**:341, 1966.
47. Folkow, B., and Neil, E.: Circulation, New York, 1971, Oxford University Press, Inc.
48. Fowlks, W. L., Havener, V., and Good, J. S.: Meridional flow from the corona ciliaris through the pararetinal zone of the rabbit vitreous, Invest. Ophthalmol. **2**:63, 1963.
49. Frayser, R., and Hickam, J. B.: Retinal vascular response to breathing increased carbon dioxide and oxygen concentrations, Invest. Ophthalmol. **3**:427, 1964.
50. Frayser, R., and Hickam, J. B.: Effect of vasodilator drugs in the retinal blood flow in man, Arch. Ophthalmol. **73**:640, 1965.
51. Friedenwald, J. S.: Retinal and choroidal arteriosclerosis. In Ridley, F., and Sorsby, A., editors: Modern trends in ophthalmology, New York, 1940, Appleton-Century-Crofts.
52. Friedman, E., and Chandra, S. R.: Choroidal blood flow. II, The effects of autonomic agents, Arch. Ophthalmol. **87**:67, 1972.
53. Friedman, E., and Chandra, S. R.: Choroidal blood flow. III, Effects of oxygen and carbon dioxide, Arch. Ophthalmol. **87**:70, 1972.
54. Friedman, E., and Oak, S. M.: Choroidal microcirculation in vivo, Bibl. Anat. **7**:129, 1965.
55. Friedman, E., and Smith, T. R.: Estimation of retinal blood flow in animals, Invest. Ophthalmol. **4**:1122, 1965.
56. Friedman, E., Kopald, H. H., and Smith, T. R.: Retinal and choroidal blood flow determined with krypton-85 in anesthetized animals, Invest. Ophthalmol. **3**:539, 1964.
57. Friedman, E., Smith, T. R., and Kuwabara, T.: Retinal microcirculation in vivo, Invest. Ophthalmol. **3**:217, 1964.
58. Fujino, T., and Hamasaki, D. I.: Effect of intraocular pressure on the electroretinogram, Arch. Ophthalmol. **78**:757, 1967.
59. Gay, A. J., and Rosenbaum, A. L.: Retinal artery pressure in asymmetric diabetic retinopathy, Arch. Ophthalmol. **75**:758, 1966.
60. Gay, A. J., Goldor, H., and Smith, M.: Chorioretinal vascular occlusions with latex spheres, Invest. Ophthalmol. **3**:647, 1964.
61. Gloster, J.: Fundus oximetry, Exp. Eye Res. **6**:187, 1967.
62. Grayson, M. C., and Laties, A. M.: Ocular localization of sodium fluorescein, Arch. Ophthalmol. **85**: 600, 1971.
63. Grotte, G.: Passage of dextran molecules across the blood-lymph barrier, Acta Chir. Scand. (suppl.): 1211, 1965.
64. Hayreh, S. S.: The ophthalmic artery. III, Branches, Br. J. Ophthalmol. **46**:212, 1962.
65. Hayreh, S. S.: Posterior drainage of the intraocular fluid from the vitreous, Exp. Eye Res. **5**:123, 1966.
66. Hayreh, S. S.: Optic disc edema in raised intracranial pressure, V, Pathogenesis, Arch. Ophthalmol. **95**: 1553, 1977.
67. Hayreh, S. S.: Pathogenesis of optic nerve damage and visual field defects. In Heilmann, L. K., and Richardson, K. T., editors: Glaucoma, Stuttgart, West Germany, 1978, Georg Thieme Verlag KG.
68. Henkind, P., and de Oliveira, F.: Development of the retinal vessels in the rat, Invest. Ophthalmol. **6**:520, 1967.
69. Hickam, J. B., and Frayser, R.: A photographic method for measuring the mean retinal circulation time using fluorescein, Invest. Ophthalmol. **4**:876, 1965.
70. Hickam, J. B., Frayser, R., and Ross, J.: A study of retinal venous blood oxygen saturation in human subjects by photographic means, Circulation **27**:375, 1963.
71. Kogure, K., et al.: Infrared absorption angiography of the fundus circulation, Arch. Ophthalmol. **83**:209, 1970.
72. Kaskel, D., Mettais, P., and Bill, A.: The effects of arterial blood pressure increments on the regional

blood flow in the eye and brain of cats. In Proceedings of the Ninth European Conference on Mirocirculation, Antwerp, Belgium, 1976, Bibl. Anat. **16:**36, 1977.

73. Kuwabara, T., and Cogan, D.: Retinal vascular patterns. IV, Mural cells of the retinal capillaries, Arch. Ophthalmol. **69:**492, 1963.
74. Landis, E. M., and Pappenheimer, J. R.: Exchange of substances through the capillary wall. In Handbook of physiology. Section 2, Circulation, Baltimore, 1963, The Williams & Wilkins Co., vol. 2, p. 961.
75. Laties, A. M.: Central retinal artery innervation, Arch. Ophthalmol. **77:**405, 1967.
76. Leber, T.: Circulations und Ernährungsverhältnisse des Auges. In Graefe, A., and Saemisch, T., editors: Handbuch der gesammten Augenheilkunde, Leipzig, Germany, 1903, Springer Verlag.
77. Lemmingson, W.: Über das Vorkommen von Vasomotion im Retinakreislauf, v. Graefes Arch. Ophthalmol. **176:**368, 1968.
78. Linnér, E.: Ascorbic acid as a test substance for measuring relative changes in the rate of plasma flow through the ciliary processes: the effect of ligation of the common carotid artery on the blood flow through the uveal tract as measured directly in a vortex vein, Acta Physiol. Scand. **26:**70, 1952.
79. Mettais, P., Kaskel, D., and Bill, A.: The effects of arterial blood pressure increments on the intraocular pressure, the intrascleral venous pressure and the facility of outflow in cats. In Proceedings of the Ninth European Conference on Microcirculation, Antwerp, Belgium, 1976, Bibl. Anat. **16:**59, 1977.
80. Michaelson, I. C.: Retinal circulation in man and animals, Springfield, Ill., 1954, Charles C Thomas, Publisher.
81. Neufeld, A. H., Jampol, L. M., and Sears, M. L.: Aspirin prevents the disruption of the blood-aqueous barrier in the rabbit eye, Nature **238:**158, 1972.
82. Niesel, P.: Messungen von experimentell erzeugten Aenderungen der Aderhautdurchblutung bei Kaninchen, Basel, Switzerland, 1962, S. Karger. AG.
83. O'Day, D. M., et al.: Ocular blood flow measurements by nuclide labelled microspheres, Arch. Ophthalmol. **86:**205, 1971.
84. Patz, A.: New role of the ophthalmologist in prevention of retrolental fibroplasia, Arch. Ophthalmol. **78:**565, 1967.
85. Perkins, E. S.: Influence of the fifth cranial nerve on the intraocular pressure of the rabbit eye, Br. J. Ophthalmol. **41:**257, 1957.
86. Rodenhäuser, J. H.: Uvea-durchblutung und Augeninnendruck, Stuttgart, West Germany, 1962, Ferdinand Enke Verlag.
87. Rohen, J.: Über das Gefässystem der Retina beim Kaninchen, Ophthalmologica **128:**307, 1954.
88. Ruskell, G. L.: Anterior communications between the intrinsic and the extrinsic arteries of the rabbit eye, Anat. Rec. **142:**147, 1962.
89. Ruskell, G. L.: Facial parasympathetic innervation of the choroidal blood vessels in monkeys, Exp. Eye Res. **12:**166, 1971.
90. Rusznyak, I., Földi, M., and Szabo, G.: Lymphatics and lymph circulation, ed. 2, New York, 1967, Pergamon Press, Inc.
91. Sears, M. L.: Miosis and intraocular pressure changes during manometry, Arch. Ophthalmol. **63:**707, 1960.
92. Shabo, A. L., and Maxwell, D. S.: The blood-aqueous barrier to tracer protein: a light and electron microscopic study of the primate ciliary process, Microvasc. Res. **4:**142, 1972.
93. Shiose, Y.: Electron microscopic studies on blood-retinal and blood-aqueous barriers, Jpn. J. Ophthalmol. **14:**73, 1971.
94. Stjernschantz, J., and Bill, A.: Effects of facial nerve stimulation on the blood flow of the eye and tongue, Acta Physiol. Scand. (in press).
95. Stjernschantz, J., and Bill, A.: Effect of intracranial stimulation of the oculomotor nerve on ocular blood flow in the monkey, cat and rabbit, Invest. Ophthalmol. **18:**99, 1979.
96. Stjernschantz, J., Alm, A., and Bill, A.: Effects of intracranial oculomotor nerve stimulation on ocular blood flow in rabbits. Modification by indomethacin, Exp. Eye Res. **23:**461, 1976.
97. Törnquist, P.: Capillary permeability in cat choroid studied with the single injection technique, II, Acta Physiol. Scand. **106:**425, 1979.
98. Törnquist, P., and Alm, A.: Retinal and choroidal contribution to retinal metabolism in vivo. A study in pigs, Acta Physiol. Scand. **106:**351, 1979.
99. Törnquist, P., Alm, A., and Bill, A.: Studies on ocular blood flow and retinal capillary permeability to sodium in pigs, Acta Physiol. Scand. **106:**343, 1979.
100. Toussaint, D., Kuwabara, T., and Cogan, D. G.: Retinal vascular patterns, Arch. Ophthalmol. **65:**575, 1961.
101. Uusitalo, R., and Palkama, A.: Evidence for the nervous control of secretion in the ciliary processes, Prog. Brain Res. **43:**513, 1971.
102. van Alphen, G. W.: On emmetropia and ametropia, Ophthalmologica **142**(suppl.), 1961.
103. Vegge, T.: An epithelial blood-aqueous barrier to horseradish peroxidase in the ciliary processes of the vervet monkey *(Cercopithecus ethiops),* Z. Zellforsch. Mikrosk. Anat. **114:**309, 1971.
104. Vegge, T.: An electron microscopic study of the permeability of iris capillaries to horseradish peroxidase in the vervet monkey *(Cercopithecus ethiops),* Z. Zellforsch. Mikrosk. Anat. **121:**74, 1971.
105. Vegge, T.: A study of the ultrastructure of the small iris vessels in the vervet monkey, Z. Zellforsch. Mikrosk. Anat. **123:**195, 1972.
106. Weigelin, E., and Lobstein, A.: Ophthalmodynamometry, New York, 1963, Hafner Publishing Co., Inc.
107. Wudka, E., and Leopold, I. H.: Experimental studies of the choroidal vessels, Arch. Ophthalmol. **55:**857, 1956.
108. Yudilevich, D. L., and Alvarez, O. A.: Water, sodium and thiourea transcapillary diffusion in the dog heart, Am. J. Physiol. **213:**308, 1967.

Chapter 7

The aqueous

MARVIN L. SEARS, M.D.

DEDICATED TO THE MEMORY OF V. E. KINSEY

Ordinarily no blood vessels are found in the cornea, lens, vitreous, or trabecular meshwork. Indeed, proportionately, the eye contains the largest avascular mass found in an organ anywhere in the body. Even those vessels found in the sclera are merely in transit. The nutrition of these avascular structures is accomplished mainly by the formation and delivery of the aqueous humor. The process is blood flow dependent, but it is largely the energy available from the cellular metabolism of the epithelia of the ciliary processes that accounts for the extraction and formation of the aqueous humor.

Interest in the aqueous humor was perhaps first kindled by knowledge of the relationship between the steady state level of intraocular pressure and the rate of formation and exit of aqueous from the eye. In ''glaucoma'' elevated pressure levels related to reduced facility of outflow of aqueous are found. Of course, it is primarily through the avascular trabecular mesh that outflow occurs. The dependence of the mesh on the aqueous for its nutrition is evident. Unfortunately the mystery of the composition of the aqueous, especially from this nutritive point of view, is only just now being explored. The same problem confronts investigators studying the substrates required for the metabolism of the lens, the corneal endothelium, and the vitreous body. Thus to understand the normal and altered physiology of several ocular functions, it is the challenge of aqueous production that must be explored.

☐ This work was supported in part by Research Grant EY 00785 from the National Eye Institute, National Institutes of Health, Research to Prevent Blindness, Inc., and the Connecticut Lions Eye Research Foundation, Inc.

CILIARY BODY

The ciliary body is a specialized structure of the uveal tract. It has at least four functions other than the formation of aqueous humor. It (1) secretes hyaluronic acid of the vitreous, (2) plays the dominant role in accommodation, (3) influences the outflow facility of aqueous by way of its smooth muscle, and (4) constitutes in large part the so-called blood-aqueous barrier.

The ciliary body extends from the ora serrata posteriorly to the scleral spur anteriorly. On sagittal section (Fig. 7-1) the ciliary body is seen as a triangular structure with its shortest side facing anteriorly. This anterior surface participates in the uveal portion of the trabecular meshwork of the angle of the anterior chamber. Also, the iris originates from its middle portion. The outer surface of the ciliary body lies against the sclera. The potential supraciliary space is continuous posteriorly with the potential suprachoroidal space. The ciliary body is held against the sclera by the intraocular pressure. The inner surface of the ciliary body can be divided into two regions; the posterior two thirds appears grossly smooth and is called the pars plana ciliaris, and the anterior third is called the pars plicata ciliaris because about 70 radiating villiform ridges project inward from it. These ridges are the ciliary processes.

The layers comprising the ciliary body are (1) ciliary muscle, (2) layer of vessels and the cili-

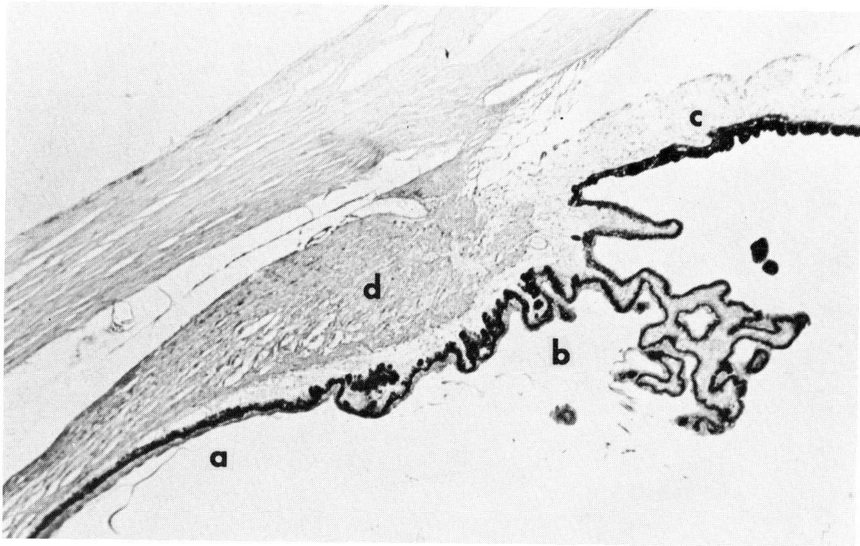

Fig. 7-1. Ciliary body. *a*, Pars plana ciliaris; *b*, plicata ciliaris; *c*, iris; *d*, ciliary muscle. (Hematoxylin and eosin stain; magnification ×10.) (Courtesy Dr. T. Farkas.)

ary processes, (3) basal lamina, (4) ciliary epithelium, and (5) internal limiting membrane.

The ciliary muscle resembles a right-angled triangle in anteroposterior section with the right angle facing inward (Fig. 7-1). The hypotenuse of the triangle runs parallel with the sclera, and the acute posterior angle points toward the choroid. The muscle that gives shape to the ciliary body is made up of flat, unstriated muscle fibers. These are organized into bundles of tightly packed muscle cells[33] rather than syncytia, with each bundle containing its own nerve supply. The bundles, with interweaving connections, tend to be organized into three masses, the meridional one of which has the scleral spur as its locus of insertion. The meridional bundles are likely the most involved in the regulation of outflow of aqueous humor.

The stroma of the ciliary body is separated from the ciliary epithelium by the forward continuation of Bruch's membrane. This basal lamina, as it approaches the ora serrata, splits into an outer elastic and an inner cuticular layer separated by a layer of avascular connective tissue.

The ciliary processes are composed of blood vessels embedded in a loose connective tissue (Fig. 7-2). The capillaries in this loose stroma are fenestrated, like those in the kidney. This anatomic feature will, as we shall see later, have some bearing on our ideas about the way in which the aqueous humor is formed.

Amazingly, a double layer of epithelial cells lines the inner surface of the basal lamina—the outer pigmented and the inner nonpigmented (Fig. 7-3). During embryogenesis of the ciliary processes the invagination of the optic cup causes these two cell layers to become apposed apex to apex. The pigmented cells represent the forward continuation of the retinal pigment epithelium. The nonpigmented cells represent the forward continuation of the retina. The pigmented cells do not have pigmented processes like the true retinal pigment epithelium; also, they are shorter and their pigment granules are darker. The dark color of the ciliary body is caused by these dark pigment granules. These cells invaginate in the anterior portion of the ciliary processes and resemble tubular glands. Indeed the entire group of processes can be thought of as an everted gland.

Electron microscopic studies[57] (Fig. 7-4) show that processes of the pigmented cells indent the characteristically thin fenestrated walls of the capillaries. There is an elaborate interdigitation between the adjacent surfaces of the

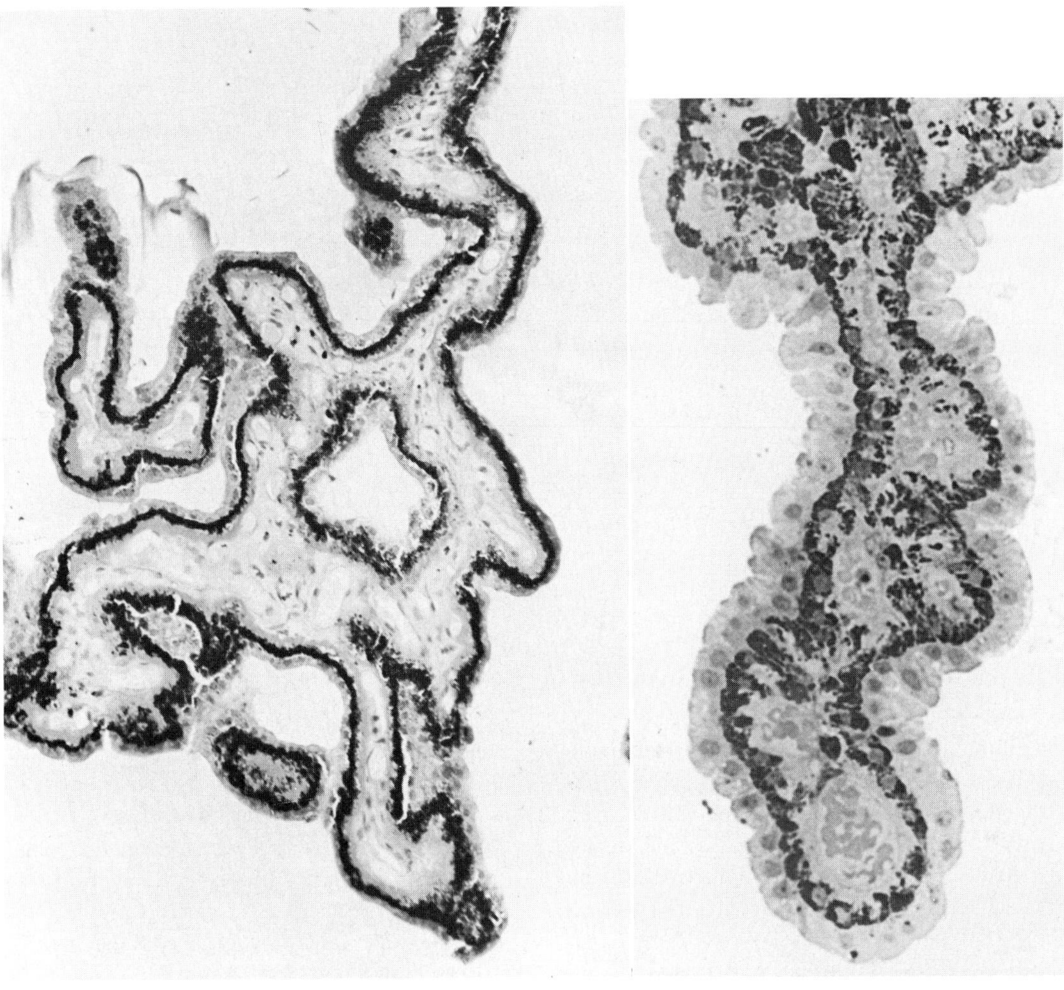

Fig. 7-2 Fig. 7-3

Fig. 7-2. Ciliary process. (Hematoxylin and eosin stain; magnification ×40.) (Courtesy Dr. T. Farkas.)
Fig. 7-3. Ciliary process in cross section, rhesus monkey. Light micrograph 2 μm thin section. (Magnification ×320.) (Courtesy T. Vegge.)

pigmented and the nonpigmented cells, which leads to a relatively firm union between the cell layers; hence spontaneous detachment of neuroepithelium does not occur over the ciliary body. Furthermore, interdigitation of cells may occur, so the intercellular clefts between the two cell layers are not lined up. Thus to reach the posterior chamber of the eye, a substance that could pass *between* the cells of the outer layer may be required to pass *through* the cells of the inner layer. Functional correlates of this morphologic arrangement have not been fully developed, but undoubtedly will bear heavily on the discrete purposes of each cell layer.

The internal limiting membrane of the retina continues forward over the bases of the nonpigmented ciliary epithelium, thus forming the internal limiting membrane of the ciliary body and lining the posterior chamber.

Recent work supports the idea that both nonpigmented and pigmented epithelial cells of the ciliary processes can regenerate after injury.[62]

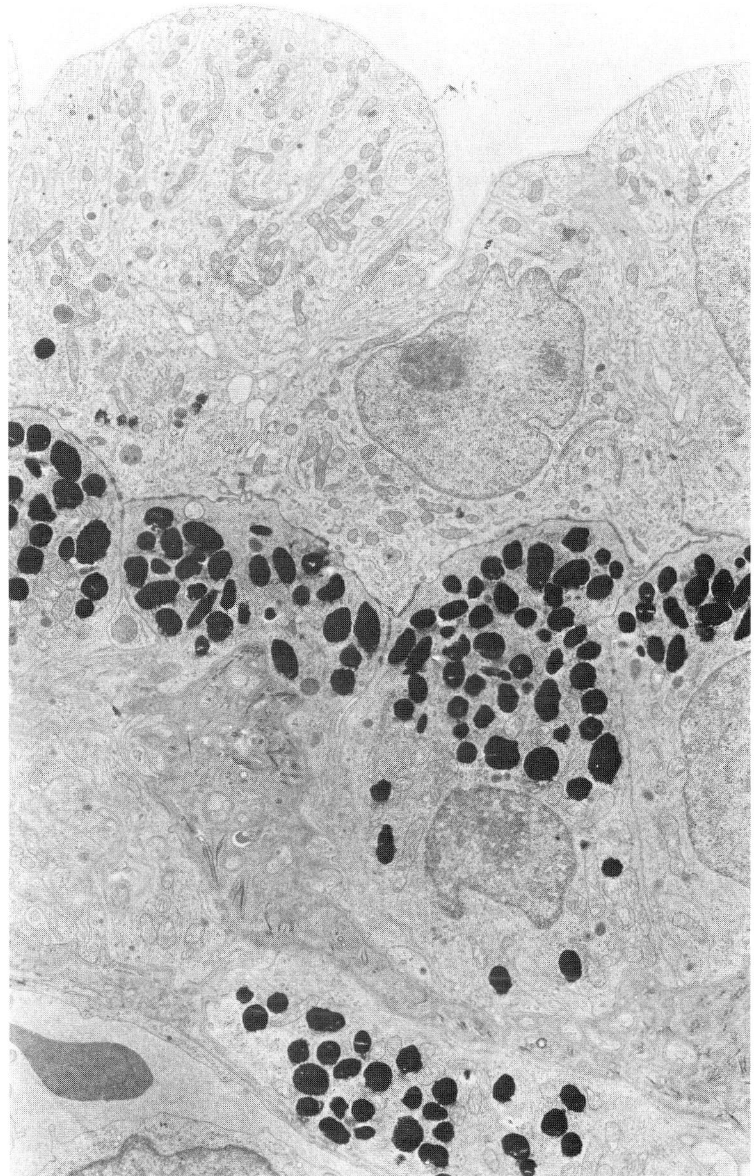

Fig. 7-4. Detail of ciliary process, vervet monkey. Section is slightly tangential to the process, grazing a pigment epithelial cell close to capillary at bottom of illustration. Section also grazes basement membrane below this cell, showing some collagen fibrils and cellular profiles in a darker mass of basement membrane material. In isotonically fixed tissue (like this) intercellular clefts of epithelium are 200 to 300 Å wide, with very few dilations, except for "ciliary channels" frequently present at apical end of intercellular clefts of nonpigmented epithelium. Note capillary in stroma beneath basement membrane. (Magnification ×7500.) (Courtesy T. Vegge.)

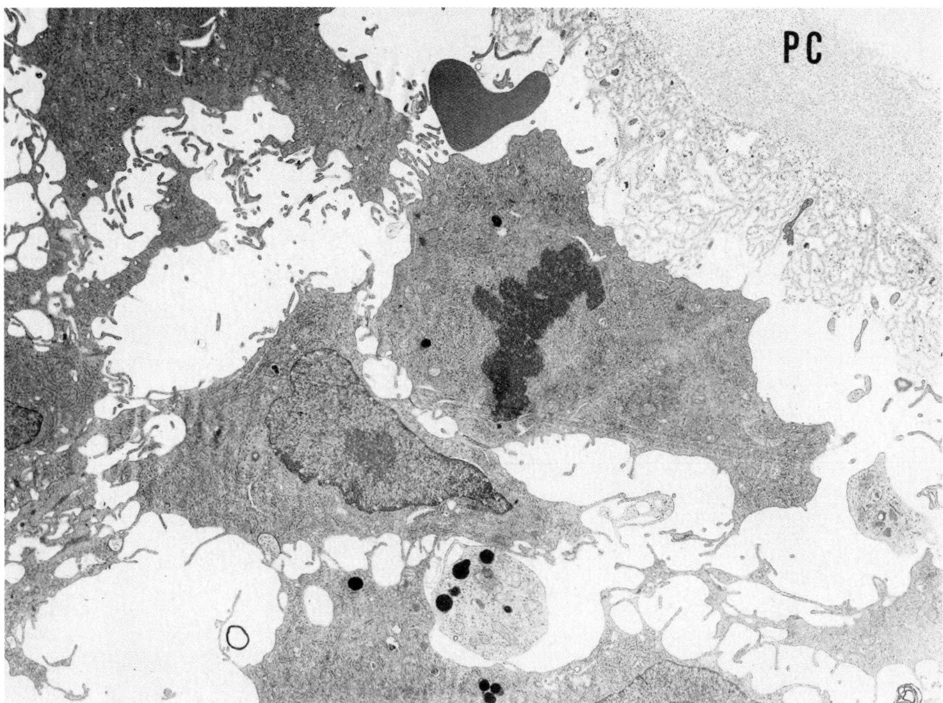

Fig. 7-5. Regeneration of nonpigmented ciliary epithelium. Electronmicrograph of mitosis within a nonpigmented epithelial cell. *PC*, Posterior chamber. (Magnification ×1300.) (Courtesy Dr. Hideaki Yamashita; from Sears, M. L.: Invest. Ophthalmol. Vis. Sci. **17**:6, 1978.)

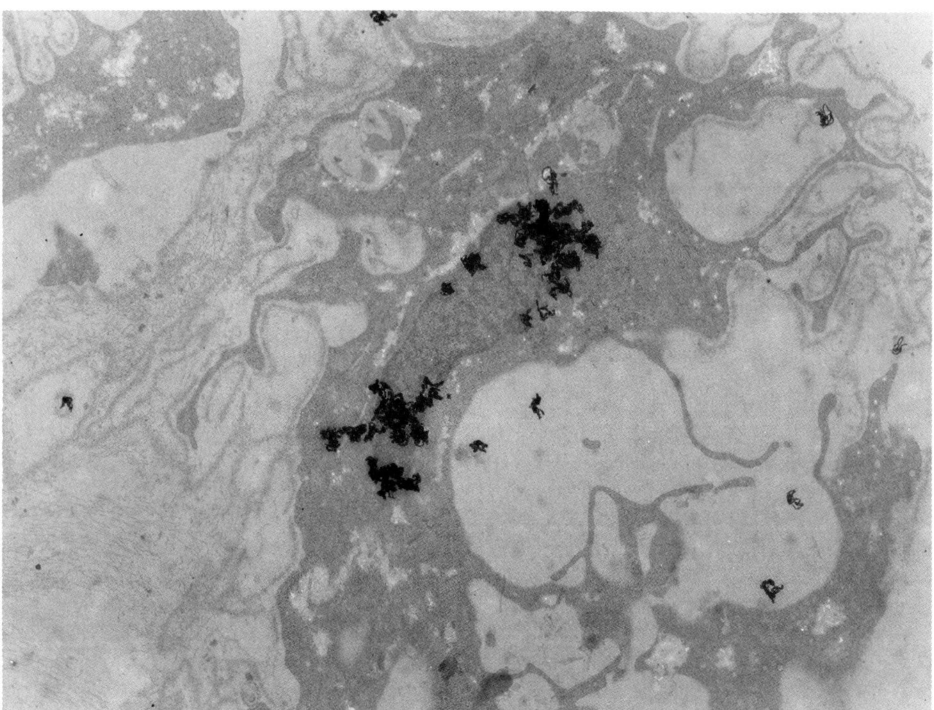

Fig. 7-6. Autoradiograph of ciliary body specimen taken seven days after epithelial injury shows incorporation of tritiated thymidine. (Courtesy Dr. Hideaki Yamashita; from Sears, M. L.: Invest. Ophthalmol. Vis. Sci. **17**:6, 1978.)

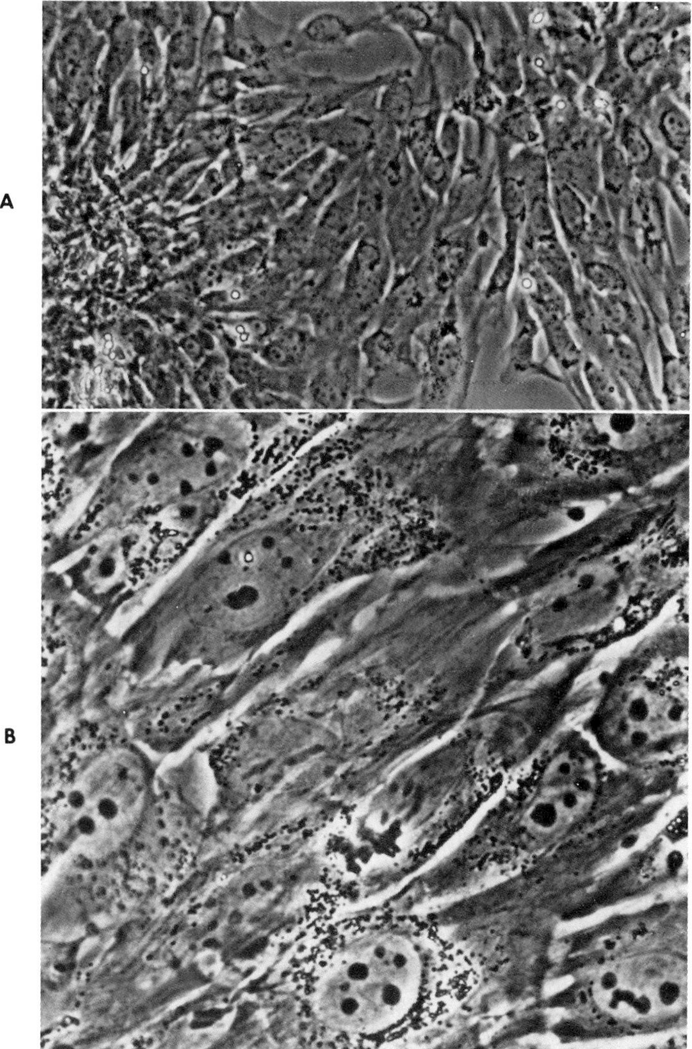

Fig. 7-7. A, Ciliary process explant from which epithelial cells have grown out in a monolayer. **B,** Higher magnification taken through a Zeiss phase microscope. Cells were viewed through a modified Rose chamber.

Mitoses are rarely, if ever, observed in normal ciliary epithelium. With colchicine treatment, however, many mitotic figures can be seen in nonpigmented epithelial cells regenerating after injury (Fig. 7-5). Furthermore, autoradiographic studies define the incorporation of tritiated thymidine and indicate that both nonpigmented and pigmented cells can undergo DNA synthesis, although the regenerative capacity of the nonpigmented layer is greater than that of the pigmented cell layer (Fig. 7-6). These studies will probably lead to an analysis of the discrete functions of the individual cell layers in tissue culture (Fig. 7-7).

Blood supply

The long posterior ciliary arteries and the anterior ciliary arteries unite in the anterior portion of the ciliary body in front of the circular portion of the ciliary muscle to form the major arterial circle of the iris. This arterial circle supplies the ciliary muscle, the ciliary processes, and the iris; it sends recurrent branches to the anterior choroid, where they anastomose with the end-branches of the short ciliary arteries.

The ciliary muscle is nourished by a large number of arterial branches, which divide dichotomously to form a dense capillary network.

The short radial ciliary arteries that supply the ciliary processes pierce the ciliary muscle. Usually each ciliary process receives its own arterial branch, which enters the ciliary process anteriorly, breaks up into many branches that anastomose with one another, and finally forms a rich capillary network.[46] This capillary network constitutes the main mass of the ciliary process. The capillary network drains into a continually anastomosing venous system that receives blood from the ciliary muscle also. This venous system runs posteriorly to drain through the vortex veins.

Nerve supply

The ciliary body is innervated by the long and short ciliary nerves that accompany the similarly named arteries. In the ciliary processes, nonmedullated nerve fibers can be found surrounding the vessels. Many of these are adrenergic and probably subserve vasomotion. Anatomic studies have demonstrated cholinergic fibers along the ciliary arteries and a few in the stroma of the processes, but the physiologic significance of these is not yet proved. Despite intensive search, no direct innervation of the ciliary epithelia themselves has been found. Scattered throughout the ciliary body are sensory nerve fibers that can be recognized by club-shaped nerve endings. It is unlikely that these are pressure-sensitive nerve endings.

THE AQUEOUS

The aqueous fills the anterior and posterior chambers of the eye. The volume of the anterior chamber aqueous of the human eye is about 0.25 ml. The posterior chamber volume is about 0.06 ml. The index of refraction of the aqueous is 1.336,* slightly lower than that of the cornea, so the cornea-aqueous interface acts as a diverging lens of low power. The density of the normal aqueous is slightly greater than that of water. The viscosity is 1.025 to 1.040 relative to water.

*Depending on temperature, concentration of solids, and the wavelength of light.

The osmotic pressure of the aqueous is probably slightly hyperosmotic to plasma by some 3 to 5 mOsm/L.*

The aqueous is a slowly flowing stream (roughly 2 μl/min in humans, or usually 1% of the anterior chamber volume per minute in most species) into which surrounding tissues can discharge waste products of metabolism. By means of a steady aqueous formation and drainage, the intraocular pressure is maintained and the globe is preserved in an optical form with the position of the refractive surfaces relative to one another.

Factors influencing composition

The composition of the aqueous will be determined by the following:

1. The physical characteristics of the membrane (barrier) separating the blood from the aqueous
2. Hemodynamic factors influencing the content of the ciliary stromal filtrate from which "aqueous" will be extracted
3. The quality of the formation process and factors, neurohumoral and others, regulating this process
4. The rate of aqueous drainage
5. Other factors diluting or altering the aqueous in its course, such as diffusional exchange across the iris
6. The contribution made by the metabolism of structures contiguous to the path of flow of aqueous

*The significance of this difference, even with accurate measurements of osmolality, is debatable. In all likelihood, water traverses the ciliary epithelium in proportion to the rate of entry of solute. Therefore a slight difference in aqueous-plasma osmolality probably does not generate aqueous formation but is a consequence of it. The osmolality of the posterior aqueous depends on the rate of ion transport and the passive influx of water that will be limited by the osmolality of the plasma. In any event, it is the value for osmolality of the fluid immediately adjacent to the epithelia, perhaps even in the intercellular clefts, that would be more enlightening than values obtained so far downstream. (See Cole[22] for full discussion.) Similar considerations apply to the values obtained for the concentration of solutes in the aqueous chambers (p. 220).

Blood-ocular barrier

The term aqueous is properly limited to the fluid found in the healthy, undisturbed eye. Any change in the state of dilation of the intraocular blood vessels, in the membranes forming the blood-ocular barrier, in the intravascular pressure, or in the composition of the blood may lead to alterations in the chemical and physical properties of the aqueous. The fluid can no longer be regarded as normal. Irritation of the eye may lead to vasodilation by axon reflexes,[24,47] and the irritative response can be reproduced in the rabbit in part by prostaglandins[1,2] and related compounds. The abnormal aqueous usually has an increased protein content of variable amount, and depending on the extent of breakdown of the various barrier structures, the composition of aqueous approaches that of plasma. It is therefore called plasmoid aqueous. Increased protein content is recognized clinically by the visibility of the slit lamp beam as it traverses the normally optically empty anterior chamber. Normally the path of the beam is invisible because the protein content is only 50 mg/100 ml, but greater amounts of protein scatter light, so the light path is seen as the aqueous flare or ray. Cells or other particles floating in the aqueous are seen as tiny bright points in the slit beam.

The most obvious chemical difference between aqueous humor and blood plasma is found in the protein concentrations of the two fluids. In the plasma of humans and rabbits there are 6 to 7 g/100 ml, but in the aqueous of rabbit there are only 50 mg/100 ml and, in humans, still less. Electrophoretic studies have shown that all the proteins of the plasma are present in the aqueous but in different proportions.[26] The proteins of aqueous likely originate from the plasma rather than by synthesis during the secretory process or as a consequence of cell turnover and production of cellular debris. The great difference in concentrations of protein and other large molecules between aqueous and blood indicates that a "barrier" is present, causing a relatively slow rate of penetration from the blood side to the aqueous side of the ciliary process. The concentration of these large molecules never rises very much in the aqueous because the fluid continuously exits.

There are at least two barriers for relatively large substances (40 Å in diameter). One is vascular and the other epithelial. At the retinal level, tight junctions between the endothelial cells of the retinal capillaries provide a vascular locus,[49] and tight junctions between the pigment epithelial cells of the retina[52] provide an additional tissue barrier against substances diffusing from the choroid via the fenestrated capillaries of that tissue. Substances diffusing from the choroid into the optic nerve may enter the vitreous through the optic papilla.[30] At the aqueous level, again, two types of permeability barriers are present. In the iris vessels of most species, tight junctions between capillary endothelial cells are found.[58] Although the reactivity of these may differ somewhat from that of retinal vessels, they are nonetheless effective in maintaining a relatively protein-free anterior aqueous. The barrier is bolstered by the posterior chamber lined by the inner layer of nonpigmented epithelial cells ringed to each other near their apices with tight junctions.[54,57] In contrast, the ciliary stromal capillaries beneath are fenestrated, like those in the choroid. Large lipid-insoluble molecules will pass to a moderate degree from fenestrated capillaries into stroma, but are markedly restricted by cells and cell junctions lining the barrier (Figs. 7-8 and 7-9).

Another sort of permeability barrier must exist for small molecules like urea. Penetration of small molecules like urea is, of course, faster than for proteins. A third type of barrier is represented by the lesser restriction to the passage of lipid-soluble molecules. Lipid-soluble molecules and many ions will pass with relative ease across the blood-ocular barrier. If unionized compounds have the same lipid solubility, the dissociation constant will determine the rate of penetration. The barrier strongly influences the steady state level of substances in the aqueous because its permeability characteristics determine their rates of penetration, and aqueous flows away all the time.

There is also a barrier to the entry of water into the eye. Water penetrates a good deal more rapidly than solutes, however. Thus the abrupt systemic administration of hypertonic solutions causes an acute transient fall in intraocular pres-

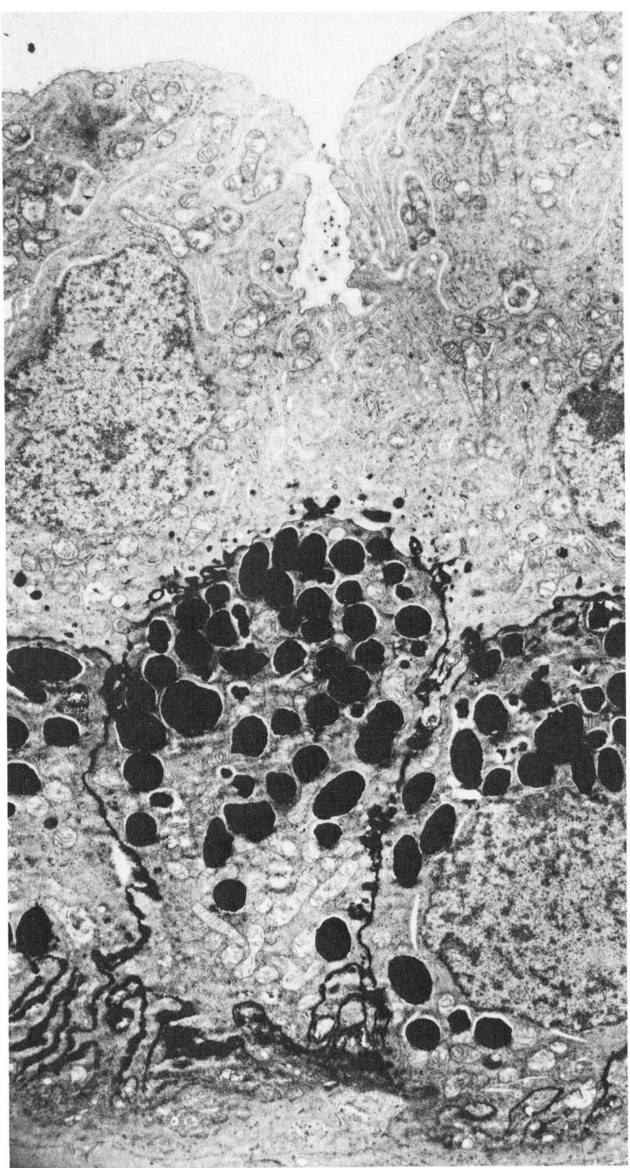

Fig. 7-8. Detail of ciliary process, vervet monkey. This animal received an intravenous injection of a protein tracer (horseradish peroxidase) 9 minutes prior to enucleation. Tissue was later reacted with appropriate substrate to create black reaction product. Tracer is present in stroma of process, in intercellular cleft of pigmented layer, and in cleft between pigmented and nonpigmented cells. A few intracellular vesicles show tracer, but no tracer has passed into lateral intercellular clefts of nonpigmented layer or into posterior chamber. (Magnification ×7500.) (Courtesy T. Vegge.)

sure by a loss of intraocular water to the blood. The duration of action of the solute will be related to the degree to which it is excluded from the ocular fluids and to other systemic factors, such as its rate of excretion by the kidneys. The response of intraocular pressure to parenteral hyperosmotic agents is useful in the preoperative preparation of patients with elevated intraocular pressure.

It is clear from this description of the blood-ocular barrier that the barrier concept must be broad to include observations derived from both

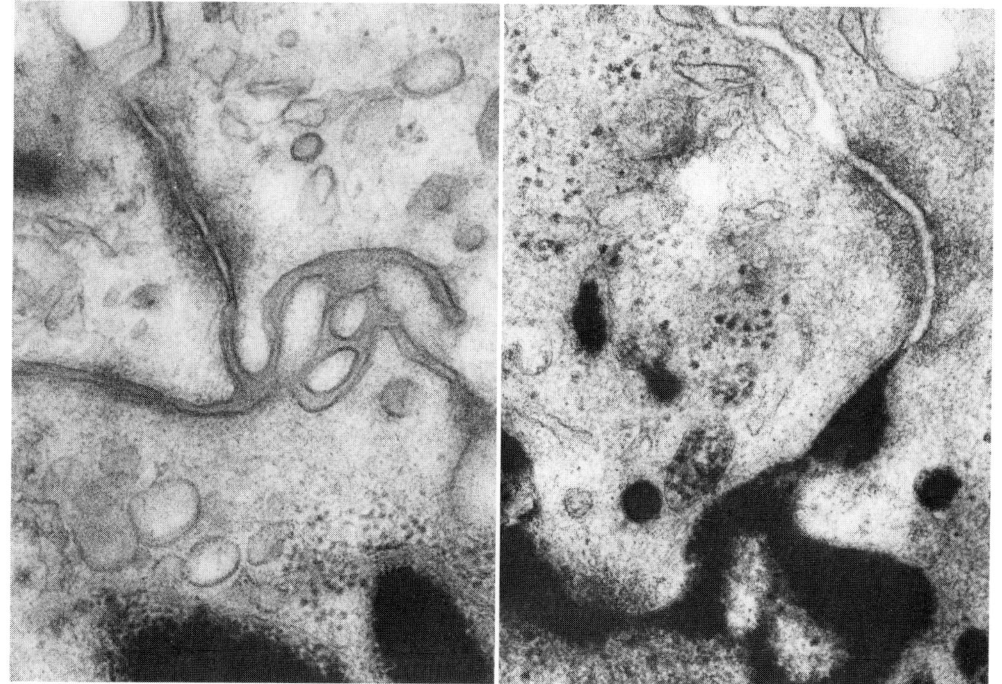

Fig. 7-9. A, Detail of junctional complex at "apical" end of intercellular cleft of nonpigmented epithelium. Tight junction seals cleft followed by adherent junction. A few microvillous projections are seen in ciliary channel. (Magnification ×60,000.) **B,** Area similar to **A,** after tracer injection. Tracer fills ciliary channel but does not pass the tight junction into lateral intercellular cleft of nonpigmented epithelium. (Magnification ×60,000.) (Courtesy T. Vegge.)

anatomic and physiologic sources. Although the true nature of the barrier is not understood, an accurate accounting will need to consider the barrier as a series of variable resistances located between the plasma and the aqueous: the capillary wall, ciliary stroma, basement membrane of the epithelia, the double epithelial layer, and the junctions between the two layers and between the adjacent epithelial cells of each layer.

Stromal pool

The first step in the formation of aqueous is the development of a plasma filtrate in the stroma of the ciliary processes. This process probably occurs by filtration through the capillary wall. Protein is to be expected in the filtrate because of the fenestrated nature of the ciliary capillaries.[13] It is on this stromal pool that the epithelia perform their work. Proponents of the hypothesis that ultrafiltration is an important factor in aqueous formation still argue that a good deal of this stromal filtrate can be hydrostatically forced across the epithelial layer. Bill,[14] however, has argued strenuously that transepithelial ultrafiltration is most unlikely to contribute to the formation of aqueous. The formation of aqueous persists in monkeys at very low arterial pressures. Also, the formation of aqueous is apparently independent of increments in blood flow through the ciliary processes. Finally, Bill[14] calculated an approximate value for the transepithelial difference in hydrostatic pressures from data for permeability of iris–ciliary process preparations and from data for the turnover of albumin in the tissue fluid of the ciliary processes. The value for this pressure difference is less than the oncotic pressure of the stromal tissue fluid, indicating, if anything, a net movement of fluid out of the eye from the posterior chamber, or opposite to the effect predicted by the ultrafiltration theory. These and other studies have indicated that the process of aqueous humor formation occurs first by filtration across the fenestrated capillaries of the ciliary stroma followed by secretion across the ciliary epithelium.[15] The site of "pseudofacility," the pressure sensitive part of aqueous humor formation, would be ultrafiltration and would be located at the vascular level within the stroma of the ciliary body.

Biophysical considerations

Before the biology of the formation process is described, it may be well to review certain of the physical principles on which these processes depend.

Diffusion. Molecules of gases and liquids are in continuous random motion. The magnitude of the motion depends on the temperature and structure of the molecules.

Diffusion is the process by which molecules of a gas or solution distribute themselves uniformly throughout the space in which they are contained. Diffusion of substances may occur across biologic membranes. The simplest case is one in which a solution is divided into two portions by a membrane permeable to both the solvent and solute, for example, water and sodium chloride. In this instance, both the salt and water molecules on one side of the membrane will interchange continuously with those on the other side (Fig. 7-10, *A*). If the concentrations of the solutions on the two sides of the membrane are initially different, exchange in both directions will proceed as before, but until equilibrium is reached, more salt will cross the membrane in the direction of higher to lower salt concentration, and more water will move in the opposite direction (Fig. 7-10, *B*). It should be pointed out that, complementary to the case of the salt, the direction of the larger water movement is also from higher concentration of water to lower concentration. The concentration of water is highest in pure water; that is, 1 L of pure water contains 1000 g of water, and therefore its molar concentration is 1000 g/18 or 55.5 M. Addition of salt reduces the amount of water present per unit volume; therefore the water is less concentrated in a salt solution. It will be apparent later why emphasis is placed on this point of view. Movement of solvent or solute across a membrane in the manner described has been expressed quantitatively (Fick's law) as follows:

$$\text{Rate of transfer} = k\,(C_{\text{side }1*} - C_{\text{side }2})$$

where

C = Concentration
k = Constant, depending on permeability of membrane, temperature, and nature of substance
$*$ = Side containing highest concentration

A semipermeable membrane is permeable to the solvent but impermeable to one or more solutes, for example, one that will permit the passage of water but not salt, or water and salt but not protein. The water will exchange across the membrane and more will move in the direction of higher concentration of water to lower concentration. The rate of net transfer will continuously decrease as the concentrations of water on the two sides approach each other.

The net transfer of water across the mem-

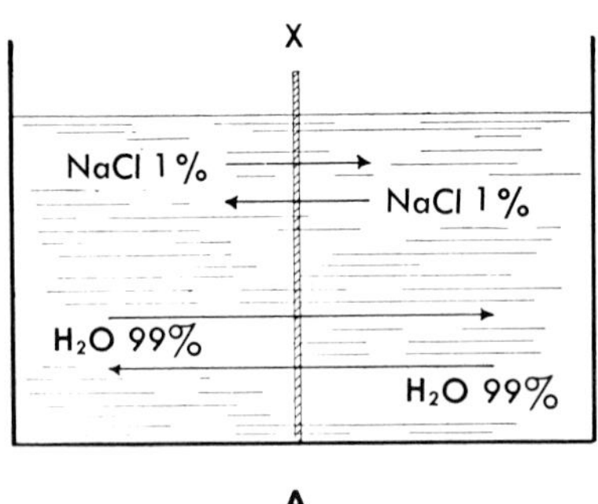

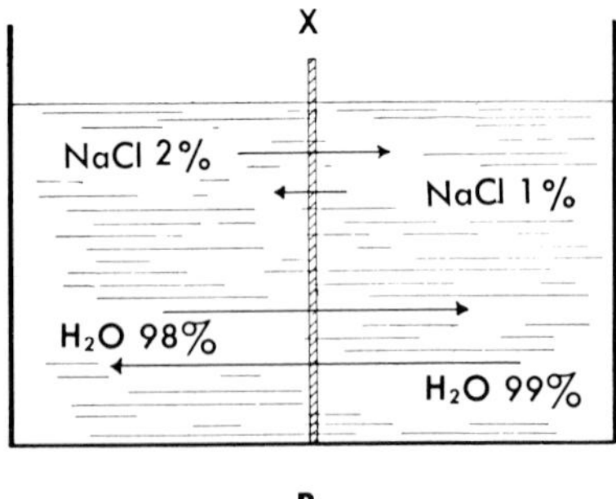

X—Membrane
Permeable to NaCl
and H_2O

Fig. 7-10. Diffusion of solutes and solutions according to gas laws.

brane can be explained on the basis that there are more molecules of water bombarding the membrane on one side than bombarding it on the other side; thus the chance of one molecule passing through the pores in one direction is correspondingly greater. This excess of water molecules going through the membrane creates a pressure known as osmotic pressure (Greek *ōsmos*, push) and the process is called osmosis.

Osmotic pressure. Osmotic pressure is measured by opposing it with sufficient hydrostatic pressure from the side of the membrane containing the higher concentration of solute so that there is no longer any net gain of water on this side.

It follows from this explanation of the source of osmotic pressure that sodium chloride, or any other solute on one side of a semipermeable membrane, does not "attract" the water from the other side, as is commonly stated.

The magnitude of the osmotic pressure depends on the number of dissolved particles to which the membrane is impermeable. In dilute solutions where particles act essentially like gas molecules, for example, in a 0.1 M solution of a nonionizing solute like glucose, a pressure of 2.24 atm (1700 mm Hg) will be developed when the solution is separated from water by a membrane that will not permit passage of the solute. If, in any solution, the solute dissociates into two ions, for example, sodium and chloride, twice the osmotic pressure is produced per mole fraction; that is, a 0.1 M solution in this case produces an osmotic pressure of 3400 mm Hg.

It is evident that relatively slight differences in concentration of salts can produce large differences in pressure; for example, an osmotic pressure of 58 mm Hg will be generated when solutions differing in concentration by only 0.01% (1.72 mmol/L) sodium chloride are separated by a membrane impermeable to the salt. Since osmotic pressure depends on the number rather than the mass of the dissolved particles, a difference in concentration of protein equal to 0.01% will generate only a small difference in osmotic pressure. If the molecular weight of the protein were 60,000 (albumin), the osmotic pressure of a 0.01% solution separated from water by a semipermeable membrane would be but 0.025 mm Hg. The difference in osmotic pressure of the plasma compared with interstitial fluids will depend, of course, on the regional differences in the plasma concentration of proteins as well as on the protein content of the tissue fluid. If tissue fluid contained essentially no protein, the difference in plasma osmotic pressure would be about 30 mm Hg, caused by a difference in protein of approximately 7%.

Dialysis and ultrafiltration. When a solution containing protein and salt is separated from pure water by a membrane permeable to salt and water but not to protein, the salt will diffuse into the water and leave the protein behind. In this case there will be a net transfer of water into the solution containing the protein. This process is called dialysis. If, on the side of the membrane containing the protein, a hydrostatic pressure is applied, for example, the blood pressure, the transfer of salt across the membrane is accelerated. The term ultrafiltration is used to describe this process. Ultrafiltration may be defined, therefore, as dialysis in the presence of hydrostatic pressure.

The Gibbs-Donnan equilibrium. If the solutions on both sides of a membrane impermeable to protein contain salt but only one solution contains protein, an unequal distribution of the ions of the salt will occur. This effect is caused by binding of positive ions (cations), like sodium, by the protein that, at the hydrogen-ion concentration of the plasma, carries a negative charge. The magnitude of the unequal distribution of ions depends on the concentration of the protein. This unequal distribution was predicted by Gibbs and confirmed later experimentally by Donnan, and is known as the Gibbs-Donnan effect, or Gibbs-Donnan equilibrium.

The following diagram may be used to illustrate the phenomenon. Two solutions, NaPr and NaCl, are separated by a membrane permeable to the sodium and chloride but not to the protein. The numbers are used to illustrate the relations to be discussed.

Side 1		Side 2
Na 7		Na 21
Pr 7		Cl 21
	Membrane	

Let us assume that these solutions have just been separated from one another by a membrane. If the membrane does not allow the protein anions to pass through it, but allows all the other ions to do so, then the following rearrangement of ions will take place:

Side 1	Side 2
Na 16	Na 12
Pr 7	Cl 12
Cl 9	

Membrane

The Gibbs-Donnan theory of equilibrium states that:

1. The product of the concentration of the sodium and chloride on one side of the membrane is equal to the product of the concentration of these ions on the other side.
2. The amount of the sodium on any one side must equal the sum of the chloride and protein ions on the same side.

Thus

$$[\mathrm{Na}]_1 \times [\mathrm{Cl}]_1 = [\mathrm{Na}]_2 \times [\mathrm{Cl}]_2$$

where the concentrations of Na and Cl are expressed in gram molecules per liter of solution. On side 1, which contains the nondiffusible substance, the sodium comes from the NaPr and the NaCl. On the other side of the membrane, Na comes only from NaCl.

It therefore follows that on side 1 the concentration of Na must exceed that of Cl because some of the Na remains with the protein in addition to that from the Cl.

On side 2, Na and Cl must be equal. Since the products of the concentrations are the same on both sides, it follows that Na on side 1, which is greater than the Cl, must be greater than the Na on side 2. The Cl on side 1 must be less than the Cl on side 2.

The Gibbs-Donnan equilibrium determines, therefore, the distribution of ions across a membrane permeable to some ions and impermeable to others. Many experiments carried out in vitro between protein and salt solutions with semipermeable membranes have confirmed this equilibrium; namely, on the side of the membrane where there is protein, the cations, for example, sodium, are greater than on the other side; and the anions, for example, chloride, are greater on the side of the membrane where there is no protein. An additional cation and anion, which may also be important in the consideration of the Gibbs-Donnan effect, are the hydrogen ion and the hydroxyl ion, respectively.

In considering influences of the Gibbs-Donnan equilibrium on distribution of ions, one must take into account that some 3 to 6 mol of chloride may be bound/mol of albumin (some 2 to 5 mEq/L in plasma) at pH 7.4[17] and that probably no sodium is bound to albumin, although a small amount may be bound to globulin.[18] Chloride had traditionally been found to be deficit in all rabbit ocular humors. However, Kinsey[37] has pointed out previous errors in methods and showed that there is an excess of chloride in anterior aqueous and vitreous and a slight deficiency in posterior aqueous. Modern knowledge of aqueous humor dynamics has been used to revalue the role of the Gibbs-Donnan equilibrium.[38]

Diffusional exchange

Dilution of the newly formed aqueous will occur both in the anterior and posterior chambers. Aqueous, just as it moves beyond the ciliary cells of origin, will be modified shortly after it is produced. The rate at which new aqueous dilutes the old is different in the two chambers. At a bulk flow of 2 μl/min, 0.033 of the volume of the posterior chamber is replaced each minute, but only 0.01 of the anterior chamber volume. Furthermore, the aqueous of both the anterior and posterior chamber is subject to diffusional exchange with the blood. The greater the exchange, the more rapidly the substance comes into equilibrium with the aqueous. Diffusion into the vitreous body is also a factor to be considered. In the instance of sodium, for example, a large fraction of the total entering from the plasma passes from the posterior chamber into the vitreous body as well as into the anterior chamber. Taking into account all these compartments requires complex mathematical analysis and has been dealt with by Kinsey and Reddy.[40] The value of these studies lies, of course, in estimating the character of the original aqueous secretate.

Exit of aqueous and solutes

The steady state or equilibrium level of a given compound in the eye, as we have seen, depends upon the factors influencing its ability to penetrate the blood-ocular barrier. For large lipid-insoluble molecules, high aqueous concentrations are never reached because the aqueous is continuously exiting. (A second reason for low levels, especially in the instance of penetration into the vitreous, is that a given compound or drug may penetrate so slowly that a true equilibrium is never reached. This is often the case for the development of equilibrium levels in vitreous where no significant ''circulation'' is found.) In addition to the drainage of aqueous, there is the rapid exit of some of the substances themselves. Indeed, systems for transport of materials out of the eye have been demonstrated for both organic[9,23] and inorganic[7] anions. The former system transports penicillin, and this outward pump may account for the low therapeutic levels achieved during drug therapy of infections.

Although the implications of the pump out system for pharmacokinetics are clear,[4] the biologic utility of such an outward bound pump is open to speculation. The demonstration of a specific carrier-mediated mechanism to prevent the intraocular accumulation of potentially toxic substances would support this concept.

Metabolites

The composition of the aqueous can be modified by metabolic processes ongoing in the cornea, lens, vitreous, and retina. For example, in the cornea, utilization of glucose and the generation of lactic acid will lower the concentration of the former and elevate the latter. It is known that glycolysis within the ciliary epithelium produces lactate. Since lactic acid is stronger than carbonic, in the presence of excessive lactate, bicarbonate levels found will be falsely low as a result of decomposition. This finding leads to the idea that the original bicarbonate concentration was somewhat higher, a suspicion confirmed.[40] In the lens, glycolytic activity will raise the levels of pyruvate and lactate. The production of lactic acid will reduce the level of anterior aqueous bicarbonate. Retinal requirements for glucose will decrease the concentration of the vitreous glucose.

Composition

Until some of the preceding factors were considered, a thorough analysis of the composition of the aqueous was not possible. Furthermore, the many striking similarities between the chemical content of plasma and aqueous encouraged simplified hypotheses. Nevertheless, new knowledge concerning the origin and nature of aqueous did originate from careful studies of the differences in the concentration of various substances in the anterior aqueous and plasma. (See Friedenwald[27] for early review.) For several reasons aqueous cannot be a dialysate of plasma; among them is the fact that the concentration of urea is lower in the aqueous than in the blood.[42]

A second important source of support for the idea that aqueous is not simply an ultrafiltrate of the plasma came from studies of ascorbic acid.[29,43] The concentration of ascorbic acid in rabbit aqueous humor is many times that in the plasma.[35] When the plasma level of ascorbic acid is increased, the concentration in the aqueous humor rises until about 3 mg/100 ml is reached in the plasma. At this plasma concentration, ascorbate in the aqueous humor reaches 50 mg/100 ml. Further increases in the plasma level cause no further rise in concentration in the aqueous humor (Fig. 7-11); transport of ascorbate into the eye is saturated. The eye apparently maintains its concentration of ascorbic acid by blood flow. Unilateral carotid ligation can cause a 17% ipsilateral reduction in the concentration of ascorbic acid.[41] Thus under desaturated conditions the concentration of ascorbic acid in the aqueous humor is some measure of the rate of blood flow through the ciliary body. Finally, observations that the isolated ciliary body can accumulate ascorbic acid against a gradient strengthened the idea of a secretory mechanism for the movement of ascorbate across the processes and into the aqueous.[8] These were among several early studies strong-

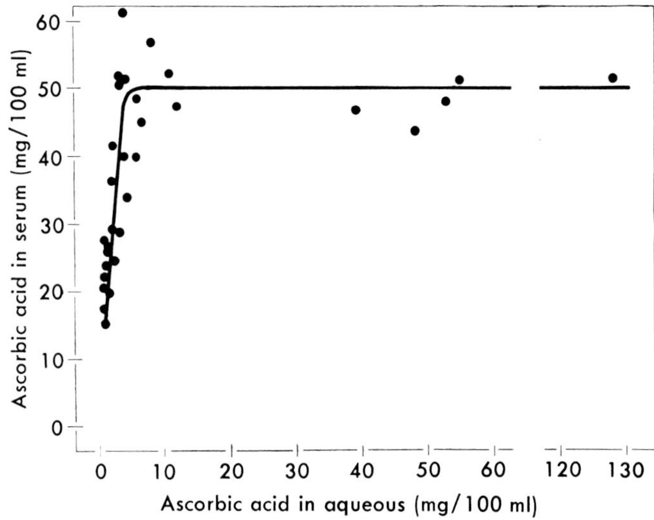

Fig. 7-11. Concentration of ascorbic acid in aqueous 60 minutes after injection of various quantities, plotted as function of concentration of blood. (Courtesy Dr. V. E. Kinsey.)

Table 7-1. Composition of plasma and aqueous humor of the anterior and posterior chambers of rabbits*

Substance	Inorganic constituents (mmol/kg H_2O)			Anterior aqueous/ plasma	Posterior aqueous/ plasma
	Anterior aqueous	Posterior aqueous	Plasma		
Sodium	146.5	144.5	—	—	—
	143.0	—	146.0	0.98	—
Potassium	4.7	—	4.8	0.98	—
Chloride	105.1	100.0	111.8	0.94	0.89
Bicarbonate	27.7	34.1	24.0	1.15	1.42
Phosphate	0.89	0.52	1.49	0.60	0.35
Hydrogen ion (pH)	7.60	7.57	7.40	—	—

*From Kinsey, V. E., and Reddy, D. V. N.: In Prince, J. H., editor: The rabbit in eye research, Springfield, Ill., 1966, Charles C Thomas, Publisher.

Table 7-2. Composition of plasma and aqueous humor of the anterior and posterior chambers of rabbits*

Substance	Organic constituents (mmol/kg H_2O)			Anterior aqueous/ plasma	Posterior aqueous/ plasma
	Anterior aqueous	Posterior aqueous	Plasma		
Ascorbate	0.96	1.30	0.02	48.0	65.0
Lactate	12.1	11.2	8.2	1.47	1.37
Glucose	6.7	7.1	8.3	0.81	0.85
Nonprotein nitrogen	13.4	13.5	17.6	0.76	0.77
Urea	6.3	5.8	7.3	0.86	0.80

*From Kinsey, V. E., and Reddy, D. V. N.: In Prince, J. H., editor: The rabbit in eye research, Springfield, Ill., 1966, Charles C Thomas, Publisher.

ly suggesting that secretory mechanisms were involved in the formation of aqueous humor.

Detailed knowledge of the formation process and of the aqueous secretate was delayed until 1953 when Kinsey[36] reached closer to the composition of the fluid actually secreted by the ciliary processes by utilizing paracentesis of the posterior chamber. Concentration differences for plasma, posterior aqueous, and anterior aqueous of rabbits are shown in Tables 7-1 and 7-2. The bicarbonate concentration in the posterior aqueous is higher than in the anterior because the newly secreted fluid has a concentration greater than plasma. Then the diffusional and metabolic modifications discussed earlier alter the newly formed aqueous. Diffusion into the vitreous and decomposition by acids in the anterior aqueous reduce the concentration in the anterior chamber. Similar considerations apply, for example, to chloride and ascorbic acid. In rabbits the newly generated posterior fluid has a relatively low concentration of chloride, and chloride from the blood diffuses into the anterior chamber. In primate aqueous the converse occurs: chloride concentrations are high and bicarbonate concentrations are low. In the case of ascorbate the gradient of diffusion for ascorbic acid from the anterior chamber to the blood reduces the concentration of anterior aqueous ascorbic acid.

The discovery of high levels of bicarbonate in the rabbit aqueous has been used to draw some valuable, if not completely accurate, conclusions. Friedenwald hypothesized that an electron pump in the ciliary processes might operate from the energy afforded by differences in the oxidation-reduction potential between the stroma and the epithelium. The cycle produces hydroxyl ions from oxygen. The reaction of these ions with carbon dioxide, catalyzed by carbonic anhydrase, would produce bicarbonate. The redox pump was also thought to account for sodium transport into the eye. The merits and objections to this idea have been summarized.[12] In any event, the use of carbonic anhydrase inhibitors causes a 50% to 60% decrease in the formation of aqueous, and these compounds have become a part of the therapy for glaucoma.[6]

The mode of action of carbonic anhydrase inhibitors in lowering intraocular pressure is not understood. In rabbits they reduce the bicarbonate concentration in the aqueous when it is high, but in primates they reduce the concentration of chloride when it is high. Somehow, as a consequence of impeding this anion transfer mechanism, *aqueous* formation is reduced. The effects are complex. For example, carbonic anhydrase inhibition slows the rate of water entry into the eye but has less effect on the ascorbate pump. Thus aqueous ascorbate concentration rises (Table 7-3). Also, metabolites

Table 7-3. Effect of acetazolamide administration* on the average steady state composition of the anterior chamber aqueous humor†

	Ion				
	Chloride	**Bicarbonate**	**Hydrogen**	**Ascorbate**	**Lactate**
Human					
Before	1.08	0.83	1.53	15	—
After	1.02	0.94	1.30	18	—
Rabbit					
Before	0.94	1.35	0.70	44	2.5
After	0.96	1.11	0.89	52	5.8
Guinea pig					
Before	0.90	1.42	0.73	17	—
After	0.93	1.19	0.83	24	—

*Dosage for humans, 500 mg orally every 6 hours; rabbit, 100 mg/kg intravenously followed by 25 mg/kg every 30 minutes; guinea pig, 70 mg/kg intraperitoneally followed by 35 mg/kg every 60 minutes. All values are for nonnephrectomized animals and are expressed as ratio aqueous to plasma concentration in mmol/kg water.

†From Becker, B.: Am. J. Ophthalmol. **47:**342, 1959.

such as lactic acid from lens glycolysis accumulate in the aqueous because the rate of water flow has decreased.

It is easier to understand the action of carbonic anhydrase inhibitors if emphasis is placed on the *rate* of entry of bicarbonate into the posterior aqueous rather than on its concentration therein. Measuring the rate of transfer of labeled ions, that is, $H^{14}CO_3^-$, from plasma to aqueous, it is possible to show that carbonic anhydrase inhibition slows the entry of bicarbonate into the posterior aqueous. As a consequence the rate of entry of water or aqueous flow may be diminished.

Mechanism of formation

The value of sampling posterior aqueous for studies of aqueous humor formation was greatly enhanced when radioactively tagged inorganic and organic compounds became available. Within certain experimental limits and assumptions, at least three approaches were now possible for the detailed study of the products secreted by the ciliary epithelia. Kinsey and Reddy[40] utilized mathematic analysis of time-concentration curves obtained by analysis of the fluids in the anterior and posterior chambers. Cole[19] collected fluid as it formed in the posterior chamber by blocking the outflow channels of the anterior chamber and by perfusion of the posterior chamber with fluids of known composition and analyzing the effluent for changes in their concentration. Finally, in the excised isolated ciliary body, there have been studies of fluxes,[32] measurements of oxidation-reduction potentials,[27] and important values obtained from intracellular epithelial recordings.[10] Another in vitro approach made use of planimetric

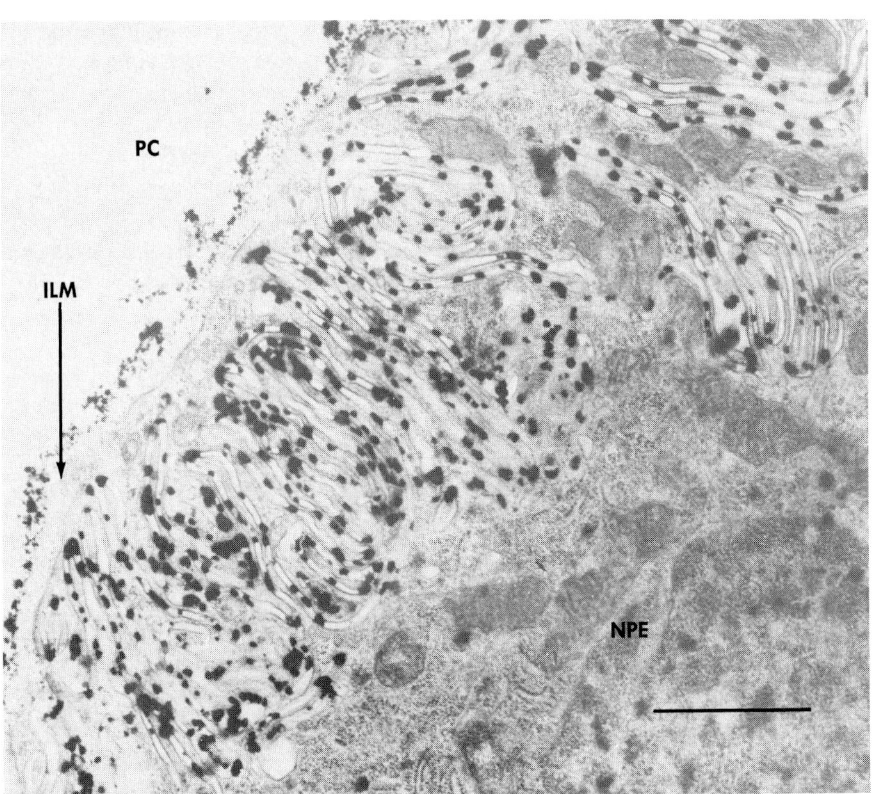

Fig. 7-12. Fresh ciliary epithelia incubated in ATP medium. Black reaction products indicate likely sites of hydrolysis of ATP. Products are dispersed as precipitates mainly along lateral and basal interdigitating cell membranes. Internal limiting membrane, *ILM;* posterior chamber, *PC;* nonpigmented epithelium, *NPE.* (Magnification ×23,900.) (From Shiose, Y., and Sears, M. L.: Invest. Ophthalmol. **5:**152, 1966.)

measurements of photographs of optical sections to estimate the rate of shrinkage of the processes.[11] Shrinkage was assumed to be a measure of transport across the processes. In this way vascular and other indirect effects on secretion were eliminated (Fig. 7-14). The isolated ciliary body has been used to study the uptake and accumulation of substances like ascorbic acid.[8] Here it is hoped accumulation of substances by the ciliary processes in vitro represents secretion of these substances in vivo.

Primary secretate

Probably 70% of the sodium found in the posterior chamber enters by a cell-mediated process. Considerably smaller amounts of chloride are similarly transported actively, but the mechanism is more obscure. The remaining fraction of sodium and chloride requires a diffusion mechanism. Bicarbonate also enters by flow and diffusion, but because of the extremely rapid diffusional exchange of carbon dioxide and its subsequent hydration, the relative proportions cannot be determined. Amino acids are transported inward across the ciliary epithelium by at least three carrier systems, one each for acidic, basic, and neutral molecules.

Evidence that secretion of the sodium ion into the posterior chamber is a key event in the formation of aqueous comes from these sources:

1. Sodium accumulates in the posterior chamber of rabbit eyes after parenteral administration of radioactive sodium.[40] (The amount of chloride that enters is insufficient to neutralize the charge on sodium. Bicarbonate together with lactate accounts for the remainder.)
2. Ouabain, an inhibitor of Na^+,K^+-activated ATPase, effectively reduces the influx of sodium and aqueous.[19]
3. Shrinkage of ciliary processes (see above) is inhibited by ouabain and requires sodium and external potassium.[11]
4. Suggestive histochemical[20] and fine structural studies[53] and biochemical analyses[51] show large concentrations of Na^+,K^+-activated ATPase along the membranes of the inner nonpigmented epithelial cells (Fig. 7-12), especially at the lateral and basal interdigitations.

Some authors have assumed that the primary event in the secretory process is the formation of hydroxyl ions. HCO_3^- as a secretate would then largely depend on its catalytic formation from OH + CO_2. In the last 15 years, however, much of the work referred to earlier (done in species other than primates) has dealt largely with identifying sodium as the primary secretate.

Metabolic requirements

The active transport of sodium by the ciliary epithelia is intimately related to the oxidation of citrate cycle substrates and derives energy

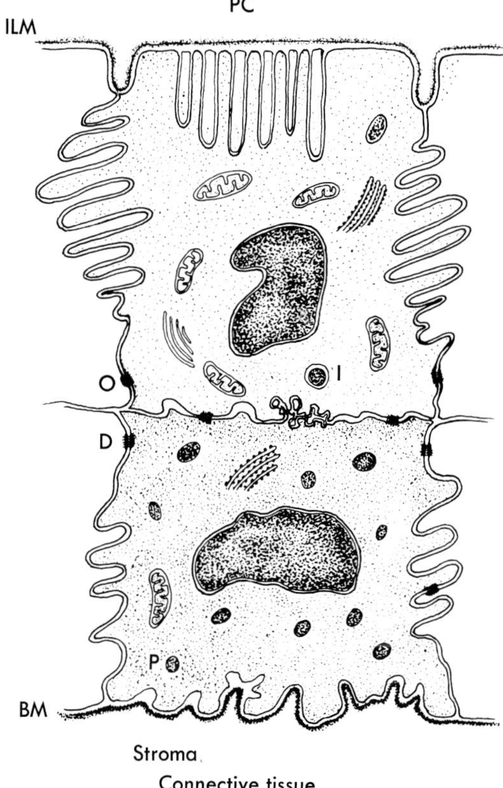

Fig. 7-13. Schematic drawing of double layer of ciliary epithelia. *PC,* Posterior chamber; *ILM,* internal limiting membrane. Note interdigitations that are lateral and basal. These probably constitute main sites of active transport. This activity, largely confined to inner cell layer, is supported by energy of cellular metabolism. In this connection note large number of mitochondria in inner cell layer. *I,* Processes interdigitating from adjacent cells in another plane of section; *O,* zonula occludens; *D,* desmosomes and gap junctions less restrictive than barrier constituted by *O; P,* pigment granules; *BM,* basement membrane for outer pigmented epithelia. Stroma is beneath as indicated.

from the hydrolysis of ATP. The origin of the energy-yielding process is probably the mitochondria of the ciliary epithelium and the citric acid–cycle enzymes located therein. Metabolic profiles of the separate epithelial layers indicate rather conclusively that the inner epithelial layer meets the requirements as the transporting layer of cells.[21,51] Thus the puzzle of the double layer of epithelia is partly solved in that the metabolism of the two layers is different; the inner nonpigmented layer energizes the active transport of sodium, while the outer pigmented cells, coupled electrically to the inner layer, are in secondary roles of permitting penetration of substrates to the active site[21,39] (Fig. 7-13) and of making pigment.

Controls

The formation of aqueous humor is subject to a wide variety of endogenous influences, most of them poorly understood. The fluctuation in the rate of aqueous formation certainly accounts for diurnal rhythms in intraocular pressure.[31,61] There is no doubt that neural or neurohumoral mechanisms mediate these changes. The idea that central mechanisms influence the rate of secretion of aqueous is important and attractive, but elusive to substantiate. The absence of nerves within the epithelial cell layers of the ciliary processes indicates that they are probably not under direct neural control, but the vessels of the ciliary processes have a very dense adrenergic innervation. A vascular receptor could conceivably mediate the rate of aqueous humor formation. It is difficult, however, to distinguish those effects related to changes in vascular tone from those caused by epithelial receptors. The secretory activity of the epithelial cell layers could also be controlled, if not by neural factors, then by humoral ones. Under humoral influence an adenyl cyclase receptor causes the secretory suppression that occurs after administration of adrenergic compounds.[25,28,45,60]

In vitro evidence for the hypothesis that ciliary epithelial secretion is regulated by an adrenergic system is as follows: Adenylate cyclase activity sensitive to catecholamines has

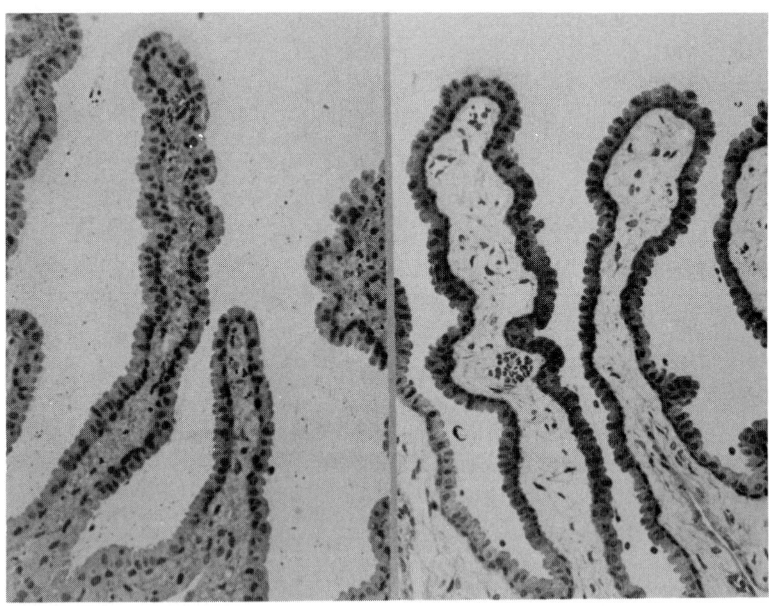

Fig. 7-14. Ciliary processes taken from right and left eyes of rabbits several hours after treatment with close arterial infusion of 2 μg cholera toxin, right, and normal saline solution, left. The pigmented and unpigmented cell layers are intact, but prominent right stromal edema indicates inhibition of net aqueous humor transport. (Courtesy Dr. Hiromu Mishima; from Aqueous Dynamics and Glaucoma Laboratory.)

been demonstrated in homogenates of the tips of the ciliary processes of rabbits[59] and has been localized to the nonpigmented ciliary epithelial cells.[55] Catecholamines accelerate the production of cyclic AMP in the rabbit ciliary body.[44] Gregory, at Yale, has shown that not only isoproterenol but also cholera toxin, the most potent stimulator of adenyl cyclase known, greatly accelerates the rate of production of cyclic AMP. Stimulation of cyclic AMP production in ciliary processes taken from six human eyes was 5 to 10 times greater than basal values. This finding indicates the presence of membrane bound adenyl cyclase in human ciliary processes. Cyclic AMP phosphodiesterase activity has been found in ciliary processes both by histochemical[50] and direct biochemical[48] assay. Beta-adrenergic receptors have been demonstrated by direct ligand-binding techniques.[16] On a molecular level, strict support for this adrenergic model requires the presence of cyclic AMP–dependent protein kinase activity in the ciliary epithelial cells, and the demonstration that specific protein phosphorylation in the cells is related to changes in the rate of aqueous formation. The in vivo response to potent stimulation of adenyl cyclase, as we have learned,[25,28,45,60] is decreased aqueous humor formation. Direct experimental measurements of net flow rates of aqueous humor, done after close arterial injection of minute quantities of cholera toxin into rabbits, confirms the idea that major reductions of aqueous humor formation occur after activation of adenyl cyclase (Fig. 7-14). How does the adenyl cyclase system work to affect a final secretate in the aqueous of the posterior chamber? It is possible that the activities of adenyl cyclase and of Na^+,K^+-activated ATPases in the ciliary epithelial cells are coupled in series, as they are in the renal tubule. The former may regulate the extraction of sodium from the stroma and outer epithelial layer lining the posterior chamber. Perhaps the protein kinase activated by intracellular cyclic AMP directly affects the permeability of the membranes of the inner nonpigmented epithelial cells (Fig. 7-15). In either case the concept of adenyl cyclase as a receptor

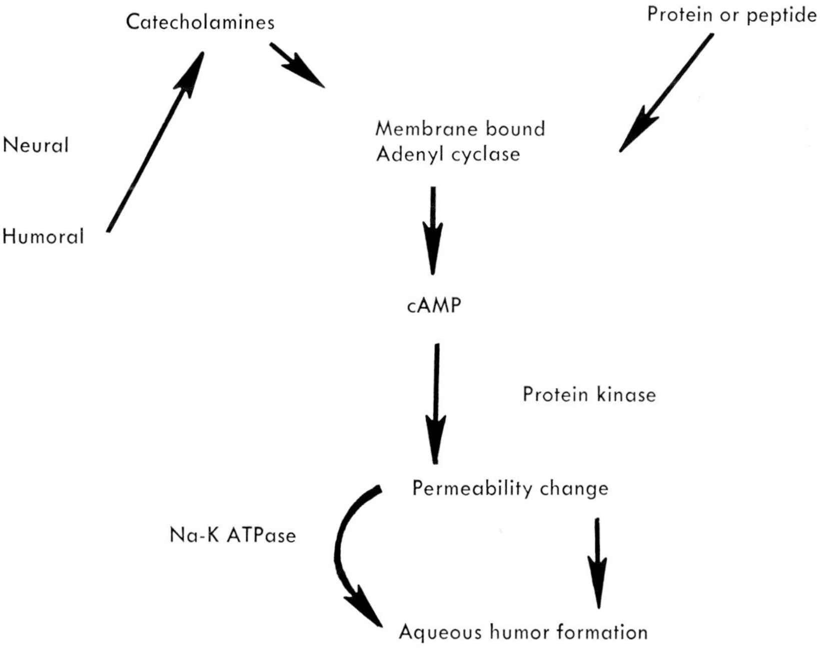

Fig. 7-15. See text for discussion.

Table 7-4. Concentration (mmol/kg H_2O) of free amino acids in plasma, aqueous, and vitreous humors of rabbits*

Amino acid	Plasma average	Posterior aqueous	Anterior aqueous	Vitreous
Alanine	0.302	0.445	0.480	0.074
Arginine	0.096	0.276	0.272	0.045
Aspartic acid	0.029	0.087	0.055	0.019
Cystine	0.054	0.008	0.030	Trace
Glutamic acid	0.178	0.401	0.295	0.169
Glycine	1.190	0.314	0.614	0.065
Histidine	0.116	0.181	0.210	0.020
Isoleucine	0.112	0.121	0.116	0.010
Leucine	0.163	0.191	0.174	0.018
Lysine	0.212	0.397	0.423	0.070
Methionine	0.014	0.026	0.023	Trace
M-Sulphone	0.011	0.005	0.010	—
M-Sulphoxide	0.027	0.026	0.034	—
Phenylalanine	0.097	0.141	0.097	0.011
Proline	0.323	0.223	0.267	0.048
Serine	0.420	0.597	0.585	0.130
Threonine	0.165	0.149	0.138	0.025
Tryptophan	0.013	0.004	0.024	Trace
Tyrosine	0.058	0.161	0.101	0.021
Valine	0.195	0.214	0.230	0.045

*From Reddy, D. V. N., Rosenberg, C., and Kinsey, V. E.: Exp. Eye Res. **1**:179, 1961.

for mediation of aqueous humor formation unifies the mode by which both experimental administration of catecholamines and endogenous humoral influences can affect aqueous humor formation (Fig. 7-15).

By a cellular mechanism not clearly understood, all beta-adrenergic blocking agents, but especially timolol, reduce intraocular pressure both in experimental animals[56] and in patients[34] with glaucoma. Timolol is the first useful topical agent in the treatment of glaucoma to become commercially available since the early 1920s, and its use emphasizes our need to understand adrenergic mechanisms in the control of intraocular pressure.

Other biological activity

Our knowledge of mechanisms fundamental to the secretory process of aqueous humor may require continued study and refinement, but the source and effect of numerous substances with biologic activity found in the aqueous are only beginning to be explored. The influences of plasma activators exerted via a disrupted blood-ocular barrier, amino acids (Table 7-4), growth and antigrowth factors, rubeosis-producing factors, trace metals, biogenic amines, acid hydrolases, prostaglandins and cyclic AMP, glycoproteins and neuropeptides like substance P provide fertile field for study. Who can forget the excitement engendered by Bárány's[3] discovery that in some species intracameral hyaluronidase can significantly decrease the resistance of aqueous outflow? What answers to old problems will be found in the aqueous solution?

ACKNOWLEDGEMENT

I am especially grateful to Mrs. Susan Fleischmann for her expert assistance in the preparation of this chapter, and to residents in ophthalmology at Yale and elsewhere for their stimulating comments and enthusiastic interest.

Marvin L. Sears

REFERENCES

1. Al-Ghadyan, A., Mead, A., and Sears, M. L.: Increased pressure after paracentesis of the rabbit eye is completely accounted for by prostaglandin synthesis and release plus pupillary block, Invest. Ophthalmol. Vis. Sci. **18**:361, 1979.
2. Ambache, N.: Properties of irin, a physiological constituent of the rabbit's iris, J. Physiol. **135**:114, 1957.
3. Bárány, E. H.: In vitro studies of the resistance to flow through the angle of the anterior chamber, Acta Soc. Med. Ups. **59**:260, 1954.
4. Bárány, E. H.: The liver-like anion transport system in rabbit kidney, uvea and choroid plexus. II, Efficiency of acidic drugs and other anions as inhibitors, Acta Physiol. Scand. **88**:491, 1973.
5. Baurmann, M.: Ueber das Ciliarfortsatzgefässsystem, Ber. Dtsch. Ophthalmol. Ges. **48**:364, 1930.
6. Becker, B.: Carbonic anhydrase and the formation of aqueous humor, Am. J. Ophthalmol. **47**:342, 1959.
7. Becker, B.: Iodide transport by the rabbit eye, Am. J. Physiol. **200**:804, 1961.
8. Becker, B.: Ascorbate transport in guinea pig eyes, Invest. Ophthalmol. **6**:410, 1967.
9. Becker, B., and Forbes, M.: Iodopyracet (diodrast) transport by the rabbit eye, Am. J. Physiol. **200**:461, 1961.
10. Berggren, L.: Intracellular potential measurements from the ciliary processes of the rabbit eye in vivo and in vitro, Acta Physiol. Scand. **48**:461, 1960.
11. Berggren, L.: Direct observation of secretory pumping in vitro of the rabbit eye ciliary processes, Invest. Ophthalmol. **3**:266, 1964.
12. Berliner, R. W.: Mechanisms of transport by membranes. In Newell, F. W., editor: Glaucoma. Transactions of the Second Conference, Dec. 3 to 5, 1956, Josiah Macy, Jr., Foundation, 1957, p. 108.

13. Bill, A.: The albumin exchange in the rabbit eye, Acta Physiol. Scand. **60:**18, 1964.
14. Bill, A.: The role of ciliary blood flow and ultrafiltration in aqueous humor formation, Exp. Eye Res. **16:**287, 1973.
15. Bill, A.: Blood circulation and fluid dynamics in the eye, Physiol. Rev. **55:**383, 1975.
16. Bromberg, B. B., Gregory, D. S., and Sears, M. L.: Beta-adrenergic receptors in ciliary processes of the rabbit, Invest. Ophthalmol. Vis. Sci. **19:**203, 1980.
17. Carr, C. W.: Studies on the bindings of small ions in protein solutions with the use of membrane electrodes. I, The binding of the chloride ion in other inorganic anions in solutions of serum albumin, Arch. Biochem. Biophys. **40:**286, 1952.
18. Carr, C. W.: Applications of membrane electrodes, Ann. N.Y. Acad. Sci. **148:**180, 1968.
19. Cole, D. F.: Effects of some metabolic inhibitors upon the formation of the aqueous humour in rabbits, Br. J. Ophthalmol. **44:**739, 1960.
20. Cole, D. F.: Location of ouabain-sensitive adenosine triphosphatase in ciliary epithelium, Exp. Eye Res. **3:**72, 1964.
21. Cole, D. F.: Aqueous humour formation, Doc. Ophthalmol. **21:**116, 1966.
22. Cole, D. F.: Aqueous and ciliary body. In Graymore, C. N., editor: Biochemistry of the eye, New York, 1970, Academic Press, Inc., p. 132.
23. Cunha-Vaz, J. G., and Maurice, D. M.: The active transport of fluorescein by the retinal vessels and the retina, J. Physiol. **191:**467, 1967.
24. Duke-Elder, P. M., and Duke-Elder, W. S.: Vascular responses of the eye, Proc. R. Soc. Lond. (Biol.) **109:**19, 1931.
25. Eakins, K. E.: The effect of intravitreous injections of norepinephrine, epinephrine and isoproterenol on the intraocular pressure and aqueous humor dynamics of rabbit eyes, J. Pharmacol. Exp. Ther. **140:**79, 1963.
26. François, J., and Rabaey, M.: Microelectrophoresis on agar of normal and pathological aqueous humor, Arch. Ophthalmol. **63:**836, 1960.
27. Friedenwald, J. S.: The formation of the intraocular fluid, Am. J. Ophthalmol. **32:**9, 1949.
28. Goldmann, H.: L'origine de l'hypertension oculaire dans le glaucome primitif, Ann. Ocul. (Paris) **184:** 1086, 1951.
29. Goldmann, H., and Buschke, W.: Blutkammerwasserschranke und Vitamin C. II, Die Abhängigkeit der Kammerwasseraskorbinsäure vom Vitamin C des blutes, Arch. Augenheilkd. **109:**314, 1935.
30. Grayson, M. C., and Laties, A. M.: Ocular localization of sodium fluorescein, Arch. Ophthalmol. **85:**600, 1971.
31. Henkind, P., Leitman, M., and Weitzman, E.: The diurnal curve in man: new observations, Invest. Ophthalmol. **12:**705, 1973.
32. Holland, M. G., and Gipson, C. C.: Chloride ion transport in the isolated ciliary body, Invest. Ophthalmol. **9:**20, 1970.
33. Ishikawa, T.: Fine structure of the human ciliary muscle, Invest. Ophthalmol. **1:**587, 1962.
34. Katz, I. M.: Beta-blockers and the eye. An overview, Ann. Ophthalmol. **10:**847, 1978.
35. Kinsey, V. E.: Transfer of ascorbic acid and related compounds across the blood-aqueous barrier, Am. J. Ophthalmol. **30:**1262, 1947.
36. Kinsey, V. E.: Comparative chemistry of aqueous humor in posterior and anterior chambers of rabbit eye, Arch. Ophthalmol. **50:**401, 1953.
37. Kinsey, V. E.: Further study of the distribution of chloride between plasma and intraocular fluids of the eye, Invest. Ophthalmol. **6:**395, 1967.
38. Kinsey, V. E.: Evaluation of the effective electrochemical potential and diffusion coefficients involved in the movement of sodium and chloride across the blood-aqueous barriers of the rabbit eye. In Glaucoma Tutzing Symposium, Basel, Switzerland, 1967, S. Karger AG.
39. Kinsey, V. E.: Ion movement in ciliary processes. In Bittar, E. E., editor: Membranes and ion transport, New York, 1971, John Wiley & Sons, Inc., vol. 3.
40. Kinsey, V. E., and Reddy, D. V. N.: Chemistry and dynamics of aqueous humor. In Prince, J. H., editor: The rabbit in eye research, Springfield, Ill., 1964, Charles C Thomas, Publisher, pp. 215-316.
41. Linnér, E.: Ascorbic acid as a test substance for measuring relative changes in the rate of plasma flow through the ciliary processes. I, The effect of unilateral ligation of the common carotid artery in rabbits on the ascorbic acid content of the aqueous humour at varying plasma levels, Acta Physiol. Scand. **26:** 130, 1952.
42. Moore, E., Scheie, H. G., and Adler, F. H.: Chemical equilibrium between blood and aqueous humor; further studies, Arch. Ophthalmol. **27:**317, 1942.
43. Müller, H. K., and Buschke, W.: Vitamin C in Linse, Kammerwasser und Blut bie normalen und pathologischem Linsenstoffwechsel, Arch. Augenheilkd. **108:**368, 1934.
44. Neufeld, A. H., and Sears, M. L.: Cyclic AMP in the ocular tissues of the rabbit, Invest. Ophthalmol. **14:** 475, 1974.
45. Ross, R. A., and Drance, S. M.: Effects of topically applied isoproterenol on aqueous dynamics in man, Arch. Ophthalmol. **83:**39, 1970.
46. Scullica, L.: Observations of uveal blood flow pattern in excised arterially perfused rabbit eyes, Am. J. Ophthalmol. **54:**1057, 1962.
47. Sears, M. L.: Miosis and intraocular pressure changes during manometry in mechanically irritated rabbit eyes studied with improved manometric techniques, Arch. Ophthalmol. **63:**707, 1960.
48. Sears, M. L.: Perspectives in glaucoma research. Friedenwald Lecture, April, 1977, Invest. Ophthalmol. Vis. Sci. **17:**6, 1978.
49. Shakib, M., and Cunha-Vaz, J. G.: Studies on the permeability of the blood-retinal barrier. IV, Junctional complexes of the retinal vessels and their role in the permeability of the blood-retinal barrier, Exp. Eye Res. **5:**229, 1966.

50. Shanta, T. R., et al.: Histochemical method for localization of cyclic 3′,5′-nucleotide phosphodiesterase, Histochemie **7:**177, 1966.
51. Shimizu, H., Riley, M. V., and Cole, D. F.: The isolation of whole cells from the ciliary epithelium together with some observations on the metabolism of the two cell types, Exp. Eye Res. **6:**141, 1967.
52. Shiose, Y.: Electron microscopic studies on blood-retinal and blood-aqueous barriers, Jpn. J. Ophthalmol. **14:**73, 1970.
53. Shiose, Y., and Sears, M. L.: Fine structural localization of nucleoside phosphatase activity in the ciliary epithelium of albino rabbits, Invest. Ophthalmol. **5:** 152, 1966.
54. Tormey, J. M.: Artifactual localization of ferritin in the ciliary epithelium in vitro, J. Cell Biol. **25:**1, 1965.
55. Tsukahara, S., and Maezawa, N.: Cytochemical localization of adenyl cyclase in the rabbit ciliary body, Exp. Eye Res. **26:**99, 1978.
56. Vareilles, P., et al.: Comparison of the effects of timolol and other adrenergic agents on intraocular pressure in the rabbit, Invest. Ophthalmol. Vis. Sci. **16:**987, 1977.
57. Vegge, T.: An epithelial blood-aqueous barrier to horseradish peroxidase in the ciliary processes of the vervet monkey *(Cercopithecus aethiops)*, Z. Zellforsch. Mikrosk. Anat. **114:**309, 1971.
58. Vegge, T., and Ringvold, A.: Ultrastructure of the wall of human iris vessels, Z. Zellforsch. Mikrosk. Anat. **94:**19, 1969.
59. Waitzman, M. B., and Woods, W. D.: Some characteristics of an adenyl cyclase preparation, Exp. Eye Res. **12:**99, 1971.
60. Weekers, R., Gilson, M., and Prijot, E.: Étude comparative des effets de diverses amines sympathicomimetiques sur la pression oculaire, Arch. Ophthalmol. (Paris) **21:**545, 1961.
61. Worthen, D. M.: Effect of pilocarpine drops on the diurnal intraocular pressure variation in patients with glaucoma, Invest. Ophthalmol. **15:**784, 1976.
62. Yamashita, H., and Sears, M.: Proof that the ciliary epithelium can regenerate, Exp. Eye Res. **27:**199, 1978.

GENERAL REFERENCES

Cole, D. F.: Aqueous and ciliary body. In Graymore, C. N., editor: Biochemistry of the eye, New York, 1970, Academic Press, Inc., Ltd.

Davson, H.: Vegetative physiology and biochemistry. In The eye, London, 1969, Academic Press, Inc., vol. 1.

Friedenwald, J. S., and Becker, B.: Aqueous humor dynamics, Arch. Ophthalmol. **54:**799, 1955.

Kinsey, V. E.: Ion movement in ciliary processes. In Bittar, E. E., editor: Membrane and ion transport, New York, 1971, John Wiley & Sons, Inc., vol. 3.

Maren, T. H.: HCO_3^- formation in aqueous humor; mechanism and relation to the treatment of glaucoma, Invest. Ophthalmol. **13:**479, 1974.

Chapter 8

Intraocular pressure

ROBERT A. MOSES, M.D.

The pressure in the normal human aqueous and vitreous is, on the average, 15.5 mm Hg. All the tissues within the leathery corneoscleral coat are subject to this pressure, which is higher than tissue pressure elsewhere in the body. Blood pressure in the uveal and retinal vessels must counterbalance or exceed intraocular pressure, or else the vessels collapse. Ocular hypertension is associated with the glaucoma disease complex. Ocular hypotension is frequently a concomitant of retinal detachment and uveitis. Prolonged hypotension may herald shrinking of the globe (phthisis bulbi). Because of these important associations, intraocular pressure has been the subject of investigation and speculation for many years.

DEFINITION OF PRESSURE

Pressure is force per unit area:

$$P = f/A$$

Pressure may be expressed as g of force/cm^2. The density of water is about 1.0, so a column of water 30 cm high exerts a pressure of about 30 g of force/cm^2. Mercury is 13.5 times as dense as water, so a column of mercury 2.2 cm or 22 mm high also develops pressure of 30 g of force/cm^2.

ORIGIN OF INTRAOCULAR PRESSURE

A slow, continuous stream of clear liquid flows from the blood in capillaries of the ciliary processes through the ciliary epithelium, into the cavity of the eye. From the ocular interior, liquid escapes through narrow channels into veins. Resistance to flow in the narrow channels causes fluid to dam up in the ocular cavity until its pressure is great enough to force it through the outflow channels at a rate equal to the inflow rate. In the ocular cavity the liquid fills the posterior and anterior chambers and is called aqueous. The pressure of the aqueous is intraocular pressure.

Three potential causes of increased intraocular pressure can immediately be discussed.

1. Stenosis or blockage of the aqueous outflow channels is by far the commonest cause of glaucoma.
2. Increase in venous pressure outside the eye is reflected back through the aqueous drainage channels. Ocular hypertension secondary to increased venous pressure in the head region occurs, for example, in obstruction of the superior vena cava by mediastinal tumor and in carotidcavernous fistula.
3. Increased production of aqueous has been difficult to identify, although it may occur in certain persons with a particular type of vascular hypertension.

In the transit of the aqueous through the eye its chemical composition is modified by interchange with surrounding tissues (ciliary epithelium, vitreous, lens, iris, and cornea), so that the fluid leaving the eye from the anterior chamber is different from that entering the posterior chamber. The chemical aspects of aqueous formation and composition are set forth in Chapter 7. The present discussion is concerned with the relation of bulk flow of aqueous to intraocular pressure.

□ This work was supported in part by Research Grant EY 00256 from the National Eye Institute, National Institutes of Health.

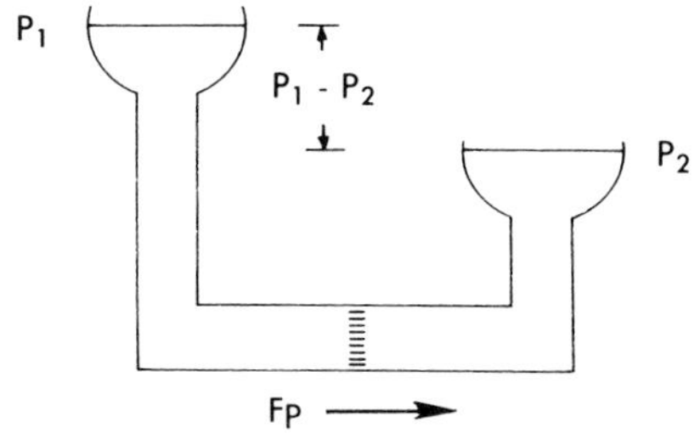

Fig. 8-1. Flow as result of pressure head. Tubing is facility of flow. Facility of large-diameter tubes is greater than facility of group of small-diameter tubes. P_1-P_2, Pressure head; F_p, rate of flow (volume per unit time) caused by pressure head.

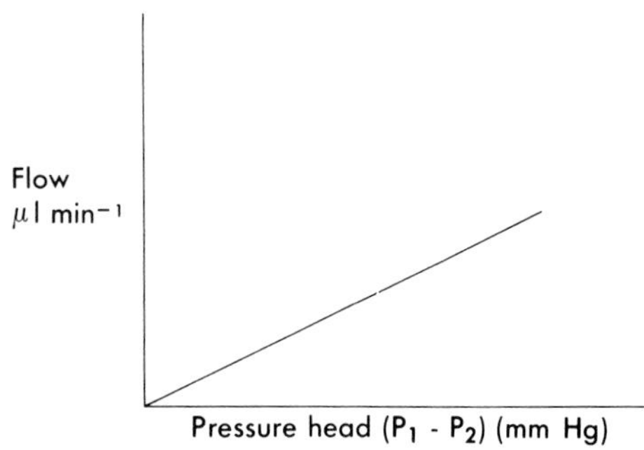

Fig. 8-2. Graphic representation of $F = (P_1 - P_2) C$. Slope of line relating pressure head and rate of flow is facility, C.

Table 8-1. Glossary

Symbol	Quantity	Units	
F	Rate of flow	Microliters per minute	μl/min or μl min^{-1}
P_1, P_2	Pressure	Millimeters of mercury	mm Hg
$P_1 - P_2 = \Delta P$	Pressure head	Millimeters of mercury	mm Hg
C	Facility of flow	Microliters per minute per millimeter of mercury	μl/min/mm Hg *or* μl min^{-1} mm Hg^{-1}
R	Resistance to flow	Millimeters of mercury minutes per microliter	mm Hg min/μl *or* mm Hg min μl^{-1}

Production of liquid flow requires that energy be expended. The energy may be provided by a pressure gradient or by a pump. The difference in pressure across a pressure gradient is called a pressure head. The rate of flow is the volume of liquid transferred in a unit of time. The rate of flow across a pressure head is restricted by the passage through which flow occurs. In ophthalmology such a flow passage is called a facility of flow.

Flow through a facility as a result of a pressure head is shown in Fig. 8-1. Fluid flows through the tubing from the higher pressure, P_1, to the lower pressure, P_2. The tubing is a facility of flow. The facility of large tubes is greater than the facility of small tubes. The relation of pressure, rate of flow, and facility of flow is given quantitatively as follows:

$$F = (P_1 - P_2)\, C$$

This equation is presented graphically in Fig. 8-2. The symbols and units in which they are expressed are given in Table 8-1.

Some authors prefer to discuss the resistance of a passage rather than the facility.

$$\text{Resistance} = 1/\text{Facility}$$

The flow system in Fig. 8-1 will constantly change as liquid from the vessel on the left moves to the one on the right until $P_1 = P_2$. In changing state the first equation will apply only for the instant at which P_1 and P_2 are measured. If the flow system is such that liquid is added to the vessel on the left as rapidly as it leaves, and a similar arrangement is made to remove fluid from the vessel on the right, the pressure head and the rate of flow through the system are constant. This is a steady state. Flow of aqueous through the eye approximates a steady state.

OSMOTIC PRESSURE

Osmotic pressure is caused by an unequal distribution of dissolved material on the two sides of a barrier through which solvent but

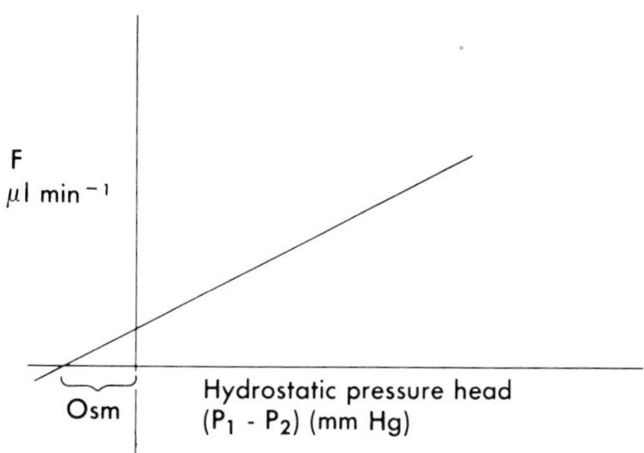

Fig. 8-3. Constant osmotic pressure difference across hydrostatic pressure head shifts facility slope, but slope itself is not changed. Osmotic pressure is greater on side 2 of pressure head.

not solute may pass. Solvent moves across the barrier from the region of higher solvent concentration to the region of lower solvent concentration. Osmotic pressure is measured by the force per unit area that must be exerted on the region of low solvent concentration (the region of the impermeable solute) to stop net movement of solvent through the barrier. Thus osmotic pressure is measured by hydrostatic pressure. In quantitative expression of flow through a semipermeable barrier the pressure head is the difference in total pressure on the two sides of the barrier.

$$P_{1\ \mathrm{total}} - P_{2\ \mathrm{total}} = (P_{1\ \mathrm{hydro}} - P_{1\ \mathrm{osm}}) - (P_{2\ \mathrm{hydro}} - P_{2\ \mathrm{osm}})$$

The movement of fluid will then be in response to the total pressure difference.

$$F = (P_{1\ \mathrm{total}} - P_{2\ \mathrm{total}})\ C$$

If the osmotic pressure difference is constant, it will shift the line relating flow to hydrostatic pressure head. Fig. 8-3 depicts the case in which osmotic pressure on side 2 is greater than that on side 1. Rate of flow is still related to pressure head by facility slope. Flow caused by hydrostatic pressure against an osmotic pressure gradient is called ultrafiltration.

The term pump may have various connotations, but in physiology it is taken to mean a mechanism that creates or controls flow of one or more substances by the expenditure of metabolic energy.

FORMATION OF AQUEOUS

Blood flows from the major arterial circle at the iris root, through the ciliary capillaries, to the uveal veins. In the ciliary process fluid passes through the capillary walls into the stroma of the process. So far as is known, this is similar to passage of fluid through capillary walls elsewhere in the body. The passage of fluid is controlled by the difference in (total) pressure inside the capillary and in the stroma and by the facility or porosity of the capillary wall.

The ciliary epithelium lies between the stroma of the ciliary processes and the posterior chamber. Fluid is moved through the ciliary epithelium to the posterior chamber, either by osmotic excess of the posterior aqueous, by a pressure gradient, or by a metabolic pump. The possibility that the aqueous of the posterior chamber is hyperosmotic to the fluid in the ciliary stroma has not been ruled out, but the existence of a metabolic pump has considerable supportive evidence.

CIRCULATION OF AQUEOUS

Aqueous humor liberated into the posterior chamber of the eye flows through the pupil into the anterior chamber. In the anterior chamber, evidence of a convection (thermal) current that moves down along the cooler cornea and up along the warmer iris may be seen, such as when intravenously injected fluorescein enters the anterior chamber through the pupil and it is seen to flow upward along the iris (Ehrlich's line). Aqueous escapes from the anterior chamber and returns to the bloodstream by two routes: (1) by passing into the anterior face of the ciliary body—the uveoscleral or unconventional outflow and (2) by leaking through the trabecular mesh into Schlemm's canal—the conventional outflow.

In the uveoscleral outflow a portion of the aqueous passing into the ciliary body is absorbed into uveal blood vessels; the remainder seeps through the sclera into orbital tissues. The uveoscleral outflow drains some 0.2 to 0.5 μl of aqueous/min at a steady rate not affected by intraocular pressure but increased by atropine and decreased by pilocarpine.

The trabecular outflow carries the greater

portion of the aqueous from the eye, approximately 1.8 to 2.5 μl/min.

The trabecular mesh consists of crisscrossing collagen cords covered by endothelium. The collagen beams branch into each other to form a three-dimensional spongelike net. Endothelium rounds out the angular crossings of the cords or beams and joins adjacent layers (lamellae) of beams, leaving a set of tortuous passages through the mesh. The mesh bridges the scleral sulcus, in which Schlemm's canal lies, spanning from the end of Descement's membrane at Schwalbe's line, well anterior to the canal, to the scleral spur and ciliary body. The mesh is a tendon of origin of the ciliary muscle, and it has been speculated that in contraction the muscle separates the lamellae of the mesh. Since the rounded apertures in adjacent lamellae do not correspond in location, narrowing of the spaces between the layers would valve the egress of aqueous, whereas spreading the lamellae, either by ciliary contraction in accommodation or by means of parasympathomimetic drugs, would open the valve (Fig. 8-4). This old and attractive hypothesis has had some experimental substantiation.[22] If such valving does not occur, the passages within the trabecular mesh would seem to be too large and too numerous to offer much resistance to the outflow of aqueous.

A thin layer of spongy tissue, the juxtacanalicular tissue, lies between the outer beams of the trabecular mesh and the single layer of endothelial cells that line Schlemm's canal. The endothelial cells are penetrated by relatively few, narrow (1.5 to 2.0 μm diameter), tortuous channels[24] (Fig. 8-5).

The course of the aqueous in the trabecular outflow is through the trabecular mesh, juxtacanalicular tissue, and canal endothelium into the canal of Schlemm and from there via veins draining the canal (collector channels) through the sclera to episcleral veins (recipient veins). Aqueous may be seen flowing in clear veins emerging from the limbus (aqueous veins) and in recipient veins, where it streams side by side with blood before mixing (laminated veins).[2] The narrow passages enumerated earlier offer resistance to outflow. The trabecular outflow behaves, to a first approximation, like a leak from a garden hose; if the pressure in the eye is artificially raised, as by pressing on the globe, the aqueous escapes more rapidly. If the passages of the leak were to remain constant in their dimensions, one would expect the *rate of flow* through the leak, F, to be linearly proportional to the *pressure head,* ΔP, driving the fluid through the leak, or as follows:

$$\Delta P = FR$$

R is the *resistance* of the leak, and since (in the eye) pressure is measured clinically in millimeters of mercury (mm Hg) and flow rate in microliters per minute (μl min^{-1}), R is expressed in mm Hg min μl^{-1}. Resistance, then, is a measure of the constriction of a passage.

The pressure effective in forcing fluid through a leak is the difference between the higher pressure on one side and the lower pressure on the other. In the case of the trabecular system, the higher pressure is intraocular pressure, P_1, and the lower pressure is episcleral venous pressure, P_e. The pressure head, $\Delta P = P_i - P_e$, produces outflow, F_{trab}, limited by resistance, R_{trab}. Thus at any particular intraocular pressure the outflow through the trabecular mesh is as follows:

$$P_i - P_e = F_{trab} R_{trab} \quad (1)$$

or the equivalent expressions:

$$F_{trab} = (P_i - P_e)/R_{trab} \quad (1a)$$

or

$$F_{trab} = (P_i - P_e)\, C_{trab} \quad (1b)$$

In steady state, when the intraocular pressure is neither rising nor falling, aqueous production rate, F_{in}, is balanced by the rate of aqueous loss through the trabecular mesh and through the uveoscleral pathway, U, as follows:

$$F_{in} = F_{trab} + U \quad (2)$$

Substituting from 1a as follows:

$$F_{in} = \frac{(P_i - P_e)}{R_{trab}} + U \quad (3)$$

The reason we specified steady state at a particular intraocular pressure is that both F_{in} and R_{trab} are sensitive to intraocular pressure.

Formation of aqueous from the tissue fluid of the ciliary processes is partly the result of ultrafiltration and largely the result of secretion. Ultrafiltration is a process that depends on pressure, in this case the net pressure difference between the tissue space and the ocular cavity. Formation of aqueous decreases slightly as intraocular pressure increases. Bárány[4] called the suppression of aqueous formation with rise of intraocular pressure "pseudofacility" because in estimation of trabecular facility, decrease in inflow of aqueous is easily mistaken for increase in outflow.

Resistance to trabecular outflow increases

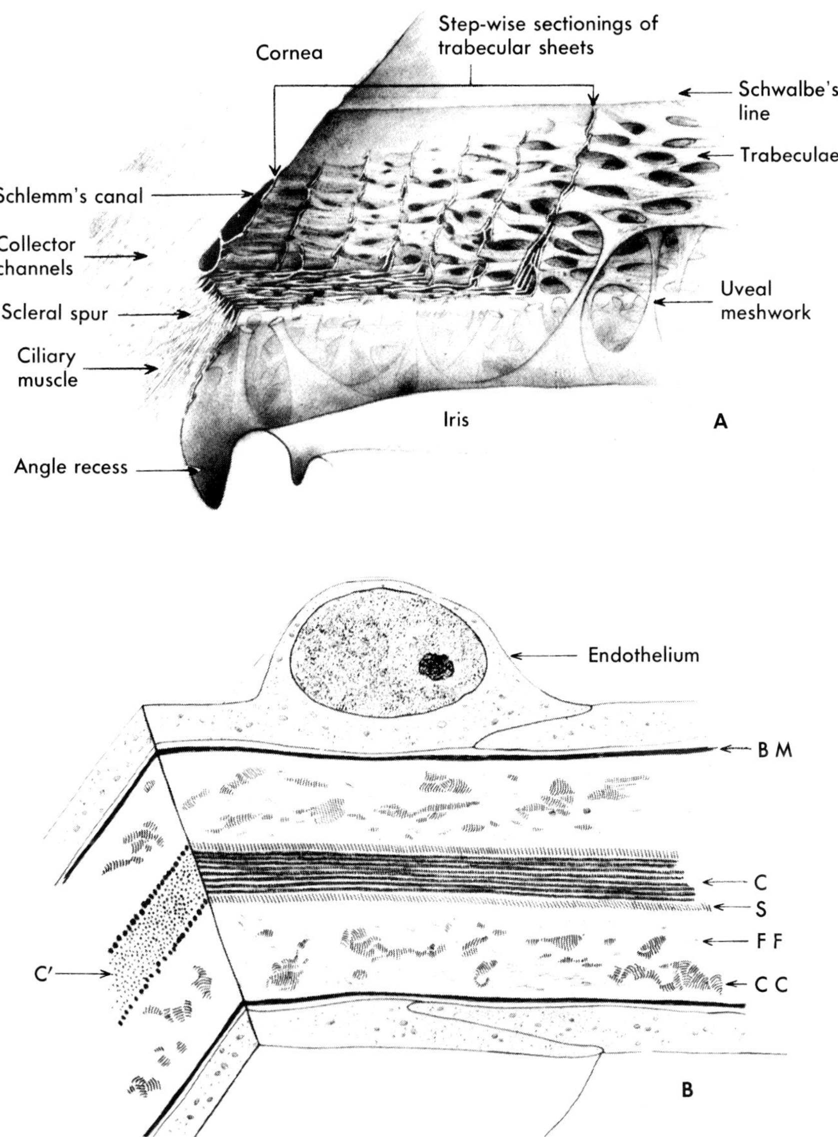

Fig. 8-4. A, Diagrammatic representation of trabecular area. Trabecular meshwork is cut away at progressively deeper levels to show size of openings and disposition of layer. Note large openings in uveal meshwork, and progressive diminution of this size as Schlemm's canal is approached. **B,** Tilted frontal and meridional representation of corneoscleral trabecular sheet. *C,* Central collagenous tissue; *S,* sheath of thick fibers; *FF* and *CC*, zone of ground substance containing irregular clumps of material and very fine fibrils; *BM,* basement membrane; *C′,* collagen fibers cut in cross section when trabeculae are sectioned in meridional plane. (From Garron, L. K., and Feeney, M. L.: Arch. Ophthalmol. **62:**966, 1959.)

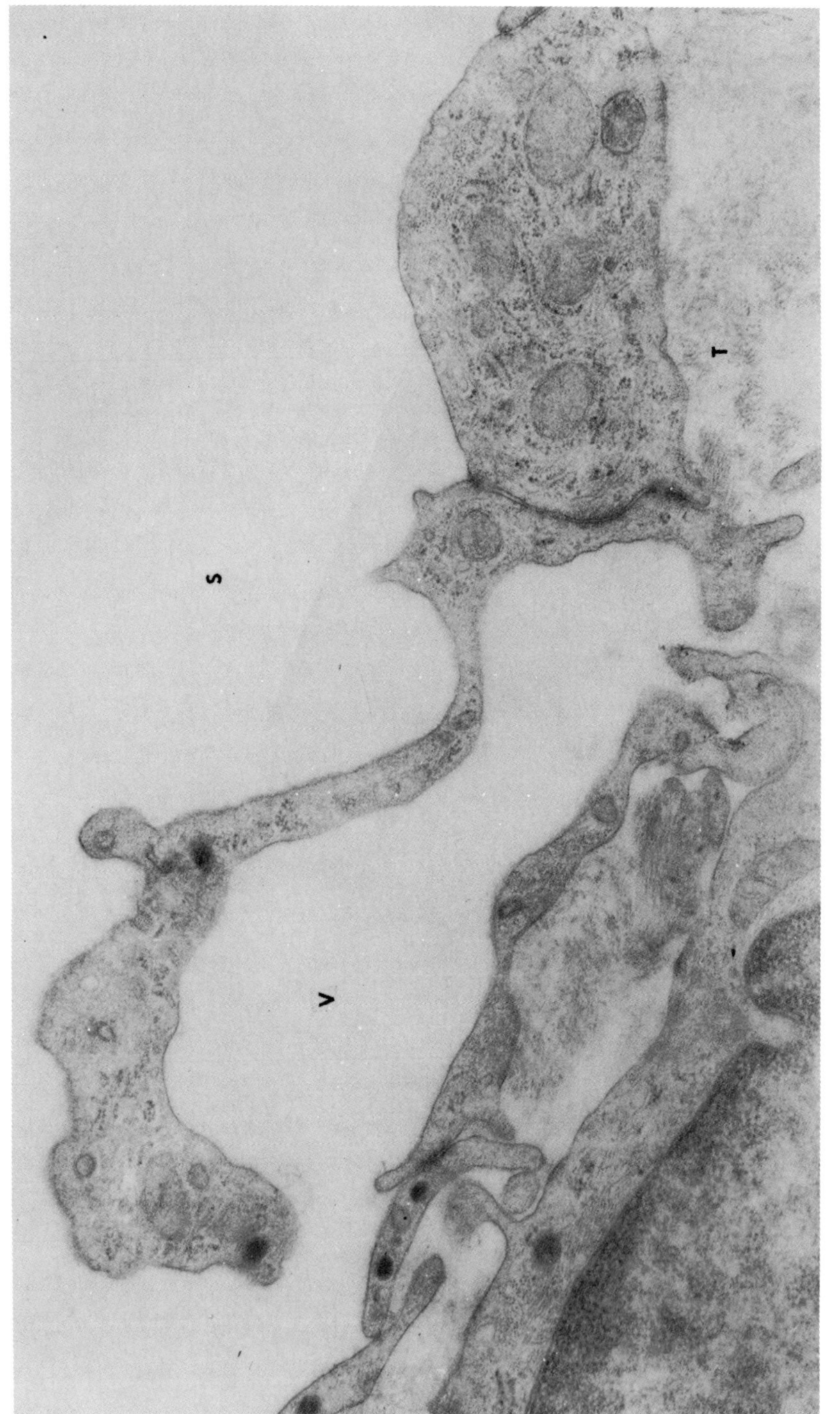

Fig. 8-5. Section through vacuole of endothelial cell lining inner wall of Schlemm's canal. (Original magnification ×32,500.) *S*, Schlemm's canal; *V*, passage through endothelial cell; *T*, trabecular space. (From Kayes, J.: Invest. Ophthalmol. **6:**381, 1967.)

with experimentally raised intraocular pressure,[42] probably because the inner wall of Schlemm's canal is bowed outward by intraocular pressure, narrowing the canal and increasing resistance to flow of aqueous along the canal to collector channels.[43] The relation between trabecular outflow resistance, R_{trab}, and P_i is given by Brubaker[10] as follows:

$$R_{trab} = R_0 + R_0Q\,(P_i - P_e) \qquad (4)$$

In this expression R_0 is the trabecular resistance when $P_i = P_e$, and "Q is the fractional change in R_0 brought about by a change in outflow pressure of 1 mm Hg."[10] Brubaker suggests that Q may be thought of as an "outflow obstruction coefficient." It has the units mm Hg^{-1}. Brubaker found the value of Q to be about 0.012 mm Hg^{-1}. He pointed out that this could be an important factor in glaucoma and in the elevated intraocular pressure incidental to tonography.

The rate of outflow from the eye is then as follows:

$$F_{out} = F_{trab} + U$$

$$F_{out} = \frac{P_i - P_e}{R_0 + R_0Q\,(P_i - P_e)} + U \qquad (5)$$

The relationship of aqueous production to intraocular pressure is less certain and at the present time an accurate quantitative expression for aqueous inflow cannot be written. In previous editions of this book I have treated aqueous production in a less comprehensive manner; however, this information can now be updated and expanded.

In Fig. 8-6, a partial graphic solution for steady state intraocular pressure, the curves rising from the lower left of the graph are the lines

$$F_{out} = \frac{(P_i - P_e)}{R_0 + R_0\,Q\,(P_i - P_e)} + U$$

and are based on an episcleral venous pressure of 9 mm Hg. The curves thus give the total rate of aqueous outflow as a function of intraocular pressure and are calculated for trabecular outflow resistances, R_0, of 3, 5, 10, and 15 minutes mm Hg μl^{-1}. In each case, the solid line is for $Q = 0.12$ mm Hg^{-1} and the interrupted line is for $Q = 0.08$ mm Hg^{-1}. The slope of an outflow curve at any point is a measure of $1/R_{trab} = C_{trab}$. Goldmann[19] views pseudofacility as the controlling factor in the

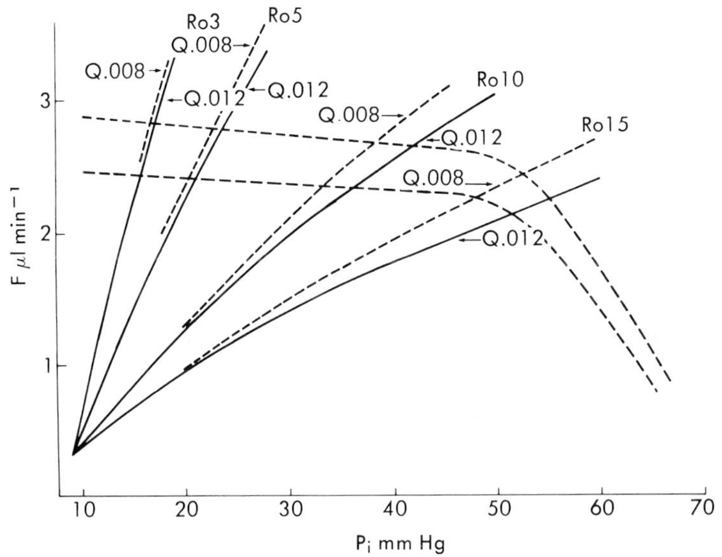

Fig. 8-6. Intraocular pressure at steady state, when rate of aqueous production equals rate of aqueous outflow. The curves rising up and to the right are total rate of outflow equal to the sum of trabecular outflow rate (based on Brubaker's relationship; see text) plus uveoscleral outflow rate, assumed to be 0.3 μl min^{-1}. Episcleral venous pressure is assumed to be 9 mm Hg. The dashed lines curving down to the right represent rates of aqueous inflow (equation unknown). The rate of aqueous formation decreases slightly as intraocular pressure increases until the region of ciliary artery blood pressure is approached, when it decreases rapidly. The intersection of inflow and outflow curves is at P_i. R_0 is trabecular resistance at $P_i = 0$. Q is increase of trabecular resistance/mm Hg.

rate of aqueous production. Bárány insists this is not so and that aqueous formation proceeds at an only slightly diminished rate as intraocular pressure increases. When intraocular pressure approaches ciliary body arterial pressure, aqueous production then falls off rapidly (personal communication). The aqueous production curves are represented by interrupted lines whose equations are not known. As the curves proceed from left to right they slope gently downward to represent some decrease of aqueous formation at increased intraocular pressure. At high intraocular pressure, the aqueous production curves slope downward more steeply as intraocular pressure approaches ciliary body arterial pressure. The aqueous formation curves of Fig. 8-6 arbitrarily give Bárány's point of view, but the true behavior of aqueous formation rate as a function of intraocular pressure is not known. The intersections of the inflow and outflow curves are at intraocular pressures where $F_{in} = F_{out}$, or steady state.

Several aspects of Fig. 8-6 are noteworthy. First, any intraocular pressure may be the result of many pairs of inflow and outflow parameters that result in $F_{in} = F_{out}$; the intraocular pressure alone does not indicate whether it is the result of rapid inflow and low resistance to outflow, or if the same pressure is the result of a lower inflow rate and a higher outflow resistance. Second, when P_i equals the assumed P_e (9 mm Hg in this example), outflow is not zero, but rather, in this case it is 0.14 μl min^{-1}, the value here assumed for U. This would imply that if inflow of aqueous ceases, intraocular pressure will drop below P_e because uveoscleral flow continues; perhaps this might explain the low pressures found in secretory failure. Third, the graph reiterates Brubaker's comment that an increase in intraocular pressure, as in tonog-

Table 8-2. Glossary

Symbol	Quantity	Units
P	Pressure	g/cm^2 mm Hg
P_i	Intraocular pressure	
P_t	Intraocular pressure during tonometry; increased from P_i by application of tonometer	
P_o	Estimate of intraocular pressure from single Schiøtz tonometer scale reading assuming average ocular rigidity	
P_e, P_v	Episcleral venous pressure	
P_a	Ciliary artery blood pressure	
F	Flow	Volume per unit time; μl/min
F_{in}	Rate of formation of aqueous	
F_{out}	Rate of discharge of aqueous	
U	Rate of discharge of aqueous through the uveoscleral route	
C	Coefficient of facility of flow	Volume per unit time per unit pressure gradient; μl/min/mm Hg; μl min^{-1} mm Hg^{-1}
C_{in}	Facility of inflow from blood to ocular cavity	
C_{out}, C_{trab}	Facility of trabecular mesh outflow	
R	Resistance to flow	mm Hg min μl^{-1}
R_{in}	Resistance to inflow of aqueous	
R_{trab}	Resistance of trabecular outflow	
R_o	Resistance of trabecular outflow when $P_i = P_e$	
Q	Fractional increase in R_o with P_i	mm Hg^{-1}
E	Ocular rigidity; measure of the resistance of the ocular coats to distention	μl^{-1}

raphy, increases R_{trab}, markedly so for high values of R and Q.

We have presented our current concepts of intraocular pressure and its parameters. These views are consistent with many known facts. For example, an increase of P_e causes an increase of P_i of slightly less magnitude; P_i rises higher in angle-closure glaucoma than it ever does in primary open-angle glaucoma because in angle closure the iris blocks uveoscleral outflow as well as the trabecular mesh; marked ciliary vasodilation, such as resulting from a large dosage of prostaglandin in a rabbit eye, can cause increased intraocular pressure and at the same time increased facility of outflow as measured by tonography. The last effect is at least partly because the rapid aqueous formation from the dilated ciliary vessels is easily suppressed by the further pressure increase occasioned by tonography, and this pseudofacility cannot be distinguished from trabecular facility by tonography alone. (However, prostaglandins also increase "true" or trabecular facility in rabbit.[11]) The total picture of aqueous dynamics is still very far from complete. Several persons have tried to apply sophisticated mathematic analysis to the aqueous flow system, sometimes with misleading results because the data employed were much too crude. While I recognize the desirability of mathematic statement, I must urge that better data are needed. A few hundred years ago, crude data clearly showed that the earth was flat!

The concepts presented here are drawn from several sources[4,9,19] (Chapter 6). For rough approximate calculations, the original Goldmann equation:

$$(P_i - P_e) = F_{in} R_{trab}$$

is useful, and F_{trab} is roughly equal to F_{in} (Table 8-2).

Changes in the trabecular mesh and canal of Schlemm with intraocular pressure

Johnstone and Grant[25] have demonstrated changes in the configuration of the trabecular mesh and inner wall of Schlemm's canal that explain the decrease of outflow facility seen with high intraocular pressure.[41] At low pressure the canal is wide, but the trabecular spaces adjoining the canal are narrow (Fig. 8-7). As ocular pressure increases, the inner wall of the canal, the site of the major resistance to outflow, is bowed outward into the canal. This has the effect of opening the intertrabecular spaces, but the net effect on resistance to out-

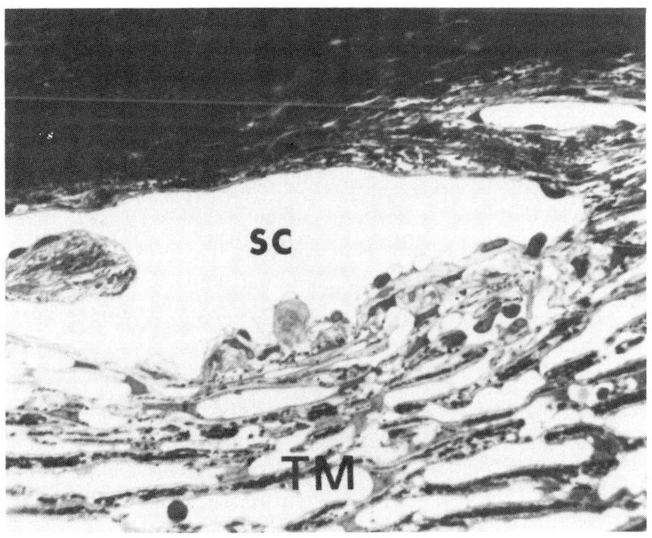

Fig. 8-7. Enucleated human eye fixed at 5 mm Hg intraocular pressure. Schlemm's canal, *SC*, remains moderately large, and there is space between lamellae of trabecular meshwork, *TM*. (From Johnstone, M. A., and Grant, W. M.: Am. J. Ophthalmol. **75**:365, 1973.)

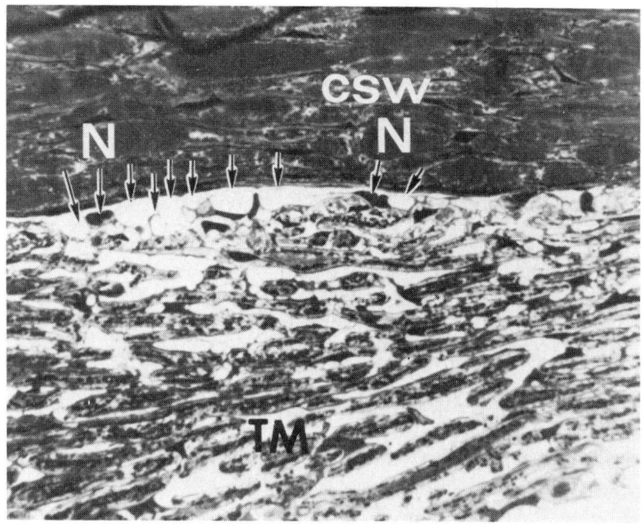

Fig. 8-8. Enucleated human eye fixed at 50 mm Hg intraocular pressure, showing nuclei, *N*, and vacuoles (arrows) of endothelial cells lining inner wall of Schlemm's canal, *SC*, in apposition to external corneoscleral wall, *CSW*. Canal has been reduced to potential space. (Magnification ×550.) (From Johnstone, M. A., and Grant, W. M.: Am. J. Ophthalmol. **75**:365, 1973.)

flow is probably small. At still higher pressure the outward-bulging inner wall comes into apposition with the rigid outer wall of the canal (Fig. 8-8). Apposition of the canal walls may close off access of aqueous to collector channels and aqueous veins.

Secondary or plasmoid aqueous

If the intraocular pressure is markedly lowered but arterial pressure is not, fluid leaks into the ciliary process stroma more rapidly than it can be pumped through the epithelium. The ciliary processes swell, indicating that the epithelium offers resistance to such rapid passage of fluid. The protein content of the rapidly formed fluid is much higher than that of normal aqueous. In cats, dogs, and rabbits this so-called secondary aqueous, or plasmoid aqueous, is often rich enough in fibrinogen to clot. In humans, secondary aqueous is rich in protein but infrequently forms a clot. The source of the protein is largely ciliary process vessels, although some may be from iris vessels.

DISTRIBUTION OF PRESSURE IN THE EYE

Intraocular pressure is borne by the cornea alone anteriorly and by the sclera and its thin uveal and fragile retinal lining posteriorly. The ciliary body and choroid are distended against the sclera by intraocular pressure. The uveal tissues support some intraocular pressure, so pressure in the potential space between uvea and sclera is about 2 mm Hg less than intraocular pressure.[40,51] When the eye is opened, as in cataract surgery, intraocular pressure drops to atmospheric pressure, while the pressure in the suprachoroidal space falls below atmospheric pressure. Under such circumstances fluid often transudes into the suprachoroidal space, producing choroidal detachment. The thicker ciliary body tends to contract to a cone, bridging the scleral shell. Hence in uveal detachment fluid always accumulates between ciliary body and sclera.

Aqueous flows from its site of formation in the posterior chamber, through the pupil, into the anterior chamber. A small pressure head from posterior chamber to anterior chamber must exist. The higher pressure of the posterior chamber transmitted to the vitreous tends to press the lens forward. The lens is restrained by its suspensory ligament, the zonule, but if the zonule is broken, the lens and iris may be carried forward or the lens may be extruded through the pupil into the anterior chamber.

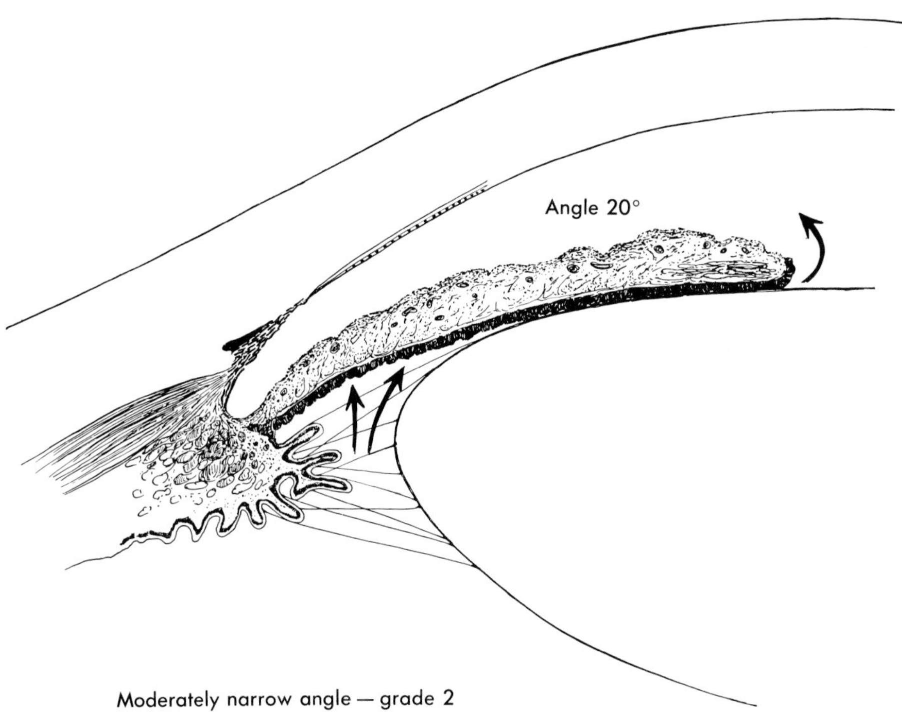

Fig. 8-9. Physiologic iris bombé. (From Kolker, A. E., and Hetherington, J., Jr.: Becker-Shaffer's diagnosis and therapy of the glaucomas, ed. 4, St. Louis, 1976, The C. V. Mosby Co.)

Angle closure

In some middle-aged persons the face of the lens is well anterior to the iris root. This may be because growth of the lens during life has thickened the lens anteroposteriorly. Also, increased equatorial diameter of the lens shortens the distance between lens equator and ciliary body, thus making the zonule fibers slack, allowing the lens to ride forward. As a result of these mechanisms or from some other cause, the iris is more tightly applied to the lens capsule than is usual. The higher posterior chamber pressure may then be manifest by a forward ballooning of the iris, physiologic iris bombé (Fig. 8-9). The anterior chamber in these persons is shallowed by the forward placement of the lens and the ballooned iris; the angle between surface of the iris and the trabecular mesh is narrow as seen with the gonioscope. In extreme cases the peripheral iris lies against the trabecular mesh and prevents egress of aqueous. If there is complete angle closure, continued formation of aqueous causes the intraocular pressure to rise sharply and to give the clinical picture of angle-closure glaucoma. However, often only part of the mesh is occluded by iris, and the pressure rise is not so dramatic and symptom producing. Partial angle closure may produce moderate ocular hypertension and slow loss of vision.

Pathologic iris bombé is seen when the iris is adherent to the lens capsule as a result of inflammation. (In ophthalmology adhesions of the iris are called synechias; adhesion of the iris to the lens is said to be a posterior synechia.) Ballooning of the iris midzone anterior to the pupil gives the iris a doughnut shape. As in physiologic iris bombé, if the trabecular mesh is occluded by the iris, intraocular pressure becomes elevated.

The surgical treatment of angle closure is to increase the facility of passage of aqueous from posterior to anterior chamber. Excision of a portion of the peripheral iris (peripheral iridectomy) allows freer forward passage of aqueous and reduces the pressure head from posterior to anterior chamber. The iris balloon collapses away from the trabecular mesh, and access of aqueous to the outflow channels is restored.

SHORT-TERM AND LONG-TERM CHANGES IN INTRAOCULAR PRESSURE

Intraocular pressure can be changed rapidly and sometimes dramatically by pressure on the eye from outside, as by forced lid closure, pressing on the eyeball with a finger, or cocontraction of the rectus muscles. These maneuvers distort the eye and require the ocular coats to stretch over the same ocular volume disposed in a shape with greater surface, as does indentation with the Schiøtz tonometer. The increased tension of the ocular coats and the concomitant increase in intraocular pressure are directly related. However, if the distortion persists—for instance, a scleral buckle for retinal detachment—the intraocular pressure, acutely raised by distortion from the equilibrium level (P_i) to a higher level ($P_i + \Delta P_i$), increases the rate of flow of aqueous from the eye and decreases the rate of inflow. The increased rate of outflow and decreased rate of inflow will in time return the intraocular pressure to the steady state level existing before the buckle was placed unless the iris has been pushed forward against the trabecular meshwork. Acute decrease of retinal and uveal vascular tone will be reflected in intraocular pressure as the intraocular vessels expand, but if this vascular dilation becomes chronic, the acute rise in intraocular pressure drains away as before. Fluctuations of blood pressure with pulse and respiration are too rapid to be completely compensated; such changes are reflected in ocular pressure. The amplitude of these blood pressure changes is damped by the distensibility of the vessel walls and the ocular coats, so arterial pulse produces a variation of 1 to 2 mm Hg in ocular pressure superimposed on the longer swings of the same magnitude associated with respiration.

Changes in intraocular pressure that occur with coughing, bending over, neck compression, and so on are generally traceable to changes in arterial blood pressure, venous blood pressure, or both. For instance, attempted expiration against the closed glottis (Valsalva maneuver) produces a marked rise in intraocular pressure of a complicated pattern. Moderate forced expiration raises intrathoracic pressure, which in turn presses on the superior vena cava. Blood backs up in the neck and head veins and raises intraocular pressure. Marked expiratory effort collapses the vena cava to the extent that blood return to the heart is impeded. Since the heart does not fill with blood, arterial pressure falls, and eye pressure falls with it. Such patterns may be quite complex and difficult to interpret.[3,48]

It is apparent that the term intraocular pressure is inadequate when a statement such as the first sentence of this chapter is made. What is meant, of course, is "mean intraocular pressure while sitting or lying quietly, breathing evenly, and without any compression of the neck vessels." The pressure in the eye may either rise or fall on change from sitting to lying. On the average it rises 2 to 3 mm Hg. Even measurement of intraocular pressure is attended by change. Repeated measurement with the Goldmann applanation tonometer usually reveals a slow fall of pressure that may amount to more than 5 mm Hg.[1,6,38] In short, intraocular pressure is not a fixed quantity. The examiner must be aware that the tense patient who holds his breath may show a high intraocular pressure, as may the man with a tight collar.

Diurnal variation in intraocular pressure

Intraocular pressure varies with the time of day; it is usually highest about the time of awakening and lowest during the evening (Fig. 8-10). There are many body functions that vary in a similar circadian pattern; blood pressure, renal activity, and secretion of adrenal steroids are a few. A clinical study[12] showed a correlation between body temperature and intraocular pressure, a relationship shown earlier experimentally by Pollack and Becker[45] in rabbits, but the pressure fall found in humans, about 20%/° C, was five times that noted in rabbits, so it is questionable if body temperature is the major factor operative in production of diurnal variation in intraocular pressure. The diurnal variation in ocular pressure is commonly much more marked in ocular hypertension than in the

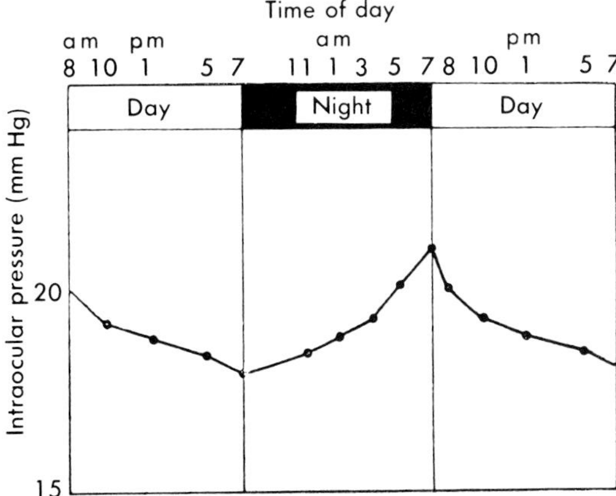

Fig. 8-10. Diurnal variation of intraocular pressure. (From Duke-Elder, W. S.: Text-book of ophthalmology, St. Louis, 1939, The C. V. Mosby Co., vol. 1.)

normal eye. There may be some diurnal change in facility of outflow,[29] but the major factor appears to be a variation in the rate of formation of aqueous. An increment in aqueous flow will pass through a normal outflow facility with little increase in pressure head, whereas the same increased rate of flow through a lesser facility will require a greater pressure head. The intimate cause of the diurnal fluctuation in the rate of aqueous flow is not known.

Intraocular pressure is always changing. We must understand that a single pressure measurement is a small sample of a continuing process and may not be representative of the long-term pressure trend. We must recognize that a patient's intraocular pressure, although normal during the afternoon, may be pathologically high in the morning. Moreover, we take our measurement—usually a visual estimate of mean pressure—under a relatively basal condition.

Several potentially confusing topics are worthy of mention. Swelling of the lens cannot cause any change in intraocular pressure unless the swollen lens has a secondary action, such as displacement of the iris against the trabecular mesh to produce angle closure or clogging of the mesh by phagocytes attracted by leakage of degraded protein from a swollen lens. The lens itself swells by taking aqueous from outside the lens capsule and accumulating it within the capsule; there is no net change in intraocular volume. A tumor cannot grow rapidly enough to raise intraocular pressure by adding volume to the eye; hemorrhage into a tumor (or elsewhere in the eye) can cause rapid pressure rise.

FREQUENCY DISTRIBUTION OF INTRAOCULAR PRESSURE IN THE POPULATION

Intraocular pressure is usually in the range of 10 to 22 mm Hg. The most frequent intraocular pressures are 15 to 16 mm Hg. The frequency distribution of intraocular pressure in the population is not a normal (Gaussian) curve, but rather is skewed to the right. This type of distribution curve is often seen in medicine; blood pressure and blood sugar distributions, for instance, are similar in that the values peak in a region compatible with continued health. To the left of the peak, or mode, the frequency decreases as if the curve were the left half of a Gaussian curve, but to the right of the mode the distribution curve trails off to a long tail. Efforts have been made to impose a normal distribution on these skewed curves and to classify those measurements to the right of the normal curve as abnormal, unhealthy, and representative of a disease state. If this is done, there is a sizable area of overlap in which it is not clear which persons should be classified as healthy and which as diseased. Without further information the individuals belonging to one or the other group cannot be identified. (For an analysis of the skew curve of intraocular pressure on the basis of hypothetical trabecular porosity, see Davanger and Holter.[13])

OCULAR HYPERTENSION

The ophthalmologist is confronted with a large number of patients who have intraocular pressure greater than the accepted safe range. Since those eyes that will manifest visual loss cannot be identified on the basis of pressure alone, intraocular pressure should be a guide to other diagnostic procedures needed. Gonioscopy is essential to the analysis of ocular hypertension, for if the hypertension is based on any degree of angle closure, prompt action is indicated to avoid increased closure and further pressure rise. Careful visual fields with small test objects are needed to detect the earliest sign of retinal dysfunction and are essential to the management of ocular hypertension. Tonography gives insight into the capability of the eye to cope with possible increases in aqueous formation. Measurement of ocular pressure every few hours during the day and at night will disclose the patient's individual pattern of pressure fluctuation. Ophthalmoscopic examination of the optic disc demonstrates how the tissues are withstanding the ocular pressure.

There is probably some intraocular pressure above which every ophthalmologist will treat the eye as glaucomatous, regardless of other findings; the conviction that very high intraocular pressure will sooner or later cause visual field loss is widespread. The lower this diagnostic limiting value is, the more patients will be placed in therapy; a greater number of patients will be classified as having glaucoma. As the upper limit of pressure thought to be consistent with continued retinal function is lowered, the number of persons who would probably never lose visual field is increased, and the percentage of therapeutic triumphs increases. On the other hand, if the diagnostic pressure limit is set high, it is advisable to repeat visual fields several times a year to detect the earliest signs of change so that therapeutic measures can be initiated promptly. At the present time, benign ocular hypertension and glaucoma cannot be separated by formula. The physician must be guided by frequent careful observation and precise measurement; the clinical course dictates the therapeutic course.

PRESSURE-VOLUME RELATIONSHIP

The relationship between pressure in the eye and volume of the ocular content is important to measurement of pressure in the eye and to measurement of facility of outflow, as we shall presently see. The difficulties imposed on a straightforward analysis by the complex shape and properties of the eye have led many investigators to adopt an empirical approach.

The usual experimental method has been to simultaneously measure pressure and volume change. For example, an eye is cannulated and connected to a pressure transducer. The manometric pressure is noted. A known volume of liquid is injected into the eye, producing a pressure rise. By repeated injection a curve relating pressure to volume change is developed (Fig. 8-11). It is to be noted that absolute volume of the eye is more difficult to measure than volume change because the volume of the eye at zero (atmospheric) pressure is ambiguous. At atmospheric pressure the eyeball may be collapsed to varying degrees without an appreciable change in intraocular pressure. However, if the curve that relates pressure to volume increment is well established, the ocular volume corresponding to atmospheric pressure can be calculated. The range of volume of individual eyes is considerable; a representative volume near zero pressure is 6.2 ml, as indicated in Fig. 8-11.

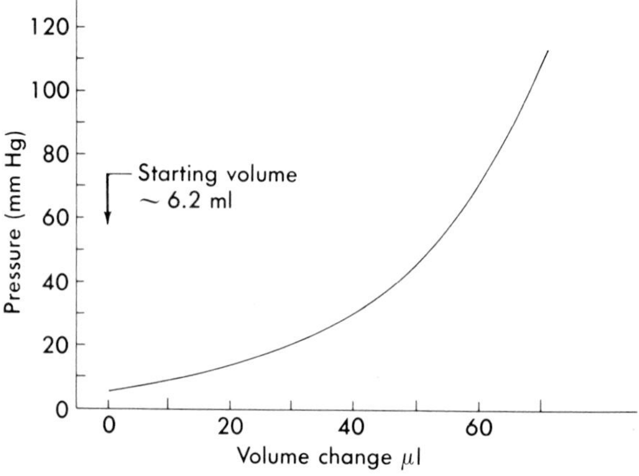

Fig. 8-11. Increase of ocular volume increases intraocular pressure.

DISTENSIBILITY OF THE EYE

The tough corneoscleral wall bounds the ocular contents. Pressure within the eye stretches the walls. This is to say that intraocular pressure develops tension in the cornea and in the sclera and its lining choroid. The relation between pressure in a spherical container and tension in the container wall is theoretically simple.

$$T = Pr/2t$$

where

T = Tension (g force/cm^2)
P = Pressure (g force/cm^2)
r = Radius of sphere (cm)
t = Thickness of vessel wall (cm)

It should be noted that although wall tension and intraocular pressure are measured in the same units, they are not the same; rather, they differ by the factor $r/2t$. It may be estimated that at 15 mm Hg intraocular pressure (approximately 20 g force/cm^2), tension in the cornea is about 16 g force/cm^2 and in the sclera at the equator of the eye is about 25 g force/cm^2. However, such calculations should not be extended to the eye as a whole until two considerations are made: (1) the shape of the eye is not truly spherical*; (2) the wall thickness varies from one region of the eye to another.

The steepness of the pressure-volume curve varies from eye to eye. This variation resisted analysis until Friedenwald showed that such curves approximate straight lines when plotted as log pressure versus volume[16] (Fig. 8-12). The steepness of slope, E, of the log pressure–volume line is a measure of the resistance of that eye to distention and is given by the change in log pressure associated with a given change of volume.

$$E = \frac{\log P_2 - \log P_1}{V_2 - V_1}$$

One of the convenient features of this relationship is that it is not necessary to know the total volume of the eye, but only the *volume change*. Thus, if the volume of the eye is changed by a known amount, $\Delta V = V_2 - V_1$, and the pressure before and after the volume change is known, then

$$E = \frac{\log P_2 - \log P_1}{\Delta V}$$

*For treatment of nonspherical vessels, consult an engineering text on strength of materials.

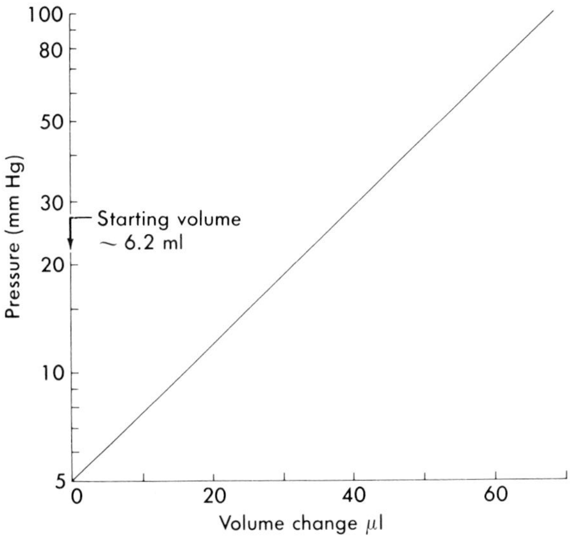

Fig. 8-12. According to Friedenwald, log intraocular pressure versus volume change is straight line; slope of line is E.

The name chosen for the distention resistance constant, E, was the coefficient of scleral rigidity, now called ocular rigidity. This is an unfortunate term because most persons associate the word rigid with resistance to bending rather than resistance to stretching. Any eye surgeon knows (sometimes sorrowfully) that the walls of the eye bend with great ease. Perhaps a name such as distensibility resistance coefficient would be preferable to ocular rigidity coefficient.

The average value of E arrived at by Friedenwald was 0.0215. Various researchers have extended Friedenwald's work. It appears that E is not truly constant for a given eye, but decreases as pressure rises[36,37,47]; other formulations have been given for the pressure-volume relationship. The values of E found by cannulation of the living human eye have been much lower.[47] Friedenwald's formula is still used in clinical ophthalmology because of the following:

1. It is the simplest formulation presented so far.
2. Much data have been accumulated in which the formula was used.
3. The value of the newer data to clinical practice has not yet become apparent.

The material on tonometry and tonography in the following discussion is subject to the criticism that it does not represent the best information available. I can only hope that inaccuracy is outweighed by simplicity.

Two qualifications of the concept of ocular rigidity must be noted, however. A rise of pressure in the living eye distends the ocular coats and forces some blood from the eye. Friedenwald was quite cognizant of the contamination of the value of E by blood loss.

The second qualification is that distention of the cornea and sclera has a time course.[37] When these materials are stretched, the molecules straighten, but this rearrangement is in a viscous material and can occur only as rapidly as viscous flow allows. This effect is called viscoelasticity. Thus when intraocular pressure is raised by rapid injection of fluid, the immediate pressure attained is decreased over a period of time by viscoelastic rearrangement of the collagen fibers of the cornea and sclera. If pressure is lowered, a similar viscoelastic retraction of the coats is seen.

MEASUREMENT OF INTRAOCULAR PRESSURE

Manometry

Intraocular pressure may be measured experimentally by cannulation of the eye. Special care must be taken not to alter the eye by rough handling. Sears devised a spring gun to introduce a needle through the leathery cornea with minimal trauma.[49] General anesthesia may alter eye pressure; for instance, most barbiturates cause intraocular pressure to decrease. Leakage of aqueous will obviously yield false results. Constriction of the neck, head-down position, and other factors will raise venous pressure and ocular pressure. In short, ocular cannulation, although direct, is not necessarily a simple means of measuring the pressure in the eye.

Tonometry

Clinical practice excludes cannulation of the eye. Resort must be made to a device that measures pressure through the eye wall. Such a device is known as a tonometer. (Ophthalmotonometer is the proper name.) The long history of tonometry is given by Draeger.[14]

Tonometers fall into two large groups: those that flatten, or applanate, the cornea and those that indent the cornea. Repeated attempts have been made to measure intraocular pressure through the sclera, but the results have been erratic because of the nonuniformity of the anterior sclera and the added thickness of the conjunctiva and ciliary body. Any device that measures pressure must in some way measure the force on an area.

Applanation tonometry

The process of applanation tonometry is based on the definition of pressure—pressure = force/area, or $P = f/A$. To a first approximation, if a segment of a flexible sphere is flattened by a plane surface, pressure in the sphere acting over the flattened area will balance the applied force, or $PA = f$. To estimate intraocular pressure, one needs to know the area flattened and the applied force (Fig. 8-13).

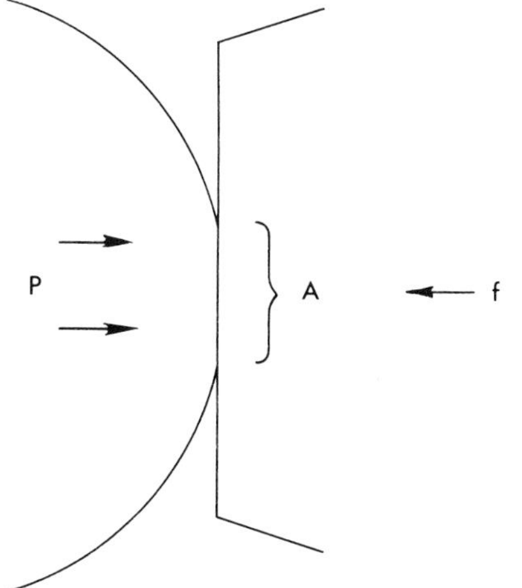

Fig. 8-13. Applanation tonometry is based on definition of pressure. Pressure = force per unit area.

The first practical tonometer was that of Maklakoff (1885). It consisted of a flat-bottomed weight. A thin coat of dye was applied to its lower surface. The weight was rested briefly on the anesthetized cornea of the supine patient. The area of contact between tonometer and cornea was estimated from the diameter of the dye disturbance on the weight. The approximate P_t* was readily calculated from imprint area and tonometer weight. A modern Maklakoff tonometer is shown in Fig. 8-14.[46]

The second applanation tonometer (Fick, 1888) employed a fixed area of flattening produced by an adjustable force. The modern version of this instrument (Goldmann, 1954) is a very accurate clinical tonometer (Fig. 8-15).[20,21]

The basic concept of applanation tonometry is very simple. On closer analysis, three complicating factors appear (Fig. 8-16).

1. The cornea, although very flexible, has some structural resistance to deformation, or a springlike quality. This force, *N*, tending to push the applanating surface away from the eye, is added to *PA*.

2. The surface of the cornea is covered with tears, which creep onto the applanating surface and pull it toward the cornea by surface tension. This force, *M*, acts in the same direction as the applied force. Estimate of intraocular pressure then is

$$PA + N = f + M$$

or

$$P = \frac{f + M - N}{A}$$

3. Flattening of a segment of the cornea displaces aqueous from the segment. In a sense this is similar to injecting a quantity of fluid into the eye. The displaced fluid finds room by further distending the globe, but the eye walls resist distension, and displacement of aqueous raises intraocular pressure.

At applanation diameters of 3 to 4 mm, the *M* and *N* forces are equal and cancel each other. Goldmann chose an applanation diameter of 3.06 mm, since at that diameter *M* (surface tension force of the tears) balances *N* (elastic force of the cornea), and at this particular diameter grams force × 10 is directly converted to mm Hg. Flattening of this small dome of cornea displaces a virtually negligible 0.5 μl of aqueous, so P_t is only 3% greater than P_i. Determination of the flattened area by a split-field device allows vernier acuity. Tears are distinguished from flattened cornea by fluorescein, which appears as a greenish rim around the blue flattened area when the field is viewed in blue light. Several handheld tonometers based on the Goldmann tonometer are available.

The calibration of the Goldmann tonometer is for human eyes. When used on the more plastic corneas of animals other than primates, other calibrations must be used.[21]

The Maklakoff tonometer is not corrected for corneal elasticity *(N)*, tear meniscus attraction *(M)*, or elevation of intraocular pressure by fluid displacement. In addition, the imprint is too large because of the tear meniscus. Nevertheless, it gives an estimate of intraocular pressure that is useful.[46] Serious defects are that rocking the tonometer on the cornea and excessive creep of tears onto the plane lead to *underestimation* of intraocular pressure. Tonometry is performed to detect increased intraocular pressure, one of the signs of glaucoma;

*P_t is pressure in the eye during application of the tonometer.

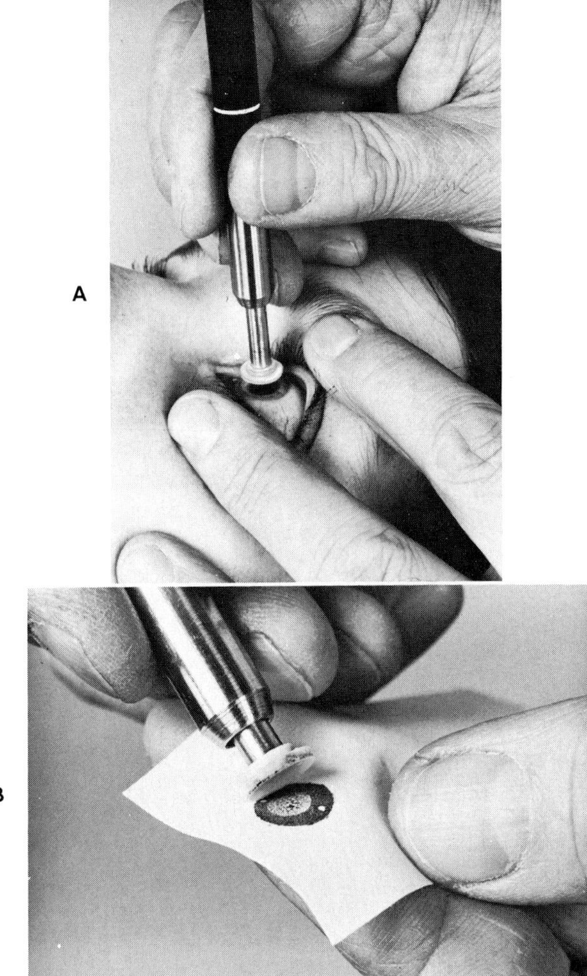

Fig. 8-14. Modern applanation tonometer of Maklakoff type, Posner-Inglima applanometer. **A,** Cornea leaves its imprint on thin film of dye spread on flat base of tonometer. **B,** Imprint is transferred to special paper. (Courtesy Dr. Adolph Posner and Mr. Richard Inglima.)

while an overestimate of intraocular pressure might cause undue alarm, an underestimate might lead to dismissal of a pathologic condition, with tragic consequences.

An applanation tonometer that has proved useful in the face of corneal edema or scarring as well as in healthy corneas is the Mackay-Marg tonometer.[39] This tonometer senses the force on a plunger set in a flat surface. When the tonometer is advanced against the cornea, the plunger bears both corneal bending force and intraocular pressure until the applanation is large enough for the footplate to bear corneal bending force. When the plunger is relieved of bending force, its recoil produces a notch in the force record. At the notch the force on the plunger is that of intraocular pressure over an area slightly larger than the plunger, and the intraocular pressure is inferred from the height of the trough above baseline (Fig. 8-17).

Krakau[27] has recently developed a tonometer that senses the vibration characteristics of the cornea. It may rest against the cornea with a force of only 0.1 g (1/15 the force of the Goldmann tonometer on an eye of normal pressure, 1/165 of the force of the Schiøtz tonometer).

Another tonometer, the noncontact tonome-

Fig. 8-15. Goldman applanation tonometer. Force on tonometer prism is adjusted to applanate cornea to standard diameter.

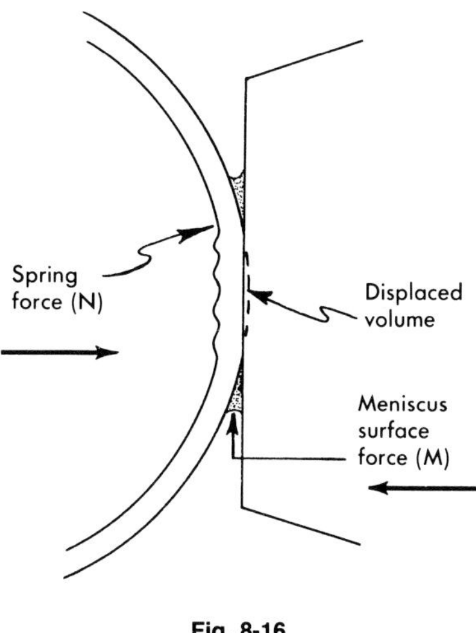

Fig. 8-16

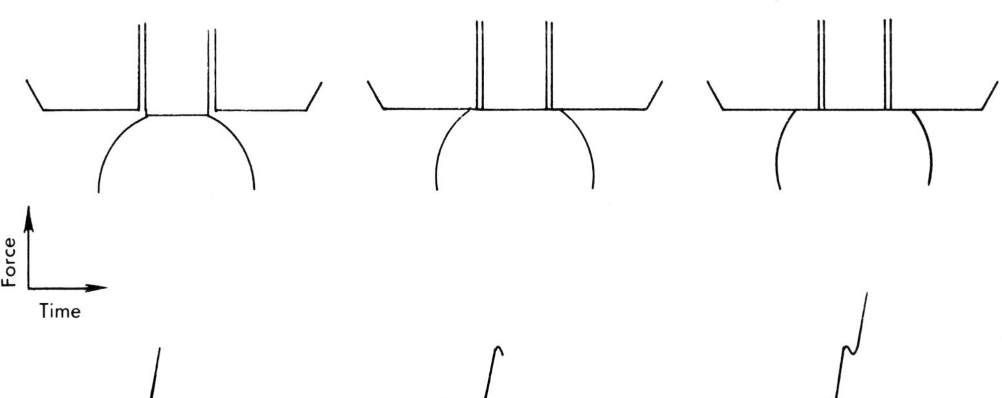

Fig. 8-17. Mackay-Marg tonometer. Left, plunger supports both intraocular pressure and spring force of cornea. Center, corneal bend is supported by footplate, and force on plunger decreases. Right, as tonometer is advanced, displaced aqueous raises intraocular pressure. Upper, tonometer probe and cornea (grossly exaggerated). Lower, record of tonometer plunger movement.

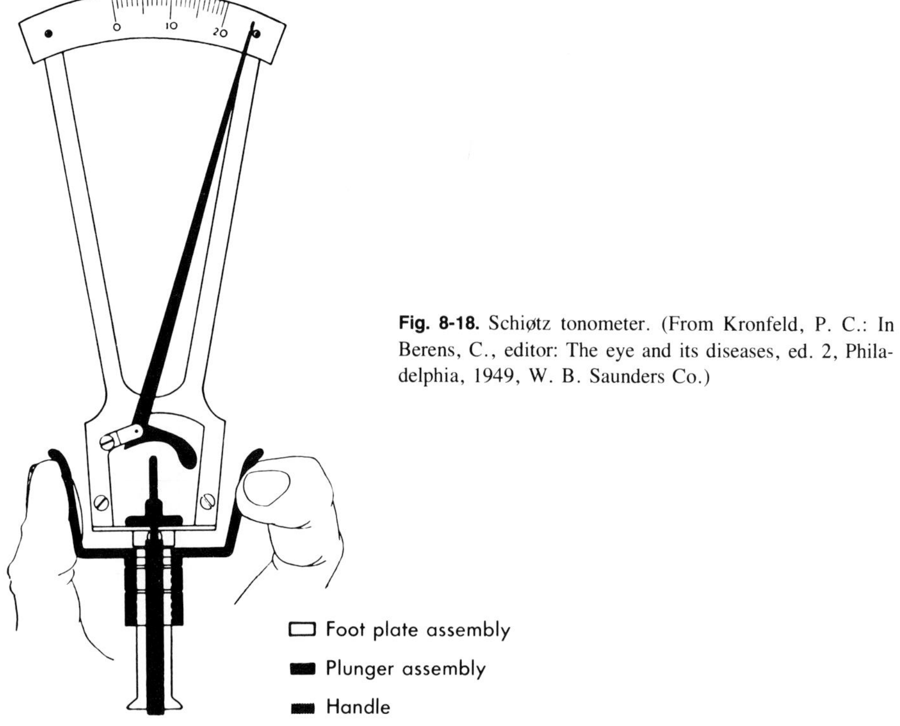

Fig. 8-18. Schiøtz tonometer. (From Kronfeld, P. C.: In Berens, C., editor: The eye and its diseases, ed. 2, Philadelphia, 1949, W. B. Saunders Co.)

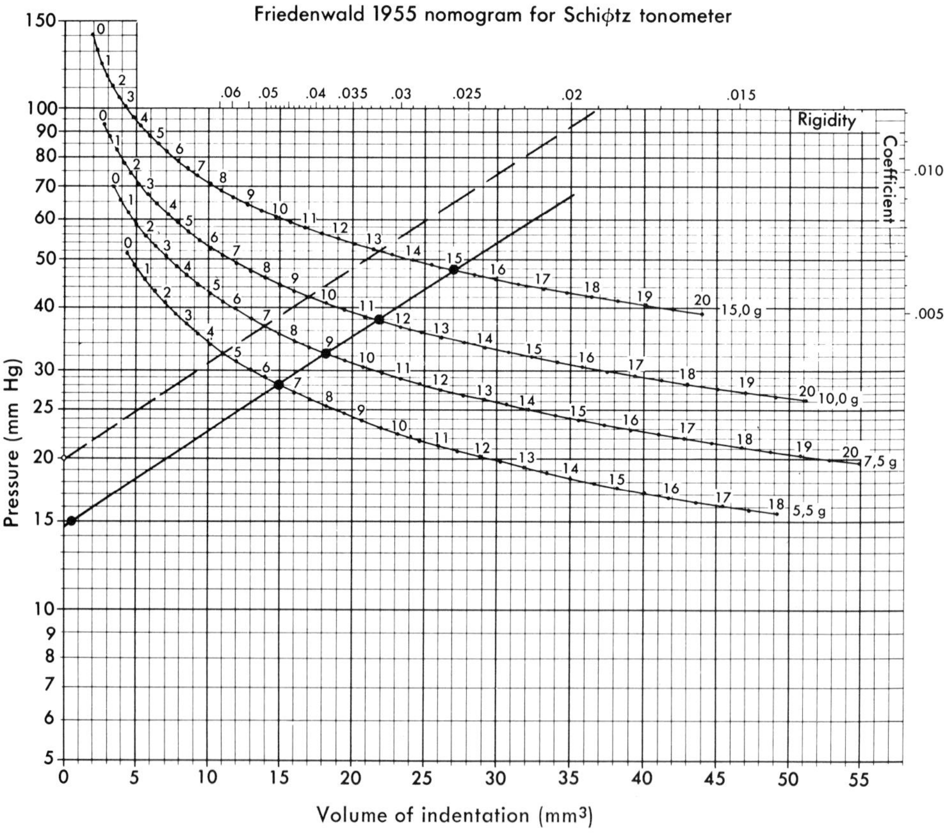

Fig. 8-19. According to Friedenwald, tonometric distortion of eye is equivalent to injection into eye. Tonometries causing different distortions should give set of points on ocular rigidity slope (solid line). Tonometries indicated are with Goldmann applanation tonometer and four different Schiøtz plunger loads. Dashed line drawn parallel to line of tonometries through origin of ocular rigidity scale (upper and right margins). Ocular rigidity indicated is 0.019.

ter,[15] uses a jet of air to flatten a small segment of cornea. The jet is rapidly increased in force, and the moment of corneal flattening is sensed optically. The air pressure of the jet at the instant of flattening is well correlated with intraocular pressure measured with the Goldmann tonometer.

Indentation tonometry

Indentation tonometers have been in use even longer than applanation tonometers. In general, a concave footplate in contact with the cornea provides a reference level, while a plunger sliding through the center of the footplate indents the cornea. The depth of indentation is a measure of pressure in the eye; the plunger indents a soft eye more deeply than a firm eye. The weight of the plunger can be set at one of several values. The most widely accepted indentation tonometer is the one devised by Schiøtz (Fig. 8-18); although it is less accurate than the modern applanation tonometer as a pressure-measuring instrument, its calibration by Friedenwald led to the development of tonography, a clinical estimate of the facility of outflow.[17] The pressure with the tonometer in place (P_t) is significantly greater than the resting pressure of the eye (P_i). Friedenwald reasoned that elevation of intraocular pressure by tonometric distortion is similar to but not identical with elevation of pressure by distention with injected fluid. If this is so, then two independent determinations of tonometric pressure and displaced volume should give the ocular rigidity slope, E. To this end the tonometric pressure corresponding to each scale indication of the Schiøtz tonometer with each plunger load was determined. The distortion volumes were likewise measured. On a semilog plot these values form the Friedenwald nomogram (Fig. 8-19). In theory, tonometer readings with different plunger loads (and applanation tonometer mea-

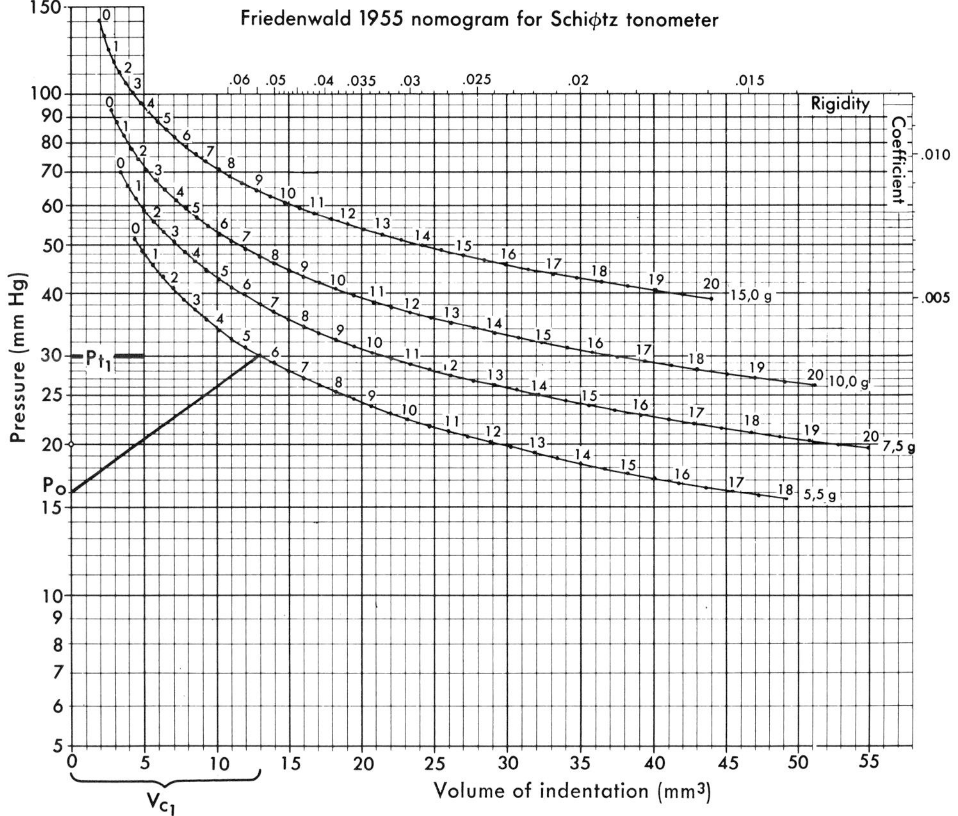

Fig. 8-20. Placement of tonometer indents cornea. Displaced volume, V_{c_1}, raises pressure along rigidity slope from P_o to P_{t_1}.

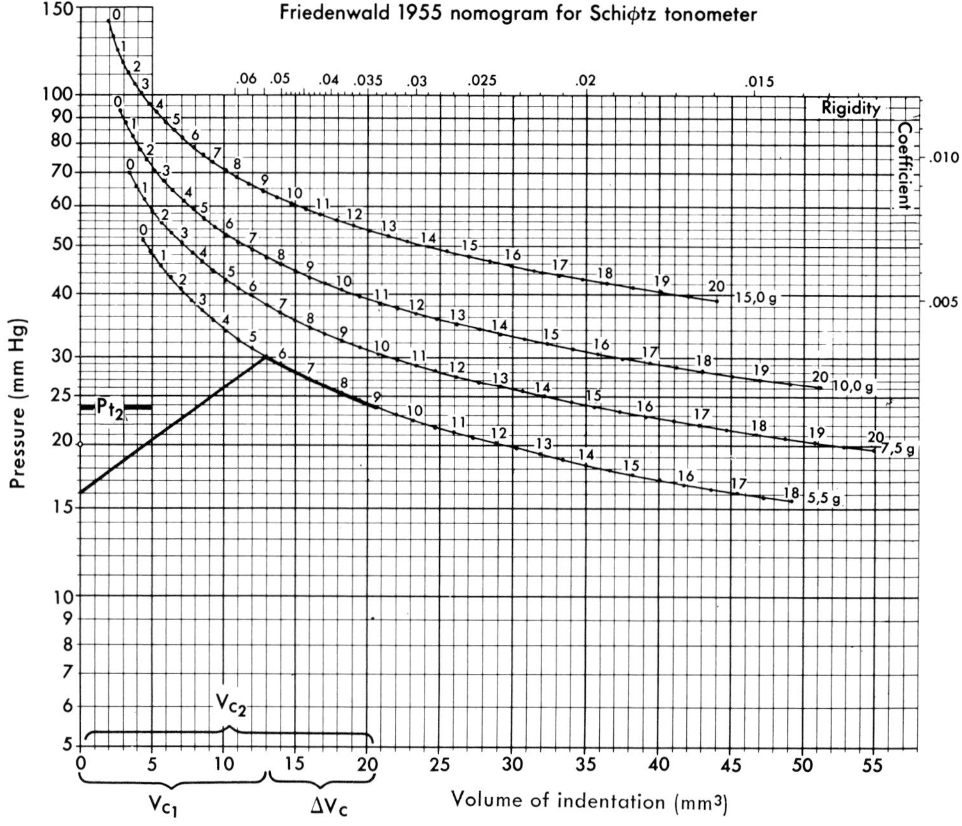

Fig. 8-21. After tonometer has rested on eye for 4 minutes, tonometer indentation has deepened from V_{c_1} to V_{c_2}. Increase in indentation volume is $\Delta V_c = V_{c_2} - V_{c_1}$.

surements) fall in a straight line on this nomogram. The line can be extrapolated to zero corneal indentation, and the pressure intercept is theoretically undisturbed intraocular pressure (P_i). In practice the two Schiøtz measurements and the applanation measurement often do not fall in line. The clinical usefulness of ocular rigidity measurements is dubious.

MEASUREMENT OF FACILITY OF OUTFLOW

Clinical measurement of the facility of outflow is accomplished by artificially raising intraocular pressure and measuring the rate of volume loss from the eye at increased pressure.

Tonography

Tonography employs an electronically recording Schiøtz tonometer and the theories and data of Friedenwald to evaluate the facility of outflow. When a Schiøtz tonometer is placed on the eye, the tonometer indents the cornea. The aqueous displaced by the indentation V_{c_1} distends the eye and raises the pressure from P_0 to P_{t_1} (Fig. 8-20). With time (usually 4 minutes), fluid is lost from the eye, the pressure falls to P_{t_2}, and the corneal indentation deepens to V_{c_2} (Fig. 8-21). The change in corneal indentation volume

$$\Delta V_c = V_{c_2} - V_{c_1}$$

is fluid lost from the eye. But the pressure drop from P_{t_1} to P_{t_2} implies that the eye is less distended. The reduction in distention ΔV_s is given by

$$\Delta V_s = (1/E)(\log P_{t_1} - \log P_{t_2})$$

Fig. 8-22 demonstrates ΔV_s graphically. The total fluid loss during the tonogram is the sum of increase in corneal indentation volume and decrease in ocular distention.

$$\Delta V_{total} = \Delta V_c + \Delta V_s$$

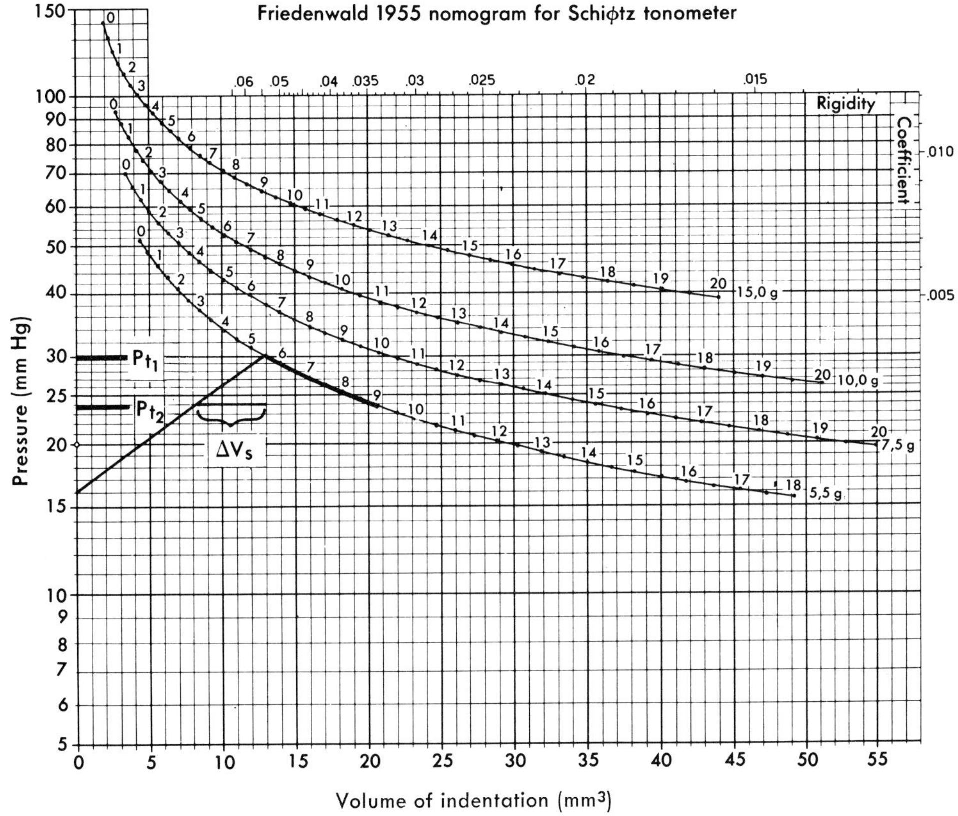

Fig. 8-22. During tonography pressure has fallen from P_{t_1} to P_{t_2}, implying that ocular coats are less distended by volume ΔV_s.

At steady state, trabecular outflow rate is

$$F_{trab} = (P_o - P_e)\, C_{trab}$$

(See Table 8-2 for symbols.) At average tonometric pressure the outflow rate is

$$F_{trab} + \Delta F_{trab} = (P_{tav} - P_e)\, C_{trab}$$

Subtraction of the steady state expression gives

$$\Delta F_{trab} = (P_{tav} - P_o)\, C_{trab}$$

Since

$$\Delta F_{trab} = \Delta V_{total}/\text{time}$$

Therefore

$$C_{trab} = \frac{\Delta V_{total}}{(P_{tav} - P_o)\, t}$$

A small correction for increase in episcleral venous pressure resulting from compression of the eye into the orbit by the tonometer is usually added.[33] When the equation is solved for C and fully expanded, it takes the form

$$C = \frac{(V_{c_2} - V_{c_1}) + (1/E)(\log P_{t_1} - \log P_{t_2})}{t\,(P_{t_{av}} - P_o - \Delta P_e)}$$

In clinical work E is given the average value 0.0215 and ΔP_e the average value 1.25 mm Hg.

In the first section of this chapter the effect of intraocular pressure on both the rate of aqueous formation and outflow facility was discussed. Increase of intraocular pressure decreases both of these. Since tonography materially increases intraocular pressure during the test, the estimate of outflow facility is only a rough estimate. The suppression of aqueous formation (pseudofacility) tends to make tonography overestimate outflow facility, while the decrease in outflow facility due to increase in intraocular pressure tends to make tonography underestimate outflow facility at steady state pressure. It is clear that we need better measures.

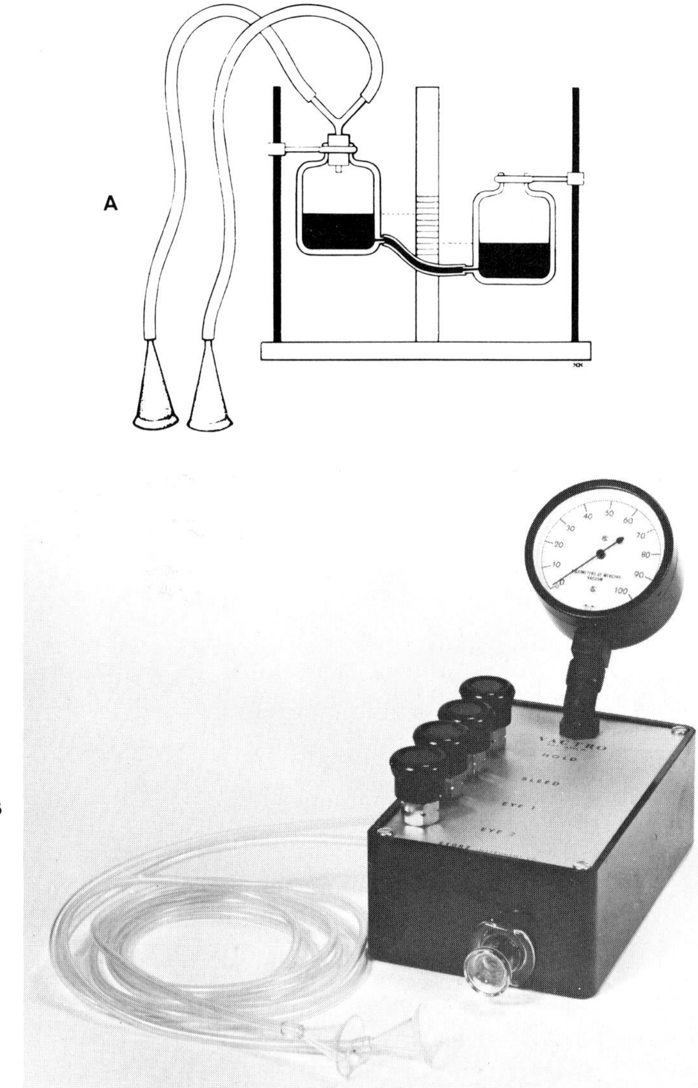

Fig. 8-23. A and **B,** Suction cup for compression of episcleral aqueous outflow channels. (From Galin, M. A., et al.: Arch Ophthalmol. **70:**202, 1963.)

Kupfer and Sanderson[31] devised an ingenious means of measuring pseudofacility; however, since this method employs tonography but does not take reduction of trabecular facility with increase in intraocular pressure into account, its accuracy is unknown.

Reasoning similar to that employed in tonography is used in experimental perfusion of the eye and is subject to many of the same pitfalls. If the perfusion is carried out at various levels of constant pressure, estimation or assumption of ocular rigidity is not necessary, but changes in aqueous formation rate and outflow facility with pressure still occur.

Suction cup

A second means of raising intraocular pressure employs the Rosengren suction cup[32] (Fig. 8-23). When the air pressure in the space between cup and cornea is lowered, the cup rim is forced against the anterior sclera, blocking the aqueous outflow veins. Newly formed aqueous impounded within the eye raises intraocular pressure. When the cup is removed, the

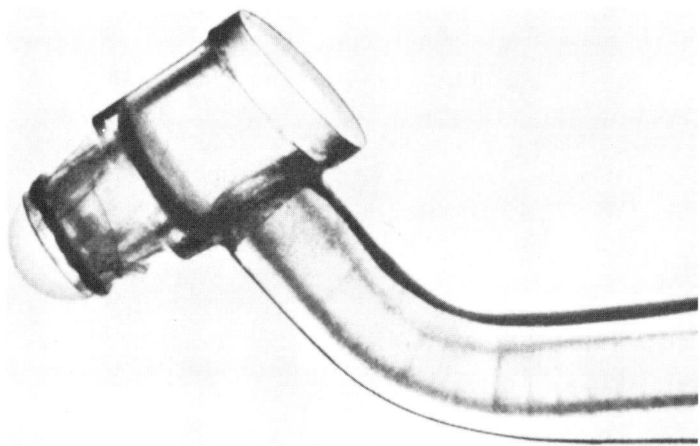

Fig. 8-24. Tambour for measuring episcleral venous pressure. (From Leith, A. B.: Br. J. Ophthalmol. **47:**271, 1963.)

rate of intraocular pressure fall is measured by tonometry, and in conjunction with the relation of pressure to volume can be interpreted in terms of facility of outflow.

MEASUREMENT OF EPISCLERAL VENOUS PRESSURE

Episcleral venous pressure can be measured clinically by means of a thin inflatable membrane and a microscope[50] (Fig. 8-24). The membrane is placed on the bulbar conjunctiva over a vein and inflated until the vein collapses. Brubaker has used the very flexible pericardial sac of a toad as the membrane and finds this gives a sensitive measure of P_e.[10]

Krakau and co-workers[28] increase the pressure of a fine jet of air on the episcleral vessel until the vessel partially collapses. The pressure of the air in the jet is thus a measure of episcleral venous pressure. The method holds much promise for clinical application.

MEASUREMENT OF FLOW

The rate of aqueous formation is about 2 to 3 μl/min at normal intraocular pressure. Fluorescein can be introduced into the anterior chamber by intravenous injection or through the cornea by iontophoresis, and the rate of dilution by newly formed aqueous can be measured,[18,23,26,35] giving a measure of the rate of formation in the intact human eye.[23] Holm[23] has observed the rate of growth of a bubble of unstained aqueous entering the anterior chamber and has calculated pupillary flow from photogrammetric measurements (Fig. 8-25). The rate of aqueous formation has also been calculated from tonographic data. The method is to

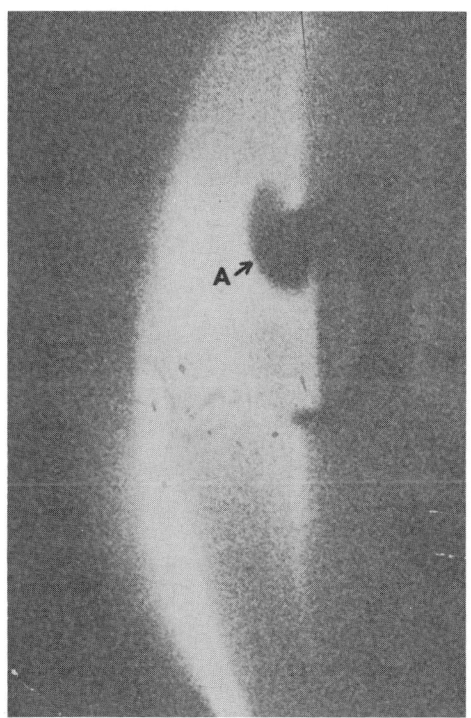

Fig. 8-25. Clear aqueous entering anterior chamber forms bubble, *A*, in stained aqueous. (From Holm, O.: Acta Ophthalmol. **46:**254, 1968.)

first evaluate facility of flow and then to substitute in the formula $F = (P_o - P_e) C$. If an assumed value of P_e is used, it is to be noted that no new information is introduced and that F is implicit in C and P_o. When P_e is assumed, this calculation is of tenuous value and is not recommended.

PHARMACOLOGY OF INTRAOCULAR PRESSURE

The miotic drugs, named for their action on the pupil, produce contraction of the ciliary muscle. The trabecular mesh, a tendon of origin of the ciliary muscle, is made more porous by ciliary muscle traction. These drugs reduce uveoscleral outflow, but the trabecular effect is so much greater that the net effect is to increase outflow facility and lower intraocular pressure.

The drugs of the parasympathomimetic group used topically in the treatment of glaucoma fall into two groups: (1) that which acts directly on the muscle (pilocarpine) and (2) cholinesterase inhibitors—physostigmine (eserine), echothiophate (Phospholine iodide), isoflurophate (DFP), demecarium bromide (Humorsol), and carbachol.

Topical epinephrine (1% to 2% epinephrine base) is a valuable drug in the treatment of open-angle glaucoma. The drug decreases the rate of production of aqueous and increases the outflow facility. The decrease in inflow is probably caused by arteriolar spasm with reduction of the pressure head $P_a - P_i$. The reason for increased trabecular facility is not clear.[5,7] Topical timolol (0.12% to 0.5%), a beta-adrenergic inhibitor also decreases aqueous production, but has no known effect on facility of outflow.

Carbonic anhydrase inhibitors decrease the rate of aqueous formation. Carbonic anhydrase is involved in transport across the ciliary epithelium. The first such agent to be used in treatment of glaucoma was acetazolamide (Diamox). It is given in the typical oral dosage of 1 g/day in divided doses. There are several other drugs of this group that are useful in the treatment of glaucoma. The carbonic anhydrase inhibitors have no effect on topical administration. Macri and co-workers[34] presented evidence of a vasoconstrictor effect of acetazolamide.

Intraocular pressure may be rapidly lowered by increasing blood osmolarity; this may be accomplished by intravenous infusion of urea or mannitol or by oral ingestion of glycerine. These agents are invaluable in treatment of acute angle-closure glaucoma. Oral glycerine is the preferred agent, and doses may be repeated as necessary until surgical measures can be instituted. Water loading, on the other hand, may lower blood osmolarity sufficiently so water is drawn into the eye. If the facility of outflow is good, there is little change of intraocular pressure, but if facility of outflow is reduced, ingestion of a quart of water may raise the intraocular pressure 8 mm Hg or more. The water-provocative test has been widely used in the diagnosis of glaucoma.

Topically administered corticosteroids have an adverse effect on open-angle glaucoma. Administration of topical steroids over a period of weeks produces a pressure rise in many otherwise normal eyes. It is necessary to stress the ocular hypertensive effect because glaucomatous field loss has occasionally been produced by excessive dosage and persistent use of these drugs. The ocular hypertension is the result of decreased trabecular outflow facility. Systemically administered corticosteroids produce a small pressure rise in some eyes, but statistical analysis is required to demonstrate the effect.[8] The steroid concentration in the eye achieved by systemic administration is much smaller than that reached by topical administration.

SURGERY FOR GLAUCOMA

There are essentially two categories of adult glaucoma: angle-closure glaucoma and open-angle glaucoma. Either may be primary or secondary to some other disease process.

Angle-closure glaucoma is the result of occlusion of the trabecular mesh by iris contact. In general, relief of this condition is best achieved by surgical means. Where iris bombé is the proximal factor, ballooning of the iris is relieved by augmenting the aqueous passageway from posterior to anterior chamber. Removal of a piece of iris (iridectomy) gives the aqueous free passage, the bombé flattens, and the angle opens unless adhesions have formed. In those cases in which angle closure is secondary, for example, to dislocated lens, the surgery must be designed to relieve the inciting condition.

In primary open-angle glaucoma unsuccessfully controlled by medical means, there are two obvious surgical approaches analogous to the pharmacologic approaches: (1) to decrease the formation of aqueous and (2) to increase the outflow facility. Formation of aqueous may be reduced by surgical injury of the ciliary body. Heat (cyclodiathermy) is not so popular for this purpose as it was, but cold (cyclocryosurgery) appears to be useful in some cases. The outflow facility may be increased by opening the inner wall of Schlemm's canal (trabeculotomy), or by bypassing the normal outflow channels entirely, creating a fistula between the anterior chamber and the tissues external to the sclera. The fistula may be created by means of a trephine, cautery, or iris inclusion in a linear incision (iridencleisis). It is not certain whether such operations function primarily by absorption of aqueous into the blood or lymph vessels or by leakage through the conjunctiva. Cyclodialysis is the creation of a communication between anterior chamber and supraciliary space; the operation appears to increase uveoscleral outflow.[44]

REFERENCES

1. Armaly, M. F., and Rubin, M.: Accommodation and applanation tonometry, Arch. Ophthalmol. **65:**415, 1961.
2. Ascher, K. W.: The aqueous veins, Springfield, Ill., 1961, Charles C Thomas, Publisher.
3. Bain, W. E. S., and Maurice, D. M.: Physiological variations in the intraocular pressure, Trans. Ophthalmol. Soc. U.K. **79:**249, 1951.
4. Bárány, E. H.: Mathematical formulation of intraocular pressure as dependent on secretion, ultrafiltration, bulk outflow, and osmotic reabsorption of fluid, Invest. Ophthalmol. Vis. Sci. **2:**584, 1963.
5. Bárány, E. H.: Topical epinephrine effects on true outflow resistance and pseudofacility in vervet monkeys studied by a new anterior chamber perfusion technique, Invest. Ophthalmol. **7:**88, 1968.
6. Bechrakis, S.: Über den spontanen Druckabfall bei Applanationstonometrie, Ophthalmologica **151:**604, 1966.
7. Becker, B.: Topical corticosteroids and intraocular pressure. In Becker, B., and Drews, R. C.: Current concepts in ophthalmology, St. Louis, 1967, The C. V. Mosby Co., vol. 1.
8. Bernstein, H. N., and Schwartz, B.: Effects of long-term systemic steroids of ocular pressure and tonographic values, Arch. Ophthalmol. **68:**742, 1962.
9. Brubaker, R. F.: Determination of episcleral venous pressure in the eye, Arch. Ophthalmol. **77:**110, 1967.
10. Brubaker, R. F.: The effect of intraocular pressure on conventional outflow resistance in the enucleated human eye, Invest. Ophthalmol. **14:**286, 1975.
11. Camras, C. B., Bito, L. Z., and Eakins, K. E.: Reduction of intraocular pressure by prostaglandins applied topically to eyes of conscious rabbits, Invest. Ophthalmol. Vis. Sci. **16:**1125, 1977.
12. Daubs, J. G.: Arrhenius plots of human aqueous flow rates, Am. J. Optom. **49**(12):1005, 1972.
13. Davanger, M., and Holter, Ö.: The statistical distribution of intraocular pressure in the population, Acta Ophthalmol. **43:**314, 1965.
14. Draeger, J.: Tonometry, Basel, Switzerland, 1966, S. Karger AG.
15. Forbes, M., Pico, G., Jr., and Grolman, B.: A noncontact applanation tonometer, Arch. Ophthalmol. **91:**134, 1974.
16. Friedenwald, J. S.: Contribution to the theory and practice of tonometry, Am. J. Ophthalmol. **20:**985, 1937.
17. Friedenwald, J. S.: Tonometer calibration, Trans. Am. Acad. Ophthalmol. Otolaryngol. **61:**108, 1957.
18. Goldmann, H.: Some remarks about fluorometric and tonographic determination of the rate of flow of aqueous, XVII Concilium Ophthalmologicum **2:**1573, 1954.
19. Goldmann, H.: On pseudofacility, Bibl. Ophthalmol. **76:**1, 1968.
20. Goldmann, H., and Schmidt, T.: Über Applanationstonometrie, Ophthalmologica **134:**221, 1957.
21. Goldmann, H., and Schmidt, T.: Weiterer Beitrag zur Applanationstonometrie, Ophthalmologica **141:** 441, 1961.
22. Harris, L. S.: Cycloplegic-induced intraocular pressure elevations, Arch. Ophthalmol. **79:**242, 1968.
23. Holm, O.: A photogrammetric method for estimation of pupillary aqueous flow in living human eye, Acta Ophthalmol. **46:**254, 1968.
24. Holmberg, Å. S. In Leydhacker, W., editor: Glaucoma. Twentieth International Congress of Ophthalmology, Tutzing Symposium, Basel, Switzerland, 1966, S. Karger AG.
25. Johnstone, M. A., and Grant, W. M.: Pressure-dependent changes in structures of the aqueous outflow system of human and monkey eyes, Am. J. Ophthalmol. **75:**365, 1973.
26. Jones, R. F., and Maurice, D. M.: New methods of measuring the rate of aqueous flow in man with fluorescein, Exp. Eye Res. **5:**208, 1966.
27. Krakau, C. E. T.: A vibration tonometer, Ophthalmol. Res. **1:**129, 1970.
28. Krakau, C. E. T., Widakowich, J., and Wilke, K.:

Measurements of the episcleral venous pressure by means of an air jet, Acta Ophthalmol. **51:**185, 1973.
29. Kronfeld, P. C. Some basic statistics of clinical tonography, Invest. Ophthalmol. **7:**319, 1968.
30. Kupfer, C., Gaasterland, D., and Ross, K.: Studies of aqueous humor dynamics in man. V, Effects of acetazolamide and isoproterenol in young and old normal volunteers, Invest. Ophthalmol. **15:**349, 1976.
31. Kupfer, C., and Sanderson, P.: Determination of pseudofacility in the eye of man, Arch. Ophthalmol. **80:**194, 1968.
32. Langham, M. F., and Maumenee, A. E.: The diagnosis and treatment of glaucoma based on a new procedure for the measurement of intraocular dynamics, Trans. Am. Acad. Ophthalmol. Otolaryngol. **68:**227, 1964.
33. Linnér, E.: Episcleral pressure during tonography, XVII Concilium Ophthalmologicum **3:**1532, 1955.
34. Macri, R. J., Dixon, R. L., and Rall, D. P.: Aqueous humor turnover rates in the cat. I, Effect of acetazolamide, Invest. Ophthalmol. **4:**927, 1965.
35. Maurice, D. M.: The use of fluorescein in ophthalmological research, Invest. Ophthalmol. **6:**464, 1967.
36. McBain, E. H.: Tonometer calibration. II, Ocular rigidity, Arch. Ophthalmol. **60:**1080, 1958.
37. McEwen, W. K., Shepherd, M., and McBain, E. H.: An electrical model of the human eye. I, The basic model, Invest. Ophthalmol. **6:**155, 1967.
38. Moses, R. A.: Repeated applanation tonometry, Ophthalmologica **142:**663, 1961.
39. Moses, R. A.: The Mackay-Marg tonometer, Trans. Am. Acad. Ophthalmol. Otolaryngol. **66:**88, 1962.
40. Moses, R. A.: Detachment of the ciliary body: anatomical and physical considerations, Invest. Ophthalmol. **4:**935, 1965.
41. Moses, R. A.: Constant pressure applanation tonography. III, The relationship of tonometric pressure to rate of loss of ocular volume, Arch. Ophthalmol **77:**181, 1967.
42. Moses, R. A.: The effect of intraocular pressure on resistance to outflow, a review, Surv. Ophthalmol. **22:**88, 1977.
43. Moses, R. A.: Circumferential flow in Schlemm's canal, Am. J. Ophthalmol. **88:**585, 1979.
44. Pederson, J. E., Gaasterland, D. E., and MacLellan, H. M.: Experimental ciliochoroidal detachment; effect on intraocular pressure and aqueous humor flow, Arch. Ophthalmol. **97:**536, 1979.
45. Pollack, I. P., and Becker, B.: The effect of hypothermia on aqueous humor dynamics. IV, Carotid artery ligation and blood flow, Am. J. Ophthalmol **51:**1039, 1961.
46. Posner, A.: A disposable applanation tonometer, Eye Ear Nose Throat Mon. **44:**70, 1965.
47. Prijot, É.: Contribution à l'étude de la tonométrie et de la tonographie en ophtalmolgie uitgeverij, The Hague, 1961, W. Junk BV.
48. Rosen, D. A., and Johnston, V. C.: Ocular pressure patterns in the Valsalva maneuver, Arch. Ophthalmol. **62:**810, 1959.
49. Sears, M. L.: Miosis and intraocular pressure changes during manometry, Arch. Ophthalmol. **63:**707, 1960.
50. Seidel, F.: Methoden zur Untersuchung des intraokularen Flüssigkeitwechsels. In Abderhalden, L., editor: Handbuch der biologischen Arbeitsmethoden, Vienna, 1927, Urban & Schwartzenberg, vol. 5.
51. Van Alphen, G. W.: On emmetropia and ametropia, Ophthalmologica **142:**47, 1961.

GENERAL REFERENCES

Kolker, A. E., and Hetherington, J., Jr.: Becker-Shaffer's diagnosis and therapy of the glaucomas, ed. 4, St. Louis, 1976, The C. V. Mosby Co.

Leydhacker, W., editor: Glaucoma. Twentieth International Congress of Ophthalmology, Tutzing Symposium, Basel, Switzerland, 1966, S. Karger AG.

Chapter 9

The vitreous

BALDER P. GLOOR, M.D.

PHYSICAL PROPERTIES

The vitreous is a clear mass that in most mammalians is physically a hydrogel. It fills the posterior cavity of the eye. The vitreous forms a semisolid support for the retina. It is probably not a tissue in the proper sense, but rather a product of the surrounding tissue, the neuroretina. It allows light to reach the retina and allows nutrients to diffuse from the ciliary body to the retina. The total weight of the vitreous in humans is 3.9 g, and its volume is about 3.9 ml.[119] The water content is extremely high, between 98%[119] and 99.7%.[134] The density of the vitreous is between 1.0053 and 1.0089.[143] The osmotic pressure determined by freezing point depression is −0.554° to −0.518° C.[96] The pH of the human vitreous is approximately 7.5.[96] The viscosity varies with age, within different parts of the vitreous, from eye to eye in the same species, and from one species to another. The value given for the intrinsic viscosity is 4.200 cm^3/g.[9,22] The refractive index is less than the value for the aqueous, the value in humans being between 1.3345[75] and 1.3348.[110] More than 90% of the visible light is transmitted through the normal vitreous. In the ultraviolet range, transparency drops rapidly between 350 and 300 nm and is zero below 300 nm. In phakic eyes ultraviolet light is absorbed by the lens to a great extent, and little reaches the vitreous. In the infrared range the transmission of the vitreous begins to drop at 800 nm, and the absorption is complete in wavelengths of over 1600 nm.[27,151]

□ This work was supported by *Fight For Sight* Postdoctoral Research Fellowship from Fight For Sight, Inc., New York, and in part by National Institutes of Health Grant NB 621 and Schweizerischer National Fonds.

EMBRYOLOGY

In the embryologic development three stages can be distinguished: (1) the primary, or hyaloid, vitreous, (2) the secondary, or definitive, vitreous, and (3) the tertiary vitreous, or the zonule of Zinn. The primary vitreous begins to form when a space opens between the lens plate and the optic vesicle. This space is filled by a protein-rich PAS-positive material. In the fifth week (6.3 mm stage) vascularized mesoderm enters the optic fissure, and in the sixth week the hyaloid artery is developed and reaches the posterior pole of the lens vesicle. In the 11 to 12 mm stage the embryonic cleft begins to close and the secondary vitreous begins to appear at the margin of the retina.

The secondary vitreous has a finer network of filaments and fewer cells than the primary vitreous. A distinct limit between the primary vitreous and the secondary vitreous in the form of a denser layer may be observed. By the ninth week the hyaloid artery and the primary vitreous are fully developed, and the secondary vitreous is also evident. The primary vitreous atrophies during development. By the seventh month the hyaloid artery no longer carries blood, and around the time of birth it is reabsorbed. The embryonic remnant of the primary vitreous, called Cloquet's canal, is still visible during the first several decades of life. In the human an optically empty space, the so-called prefoveolar channel, develops in the vitreous in front of the differentiating macula. The channel disappears during the first year of life, leaving an interruption in the vitreous cortex, the prefoveal hole.[53] The examination of optical sections of the vitreous in developing sheep eyes, photocoagulated several months previously, showed a disturbed growth above the destroyed retinal areas. The shape of the resulting deficiencies of the vitreous structure in form of holes suggests that the vitreous is produced by retinal elements and develops by appositional growth.[55] The electron microscopic findings of Witschel,[152] that glial cells produce collagen which is similar to vitreal collagen throughout life, also supports the view that the definitive vitreous is of neuroectodermal origin.

How the regression of the primary vitreous is induced is unknown. Autolytic processes[68] are responsible for the dissolution of its vessels, not immigrating macrophages as proposed by Jack.[82] Autolytic vacuoles appearing in large amounts in the cells of the vessel walls introduce the process of dissolution. These morphologic changes in the regressing hyaloid vessels are quite similar to the changes observed in developing retinal vessels following hyperoxia.[4-6] The tertiary vitreous, or the zonula, begins to form in the sixth month (170 mm stage). The development of zonular fibers is not yet clearly understood. Investigations with labeled amino acids such as proline, glycine, methionine, and cystine suggest that zonules are products of the ciliary epithelium.[70] In the chicken the zonular fibrils appear first as loose fibrils in the vitreous space between the lens and the ciliary body. A short time later they become arranged in dense bundles inserted at the basement membrane of the ciliary epithelium and at the lens capsule.[30]

Chemical development

The knowledge of the chemical development of the vitreous lags behind morphologic information. Development is not fully terminated until the eye reaches adult size.[9,14,16] The embryonal vitreous, and in some species also the young vitreous, contains a much higher concentration of soluble proteins than does the adult vitreous.[19] In humans[22] and in cattle[14] the level does not change after birth. The concentration of collagen and hyaluronic acid is low in the early stages of development. The collagen content increases slowly after birth, proportionately to the augmentation of the whole globe. The concentration in the different parts of the vitreous (anterior, posterior, cortical, and central) remains quite stable.[14] The rate of converting glucose into glucosamine was found to be higher in the vitreous of young calves than of adult animals,[21] and the hyaluronic acid content reaches its adult level at different rates in different species—in the chicken, as early as the end of the embryonic development,[17] in rabbits and cattle much later,[9,14,19] and in humans at the age of about 13 years.[22]

Developmental anomalies[36]

Smaller or larger remnants of the primary vitreous may be found in the form of a remaining dot on the posterior lens capsule (Mittendorf's dot) or as a persistent hyaloid artery. This condition is often associated with falciform retinal folds and congenital retinal septa.[8] Persistence of the primary vitreous in a more extensive form is a serious condition commonly associated with persistence of hyaloid vessels and fibrovascular tissue behind the lens.[120] Most cases of persistent primary vitreous are unilateral, and the condition is more frequent in premature infants.

The differential diagnosis between a persisting primary vitreous and a tumor (retinoblastoma) may be extremely difficult. The condition also has to be differentiated from retrolental fibroplasia, which always occurs bilaterally, usually after a newborn infant has been kept in an incubator under increased oxygen tension.

Partial or total failure of the development of the tertiary vitreous results in a local or total absence of the zonular fibers with subluxation or ectopia lentis. It occurs in three conditions:

1. As an isolated anomaly
2. In association with ectopia pupillae and other ocular anomalies
3. As a part of systemic syndromes with other, particularly mesodermal, anomalies (Marfan's syndrome and Marchesani's syndrome)

ANATOMY OF THE VITREOUS

The vitreous is adjacent to the lens, the zonular fibers, the pars plana of the ciliary body, the retina, and the optic disc. Its shape is not only determined by the tissue surrounding it but also by its own framework, elasticity, and turgescence. It is firmly attached to its neighboring tissues at the peripheral retina and at the pars plana of the ciliary body, the vitreous base,[126] and around the optic disc[109] (Fig. 9-1). A firm attachment may be found in some cases around the macula.[63,129] Adherences also exist around the retinal vessels.[64,123] Eisner[52] was able to demonstrate very clearly the structures of the secondary vitreous in autopsy eyes with a xenon arc slit lamp and a stereoscopic microscope with a 10 to 16 times magnification. These findings can be correlated to the histologic findings of Szent-Györgyi[138] (1917) and to the structures of the vitreous observed biomicroscopically during life. Three main zones of the vitreous can be differentiated: the cortical layer, the intermediate zone, and the central channel and in newborn children an additional channel, the prefoveolar.[54] These zones are marked off by funnel-shaped tracts, which diverge from the posterior pole, traverse the vitreous cavity, and insert anteriorly on circular belts. The densest of them, the preretinal tract, demarcates the cortical zone from the intermediate zone; the most central one, the hyaloid tract, inserts at the lens near or at Wieger's ligament. Other similar dense membranelles, the coronary tract and the median tract, extend to the coronary ligament and the medial ligament, respectively (Figs. 9-1 and 9-2). The vitreoretinal juncture (internal limiting lamina[60]) consists of a basement membrane, delimiting the internal layer of retinal glial cells (Müller cells), and a layer of dense collagen fibrils.[30,31,57,58,78] Between the two layers is an electron-lucent space, about 400 Å wide, across which delicate fibrils extend.[60] The vitreoretinal juncture shows marked topographic variations. It is thinnest in the basal zone (510 Å) and thickens in the equatorial zone (sixfold) and posterior zone (37-fold).[60] The fovea shows marked thinning of the internal limiting lamina.[60] The condensation of collagen fibrils to a membranelike belt is a common phenomenon in all border relations, in which connective tissue comes in contact with other tissues.[65] In the region of the vitreous base the vitreous fibrils are strongly interwoven in the basement membrane of the ciliary epithelium.[78]

A clinically important structure is the posterior vitreous membrane (posterior hyaloid) as seen biomicroscopically after vitreous detachment. It corresponds to the cortical vitreous and is not a basement membrane.

A population of uniform cells[17,139] is trapped in the collagen network of the outermost part of the vitreous, comprising a layer about 100 μm thick. These cells are more numerous in the region adjacent to the ciliary body and the pars plana and less frequent in the region adjacent to the retina. They are of mesodermal origin and can be differentiated from fibrocytes by their morphologic and histochemical proper-

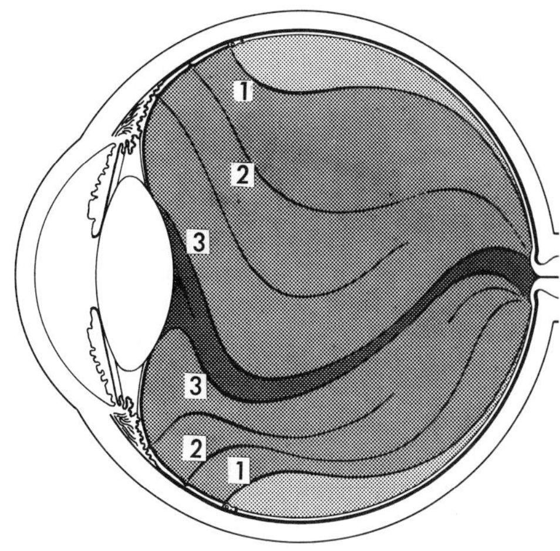

Fig. 9-1. Schematic drawing of intravitreal membranelles (according to slit-lamp examinations of eyes obtained at autopsy). The vitreous body is divided into three zones: externally, as far as the retina extends, there is a relatively thick vitreous cortex (whitish gray here). It has holes at characteristic locations: in front of papilla, in region of fovea centralis, in front of vessels, and in front of anomalies of ora serrata region (enclosed ora bays, meridional folds, zonular traction tufts). Intermediate zone (light gray here) contains vitreal tracts; membranelles that form funnels packed into one another and diverge from region of papilla anteriorly. Central channel is space (dark gray here) delimited by hyaloid tract. It is closed off anteriorly by retrolental section of anterior vitreous membrane. It contains no typical tracts but only irregularly arranged vitreous fibers, part of which are residual of Cloquet's canal. The outermost vitreous tract, the preretinal tract, *1,* separates intermediary substance from vitreous cortex. The innermost tract, the hyaloid tract, *3,* inserts at edge of lens. Between them extends median tract, *2,* to median ligament of pars plana, and coronary tract to coronary ligament.[53] (Modified from Eisner, G.: Biomicroscopy of the peripheral fundus, Berlin, 1973, Springer-Verlag.)

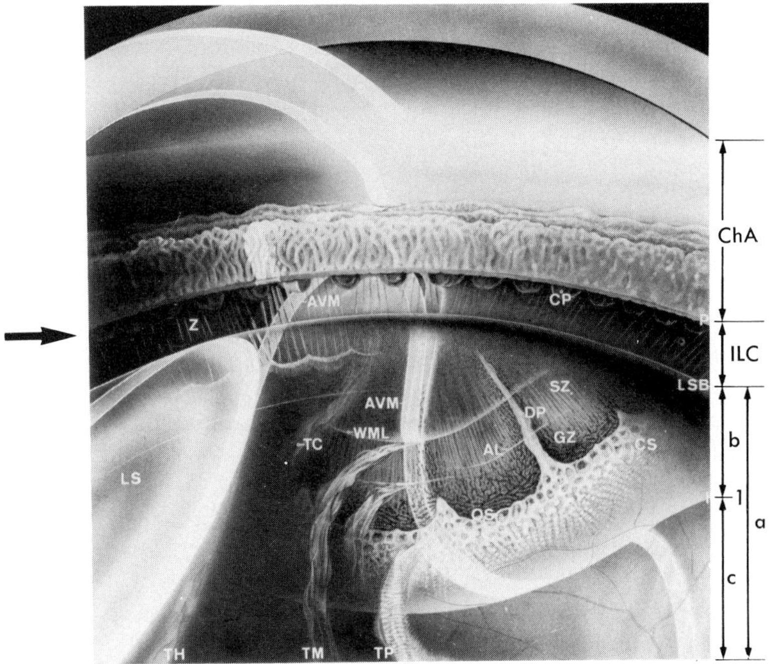

Fig. 9-2. Normal biomicroscopic appearance of indented peripheral fundus. Semischematic drawing of the picture in the mirror of the contact lens. Notice the oraparallel circular tracts; the median ligament, white midline, *WML;* and an accessory line, *AL. AVM,* Anterior vitreous membrane; *CP,* ciliary process; *CS,* cystoid spaces, cystoid degeneration; *CHA,* anterior chamber; *DP,* dentate processes of the ora serrata; *GZ,* granular zone of the pars plana; *1,* top of indented protuberance; *ILC,* iridolenticular cleft; *LSB,* lens border; *LS,* lens; *OS,* ora serrata; *P,* pupillary border; *SZ,* striate zone of the pars plana; *TC,* tractus coronarius (coronary tract); *TH,* tractus hyaloideus (hyaloid tract); *TM,* tractus medianus (median tract); *TP,* tractus preretinalis (preretinal tract); *Z,* zonula; *a,* region seen translentally; *b,* indented area; *c,* nonindented area.[53] (Modified from Eisner, G.: Biomicroscopy of the peripheral fundus, Berlin, 1973, Springer-Verlag.)

ties.[15,139] Balazs and co-workers[15] believed themselves justified in giving these cells a proper name and called them hyalocytes, but labeling experiments with [^{3}H]thymidine suggest that the cortical vitreous cells are continually immigrating hematogenous monocytes and the special name should be dropped. In the vitreous these cells still show some mitotic activity; their lifetime is short, their half-life being less than 1 week.[67,69] It is suggested that hyalocytes are the metabolic center of the vitreous,[10,102,105] but it is very difficult to believe that immigrating macrophages could be responsible for the construction of such a regular structure as the vitreous. A few fibrocytes are restricted to the area around the optic disc.

The anterior vitreous is attached to the lens approximately 1 mm behind the equator (Fig. 9-1). The attachment forms a circle on the posterior surface of the lens (Wieger's ligament, or hyaloideocapsular ligament). The zonular fibers insert into the lens capsule anterior to the hyaloideocapsular ligament. The depression formed by the lens in the vitreous is the patellar fossa. The hyaloideocapsular ligament passes centrally into the membrana plicata. The plicata opens to a tube, which runs backward toward the disc; this is, as mentioned, Cloquet's canal, the remnant of the primary vitreous. Around Cloquet's canal are alternating layers of optically dense and transparent substances. Biomicroscopically these structures appear as bands or lamellae. They seem to run toward the vitreous base, where the vitreous structure becomes denser biomicroscopically as well as histologically. Cloquet's canal ends around the optic disc as a circular glial adhesion to the retina 3 to 4 mm in diameter (the area of Martegiani, or posterior vitreous base). After vitreous detachment this appears as the so-called papillary ring. The anterior opening of Cloquet's canal behind the lens varies from 2.5 to 6.0 mm in diameter. The anterior limit of the vitreous is not the plicata, but a normally indiscernible anterior vitreous border line that becomes apparent after intracapsular cataract extraction or if the potential space between lens and this membrane opens in an anterior vitreous detachment (Berger's space, or retrolental space). The virtual space between zonula and vitreous is called Petit's canal, or the retrozonular space.

With histologic techniques using connective tissue and acid mucopolysaccharide stains, for example, Masson's trichrome, Alcian blue–PAS, a networklike fibrillar structure of the vitreous can be shown at its base radiating toward Cloquet's canal and toward the posterior part. The elucidation of the structure of the central parts of the vitreous poses several problems regarding the fixation techniques.[138] The difficulties are responsible for the gap between biomicroscopy and histology of the vitreous. Whether the histologically apparent fibrillar structure really exists, or whether it is a fixation artifact cannot be resolved, even with the aid of phase-contrast microscopy.[19,73,131] Fresh preparations show a complex structure of fibers of different sizes in the range of 1.1 to 1.3 μm, but there are still critics who claim that these are artificial condensations.[125]

MOLECULAR STRUCTURE OF THE VITREOUS

Collagen and hyaluronic acid are responsible for the molecular structure. Collagen mainly makes up what is called the framework of the vitreous. This is the part that remains after filtration, after the vitreous is washed out several times, or after both procedures. Hyaluronic acid, on the other hand, is found in the soluble part, the vitreous humor.

Young[153] and Mörner[95] first suggested that the fibrous protein in the vitreous (vitrosin) is collagen; this was confirmed by Pirie and associates.[114] The substance contains 18% glycine, 8.4% [illegible]ne, 15.4% hydroxyproline, and small am[illegible] of cystine.[154] Hydroxyproline is specific for collagen. To determine the collagen content of a specimen the hydroxyproline content is analyzed. Vitrosin also has the same x-ray diffraction properties as collagen. In electron-microscopic pictures the width of the elementary fibrils is reported to be 10 to 15 nm[74] and 25 nm.[91] A cross striation of 64 nm was observed.[74] The shrinkage temperature of the fibrils is that of collagen, around 60°[32] to 65° C,[9] between a pH of 3.0 and 9.5. Vitrosin differs from normal collagen in that 4% to 9% of its weight is a complex polysaccharide that cannot be separated from it. The polyacrylamide gel electrophoretic pattern of pepsin-solubilized vitreous collagen, as well as the CNBr-derived peptides of pepsin-solubilized vitreous body collagen, revealed no common identity with types I, II, III, or IV collagen.[130] This again suggests that the vitreous collagen is very specific and probably not the product of connective tissue cells, but rather of the neuroectoderm. The large, coarse fibrils of the vitreous, as shown by phase-contrast microscopy, are broken down in some degree (20%) by trypsin.[20] Other enzymes such as chymotrypsin, emulsin, and elastase have some effect, but collagenase is most powerful and destroys 60% to 93% of the framework.[131,136] The collagen content is highest where the vitreous is a gel, that is, in the cortical vitreous and near the ciliary body, and lowest in the central part. Large differences exist between the different species. The collagen content varies from 25 mg/100 mg in the owl monkey, in which the vitreous is mostly fluid, to 286 mg/100 ml in human beings, in whom the vitreous is normally a gel (Table 9-1).

The unit of hyaluronic acid consists of one molecule of glucuronic acid and one of acetylglucosamine. A small amount of galactosamine is also present. Hexosamine is predominant in the embryo; later the ratio between hexosamine and glucuronic acid becomes 1:1. Determination of hyaluronic acid content is now on the basis of glucuronic acid only, because glucosamine may be found in the vitreous not incorporated in hyaluronic acid.[132] The molecular weight of hyaluronic acid in the vitreous is lower than in other tissues (umbilical cord, synovial fluid, connective tissue) and varies be-

Table 9-1. Mucopolysaccharide and collagen content of aqueous humor and vitreous body of various animal species ($\mu g/ml$)*

Species	Aqueous humor (mucopolysaccharide)	Vitreous body	
		Mucopolysaccharide	Collagen
Squid *(Loligo pealii)*		230	3.3
Frog *(Rana catesbeiana)*		<1	182
Carp *(Cyprinus carpio)*	591	575	690
Tuna *(Thunnus thynnus)*	880	660	217
Chicken *(Gallus gallus)*	21	21 (17)†	209 (1.5)†
Barred owl *(Strix varia varia)*	262‡ 15§	35	90
Rabbit *(Lepus cuniculus)*	<3	31	104
Guinea pig *(Cavia porcellus)*		37	134
Steer *(Bos taurus)*	<2	710	57
Owl monkey *(Aotus trivirgatus)*	<2	423	25
Human being		240	286

*From Balazs, E. A.: Physiology of the vitreous body. In Schepens, C. L., editor: Importance of the vitreous body in retina surgery with special emphasis on reoperations, St. Louis, 1960, The C. V. Mosby Co.
†In liquid portion of vitreous.
‡Next to corneal endothelium.
§Next to lens.

tween 61,000 and 381,000[9]; the length varies between 160 and 350 nm (polymerization effect, see p. 267). The highest concentration of hyaluronic acid is, like collagen, found in the cortical vitreous layers. The concentration varies also with the species and is highest in cattle and owl monkeys and lowest in rabbits and birds (Table 9-1). The hyaluronic acid is responsible for the viscostiy of the vitreous.

In its specific organization the collagen-hyaluronic network is responsible for the structure, the volume, and the distribution of the cells and the transparency of the vitreous.[10] As in all other connective tissues, the vitreous has two compartments; the solid compartment is the vitreous gel, which contains collagen fibrils, and the liquid compartment is the vitreous sol, which does not contain collagen fibrils. The cortical vitreous is always present in a solid or a gel state. The remainder of the vitreous is liquid in some animals, gel in others. In humans the vitreous is normally present in a gel state, except in older people and in those with pathologic conditions. The collagen network is made up of crossed rods, which give it its plasticity. The spongelike hyaluronic acid network is entangled in the collagen network and provides the viscoelasticity. The two elements interact in a manner that stabilizes the system; the hyaluronic acid molecules prevent the collagen network of the gel from separating into solid and liquid phases if mechanically agitated or from collapsing if submitted to centrifugal forces. The molecular weight of the hyaluronic acid plays an important role in the stabilization effect. The rigidity of the vitreous gel is greatest in the cortical layer and at the vitreous base, where the collagen content is highest. The viscosity depends on the concentration and size of the hyaluronic acid. The elasticity of the vitreous gel depends on the concentration and molecular qualities of the collagen network and the hyaluronic acid molecules.

Changes in vitreous volume

The volume of the vitreous is altered if the electrochemical characteristics of the hyaluronic acid molecules, which have a high negative electrostatic potential, are changed.[10] As long as these negative charges are neutralized by Na^+ ions and probably proteins, the volume of the hyaluronic acid is restricted. If the neutralizing substances are removed, the negatively charged hyaluronic acid elements are free to repel one another, and the system will expand. With the addition of positively charged molecules, such as protamine sulfate or cationic dyes, the entire system will collapse. Also, all processes that shrink the collagen network, such as heat, radiation, and freezing and tha[illegible] decrease the volume of the vitreous gel.[illegible]

The collagen network has a restricted space

Table 9-2. Chemical composition of the vitreous*

Constituent	Cattle	Rabbits	Humans
Inorganic constituents (mmol/kg H_2O)			
Sodium	130.5	133.9 to 152.2	137.0
Potassium	7.7	5.1 to 10.2	3.8
Calcium	3.9	1.5	—
Magnesium	0.8	—	—
Chloride	115.6	104.3	112.8
Phosphate	0.92 to 0.42	0.40	—
Sulfate	1.2	—	—
Bicarbonate water	19.6 to 32.4 mEq/kg H_2O	19.6 to 32.4 mEq/kg H_2O	19.6 to 32.4 mEq/kg H_2O
Water and organic constituents (mg/100 ml H_2O)			
Water	99,000	99,000	99,000
Total nitrogen	22	13	23.5
Nonprotein nitrogen	10.4	18 to 25	
Amino acid nitrogen	3.0		
Peptide nitrogen	1.2		
Urea	14.2		
Uric acid	2.8		
Creatine	1.6		
Creatinine	1		
Total protein nitrogen	11.5		
Proteins	40 to 70		40
Lipids	0.6		
Glucose	55 to 62	55 to 80	30 to 70
Lactic acid	14.8	65	70
Pyruvic acid	1.2	5 to 6	7.3
Citrate	13.7	1.8	1.9
Ascorbate		8 to 15	

*From Nordmann, J.: Biologie du corps vitre. Chapter III, Chimie. In Brini, A., et al., editors: Biologie et chirurgie du corps vitre, Paris, 1968, Masson & Cie., Editeurs.

for large molecules. This space is mainly occupied by the hyaluronic acid molecules; the space for other large molecules is extremely limited[98] (effect of excluded volume[10]). This may explain some peculiarities in the exchange of substances into and out of the vitreous.

The double network also acts as a molecular sieve, slowing down the movement of large molecules and cells. Thus normal vitreous cells are restricted to its cortical layers.

The exclusion of proteins, other large molecules, and cells is essential to the transparency of the vitreous.[10]

CHEMICAL COMPOSITION OF THE VITREOUS AND EXCHANGE OF SUBSTANCES

A survey of the content of different substances in the vitreous of cattle, rabbits, and humans, as compiled by Nordmann,[96] is presented in Table 9-2. Besides the substances mentioned, traces of ash (SiO_2, FeO_3, Al_2O_3, MgO) and of metals (strontium, barium, aluminum, molybdenum, manganese, iron, nickel, copper, zinc, and lead[92]) can be found. No large differences exist between the three species except for the lower chloride content in the rabbit. Most of the analyses carried out on vitreous have been performed in the whole vitreous body. Since the fibrillar component, the already mentioned collagen framework, is so small compared to the volume of the fluid, the resulting analysis can be considered to represent almost entirely the chemistry of the vitreous humor. Except for the content of collagen and hyaluronic acid, the composition of the vitreous humor has many similarities to that of the aqueous (Table 9-3). Maurice[92] suggests that there exists a free diffusion across the surface of separation between the vitreous and the aqueous

Table 9-3. Concentration of various substances in the intraocular fluids of the rabbit eye (weighted averages in mmol/kg H_2O)*

Substance	Posterior aqueous	Anterior aqueous	Plasma	Vitreous
Sodium	136.00	138.00	143.00	134.00
Potassium		5.00	5.60	9.50
Chloride	96.50	101.00	97†	104.70
Total CO_2	37.50	30.20	20.60	26.00
Phosphate	0.58	0.89	2.04	0.40
Ascorbate	1.38	1.11	0.04	0.46
Lactate	9.90	9.30	10.30	12.00
Glucose	5.60	5.40	5.70	3.00
NPN	23.00	25.00	34.00	17.00

*From Reddy, D. V. N., and Kinsey, V. E.: Arch. Ophthalmol. **63:**715, 1960.
†Value modified in accordance with later data published by Kinsey, V. E.: Invest. Ophthalmol. **6:**395, 1967.

and throughout the vitreous itself. Reddy and Kinsey[115] proposed that the vitreous acts as a "source" of substances that are in lower concentration in the posterior aqueous, such as chloride in the rabbit, and as a "sink" for substances that are in higher concentration in the posterior aqueous, such as carbon dioxide. But some substances do not follow these rules, and considerable differences exist in their concentration in relation to the aqueous. The content of glucose, its metabolic derivatives (such as lactate and pyruvate), and the content of phosphate are also affected by the metabolism of the lens and the retina.[1,2] Therefore there are three avenues of interchange: across the boundary between aqueous and vitreous, across the retina, and with the lens. Many of the substances use more than one path. Labeled substances reach a steady state more slowly in the vitreous than in the aqueous.[25,47,115] The relatively small surface and large volume of the vitreous compared to the large surface and small volume of the posterior chamber and also the peculiar structure of the cortical layer of the vitreous may be responsible (molecular sieve effect[10]). In physiologic conditions, as well as after administration of different substances, very different concentration gradients between plasma, aqueous, and vitreous humor are reached from substance to substance. An equilibrium in the ratio of the concentration in the vitreous to the concentration in the aqueous (C_V/C_A of 1.0) is possible for several substances such as chloramphenicol.[26] The level of other substances in the vitreous, such as urea, always remains below the level in the aqueous (excluded volume effect, according to Balazs[10]).

Specific differences between the composition of plasma, aqueous, and vitreous humor suggest a blood-retinal or a blood-vitreal barrier with selective properties similar to the blood-brain barrier.* The permeability depends on the size of the molecules and, in part, on their lipid solubility.[25,44] Very small amounts of protein pass this barrier. The protein content of the vitreous is quite characteristic and differs considerably from that of the plasma and aqueous, which suggests that the vitreous must be regarded as a separate compartment.[9] The barrier between the vitreous and adjacent tissues was localized by Rodriquez-Peralta[124] in the vascular endothelium of the retinal vessels, in the pigment epithelium of the retina, in the inner layer of the ciliary epithelium, and in the pigment epithelium of the ciliary body.[39] Important facts suggest a pump at the level of the retinal vessels,[71,133] the retinal and ciliary pigment epithelium, or at both levels,[61] which moves organic anions from the vitreous into the bloodstream. Fluorescein, for instance, does not pass into the vitreous, but is quickly moved out if injected into it. This process can be blocked by closing retinal vessels with diathermy and by metabolic and competitive inhibitors, which also block the organic anion transport in the kidney. A relationship may exist between this pump and a constant flow

*References 25, 26, 37-40, 45, 106, 107, and 124.

from the ciliary body through the vitreous toward the retina, and also between this pump and the relatively high exchange rate of the water in the vitreous. This flow may assist in maintaining a normal adherence between retina and pigment epithelium.[39]

Oxygen

The oxygen tension in the vitreous as measured in the cat's eye[89] depends strictly on the oxygen tension of the inspired air. When the animal was submitted to moderate hypoxia (15% oxygen in the inspired gas), the oxygen tension in the vitreous fell from the normal 53 mm Hg to 28 mm Hg. With a rise in oxygen tension of the inspired air to 600 mm Hg, the tension in the vitreous reached 160 mm Hg. It is supposed that the oxygen exchange takes place through the vessels surrounding the vitreous and that no exchange takes place through the posterior chamber.[96]

Water

The water movement appears surprisingly high. Tritiated water was injected into the peritoneum, and its content in the vitreous was determined at different times. It was calculated that half the water of the vitreous was replaced every 10 to 15 minutes (in the aqueous, every 2.5 to 3.0 minutes).[33,86] This corresponds to a total movement of water in and out of approximately 85 mm^3/min. A flow from the ciliary body toward the posterior pole in the rabbit was traced by Fowlks[62]; he also investigated a pump mechanism out of the vitreous pointing in this direction.

Sodium

Sodium enters the vitreous at its base from the ciliary body and the posterior chamber.[85,146] Passage across the retina seems to be less important. Penetration into the vitreous, determined with ^{24}Na, is slower than into the posterior chamber. In the posterior chamber, sodium reaches a steady state level after 40 minutes, whereas in the most anterior part of the vitreous, steady state is achieved after 3 hours. In the posterior parts of the vitreous no labeled sodium was found after 3 hours. This demonstrates a lack of mixing in different parts of the vitreous[85] (Fig. 9-3). After intravitreal injection, 90% of the labeled sodium leaves the vitreous in 24 hours, mainly via the aqueous; labeled sodium appears in the aqueous after 1 hour. An equal level of labeled sodium is reached in aqueous and vitreous after about 6 hours. Direct passage of sodium toward the blood seems to be limited by a membrane of quite low permeability.[115,146,150]

Potassium

According to Bito and Davson,[23] an active carrier mechanism in the ciliary epithelium brings the potassium concentration in the posterior chamber to a higher level than in the plasma (in the dog 5.64 mEq/kg H_2O in the posterior aqueous, 3.9 mEq/kg H_2O in the plasma). The level in the anterior chamber aqueous is considerably lower (4.9 mEq/kg H_2O), which may be explained by a pickup by the anterior surface of the lens. The concentration in the vitreous is higher than in the anterior chamber (5.35 mEq/kg H_2O). Furthermore, a difference between anterior and posterior parts of the vitreous was demonstrated (5.28 mEq/kg H_2O versus 5.01 mEq/kg H_2O). Therefore not only a diffusion from the posterior chamber across the hyaloid seems to be responsible for the relative high potassium concentration in the vitreous, but also the active accumulation of potassium by the anterior surface of the lens associated with a passive diffusion from its posterior surface into the vitreous.[23] Potassium leaves the vitreous mainly across the retina.

Calcium

The calcium content of the vitreous and aqueous is equal to that of the serum.[83]

Chloride

In the cat and in the rabbit, chloride is especially low in the aqueous. In these animals the low chloride content in the aqueous is associated in some way with a high bicarbonate concentration. In humans and other primates the distribution ratio of chloride and bicarbonate is the reverse of that in the cat and rabbit, but in primates there is also a gradient of chloride concentration from the vitreous to the aqueous. The chloride exchange differs from

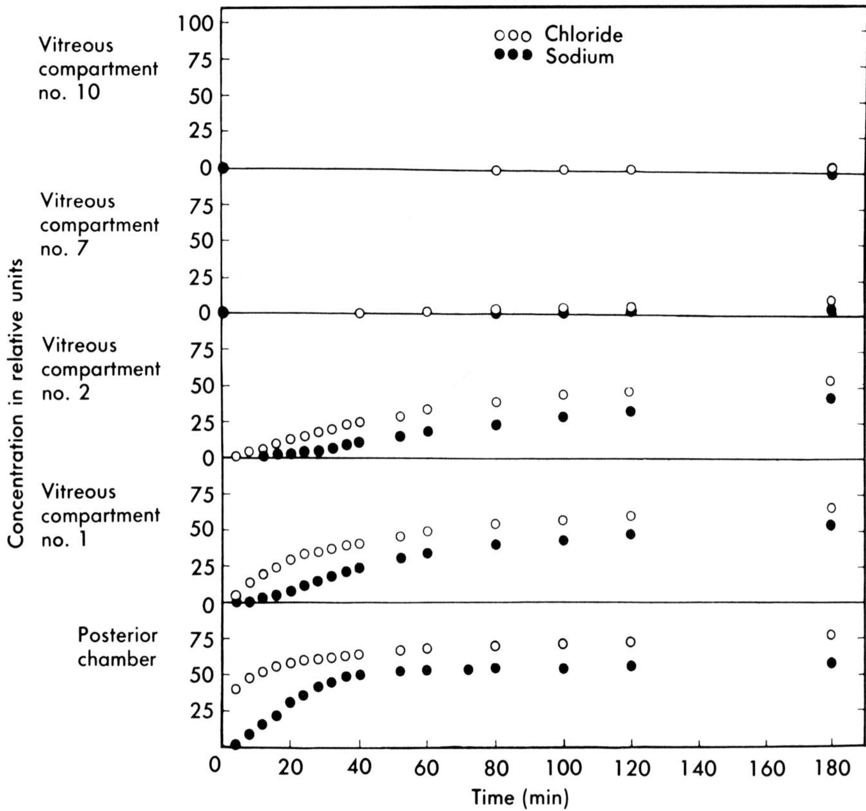

Fig. 9-3. Concentration of ^{24}Na and ^{36}Cl in posterior chamber and several compartments of vitreous model as determined by analog computer. (After Kinsey, V. E., and Reddy, D. V. N.: Doc. Ophthalmol. **13:**13, 1959.)

Table 9-4. Glucose content of blood and aqueous and vitreous humors*

Eye number	Blood	Aqueous humor	Anterior vitreous	Posterior vitreous	Difference
1†	284	86	63	53	−10
2†	284	86	81	64	−17
3	108	99	71	44	−27
4	104	73	63	70	7
5	104	83	63	52	−11
6	118	96	65	67	2
7	118	127	64	47	−17
8†	281	158	129	106	−23
9	127	65	42	35	− 7
10	150	—	68	57	−11
11	150	114	88	51	−37

*From Adler, F.: Trans. Am. Ophthalmol. Soc. **28:**307, 1930.

†These animals were anesthetized with ether and chloroform, hence the high glucose values in blood.

that of sodium as determined in the rabbit.[85] (Fig. 9-3). It is more rapid; half takes place across the retina, the other half between posterior chamber and vitreous. Why a higher concentration of chloride is reached in the vitreous than in the posterior chamber and why the vitreous loses chloride again toward the posterior chamber is not yet clearly understood.

Phosphate

The low level of phosphate in the vitreous compared to that in the aqueous attracts attention. First of all, much less phosphate penetrates into the vitreous than into the anterior chamber.[47,107] As demonstrated by autoradiographic techniques, the portal of entrance is the anterior vitreous bordering the ciliary body.[35,107,144,145] Phosphate then diffuses slowly toward the posterior vitreous. The metabolic use of the phosphate by the retina is principally responsible for its low content in the vitreous.[43] If the metabolism of the retina is altered by experimental retinal degeneration, the phosphate content rises in the vitreous,[48] and if labeled phosphate is injected into the vitreous, the isotopes are fixed in the lipids and nucleoproteins of the retina, the choroid, and the iris.[141] The lens also utilizes phosphate of the aqueous, but it returns it again and is therefore not responsible for the low content of this substance in the vitreous.

Bicarbonate

Bicarbonate is secreted by the ciliary body. The concentrations of bicarbonate and of chloride are different in different species. Low bicarbonate levels are combined with high chloride concentrations in horses, oxen, sheep, and primates; the proportions are reversed in cats and rabbits. The bicarbonate concentration is considerably lower in the vitreous than in the aqueous and lower in the posterior vitreous than in the anterior vitreous.[46] The bicarbonate content of the vitreous is increased by damage to the retinal metabolism.[7] The low content of bicarbonate in the posterior vitreous is normally caused by the presence of lactate and pyruvate, which reduce the pH and prevent the transformation of carbon dioxide into bicarbonate.

Glucose and derivatives of glucose metabolism

Glucose diffuses into the vitreous across all its surrounding tissues, of which the main part is the retina. Its penetration into the vitreous is slower than into the aqueous, and the concentration reached is lower than would be expected if simple diffusion were to take place (Table 9-4). The concentration is also lower in the posterior vitreous than in the anterior vitreous; this is attributed to the use of glucose by the retina. The concentration of glucose rises after inhibition of the metabolism of the retina by cutting the optic nerve[1] or by inhibition with sodium iodate.[127]

Determinations of the penetration rate of glucose, galactose, 3-methylglucose, xylose, and sucrose into the vitreous after intravenous injection demonstrate not only a lower penetration rate into the vitreous than into the aqueous but also differences between the particular saccharides. This is one of several facts that suggest the existence of a selective blood-vitreal barrier. In species with an aerobic glucose metabolism (mammals, birds) the lactate and pyruvate contents of the vitreous are higher than in the plasma or the aqueous (ratio between lactate in the vitreous to lactate in the plasma is 2.6:4.3), whereas in species with anaerobic glycolysis (amphibians, fishes, cephalopods) this ratio is 1:1.[50]

Proteins and amino acids

The blood-vitreal barrier is quite efficient in dealing with soluble proteins. As long as the vessel walls are not damaged (by inflammatory processes, diabetes, or central vein occlusion), the penetration rate of blood proteins and hemoglobin is very low. The concentration of soluble protein is higher in the cortical layer close to the retina than in the central and anterior parts (96 mg/100 ml, 61.9 mg/100 ml, and 37.9 mg/100 ml).[9] This supports the theory that the soluble proteins originate from the retinal vessels. The concentration is similar to that in the aqueous, but the distribution of the components is different. The albumin-globulin ratio is much lower than in the plasma (0.46 to 0.75 instead of 2).[12,142] The globulin content is very low. The content of glycoproteins is relatively high (0.33% to 0.50% of the protein fraction).[132] The carbohydrate component of the glycoprotein contains a surprisingly high amount of sialic acid. The ratio of sialic acid to protein nitrogen is about three times higher than in the serum, the synovial fluid, and the aqueous humor. According to Balazs,[9] this glycoprotein, as well as the specific distribution of the proteins, indicates that the vitreous body is a compartment well separated from the anterior chamber as well as from the intercellular substance of the blood vessel bed of the retina. The sialic acid (glycoproteins) may be responsible for the fact that the proteins of the vitreous are resistant to trypsin.[12]

Proteins leave the vitreous by way of the aqueous,[34,93] and a small amount exits through the retina.[49]

Free amino acids reach a higher level in the aqueous than in the plasma, but a much lower level in the vitreous. The vitreous concentration is only about one fifth that in the plasma.[117] Glutaminic acid, with a level similar to that in the serum, is an exception. This is attributed to a liberation of this substance in retinal metabolism.[114] The concentration gradient of other amino acids existing between aqueous and vitreous, but mainly the gradient existing between anterior and posterior vitreous, suggests utilization by the retina, the cellular elements of the cortical layer of the vitreous, or both.[24] Furthermore, an active transport of amino acids from the vitreous into the blood by way of the retinal circulation[116,117] or by an active transport mechanism located in the pigment epithelium may exist.[24,118]

Urea

Adler[2] pointed out that the urea content in aqueous and vitreous is lower than in the

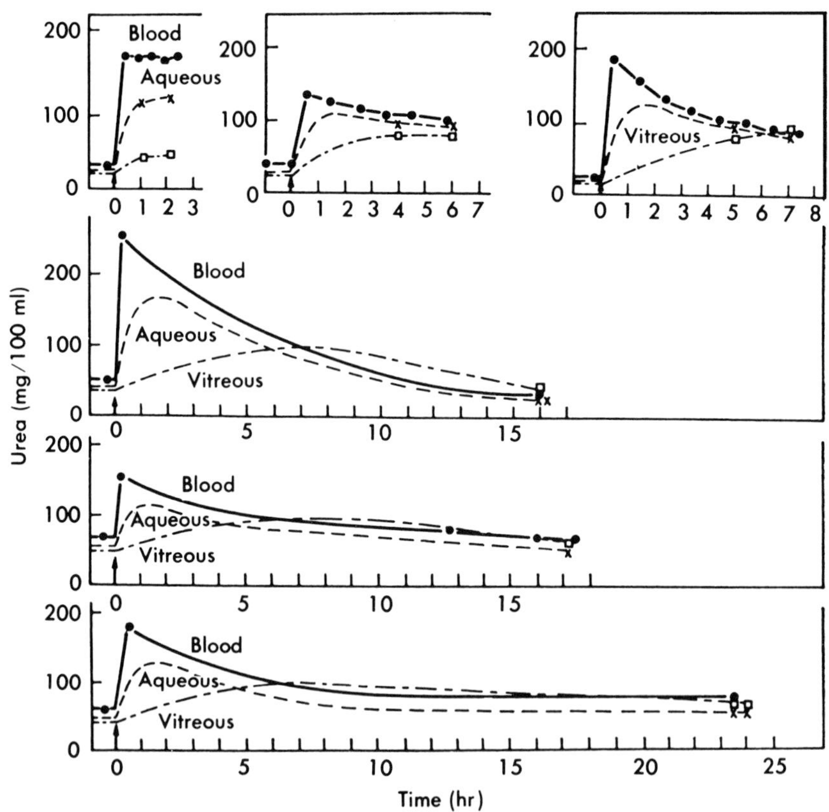

Fig. 9-4. Urea concentration in blood, aqueous humor, and vitreous body after intravenous infusion of 1 g of urea in rabbits. Graphs are assembled from infusions in six rabbits. Urea blood level estimated at regular intervals in each case. (After Bleeker, G. M., van Haeringen, N. J., and Glasius, E.: Exp. Eye Res. **7**:30, 1968.)

plasma. In recent investigations involving the rabbit the ratio of the steady state of the concentration in the plasma to that in the aqueous and vitreous was always around 5:4:3.7 and is not related to large variation in the blood concentration from animal to animal.[25] It is surprising that urea, with a low molecular weight of 60, which usually penetrates tissues very well, enters the aqueous slowly, and the vitreous even more slowly. On the other hand, the concentration of the urea in the vitreous exceeds that of other slowly penetrating substances.[96]

After intravenous infusion of urea the ratio of plasma urea to aqueous urea remains unchanged but becomes changed for vitreous in relation to aqueous and blood (Fig. 9-4). In the blood a peak is reached suddenly, and the rise in the aqueous is a little slower; in the vitreous the urea content lags far behind, but exceeds that of the aqueous and blood after about 7 hours. The concentration rises first in the cortical vitreous; the central part follows later.

In the interpretation of these facts, Bleeker and associates suggest two conclusions.

1. In acute blood urea rise the lag in urea content of the vitreous is the result of the difference of the ratio between area of membrane and volume of fluid, which for the vitreous is very small.
2. The slow penetration into the vitreous body may be caused by diffusional difficulties inherent in the physiochemical properties of the gel.[10]

Intravitreally injected urea ^{14}C leaves the vitreous by passage into the posterior chamber after equilibration between the two fluid compartments is reached.[87] The peculiarities of the urea exchange not fully explained are also interesting in that the structural analog thiourate (ethylmethylpropylthiourate) enters the vitreous quickly, and the concentration ratio (C_A/C_V) is about 1:1. The higher penetration rate is attributed to the better liposolubility of thiourate.[44]

Ascorbate

The content of ascorbate in the vitreous increases during embryonal development; at birth it reaches a level that remains practically steady during life. The ascorbate content of the vitreous is higher than that of the plasma and lower than that of the posterior chamber. The values given for the rabbit differ between 0.46 mmol/kg H_2O and 0.85 mmol/kg H_2O.[4,92]

A significant drop from 1.0 mmol/kg H_2O to 0.6 mmol/kg H_2O was observed in the guinea pig after lens extraction.[77] An interesting fact is that photosensitized oxidative systems such as "ascorbic acid-Cu^{2+}-O_2-light" can depolymerize hyaluronic acid (see p. 260). This may explain the relatively low average molecular weight and the high polydispersity of the hyaluronic acid present in the vitreous.[135] The biologic role of such a depolymerizing system in the vitreous body is not yet clear.

Other substances

The concentration and penetration of different sugars, as well as the specific distribution of proteins in the vitreous, already suggested the existence of a selective retinovitreal barrier. The investigation of foreign substances such as dyes furnishes some more information about the function of this barrier as well as about the direction in which some of the exchange mechanisms are working. The existence of a selective barrier of the vitreous is probable when the ratio of the concentration in the vitreous to the concentration in the aqueous (C_V/C_A) of a substance is significantly less than 1. Some of these substances are of great practical value, such as diagnostically used dyes like fluorescein as well as medicaments like sulfonamides and antibiotics. The penetration rates of many such substances as investigated by Bleeker and co-workers[26] are shown in Table 9-5. The sequence of the substances is arranged for the purpose of this chapter. The relative ratio of plasma to aqueous to vitreous and the ratio of concentration in the vitreous to concentration in the aqueous (C_V/C_A) are given. The molecular weights of most of the substances are in a similar range.

Dyes

Fluorescein has become an important tool for diagnostic evaluation of every type of barrier in the eye under physiologic and patho-

Table 9-5. Penetration of various compounds from the blood into the aqueous humor and the vitreous body of the rabbit eye*

Compound	Lipoid solubility	Molecular weight	Blood	Concentration ratio		
				Aqueous	Vitreous	C_V/C_A
Dyes						
Fluorescein	0.18	332	1	0.030	<0.005	<0.17
Phenolsulfonphthalein	0.0008	354	1	0.05	<0.005	<0.1
Rhodamine B	23	428	1	0.40	0.40	1.0
Antibiotics						
Penicillin	0.0004	334	1	0.15	0.002	0.01
Phenoxyethylpenicillin	0.0004	365	1	0.02	<0.002	0.1
Dimethoxyphenylpenicillin	0.0004	380	1	<0.02	<0.002	<0.1
Diethylaminoethylpenicillin (penethamate)	0.025	443	1	0.1	0.1	1.0
p-Aminohippuric acid	0.004	194	1	0.38	0.02	0.05
p-Aminohippuric ethylester	0.33	223	1	0.46	0.27	0.59
Streptomycin	Negligible	581	1	<0.05	<0.05	—
Chlortetracycline	Negligible	423	1	<0.05	<0.05	—
Oxytetracycline	Negligible	441	1	<0.05	<0.05	—
Chloramphenicol	4	323	1	0.4	0.4	1.0
Other substances						
Acetazolamide	0.06	222	1	—	Negligible	—
Urea	0.0006	60	1	0.80	0.74	0.92
Glucose	<0.0004	180	1	0.9	0.5	0.6
Sucrose	<0.0004	342	1	Moderate	Negligible	—

*After Bleeker, G. M., et al.: Exp. Eye Res. **7**:37, 1968.

logic conditions.[94] Normally, as long as the blood-retinal barrier is intact, fluorescein does not enter into the vitreous space (concentration less than 0.005 in the blood).[94] A second barrier is the cortical layer of the formed vitreous. This may become evident in cases of vitreous detachment, in which a stainable liquid vitreous can be differentiated from a formed vitreous. On the other hand, the vitreoretinal barrier is broken down in central vein occlusion. Alteration of the blood-retinal barrier, as measured by fluorophotometry, is a very early sign of diabetic retinopathy.[41,147] Fluorescein, if injected into the vitreous, is transported out of the vitreous by a pump mechanism.[39] Compared with other dyes of similar chemical constitution (xanthine derivates), phenolsulfonphthalein shows about the same characteristics as fluorescein (Table 9-5), but rhodamine B penetrates easily and reaches a value of 0.4 and a C_V/C_A ratio of 1 : 1. This substance shows a much higher liposolubility than related substances do, but if the liposolubility of fluorescein is compared with that of other substances such as para-aminobenzoic acid, which gets into the vitreous relatively well, we would also expect a better permeability of fluorescein.

Antibiotics

Antibiotics of the penicillin group, as well as related substances such as *p*-aminohippuric acid and *p*-aminohippuric ethylester, are members of an interesting larger group including radiopaque components such as sodium iodomethamate (Iodoxyl), iodopyracet, iodohippurate (Hippuran), and others. They are eliminated from the blood by a carrier mechanism in the tubular epithelium of the kidney. This mechanism can be obstructed by administration of probenecid. A similar process could be shown in the eye for other substances of this group, iodopyracet[61] and fluorescein.[39] These substances do not penetrate into the vitreous but are carried out rapidly if their passage is not blocked by probenecid. This pump working only in one direction may possibly be located in the pigment epithelium. Similarities between pigment epithelium and tubular epithelium of the kidney have already been suggested by the investigation of Koyanagi and Kinukawa[88] regarding intoxication with sodium iodate and sodium iodacetate, sublimate, naphthan, and other substances. The deficit of iodate in intraocular fluids is also a consequence of an active transport system out of the eye.[18] This transport system is inhibited by perchlorate, thiocyanate, and fluoroborate but not by organic anions such as iodopyracet, probenecid, and fluorescein; therefore these otherwise analogous transport mechanisms seem to be independent of each other.[18]

Penetration of the blood-vitreal barrier from blood to vitreous is in some respects parallel to liposolubility. Esterification of *p*-aminohippuric acid and penicillin increases the liposolubility and allows a sizable concentration to develop in the vitreous.[26] These penicillins are therefore desirable for therapy of intravitreal infections. The permeability of antibiotics such as streptomycin, oxytetracycline, and chlortetracycline, which all show a negligible liposolubility, is so low that they are useless for therapeutic purposes against vitreous infections (Table 9-5). Permeability of chloramphenicol into the vitreous is esceptionally high; this is similar to the high permeability of this substance into the cerebrospinal fluid. It shows a high liposolubility, penetrates as well in the aqueous as in the vitreous, and reaches therapeutically useful concentrations in a short time (Table 9-5).

Permeability of antibiotics into the vitreous may be enhanced by cryocoagulation or photocoagulation.[122] The low permeability of most antibiotics into the vitreous is now overcome by intravitreal injection of these drugs.[112] At the present time the following substances are used: gentamicin (0.3 mg), tobramycin (0.5 mg), methicillin (2 mg), cephaloridine (0.25 mg), carbenicillin (2 mg). Higher dosages imply a high risk for the retina.

• • •

From scrutiny of the data in Table 9-5, we must conclude that liposolubility is not the only parameter affecting permeability of the vitreoretinal barrier. On one hand, a liposolubility of 0.004 seems to be the limit of penetration; on the other hand, acetazolamide, which has a liposolubility of 0.06 does not penetrate at all, and glucose, urea, and sucrose also follow other laws.

FLOW IN THE VITREOUS

A flow of fluid through the ciliary body backward toward the optic disc exists in rabbits[62,76] and mice,[66] but under normal conditions it has never been proved in humans or monkeys.[66,76]

METABOLISM

Fischer-von Bunau and Fischer[59] could find no metabolic activities of the vitreous that use oxygen and produce carbon dioxide, but these experiments do not negate metabolism of the cortical tissue layers. It was suggested that, beside their phagocytotic activities, the cortical vitreous cells are involved in the production of aminoglycans.[102-105] That the concentration of hyaluronic acid is highest in the cortical vitreous, where the hyalocytes are localized, is considered a hint of their specific metabolic activity. If the hyaluronic acid is depolymerized by intravitreal injection of hyaluronidase in the rabbit, the hyaluronic acid is reformed after 6 weeks.[113] If the liquid vitreous of the owl monkey is replaced by collagen gel and hyaluronic acid, a liquid humor similar to that present before replacement is found after 6 to 10 months.[137] If samples of cortical vitreous are incubated with [^{14}C]glucose the amount of newly formed ^{14}C-containing glucosamine is proportional to the number of hyalocytes present.[9] An enzyme system with the property of transforming glucose to glucosamine and transferring the latter to hyaluronic acid of high molecular weight has been demonstrated in hyalocytes.[102,103] However, these experiments demonstrate only that these cells possess, like any hematogenous macrophage, an enzyme system able to induce the synthesis of aminoglycans, but not that they are really responsible for the synthesis of the proper aminoglycans of the vitreous. Other experiments showed that the central acellular regions of the vitreous were nearly as active as the peripheral regions in incorporating labeled glucose into nondialyzable polymers.[21] In aphakic eyes the content of hyaluronic acid is markedly lower than in phakic eyes, probably because the aminoglycans escape into the aqueous.[97]

VITREOUS SUBSTITUTION

Substitution of a nontoxic, clear substance for the vitreous is desirable for two purposes: first, the replacement of hazy vitreous by a clear fluid may be necessary after repeated hemorrhages and inflammatory processes; second, a mechanical means to push a detached retina back toward the pigment epithelium is indicated in cases of massive preretinal retraction. Balanced physiologic saline solution and air are commonly used in retinal detachment surgery. Human vitreous has been transplanted with some degree of success. Bacteriologic problems are not yet solved. Good results have been reported following the use of lyophilized vitreous,[110,111] pure hyaluronic acid,[80,149] a biologically active fraction of sodium hyaluronate (Healon),[11] and polygeline.[100,101] The use of collagen is still in the experimental stage.[13,51] Silicone oil has become a useful tool in vitreoretinal surgery for the management of selected cases of retinal detachment with massive preretinal retraction,[3,36,99,148] but the last word in vitreous replacement has not yet been spoken.

PATHOLOGY

Biomicroscopy of the fundus periphery and pars plana[72,81] and stereoscopic binocular indirect ophthalmoscopy combined with indentation[56] have produced considerable advances in the evaluation of morphologic changes of the vitreous, mainly in relation to retinal pathology. The knowledge of the physiopathologic and chemical changes lags far behind. More knowledge of the metabolism of the vitreous would be very important because the degenerative changes occurring in it and at the vitreoretinal juncture during lifetime do not only appear to be essentially irreversible but are responsible for what finally leads to the detachment of the vitreous. Degenerative changes of the vitreous are the main cause of retinal detachments because partial detachment of the vitreous with formation of a retinal tear and lasting vitreous pull are the prerequisites of most retinal detachments. Changes in the vitreous must be evaluated with regard to possible alteration in the surrounding tissues—choroid, retina, retinal vessels, ciliary body, zonula, and lens. No isolated vitreous disease really exists, but there are distinct clinical entities in which morphologic vitreous changes are prominent.

Vitreous detachments

Two forms of vitreous detachment may be differentiated: acute complete posterior vitreous detachment with collapse, and simple posterior vitreous detachment with retraction.[81] The former is common in elderly persons; in several studies it was found in 65% to 75% of patients over 65 years of age.[36,128] As a forerunner of complete posterior vitreous detachment with collapse, a threadlike demarcation of the vitreous fibrils begins in the mid-forties (earlier in myopes), followed by the formation of liquid-filled pockets, so-called syneresis pockets or cavities. The cavities enlarge and become confluent. At the same time a degenerative process in the cortical vitreous takes place[65]; then the slowly progressing degenerative process enters an acute phase. The thin remaining coat of cortical vitreous suddenly breaks, and the collected fluid in the cavities empties itself into space between vitreous and retina within a few hours.[90,128] This creates very disturbing symptoms in the form of lightning flashes and suddenly appearing floaters described as spiders, flies, rings, or hooks jumping around the visual field. Biomicroscopically, a condensed, collapsed vitreous is found behind the lens and at the bottom of the vitreous cavity. The formed vitreous is delimited from a large optically empty retrovitreal space by the detached vitreous membrane, corresponding to the cortical vitreous. Toward the posterior pole the membrane is less easily seen, but if there is a ring-shaped or hook-shaped opacity near the optic disc, the diagnosis of a vitreous detachment is assured. The ring corresponds to the glial tissue, which surrounded the optic disc. The patient who has the symptoms of acute posterior vitreous detachment with collapse must be examined carefully and repeatedly because retinal detachment may follow due to formation of retinal tears by vitreous pull at places where the vitreous does not detach easily from the retina. When the retina is torn, hemorrhage into the vitreous space may occur. A horseshoe-shaped tear originating with an acute vitreous detachment is an indication for prophylactic retinal surgery to prevent further retinal detachment. Traction on the macula with disturbance of the visual acuity is another complication; it is rare and temporary in most cases.

In simple vitreous detachment without collapse the vitreous detached from the retina remains spherical. Only a small sickle-shaped space between cortical vitreous gel and the retina appears. A retraction of the collagen framework seems to occur in this form of vitreous detachment, which is often observed after inflammatory processes and in diabetic retinopathy. In diabetic retinopathy with superficial vascular proliferation it may be especially disastrous. Retraction of the vitreous easily tears the newly formed vessels and causes a retrovitreal hemorrhage, which starts a vicious circle; with the hemorrhage, new vessel proliferation and more vitreous retraction are promoted, and therefore more hemorrhages. The shrinking vitreous, strongly adherent to the retina in some places, may produce a retinal detachment. This type of vitreous detachment usually develops around the posterior pole and is not preceded by the formation of syneresis cavities. The surgical therapy of these detachments poses special problems. As long as the vitreous is not detached, vessels growing into the formed vitreous could not be observed,[42,79,140] confirming again that the cortical vitreous layer acts as a barrier.[10]

Vitreous opacities

Cellular infiltrations of the vitreous are common in inflammatory processes of the surrounding tissues. Opacities can become especially dense and membranelike in peripheral uveitis. A degeneration of the framework usually occurs in inflammation. Formation of coarse fibrils is frequent. Fibroplastic processes are common after perforating injuries and repeated hemorrhages. As long as the blood does not penetrate in the formed vitreous but remains near the retina, it is reabsorbed quite easily. A fibrinolytic activity in the vitreous is doubtful,[28,84] but a preactivator susceptible to streptokinase, which acts on the transformation of profibrinolysin into fibrinolysin could be proved in small concentrations.[29,121]

Very conspicuous to the examiner are the white dots suspended in formed vitreous in asteroid hyalosis; not so for the patient, who does not complain of any symptoms. The deposits are calcium soaps, which are bilateral in most cases and appear in otherwise normal vitreous at an average age of 60 years. Somewhat similar deposits, composed primarily of cholesterol, are called synchysis scintillans. These flashing crystals form as a sequel to ocular diseases such as inflammation or hemorrhage. The cholesterol crystals tend to sink in the liquified vitreous of the damaged eye, only to swirl upward with sudden eye movements.

REFERENCES

1. Adler, F. H.: An investigation of the sugar content of the ocular fluids under normal and subnormal conditions and the glycolytic activity of the tissues of the eye, Trans. Am. Ophthalmol. Soc. **28:**301, 1930.
2. Adler, F. H.: The metabolism of the retina, further notes, Arch. Ophthalmol. **6:**901, 1931.
3. Armaly, M. F.: Ocular tolerances to silicones: replacement of aqueous and vitreous by silicone fluids, Arch. Ophthalmol. **68:**390, 1962.
4. Ashton, N., and Pedler, C.: Studies on developing retinal vessels. IX, Reaction of endothelial cells to oxygen, Br. J. Ophthalmol. **46:**257, 1962.
5. Ashton, N., Tripathi, B., and Knight, G.: Effect of oxygen on the developing retinal vessels of the rabbit. I, Anatomy and development of the retinal vessels of the rabbit, Exp. Eye Res. **14:**214, 1972.
6. Ashton, N., Tripathi, B., and Knight, G.: Effect of oxygen on the developing retinal vessels of the rabbit. II, In vivo experiments, Exp. Eye Res. **14:**221, 1972.
7. Auricchio, G., and Ambrosio, A.: Sul compartamento della riserva alcalina dei liquidi endoculari in condizioni normali e pathologiche, Boll. Soc. Ital. Biol. Sper. **29:**1172, 1953.
8. Badtke, G.: Ueber seltene Duplikaturenbildungen in der embryonalen Netzhaut, Beitrag zur Genese der Ablatio falciformis congenita, v. Graefes Arch. Ophthalmol. **155:**266, 1954.
9. Balazs, E. A.: Physiology of the vitreous body. In Schepens, C. L., editor: Importance of the vitreous body in retina surgery with special emphasis on reoperations, St. Louis, 1960, The C. V. Mosby Co.

10. Balazs, E. A.: The molecular biology of the vitreous. In McPherson, A., editor: New and controversial aspects of retinal detachment, New York, 1968, Harper & Row, Publishers.
11. Balazs, E. A.: The biological effect of Healon in the vitreous. Club Jules Gonin, read at the Eighth Biannual Meeting, Miami, Fla., 1972.
12. Balazs, E. A., and Sundblad, L.: Studies on the structure of the vitreous body. V, Soluble protein content, J. Biol. Chem. **235:**1973, 1960.
13. Balazs, E. A., and Sweeney, D. B.: The replacement of the vitreous body in the monkey by reconstituted vitreous and by hyaluronic acid in surgery of retinal vascular diseases, Mod. Probl. Ophthalmol. **4:**230, 1966.
14. Balazs, E. A., Laurent, T. C., and Laurent, U. B. G.: Studies on the structure of the vitreous body. VI, Biochemical changes during development, J. Biol. Chem. **234:**422, 1959.
15. Balazs, E. A., Toth, L. Z. J., and Mitchell, A. P.: Studies on the vitreous body. XII, Cytological and hostochemical studies on the cortical tissue layer, Exp. Eye Res. **3:**57, 1963.
16. Balazs, E. A., et al.: Studies on the structure of the vitreous body. VIII, Comparative biochemistry, Arch. Biochem. Biophys. **81:**464, 1959.
17. Balazs, E. A., et al.: Cytological and biochemical studies on the developing chicken vitreous, Exp. Eye Res. **4:**237, 1965.
18. Becker, B.: Iodide transport by the rabbit eye, Am. J. Physiol. **200:**804, 1961.
19. Bembridge, B. A., and Pirie, A.: Biochemical and histological changes in developing rabbit eyes, Br. J. Ophthalmol. **35:**784, 1951.
20. Bembridge, B. A., Crawford, G. N. C., and Pirie, A.: Phase contrast microscopy of the animal vitreous body, Br. J. Ophthalmol. **36:**131, 1952.
21. Berman, E. R., and Gombos, G. M.: Studies on the incorporation of U-^{14}C-glucose into vitreous polymers in vitro and in vivo, Invest. Ophthalmol. Vis. Sci. **8:**521, 1969.
22. Berman, E. R., and Michaelson, I. C.: The chemical composition of the vitreous body as related to age and myopia, Exp. Eye Res. **3:**9, 1964.
23. Bito, L. Z., and Davson, H.: Steady state concentrations of potassium in ocular fluids, Exp. Eye Res. **3:**283, 1964.
24. Bito, L. Z., et al.: The relationship between the concentrations of amino acids in the ocular fluids and blood plasma of dogs, Exp. Eye Res. **4:**374, 1965.
25. Bleeker, G. M., van Haeringen, N. J., and Glasius, E.: Urea and the vitreous barrier of the eye, Exp. Eye Res. **7:**30, 1968.
26. Bleeker, G. M., et al.: Selective properties of the vitreous barrier, Exp. Eye Res. **7:**37, 1968.
27. Boettner, E. A., and Wolter, J. R.: Transmission of the ocular media, Invest. Ophthalmol. **1:**776, 1962.
28. Borello, C.: Ricerche sul sistema fibrinolitico nel corpo vitreo, Rass. Ital. Ottalmol. **29:**309, 1960.
29. Brégeat, P., Regnault, F., and Perdriel, G.: Etude expérimentale de l'évolution des hémorrhagies intravitréennes, Bull. Soc. Ophthalmol. Fr. **79:**437, 1966.
30. Brini, A., Porte, A., and Stoeckel, M. E.: Embryologie du vitré. In Brini, A., et al., editors: Biologie et chirurgie du corps vitré, Paris, 1968, Masson & Cie., Editeurs.
31. Brini, A., Porte, A., and Stoeckel, M. E.: Morphologie et structure du vitré. In Brini, A., et al., editors: Biologie et chirurgie du corps vitré, Paris, 1968, Masson & Cie., Editeurs.
32. Brunish, R.: Biochemical studies on vitreous stability, Am. J. Ophthalmol. **41:**1024, 1956.
33. Cagianut, B., and Verrey, F.: Essai de dépistage du métabolisme hydrique dans les milieux transparants de l'oeil humain par injection d'eau lourde dans la chambre antérieure, Ann. Ocul. **182:**649, 1949.
34. Castren, J. A., and Laamanen, A.: The resorption speed of intravitreally injected traced proteins, Acta Ophthalmol. **41:**688, 692, 1963.
35. Christiansson, J., and Palm, E.: The exchange of substances in the anterior part of the vitreous body, bordering upon the lens, Acta Ophthalmol. **32:**197, 1954.
36. Cibis, P. A.: Vitreoretinal pathology and surgery in retinal detachment, St. Louis, 1965, The C. V. Mosby Co.
37. Cunha-Vaz, J. G.: Studies on the permeability of the blood retinal barrier. II, Breakdown of the blood-retinal barrier by injury. Br. J. Ophthalmol. **50:**454, 1966.
38. Cunha-Vaz, J. G.: Studies on the permeability of the blood retinal barrier. III, Breakdown of the blood-retinal barrier by circulatory disturbances: a topographical study of the vascular tree, Br. J. Ophthalmol. **50:**505, 1966.
39. Cunha-Vaz, J. G., and Maurice, D. M.: The active transport of fluorescein by the retinal vessels and the retina, J. Physiol. (Lond) **191:**467, 1967.
40. Cunha-Vaz, J. G., Shakib, M., and Ashton, N.: Studies on the permeability of the blood-retinal barrier. I, On the existence, development and site of a blood-retinal barrier, Br. J. Ophthalmol. **50:**441, 1966.
41. Cunha-Vaz, J., et al.: Early breakdown of the blood-retinal barrier in diabetes, Br. J. Ophthalmol. **59:**649, 1975.
42. Davis, M. D.: Vitreous contraction in proliferative diabetic retinopathy, Arch. Ophthalmol. **74:**741, 1965.
43. Davson, H.: Physiology of the ocular and cerebrospinal fluids, Boston, 1956, Little, Brown & Co.
44. Davson, H.: The intraocular fluids. In Davson, H., editor: The eye. Vegetative physiology and biochemistry. New York, 1962, Academic Press, Inc., vol. 1.
45. Davson, H., and Duke-Elder, W. S.: The distribution between the intraocular fluids and blood plasma, and the kinetics of penetration of various sugars into these fluids, J. Physiol. (Lond) **107:**141, 1948.
46. Davson, H., and Luck, C. D.: A comparative study

of the total carbon dioxide in the ocular fluids, cerebrospinal fluid and plasma of some mammalian species, J. Physiol. (Lond) **132**:454, 1956.

47. Davson, H., et al.: The penetration of some electrolytes and non-electrolytes into the aqueous humour and the vitreous body of the cat, J. Physiol. (Lond) **108**:203, 1949.
48. De Rosa, L.: Il fosforo del vitreo nel corso della degenerazione pigmentaria sperimentale al iodato da sodio, Rass. Ital. Ottalmol. **24**:114, 1955.
49. De Rosa, L., and De Vincentiis, M.: L'azoto del vitreo nella degenerazione pigmentaria sperimentale della retina, Rass. Ital. Ottalmol. **22**:481, 1953.
50. De Vincentiis, M.: Ulteriori osservazioni sul contenuto di acido piruvico e acido lattico nel vitreo, Boll. Soc. Ital. Biol. Sper. **27**:309, 1951.
51. Dunn, M. W., et al.: Studies in collagen implants in the vitreous, Surg. Forum **19**:492, 1968.
52. Eisner, G.: Autoptische Spaltlampenuntersuchung des Glaskörpers. I to III, v. Graefes Arch. Ophthalmol. **182**:1, 1971.
53. Eisner, G.: Biomicroscopy of the peripheral fundus, Surv. Ophthalmol. **17**:1, 1972.
54. Eisner, G.: Autoptische Spaltlampenuntersuchung des Glaskörpes. V, v. Graefes Arch. Ophthalmol. **187**: 5, 1973.
55. Eisner, G.: Lichtkoagulation und Glaskörperbildung: zur Frage der Glaskörperentstehung, v. Graefes Arch. Ophthalmol. **206**:33, 1978.
56. Favre, M., and Goldmann, H.: Zur Genese der hinteren Glaskörperabhebung, Ophthalmologica **132**:87, 1956.
57. Fine, B. S.: Retinal structure: light and electron-microscopic observations. In McPherson, A., editor: New and controversial aspects of retinal detachment, New York, 1968, Harper & Row, Publishers.
58. Fine, B. S., and Tousimis, A. J.: The structure of the vitreous body and the suspensory ligaments of the lens, Arch. Ophthalmol. **65**:95, 1961.
59. Fischer-von Bünau, H., and Fischer, F. P.: Hat der Glaskörper einen Stoffwechsel? Arch. Augenheilkd. **106**:463, 1932.
60. Foos, R. Y.: Vitreoretinal juncture: topographical variations, Invest. Ophthalmol. **11**:801, 1972.
61. Forbes, M., and Becker, B.: The transport of organic ions by the rabbit eye. II, In vivo transport of iodopyracet, Am. J. Ophthalmol. **50**:867, 1960.
62. Fowlks, W. L.: Meridional flow from the corona ciliaris through the pararetinal zone of the rabbit vitreous, Invest. Ophthalmol. **2**:63, 1963.
63. Gärtner, J.: Klinische Beobachtungen über Glaskörperadhärenzen am hinteren Augenpol, Klin. Monatsbl. Augenheilkd. **140**:161, 1962.
64. Gärtner, J.: Klinische Beobachtungen über den Zusammenhang der Glaskörpergrenzmembran mit Glaskörpergerüst und Netzhautgefässen in der Ora-Aequatorgegend, Klin. Monatsbl. Augenheilkd. **140**:524, 1962.
65. Gärtner, J.: Histologische Beobachtungen über das Verhalten der vitreoretinalen Grenzschicht bei Glaskörperabhebung, Klin. Monatsbl. Augenheilkd. **142**:769, 1963.
66. Gärtner, J.: Die Bedeutung des perivasculären Raums der Zentralgefässe für den vitreopapillären Stofftransport beim normalen Mäuseauge und beim menschlichen Auge unter pathologischen Bedingungen, v. Graefes Arch. Ophthalmol. **175**:13, 1968.
67. Gloor, B. P.: Mitotic activity in the cortical vitreous cells (hyalocytes) after photocoagulation, Invest. Ophthalmol. **8**:633, 1969.
68. Gloor, B. P.: Zur Entwicklung des Glaskörpers und der Zonula. II, Glaskörperzellen während Entwicklung und Rückbildung der Vasa hyaloidea und der Tunica vasculosa lentis, v. Graefes Arch. Ophthalmol. **186**:311, 1973.
69. Gloor, B. P.: Zur Entwicklung des Glaskörpers und der Zonula. III, Herkunft, Lebenszeit und Ersatz der Glaskörperzellen beim Kaninchen (autoradiographische Untersuchungen mit ^{3}H-thymidin), v. Graefes Arch. Ophthalmol. **187**:21, 1973.
70. Gloor, B. P.: Zur Entwicklung des Glaskörpers und der Zonula. VI, Autoradiographische Untersuchungen zur Entwicklung der Zonula der Maus mit ^{3}H-markierten Aminosäuren und ^{3}H-Glucose, v. Graefes Arch. Ophthalmol. **169**:105, 1974.
71. Gloor, B. P., Rokos, L., and Leuenberger, R.: Abtransport von ^{131}J-Hippursäure und ^{22}Na aus dem Retroretinal-Raum bei experimenteller arhegmatogener Netzhautablösung, v. Graefes Arch. Ophthalmol. **197**:107, 1975.
72. Goldmann, H.: Zur Technik der Spaltlampenmikroskopie, Ophthalmologica **96**:90, 1938.
73. Grignolo, A.: Fibrous components of the vitreous body, Arch. Ophthalmol. **47**:760, 1952.
74. Gross, J., Matoltsy, A. G., and Cohen, C.: Vitrosin: a member of the collagen class, J. Biophys. Biochem. Cytol. **1**:215, 1955.
75. Guggenheim, I., and Franceschetti, A.: Refraktometrische Untersuchungen des Glaskörpers von Kaninchen und Mensch unter physiologischen und pathologischen Bedingungen, Arch. Augenheilkd. **98**:448, 1928.
76. Hayreh, S. S.: Posterior drainage of the intraocular fluid from the vitreous, Exp. Eye Res. **5**:123, 1966.
77. Heath, H., et al.: Biochemical changes in aphakia, Vision Res. **1**:274, 1961.
78. Hogan, M. J.: The vitreous, its structure and relation to the ciliary body and retina, Invest. Ophthalmol. **2**:418, 1963.
79. Hruby, K.: Klinische Untersuchungen über das Verhalten des Glaskörpers bei Retinitis proliferans, v. Graefes Arch. Ophthalmol. **144**:435, 1942.
80. Hruby, K.: Hyaluronsäure als Glaskörperersatz bei Netzhautablösung, Klin. Monatsbl. Augenheilkd. **138**:484, 1961.
81. Hruby, K.: Slitlamp examination of vitreous and retina (Posner, A., translator), Baltimore, 1967, The Williams & Wilkins Co.
82. Jack, L. R.: Regression of the hyaloid vascular system, Am. J. Ophthalmol. **74**:261, 1972.

83. Jess, A.: Der Glaskörper und seine Erkrankungen, Kurzes Hdb. Ophthalmol. **5**:325, 1930.
84. Kimura, S.: Fibrinolysis in the eye. Report: fibrinolytic enzyme system of the vitreous and the retina with choroid, Acta Soc. Ophthalmol. **67**:852, 1963.
85. Kinsey, V. E., and Reddy, D. V. N.: An estimate of the ionic composition of the fluid secreted into the posterior chamber inferred from a study of aqueous humor dynamics, Doc. Ophthalmol. **13**:7, 1959.
86. Kinsey, V. E., Grant, M., and Cogan, D. G.: Water movement and the eye, Arch. Ophthalmol. **27**:242, 1942.
87. Kinsey, V. E., Reddy, D. V. N., and Skrentny, B. A.: Intraocular transport of C^{14} labeled urea and the influence of Diamox on its role of accumulation in aqueous humors, Am. J. Ophthalmol. **50**:1130, 1960.
88. Koyanagi, Y., and Kinukawa, C.: Gleichartige Veränderungen des retinalen Pigmentepithels und des renalen Tubulusepithels bei verschiedenen Vergiftungen: Beiträge zur Pathogenese der Retinitis albuminurica, v. Graefes Arch. Ophthalmol. **137**:261, 1937.
89. Krause, A. C., and Goren, D. V. N.: The effects of hypoxia and hyperoxia upon the oxygen tension in the vitreous humor of the cat, Am. J. Ophthalmol. **42**: 647, 1956.
90. Linder, B.: Acute posterior vitreous detachment and its retinal complications, Acta Ophthalmol. **87**:306, 1966.
91. Maltoltsy, Y.: A study on the structural protein of the vitreous body, J. Gen. Physiol. **36**:29, 1952.
92. Maurice, D. M.: The exchange of sodium between the vitreous body and the blood and the aqueous humour, J. Physiol. (Lond) **137**:110, 1957.
93. Maurice, D. M.: Protein dynamics in the eye studied with labeled proteins, Am. J. Ophthalmol. **47**:361, 1959.
94. Maurice, D. M.: The use of fluorescein in ophthalmological research, Invest. Ophthalmol. **6**:464, 1967.
95. Mörner, C. T.: Untersuchung der Proteinsubstanzen in den lichtbrechenden Medien des Auges, Z. Physiol. Chem. **18**:233, 1894.
96. Nordmann, J.: Biologie due corps vitré: chimie, physique et physico-chimie, physiologie, bacteriologie, immunologie, physiopathologie. In Brini, A., et al., editors: Biologie et chirurgie du corps vitré, Paris, 1968, Masson & Cie., Editeurs.
97. Oesterlin, S.: Changes in the macromolecular composition of the vitreous produced by removal of the lens, Excerta Medica International Congress Series, no. 222, Ophthalmology, Amsterdam, 1970, pp. 1620-1923. See also Österlin.
98. Ogston, A. G., and Phelps, C. F.: The partition of solutes between buffer solutions and solutions containing hyaluronic acid, Biochem. J. **78**:827, 1961.
99. Okun, E., Aribes, N., and Johnston, G. P.: Therapy of retinal detachment complicated by massive preretinal fibroplasia: long-term follow-up of patients treated with intravitreal liquid silicone, Mod. Probl. Ophthalmol. **8**:563, 1969.
100. Oosterhuis, J. A.: Polygeline as a vitreous substitute. II, Clinical results, Arch. Ophthalmol. **76**:374, 1966.
101. Oosterhuis, J. A., et al.: Polygeline as a vitreous substitute. I, Observations in rabbits, Arch. Ophthalmol. **76**:258, 1966.
102. Österlin, S. E.: The synthesis of hyaluronic acid in vitreous. III, In vivo metabolism in the owl monkey, Exp. Eye Res. **7**:524, 1968.
103. Österlin, S. E., and Jacobson, B.: The metabolism of the vitreous. In Dardenne, M. U., and Nordmann, J., editors: Biochemistry of the eye. Twentieth International Congress of Ophthalmology, Symposium, Tutzing Castle, Basel, 1968, S. Karger AG.
104. Österlin, S. E., and Jacobson, B.: The synthesis of hyaluronic acid in vitreous. I, Soluble and particulate transferases in hyalocytes, Exp. Eye Res. **7**:497, 1968.
105. Österlin, S. E., and Jacobson, B.: The synthesis of hyaluronic acid in vitreous. II, The presence of soluble transferase and nucleotide sugar in the acellular vitreous gel, Exp. Eye Res. **7**:511, 1968. See also Oesterlin.
106. Palm, E.: On the occurrence in the retina of conditions corresponding to the "blood-brain barrier," Acta Ophthalmol. **25**:29, 1947.
107. Palm, E.: On the phosphate exchange between the blood and the eye: experiments on the entrance of radioactive phosphate into the aqueous humour, the anterior uvea and the lens, Acta Ophthalmol. **32**:1, 1948.
108. Palm, E.: The exchange of phosphate between the blood and the eye, studied by means of radio-autographs, Acta Ophthalmol. **27**:267, 1949.
109. Pau, M.: Zur Entwicklung der Glaskörperstrukturen und der Zonula, Ophthalmologica **134**:320, 1957.
110. Paufique, L., and Moreau, P. G.: Les greffes de vitré lyophilisé, Ann. Ocul. **186**:873, 1953.
111. Paufique, L., Fayet, M. T., and Ravault, M.: Etude comparative du vitré humain normal et lyophilisé, Ann. Ocul. **192**:241, 1959.
112. Peyman, G. A., and Sanders, D. R.: Advances in uveal surgery, vitreous surgery, and the treatment of endophthalmitis, New York, Appleton-Century-Crofts, p. 179.
113. Pirie, A.: The effect of hyaluronidase injection on the vitreous humour of the rabbit, Br. J. Ophthalmol. **33**:678, 1949.
114. Pirie, A., Schmidt, Y., and Waters, J. W.: Ox vitreous humor. I, The residual protein, Br. J. Ophthalmol. **32**:321, 1948.
115. Reddy, D. V. N., and Kinsey, V. E.: Composition of the vitreous humor in relation to that of plasma and aqueous humors, Arch. Ophthalmol. **63**:715, 1960.
116. Reddy, D. V. N., and Kinsey, V. E.: Transport of alpha aminoisobutyric acid into ocular fluids and lens, Invest. Ophthalmol. **1**:41, 1962.
117. Reddy, D. V. N., Rosenberg, C., and Kinsey, V. E.: Steady state distribution of free amino acids in the aqueous humours, vitreous body and plasma of the rabbit, Exp. Eye Res. **1**:175, 1961.

118. Reddy, V. N., Chakrapani, B., and Lim, C. P.: Blood vitreous barrier to amino acids, Exp. Eye Res. **25:**543, 1977.

119. Redslob, E.: Le corps vitré. Son dévelopment, sa structure, ses propriétes physicochimiques, Paris, 1932, Masson & Cie., Editeurs.

120. Reese, A. B.: Persistent hyperplastic primary vitreous, Am. J. Ophthalmol. **40:**317, 331, 1955.

121. Regnault, F.: L'hémorragie intravitréenne, Paris, 1966, Polycopie Gitanne.

122. Röber, H., et al.: Concentrations of ampicillin in the vitreous after cryocoagulation, v. Graefes Arch. Ophthalmol. **204:**275, 1977.

123. Rieger, H.: Zur Histologie der Glaskörperabhebung, Teil II. Ueber die Beziehungen des abgehobenen Glaskörpers zur Netzhaut, v. Graefes Arch. Ophthalmol. **146:**447, 1944.

124. Rodriguez-Peralta, J., and Lorenzo, A.: Hematic and fluid barriers of the retina and the vitreous body, J. Comp. Neurol. **132:**109, 1968.

125. Rossi, A.: Structure of the vitreous body, observations by means of the phase contrast microscope, Br. J. Ophthalmol. **37:**343, 1953.

126. Salzmann, M.: The anatomy and histology of the human eyeball (Brown, E. V. L., translator), Chicago, 1912, University of Chicago Press.

127. Santoni, A.: Ricerche sul indice di refrazione e sul contenuto in glucosio del vitreo nel corso della degenerazione corioretinica spermentale da iodato da sodio, Ann. Ottalmol. **74:**590, 1948.

128. Schepens, C. L.: Diagnostic and prognostic factors as found in preoperative examination, Trans. Am. Acad. Ophthalmol. Otolaryngol. **56:**298, 1952.

129. Schepens, C. L.: Clinical aspects of pathologic changes in the vitreous body, Am. J. Ophthalmol. **38:**8, 1954.

130. Schmut, O.: The organization of tissues of the eye by different collagen types, v. Graefes Arch. Ophthalmol. **207:**189, 1978.

131. Schwarz, W., and Schuchardt, E.: Die Struktur des frischen Glaskörpers, Z. Zellforsch. Mikrosk. Anat. **35:**293, 1950.

132. Schweer, G.: Glaskörper und Hyaluronsäure-Hyaluronidase System, Sammlung zwangloser Abhandlungen aus dem Gebiete der Augenheilkunde, Stuttgart, West Germany, 1962, Georg Thieme Verlag KG.

133. Stürm, R., and Rokos, L.: Einfluss von Netzhautgefässverschlüssen auf den Abtransport von ^{22}Na und o-^{131}J-Hippursäure aus dem Raume zwischen Netzhaut und Pigmentepithel (Katzenversuche), v. Graefes Arch. Ophthalmol. **197:**119, 1975.

134. Süllmann, H.: Chemie des Auges, Tabul. Biol. **22:**79, 1951.

135. Sundblad, L., and Balazs, E. A.: Chemical and physical changes of glycosamino-glycans and glycoproteins caused by oxidation-reduction systems and radiation. In Balazs, E. A., and Jeanloz, R. W., editors: The amino sugars: the chemistry and biology of components containing amino sugars. Metabolism and interactions, vol. 2B, New York, 1966, Academic Press, Inc.

136. Suran, A. A., and McEwen, W. K.: Residual proteins of the vitreous, Am. J. Ophthalmol. **42:**338, 1956.

137. Sweeney, D. B., and Balazs, E. A.: Fate of collagen and hyaluronic acid gels and solutions into the vitreous of the owl monkey, Invest. Ophthalmol. **3:** 473, 1964.

138. Szent-Györgyi, A.: Untersuchungen über den Bau des Glaskörpers des Menschen, Arch. Mikrobiol. **89:**324, 1917.

139. Szirmai, J. A., and Balazs, E. A.: Studies on the structure of the vitreous body. III, Cells in the cortical layer, Arch. Ophthalmol. **59:**34, 1958.

140. Tolentino, F. I., Lee, D. F., and Schepens, C. L.: Biomicroscopic study of vitreous cavity in diabetic retinopathy, Arch. Ophthalmol. **75:**238, 1966.

141. Tronche, P., et al.: Répartition et métabolisme du radiophosphore 32 injecté dans le gel vitréen, Ann. Ocul. **196:**159, 1963.

142. Vilstrup, G., and Kornerup, V.: Protein fractions in corpus vitreum examined by paper electrophoresis, Acta Ophthalmol. **33:**13, 1955.

143. Visser-Heerema, J.: Ueber das spezifische Gewicht der bei der Operation von Netzhautablösungen gewonnenen Flüssigkeit, Arch. Augenheilkd. **109:**543, 1936.

144. von Sallmann, L., and Dillon, B.: Studies of the eye with radioiodine autographs, Am. J. Ophthalmol. **33:**429, 1950.

145. von Sallmann, L., and Locke, B. D.: Experimental study on early lens changes after roentgen irradiation: exchanges and penetration of radioactive indicators (Na^{24}, K^{42}, I^{131}, P^{32}) in normal and irradiated lenses of rabbits, Arch. Ophthalmol. **45:**431, 1952.

146. von Sallmann, L., Evans, T. C., and Dillon, B.: Studies of the eye with radiosodium autographs, Arch. Ophthalmol. **41:**611, 1949.

147. Waltman, S., et al.: Alteration of the blood-retinal barrier in experimental diabetes mellitus, Arch. Ophthalmol. **96:**878, 1978.

148. Watzke, R. C.: Silicone retinopoiesis for retinal detachment: a long-term clinical evaluation, Arch. Ophthalmol. **77:**185, 1967.

149. Widder, W.: Hyaluronsäure als Glaskörperimplantat bei Netzhautablösung, v. Graefes Arch. Ophthalmol. **162:**416, 1960.

150. Widder, W.: Tierversuche über die Verweildauer verschiedener Glaskörperimplantate, v. Graefes Arch. Ophthalmol. **164:**550, 1962.

151. Wiesinger, A. M., et al.: The transmission of light through the ocular media of the rabbit eye, Am. J. Ophthalmol. **42:**907, 1956.

152. Witschel, H.: Reaktions- und Reparationsvorgänge an der menschlichen Netzhautglia. In Naumann, G., and Gloor, B., editors: Wundheilung des Auges und ihre Komplikationen, Symposium der DOG, Tübingen, West Germany, 1979, publ. no. 30.03.-01.04, Munich, West Germany, 1980, J. F. Bergmann.

153. Young, R. A.: The ground substance of connective tissue, J. Physiol. (Lond) **16:**325, 1894.
154. Young, R. G., and Williams, H. H.: Biochemistry of the eye: gelatinous protein of vitreous body, Arch. Ophthalmol. **51:**593, 1954.

GENERAL REFERENCES

Balazs, E. A.: Die Mikrostruktur und Chemie des Glaskörpers, Ber. Dtsch. Ophthalmol. Ges. **68:**536, 1968.

Balazs, E. A.: The molecular biology of the vitreous. In McPherson, A., editor: New and controversial aspects of retinal detachment, New York, 1968, Harper & Row, Publishers.

Berman, E. R., and Voaden, M.: The vitreous body. In Graymore, C. N.: Biochemistry of the eye, New York, 1970, Academic Press, Inc., pp. 373-471.

Brini, A., et al., editors: Biologie et chirurgie due corps vitré, Rapport Societe Française Ophtalmologie, Paris, 1968, Masson & Cie., Editeurs.

Busacca, A.: Biomicroscopie et histopathologie du fond de l'oeil, Zurich, Switzerland, 1967, Schweizer Verlagshaus AG.

Cibis, P. A.: Vitreoretinal pathology and surgery in retinal detachment, St. Louis, 1965, The C. V. Mosby Co.

Eisner, G.: Biomicroscopy of the peripheral fundus, Survey Ophthalmol. **17:**1, 1972.

Eisner, G.: Biomicroscopy of the peripheral fundus, Berlin, 1973, Springer-Verlag.

Goldmann, H.: Senescenz des Glaskörpers, Ophthalmologica **143:**253, 1962.

Goldmann, H.: Biomikroskopie des normalen menschlichen Glaskörpers während des Lebens, Ber. Dtsch. Ophthalmol. Ges. **23:**15, 1968.

Hruby, K.: Slitlamp examination of vitreous and retina, Baltimore, 1967, The Williams & Wilkins Co.

Kinsey, V. E., and Reddy, D. V. N.: Chemistry and dynamics of aqueous humor. In Prince, J. H., editor: The rabbit in eye research, Springfield, Ill., 1964, Charles C Thomas, Publisher.

Chapter 10

The lens

EDWARD COTLIER, M.D.

ANATOMY OF THE LENS

The crystalline lens, which is positioned behind the iris, refracts the light entering the eye through the pupil and focuses it on the retina. The main functions of the lens are as follows:

1. To maintain its own clarity
2. To provide refractive power by contributing to the optical system of the eye
3. To provide accommodation, which allows the eye to clearly focus objects placed within a 6 m range
4. To absorb ultraviolet light

The absence of the lens, aphakia, results in loss of approximately 20 diopters of refractive power.

The lens is avascular but obtains nutrition from the surrounding fluids: the aqueous and the vitreous. Glucose from these fluids provides the lens with chemical energy required to continue growth and maintain transparency.

The lens has the shape of a flattened globe (Fig. 10-1); its various parts are given geodesic terms. The most anterior part of the lens is the anterior pole, the peripheral area is the equator, and the most posterior area, the posterior pole.

The anatomy of the lens is simple. The lens is entirely surrounded by a capsule, and under the anterior capsule is a layer of epithelial cells. The epithelial cells reproduce; near the lens equator the cells elongate toward the poles and become lens fibers. The lens fibers stretch in arcuate fashion from the anterior to the posterior pole of the lens. On cross section, they have a hexagonal shape. Lens fibers are formed throughout life, and new fibers cover the old.

The lens is suspended from the ciliary process by the zonules (Fig. 10-1). When the ciliary muscle contracts and the zonules relax, the lens becomes thicker and more convex, increasing its refractive power. The lens becomes thinner following relaxation of the ciliary muscle. If the zonules are broken, the lens shifts position (dislocates). During a gross dissection, after the capsule is removed the superficial, or cortex, fibers (Fig. 10-1) are easily separated, whereas the fibers in the center of the lens are glued together and form a compact mass known as the nucleus (Fig. 10-1). For humans slit-lamp examination reveals zones of optical discontinuity in the cortex and the nucleus corresponding to periods of lens growth throughout life (Fig. 10-2). Any alteration in the optical homogeneity of the lens or decrease in its transparency is known as a cataract and can be documented by photography and by densitometric measurements (Fig. 10-3). The young lens presents a uniform dark-blue fluorescence that increases with age.[84] The nucleus of the lens after age 30 fluoresces yellow green. Human senile cataracts begin with vacuoles (optically clear spaces) in the lens cortex and small opacities in the cortex or with changes in color and opacities in the nucleus (Fig. 10-3). Such cataracts are known as "immature" cataracts. If the cataract continues to progress, the entire lens cortex becomes opaque and the lens becomes white. Such cataracts are known as "mature" cataracts.

Capsule

An elastic capsule surrounds the lens and maintains its structural integrity (Fig. 10-4). The capsule is secreted at the embryonic stage by the lens epithelium. The capsule is 2 to 20 μm thick (thinnest at the poles and equator)

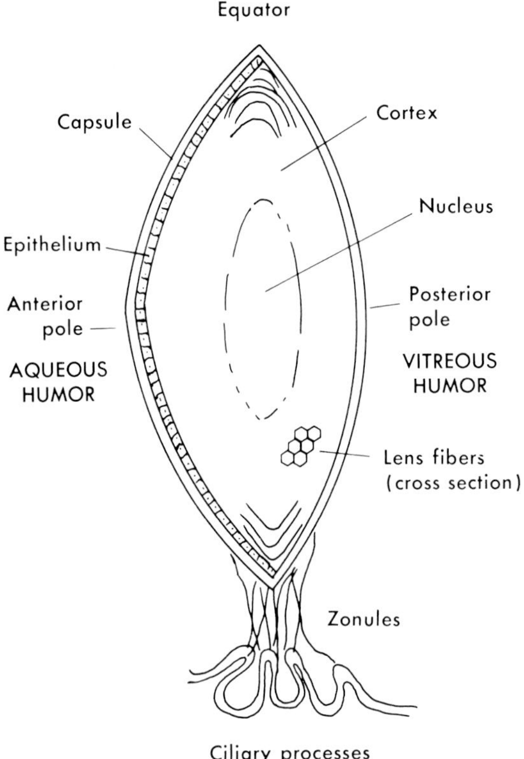

Fig. 10-1. Crystalline lens and zonular ligaments.

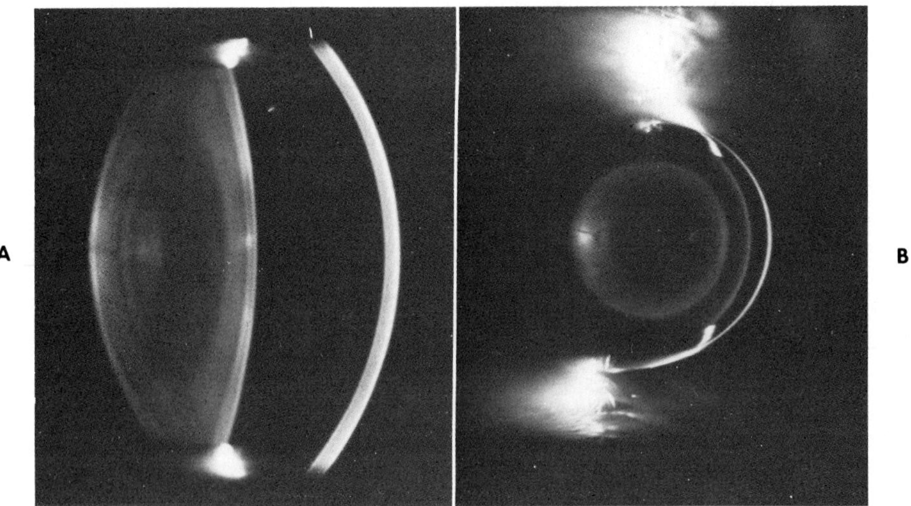

Fig. 10-2. Zones of discontinuity in the human lens **A,** as compared to homogeneous lens of fat, **B.** Obtained with Bonn camera.[47] (Courtesy Dragomirescu et al.)

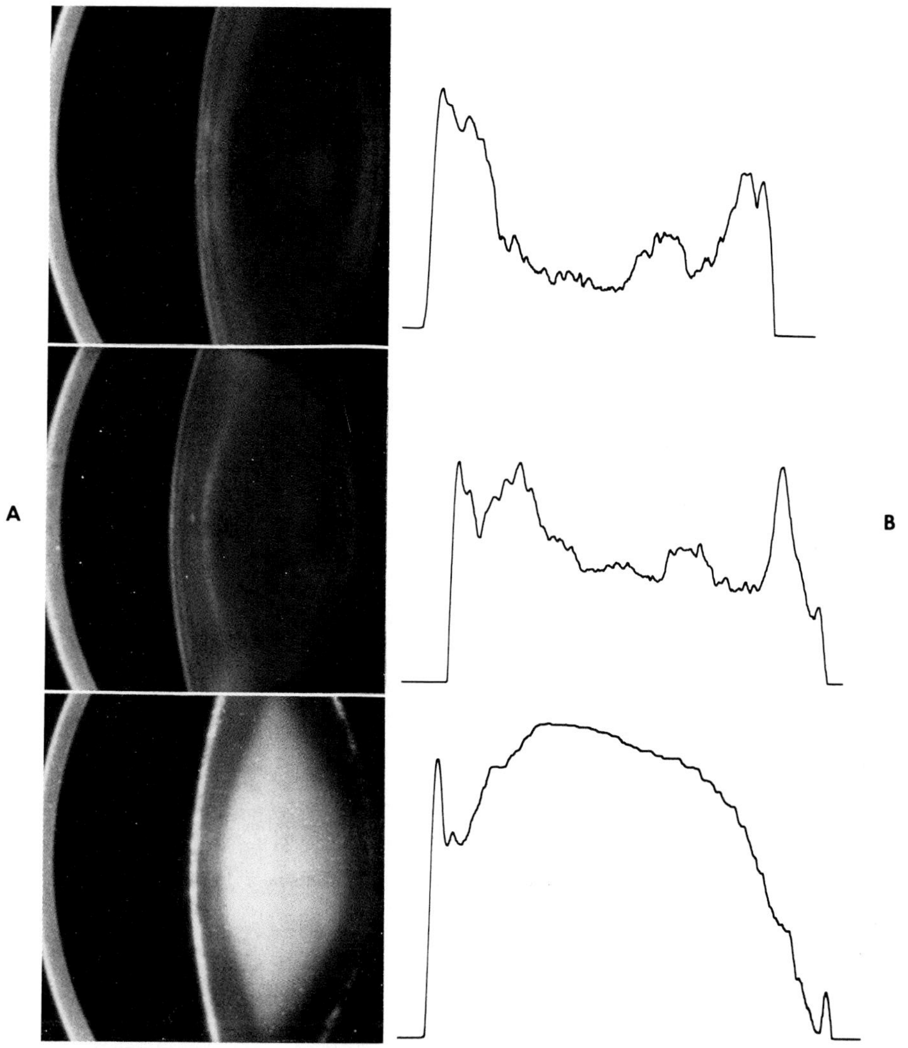

A

B

Fig. 10-3. Photographs **A,** and densitometric recordings **B,** of human lenses without opacities (top), supranuclear and early nuclear opacities (center), and dense nuclear opacity (bottom).[47] (Courtesy Dragomirescu et al.)

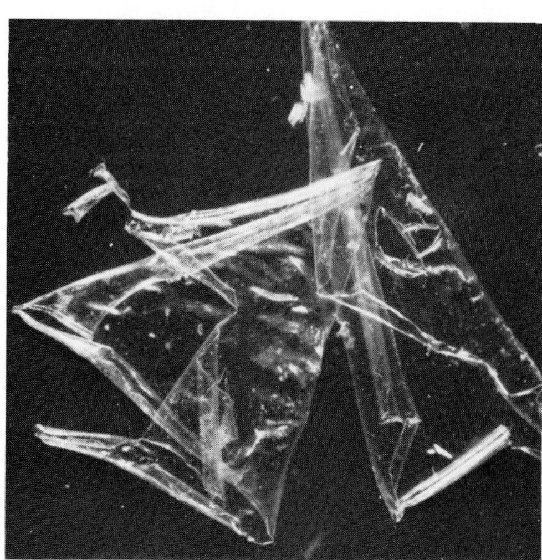

Fig. 10-4. Lens capsule (bovine) stripped of lens epithelium and curled onto itself. (Magnification ×9.5.) (From Fukushi, S., and Spiro, R. G.: J. Biol. Chem. **244**[8]:2041, 1969.)

and homogeneous by light microscopy. With the electron microscope many lamellae, measuring 300 to 400 Å in thickness are found. Each lamella contains fine filaments.[68] The capsule is made up of collagen-like glycoprotein material[117] and is digested by collagenase.[6] The lens capsule is considered a model basement membrane. As with other basement membranes, the lens capsule stains intensely with the PAS (periodic acid–Schiff) reagent. There are chemical and antigenic similarities between the lens capsule and basement membranes of the kidney glomeruli, blood vessels, spleen, and lungs. All these membranes stain with fluorescent antibody to kidney glomeruli.[105]

The lens capsule contains enzymes,[138] ATP, and glycolytic intermediates,[44] but cannot be considered to have independent metabolism. The lens capsule depends on contact with lens epithelium and fibers for metabolic supplies.

The capsule restricts penetration into the lens by various molecules based on size, charge, and lipoid solubility.[56] In vitro removal of the lens capsule with collagenase does not affect ion or amino acid transport into the lens.[6]

The capsule is tough and resistant to traction. The total surgical extraction of the human adult cataractous lens can be accomplished by traction either after grasping the lens capsule with forceps or by freezing a cryoextractor instrument to its surface.[86] This is known as intracapsular lens extraction. If the capsule breaks, the lens extraction is said to be extracapsular. In diabetic patients the lens capsule is more fragile, and extracapsular lens extractions are more frequent.[15] Children and young adults have attachments between the posterior lens capsule and the anterior hyaloid of the vitreous, and thus intracapsular extraction of the lens is not indicated.

The capsule appears to be an anatomic structure unchanged through life, but the presence of glycosidases in the capsule suggests a slow degradation of its glycoprotein, and thus chemical turnover of its constituent materials.[17]

Abnormal formation of capsulelike material is found in pseudoexfoliation of the lens capsule. Dandrufflike material accumulates on the lens capsule, the ciliary processes, and the posterior iris layer.[3,50] Open-angle glaucoma is often found in pseudoexfoliation of the lens capsule. Lamellar splitting of the capsule, known as true exfoliation of the lens capsule, is mainly found among glassblowers. Small cuts or rents of the capsule may result in localized opacities of the underlying lens cortex, more extensive tears in complete cataracts.

Epithelium

The cuboidal cells of the lens epithelium form a monolayer. The epithelial cells are firmly attached to the anterior capsule and loosely attached to the underlying fibers. This anatomic arrangement allows dissection of the anterior lens capsule and flat mount preparations in which the entire epithelium can be stained and visualized.[135] The adult rabbit lens contains approximately 800,000 epithelial cells, of which 100 cells are in mitosis. In a flat-mount preparation (Fig. 10-5) three areas are recognized: central, preequatorial, and equatorial. Cells are more densely packed in the preequatorial and equatorial areas. Most of the mitotic cells are found in the preequatorial region. The rate of cells entering mitosis in the young rat is approximately 100 cells/hour. The number of mitotic cells decreases with age.[25] Cells in mitosis in the preequatorial area of the lens are extremely susceptible to radiation.

The epithelium is the area of the lens with the highest metabolic rate. Glucose and oxygen are utilized by the lens epithelium. The content of ATP and enzymes is the highest in this area.[26,44,138] Chemical energy is needed for the epithelium to transport carbohydrates, electrolytes, and amino acids into the lens. Energy is also needed for protein synthesis in the newly formed lens fibers at the lens equator.

Harding and co-workers[63] found that mechanical trauma to the epithelium results in mitotic stimulation with increased incorporation of thymidine, a DNA precursor. "Repair" epithelial cells are elongated, resemble fibroblasts, and can pile up to 10 layers thick under the lens capsule. Heaping up of epithelial cells is found in the anterior pole of the lens in many human and experimental cataracts.

Lens fibers

The fibers make up the bulk of the lens cortex and nucleus. The 4-month-old rabbit lens cortex alone contains approximately 800,000 lens

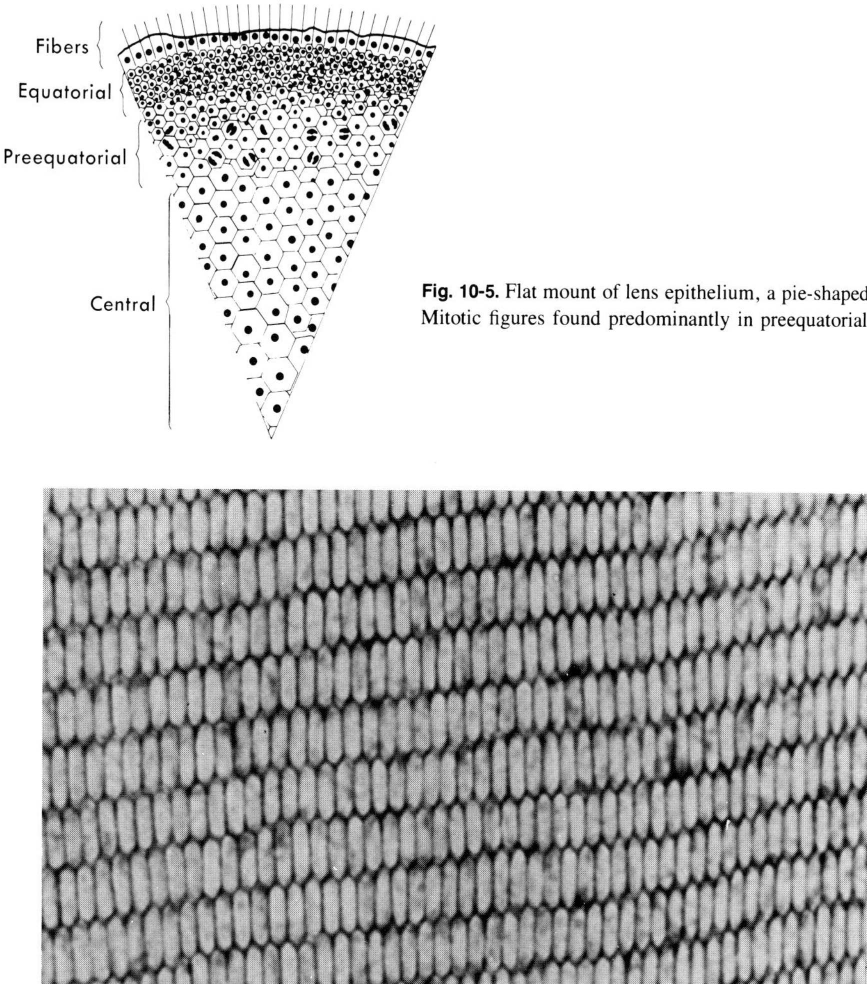

Fig. 10-5. Flat mount of lens epithelium, a pie-shaped area. Mitotic figures found predominantly in preequatorial area.

Fig. 10-6. Hexagonal appearance of rabbit cortical lens fibers in cross section. (Magnification ×100.)

fibers. Each fiber, hexagonal in cross section, represents an elongated cell with a membrane (Fig. 10-6). The areas in which the fibers meet anteriorly and posteriorly are the sutures of the lens. These are the two Y-shaped sutures in human lenses. Only a vertical anterior and a horizontal posterior suture are found in the rabbit lens. The lens fibers lose their nuclei as they grow old; nuclei are found near the equator in the younger fibers. The membranes of the fibers have side digitations that result in fiber interlocking (Fig. 10-7). Such a system results in the exquisite plasticity needed for the fibers to passively change shape during accommodation.

The spaces between the lens fibers, the extracellular spaces of the lens, are very small, accounting for only about 5% of the lens volume. These spaces are enlarged in human and experimental cataracts. Fluid collection between the fibers and destruction of many lens fibers result in the vacuoles or clefts seen by slit-lamp observation. The spaces between the fibers cannot be greatly increased unless the membranes of the lens fibers are broken. Thus the biochemistry of lens fiber membranes is of great significance for an understanding of membrane rupture and cataract formation. The membranes of lens fibers are made of glycoproteins, glycolipids, phospholipids, acidic mucopolysac-

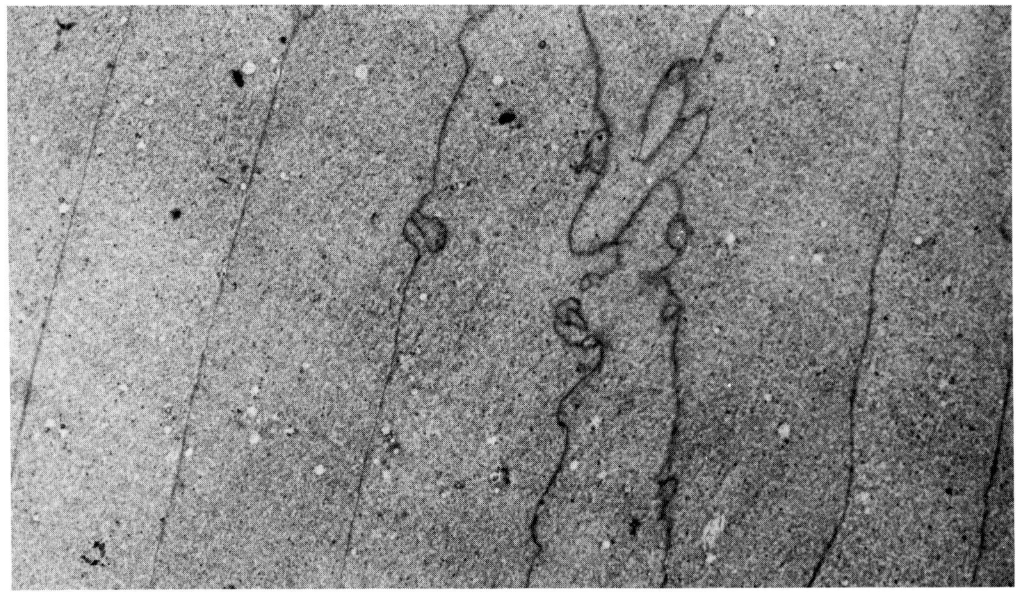

Fig. 10-7. Fibers in equatorial region of rabbit lens showing interdigitations among them. (Magnification ×3000.) (Courtesy Dr. George M. Wyhinny.)

charides, and sulfhydryl-protein.* Ribonucleic acid (RNA) attached to fiber membranes was found by Dische[45] and by Bloemendal[9] and co-workers. The chemical composition of the membranes is similar to that of other cell membranes, but specific intrinsic glycoprotein subunits of 26,000 daltons are present.[9,14] The membranes of the lens are broken by anionic and cationic surface-acting detergents.[107] The membranes are synthesized by transferases, enzymes assembling specific amino acids, sialic acid, or fatty acids, or a combination of these. The membranes of the fibers are degraded by glycosidases, lipases, and proteases.

Zonules

The zonules are thin, delicate filaments maintaining the lens suspended in position. The zonules stretch from the ciliary epithelium and peripheral retina to the lens capsule and form a thin pericapsular membrane at the lens equator. The large zonular fibers represent bundles of many smaller fibrils, 80 to 120 Å wide and having a regular periodicity of 110 to 180 Å. Chemically, the zonules are made of the following:

1. Collagen-like glycoprotein staining positive with PAS (amino acid analysis of the zonules reveals no hydroxyproline, an amino acid found in collagen)
2. Acidic mucopolysaccharide, probably chondroitin sulfate, that stains positive with Alcian blue[137]

The zonular fibers are broken by α-chymotrypsin, a proteolytic enzyme. This enzyme is often used to facilitate the delivery of the lens during cataract surgery. However, fragments of zonules broken by α-chymotrypsin can plug the trabecular meshwork and induce transient glaucoma in both humans and monkeys.[83]

The zonules contain 7% cysteine, and failure to convert the amino acid homocysteine into cysteine results in broken zonules and dislocated lenses in homocystinuria. Homocystinuria is an inborn error of metabolism in which the enzyme cystathionine synthetase is absent and excessive homocystine is excreted in urine. These patients have dislocated lenses and fair hair and can suffer thrombotic closure of vessels and mild mental retardation. The disease is inherited in an autosomal recessive fashion. Patients with Marfan's syndrome also have dislocated lenses, but the homocystine urine test is

*See references 9, 12, 33, 36, 43, and 112.

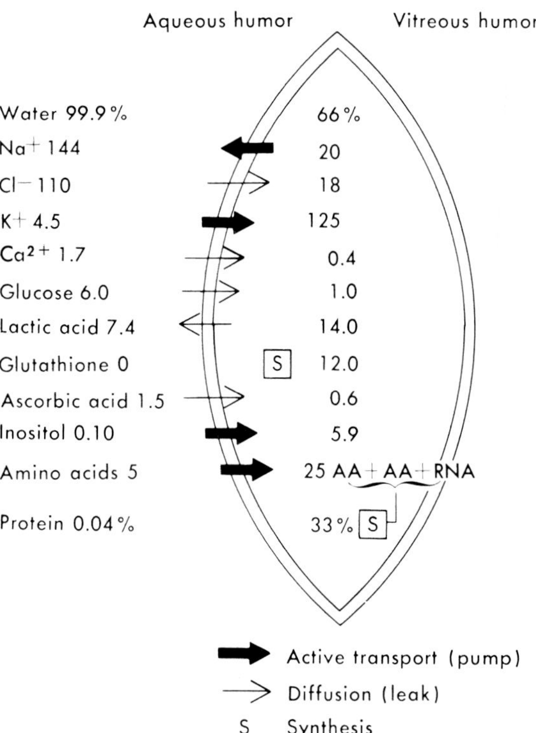

Fig. 10-8. Chemical composition of aqueous humor and lens. Water and protein are expressed in percent of lens weight. Na^+, Cl^-, K^+, and Ca^{2+} ions in μEq/ml lens water. Other compounds are expressed in μmol/g of lens weight or μmol/ml of aqueous humor. *AA*, Amino acids; *RNA*, ribonucleic acid.

negative. Marfan's syndrome is characterized by autosomal dominant inheritance. The patients often develop dissecting aortic aneurysm. Arachnodactyly (spider fingers) and dolichostenomelia (long, thin extremities) are found in both homocystinuric homozygotes and Marfan's heterozygotes.[29]

BIOCHEMISTRY OF THE LENS

Lens culture

The animal lens can be removed with ease by freeing the vitreous attachments to the posterior pole and cutting the zonules. For short-term culture the isolated lens is placed in a culture flask containing an isotonic oxygenated balanced salt solution with glucose, such as Tyrode's solution or Krebs-Ringer's solution.[5] TC199, a tissue culture medium containing salts, glucose, amino acids, and vitamins, is used for long-term lens culture of days or weeks.[82] Most metabolic functions of the lens such as glucose utilization, transparency, and cation and amino acid transport can be evaluated while the lens is in culture. Chemicals or drugs harmful or beneficial to the lens in culture can be tested by comparing the treated with the control contralateral lens of the same animal.

Chemistry of the normal lens

The lens is a dehydrated organ. The adult human or animal (rabbit) lens contains approximately 66% water and 33% protein. The lens cortex is more hydrated than the lens nucleus. Lens dehydration is maintained by an active Na^+ ion–water pump that resides within the membranes of cells in the lens epithelium and each lens fiber (Fig. 10-8).

Electric potential of the lens

The inside of the lens is electronegative. Potentials of −64 mV to −78 mV are recorded across the intact lens capsule and of −23 mV in the lens fibers.[99] There is a −23 mV difference between the anterior and posterior surfaces of the lens.[81] Thus the flow of electrolytes into the lens is directed by an electrical gradient.

The lens as an osmometer

The capsule confers on the lens properties of an intact cell, such as swelling in hypotonic media and dehydration in hypertonic media (Fig. 10-9). The osmolarity of the rabbit lens is 302 mOsm and equals the osmolarity of the aqueous.[35] The cations (sodium and potassium), which equal 145 mEq/L, and the anions (chloride, bicarbonate, sulfate, ascorbate, and glutathione), which equal 50 to 60 mEq/L, contribute to lens osmolarity. An anionic deficit of 90 mEq/L is probably made by acidic groups of lens proteins and glycoproteins.

Water increase in the lens breaks the lens fibers' membranes and results in microscopic vacuoles later seen by slit-lamp examination. The water equilibrium between the lens and the surrounding fluids is disrupted if the concentration of osmotically active compounds ($Na^+ + K^+$ or others) increases inside the lens. For example, poisoning the lens with iodoacetate increases the total levels of $Na^+ + K^+$ ions

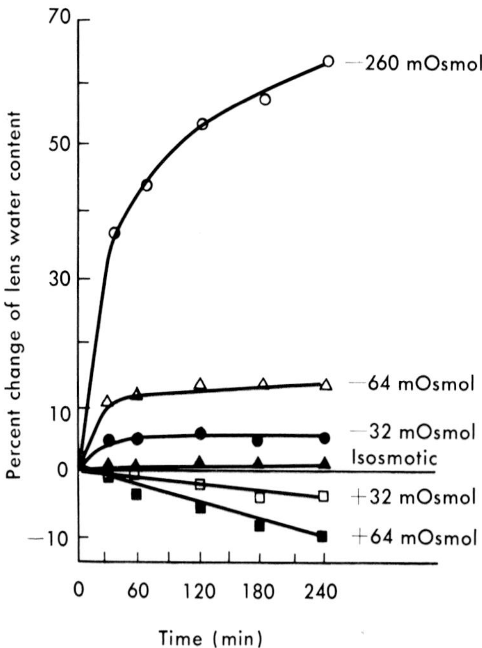

Fig. 10-9. Water gains by rabbit lenses incubated in hyposmotic media (−32, −64, and −260 mOsmol) and water losses by lenses incubated in hyperosmotic media (+32 and +64 mOsmol).

with resultant lens swelling. Increase in $Na^+ + K^+$ levels also follows lens exposure to surface-active detergents or antibiotics.[31] When retained inside the lens, abnormal products of galactose metabolism such as galactitol (dulcitol) can exert osmotic effects and result in water influx and lens swelling.[75,77]

Lens protein

The protein content of the lens, 33% of the total weight, is higher than that of any other organ in the body (e.g., brain 10%, muscle 18%). The perfect physicochemical arrangement of the lens proteins, living in an optimum environment of water, electrolytes, and sulfhydryl (SH^-) groups, gives transparency to the lens.

Protein fractions

When the lens is homogenized in a water or saline solution and the homogenate is spun down at low centrifugal forces, most proteins (85%) remain in suspension (soluble proteins), but some (15%) precipitate (insoluble protein). The insoluble proteins are attached to or constitute the membranes of the lens fibers. Urea washings remove protein attached to lens membranes. This fraction, also known as albuminoid, increases with age and during experimental or human senile cataracts. Most of the urea-insoluble protein fraction becomes partially soluble after treatment with high molar guanidine or digestion with proteases and aminopeptidases.[41,62,96] Most insoluble proteins are found in the lens nucleus, whereas most soluble proteins are in the lens cortex.[62]

The water-soluble proteins of the lens (85%) are divided into three main groups: α-crystallin (15%), β-crystallin (55%) and γ-crystallin (15%). On the basis of their electrophoretic mobility toward the anode, α-crystallin is fastest, β-crystallin is intermediate, and γ-crystallin is slowest. The molecular weight of the crystallins (in daltons) is α-crystallin 1,000,000, β-crystallin 50,000 to 200,000, and γ-crystallin 20,000.[62,114] Extensive chemical and physicochemical studies of each crystallin have been made. α-Crystallin and β-crystallin are made of subunits, and aggregation or separation of these subunits determines the physicochemical characteristics of each crystallin.

In the bovine lens Spector and co-workers[115,116] found progressive aggregation of α-crystallin with age such that giant macromolecules of 50 to 200 million daltons are formed—high molecular weight (HMW) α-crystallin. From light-scattering theory and studies of lens transparency Jedziniak and Benedek[69] and Spector[116] conclude that HMW α-crystallin could cause lens opacification if present in sufficient concentration. The protein subunits are assembled by the alignment of amino acids through RNA[38] as specified by the genetic code (DNA) (Fig. 10-10). Lens proteins are degraded by proteases and aminopeptidases.[113]

- Lens proteins, 100%
  - Soluble, 85% (crystallins)
    - α-crystallin 15%
    - β-crystallin 55%
    - γ-crystallin 15%
    - (in birds, δ-crystallin)
  - Insoluble, 15% (albuminoid)

Immunology

An animal immunized by injections of lens protein will only develop serum antibodies against lens protein, regardless of the animal species from which it is obtained. Lens proteins are organ specific but not species specific. In the normal lens the membrane of the lens fibers and lens capsule do not allow the passage of protein molecules from the lens to the aqueous humor. When a mature cataract develops, the membranes of the lens fibers are lysed, the capsule becomes more permeable, and protein can leak out of the lens. Lens protein in the anterior chamber can act as an antigen and induce anterior chamber and serum antibodies. The result is an inflammation of the uveal tissues known as lens-induced uveitis, phacogenic uveitis, or phacoanaphylactic uveitis.[67]

Following extracapsular extraction of the lens, lens material remaining in the eye is rapidly hydrolyzed and leaves the eye in a matter of days or weeks. Extracapsular lens extrac-

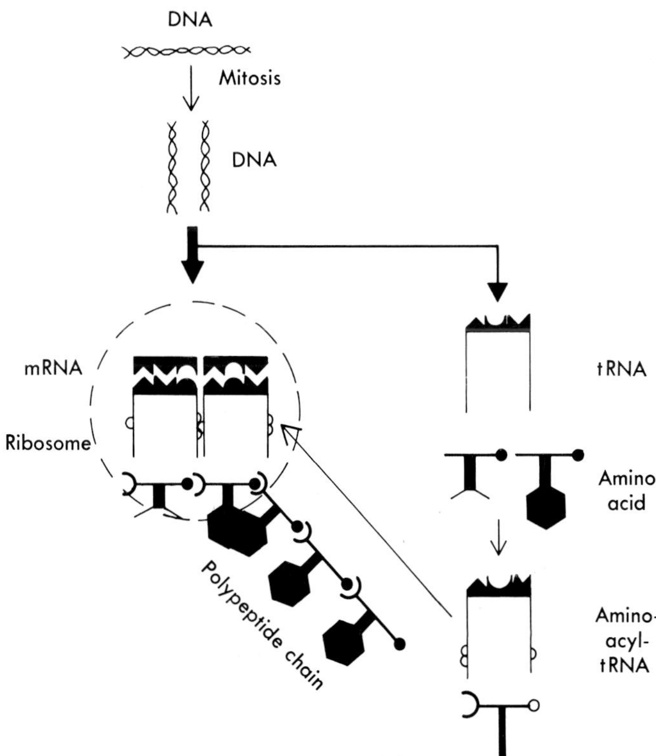

Fig. 10-10. Biochemistry of protein synthesis by the lens. Replication of DNA and transfer of genetic information to tRNA (transfer RNA), which assembles amino acids into polypeptide chain through mRNA (messenger RNA) in ribosome.

tion or rupture of the lens by accidental trauma results in high titers of serum antilens antibodies. Wirostko and Spalter[136] found hemagglutination antilens titers of 32 to 256 in human serum following unplanned extracapsular lens extractions. In one patient the serum antilens titer rose to 4096, 3 weeks following extracapsular lens extraction. Agar immunodiffusion of the patient's serum globulins showed reaction against lens and also uveal tissues. Antilens antibodies may produce uveitis in the second eye if mature cataract develops or another extracapsular extraction takes place. Occasionally, degraded lens protein leaked through the capsule into the aqueous humor is engulfed by macrophages, which can plug up the trabecular meshwork, thus blocking aqueous humor outflow and producing increased intraocular pressure—phacolytic glaucoma.[67]

François and co-workers[54] and Charlton and van Heyningen[18] found that the small molecular weight proteins (γ-crystallin and some β-crystallin) are markedly decreased in immature human senile cataracts. This material may leak into the anterior chamber but does not seem to elicit a primary immune response. α-Crystallin, as detected in the anterior chamber by radioimmunoassay, is increased in eyes with heterochromic uveitis and in eyes with hypermature senile cataracts, compared to normal eyes.[106] To initiate intraocular sensitization, the rupture and release of (1) antigenic membrane material (glycoproteins, etc.), (2) chemically denatured lens protein, or (3) large amounts of lens protein seems to be required.

Water and electrolyte transport

Na^+ and water extrusion

The lens is dehydrated and has higher levels of K^+ ions and lower levels of Na^+ ions, Cl^- ions, and water than the aqueous and vitreous have (Fig. 10-8). The electrolyte and water content of the lens resembles that of an intact cell, whereas the Na^+, Cl^-, and K^+ ion and water content of aqueous and vitreous is similar to that in plasma or extracellular fluids. To maintain electrolyte and water gradients against the surrounding fluids, the lens generates chemical and electrical energy. Chemical energy to extrude Na^+ ions and water is provided by ATP through glucose metabolism.

To maintain Na^+ ions and water out of the lens the following mechanisms are operative:

1. Na^+ ion extrusion "pump" coupled to K^+ ion intake. This mechanism depends on ATP breakdown and is regulated by the enzyme Na^+,K^+-ATPase. Na^+,K^+-ATPase will transport one Na^+ equivalent out of the lens and one K^+ equivalent into the lens. Na^+,K^+-ATPase is inhibited by ouabain, the cardiac glycoside.[10] Chelation of free sulfhydryl groups, probably attached to the Na^+,K^+-ATPase, also decreases the exchange of K^+ and Na^+ ions.[52] Furthermore, the K^+-Na^+ ionic balance of the lens is affected by anticholinesterase agents such as echothiophate iodide (Phospholine Iodide), if present in high concentrations in culture media.[93] Topical administration of Phospholine Iodide to iridectomized monkeys results in anterior cortical and subcapsular lens opacities.[72] By contrast, pilocarpine does not induce cataracts or have adverse effects on lens metabolism. These drugs may act directly on the membrane ATPases or indirectly through the cholinesterase system.

2. Na^+ ion extrusion "pump" depending on metabolism. If a lens is cultured in vitro without glucose, there is a progressive increase in Na^+ ions and water. Similar effects result from lens exposure to metabolic poisons such as iodoacetate or fluoride.[64,78]

3. Na^+ ion extrusion "pump" depending on the physicochemical integrity of lens fiber membranes. The surface membranes of the lens fibers maintain Na^+ ions and water out of the lens. Surface active agents (antibiotics, detergents, lysophospholipids) disrupt the physicochemical integrity of the membrane and the Na^+ extrusion pump with subsequent gain of Na^+ ions and water by the lens. Lens swelling and, eventually, complete loss of lens transparency follow.[31,37,107]

K^+ and amino acid concentration

The concentration of K^+ ion inside the lens is 25 times higher than in aqueous or vitreous (Fig. 10-8). The concentration of amino acids in the lens is two to six times higher than in aqueous (Figs. 10-8 and 10-11).[103] The levels of inositol in the lens are 95 times higher than those in aqueous (Fig. 10-8).[27] Thus K^+ ion, inositol, and amino acids are retained inside the lens.

Studies with isotopes in animal lenses indicate the lens transports and accumulates K^+ ion, amino acids, and inositol against the concentration gradient, while a slow efflux toward the outside takes place. The active transport of these compounds is known as the "pump" component, and the efflux as the "leak" (permeability) component.[80] Some investigators postulate that the lens acts as a giant pump-and-leak unit similar to red blood cells, bac-

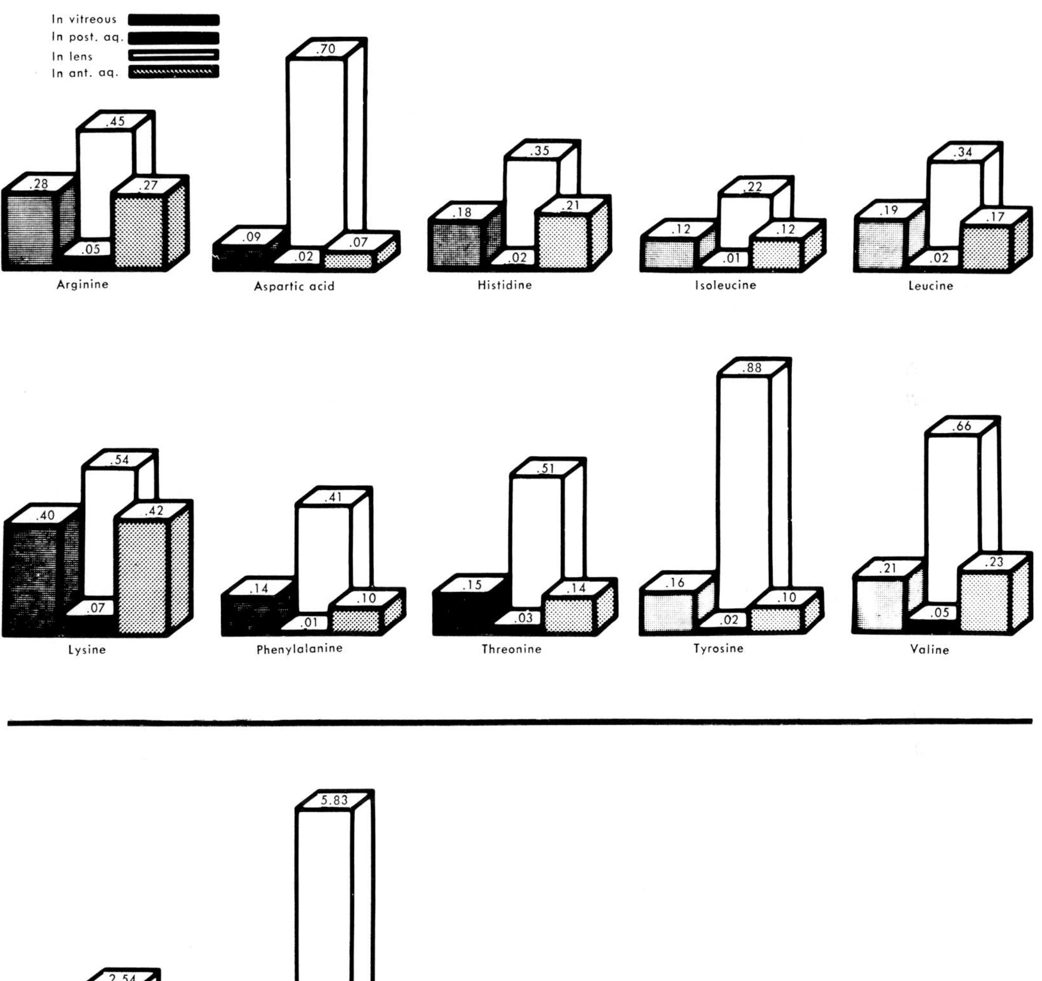

Fig. 10-11. Concentration of free amino acids in rabbit lens. Note higher concentration in lens as compared to posterior or anterior aqueous humor and vitreous humor. (From Reddy, V. N., and Kinsey, V. E.: Invest. Ophthalmol. **1:**635, 1962.)

teria, or other cells. This pump-and-leak system would result from the addition of each pump-and-leak membrane located in the epithelial cells and in each lens fiber. The pump is more active in the epithelial cells than in lens fibers.[64,80] The fibers, however, retain the K^+ ions, amino acids, and inositol inside the lens and against the concentration gradient. Electrical coupling between fibers exists. Another interpretation of the lens fibers as a transport unit is that they represent a syncytium similar to that of heart muscle fiber.[99] These specialized pumps are abolished totally or in part by metabolic inhibitors, by ouabain, or by lack of Na^+ ions in the media.[34,64,78] For practical experimental purposes the ion Rb^+ is used in lieu of K^+. α-Aminoisobutyric acid is used as a test substance, as this amino acid is not metabolized by the lens nor incorporated into lens protein.

Following chemical or anatomic damage to the membranes of the lens epithelium or fibers, there is increased leak out (permeability) of K^+, inositol, and amino acids from the lens. Low Ca^{2+} ion concentration in serum or aqueous, such as is found in hypoparathyroidism, increases the leak out from the lens of K^+, amino acids, and inositol.[6,27,127] Corticosteroids increase the leak out of K^+ ions and increase the entrance of Na^+ ions into the lens.[6,64]

Amino acids in the lens

Amino acids are actively transported into the lens. Once in the lens, free amino acids are incorporated into RNA[38] to form lens protein, can be metabolized with formation of CO_2 (Fig. 10-12), or can efflux from the lens. The turnover of free amino acids in the lens is very rapid, the renewal rate for lysine being 16% of the total in the lens per hour.[28]

The transport of tryptophan (Fig. 10-8) by the lens is the most active among all amino acids and requires metabolic energy. Tryptophan transport, however, unlike the transport of most amino acids, is not significantly decreased by inhibition of Na^+,K^+-ATPase or by decreased levels of Na^+ ions in the culture media.[34] In my laboratories Schweitzer (unpublished data) found active lens transport of kynurenine. Kynurenine-incubated calf or rabbit lenses acquire a uniform yellowish tint with complete preservation of transparency similar to the color of clear lenses in humans. The absorption of light at 360 nm indicated kynurenine may be a natural filter for ultraviolet light in the human eye, thus protecting the retina.[132] With age, the gap between the ultraviolet absorption by the corneal epithelium (absorption maximum 290 nm) and the lens (absorption maximum 360 nm) may be closed by the progressive accumu-

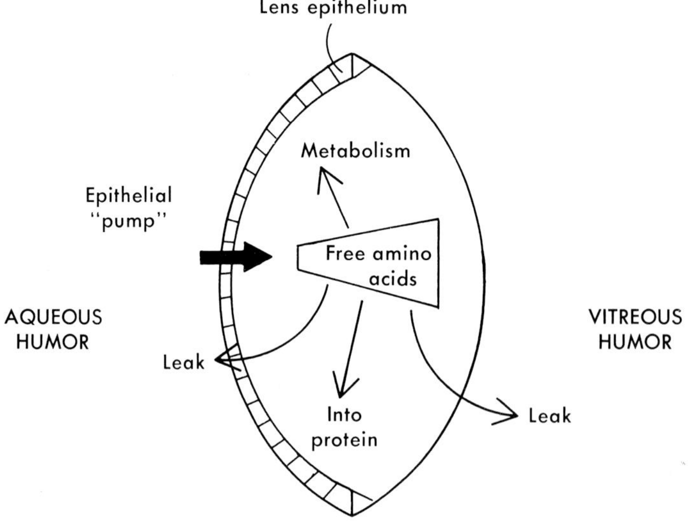

Fig. 10-12. Amino acids are actively transported into the lens mainly by an epithelial pump, and leak out of the lens through anterior and posterior surfaces. Free amino acids are metabolized or incorporated into protein to form new lens fibers.

lation in the lens of formylkynurenine (absorption maximum 325 nm) and anthranilic acid (absorption maximum 320 nm).

Glutathione, sulfhydryl proteins

Glutathione, a polypeptide, is synthesized in the lens. Glutathione is made of three amino acids: glycine, cysteine, and glutamic acid. The levels of glutathione in the lens are high, and most of lens glutathione is in the reduced form (GSH). Only 6.8% of all lens glutathione is in the oxidized form (GSSG).[102] GSH and GSSG are in equilibrium.

$$2\ GSH + \tfrac{1}{2}O_2 \rightleftarrows GSSG + H_2O$$

GSH lens concentrations (in micromoles per gram) are 2.80 in the monkey, 6.70 in the rat, 12.0 in the rabbit. GSH levels are much higher in the cortex than in the nucleus of the lens. In the human lens, GSH levels decrease slightly with age. Another polypeptide, ophthalmic acid, is found in the lens in concentrations of $^1/_{10}$ to $^1/_{100}$ those of GSH.[102]

The pentose shunt of glucose metabolism is active in the lens (see glucose metabolism below). The shunt generates NADPH (reduced nicotinamide-adenine dinucleotide phosphate) that maintains glutathione in the reduced state by the following reaction catalyzed by glutathione reductase.[75]

$$GSSG + NADPH + H^+ \leftrightharpoons 2\ GSH + NADP$$

Lens proteins contain reduced sulfhydryl groups (PSH) and oxidized disulfide groups (PSSP), maintaining high levels of GSH, as shown in the following reaction[62,75]:

$$2\ PSH + GSSG \rightarrow 2\ GSH + PSSP$$

Thus decreased GSH or increased GSSG will result in PSH oxidation and alterations in protein linkages, their solubility, and their transparency.[61,62] Sippel[112] localized PSH and PSSG within the membranes of the lens fibers.

The main roles of lens GSH are:

1. To preserve the physicochemical equilibrium of lens proteins by maintaining high levels of reduced sulfhydryl (SH^-) groups
2. To maintain transport "pumps" and the molecular integrity of lens fiber membranes

SH^- groups are needed for the enzyme Na^+, K^+-ATPase involved in lens cation transport. In addition, a role of GSH in membrane function and transport was uncovered by Epstein and Kinoshita,[52] who found that diamine (a GSH oxidant) reduced the activity of the ^{86}Rb pump and increased the ^{86}Rb efflux from the lens.

In human senile cataracts and all tested experimental cataracts there is loss of lens GSH. GSH decreases in lens are always associated with or precede disruption of the membranes of the lens fibers.

Lipids: transport, structure, and synthesis

The lipids of the human lens are unique and differ markedly from those of other animal species.[12] Lipids represent 3% to 5% of the dry weight of the lens.[53] In the human lens cholesterol represents approximately 50% of lipids, whereas phospholipids account for 45% and glycosphingolipids and ceramides for 5%. The lipids are major components of the lens fiber membranes, and either decrease in their synthesis or impaired degradation brings about membrane damage and lens opacities. Harris and Gruber[65] found that preventing cholesterol synthesis in triparanol-fed rats increases Na^+ ions and causes lens hydration. Cataracts also develop in humans treated with anticholesterolemic agents such as triparanol. On the other hand, cataracts are present in cholesterolemic xanthomatosis, an inborn error of cholesterol degradation.[120] Although fatty acids are rapidly transported into the rabbit lens and incorporated into lens phospholipids, cholesterol is not transported into the lens[13] and is derived primarily from acetate synthesis.[32] Esterification of cholesterol takes place in the human lens, where 25% of total cholesterol is in the ester form.[33,47]

Among phospholipids the human lens is particularly rich in sphingomyelin[12] and both sphingomyelin and its precursor ceramides are increased in senile cataracts.[36,122] Tao and Shen[123] found that ceramide synthesis proceeds via fatty acid and sphingosine. Its degradative enzyme, ceramidase, is present in the human lens. Sphingomyelin is degraded by sphingomyelinase, which is somewhat decreased in senile cataract,[32] but not absent, as in Niemann-Pick's disease. In the latter disorder, myelin-like figures are found in the lens by transmission electron microscopy, but distinct sutural opacities are rarely seen by ophthalmoscopy.[104] Lens abnormalities, however, are found in rat experimental models of Niemann-Pick's disease.[90]

The cholesterol-phospholipid ratio of human lens fiber membranes is the highest among cell or organelle membranes, thus conferring the lens resistance to deformation.[36] Lipids, as structural components of lens fiber membranes, are intimately associated with insoluble lens proteins.[9,36] The increased insolubility of the proteins with age or during cataract formation may be due to derangements in the stereochemical arrangement between lipids and proteins in the membrane and the soluble proteins inside the fibers.[33]

Ascorbic acid

In rabbits and guinea pigs ascorbic acid concentration is lower in the lens than in the aqueous. However, in other animal species, including humans, ascorbic acid values are higher in the lens than in the aqueous.[128] The role of ascorbic acid in the lens is not clear, but it could participate in oxidation-reduction reactions alone or coupled to glutathione.

Glucose metabolism

Glucose from the aqueous and vitreous diffuses into the lens and is rapidly metabolized through four main pathways (Table 10-1)[76,79]:

1. The glycolytic pathway
2. The Krebs (oxidative) cycle
3. The hexose monophosphate (pentose) shunt
4. The sorbitol pathway

End products of glucose metabolism are lactic acid, carbon dioxide, and water. Lactic acid from the lens diffuses to the aqueous and is eliminated via this circulating fluid.

Glucose metabolism generates adenosine triphosphate. ATP breakdown is required for active transport of ions and amino acids, maintenance of lens dehydration and lens transparency, and for continuous protein and GSH synthesis. In vitro the rabbit lens metabolizes 1 μmol of glucose/lens/hour. The pentose shunt does not generate ATP. However, it forms pentoses needed for ribonucleic acid synthesis. In addition, NADPH generated from the shunt is utilized to maintain lens glutathione in the reduced state. The pentose shunt is extremely active in the lens, a feature in common with red blood cells.[76] In addition, an active mechanism for pyruvate decarboxylation exists in the lens, which results in formation of carbon dioxide and acetaldehyde. The latter is metabolized through lens aldehyde dehydrogenase.[32] The carbon dioxide combination with water to form bicarbonate (CO_3H^-) is particularly active in the lens nucleus, where the levels of carbonic anhydrase exceed those in lens cortex.[16] However, carbonic anhydrase inhibitors do not produce cataracts.

The enzymes hexokinase and phosphofructokinase regulate the rate of glucose metabolism by the lens, whereas oxygen is not es-

Table 10-1. Glucose metabolism by lens

Pathway	Main intermediates	End products	Glucose through pathway (%)	Mol ATP gained/mol glucose metabolized
Glycolytic	Glucose 6-phosphate; fructose, 1,6-diphosphate; pyruvic acid	Lactic acid	80	2
Krebs cycle (oxidative) $+O_2$	Tricarboxylic acid	CO_2; H_2O	5	36
Pentose shunt	Pentoses	CO_2; NADPH	15	—
Sorbitol pathway	Sorbitol; fructose	Lactic acid*	Unknown	2

*Some lactic acid attributed to glucose metabolism by the glycolytic pathway may actually be formed through the sorbitol pathway.

sential for glucose metabolism.[19,20] Conversion of glucose to amino acids such as glutamic acid, aspartic acid, glycine, and others may account for 6% to 8% of glucose metabolism is the rat lens,[60] but not in the rabbit lens.[102] Hockwin and Korte[66] find oxygen consumption by the lens to be minimal: 0.5 μatom/g lens/hour. The Krebs cycle requires oxygen, and it is very inactive in the lens as there is a paucity of mitochondria and oxidative enzymes. In culture the lens can maintain all metabolic functions if provided with glucose only (but not if provided with oxygen only). If the lens is deprived of glucose, it will utilize its own endogenous energy reserves: ATP glucose, sorbitol, and fructose.[30,78]

When deprived of glucose, lenses in culture will gain water and lose transparency. Cataracts can develop in infantile hypoglycemia, a group of diseases in which low plasma glucose levels are present.[92] However, evidence to support the widely circulated concept that human senile cataracts may be caused by decreased glucose metabolism or lens anoxia is not available.

The levels of glucose are higher in the aqueous (5.5 μmol/ml) that in the lens (1 μmol/ml), and glucose diffuses readily into the lens. Transport of glucose into the lens is not affected by ouabain, absence of sodium or calcium ions in the incubation media, insulin, or epinephrine. However, other sugars or phloretin can inhibit lens glucose transport.[51] Thus the transport of glucose into the lens resembles that of the mature mammalian erythrocyte.

CATARACT

Human senile cataracts

The human lens normally undergoes changes with age: it slowly increases in size as new lens fibers develop throughout life; older lens fibers in the depths of the lens become dehydrated, compacted, and "sclerosed"; a yellow-brown pigment accumulates. The increase in the optical density of the nucleus tends to increase the refractive power of the lens so that less hyperopic spectacle correction may be needed in old age. The yellow-brown pigment may become so dense as to constitute nuclear sclerosis and later brunescent cataract. Cortical cataract, however, is the development of vacuoles and water clefts in the lens cortex that tend to increase in extent and in the advanced state give the lens a pearllike appearance. Approximately 60% of human beings have some alteration in lens transparency after 65 years of age. Progression of lens changes differs among individuals, and lens opacities can cause visual deficits in a shorter or longer period of time.

Cataract can be defined in terms of abnormal morphology or biochemistry, decreased light transmission, optical aberrations, decreased visual acuity; or all of these parameters. Morphologically, a cataract is any disturbance of the optical homogeneity of the lens. This definition includes vacuoles, water clefts, dense areas reflecting or refracting light, and punctate microscopic dots. The definition is narrow inasmuch as small congenital or developmental punctate cortical lens opacities may be seen in 10% to 20% of humans. Biochemical abnormalities may herald morphologic changes of the lens, as found in the experimental cataracts induced by galactose or diabetes. In humans, decreased light transmission is accompanied by opacities visible through an ophthalmoscope or slit lamp. Abnormalities of lens refraction, lenticular astigmatism, or optical aberrations due to increased refractive index of certain lens areas, however, may be detected only by retinoscopy. Documentation of differences in refractive index among various areas of the normal lens and of lens opacities is best accomplished through slit-lamp photography in vivo or stereophotography in vitro (Fig. 10-3). Opacities may extend toward the central area of the lens that is traversed by light and thus induce minor or major distortions in visual images and decreased vision. In humans the term cataracts carries the connotation of a visual disorder eventually leading to surgery and, thus, should be used with caution. Therefore, if your patients are those with minimal visual disabilities and slow-forming senile cataracts, the term "lens opacities" should be used. An all-encompassing biochemical interpretation of human senile cataract is not available. Current concepts on chemical lesions associated with certain morphologic changes of human cataracts will be discussed.

Cortical vacuoles, water clefts, and cortical lens opacities

The association of lens hydration and Na^+ ion imbalances in the senile cataract is not as clear-cut as in the various animal models. Nevertheless, several investigators found increased Na^+ ions levels in senile cataracts and more so in the advanced cataracts.[48,91,129,130] Although increased hydration is found in mature cataracts, the water content of incipient cataracts is not increased and may even be decreased.[32,130] During senile cataract formation, there is progressive loss of glutathione, inositol, and soluble protein, most likely due to abnormal transport and increased permeability.[128,130] Increased free amino acid levels in senile cataract may indicate accelerated proteolysis.[4] However, the plasma/aqueous humor levels of amino acids is not changed in senile cataracts.[110] An adverse environment, including gains in Ca^{2+} ions may trigger abnormalities in the molecular organization of lens proteins that eventually lead to protein and lipid aggregation and insolubilization.[69] Opacification in animal lenses can be explained on the basis of protein aggregation, a reversible or irreversible process.[115,116] Cortical opacities indicate irreversible protein or protein-lipid aggregation, and such aggregates may also be related to the changes in color to dark yellow or brown that take place in the nucleus of the lens. The insoluble protein of cataract, however, does not differ significantly in amino acid composition as compared to clear lenses.[21,121] Certain insoluble protein fractions of cataracts, however, are particularly rich in amino acids, cholesterol, and sphingomyelin. There is a moderate increase of these lipids and of ceramide in senile cataract, which indicates some similarities to the arteriosclerotic plaque.[36]

Yellowing of the lens nucleus and nuclear sclerosis

The nucleus of the lens becomes more compact and resists mechanical disruption with aging. In senile cataracts, changes in color to dark yellow, then to yellow-brown or brown, and hardening of the lens nucleus parallel decreased transparency.

Because of past failure to uncover specific changes in amino acid composition in yellow-brown cataracts, many theories on the pigment formation have been proposed. The majority of these theories revolve around the following:

1. Oxidation or photoxidation of tryptophan and/or tyrosine in lens proteins[97,140]
2. Formation of sugar–amino acid bonds with subsequently ''browning'' of lens proteins[109,119]
3. Formation of lipid peroxidation products[8,36]

Oxidation or photoxidation of tyrosine or tryptophan can be induced by ultraviolet light or enzymes. Exposure of intact lenses or lens proteins to the photoxidation products results in yellowing or browning of the protein. However, only the cortex area of animal lenses cultured in oxidized tyrosine or tryptophan acquires a brown color, whereas the nucleus remains clear and colorless. Furthermore, decreased tryptophan levels in cataracts should be found if photoxidation occurs. Actually, the opposite is found.[22,32,121] By contrast, total yellowing or browning of the lens results from exposure to glyceraldehyde or glycoaldehyde, which is caused by formation of sugar-amino acid bonds.[109] We found that acetaldehyde induces selected browning in the nucleus of the calf lens incubated for periods of weeks.[32]

Among repeatedly confirmed findings in the nuclear yellow or brown cataracts are the progressive decreases in reactive SH^- groups[61,62,85] and the presence of a fluorescent material in the insoluble lens proteins.[57,96] Further characterization of human lens fluorogens has been accomplished by isolation of fluorescent polypeptides by Garcia-Castaneiras and co-workers.[57] Angi, in our laboratories, isolated a similar fluorescent peptide rich in water-soluble amino acids—mainly aspartic acid and glutamic acid.[32] This yellow fluorescent polypeptide increased in concentration with progression of lens density or color in the senile cataracts and differed in amino acid composition from other soluble or insoluble proteins or peptides from the human lens. Previously, Van Heyningen[131] found kynurenine derivatives in the human lens, but their concentrations were not augmented in cataracts. Dillon and co-workers[42] have described β-carbolines as major fluorogens in human cataracts, but we have failed to detect them in human lens protein in well-controlled conditions.[32]

Because of the progressive increases in tryptophan and the FKN-kynurenine complex, a derangement of tryptophan metabolism probably occurs in the most advanced senile cataracts. We found the human and calf lenses, however, are unable to significantly metabolize tryptophan and thus the only route for tryptophan or kynurenine penetration is active transport into the lens. Nonenzymatic oxidation of tryptophan to FKN or kynurenine may also occur, but it is probably minimal due to the low oxygen content of lens. Pirie,[97] however, claims light could photoxidize lens tryptophan in situ with formation of formyl kynurenine.

Role of ultraviolet light and temperature in cataractogenesis

The susceptibility of the lens epithelium to ultraviolet light was first demonstrated by Duke-Elder.[48] Others have shown that ultraviolet irradiation induces changes in human or animal lens proteins either alone or in combination with kynurenine peptides with significant destruction of lens tryptophan and formation of β-carboline fluorescent pigments.[11,58,97,141] Epidemiologic evidence indicates ultraviolet light may contribute to cataractogenesis and early presbyopia.[26,101] The incidence of cataracts is significantly higher in certain areas of India and among Australian aborigines with high ultraviolet radiation exposures.[26,125] In at least one case Lerman[88] found a direct relationship between exposure to ultraviolet light and posterior subcapsular or cortical lens opacities. Girgus and co-workers,[58] however, have shown independence of lens pigmentation from ultraviolet light exposure in humans. Furthermore, senile cataracts develop in all climates, including areas with year-round cloudiness, and in individuals who have undergone minimal sunlight exposure throughout life.

The effect of sunlight elevating the lens and posterior chamber temperature must also be considered. Summer midday sunlight (10,000 lux) in New Haven, Connecticut, raises the temperature of the rabbit lens and posterior chamber by 3 ° to 6 ° C in pigmented rabbits but not in albino rabbits.[32] Increased temperature reduces the rate of mitotic activity in the cultured rabbit lens.[134]

Mature cataract

The complete opacification of the lens results from the complete disruption of lens chemistry. In mature cataracts, great increases in Na^+ ions and water are accompanied by losses of K^+ ions, free amino acids, glutathione, inositol, and soluble lens proteins. Marked increases in Ca^{2+} ions and the insoluble lens proteins are often found in mature cataracts.

Experimental cataract

Chemical changes

Chemical changes occurring during the development of experimental cataracts include:

1. Gain in Na^+ ions and water
2. Loss of K^+ ions, amino acids, and inositol
3. Loss of glutathione
4. Increase in Ca^{2+} ions
5. Decrease in soluble lens protein and increase in insoluble lens protein
6. Decreased activities of most enzymes; increased activity of proteolytic enzymes and glycosidases
7. Decreased ATP content

Damage to the membranes of the lens fibers represents the initial pathogenesis of most experimental cataracts. As the fibers become unable to extrude Na^+ ions, there are sizeable gains of this ion, which, added to other intracellular cations (mainly to K^+), increase intracellular osmolarity with subsequent water gains. This pattern follows the model of overhydration accepted for most cells.

Abnormal lens permeability is also found in x ray–, triparanol-, or aminotriazole-induced cataracts. Confluent fiber membrane damage produces the small vacuoles or clefts seen by slit-lamp examination. Concomitant lens losses of K^+ ions, amino acids, inositol, and glutathione occur, as the leak out of these compounds is greater than the total pumped into the lens. The progressive loss of these compounds coupled to Na^+ ion and Ca^{2+} ion gains sets up an adverse physiocochemical environment for the lens proteins. The synthesis of lens proteins may be totally or partially decreased early in cataract development. The increased permeability of the lens fibers allows loss of soluble lens proteins. Soluble lens proteins are also converted in increased amounts into insoluble protein. At this stage of cataractogenesis, decreased activities of glycolytic and other enzymes are found. However, the hydrolytic glycosidases that break down fiber membranes and proteases that cleave lens proteins may increase in activity. The generalized disarray of lens metabolism is accompanied by ATP loss. The end result is a totally opaque lens.

Regional sensitivity of the lens

An injury to the lens, physical or chemical, may involve only a region of the tissue, and a partial cataract may develop. The magnitude of the injury will determine whether this regional cataract will progress further and become total or will be "repaired" with growth of new, clear lens fibers on top of it (zonular cataract). Thus the age of the injury can be derived from the depth of an opacity within the lens stroma. The regional sensitivity of the lens has been investigated by the following:

1. Shielding of lens quadrants prior to irradiation with x rays
2. Localized mechanical trauma
3. Injection of surface-active detergents into the vitreous cavity
4. Discontinuous administration of a toxic agent (e.g., galactose) or discontinuous therapy for a cataractogenic disease (e.g., insulin for diabetes, calcium for hypoparathyroidism)

Peripheral cataracts. Mitotic activity and the transformation of epithelial cells into lens fibers and protein synthesis occur at the lens equator. Furthermore, active transport of K^+ and inositol and glutathione synthesis are found at the lens equator. Damage to the lens equator by trauma or x rays results in lens vacuolation localized initially only to this area. Vacuolation in the experimental carbohydrate cataracts produced by diabetes and galactose also begins at the lens equator.

Posterior subcapsular cataracts. Lens fibers extend in an arcuate fashion and form the posterior subcapsular area. Metabolic exchanges occur between the posterior lens area and the vitreous cavity. Toxic agents in the vitreous cavity cause vacuolation in the posterior subcapsular area. Such lesions are readily obtained after intravitreal injection of surface-active detergents or antibiotics. In other experimental cataracts (traumatic, x ray), injury to the lens equator with posterior migration of dead nuclei causes secondary posterior subcapsular cataracts.

Central anterior cataracts. Cataracts of the central anterior area result from proliferation of the central epithelial cells heaping up in layers and forming a pyramid, with the apex toward the center of the lens and the base resting on the capsule.

Nuclear cataracts. A massive continuous injury to the lens causes a total cataract. However, if the toxic agents are removed and the lens allowed to recover, new lens fibers cover the opaque tissue. The residual opacity is a zonular or apparently nuclear cataract.

Pie-shaped lesions. Only a quadrant or several clock hours of the lens are damaged. These cataracts can be produced by localized mechanical injury or by x-ray exposure to a quadrant of the lens. The quadrantic opacities are generally well repaired, and residual lesions remain as zonular cataracts.

Cataract production

Many agents, including trauma, chemicals, radiation, electricity, viruses, and vitamin or amino acids deficiencies can induce cataracts in humans or in experimental animals. Cataracts may be associated with skin, central nervous system, and skeletal diseases, chromosomal abnormalities, and other ocular disease or malformation. Our list is by no means exhaustive. A few cataractogenic agents and conditions have been subjected to extensive investigation, and of these the cataracts produced by ionizing radiation, diabetes, galactosemia, and surface-active agents have been selected for review.

Radiation. Ionizing radiation (x rays, gamma rays, neutrons, beta rays) induces lens vacuoles and lens opacities. Cataracts follow excessive irradiation of the unprotected human lens. Lead shields can provide adequate protection to the lens and are used when ocular and periocular tumors are irradiated. Occasionally, scattered radiation produces lens injuries. Atomic explosions cause lens damage.

The morphology of x-ray cataracts in the rabbit has been detailed by Cogan and Donaldson.[23,24] Initially, vacuoles appear in the equatorial and posterior subcapsular areas. Later, cortical opacities are found. Cataracts induced by x rays appear to result from a combination of several mechanisms.

1. Radiation effects on the permeability of the lens. Lenses exposed to radiation are more leaky to K^+, inositol, and glutathione. These effects of x rays may occur at the level of the membranes of the lens fibers. X rays are known to increase the permeability of other cells, most notably red blood cells and bacteria.
2. Radiation effects on the synthesis of lens proteins. Decreased incorporation of amino acids into proteins with decreased levels of soluble lens protein is produced by experimental x-ray irradiation of rats.[46]
3. Arrest of cells entering mitosis and cell damage in the lens epithelium. These findings were well documented by the pioneer studies of von Sallmann.[135]

Cataractogenic dosages of x rays vary with animal species and age. Lenses of younger animals are susceptible to dosages of radiation that have little or no effects in adult animals. Standard cataractogenic dosages in rabbits range from 1000 to 4000 roentgens delivered to the lens area. Mature cataracts are induced with the higher dosages only. In humans, however, cataracts can result from irradiations with 300 roentgens or higher.

The animal lens can be protected against radiation if thiourea, glutathione, or cysteine is given prior to irradiation.[135] A high concentration of free sulfhydryl (SH^-) groups in the lens or aqueous humor may provide radioresistance.

Diabetic cataract. Increased levels of glucose in the aqueous and lens are found in patients with diabetes mellitus. In general, the glucose concentration in the aqueous parallels the concentrations in the plasma. From the aqueous, glucose diffuses rapidly into the lens. The lens metabolizes glucose through the four main pathways described on p. 290.[76,79] Basically, the initial rate of glucose metabolism by the lens is determined by hexokinase, a regulatory enzyme. In diabetes excessive glucose in the lens (over 200 mg/100 ml) saturates hexokinase. Glucose then piles up and is converted in part into sorbitol and fructose, as shown below (= indicates saturation).[79]

Aldose reductase catalyzes the glucose-to-sorbitol reaction. In this reaction NADPH is reduced to NADP. A major source of NADPH in the lens is the pentose-phosphate shunt.[75] High levels of sorbitol and fructose were found in diabetic animal lenses by Kuck[87] and in human diabetic lenses by Pirie and van Heyningen.[98] The excess of sorbitol is also found in rabbit lenses cultured in vitro in high-glucose media.[79]

The effects of excessive sorbitol and fructose in the diabetic lens result in increased hydration and Na^+ ions gains with subsequent loss of K^+ ions, amino acids, and inositol.[77] The swelling of the lens has led Kinoshita[77] to postulate that high levels of sorbitol in the diabetic lens draw water, thus rupturing lens fibers and causing vacuolation in lens cortex.

Varma and Kinoshita[133] succeeded in slow-

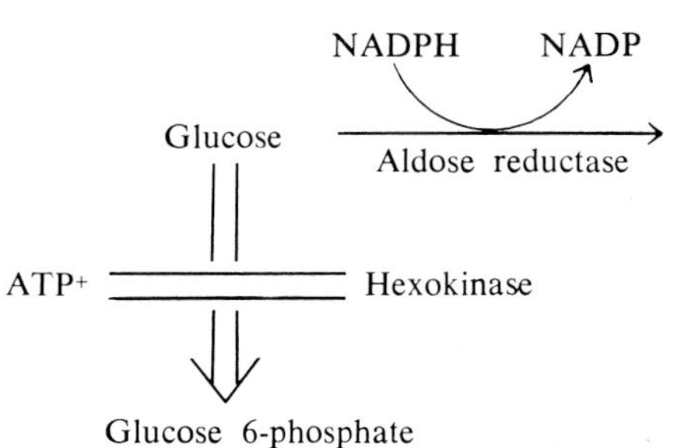

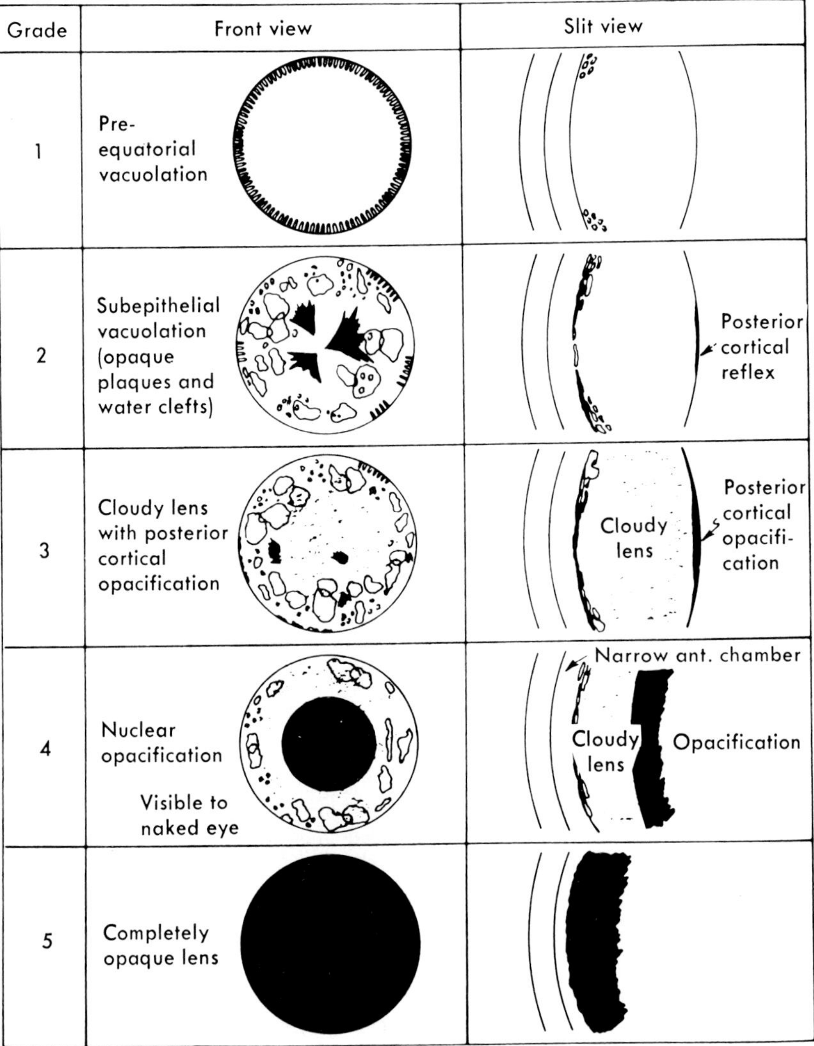

Fig. 10-13. Cataract stages in diabetes or galactosemia in rat.

ing down the development of diabetic cataracts in animals by administration of flavonoids, which decrease sorbitol formation by inhibition of aldose reductase. Furthermore, Chylack and Cheng[20] inhibited sorbitol formation in human lenses incubated in media containing high glucose levels by adding aldose reductase inhibitors. The progression of diabetic cataracts in alloxan-diabetic rats is shown in Fig. 10-13. The rate of cataract progression depends on the glucose levels in serum; the higher the glycemia, the faster the cataracts become mature.[95] However, cataracts in rats or rabbits can be prevented if the blood sugar levels are lowered by insulin or other hypoglycemic agents.

The glycemia of the majority of diabetics is well controlled prior to surgery, and thus the sorbitol and fructose cataract levels are minimally elevated in some patients and within normal limits in others.[98] However, it would appear the accumulation of sugar alcohol provides a ready explanation for the osmotic hydration that occurs in acute-onset high-glycemia diabetes. In humans refractive errors, mainly acute myopia, caused by increased lens refraction are reversible following a lowering of glycemia to normal levels. In rabbits, however, experimental diabetes is accompanied by hyperopia, which can be reversed by aldose reductase inhibitors.

In humans diabetic cataracts are rarely seen in juveniles but may appear suddenly in adults with uncontrolled diabetes mellitus. Vacuoles or clefts in the lens cortex may "heal," leaving small, snowflake cortical opacities. Snowflake cortical lens opacities, also known as "metabolic cataracts," are found in diabetes, hypoparathyroidism, and myotonic dystrophy. Diabetes, overt or borderline, is prevalent among older patients with senile cataracts. From 10% to 15% of patients admitted to hospitals for senile cataract surgery in Oxford, England, had abnormal plasma glucose levels or overt diabetes mellitus.[15]

Although sugar alcohols accumulate in both human and experimental cataracts, it would appear other mechanisms are involved in the slight acceleration of senile cataract formation experienced by adult-onset overt or borderline diabetics. The known damage to basement membrane in these patients may be the cause of the decreased fragility of their lens capsule.[15] Furthermore, excessive red blood cell glycosylation is found in uncontrolled diabetics, and it may have an effect on the lens as well. Recently, Stevens and co-workers[119] found evidence of increased protein glycosylation by animal lenses incubated in high-glucose media. The formation of sugar–amino acid (lysine) bonds may result in protein aggregation and decreased transparency. Ansari and co-workers,[2] however, have failed to link lens glycosylation to formation of S-S bonds or the aggregation of lens proteins.

Galactosemic cataract. Galactosemic cataracts result from an excess of galactose, which along with glucose results from the hydrolysis of lactose in the serum and aqueous humor. Galactosemia in humans is the result of an inborn error of metabolism that leads to excessive galactose in serum when milk is ingested. Two clinical forms of galactosemia are recognized[111]: (1) classic galactosemia in which the enzyme galactose 1-phosphate uridyl transferase is absent and (2) galactokinase deficiency.

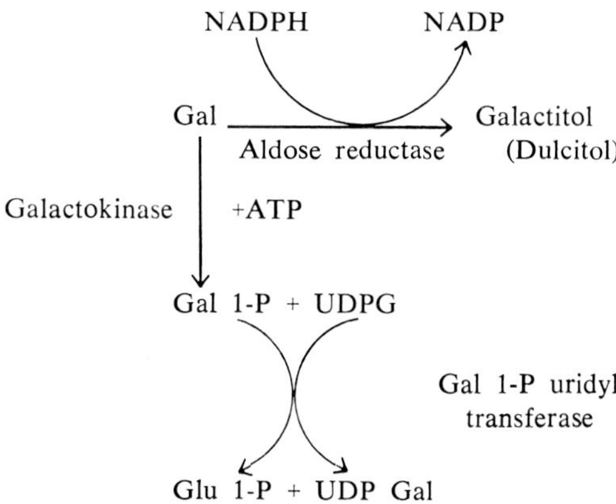

The pathway of galactose metabolism in human tissue is shown above.

Excessive galactose in the serum and the lens results in a shift toward the formation of galactitol. Excessive galactitol is excreted in urine by homozygotes for classic galactosemia or galactokinase deficiency. The accumulation of galactitol has deleterious effects on the lens and leads to cataract formation. These cataracts are easily produced in rats or hamsters fed excessive amounts (25% to 60%) of galactose in the diet. The morphologic changes of galactose cataracts in rats are shown in Fig. 10-13 and are similar to those found in diabetes. Indeed, it is likely that both diabetic and galactosemic cataracts result from the excessive amounts of sugar alcohols drawing water into the lens. The increased water content of galactosemic rat lenses parallels the galactitol accumulations as shown in Fig. 10-14. Water gains by the lens fibers rupture their membranes, with resultant loss of K^+, amino acids, and inositol. The biochemical changes that follow are those common to all other experimental cataracts and are discussed above.

Cataract due to plasma phospholipids in uveitis and after intraocular surgery. The preservation of an intact blood-aqueous barrier is fundamental for lens viability and transparency. In uveitis or following anterior segment eye surgery the blood-aqueous barrier is broken with influx of plasma components into the aqueous humor. Lens vacuoles or lens opacities develop in human or animal eyes with prolonged breakdown of the blood-aqueous barrier. Furthermore, vitrectomy surgery or break-

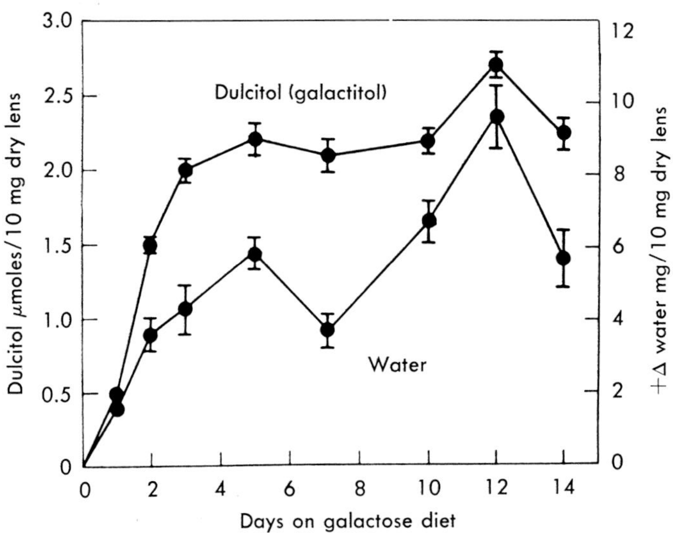

Fig. 10-14. Accumulation of dulcitol (galactitol) and water by the lens of young rats fed a high-galactose diet. (Data from Kinoshita.[77])

down of the blood-vitreous barrier can lead to posterior cortical or subcapsular cataracts. Plasma proteins do not appear to be toxic to the lens. However, plasma phospholipids, specifically, lysophosphatidyl choline (LPC) can induce cataracts.[37] In the intact rabbit or human eye the concentration of LPC is only 1/10 to 1/50 that in the plasma. In uveitis the levels of LPC increase to cataractogenic levels (10μg/ml or higher). LPC lyses membranes of lens fibers either directly or through the marked increases in lens Na^+ ions and water that it induced. LPC penetrates readily into the cultured lens, depending on the protein concentration in the incubating media,[37] inducing hydration and increased efflux of K^+ and inositol.[70] At higher protein media concentrations, LPC penetration into the lens decreases. LPC appears to be the major cataractogenic factor in uveitis, depending on time of exposure and aqueous humor plasma levels. Phospholipase A, which is also cataractogenic, is found in leukocytes. The enzyme hydrolyzes phosphatidyl choline with formation of LPC.

$$\text{Phosphatidyl choline (PC)} \xrightarrow{\text{Phospholipase A}} \text{LPC} + \text{Fatty acid}$$

Penetration of phospholipase A into the lens hydrolyzes PC in the membranes of lens fibers and results in direct formation of LPC. Products of inflammation such as lymphokines or hydrolytic enzymes derived from leukocytes may also contribute to lens membrane damage in eyes with uveitis, but their role remains to be studied.

Cataract with storage of interstitial glycoprotein (oligosaccharide) or glycolipid. The chemical composition of the interstitial substance between lens fibers is unknown. The space among fibers, however, is large enough to be optically distinct as a suture or area of junction in most mammalian lenses. Increased density of the sutures or deposits in the perisutural areas is found in a series of glycoprotein or glycolipid storage diseases, indicating such substances are normally metabolized by the lens. A variety of lysosomal hydrolases including α-mannosidase, α-galactosidase, sialidase, and sphingomyelinase occur in lens and lens fiber membrane.[17,32] Among persons with inborn errors of glycolipid metabolism, hemizygotes and heterozygotes for Fabry's disease show increased density of the sutures.[55] Patients with Fabry's disease store ceramide trihexoside in their tissues due to deficiency of α-galactosidase.[29] In mannosidosis, an error of glycoprotein (oligosaccharide) metabolism, sutural lens opacities are present due to deficient α-mannosidase,[94] whereas punctate lens opacities are found in type II sialidosis with deficient sialidase (α-neuraminidase).[89] Pathways for sialic acid synthesis and degradation exist in the mammalian lens.[73,126] Patients with sialidosis present corneal opacities, cherry-red spots, and Hurler's syndrome–like facies, an indication of the generalized disturbance of the oligosaccharide metabolism. Massive sialic acid oligosacchariduria is found.[89]

Cataract and oxygen or peroxide toxicity. Superoxide anion free radical (O_2^-) or its derivatives peroxide (H_2O_2), singlet oxygen ($'O_2$), and OH induce oxidative damage to a variety of cells. The lens is highly susceptible to these radicals, and mice exposed to hyperbaric oxygen develop cataracts. Peroxide (H_2O_2) is catalyzed through catalase and peroxidase and synthesized through superoxide dismutase (SOD).

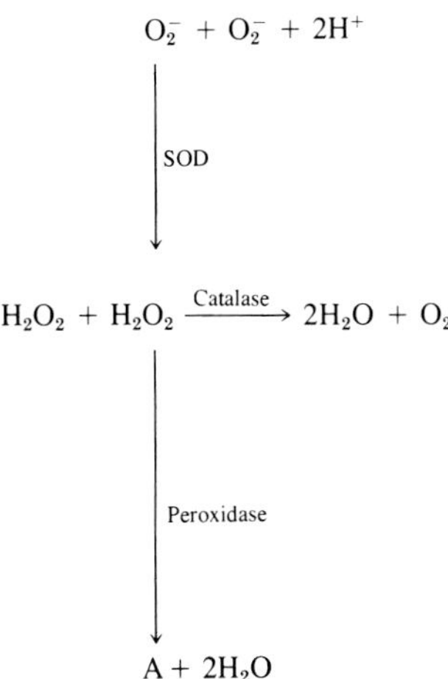

Bhuyan and Bhuyan[7] have shown that cataracts result from catalase inhibition with 3-aminotrizole fed to rabbits or added to lens culture media. Peroxide inhibits ^{86}Rb lens transport and induces lens membrane damage. The accumulation of peroxide has not been found in human cataracts, and human disorders with deficient leukocyte peroxidase are not accompanied by cataracts. It is possible, however, that peroxidation of lipids may take place in the lens. Recently, Bhuyan and co-workers[8] found malonaldehyde in human lens (0.91 nmol/g) and increased levels of malonaldehyde in senile cortical cataract (4.07 nmol/g), an indication of lipid peroxidation in the tissue. The theory of lipid peroxidation is based on the premise that peroxide oxidation of polyunsaturated fatty acids such as arachidonic acid or phosphatidyl ethanolamine originates short-chain aldehydes (such as malonaldehyde), which could then bind to reactive amino groups or sulfhydryl groups.[124]

Lens lipids, however, are not particularly rich in these fatty acids and phospholipids.[12,33] The proportion of arachidonic acid decreases in senile cataract.[118] The above-mentioned contradictions cast some doubts as to whether peroxidation represents a major mechanism in human cataract formation. Furthermore, if high levels of peroxide formation in aqueous humor were to take place, it would be highly toxic to the corneal endothelium.

REFERENCES

1. Anderson, E. I., and Spector, A.: The state of sulfhydryl groups in normal and cataractous human lens proteins. I, Nuclear region, Exp. Eye Res. **26:**407, 1978.
2. Ansari, N. H., Awasthi, C., and Srivastava, S. K.: Role of glycosylation of human lens crystallin in protein disulfide and cataract formation, Invest. Ophthalmol. (ARVO suppl.) p. 98, April 1979.
3. Ashton, N., et al.: Electron microscopic study of pseudoexfoliation of the lens capsule. I, Lens capsule and zonular fibers, Invest. Ophthalmol. **4:**141, 1968.
4. Barber, G. W.: Free amino acids in senile cataractous lenses: possible osmotic etiology, Invest. Ophthalmol. **7:**564, 1968.
5. Becker, B.: Accumulation of rubidium-86 by the rabbit lens, Invest. Ophthalmol. **1:**502, 1962.
6. Becker, B., and Cotlier, E.: The efflux of ^{86}rubidium from the rabbit lens, Invest. Ophthalmol. **4:**117, 1965.
7. Bhuyan, K. C., and Bhuyan, D. K.: Superoxide dismutase of the eye. Relative functions of superoxide dismutase and catalase in protecting the ocular lens from oxidative damage, Biochim. Biophys. Acta **542:**28, 1978.
8. Bhuyan, K. C., Bhuyan, D. K., and Podos, S. M.: Lipid peroxidation and membrane damage in the pathogenesis of cataract, Invest. Ophthalmol. (ARVO suppl.) p. 97, April 1979.
9. Bloemendal, H., et al.: The plasma membranes of eye lens fibers. Biochemical and structural characterization, Cell Differ. **1:**91, 1972.
10. Bonting, S. J.: Na^+-K^+ activated adenosine-triphosphatase and active cation transport in the lens, Invest. Ophthalmol. **4:**723, 1965.
11. Borkman, F. R., Dalrymple, A., and Lerman, S.: Ultraviolet action spectrum for fluorogen production in the ocular lens, Photochem. Photobiol. **26:**129, 1977.
12. Broekhuyse, R. M.: Phospholipids in tissues of the eye. III, Composition and metabolism of phospholipids in human lens in relation to age and cataract formation, Biochim. Biophys. Acta **187:**354, 1969.

13. Broekhuyse, R. M., and Bogemann, B.: Lipids in tissues of the eye. XVI, Uptake of lipids by the rabbit lens in vitro, Exp. Eye Res. **26**:567, 1978.
14. Broekhuyse, R. M., and Kuhlmann, E. D.: Lens membranes. I, Composition of urea-treated plasma membranes from calf lens, letter to the editor, Exp. Eye Res. **19**:297, 1974.
15. Caird, F. I., Pirie, A., and Ramsell, T. F.: Diabetes and the eye, Oxford, 1968, Blackwell Scientific Publications, Ltd.
16. Canade, M. R., and Bonting, S. L.: Carbonic anhydrase distribution in rabbit lens, Exp. Eye Res. **4**:283, 1965.
17. Carlin, R., and Cotlier, E.: Glycosidases of the crystalline lens, Invest. Ophthalmol. **10**:887, 1971.
18. Charlton, J., and van Heyningen, R.: An investigation into the loss of proteins of low molecular size from the lens in senile cataract, Exp. Eye Res. **7**:47, 1968.
19. Chylack, L. T., Jr.: Mechanism of "hypoglycemic" cataract formation in the rat lens. I, The role of hexokinase instability, Invest. Ophthalmol. **14**:746, 1975.
20. Chylack, L. T., Jr., and Cheng, H.: Sugar metabolism in the crystalline lens, Survey Ophthalmol. **23**: 26, 1978.
21. Clark, R., Zigman, S., and Lerman, S.: Studies on the structural proteins of the human lens, Exp. Eye Res. **8**:172, 1969.
22. Ciusa, W., Cristini, G., and Barbiroli, G.: The transmethylation processes in the aqueous humour and in the lens, Exp. Eye Res. **3**:169, 1964.
23. Cogan, D. G., and Donaldson, D. D.: Experimental radiation cataracts. I, Cataracts in the rabbit following single X-ray exposure, Arch. Ophthalmol. **45**: 508, 1951.
24. Cogan, D. G., et al.: Experimental radiation cataract. III, Further experimental studies on X-ray and neutron irradiation of the lens, Arch. Ophthalmol. **50**:547, 1953.
25. Cotlier, E.: The mitotic cycle of the lens epithelium: effect of age and galactose, Arch. Ophthalmol. **68**:801, 1962.
26. Cotlier, E.: Histo- and cytochemical localization of dehydrogenases in rabbit lens epithelium, J. Histochem. **12**:419, 1964.
27. Cotlier, E.: Myo-inositol: active transport by the crystalline lens, Invest. Ophthalmol. **9**:681, 1970.
28. Cotlier, E.: Lysine transport and protein incorporation by the lens, Biochim. Biophys. Acta **241**:798, 1971.
29. Cotlier, E.: Biochemical detection of inborn errors of metabolism affecting the eye, Trans. Am. Acad. Ophthalmol. Otolaryngol. **76**:1167, 1972.
30. Cotlier, E.: Glycolytic intermediates in the lens: effect of external energy deprivation and osmolarity, Life Sci. **11**:733, 1972.
31. Cotlier, E.: Rb^{86} transport, water, and cation concentrations in lenses exposed to polymyxin B sulfate, Exp. Eye Res. **15**:711, 1973.
32. Cotlier, E., and Angi, M.: Unpublished observations.
33. Cotlier, E., and Apple, D.: Cataracts induced by the polypeptide antibiotic polymyxin B sulfate, Exp. Eye Res. **16**:69, 1973.
34. Cotlier, E., and Beaty, C.: The role of Na^+ ions in the transport of α-aminoisobutyric and other amino acids into the lens, Invest. Ophthalmol. **6**:64, 1967.
35. Cotlier, E., Kwan, B., and Beaty, C.: The lens as an osmometer and the effects of medium osmolarity on water transport, Rb^{86} transport by the lens, Biochim. Biophys. Acta **150**:705, 1968.
36. Cotlier, E., Obara, Y., and Toftness, B.: Cholesterol and phospholipids in protein fractions of human lens and senile cataract, Biochim. Biophys. Acta **530**:267, 1978.
37. Cotlier, E., et al.: Lysophosphatidyl choline and cataracts in uveitis, Arch. Ophthalmol. **94**:1159, 1976.
38. Devi, A., Friel, R., and Lerman, S.: The RNA content and the amino acid-RNA incorporating system in the lenses of various species, Arch. Ophthalmol. **65**:855, 1961.
39. Dhir, S. P., Detels, R., and Alexander, E. R.: The role of environmental factors in cataract, pterygium, and trachoma, Am. J. Ophthalmol. **64**:128, 1967.
40. Dilley, K. J.: Loss of tryptophan associated with photo-polymerization and yellowing of proteins exposed to light over 300 nm., Biochem. J. **133**:821, 1973.
41. Dilley, K. J., and Pirie, A.: Changes to the proteins of the human lens nucleus in cataract, Exp. Eye Res. **19**:59, 1974.
42. Dillon, J., Spector, A., and Nakanishi, K.: Identification of β carbolines isolated from fluorescent human lens proteins, Nature **29**:422, 1976.
43. Dische, Z.: The glycoproteins and glycolipoproteins of the bovine lens and their relation to albuminoid, Invest. Ophthalmol. **4**:759, 1965.
44. Dische, Z., and Ehrlich, G.: The breakdown of glucose and its phosphoric esters in the bovine lens capsule. Part 2, Am. J. Ophthalmol. **39**(2):99, 1955.
45. Dische, Z., Devi, A., and Zelmenis, G.: Influence of age and cataract formation on the ribonucleic acid of the lens, Am. J. Ophthalmol. **51**:993, 1961.
46. Dische, Z., et al.: Changes in proteins and protein synthesis. In tryptophan deficiency and radiation cataracts of rats. Part 2, Am. J. Ophthalmol. **47**(1): 368, 1959.
47. Dragomirescu, V., et al.: Development of a new equipment for rotating slit image photography according to Scheimpflug's principle, Interdisc. Top. Gerontol. **13**:1, 1978.
48. Duke-Elder, W. S.: The pathological action of light upon the eye. Part II, Action upon the lens: theory of the genesis of cataract, Lancet **1**:1188, 1926.
49. Duncan, G., and Bushell, A. R.: Ion analyses of human cataractous lenses, Exp. Eye Res. **20**:223, 1975.
50. Dvorak-Theobald, G.: Pseudo-exfoliation of the lens capsule: relation to "true" exfoliation of the lens

capsule as reported in the literature and role in the production of glaucoma capsulocuticulare, Am. J. Ophthalmol. **37:**1, 1954.

51. Elbrink, J., and Bihler, I.: Characteristics of the membrane transport of sugars in the lens of the eye, Biochim. Biophys. Acta **282:**337, 1972.
52. Epstein, D., and Kinoshita, J. H.: The effect of diamide on lens glutathione and lens membrane function, Invest. Ophthalmol. **9:**629, 1970.
53. Feldman, G. L.: Lipids of the human lens. In Dardenne, M. U., and Nordmann, J., editors: Biochemistry of the eye. Symposium, Tutzing Castle, 1966, Basel, Switzerland, 1968, S. Karger AG., pp. 348-358.
54. François, J., Rabaey, M., and Stockmans, L.: Gel filtration of the soluble proteins from normal and cataractous human lens, Exp. Eye Res. **4:**312, 1965.
55. Franceschetti, A.: Fabry disease: ocular manifestations, original article series (Bergsma, D., Bron, A. J., and Cotlier, E., editors), Birth Defects **12**(3): 195, 1976.
56. Friedenwald, J. S.: Permeability of the lens capsule with special reference to the etiology of senile cataract, Arch. Ophthalmol. **3:**182, 1930.
57. Garcia-Castineiras, S., Dillon, J., and Spector, A.: Non-tryptophan fluorescence associated with human lens protein; apparent complexity and isolation of bityrosine and anthranilic acid, Exp. Eye Res. **26:** 461, 1978.
58. Girgus, J. S., Coren, S., and Porac, C.: Independence of in vivo human lens pigmentation from U.V. light exposure, Vision Res. **17:**749, 1977.
59. Goldmann, H.: Senile changes of the lens and the vitreous, Am. J. Ophthalmol. **57:**1, 1964.
60. Hammar, H.: The formation of amino acids in vitro from glucose-U-^{14}C in the eye lens of rats and the influence of sodium fluoride and alloxan diabetes, Acta Ophthalmol. **14:**1, 1965.
61. Harding, J. J.: Free and protein-bound glutathione in normal and cataractous human lenses, Biochem. J. **117:**957, 1970.
62. Harding, J. J., and Dilley, K. J.: Structural proteins of the mammalian lens: a review with emphasis on changes in development, aging, and cataract, Exp. Eye Res. **22:**1, 1976.
63. Harding, C. V., Donn, A., and Srinivasan, B. D.: Incorporation of thymidine by injured lens epithelium, Exp. Cell Res. **18:**582, 1959.
64. Harris, J. E., and Becker, B.: Cation transport of the lens, Invest. Ophthalmol. **4:**709, 1965.
65. Harris, J. E., and Gruber, L.: Reversal of triparanol-induced cataracts in the rat, Invest. Ophthalmol. **11:** 608, 1972.
66. Hockwin, O., and Korte, I.: Biochemistry of the various catabolic pathways of the lens. In Dardenne, M. U., and Nordmann, J., editors: Biochemistry of the eye. Symposium, Tutzing Castle, 1966, Basel, Switzerland, 1968, S. Karger AG, pp. 216-225.
67. Hogan, M. J., and Zimmerman, L. E.: Diseases of the lens. In Hogan, M. J., and Zimmerman, L. E., editors: Ophthalmic pathology, ed. 2, Philadelphia, 1962, W. B. Saunders Co.
68. Hogan, M. J., Alvarado, J. A., and Weddell, J. E.: Lens. In Histology of the human eye, Philadelphia, 1971, W. B. Saunders Co.
69. Jedziniak, J. A., et al.: On the presence and mechanism of formation of heavy molecular weight aggregates in human normal and cataractous lenses, Exp. Eye Res. **15:**185, 1973.
70. Kador, P. F., and Kinoshita, J. H.: Phospholipid effects on the rat lens transport systems, Exp. Eye Res. **26:**657, 1978.
71. Kahn, H. A., et al.: The Framingham eye study. II, Association of ophthalmic pathology with single variables previously measured in the Framingham heart study, Am. J. Epidemiol. **106:**33, 1977.
72. Kaufman, P. L., and Axelsson, V.: Induction of subcapsular cataract in aniridic vervet monkey by echothiophate, Invest. Ophthalmol. **14:**863, 1975.
73. Kean, K. L.: Sialic acid activating enzyme in ocular tissue, Exp. Eye Res. **8:**44, 1969.
74. Kern, H. L.: Transport of organic solutes in the lens. In Zadunaisky, J., and Davson, H., editors: Current topics in eye research, New York, 1979, Academic Press, Inc.
75. Kinoshita, J. H.: Selected topics in ophthalmic biochemistry, Arch. Ophthalmol. **72:**554, 1964.
76. Kinoshita, J. H.: Pathways of glucose metabolism in the lens, Invest. Ophthalmol. **4:**619, 1965.
77. Kinoshita, J. H.: Cataracts in galactosemia, Invest. Ophthalmol. **4:**786, 1965.
78. Kinoshita, J. H., Merola, L. A., and Kern, H.: Factors affecting the cation transport of calf lens, Biochim. Biophys. Acta **47:**458, 1961.
79. Kinoshita, J. H., et al.: Factors affecting the formation of sugar alcohols in the ocular lens, Biochim. Biophys. Acta **74:**340, 1963.
80. Kinsey, V. E.: The pump-leak concept of transport in the ocular lens. In Rohen, J. W., editor: Eye structure II. Symposium, Stuttgart, West Germany, 1965, FK Schattauer Verlag, pp. 383-394.
81. Kinsey, V. E., and McGrady, A. V.: Studies on the crystalline lens. XVII, Bioelectric potentials as affected by cationic compositions of bathing medium, Invest. Ophthalmol. **4:**282, 1971.
82. Kinsey, V. E., et al.: Studies on the crystalline lens. Mitotic activity in the epithelia of lenses cultured in various media, Am. J. Ophthalmol. **40:**216, 1955.
83. Kirsch, R. E.: Glaucoma following cataract extraction associated with use of alpha-chymotrypsin, Arch. Ophthalmol. **72:**612, 1964.
84. Klang, G.: Measurements and studies of the fluorescence of the human lens in vivo, Acta Ophthalmol. **31**(suppl.)1, 1948.
85. Kramps, J. A., Hoenders, H. J., and Wollensak, J.: Increase of nondisulfide cross-links during progress of nuclear cataract, Exp. Eye Res. **27:**731, 1978.
86. Krwawicz, T.: Intracapsular extraction of intumescent cataract by application of low temperature, Br. J. Ophthalmol. **45:**279, 1961.

87. Kuck, J. F. R., Jr.: The formation of fructose in the ocular lens, Arch. Ophthalmol. **65**:840, 1961.
88. Lerman, S.: Ultraviolet radiation and human cataractogenesis, Invest. Ophthalmol. (ARVO suppl.), p. 128, April 1979.
89. Loewen, J. A., and O'Brien, J. S.: Sialidosis: a review of human neuraminidase deficiency, Am. J. Hum. Genet. **31**:1, 1979.
90. Lullmann, H., Lullmann-Rauch, R., and Brady, R.: Niemann-Pick disease experimental model, Science **200**:568, 1978.
91. Maraini, G., and Orsoni, J. G.: Membrane water permeability in normal and cataractous human lenses, Exp. Eye Res. **20**:121, 1975.
92. Merin, S., and Crawford, J. S.: Hypoglycemia and infantile cataract, Arch. Ophthalmol. **86**:495, 1971.
93. Michon, J., and Kinoshita, J. H.: Experimental miotic cataract. I, Effects of miotics on lens structure, cation content, and hydration, Arch. Ophthalmol. **79**:79, 1968.
94. Murphree, A. L., et al.: Cataract in mannosidosis, original article series (Bergsma, D., Bron, A. J., and Cotlier, E., editors), Birth Defects **12**(3):319, 1976.
95. Patterson, J. W.: Development of diabetic cataracts. Part 2, Am. J. Ophthalmol. **35**(5):68, 1952.
96. Pirie, A.: Color and solubility of the proteins of human cataracts, Invest. Ophthalmol. **7**:634, 1968.
97. Pirie, A.: Formation of N-formylkynurenine in proteins from lens and other sources by exposure to sunlight, Biochem. J. **125**:203, 1971.
98. Pirie, A., and van Heyningen, R.: The effect of diabetes on the content of sorbitol, glucose, fructose and inositol in the human lens, Exp. Eye Res. **3**:124, 1964.
99. Rae, J. L.: The electrophysiology of the crystalline lens (Zadunaisky, J., and Davson, H., editors), Curr. Top. Eye Res. **1**:37, 1979.
100. Rambo, V. C.: Further notes on the varying ages at which different peoples develop presbyopia, Am. J. Ophthalmol. **36**:709, 1953.
101. Rambo, V. C.: A study of the accommodation of the people of India (with further notes on the development of presbyopia at different ages in different people), Am. J. Ophthalmol. **49**:993, 1960.
102. Reddy, D. V. N.: Metabolism of glutathione in the lens, Exp. Eye Res. **11**:310, 1971.
103. Reddy, D. V. N., and Kinsey, V. E.: Studies on the crystalline lens. Quantitative analysis of free amino acids and related compounds, Invest. Ophthalmol. **1**:635, 1962.
104. Robb, R. M., and Kuwabara, T.: The ocular pathology of Type A Niemann-Pick disease. A light and electron microscopic study, Invest. Ophthalmol. **12**:366, 1973.
105. Roberts, D. St. C.: Studies on the antigenic structure of the eye using the fluorescent antibody technique, Br. J. Ophthalmol. **41**:338, 1957.
106. Sandberg, H. O.: The alpha-crystallin content of aqueous humor in cortical, nuclear, and complicated cataracts, Exp. Eye Res. **22**:75, 1976.
107. Sanders, D., et al.: Cataracts induced by surface active agents, Exp. Eye Res. **19**:35, 1974.
108. Satoh, K., Bando, M., and Nakajima, A.: Fluorescence in the human lens, Exp. Eye Res. **16**:167, 1973.
109. Scharff, T. G., and Montgomery, O. L.: Pigmentation and enzymic inhibition by carbonyl compounds in the rabbit lens, Biochem. Pharmacol. **21**:1799, 1972.
110. Schonheyder, F., Ehlers, N., and Hust, B.: Remarks on the aqueous humor/plasma ratios for amino acids and related compounds in patients with various chronic ocular disorders, Acta Ophthalmol. **53**:627, 1975.
111. Segal, S.: Disorders of galactose metabolism. In Stanbury, J. B., et al., editors: The metabolic basis of inherited disease, ed. 3, New York, 1972, McGraw-Hill Book Co., pp. 174-195.
112. Sippel, T. O.: Histochemistry of protein sulfhydryl and disulfide groups in the lens, Exp. Eye Res. **11**:383, 1971.
113. Spector, A.: A study of peptidase and esterase activity in calf lens, Exp. Eye Res. **1**:330, 1962.
114. Spector, A.: Methods of isolation of alpha, beta and gamma crystallins and their subgroups, Invest. Ophthalmol. **3**(3):182, 1964.
115. Spector, A.: Aggregation of alpha-crystallin and its possible relationship to cataract formation, Isr. J. Med. Sci. **8**:1577, 1972.
116. Spector, A., Roy, D., and Stauffer, J.: Isolation and characterization of an age-dependent polypeptide from human lens with non-tryptophan fluorescence, Exp. Eye Res. **21**:9, 1975.
117. Spiro, R. G., and Fukushi, S.: The lens capsule: studies on the carbohydrate units, J. Biol. Chem. **244**(8):2049, 1969.
118. Sternberg, P., Cotlier, E., and Obara, Y.: Lysophosphatidyl ethanolamine effects on rabbit lens and levels in human lenses and senile cataracts, Exp. Eye Res. **26**:729, 1978.
119. Stevens, V. J., et al.: Diabetic cataract formation: potential role of glycosylation of lens crystallins, Proc. Natl. Acad. Sci. USA **75**:2918, 1978.
120. Swartz, M., Salem, G., and Burman, K. D.: Cerebrotendinous xanthomatosis: a treatable lethal disease with cataracts, Invest. Ophthalmol. (ARVO suppl.), p. 71, April 1979.
121. Takemoto, L., and Azari, P.: Amino acid composition of normal and cataractous human lens proteins, Exp. Eye Res. **23**:1, 1976.
122. Tao, V. P., and Cotlier, E.: Ceramides of human normal and cataractous lenses, Biochim. Biophys. Acta **409**:329, 1975.
123. Tao, V., and Shen, Y. W.: Occurrence of a ceramide synthetase in human cataract and pig lens capsule, Invest. Ophthalmol. (ARVO suppl.), p. 97, April 1979.
124. Tappel, A. L.: Lipid peroxidation and fluorescent molecular damage to membranes. In Trump, B. F., and Arstila, A. U., editors: Pathobiology of cell

membranes, New York, 1975, Academic Press, Inc., vol. I.

125. Taylor, H. R.: Environmental factors associated with senile nuclear sclerosis and pseudoexfoliation of the lens, Invest. Ophthalmol. (ARVO suppl.), p. 129, April 1979.

126. Tulsiani, D. R. P., Nordquist, R. E., and Carubelli, R.: The neuraminidases of rat eyes, Exp. Eye Res. **15:**93, 1973.

127. Thoft, R. A., and Kinoshita, J. H.: The effect of calcium on rat lens permeability, Invest. Ophthalmol. **4:**122, 1965.

128. Van Heyningen, R.: The lens: metabolism and cataract. In Davson, H., editor: The eye, New York, 1969, Academic Press, Inc., pp. 423-434.

129. Van Heyningen, R.: The human lens. I, A comparison of cataracts extracted in Oxford (England) and Shikarpur (W. Pakistan), Exp. Eye Res. **13:**136, 1972.

130. Van Heyningen, R.: The human lens. II, Some observations on cataracts removed in Oxford, England, Exp. Eye Res. **13:**148, 1972.

131. Van Heyningen, R.: Assay of fluorescent glucosides in the human lens, Exp. Eye Res. **15:**121, 1973.

132. Van Heyningen, R.: The glucoside of 3-hydroxykynurenine and other fluorescent compounds in the human lens. In The human lens, in relation to cataract. Ciba Foundation Symposium 19 (new series), 151-71, Amsterdam, 1973, BV Associated Scientific Publishers.

133. Varma, S. D., and Kinoshita, J. H.: Inhibition of lens aldose reductase by flavonoids—their possible role in the prevention of diabetic cataracts, Biochem. Pharmacol. **25:**2505, 1976.

134. Voaden, M. J.: Effect of temperature on cell division in the cultural rabbit lens, Exp. Eye Res. **11:**7, 1971.

135. Von Sallmann, L.: Experimental studies on early lens changes after roentgen irradiation, Arch. Ophthalmol. **47:**305, 1952.

136. Wirostko, E., and Spalter, H. L.: Lens-induced uveitis, Arch. Ophthalmol. **78:**1, 1967.

137. Wollensak, J.: Zonula zinnii, Fortschr. Augenheilkd. **16:**240, 1965.

138. Wortman, B., and Becker, B.: Enzymatic activities of the lens. A preliminary study. Part 2, Am. J. Ophthalmol. **42**(4):342, 1956.

139. Zigler, J. S., Jr., et al.: Studies on brunescent cataracts. I, Analysis of free and protein-bound amino acids, Ophthalmic Res. **8:**379, 1976.

140. Zigman, S.: Eye lens color: formation and function, Science **171:**807, 1971.

141. Zigman, S., Schultz, J., and Yulo, T.: Possible role of near UV light in the cataractous process, Exp. Eye Res. **15:**255, 1973.

Chapter 11

Accommodation

ROBERT A. MOSES, M.D.

The normal eye is so constructed that when it is at rest, rays of light coming from a distant object are focused on the retina. The refractive indices of the various ocular media, the curvatures of the refracting surfaces, and the position of the retina are such that the rays of light entering the eye parallel to its axis are focused on the sensitive outer layer of the retina, the rods and cones. That is to say that the normal eye is emmetropic, and that an emmetropic eye is one in which the retina coincides with the posterior principal focus of the optical system when the muscular activity controlling focusing is at a physiologic minimum (Fig. 11-1, *A*).

The eye can adjust to bring near objects into focus. As we will see, accommodation is accomplished by causing the lens to become more convex (Fig. 11-1, *B*). When the eye is adjusted for near vision, it is said to be accommodated; in distance vision accommodation is relaxed.

DEPTH OF FIELD AND DEPTH OF FOCUS

Depth of field refers to the range of distance from the eye in which an object appears clear without change of accommodation. Depth of focus is the range at the retina in which an optical image may move without impairment of clarity. If an object is accurately focused monocularly, other objects somewhat nearer and farther away are also seen clearly without any change in accommodation. Still nearer or more distant objects are blurred. The depth of the clear zone is the depth of the field. Depth of field reduces the necessity for precise accommodation; indeed, the accommodative mechanism is usually exerted only the minimum required for clear vision.

Part of the explanation of the range of clear vision surrounding the best focus of the optical system of the eye is based on the fact that the retinal receptors are discrete units of finite diameter. Fig. 11-2, *A*, represents image formation by an eye with a fixed degree of accommodation. Each point of the object is brought to a point focus on the retina. If the object is brought closer to the fixed focus lens system, the image or sharp focus recedes into the retina. The retinal receptors then intercept the light rays before they achieve sharp focus (Fig. 11-2, *B*). Each point of the object is now represented on the retina by a blur circle. Similarly, if the object is moved away from the eye, the image falls in front of the retina; points of the object will be represented by blur circles of light diverging past the focal point (Fig. 11-2, *C*). Whether a point of light falls on a retinal cone or whether a circle of light fills its aperture, the cone responds by sending a message to its bipolar and ganglion cell chain. This latitude, shown schematically in Fig. 11-3, allows the image on the retina to be somewhat underfocused or overfocused without impairing the quality of the message transmitted centrally through the optic nerve.

It must be understood that this is an oversimplification, as true point focus does not exist in a real optical system (p. 587).

The range through which the optical image may move into or in front of the retina without impairing the clarity of observation is the depth

□ This work was supported in part by Research Grant EY 00256 from the National Eye Institute, National Institutes of Health.

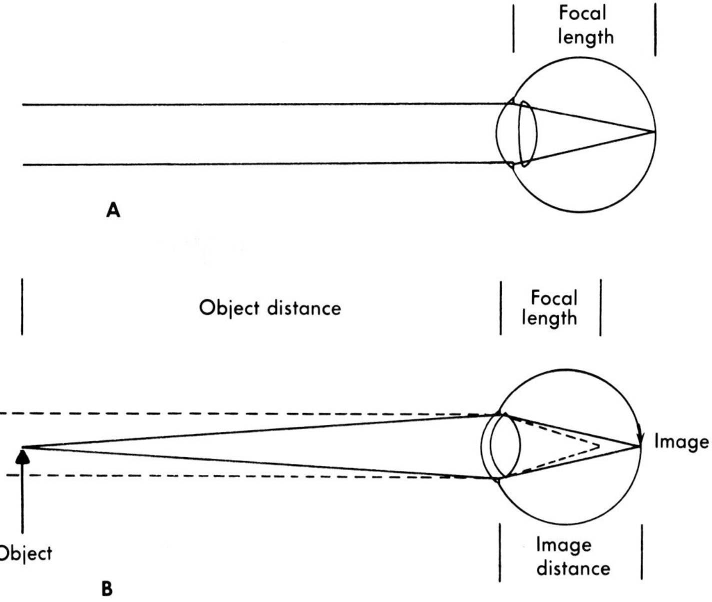

Fig. 11-1. A, Emmetropic eye at rest brings parallel rays of light from far distant object to focus on retina. **B,** Accommodation shortens focal length of optical system of eye so that near object is brought to conjugate focus on retina.

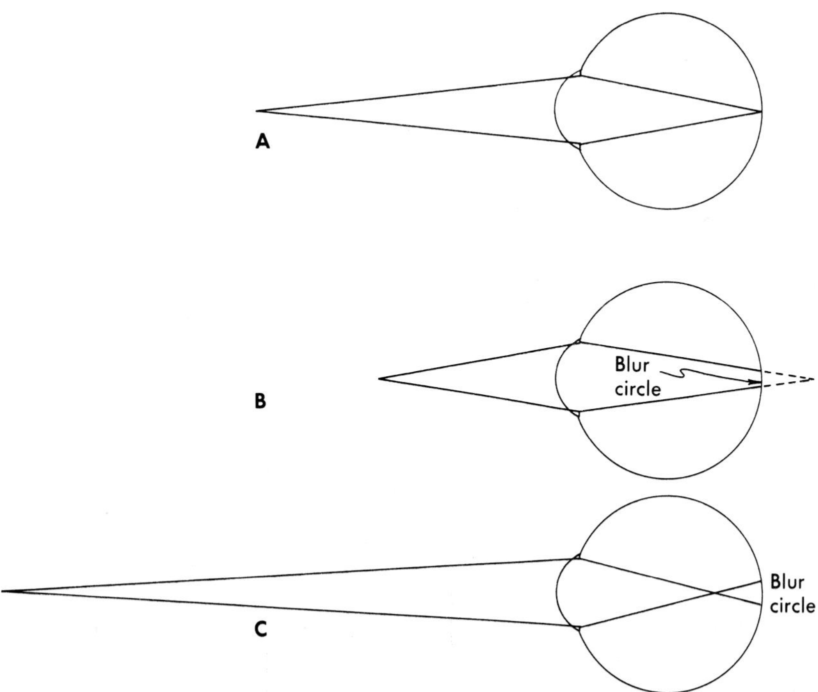

Fig. 11-2. A, Eye is accommodated so that each point of object is represented by point image on retina. Objects closer, **B,** or more distant, **C,** than distance of accommodation are represented on retina by blur circles.

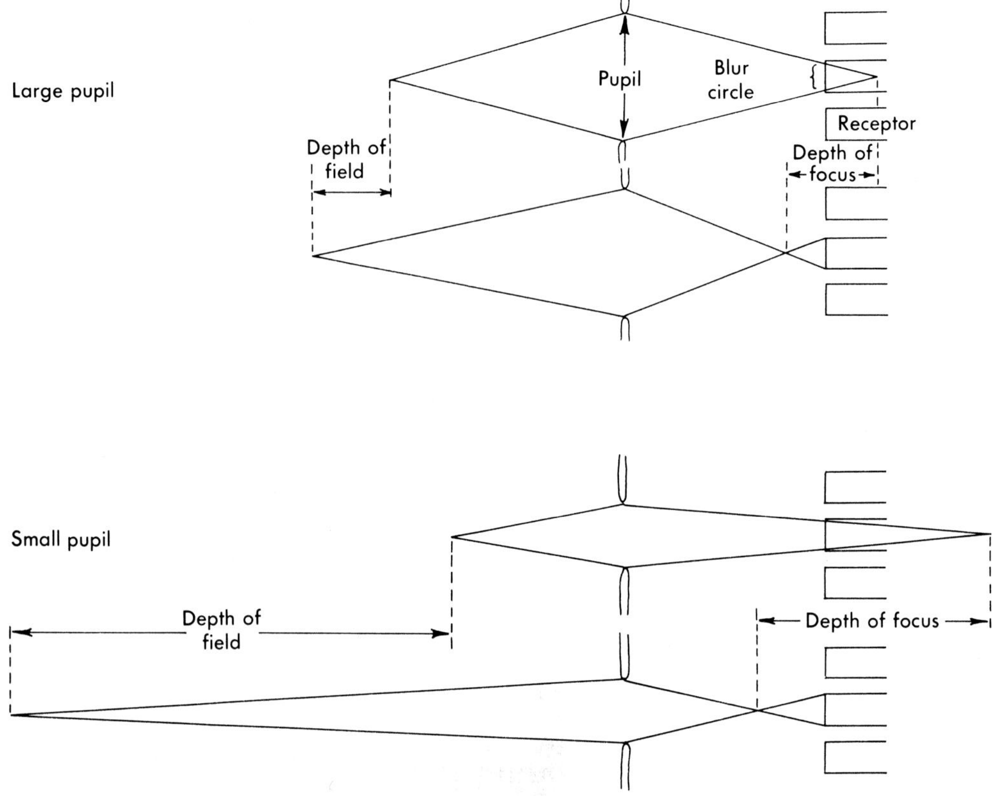

Fig. 11-3. As long as light from object point falls on one receptor, only one receptor is stimulated. Range through which object point may be moved so that blur circle does not exceed receptor diameter is geometric depth of field. Size of blur circle produced on retina is proportional to size of pupil. Depth of field is inversely proportional to pupil size.

of focus of the eye. Depth of field and depth of focus are related to each other by the formula for the distance of object and image from a lens.

If p is the distance of the object from the lens, q is the distance of the image from the lens, and f is the focal length of the lens, then

$$1/p + 1/q = 1/f^*$$

When the distances are given in meters, $1/p$ m $= P$ diopters, so that the same formula may be written in diopters as

$$P + Q = F$$

The depth of field (and of focus) is markedly influenced by the diameter of the pupil (Fig. 11-3). A larger pupil permits a cone of light with a more obtuse apex to reach the retina and shallows the field. Conversely, a small pupil, or the pinhole disc often used in visual testing, overrides minor refractive errors by increasing the depth of field of the eye. The narrow pencils of inaccurately focused rays permitted to enter the eye through the pinhole produce small blur circles, and small blur circles are restricted to few cones at the fovea. Consequently the image on the retina of the poorly focused eye is improved.

An estimate of the depth of focus for a given pupil size is easily obtained from the geometry in Fig. 11-3. Table 11-1, derived from the results of such calculations, shows the effect of pupil size on depth of field. Although the image of a point source is not a point on the retina but is blurred by a diffraction pattern formed by bending of light at the edge of the pupil,

*In this formulation all distances from the lens are taken as positive; other sign conventions are also in common use.

Table 11-1. Geometric estimate of depth of field

Pupil diameter	Depth of field (diopters)	Accurate focus	
		1 m	17 m
2 mm	± 0.06 D	0.94 to 1.06 m	8.5 to ∞ m
4 mm	± 0.03 D	0.97 to 1.03 m	11 to 33 m

and although the image suffers from abberations of the lens system, the illustration points out that the "grain" of the retina allows a certain leeway in focus.

It must not be supposed that the pinhole, or pupil, can be made smaller and smaller and give better and better vision to the out-of-focus eye. On clinical testing, diffraction disturbances overtake the benefit of the small aperture at pinhole diameters of less than 0.75 mm.[33]

Even if the emmetropic eye is sharply focused, pupil diameters less than 2.4 mm reduce visual acuity because of diffraction. Both pinhole disc and small pupil curtail the amount of light admitted to the eye, which in itself may decrease visual acuity. Therefore the net result of narrowing the pupil or use of a pinhole in ametropia* is a balance between increased depth of field, increased diffraction, and decreased retinal illumination.

On the other hand, vision of the emmetropic eye with a very large pupil is not so afflicted by shallow depth of field as one might suppose, since the peripheral rays that strike the retina most obliquely are less effective in stimulating vision (Stiles-Crawford effect).

Since retinal images are never truly sharp, the visual system is constantly processing somewhat blurred images, and tends to tolerate blur up to the point at which essential detail is obscured. This tolerance of blur therefore extends the apparent depth of field considerably, so that the eye may be as much as ±0.25 D out of focus without stimulating accommodative change.

Depth of field has repeatedly been mistaken for accommodation. It is often incorrectly stated that the eye has "residual accommodation" after application of cycloplegic drugs such as atropine. It has even been stated that the aphakic eye* has a small amount of accommodation, an obvious impossibility. That the aphakic eye has less depth of field than the phakic eye is related to the length of the optical system.

In aphakia corrected with a spectacle lens the lens-cornea combination has a focal length considerably longer than that of the normal eye, and as the focal length of the system increases, the depth of field decreases. It may be well to note that in refraction for glasses the strongest convex lens or weakest concave lens which will allow for clear vision at 6 m throws the entire depth of field between the test chart and the patient. When the patient attempts to gaze at distant objects he may well observe them as blurred by 0.17 D of artificial myopia. The old adage of "force plus lens" in refraction is very dubious.

EXISTENCE OF ACCOMMODATION

It is common experience that when one observes a spot on a windowpane a foot or two away, the spot appears clear while objects outside the window are blurred. When attention is shifted to an object in the background, the spot on the windowpane blurs and may even disappear. Christoph Scheiner (1619) elaborated this simple observation into an elegant demonstration of accommodation (Fig. 11-4). A card with two pinholes, their separation being less than the diameter of the pupil, is placed close before one eye; the other eye is occluded. The subject is aware of two intersecting circular fields. While a distant object is observed, a pin is introduced a foot or so from the eye in the axis of the intersection field. This pinpoint appears double as long as the distant object is seen

*Ametropic—not emmetropic.

*Without lens, as after cataract extraction.

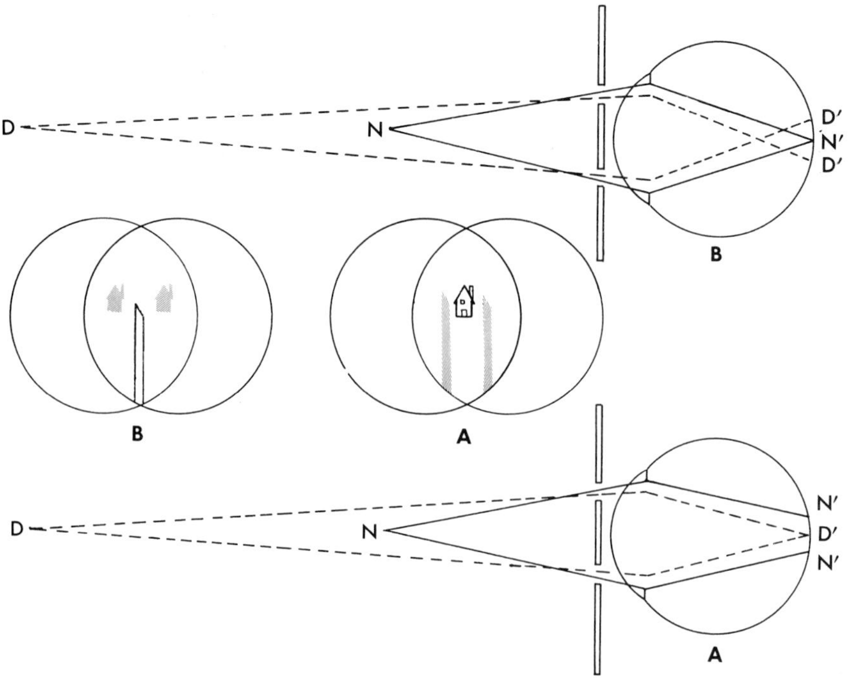

Fig. 11-4. Scheiner's two-pinhole demonstration of accommodation. **A,** Unaccommodated eye sees far object single and near object double. **B,** Accommodated eye sees near object single and far object double.

clearly (Fig. 11-4, *A*). When attention is shifted to the pinpoint, the distant object blurs and doubles while the pin becomes single and clear (Fig. 11-4, *B*). (Older readers who have little or no accommodation may look through their bifocals to make the near object single and clear.)

Scheiner's experiment established that the young eye has a mechanism for changing focus. Four focusing devices can be conceived: the cornea might change its curvature, the eye might change in length, the lens might change in axial position, or the lens might change its shape. We know that the lens changes shape in the human eye, but the demonstrations that the other mechanisms are not involved are also worth mentioning.

Accommodation persists when the cornea is eliminated as a refracting surface. This was demonstrated by Thomas Young (1801), who applied a contact lens (the first contact lens) to his eye. Young also proved that the accommodation is not accompanied by a change in length of the eye. He noted that pressure on the posterior part of the globe by a blunt object produces a blind area (scotoma), and that the size of this scotoma increases with increasing pressure. This is easily demonstrated: occlude one eye, look to the occluded side, and press over the lateral canthus just anterior to the lateral orbital margin. A scotoma appears in the image of the nose. Young constructed a device consisting of a ring that encircled his cornea connected to a key that curved over his globe and rested on the eye near the posterior pole. The pressure of the key produced a scotoma. Young was able to accommodate normally despite the iron frame on his eye. If the eye had lengthened during accommodation, the bar would have pressed more deeply and the scotoma would have enlarged; this did not happen.

The position of the lens and the changes in its curvature have been studied by observation of reflections from its surfaces. The reflections from the surfaces of the refracting media of the eye were described by Purkinje and bear his name (Fig. 11-5). The first image is from re-

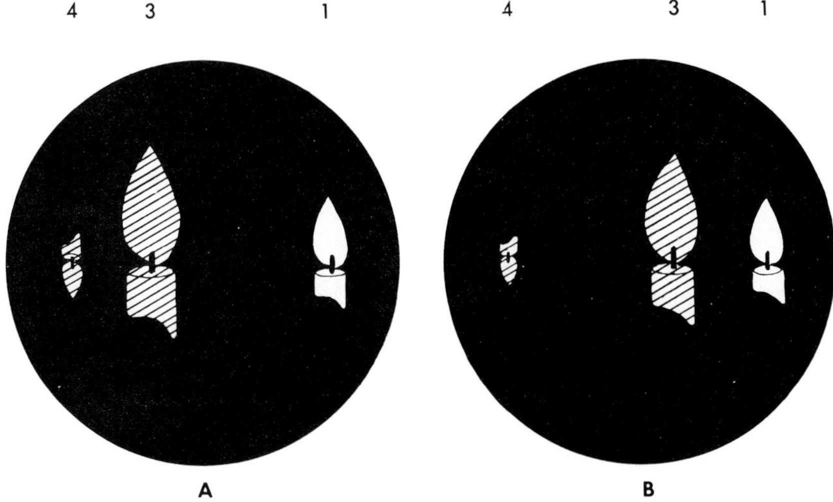

Fig. 11-5. Catoptric images from refracting surfaces of eye (Purkinje). **A,** When eye is at rest. **B,** During active accommodation. Reflections are from anterior surface of cornea, *1;* anterior surface of lens, *3;* and posterior surface of lens, *4*.

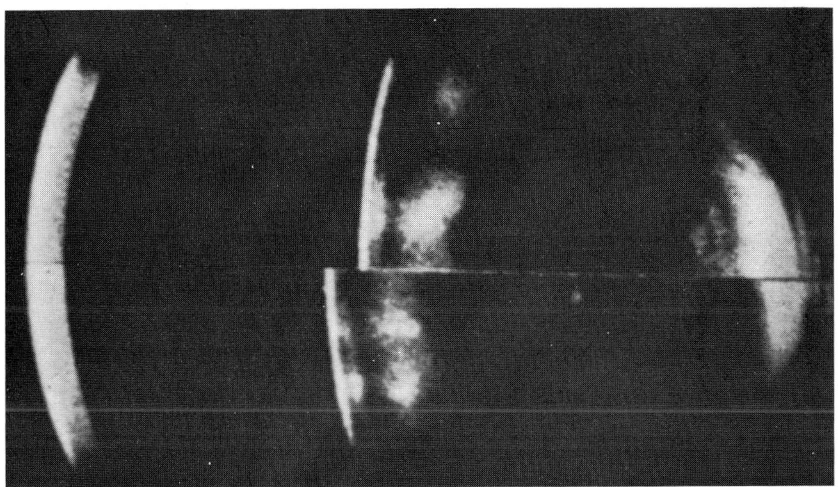

Fig. 11-6. Side-by-side comparison of photographs of same lens with (below) and without (above) accommodation. Note change in thickness and anterior curvature of lens, change in depth of anterior chamber, and smaller strip of lens photographed because of contraction of pupil under accommodative stimulus. (From Patnaik, B. A.: Invest. Ophthalmol. **6:**601, 1967.)

flection at the anterior surface of the cornea. It is bright and upright, and it does not change size with accommodation, since the cornea does not change curvature. The second Purkinje image, formed by the posterior surface of the cornea, is very faint and upright and is also fixed in size. The third upright Purkinje image is formed by the anterior surface of the lens, and the fourth inverted image is formed by the posterior surface of the lens. During accommodation the third image becomes markedly smaller, and the fourth image somewhat smaller. The Purkinje images were carefully studied by Helmholtz.[17] The size and position of the images allow calculation that during accommodation the posterior pole of the lens remains virtually fixed in position, while the anterior pole moves forward. Accommodation causes the posterior lens surface to become somewhat more curved, while the central portion of the anterior surface becomes much more convex.

Scheiner's experiment establishes that the eye has a mechanism for changing, or accommodating, its focus. Young demonstrated that this change does not occur in the cornea or in the length of the eyeball. All evidence indicates that the dioptric change in accommodation results from a change in the form of the lens.

The work of these pioneers has been vastly elaborated through the years, and new methods of measurement have been introduced. For instance, the length of the eyeball may be measured by x rays[31] and by ultrasonography[28]; the changes in the lens can be recorded by slit-lamp photography (Fig. 11-6).

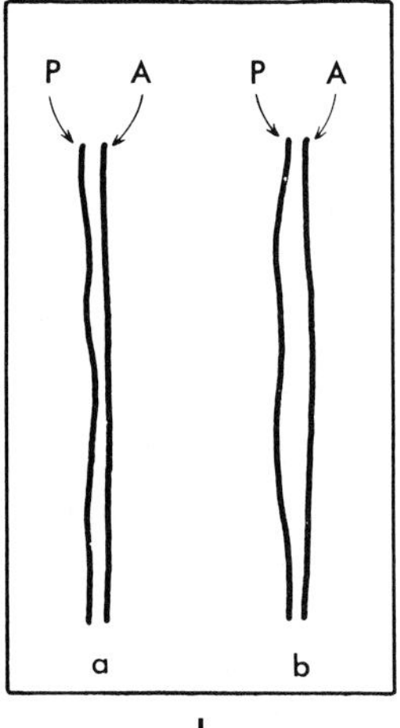

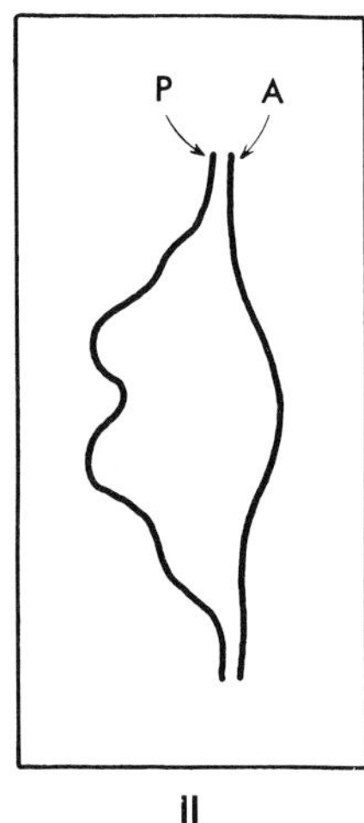

Fig. 11-7. **I,** Normal state of capsule at rest. Posterior capsule, *P*, is more mobile than anterior capsule, *A*, and sometimes floats forward almost to touch anterior capsule, *a*, and at other times floats backward, *b*. **II,** Capsule under influence of eserine. It has now become very lax. (After Graves.)

CHANGES IN ACCOMMODATION

All observers agree that the following changes take place in the eye during accommodation.

1. The pupil contracts during accommodation and convergence. This is a synkinesis and not a true reflex, for it does not depend on either accommodation or convergence alone for its appearance. If convergence is prevented by the interposition of suitable base-in prisms before each eye, contraction of the pupil will take place during accommodation alone. If accommodation is prevented by placing a suitable convex lens in front of each eye and the person converges to the near point, the pupil will likewise contract.

2. The anterior pole of the lens moves forward, carrying the iris with it. Hence the anterior chamber becomes slightly shallower in the center as the anterior pole of the lens approaches the back surface of the cornea. The posterior pole does not change its position to any extent.

3. The anterior surface of the lens becomes more convex. The posterior surface increases its curvature slightly.

4. Since the posterior pole remains fixed and the anterior pole moves forward, the thickness of the lens at the center increases.

5. As the lens increases in axial thickness, it diminishes in diameter.

6. Changes occur in the tension of the lens capsule. In a unique case of aphakia following trauma in a 30-year-old man, Graves observed the anterior and posterior capsules of the lens with a slit lamp and studied the changes that accommodation and various drugs produced in them.[15] From his drawings (Fig. 11-7) it seems indisputable that both during normal accommodation and following the administration of a drug such as physostigmine (eserine), which stimulates contraction of the ciliary muscle, the anterior capsule becomes slack and separates from the posterior capsule.

7. During accommodation the lens sinks in the direction of gravity. Hess first called attention to this fact by observing the apparent upward displacement of the entoptic image of an opacity in his own lens when he accommodated.[10] Fincham confirmed this by finding the distance from the anterior surface of the lens to the cornea in the accommodated state to be 0.2 mm less when the head was held forward parallel to the floor than when the head was held erect, showing the effect of gravity on

Table 11-2. Radii of curvature of lens

Surface	Radius of curvature (mm)
Anterior surface	11.0 (8.4 to 13.8)
Posterior surface	5.7
Anterior surface of core lens	4.6 to 7.5
Posterior surface of core lens	7.9

the position of the lens during accommodation.[10] No change in this distance was observable with a change in position of the head when the eye was not fully accommodated. Little doubt exists, therefore, that the lens is displaced in the direction of gravity during maximum accommodation.

Changes within the lens substance create a change in refractive power of the lens, in addition to the changes in curvature of the surface when the ciliary muscle contracts. These internal changes are brought about by changes in curvature of the various portions of the lens having different indices of refraction. The total dioptric power of the lens has been found to be greater during maximal accommodation than could be accounted for on the basis of changes in the curvature of the surface. Gullstrand suggested that the changes in the isoindical surfaces made up this deficit.[17]

These changes have been documented by slit-lamp photography as in Fig. 11-6. Brown[3] has by this means studied the effect of accommodation on movement of the lens poles, shape of the lens surfaces, and nucleus and depth of the anterior chamber in subjects of different ages.

The radii of curvature of the anterior and posterior surfaces of the lens and those of the central layers of the lens, which are considered to form a unit or core lens, are given in Table 11-2.

MECHANISM OF ACCOMMODATION

The young lens is a soft, easily molded material contained in an elastic capsule. The lens capsule tends to mold the lens to a spherical shape, but this is opposed by the tension of the zonular fibers that suspend the lens from the ciliary body. Traction of the zonule on the lens capsule flattens the lens. The ciliary body is stretched backward and outward along the sclera by the elastic choroid. Contraction of the ciliary muscle pulls the choroid forward and the ciliary attachment of the zonule inward toward the lens, reducing the tension in the zonule and allowing the lens to increase in convexity. As the index of refraction of the lens (about 1.390) is greater than the indices of aqueous and vitreous (1.334), increase in convexity of the lens increases its dioptric power and allows near objects to be imaged clearly on the retina.

This statement of the basic mechanism of accommodation can be attributed principally to Helmholtz.[17] Emphasis on the importance of the lens capsule was added later by Fincham.[10] The mechanism of accommodation may be divided into two parts—active contraction of the ciliary muscle, and passive change of shape of the lens.

The ciliary muscle

The ciliary muscle is a ring of smooth muscle lying against the inner surface of the anterior portion of the sclera. Some fibers of the muscle take origin from the anterior sclera at the scleral spur, while the tendons of some fibers bridge or penetrate the spur to take origin from the trabecular mesh. The scleral spur is a ring of collagenous fibers attached to the inner sclera just posterior to Schlemm's canal. It serves as a pulley for the outer ciliary muscle tendons to prevent canal collapse.[26] The muscle fibers originating from the spur and mesh radiate from their origin, the outer ones passing to the anterior choroid (meridional or longitudinal group), the inner ones fanning toward the axis of the eye (Fig. 11-8). The innermost fibers of the ciliary muscle do not originate from the sclera but form a sphincter lying in the most axial portion of the muscle. The exact disposition of the muscle fibers is not so clearly divided as has just been described or as is stated in many texts, but is probably a mesh, as illustrated by Fincham[10] (Fig. 11-9).

The action of the ciliary muscle on the ciliary zonular attachment can only be to move the attachment forward and inward (axipetal). The action of the ciliary muscle on the choroid can only be to stretch the choroid forward along the sclera. These actions are consistent with the

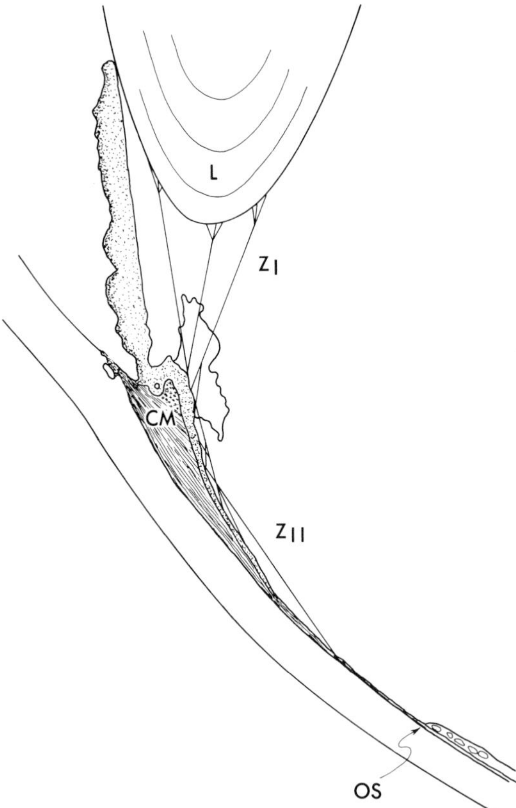

Fig. 11-8. Meridional section. *CM*, Ciliary muscle; *L*, lens; Z_I, zonules between lens and ciliary body; Z_{II}, zonules attached to ciliary body but not to lens; *OS*, ora serrata.

simple principles that a muscle fiber can only contract or not contract (relax), that a muscle can shift its bulk but not change its volume, and that the contents of the eye are incompressible.

At one time it was speculated that the choroid was too tender a structure to be pulled on by the ciliary muscle.[18] The choroid is thin but is actually as elastic as rubber over a limited range.[24] Attachment of the ciliary body and anterior choroid to the sclera is relatively strong only at the scleral spur and where vessels, the anterior ciliary arteries and veins, and the vortex veins perforate. Posterior to the equator the choroid is connected to the sclera by increasing numbers of fine lamellae. Each lamella is delicate and easily torn, but in aggregate the lamellae form an elastic but strong union between the choroid and sclera. Thus it is that the

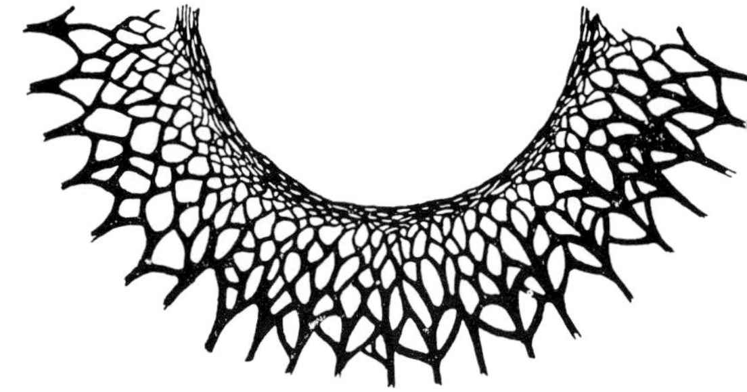

Fig. 11-9. Internal part of ciliary muscle. (From Fincham, E. F.: Br. J. Ophthalmol. Monogr. [suppl. VIII], 1937.)

ciliary muscle, attached to the sclera at its anterior end, can slide itself forward along the concavity of the sclera, and in doing so can thicken the muscle mass inward toward the axis of the eye. The ciliary sphincter adds to this motion. The choroid is stretched forward by the ciliary muscle contraction.

It would be anticipated that if the ciliary body could be visualized from in front, the ciliary processes would be seen to move axipetally during accommodation. This has been observed in a patient with traumatic aniridia and in persons with surgical iridectomy.[4,10]

The forward movement of the ora serrata with accommodation has been measured by two methods.[25] The differentiated retina ends abuptly in a scalloped margin, the ora serrata, a short distance anterior to the equator of the globe. Bands of pigment density or rarification of this region may easily be seen as dark or light stripes on the sclera when the eye is transilluminated. Photographs of the position of such bands show that the ora serrata moves forward about 0.05 mm with each diopter of accommodation (Fig. 11-10). In the second method the visual field is mapped using diascleral illumination (Fig. 11-11). With this technique it may be shown that the margin of the sensitive retina* moves forward with accommodation. The experimental method is difficult, and the scatter in measurements is wide, but the results are consistent with those obtained by the transillumination method.

*The retina is not sensitive to light in its periphery, particularly on the temporal side where there are several millimeters of histologically normal retina posterior to the ora that are not represented in the visual field.

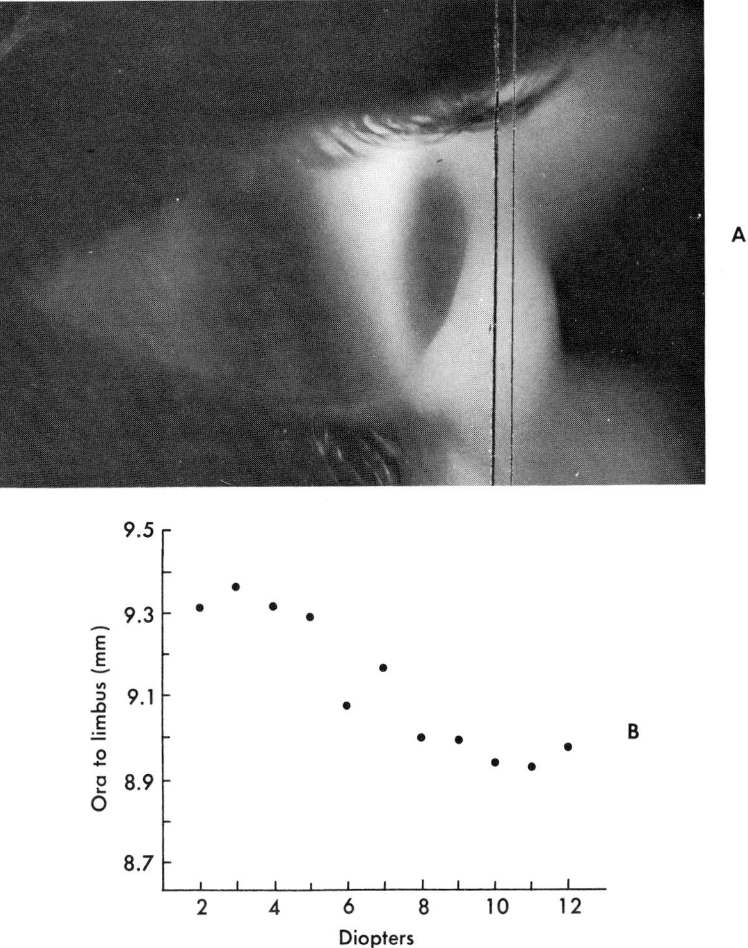

Fig. 11-10. A, Ciliary body shadow cast on temporal sclera by light on nasal sclera. **B,** Movement of pigment bands near ora serrata with accommodation in 17-year-old subject. Measurements from photographs of transilluminated eye.

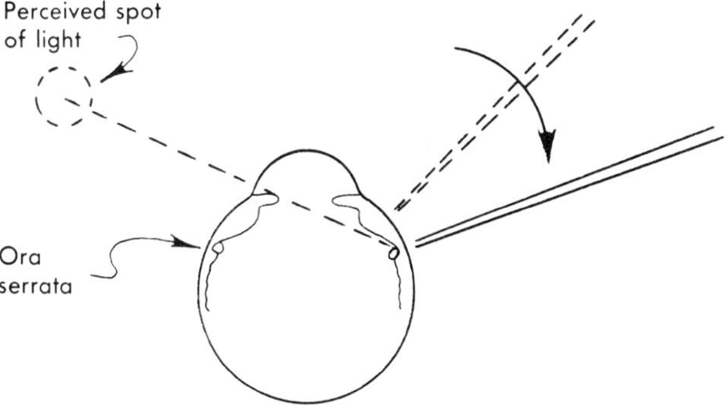

Fig. 11-11. Anterior end of sensitive retina located by diascleral visual field mapping.

It may be that the continual stretching of the retina in accommodation leads to its peripheral cystoid degeneration*; it is accepted that the intense spasm of the ciliary muscle caused by strong cholinesterase inhibitors such as diisopropyl fluorophosphate (DFP) may be followed by retinal breaks near the ora serrata.

The elastic choroid and ciliary body are distended against the sclera by intraocular pressure. The pressure in the potential space between uvea and sclera is less than intraocular pressure because of the elasticity of the choroid and ciliary body.[24,35] When intraocular pressure is lowered to atmospheric pressure, as in cataract surgery, the elastic uvea contracting away from the domed sclera causes the pressure in the suprachoroidal space to fall below atmospheric pressure. Fluid transuded into the low-pressure region is seen clinically as a separation or detachment of the choroid. The ciliary body, thicker than the choroid, stretches as a cone from the scleral spur across the transudate-filled spherical dome of the sclera. Fluid between the uvea and sclera may therefore be tapped through an incision over the ciliary body.

Separation of the ciliary body from the sclera by transudate, although the result of low intraocular pressure, appears to interfere with formation of aqueous humor, and occasionally a circular and serious situation arises following intraocular surgery in which the anterior chamber will not reform in the presence of ciliary detachment.[7]

The zonule

The great number of fine, acellular fibers that constitute the zonule (of Zinn) may be functionally divided into two groups (Fig. 11-8). The first group connects ciliary body and lens and transmits the ciliary muscle posture to the lens capsule (Fig. 11-12, *A*). These are arranged into three subgroups, which attach to the anterior lens capsule, the posterior lens capsule, and the lens equator, respectively. The anterior and posterior subgroups are by far more numerous than the equatorial zonules. I am convinced from my own studies that these suspensory fibers of the zonule do not cross each other to any major extent and that the appearance of crossing is the result of oblique sectioning.

Near their capsular termination the fibers split into many finer fibrils, which attach to the capsule over a relatively large area (Fig. 11-13). The ciliary attachment is likewise spread over a number of epithelial cells. Fibrils branching from the principal fibers to insert more posteriorly, in addition to fibers of the interciliary group, give the illusion of an arch under low magnification, but are actually a large number of straight segments (Fig. 11-14). The zonular fibers suspending the lens attach to the ciliary body at the sides of the ciliary processes and in the valleys between the processes. The tips of the ciliary process float freely in the aqueous.

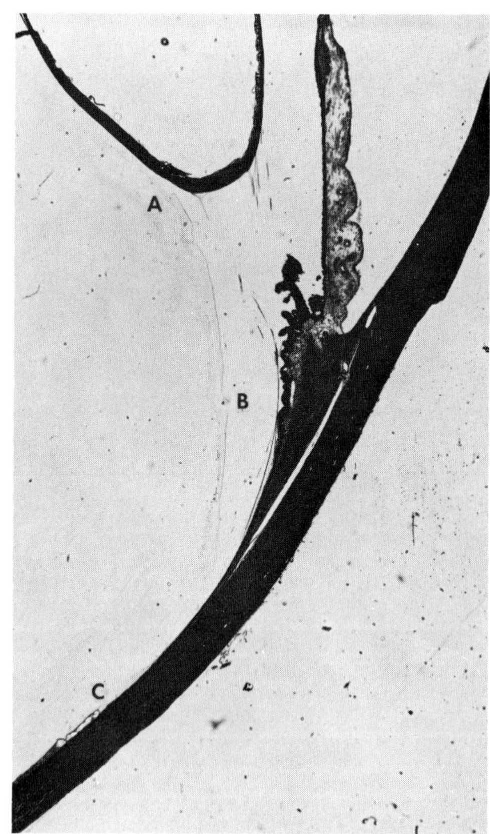

Fig. 11-12. Zonules. *A*, *B*, and *C* magnified in Figs. 11-13, 11-14, and 11-16.

*Lacunae develop in the substance of the peripheral retina, increasing in extent with age.[20]

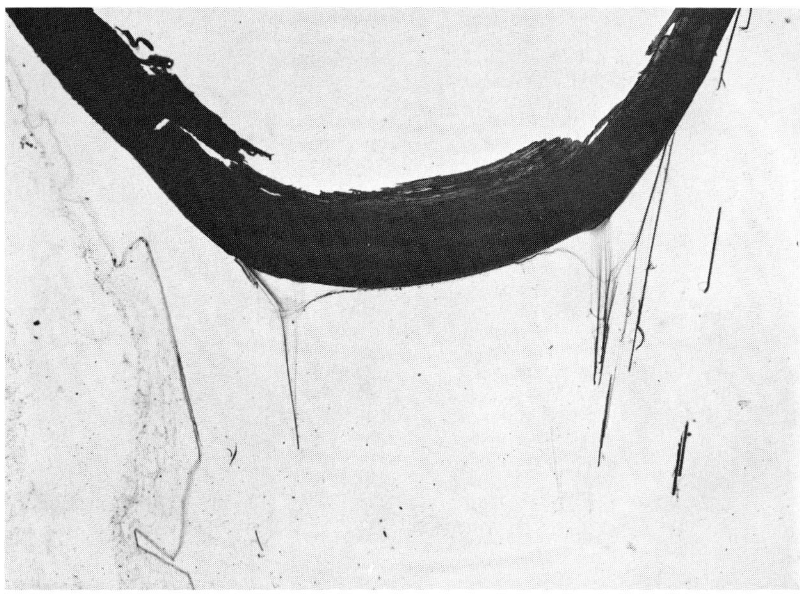

Fig. 11-13. Zonular attachments to lens. (Magnified region *A* from Fig. 11-12.)

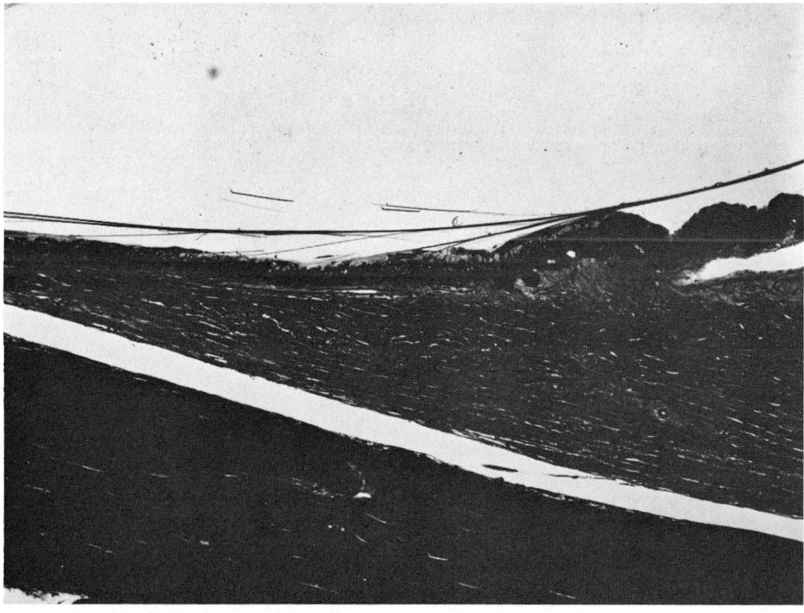

Fig. 11-14. Zonules running in meridional direction from one part of ciliary body to another part. (Magnified region *B* from Fig. 11-12.)

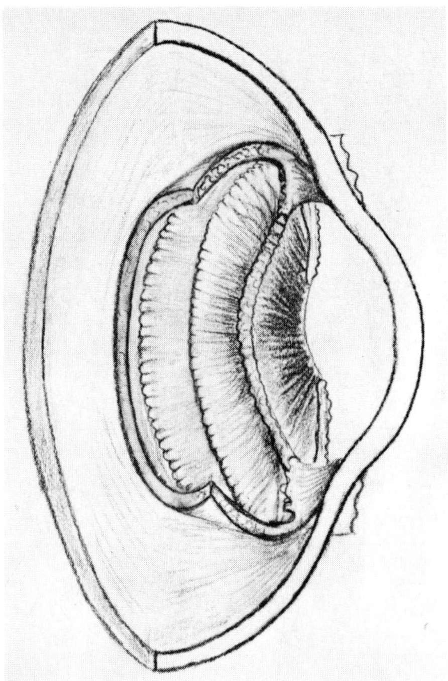

Fig. 11-15. Meridional section of anterior segment of fresh monkey eye showing marked separation of ciliary body. (From Moses, R. A.: Invest. Ophthalmol. **4**:939, 1965.)

The zonular fibers of the second group do not attach to the lens. They may be divided into two major subgroups.

1. Many fibers bridge from one place on the ciliary body to another. Their direction is principally meridional (Figs. 11-8 and 11-14). These fibers form a dense mesh on the inner surface of the ciliary body, which transmits the elastic traction of the choroid to the ciliary attachment of the suspensory fibers when the ciliary muscle is relaxed. The elasticity of the interciliary fibers tends to curl the ciliary body inward and retract the cut edge of the choroid forward if the enucleated eye is sectioned frontally near the ora serrata (Fig. 11-15).

2. Many fine zonular fibers extend from the pars plana of the ciliary body into the vitreous (Fig. 11-16) to form the vitreous base. Whether these fibers more properly belong to the vitreous than to the zonule seems to be an academic question. Other zonular fibers extending from the lens into the vitreous have been observed; their numbers are relatively few, and their functional significance is unknown. The zonule as seen by the scanning electron microscopy is discussed by Rohen.[30]

Obviously the zonules are important, since they transmit ciliary and choroidal forces to the lens capsule for regulation of the convexity of the lens. Moreover the suspensory zonules restrain the lens from prolapsing. If the lens is not restrained, the small excess of pressure of the posterior chamber aqueous and the vitreous can press the lens into the pupil and carry the lens and the iris forward as aqueous is lost from the anterior chamber through the trabecular mesh. On occasion an unattached lens will extrude through the pupil into the anterior chamber, where it may block the pupil from in front. Thus, whether the loose lens lies behind the pupil or anterior to it, pupillary block will move the iris forward in the stream of aqueous until the iris periphery contacts the trabecular mesh, shutting off egress of aqueous. When the angle of the anterior chamber is thus closed, intraocular pressure rises rapidly.

Little can be stated about the metabolism of the zonule. The fibers show a suggestion of banding under the electron microscope and are presumably a form of collagen. They may be dissolved by α-chymotrypsin, a fact often used to advantage in cataract surgery. Their embryonic origin is uncertain. In the young a few

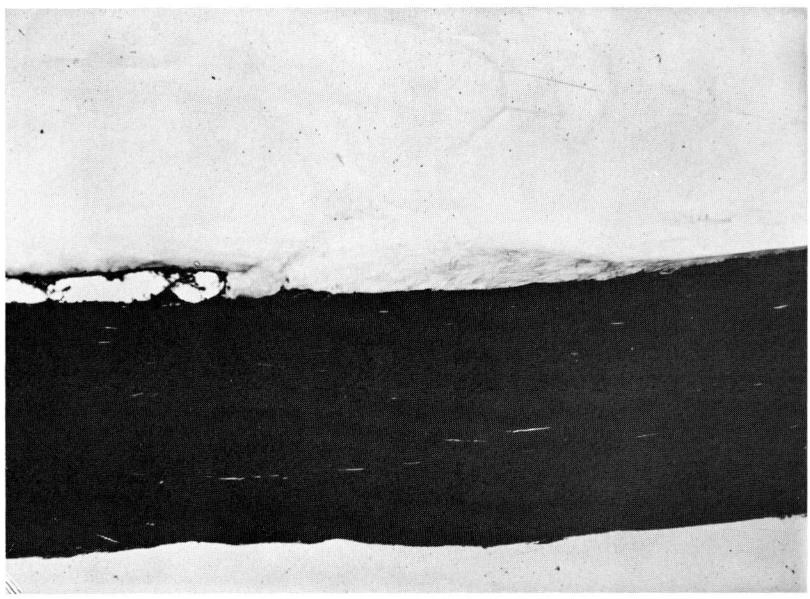

Fig. 11-16. Vitreous base. (Magnified region *C* from Fig. 11-12.)

cells may be seen clinging to the zonular fibers, but in the normal adult no zonular cells are found. If acellularity implies absence of metabolism, then these fibers must last a lifetime. It would also be implied that after the zonules are formed, if the diameter of the ciliary ring grows more rapidly than the diameter of the lens, the zonular tension will be increased and the lens thereby flattened. (The interplay of the various structure of the eye in growth is the subject of a treatise by Van Alphen.[35]) On the other hand, the lens continues to grow throughout life. In time, as the lens equator slowly expands toward the ciliary ring, the lens-ciliary zonules become less tense and allow the lens to ride forward into the pupil. This may be one of the causes of shallowing of the anterior chamber and narrowing of the chamber angle, which predispose to angle-closure glaucoma in middle life.

The capsular mechanism

The shape of the lens depends on the elasticity of its capsule, a relatively thick basement membrane of the lens cells, and on the traction on the capsule by the zonules. The lens surfaces are not perfectly spherical in contour; the anterior surface, in particular, is more convex centrally during accommodation. Fincham[10] suggested that variations in thickness of the lens capsule accounted for the local variations in curvature. He proposed that during accommodation the thicker ring of anterior capsule surrounding the central region contracts under the lessened zonular traction, while the thinner central capsule bulges forward in a more pronounced fashion. The physiologic anterior lenticonus thus formed has a short radius of curvature and high refraction. The relative thickness of the lens capsule at various locations is shown in Fig. 11-17, *B*. (It must be emphasized that the thickness differences in the figure are grossly exaggerated for purposes of illustration.) O'Neill and Doyle[29] have analyzed the lens as an elastic bag (the capsule) containing viscous fluid (the lens cortex). They point out that the contents of the lens capsule must be under pressure slightly greater than intraocular pressure, and this pressure must be greatest in the unaccommodated state. Using data available in the literature and their own measurements of the elasticity of the lens capsule of the cat, these authors show that the shape of the lens surfaces can be accurately computed.

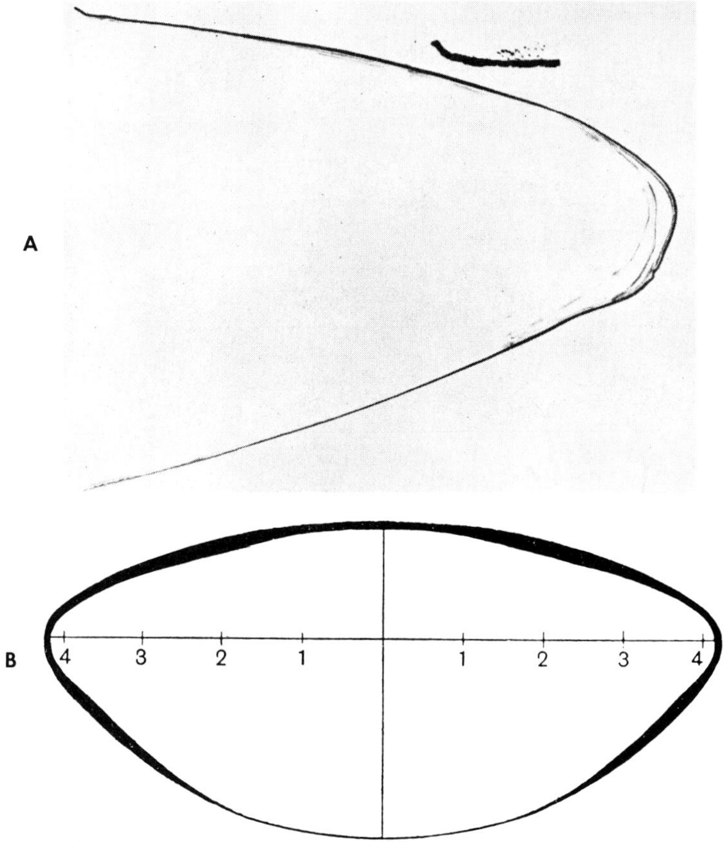

Fig. 11-17. A, Human lens capsule. **B,** Diagram of lens capsule (thickness magnified). (From Fincham, E. F.: Br. J. Ophthalmol. Monogr. [suppl. VIII], 1937.)

It was noted that in extreme accommodative effort the lens becomes displaced in the direction of gravity.[10] This is taken to indicate that the ciliary ring has contracted to such an extreme that the zonules are relaxed and the lens sinks. Since convergence tends to stimulate accommodative effort, even in the presbyopic subject,[27] reading in the usual head-forward position may allow the lens to sink against the pupil and increase the physiologic pupillary block. This may result in angle-closure ocular hypertension (glaucoma) in predisposed (narrow-angle) persons.[19,21]

Many other theories of accommodation have been proposed, but at this time the Helmholtz theory as modified by Fincham appears to be the most probable. The theory of Tscherning, in which forward movement of the vitreous base against the posterior lens periphery alters the lens shape, has had some strong advocates until relatively recent times.[23]

STIMULUS TO ACCOMMODATION

The stimulus to accommodation could be related to blur, chromatic aberration, or awareness of proximity. In experimental situations in which the apparent proximity of the target is constant and only the vergence of light rays entering the eye is varied (and binocular clues are eliminated by occluding one eye), a blurred image on the retina stimulates a change in accommodation. Stark and Takahashi[32] found that the initial alteration of accommodation was appropriate in only half and inappropriate in half of their subject trials, suggesting that the initial accommodative adjustment was based on trial and error. These authors were careful to eliminate clues to target distance such as are obtained from target size, translation across the field of vision, and binocular viewing. If their results bear the test of time, the old problem of the stimulus to accommodation may be solved with the answer—blur—although Fincham's results showing lack of accommodative response to blur when the target was illuminated by monochromatic light are not explained.[11]

Reaction time

The accommodative response follows the presentation of an accommodative stimulus after a reaction time that averages about 0.36 second.[5] When the response is carried out in a single sweep, the time required to make the adjustment is independent of the amplitude of the adjustment and averages 0.64 second for far-to-near accommodation and 0.56 second for near-to-far accommodation. The reaction time for an accommodative response is considerably longer than that for the contraction of the pupil to light, 0.26 to 0.30 second,[22] or for eye movements, with a minimum of 0.12 second. That the reaction time of the pupil response to near is longer than the reaction time of the pupil to light is probably because in the near response of the pupil there is a longer central component. The latency of a convergence response is about 0.20 second.[36]

Other evidence suggests that in the case of accommodation the accommodation itself can be halted during its progression and that some continuous monitoring system must be present which guides the accommodation mechanism once it has begun to act, up to a point at which the satisfactory amount of accommodation has been produced. This can be contrasted with the behavior of the extraocular muscle system in which pulse stimuli, if they yield a response, produce saccadic movements that cannot be modified during their progress.

AMPLITUDE OF ACCOMMODATION

The amount that the eye can alter its refraction is greatest in childhood and slowly decreases until it is lost in middle age. The amplitude or range of accommodation is usually measured by first placing convex lenses before the eye until a target at a convenient distance just begins to blur. It is then assumed that accommodation is relaxed. The convex lens is reduced, or, alternatively, the target is brought closer to the subject until the target again begins to blur. The range between far blur and near blur or maximum plus (convex lens) to blur and maximum minus (concave lens) to blur is the range of accommodation in diopters; this range is depicted as a function of age in Fig. 11-18. As was pointed out previously, part of the apparent range of accommodation is depth of field and part is tolerance of blur. When more critical tests are used, the range of accommodation is found to be smaller[16] (Fig. 11-19).

The amplitude of accommodation depends somewhat on various other factors such as the background luminance of the object of regard. As the level of luminance is diminished, the amplitude of accommodation is lessened.[1] This is in no way related to any change in the accommodative-convergence/accommodation (AC/A) ratio, however, which does not change significantly with decreasing retinal illuminance.[2]

ACCOMMODATION AND REFRACTION

It has been assumed that accommodation can be kept in abeyance when a manifest refraction is being done in patients under the presbyopic age by the method of fogging (inserting plus lenses in the trial frame sufficiently strong to blur distance vision). This, in effect, brings the conjugate focus of the test letters well in front of the retina, and hence any accommodative effort would blur the patient's vision still further. Evidence has been produced, however, that this is not always the case and that fogging may stimulate accommodation instead of relaxing it maximally or allowing it to remain passive.[14]

Once the eye has accommodated accurately on a target, small fluctuations in the accommodation take place. This is not surprising, since most motor systems show similar variations in tone. Thus the pupil shows constant fluctuations in size at any constant illumination, and the eyeball shows constant micronystagmus during steady fixation. The fluctuations in ac-

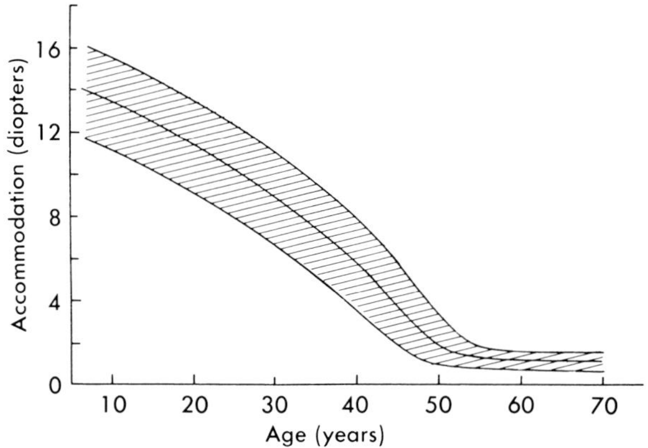

Fig. 11-18. Amplitude of accommodation, showing range found in population and mean. (From Duane, A.: Arch. Ophthalmol. **54:**568, 1925.)

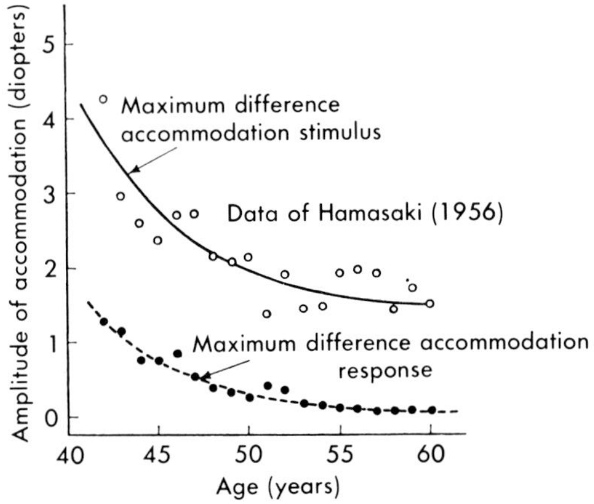

Fig. 11-19. Mean values for maximum amount dioptric power of eye can increase (filled circles) for various ages, according to data of Hamasaki, Ong, and Marg. Open circles represent near-point determinations by push-up method in same population sample (From Hamasaki, D., Ong, J., and Marg, E.: Am. J. Optom. **33:**3, 1956.)

commodation disappear when the subject views a target at infinity or when a cycloplegic is instilled in the eye, paralyzing accommodation.[6]

RELATION BETWEEN ACCOMMODATION AND CONVERGENCE

In normal binocular vision, accommodation and convergence operate in unison. The chief stimulus to accommodation is a change in the vergence of light striking the retina. If a distant object is in focus on the foveas, as it approaches the eyes the pencils of rays entering the pupils become more divergent. This stimulates accommodation. If the object approaches the individual on a line midway between the two eyes, the image moves away from each fovea in a temporal direction. This temporal displacement of the retinal images relative to each other provides the stimulus for convergence of the visual axes. The basic stimuli to the *binocular* adjustment for near vision are therefore (1) change in the vergence of the light reaching each fovea and (2) temporal disparity of the two images relative to the two foveas.

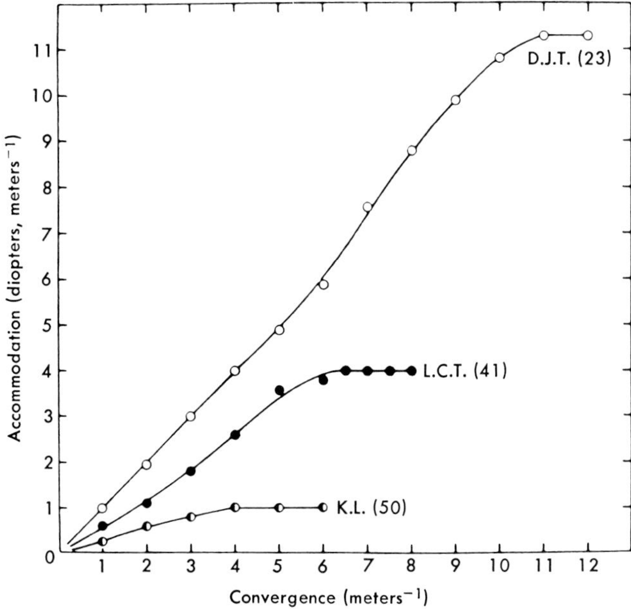

Fig. 11-20. Accommodation induced by convergence. Results on three subjects of different ages. Maximum and rate of increase of accommodation relative to convergence become less with age. Convergence is shown in meters^{-1}, as comparable with diopters of accommodation. (From Fincham, E. F.: J. Physiol. [Lond] **128:**105, 1955.)

The relationship of these two functions is not strictly fixed, although under normal conditions a unit change in one is accompanied by a unit change in the second. Thus 1 D of accommodation is ordinarily accompanied by 1 meter angle of convergence.* But it has been shown by most investigators that some latitude of action is possible between accommodation and convergence. There is an amplitude of accommodation corresponding to each convergence value, and convergence can vary somewhat relative to a fixed condition of accommodation. To show the amplitudes of either accommodation or convergence, each function in turn must be held in a fixed state of action while the other is varied. Provision must be made in the experimental setup to change either the vergence of the light or the temporal disparity of the image, but not both at the same time. Such measurements will tell us then how far either function, being unstimulated, responds in sympathy to stimulated changes in the other.

*A meter angle is the angle formed by each visual axis with a line perpendicular to the interpupillary line.

The change in convergence produced by a change in accommodation is known as the accommodative convergence. This was considered on p. 145. The change in accommodation produced by a change in convergence will be considered here. Until recently it has been debated whether convergence alone, without any accompanying change in the vergence of light on the retina, was able to change accommodation. In 1956 Christoferson and Ogle[8] stated that the difficulties of measurement were such that they prevented any investigation as to what extent convergence-accommodation exists. Fincham and Walton,[13] however, have succeeded in measuring the controlling effect of convergence on accommodation in the absence of the effects of light vergence (Fig. 11-20). They find that in young subjects, up to the age of 24 years, the convergence-induced accommodation is equal to the convergence, but above that age it gradually diminishes. Accommodation is induced by convergence as an unconditioned reflex. It is probable that the diminution of the convergence-induced accommodation with age is due to the progressive

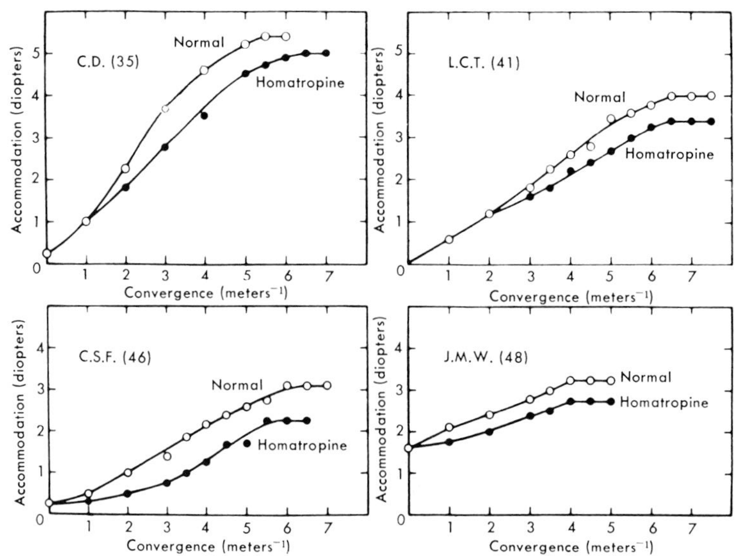

Fig. 11-21. Effect of slight paresis of ciliary muscle on accommodation induced by convergence. (From Fincham, E. F.: J. Physiol. [Lond] **128:**107, 1955.)

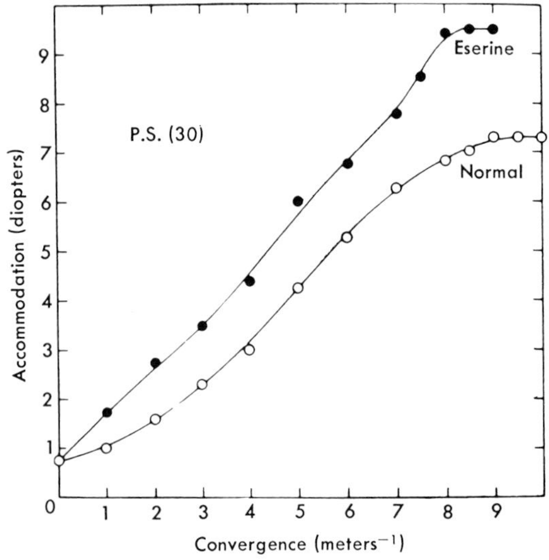

Fig. 11-22. Effect of eserine on accommodation induced by convergnce. (From Fincham, E. F.: J. Physiol. [Lond] **128:**108, 1955.)

sclerosis of the crystalline lens causing a reduced response to a given nerve discharge and contraction of the ciliary muscle rather than to a change in the relative innervations to the two functions.

Accommodation can be produced, therefore, by convergence alone, and up to 24 years of age the physiologic maximum can be produced by this stimulus. In older persons the rate of increase of accommodation with increasing convergence is reduced. Fincham has produced evidence which shows that this is caused by the need for greater ciliary force to produce a given amount of accommodation.[12] Furthermore, he found that artificial paresis of the ciliary muscle with homatropine sufficient to produce the slightest reduction in accommodation relative to convergence also causes the maximum of accommodation to be reduced at all ages (Fig. 11-21). Therefore maximal accommodation requires the maximal force of the ciliary muscle. Miotics enhance the contraction of the ciliary muscle accompanying convergence. This results in an increase of accommodation relative to convergence, and the maximal accommodation is increased above the physiologic maximum (Fig. 11-22). As the lens substance becomes firmer with age, greater force from the elastic capsule will be required to produce a given change in curvature. This force can only be applied by greater contraction of the ciliary muscle. Thus we are unable to maintain our full accommodation for more than a short time, and the presbyope requires optical aid although his near point may not lie outside his working distance. The fact that the presbyope has to contract his ciliary muscle to a great extent to produce even minimal accommodation naturally gives rise to fatigue, headaches, and all the symptoms of asthenopia.

The loss of accommodation with age is fairly uniform. In middle life when near vision becomes difficult or impossible the condition is called presbyopia.

PRESBYOPIA

The amplitude of accommodation gradually diminishes with age. At the age of 8 years the dioptric power of the eye can be raised by accommodation approximately 14 D; at the age of 20 years this has fallen to 11 D. At the age of 30 years it falls to 9 D, and at the age of 50 years it is less than 2 D. The nearest point for which the eye can accommodate so that a clear image is formed on the retina is called the near point or the punctum proximum. If the eye is emmetropic or is made so with the proper corrective lenses, the near point will vary with the age of a person, as just stated. It is closest to the eye in the young and recedes gradually until about the age of 45 years, when a much more rapid recession occurs. Further loss of the accommodative power, with recession of the near point, continues unabated until about the age of 60 years, by which time all the accommodation has been lost.

By the time the near point has become so far removed that the subject cannot read fine print, the eye is said to have become presbyopic. Presbyopia is the normal recession of the near point due to age and usually begins around the age of 46 years. Convex glasses must then be prescribed as a substitute for the decrease in the accommodative power of the lens.

It is generally believed that presbyopia is the result of sclerosis or hardening of the nucleus of the lens so that the forces which normally deform the soft lens during youth (capsular elasticity) are now no longer effective. It has also been suggested that the decrease in plasticity of the cortex is the result of a loss of water content of the lens with age. Fincham found the surfaces of lenses that had been removed from the eye to be less strongly curved in older people than in children, which implies that the elasticity of the capsule is no longer able to deform the senile lens.

Two explanations of presbyopia have been offered: (1) that the lens nucleus grows and stiffens with age and the moldable cortex is reduced and (2) that the ciliary muscle weakens with age and can no longer adequately relax the zonules. Clinical experience with lenses dislocated by trauma or removed in cataract surgery confirms the findings of Fincham. The usual senile cataract is relatively discus-shaped when removed from the eye, its shape being maintained by the firm lens substance. The liquefied hypermature cataract is globular when extracted, rounded up by the elastic capsule.

Measurements of ciliary body movement in accommodation appear to show that the same movement is required per diopter of accommodation at all ages tested, but the results must be confirmed.

PHARMACOLOGY

The ciliary muscle is under the control of the parasympathetic nervous system and shows typical responses to the drugs that act on this system. Pilocarpine acts directly at the myoneural junction to cause muscle contraction and hence accommodative spasm. Cholinesterase inhibitors such as physostigmine (eserine), diisopropyl fluorophosphate, and echothiophate iodide allow acetylcholine to accumulate at the parasympathetic third nerve endings and produce accommodative spasm. Atropine blocks the action of acetylcholine and pilocarpine at the muscle, and it and other drugs that paralyze the ciliary muscle are called cycloplegics. A drug that produces rapid cycloplegia but has a relatively short action is tropicamide (Mydriacyl). Cyclopentolate (Cyclogyl) is a blocker with longer action.

Patients who develop paralysis of the sympathetic supply to the head (Horner's syndrome) have a shift of refraction, becoming slightly more myopic or less hyperopic.[9] The amplitude of accommodation remains unchanged. Conversely, sympathomimetic drugs such as 10% phenylephrine (Neo-Synephrine) cause the eye to become slightly hyperopic. Tornqvist[34] has shown that adrenergic activity in the ciliary body is responsible for these effects. That is, sympathetic activity inhibits the ciliary muscle (beta-adrenergic mechanism); sympathetic denervation or beta-adrenergic blockade with propranolol removes the inhibitory activity, and the accommodation of the eye is increased. Drugs with beta-adrenergic activity (epinephrine) inhibit the ciliary muscle with a consequent shift of the refraction toward hyperopia.

REFERENCES

1. Alpern, M., and David, H.: The effects of illuminance quantity on the accommodation of the eyes, Indust. Med. **27:**551, 1958.
2. Alpern, M., and Larson, B.: Vergence and accommodation. IV, Effect of luminance quantity on the AC/A, Am. J. Ophthalmol. **49:**1140, 1960.
3. Brown, N.: The change in shape and internal form of the lens of the eye on accommodation, Exp. Eye Res. **15:**441, 1973.
4. Busacca, A.: Éléments de gonioscopie, São Paulo, Brazil, 1945, Tipografia Rossolillo.
5. Campbell, F., and Westheimer, G.: Dynamics of accommodation response of the human eye, J. Physiol. (Lond) **151:**285, 1960.
6. Campbell, F., Robson, J., and Westheimer, G.: Fluctuations in accommodation under steady viewing conditions, J. Physiol. (Lond) **145:**579, 1959.
7. Chandler, P. A., and Maumenee, A. E.: A major cause of hypotony, Trans. Am. Acad. Ophthalmol. Otolaryngol. **65:**563, 1961.
8. Christoferson, K. W., and Ogle, K. N.: Effect of homatropine on accommodation-convergence association, Arch. Ophthalmol. **55:**779, 1956.
9. Cogan, D. C.: Accommodation and the autonomic nervous system, Arch. Ophthalmol. **18:**739, 1937.
10. Fincham, E.: The mechanism of accommodation, Br. J. Ophthalmol. **8:**1, 1937.
11. Fincham, E.: The accommodation reflex and its stimulus, Br. J. Ophthalmol. **35:**381, 1951.
12. Fincham, E.: The proportion of ciliary muscular force required for accommodation, J. Physiol. (Lond) **128:**99, 1955.
13. Fincham, E., and Walton, J.: The reciprocal actions of accommodation and convergence, J. Physiol. (Lond) **137:**488, 1957.
14. Flom, M.: Variations in convergence and accommodation induced by successive spherical lens additions with distance fixation, Am. J. Optom. **32:**176, 1955.
15. Graves, B.: The response of the lens capsules in the act of accommodation, Trans. Am. Ophthalmol. Soc. **23:**184, 1925.
16. Hamasaki, D., Ong, J., and Marg, E.: The amplitude of accommodation in presbyopia, Am. J. Optom. **33:**3, 1956.
17. Helmholtz, H.: Treatise on physiologic optics, New York, 1962, Dover Publications, Inc., vol. 1.
18. Henderson, T.: The anatomy and physiology of accommodation in mammalia, Trans. Ophthalmol. Soc. U.K. **46:**280, 1926.
19. Higgitt, A. C., and Smith, R.: Reading test in glaucoma, Br. J. Ophthalmol. **39:**103, 1955.
20. Hogan, M. J., and Zimmerman, L. E.: Ophthalmic pathology, Philadelphia, 1962, W. B. Saunders Co.
21. Hyams, S. W., Friedman, Z., and Newmann, E.: Elevated intraocular pressure in the prone position, Am. J. Ophthalmol. **66:**661, 1968.
22. Lowenstein, O., and Lowenfeld, I.: Electronic pupillography, Arch. Ophthalmol. **59:**352, 1958.
23. Luedde, W. H.: Accommodation. In Berens, C., editor: The eye and its diseases, Philadelphia, 1949, W. B. Saunders Co.
24. Moses, R. A.: Detachment of the ciliary body—anatomical and physical considerations, Invest. Ophthalmol. **4:**935, 1965.
25. Moses, R. A.: Unpublished data.
26. Moses, R. A., and Grodzki, W. J., Jr.: The scleral spur and scleral roll. Invest. Ophthalmol. Vis. Sci. **16:**925, 1977.
27. Ogle, K. N.: The accommodative convergence, accommodation ratio and its relation to the correction of refractive error, Trans. Am. Acad. Ophthalmol. Otolaryngol. **70:**322, 1966.
28. Oksala, A.: Ten years experiences in clinical ultrasound investigation, Acta Ophthalmol. **45:**489, 1967.
29. O'Neill, W. D., and Doyle, J. M.: A thin shell deformation analysis of the human eye, Vision Res. **8:**193, 1968.
30. Rohen, J. W.: Scanning electron microscopic studies of the zonular apparatus in human and monkey eyes, Invest. Ophthalmol. Vis. Sci. **18:**133, 1979.
31. Sorsby, A., and O'Connor, A. D.: Measurement of the diameters of the living eye by means of x-rays, Nature **156:**779, 1945.
32. Stark, L., and Takahashi, Y.: Absence of an odd-

error signal mechanism in human accommodation, IEEE Trans. Biomed. Eng. **12:**138, 1965.

33. Takahashi, E., cited by Westheimer, G.: Visual acuity, Ann. Rev. Psychol. **16:**359, 1965.
34. Tornqvist, G.: Effect of cervical sympathetic stimulation on accommodation in monkeys, Acta Physiol. Scand. **67:**363, 1966.
35. Van Alphen, G. W.: On emmetropia and ametropia, Ophthalmologica **142:**310, 1961.
36. Westheimer, G., and Mitchell, A.: Eye movement responses to convergent stimuli, Arch. Ophthalmol. **55:**848, 1956.

GENERAL REFERENCE

Alpern, M.: Muscular mechanisms. In Davson, H., editor: The eye, New York, 1962, Academic Press, Inc., vol. 3.

Chapter 12

The pupil

H. STANLEY THOMPSON, M.D.

This chapter is chiefly concerned with the application of pupillary signs to clinical diagnosis. After a brief review of pupillary neuroanatomy, physiology, and pharmacology, the reader is asked to consider afferent and efferent pupillary defects, tonic pupils, light-near dissociation, anisocoria, and Horner's syndrome.

ANATOMY

Muscles of the iris

The iris contains two muscles, the sphincter pupillae and the dilator pupillae. They are among the very few muscles of the body derived from the neural ectoderm. At about the beginning of the fourth month of gestation, the rim of the optic cup differentiates into a circular band of muscle, the sphincter pupillae (Fig. 12-1). When fully formed, the sphincter is typical smooth muscle and lies anterior to the neuroectodermal pigment epithelium in the iris stroma. The dilator muscle arises from the outer layer of the primitive optic cup at about the seventh month by elongation of the cells in a radial direction. The myoid elongations of the dilator cells lie anterior to their pigmented cell bodies; the cell bodies remain in apposition to the pigmented inner wall of the optic cup. As can be seen in Fig. 12-1, the inner, retinal layer of the optic cup becomes the single, deeply pigmented cell layer of the posterior iris epithelium, while the outer layer of the optic cup, corresponding to the pigment epithelium of the retina, gives rise to a pigmented myoepithelium, the dilator of the iris.

The excursion of the pupil can be extraordinarily large. When maximally contracted, the diameter of the human pupil may be less than 1 mm; when maximally dilated, it may be more than 9 mm. The iris sphincter shortens about 87% of its length, a change seldom found in other smooth or striated muscles in the body. It is generally believed that those muscles which move their insertions through large distances have long fibers.

The sphincter muscle does not move freely in the iris like a pajama string; its fibers are intimately connected with the underlying stroma and with the dilator muscle to such an extent that constriction of the pupil is possible even after a section of the sphincter has been cut—that is, following sphincterotomy or sector iridectomy.

Innervation of the iris muscles

The muscles of the iris are innervated by the autonomic nervous system. The sympathetic innervation of the dilator muscle is shown in Fig. 12-16 and will be described on p. 348, where Horner's syndrome is discussed.

Afferent pathway of the light reflex. There can be little question that the receptors for the pupillary light reflex are identical to those concerned with vision. The question arises as to whether two separate sets of fibers convey the pupillary and visual impulses, or whether these impulses reach the brain by the same fiber. Each optic nerve is composed of both large and small fibers. It has been assumed by some authors that the thick fibers conduct the afferent pupillary impulses and the smaller fibers the visual impulses. No physiologic or anatomic evidence is available to decide this point. The pupillary fibers may, on the other hand, be collaterals given off by the visual fibers, and the same optic nerve fiber may conduct both visual and afferent pupillary impulses, which would separate from each other before reaching their appropriate cell terminations.

As far as is known, the pupillary fibers follow the course of the visual fibers in every respect and undergo semidecussation in the chiasm. They travel with the visual fibers in the optic tract. Near the lateral geniculate body they leave the optic tract to enter the brachium of the superior colliculus and proceed to the pretectal region. Here the pupillary light reflex fibers synapse with cells of the pretectal nucleus whose axons connect to the cells of the Edinger-Westphal group of the oculomotor nucleus.

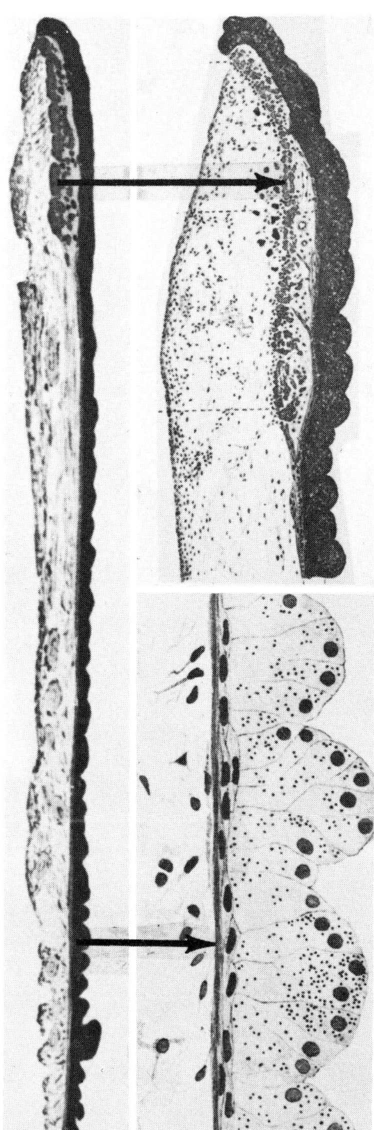

Fig. 12-1. Cross section of iris. Upper arrow points to sphincter muscle drawn in higher magnification; lower arrow points to dilator muscle of bleached specimen drawn in higher magnification. (From Saltzmann, M.: Anatomy and histology of the human eyeball, Chicago, 1912, The University of Chicago Press.)

In humans the intercalated neurons from each side of the pretectum are distributed equally to the oculomotor nuclei of both sides. The intercalated fibers destined for the opposite third nerve parasympathetic nucleus (of Edinger-Westphal) cross dorsal to the cerebral aqueduct in the posterior commissure; then, in company with axons arising on that side, they arch ventrally to reach the Edinger-Westphal nucleus. Thus each optic tract carries pupillary fibers from both eyes (Fig. 12-2). The pupillary fibers synapse with intercalated neurons that distribute the impulses of each tract to both third nerve nuclei.

Pupilloconstrictor responses are obtained from stimulation of the optic chiasm, the optic tract on the lateral surface of the brain stem and ventral to the lateral geniculate body, the brachium of the superior colliculus, the pretectal region, the posterior commissure, and fibers arching ventrally around the central gray matter at the level of transition between the third ventricle and cerebral aqueduct (Fig. 12-2).

Efferent pathway to sphincter and ciliary muscle. The intraocular muscles are controlled by the Edinger-Westphal nucleus, which is part of the oculomotor nucleus (see Chapter 5). The third nerve leaves the midbrain in the interpeduncular fossa carrying near its epineurium the small caliber fibers that serve the intraocular muscles. The synapse in this typical two-neuron parasympathetic outflow path is in the ciliary ganglion, deep in the muscle cone of the orbit (Fig. 12-3). The postganglionic fibers follow the short ciliary nerves to the anterior segment of the globe.

PHYSIOLOGY

The normal pupil

Normally, the pupil is placed not quite at the center of the iris, but rather slightly nasally and

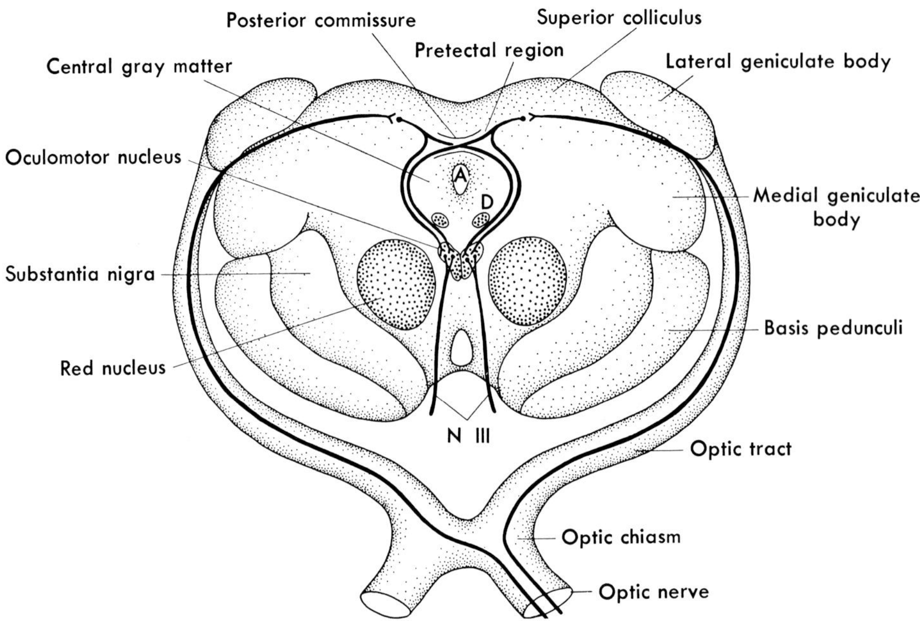

Fig. 12-2. Diagram of path for pupillary light reflex. *A,* Aqueduct; *D,* nucleus of Darkschewitsch; *N III,* oculomotor nerve. (After Ranson.)

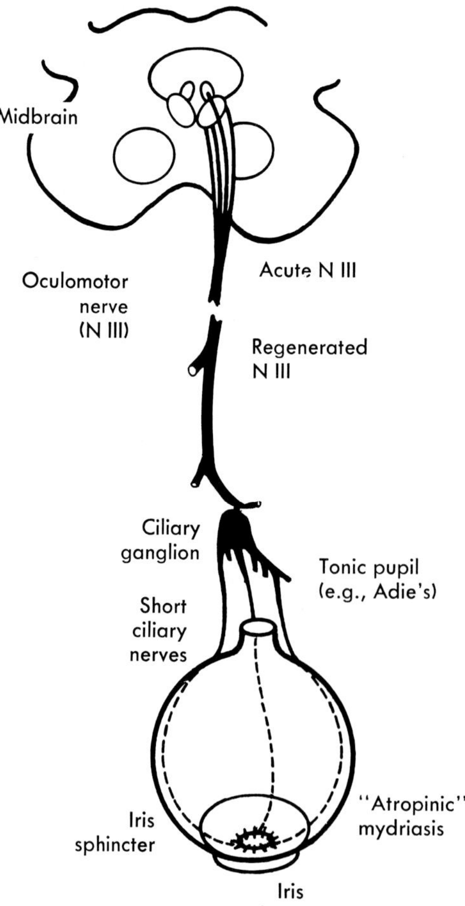

Fig. 12-3. Innervation of the iris sphincter, from the Edinger-Westphal nucleus via oculomotor nerve, ciliary ganglion, and short ciliary nerves, with some of the causes of a fixed dilated pupil.

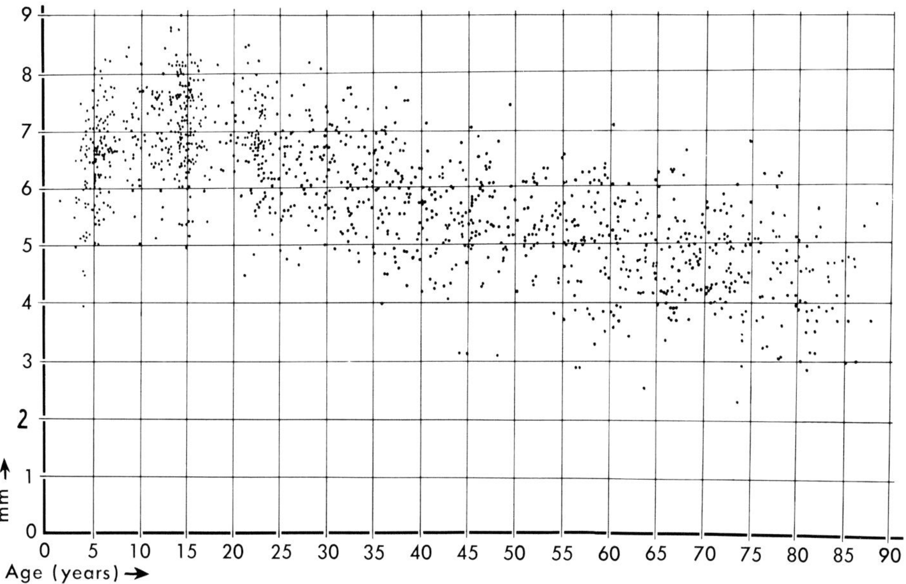

Fig. 12-4. Pupillary size in darkness at various ages. 1263 subjects chosen at random from a population survey. Each point represents the average diameter of the two pupils taken together. The ordinate shows horizontal diameters (in millimeters); the abscissa shows age in years. Note the wide scatter among individuals and the obvious age trend. (From Loewenfeld, I. E.: Pupillary changes related to age. In Thompson, H. S., editor: Topics in neuro-ophthalmology, © [1979], The Williams & Wilkins Co., Baltimore.)

inferiorly. The size of the pupil varies with age. Sympathetic innervation reaches the dilator muscle at full term, as evidenced by the fact that the pupils of premature infants do not dilate to hydroxyamphetamine but do dilate to phenylephrine.[10] The miosis of the newborn is due in part to the fact that the baby is asleep or sleepy 22 hours out of every 24. In adolescence the pupils are at their largest. They then become steadily smaller until about age 60, when the size levels off (Fig. 12-4).[15]

Pupillary unrest and hippus

A healthy iris is moving most of the time, even when illumination and accommodation are constant. This physiologic pupillary unrest is presumed to be due to fluctuations in the activity of the sympathetic and parasympathetic innervation of the iris muscles, which have reached an unsteady sort of equilibrium. The movements of the iris are largest in moderately bright light, and the frequency of the oscillation increases with the light intensity (Fig. 12-5). This pupillary instability is most apparent in young people and may be unhesitatingly accepted as a normal phenomenon and called "physiologic pupillary unrest" or "hippus." The word hippus has a long and interesting history. For the last 200 years it has meant a pupillary unrest "of abnormal degree," a tremor of the iris muscles. It was thought during the nineteenth century to be a sign of various neurologic diseases, but it has gradually become accepted as a normal phenomenon.[36]

Simple anisocoria

A small amount of pupillary inequality is common. Roughly a quarter of the normal population has an anisocoria of 0.4 mm or more. This figure increases with the age of the subjects; one fifth of the population under 17 and about one third of the subjects over 60 years of age have this amount of anisocoria.[14] It is interesting that this pupillary inequality is not constant in a given individual; it may increase or decrease or even reverse sides within days or hours. It has been called a "see-saw" anisocoria when these changes are dramatic.

This is an anisocoria that decreases in bright light; it has not been associated with any dis-

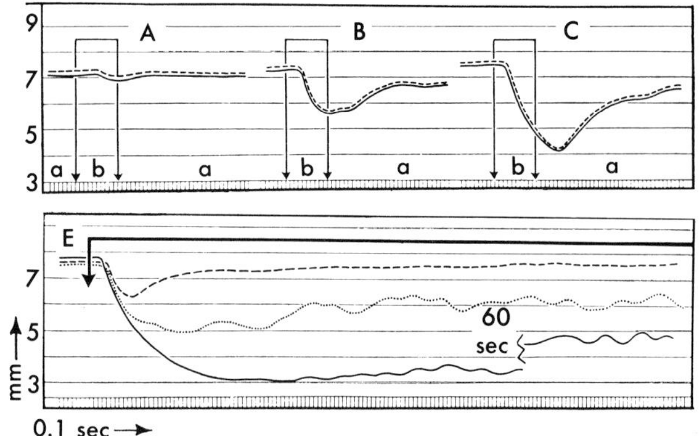

Fig. 12-5. *A, B,* and *C,* Dark-adapted normal subject. Light flashes, *b,* of increasing intensity produce increasing pupillary constriction. Latent period decreases with intensity of flash. Right eye (solid line) was stimulated; left eye (broken line) remained in darkness. Reactions are equal on two sides. *E,* Reaction of pupil to prolonged light of different intensities. (From Lowenstein, O., and Loewenfeld, I. E.: In Davson, H., editor: The eye, New York, 1962, Academic Press, Inc., vol 3.)

ease process, it produces no symptoms, it is common, it is benign. For these reasons it has been called a "simple" anisocoria. The light reactions are normal in both eyes and no further work-up is called for.

Tournay's phenomenon

In 1917 Tournay said that it was normal to have unequal pupils in lateral gaze—with the larger pupil in the abducting eye. It has recently been established that this is not the case;[18,29] almost everyone has equal pupils in all directions of gaze. Lateral gaze induced an anisocoria of more than 0.5 mm in less than 5% of a population studied.[18] In these rare cases it is interesting to speculate that there might be an anomalous co-innervation of the iris sphincter and the medial rectus muscle, so both muscles are inhibited in abduction.

Pupillary reaction to light

The pupil seems easy to observe, but it is also easy to introduce artifacts into the observation and to draw false conclusions because of these artifacts. Through the years a vast and confusing literature on the pupil has accumulated. Lowenstein and Loewenfeld[12,21] have sorted through the maze of conflicting opinion and have tested the evidence with their own observations. Their development and extensive utilization of measurement of the human pupil by infrared light has been helpful to our understanding of pupillary response in health and disease. Much of the following section has been drawn from their work.

Effect and intensity of the stimulus

When the dark-adapted eye is exposed to light flashes of short duration, the pupillary threshold is found to be very low. With the use of appropriate recording techniques, small but distinct pupillary reactions usually can be obtained well within the first log unit of stimulus luminance above the subject's scotopic visual threshold, except for stimuli shorter than about 100 milliseconds, inasmuch as intensity and duration are interchangeable for very short flashes.

When the intensity of the light is increased over a range of approximately 3 log units, the pupillary contractions become more constant and more extensive. Throughout this low-intensity range of luminance the responses are, however, typically shallow; the contraction is preceded by a long latent period, and it is slow, inextensive, and of short duration. When the light intensity is further increased, the reflexes begin to grow markedly in amplitude, speed of movement, and duration of contraction until maximal values are reached at about 7 to 9 log units above the scotopic visual threshold. This sudden increase in effectiveness of the light stimuli occurs because the cone threshold has been exceeded.

Very powerful light flashes fail to add further to the amplitude and speed, and they do not reduce the latent period of the reactions, but they greatly prolong the contraction; after such stimuli the pupil may remain in spastic miosis for several seconds. There is no doubt that the afterimage contributes to this delayed redilation.

Modifying effects of fatigue and emotional excitement

Fatigue and emotional excitement are so much a part of everyday life and their modifying influence on pupillary diameter and reactions is so profound that their effects must be well understood and constantly borne in mind when considering the pupillary reaction to light.

The light reflex is not independent of the subject's level of consciousness. While the subject is alert, the central synapse of the pupillomotor reflex arc in the Edinger-Westphal nucleus is subject to supranuclear inhibitory influences. Simultaneously, hypothalamic discharges are brought into play by sensory or emotional stimuli provided by the environment or, at least in humans, by spontaneous thoughts or emotions; they travel via the brain stem, cervical cord, and peripheral sympathetic chain to the dilator muscle of the iris. Under the influence of these mechanisms the pupil in healthy alert subjects is relatively large and quiet in darkness, and this condition may be maintained for long periods of time. But when the subject becomes tired, the pupils gradually become smaller and begin to oscillate. In the moments immediately preceding sleep, the pupils are quite small, but a psychosensory stimulus such as a sudden sound will restore the waking condition and redilate the pupils.

As the subject drifts toward sleep, cerebral and diencephalic centers cease to function in an orderly sequence. Supranuclear inhibition of the Edinger-Westphal nucleus decreases, and sympathetic activity is gradually lost. The consequent relative preponderance of parasympathetic outflow results in miosis.[22] At the moment of spontaneous or reactive awakening, the sympathetic activity and supranuclear inhibition of the sphincter nucleus cooperate in dilating the pupil.[11]

Pupillary light reflexes are superimposed on this constantly shifting equilibrium of autonomic innervation of the iris, which can be further modified by humoral adrenergic mechanisms and influenced by the mechanical limitations of the iris.

Latent period of the light reflex

The latent period of the light reaction is relatively long, as is to be expected in a reflex with a smooth muscle effector. With a very bright light the latent period can be approximately 0.2 second, and as the stimulus is dimmed the latent period is prolonged until it may approach 0.5 second.[23] Because the response is slow, contraction to light flickering at the low rate of 5 Hz is fused into a steady contraction.

Spectral sensitivity of the light reaction

The pupillomotor effectiveness of a colored light simulus is related to its apparent brightness; for each color, the threshold of pupillary reactions is almost as low as the corresponding visual threshold. This is true for all areas of the retina and in the dark-adapted as well as the light-adapted eye. In other words, the Purkinje shift and the rod-cone break can also be seen in dark-adapted pupillary responses. (These visual phenomena are discussed in Chapters 19 and 22.)

Pupillomotor sensitivity of various parts of the retina

The retinal periphery is far less efficient than the fovea for the production of extensive pupillary contractions, and if the amplitude of the light reaction were used as a measure of retinal sensitivity, the pupillomotor sensitivity of the fovea would seem to be far greater than that of the periphery. However, with light stimuli of low intensity, the periphery of the retina has a much lower pupillomotor threshold and is thus much more sensitive than the fovea, although the light reactions produced by stimulating the periphery are of low amplitude. The pupillary sensitivity of the retina is thus remarkably similar to the visual sensitivity.

Consensual light reflex

The stimulation of one retina by light produces a contraction of the pupil in the opposite eye in all animals in which there is partial decussation of both the optic nerve fibers in the chiasm and the pupillary fibers in the midbrain. In primates the hemidecussation in the dorsal midbrain seems to be functionally 50-50. That is, any pupillomotor stimuli arriving in the midbrain are divided equally to the two eyes.

Consensual deficits

It has been suggested by careful observers that in humans this wiring is not as precise as it seems and that a large number of normal subjects will have a slightly smaller consensual reaction than direct reaction. However, most of these differences are of the order of 0.1 mm, and clinically visible consensual deficits (0.4 mm or more) are rare (perhaps 5% of the population). Unilateral consensual deficits are more common than bilateral consensual deficits ("alternating contraction of anisocoria" of Lowenstein).[19]

Animals phylogenetically lower than primates normally have a bilateral consensual deficit because the pupillary pathways in midbrain and chiasm are more crossed than uncrossed. An owl illuminated from the left will have an anisocoria, with the left pupil smaller than the right. This situation is rare in humans and in monkeys, and when present it is seldom recognized. An impaired consensual pupillary reaction is not of any known clinical significance except as a potential source of confusion when looking carefully at the pupillary light reactions (Fig. 12-6). If you are going to look carefully, it is worth knowing about.[32] A consensual deficit can be readily seen in flash photographs taken with one eye in bright light and the other in darkness.

Accommodation-convergence reaction (near reflex)

When a person is requested to direct his eyes to an object held close to the face, his pupils contract. This contraction is independent of any change in illumination and depends on an association between the sphincter pupillae and the ciliary muscle and the medial recti. It is not a true reflex but an associated movement, or synkinesis. It may be said that the impulse to look at an object close at hand should associate these three mechanisms, for they serve a common purpose. The medial recti contract so that the image of an object close to the eyes will be thrown on both foveas. Contraction of the ciliary muscles takes place so that the image

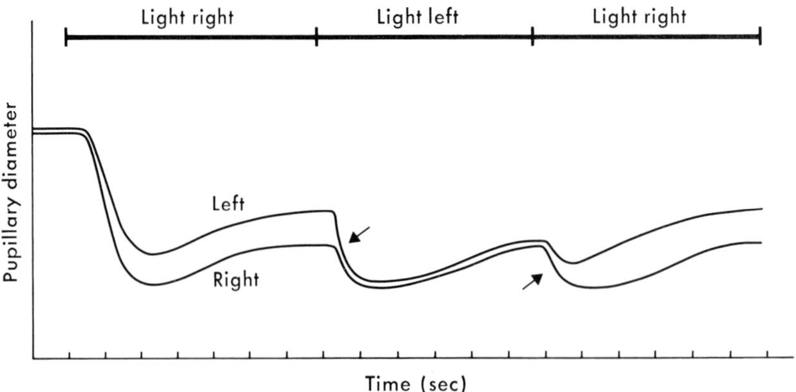

Fig. 12-6. Pupillographic tracing of a patient with a left consensual deficit. Notice that an anisocoria is present only when the right eye is stimulated and the left eye is in darkness. Since in the ordinary swinging light test you look only at the illuminated eye, you may not be aware that you are producing an anisocoria; all you see is the larger initial constriction in the eye with the impaired consensual reaction (arrows). A unilateral consensual deficit presents itself clinically during the swinging light test as asymmetric initial pupillary constriction in the two eyes without the usual accompanying asymmetric pupillary escape seen in relative afferent defects. (From Thompson, H. S.: Trans. Ophthalmol. Soc. U.K. **96:**377, 1976.)

will be in focus on each fovea, and contraction of the pupil takes place, probably as a further optical aid to increase the depth of field. All three components depend on supranuclear connections; it is evident that the motivation for clear, single binocular vision requires cortical processing of the messages from the retinal receptors. The parasympathetic discharges to the sphincter are markedly influenced by activity in the cortex. It is fair to say that, although the near reaction has been the subject of much investigation, we do not yet know the intimate details of the cortical anatomy and physiology mechanisms involved. Since this pupillary reaction is a synkinesis that depends on the nearness of the object of regard, it is best termed the near-point reaction, or near reaction of the pupil.

The contraction of the pupil "at near" does not depend on either the accommodation or the convergence any more than does the convergence or accommodation depend on others of the triad. All three must be looked on as being associated in a common function. Normally they work together, but one may be dissociated from the others. Thus it is possible to have contraction of the pupil with convergence of the visual axes without accommodation by placing plus lenses in front of each eye to take the place of the accommodation needed for the near point. Likewise, convergence can be prevented by placing base-in prisms in front of each eye. When a person is asked to read fine print under both these circumstances, the pupil will contract. The near pupillary contraction exists in uncorrected myopic persons and in very old people who have lost all accommodation. The constriction of the pupil takes place equally in both eyes, although one is covered. This is true even though the vision in one eye is considerably impaired.

When a near effort is difficult to obtain, try the "orbicularis reflex."[21] The patient is asked to read the chart at distance while trying to close his eyes. One eye is held open by the examiner. This often brings out an involuntary "near reaction."

PHARMACOLOGY OF THE PUPIL

The parasympathetic and sympathetic neural impulses to the iris muscles can be modified by drugs at the synapses and at the effector sites because it is at these locations that the transmission of the impulse depends on chemical mediators. Drugs can also have central actions that affect the pupils. These drugs and their various modes of action are summarized in Fig. 12-7 and in the following paragraphs, where they are grouped according to the site and mechanism of their action.

A few cautionary words should first be said about the interpretation of pupillary responses to topically instilled drugs. There are large interindividual differences in the responsiveness of the iris to topical drugs. This becomes most evident when weak concentrations are used. For example, 0.25% pilocarpine will produce a minimal constriction in some patients and an intense miosis in others. This means that the most secure clinical judgments stem from comparisons with the action of the drug on the other, normal eye.

It should also be remembered that the general status of the patient will influence the size of the pupils. If the patient becomes uncomfortable or anxious while waiting for the drug to act, both pupils may dilate. If the patient becomes drowsy, both pupils will constrict. Thus, if a judgment is to be made about the dilation or contraction of the pupil in response to a drug placed in the conjunctival sac, one pupil should be used as a control whenever possible.

If only one eye is involved, the drug should be put in both eyes so that the response of the normal and abnormal eye can be compared. When the condition is bilateral, no such comparisons are possible, but an attempt should be made to make sure that the observed response is indeed caused by the instilled drug. Thus in bilateral cases the drop should be put in one eye only so that the responses of the medicated and unmedicated eyes can be compared.

Parasympatholytic (anticholinergic) drugs

The belladonna alkaloids occur naturally. They can be found in various proportions in deadly nightshade *(Atropa belladonna),* henbane *(Hyoscyamus niger),* and jimsonweed *(Datura stramonium)*. Potions made from these plants were the tools of professional poisoners in ancient times. The word "belladonna"

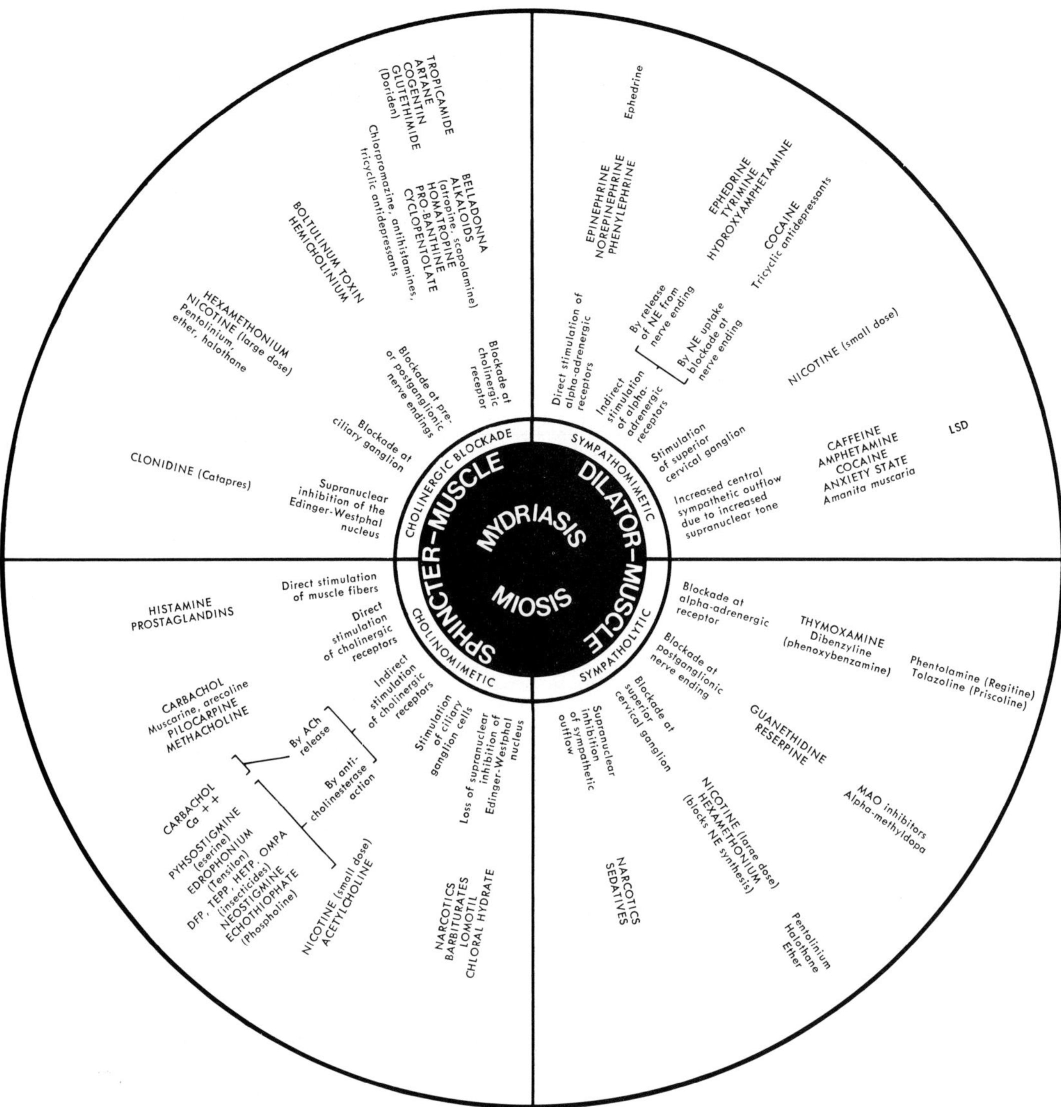

Fig. 12-7. Drugs that affect the size of the pupil and their mechanism of action.

(''beautiful lady'') was derived from the cosmetic use of these substances as mydriatics in sixteenth-century Venice. Youth and excitement have always dilated the pupils, and age and boredom still constrict them. The mischief caused by the ubiquitous jimsonweed is typical of this group of plants.[30] Jimsonweed has been used as a poison, has been taken as a hallucinogen, and has caused accidental illness and death, and it can cause an alarming accidental mydriasis. These solaneceous plants, which are related to the tomato, potato, and eggplant, are still cultivated for medicinal purposes.

Atropine and scopolamine block parasympathetic activity by competing with acetylcholine at the effector cell (i.e., iris sphincter), thus preventing depolarization. After conjunctival atropine (1%), mydriasis begins within about 10 minutes and is fully developed at 35 to 45 minutes; cycloplegia is complete within 1 hour. The pupil may stay dilated for several days, but accommodation usually returns in 48 hours.

Table 12-1. Response of the pupil to miotic drugs—parasympathetic defect*

Topically applied drug	Normal	Third nerve lesion, e.g., aneurysm (preganglionic)	"Adie's" tonic pupil (postganglionic)	Atropinized pupil (pharmacologic blockade)
Mecholyl 2.5%	O No miosis	O → + No miosis	O → + → +++ Mild-marked miosis	O No miosis
Pilocarpine 0.125%	O → + Minimal miosis	+ Minimal miosis	++ → +++ Moderate → marked miosis	O No miosis
Pilocarpine 0.25%	+ Mild miosis	+ → ++ Mild miosis	+++ → ++++ Marked miosis	O No miosis
Pilocarpine 0.50%	++ Moderate miosis	++ Moderate miosis	++++ Intense miosis	O No miosis
Pilocarpine 1.0%	+++ Marked miosis	+++ Marked miosis	++++ Intense miosis	O No miosis

*From Thompson, H. S.: Bristol Med. Chir. J. **90:**37, 1976.

Scopolamine (0.2%) causes mydriasis that lasts, in an uninflamed eye, for about 2 days; it is a less effective cycloplegic than atropine.

Homatropine and eucatropine (Euphthalmine) are synthetic anticholinergic drugs that were introduced into ophthalmology late in the nineteenth century. Homatropine (2% to 5%) causes mydriasis that lasts for 1 to 2 days, but repeated applications are necessary for effective cycloplegia. Eucatropine mydriasis (2% to 10%) lasts for less than 12 hours and is accompanied by only slight cycloplegia.

Tropicamide (Mydriacyl) and cyclopentolate (Cyclogyl) are more recently developed synthetic parasympatholytics with a short duration of action. Tropicamide (1%) is an effective, short-acting mydriatic (3 to 6 hours), which results in only a very transient paresis of accommodation. Compared with tropicamide, cyclopentolate (1%) seems to be a more effective cycloplegic and a slightly less effective mydriatic, especially in dark eyes; accommodation takes about half a day to return and the pupil still may not be working perfectly after more than 24 hours.

Botulinum toxin blocks the release of acetylcholine, and hemicholinium interferes with the synthesis of acetylcholine both at the preganglionic and at the postganglionic nerve endings, thus interrupting the parasympathetic pathway in two places. The outflow of sympathetic impulses is also interrupted by systemic doses of these drugs, since the chemical mediator in sympathetic ganglia is also acetylcholine.

Parasympathomimetic (cholinergic) drugs

Pilocarpine and methacholine (Mecholyl) are structurally similar to acetylcholine and are capable of depolarizing the effector cell, thus causing miosis and spasm of accommodation. Mecholyl is still sometimes used in a weak (2.5%) solution to test for cholinergic supersensitivity of the sphincter muscle. It is being replaced by weak pilocarpine (0.1%) (Table 12-1).

Carbachol (carbamylcholine, Doryl) acts chiefly at the postganglionic cholinergic nerve ending to release the stores of acetylcholine. There is also some direct action of carbachol on the effector cell. A 1.5% solution causes intense miosis, but the drug does not penetrate the cornea easily and is therefore usually mixed with a wetting agent (1:3500 benzalkonium chloride).

Acetylcholine is liberated at the cholinergic nerve endings by the neural action potential and is promptly hydrolyzed and inactivated by cholinesterase. Cholinesterase, in turn, can be inactivated by any one of many anticholinesterase drugs. These drugs either block the action of cholinesterase or deplete the stores of the enzyme in the tissue. They are thus able to potentiate the action of the chemical mediator by preventing its destruction by cholinesterase. It follows from their mode of action that these drugs will lose their cholinergic activity after the nerve supply has been completely blocked.

Along the Calabar coast of West Africa the native tribes once conducted trials "by ordeal" using a poison prepared from the bean of the plant *Physostigma venenosum*. The local name for this big bean was the "esere nut." If justice was ever served at these trials it was surely for the wrong reasons. The guilty man can be imagined, lingering at the poison cup, sipping slowly, aware that this was the end; he might thus take the time to absorb a fatal dose. However, the innocent man, knowing that the potion could not harm him, would cheerfully quaff it down, and his stomach would respond to this assault by vomiting it all back up.

A paper published by Argyll Robertson helped to introduce the miotic Calabar extract into clinical ophthalmology. Physostigmine (eserine) was first used to stretch out the iris so that a peripheral iridectomy could be done more easily.[9]

The organic phosphate esters (echothiophate [Phospholine], isoflurophate [diisopropyl fluorophosphate—DFP], tetraethyl pyrophosphate, hexaethyltetraphosphate, parathion), many of which are in widespread use as insecticides, cause a much longer lasting miosis than the other anticholinesterases, but even this potent effect, thought to be due to interference with cholinesterase synthesis, can be reversed by pralidoxime chloride (P-2-AM).

Sympathomimetic drugs

Epinephrine (Adrenalin) is capable of direct stimulation of the receptor sites of the dilator cell. When applied to the conjunctiva, the 1:1000 solution does not penetrate into the normal eye in sufficient quantity to have an obvious adrenergic effect. If, however, the receptors have been made supersensitive by previous denervation, this concentration of epinephrine usually dilates the pupil. Phenylephrine (Neo-Synephrine) in the 10% solution has a powerful mydriatic effect. Its action is almost exclusively a direct alpha stimulation of the effector cell. The pupil recovers in 8 hours and shows a "rebound miosis" lasting several days.[5] Both 5% and 2.5% solutions are now commonly used. Ephedrine acts chiefly by releasing endogenous norepinephrine from the nerve ending, but it also has a definite direct stimulating effect on the dilator cells. Tyramine and hydroxyamphetamine act adrenergically by releasing norepinephrine from the stores in the postganglionic nerve endings; as far as is known this is their only effective mechanism. Cocaine (5% to 10%) is applied to the conjunctiva as a topical anesthetic, as a mydriatic, and as a test for Horner's syndrome (Table 12-2). Its mydriatic effect is the result of an accumulation of norepinephrine at the receptor sites of the dilator cells. The amount of transmitter substance builds up at the neuroeffector junction because cocaine prevents the reuptake of the norepinephrine back into the cytoplasm of the nerve ending. The action of cocaine is thus analogous to the action of the anticholinesterases at the cholinergic junction in that it interferes with the mechanism for prompt disposition of the chemical mediator.

Cocaine itself has no direct action on the effector cell nor does it serve to release norepinephrine from the nerve ending. Furthermore, it does not block the physiologic release of norepinephrine from the stores in the nerve ending. Since the reuptake mechanism is blocked, the norepinephrine accumulates at the junction until it throws the muscle cell of the iris dilator into spasm. If the nerve action potential in the sympathetic pathway is interrupted, as in Horner's syndrome, the transmitter substance will not accumulate and the pupil will not dilate. The duration of cocaine mydriasis is quite variable; it may last more than 24 hours. It does not show "rebound miosis."[5]

Sympatholytic drugs

Thymoxamine is a clean alpha-adrenergic blocker that will reverse phenylephrine mydriasis by occupying the alpha-receptor sites on the iris dilator muscle. The other drugs producing alpha-receptor blockade (dibenzyline [phenoxybenzamine], phentolamine [Regitine], tolazoline [Priscoline]) are less precise in their modes of action and are no longer used in clinical ophthalmology.

Guanethidine (Ismelin) and reserpine interfere with the normal release of norepinephrine from the nerve ending and deplete the norepinephrine stores. When applied to the eye, they cause Horner's syndrome complete with ptosis, miosis, and supersensitivity to adrenergic drugs.

Other drugs that affect the pupil

Histamine is a substance that can be obtained as an extract of various tissues. It acts as a protoplasmic irritant and affects the sphincter fibers directly; it will constrict the pupil of a thoroughly atropinized eye.

The chief pupillary action of *morphine* is to cut off cortical inhibition of the Edinger-Westphal nuclei with resultant miosis. It may also have a direct central stimulating action on the sphincter nucleus. Topical morphine, however, even in strong solutions (5%) has a minimal miotic effect on the pupil.

Nalorphine (Nalline) and levallorphan (Lorfan) are antinarcotic agents that, when given subcutaneously, reverse the miotic action of morphine.

During the induction of anesthesia the patient may be in an excited state and the pupils are often dilated. As the anesthesia deepens, supranuclear inhibition of the sphincter nuclei is cut off and the pupils become small. If the anesthesia becomes dangerously deep and begins to encroach on the midbrain, the pupils become dilated and fail to react to light.

The concentration of *calcium* and *magnesium* ions in the blood may affect the pupil. Calcium facilitates the release of acetylcholine, and when calcium levels are abnormally low, the amount of acetylcholine liberated by each nerve impulse drops below the level needed to produce a postsynaptic potential, thus effectively blocking synaptic transmission. Magnesium has an opposite effect; thus a high concentration of magnesium can block transmission and may dilate the pupil.

Topical dexamethasone (Decadron) not only tends to increase the intraocular pressure, but at the same time acts as a mild mydriatic. The reaction of the pupil to light does not, however, appear to be blocked. The mechanism of this mydriasis is not yet understood.

Maximal mydriasis

Full mydriasis can be obtained with eyedrops by using cocaine 10%, cyclopentolate 1%, and phenylephrine 10%, in that order, waiting only 3 minutes between drops. The peak effect is in 1 hour. The cocaine increases the corneal penetration of the other drugs, adds to the tightening of the dilator muscle, numbs the eye, and diminishes reflex tearing that might dilute the other drops. The cyclopentolate takes the sphincter out of action at 30 minutes so that when a high concentration of phenylephrine hits the dilator muscle 20 to 30 minutes later, a sudden and maximal mydriasis occurs.

Iris pigment and pupillary response to drugs

In general, the more pigment in the iris, the more slowly the drug takes effect and the longer its action lingers. This is probably due to the drug being bound to iris melanin and then slowly released. It should be noted that there are wide individual differences in pupillary responses to topical drugs. There is probably a greater range of responses among blue eyes than there is between the average response of blue eyes and the average response of dark-brown eyes. Some of these individual differences are due to corneal penetration of the drug.

AFFERENT PUPILLARY DEFECTS

The pupil of an eye that is blind from retinal or optic nerve disease will fail to react directly to light, but will constrict consensually when the other, healthy, eye is stimulated. The blind eye is said to show a "relative afferent pupillary defect" because the pupillomotor stimulus reaching the brain from the blind eye is diminished relative to the seeing eye.

Swinging light test

Relative afferent pupillary defects are best seen by swinging a hand-held light from one eye to the other. Understanding and making use of this sign is of the greatest clinical importance. The pupils are clinically more useful as indicators of optic nerve disease than they are as indicators of the integrity of iris innervation.

The patient looks into the distance while the

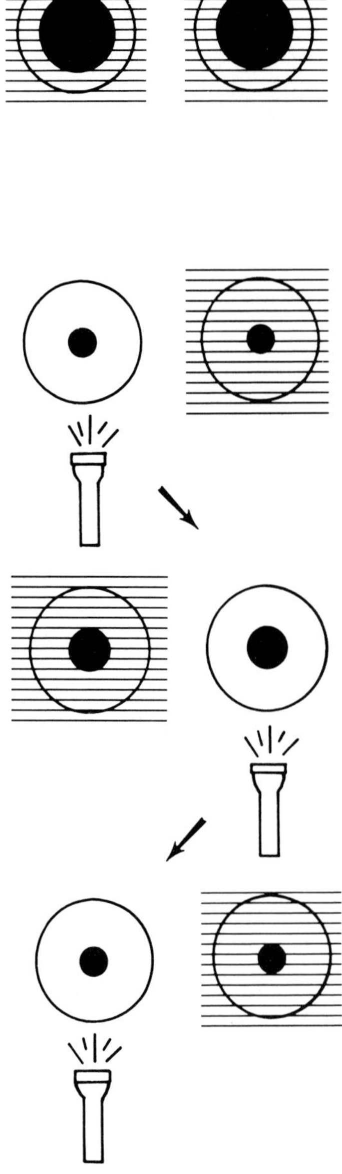

Fig. 12-8. "Swinging light test" done in dim light in a patient with a moderate relative afferent pupillary defect.

examiner shines a bright hand-held light first into one eye and then the other. The light is quickly shifted across to the other eye every 2 to 4 seconds. As the light shifts from the "good" eye to the "bad" eye, the direct light stimulus is no longer sufficient to keep the pupils small, so they both dilate (Fig. 12-8). This test is quick and easy to do, and it will reveal subtle differences in the photomotor input of the two eyes. For example, a definite pupillary dilation is usually seen in an eye that has recovered to 20/20 after an attack of optic neuritis.

Cautionary notes and hints

Only one visible and working iris sphincter is required to test for afferent defects. The test can be done in the presence of a third nerve palsy, corneal opacity, or posterior synechia by observing the direct and consensual responses of the intact pupil as each eye is alternately stimulated (Fig. 12-9).

If a bright hand-held light is used in a darkened room, the pupillary excursions are amplified; this makes the test more sensitive. The light should not be left longer on one eye than on the other because this would tend to bleach one retina more than the other. This tends to create a small afferent defect in the eye in which the examiner is expecting a defect.

An anisocoria in bright light that is larger than 2 mm (i.e., a very large anisocoria) will tend to produce a small afferent defect because one retina is being shaded by the iris more completely than the other.

In a patient with a profound suppression amblyopia there is often a small relative afferent pupillary defect in the amblyopic eye. A patient with half as much loss of acuity due to retrobulbar neuritis would have a dramatic afferent defect.

A certain number of normal, healthy people (perhaps 5% to 10%) will have a clinically visible *consensual deficit,* such that the direct reaction is larger than the consensual reaction. This produces an anisocoria in light when none was present in darkness (p. 332). A unilateral consensual deficit can be a source of confusion when looking carefully for a relative afferent pupillary defect.[32]

As a rule, a cataract does not cause an afferent pupillary defect. If an afferent pupillary defect is present in a cataractous eye, there is almost certainly something else going on back there, and the prognosis for sharp aphakic vision is not good.[4,32] It is not fully understood why this is such a good clinical rule. Surely a

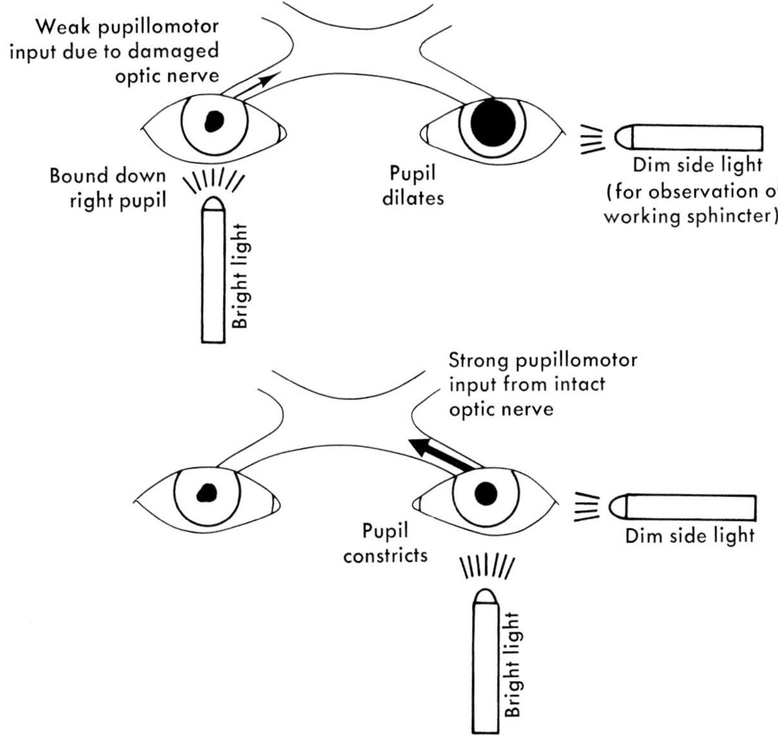

Fig. 12-9. Demonstrating an afferent pupil defect when one eye has an immobile pupil. When one pupil *(OD)* is immobilized by posterior synechiae, it is still possible to demonstrate an afferent defect by observing the direct and consensual response of the intact pupil. The side light must be kept very dim or it will induce an afferent defect.

brunescent lens must block most of the light entering the pupil. Is it because the shaded retina is more sensitive? Does the cataract scatter the light onto more retinal elements?

Measuring afferent pupillary defects

It is possible simply to record the afferent defect as present or absent for each eye without having to mumble about "direct and consensual responses." The great advantages of the swinging light test are that it automatically checks the consensual response and that it can be quickly and simply recorded.

An effort is sometimes made to refine the test by estimating the amount of relative input defect and making a comment in the patient's chart, for example, "obvious afferent defect" or "mini-afferent." Some careful observers have tried to judge the relative defect on a scale of 1+ to 4+, but two competent observers may score the afferent defect differently. It is possible to quantify the relative afferent pupil defect by putting neutral density filters over the good eye until the defect disappears.

The stimulus to the good eye must be dimmed by a certain number of log units before a balance point is reached at which no relative afferent defect can be seen. This number (in log units of neutral density filter) is a *measure* of the relative pupillomotor input deficit in the bad eye.

When making these measurements it is important to reach a decision about the afferent defect promptly (within two or three swings), then take the filter away and flash both retinas with the light to balance the retinal bleach before proceeding with the next filter. If this technique is not used, it is very difficult to get a reproducible end point. The balance point keeps slipping away; thus more and more filter density is required. This is because the better eye is shaded by the filters and is becoming relatively dark adapted while the retina of the eye with the afferent defect is becoming progressively more bleached. This tends to exaggerate the difference in pupillomotor input of the two eyes. These measurements provide an objective indicator of the amount of retina or optic nerve dysfunction relative to the other eye.

Afferent defects in tract lesions

Consider the patient with (1) an incongruous homonymous field loss, (2) nothing to suggest superimposed optic nerve or retinal disease, and (3) an afferent pupillary defect in the eye with the greater field loss. In this patient the field loss is more likely to be due to a tract lesion than to a suprageniculate lesion because, as would be expected in a tract lesion, the pupillary fibers seem to be damaged to the same extent that the visual fibers are damaged. The patient with a complete tract lesion, a total homonymous hemianopia with 20/20 vision in each eye, and a band pallor of the contralateral disc fits into this description, since a temporal field has been lost in one eye and a nasal field in the other. These patients also have a small afferent defect in the eye with the greater field loss.

The other pupillary signs of tract lesions (Behr's sign and Wernicke's pupillary hemiakinesia) cannot be counted on to be of clinical value.

Pupil cycle time

A small beam or slit of light focused at the pupillary margin induces regular, persistent oscillations of the pupil. These oscillations can be timed with a stopwatch. The period of the average complete cycle is called the "pupil cycle time" (Fig. 12-10).

The pupil cycle time is the time required for the constriction and redilation of the pupil when the pupillary system chooses its own rate of oscillation. There are many factors that could slow down the pupillary reflex arc: the speed, frequency, and intensity of the afferent nerve impulses reaching the midbrain; the synaptic delays; the efferent nerve delays; and any structural slowness of the iris musculature itself.[25] The pupil cycle time tends to be prolonged in optic neuritis and in compressive optic neuropathy.

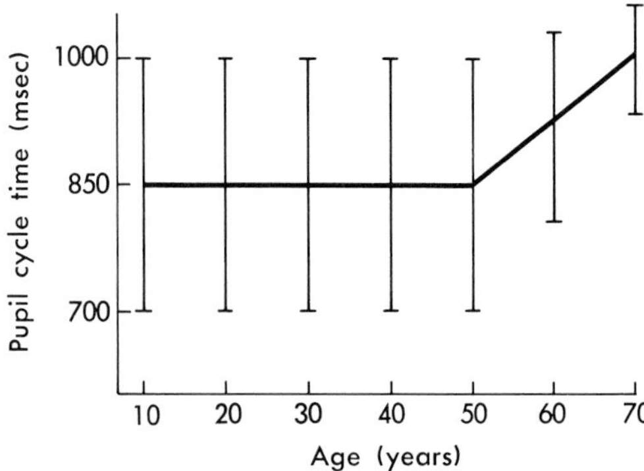

Fig. 12-10. Age and pupil cycle time. An approximation based on the combined data of Miller[25] and Zahoruk and Thompson (unpublished data, 1978). In 100 normal subjects the mean pupil cycle time was 850 milliseconds. This did not change until approximately age 50, after which it lengthened by 7.5 milliseconds per year. Under age 50, 95% of the measurements fell between 700 and 1000 milliseconds. In a healthy young subject the pupil cycle time can be measured repeatedly with a standard deviation of 15 to 20 milliseconds, but with age or with afferent or efferent disease the cycling is less certain, the measurements become less precise, and the standard deviation may extend to 40 to 50 milliseconds.

EFFERENT PUPILLARY DEFECTS

When the parasympathetic innervation to the iris sphincter is impaired, the pupil reacts poorly to light and may appear to be fixed in the dilated position.

"Fixed" dilated pupil

The differential diagnosis of a "fixed" dilated pupil is summarized diagrammatically in Fig. 12-11 and below:

I. Midbrain damage—vascular accidents, tumors, degenerative and infectious diseases, etc.
   A. Dorsal midbrain (the Edinger-Westphal nucleus and its connections)
      1. Relatively rare
      2. Usually involves both pupils
      3. Pupillary near-vision reaction often retained
      4. Often associated with supranuclear vertical gaze palsy
   B. Ventral midbrain (fascicular portion of third nerve)
      1. Often associated with other neurologic

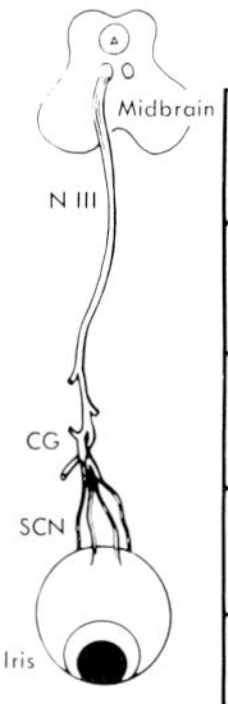

Pupil with poor light reaction

		Slit lamp signs					
	Sensitivity to cholinergic substances	Pupil immobile: "fixed" to light	Segmental palsy of iris sphincter	Segmental distortion of stroma: "stromal streaming"	Segmental pupil contraction with eye movement (Czarnecki)	Torn sphincter or pupil margin loose pigment, iris atrophy transillumination	Diplopia, ptosis, or EOM defect
Acute third nerve palsy or paresis	Normal sensitivity	Variable	Variable	Rare	Never	Never	Very common
Regenerated oculomotor palsy or paresis	Often moderately supersensitive	Variable	Common	Common	Common	Never	Very common
Tonic pupil (e.g., Adie's syndrome)	Supersensitive	(10%)	Very common	Very common	Never	Never	Never
"Atropinic" mydriasis	Profoundly undersensitive	Common	Never	Never	Never	Never	Never
Iris damage (traumatic iridoplegia and iris ischemia)	Undersensitive	Variable	Variable	Common	Never	Very common	With associated orbital injury

Midbrain
N III
CG
SCN
Iris

Fig. 12-11. Differential diagnosis of the pupil with a poor light reaction, with emphasis on the slit-lamp signs. Common—frequently seen, perhaps more than 50% of patients. Very common—expected, usually seen. Variable—depends on the extent of damage. Rare—I have never seen it, but do not want to swear it is impossible. Never—not only have I never seen it, but given the nature of the problem, it is not to be expected.

deficits, for example, Nothnagel's, Benedikt's, or Weber's syndromes
2. Unlikely to spare the extraocular components of the third nerve

II. Damage to the third nerve (from interpeduncular fossa to ciliary ganglion)
   A. Basal aneurysms
   B. Supratentorial space-occupying masses
      1. Cause displacement of the brain stem or transtentorial herniation of the uncus
      2. Usually stuporous or comatose
   C. Basal meningitis—often causes bilateral internal ophthalmoplegia
   D. Ischemic oculomotor palsy ("diabetic ophthalmoplegia")—usually spares the pupillary fibers
   E. Parasellar tumor (e.g., pituitary adenoma, meningioma, craniopharyngioma, nasopharyngeal carcinoma, or distant metastases)
   F. Parasellar inflammation (e.g., "Tolosa-Hunt" syndrome, temporal arteritis, herpes zoster)

III. Damage to the ciliary ganglion or short ciliary nerves—results in "tonic pupil"
   A. Local tonic pupil
      1. Viral ciliary ganglionitis (e.g., herpes zoster)
      2. Orbital or choroidal trauma, tumor, or injury
      3. Blunt trauma to the globe; may injure branches of the short ciliary nerves at the iris root (traumatic iridoplegia)
   B. Neuropathic tonic pupil—part of the picture of peripheral neuropathy
   C. Idiopathic tonic pupil with benign areflexia (Adie's syndrome)

IV. Damage to the iris
   A. Degenerative or inflammatory diseases of the iris
   B. Posterior synechiae
   C. Acute rise of intraocular pressure (hypoxia of sphincter)
   D. Blunt injury to the globe with sphincter damage (traumatic iridoplegia)

V. Drug-induced mydriasis
   A. Anticholinergic blockage at the sphincter muscle (e.g., atropine, etc.)
   B. Adrenergic stimulation of the dilator muscle (e.g., phenylephrine, cocaine, hydroxyamphetamine)

Of all the possible causes of a poor light reaction in one eye, Adie's syndrome is probably the most common. About 90% of Adie's pupils have a sector of residual light reaction when examined at the slit lamp. However, a patient who walks into an ophthalmologist's office with a dilated and absolutely unresponsive pupil of recent onset probably has been exposed to an atropinic drug; intracranial aneurysm and Adie's tonic pupil are less likely possibilities. Nurses and pharmacists who handle these drugs may get them on their hands and inadvertently rub them into the eye. Farmers and others who work outside may be exposed to plants that are rich in belladonna alkaloids, and an occasional patient will deliberately use a mydriatic drop or ointment and then deny it. There is a simple way for the ophthalmologist to recognize the pupil dilated because of pharmacologic blockade and to distinguish it from a paralytic mydriasis such as a partial third nerve palsy or Adie's tonic pupil. One drop of a weak pilocarpine solution (0.5% to 1.0%) will contract the normal pupil. It will also contract the pupil in cases of paralytic mydriasis due to preganglionic or postganglionic nerve damage. In fact, an interruption of the parasympathetic innervation of the iris tends to cause supersensitivity of the sphincter to cholinergic drugs. However, when the iris sphincter has been blocked by atropine or other anticholingeric drugs, one drop of a weak pilocarpine solution will not constrict the pupil. Atropinic drugs are successful competitive antagonists of pilocarpine; they use up the available cholinergic receptor sites on the sphincter muscle cell. The pilocarpine test clearly distinguishes parasympathetic denervation from pharmacologic blockade because denervation can only increase the sensitivity of the iris to pilocarpine, whereas atropine drugs decrease pilocarpine miosis.[37]

The Edinger-Westphal nucleus is at the rostral end of the oculomotor group, and the parasympathetic fibers for the pupil and accommodation are at the upper edge of the third nerve as it emerges into the interpeduncular fossa. These smaller fibers stay at the periphery of the bundle and gradually move to a more medial position. When the third nerve enters the cavernous sinus, the pupil fibers have worked their way around to the lower edge of the nerve, so that when the nerve separates into superior and inferior divisions, the pupil fibers are always in the lower bundle.

Basal meningitis, especially tubercular or syphilitic, was once a common cause of bilateral internal ophthalmoplegia. This was considered to be due to the peripheral and hence vulnerable location of the parasympathetic fibers in the subarachnoid course of the oculomotor nerve. The third nerve can be stretched or compressed by an adjacent aneurysm, and this usually produces a complete oculomotor palsy but occasionally the first sign is a dilated pupil—usually soon followed by ptosis or diplopia.

In the obtunded or comatose patient, unilateral pupillary dilation is an important sign of transtentorial herniation of the uncus. This is also due to stretching and compression of the third nerve. Again, the pupillary fibers may be damaged early but are soon followed by impairment of fibers serving the extraocular muscles.

Aberrant regeneration of the third nerve

After a third nerve palsy due to an aneurysm, the fibers start to grow again, and in 6 to 8 weeks there appears to be some recovery. The lid lifts, but the pupillary light reaction seldom recovers,[8] and there is usually a residual diplopia in various directions of gaze.

Nerves, like the oculomotor nerve and the facial nerve, which carry fibers bound for several different destinations in a single bundle, are particularly subject to aberrant regeneration. If there has been scar formation or structural damage to the nerve, the signs of aberrant regeneration are to be expected.

An ischemic oculomotor palsy (a pupil-sparing third nerve palsy) usually has recovered completely in 2 months without signs of aberrant regeneration. In fact, if the following signs appear as the nerve is recovering, the possibility of aneurysm should be reconsidered.

1. Inappropriate lid retraction on adduction or depression (pseudo–von Graefe's sign)
2. Pupil constriction on adduction, elevation, or depression, especially up and in
3. Segmental constriction of parts of the iris sphincter with eye movements, a slit-lamp sign.[2]

It is common for fibers destined for the extraocular muscles to innervate the pupil during the recovery of a third nerve palsy. This results in a pupillary constriction when an effort is made to turn the eye in one of the directions served by the third nerve. Because the pupil fails to react to light but may constrict on convergence, this has been called a "pseudo–Argyll Robertson pupil," but the pupil in such a case will constrict with any adduction movement of the eye.

With the magnification provided by the slit lamp it can be seen that this aberrant reinnervation of the iris sphincter is sometimes segmental and of a degree insufficient to produce an obvious pupillary constriction.[2]

Cyclic oculomotor spasms

Patients with a congenital oculomotor palsy will sometimes show an intermittent spasm of all the components of the third nerve. These cyclic spasms on a background of paresis ("cyclic oculomotor palsy") may well be a mixture of aberrant regeneration of the peripheral nerve and an abnormality of the oculomotor nucleus that results from damge to the immature nervous system.[17]

Iris damage

Blunt injury to the globe will cause a traumatic iridoplegia. There seem to be several factors at work: (1) the chamber angle can be recessed, tearing the branches of the short ciliary nerves that serve the iris muscles; (2) the sphincter muscle itself can be injured, so it will not constrict to pilocarpine; (3) the short ciliary nerves may be damaged as the choroid is ruptured. These injuries may produce a segmental palsy of the sphincter, a light-near dissociation, and an undersensitivity or oversensitivity of the sphincter to pilocarpine, depending on the exact nature of the damage (Fig. 12-11). An attack of acute angle-closure glaucoma often produces a pupil in middilation that fails to react to light or to pilocarpine. This is due to hypoxia of the iris muscles.

TONIC PUPIL

Damage to the ciliary ganglion or short ciliary nerves produces a very characteristic combination of signs:

1. A poor pupillary reaction to light, which, at the slit lamp, can be seen to be a regional palsy of the iris sphincter
2. Accommodative paresis
3. Cholinergic supersensitivity of the denervated muscles
4. Often a pupillary response to near vision that is unusually strong and tonic

Patients who show these signs are said to have "tonic pupils." That is, a "tonic pupil" is any postganglionic, parasympathetic denervation of the intraocular muscles. This can, of course, occur due to a local infection or injury (e.g., chickenpox, orbital surgery), or it may be part of a widespread peripheral neuropathy (e.g., diabetes, alcoholism, etc.).

ADIE'S SYNDROME

Adie's syndrome is a form of tonic pupil in which no local cause for the denervation is evident and there is no peripheral neuropathy to account for the tendon areflexia.

The typical patient with Adie's syndrome is a woman, aged 20 to 40, with one pupil reacting poorly to light. Less than 30% of the patients are men. The mean age of onset is 32 years for both men and women. When first seen, only 10% of the patients have both eyes involved.[33]

Light reaction in Adie's syndrome

When examined at the slit lamp, some response of the sphincter to light can be seen in 90% of the affected eyes. This residual reaction is always a segmental contraction of the iris sphincter. Most of these pupils have lost more than half of the sphincter function and continue to lose more with the passage of time. The loss of light reaction in the denervated pupils seems to occur randomly around the sphincter without any clear predilection for one quadrant.[34]

Cholinergic supersensitivity in Adie's syndrome

In 1905, Markus[2] suggested that a very weak solution of eserine could be used as a diagnostic test to localize the lesion causing a denervation of the iris sphincter; a good constriction put the lesion in the ciliary ganglion or in the short ciliary nerves. This recommendation was based on the work of Anderson, who had shown differences in the behavior of the sphincter depending on the location of the lesion. But it was not until Cannon and Rosenblueth had sorted out all these observations into a "law of denervation supersensitivity" that these phenomena were understood. Scheie and Adler, in 1940, clearly showed that Adie's tonic pupils were supersensitive to cholinergic substances. They chose a concentration (2.5%) of methacholine hydrochloride (Mecholyl) that would not contract a normal pupil, and they showed that most Adie's pupils constricted to this cholinergic stimulus, some quite dramatically.

It was soon recognized that there were some tonic pupils that failed to constrict with 2.5% methacholine. At first this was blamed on an instability of the methacholine solution, and it was recommended that a fresh solution be prepared for each use. It was finally concluded that an aqueous solution of methacholine hydrochloride is reasonably stable, but that there are large interindividual variations in the ability of the drug to penetrate a normal cornea. In addition, there appear to be large variations from one tonic pupil to another in the sensitivity of the denervated iris sphincter.

Methacholine has not been available as a commercial eyedrop for a number of years and has become increasingly difficult to find. This has prompted a search for substitutes, and the most readily available one is pilocarpine, which is known to be stable and is cheap and always at hand.

Pilocarpine 0.125% (⅛%) constricts most normal pupils slightly, with a degree of miosis differing among individuals from just noticeable to about 2 mm (mean = 1 mm), whereas methacholine 2.5% has no consistent miotic effect. There are some advantages to using a concentration of the cholinergic substance sufficient to bring the normal pupil down a small amount. If neither pupil constricts to methacholine 2.5%, as often happens, it could be because the drug has not penetrated the patient's cornea, in which case the absence of a cholinergic supersensitivity has not been demonstrated. However, when the normal pupil shows a drug effect and constricts slightly and the Adie's pupil does not constrict more than the normal pupil, it can be said confidently that the affected sphincter is not supersensitive.

Pilocarpine 0.125% will demonstrate a supersensitivity in 80% of Adie's pupils, whereas Methacholine 2.5% reveals the supersensitivity in only 67% of them. Pilocarpine 0.1% or 0.0625% will also serve to demonstrate supersensitivity of the sphincter. These concentrations of pilocarpine are not only adequate substitutes for Methacholine 2.5%, they are also more sensitive testing substances in unilateral cases simply because they have a slightly stronger miotic action.[1]

Near reaction in Adie's syndrome

The near response of the pupil generally exceeds the light reaction in Adie's syndrome. The near response is slow and steady, and on looking back into the distance it tends to hold the contraction for a few seconds (i.e., it is "tonic"). The reasons for this behavior are not fully understood. The slowness of the tonic pupil might be due to the diffusion of acetylcholine through the aqueous to the supersensitive receptors of the iris sphincter, but the light-near dissociation cannot be so easily explained.

Tonicity. The tonic behavior of an Adie's pupil might be due to the diffusion of acetylcholine released from remaining or regenerated nerve endings through the aqueous to the nearby supersensitive receptors of the iris sphincter.[3,16,28,39] This behavior might also be due to the lack of acetylcholinesterase in the collateral sprouts of regenerating nerves.

Light-near dissociation. The near reaction is not spared; it is restored. Its strength is best explained by postulating aberrant regeneration of fibers that were originally destined for the ciliary muscle into the iris sphincter, so with every effort to focus the eye on a near object, impulses spill into the sphincter, constricting the pupil. Accommodative fibers in the short ciliary nerves outnumber sphincter fibers by 30 to 1. This means that the ciliary muscle will probably receive appropriate reinnervation, but the odds against the iris sphincter receiving the right fibers are very high. Thus with random regeneration of fibers the power of accommodation is likely to recover, while the light reaction does not; at the same time the sphincter is likely to be served by accommodative impulses that constrict the pupil firmly with every near effort.[16]

Tendon reflexes in Adie's syndrome

About 90% of the patients with Adie's tonic pupil have diminished or absent tendon reflexes. This is a "benign areflexia," in that it is not due to a sensory or motor deficit; there is no peripheral neuropathy. The reflex impairment is widespread, affecting the arms as well as the legs, and is bilaterally symmetric in half the cases. In the patients with asymmetric tendon reflex loss it cannot be predicted from the distribution of the defect which eye is most likely to have the tonic pupil.[35]

Corneal sensation in Adie's syndrome

Many patients with Adie's syndrome have a mild regional impairment of corneal sensation when tested with a corneal esthesiometer.[27] This is thought to be due to damage to the sensory afferent nerves from the cornea, which travel with the short ciliary nerves and pass through the ciliary ganglion.

Natural history of Adie's syndrome

The onset of the condition seems to be abrupt and is marked by new symptoms of anisocoria or blurred near vision, yet patients will occasionally say that their tendon reflexes have been weak for years and that physicians have always had trouble eliciting knee jerks. The accommodative paresis soon starts to recover; reading glasses that have been prescribed are quickly discarded. The dilated pupil slowly comes down until the anisocoria is hardly noticeable. In darkness the tonic pupil does not dilate fully and is now the smaller pupil. Tonicity of accommodation sometimes develops.

There is a strong tendency for the second eye to be affected by a similar process some years later, but the onset is seldom noticed by the patient. This may be because near vision is already disturbed and presbyopia is approaching. Eventually both pupils are small and react poorly to light. The characteristic tonic near response, light-near dissociation, and cholinergic supersensitivity are still present, but are much harder to demonstrate because both pupils are involved and both are miotic. Fig. 12-12 plots the fluctuating anisocoria in just such a typical case.

Many patients believe that they have recovered from their Adie's syndrome because after 2 years accommodation has improved and the pupil is no longer dilated. But the light reaction has not recovered; in fact, it is likely to have lost further segments of the sphincter. Their tendon reflexes are just as bad, and the second eye may now be involved. The patient's have actually lost more function through progressive loss of ganglion cells than they have regained through regeneration of nerve fibers.

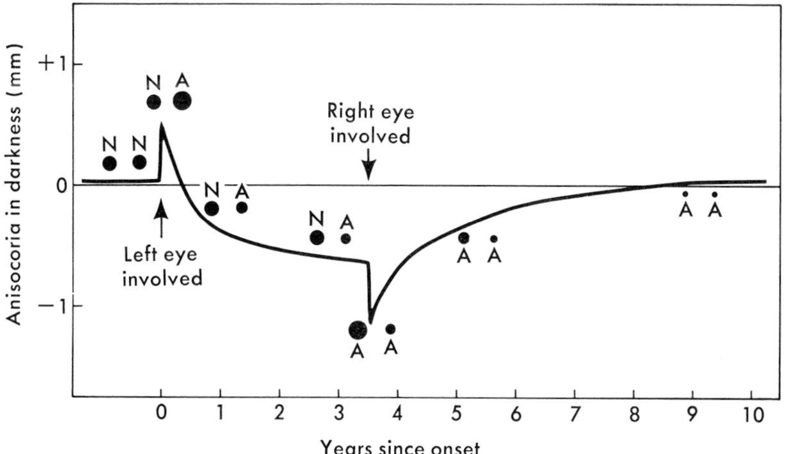

Fig. 12-12. Changing anisocoria of Adie's syndrome. First one pupil is involved and then the other. Both pupils gradually become smaller and smaller until it is hard to recognize them as Adie's tonic pupils. *N,* Normal; *A,* Adie's. (From Thompson, H. S., editor: Topics in neuro-ophthalmology, © [1979] The Williams & Wilkins Co., Baltimore.)

Etiology of Adie's syndrome

It has been suggested that a viral ciliary ganglionitis is the cause of the denervated intraocular muscles, but no explanation for the loss of tendon reflexes has been offered. A search was made for viral antibodies in the serum of patients with Adie's syndrome, but nothing much was found.[24] In results from an autopsied patient, Harriman[7] found that the ciliary ganglion on the affected side seemed to be suffering from an indolent degenerative process such as might occur with a "slow virus" infection.

PUPILLARY LIGHT-NEAR DISSOCIATION

In 1921,[13] Wilson encouraged the use of the phrase "Argyll Robertson's pupil" in a general sense, meaning any pupil that reacted better to near stimuli than to light. This resulted in the proliferation of awkward phrases like "pseudo–Argyll Robertson pupil" and "Argyll Robertson–like pupil." When the term "light-near dissociation" appeared, it was gratefully accepted as meaning any situation in which the pupillary *near* reaction was greater than the *light* reaction.

Recognizing a light-near dissociation

An adequate near effort is seldom made in dim light; the patient needs to see what he is looking at before he tries to get it in focus. The near response should therefore be checked in good room light and repeated several times. If the near reaction is elicited by having the patient look at a near object while a bright light is shining in the eye, the near reaction will always seem to be greater than the light reaction, since the two responses are added together. To test for a light-near dissociation, first see how small the pupil becomes with a very bright light, then with side lighting, or in room light have the patient look closely at a near object, such as his own thumbnail. If the near response makes the pupil smaller than the bright light does, the eye is showing a "light-near dissociation."

Causes of light-near dissociation

If we assume first that all light-near dissociations are due to some failure of the light reaction which somehow spares the near reaction, and second that there are no hyperreactive near responses, one trip around the light-reflex pathway should identify all the ways in which the light reaction can be impaired, and at the same time it should provide us with a classification of all the light-near dissociations (Fig. 12-13).

Any lesion that impairs the light reaction but allows the near reaction to be spared (or restored) should cause a light-near dissociation.

1. The lesion is in the retina. If both eyes are blind, for example, from diabetic retinopathy or old retinal detachment, there will be no light perception and no pupillary response to light, but if the iris and the sphincter innervation remain intact, there may be a good near response.

2. The lesion is in the anterior visual path (optic nerves, tracts, or chiasm). Once again, if both eyes are blind from optic nerve disease (e.g., temporal arteritis), there will be no light reaction, but the near response should be preserved—if the patient can remember how to cross his eyes.

3. The lesion is in the midbrain. The light reflex path can be interrupted here in the pretectal area without damaging the more ventrally located input for the near reflex. Tumors are the most common cause for this kind of light-near dissociation. Pinealomas classically cause Parinaud's syndrome, which includes poor light reaction and light-near dissociation; this is also called the sylvian aqueduct syndrome. Vascular disease in this area can cause a light-near dissociation, for example, occlusion of a posterior choroidal artery. Encephalitis and demyelinization are rare causes of light-near dissociation.

4. Argyll Robertson's pupils deserve a category of their own, although the defect is most likely a variation of the midbrain defects listed above.

5. The lesion is in the third nerve. How is it possible to damage the light reflex and spare the pupillary near reaction with a lesion in the outflow pathway when the fibers serving the two functions are identical? It is not possible. The near reaction is not actually spared; it appears to be restored in some cases because of aberrant regeneration of medial rectus innervation into the sphincter innervation pathway. This has been called the "pseudo–Argyll Robertson

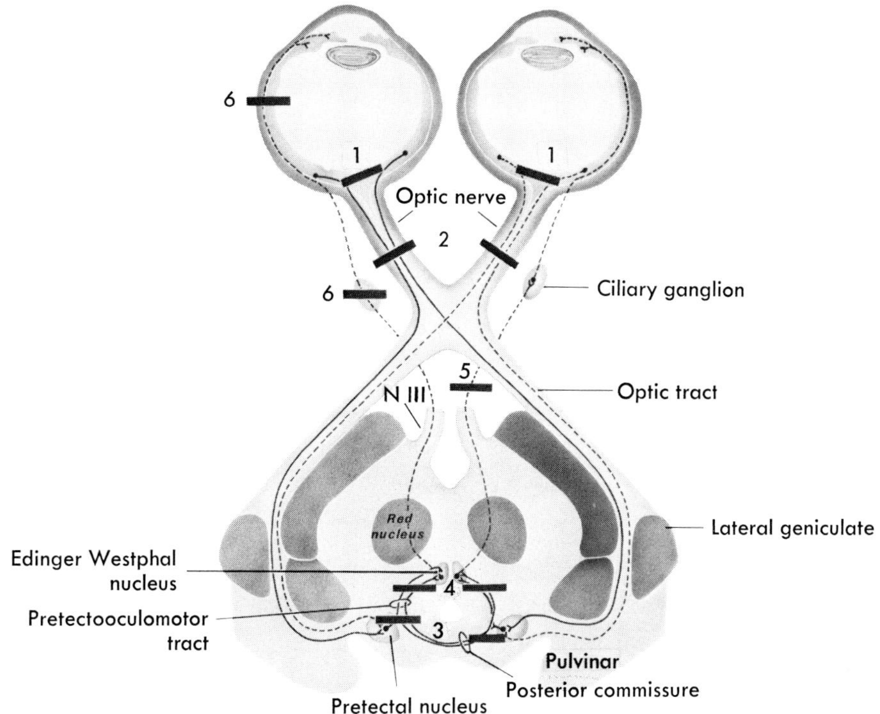

Fig. 12-13. Possible sites of isolated damage to pupillary light reflex pathway, which may result in light-near dissociation. *1* and *2*, Lesion in retina or optic nerve (e.g., optic neuritis). *3*, Lesion in midbrain (e.g., pinealoma). *4*, Lesion in intercalated neuron (e.g., Argyll Robertson's pupils). *5*, Lesion in third nerve (e.g., aneurysm). *6*, Lesion in ciliary ganglion or short ciliary nerves (e.g., tonic pupil). The "near" input to the Edinger-Westphal nuclei is thought to approach from the ventral side because it is spared in Parinaud's syndrome.

pupil associated with aberrant regeneration of the third nerve." This is a false light-near dissociation; the pupil (which fails to react to light because of the third nerve damage) constricts not only on convergence, but also any time the medial rectus is innervated, for example, with horizontal gaze (p. 342).

6. The lesion is in the ciliary ganglion or short ciliary nerves. Why do Adie's tonic pupils almost always show a light-near dissociation, when the "light" and "near" innervation of the sphincter follow an identical final common pathway? This is another example of aberrant regeneration in a mixed nerve. This time it is the accommodative impulses that find their way into the innervation of the iris sphincter; thus every time an effort is made to focus the lens, ciliary muscle innervation spills over into the sphincter and constricts the pupil.

7. A pupillary light-near dissociation has been observed with various peripheral neuropathies (amyloidosis, diabetes, alcoholism, Charcot-Marie-Tooth's disease, Dejerine-Sottas' disease). The mechanism of this light-near dissociation has not been established, but "aberrant regeneration" seems to be a good guess.

Argyll Robertson pupils

In 1869, Douglas Argyll Robertson published two papers in which he described a group of patients with tabes dorsalis who showed the following characteristics:

1. The retina was sensitive to light.
2. The pupil did not respond on exposure to light.
3. There was pupillary contraction during act of accommodation for near objects.
4. The pupils contracted further with extracts of the Calabar bean (physostigmine), but they dilated poorly with atropine.
5. The pupils were very small.

This curious finding of pupils that failed to respond to light despite good vision, yet contracted well on accommodation to near stimuli attracted immediate attention and soon became the most famous of all pupillary phenomena. One reason for the great interest in the Argyll Robertson syndrome was the fact that it was found in a high percentage of patients suffering from tabes dorsalis, general paresis, and "lues cerebri" and in very few other diseases. About 30 years after Argyll Robertson's publications, when Wassermann helped establish that these three conditions were actually only different manifestations of central nervous system syphilis, the Argyll Robertson pupil became widely accepted as being virtually pathognomonic of neurosyphilis.

After all these years, the pathogenesis of this kind of light-near dissociation is still not clearly understood. The reader is referred to Loewenfeld's monograph for an authoritative discussion of this question. She concludes that the classic Argyll Robertson pupil is the result of neuronal damage in the rostral midbrain near the sylvian aqueduct, interfering with the light reflex fibers and the supranuclear inhibitory fibers as they approach the Edinger-Westphal nucleus.[13]

Argyll Robertson pupils today

There is no doubt that neurosyphilis still produces pupillary abnormalities, including in many cases small, poorly reactive pupils with a light-near dissociation (Argyll Robertson's pupils). But there has always been a large group of patients with peripheral neuropathies of all kinds that tend to denervate both iris sphincters as part of the generalized neuropathy. Denervated sphincters eventually become miotic. Even Adie's tonic pupils tend to become bilateral and miotic with time.

Flagrant neurosyphilis (tabes dorsalis and general paresis) has become uncommon, and therefore a smaller percentage of miotic, light-palsied pupils can be proved to be related to syphilis. In making the diagnosis of syphilis, more weight is now put on the serologic, rather than the clinical, signs. A quick look at the pupils used to give the right answer 90% of the time. Nowadays, if we say "syphilis" when we see "Argyll Robertson pupils," we are often wrong; the patient is just as likely to have a diabetic or an alcoholic neuropathy. Argyll Robertson's pupils are not quite as firm a diagnostic sign as they used to be, so we are not as careful to look for them as we used to be. If we do not look for something, we seldom see it.

I suspect that the same percentage of patients with neurosyphilis have Argyll Robertson's pupils as ever did, but far fewer patients with Argyll Robertson's pupils have neurosyphilis.

Argyll Robertson's pupils have not vanished —we just do not bother to look for them anymore. This is too bad, because Argyll Robertson's pupils are still strongly suggestive of neurosyphilis. Any patient with both pupils abnormal (small, large, irregular, or reacting poorly to light) should at least have a serum fluorescent treponemal antibody test.

HORNER'S SYNDROME

Horner's syndrome is due to ipsilateral interruption of the sympathetic outflow to the head and neck. Bilateral Horner's syndrome is very hard to diagnose because most of the signs are recognized only by comparison with the normal eye on the other side.

Horner's syndrome can result from lesions in the brain stem or spinal cord, damage at the apex of the lung or in the supraclavicular space, or interruption of the fibers in the carotid plexus along the internal carotid artery all the way up to the cavernous sinus. Any defect in this long pathway produces the same kind of clinical picture; thus the combination of miosis and ptosis is by itself of little localizing value. However, the lesion causing Horner's syndrome often affects neighboring structures, producing signs and symptoms that suggest the location of the problem.

J. F. Horner (1834 to 1886) was a Swiss ophthalmologist. In 1869, he described a patient who showed a ptosis, miosis, and facial anhidrosis and reminded his readers that this combination of signs pointed to damage to the ipsilateral cervical sympathetic pathway. The features of sympathetic denervation and irritation had been thoroughly described many times during the preceding 140 years by Pourfour du

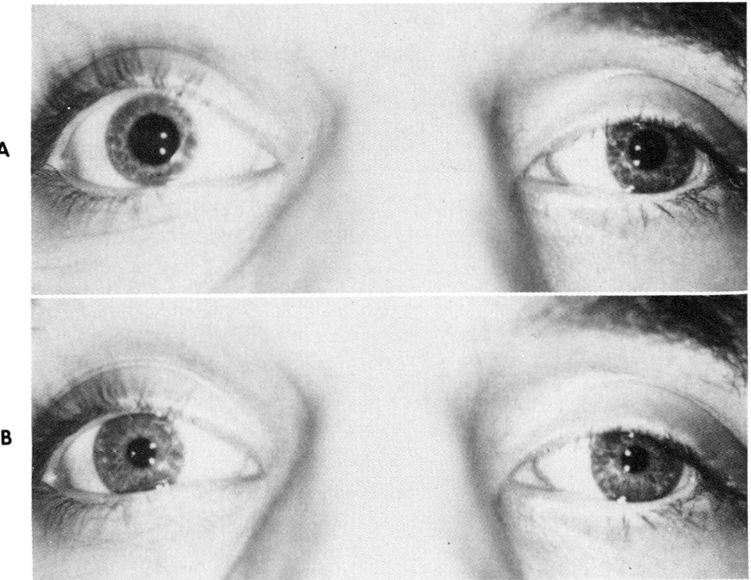

Fig. 12-14. Horner's syndrome in darkness and in light. **A,** taken in darkness shows more anisocoria than **B,** which was taken in bright light.

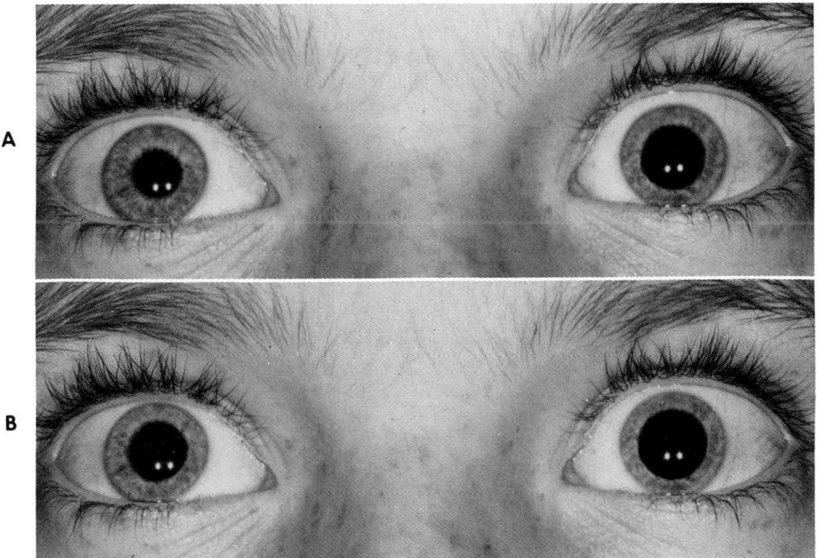

Fig. 12-15. "Dilation lag" in Horner's syndrome. **A** was taken in darkness 5 seconds after the lights went out. **B** was taken after 15 seconds of darkness. The Horner's pupil (the patient's right eye) is slow to dilate because it lacks dilator tone.

Petit, 1727; Claude Bernard, 1852; Brown-Sequard, 1854; John W. Ogle, 1858; S. Weir Mitchell, 1864; and many others (see historical review by Bonnet[1]). However, it was Horner's name that stuck. Modern usage seems to allow Horner's eponym for any oculosympathetic paresis, wherever the lesion, even though Horner's case was presumably a neck lesion. Thus we speak of "central Horner's syndrome" when the lesion is in the cord or brain stem and "postganglionic Horner's syndrome" when the lesion is in the head, even though the anhidrosis of Horner's clinical triad is missing. The term "Horner's syndrome" will be used in this way throughout this section. I admit that this is both careless and historically unjust.

Clinical signs of Horner's syndrome

Ptosis. There is a moderate droop of the upper lid because of the paralysis of Mueller's muscle. When the patient's eyes are wide open, this ptosis almost disappears.

There are also sympathetically innervated retractor fibers in the lower lid. In Horner's syndrome the lower lid rises slightly ("upside-down ptosis"), contributing to the narrowing of the palpebral fissure and to the apparent enophthalmos.

Miosis. There is a moderate decrease in pupil size because of paralysis of the dilator muscle. In bright light both pupils are nearly equal. The anisocoria is much more evident in dim light because the sphincter relaxes in both eyes, leaving a tight dilator muscle in the normal eye and a flaccid one in the affected eye. Self-developing flash photographs taken in darkness and in bright light help to quickly confirm this observation (Fig. 12-14).

Facial anhidrosis. Impairment of sweating on the ipsilateral face and neck is characteristic of preganglionic Horner's syndrome, but in an air-conditioned office this may not be a helpful sign. The sudomotor fibers to the face follow the branches of the *external* carotid artery; thus a Horner's syndrome caused by damage to the postganglionic fibers along the *internal* carotid plexus is not associated with any anhidrosis of the face, except in a small patch on the forehead where sudomotor fibers may reach the face via the supraorbital artery or nerve.

Ocular hypotony and conjunctival hyperemia. Ocular hypotony and conjunctival hyperemia are transient signs in acute Horner's syndrome and cannot be depended on to be present after the first few weeks.

Dilation lag. When the lights are turned out, the Horner's pupil dilates more slowly than the normal pupil does because it lacks the pull of the dilator muscle. This characteristic behavior of the sympathetically denervated iris can be seen clinically, and it can be documented with self-developing flash photographs taken 5 seconds after the lights go out and again after 15 seconds of darkness.[26] In Horner's syndrome there will be more anisocoria in the 5-second photograph than in the 15-second photograph (Fig. 12-15). This is a useful and highly specific sign for Horner's syndrome, and the diagnosis can often be made with certainty at this point without proceeding to the cocaine test.

Cocaine test

Cocaine will only dilate the pupil when the sympathetic pathway is intact and norepinephrine is being released from the nerve endings in the dilator muscle. A drop of a 5% or 10% solution is placed in each eye and repeated a minute or two later. Ideally, three photographs should be taken: before instillation at 30 minutes, and at 60 minutes. The anisocoria of Horner's syndrome will increase after cocaine instillation because the Horner's pupil dilates less than the normal pupil. All Horner's pupils, no matter where the defect in the pathway is located, will dilate poorly to cocaine. This means that cocaine helps to confirm the diagnosis and establishes that the miosis is caused by a sympathetic innervation deficit, but it does not help much in localizing the lesion.

Localizing the lesion

When it has been determined that the patient does have Horner's syndrome, the next step is to find out where the lesion is in the sympathetic pathway. The pathway is long and complex, but it is usually thought of as falling into three divisions, each separated by a synapse (Fig. 12-16).

The *central neuron* is in the brain stem and

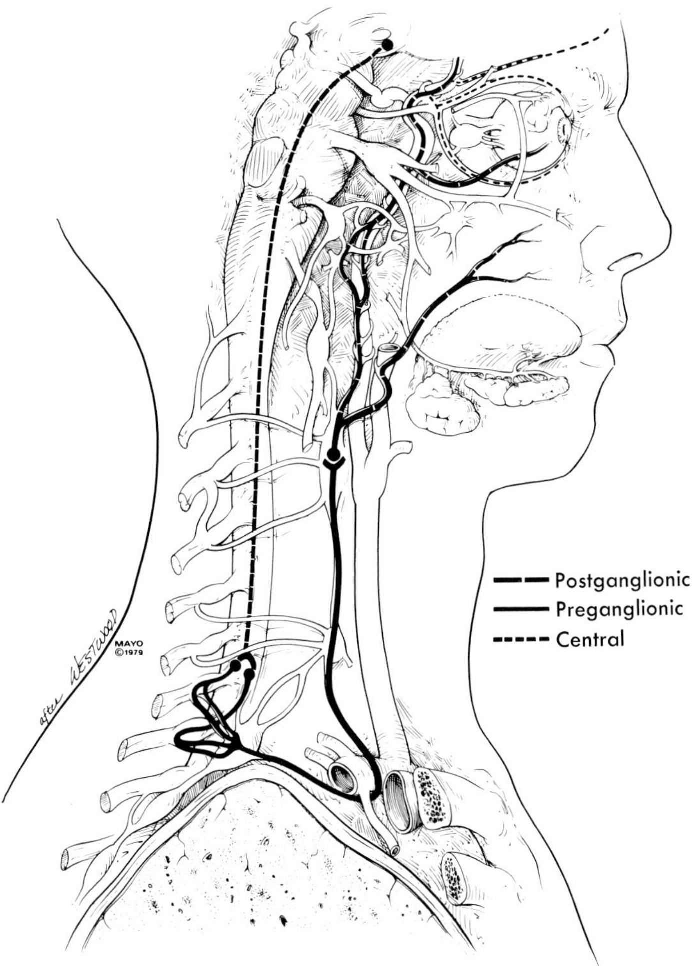

Fig. 12-16. Sympathetic innervation of the eye. (Modified from Maloney, W. F., Young, B. R., and Moyer, N. J.: Horner's pupils, Am. J. Ophthalmol. **90:**395, 1980.

cervical cord (from hypothalamus to the ciliospinal center of Budge at C8-T2). The *preganglionic neuron* is in the chest and in the neck (from cervical cord via stellate ganglion at the pulmonary apex to the superior cervical ganglion at the carotid bifurcation). The *postganglionic neuron* penetrates the base of the skull and passes through the cavernous sinus to enter the orbit (from the superior cervical ganglion at the angle of the jaw, via the carotid plexus, the cavernous sinus, and the long ciliary nerves to the iris).

It is necessary and important to understand this neuroanatomy; without it the differential diagnosis of Horner's syndrome is meaningless. It is into these three categories (central, preganglionic, and postganglionic) that Horner's syndrome can be divided by making use of associated signs and symptoms and the hydroxyamphetamine test.

Associated signs and symptoms

A patient with a *central* neuron lesion often has had the sudden onset of vertigo and sensory deficits caused by a medullary infarction, or the signs may point to cervical cord disease. The sweating deficit extends over more of the body than just the head and neck.

The patients with an isolated *preganglionic* Horner's syndrome may have a lung or breast

Table 12-2. Adrenergic mydriasis in Horner's syndrome—sympathetic defect*

Drug	Normal	Central lesion	Preganglionic lesion	Postganglionic lesion
Cocaine 5% to 10% (2 drops)	Mydriasis	Impaired dilation	No dilation†	No dilation†
Hydroxyamphetamine HBr 1% (2 drops)	Mydriasis	Normal dilation; pupils become equal	At least normal dilation; Horner's pupil may become slightly larger	No dilation†
Supersensitivity tests		Least dilation	Some dilation	Most dilation
Epinephrine 1:1000 (several drops)	No dilation	No dilation	Moderate dilation	Dramatic dilation
Phenylephrine 1% (2 drops)	Slight dilation	Slight dilation	Slight to moderate dilation	Moderate to dramatic dilation

*Thompson, H. S.: Bristol Med. Chir. J. **90:**37, 1976.
†Partial dilation suggests a partial defect.

malignancy that has spread to the thoracic outlet. There may be a history of surgery or injury to the neck, chest, or cervical spine. The anhidrosis involves the face and neck.

The patient with a *postganglionic* lesion most commonly has an ipsilateral vascular headache syndrome. There may have been a skull fracture or intraoral or retroparotid trauma; anything that gets the carotid artery into trouble may damage the carotid plexus. Occasionally a tumor of the middle cranial fossa or the cavernous sinus will involve some parasellar cranial nerves and affect the sympathetic fibers to the eye. Isolated Horner's syndrome due to orbital lesions is rare.

Although the history and physical examination can tell much about the location of the lesion in Horner's syndrome, sometimes despite our best efforts there are no clues to help us decide whether the lesion causing the Horner's syndrome is in the chest or in the head. It is at this point that the hydroxyamphetamine test can be extremely helpful.

Hydroxyamphetamine test

Hydroxyamphetamine hydrobromide 1% (Paredrine) is a commercially available adrenergic mydriatic that acts by releasing norepinephrine from the nerve endings in the dilator muscle. If there are no nerve endings because there are no postganglionic nerves, the drug will have no effect. This is the best drug test for identifying a postganglionic lesion. There is no drug test which will clearly separate central from preganglionic lesions.[31]

Drug tests making use of the principle of denervation supersensitivity (epinephrine 1:1000, phenylephrine 1%) are also helpful in identifying postganglionic lesions, but the drugs must be placed on strictly untouched corneas so that approximately the same dose reaches each iris. If this precaution is not taken, the tests are useless. The hydroxyamphetamine test does not suffer from these limitations, since hydroxyamphetamine mydriasis is limited by the amount of norepinephrine available in the nerve endings for release rather than by the amount of the drug that penetrates the cornea (Table 12-2).

Table 12-3. The diagnosis of acquired Horner's syndrome—Is it a Horner's syndrome?*

History and physical examination	Pupillary drug test
Ptosis Miosis More anisocoria in darkness Facial anhidrosis Upside-down ptosis Pupillary dilation lag Hyperemia of conjunctiva Apparent enophthalmus	Cocaine 5%—all Horner's pupils dilate poorly to cocaine (compared to the full dilation of the normal pupil)

*From Grimson, B. S., and Thompson, H. S.: Horner's syndrome, overall view of 120 cases. In Thompson, H. S., editor: Topics in neuro-ophthalmology, © (1979), The Williams & Wilkins Co., Baltimore.

Table 12-4. The diagnosis of acquired Horner's syndrome—what kind of Horner's syndrome is it?*

History	Physical examination	Pupillary drug tests
Central		
Sensory deficits Vertigo	Brain stem signs Cervical cord signs Syringomyelia	Hydroxyamphetamine in both eyes; both pupils dilate Phenylephrine in both eyes; both pupils dilate
Preganglionic		
Lung or breast tumor Thoracotomy Neck injury or surgery Brachial plexus injury	Facial anhidrosis Brachial plexus palsy Pancoast's syndrome	Hydroxyamphetamine in both eyes; both pupils dilate Phenylephrine in both eyes; both pupils dilate
Postganglionic		
Ipsilateral vascular headache Head trauma Intraoral or retroparotid trauma	Normal facial sweating Parasellar cranial nerve involvement Signs of internal carotid artery (contralateral hemiplegia, etc.)	Hydroxyamphetamine 1%—little or no dilation of Horner's pupil; full dilation of normal pupil Phenylephrine 1%—Horner's pupil dilates more than the normal pupil does (denervation supersensitivity)

*From Grimson, B. S., and Thompson, H. S.: Horner's syndrome, overall view of 120 cases. In Thompson, H. S., editor: Topics in neuro-ophthalmology, © (1979), The Williams & Wilkins Co., Baltimore.

As can be seen in Tables 12-3 and 12-4, the diagnosis of Horner's syndrome is divided into two stages: (1) the recognition of the sympathetic deficit and (2) the localization of the lesion. The second step is of vital importance in the management of the patient with Horner's syndrome. In the patient with an isolated preganglionic lesion there may be an occult malignancy; in Pancoast's syndrome there is an associated pain in the arm. If the lesion is postganglionic, it is most likely to be due to a benign vascular headache syndrome affecting the internal carotid artery.

Raeder's syndrome

The term "Raeder's syndrome" has been applied to many types of painful postganglionic Horner's syndrome, thus blurring its meaning and weakening its clinical usefulness. Most of Raeder's own cases had a parasellar syndrome of multiple cranial nerve palsies, but Raeder's name has been attached to the benign syndrome of cluster headaches with Horner's syndrome. Cluster headaches are easily recognizable by history and are not associated with cranial nerve palsies. "Raeder's paratrigeminal syndrome" is a term that should probably be limited to the occasional middle fossa mass which produces trigeminal nerve involvement with pain and a postganglionic Horner's syndrome.[6]

Congenital Horner's syndrome

When the sympathetic ocular innervation is interrupted early in life, the pigment of the iris stroma fails to develop. This produces an iris heterochromia. It is unusual nowadays to see a Klumpke's palsy associated with congenital Horner's syndrome, but birth trauma is still a significant cause (stretching of the pathway, forceps injury, skull fracture). Many of these children have sweating and vasomotor abnormalities of the entire side of the face, yet the pupil fails to dilate to hydroxyamphetamine (suggesting a loss of postganglionic fibers). This is probably due to a transynaptic dysgenesis of the postganglionic neuron following a preganglionic injury.[38]

Sympathetic irritation

When the sympathetic outflow suddenly fires, the pupil dilates, the lid lifts, and the conjunctiva blanches. This has been called the Claude Bernard syndrome. It may occur intermittently against a background of ipsilateral

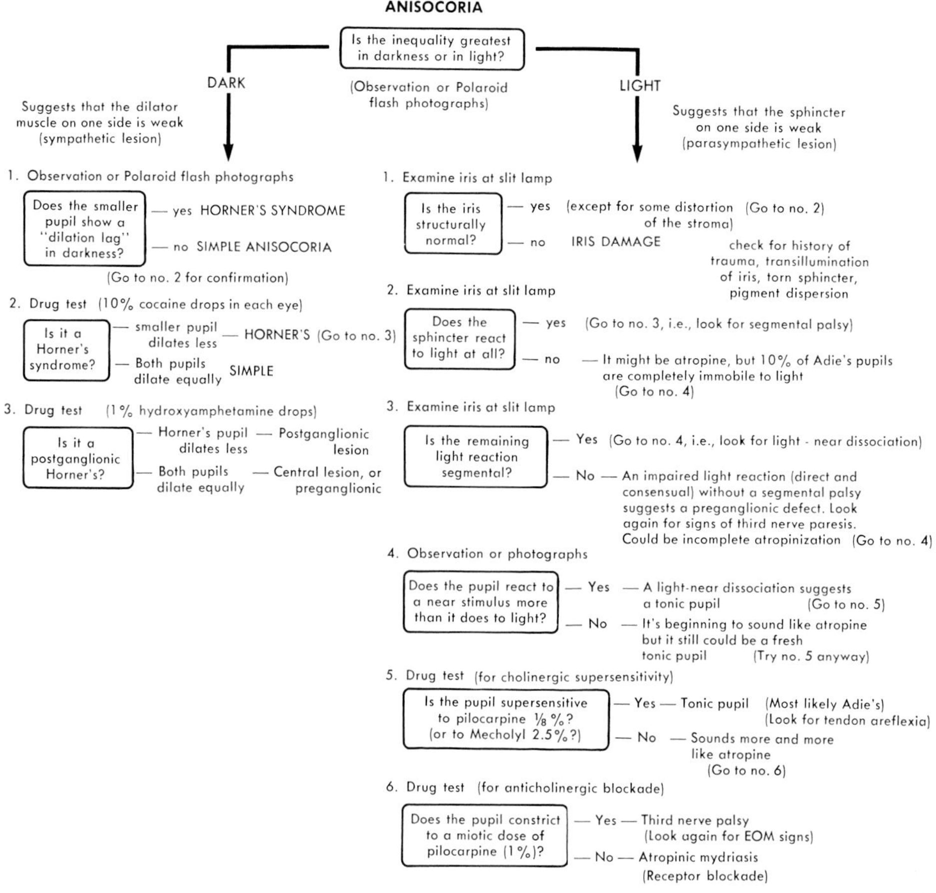

Fig. 12-17. Flow chart for analyzing anisocoria.

Horner's syndrome in cervical cord disease as in the well-documented case of Lowenstein and Levine.[20] This can give the appearance of an alternating Horner's syndrome on the other side.

Intermittent spasm of a dilator segment may occur with or without Horner's syndrome, resulting in a "springing pupil" ("amoeboid movements of the pupil"). Intermittent Horner's syndrome in an otherwise normal eye occurs during cluster headaches.

ANISOCORIA

When you see a patient with unequal pupils you naturally suspect a denervation of one of the iris muscles. If both pupils react reasonably well to light you have gone a long way toward ruling out a sphincter denervation and your attention turns to the smaller pupil. However, sometimes it is hard to be sure whether one sphincter is slightly weaker than the other. At this point a camera that takes self-developing flash photographs is very useful: take one photograph after 5 seconds of darkness and one in bright light; the weaker sphincter falls behind, so in bright light there is more anisocoria than in darkness. This test is good only for fresh denervations. The most common exception is an Adie's tonic pupil that has grown small with age, dilates poorly in darkness, and still constricts poorly in light.

Having decided that sphincter function is not impaired, the next problem is to distinguish between simple anisocoria (p. 329) and Horner's syndrome (p. 348). The patient with unequal pupils brings to his physician a problem in differential diagnosis. There are so many factors that might produce an anisocoria that it is sometimes hard to think of them in a logical order. The chart (Fig. 12-17) summarizes the steps are helpful in analyzing an anisocoria.

REFERENCES

1. Bonnet, P.: L'historique du syndrome de Claude Bernard. Le syndrome paralytique du sympathique cervical, Arch. Ophtal. **17:**121, 1957.
2. Bourgon, P., Pilley, S. F. J., and Thompson, H. S.: Cholinergic supersensitivity of the iris sphincter in Adie's tonic pupil, Am. J. Ophthalmol. **85:**373, 1978.
3. Czarnecki, J. S. C., and Thompson, H. S.: The iris sphincter in aberrant regeneration of the third nerve, Arch. Ophthalmol. **96:**1606, 1978.
4. DeLacey, P. H., editor and translator: On the doctrines of Hippocrates and Plato by Galen of Pergamum, (in press), Corpus Medicorum Graecorum.
5. Gillum, W. N.: Sympathetic stimulators and blockers, Ophthalmic Semin. **2:**283, 1977.
6. Grimson, B. S., and Thompson, H. S.: Raeder's syndrome, a clinical review, Surv. Ophthalmol. **24:**199, 1980.
7. Harriman, D. G. F.: Pathological aspects of Adie's syndrome, Adv. Ophthalmol. **23:**55, 1970.
8. Hepler, R. S., and Cantu, R. C.: Internal carotid aneurysms and third nerve palsies: ocular status of survivors, Surg. Forum **17:**436, 1966.
9. Kronfeld, P. C.: Eserine and pilocarpine: our 100 year old allies, Surv. Ophthalmol. **14:**479, 1970.
10. Lind, N., et al.: Adrenergic neurone and receptor activity in the iris of the neonate, Pediatrics **47:**105, 1971.
11. Loewenfeld, I. E.: Mechanisms of reflex dilatation of the pupil. Historical review and experimental analysis, Doc. Ophthalmol. **12:**185, 1958.
12. Loewenfeld, I. E.: Pupillary movements associated with light and near vision: an experimental review of the literature. In Whitcomb, M., editor: Recent developments in vision research, Publ. no. 1272, Washington, D.C., 1966, National Research Council, National Academy of Sciences, pp. 17-105.
13. Loewenfeld, I. E.: The Argyll Robertson pupil, 1869-1969. A critical survey of the literature. In Schwartz, B., editor: Syphilis and the eye, Baltimore, 1970, The Williams & Wilkins Co., (reprinted from Surv. Ophthalmol. **14:**199, 1969).
14. Loewenfeld, I. E.: "Simple, central" anisocoria: a common condition, seldom recognized, Trans. Am. Acad. Ophthalmol. Otolaryngol. **83:**832, 1977.
15. Loewenfeld, I. E.: Pupillary changes related to age. In Thompson, H. S., editor: Topics in neuro-ophthalmology, Baltimore, 1979, The Williams & Wilkins Co.
16. Loewenfeld, I. E., and Thompson, H. S.: The tonic pupil: a reevaluation, Am. J. Ophthalmol. **63:**46, 1967.
17. Loewenfeld, I. E., and Thompson, H. S.: Oculomotor paresis with cyclic spasms. A critical review of the literature and a new case, Surv. Ophthalmol. **20:**81, 1975.
18. Loewenfeld, I. E., Friedlaender, R. P., and McKinnon, P. F.: Pupillary inequality associated with lateral gaze (Tournay's phenomenon), Am. J. Ophthalmol. **78:**449, 1974.
19. Lowenstein, O.: Alternating contraction anisocoria, Arch. Neurol. Psychiatry **72:**742, 1954.
20. Lowenstein, O., and Levine, A. S.: Pupillographic studies. V, Periodic sympathetic spasm and relaxation and role of sympathetic nervous system in pupillary innervation, Arch. Ophthalmol. **31:**74, 1944.
21. Lowenstein, O., and Loewenfeld, I. E.: The pupil. In Davson, H., editor: The eye, ed. 2. Muscular mechanisms, New York, 1969, Academic Press, Inc., vol. 3.
22. Lowenstein, O., Feinberg, R., and Loewenfeld, I. E.: Pupillary movements during acute and chronic fatigue: a new test for the objective evaluation of tiredness, Invest. Ophthalmol. **2:**138, 1963.
23. Lowenstein, O., Kawabata, H., and Loewenfeld, I. E.: The pupil as an indicator of retinal activity, Am. J. Ophthalmol. **57:**569, 1964.
24. Meek, E. S., and Thompson, H. S.: Serum antibodies in Adie's syndrome. In Thompson, H. S., editor: Topics in neuro-ophthalmology, Baltimore, 1979, The Williams & Wilkins Co.
25. Miller, S. D., and Thompson, H. S.: Edge-light pupil cycle time, Br. J. Ophthalmol. **62:**495, 1978.
26. Pilley, S. J. F., and Thompson, H. S.: Pupillary dilatation lag in Horner's syndrome, Br. J. Ophthalmol. **50:**731, 1975.
27. Purcell, J. J., Krachmer, J. H., and Thompson, H. S.: Corneal sensation in Adie's syndrome, Am. J. Ophthalmol. **84:**496, 1977.
28. Scheie, H. G.: Site of disturbance in Adie's syndrome, Arch. Ophthalmol. **24:**225, 1940.
29. Sharpe, J. A., and Glaser, J. S.: Tournay's phenomenon, a reappraisal of anisocoria in latreral gaze, Am. J. Ophthalmol. **77:**250, 1974.
30. Thompson, H. S.: Cornpicker's pupil: jimson weed mydriasis, J. Iowa Med. Soc. **71:**475, 1971.
31. Thompson, H. S.: Diagnostic pupillary drug tests. In Blodi, F. C., editor: Current concepts in ophthalmology, St. Louis, 1972, The C. V. Mosby Co., vol. 3, chapter 6.
32. Thompson, H. S.: Pupillary signs in the diagnosis of optic nerve disease, Trans. Ophthalmol. Soc. U.K. **96:**377, 1976.
33. Thompson, H. S.: Adie's syndrome: some new observations, Trans. Am. Ophthalmol. Soc. **75:**587, 1977.
34. Thompson, H. S.: Segmental palsy of the iris sphincter in Adie's syndrome, Arch. Ophthalmol. **96:**1615, 1978.
35. Thompson, H. S., Bourgon, P., and Van Allen, M. W.: The tendon reflexes in Adie's syndrome. In Thompson, H. S., editor: Topics in neuro-ophthalmology, Baltimore, 1979, The Williams & Wilkins Co.

36. Thompson, H. S., Franceschetti, A. T., and Thompson, P. M.: Hippus, Am. J. Ophthalmol. **71:**1116, 1971.
37. Thompson, H. S., Newsome, D. A., and Loewenfeld, I. E.: The fixed dilated pupil, sudden iridoplegia or mydriatic drops? A simple diagnostic test. Arch. Ophthalmol. **86:**21, 1971.
38. Thompson, H. S., Hurtwitz, J., and Czarnecki, J. S. C.: Aberrant regeneration and the tonic pupil. In Glaser, J. S., editor: Neuro-ophthalmology, Symposium of the University of Miami and the Bascom Palmer Eye Institute, St. Louis, 1980, The C. V. Mosby Co., vol. X.
39. Weinstein, J. M., Zweifel, T. J., and Thompson, H. S.: Congenital Horner's syndrome, Arch. Ophthalmol. (in press), 1980.
40. Wirtschafter, J. D., Volk, C. R., and Sawchuk, R. J.: Transaqueous diffusion of acetylcholine to denervated iris sphincter muscle: a mechanism for the tonic pupil syndrome (Adie's syndrome), Ann. Neurol. **4:**1, 1978.

Chapter 13

Energy relationships

ROBERT A. MOSES, M.D.

VISION

The sensation of vision, like other sensations, cannot be defined in objective terms; there is no means of explaining brightness or color to the congenitally blind. Vision is a response to a physical stimulus. Some responses of the visual system may be detected by instruments as objective changes. The bleaching of retinal pigment, the electroretinogram, and potentials evoked at the occiput are examples. One may predict that more of the visual process will yield to objective investigation as techniques are devised and improved. However, a large body of valuable information has been and is being obtained by using the subject as if he were an instrument, and one of the most sensitive applications of instruments is to detect if two things equate or differ. Familiar examples of the "null" or no difference principle are the chemical balance and the Wheatstone bridge. In the study of vision the analogous question to the subject is, "Do you perceive a difference between these two fields?" The end point is when the fields appear the same. Such a technique is used, for example, in color matching. Often the test is shortened by using the subject as a servomechanism, and he is requested to adjust the fields until they appear the same. A second piece of information that may be obtained from this technique is the sensitivity of the detector, how different the fields must be to produce a difference that is just noticeable. This is the question we ask in measuring the extent of the visual field for a given test object; is there a perceptible difference between the object and the blackground? Thus the isopter for a particular test object is the limit of the area of visual field in which the test object is perceived as different from the background. These concepts have been applied to a number of visual functions. Some functions of the human visual system are remarkably sensitive and rival the most elaborate equipment available. For example, the dark-adapted human eye can detect a flash of light when the retina absorbs a dozen photons. The best photographic film is not so sensitive, and the best photomultiplier is only slightly more sensitive.

□ This work was supported in part by Research Grant EY 00256 from the National Eye Institute, National Institutes of Health.

Stimulus to vision

Stimuli to visual sensation may be divided in a purely physical sense into two types: inadequate and adequate.

Inadequate stimuli. Inadequate stimuli produce glowing sensations called phosphenes. The pressure phosphene appears as a patch with contrasting border when pressure is applied to the sclera, as in mechanical stimulation. The appearance of the phosphene depends on whether it is observed in the light or in the dark; it is described somewhat differently by different subjects (p. 568). The movement phosphene is seen in the dark. On rapid movements of the eyes a light circle corresponding to the optic disc and a larger circle corresponding to the insertions of the rectus muscles appear. These flashes are the result of distortion of the retina, by the inertial lag of the optic nerve and vitreous and the tug of the extraocular muscles (p. 569). The accommodation phosphene is seen in the dark as a peripheral light when the ciliary muscle contracts during accommodation or in the light as a transient graying on a bright field.

Electrical phosphenes are observed with the passage of weak electric currents through the eye. Potts and co-workers[4] have shown that the phosphene arises proximally to the receptors and may be used to study the integrity of the conduction pathway.

Radiation phosphenes appear with the passage of x rays or other ionizing radiation through the retina.

Adequate stimuli. Adequate stimuli to vision are defined as electromagnetic radiations in the range of approximately 400 to 750 nm.

ELECTROMAGNETIC RADIATION

There are two ways of recognizing and evaluating electromagnetic radiation: (1) as a physical entity and (2) by its physiologic and psychophysical effects. From the physical viewpoint, electromagnetic radiation is characterized by frequency (or alternatively wavelength in a given medium) and energy content. From a physiologic viewpoint appropriate radiation may produce color sensation and the sensation of brightness. The brightness observed is not a simple function of the energy content. Wavelengths subjectively identified as green are much more efficient in stimulating visual sensation than are wavelengths producing the sensations of blue or red, even though the unit quantity, or quantum, of "green" radiation contains less energy than a quantum of "blue" radiation.

Electromagnetic radiation occurs when the distribution of electrical charge in a system changes. Electromagnetic radiation arises around a conductor carrying alternating current or pulsating direct current; radio waves are generated by such means. Oscillation of the electrons of atoms produces radiation varying from the infrared through the visible spectrum and the ultraviolet. The abrupt deceleration of rapidly moving electrons produces x rays. Pulsating nuclear charges produce gamma rays.

The different forms of electromagnetic radiation have the following factors in common.

1. They all have properties that may be thought of as waves.
2. The radiation is emitted in discrete units, each of which is called a quantum.
3. The radiation is propagated in a vacuum, at the velocity of 3×10^{10} cm/sec (186,000 miles per second). The velocity of electromagnetic radiation in vacuum is symbolized by the letter c.

The characteristic that distinguishes the different electromagnetic radiations is the number of vibrations per second, or frequency, symbolized by the Greek letter ν. Frequency is given in cycles per second, or hertz (Hz). Radio waves are of low frequency; light waves are of higher frequency (Fig. 13-1). The frequency of an electromagnetic wave is constant for that wave, regardless of whether it is traveling in vacuum or through a dense medium such as glass.

Only a small portion of the electromagnetic radiation range is capable of exciting visual sensation. Electromagnetic radiation that produces visual sensation is called light. (Ionizing radiations, such as x rays, produce visual sensation but are not bent [focused or refracted] by the eye and are classed as inadequate stimuli.) A quantum of light is often called a photon.

Although the frequency of radiation does not vary with the density of the medium, the velocity of the radiation is less in dense media. The ratio of the velocity in vacuum, c, to the velocity in a medium, v_m, is the index of refraction, n_m, of the medium.

$$n_m = c/v_m$$

The velocity of radiation in air is nearly the same as that in vacuum, and the index of refraction of air is taken as 1. Since the velocity in air will always be greater than the velocity in a denser medium, the index of refraction of substances such as water and glass will always be greater than 1.

Radiation in the visible range is conventionally described in terms of wavelength, which is related to frequency in the following way. If radiation frequency ν cycles per second is propagated in a particular medium with a veloc-

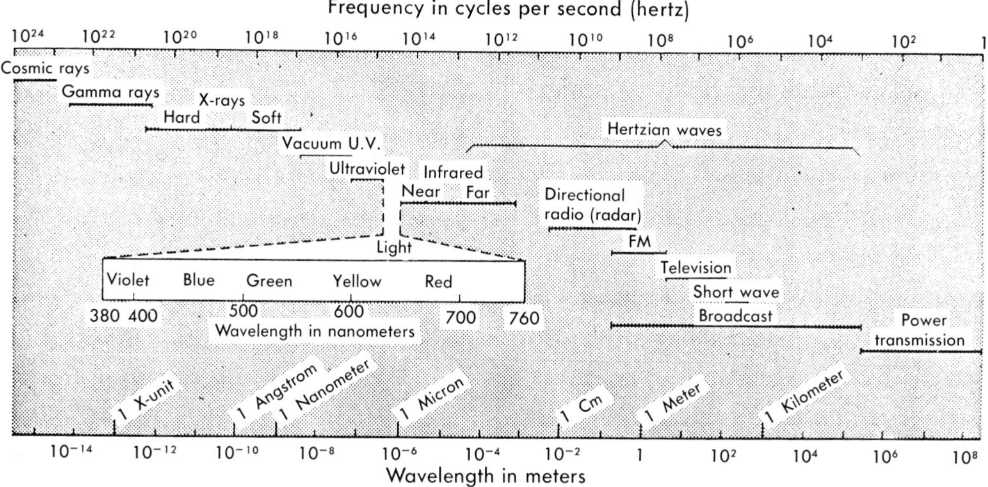

Fig. 13-1. Spectrum of electromagnetic radiation. (From Illuminating Engineering Society: IES lighting handbook, ed. 4, New York, 1966, The Society.)

ity, v_m, in 1 second ν waves, or cycles, will pass a given point in the medium. During this second the first wave will travel v cm from the point. There will then be v cm for ν cycles or v/ν cm for one cycle. The distance occupied by one cycle is the wavelength λ.

$$\lambda = v/\nu$$

It is to be noted that the wavelength varies with the density of the medium in which it is traveling. For example, light of 510 nm* wavelength in air is about 382 nm at the retina, as we will now show.

In vacuum (or for our purposes, air), radiation of any wavelength, λ_a, and frequency, ν, travels at a constant velocity, c.

$$\lambda_a = c/\nu \quad \text{or} \quad \lambda_a \nu = c$$

In a denser medium the velocity of propagation is less, v_m. Frequency remains constant, so wavelength in the medium, λ_m, is less than in air:

$$\lambda_m \nu = v_m$$

The change in wavelength is then proportional to the change in velocity

$$\frac{\lambda_a \nu}{\lambda_m \nu} = \frac{\lambda_a}{\lambda_m} = \frac{c}{v_m}$$

*1 nanometer is 10^{-9} m (0.000,000,001 m). The wavelength of light is often given in Ångstrom units (1 nm = 10 Å) or in millimicrons (1 nm = 1 mμ).

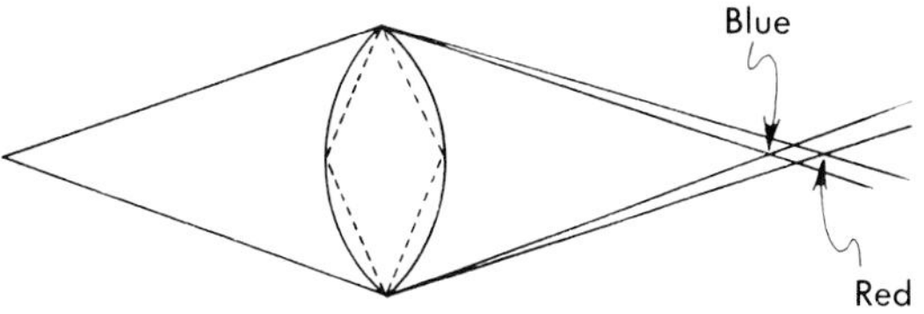

Fig. 13-2. Chromatic aberration. Distance of image from lens is proportional to wavelength. Prismatic effect of lens is suggested by dotted lines.

However, $c/v_m = n$, the index of refraction of the medium, so

$$\lambda_a/\lambda_m = n_m \quad \text{or} \quad \lambda_m = \lambda_a/n_m$$

The wavelength in a medium is the wavelength in air divided by the refractive index of the medium.

The refractive index of vitreous is about that of water, 1.336, so green light of 510 nm in air will have a wavelength of

$$\lambda_{vit} = \frac{510 \text{ nm}}{1.336} = 382 \text{ nm}$$

in vitreous.

I have just spoken loosely of the refractive index. More accurately, each medium has a refractive index for each wavelength. Short-wavelength radiation is retarded more by media than is long-wavelength radiation. Thus the refractive index for blue light is greater than the index for red light. A simple convex lens brings blue light to a focus closer to the lens than it brings the focus for red light (Fig. 13-2). This phenomenon is called chromatic aberration. Use is made of the chromatic aberration of the eye in the practice of refraction in the familiar red-green (duochrome) test. When the test letters illuminated with green light and those illuminated with red light are seen equally blurred, the green image is on the anterior side of the point of detection in the retina and the red image is on the posterior side.

Energy of light

Electromagnetic radiation is emitted in discrete packets of energy called quanta or photons.* The energy, U, of a photon is directly proportional to the frequency, ν.

$$U = h\nu$$

The proportionality constant, h, is Planck's constant, which has a value of 6.62×10^{-27} erg sec.

Since wavelength of radiation in air, λ, is related to frequency, ν, by

$$\nu = c/\lambda$$

the energy of quanta can also be calculated on the basis of wavelength.

$$U = hc/\lambda$$

Pirenne has given an illustration of the different energy values of quanta from different parts of the spectrum.[2] A metal such as sodium will emit electrons if radiation from the violet end of the spectrum is played on it. The number of electrons emitted depends, however, on the wavelength of the radiation acting on the sodium. The shorter the wavelength, the higher the energy of each quantum; therefore less intense radiation is needed to produce an emission of electrons. As the wavelength becomes longer, it is increasingly necessary to augment the energy in the radiation used because each quantum has less energy until finally no emission of electrons takes place, no matter how intense the radiation.

> As a certain definite amount of energy is necessary to expel an electron, the light quantum interacting with the metal . . . must have sufficient energy. . . . If the wave length is longer than the critical value, none of the quanta will have the requisite energy to expel an electron. Each quantum acts independently.*

Since the energy of the photon is inversely proportional to wavelength in a given medium, the energy of a photon of wavelength 400 nm (violet) is twice as great as that of a photon of wavelength 800 nm (red). The energy of a photon can be calculated by means of the preceding formula $U = hc/\lambda$; for instance, the energy of a quantum of radiation of 500 nm:

$$U = \frac{(6.62 \times 10^{-27}\ \text{erg sec})\,(3 \times 10^{10}\ \text{cm/sec})}{5 \times 10^{-5}\ \text{cm}}$$

$$= 3.97 \times 10^{-12}\ \text{erg}$$

where

$\lambda = 500\ \text{nm} = 5 \times 10^{-5}$ cm
$h = 6.62 \times 10^{-27}$ erg sec
$c = 3 \times 10^{10}$ cm/sec

From a practical standpoint, when one is dealing with quantities as minute as 1 or 2 quanta, it is not physically possible to produce or to deliver to some predetermined spot the exact number of quanta desired. It is impossible to set up a source of radiation that can always emit a constant number of quanta in unit of time. It is possible to make a source that delivers several thousand quanta per millisecond with a relatively small variability in this number on a percentage basis, but, when a source that will deliver exactly 6 quanta per millisecond is desired, the variability of this number on a percentage basis will be necessarily enormous. At one time the source may deliver exactly 6 quanta, at another 3 quanta, and at still another 8 quanta. This must be kept in mind in dealing with experiments in which the eye is exposed to minimal quantities of light for the purpose of determining its threshold.

RADIOMETRY

A few quanta per second over a long period of time or a short burst of quanta could contain

---

*Do not confuse this term with the older term used to describe retinal illuminance, now termed the troland.

*From Pirenne, M. H.: Vision and the eye, London, 1967, Chapman & Hall, Ltd., p. 88.

the same energy, so the energy delivered in a specified time interval must be stated. The watt, a practical unit of power, is 10^7 ergs/sec.

The power of a beam of light may be measured with several types of instruments. Typically the radiation is absorbed in a black chamber, converting the radiation to heat. The heat produced per unit time, measured with a thermocouple, is a direct measure of the power in the beam. Such a device is nonselective: it measures the total power, but not the distribution among the various wavelengths. For the distribution of power in a radiation beam to be measured, the beam is first spread into a spectrum by a prism or grating, and the power of successive narrow regions or bands of the spectrum is measured. Photocells of various types are widely used in radiometry. It must be emphasized that to speak with any certainty about the power of a radiation beam as measured by a photocell, one must know the response of the photocell as a function of wavelength, since the cell may be very sensitive to some wavelengths and not respond to others.

It is usually not enough to know the total power available and its wavelength distribution; one must also know how the power is delivered to a given area, or watts per square meter. The power per unit area is the energy density.

The physical specification of a light source, whether it be a primary source, such as an electric lamp, or a secondary source, such as an

Fig. 13-3. Reflection from imperfectly diffusing screen.

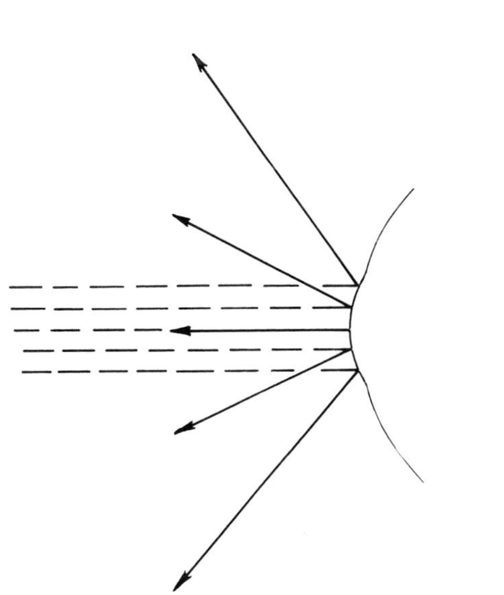

Fig. 13-4. Reflection of light parallel to optic axis from cornea. Peripheral rays are more reflected than axial rays because of greater angle of incidence.

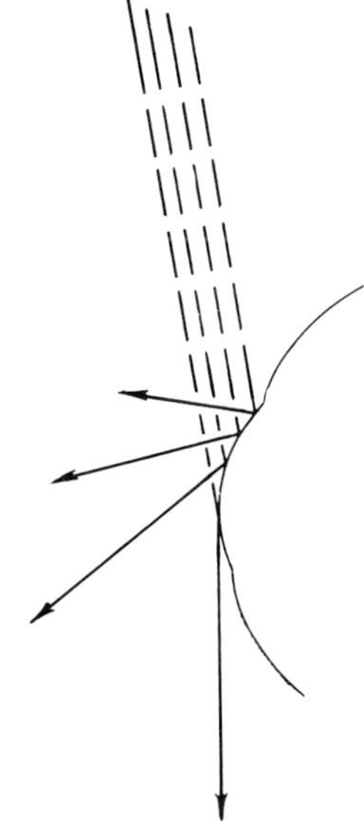

Fig. 13-5. Reflection of light obliquely incident to cornea. At grazing incidence, large proportion of light is reflected.

illuminated object, must give the area of the source, the energy density of the source, the spectral distribution of the energy, and the directional distribution of the radiation.

The factors mentioned in this section are of practical importance in the practice of ophthalmology. For instance, when test letters are projected on a screen for visual acuity testing, the screen becomes a secondary source of radiation; the light that enters the patient's eye diverges from the screen. It is of little use to place a light-measuring instrument on the screen facing the projector, as this registers only the amount of light falling on the screen, not the amount reflected from it (Fig. 13-3). Since most screens are highly directional diffuse reflectors, it is necessary to measure the light emanating from the screen in the direction of the patient.

The spectral distribution of the radiation reflected from the screen could be of some consequence. If the projector produced ultraviolet radiation, a photographic light meter would indicate more "light" reflected from the screen than the eye can use, since the light meter (and photographic film) is sensitive to ultraviolet radiation, but the "transparent" media of the eye, particularly the lens, are opaque to the ultraviolet.

A second example of importance is in use of the tangent screen and perimeter in the measurement of the visual field. Here the subject is required merely to indicate if an object is perceived against a background. It is of little interest to know the characteristics of the light falling on the test field; what we wish to know is the amount of light reflected from the test objects and from the background toward the patient. Clearly a soiled test object will reflect less light and therefore will be a weaker stimulus than a clean test object. If a flat test object is used, when it is tilted, the amount of light reflected toward the subject is less, due to directional reflectance. Moreover, its effective area is smaller. For both reasons the tilted flat test object will be a weaker stimulus than its nominal value.

In the discussion of visual acuity (Chapter 18) it will be found that the relation of the light reflecting from the object to the light from the background is very important.

A third example stresses the importance of spectral distribution. Various colored objects are used in testing the patient's ability to discriminate colors. Unless these standardized objects are presented in light of standard spectral composition, the tests become much less meaningful and may be quite misleading. It is common experience that two spools of thread that appear the same by incandescent electric light may be an obvious mismatch by sunlight. Incandescent light is rich in the yellow and relatively deficient in the blue.

The total amount of energy is an area depends on the extent of the area, the energy density, and the duration of exposure to the radiation. The energy may be absorbed, transmitted, or reflected. The development of photocoagulators based on xenon arcs and lasers has required careful consideration of these factors. The radiant energy of the photocoagulator passes through the transparent eye media but is absorbed by pigmented tissue. The absorbed energy is reemitted as heat. Surgical burns may be produced by the xenon arc device used for a fraction of a second or by the more intense laser beam in a much shorter time.

Reflection in the eye occurs where the refractive index changes abruptly. Reflection increases with the angle of incidence (Fig. 13-4). The index difference is greatest at the air-cornea interface, and the greatest amount of reflection occurs here, but reflections from the posterior surface of the cornea, the surfaces of the lens, and the surface of the retina contribute to the light loss. At normal incidence the intensity of the reflected beam is

$$I_r = I\left(\frac{n_1 - n}{n_1 + n}\right)^2$$

where

I = Intensity of incident beam
I_r = Intensity of reflected beam
n = Index of refraction of surrounding medium (for air, $n = 1$)
n_1 = Index of refraction of reflecting medium (for cornea, $n_1 = 1.336$)

At the anterior surface of the cornea

$$I_r = I\left(\frac{1.336 - 1}{1.336 + 1}\right)^2$$

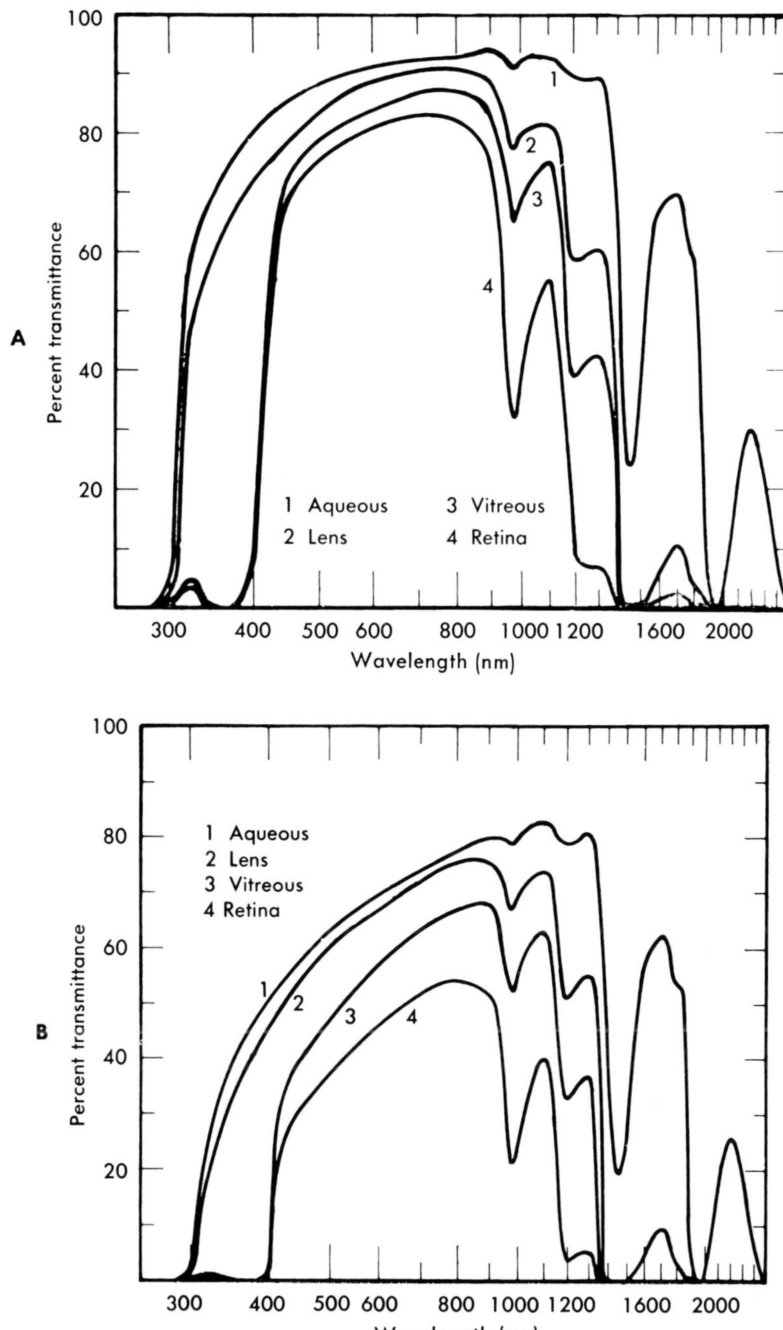

Fig. 13-6. Transmission of visible light and near infrared by ocular media. **A,** Total transmittance through entire eye. **B,** Direct transmittance through entire eye. (From Boettner, E. A., and Wolter, J. R.: Invest. Ophthalmol. **1:**776, 1962.)

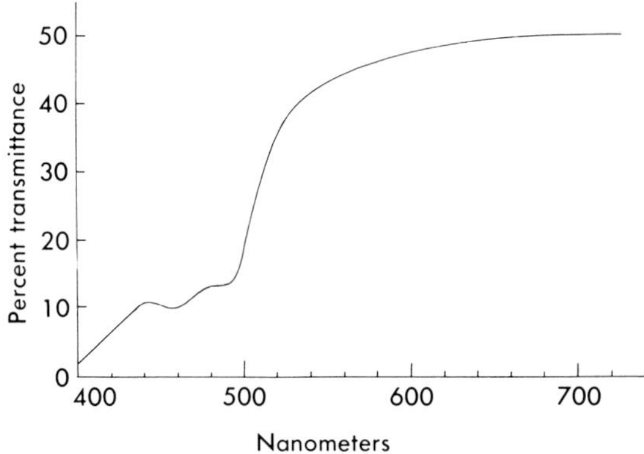

Fig. 13-7. Transmission of light of visible range by ocular media. Low transmission of short wavelengths is due largely to lens. Irregular dip in blue region (440 to 500 nm) is due to absorption by yellow macular pigment, xanthophyll. (From Wyszecki, G., and Stiles, W. S.: Color science, New York, 1967, John Wiley & Sons, Inc.)

Thus I_r is 0.02, or 2% of the light incident normal to the cornea is reflected. The intensity of the reflected light increases with departure from perpendicular incidence until, at grazing incidence, much of the light is reflected. The illumination of the retina by a test object presented in the periphery of the visual field is considerably reduced (Fig. 13-5), not only by the obliquity of the pupil, but also by reflection from the surfaces of the cornea and lens.

The lens is responsible for much of the radiation absorption in the ocular media. Its absorption is greatest at short wavelengths; the aphakic eye can be stimulated by ultraviolet light, which is invisible to the normal eye because it is blocked from the retina by the lens.

Absorption by the lens, particularly in the blue part of the spectrum, increases with age as a brown protein accumulates.[3]

Considerable short-wavelength radiation is absorbed by the yellow macular pigment, xanthophyll, which lies in crystals between the elongated cone fibers of the macula (Henle's fiber layer). The partially oriented xanthophyll crystals polarize light, so sheaflike figures are seen entoptically in a uniform field of polarized light.

The polarization figures, Haidinger's brushes, have two clinical applications.

1. When macular function is impaired, as in central serous retinopathy, the brushes are no longer seen.
2. The brushes form a subjective marker of the macula and are used to indicate to the orthoptic patient the straight-ahead or central region of the visual field.

Since the brushes rapidly fade because of local adaptation if the polarization plane is fixed, a rotating polarizer is used in clinical application. A blue filter enhances the effect.

Scattering produces light loss in the normal eye. The difference between total transmission (Fig. 13-6, *A*) and direct transmission (Fig. 13-6, *B*) is the result of scatter.[1,5] The net light actually reaching the receptors is shown in Fig. 13-7. The optical section of the cornea and other media seen with the slit lamp is produced by scattered light.

Scattering of light by submicroscopic particles, the Tyndall effect, is ordinarily not seen in the aqueous; visibility of the slit-lamp beam in the aqueous is caused by excessive protein and is a sign of disease. Larger particles, such as cells in the aqueous, appear as points of light.

LUMINOSITY

Radiation of the visual spectrum can be characterized physically by the wavelength of the radiation and the energy content or subjectively by the color and the sensation of brightness it produces. It has long been known that green light is more efficient than other spectral colors in producing visual sensation; it requires fewer

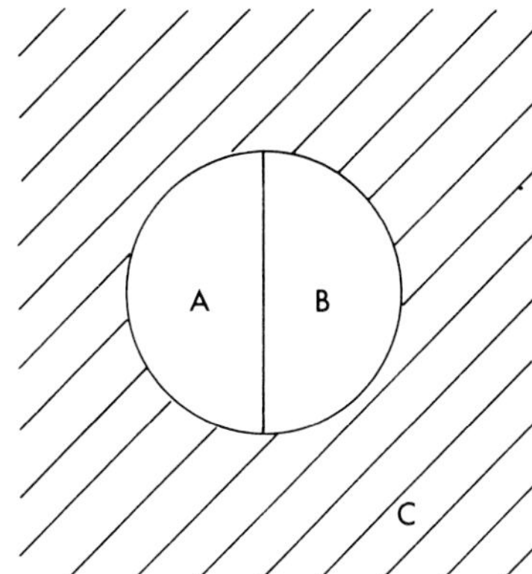

Fig. 13-8. Bipartite field. *A* and *B*, fields presented for comparison. Surround, *C*, may be dark or of other specified characteristics.

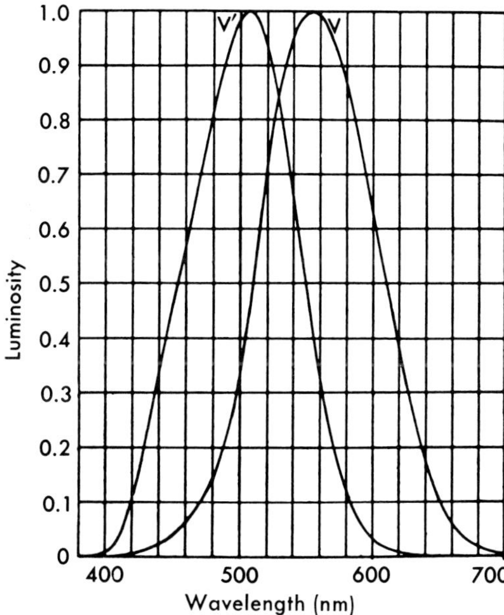

Fig. 13-9. Photopic, *V*, and scotopic, *V'*, luminous efficiency curves. Most efficient wavelength in light-adapted eye is 555 nm, in dark-adapted eye is 510 nm. Efficiency of other wavelengths given as decimal fraction of most efficient wavelength.

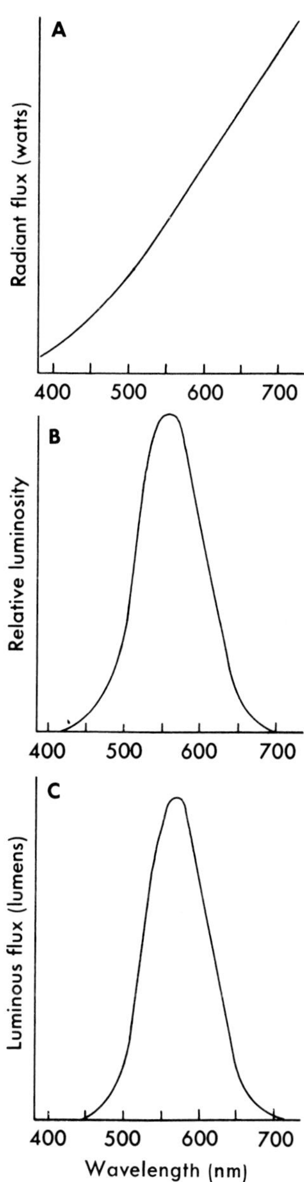

Fig. 13-10. Evaluation of spectroradiometric data in terms of luminosity. **A,** Power distribution in incandescent light (standard illuminant). **B,** Standard photopic luminous efficiency curve. Luminous flux of light source is calculated by multiplying power ordinates of **A** by efficiency ordinates of **B,** giving curve **C.** Area under curve **A** is watts; area under curve **C** is lumens.

watts of green light falling on a retinal area to appear a given brightness than for any other color. The most efficient wavelength of green depends on whether the determination is made in the presence of relatively bright surroundings, or whether the test is made in the dark after the eye has become adjusted to the dark. When the eye is accustomed to light, it is said to be light adapted and to have photopic vision. The dark-adapted eye is adjusted for scotopic vision.

One way of measuring the relative efficiency of the various spectral wavelengths is to illuminate half of a screen with light of one wavelength, the other half with light of a different wavelength. The intensity of one field is adjusted until the apparent brightness or luminosity of the two fields is equal (Fig. 13-8). This process is called heterochromatic brightness matching. The number of watts per unit area of each light required to make the two halves appear equally bright is a measure of the relative efficiency of the wavelengths to produce subjective brightness. When the most efficient wavelength is found, it is given the value of 1.0, and the efficiency of the other wavelengths is rated as a fraction of the most efficient wavelength. If this assessment is made on the light-adapted eye, the most efficient wavelength is 555 nm; under scotopic conditions the efficiency curve shifts and the most efficient wavelength is 510 nm (Fig. 13-9). The shift in the efficiency or luminosity curve was first described by Purkinje. Different subjects and different conditions produce somewhat varying luminosity curves. It is important to illuminating engineers, paint manufacturers, dye manufacturers, and others, as well as ophthalmologists, to have color standards and standard curves giving the visual efficiency of the wavelengths of the spectrum. To this end the International Committee on Illumination* has agreed on the "standard observer" photopic and scotopic luminosity curves seen in Fig. 13-9.

The existence of these standard values of relative efficiency allows us to weight each narrow wave band of a given radiation in accordance with its ability to excite visual sensation. The radiant power of a band times its efficiency gives the luminous flux density of the band. The sum of the luminous flux densities of the individual spectral bands is the luminous flux density of the particular illuminant as a whole. Fig. 13-10 shows how any spectroradiometric curve can be evaluated with respect to the eye. Curve *A* represents standard illuminant *A*,* and curve *B* is the photopic luminosity curve. Curve *C* is obtained by multiplying the ordinates of *A* and *B* at each wavelength. The area under curve *A* is expressed in watts. The area under curve *C* is expressed in lumens or equivalent units. The link between the physical measurements of radiometry and the effect of the radiation as a stimulus to vision is the standard observer. Luminous flux, *F*, is measured in lumens. One lumen of light of wavelength 555 nm is 1/685 watt. Various units derived from this definition are given in the Appendix to this chapter. It must be remembered that the standard observer is as much a fiction as is the average man. While many persons respond to light in a manner reasonably well described by the standard curves, many do not, and prominent among these are the color deficient.

*In French, CIE, Commission Internationale de l'Eclairage.

*Incandescent source at color temperature 2854° K.

REFERENCES

1. Boettner, E. A., and Wolter, J. R.: Transmission of the ocular media, Invest. Ophthalmol. **1**:776, 1962.
2. Pirenne, M. H.: Vision and the eye, ed. 2, London, 1967, Science Paperbacks and Chapman & Hall Ltd.
3. Pirie, A.: Color and solubility of the proteins of human cataracts, Invest. Ophthalmol. **7**:634, 1968.
4. Potts, A. M., Inoue, J., and Buffum, D.: The electrically evoked response of the visual system (EER), Invest. Ophthalmol. **7**:269, 1968.
5. Wyszecki, G., and Stiles, W. S.: Color science, New York, 1967, John Wiley & Sons, Inc.

GENERAL REFERENCES

Enoch, J. M.: Physiology, vision. In Sorsby, A., editor: Modern ophthalmology, New York, 1963, Appleton-Century-Crofts, vol. 1.

Graham, C. H., editor: Vision and visual perception, New York, 1965, John Wiley & Sons, Inc.

Padgham, C. A.: Subjective limitations on physical measurements, Philadelphia, 1965, Franklin Publishing Co.

Rubin, M., and Walls, G.: Studies in physiological optics, Springfield, Ill., 1965, Charles C Thomas, Publisher.

APPENDIX TO CHAPTER 13

The amount of radiation reaching the retina from a given source may be approximated as follows. Unit area of source emits W watts power, which is to say that the emissive power of the source is W watts/cm^2. At the distance of the eye, d, the power is spread over a spherical surface (Fig. 13-11). The pupil of the eye is a portion of the sphere. The proportion of the radiant power entering the pupil from unit area of source is

$$W \times \frac{A_{\text{pupil}}}{A_{\text{sphere}}}$$

The power in the pupil coming from unit area of source is focused on an area of retina, A_{ret} (Fig. 13-12).

$$\frac{A_{\text{source}}}{d^2} = \frac{A_{\text{ret}}}{f^2}$$

in which f is the posterior focal length of the eye.

The area of the retinal image is

$$A_{\text{ret}} = \frac{A_{\text{source}} \times f^2}{d^2}$$

The radiant flux falling on the pupil from 1 cm^2 of the source

$$W \times \frac{A_{\text{pupil}}}{A_{\text{sphere}}}$$

is distributed over retinal area (f^2/d^2) cm^2. The radiant power per unit area of retinal image is

$$H = \frac{W \times (A_{\text{pupil}}/A_{\text{sphere}})}{f^2/d^2 \text{ watts/cm}^2}$$

The area of the pupil is πr^2. The area of the spherical surface, including the pupil, is $4\pi d^2$.

The radiant power per unit area of retinal image then reduces to

$$H = \frac{W \times r^2_{\text{pupil}}}{4f^2}$$

The result is surprising at first; the amount of energy per second reaching unit area of retina is independent of the distance of the eye from the light source. However, consider the fact that, as the eye recedes from the light source and the light that enters the pupil diminishes as the square of the distance, the area of the retinal image also diminishes as the square of the distance.

Assumptions in this first-order estimate include the following.

1. All the radiation is transmitted; none is reflected and none is absorbed or scattered.
2. A sharp image is formed on the retina.
3. The source is uniformly emitting.

RADIOMETRIC TERMS

radiant flux or radiant power *(P)* Rate of flow of electromagnetic radiation.

radiant intensity *(J)* Radiant flux within unit solid angle around a point on the radiation source.

irradiance *(H)* Power of radiant energy incident per unit area of surface.

PHOTOMETRIC TERMS

luminous flux *(F)* Rate of flow of radiant electromagnetic energy evaluated according to its ability to produce visual sensation in the average human eye.

luminous intensity *(I)* Luminous flux emitted per unit solid angle in a given direction.

illuminance *(E)* Luminous flux incidence on unit area of a surface.

luminance *(L)* Luminous intensity per unit apparent area of a surface source of light (either emitting or reflecting).

Radiometric equivalents

Energy: 1 joule = 10^7 ergs
Power: 1 watt = 1 joule/sec

Photometric equivalents

Luminous flux (F): 1 lm = 1/685 watt at λ = 555 nm
Luminous intensity (I): 1 lm/ω = 1 candela (cd)*
Illuminance (E):
1 lm/m^2 = 1 lux (lx)
Luminance (L):
1/lm/ω/m^2 = 0.3142 millilambert (mL)
1/lm/ω/m^2 = 0.2919 foot-lambert (ft-L)

For an extended source, emittance of
1 lm/cm^2 = 1 lambert (L)
1 mL = 10 apostilbs
1 mL = 3.183 c/m^2
1 mL = 3.183 lm/steradian/m^2

*The candela is 1/60 of the luminous intensity of 1 cm^2 of a black-body radiator at the temperature of the freezing point of platinum.

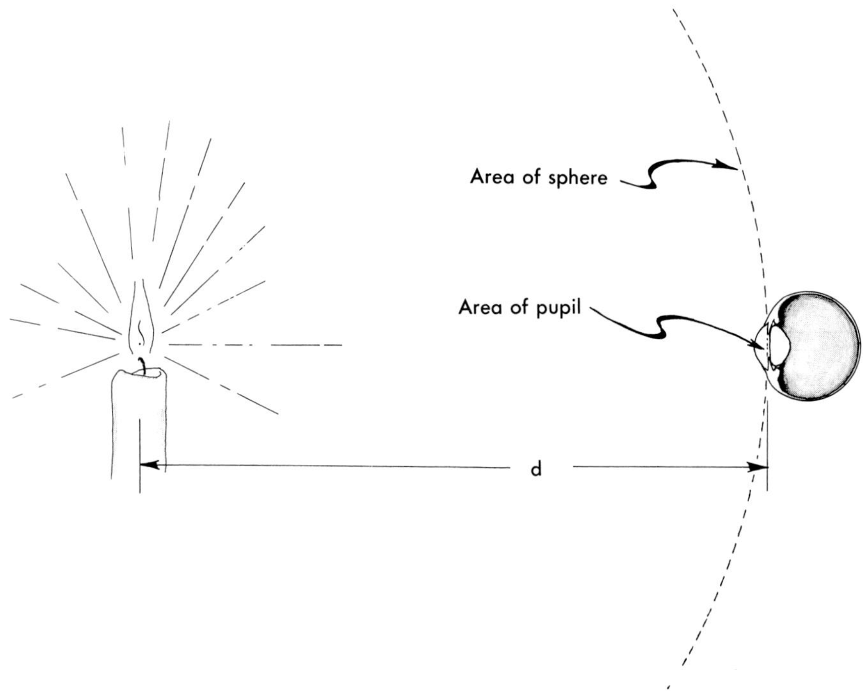

Fig. 13-11. Pupil as part of sphere around source.

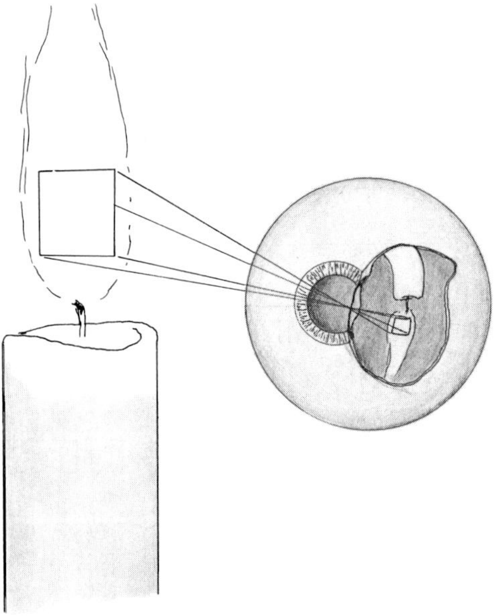

Fig. 13-12. Projection of unit area of source on retina.

Table 13-1. Comparison of photometric and radiometric terms

Radiometric term	Symbol	Units	Photometric term	Symbol	Units
Radiant energy	U	joule	—	—	—
Radiant flux	P	watt	Luminous flux	F	lumen (lm)
Radiant intensity	J	watt/ω	Luminous intensity	I	lm/ω
Irradiance	H	watt/m^2	Illuminance	E	lm/m^2
Radiance	N	watt/ω/m^2	Luminance	L	lm/ω/m^2

Table 13-2. Symbols

Symbol	Term	Definition	Symbol	Term	Definition
ω	Steradian, unit solid angle	$1/4\pi$ sphere	n	Index of refraction	$n = c/v_m$
c	Velocity of light in vacuum	2.99776×10^{10} cm/sec	U	Energy	joule = 10^7 ergs
v_m	Velocity in medium (m)	cm/sec	h	Planck's constant	6.624×10^{-27} erg sec
ν	Frequency	cps or Hz	V_λ	Relative photopic luminous efficiency	—
λ	Wavelength	nm	V_λ'	Relative scoptic luminous efficiency	—

Retinal illuminance $(L \times S)$

*Troland** = luminance of 1 c/m^2 on surface viewed through artificial pupil of area S = 1 mm^2

Conversion of radiometric to photopic (photometric) values

lm = 1/685 watt at λ = 555 nm

$$F = 685 \sum_{o}^{\infty} P_\lambda V_\lambda \Delta_\lambda \text{ lm}$$

*Photopic–uncorrected for Stiles-Crawford effect.

APPENDIX REFERENCES

Committee on Colorimetry of the Optical Society of America: The science of color, Washington, D.C., 1966, The Optical Society of America.

Graham, C. H., editor: Vision and visual perception, New York, 1965, John Wiley & Sons, Inc.

Padgham, C. A.: Subjective limitations on physical measurements, Philadelphia, 1965, Franklin Publishing Co.

Wyszecki, G., and Stiles, W. S.: Color science, New York, 1967, John Wiley & Sons, Inc.

Chapter 14

The retina and optic nerve

ADOLPH I. COHEN, Ph.D.

DEVELOPMENT OF THE RETINA

The eye is an externalized portion of the brain. The neural retina is, in fact, a derivative of and an extension of the diencephalon. The optic nerve, which connects the retina with higher visual centers, is structurally and functionally a tract of the central nervous system, rather than a peripheral nerve. The layers of collagen bundles within the cornea and sclera are continuous with those of the dural investment of the optic nerve and via that connection with that of the brain. The clear fluid within the eye most closely resembles the cerebrospinal fluid of the brain, rather than any other body fluid. The blood vessels of the retina are identical with those of the brain proper.

In the human, formation of the eye[48] begins with lateral outpouchings of the prosencephalon (the precursor of the telencephalon and diencephalon) during the third week of development (2.6 mm stage). These enlarge to form primary optic vesicles whose lateral aspects invaginate during the fourth week of development to form double-walled, optic cups. These continue to enlarge, largely on their dorsal and lateral aspects, to form more mature double-walled optic cups whose concavities are directed lateroventrally. As the optic vesicles are outpouchings of the hollow neural tube, the cavity within each optic vesicle is initially continuous with that of the neural tube. The neural tube's system of connected cavities develops into the hollow, fluid-filled ventricular system of the brain. Each optic vesicle's lumen becomes cut off from the lumen of the neural tube because of total obliteration of the lumen within the optic stalk that connects the optic vesicle to the neural tube. However, an important residuum of this cavity persists between the two walls of the optic cup in the adult vertebrate, between the retina and the pigment epithelium. Clinicians refer to this space in the human eye as the subretinal space, and it is this space that enlarges in detachment of the retina, but it is, in fact, the ocular ventricle.

The fates of the two walls of the embryonic optic cup in the vertebrate eye are as follows: the outer wall, closest to the brain proper, remains as a sheet of cells, one-cell thick, and becomes the pigment epithelium. The invaginated inner wall, facing the future vitreal chamber of the eye, becomes several structures. Over most of its central extent it becomes the neural retina, a relatively thick multicellular tissue, but the peripheral extent of this layer remains one-cell thick. This peripheral, one-cell thick extension from the retina becomes physically attached to the corresponding peripheral portion of the pigment epithelium, also one-cell thick, and this double epithelium forms a peripheral extension of the double-walled optic cup. This persists in the adult eye as the two-layered epithelium of the ciliary body and iris. Between these adherent layers of pigment epithelium and nonneural epithelium the ventricular space is essentially obliterated.

Because of the invagination process during development and the spherical organization of the eye, the retinal surface facing the vitreal chamber is its inner surface, that facing the

☐ This work was supported in part by Research Grant EY 00258-16 from the National Eye Institute, National Institutes of Health.

ocular ventricle and pigment epithelium, its outer surface. Had the invagination not occurred, the inner or vitreal surface of the retina would have faced the outside of a sphere. This reveals its homology to the outer or pial surface of the brain. Similarly, the outer surface of the retina, in the absence of invagination, would have faced the cavity of a hollow sphere. Thus its homology is to the inner or ventricular surface of the brain. In an anterior facing eye, as in the human, the inner layer of the optic cup is sometimes designated the anterior layer; the outer layer, the posterior layer.

Paralleling the situation at the pial surface of the brain, all the cell membrane at the inner surface of the retina belongs to a mosaic of processes of glial cells, and the surface is covered by a basal lamina. This surface with its covering (Fig. 14-1) constitutes the vaguely defined "inner limiting membrane" of the retina.[35]

The outer surface of the retina and the inner surface of the pigment epithelium face the ocular ventricle. The margins of all cells facing *brain* ventricles are joined by presumptively adhesive structures called terminal bars. These form a sort of collar around each cell at the surface, and the proteins of these attachment specializations stain intensely with certain procedures. The same cytologic description holds true for the cells facing the ocular ventricle. In thick sections of retina this line of intense stain, because of terminal bars at the outer or ventricular surface of the retina, was originally interpreted as representing a membrane, hence the term "external limiting membrane" of the retina (Fig. 14-2). It forms a useful marker for the retinal surface at the ventricle.[10,12] Many terminal bars consist of a junctional complex: a *zonula occludens* or tight junction, a *zonula adherens* or adhesive junction, and a *zonula communicans* or gap junction. However, the

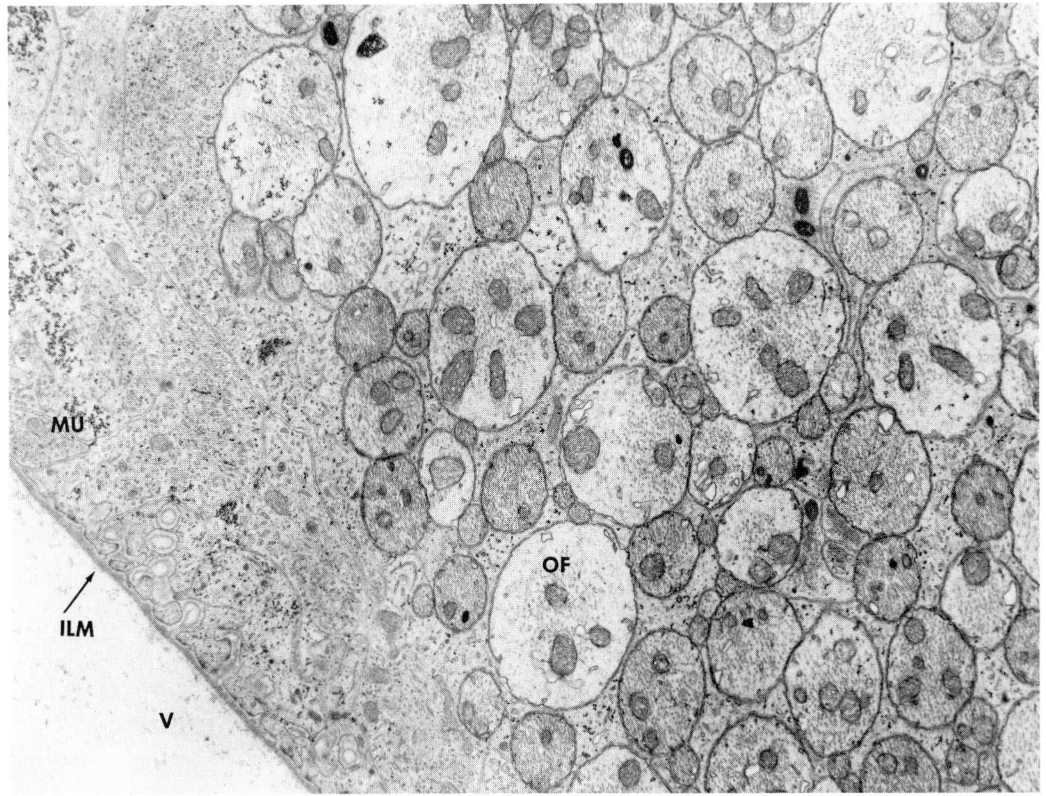

Fig. 14-1. Vitreal surface of retina of rhesus monkey. Note vitreous, *V;* inner limiting membrane, *ILM;* foot processes of glial cells of Müller, *MU;* optic fibers, *OF*. (Magnification ×14,000.)

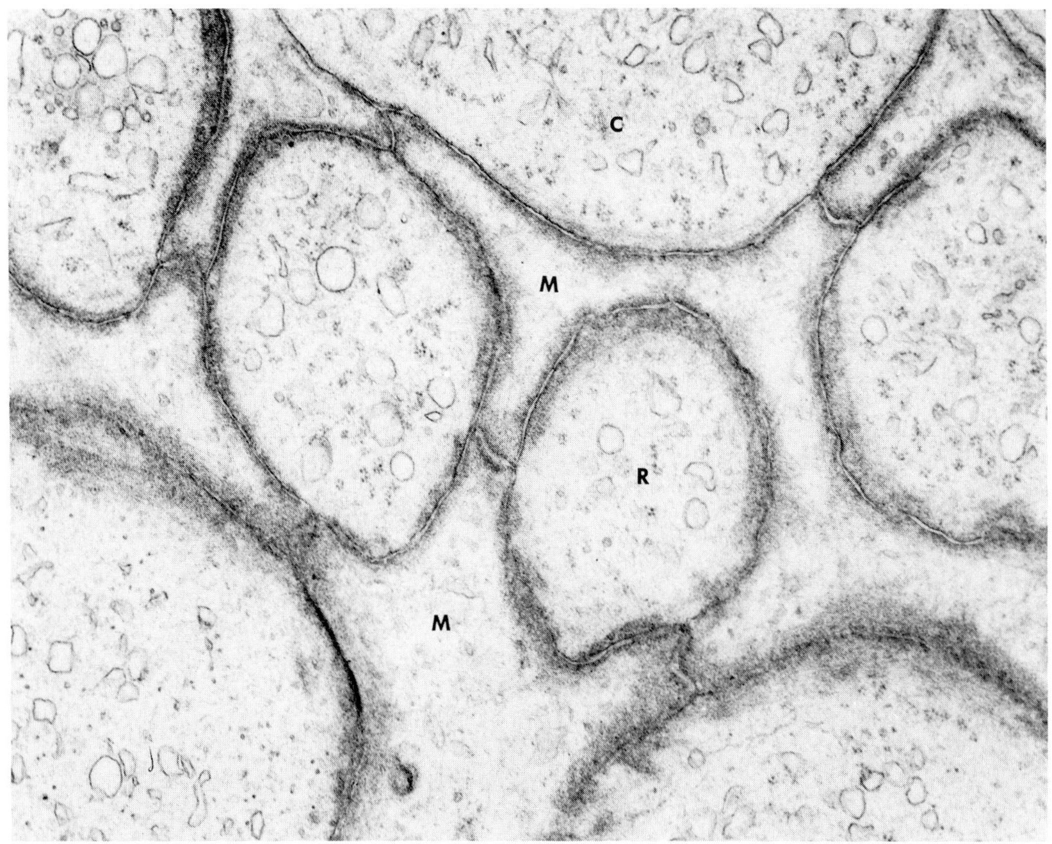

Fig. 14-2. Tangential section of human retina at level of external limiting membrane. Note glial cells of Müller, *M;* separating rods, *R;* cones, *C.* (Magnification ×26,400.)

external limiting membrane solely consists of a *zonula adherens.*

PIGMENT EPITHELIUM

The outer layer of the optic cup or pigment epithelium (PE) is the homolog of the epithelium of choroid plexus of the brain. In regions of the brain where choroid plexus forms, the wall of the neural tube remains one-cell thick, and this sheet of cells is invaginated into a brain ventricle before ingrowing tufts of capillaries. The epithelium remains intact and persistently intervenes between the capillaries and the ventricular volume. The capillaries of choroid plexus exhibit fenestrated endothelial cells, and various molecules introduced into the blood readily leak out of the capillaries, apparently through these fenestrae. The choroid plexus epithelium, however, forms an effective barrier for these substances. Included in the terminal bar complex on the ventricular side of plexus epithelium, joining these epithelial cells to one another, is a tight or occluded intercellular zone that largely restricts the ability of materials to enter or leave the ventricle by passage *between* these cells. Thus to enter or leave a ventricle via choroid plexus, materials must largely pass through the cells of the epithelium itself, suggesting that this is a site for control and selectivity.

Similarly, although the pigment epithelium (Fig. 14-3) has some specialized functions, which are described in connection with the photoreceptors, these are superimposed on its persistent choroid plexus functions. The pigment epithelium is backed by fenestrated (Fig. 14-4) choroid capillaries.[55] Included in the terminal bar system girdling the inner aspect of

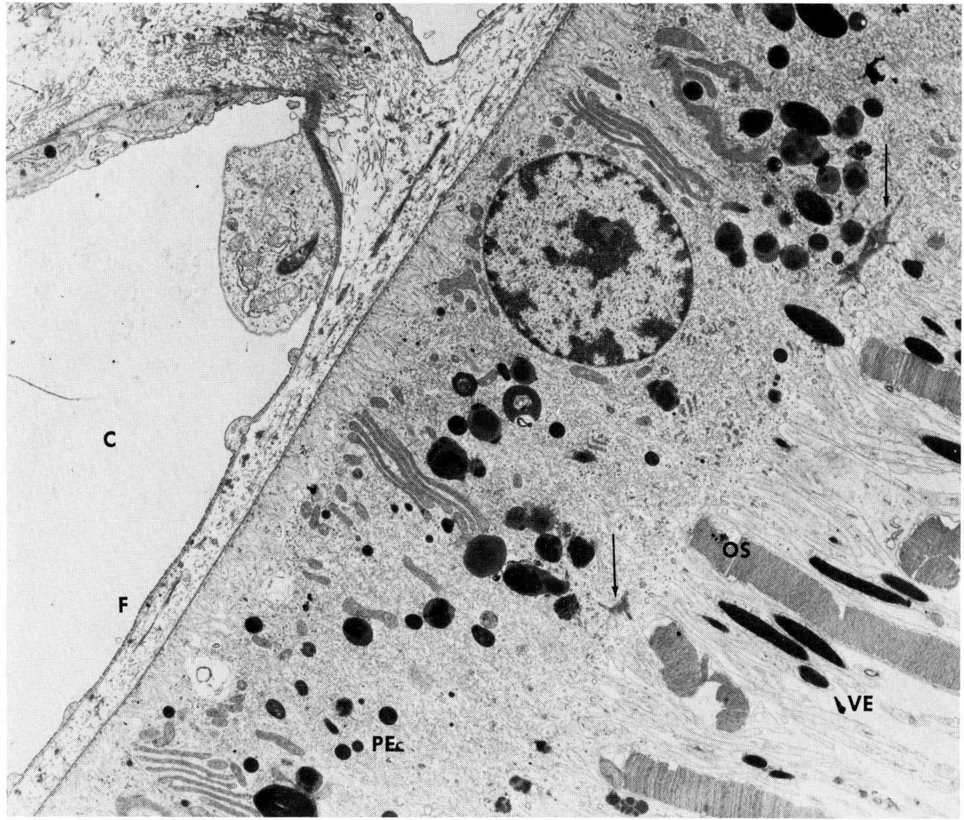

Fig. 14-3. Pigment epithelium, *PE*, of rhesus monkey. Note ventricle, *VE;* outer segments, *OS;* choroid capillary, *C*, with its fenestrated inner wall, *F*. Arrows indicate terminal bars. (Magnification ×4250.)

the cells, facing the ocular ventricle, is an occluded zone (Fig. 14-5) that resists intercellular movements of molecules across the epithelium, thus facilitating control of transport by the epithelium cells. The layer of cells of the pigment epithelium, thus joined by occluded junctions, forms a barrier of considerable electrical resistance known as the "R" membrane.[8] In addition to occluded junctions, ther terminal bars between cells of the pigment epithelium include extensive gap junctions, which probably electrically couple the epithelial cells to one another. The functional significance of the arrangement is still unclear.

CELLULAR ORGANIZATION OF THE RETINA

The retina develops from a primitive, pseudostratified neuroepithelium, whose tall, thin cells run the full retinal thickness, reaching both inner and outer surfaces. Within the cells of this epithelium, nuclei migrate toward the ventricle, and nuclear and cell division occur at this level. Before or following these events the cells' nuclei are located away from the ventricle, where they form a single thick band. Eventually certain cells cease dividing and differentiate into one of the varieties of nerve or glial cells of the mature retina. These lose their connections with the retinal surfaces. The first mature cells to be recognizable are the ganglion cells of the retina, which form the nerve cell layer closest to the vitreal chamber. The sequence of appearance of other cell varieties is less precise. The sequence of appearance of recognizable synapses, however, does not necessarily correspond with the sequence of appearance of cell types, for synapses involv-

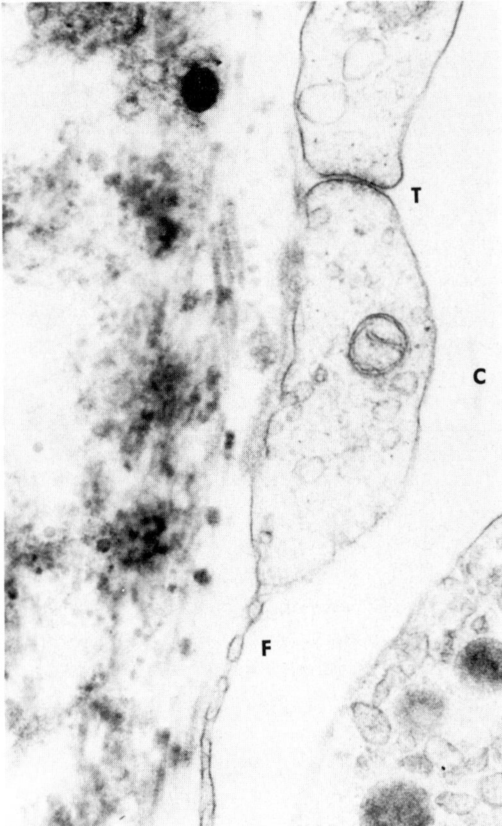

Fig. 14-4. Inner wall of human choroid capillary, *C*, showing tight junction, *T*, between endothelial cells and fenestrations, *F*, in endothelial cell wall. (Magnification ×50,000.) (From Moyer, F. H.: In Straatsma, B. R., et al., editors: The retina: morphology, function, and clinical characteristics, U.C.L.A. Forum Med. Sci. no. 8, Los Angeles, 1969, University of California Press. Reprinted by permission of the Regents of the University of California.)

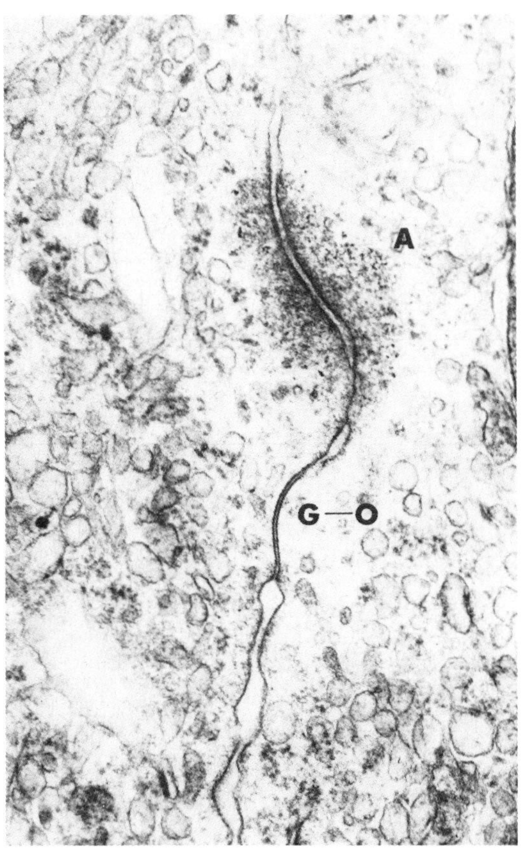

Fig. 14-5. Terminal bar of human pigment epithelium showing an adherent junction, *A*, and a zone, *G-O*, facing the retina where freeze-fracturing reveals gap junctions between strands of a zonula occludens (tight junction). (Magnification ×48,000.)

ing the photoreceptors are the first to appear and reach their mature organization and numbers.

Eventually the retina reaches its well-known layered appearance (Fig. 14-6). The layers of cell nuclei are as follows:

1. The outer nuclear layer (ONL), which contains the cell bodies of the photoreceptors
2. The inner nuclear layer (INL), which contains the cell bodies of horizontal neurons, bipolar neurons, amacrine neurons, and those of the glial cells of Müller
3. The ganglion cell layer (GL), which contains the cell bodies of the ganglion cells and those of some astroglial cells

Between the outer and inner nuclear layers is the outer synaptic layer (OSL), also called the outer plexiform layer (OPL). Here take place synaptic interactions of photoreceptors, horizontal cells, and bipolar neurons. Between the inner nuclear layer and the ganglion cell layer is the inner synaptic layer (ISL), also called the inner plexiform layer (IPL). Here take place synaptic interactions involving bipolar neurons, amacrine cells, ganglion cells, and in certain retinas, such as those of birds, interactions with terminals of centrifugal fibers to the retina from extraretinal sources.

Recently, interplexiform neurons have been identified in a number of species, including the

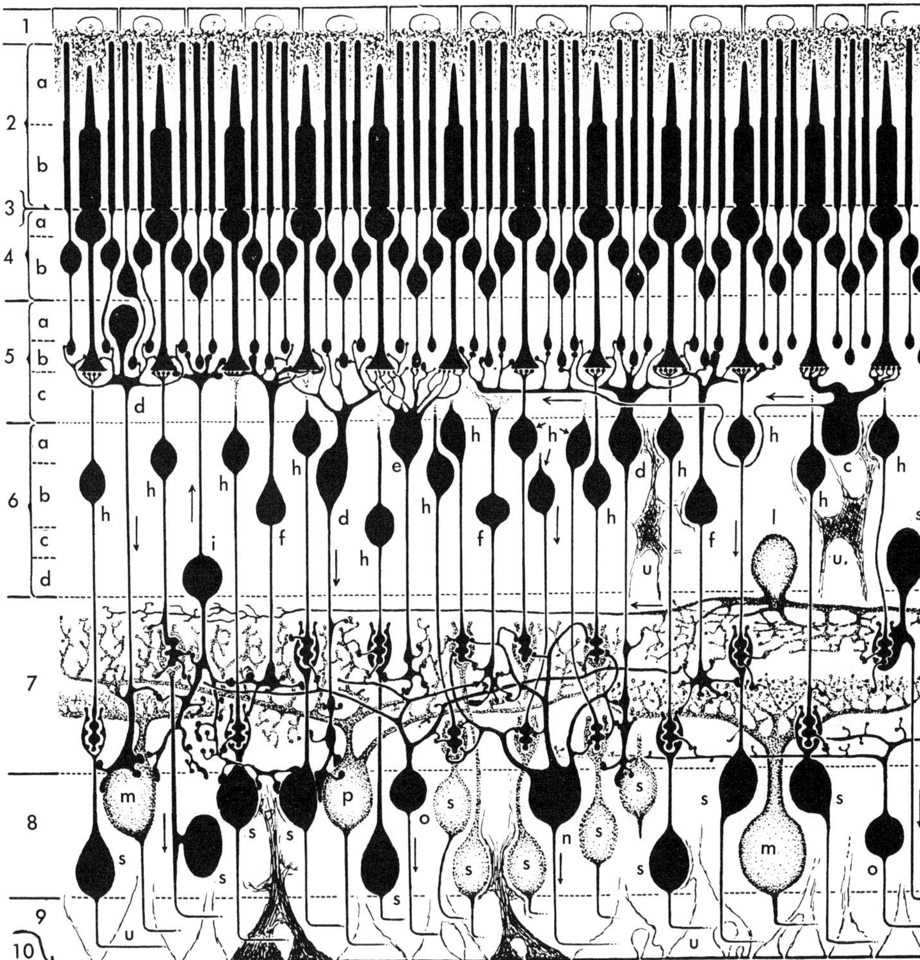

Fig. 14-6. Scheme of structures of primate retina as revealed by method of Golgi. Layers and zones are indicated as follows: *1,* Pigmented epithelium; *2a,* outer segments of photoreceptors; *2b,* inner segments of photoreceptors; *3,* external limiting membrane; *4a, b,* photoreceptor nuclei; *5a,* photoreceptor axons; *5b,* photoreceptor terminals; *5a, b, c,* outer synaptic layer; *6a, b, c, d,* outer nuclear layer; *7,* inner synaptic layer; *8,* ganglion cell somata; *9,* optic fiber layer; *10,* inner limiting membrane. Nerve cells are indicated as follows: *c,* Horizontal neuron; *d, e, f, h, i,* various types of bipolar neurons; *l,* amacrine neuron; *m, n, o, p, s,* various ganglion cells; *u,* glial cells of Müller. (From Polyak, S.: The retina, Chicago, 1941, University of Chicago Press.)

goldfish, cat, and cebus monkey. These neurons have their cell bodies in the INL and processes in both synaptic layers. Cytologic examination suggests that they convey information from the inner to the outer synaptic layer, and histochemical analysis suggests that there may be two or more varieties of these cells.[17]

The expanded processes of glial cells of Müller fully occupy the retinal face on the vitreous chamber. Between these processes and the ganglion cells lies the layer of optic fibers (OFL). The optic fibers consist of the axons of the ganglion cells and, in most species, including humans, are unmyelinated while within the retina. In most species these course across the retina to a central point, the optic nerve head or optic papilla, where they turn and leave the globe as the optic nerve. In some species the exit of the optic nerve is via a band or stripe, and the collection of axons into an optic nerve occurs just outside the eye.

The glial cells of Müller fall into a class of

cells termed "ependymoglia." They retain the primitive condition in extending, in their height, the full thickness of the retina, and the dual name indicates that they are glial cells and are also fulfilling a role corresponding to that of the ependymal cells that line other ventricular cavities of the brain. As already noted, at the inner retinal surface their expanded "feet" form a complete mosaic occupying the vitreal face of the retina. At the outer surface of the retina, where the Müller cells reach the surface, they separate the individual photoreceptors from one another, and microvillous processes of the Müller cells form so-called "fiber baskets" about the bases of the portions of the photoreceptors protruding into the ventricle. Within the retina, Müller cell processes fill in almost all volumes not occupied by nerve cells, relatively rare astroglia, or blood vessels. Most retinas, including primate retinas, lack oligodendroglial cells.

There appears to be no physical barrier to the diffusion of materials from the vitreous through the retina into the ocular ventricle. The terminal bars linking the photoreceptors and Müller cells at the level of the external limiting membrane lack occluded zones, although small gap junctions are sometimes seen between adjacent Müller cells at this level. This again parallels the situation in the brain in that, apart from the epithelium of the choroid plexus, the cells facing the ventricles seem to lack complete circumferentially occluding junctions. Substances injected into brain ventricles readily enter the tissue of the brain.

BLOOD SUPPLY OF THE RETINA

During the development of the optic cup a vertical cleft forms in the ventral aspect of the cup and optic stalk. In the retina this extends dorsally to the site of the future nerve head.[48] Through this cleft run blood vessels that gain access to the inner surface of the developing retina. Branches of these vessels eventually come to supply the inner retina. The cleft through which they reach the inner surface progressively heals, uniting the divisions of retina it formerly separated, and the penetrating main blood vessels become restricted to a residue of the cleft in the center of the optic nerve head. In most animals the capillaries derived from the branches of arteries originating at the nerve head tunnel into and supply the inner two thirds of the retina, the choroidal circulation apparently sufficing to supply the remaining outer retina via regulated transport across the pigment epithelium. This is the situation in humans. However, in some animals, like the rabbit, blood vessels are both superficial (i.e., on the vitreal surface) and confined to rather limited regions of the retina, whereas in others, such as the flying squirrel, capillaries penetrate the full thickness of all retinal regions, even lying just below the outer retinal margin denoted by the external limiting membrane. The junctions of endothelial cells in retinal vessels are tight or occluded. Thus to enter or leave the retina, most substances require active transport across the endothelial cells. The capillaries are surrounded by a basal lamella and by intermittent pericytes (mural cells),[35] which are also covered by a basal lamina.

PHOTORECEPTORS

The photoreceptors of the retina[12] are cells whose structure has suggested to many that they are evolutionarily related to the ciliated ependymal cells that line brain ventricles. The cells are bipolar in form (Fig. 14-6). Their cell bodies lie in the outer nuclear layer, with those of cones tending to be larger and lying closer to the outer retinal surface than those of rods. Their synaptic terminals lie at the outer aspect of the outer synaptic layer. From the cell body an elongated portion of the cell protrudes into the ventricular space. This is structurally divided into two portions, linked by a ciliary derivative. There is an outer segment (OS), whose apex reaches the inner aspect of the pigment epithelium, and an inner segment (IS), which connects to the cell body within the retina (Fig. 14-7).

Slightly below the apex of the inner segment, and lying eccentrically, is a basal body from which arises a ciliary stalk. This stalk links the inner to the outer segment, and the nine pairs of microtubules within the stalk (which lacks the central pair of microtubules seen in motile cilia) continue as a sort of backbone, eccentrically within the outer segment.

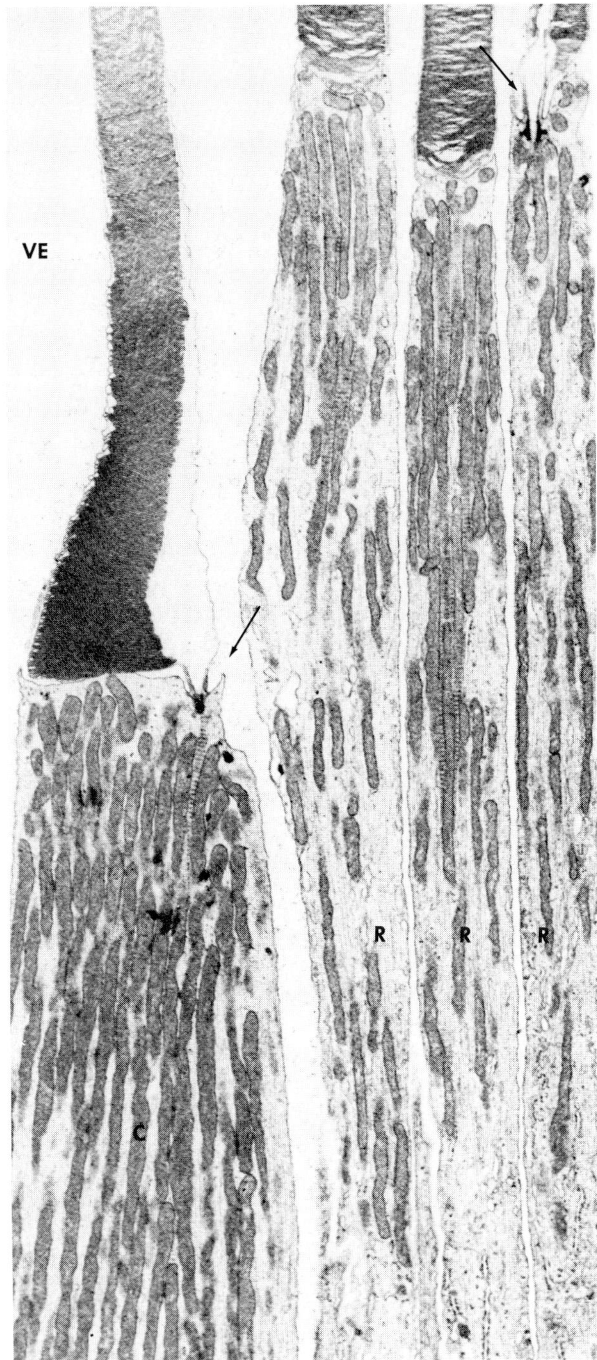

Fig. 14-7. Ventricular portions of rods, *R*, and cones, *C*, of rhesus monkey. Outer segments (above) are linked to inner segments (below) by ciliary connective (arrows), and these portions of photoreceptors are surrounded by ventricular space, *VE*. (Magnification ×7700.)

At the apex of the inner segment is a large collection of mitochondria, and this region of the cell is sometimes known as the ellipsoid. The remainder of the inner segment has more usual cell organelles including ribosomes and a Golgi region. In certain species the apical portion of the inner segment may have a colored oil droplet or other specialized organelles. In some species, but not in mammals, the ventricular portions of the cell may extend or contract in the radial direction in response to light. In these species the name myoid is sometimes applied to the portion of the inner segment closest to the retinal surface at the external limiting membrane, since this is where the active process responsible for the translational movement is thought likely to occur.

Before turning to the outer segments, the apparent site of initiation of visual signals, one must remark on the two varieties of photoreceptors, the rods and cones. As the names imply, these are morphologic terms. Originally the morphologic distinction was that in rods the inner and outer segments were both cylindric and of similar diameter, whereas in cones the outer segment was conical and the inner segment of clearly greater diameter than the outer. In current usage, photoreceptors are often called cones on morphologic grounds if the inner segment is of considerably greater diameter than a short outer segment, whether the latter is clearly conical or not.

The importance of the distinction in shapes was first recognized by Schultze, who in 1866 pointed to the correlation that rods tended to be the numerically dominant photoreceptor variety in animals whose activity extended into the range of twilight and night, whereas strictly diurnal animals tended to be very rich in cones. This correlation of the period of animal activity and shape of the dominant variety of photoreceptor hints at something related to the optical properties of the photoreceptors. Photoreceptor diameters are of the same order of magnitude as the wavelengths of light in water. But just how the shape is of functional significance in the two illumination ranges of the diurnal cycle is not altogether clear. However, the evidence is rather overwhelming that the presence of rods or cones is a very useful probabilistic indicator for the presence of other photoreceptor and visual properties—the general spectral area of the absorption maxima of visual pigments, ability to follow flickering light, certain ultrastructural details of the photoreceptors, frequency of certain retinal neuronal classes, and so on. Presumably these all relate to a heightened efficiency for the visual tasks of the animal. The shape of the cells is, however, by no means a certain indicator for the presence of any particular property, and in a relatively few species the decision as to whether a given receptor should be called a rod or cone on morphologic grounds is not readily made.

The outer segments of both rods and cones contain many double-membrane discs or flattened saccules (Fig. 14-8). At the inner or vitread end of the outer segments of rods, one observes that the plasma membrane is infolded to form hollow discs whose lumina are continuous with extracellular space. But in rods, apart from this small zone, one observes that most of the discs have apparently become isolated from the cell membrane (Figs. 14-8, *A* and *B*, and 14-9, *A*), and from each other, and the disc lumina are thus isolated from extracellular space. The individual discs seem completely and uniformly separated from each other, and some special techniques have suggested the presence of a gel between the discs. In cones, on the other hand, one observes in inframammalian species that virtually all the discs in the outer segments retain their connection with the cell membrane, and therefore the disc lumina retain their connections with extracellular space. In mammalian cones, however, while the discs in the inner third of the outer segment have their membrane confluent with the cell membrane (Figs. 14-8, *C* and *D*, and 14-9, *B*), in the outer two thirds the cone discs seem isolated. Occasionally special techniques reveal persistent connections of distal mammalian discs of cones to the plasma membrane.

The discs are of great importance because the visual pigments, which capture the photons to begin the visual process, appear to be built into the discs. Although visual pigment molecules are also present in the plasma membrane of the outer segment, they constitute a small fraction of those in disc membranes and are

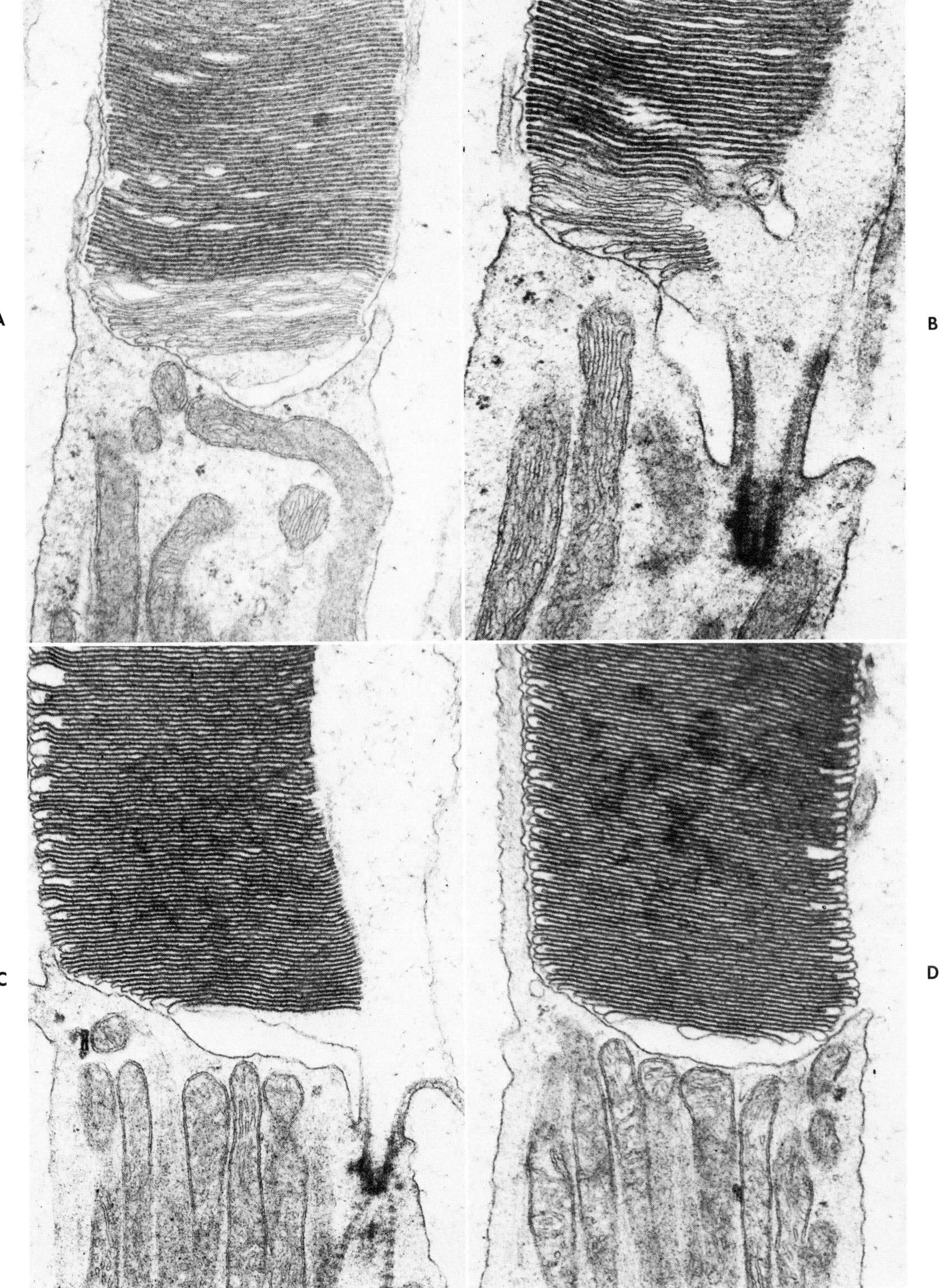

Fig. 14-8. Bases of outer segments of rhesus rods, **A** and **B,** and cones, **C** and **D. A** and **C** are cut at right angles to **B** and **D.** Note plasma membrane infolding to form saccules in **A** (magnification ×37,000) and total separation of outer and inner segments in **B** (magnification ×30,000). Similar phenomena are observed in **C** (magnification ×22,500) and **D** (magnification ×25,000) except that almost all saccules are seen to connect with the plasma membrane.

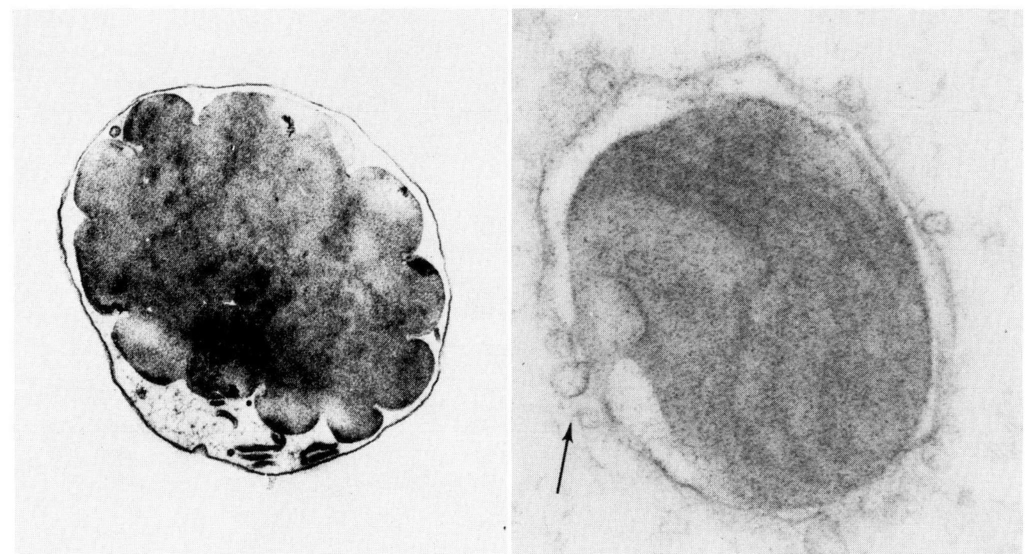

Fig. 14-9. A, Cross section of human rod outer segment. (Magnification ×40,000.) **B,** Cross section of distal portion of outer segment of rhesus cone. Note plasma membrane infolding at arrow to form disc. (Magnification ×54,500.) (**A** from Cohen, A. I.: Anat. Rec. **152:**63, 1965.)

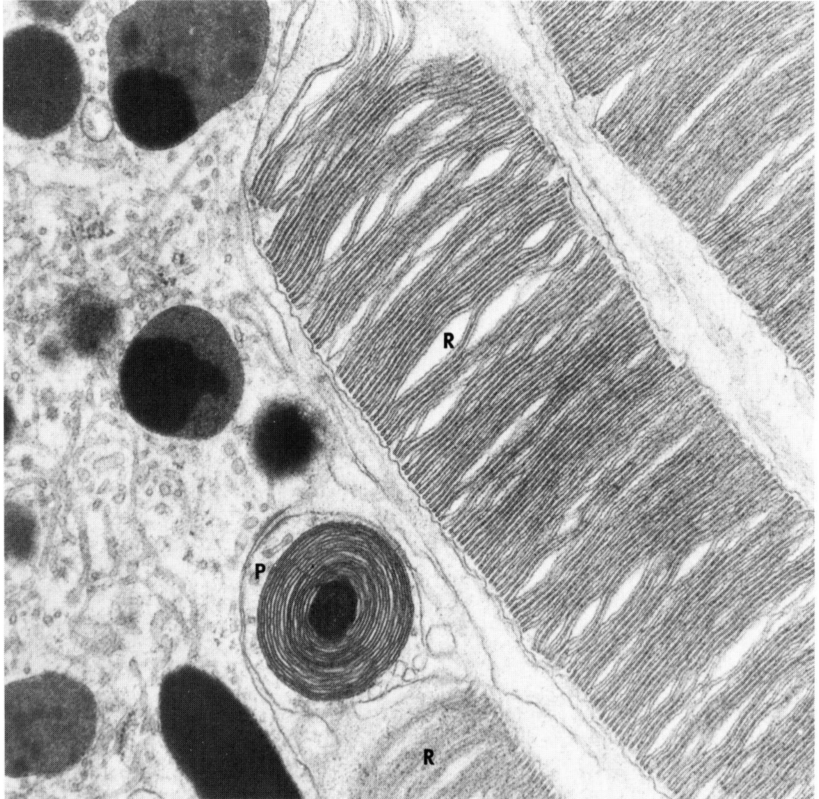

Fig. 14-10. Rod outer segments, *R,* abutting on pigmented epithelium of rhesus monkey. Note phagosome, *P,* forming at rod tip. (Magnification ×29,000.)

not optimally oriented for interaction with light. The visual pigments are rather insoluble, so that detergents must be employed to bring them into solution. This suggests that they are membrane proteins. Also supporting this view are studies with polarized light which indicate that the light-capturing or chromophore portions of the visual pigment molecules lie in planes essentially parallel to the planes of the discs. Visual pigment molecules may constitute in excess of 50% of the protein of the outer segments. The pigments are combinations of the aldehyde of vitamin A and various proteins. Their chemistry is discussed in Chapter 15.

Outer segments are capable of regeneration. If outer segments of rods or cones are destroyed via experimental retinal detachments, they may regenerate on reattaching the retina. Similarly, rod outer segments may break down because of vitamin A deficiency, but if vitamin A refeeding is initiated before the whole cell deteriorates, the outer segments may reform.

In 1961, Droz noted that radiolabeled protein from the inner segment was moving into the outer segment. Putting this observation together with observations of possibly forming saccules at the base of rod outer segments and observations of fragments of outer segments in cells of the pigment epithelium, it was suggested[3,10] that rod saccules might be in continuous formation. That this was indeed the case was proved by the elegant work of Young.[84] By supplying a pulse of a radioactive amino acid, he was able to show that amino acids were incorporated into rod discs and into rhodopsin, the rod pigment, and the pulse could be followed as a band of radioactivity, detected in autoradiographs, both at the levels of light and electron microscopy. With time this band of radioactive protein was located at increasingly sclerad levels of outer segments, and eventually in fragments of outer segments ingested (phagosomes) by the cells of the pigment epithelium[90] (Fig 14-10).

In cones of both frogs[85] and monkeys,[86] however, no banding was seen, only a diffuse labeling of the outer segments. This appears to be due to the confluence of the membrane of cone discs with the plasma membrane, which permits the mixing and randomization of newly added molecules of photopigment by diffusion. Cone discs are also shed,[34,73] and an interesting distinction from rods has been observed. Whereas the shedding of rod discs largely occurs shortly after morning when lights come on in an animal room,[5,46] the shedding of cone discs peaks after lights go out.[60,88,89] These results suggest that some rod and cone discs are replaced each day and further suggest that certain pathologic conditions may be explained by an imbalance in the production and destruction of discs. For example, in the RCS strain of rats in which an inherited receptor dystrophy occurs, Dowling and Sidman[18] observed accumulating discs and rhodopsin in the ocular ventricle during the onset of the pathologic process. However, this is by no means the typical inherited receptor dystrophy. In another receptor dystrophy, which occurs in mice, no membranes accumulate in this area. Both inherited lesions somewhat resemble the human disease called retinitis pigmentosa.

The functional significance of disc renewal and of the temporal differences in phagosome formation in rods and cones is unclear. There is a possibility that in the course of a normal exposure to light the radiation denatures a certain fraction of the protein of visual pigments. Possibly it is unsaturated lipid of the discs that is broken down. It has been shown in certain species that rods are surprisingly sensitive to damage by a prolonged exposure to light.[41,57]

RECEPTOR OUTER SEGMENT-PIGMENT EPITHELIUM RELATIONS

Pigment epithelial cells are implicated in the ocular transport of vitamin A and its derivatives, and in most species the close approximation of the photoreceptor outer segments to the pigment epithelial cells or their processes is important for the most efficient regeneration of visual pigment. The regeneration of visual pigment is one factor in dark adaptation after the significant bleaching of such pigment. However, when substantial levels of visual pigment are bleached in isolated retinas free of pigment epithelium, dark adaption not only produces a fairly rapid return of cone-initiated electrophysiologic response, but also a much slower (tens of minutes) return of rod response.

Where the apices of receptor outer segments approach the pigment epithelium, they are well draped by long microvillous and/or sleevelike processes extending from these cells. Within the inner aspect of the cells of the pigment epithelium, and within their processes, are found numerous melanosomes. In some species, but not noticeably in mammals, these cytoplasmic granules move into and out of these processes in response to the intensity of light. In a state of light adaptation, when these particles are in pigment cell processes between outer segments, they obviously minimize the scattering of light from one receptor to another. However, it should be noted that by far the greatest amount of melanin pigment in the eye lies in numerous melanophores in the uveal coat. The bulk of absorption of stray light in the eye thus occurs in the latter, but some light is backscattered out of the eye. This is what permits the use of the ophthalmoscope and certain sophisticated photometric devices used in studying the dynamics of photopigment bleaching in the living eye.

The phenomenon of detachment of the retina consists of the physical separation of the retina from its close approximation to the pigment epithelium. There is, however, no convincing evidence for specialized membrane junctions between the photoreceptors and the latter cells. All the parameters that contribute to attachment are probably not known and many are not well understood, and the importance of different parameters may vary in different species. These parameters include the following:

1. Factors regulating the volume of the fluid in the ocular ventricle
2. An acid mucopolysaccharide, known to be present in the fluid of the ocular ventricle and which could contribute to its viscosity or to the cohesion of neighboring membranes
3. A barb action of the elongated melanosomes within the long microvilli from the pigment epithelium.

It is well known that retinas are easily removed from some eyes (elderly human) but that others resist with such force that outer segments are torn from the photoreceptors and left behind adherent to the pigment epithelium (many squirrels). An hour's dark adaptation facilitates the removal of some retinas (frog), but light adaptation or longer dark adaptation renders this difficult. Punctate retinal perforations in monkeys cause a rapidly spreading detachment in a retina otherwise removed with difficulty. When photoreceptors vanish in mice with inherited receptor dystrophy, after a time the remaining retina is detached with more difficulty than before.

The pigment epithelium also has phagocytic functions. As noted earlier, portions of the tips of outer segments are ingested and digested within the pigment epithelium.[34,89,90] The membrane of the outer segment heals over. It is not yet clear how this terminal fragmentation is accomplished. The morphologic evidence in one study[70] pointed to the active participation of the epithelial cells, as processes of the pigment epithelium gave the appearance of pinching off the tips of the outer segments. However, in another study,[87] the morphologic evidence pointed to an independent mechanism within the receptor outer segments, as fragments seemed to be forming without the participation of the pigment epithelium.

Another possible relation of the pigment epithelium to the outer segments has recently come to light. Laties and co-workers[45] noted that the relationship of the axis formed by the aligned inner and outer segments to the plane of the retina at the external limiting membrane was often not perpendicular. Indeed, this proved to vary progressively from the perpendicular as one examined retina farther and farther from that lying on the optic axis of the eye. An analysis of the optical aspects of the phenomenon by Laties and Enoch[44] revealed that the receptor axes are so tipped as to orient them to the exit pupil of the eye rather than to the center of the ocular sphere. This maximizes the ability of any particular photoreceptor to capture light.

There is good evidence that during the act of accommodation the contraction of the ciliary muscle drags the choroid and pigment epithelium forward (Chapter 11) and the retina likewise, to some extent. This shifting is unequal because of the "tacking down" of the retina at the nerve head and possibly through the resistance of uveal vessels. However, this shearing could alter the orientation of receptor outer segments.

Finally, it has been known since the last century that the regeneration of bleached photopigment was facilitated by and often required the presence of the pigment epithelium in close proximity to the outer segments. This suggests some metabolic interactions of the two cell types.

DISTRIBUTION OF PHOTORECEPTORS AND OTHER NEURONS WITHIN THE RETINA: CENTRAL REGIONS AND FOVEAS

Almost all retinas contain more than one variety of photoreceptor. Claims of the existence of "all-rod" or "all-cone" retinas notwithstanding, electron-microscopic examinations have revealed some cones in many "all-rod" retinas, two sharply distinguished classes of photoreceptors based on differences in synaptic terminals in other "all-rod" retinas, some rods in "all-cone" retinas, and so on. In any event the existence of photoreceptors with differing visual pigments within the class of rods or among cones[12] in some retinas raises questions as to how different types of photoreceptors are distributed in retinas.

Retinal regions are usually functionally specialized. This specialization may be inferred from the differences in the local relative concentrations of the varieties of receptors and other cells. One functional subdivision commonly considered is a distinction between regions with an organizational bias for detecting gross form and movements versus regions organizationally biased for inspecting detail. Regions biased for inspecting detail are richer in cones by virtue of containing thinner cones and more of them per unit area than elsewhere, and more ganglion cells per unit area as well. Such a region is termed a "central" region and is usually a circular patch of retina, as in the human, cat, and bird, but in some animals, such as ground squirrels, there is a central band running horizontally across the retina. This is sometimes called a visual streak. A more precise definition of "central" region, which makes the remainder of the retina "peripheral," includes physiologic and psychophysical, as well as neuroanatomic distinctions. It is important to note that in terms of area, "central" regions may occupy a rather small portion of the total area of retinas. Thus in a conventional, nonphysiologic sense, much of the functional periphery of the retina is still physically rather central in the fundus. Physiologically central regions tend to be free of major blood vessels and in certain retinas, even capillaries. In the human the extent of the cone-rich area is about 1 cm in diameter, and it tends to be demarked by the presence of a yellow, nonphotolabile carotenoid in some cell as yet unidentified, which gives the region the name, *macula lutea*. It is therefore referred to as the macular region. The intensity of the yellow pigment, which has some effect on color perception, varies considerably from individual to individual.

In some species, including humans, the center of the cone-rich region contains a pit or fovea. In the human the full depression occupies about 5 degrees of arc or about 1.5 mm on the retina (Fig. 14-11). In the center of the fovea is a region of 54 minutes of arc or about 260 μm in extent on the retina, and in this region the only photoreceptor type present, in terms of form, are cones. This region is sometimes called the central fovea or foveola. Although a general definition of "central" fitting all vertebrates would make the macula the central region of the human retina, most physiologists confine "central" to the fixation center or foveola in the human and other foveate primates. Cones in this region have the finest diameters of the retinal cones (1.5 μm), and this is the region of the highest concentration of cones in the retina. Foveas occur in other primates and in certain fish and birds. Certain predatory birds may have a fovea in each of two well-separated central retinal regions. Functionally the fovea is the position of the retina to which, by turning the eyeball, a person brings the image of whatever is of greatest psychologic interest in his visual field. To perform an inspection of a static object, the eye, by small movements (saccades), moves successively to the foveal position the elements in the image of the visual field essential to discrimination and evaluation, such as borders of contrast.

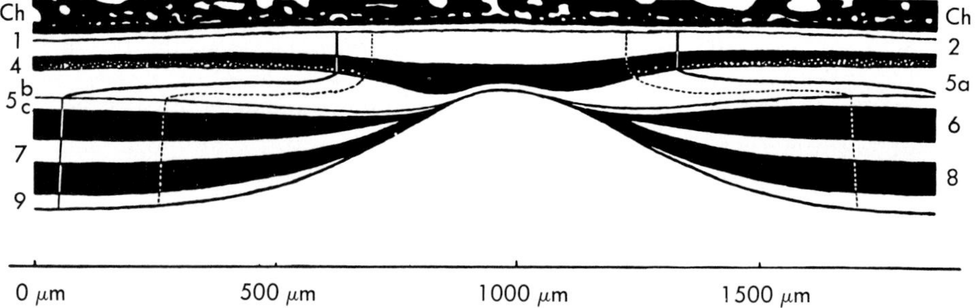

Fig. 14-11. Semidiagrammatic representation of changes in relative thickness and position of retinal layers brought about by foveal excavation in adult human eye; peculiar topical functional relationship of photoreceptor layer, *2,* and deeper layers, *4* to *9,* caused by latter's displacement owing to formation of fovea. Broken lines indicate rodless territory; white dots, rod cell nuclei; solid lines, region of high photoreceptor (rods plus cones) density. (From Polyak, S.: The retina, Chicago, 1941, University of Chicago Press.)

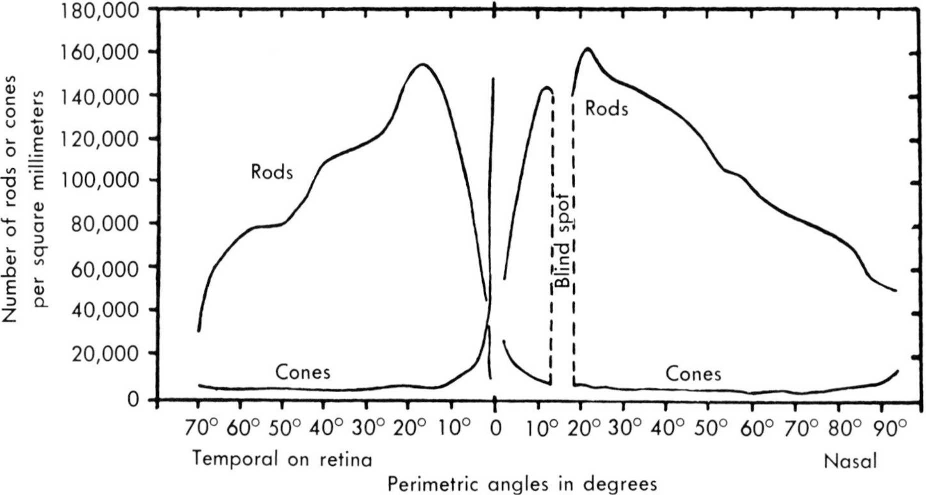

Fig. 14-12. Distribution of rods and cones in human retina. Instead of retinal distances, Østerberg's values for corresponding perimetric angles are given. Although approximate only, especially at higher angles, such values are more useful in practice than distances on the retina. Note distribution of rods and cones on nasal side in and near fovea, not given on this graph, would be approximately the same as distribution on temporal side of retina, which is seen on left of vertical passing through 0 degree on angle scale. (After Østerberg; from Pirenne, M.: Vision and the eye, London, 1948, The Pilot Press.)

Anatomically, the retina in the central fovea entirely consists of the outer and inner segments of the photoreceptors, the photoreceptor cell bodies, and the intervening glial cell processes. The axons of the photoreceptors, the so-called Henle fibers, are swept horizontally and leave the foveal area. The terminals of foveal cones, the horizontal neurons and bipolar neurons with which they interact, as well as those amacrine cells and ganglion cells that receive information from the foveal cones, are all in the parafovea, the region surrounding the fovea. Although qualitatively it is easy to imagine that this arrangement minimizes possible disturbing effects in the light path for this critical area, no satisfactory quantitative statement of this advantage has been forthcoming, probably because the extent to which tissue in front of the fovea would scatter light is hard to estimate. The foveola is surrounded by a parafoveal region, and this by a perifoveal region. The circle defined by the limits of the parafovea has a diameter of 2500 μm, that defined by the limits of the perifovea, a diameter of 5500 μm. Beyond the perifovea, defining its peripheral margin, the layer of ganglion cell bodies is never more than one cell body deep.

In the human retina, most of our knowledge of the distribution of rods and cones in a horizontal section across the retina (Fig. 14-12) is due to a quantitative study by Østerberg.[61] No comparable vertical analysis exists, and Østerberg's study was of a single normal retina. Nevertheless the fit of the data to many psychophysical studies suggests that the variance among humans with regard to the distribution of photoreceptors cannot be too great. The main points to note are as follows. The human retina has about 100 million rods and 6 million cones. The region of the nerve head has no retina or photoreceptors and therefore represents a blind spot in the subject's visual field. The cone concentration peaks sharply at about 150,000 cones/mm^2 at the central fovea, and then falls off abruptly to a low level, some 4000 to 5000 cones/mm^2, but essentially remains at that level across the entire remaining retina. The rod concentration, also the ratio of rods to cones, peaks at 20 degrees off the fovea, the angle being taken from the second optical nodal point. This region of highest rod density is approximately 6 mm from the fovea, and the rod density here is about 150,000 rods/mm^2. It then falls off less abruptly than the cone population, reaching about 30,000 to 40,000 rods/mm^2 at the extreme periphery of the retina. The area for useful color vision in humans has a diameter of about 9 mm, centered on the fovea. Three types of primate cones with pigments showing peak absorptions either in the yellow, green, or blue have been identified by spectrophotometry (Chapter 19). Some evidence suggests that the very center of the rod-free fovea may be deficient in cones with a peak absorption in the blue, but apart from this there is little information on the distribution of cone varieties in the human retina. Color vision also occurs in some nonprimate mammals, birds, reptiles, amphibians, and fish. However, many retinas exhibit cones in animals where there is little or no evidence of color vision. Thus the more general function of cones probably relates to vision at higher levels of light.

In the interpretation of analyses of the inner nuclear layer in the literature one may be misled by an unhappy tendency to refer to this layer as "the bipolar layer." Although it certainly contains the somal regions of these neurons, it also contains those of horizontal and amacrine neurons, possible interplexiform neurons, rare "displaced" ganglion cells, and the somal regions of the glial cells of Müller. If cell counts were to be made of this layer by measuring the concentrations of cell nuclei across the retina, they would lump these cell varieties together on an unwarranted assumption that their relative concentrations are the same across the retina. Not only would one like to have individual data for the frequency of these individual cell varieties, but as there are different types of horizontal, bipolar, and amacrine cells, one would ultimately want to have figures for their individual distributions. Although technical problems at the moment make it difficult to obtain such data, it is important to recognize the existence of the problem.

The situation for the distribution of ganglion cells is somewhat better because, apart from some astroglial nuclei, the neuronal nuclei in this region belong only to ganglion cells and to "displaced" amacrines. However, there are several varieties of ganglion cells in terms of size and distribution of processes. Some information does exist as to their distribution; thus it is fair to state that the macula region in the human retina is rich in small ganglion cells and that, by comparison to the concentration of cones in this region, it seems likely that there are enough small ganglion cells to permit the consideration that each could receive information via intermediate cells from a rather small population of cones. A chain of information transmission in which the ratio of receptors connected via intermediates to ganglion cells approaches 1 : 1 is what one might idealize for a region of high detail discrimination. In contrast to low information convergence in such regions, in other retinal regions there is a high ratio of rods to ganglion cells and, as expected, a high sensitivity to detecting light but poor form discrimination.

SYNAPTIC CONNECTIONS OF THE RETINA

Receptor terminals fall into two morphologic categories,[12] spherules and pedicles. As these terms imply, spherules are small and round and pedicles are large and have flat bases facing the rest of the outer synaptic layer. In humans and in those other mammals having clearly defined rods and cones, rods end in spherules and cones in pedicles, but in many nonmammalian species both rods and cones may terminate in pedicles, with those of rods usually being smaller. In some mammals, such as gray squirrels, all the terminals seem to be pedicles, and in others, for example, mice, in which almost all receptors seem rodlike, some receptors ending in pedicles are found among those ending in spherules.

Processes of horizontal cells and bipolar neurons are deeply invaginated within rod spherules (Fig. 14-13) but only superficially invaginated into the bases of pedicles (Fig. 14-14). The typical appearance of invaginated synaptic processes has a pair of horizontal cell processes flanking a bipolar dendrite lying between but slightly vitread to them (triads). Lying within the receptor terminal, in a plane that passes between the contacting horizontal processes, is a membranous lunate plate, or synaptic lamella. This is often called a synaptic ribbon because of its appearance in sections. Underlying the lunate plate, paralleling its concavity, is the so-called arciform density. The receptor terminal is full of synaptic vesicles, and a halo of these surrounds the lamella, whose function is unknown. Cone terminals also have superficial, noninvaginated synaptic contacts on their base and, in addition, send out processes ending on nearby cone pedicles and on nearby spherules. Recent evidence suggests that these interreceptor contacts[65] resemble so-called "gap" junctions, which in some neuronal junctions permit

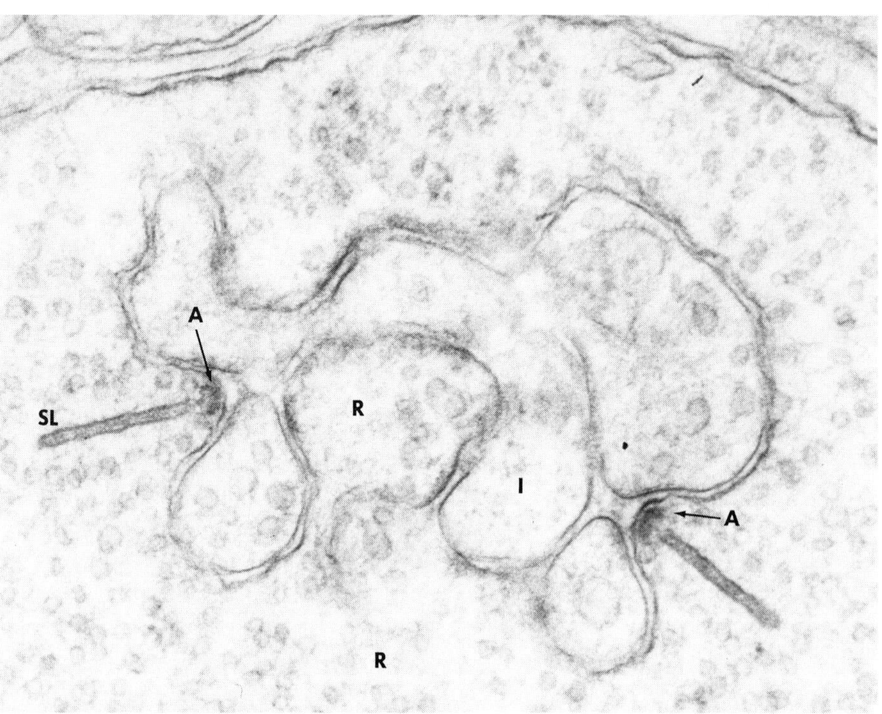

Fig. 14-13. Synaptic processes invaginated within a human rod spherule, *R*. Note synaptic lamella, *SL;* arciform densities beneath lamellae, *A* (arrows); invaginated neurites, *I*. (Magnification ×74,500.)

the electrotonic spread of current between the cells they join. In addition, minute gap junctions may be present between short laterally directed processes of photoreceptors occurring just sclerad to the external-limiting membrane.

A recent summary of retinal neuronal connections is that of Stell,[74] and a generalized diagram of synaptic connections in the vertebrate retina is given in Fig. 14-15. However, the diagram omits the interplexiform neuron, which appears to receive inputs in the IPL and to project to the OPL. Horizontal cells occur in the outer portion of the inner nuclear layer and are neurons whose processes are disposed in a manner suggesting a role in the horizontal integration of retinal activity. Their activity has been studied in very few species. In fish there may be three layers of these cells, and these are further distinguished by their large size and by the presence of putative electrotonic junctions that join cells in the same layer to one another and may permit the spread of current from one cell to another. Some physiologic evidence in turtles also suggests such linkage of horizontal cells. In most species there seem to be at least two varieties of these cells. Sometimes only one type has a well-defined axon, but this may be a technical problem. Mammalian horizontal cells are conventionally neuronal in appearance, and there is no anatomic evidence suggesting that they are electrotonically linked. In primates, horizontal cell somas have dendrites that clearly have been traced to the lateral portions of the triads in cones, and similar lateral processes in rod spherules originate in horizontal cells. The processes in rod spherules appear to be the axon terminals of cells whose dendrites enter cones. A second type of horizontal cell seems to be purely involved with cones.

Turning to bipolars, there is considerable variation in different species as to form and

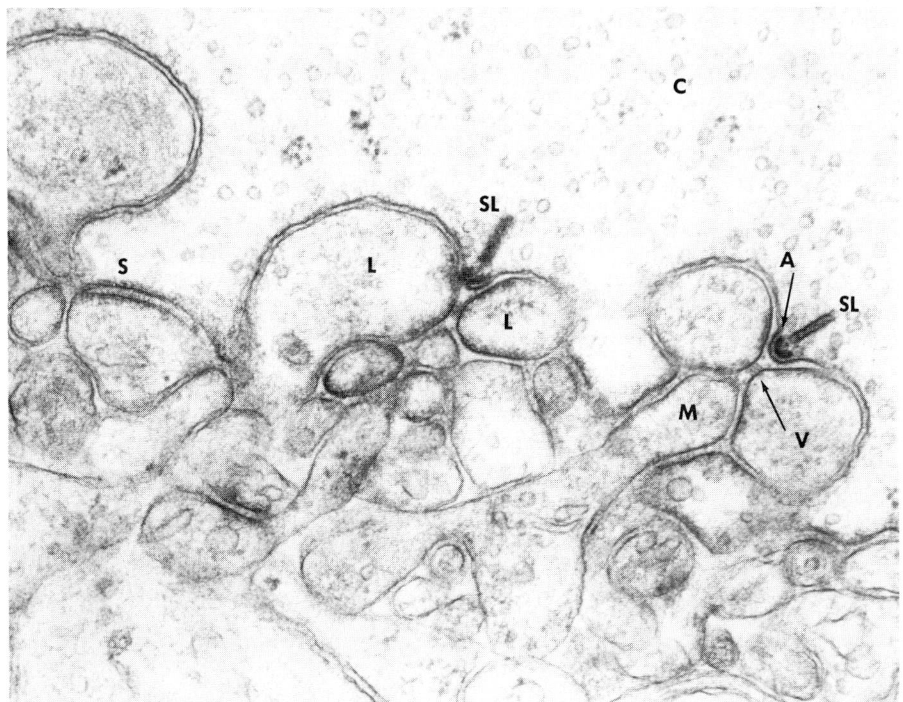

Fig. 14-14. Synaptic region of pedicle of cone, *C*, of rhesus monkey. Note synaptic lamellae, *SL*, and underlying arciform density, *A*. Synapses may be partially invaginated triads that involve two lateral, *L*, and one medial, *M*, component, probably of horizontal cell and bipolar cell origin, respectively, or superficial synapses, *S*, probably from another type of bipolar neuron. Also note small circular profiles in synaptic cleft, *V* (arrow). (Magnification ×60,000.)

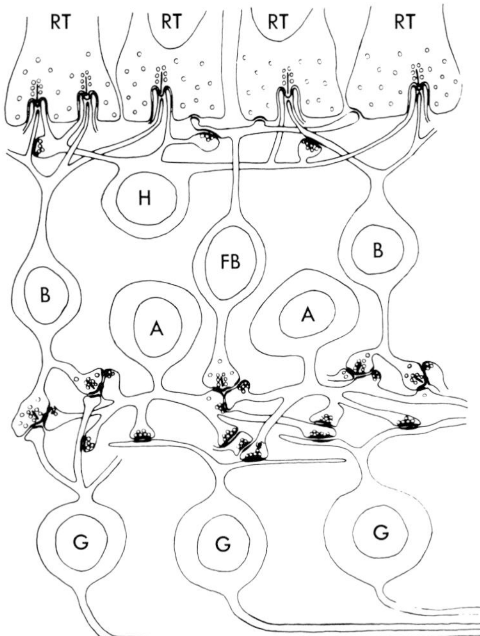

Fig. 14-15. Summary diagram of arrangements of synaptic contacts found in vertebrate retinas. In outer synaptic layer, processes from bipolar, *B*, and horizontal cells, *H*, penetrate into invaginations in receptor terminals, *RT*, and terminate near synaptic ribbons, lamellae, of receptor. Processes of flat bipolar cells, *FB*, make superficial contacts on bases of some receptor terminals. Horizontal cells make conventional synaptic contacts onto bipolar dendrites and other horizontal cell processes (not shown). Since horizontal cells usually extend farther laterally in outer synaptic layer than do bipolar dendrites, distant receptors can presumably influence bipolar cells via horizontal cells. In inner synaptic layer, two basic synaptic pathways are suggested. Bipolar terminals may contact one ganglion cell dendrite and one amacrine process at ribbon synapses (left side of diagram) or two amacrine cell, *A*, processes (right side of diagram). When latter arrangement predominates in a retina, numerous conventional synapses between amacrine processes (serial synapses) are observed, and ganglion cells, *G*, are contacted mainly by amacrine processes (right side of diagram). Amacrine processes in all retinas make synapses of conventional type back onto bipolar terminals (reciprocal synapses). (From Dowling, J. E.: Invest. Ophthalmol. **9**:655, 1970).

connectivity. In primates, cone pedicles may be in synaptic contact with three varieties of cone bipolars. One of these, the invaginating, midget bipolar, has its dendrites enter the triads, where it forms the central elements. The others, flat, midget bipolars and diffuse bipolars, have their dendrites end superficially on the pedicle base. All cones are contacted by at least one invaginating, midget bipolar and commonly by other cone bipolars, but the contacts of one midget bipolar seem confined to one cone. There are, in addition, rod bipolars whose dendrites synapse within the triads of a group of spherules. Moreover, there is an overlap of synaptic fields of adjacent rod bipolars. In other species, certain fish, for example, single bipolars contacting both rods and cones have been reported. In certain fish central processes of some triads have proved to originate in a particular class of horizontal cells; thus central triad processes are not invariably of bipolar origin.

Despite the fact that in higher primates rods outnumber cones by about 16 or 17 to 1, there is a greater convergence of rods than of cones onto bipolar neurons, and this, combined with the saturation of rod signalling in these species at modest light levels, means that under many situations cone-initiated activity can dominate neuronal signalling even where rods locally outnumber cones in the retina. Although the light levels at which rod signalling saturates may vary among species, cone signalling can sometimes dominate retinal activity even where cones are but 2% to 3% of the photoreceptors.

Before considering the terminations of these bipolar varieties in the IPL, since both amacrine cell processes and dendrites of ganglion cells are involved in these junctions, one must consider the disposition of the two latter cell varieties. By an amacrine cell is meant a neuron with no morphologically definable axon. In most cases the somal regions of these cells lie at the inner aspect of the inner nuclear layer. From the soma of the cell an inwardly directed process arises that simply branches and rebranches into finer processes. If these occur at a fairly precise horizontal level, the cell is said to show a stratified disposition of processes and may be unistratified or bistratified, depending on whether the processes spread on one or two levels. Stratified amacrine cells tend to be moderately branched. Amacrine cells may also exhibit a diffuse or nonstratified but highly branched dispersion of their processes with the horizontal extent of the volume of spreading being wide or narrow. Another variety is "displaced" in the sense that its soma occurs in the layer of ganglion cells rather than in the inner aspect of the inner nuclear layer, and its principal process is then directed outwardly. Some amacrine cells possess large dense core vesicles in their cytoplasm and processes, and these probably correspond to cells that histochemical studies reveal to contain dopamine. Some of the recently described interplexiform neurons of the fish and cebus monkey (a new-world monkey) have OPL terminals containing dopamine.[17]

The ganglion cells, large or small, may also be classified by how their dendritic expanses are distributed—unistratified, bistratified, multistratified, or diffuse—and like the amacrine cells, some ganglion cells are ''displaced'' by having their somas occur near amacrine somas in the inner nuclear layer. In one species, the axons of these ''displaced'' ganglion cells have been reported to terminate in a distinctive region of the optic tectum. Ganglion cells may be small cells of the midget variety, which sometimes contact but one midget bipolar. Thus a color-coded path could exist from a single cone to this midget ganglion cell via a midget bipolar. Of the larger ganglion cells, there is something of a spectrum, in that from one or more levels of fairly precise stratifications of dendrites, one can go to a single, thicker stratification of dendrites (diffusely stratified) and to a partly globular disposition of dendrites (diffuse).

In birds it is quite certain that centrifugal fibers from the isthmooptic nucleus of the midbrain enter the retina via the optic nerve and synapse on certain amacrine cells. Less secure evidence for centrifugal fibers exists for some amphibians and fish, but evidence for the existence of such a system in mammals is exceedingly thin. It has also been claimed that there are associational ganglion cells whose axons never leave the retina but terminate in the inner synaptic layer.

The analysis of synaptic connections in the inner synaptic layer is facilitated by the fact that the terminals of bipolar neurons somewhat resemble those of photoreceptors. The bipolar axon branches to some extent, and the branches end in large ''bags'' filled with synaptic vesicles. A number of synaptic ribbons, strongly resembling those in photoreceptors, occur at various points in each bipolar terminal perpendicular to and against the cell membrane. In a given section the plane of each lamella or ribbon within the terminal passes between two, closely opposed, contacting processes on the adjacent outer surface of the terminal. Such an arrangement is termed a dyad; however, serial reconstructions of a retinal *volume* may suggest more than two postsynaptic elements in the area. The two contacting processes in a dyad may both contain synaptic vesicles and therefore may both originate in amacrine cells, or one may contain vesicles but the other be free of them and be identifiable as a dendrite originating in a ganglion cell (Fig. 14-16). Cytologic details suggest that these double contacts are postsynaptic to the bipolar cell. Often a single vesicle-filled process is seen to make contact with a bipolar terminal, and the synaptic structure suggests that, in this case, the bipolar is postsynaptic. Since this presynaptic amacrine process may sometimes branch off a process about to terminate in a dyad, or a presynaptic contact may even be seen *within* a postsynaptic amacrine element of a dyad, the possibility of inhibitory feedback of stimulation via a reciprocal synapse arrangement is suggested. Amacrine-to-amacrine synapses are frequently seen in the inner synaptic layer, and in certain species serial chains of interamacrine synapses are seen. Four or five processes may be involved in such a chain, but it is not clear how many cells are involved. Short serial chains may sometimes be seen in the peripheral retina of primates. Finally, processes of amacrine cells are often seen to terminate on the outer aspects of the somas of ganglion cells and on their principal dendrites.

A statistical analysis of the frequency of amacrine-amacrine versus amacrine-ganglion dyads and other amacrine synapses per given area of the inner synaptic layer in sections supports the impression obtained from Golgi and methylene blue studies that retinas proportionally richer in amacrine cells (pigeons, rabbits) have a proportionally greater extent of the inner synaptic layer volume that is devoted to amacrine cell processes and their synapses. This seems to correlate with the presence of more complex data-processing operations within these retinas.

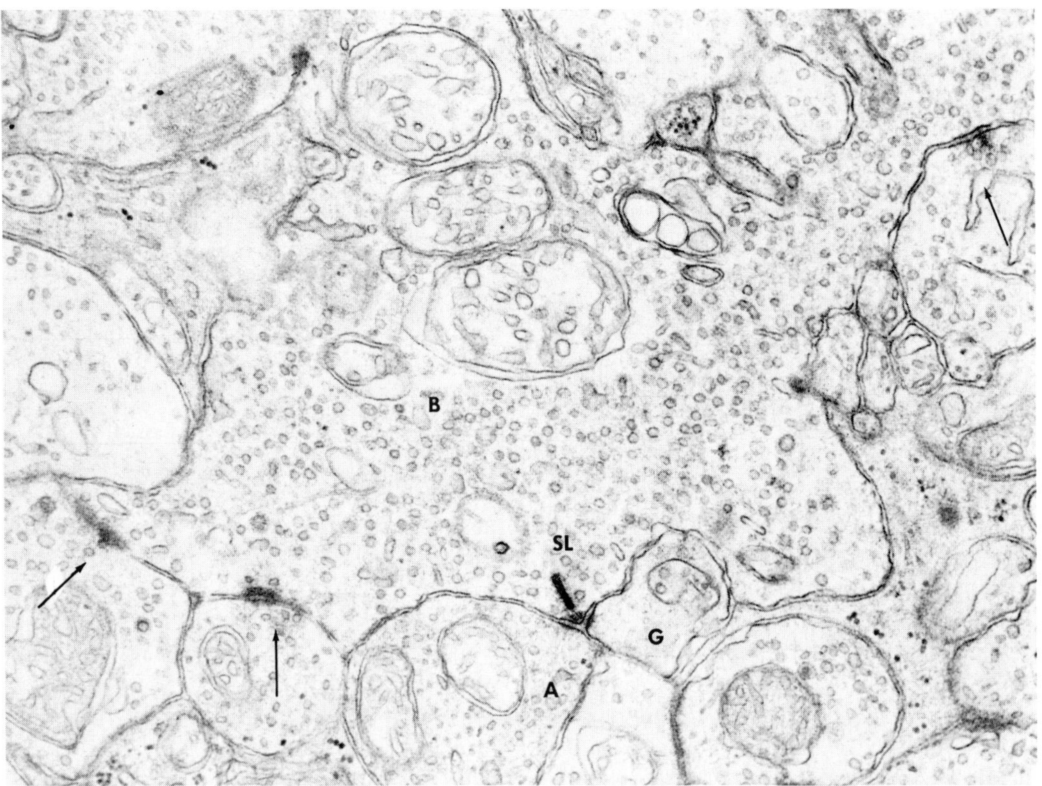

Fig. 14-16. Portion of bipolar terminal, *B*, of retina of rhesus monkey. Synaptic lamella, *SL*, indicates "dyad" synapse with adjacent postsynaptic processes from an amacrine, *A*, and ganglion, *G*, cell. Amacrine cell contacts (arrows) that are presynaptic to the bipolar are also evident. (Magnification ×26,000.)

An impression also coming from such studies is that in some species there may possibly exist a class of ganglion cells whose synaptic contacts are largely or even exclusively with amacrine cells. In the cat rod signals may only reach ganglion cells via such a path, lacking direct bipolar to ganglion cell input. There is no evidence that the primate retina falls into such a group, although it is likely that different varieties of primate ganglion cells differ both as to the kinds of amacrine and bipolar cells with which they interact as well as in the frequency of contacts they make with such cells.

RETINAL SYNAPTIC MECHANISMS AND PUTATIVE CHEMICAL NEUROTRANSMITTERS

Before entering into a discussion of the information obtained from studying the electrical activity of retinal cells, it is worth summarizing what is known of synaptic mechanisms in the retina. The photoreceptors have terminals rich in synaptic vesicles. This suggests that these cells communicate with postsynaptic cells via a chemical transmitter, but there is no direct evidence for what this transmitter might be. However, the deep invagination of processes of second-order cells within spherules, and to a lesser extent within pedicles, suggests that there could be some restriction of the diffusion of current and a degree of electrical interaction either between the receptors and second order neurons, or between processes of second-order neurons within the invaginations. Interreceptor contacts between cones, and between cones and rods, have frequently been noted and are said to consist of "gap" junctions.[65] Such structures might permit electrical interactions of these cells, and electrophysiologic evidence for the direct interaction of some turtle cones has recently been advanced.[6] Although horizontal cells in fish and some species in other classes are clearly electrically coupled to one another, there are no obvious junctions of this type between mammalian horizontal cells. However, outer plexiform layer contacts between primate horizontal cells, previously called desmosomal, have recently been shown to have some characteristics of "gap" junctions.[65] Lateral triad processes within spherules have a probable horizontal cell origin and often possess vesicles, so perhaps chemical transmission is involved at these sites. Horizontal cells in some species (rabbits, cats, mice) make relatively sparse synapses on bipolar dendrites in the outer synaptic layer, and these, too, have presynaptic vesicles. The outer synaptic layer in most species tends not to stain intensely with techniques for specific acetylcholine esterase, but weak staining for nonspecific esterase is sometimes seen. Conventional wisdom has held that such observations made it unlikely that acetylcholine is a transmitter in this region. However, Lam[42] has recently found a capacity for the synthesis of acetylcholine in photoreceptors of the turtle. Other speculative candidates for photoreceptor transmitters include the amino acids glutamate and aspartate, as these are the active candidates being investigated in primary sensory cells other than photoreceptors. Thus far, aspartate seems to be the best mimic of the receptor transmitter, and D-α-aminoadipate, a potent inhibitor of high-affinity aspartate uptake, mimics light action to some extent.

The inner synaptic layer stains heavily for acetylcholine esterase activity. It is therefore likely that at least some cell type in this region utilizes acetylcholine as a transmitter, and some inhibitors of cholinergic synapses have been found to inhibit the firing of ganglion cells. Current evidence implicates an amacrine cell as the source of acetylcholine in the inner retina. There is also evidence that some ganglion cells can synthesize acetylcholine—which presumably would be employed at synapses outside the eye. Approximately 1:30 amacrine cells fluoresce when retinas are processed by the technique of Falck and Hillarp. The material responsible for this result seems to be dopamine, a potential transmitter. Retinas take up dopamine and then release it in response to light flashes.[38] The action of a transmitter—excitation or inhibition—is a parameter both of the nature of the agent and of membrane mechanisms determining the response of a particular cell type to the agent. For example, the action of acetylcholine in skeletal muscle is excitatory, but it is inhibitory in its role in cardiac physiology. There is some evidence that retinal dopamine may play an inhibitory role. In addition, both glycine and γ-aminobutyric acid (GABA) are present in high concentrations in the inner retina, and studies on how inhibitors of glycine and GABA affect ganglion cells suggest that these amino acids may be inhibitory, if indeed they are transmitters. Autoradiography suggests that glycine and GABA are concentrated by different groups of amacrine cells, and some ganglion cells also take up GABA.[20] No clues exist as to the transmitter in the bipolar terminal, but its physical resemblance to the photoreceptor terminal might suggest the utilization of the same agent. The retina is also exceedingly rich in taurine, which occasionally has been suspected of transmitter or neuromodulator activity. However, while substantial throughout the retina, endogenous levels of taurine are highest in the cell bodies of photoreceptors in all species studied. In the cat taurine is an essential amino acid, and in this species a taurine deficiency results in the destruction of the photoreceptors. On the other hand, taurine has exhibited an inhibitory action at synapses of the inner retina, but this may be based on its chemical resemblance to glycine. When transmitter agents are released, it is obviously desirable to terminate their presence after they have carried out their signaling function. This can be done by destroying them chemically or by sweeping them back up into the cell that released them or into some other cell. The glial cell of Müller, for example, could carry out such a sequestering function, since it has been recently shown to take up glutamate to a greater extent than neighboring cells.

ELECTRICAL ACTIVITY AND INFORMATION PROCESSING BY RETINAL NEURONS

To attempt to discern how the retina processes visual information, one ideally studies the activity of specific cells of different types in specific retinal areas and relates their responses to specific spatial and temporal stimulus patterns in the visual field of the animal. A recent summary of this approach is given by Dowling.[16] The electrical activity of individual cells can be recorded by intracellular electrodes and sometimes by extracellular electrodes while an animal is held in a frame with its eyes immobilized and with the stimuli projected onto a screen whose surface includes the visual field of the animal. Any portion of the receptor mosaic where appropriate stimuli from the visual field evoke or modify responses from the cell under investigation is referred to as part of the *receptive field* of that cell. This is a physiologic concept. The cell in question may be at any level in the chain of nerve cells processing visual information. It is also important to realize that there is a considerable amount of overlapping of receptive fields of cells near each other in the retina.

Within a cell's receptive field, one sometimes distinguishes physiologically distinct regions, such as a "center" and a "surround." When a small spot of light, at an intensity above background, is first positioned on the center and then on the surround, opposite responses are often elicited, with a diminished or absent response if the spot of light is expanded to simultaneously stimulate both center and surround. One can also see center-surround effects with spots of "darkness," the illumination of the dark spot being well below that of the background. The spatial dimensions of receptive field centers are one determinant of spatial resolution, the smaller the center the smaller the possible spatial resolution.

Depending on the parameters being studied, the stimuli presented in receptive field analysis may have to be specified with reference to size, shape and orientation, intensity, contrast with background, hue, level and hue of background illumination, whether the stimulus is steady or in motion, and, if in motion, the direction and speed of motion.

One can, of course, obtain valuable information from the gross electrical responses of the retina. Electrodes across the eye "see" a summation of the various individual cell responses as attenuated by various geometries of the cells within the retina and by variations in the resistances in the extracellular pathways in the eye. The techniques can be refined by presenting stimuli to specific regions of the retina while supplying a steady background illumination to minimize responses from scattered light and by refining the data through computer averaging techniques. It is also possible to project gratings or other patterns directly on the retina, sometimes bypassing the optics of the eye. The results of such studies have provided much of the background for the advances in the study of information processing, and since they do not usually involve entering the eye, they can be carried out on intact animals or human subjects. These methods have also yielded valuable clinical information. Such gross electrical studies of the eye are described in Chapter 17. In addition, a duplex retina with its population of rods and cones modifies the signals reaching ganglion cells as a function of its adaptational state, that is to say, when it is dark adapted to a lower level of illumination, or when it is light adapted to a more intense illumination. Altering the adaptational level involves both photochemical and electrochemical changes in the receptors and probably at subsequent retinal processing levels. There are treated in Chapter 22.

The most elegant methods for studying information processing involve intracellular recordings because of the certainty that information is being derived from single cells and because it is often possible to mark the cell from which the record was obtained by injecting dyes into the cell and then processing the tissue for histology. In the case of the fluorescent dye, Procion yellow, the dye diffuses throughout the entire cell, and when processed tissue is observed by fluorescence microscopy, the form of the entire cell may be observed with a clarity matching that given by the Golgi method (Fig. 14-17). Cells may also be marked by injecting the enzyme horseradish peroxidase through the recording electrode. Much of the activity of this enzyme survives fixation, and an electron-dense product can be generated to permit locating the cell and its processes in the electron microscope. Thus the electrical record may be associated with a particular cell type.

However, intracellular recording is greatly

facilitated by having large cells available for penetration, and for this reason most of the intracellular data comes from amphibians or fish.

Extracellular records of the activity of single cells are most easily obtained from large ganglion cells at the retinal face. It proves to be relatively difficult to obtain records from small ganglion cells. Thus it is important to remember in evaluating schemes of retinal data processing that they may be based on an unrepresentative sampling. Certain types or sizes of cells may less frequently or even never have been successfully observed by either the intracellular or extracellular methods.

The electrical activity of most retinal cells does not include "spikes," that is, propagated all-or-none discharges. The cell dimensions are such that disturbances spreading electrotonically from an active locus of the cell are apparently adequate for activating synaptic mechanisms elsewhere in the same cell. However, some amacrine cells show some spiking discharges riding on depolarizations, and the ganglion cells, whose axons travel considerable distances, have conventional all-or-none behavior with propagated discharges.

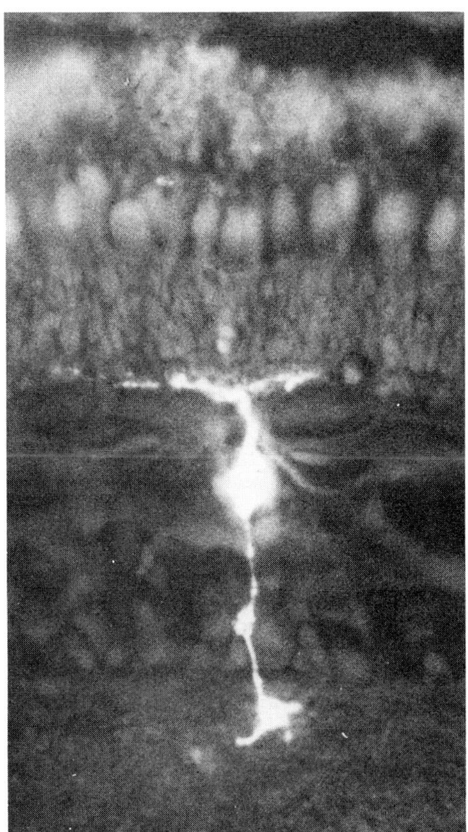

Fig. 14-17. Fluorescent bipolar cell in section of goldfish retina fixed after injection of Procion yellow subsequent to electrical recording. No magnification given. (From Kaneko, A.: J. Physiol. [Lond] **207:**623, 1970.)

It cannot be too strongly emphasized, in considering the following brief summation of the electrical activity of retinal cell types and retinal information processing, that the intracellular data are derived from a few species and from a field in which rapidly appearing reports of research sometimes bring rapid remodelings of working hypotheses. The description of electrical activity will begin with the receptors and follow the visual pathway, but it should be appreciated that, historically, the earliest cells studied were the ganglion cells of the retina, and that much of the basic concept of receptive fields and their functional organization originated with studies of ganglion cell activity.

Before turning to what is known of the electrical responses of various retinal neurons, there are two important statements that can be made about the visual process. First, all evidence points to a functional organization in the retina and higher visual system that is relativistic and directed at discerning local contrasts that establish borders between areal elements in the complex image of the visual field, rather than mechanisms for assaying the absolute levels of light in local areas. Indeed, it is the case that a retinal locus receiving an image of an area perceived as "black" at a high level of illumination may actually be receiving a greater absolute quantity of light than a retinal locus receiving an image of an area perceived as "white" at a dim illumination if, in the former instance, the "black" area is *relatively* receiving much *less* light than its general surround, and in the latter instance, if the "white" area is *relatively* receiving much *more* light than its surround. The second important point is that the neural networks of the visual apparatus are more keyed to detecting fluctuations in the retinal image due to changes in local relative intensity than to detecting steady displays. One source of this fluctuation is movement of the image of the visual field on the retina. The latter fact raises an important point regarding movements of the eye. In their normal function eyes cannot hold their positions for more than a second or two, and in the case of humans, when objects in the visual field are intentionally fixed on the fovea, they drift off, and continual corrections of eye position must be made to restore the intended positioning. Moreover, if by elaborate optical means the image of the visual field is made to hold its position on the retina despite eye movements, the image fades and is no longer seen by the observer. This explains why the shadows of the blood vessels of the retina are not in constant view in superimposition on the field of vision, for by having a fixed relation to the retina and the pathway of light, they are "adapted" out of the perceived image. The important point is, therefore, that a normal fine instability of the eye contributes to the normal visual process.

One of the major surprises in the field of retinal physiology was the discovery by a number of investigators[7,37,77] that vertebrate photoreceptors hyperpolarize when exposed to flashes of light. Not only was this the reverse of the well-known behavior of invertebrate photoreceptors, but it also seemed contrary to the conventional wisdom that nerve cells excite other nerve cells by releasing an appropriate quantity of transmitter agent when their terminals are depolarized, and it is certainly the case that photoreceptors signal second-order cells. One may consider that the vertebrate photoreceptors are normally neither in absolute darkness nor in intense light. They signal shifts toward darkness by depolarizing, and proportionally increasing their liberations of a chemical transmitter (neurohumor) and shifts toward light by hyperpolarizing and proportionally reducing the release of neurohumor.

It has been known by deduction from the findings of Hecht, Shlaer, and Pirenne[33] that single rods may be excited by a single quantum of light. What these workers actually found is that at the absolute threshold for vision a flash bringing small numbers of quanta to the photoreceptors had some of these quanta captured by a group of rods of such a number that made it unlikely that any individual rod within the group caught more than one. One must concede that for a single rod to be excited by the capture of a single photon represents an exquisite sensitivity. If photoreceptors are exquisitely sensitive near their absolute threshold in total darkness, yet maximally releasing transmitter in this state, second-order cells are detecting a reduction in the amount of transmitter released against the largest possible background.

Is there evidence pointing to whether photoreceptors are diminishing or increasing their putative release of a putatively excitatory neurohumor in response to light? Such evidence has been sought in the electrical behavior of the second-order cells, the horizontal and bipolar cells, and seems to support the maximum release of transmitter in the dark. Another indication that photoreceptor transmitters are turning over maximally in the dark has been seen when comparing their terminals following exposure to the enzyme horseradish peroxidase in either the light or dark. The enzyme is taken up passively in vesicles that are known to be forming from terminal plasma membrane during synaptic activity. The retinas are fixed for electron microscopy. Some enzyme activity survives fixation and can be used to generate an electron-dense product that appears in recently formed terminal vesicles. Since peroxidase-containing vesicles prove to be much more numerous in photoreceptor terminals of dark retinas, this indicates a more vigorous synaptic activity in the dark.[66,67]

Studies with extracellular electrodes on retinal slices showed that in the dark a current is flowing into the outer segment from the rest of the photoreceptor. The effect of a flash of light is to diminish this dark current,[62] and a number of studies indicate that this diminution is largely achieved by decreasing the conductance for sodium ion across the plasma membrane of the outer segment. The extent of reduction of the dark current by light quanta seems adequate to explain the required modulation of transmitter output for receptor signaling. The light effect can be mimicked to a large extent by manipulating the concentration of external calcium ions, leading to the suggestion that calcium ions may normally be involved in the control of sodium current in the outer segment. Large diminutions in the concentration of guanosine 3′, 5′-cyclic monophosphate (cGMP) in rod outer segments also accompany light flashes and suggest that the conductance control mechanism could be rather complicated.

Bader and co-workers,[2] recording from single rods of a salamander, found a dark membrane potential of about −45 mV, which could be driven to −75 mV by a bright flash. At bright flash intensities a sharp transient hyperpolarization was seen initially, which then fell to a plateau of lesser hyperpolarization. Control of the membrane potential by injecting current showed that the reversal or blocking potential for the light response was about 0 mV.

Although the alteration by light of a membrane conductance to sodium in the outer segment largely explains potential changes in receptors, certain complexities seem to require the existence of voltage controlled conductances, possibly to potassium and calcium, in as yet unknown locations in the cell, and Fain and co-workers[22,23] have presented evidence to support this view. Werblin[81] has presented evidence for a voltage-induced regenerative hyperpolarization in rods. If present in cones, such a property could help explain how adequate signals reach the distant terminals of foveal cones.

Intracellular recordings from amphibian rods and cones by Norman and Werblin[58] have revealed differences that to a first approximation are likely to be reflected in the differential behavior of mammalian rods and cones, in which such recordings are not yet feasible. Both rods and cones hyperpolarize to light flashes, but rods recover more slowly than do cones from bright flashes presented against a dark background, and rods have a lower absolute threshold to flashes presented against a dark background. The magnitude of the response of both rods and cones is graded over 3 log units of intensity of light flashes presented against a dark background. With brighter flashes both show amplitude saturation (although the rise times to reach these amplitude maxima continue to decrease with increasing flash intensities). However, when light flashes are presented against illuminated backgrounds, the threshold for the response of cones is raised in relation to the background intensity, but they continue to respond over 3 log units of intensity above background; that is, their response curve is simply shifted. On the other hand, the response of rods to flashes against illuminated backgrounds is also elevated in threshold but is very poorly responsive to increasing flash intensity; that is, their response is essentially saturated.

Isolated whole individual rods of the tiger salamander show time constants of voltage response to injected current that are 170 times those of intact rods in the retina. This supports the view that the frequently observed gap junctions between receptors allow spread of current.

The more rapid recovery of cones than rods to light flashes has often been held to partly account for the ability of cone-initiated vision to more readily follow higher flicker frequencies of bright (photopic) light than can rod vision in the dim (scotopic) light range. However, under certain circumstances it has been shown that the rods of the skate,[29] and possibly those of humans,[13] can adapt and begin to follow fast flicker.

A further aspect of receptor physiology has been discerned by the use of potassium-sensitive electrodes. These reveal that the onset of light causes a decrease in extracellular potassium in the ventricular space about the outer and inner limbs of the photoreceptors. This produces a greater hyperpolarization of the apical (ventricle-facing) aspect of the pigment epithelial cells than of their basal aspect and creates a transepithelial potential, which in rod-dominated retinas persists beyond the termination of the flash. The effect is of short duration and hard to detect in cone-dominated retinas, probably due to temporal differences in the recovery processes of rods and cones. This potential, variably attenuated by some opposing retinal potentials in differenet species, largely accounts for the c wave of the electroretinogram (ERG).

The earliest records of potentials from what we now know to be horizontal cells were obtained from fish and were termed ''S'' (slow) potentials. These fell into two groupings—one sensitive to color (''C'' type) in that the cell's membrane potential shifted in the hyperpolarizing or depolarizing direction as a function of wavelength, and another ("L" type) whose membrane potential when disturbed by luminosity of any wavelength always was displaced in the hyperpolarizing direction. In the mud puppy, an amphibian with small eyes but with extraordinarily large retinal cells, Werblin and Dowling[82] were able to record from all principal types of retinal neurons and then mark the cells (Fig. 14-18). Horizontal cells, when studied in *Necturus*, behaved like receptors in that they hyperpolarized to all wavelengths to which they reacted (i.e., there were no records from C-type horizontal cells) and regardless of the spatial distribution of light within their receptive field. A number of intracellular recordings have been obtained from L-type horizontal cells of the cat,[71,72] and they likewise show a large receptive field, within which an enlarged stimulus only brings an enhanced

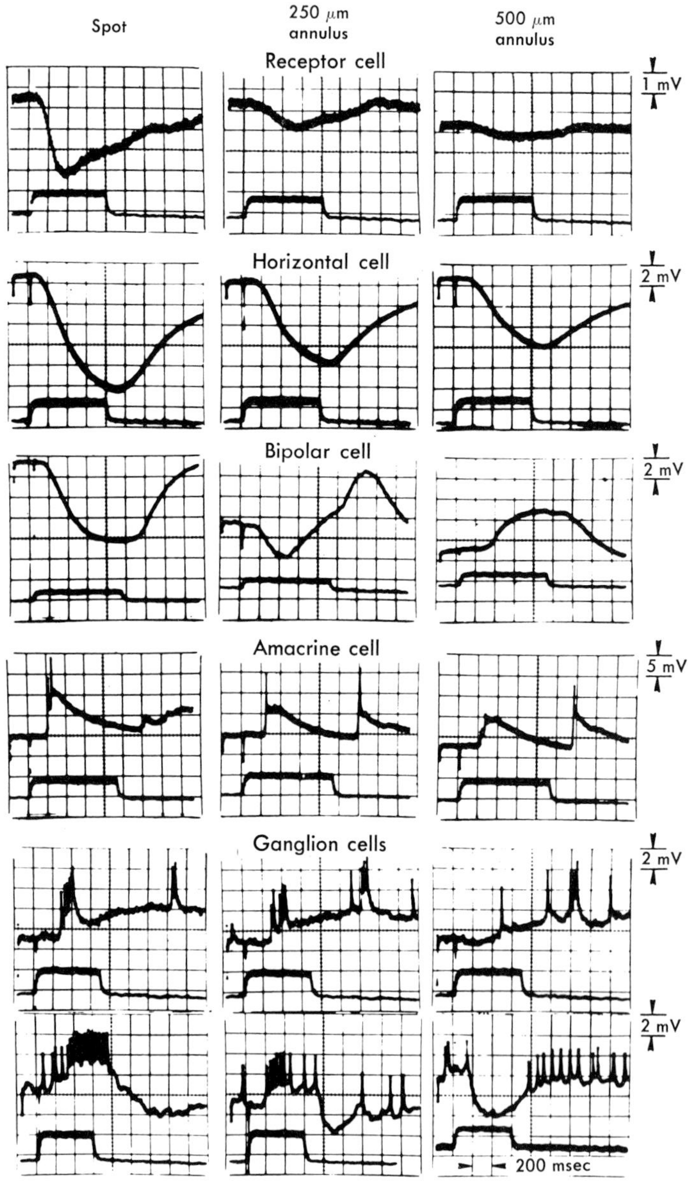

Fig. 14-18. For legend see opposite page.

response that is always a hyperpolarization to light. However, to return to *Necturus*, bipolar cells exhibited a behavior that was influenced by the spatial distribution of light within their receptive field. This field could be subdivided into a central and peripheral region. When a spot of light fell solely in the central region of their receptive field, the membrane potentials of many of the bipolar cells shifted in the depolarizing direction, the remainder in the hyperpolarizing direction.

These cells have come to be defined as depolarizing and hyperpolarizing bipolars on the basis of this response to light on their centers. These same bipolar classes can be viewed in another way. In one class light on their centers causes depolarization, while in the other darkness on their centers causes depolarization. When a test spot of light (or darkness) is enlarged to include the surround of the receptive field, the previous center response is antagonized, thus contrast between a center and its surround enhances the center response in either class.

The likely anatomic basis for the center of this receptive field is the population of photoreceptors with which these bipolars are in direct synaptic contact, whereas the surround may represent another population of photoreceptors by which the bipolars are indirectly influenced, via horizontal neurons. In *Necturus* an annulus of light falling solely on the periphery only exerted the effect on the center when the latter was also illuminated, but in most retinas an effect of an annulus of light on a center is seen without center illumination and gives the opposite effect to center illumination.

Since some types of bipolars possess dendrites that exclusively end within triads of cones (invaginating bipolars) and the dendrites of others end superficially on cones (flat bipolars), a suggestion has been advanced that these classes may be respectivley identical with depolarizing and hyperpolarizing bipolars. However, some studies by Lasansky[43] in the salamander do not support this view. When the membrane polarities of a presynaptic and postsynaptic cell shift in the same direction, the synapse is sometimes called "sign-conserving"; when they shift in opposite directions, the synapse is termed "sign-reversing." As all the dendrites of all rod bipolars of the cat are invaginated, only one bipolar class should then be associated with rods, and consistency would require that their synapse be sign-reversing. However, scotopic stimuli in the cat appear to only activate hyperpolarizing bipolars.[56] How these relate synaptically to rod-driven "on" and "off" center ganglion cells *(vide infra)* remains to be determined. The two bipolar classes (depolarizing versus hyperpolarizing) likely differ in responsiveness to the same photoreceptor neurohumor, a possibility favored by the observation that only depolariz-

Fig. 14-18. Recordings show major response types in *Necturus* retina and difference in response of a given cell type to a spot and to anuli of 250 and 500 μm radii. Receptors have relatively narrow receptive fields, so that anular stimulation evokes very little response. Small potentials recorded on anular stimulation were probably due to scattered light. Horizontal cell response over broader region of retina, so that anular stimulation with same energy as that of the spot (left column) does not reduce response significantly (right columns). Bipolar cell responds by hyperpolarization when center of its receptive field is illuminated (left column). With central illumination maintained (right trace; note lowered baseline of recording and elevated baseline of stimulus trace in records) anular illumination antagonizes sustained polarization elicited by central illumination, and response of opposite polarity is observed. In middle column, anulus was so small that it stimulated center and periphery of field simultaneously. Amacrine cell was stimulated under same conditions as was bipolar cell and gave transient responses at both onset and cessation of illumination. Its receptive field was somewhat concentrically organized, given a larger on response to spot illumination and a larger off response to anular illumination of 500 μm radius. When an anulus of 250 μm radius, cell responded with large responses at both on and off. Ganglion cell shown in upper row was of transient type and gave bursts at both on and off. Its receptive field organization was similar to that of amacrine cell illustrated. Ganglion cell shown in lower row was of sustained type. It gave a maintained discharge of impulses with spot illumination. With central illumination maintained, large anular illumination (right column) inhibited impulse firing for duration of stimulus. Smaller anulus (middle column) elicited brief depolarization and discharge of impulses at on, and brief hyperpolarization and inhibition of impulses at off. (From Werblin, F. S., and Dowling, J. E.: J. Neurophysiol. **32**:339, 1969.)

ing bipolars lose their responsiveness in chloride-free media.[50] Hyperpolarizing bipolars only lose their antagonistic surround under these circumstances. However, the possibility of two different neurohumors has not been eliminated. The two classes may function in detecting relatively bright centers in darker surrounds, or the converse.

Intracellular recordings from cells subsequently identified as amacrine cells by anatomic means have only been accomplished in the amphibian, *Necturus,* and in some fish. In *Necturus* amacrine cells proved to be the first cells in the visual pathway to exhibit a behavior other than a slow, graded, and sustained response. To a flashing, positionally static light spot they usually exhibited transient depolarizations at both the onset and termination of the stimulus. In addition, a few spikes were seen accompanying these depolarizations. In some fish, however, in addition to amacrines giving transient "on-off" responses, many showed sustained depolarizations or hyperpolarizations. In *Necturus*[80] a moving light spot sometimes indicated amacrines that possessed a center-surround mechanism in that as the spot entered the receptive field of some cells, there would be hyperpolarization, the depolarization, and the reverse effects occurred on leaving the field. Sometimes a response might only occur on entering (on center) or on leaving (off center) the receptive field, but not both. However, half the amacrine cells studied by Toyoda[76a] in the carp had a center-surround organization with on or off centers, responding to static spots and annuli of light. Amacrine effects on *Necturus* bipolars could be evoked by stimuli at distances suggesting a chain of amacrine cells bringing this influence to the bipolar being studied. This may relate to the serial synapses seen anatomically. The transient responses of some amacrine cells could be an intrinsic cell character or could result from an inhibitory feedback onto the same bipolar exciting the amacrine cell at a dyad synapse.

Recordings from single ganglion cells can be obtained extracellularly at the retinal face, or from single, surviving active optic nerve axons in small bundles teased from the nerve, or by intracellular recordings. It is possible to classify ganglion cells by a number of physiologic as well as anatomic criteria. Ganglion cells fire "all-or-none" spike discharges. In retinas of primates and certain mammals (e.g., cat) many ganglion cells have a low, sustained firing rate in total darkness. This rate may either be depressed by a spot of light placed in the center of their receptive field, in which case the ganglion cells are termed "off" cells, or it may be enhanced by the spot of light, in which case they are termed "on" cells. One can also view "on" cells as having enhanced firing to center illumination above that of the surround, while "off" cells show an enhanced firing to center illumination below that of the surround. In cat retinas and in those of some fish the dendrites of "off" center ganglion cells seem to branch in the outer third of the IPL and receive input from hyperpolarizing bipolars, while those of "on" center ganglion cells seem to branch in the vitread two thirds of the IPL and receive information from depolarizing bipolars. However, it is not clear whether this anatomy holds for all mammals. Another independent ganglion cell dichotomy is based on whether the response is sustained as long as the spot of light or darkness persists or whether it is transient—that is, appearing only after the onset or offset of the stimulus spot. The two classes just described are compared as "tonic or sustained" versus "transient or phasic," but the distinction proves difficult for some cells. An experimental classification with fewer ambiguities is that of "X" and "Y" cells.[21] These roughly correspond to the sustained and phasic types, respectively. An "X" cell maintains a steady firing rate while light and dark elements (e.g., grating bars or a checkerboard pattern) in the center of its receptive field are moved or rearranged, providing that the net illumination is unchanged, whereas the "Y" cells respond to each rearrangement. In addition, the centers of "X" cells tend to be more sharply delimited from the surrounds than those of "Y" cells. The "briskness" or "sluggishness" of a response is another independent dichotomy that has been found useful in classifying the reponse of ganglion cells. In addition, as originally noted by Hartline (reviewed by Kuffler[39]), some ganglion cells, termed "on-off" cells, show phasic responses both at the onset and offset of center illumination, while other phasic cells only respond at "on" or at "off." This behavior is somewhat influenced by the size of the stimulating spot, the dual response sometimes disappearing when a diffuse spot is reduced in size. In addition, phasic or sustained "on" center responses tend to be suppressed by chloride-free media,[51] suggesting a direct or indirect association with depolarizing bipolars, since the response of these cells is suppressed by the same medium, as noted above. Finally, a ganglion cell type named the "W" cell has been described.[76] These are quite numerous, constitute the smallest ganglion cells, and have the slowest conductance rates.[75] They may be brisk or sluggish, may lack a surround influence, and the more complex types may respond to nonuniformity in a field.

The destinations of axons of ganglion cells of the rhesus monkey have been investigated anatomically with the horseradish peroxidase tracing technique by Bunt and co-workers[9] and electrophysiologically by Schiller and Malpeli.[68,69] The latter find that the fast-conducting axons from the retina tend to end in the magnocellular layers of the dorsal lateral geniculate nucleus (DLGN), while axons of medium conduction velocity project to the parvocellular layers of the DLGN. The majority of cells projecting to the superior colliculus were phasic and not color opponent (below). The conclusions from the peroxidase studies differed in that *all* rhesus ganglion cells projected to the parvocellular layers of the DLGN, but only the largest ganglion cells of the peripheral retina and about 25% of the ganglion cells of the parafovea also projected to the magnocellular layers. Scattered ganglion cells of a wide range of sizes also projected to the superior colliculus. If peroxidase injected in the parvocellular DLGN can label *all* retinal neurons, it follows that the labeling of some retinal ganglion cells via peroxidase injected into the magnocellular layers or colliculus must be occurring via collaterals. This situation in the monkey appears to differ from that in the cat, in which the "W" cells tend to project to the superior colliculus and to some large cells of the DLGN, "X" cells appear to project only to the DLGN, and "Y" cells seem to project to the DLGN and SC.

Early studies of the electrical discharges of ganglion cells were those on the eel by Adrian and Matthews, on the frog by Hartline, and on a variety of animals by Granit and various coworkers. Later studies were performed by Kuffler,[39] who reviews the earlier work. Their findings led to some important generalizations. Kuffler had paid critical attention to the correlation of the behavior of these cells with the position and sizes of the spots of light and discerned that within the receptive field of a ganglion cell there was a central region whose response to light tended to be antagonized by light positioned in its surround. Thus some of the brevity of on firing in a phasic cell could be attributed to an inhibitory effect of a surround. A later report by Barlow, Fitzhugh, and Kuffler[4] showed that adaptational effects played a role in defining these center-surround mechanisms, the antagonistic surround effect often being greatly diminished in the dark-adapted state. Kuffler[39] considered it likely that ganglion cells with sustained firing to light (tonic cells) had a fairly direct and uncomplicated input from bipolars, whereas those ganglion cells exhibiting transient responses to light (phasic cells) had a more complex unput. This early view seems well supported by Werblin and Dowling's[82] interpretations based on their intracellular recordings of ganglion cell activity in *Necturus*. They found one type of ganglion cell with a receptive field organization quite similar to those of bipolars. Central illumination produced a sustained depolarization and sustained firing of this cell that could be inhibited in a sustained fashion by illumination of the surround. The second type of ganglion cell gave a phasic discharge, and many ganglion cells of this type fired phasically to moving stimuli passing through their receptive fields. Werblin and Dowling suggest that the important source of input to the phasic cells are amacrine cells, whereas bipolars are the important input to ganglion cells exhibiting tonic responses.

In mammalian retinas there is a distinct tendency for the mean size of the receptive fields of ganglion cells to increase with distance from a central area or fovea. However, in any given retinal region there is a wide spread of sizes of the receptive fields of ganglion cells.

In species with color vision the center-surround mechanism in ganglion cells may be color sensitive and include opponent wavelength effects in addition to the spatial effects. For example, a center excited by red light may be inhibited by a red surround, but not by a green surround, or a green center may excite and a red surround excite. Green and blue may also be antagonists. In addition, some ganglion cells, while exhibiting color opponency effects, may not show a center-surround mechanism. For example, blue light may inhibit and green light increase the spontaneous firing rate on illumination of the identical area. Some ganglion cells excited by cones show center-surround effects without wavelength preferences.

In the primate retina, Gouras[27] has also observed two distinct types of on-center ganglion cells. One received signals from two cone varieties, both of which excited in the center and inhibited in the surround. This broad-band type responded phasically to maintained stimuli of wavelengths to which these cones responded. The other on-center ganglion cell responded only to a single cone mechanism, that is, either giving peak response in the blue, green, or yellow and discharged tonically to a sustained stimulus. Conduction velocities of axons of the phasic-responding cells were faster than those measured in tonic-responding cells, and thus phasic-responding cells are likely to have larger axons. Schiller and Malpeli[69] have found ganglion cells in the rhesus monkey that show "color opponency." They defined this operationally as exhibited when the receptive field center and surround produced different responses to different wavelengths or when the surround produced a wavelength specific inhibition of the center. In contrast to "broad-band" ganglion cells (BBGC), these color-opponent ganglion cells (COGC) were of the "sustained" type. The "on" COGC projected to layers 5 and 6 of the dorsal lateral geniculate nucleus (DLGN), whereas the "off" COGC projected to layers 3 and 4. The transient BBGC mostly projected to layers 1 and 2, but some projected to the colliculus.

DeValois has pointed out that when the luminance of center and surround are both increased in color-opponent systems, antagonism persists, but if instead of a luminance increase there is a spectral shift, one can obtain synergy. For example, if a center is being stimulated by a long wavelength and the surround by a middle wavelength, a color shift to wavelengths in the shorter part of the spectrum can both undo the antagonism of the surround and enhance the activity of the center.

Hubel and Wiesel[36] had earlier observed ganglion cells in the retina of the spider monkey that were excited by certain wavelengths and inhibited by others, and DeValois and coworkers[15] had observed cells in the lateral geniculate nucleus that were similarly excited or inhibited by different wavelengths. This had raised the question as to whether spatial and chromatic information might sometimes move to the lateral geniculate via the same channel as had been observed earlier in goldfish by Wagner and associates.[79] The central processing of information is discussed in Chapter 16.

Confirming earlier studies on cats, Gouras and Link[28] have also noted that certain perifoveal ganglion cells responded to rod-initiated stimuli in dim light but switched to cone-initiated inputs when cone thresholds were exceeded. Signals above the threshold for cone-initiated firing of these ganglion cells took less time to reach the ganglion cells than did those from rods when tested just above rod threshold. Using different stimulating wavelengths, Gouras observed that in a moderate state of dark adaptation the same cell could send either rod- or cone-initiated signals but not simultaneously. Thus the receptive field of some perifoveal ganglion cells in the monkey appears to be organized into two superimposed fields. In recording from optic nerve inputs at the level of the monkey lateral geniculate nucleus, Wiesel and Hubel[83] confirmed the existence of these dual-function ganglion cells, but found ganglion cells with a sole cone input to be in the majority, even at 10 degrees off the fovea, and found no sign of ganglion cells with a sole-rod input.

One of the most seminal investigations in visual data processing was that of Lettvin,[47] and co-workers. This dealt with the presentation in the visual fields of frogs of complex stimuli rather than simple spots of light. Recording from axons of retinal ganglion cells near their termination in the optic tectum of the frog, they discovered cells that were preferentially activated by one of a variety of such complex stimuli as straight or curved, black and white borders, moving spots of light, and spots of light with changing intensity. The frog has some unmyelinated fibers in its optic nerve, and Lettvin and colleagues associated sustained activity in the presence of a static, contrasting border with such fibers. Other unmyelinated fibers were active when small spots were moving across a complex patterned background. Some myelinated fibers were active when moving edges passed through the receptive field of their ganglion cells, and still others responded to dimming spots.

Thus one line of investigation has become the search for the presence of nerve cells in the visual system that solely respond to stimuli of particular types, such as moving spots, borders, or bars of contrast (black on white or white on black), with attention paid to the size of spots, width of bars, and preferential responses to particular directions of movement, wavelength sensitivity, and the presence of surrounds with opponent mechanisms.

It was discerned that at the retinal level there were some species in which ganglion cells exhibited only a fairly simple, center-surround mechanism of circular configuration. In the same animals, cells specifically responsive to more complex inputs were only discerned at such higher levels as the visual cortex and superior colliculus. Retinas in this group include those of the cat, monkeys, and presumably humans. Other animals, such as rabbits, ground squirrels, and pigeons, had retinas in which additional categories of ganglion cells could be found that responded best to borders or bars and to motions in particular directions. It was apparent that the presence of ganglion cells with such properties correlated with the presence of a relatively greater proportion of amacrine cells in the inner retina. The percentage of ganglion cells that exhibit phasic behavior tends to be greater in the complex retinas. As noted earlier, the various classes of amacrine cells with their unique transmitters and connections play a role in creating the field properties of the ganglion cells.

One may imagine populations of specific, first-order cells connected to some second-order cells, and populations of specific, second-order cells connected to some third-order cell, and so on to achieve a high degree of specificity of response in some nth-order cell. Clearly some vertebrates carry out more stages of the data processing in the retina than do others, and operations may vary with retinal region, that is, a particular region of a retina may be more complex than another.

It should then be clear from the preceding discussion that the retina is organized to permit the lateral interaction of nerve cell networks and that this can often take the form of lateral inhibition. Certain perceptive phenomena may therefore be explicable on the basis of processing at the retinal level. Perhaps the most interesting of these is the phenomenon of Mach bands, or the enhancement of contrast at borders. Adjacent to the very border between black and white areas, the black appears somewhat blacker than the remaining black area, and the white appears somewhat whiter than the remaining white area. The basis of this observation might be as follows. Whereas most portions of the white area as imaged on the retina are surrounded by other white areas and therefore receive uniform lateral inhibitory inputs from all directions, the white region bordering on the black zone receives minimal lateral inhibition from the black direction. It therefore appears whiter. Conversely, while the image of a black region on the retina when surrounded by black regions has minimal inhibitory inputs from any part of its surround, that adjacent to the white zone receives inhibition from across this border, and the adjacent black therefore appears blacker. A thorough treatment of such problems is given in a delightful monograph by Ratliff.[64]

ROLE OF GLIA

The electrical activities of retinal cells are doubtlessly accompanied by movements of ions across membranes. Kuffler and co-workers[40] having noted that astroglia in the optic nerve of the frog exhibited slow potential changes as potassium accumulated about them during optic nerve discharges, Miller and Dowling[52] were led to investigate potential changes in the principal glial cell of the retina, the Müller cell, during retinal activity. In *Necturus* they found that the responses of the retinal Müller cells were indeed highly similar to those observed by Kuffler and associates in the optic nerve. Miller and Dowling made the reasonable suggestion that the b wave of the ERG (Chapter 17) may largely reflect potential changes in the Müller cells. Based on further studies by a number of investigators, the current detailed hypothesis contains the following elements. The membrane potential of the Müller cell may reflect its behavior as a potassium electrode. Following a light flash, the level of extracellular potassium falls near the receptors[59] and the apices of the Müller cells, a factor probably underlying a Müller cell contribution to PIII of the ERG, but then increases at the OPL level of the retina and more slowly at a deeper level. The source of this increased extracellular potassium appears to be the activity of depolarizing bipolars in the outer retina and depolarizing third-order neurons in the inner retina, and these effects of locally increased extracellular potassium on Müller cells seem responsible for the appearance of local current sinks in these cells, which lead to radial currents. Temporal and magnitude differences in these radial Müller cell currents are reflected in the changing transretinal potentials representing the initial and sustained portions of the b wave. Sustained increases in extracellular potassium

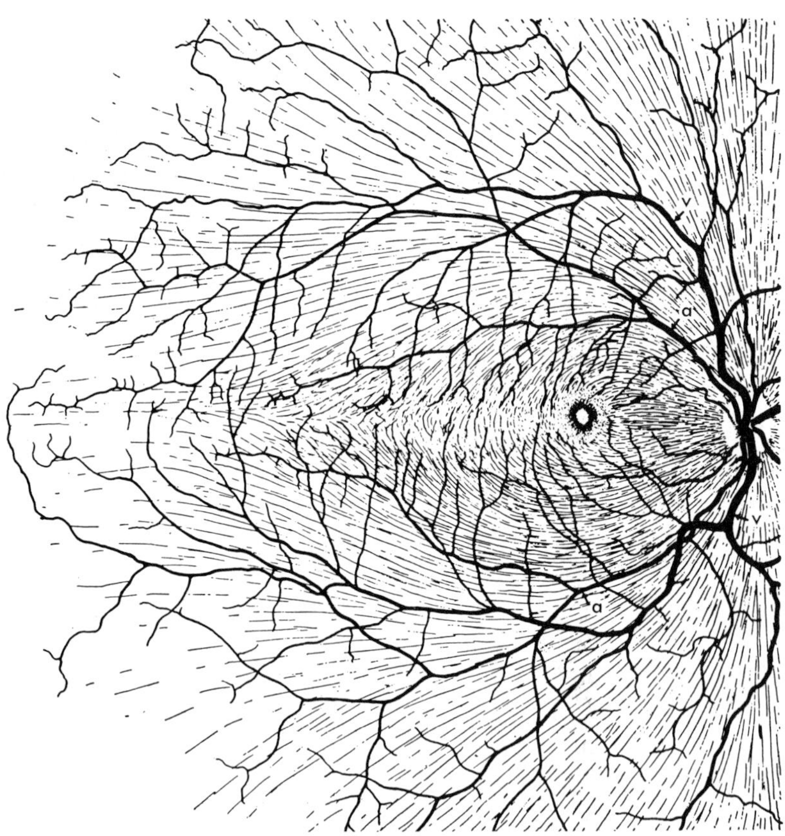

Fig. 14-19. Distribution of optic nerve fibers (thin lines) and blood vessels in retina of adult rhesus monkey as seen in whole-mount preparation stained with methylene dye during life. *a,* Arteries; *v,* veins. Nerve head is to extreme right; white spot surrounded by dark ring represents foveal floor and sloping sides. (From Polyak, S.: The retina, Chicago, 1941, University of Chicago Press.)

in the inner retina consequent to backgrounds of light that do little bleaching have also been suggested as a source of "network adaptation."

Glia undoubtedly have phagocytic functions in pathologic states and certainly functions that are still unknown. Because of differences in the vegetative metabolism of retinas of different species, partly due to differences in vascularity, one is handicapped in finding animal retinas whose glia may be useful in predicting the spectrum of functions of glia in primate retinas.

OPTIC NERVE

The axons of the retinal ganglion cells sweep across the inner face of the retina, just below the surface mosaic of expanded foot processes of the glial cells of Müller, and head toward the nerve head, where they turn sharply, penetrate the retina, and leave the eye (Fig. 14-19).

In their retinal course, as noted earlier, the foveal region is avoided and axons of macular origin, originating in the parafovea, sweep in arcs above and below the foveal pit. An exception to this, of course, are macular axons originating between the fovea and the nerve head. These enjoy a straight route to the nerve head. They are joined by some axons sweeping around the foveal area and form the important papillomacular bundle of retinal axons. Various defects in the perceived visual field, called scotomas, can reflect damage to particular groupings of retinal axons (Figs. 14-20 and 14-21).

In humans there are approximately one million axons leaving the eye. In their course across the retina they are unmyelinated and accumulate in fascicles of a few hundred axons separated by Müller cell processes, but there is a constant interchange of axons between adjoining fascicles. Within the fascicles the naked axons are often contiguous to one another, which suggests that the external currents from

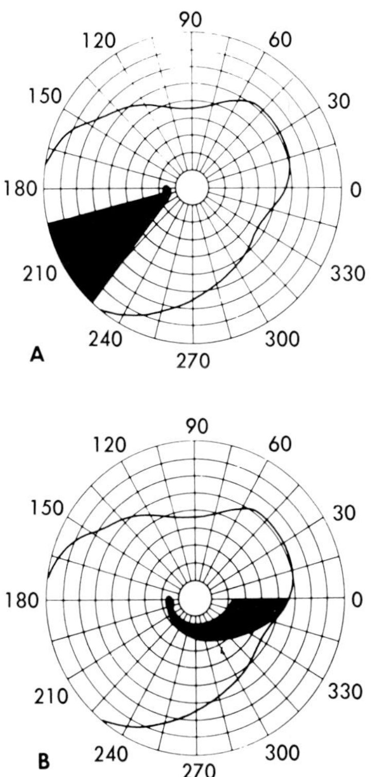

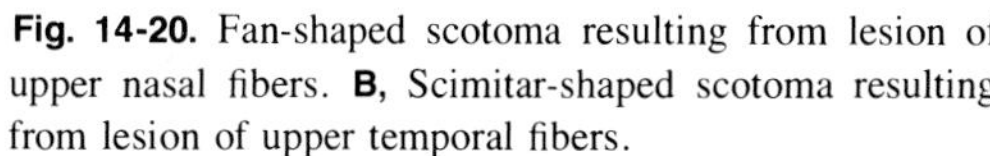
Fig. 14-20. Fan-shaped scotoma resulting from lesion of upper nasal fibers. **B,** Scimitar-shaped scotoma resulting from lesion of upper temporal fibers.

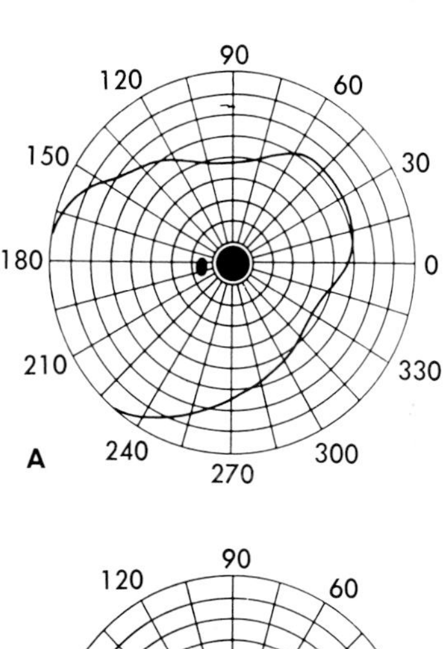

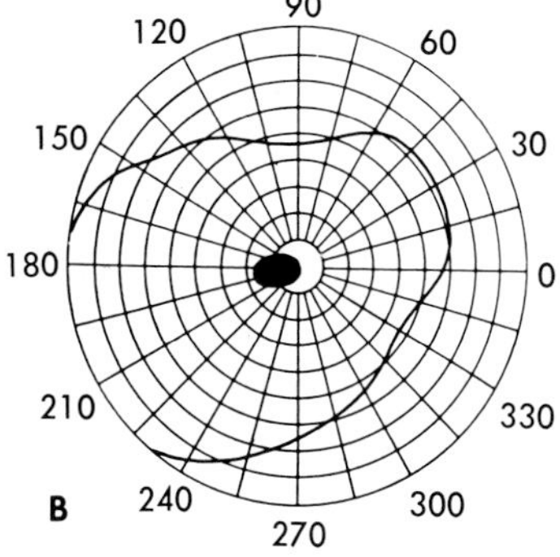

Fig. 14-21. A, Central scotoma. **B,** Cecocentral scotoma resulting from lesion of macular fibers.

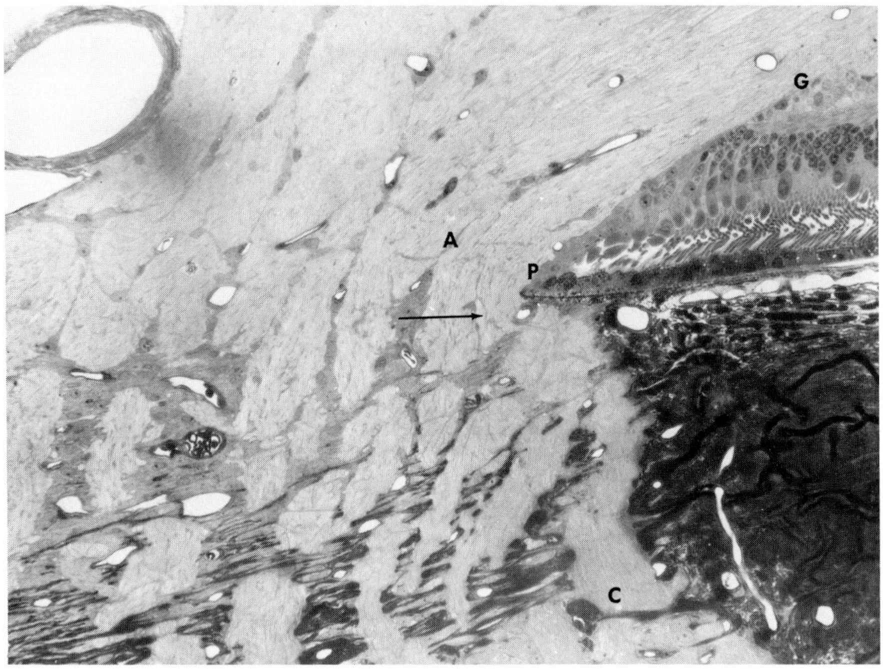

Fig. 14-22. Optic fibers exiting from retina of rhesus monkey. Arrow indicates level of choroid capillaries. There is no barrier between pial surface of forming optic nerve and these vessels. Locations of retinal ganglion cell levels, *G*, their axons, *A*, and pigmented epithelium, *P*, are indicated. Note that collagen bundles of lamina cribrosa, *C*, are confluent with those of uveal region. (Magnification ×270.)

electrical activity in any single axon cannot typically cause firing of adjacent axons, since this might confuse visual data processing. This also points up the fact that the principal insulatory role of the myelin sheaths, which these axons acquire after leaving the eye, is to provide a basis for saltatory conduction. This increases the rate of conduction of impulses in these axons when outside the eye. Thus for their retinal extent, these fibers have a different conduction rate from that which they possess when in the nerve.

In leaving the eye the retina and the pigment epithelium are penetrated. In sweeping past the pigment epithelium, with its backing of fenestrated choroid capillaries, the nerve fibers pass what is at least a theoretical weak point in the blood retinal barrier, since molecules leaving the choroid capillaries and diffusing in the horizontal direction could reach the pial surface of the optic nerve (Fig. 14-22).

The naked axons pass through the cribriform plate of the eye, and shortly thereafter they are seen to be myelinated. In the region of the nerve head, fibrous astrocytes begin to appear in the retina. These glial cells owe their name to numerous fine protoplasmic processes that extend from these cells in all directions. Fibrous astrocytes constitute the principal variety of glial cell of the optic nerve. In addition, oligodendroglia appear in the nerve where the myelination begins, this apparently being their principal function; the retina itself, in many species including primates, is devoid of this type of glial cell. In humans, myelination of the optic nerve begins in the fifth month of fetal life.

The human optic nerve[11] has the form of a solid, cylindric mass that has been extensively penetrated by many tunnels. The external surface of the cylinder and the internal surface of all the interlacing tunnels within it are coated by a basal lamina, beneath which the cellular surface consists of a mosaic of processes of the fibrous astrocytes. Thus an astroglial surface, covered by a basal lamina, constitutes the external and internal surfaces of the neural tissue of the optic nerve, just as it does in brain tissue elsewhere, except at surfaces of brain ventricles. The tunnels are occupied by mesodermal tissue such as capillaries and connective tissue elements including bundles of collagen fibers. These mesodermal gaps seem to subdivide the mass of the optic nerve when it is seen in sections, but in fact there is no interruption of the three-dimensional continuity of the neural mass of axons and glial cells.

Most of the collagen bundles of the sclera continue onto the optic nerve as the collagen bundles of its dural sheath. However, in the formation of the lamina cribrosa, some of these bundles pass across the optic nerve by means of the aforementioned system of tunnels occupied by mesodermal derivatives. The latter bundles thus complete the collagenous jacketing of the ocular sphere and clearly play a role in resisting the pressure within the eye, but the cribriform region is a potential weak point of this anatomic arrangement.

In humans the central retinal artery enters the optic nerve while within the orbit and eventually runs through one of the larger tunnels within the nerve together with several veins. It may branch while within this tunnel. Details of the blood supply of the nerve have been described by Hayreh[31,32] and by François and Neetens.[24,25] These authors disagree on a number of points. Sympathetic nerve fibers accompany the arteries and typically end in proximity to them while in the cribriform zone. Sympathetic nerves and their terminals have but rarely been noted by electron microscopists as existing alongside arteries and arterioles within the human retina, nor has their catecholamine transmitter been noted much beyond the nerve head after histochemical procedures, but a thorough investigation of this question remains to be done. Peripheral nerves, however, supply the dural sheath of the optic nerve, as they supply other dural regions. The pain in retrobulbar neuritis arises from the sheath, not the neural tissue, of the optic nerve.

PRODUCTION OF CUPPING IN THE OPTIC NERVE

It was formerly believed that the so-called glaucomatous cupping that occurs in chronic simple glaucoma of the open-angle type was caused solely by the elevated intraocular pressure that pushed back the lamina cribrosa. It was reasoned that this was the weakest spot in the tunics of the eye, thus it was the first to yield to the elevated intraocular pressure. It is generally recognized, however, that there is no direct relationship between the degree of elevation of the pressure and the amount of cupping.

Goldman[26] has pointed out that the cupping is not caused by the increased pressure primarily, but rather by the cytopathologic changes that occur simultaneously with the rise of pressure in the posterior segment of eyes in primary open-angle glaucoma. Measurements of the distance between the retinal surface and the lamina cribrosa in normal patients and in patients with chronic simple glaucoma who have a fully developed glaucomatous excavation but still retain visual function show no differences. If anything, the distance is less in patients with glaucoma. This indicates that the glaucomatous cupping is not caused by the lamina cribrosa bending backward. Bowing of the lamina is really a late effect and occurs frequently in patients with absolute glaucoma. The cupping, according to Goldmann, is caused by the changes that occur behind the lamina cribrosa—a fact pointed out first by Schnabel and confirmed by Elschnig and Fuchs. For a long period the lamina cribrosa does not participate in the glaucomatous excavation, but the cupping is caused by the disappearance of the nerve and supporting glial tissue because of a peculiar type of change that Schnabel called cavernous optic atrophy. Therefore the cause of galucomatous excavation is the disappearance of the supporting glial tissue. In primary atrophy, on the contrary, this supporting tissue remains practically unchanged so that no cup is produced.

The duration of the pressure, rather than the height of the pressure, is the chief factor in determining the extent of the cupping. An acute rise in pressure, lasting a week or more, seldom gives rise to any cupping of the optic disc, whereas pressure slightly above normal that continues for months and years eventually gives rise to a deep glaucomatous cup and atrophy of the nerve fibers.

DISTRIBUTION OF AXONS WITHIN THE OPTIC NERVE

It is of considerable significance for work with human patients to know something of the distribution of optic nerve fibers within the optic nerve and at the optic chiasm. Experimental studies with monkeys give results that fit clinical observations in suggesting that axons originating in the periphery of the retina tend to lie peripherally within the nerve and those from the central retina eventually lie centrally. As the papillomacular bundle enters the temporal edge of the optic nerve head this is its initial position, and its fibers gradually shift to a central position farther back in the nerve. Because there is less convergence of visual information in the central retina, the greater concentration of ganglion cells of the central retina yields more axons per unit of area of central retina than the sparser ganglion cells yield axons per unit of area of peripheral retina. These central axons constitute a greater fraction of the optic nerve than expected from a simple ratio of the retinal area from which they originate to the total retinal area. This has recently been demonstrated in an elegant quantitative fashion by Potts and co-workers.[63] The extramacular temporal fibers enter the nerve both above and below the papillomacular bundle, and the nasal fibers tend to be concentrated in the nasal side of the nerve. As one proceeds toward the chiasm, a tendency for the central movement within the nerve of the macular fibers is accompanied by some 45 degrees of rotation of the locations of the retinal quadrants represented by the peripheral fibers, thus as the nerve enters the chiasm, the inferior nasal fibers that entered on the inferior nasal side of the nerve head now tend to be concentrated on the whole of the inferior side of the nerve. Hence they are found in the inferior chiasm, and fibers from the superior retina tend to occupy the superior portion of the chiasm. However, in the prechiasmatic nerve there are no sharp segregations of fibers from the macula and other retinal regions, and there is considerable intermingling of fibers of different origin. Another parameter affecting the position of axons in the optic nerve may be the sequence of axonal sprouting from retinal ganglion cells in different retinal areas during retinal differentiation and development.

The axons originating from the temporal retina in humans are not destined to cross, and they lie in the temporal portion of the chiasm, while those originating from the nasal retina are found in the nasal part of the chiasm. In certain species, axons having a hypothalamic or accessory optic tract destination leave at the chiasm. The existence of these fibers has not been established for humans, but there is a reasonable probability of their existence.

The chiasm is richly supplied with blood, chiefly by branches of the internal carotid and the anterior cerebral and anterior communicating arteries.

As extensions of the ganglion cells of the retina, the axons of the optic nerve depend on activities within the somas of these cells for many biochemical needs. Advantage has been taken of this situation to study the ways in which materials are transported along processes in nerve cells. If a radioactive amino acid or some species of radioactive precursor of some other large molecule is injected into the vitreous chamber of the eye of a living animal, it easily passes into the retina. Here these precursors are taken up by the ganglion cells and employed in their synthetic activities. At intervals thereafter, animals may be sacrificed and the movement of radioactivity down the optic nerve observed by autoradiography or other means. Such studies have shown that some kinds of molecules are transported slowly, at about 1 to 10 mm/day, whereas another fraction is transported more rapidly at some 200 to 400 mm/day. The latter fraction includes materials rapidly transported to the terminals of the axons. Such studies provide tools for studying the effects of pathologic states of the retinal ganglion cells or their intraretinal or extraretinal axons on axonal transport. Recent experimental studies[1] support the view that elevated intraocular pressure directly or indirectly interferes with the normal movement of tagged molecules along axons of the retinal ganglion cells and that the lamina cribrosa is a site where blockage seems to occur.

When the optic nerve head becomes congested and swollen, a so-called choked disc is said to exist. In almost all cases of edema of central nervous tissue, microscopy reveals that there is a swelling of glial cells with minimal enlargement of the extracellular space, but this does not appear to be true for papilledema. Early speculations on the origins of papilledema solely assumed an expansion of the extracellular volume. This was attributed to venous stasis caused by direct or indirect pressure on the cavernous sinus (an early theory of von Graefe) or by cerebrospinal fluid (CSF) pressure on the long extent of the central retinal vein in the subarachnoid space of the optic nerve. These hypotheses have likewise been shown to be invalid. It now seems likely that when the cerebrospinal fluid (CSF) pressure in the subarachnoid space of the sheath of the optic nerve increases relative to intraocular pressure, it tends to interfere with postocular axoplasmic flow in the axons of the retinal ganglion cells. Consequent to this there ensues a discal swelling that is largely based on the intraocular pooling of ganglion cell axoplasm and engorgement of the axons in the region proximal to where these ganglion cell axons first enter an area of elevated pressure. To a lesser extent there is some expansion of interstitial space, but there is no evidence of glial swelling. This view follows from electron-microscopic observations of the edematous nerve head[53,54,78] either consequent to artifically elevated cranial CSF pressure or consequent to ocular hypotony, both procedures establishing greater extraocular than intraocular pressure.

As discharges sweep along the masses of axons in the optic nerve, these are accompanied by movements of ions into and out of the axons. As already mentioned, Kuffler and coworkers[40] placed electrodes in the astroglia of the frog optic nerve and noted that during such discharges the liberation of extracellular potassium from the axons produced a slow depolarization of the glial cells. They also discerned that these glial cells were electrically coupled to one another.

As noted earlier, the principal destinations of the optic nerve fibers are terminals in the dorsal lateral geniculate nuclei and to the superior colliculus. In humans and the higher primates a meridian may be imagined as passing vertically through the fovea, dividing the retina into temporal and nasal sectors. Axons originating in ganglion cells nasal to this meridian pass to the geniculate and colliculus of the opposite side of the body, whereas axons of ganglion cells temporal to this meridian pass to the ipsilateral centers. In lower vertebrates the general tendency is for the vast majority of axons to cross to the other side of the body. In these animals, when ipsilateral fibers are present, they come from the temporal retina but represent a specific class or classes of ganglion cells with other temporal ganglion cells still projecting contralaterally.

It has recently been noted by Guillery and associates[30] that in mammals homozygous for albinism or for certain other allelic genes in the albino series, such as those producing the Siamese pattern in cats, there is misrouting of many axons originating in the temporal retina. These aberrant axons cross in the chiasm instead of remaining ipsilateral and thus insert in geniculate positions in competition with normal axons. Thus the known fixation difficulties and poor vision of human albinos, usually solely attributed to light scattering within the eye, might prove, in part, to originate in similar axonal misroutings since the phenomenon seems rather general in albino animals, and indeed leads to neuroanatomic abnormalities at the level of the visual cortex.

DESTINATION OF OPTIC NERVE FIBERS

The dorsal lateral geniculate nucleus and the superior colliculus are the principal sites for the termination of human optic nerve fibers. In fish, amphibians, and birds the optic tectum is the main terminus. This ventricle-containing midbrain structure is related to the superior colliculus. The dorsal lateral geniculate terminus seems to have originated as a secondary center, which became more important in the course of mammalian and primate evolution.

Other brain regions that have been reported

to receive retinal inputs in mammals are the ventral lateral geniculate nucleus, the pretectal region, the hypothalamus, and certain midbrain nuclei belonging to the accessory optic tract system.[19] The primate retina clearly has some pretectal projection, and it probably has projections to some or all of these other centers as well. Fibers of specific functions may have specific destinations. Thus, in ground squirrels, Michael[49] found a fairly total separation of color-coded and non–color-coded ganglion cells. The former have a tonic firing pattern and project to the lateral geniculate nucleus. The latter have a phasic firing pattern and project to the superior colliculus.

From time to time the possibility is raised of the existence of efferent fibers to the human retina, and there is no doubt that such systems exist in some vertebrates. The most thoroughly documented case is in the pigeon, in which a midbrain nucleus, the isthmooptic nucleus, was shown by Cowan and Powell[14a] to project to the retina, ending on cell processes in the outer portion of the inner synaptic layer. A fairly precise topographic arrangement consisting of projections from specific regions of the pigeon retina to specific regions of the pigeon tectum, then to specific regions of the isthmooptic nucleus and finally back to the original retinal loci were discerned.[14] Similar evidence seems to be developing for certain snakes and turtles. Highly suggestive evidence also exists for efferents in frogs and certain teleosts, but the evidence for such a system in the mammal is still rather weak.

REFERENCES

1. Anderson, D. R., and Hendrickson, A.: Effect of intraocular pressure on rapid transport in monkey optic nerve, Invest. Ophthalmol. **13:**771, 1974.
2. Bader, C. R., MacLeish, P. R., and Schwartz, E. A.: Responses to light of solitary rod photoreceptors isolated from tiger salamander retina, Proc. Natl. Acad. Sci. U.S.A. **75:**3507, 1978.
3. Bairati, A., Jr., and Orzalesi, N.: The ultrastructure of the pigment epithelium and of the photoreceptor-pigment epithelium junction in the human retina, J. Ultrastruct. Res. **9:**484, 1963.
4. Barlow, H. B., FitzHugh, R., and Kuffler, S. W.: Change of organization in the receptive fields of the cat's retina during dark adaptation, J. Physiol. (Lond) **137:**338, 1957.
5. Basinger, S., Hoffman, R., and Matthes, M.: Photoreceptor shedding is initiated by light in the frog retina, Science **194:**1074, 1978.
6. Baylor, D. A., Fuortes, M. G. F., and O'Bryan, P. M.: Receptive fields of cones in the retina of the turtle, J. Physiol. (Lond) **214:**265, 1971.
7. Bortoff, A.: Localization of slow potential responses in the *Necturus* retina, Vison. Res. **4:**627, 1964.
8. Brindley, G. S.: Physiology of the retina and visual pathway, ed. 2, Baltimore, 1970, The Williams & Wilkins Co., p. 315.
9. Bunt, A. H., et al.: Monkey retinal ganglion cells: morphometric analysis and tracing of axonal projections, with a consideration of the peroxidase technique, J. Comp. Neurol. **164:**265, 1975.
10. Cohen, A. I.: Vertebrate retinal cells and their organization, Biol. Rev. **38:**427, 1963.
11. Cohen, A. I.: Ultrastructural aspects of the human optic nerve, Invest. Ophthalmol. **6:**294, 1967.
12. Cohen, A. I.: Rods and cones. In Fuortes, M. G. F., editor: Handbook of sensory physiology, Berlin, 1972, Springer-Verlag, vol. 7/2, p. 63.
13. Conner, J. D., and MacLeod, D. I. A.: Rod photoreceptors detect rapid flicker, Science **195:**698, 1977.
14. Cowan, W. M.: Centrifugal fibers to the avaian retina, Br. Med. Bull. **26:**112, 1970.

14a. Cowan, W. M., and Powell, T. P. S.: Centrifugal fibres in the avian visual system, Proc. R. Soc. Lond. (Biol) **158:**232, 1963.

15. DeValois, R. L., et al.: Response of single cells in monkey lateral geniculate nucleus to monochromatic light, Science **127:**238, 1958.
16. Dowling, J. E.: Organization of vertebrate retinas, Invest. Ophthalmol. **9:**655, 1970.
17. Dowling, J. E., and Ehinger, B.: Synaptic organization of the amine-containing interplexiform cells of the goldfish and cebus monkey retinas, Science **188:**270, 1975.
18. Dowling, J. E., and Sidman, R. L.: Inherited retinal dystrophy in the rat, J. Cell Biol. **14:**73, 1962.
19. Ebbesson, S. O. E.: On the organization of central visual pathways in vertebrates, Brain Behav. Evol. **3:** 178, 1970.
20. Ehinger, B.: Cellular location of the uptake of some amino acids into the rabbit retina, Brain Res. **46:**293, 1972.
21. Enroth-Cugell, C., and Robson, J. G.: The contrast sensitivity of retinal ganglion cells of the cat, J. Physiol. (Lond) **187:**517, 1966.
22. Fain, G. L., Quandt, F. N., and Gershenfeld, H. M.: Calcium-dependent regenerative responses in rods, Nature **269:**707, 1977.
23. Fain, G. L., et al.: Contribution of a caesium-sensitive conductance increase to the rod photoresponse, Nature **272:**467, 1978.
24. François, J., and Neetens, A.: Vascularization of the optic pathway; lamina cribrosa and optic nerve, Br. J. Ophthalmol. **38:**472, 1954.
25. François, J., and Neetens, A.: Central retinal artery and central optic nerve artery, Br. J. Ophthalmol. **47:**21, 1963.

26. Goldmann, H.: The glaucoma problem. In Newell, F., editor: Glaucoma. Transactions of the Second Conference, New York, 1957, Josiah Macy, Jr., Foundation.
27. Gouras, P.: Identification of cone mechanisms in monkey ganglion cells, J. Physiol. (Lond) **199:**533, 1968.
28. Gouras, P., and Link, K.: Rod and cone interaction in dark-adapted monkey ganglion cells, J. Physiol. (Lond) **184:**499, 1966.
29. Green, D. G., and Siegal, I. M.: Double branched flicker fusion curves from the all-rod skate retina, Science **188:**1120, 1975.
30. Guillery, R. W., Amorn, C. S., and Eighmy, B. B.: Mutants with abnormal visual pathways: an explanation of anomalous geniculate laminae, Science **174:** 831, 1971.
31. Hayreh, S.: The ophthalmic artery. III, Branches, Br. J. Ophthalmol. **46:**212, 1962.
32. Hayreh, S.: The central artery of the retina: its role in the blood supply of the optic nerve, Br. J. Ophthalmol. **47:**651, 1963.
33. Hecht, S., Shlaer, S., and Pirenne, M. H.: Energy, quanta, and vision, J. Gen. Physiol. **25:**819, 1942.
34. Hogan, M. J.: Role of the retinal pigment epithelium in macular disease, Trans. Am. Acad. Ophthalmol. Otolaryngol. **76:**64, 1972.
35. Hogan, M. J. Alvarado, J. A., and Weddell, J. E.: Histology of the human eye, Philadelphia, 1971, W. B. Saunders Co., p. 687.
36. Hubel, D. H., and Wiesel, T. N.: Receptive fields of optic nerve fibers in the retina of the spider monkey, J. Physiol. (Lond) **154:**572, 1960.
37. Kaneko, A., and Hashimoto, H.: Recording site of single cone response determined by an electrode marking technique, Vision Res. **7:**847, 1967.
38. Kramer, S. G.: Dopamine: a retinal neurotransmitter. I, Retinal uptake, storage, and light-stimulated release of H3-dopamine in vivo, Invest. Ophthalmol. **10:** 438, 1971.
39. Kuffler, S. W.: Discharge patterns and functional organization of mammalian retina, J. Neurophysiol. **16:**37, 1953.
40. Kuffler, S. W., Nicholls, J. G., and Orkand, R. K.: Physiological properties of glial cells in the central nervous sytem of amphibia, J. Neurophysiol. **29:**768, 1966.
41. Kuwabara, T., and Gorn, R. A.: Retinal damage by visible light, Arch. Ophthalmol. **79:**69, 1968.
42. Lam, D. M. K.: Biosynthesis of acetylcholine in turtle photoreceptors, Proc. Natl. Acad. Sci. U.S.A. **69:** 1987, 1972.
43. Lasansky, A.: Organization of the outer synaptic layer in the retina of larval tiger salamander, Philos. Trans. R. Soc. Lond. Biol. **265:**471, 1976.
44. Laties, A., and Enoch, J.: An analysis of retinal receptor orientation. I, Angular relationship of neighboring photoreceptors, Invest. Ophthalmol. **10:**69, 1971.
45. Laties, A., Liebman, P., and Campbell, C.: Photoreceptor orientation in the primate eye, Nature **218:**172, 1968,
46. LaVail, M. M.: Rod outer segment disk shedding in rat retina, Science **194:**1071, 1976.
47. Lettvin, J. Y., et al.: What the frog's eye tells the frog's brain, Proc. Inst. Radio Eng. **47:**1940, 1959.
48. Mann, I.: The development of the human eye, New York, 1950, Grune & Stratton, Inc., p. 312.
49. Michael, C. R.: Integration of retinal and cortical information in the superior colliculus of the ground squirrel, Brain Behav. Evol. **3:**205, 1970.
50. Miller, R. F., and Dacheux, R. F.: Synaptic organization and ionic basis of on and off channels in Mudpuppy retina. I, Intracellular analysis of chloride-sensitive electrogenic properties of receptors, horizontal cells, bipolar cells, and amacrine cells, J. Gen. Physiol **67:**639, 1976.
51. Miller, R. F., and Dacheux, R. F.: Synaptic organization and ionic basis of on and off channels in Mudpuppy retina. II, Chloride-dependent ganglion cell mechanisms, J. Gen. Physiol. **67:**661, 1976.
52. Miller, R. F., and Dowling, J. E.: Intracellular responses of the Müller (glial) cells of the mudpuppy retina: their relation to b-wave of the electroretinogram, J. Neurophysiol. **33:**323, 1970.
53. Minckler, D. S., and Bunt, A. H.: Axoplasmic transport in ocular hypotony and papilledema in the monkey, Arch. Ophthalmol. **95:**1430, 1977.
54. Minckler, D. S., Tso, M. O. M., and Zimmerman, L. E.: A light microscopic, autoradiographic study of axoplasmic transport in the optic nerve head during ocular hypotony, increased ocular pressure, and papilledema, Am. J. Ophthalmol. **82:**741, 1976.
55. Moyer, F. H.: Development, structure, and function of the retinal pigmented epithelium. In Straatsma, B. R., et al., editors: The retina: morphology, function and clinical characteristics, U.C.L.A. Forum Med. Sci. no. 8, 1969.
56. Nelson, R., et al.: Neural responses in the rod and cone system of the cat retina: intracellular records and Procion stains, Invest. Ophthalmol. **15:**946, 1976.
57. Noell, W. K., et al.: Retinal damage by light in rats, Invest. Ophthalmol. **5:**450, 1966.
58. Norman, R. A., and Werblin, F. S.: Control of retinal sensitivity. I, Adaptation in rods and cones, J. Gen. Physiol. **67:**37, 1974.
59. Oakley, B., II, and Green, D. G.: Correlation of light-induced changes in retinal extracellular potassium concentration with c-wave of the electroretinogram, J. Neurophysiol. **39:**1117, 1976.
60. O'Day, W. T., and Young, R. W.: Rhythmic daily shedding of outer segment membranes by visual cells in the goldfish, J. Cell Biol. **76:**593, 1978.
61. Østerberg, G.: Topography of the layer of rods and cones in the human retina, Acta Ophthalmol. **6**(suppl.):8, 1935.
62. Penn, R. D., and Hagins, W. A.: Signal transmission along retinal rods and the origin of the electroretinographic a-wave, Nature (Lond) **223:**201, 1969.
63. Potts, A. M., et al.: Morphology of the primate optic nerve. III, Fiber characteristics of the foveal outflow, Invest. Ophthalmol. **11:**1004, 1972.

64. Ratliff, F.: Mach bands: quantitative studies on neural networks in the retina, San Francisco, 1965, Holden-Day, Inc.
65. Raviola, E., and Gilula, N. B.: Intramembrane organization of specialized contacts in the outer plexiform layer of the retina: a freeze-fracture study in monkey and rabbits, J. Cell Biol. **65:**192, 1975.
66. Ripps, H., Shakib, M., and MacDonald, E. D.: Peroxidase uptake by photoreceptor terminals of the skate retina, J. Cell Biol. **70:**86, 1978.
67. Schacher, S. M., Holtzman, E., and Hood, D. C.: Uptake of horseradish peroxidase by frog photoreceptor synapses in the dark and the light, Nature (Lond) **249:** 261, 1974.
68. Schiller, P. H., and Malpeli, J. G.: Properties and tectal projections of monkey retinal ganglion cells, J. Neurophysiol. **40:**428, 1977.
69. Schiller, P. H., and Malpeli, J. G.: Functional specificity of lateral geniculate nucleus laminae in the rhesus monkey, J. Neurophysiol. **41:**788, 1978.
70. Spitznas, M., and Hogan, M. J.: Outer segments of photoreceptors and the retinal pigment epithelium, Arch. Ophthalmol. **84:**810, 1970.
71. Steinberg, R. H.: Rod and cone contributions to S-potentials from the cat retina, Vision Res. **9:**1319, 1969.
72. Steinberg, R. H., and Schmidt, R.: Identification of horizontal cells as S-potential generators in the cat retina by intracellular dye injection, Vision Res. **10:** 817, 1970.
73. Steinberg, R. H., Wood, I., and Hogan, M. J.: Pigment epithelial ensheathment and phagocytosis of extrafoveal cones in human retina, Philos. Trans. R. Soc. Lond. (Biol) **277:**459, 1978.
74. Stell, W. K.: The morphological organization of the vertebrate retina. In Fuortes, M. G. F., editor: Handbook of sensory physiology, Berlin, 1972, Springer-Verlag, vol. 7/2.
75. Stone, J., and Fukuda, Y.: The naso-temporal division of the cat's retina re-examined in terms of Y, X and W cells, J. Comp. Neurol. **155:**377, 1974.
76. Stone, J., and Hoffman, K.: Very slow-conducting ganglion cells in the cat's retina: a major new functional type? Brain Res. **43:**610, 1972.
76a. Toyoda, J. I., Hashimoto, H., and Ohtsu, K.: Bipolar-amacrine transmission in the carp retina, Vision Res. **13:**295, 1973.
77. Tomita, T.: Electrophysiological study of the mechanisms subserving color coding in the fish retina, Symp. Quant. Biol. **30:**559, 1965.
78. Tso, M. O. M., and Hayreh, S. S.: Optic disc edema in raised intracranial pressure. IV, Axoplasmic transport in experimental papilledema, Arch. Ophthalmol. **95:**1458, 1977.
79. Wagner, H. G., MacNichol, E. F., Jr., and Wolbarscht, M. L.: The response properties of single ganglion cells in the goldfish retina (part 2), J. Gen. Physiol. **43:**45, 1960.
80. Werblin, F. S.: Response of retinal cells to moving spots: intracellular recording in *Necturus maculosis,* J. Neurophysiol. **32:**342, 1970.
81. Werblin, F. S.: Regenerative hyperpolarization in rods, J. Physiol. **244:**53, 1975.
82. Werblin, F. S., and Dowling, J. E.: Organization of the retina of the mudpuppy, *Necturus maculosis.* II, Intracellular recording, J. Neurophysiol. **32:**315, 1969.
83. Wiesel, T. N., and Hubel, D. H.: Spatial and chromatic interactions in the lateral geniculate body of the rhesus monkey, J. Neurophysiol. **29:**1115, 1966.
84. Young, R. W.: The renewal of the photoreceptor cell outer segments, J. Cell Biol. **33:**61, 1967.
85. Young, R. W.: A difference between rods and cones in the renewal of outer segment protein, Invest. Ophthalmol. **8:**222, 1969.
86. Young, R. W.: The renewal of rod and cone outer segments in the rhesus monkey, J. Cell Biol. **49:**303, 1971.
87. Young, R. W.: Shedding of discs from rod outer segments in the rhesus monkey, J. Ultrastruc. Res. **34:** 190, 1971.
88. Young, R. W.: The daily rhythm of shedding and degradation of cone outer segment membranes in the lizard retina, J. Ultrastruct. Res. **61:**172, 1977.
89. Young, R. W.: The daily rhythm of shedding and degradation of rod and cone outer segment membranes in the chick retina, Invest. Ophthalmol. Vis. Sci. **17:**105, 1978.
90. Young, R. W., and Bok, D.: Participation of the retinal pigment epithelium in the rod outer segment process, J. Cell Biol. **42:**392, 1969.

Chapter 15

Metabolism and photochemistry in the retina

SIDNEY FUTTERMAN, Ph.D.
JOHN C. SAARI, Ph.D.

Visual excitation is the unique and important function of the retina. Associated with this activity are substantial demands for energy, which are superimposed on the ATP requirements of other, more common, physiologic activities. These include the biosynthesis of cellular constituents that must be replaced as a result of normal "wear and tear" and the active transport of cations to keep the intracellular concentration of potassium high and that of sodium low in the face of the natural leakiness of cell membranes, axonal transport, and rod photoreceptor renewal. As is the case in the brain,[57] the ATP produced in the retina is derived almost exclusively from the metabolism of glucose.

CARBOHYDRATE METABOLISM

It is generally thought that the glucose and oxygen metabolized by the inner layers of the retina are largely supplied by way of the retinal capillaries, but that the photoreceptor cells (p. 376) receive glucose and oxygen from the choroidal circulation. As in the brain, a limited quantity of glycogen is present in retinal tissue, although localized primarily in Müller's cells.[36] It is somewhat less abundant in the highly vascularized rat, mouse, and human retinas than in the poorly vascularized retinas of rabbit and guinea pig. The Müller's cells also appear to contain glucose-6-phosphatase activity,[38] and it seems likely that these glial cells may have the capability of converting their glycogen to glucose and releasing the glucose for uptake and utilization by neighboring neuronal cells.

Early studies revealed an unusually active metabolism for retinal tissue. Rat retinas consumed oxygen (QO_2 = ca 30 μl/min/mg) at a rate substantially higher than that of slices of any other tissues.[62] Nearly two decades elapsed before it was well established that the combustion of pyruvate to carbon dioxide and water in mitochondria consumed most of the oxygen utilized by living cells and produced most of the ATP. A tight coupling of the oxidation steps in the mitochondria to ATP production makes the tricarboxylic acid cycle dependent on the supply of ADP and inorganic phosphate. Physiologic processes that consume the energy of ATP liberate the ADP and inorganic phosphate required by the mitochondria and thereby govern the rate of mitochondrial oxygen consumption. It has been suspected for some time that ATP supplied by the strikingly numerous ellipsoid mitochondria must serve some special role in photoreception.

Despite the high rate of respiration in retinal tissue, the oxidation of pyruvate in the mitochondria does not keep pace with pyruvate production from glucose.[62] The excess pyruvate is converted to lactic acid. Many other tissues, including muscle, heart, kidney, and liver show lactate accumulation only under conditions sufficiently anaerobic to reduce mitochondrial

□ This work was supported in part by Research Grants EY 00343, EY 02317 from the National Eye Institute, National Institutes of Health.

metabolism and release them from control by the Pasteur effect (inhibition of glycolysis by the presence of abundant oxygen). Retina, in common with embryonic tissues, some tumor tissues, leukocytes, and brain, shows a net accumulation of lactate even when it is adequately supplied with oxygen.

Brain cortex has about half the rate of respiration and produces lactate at about only 15% aerobically and 35% anaerobically of that of the adult retina. It has been calculated that about half of the in vitro respiration and lactate production in adult rabbit retina is attributable to the photoreceptor cells.[10] Laborious quantitative microanalyses of the distribution of the activities of predominantly mitochondrial enzymes or primarily cytoplasmic enzymes of glycolysis at various depths in the retina of

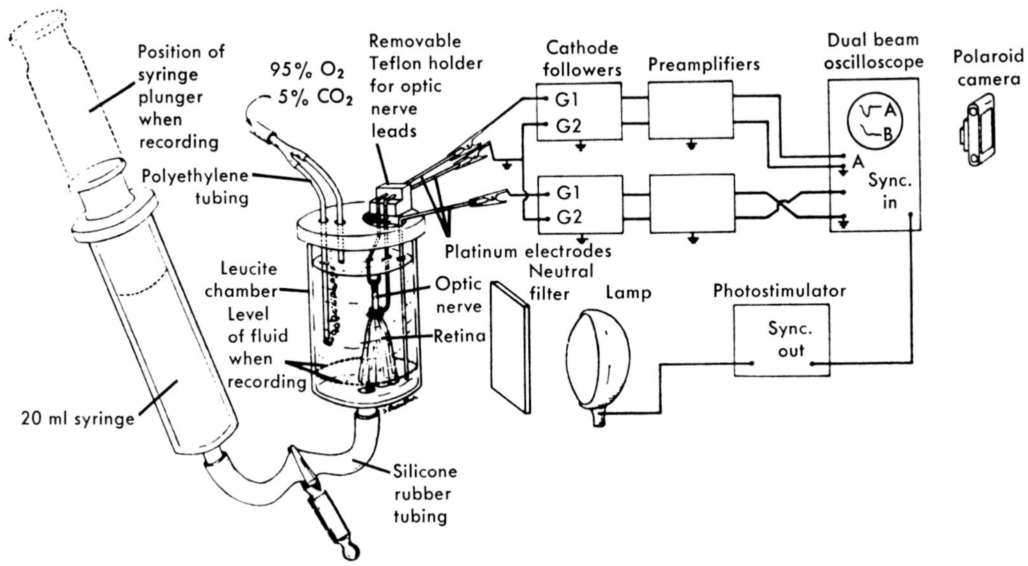

Fig. 15-1. Incubating and recording apparatus. Diagram shows retina suspended by optic nerve in medium that can be withdrawn from chamber by means of syringe when electrical recording is to be made. Gas mixture, saturated with water vapor, is admitted into chamber both above and below surface of medium. (With permission of Ames, A., III: Studies with isolated retina. In Graymore, C. N., editor: Biochemistry of the retina, London, 1965, copyright by Academic Press, Inc. [London] Ltd.)

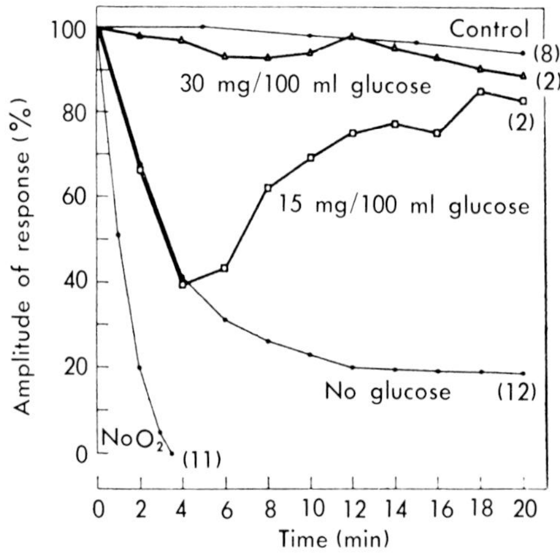

Fig. 15-2. Amplitude of light-evoked compound action potentials recorded from optic nerve and plotted relative to control response recorded at zero time. Various curves show change in amplitude during incubation in control medium and in media containing normal glucose but no O_2; normal O_2 but no glucose; normal O_2 and two intermediate levels of glucose as shown. (With permission of Ames, A., III: Studies with isolated retina. In Graymore, C. N., editor: Biochemistry of the retina, London, 1965, copyright by Academic Press, Inc. [London] Ltd.)

rabbit and monkey have been made.[42] These, together with early electron-microscopic examinations, revealed the dense accumulation of mitochondria in the ellipsoid region of photoreceptor cells. Since that time, it has been apparent that the mitochondria-rich ellipsoid region of photoreceptor cells must account for a major share of the oxygen consumption and energy production of the retina.

When conditions were achieved in which the excitability of retinal tissue could be maintained in vitro, it became possible to determine directly some of the metabolic requirements for visual excitation.[1,55] A preparation of rabbit retinal tissue and a portion of attached optic nerve with appropriately placed electrodes was suspended in a suitable medium that could be partially withdrawn when electrical measurements were to be taken (Fig. 15-1). The amplitude of what would appear to be the counterpart of the a wave of the electroretinogram (ERG) in vivo was monitored.[1] In the absence of glucose in the external medium, endogenous energy stores were insufficient to sustain retinal function, and within approximately 10 minutes the amplitude of the light-evoked electrical response fell to a fraction of the initial amplitude (Fig. 15-2). The loss in electrical response was even more rapid when the preparation was made anaerobic by substituting nitrogen as the gas phase. After deprivation, virtually complete recovery of the amplitude of the electrical response was generally observed, providing the period of deprivation had been of relatively brief duration. These studies indicate that the retina does not possess a large store of glycogen and that it cannot function anaerobically even when abundantly supplied with glucose. However, the ability of the retina to function in an external medium with a concentration of glucose only one third of that normally found circulating in plasma provides support for the concept that the retina, again like brain and unlike muscle or adipose tissue, does not require insulin for glucose uptake.

Under exposure to centrifugal forces that produce blackout or under conditions of oxygen deficiency, loss of peripheral visual field occurs with preservation of central vision. It would appear that circulation in the peripheral capillaries can be compromised under circumstances in which the central retina remains adequately supplied with oxygen.

Although many early investigators have looked for changes in the metabolism of the retina caused by the influence of light, their findings have been contradictory. Recent biophysical studies may provide a more rational basis on which to explore light-induced changes in retinal metabolism. In any case, they appear to account for the relatively high basal rate of metabolism of retinal tissue.

Vertebrate visual cells have a steady current flow along the length of the photoreceptor organelle.[23] Although external sodium ions leak into the dark-adapted outer segment steadily, there is no local extrusion of sodium ions because the Na^+,K^+-ATPase that serves as a pump mechanism in biologic membranes is lacking in the external membrane of the outer segment. The pump mechanism does operate in the inner segment, and sodium ions entering the outer segment flow into the inner segment where they are extruded from the cell. The dark current of the visual cell thus arises from this leak-flow-pump system. It is likely that a major function of the ellipsoid mitochondria is to provide ATP for the cation pumping in the visual cell associated with the maintenance of the dark current. In the overall process, one molecule of oxygen consumed in a mitochondrion would lead to the production of about six high-energy phosphate bonds in the form of ATP, which, in turn, would support the extrusion of about 18 sodium ions from the visual cell. It has been estimated that the dark-current flow rate of approximately 10 sodium ions/sec/μm^2 is sufficient to involve the complete turnover of all sodium ions in a rod cell in 45 seconds and to maintain the visual cell membrane potential reduced by approximately 20 mV below that seen when sodium channels are blocked.

NEUROTRANSMITTERS IN THE RETINA

The retina appears to contain pathways for the synthesis and degradation of the various neurotransmitters found in the brain. The presence of acetylcholine esterase in the plexiform layers was demonstrated histochemically,[32] and electrophysiologic evidence indicates that acetylcholine may be the neurotransmitter at the inner molecular layer bipolar cell synapse. The uptake of 5-hydroxytryptophan by retinal tissue and its conversion to serotonin has been demonstrated,[50] but as yet no functional role can be assigned to serotonin in the retina. Norepinephrine is virtually undetectable in retinal tissue. About 10% of the amacrine cells appear to contain dopamine, which is now thought to be an inhibitory retinal neurotransmitter. Exposure of the dark-adapted retina to light has been found to increase the rate of dopamine release from the tissue.[33] γ-Aminobutyric acid, another inhibitory neurotransmitter, is relatively abundant in the retina, but its localization and role have not been established. γ-Aminobutyric acid has been observed to depress the responses of retinal ganglion cells to repetitive light flashes, while its precursor, glutamic acid, appears to elicit the opposite response.[47] Light stimulation increases the concentration of γ-aminobutyric acid in the retina.[37] Glycine, suspected as an inhibitory transmitter in some regions of the central nervous system, is abundant in the inner layers of the retina and is concentrated in certain amacrine cells.[15]

Photoreceptors are thought to release a neurotransmitter continuously in the dark. When light hyperpolarizes the photoreceptor cell, the release of transmitter is decreased. Horizontal cells receiving input from cones in the carp retina and known to undergo hyperpolarization in the light were effectively depolarized by the amino acid, L-aspartate.[64] In high concentration L-aspartate can block light-induced electrical responses of the vertebrate retina from all cells other than photoreceptors.[56] This observation has been made in a variety of species and suggests that L-aspartate may be the transmitter for rods as well as cones.

TOXIC CHEMICALS

Studies first reported in the early 1950s reveal that intravenous injection of sodium azide produces a rapid transitory change in the direct current (DC) potential across the retina, amounting to about 20 mV. This change occurs both in dark- and light-adapted animals as well as in animals in which the photoreceptor cells have previously been destroyed by the administration of sodium iodoacetate.[47] In animals that have previously received sodium iodate, the simultaneous destruction of the cells of the pigment epithelium and abolition of the c wave of the ERG establish that the c wave has its origin in the pigment epithelium layer and that the resting transretinal DC voltage depends on an intact pigment epithelial layer.

NUTRITION AND THE RETINA

The eye requires a continuous supply of vitamin A (retinol). It is as yet the only organ in which this fat-soluble vitamin is known to play an important and reasonably well-defined role. Although it has been estimated that less than 0.01% of the total quantity of vitamin A in the human body occurs in the eye,[45] there it is localized in the retina where it is essential for vision. For this reason the metabolism of vitamin A in the retina has been studied intensively.

The effects of vitamin A deficiency include degeneration of the retina, corneal changes, and weight loss or cessation of growth. The discovery that rats deprived of retinol could be kept alive and healthy when deficient diets were supplemented with retinoic acid[58] provided a way in which to study the selective requirement of the retina for retinol. Under the usual laboratory conditions, weanling rats receiving retinoic acid in place of retinol suffered a severe and selective degeneration of the visual cells as reflected by progressive elevation of the threshold of the ERG, but in every other respect appeared to grow normally.[14] A further discovery[48] that retinol-deficient rats maintained in complete darkness incur no loss of visual cells or elevation of the ERG threshold is surprising. One reasonable explanation is that the salvage pathway for reutilization of the retinal (previously known as retinene or vitamin A aldehyde) present in the distal ends of photoreceptors that are phagocytized by cells of the underlying pigment epithelium must be remarkably efficient. In any case, it is now apparent that there must be some exposure of the retina to light for the progressive degeneration of the photoreceptor cells associated with vitamin A deficiency to occur.

Taurine is present in high concentration in

the retina,[49] and light stimulates its release from rod outer segments.[54] The role of taurine is not clear, but it may be required for intracellular osmotic regulation in photoreceptor cells. The essentiality of taurine is indicated by the degeneration of photoreceptor cells in cats fed a diet deficient in taurine.[27] In other species the biosynthesis of taurine is sufficiently active that no dietary requirement is apparent.

Reducing agents, particularly α-tocopherol (vitamin E), appear to be required to protect the pigment epithelium from the accelerated accumulation of lipofuscin, the so-called "aging pigment."[31] In monkeys a deficiency of α-tocopherol produces macular degeneration with disruption of cone photoreceptor outer segments.[26] It is thought that α-tocopherol may serve to inhibit peroxidation of polyunsaturated fatty acids and vitamin A in photoreceptors.

VISUAL PIGMENTS

Visual pigments are distinguished by their spectral properties. The light-sensitive pigments of the human retina[61] include the well-known rod pigment employed for dim light vision, rhodopsin (Fig. 15-3), and three more recently discovered cone pigments (Fig. 15-4) responsible for color discrimination and normal daylight vision. Retinal rod pigments are more abundant and stable than cone pigments. As yet, virtually nothing is known about the chemistry of the cone pigments, whereas a great deal is known about rhodopsin. Most visual pigments, including rhodopsin, which is widely distributed among vertebrates and invertebrates, consist of an apoprotein, opsin, to which is attached the chromophore molecule 11-*cis*-retinal (Fig. 15-5). In rhodopsin and probably all other visual pigments the chromophore is attached in the conjugate acid form[16] (protonated Schiff base or immonium salt) of a Schiff base linkage to a lysine residue of the apoprotein (Fig. 15-6).[5]

Comparative studies, which may ultimately reveal some of the influences that shaped the evolution of human visual pigments, indicate that some fish and amphibians contain as rod photoreceptor pigment porphyropsin,[60] in

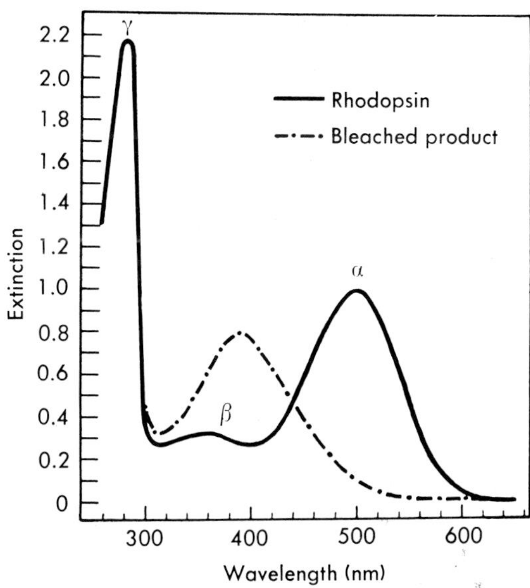

Fig. 15-3. Absorption spectrum of rhodopsin from bullfrog retinas in solution in 2% aqueous digitonin at pH 5.6 showing three absorption bands. Bleaching replaces α and β bands with absorption spectrum of retinal, but alters protein γ band only very slightly. (From Wald, G.: J. Opt. Soc. Am. **41**:949, 1951.)

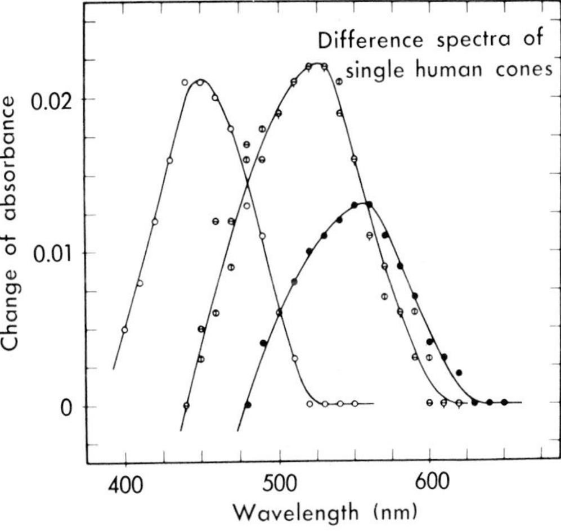

Fig. 15-4. Difference spectra of visual pigments in single cones of human parafovea. In each case, absorption spectrum was recorded in dark from 650 to 380 nm, then again after bleaching with flash of yellow light. Differences between these spectra are shown. They apparently represent one blue-sensitive cone with λmax about 450 nm, two green-sensitive cones with λmax about 525 nm, and red-sensitive cone with λmax about 555 nm. (From Wald, G., and Brown, P. K.: Cold Spring Harbor Symp. Quant. Biol. **30**:345, 1965.)

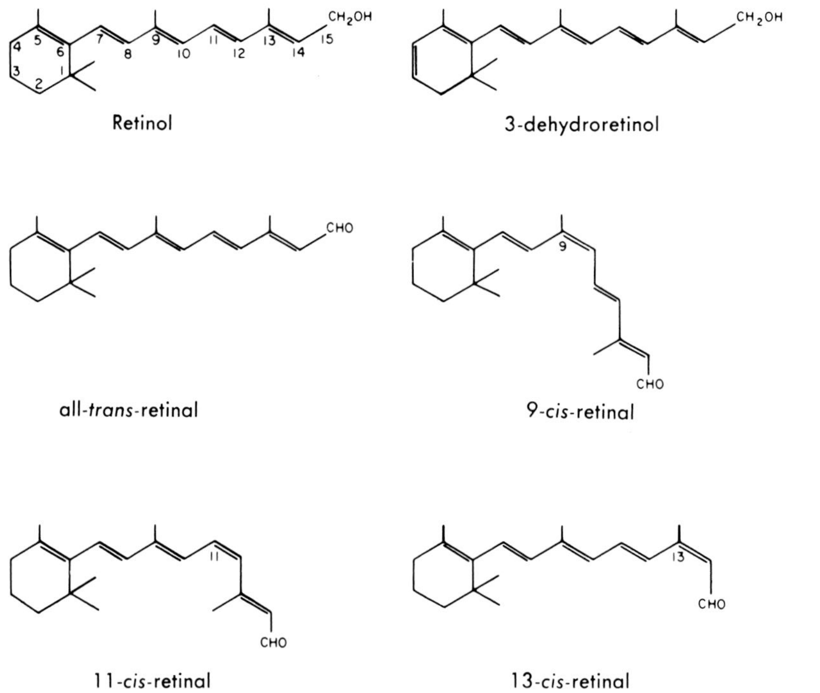

Retinol

3-dehydroretinol

all-*trans*-retinal

9-*cis*-retinal

11-*cis*-retinal

13-*cis*-retinal

Fig. 15-5. Structures of all-*trans*-retinol, all-*trans*-3-dehydroretinol, and all-*trans* and mono *cis* geometric isomers of retinal.

$$\underset{\text{11-}cis\text{-retinal}}{C_{19}H_{27}CHO + H_2N\text{-opsin}} \longrightarrow \underset{\substack{\text{Schiff base}\\ \text{(N-retinylideneopsin)}}}{C_{19}H_{27}CH{=}N\text{-opsin} + H_2O} \longrightarrow \underset{\text{Conjugate acid}}{C_{19}H_{27}CH{=}\overset{+}{\underset{H}{N}}\text{-opsin}}$$

Fig. 15-6. Linkage of chromophore to protein in rhodopsin.

which the chromophore is 11-*cis*-3-dehydroretinal. The apoprotein of porphyropsin is seemingly identical to that of rhodopsin, but the spectra of the porphyropsins are shifted toward longer wavelengths by about 20 nm. The retinas of these species are apparently able to convert retinal to 3-dehydroretinal, and this capability in some species may be regulated largely by the properties of the light available in the natural environment and in other species may be under hormonal control.[6] In some freshwater fish the rods contain variable amounts of both rhodopsin and porphyropsin formed from a single apoprotein combining indiscriminately with either chromophore. It was recognized that the most widely studied cone pigment, chicken iodopsin, could form a new pigment with hitherto undescribed properties on replacement of its 11-*cis*-retinal chromophore with 11-*cis*-3-dehydroretinal. The "synthetic" pigment was made and named cyanopsin.[60] More recently cyanopsin-type cone pigments have been identified in freshwater tortoise and tadpole retinas.[40]

It has been found that the curve reflecting the spectral sensitivity of the dark-adapted human matches almost perfectly the absorption spectrum of rhodopsin (Fig. 15-7) with peak sensitivity occurring at 500 nm, providing the appropriate corrections are made for partial absorption of shorter wavelengths by the slightly yellow human lens.

The photopic luminosity curve of normal trichromats (non–color-blind persons) matches reasonably well the composite absorption spec-

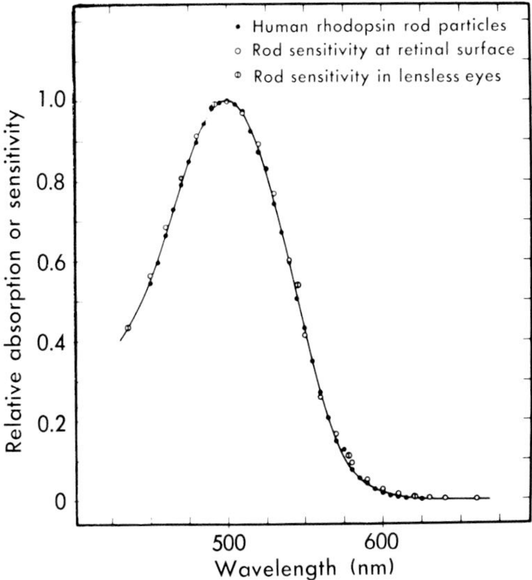

Fig. 15-7. Absorption spectrum of human rhodopsin, measured in suspension of rod outer segments, compared with spectral sensitivity of human rod vision, measured as at retinal surface. The latter data involve either average scotopic luminosity corrected for ocular transmission, or uncorrected measurements of spectral sensitivity of rod vision in aphakic (lensless) eye. (From Wald, G., and Brown, P. K.: Cold Spring Harbor Symp. Quant. Biol. **30:** 345, 1965.)

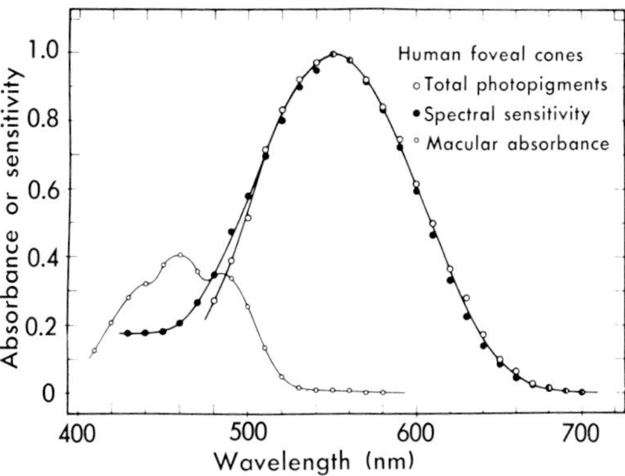

Fig. 15-8. Difference spectrum of total photopigments of human fovea (average of five), compared with spectral sensitivity of foveal vision, measured as at level of cones. To obtain the latter function, average photopic luminosity curve was converted to quantum basis and corrected for ocular and macular transmission. Spectrum of a human macula is shown at left. Corrected luminosity curve agrees with difference spectrum of foveal photopigments down to about 510 nm. Below this wavelength difference spectrum falls off, owing to formation of colored products of bleaching. (From Wald, G., and Brown, P. K.: Cold Spring Harbor Symp. Quant. Biol. **30:**345, 1965.)

trum representing the three cone pigments (Fig. 15-8).

The gene loci, one specifying the red sensitive cone pigment and another the green sensitive cone pigment, are on the X chromosome. The majority of color-blind individuals lack one of the three color mechanisms and are called dichromats. Approximately 1% of males are protanopes lacking the red-sensitive pigment and 2% are deuteranopes lacking the green-sensitive pigment.[61] Tritanopes, lacking the blue-sensitive receptor, are apparently very rare.

A detailed picture of the structure and orientation is beginning to emerge for cattle rhodopsin, the most thoroughly studied visual pigment (Fig. 15-9). All work with this material is customarily done under red light to avoid bleaching of the pigment. The protein is asymmetrically oriented within and appears to span the phospholipid bilayer of the disc membrane.[18] Glycosylation sites (at least two[25]) are thought to be localized to the intradiscal side of the membrane.[51] Phosphorylation of rhodopsin by the light-activated protein kinase specific for the visual pigment has been shown to occur in vivo and in vitro.[35] Although the time course of the reaction appears to be too slow to allow its participation in visual transduction, modification of rhodopsin may feature in the control of the sensitivity of the photoreceptor cell.[44] A considerable amount of the rhodopsin molecule is within the membrane itself, consistent with x-ray[9] and neutron diffraction[53] analyses and thus interacts directly with the phospholipids of the membrane. Although the spectral properties of rhodopsin are not altered by removal of essentially all its phospholipid,[4] particulate preparations of delipidated rhodopsin appear to undergo irreversible denaturation on bleaching. However, if phospholipid is restored to such preparations before bleaching, they are capable of regenerating photosensitive pigment in the dark in the presence of an appropriate chromophore molecule.

Particulate rhodopsin from photoreceptor

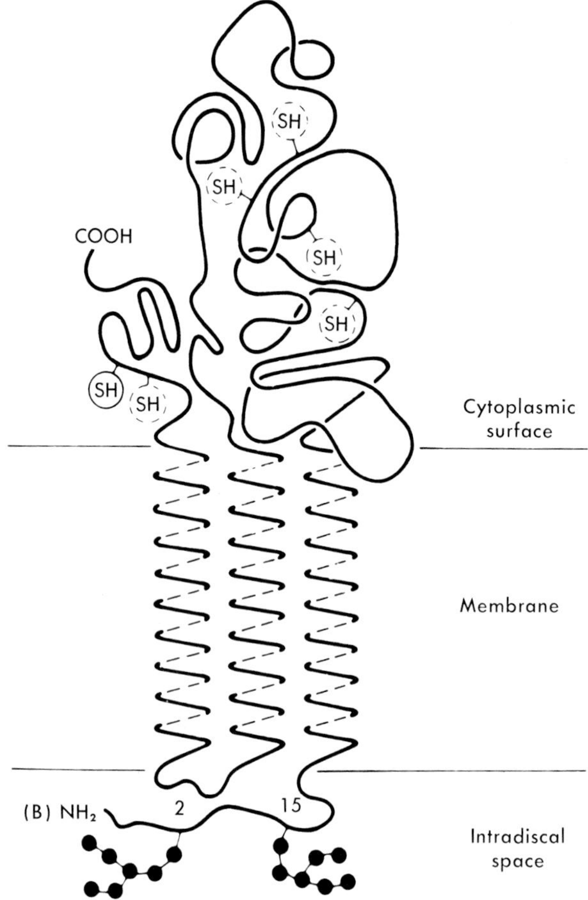

Fig. 15-9. Model for the organization of rhodopsin in disc membranes. Only enough of the polypeptide chain is shown to illustrate the major topological features. Sugar residues (filled circles) and the blocked amino terminus [(B)NH_2—] are thought to be localized to the intradiscal surface of the membrane, while protein sulfhydryl groups (—SH), the carboxyl terminus (—COOH), and a considerable amount of the polypeptide chain are localized to the cytoplasmic surface. The placement of 11-*cis*-retinal, not indicated in the figure, is uncertain. (From Fung, B. K.-K., and Hubbell, W. L.: Biochemistry **17:**4413, 1978; reproduced with permission, copyright 1978 by the American Chemical Society.)

membrane preparations can be brought into aqueous solution with the aid of a wide variety of detergents. Such solutions have made possible detailed spectral studies of the bleaching of rhodopsin.

On exposure of the eye to light of wavelengths between 400 and 800 nm, the visual pigments in the retina absorb light, and visual sensations occur. The lower wavelength limit for visual excitation is imposed by the transmission of the ocular media, especially the lens. In solution ultraviolet light in the wavelength region of 280 nm, which is absorbed by the aromatic amino acids, effectively bleaches rhodopsin[34] and presumably all other visual pigments. The first step in the process is a photochemical reaction triggered by the energy present in a photon of light absorbed by a visual pigment molecule located in the projecting outer segment of a retinal rod or cone photoreceptor cell.

BLEACHING OF VISUAL PIGMENT

When the dark-adapted retina is exposed to light, the visual pigments of the rod and cone cells absorb light, bleach, and after a time

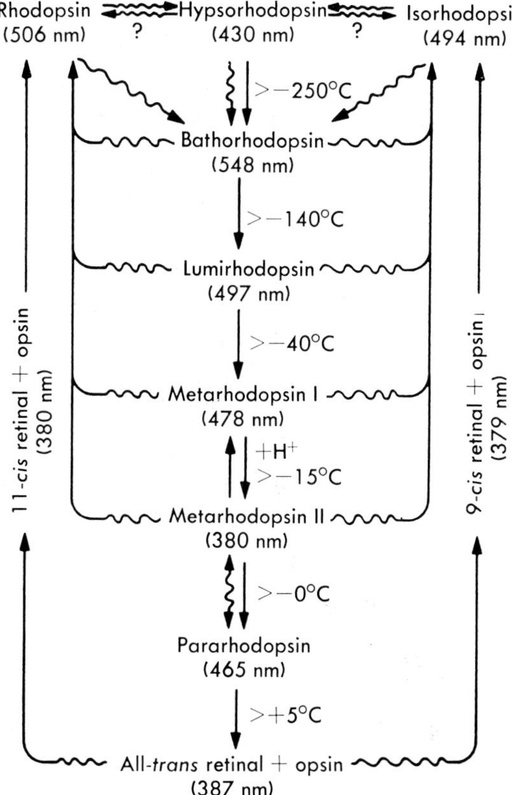

Fig. 15-10. Stages in bleaching of cattle rhodopsin. Photoreactions are symbolized by wavy lines, thermal reactions by straight ones. λ_{max} for rhodopsin, hypsorhodopsin, isorhodopsin, and bathorhodopsin at −268° C; lumirhodopsin and metarhodopsin I at −65° C; other pigments at near 0° C. (From Yoshizawa, T.: In Dartnall, H. J. A., editor: Handbook of sensory physiology, New York, 1972, Springer-Verlag New York, Inc., vol. 7/1.)

release the all-*trans* isomer of retinal from the apoprotein, opsin. In bleaching, rhodopsin passes through a series of short-lived intermediates before losing its red color. Interest in elucidating the intermediate steps in this process has been spurred by the anticipation that clues would be uncovered to the way in which visual pigment bleaching is geared to the mechanism for eliciting subsequent electrical activity.

This goal has not be realized, but much important information has been obtained. The assumption underlying this approach was that individual reactions could be arrested at appropriately low temperatures, and the spectral properties of reaction mixtures at low temperatures would reveal the fleeting intermediates that were in the past virtually undetectable at room temperature or in situ in the retina.

After a molecule of rhodopsin absorbs 1 quantum of visible light at the lowest temperature yet attainable for spectral studies, that of liquid helium, the geometric isomerization of the chromophore from the 11-*cis* to the all-*trans* isomer occurs.[66] When the irradiation is carried out with light of wavelengths greater than 520 nm, the first intermediate observed is hypsorhodopsin (Fig. 15-10). With 437 nm light the first intermediate is bathorhodopsin, which was previously named prelumirhodopsin when it was detected as the earliest intermediate produced by irradiating rhodopsin at the slightly higher temperature of liquid nitrogen. Bathorhodopsin has even been detected as a very fleeting intermediate at room temperature, forming in about 6 picoseconds after a rhodopsin solution is exposed to a flash of light.[8]

Only the first step in the bleaching sequence requires input of light. All subsequent reactions can proceed in the dark and are thought to involve secondary changes in the conformation of the chromophore and protein. Bathorhodopsin changes to lumirhodopsin when the temperature is permitted to rise above −140° C. Above −40° C lumirhodopsin is converted to metarhodopsin I, which in turn can change into metarhodopsin II above −15° C. However, an additional quantum of light absorbed can convert each of the intermediates down to the metarhodopsin II stage back to the native rhodopsin or to the artificial pigment isorhodopsin in which the chromophore is 9-*cis*-retinal.

Pararhodopsin appears to be a relatively long-lived intermediate in the intact retina. On exposure to bright light the rose color of the dark-adapted retina can be observed at pH 7.4 to change within seconds to the orange of pararhodopsin and then fade gradually in the course of several minutes to the straw color of the fully bleached retina. In the final step, hydrolysis of pararhodopsin occurs, releasing all-*trans*-retinal.

Except for the fact that no hypsointermediate has been detected in the bleaching of any pigments other than cattle rhodopsin, cor-

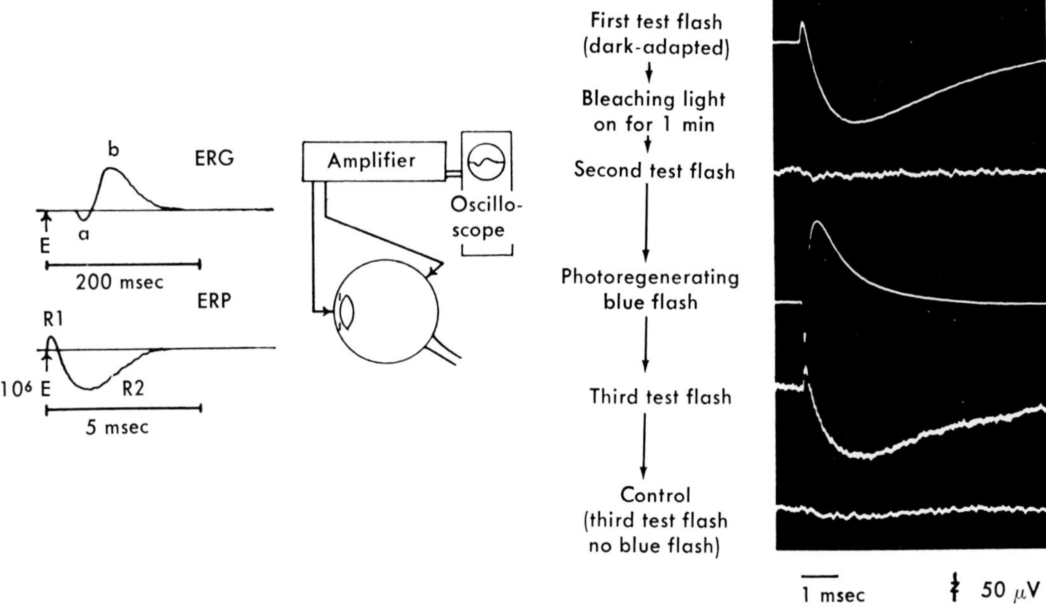

Fig. 15-11. Left, diagram to show essential hookup for observing electroretinogram (ERG) or early receptor potential (ERP) in dark-adapted vertebrate eye. Each of these responses is a biphasic fluctuation of potential, involving cornea-positive (upward) and cornea-negative (downward) components. Unlike the ERG, the ERP has no measurable latency. For both types of response to be comparable in amplitude, flash that stimulates ERP must be approximately 1 million times more intense. Right, photoregeneration of ERP in eye of albino rat. Test flash and bleaching light were composed of long wavelengths primarily absorbed by rhodopsin. Blue photoregenerating flash contained wavelengths absorbed by longer lived intermediates of bleaching (apparently mainly metarhodopsin II and pararhodopsin). Control trace was obtained from second eye subjected to same bleaching light and test flashes, but without interpolated blue flash, 27° C. First test flash yields normal ERP. Bleaching light having removed rhodopsin, second test flash yields nothing, even though amplification is increased. Blue flash, photoregenerating rhodopsin, yields ERP of reversed polarity. Third test flash, yielding again a normal ERP (high gain), shows that blue flash did regenerate rhodopsin. Control flash shows that without interpolated blue flash, no response is obtained (high gain). (From Cone, R. A.: Science **155:**1128, 1967, copyright 1967 by the American Association for the Advancement of Science.)

responding batho, lumi, and meta intermediates occur in the bleaching of chicken iodopsin and carp porphyropsin (Table 15-1).

In the intact photoreceptor organelle, absorption of light by rhodopsin is followed by an electrical response, the early receptor potential[11] (ERP), and rotation of the "light-struck" pigment molecules in the membrane in which they are situated.[7,12] The first two effects have been shown to occur with negligible latency after a light flash. The early receptor potential can be observed in preparations of retinal tissue warmed as high as 48° C and even after glutaraldehyde fixation of the tissue. After elicitation of the ERP with a brilliant flash of light (Fig. 15-11), a second blue flash acting on the bleached intermediates will produce photoregeneration of rhodopsin and at the same time will produce an accompanying receptor potential similar in form to the ERP but with reversed polarity. The ERP has been interpreted as reflecting an intramolecular dislocation or displacement of charged groups. The bleaching of rhodopsin has been shown to be accompanied by the uptake of a proton.[5,43] The importance of these effects in the overall process of visual excitation is not yet clear. In any case, there is now strong suggestive evidence that absorption of light by rhodopsin reduces the dark-current flow within about 1 millisecond.

The detailed molecular events in the overall process of visual excitation are still not under-

Table 15-1. Absorption properties of visual pigments and their intermediates*†

Visual pigment	Measured at ° C	Absorption property	Cattle rhodopsin	Squid rhodopsin	Chicken iodopsin	Carp porphyropsin
11-*cis*	Room temperature	λ_{max}(nm)	498	493	562	522
		ϵ_{max}	40,600	40,600	40,600	30,000
	−195	λ_{max}	505	498	575	528
		$\epsilon_{-195}/\epsilon_{RT}$	1.14	1.15	—	1.06
Iso (9-*cis*)	Room temperature	λ_{max}(nm)	485	473	510	506
		ϵ_{max}	43,000	47,700	—	30,100
	−195	λ_{max}(nm)	491	481	550	514
		$\epsilon_{-155}/\epsilon_{RT}$	1.13	1.12	—	1.10
Hypso (all-*trans*)	−269	λ_{max}(nm)	430	—	—	—
		ϵ_H/ϵ_R	0.91	—	—	—
Batho (all-*trans*)	−195	λ_{max}(nm)	543	543	640	592
		ϵ_B/ϵ_R	1.13	1.39	1.50	1.48
Lumi (all-*trans*)	−65	λ_{max}(nm)	497	530‡	518	542‡
		ϵ_L/ϵ_R	1.15	1.03‡	—	1.16‡
Meta I (all-*trans*)	−65	λ_{max}(nm)	481	500§	495‖	509‡
		ϵ_{MI}/ϵ_R	1.06	1.36§	—	1.07‡
Meta II (all-*trans*)	Above 0	λ_{max}(nm)	380	380¶	380‖	408
		$\epsilon_{MII}/\epsilon_R$	1.00	1.50¶	—	1.11

*From Yoshizawa, T.: The behavior of visual pigments at low temperature. In Dartnall, H. J. A., editor: Handbook of sensory physiology, New York, 1972, Springer-Verlag New York, Inc., vol. 7/1.

†NOTES: $\epsilon_{-195}/\epsilon_{RT}$ is ratio of maximum absorbance of the pigment at −195° C to that at room temperature. ϵ_H/ϵ_R, ϵ_B/ϵ_R, ϵ_L/ϵ_R, ϵ_{MI}/ϵ_R, and $\epsilon_{MII}/\epsilon_R$ are the ratios of the maximum absorbances of hypso, batho, lumi, metarhodopsin I and II, respectively, to that of rhodopsin at the temperature shown in the second column.

‡Measured at about −190° C.

§Acid metarhodopsin.

‖Measured at −38° C.

¶Alkaline metarhodopsin.

stood. Because the disc membrane in the outer segment, containing the bulk of the rhodopsin, is physically separated from the plasma membrane, where the permeability changes occur, it is clear that a chemical signal (termed an internal transmitter) is required. Such a model can also explain the enormous amplification (~10^6) observed during visual transduction if many internal transmitter molecules are released by the bleaching of a single rhodopsin molecule. A prime candidate for the internal transmitter is the calcium ion. In the hypothetical scheme proposed by Hagins[23] (Fig. 15-12) the absorption of light by rhodopsin triggers a release of calcium ions (~1000 per rhodopsin bleached) into the space external to the discs, where subsequent attachment of the calcium to sites on the plasma membrane of the photoreceptor organelle closes off sodium channels. The decreased flow of sodium into the photoreceptor organelle hyperpolarizes the visual cell, altering the membrane potential at the synapse. Although the internal transmitter hypothesis leads to several experimentally testable predictions, the hypothesis has not yet been fully confirmed. Hyperpolarization can be induced in the dark by increasing the concentration of calcium ion in the external fluid bathing the photoreceptor organelles,[23] and after the plasma membrane of the photoreceptor organelle has been made selectively permeable to calcium ion by the addition of a calcium ionophore, much lower concentrations of externally applied calcium are effective.[24] However, even though isolated rod outer segments appear to contain adequate stores of calcium, no light-induced release of sufficient amounts of the ion has been detected from isolated rods or from intact retina and a calcium pump (Ca^+-ATPase), necessary to restore the original conditions after cessation of the bleaching light, has not yet been characterized.[39]

Cyclic nucleotides remain as additional can-

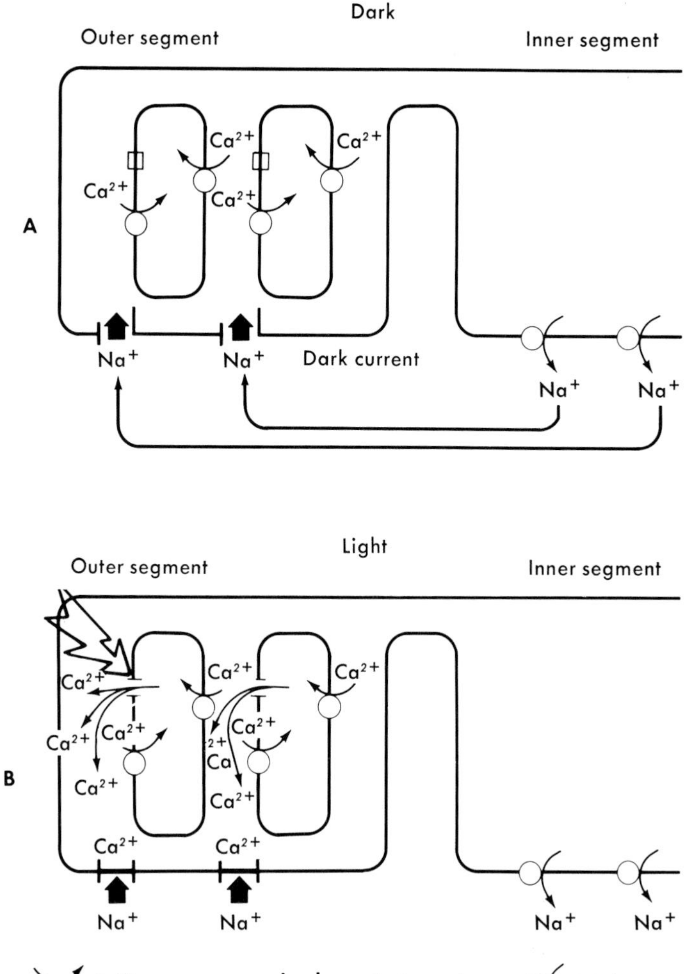

Fig. 15-12. Proposed mechanism for excitation in vertebrate rods and cones. **A,** Rod/cone in darkness. **B,** Rod/cone in light. (Modified from Hagins, W. A.: Annu. Rev. Biophys. Bioeng. **1**:131, 1972.)

didates for the internal transmitter. At present, guanosine 3′, 5′-cyclic monophosphate (cGMP) is the only substances for which a light-induced alteration of concentration has been demonstrated within the time constraint, evident from electrophysiologic experiments, of about 1 millisecond.[39, 65] Alterations in the concentration of this nucleotide are mediated by a light-activated phosphodiesterase specific for cGMP. The action spectrum for phosphodiesterase activation corresponds to the absorption spectrum of rhodopsin, although the visual pigment itself has no known enzymatic activity.[3] The high gain associated with this process as well as the rapid time course suggest a role in the visual transduction process. Iontophoretic injection of cGMP into a single rod outer segment of the toad retina produced a marked latency in the light-stimulated hyperpolarization consistent with the hypothesis that hydrolysis of cGMP initiates the phototransduction process.[46]

In the presence of continuous bleaching light, hyperpolarization is sustained; pumping action in the photoreceptors must then be diminished, but, surprisingly, the rate of oxygen consumption by the retina does not appear to be reduced.[55] Light-induced turnover of the nucleotides of the outer segment may account for this observation.

Free all-*trans*-retinal does not accumulate to any extent in the outer segments of the visual cell during light adaptation. Retinal is known to form Schiff bases readily with free amino groups of proteins and phospholipids. If this aldehyde were permitted to accumulate during the course of light adaptation, it would probably inhibit neighboring enzymes by combining with their free amino groups and thereby act as a toxic agent in the retina. Instead, retinal is rapidly reduced to retinol with TPNH (NADPH) by the enzyme retinal reductase. This enzyme, unlike soluble liver alcohol dehydrogenase, is associated with the insoluble photoreceptor membrane protein and shows a preference for long-chain aldehydes. In addition to retinal reductase, visual cell outer segments contain some enzymes representing the glycolytic and pentose cycle pathways. The latter pathway produces TPNH and under conditions of constant illumination proceeds slowly in the retina because of the slow rate at which TPNH is oxidized to TPN. However, the pentose cycle dehydrogenase reactions that produce TPNH are coupled to the retinal reductase reaction that utilizes TPNH. In this manner during light adaptation the retinaldehyde released by the bleaching of rhodopsin accelerates the rate of metabolism of glucose by way of the pentose cycle dehydrogenase reactions. TPNH generated by the pentose cycle dehydrogenases appears to be the exclusive reducing agent for retinaldehyde in the photoreceptor organelle.[22]

Much of the retinol produced during light adaptation diffuses out of the photoreceptor organelle and is esterified in microsomes (endoplasmic reticulum).[2] At the present time it is not known whether the esterification occurs in the retina or in the pigment epithelium. In any case, following prolonged adaptation to relatively intense light the bulk of the retinal released from rhodopsin accumulates as retinyl esters in the pigment epithelium.[13] The esterification step would appear to facilitate the retention of retinol in the eye and perhaps also to protect the retina from sustained exposure to high concentrations of retinol, which are well known to affect adversely the stability of lipoprotein membranes. In this esterification a spectrum of saturated and monounsaturated fatty acids are transferred to retinol from an as yet unidentified donor in microsomes.

The role of nucleotides in the photoreceptor cell has become increasingly exciting, if enigmatic, as the list of nucleotide-metabolizing enzymes present in the outer segment has expanded to include a light-sensitive opsin kinase,[35] a cGMP-stimulated protein kinase, a light-activated GTPase,[63] a guanylate cyclase,[59] and a light-stimulated phosphodiesterase specific for cGMP.[3] The high concentration of nucleotides in the outer segment, the rapid alterations in concentration induced by light,[39] and the possible involvement of nucleotide metabolism in degenerative retinal disease[41] all point toward an important role for these molecules in the photoreceptor cell.

VISUAL CYCLE

When the intensity of background illumination remains relatively constant, it is obvious that the rates of visual pigment bleaching and regeneration are in balance. When the level of illumination decreases, dark adaptation occurs and the concentration of rhodopsin in photoreceptors increases. These observations were made in early studies of light and dark adaptation and clearly revealed the occurrence of a visual cycle (Fig. 15-13).[60]

The individual steps associated with visual pigment regeneration during dark adaptation are not yet well understood. It is clear that geometric isomerization is a key reaction uniquely associated with dark adaptation. Early evidence for the existence of an enzyme catalyzing the interconversion of the all-*trans* and 11-*cis* isomers of retinal was obtained,[29] but the enzyme does not seem sufficiently active to account for the physiologic process. More recent studies[19,20] on the catalysis of the geometric isomerization of all-*trans*-retinal by dihydroflavins have shown production of the 9-*cis* and 13-*cis* isomers, but no 11-*cis*-retinal. The final step in rhodopsin regeneration involves a spontaneous reaction between 11-*cis*-retinal and the apoprotein to establish both the Schiff base linkage and all the required conformational changes to restore the original spectral properties of pigment.

In the metabolism of retinol associated with

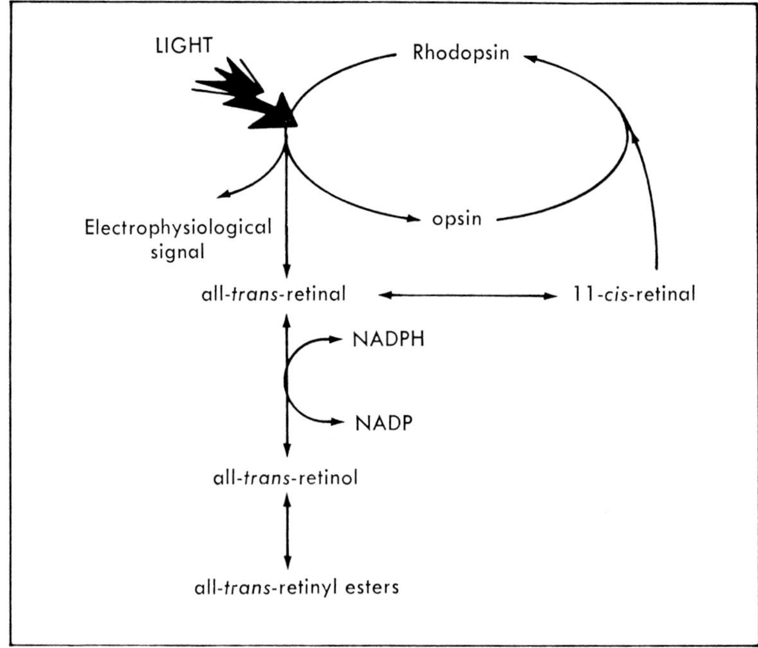

Fig. 15-13. Diagram of rhodopsin system, showing isomerization cycle. Bleaching of rhodopsin by light ends in mixture of opsin and all-*trans*-retinal. The latter must be isomerized to 11-*cis* before it can regenerate rhodopsin. While that is happening, much of it is reduced to all-*trans*–vitamin A, most of which in turn is esterified. These products must be isomerized to their 11-*cis* configuration before engaging in resynthesis of visual pigments.

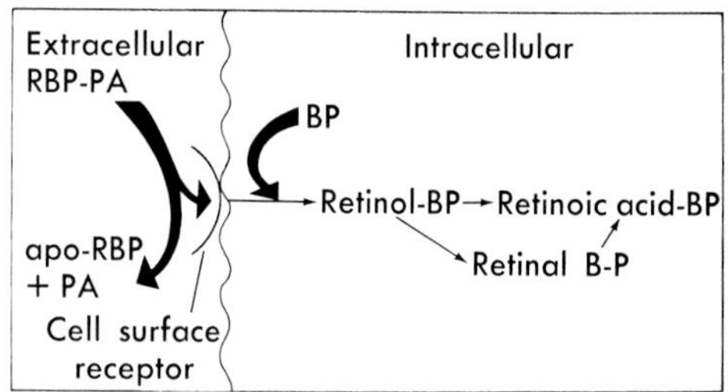

Fig. 15-14. Specific uptake of serum vitamin A appears to depend on the presence of receptors on the cell surface that recognize and interact with the serum retinol-binding protein-prealbumin complex. In a poorly understood process, the protein components are released from the receptor while retinol enters the cell, where it is picked up by a cellular retinol-binding protein. Cellular binding proteins are also present within the photoreceptor cell and can interact with retinal and retinoic acid. *PA*, Prealbumin; *RBP*, serum retinol-binding protein; *apo-RBP*, RBP, from which the retinol has been removed; *BP*, binding protein.

light and dark adaptation much of the retinol is recycled, but it is apparent that a small fraction must be lost through diffusion of retinol from the retina, peroxidative degradation, or possibly even through the enzymatic oxidation of retinal to retinoic acid. Retinol is available from the blood, where it is transported in plasma in combination with a specific binding protein[30] that protects the rather labile retinol molecule from enzymatic or free radical attack.

The retinal pigment epithelium situated between the choroidal vasculature and the neural retina serves as a barrier to the free diffusion of large molecules from the blood stream to the photoreceptor cells. Retinol, bound initially to the serum retinol-binding protein[30] within the choroid, must traverse this single layer of epithelial cells to gain access to the photoreceptors, the primary site of vitamin A function in the visual system. Uptake of retinol from the serum may depend on receptors located on the basal surface of the pigment epithelium that recognize the serum retinol-binding protein-retinol complex.[28] An intracellular retinol-binding protein from retina and pigment epithelium has been characterized[52] and may serve as an acceptor for serum retinol and also as a facilitator of the transfer of retinol between photoreceptor cells and retinal pigment epithelium. A recently discovered binding protein shows preference for the 11-*cis* isomer of retinal[21] and appears to occur uniquely in the retina in relatively large amounts (i.e., 1 mol protein for each 20 mol rhodopsin). The properties of this binding protein suggest that it serves a unique role in the retina, perhaps as a template on which all-*trans*-retinal can be isomerized to the 11-*cis* configuration. A third cellular binding protein specific for retinoic acid has also recently been discovered in extracts of retina.[52] Retinoic acid is thought to function in the maintenance and differentiation of epithelial tissues and has no known function in vision.[14,58] Fig. 15-14 depicts a plausible scheme for the interaction of these cellular binding proteins with retinol delivered to the cell surface of a target cell by the serum retinol-binding protein.

REFERENCES

1. Ames, A., III: Studies of morphology, chemistry and function in isolated retina. In Graymore, C. N., editor: Biochemistry of the retina, New York, 1965, Academic Press, Inc., p. 22.
2. Andrews, J. S., and Futterman, S.: Metabolism of the retina. V, The role of microsomes in vitamin A esterification in the visual cycle, J. Biol. Chem. **239:** 4073, 1964.
3. Bitensky, M. W., et al.: Activation of photoreceptor disk membrane phosphodiesterase by light and ATP, Adv. Cyclic Nucleotide Res. **5:**213, 1975.
4. Borggreven, J. M. P. M., et al.: Biochemical aspects of the visual process. XIII, The role of phospholipids in cattle rhodopsin studied with phospholipase C., Arch. Biochem. Biophys. **145:**290, 1971.
5. Bownds, D.: Site of attachment of retinal in rhodopsin, Nature **216:**1178, 1967.
6. Bridges, C. D. B.: The rhodopsin-porphyropsin visual system. In Dartnall, H. F. A., editor: Handbook of sensory physiology, New York, 1972, Springer-Verlag New York Inc., vol. 7/1, p. 417.
7. Brown, P. K.: Rhodopsin rotates in the visual receptor membrane, Nature **23:**35, 1972.
8. Busch, G. E., et al.: Formation and decay of prelumirhodopsin at room temperatures, Proc. Natl. Acad. Sci. U.S.A. **69:**2802, 1972.
9. Chabre, M.: X-ray diffraction studies of retinal rods. I, Structure of the disc membrane, effect of illumination, Biochim. Biophys. Acta **382:**322, 1975.
10. Cohen, L. H., and Noell, W. K.: Glucose metabolism of rabbit retina before and after development of visual function, J. Neurochem. **5:**253, 1960.
11. Cone, R. A.: Early receptor potential: photoreversible charge displacement in rhodopsin, Science **155:**1128, 1967.
12. Cone, R. A.: Rotational diffusion of rhodopsin in the visual receptor membrane, Nature **236:**39, 1972.
13. Dowling, J. E.: Chemistry of visual adaptation in the rat, Nature **188:**114, 1960.
14. Dowling, J. E., and Wald, G.: The biological function of vitamin A acid, Proc. Natl. Acad. Sci. U.S.A. **46:** 587, 1960.
15. Ehinger, B.: Cellular location of the uptake of some amino acids into the rabbit retina, Brain Res. **46:**297, 1972.
16. Erickson, J. O., and Blatz, P. E.: N-retinylidene-1-amino-2-propanol: a Schiff base analog for rhodopsin, Vision Res. **8:**1367, 1968.
17. Falk, G., and Fatt, P.: Rapid hydrogen ion uptake of rod outer segments and rhodopsin solutions on illumination, J. Physiol. (Lond) **183:**211, 1966.
18. Fung, B. K.-K., and Hubbell, W. L.: Organization of rhodopsin in photoreceptor membranes. 2, Transmembrane organization of bovine rhodopsin: evidence from proteolysis and lactoperoxidase-catalyzed iodination of native and reconstituted membranes, Biochemistry **17:**4403, 1978.
19. Futterman, S., and Rollins, M. H.: A catalytic role of dihydroflavin in the geometrical isomerization of all-*trans*-retinal. In Langer, H., editor: Biochemistry and physiology of the visual pigments, New York, 1973, Springer-Verlag New York, Inc.
20. Futterman, S., and Rollins, M. H.: Evidence for the involvement of a reduced flavin isomerization catalyst in the regeneration of bleached rhodopsin, Invest. Ophthalmol. **12:**234, 1973.
21. Futterman, S., Saari, J. C., and Blair, S.: Occurrence of a binding protein for 11-*cis*-retinal in retina, J. Biol. Chem. **252:**3267, 1977.
22. Futterman, S., et al.: Metabolism of glucose and reduction of retinaldehyde in retinal photoreceptors, J. Neurochem. **17:**149, 1970.
23. Hagins, W. A.: The visual process: excitatory mechanisms in the primary receptor cells, Annu. Rev. Biophys. Bioeng. **1:**131, 1972.
24. Hagins, W. A., and Yoshikami, S.: A role for Ca^{2+} in excitation of retinal rods and cones, Exp. Eye Res. **18:** 299, 1974.
25. Hargrave, P. A.: The amino-terminal tryptic peptide of bovine rhodopsin. A glycopeptide containing two sites of oligosaccharide attachment, Biochim. Biophys. Acta **492:**83, 1977.
26. Hayes, K. C.: Retinal degeneration in monkeys induced by deficiencies of vitamin A or E, Invest. Ophthalmol. **13:**499, 1974.
27. Hayes, K. C., Carey, R. E., and Schmidt, S. Y.: Retinal degeneration associated with taurine deficiency in the cat, Science **188:**949, 1975.

28. Heller, J., and Bok, D.: A specific receptor for retinol binding protein as detected by the binding of human and bovine retinol binding protein to pigment epithelial cells, Am. J. Ophthalmol. **81**:93, 1976.
29. Hubbard, R.: Retinene isomerase, J. Gen. Physiol. **39**: 935, 1956.
30. Kanai, M., Raz, A., and Goodman, DeW. S.: Retinol-binding protein: the transport protein for vitamin A in human plasma, J. Clin. Invest. **47**:2025, 1968.
31. Katz, M. L., Spoon, W. L., and Dratz, E. A.: Fluorescent pigment accumulation in retinal pigment epithelium of antioxidant-deficient rats, Invest. Ophthalmol. Visual Sci. **17**:1049, 1978.
32. Koelle, G., and Friedenwald, J.: A histochemical method for localizing choline esterase activity, Proc. Soc. Exp. Biol. Med. **70**:617, 1949.
33. Kramer, S. G.: Dopamine: a retinal neuro-transmitter. I, Retinal uptake, storage, and light-stimulated release of H^3-dopamine *in vivo,* Invest. Ophthalmol. **10**:438, 1971.
34. Kropf, A.: Intramolecular energy transfer in rhodopsin, Vision Res. **7**:811, 1967.
35. Kuhn, H.: Light-dependent phosphorylation of rhodopsin in living frogs, Nature **250**:588, 1974.
36. Kuwabara, T., and Cogan, D.: Retinal glycogen, Arch. Ophthalmol. **66**:680, 1961.
37. Lam, D. M. K.: The biosynthesis and content of gamma-aminobutryic acid in the goldfish retina, J. Cell Biol. **54**:225, 1972.
38. Lessell, S., and Kuwabara, T.: Phosphatase histochemistry of the eye, Arch. Ophthalmol. **71**:851, 1964.
39. Liebman, P. A.: Rod disk calcium movement and transduction: a poorly illuminated story, Ann. N.Y. Acad. Sci. **307**:642, 1978.
40. Liebman, P. A., and Entine, G.: Cyanopsin, a visual pigment of retinal origin, Nature **216**:501, 1967.
41. Lolley, R. N., et al.: Cyclic GMP accumulation causes degeneration of photoreceptor cells: simulation of an inherited disease, Science **196**:664, 1977.
42. Lowry, O., Roberts, N., and Lewis, C.: The quantitative histochemistry of the retina, J. Biol. Chem. **220**: 879, 1956.
43. McConnell, D. G., Rafferty, C. N., and Dilley, R. A.: The light-induced proton uptake in bovine retinal outer segment fragments, J. Biol. Chem. **243**:5820, 1968.
44. Miller, J. A., Brodie, A. E., and Bownds, D. E.: Light-activated rhodopsin phosphorylation may control light sensitivity in isolated rod outer segments, F.E.B.S. Lett. **59**:20, 1975.
45. Moore, T.: Systemic action of vitamin A, Exp. Eye Res. **3**:305, 1964.
46. Nicol, G. D., and Miller, W. H.: Cyclic GMP injected into retinal rod outer segments increases latency and amplitude of response to illumination, Proc. Natl. Acad. Sci. U.S.A. **75**:5217, 1978.
47. Noell, W. K.: The visual cell: electric and metabolic manifestations of its life processes, Am. J. Ophthalmol. **48**:347, 1959.
48. Noell, W. K., Delmelle, M. C., and Albrecht, R.: Vitamin A deficiency effect on retina: dependence on light, Science **172**:72, 1971.
49. Orr, H. T., Cohen, A. I., and Lowry, O. H.: The distribution of taurine in the vertebrate retina. J. Neurochem. **26**:609, 1976.
50. O'Steen, W. K.: Retinal and optic nerve serotonin and retinal degeneration as influenced by photoperiod, Exp. Neurol. **27**:194, 1970.
51. Röhlich, P.: Photoreceptor membrane carbohydrate on the intradiscal surface of retinal rod discs, Nature **263**: 789, 1976.
52. Saari, J. C., Futterman, S., and Bredberg, L.: Cellular retinol- and retinoic acid–binding proteins of bovine retina, J. Biol. Chem. **253**:6432, 1978.
53. Saibil, H., Chabre, M., and Worcester, D.: Neutron diffraction studies of retinal rod outer segment membranes, Nature **262**:266, 1976.
54. Salceda, R., López-Colomé, A. M., and Pasantes-Morales, H.: Light-stimulated release of [^{35}S]taurine from frog retinal rod outer segments, Brain Res. **135**: 186, 1977.
55. Sickel, W.: Retinal metabolism in dark and light. In Fourtes, M. G. F., editor: Handbook of sensory physiology, Berlin, 1972, Springer-Verlag, vol. 7/2, p. 667.
56. Sillman, A. J., Ito, H., and Tomita, T.: Studies of the major receptor potential of the isolated frog retina, Vision Res. **9**:1435, 1969.
57. Sokoloff, L.: Circulation and energy metabolism of the brain. In Albers, R. W., et al., editors: Basic neurochemistry, Boston, 1972, Little, Brown & Co., p. 299.
58. Van Dorp, D. A., and Arens, J. F.: Biological activity of vitamin A acid, Nature **158**:60, 1946.
59. Virmaux, N., Nullans, G., and Goridis, C.: Guanylate cyclase in vertebrate retina: evidence for specific association with rod outer segments, J. Neurochem. **26**: 233, 1976.
60. Wald, G.: Molecular basis of visual excitation, Science **162**:230, 1968.
61. Wald, G., and Brown, P. K.: Human color vision and color blindness, Cold Spring Harbor Symp. Quant. Biol. **30**:345, 1965.
62. Warburg, O.: Uber die Klassifizierung tierischer Gewebe nach ihrem Stoffwechsel, Biochem. Z. **184**: 484, 1927.
63. Wheeler, G. L., Matuo, Y., and Bitensky, M. W.: Light-activated GTPase in vertebrate photoreceptors, Nature **269**:822, 1977.
64. Wu, S. M., and Dowling, J. E.: L-aspartate: evidence for a role in cone photoreceptor synaptic transmission in the carp retina, Proc. Natl. Acad. Sci. U.S.A. **75**: 5205, 1978.
65. Yee, R., and Liebman, P. A.: Light-activated phosphodiesterase of the rod outer segment. Kinetics and parameters of activation and deactivation, J. Biol. Chem. **253**:8902, 1978.
66. Yoshizawa, T.: The behavior of visual pigments at low temperatures. In Dartnall, H. F. A., editor: Handbook of sensory physiology, New York, 1972, Springer-Verlag New York, Inc., vol. 7/1, p. 146.

Chapter 16

Anatomy and physiology of central visual pathways

ALAN L. PEARLMAN, M.D.

VISUAL PATHWAYS

The examination of the visual fields plays an important role in the clinical localization of lesions of the nervous system; knowledge of the neuroanatomy of the visual pathways is essential for the interpretation of visual field defects. This chapter will briefly review the major anatomic features of the system as they relate to clinical localization and then consider the structure and function of the lateral geniculate and striate cortex in more detail. The discussion will focus chiefly on studies of higher primates (monkey and human) with reference to studies of cats in areas where comparable work is not available for primates.

One important feature of the visual system is the fact that the system is organized in a retinotopic fashion at each level. Retinotopic simply means that a given retinal point projects to a given region of the lateral geniculate, and then to a given region of the visual cortex. It means further than an orderly map of the projection of the retina onto the lateral geniculate and cortex can be described. The retinotopic organization is quite important in the interpretation of clinical visual fields, but the concept of point-to-point representation must be expanded when we consider the fine details of neuroanatomic connections and the physiologically determined receptive fields of cells at various levels. As we have already seen (Chapter 14), the signals from many retinal cones may ultimately converge on a single ganglion cell, and the signals from a single cone may influence more than one ganglion cell. The same principles of convergence and divergence apply to the connections at the geniculate and cortex, while the principle of retinotopic organization is maintained in a general way.

Retinal ganglion cell axons form the optic nerve and project to several different sites in the central nervous system. These sites include the suprachiasmatic nuclei of the hypothalamus, the accessory optic nuclei, the pretectum, the superior colliculus, and the dorsal and ventral lateral geniculate nuclei.[93] This chapter will concentrate on the pathway from the retina to the lateral geniculate nucleus and visual cortex, since it is important in primate visual perception and in the clinical localization of lesions in humans. The pathway from the retina to the pretectum is important in the pupillary reflex (Chapter 12); the pretectal nuclei, the accessory optic nuclei, and the superior colliculus are involved in the control of eye movements (Chapter 5).

Optic nerve, chiasm, and tract

Many important features of the optic nerve have already been covered in Chapter 14. In this discussion the course of the axons of ganglion cells from various parts of the retina on their way to the lateral geniculate nucleus will be described. As these axons group together and leave the eye, they are called the optic nerve; the region where the two optic nerves

□ This work was supported in part by Research Grant EY00621 from the National Eye Institute, National Institutes of Health.

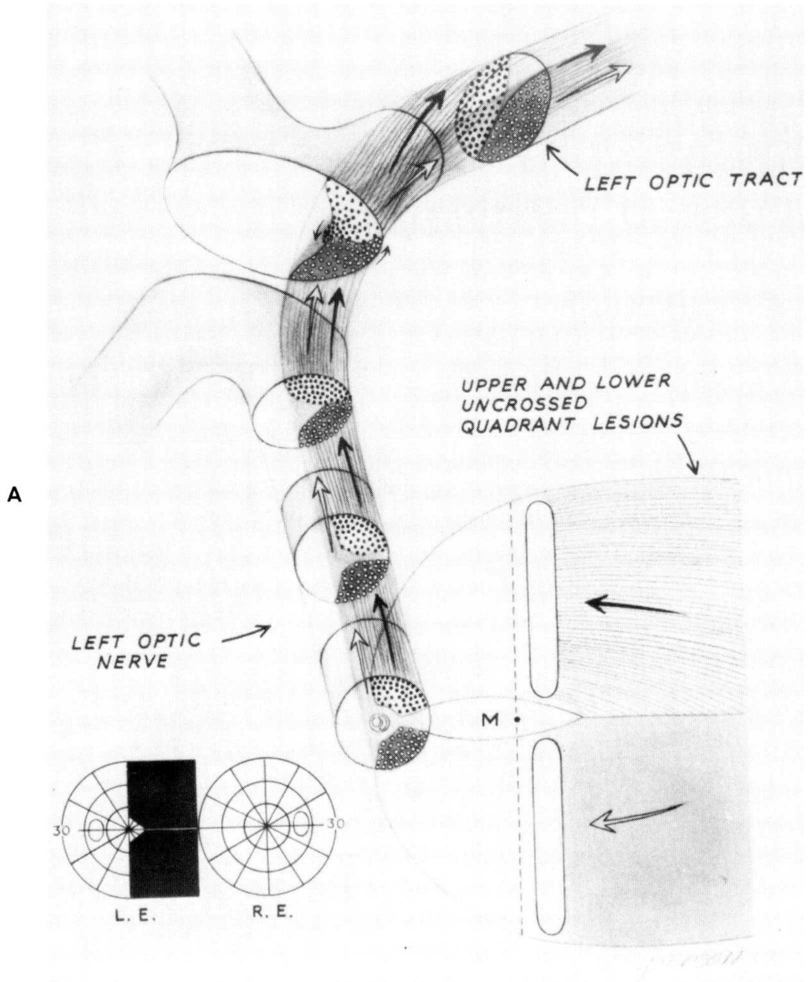

Fig. 16-1. A, Course of uncrossed retinal ganglion cell axons through optic nerve, chiasm, and tract of monkey. Left retina is represented below on right. Vertical white bars are lesions made by photocoagulator; macula, *M*, has not been damaged. Hypothetical visual field defect produced by these lesions is shown in lower left.

come together and fibers from the nasal side of each retina cross to the other side of the brain is called the optic chiasm; the remainder of the pathway from chiasm to lateral geniculate is called the optic tract. There have been a number of careful studies of this clinically important bit of neuroanatomy; each makes use of the technique of producing small lesions in the retina and tracing the degenerating axons by histologic methods.[11,45,46,92] These studies were done on primates; the degree to which the fiber anatomy of nerve, chiasm, and tract corresponds to that of humans is not precisely known, but the similarities are great.

As a consequence of the inversion produced by the optics of the eye, the superior visual field is projected onto the inferior retina, and the temporal visual field is projected onto the nasal retina. At the optic disc the distribution of retinal fibers corresponds roughly to the origin of the fibers in the retina. Fibers from upper retinal quadrants enter the disc above the horizontal meridian, and those from the lower quadrants enter below the horizontal meridian; nasal and temporal fibers are on their respective sides. The macular fibers enter the disc on the temporal aspect, occupying a third or a quarter of the disc.

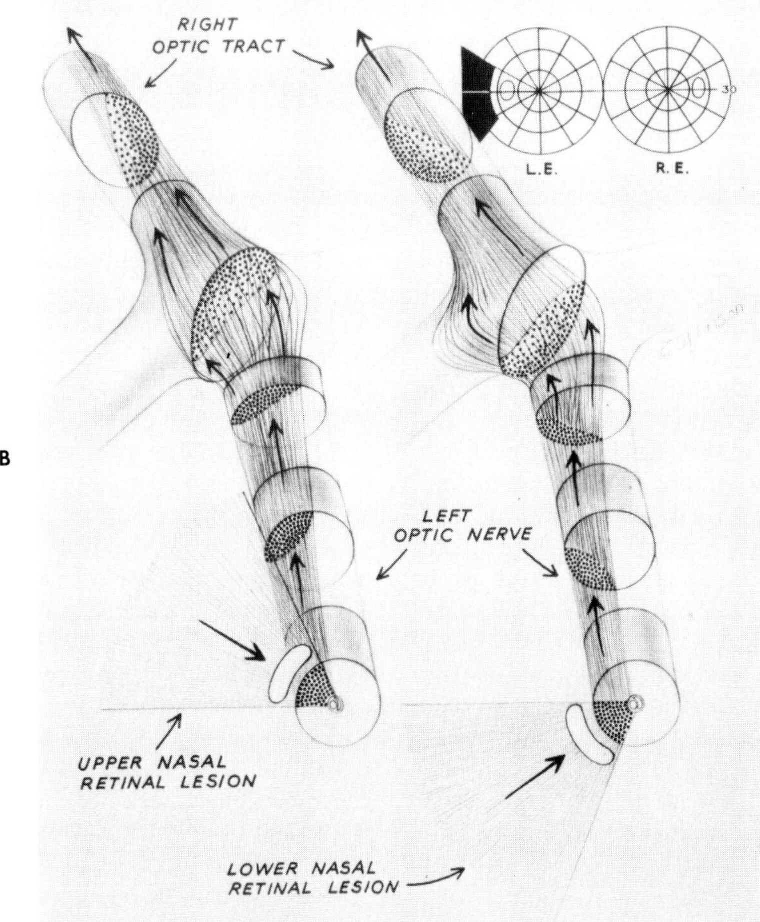

Fig. 16-1, cont'd. B, Course of crossed retinal ganglion cell axons. Photocoagulator lesions in left retina are indicated by white crescents in retinal diagrams at bottom of figure; hypothetical visual field defects produced by these lesions are shown at upper right. (From Hoyt, W. F., and Luis, O.: Arch. Ophthalmol. **68:**94, 1962.)

Ganglion cell axons that arise from the retina temporal to a vertical line through the fovea do not cross in the optic chiasm. Fibers from upper temporal retinal quadrants remain in the upper lateral aspects of the optic nerve and chiasm, but then swing medially as they leave the chiasm to become located in the medial aspect of the optic tract (Fig. 16-1, *A*). Fibers from lower temporal retinal quadrants travel in the lower lateral aspect of the optic nerve and chiasm, and then in the lateral aspect of the optic tract (Fig. 16-1, *A*).[45]

Fibers in the optic nerve that arise from ganglion cells in the nasal half of each retina also travel together in groups from upper and lower retinal quadrants. Fibers from the upper nasal retina on the left travel in the upper medial aspect of the optic nerve, cross in the upper posterior aspect of the chiasm, and then achieve a position in the medial aspect of the right optic tract (Fig. 16-1, *B*). Here they are grouped with the uncrossed fibers from the upper temporal quadrant of the right eye. Fibers from the left lower nasal retinal quadrant travel in the lower medial portions of the optic nerve and cross in the anterior and lower aspects of the chiasm. Some of these fibers loop forward into the right optic nerve as it joins the chiasm, and then pass backward again to enter the right optic tract (Fig. 16-1, *B*). The left lower nasal retinal fibers achieve a position in the lateral aspect of the right optic tract, along with the uncrossed lower temporal fibers of the right retina.[45]

Thus far it has not been possible to make lesions of the macula that involve only medial or lateral, inferior or superior aspects. Therefore it is not clear whether fibers from the respective macular quadrants remain grouped together in the same way that fibers from peripheral retinal quadrants do. Macular fibers enter the optic nerve in a group on the temporal aspect, but as they course toward the chiasm, they spread out rapidly and are soon found throughout the entire optic nerve. In the chiasm macular fibers remain diffusely spread among peripheral fibers, occupying all parts of the chiasm except the anterior-inferior and posterior-inferior aspects. Macular fibers are also spread throughout the superior aspect of the optic tract, mixing medially and laterally with peripheral fibers from both upper and lower retinal quadrants.[46] Thus the retinotopic concept does not apply to macular fibers between the retina and lateral geniculate; macular fibers are not confined to a single bundle at any point except the distal optic nerve.

Lesions of infrageniculate pathway

In considering the effects of lesions in the visual pathways, it should be kept in mind that clinical cases seldom correspond exactly to textbook diagrams, for several reasons. Most pathologic processes affecting vision are not discrete knife cuts, but rather are naturally occurring processes like tumors and vascular occlusions that do not produce lesions with sharp boundaries. In addition, as more detailed anatomic evidence becomes available, textbook pictures have to be altered to conform to the real, and usually more complex, anatomic situation.[44] The following will serve as an introduction to the consideration of visual field defects and their localization. Greater detail is available in several sources.[35,39,44,112]

Fibers from the nasal portions of the two eyes cross in the optic chiasm; those from the temporal portions of the retinas remain on the same side. In this way information about the right side of visual space, coming from both eyes, is joined together in the left hemisphere. The chiasm thus serves an important role in visual function; it also is the neuroanatomic foundation for much of the topical localization of lesions of the visual pathways. The first step in such a localization is usually to determine whether the lesion is in front of, at, or behind the optic chiasm.

Optic nerve lesions. Lesions of the optic nerve characteristically are recognized by the fact that the vision loss is in one eye and the other eye is normal (Fig. 16-2, lesion 1). One of the most common optic nerve lesions results from demyelination in the optic nerve. For some reason the macular fibers are more affected by this process, which is distressingly frequent in multiple sclerosis, and the resultant visual field loss is a central scotoma. Compression of the optic nerve, either within the orbit or in its intracranial course, may also produce a central scotoma. Since the nasal and temporal retinal fibers are grouped separately as they approach the optic chiasm, small lesions in this region can produce loss of vision in half the visual field (hemianopic field defect) of the involved eye.

When the field defects occur in both eyes, one must decide whether there is a single lesion in or behind the optic chiasm, or whether there are lesions in both optic nerves. The nature of the field defect is critical in this distinction. Field defects characteristic of optic nerve involvement (i.e., central scotomas) occurring bilaterally indicate bilateral nerve lesions. Lesions at or behind the chiasm also produce characteristic field defects, which will be described below.

Optic chiasm lesions. The optic chiasm is frequently compressed by tumors arising in the pituitary just beneath it or from the adjacent dura mater. The field defect most characteristic of chiasmal lesions is loss of vision in both temporal fields (bitemporal hemianopia, Fig. 16-2, lesion 2), either peripherally, centrally, or both. When the chiasm is involved anteriorly, at the junction with the optic nerve, there may be total loss of vision in one eye and a temporal hemianopia in the other eye (Fig. 16-2, lesion

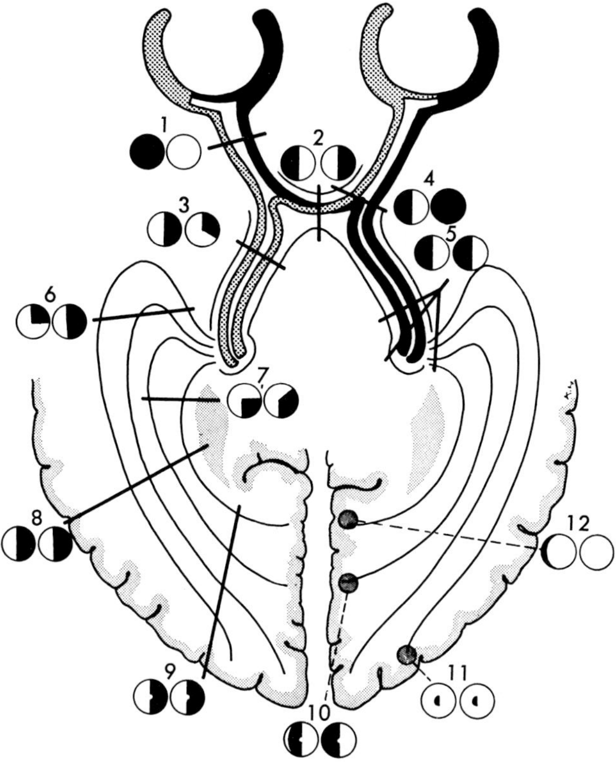

Fig. 16-2. Diagrammatic representation of primary visual pathway, showing in schematic fashion abnormalities in visual fields produced by discrete and total interruption of nerve fibers at various levels. Visual fields are shown as viewed by patient, with black areas indicating areas of absent vision. Explanations of field defect resulting from a lesion at each site are given in text. (From Harrington, D. O.: The visual fields: a textbook and atlas of clinical perimetry, ed. 4, St. Louis, 1976, The C. V. Mosby Co.)

4), the former from involvement of the optic nerve and the latter from involvement of the crossing fibers in the anterior aspect of the chiasm or as they loop forward into the optic nerve.

Optic tract lesions. Lesions of the visual pathways behind the chiasm, beginning with those in the optic tract, produce field defects in the opposite hemifield of the two eyes; a lesion of the right optic tract (Fig. 16-2, lesion 5) produces field loss in the left hemifield of both eyes (left homonymous hemianopia). Since any postchiasmatic lesion may produce homonymous hemianopia, one must therefore depend on the details of the hemianopic defect and the signs arising from involvement of neighboring parts of the nervous system for localization. Although lesions of the optic tract are generally said to produce incongruous field defects (defects that involve the homonymous hemifield of the two eyes but are not identical, as in Fig. 16-2, lesion 3), they may also be congruous. The incongruity is thought to arise in tract lesions because the nerve fibers from homonymous parts of the two retinas are not as closely associated in the tract as they will be in more posterior parts of the pathway.

In some instances an optic tract lesion producing homonymous hemianopia may be distinguished from a more posterior lesion with the same field loss by the hemianopic pupillary reaction. Just before the optic tract reaches the lateral geniculate nucleus, fibers of the afferent limb of the pupillary response leave the optic tract for the pretectal region of the midbrain. A lesion interrupting these fibers should result in an absent pupillary response when the hemianopic field is illuminated. Since it is difficult to illuminate only the hemianopic retina without scattering light to the normal hemiretina, one cannot conclude that the optic tract is intact if the pupillary reaction is observed with illumination of the hemianopic field.

Lateral geniculate lesions

Optic tract fibers from upper retinal quadrants terminate medially in the lateral geniculate nucleus, and inferior retinal quadrants project laterally. The macular fibers terminate posteriorly in a substantial portion of the nucleus. It is rare to be able to implicate the lateral geniculate body in the clinical localization of a hemianopic field defect. Lesions of the terminal optic tract, the lateral geniculate, or the initial part of the optic radiation might all produce a similar defect (Fig. 16-2, lesion 5). The geniculate is more often implicated when a homonymous hemianopia is associated with signs indicating involvement of neighboring structures. For example, a profound sensory loss on the left side of the body associated with a left homonymous hemianopia implies a lesion of the right thalamus including the lateral geniculate. Wild, persistent, flailing movements (hemiballismus) of the limbs on the left and a left homonymous hemianopia are associated with a lesion of the right subthalamic nucleus and the right lateral geniculate.

Lesions of optic radiations

The axons of lateral geniculate cells project to the primary visual cortex (also called calcarine or striate cortex) as the optic radiations. The fibers pass through the posterior part of the internal capsule, spreading out to form a large fan that sweeps around the lateral wall and the posterior horn of the lateral ventricle (Fig. 16-3).

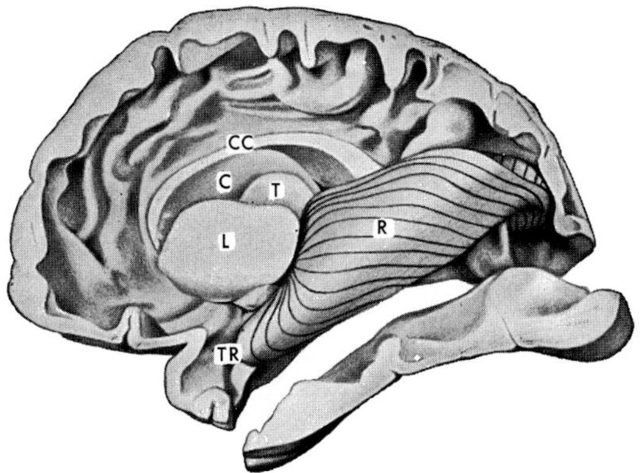

Fig. 16-3. Internal view of left hemisphere from lateral aspect. Optic radiations are seen as they leave lateral geniculate and sweep around lateral aspect of lateral ventricle and then terminate in striate cortex on medial aspect of hemisphere. Inferior radiation fibers sweep around temporal horn of lateral ventricle; the most anterior of these temporal radiation fibers are indicated by *TR*. Other structures shown diagrammatically for reference are caudate nucleus, *C;* corpus callosum, *CC;* lentiform nucleus, *L;* and thalamus, *T*. (From Duke-Elder, S.: Text-book of ophthalmology, St. Louis, 1946, The C. V. Mosby Co., vol. 1.)

Fibers arising from the medial aspect of the lateral geniculate, representing the superior retinal quadrant, run almost directly posteriorly to the striate cortex. Fibers from the lateral aspect of the geniculate, representing inferior retinal quadrants, swing anteriorly into the temporal lobe before turning posteriorly toward the striate cortex. This looping forward (called Meyer's loop) of the radiation fibers representing inferior retinal quadrants is an important feature in localizing lesions of the visual radiations. Loss of vision in the right superior visual fields (homonymous quadrantanopia) (Fig. 16-2, lesion 6) indicates a lesion in the left temporal lobe, interrupting the fibers of Meyer's loop. Conversely, a lesion in the left parietal lobe, interrupting only the radiation fibers that do not loop forward in the temporal lobe, will produce a right inferior homonymous quadrantanopia (Fig. 16-2, lesion 7). Temporal and parietal lobe lesions in the dominant (in most people the left) hemisphere also produce various types of fluent aphasia. There are many other clinical features that point to the parietal lobe; among these are difficulty in drawing and calculating, and the absence of optokinetic nystagmus when the striped stimulus is moved toward the side of the lesion.

Lesions of visual cortex

The primary visual (striate) cortex is located on the medial aspect of the cerebral hemisphere; much of it is buried deep in the calcarine fissure (Fig. 16-4). Optic radiation fibers representing superior retinal quadrants terminate on the superior lip of the calcarine fis-

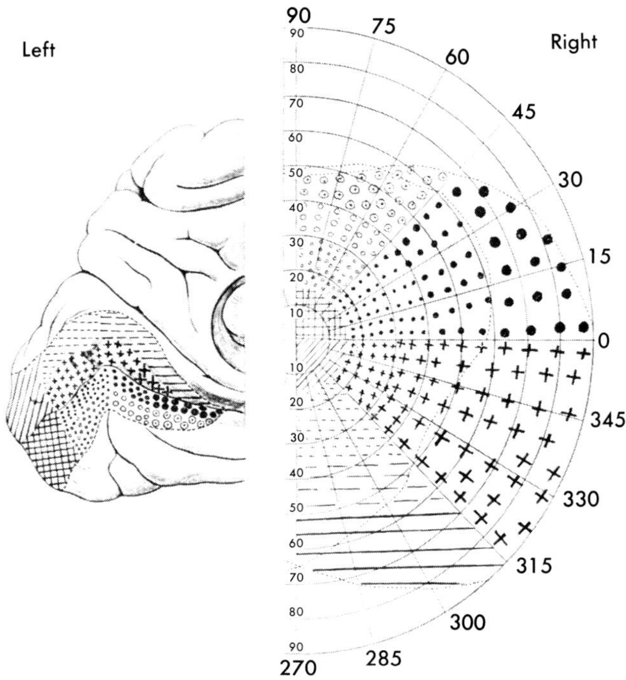

Fig. 16-4. Diagrammatic representation of right visual field on calcarine fissure of left hemisphere in human. Calcarine fissure is shown as widely opened. Symbols on cortex indicate representation of that part of visual field containing the same symbols. (From Duke-Elder, S.: Text-book of ophthalmology, St. Louis, 1946, The C. V. Mosby Co., vol. 1.)

sure, and those representing inferior retinal quadrants end in the inferior lip. The macula is represented posteriorly in a large region that extends for a variable distance onto the posterolateral aspect of the occipital pole. Retinal areas close to the macula are represented more anteriorly on the medial aspect of the hemisphere; the more peripheral the retinal area, the farther away from the occipital pole will its representation be found (Fig. 16-4). The representation of the most peripheral nasal retina of the opposite side is at the extreme anterior end of the calcarine fissure. This segment is the only cortical segment that is monocular; it represents the extreme temporal aspect of the visual field, called the temporal crescent.

Lesions of visual cortex on one side result in homonymous hemianopia, as do lesions in the visual radiations near the cortex (Fig. 16-2, lesion 8). As opposed to lesions in optic chiasm or tract, the field defect is usually quite congruous, since the inputs from the two eyes terminate in the same small cortical region. Frequently, a lesion of optic radiation or cortex will produce a homonymous hemianopia that appears to spare the macular regions of the field. This phenomenon is called macular sparing; it usually indicates a zone of preserved vision of up to 5 degrees around the fixation point (Fig. 16-2, lesion 9). There is no single, completely satisfactory explanation for the phenomenon, although its presence is helpful in localization. In many instances what is called "macular sparing" is simply an artifact produced by poor fixation during visual field testing. True macular sparing in the presence of a homonymous field defect suggests an occipital lesion; its absence does not necessarily rule out such a lesion.

Macular sparing may in some cases simply be a reflection of the anatomic fact that central vision is represented in an area of cortex that is quite large relative to more peripheral parts of the visual field. In the monkey, for example, about one third of the surface of the primary visual cortex is devoted to the central 5 degrees of the visual field; the central 1 degree of the visual field occupies 7% of the area of the visual cortex.[18] Thus a lesion of occipital cortex that is not complete is more likely to spare some or all of the macular representation. Since the cortical representation of the macula extends onto the posterolateral aspect of the occipital pole, a portion of it may fall into the border zone between posterior cerebral and middle cerebral artery territories. An occlusion of the posterior cerebral artery would therefore produce a hemianopic field defect with macular sparing if a portion of the macular representation received its blood supply from the middle cerebral artery. For many years the observation of macular sparing was used to argue for the possibility that each macula was in fact represented bilaterally in the cortex. Although there is clear evidence that the parts of the cortex in the two hemispheres representing the midline of the visual field are connected by the corpus callosum,[53] there is no solid anatomic evidence that the macula projects to both sides at any point in front of the visual cortex. The corpus callosum connection would not explain macular sparing, since a cortical lesion on one side would also eliminate its projection to the other side.

A lesion at the tip of the occipital pole may produce a congruent homonymous hemianopic field defect that involves only central or macular vision (Fig. 16-2, lesion 11). A lesion at the anterior end of the calcarine fissure causes visual field loss in the temporal crescent of the contralateral eye (Fig. 16-2, lesion 12). This is the only place behind the chiasm that a lesion producing a monocular field loss can be situated. Conversely, the temporal crescent is sometimes spared if the pathologic process involves the occipital pole but does not extend into the anterior end of the calcarine cortex (Fig. 16-2, lesion 10).

Total blindness may result from complete destruction of the calcarine cortex bilaterally. It may be distinguished from blindness caused by lesions in or in front of the chiasm by the preservation of pupillary reflexes and the normal funduscopic examination in patients with cortical blindness. Blindness with normal pupillary responses is almost always the result of bilateral occipital cortical lesions rather than bilateral geniculate or radiation lesions, since the visual cortices on the two sides are close to one another and thus subject to simultaneous damage by the same process (often trauma). They also both receive their blood supply from branches of a single artery (the basilar) that is commonly occluded by thrombosis or embolization.

STRUCTURE AND FUNCTION OF LATERAL GENICULATE NUCLEUS

Each major sensory system has an important set of synaptic connections in the thalamus; the thalamic relay in the visual system is the lateral geniculate nucleus. The term "relay" is not meant to imply that the function of the nucleus is simply to pass information along in unaltered form. As we will see, very complex synaptic interactions take place between incoming optic tract fibers and geniculate cells, and a number of as yet incompletely understood physiologic transformations result. In many animals a ventral and a dorsal nucleus of the lateral geniculate body may be defined. The principal thalamocortical relay is the dorsal nucleus; in monkey and human the ventral nucleus is probably represented by the rather small pregeniculate nucleus. In this section the term dorsal nucleus will refer only to the lateral geniculate nucleus.

Retinotopic organization

The most striking anatomic feature of the lateral geniculate nucleus is its laminated appearance in histologic sections (Fig. 16-5). In the posterior aspect of the lateral geniculate of rhesus monkey there are six distinct layers of cell bodies, separated from each other by narrow, almost cell-free, zones. The four dorsal layers contain relatively small cells and are called parvocellular layers. The two ventral layers contain larger cells and are therefore called the magnocellular layers. The four parvocellular layers of the posterior aspect of the lateral geniculate are combined anteriorly to form two layers. The posterior, six-layered segment of the lateral geniculate contains the representation of the central 15 to 20 degrees of the contralateral hemifield, with the remainder of

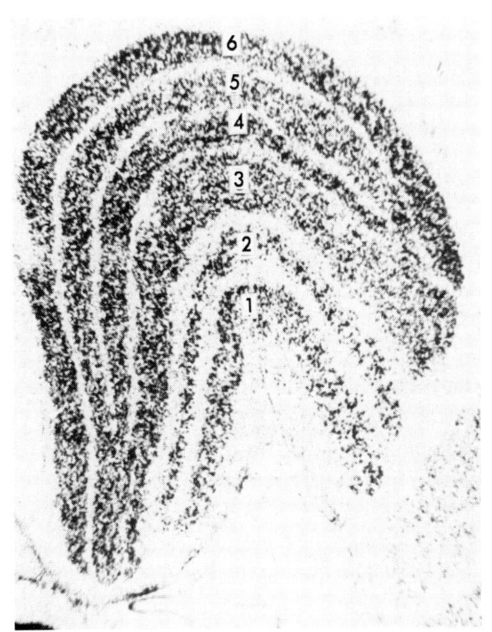

Fig. 16-5. Coronal section of rhesus monkey lateral geniculate, stained to show neuronal cell bodies. Layers are numbered from ventral to dorsal, and medial aspect is on right.

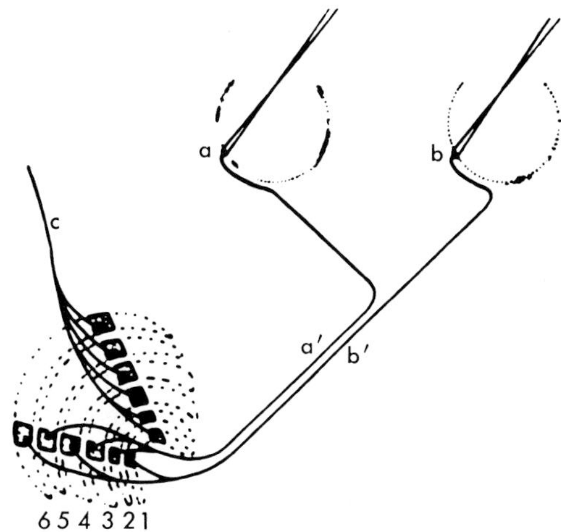

Fig. 16-6. Diagrammatic representation of lines of projection in rhesus monkey lateral geniculate. Fibers from corresponding parts of two retinas, *a* and *b*, terminate in corresponding parts of geniculate laminae. Uncrossed fibers, *a'*, terminate in layers *2*, *3*, and *5*; crossed fibers, *b'*, terminate in layers *1*, *4*, and *6*. *a'* and *b'* represent groups of fibers from retinal points *a* and *b*; individual ganglion cell fiber terminates in only one geniculate layer. Geniculate fibers, *c*, from corresponding parts of the six laminae project to small region of visual cortex. (From Le Gros Clark, W.: Trans. Ophthalmol. Soc. U.K. **62:**237, 1942.)

the hemifield represented in the more anterior, four-layered segment. The layers are conventionally numbered from ventral to dorsal (Fig. 16-5). Parvocellular layers 6 and 4 receive input only from the contralateral eye; they fuse to become a single dorsal parvocellular layer in the anterior aspect of the geniculate. Similarly, parvocellular layers 5 and 3 receive input only from the ipsilateral eye and fuse anteriorly to form the single ventral parvocellular layer. Thus the parvocellular layers contain two complete representations of the contralateral hemifield; one from the contralateral eye in the dorsal parvocellular layer and its posterior subdivisions, layers 6 and 4, and the other in the ventral parvocellular layer and its posterior subdivisions, layers 5 and 3. The two magnocellular layers are not subdivided; each contains a representation of the contralateral field, layer 2 from the ipsilateral eye and layer 1 from the contralateral eye.[11,56,73,92] The human lateral geniculate is similar in basic structure, but there is a considerable degree of variability from one brain to the next in the details of the laminar pattern.[40]

Each geniculate layer thus contains its own retinotopic map of the contralateral visual field. The visual field maps are in rather precise register: an electrode passing from one layer to the next will record from cells representing corresponding parts of the two retinas (Fig. 16-6). The straight line connecting the points in each layer that represent the same point in the visual field is called a line of projection[6] and has been likened to one of the toothpicks in a club sandwich.[111] Cells in a given line of projection in the geniculate send their axons to a restricted region of the visual cortex. Lines of projection may therefore also be defined by making small lesions in the cortex and observing the retrograde cell degeneration that takes place in each of the geniculate laminae. Lines of projection run roughly perpendicular to all the layers. Since the extreme temporal crescent of the visual field is monocular, it is represented only in the geniculate laminae receiving input from the contralateral eye. In monkey and human these layers extend anteriorly slightly beyond layers receiving ipsilateral input; the temporal crescent is represented in these extensions. Layers receiving input from the contralateral eye also have a small, cell-free discontinuity corresponding to the blind spot (optic disc) of the contralateral nasal retina.[64]

Synaptic interconnections

The synaptic interconnections in the lateral geniculate have proved to be much more complex than earlier light-microscopic studies had indicated. The electron microscope has uncovered many interesting anatomic features, but the problem of relating electron-microscope findings to light-microscope observations is incompletely settled.[36] There are many similarities in the fine structure of the geniculate of cat and monkey, although the former has received somewhat more attention than the latter. The following generalizations apply to both animals except where specifically noted.

The basic elements within the geniculate are the axons of retinal ganglion cells (retinogeniculate axons), the geniculate cells that project to the cortex (geniculocortical cells), the geniculate cells that do not send axons out of the geniculate (interneurons), and afferent axons returning from the visual cortex (corticogeniculate axons).

The retinogeniculate axons end within the cellular layers of the geniculate, not in the interlaminar zones. Each incoming fiber distributes its terminals within one layer rather than overlapping among several layers.[83] The fibers terminate on both geniculocortical cells and interneurons, ending on dendritic segments near the cell body in the cat, and sometimes on the cell body itself.

Geniculocortical projection cells may be divided into two classes in the cat. Class 1 cells have the largest cell bodies and thick straight dendrites with only occasional spines. Class 2 cells have cell bodies that are intermediate in size with fine dendrites that are often curved. These dendrites have prominent, grapelike appendages near their branch points. Interneurons

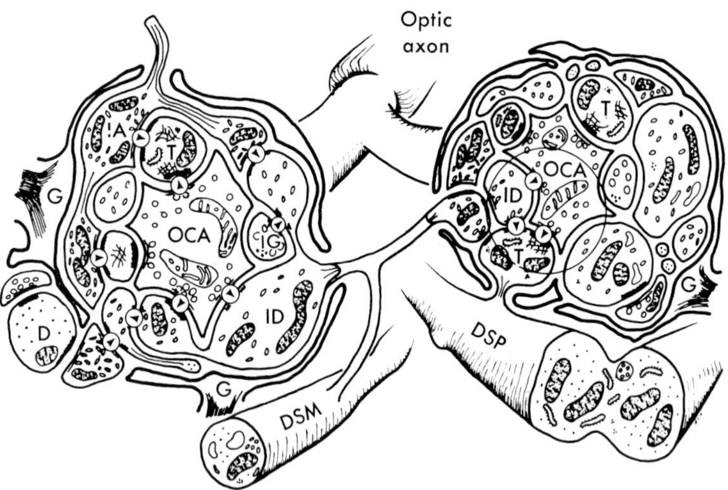

Fig. 16-7. Semidiagrammatic drawing of two synaptic glomeruli in lateral geniculate nucleus of cat. Optic axon is shown dividing and entering the two adjacent glomeruli, where it forms central element, *OCA*. Spinous dendrites, *DSP*, of geniculocortical cells contribute thorns, *T*, that enter glomeruli and receive direct input from optic axons. Smooth dendrites, *DSM*, of interneurons have multilobed dendritic terminals that may enter adjacent glomeruli; these interneuronal dendrites, *ID*, form most of profiles within glomerulus. Interneuronal dendrites are frequently interposed between optic axons and geniculocortical dendrites in a triad (circled region on right). Interneuronal axons, *IA*, form clawlike terminations at periphery of glomerulus, ending on interneuronal dendrites and geniculocortical dendrites. Glomerulus is defined by sheath of astroglial processes, *G*, that surrounds it. (From Famiglietti, E. V., and Peters, A.: J. Comp. Neurol. **144:**285, 1972.)

are the smallest cells in the nucleus; they have fine, wavy dendrites with complex stalked appendages along the dendrites. When the axons of interneurons stain, they are seen to ramify extensively near the cell body, but remain within the nucleus.[28,36,83]

An important morphologic element in the cat lateral geniculate is the synaptic glomerulus, a zone of closely packed neuronal profiles almost completely surrounded by a glial capsule (Fig. 16-7).[28,36,101] Each glomerulus characteristically contains a central retinogeniculate axon (OCA), the dendritic thorns (T) of geniculocortical relay cells, and the processes of interneurons (IA and ID). Careful serial reconstructions of electron micrographs indicate that most of the processes of interneurons within the glomerulus are extensions of dendrites (ID),[28] and some are interneuronal axons (IA). Optic nerve axons are presynaptic to the thorns of geniculocortical cell dendrites and also to adjacent interneuronal dendrites. The interneuronal dendrites is then frequently presynaptic to the dendritic thorn of the geniculocortical cell; the relationship of these three processes is called a triad (circled region in the glomerulus on the right). The interneuronal dendrite also receives input from other interneuronal dendritic processes, thus forming serial synapses between processes of the same type. The processes of the interneuronal dendrite have been likened to those of the amacrine cell in the retina, since, like the amacrine cells, they are the internuncial process in a triadic relationship, and they form serial synapses. Interneuronal dendrites also connect nearby synaptic glomeruli. Interneuronal axons (IA) play a relatively small role in the synaptic glomerulus, terminating on thorns of geniculocortical cells and on interneuronal dendrites. More prominently, they terminate on the cell body and axon initial segment of the geniculocortical cells. Although the monkey lateral geniculate does not contain glial-enclosed synaptic glomeruli, the neuronal processes seen on electron microscopy are much the same as those of the cat, and the interconnections between processes are also similar.[36,66,84]

Fibers returning from the cortex to the lateral geniculate arise from striate cortex (area 17) in the monkey and cat. There are also projections to the lateral geniculate from extrastriate visual regions (areas 18 and 19) in the cat that have not yet been demonstrated in the monkey.[12,34,72,104] The fine corticogeniculate axons pass into the geniculate along lines of projection and synapse in all geniculate laminae and in the interlaminar zones; a given fiber may terminate in more than one lamina. The fibers end on fine peripheral dendrites of both geniculocortical projection cells and interneurons, but do not form a part of the synaptic glomeruli.[28,36]

Physiology of lateral geniculate neurons

Receptive field organization

Lateral geniculate neurons, like ganglion cells in the retina, respond when visual stimuli appear in a particular part of the visual field. That part of the visual field that influences the firing of the neuron under study is called the cell's receptive field. The receptive fields of neurons in the lateral geniculate are similar to the receptive fields of retinal ganglion cells (Chapter 14). Many of the receptive fields in the lateral geniculate of cat and monkey consist of circular central regions that produce a response either when a spot of light is turned on or when it is turned off, and an inhibitory surround that produces a response opposite to that of the central region and antagonizes the response from the center when both areas are illuminated simultaneously.[26,48,74,95] Thus an on-center lateral geniculate cell will respond with a burst of action potentials when a spot of light is turned on in its receptive field center (Fig. 16-8, *A* and *B*); the cell will also respond when an annulus of light illuminating only the surround region is turned off (Fig. 16-8, *D*). Stimuli illuminating both center and surround at the same time are less effective than those illuminating either center or surround alone (Fig. 16-8, *C*); the two regions are therefore said to be antagonistic.[48]

Simultaneous recordings from retinal ganglion cells and lateral geniculate cells in the cat have demonstrated that the excitatory input to an individual geniculate cell comes from only a very few ganglion cells with receptive

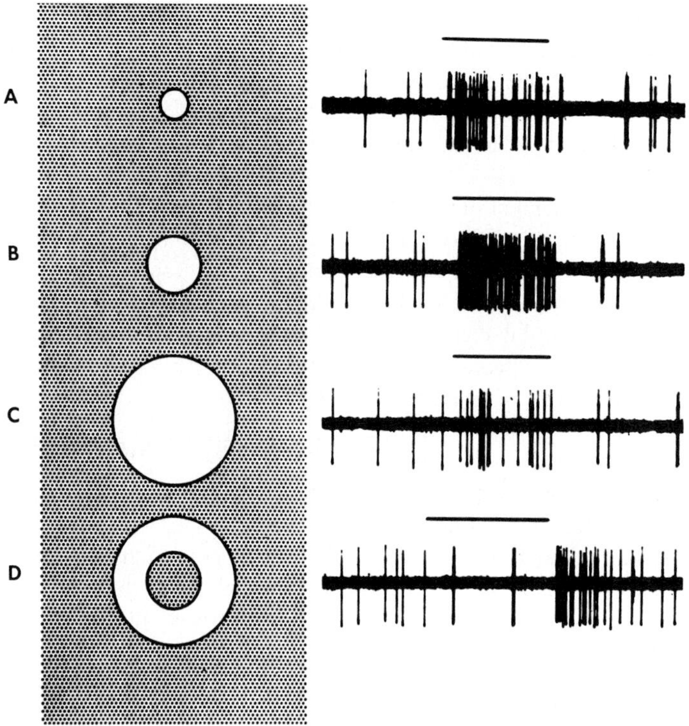

Fig. 16-8. Responses from on-center lateral geniculate neuron in the cat. Light stimuli are shown on left, and action potentials produced by stimulus, as recorded with extracellular microelectrode, are shown on right. Straight horizontal lines above responses indicate when stimulus is on; stimulus duration is 1 second. Stimuli *A*, *B*, and *C* are spots 1, 2, and 14 degrees in diameter centered in receptive field. Stimulus *D* is an anulus with inner diameter of 2 degrees and outer diameter of 14 degrees. (From Hubel, D. H., and Wiesel, T. N.: J. Physiol. [Lond] **155:**385, 1961.)

field centers slightly offset from each other. In some cases only one ganglion cell may account for all the excitatory input to a geniculate cell; for most cells the number is three or less. The excitatory inputs to on-center geniculate cells are all from on-center ganglion cells, and those to off-center geniculate cells come exclusively from off-center ganglion cells.[16,69]

X, Y, and W neurons

Like retinal ganglion cells (Chapter 14), the neurons of the lateral geniculate may be divided into several classes based on their response characteristics to particular stimuli; there is also a correlation between the response characteristics of a particular neuron and the conduction velocity of its axon. As more information is obtained, more types of receptive field have been defined, and there is some controversy as to just how retinal ganglion cells and lateral geniculate neurons should be classified.[93] Nevertheless, certain generalities are possible. Lateral geniculate X cells usually respond in a sustained or tonic manner to the appropriate stimulus presented in the receptive field center (Fig. 16-9, *A* and *C*).[15,43] X cells also have linear spatial summation within the receptive field center, while Y cells have nonlinear spatial summation.[27] Most Y cells also respond in a transient or phasic fashion to a spot in the receptive field center (Fig. 16-9, *B* and *D*).[15,43] Both X and Y cells may be either on center or off center. Lateral geniculate Y cells receive input, for the most part, from Y-type retinal ganglion cells.[15] These retinal ganglion cells have relatively large receptive fields and rapidly conducting axons (30 to 40 m/sec). The Y geniculate neurons on which they ter-

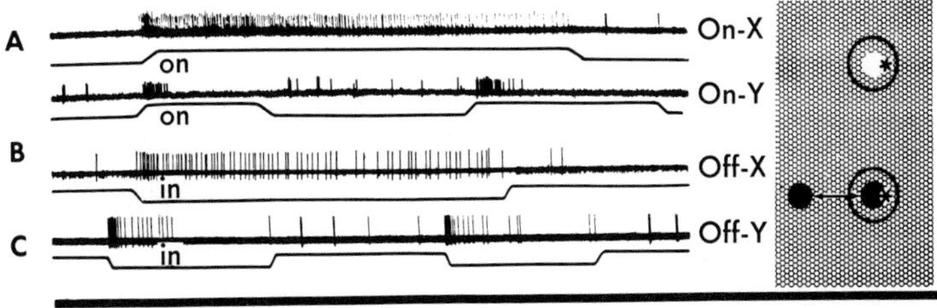

Fig. 16-9. Sustained and transient lateral ganiculate neurons. Line *A* shows action potentials produced by single on-center cell with sustained (or X-type) response to spot of light turned on in receptive field center. Turning on of stimulus is indicated by upward deflection in horizontal line just below line *A*. Stimulus and receptive field center are shown diagrammatically to right of figure. Line *C* is also from sustained cell, but with an off-center rather than an on-center, so that appropriate stimulus is introduction of small black spot. Lines *B* and *D* are taken from transient (or Y-type) geniculate neurons, line *B* from on-center cell, and line *D* from off-center cell. (From Hoffmann, K., Stone, J., and Sherman, S. M.: J. Neurophysiol. **35:**518, 1972.)

minate in turn have axons that conduct at rapid rates to the striate cortex. Conversely, X-type ganglion cells have smaller receptive fields and slower (17 to 20 m/sec) conducting axons, which terminate on geniculate X cells projecting to the striate cortex with axons that conduct at relatively slower rates.[15,16,20,31,43]

The smallest retinal ganglion cells have small axons and therefore the slowest conduction velocities. These cells, by convention grouped together as W cells, have been most clearly defined in the cat retina, but they are not a functionally uniform class. Many W cells do not have center-surround receptive field arrangements; some are direction selective, some respond with increased or decreased firing to a contrast anywhere within the receptive field, and others are color coded.[93] The W cells of cat lateral geniculate are also small and have receptive field properties that are similar to retinal W cells; they have fine, slowly conducting axons that project to the cerebral cortex.[118]

Thus far it would appear that the response properties of geniculate neurons are much like those of the retinal ganglion cells that supply their input. Yet the complex geniculate anatomy described earlier provides for action on geniculocortical relay neurons by interneurons and by axons returning from the cortex; the output of geniculocortical neurons might therefore be expected to be different from the input from ganglion cells because of these additional connections. The most readily apparent way in which geniculate cell output differs from input is in the relative effectiveness of the antagonistic surround of the receptive field.[48] When the ganglion cell input and geniculate cell output are recorded simultaneously, the geniculate neuron is usually substantially less sensitive than its primary excitatory input if the stimulus spot is larger than the receptive field center.[48,69] The manner in which cells are interconnected to account for this property is not established, although it seems probable that the added peripheral suppression is mediated by geniculate interneurons.[69]

The second major difference between retinal ganglion cells and geniculocortical relay neurons is that the latter show some evidence of input from both eyes. Although the principal input to geniculate cells is from the eye providing direct input to the lateral geniculate layer in which the cell is located, there is evidence for a weak and usually inhibitory input from the corresponding part of the other retina in the cat.[94,98] Since the direct optic nerve input to a given layer of the geniculate comes from only one eye, the input from the other eye must be arriving via another source. This inhibitory influence is not carried exclusively by the return pathway to the geniculate from cortical visual areas, since it survives ablation of these areas.[94] The binocular effects in the geniculate must therefore be mediated by connections between the geniculate laminae. The possibilities for such interlaminar connections in the cat include the dendrites of some geniculocortical cells that cross laminar borders and thus receive input from both eyes, and the axons of some interneurons that also cross laminar borders.[83,102]

Opponent color cells

Most cells in the lateral geniculate of the cat have the same spectral sensitivity in both the center and the surround of their receptive field,[20] but rare cells that have color-specific response properties are encountered in the most ventral layer of the geniculate.[21,88]

The rhesus monkey, whose capabilities for color vision are much like those of the human, has many cells in the retina and lateral geniculate that respond in different ways to different wavelengths.[19,22,26,95,116] The receptive field type most frequently encountered in the parvocellular dorsal layers of the monkey geniculate has a concentric center-surround arrangement, but the center and surround have different spectral sensitivities. Thus the center may produce an on response to a green spot and the surround produce an off response to a red annulus. Another type of cell in the parvocellular geniculate layers has a circular field with no concentric center-surround arrangement. These cells have opponent color responses throughout the field; an example would be a cell responding at on to green spots and at off to blue spots anywhere within the receptive field. There are also cells in the monkey lateral geniculate that are not color coded; these cells have center surround arrangements, but the responses do not differ when the color of the stimulus is changed. The subject of opponent color cells is covered in more detail in Chapter 19.

Functional specificity of geniculate laminae

Although cells in the individual laminae of the cat lateral geniculate differ in their receptive field properties from cells in other laminae, the segregation of cells by lamina appears to be more distinct in the rhesus monkey. In the six-layered portion of the monkey geniculate, which represents the central 18 to 20 degrees of the visual field, cells in the four parvocellular layers are predominately color opponent. They have been called X-like because of their similarity to X cells of the cat lateral geniculate. They receive input from optic tract fibers with conduction velocities in the midrange, have relatively sustained responses, and project to the cortex with axons that conduct less rapidly than cells in the magnocellular layers. In addition, cells in the dorsal pair of parvocellular layers (Fig. 16-5, layers 6 and 5) are mostly on center, while those in the ventral pair (layers 4 and 3) are predominantly off center. Cells in the two magnocellular layers (layers 2 and 1) are not color coded; they have concentric, center-surround receptive fields that are Y-like, receiving input from rapidly conducting retinal ganglion cells, responding transiently, and projecting axons.[26,74,95] Thus there is clear evidence that the anatomic segregation of cells into laminae and the functional properties of individual cells are closely correlated. To summarize, information from the two eyes is kept separate in the lateral geniculate nucleus; the contralateral hemifield of each eye is represented in a parvocellular layer that is split into two posterior subdivisions, and in a magnocellular layer. Parvocellular layers contain predominantly color-coded X cells, magnocellular layers contain non-color–coded Y cells. In addition, on-center and off-center cells are segregated in the two posterior parvocellular layers representing each eye. Although the functional consequences of this careful anatomic segregation is not yet understood, there is evidence that it is maintained in subsequent processing; axons from parvocellular and magnocellular laminae terminate in different layers of the striate cortex, as will be described in the next section.

STRUCTURE AND FUNCTION OF STRIATE CORTEX

The cell bodies of the cerebral cortex are arranged in layers, separated by relatively cell-free zones containing primarily the dendritic and axonal processes of the cells. On the basis of differences in the appearance of these layers, the cortex has been divided into cytoarchitectonic areas, the most distinctive of which is the region in the occipital lobe that receives the fibers from the lateral geniculate nucleus. This region is characterized by a thick band of dendrites and axons about midway between the surface and depth of the cortex that is visible to the unaided eye in fresh or fixed brain; the band is called the stria of Gennari after its discoverer, and the cortical area therefore is the striate cortex. The striate cortex is area 17 in the most widely used system of designating cytoarchitectonic areas.[10] The striate cortex in humans is located on both sides of the calcarine fissure on the medial aspect of the occipital lobe, extending for a short distance onto the occipital pole (Fig. 16-4).

In the rhesus monkey, area 17 is the only cortical region receiving direct input from the lateral geniculate nucleus.[32,117] Although area 17 receives the lateral geniculate projection and contains the distinctive stria of Gennari, the incoming geniculate fibers contribute only slightly to the stria itself.[14,57,70] The stria, which is shown as layer IVB in Fig. 16-10, is largely composed of interconnecting axons arising from neurons in laminae above and below the stria.[70] The system for numbering the layers of the cortex varies somewhat from author to author, depending on the criteria adapted for including a particular laminar subdivision with other subdivisions. The system shown in Fig. 16-10 for the monkey cortex is modified from that of Brodmann.[9,70,92]

Geniculate afferents

Afferents from the geniculate terminate for the most part in layer IV in monkey striate cortex, while a few reach layer I.[32,57,70] There are interesting differences in the level of terminations of the fibers from the various layers of the monkey lateral geniculate. Fibers arising in the parvocellular layers of the geniculate terminate in a broad and dense band in layer IVC β and also in a narrow, less dense band just above the stria of Gennari in layer IVA. Fibers arising from the two ventral magnocellular geniculate layers terminate in lamina IVCα.[57] This difference in terminations of the magnocellular and parvocellular layers takes on added

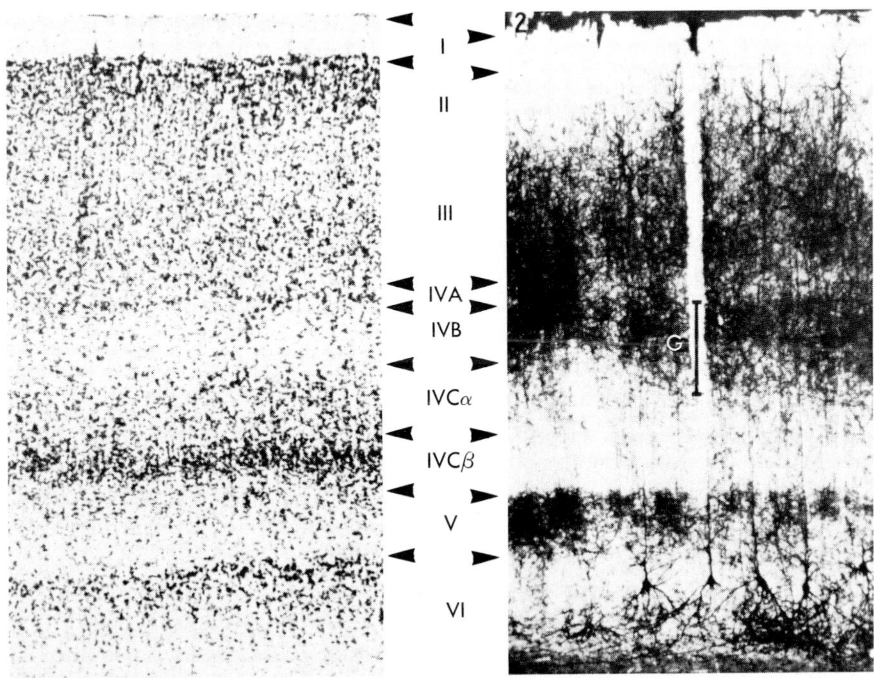

Fig. 16-10. A, Nissl-stained section of area 17 of monkey, *Macaca mulatta,* with pial surface at top. This method predominantly stains nuclei of cortical cells. Numbering system in center identifies cortical laminae of both sections. **B,** Section prepared by the Golgi rapid method, which stains cell bodies and many processes of cortical neurons. Position of stria of Gennari is indicated by *G*. Several large, solitary pyramidal cells (of Meynert) are evident in layer *VI*. (From Lund, J. S.: J. Comp. Neurol. **147**:455, 1973.)

significance in view of the differences in receptive field types in magnocellular and parvocellular layers described earlier in the section on geniculate physiology.

Stellate and pyramidal cells

Staining techniques that allow the visualization of the processes of neurons demonstrate the highly branched dendritic trees of cortical neurons. On the basis of the arrangement of the dendritic branches, cortical neurons have been grouped into two main varieties, pyramidal cells and stellate cells. Pyramidal cells (for example, the large cells in layer VI of Fig. 16-10) have a cell body shaped something like a pyramid. The apex of the pyramid points toward the cortical surface; a single dendrite arises from it and is called the apical dendrite. Several dendrites characteristically arise from the base of the pyramid, spreading in a radial fashion.

The axon of a pyramidal cell usually descends to the subcortical white matter and projects either to another cortical site or to a subcortical target such as the lateral geniculate or superior colliculus. The axon of a stellate cell, on the other hand, usually ramifies locally to form intracortical connections. Stellate cell dendrites arise around the circumference of the cell body in a star-shaped pattern (Fig. 16-11). Stellate cells have been subdivided into those that have small spines arising from the dendrites (Fig. 16-11, *A*) and those that do not (Fig. 16-11, *B*).[67,70] The spinous stellate cell is restricted to cortical layer IV,[70] where most of the lateral geniculate fibers terminate, and these terminations are predominately on dendritic spines, many of which arise from dendrites of stellate cells.[32] There are also distinctive physiologic differences between the cells of layer IV and the cells of other layers, as we will see when we consider the receptive fields of cortical neurons.

Lamination of afferents and efferents

The orderly, laminar arrangement of cell bodies that characterizes cerebral cortex takes on added significance when one considers the arrangement of the afferent and efferent connections of striate cortex. Cells of a given layer project to specific targets; afferent fibers arriving in striate cortex terminate in specific layers. As is the case in many parts of the nervous system, connections between striate cortex and other cortical and subcortical regions are often reciprocal.

Subcortical connections. As indicated earlier, the specific thalamic afferents from the monkey lateral geniculate terminate primarily in layer IV; there is also a sparse projection to layer I.[57,70] The reciprocal connections, returning from striate cortex to the lateral geniculate, arise from the pyramidal cells of layer VI. The distinction between magnocellular and parvocellular geniculate layers is again evident; cells in the upper part of cortical layer VI project primarily to the magnocellular layers, while the projection to parvocellular layers arises mainly from cells in the lower aspect of layer VI.[72]

Neurons of layer V project to the superior colliculus and to the pulvinar, a complex group of nuclei in the thalamus.[72] As mentioned earlier, the superior colliculus receives a major projection from the retina; it projects in turn to distinct regions of the pulvinar. Parts of the pulvinar are reciprocally connected with striate cortex,[85] although the major visual system connections of the pulvinar are with extrastriate regions. Axons from pulvinar to striate cortex terminate in layer I, among the dendrites of the layer V cells[71] whose axons project back to pulvinar and to superior colliculus. Thus for the most part cells of striate cortex that project to subcortical targets are located in deep cortical layers (V and VI).

Corticocortical connections. Striate cortex is extensively interconnected with other cortical regions; the adjacent cortical zones that are important in the visual process are called extrastriate visual regions and will be discussed later in this chapter. Pyramidal cells of striate cortex that project to extrastriate visual regions are primarily in the superficial cortical layers, II and III. The cells that project, via the corpus callosum, to the contralateral striate cortex also arise in layer III. Reciprocal connections from extrastriate visual regions and from contra-

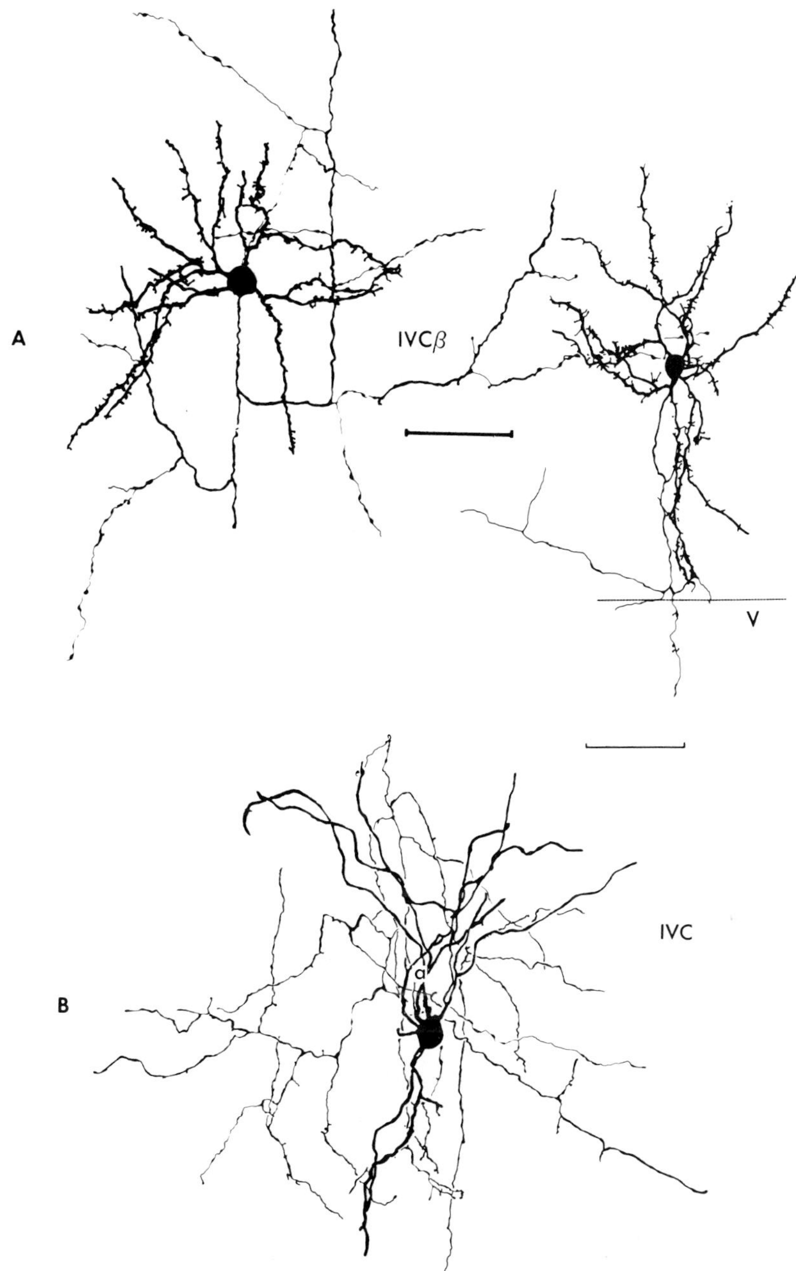

Fig. 16-11. A, Two stellate neurons from layer *IVC* of area 17 of monkey. Both have numerous spines arising from dendrites; those on left are coarse and arise from stout dendrites, while those on cell on right are fine and arise from slender dendrites. All stellate cells with spinous dendrites are found in layer *IV* of cortex. (horizontal bar = 50 μm.) **B,** Stellate neuron from layer *IVC* of area 17 of monkey that does not have spines arising from its dendrites. Such smooth stellate cells are found throughout cortex. (horizontal bar = 50 μm.) (From Lund, J. S.: J. Comp. Neurol. **147:**455, 1973.)

lateral striate cortex are made by fibers that terminate predominantly in layers II and III of striate cortex.

Physiology of striate cortex

Because of the enormous complexity of the interactions between the highly branched processes of cortical neurons, it is not possible with presently available information to construct a circuit diagram for the visual cortex. The analysis of receptive fields of cortical cells has led to a number of significant observations and to an interesting model for how cortical cells interact. Microelectrode recording and receptive field mapping has been used extensively in the study of the physiology of neurons in the visual cortex. Cortical cells studied in this way may be grouped into several major types on the basis of certain receptive field properties. The names given to these types, *concentric, simple, complex,* and *hypercomplex,* to some extent imply that the cell types are connected to each other in a serial, or hierarchical order. Hubel and Wiesel,[47,49,51,54] who first studied receptive fields of visual cortex and named these receptive field types, have suggested a hierarchical model for cell interconnections. We will first consider the basic receptive field properties and then review the evidence for the model.

Concentric cells. Many of the cells recorded in layer IV of the striate cortex have receptive field characteristics that are quite similar to those of the lateral geniculate cells.[24,54,76,96] These fields are arranged in concentric center-surround fashion and receive input from only one eye; their similarity to lateral geniculate cells indicates that very little transformation of the lateral geniculate input has occurred. The location of these cells in layer IV, where the bulk of the lateral geniculate afferents terminate, together with their geniculate-like receptive fields, strongly suggests that they receive direct geniculate input.

Simple cells. Simple cells in area 17 have receptive fields that can be mapped with small spots of light into on areas and off areas. These on and off regions, like those of cells in the lateral geniculate, are mutually antagonistic; unlike the lateral geniculate, however, receptive fields of simple cortical neurons are ar-

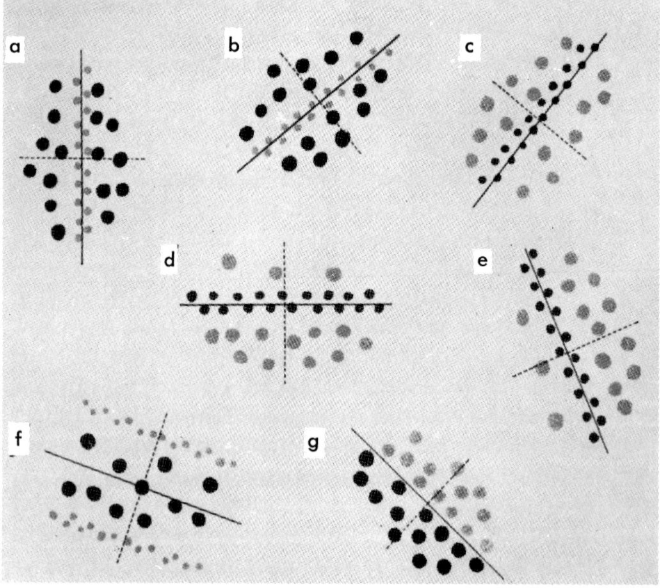

Fig. 16-12. Receptive fields of simple cortical cells from area 17 of cat. On regions are represented by black dots and off regions by gray dots. Axis of orientation of receptive field is shown by continuous straight line. (From The visual cortex of the brain by D. Hubel, Sci. Am. **209:**58, 1963, copyright © 1963 by Scientific American, Inc. All rights reserved.)

ranged in parallel bands of on and off regions, rather than the concentric center-surround arrangement of geniculate cells (Fig. 16-12). Receptive fields of simple cells often have a central band that is either an on region or an off region, with parallel flanking regions on two sides that are the opposite (Fig. 16-12, *a* to *f*). The flanking regions are not always equal in size (Fig. 16-12, *d* and e), and sometimes there is only one (Fig. 16-12, *g*). A bar of light turned on or off within one of these parallel bands is an exellent stimulus; when the bar is rotated so that it falls on mutually antagonistic regions, the stimulus is much less effective in producing a response (Fig. 16-13). The stimulus must therefore be properly oriented; the axis of orientation of the receptive field may be vertical, horizontal, or oblique. Thus far there has been no indication that any particular axis of orientation is represented more frequently than any other axis.[49,54] A moving light or dark bar of the proper orientation is frequently a very effective stimulus for these cells; in many instances it is more effective than a stationary bar turned on and off. The direction of movement can also be quite important; a given simple cell with vertical orientation may respond well to a vertical bar moved across its receptive field from left to right, but poorly or not at all to the same bar moved from right to left. The speed of stimulus movement may also be critical, with different cells responding best to different speeds.[47,49,54]

In the monkey the majority of simple cells are found in layer IV, but they are also encountered in layers superficial and deep to layer IV.

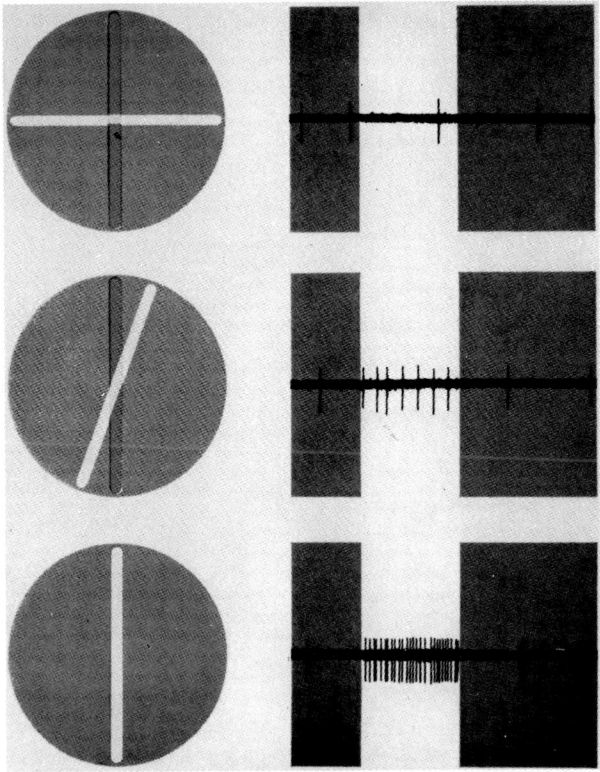

Fig. 16-13. Simple cortical cell with vertical preferred axis of orientation. Stimulus is white bar of varying orientation shown on representation of receptive field at left. When stimulus is horizontal (top), it falls on excitatory and inhibitory regions, and there is no response evident in the recording from cell shown at right. A tilt slightly off vertical, seen in middle row, gives weak response. When stimulus is vertical (bottom), stimulus is entirely in excitatory region of field, and response is vigorous. (From The visual cortex of the brain by D. Hubel, Sci. Am. **209:**58, 1963, copyright © 1963 by Scientific American, Inc. All rights reserved.)

Those in layer IV are often monocular and have smaller receptive fields than those in other layers, indicating that they are not far removed from the geniculate input.[54,96]

Complex cells. Complex cells are distinguished from simple cells on the basis of the observation that it is not possible to map out distinct antagonistic on and off regions in the receptive fields of complex cells.[49] Instead, complex cells give the same response to a line stimulus anywhere within the receptive field, and respond to a moving stimulus throughout the field. Complex cells otherwise have many properties similar to those of simple cells; they respond best to light or dark bars of a specific orientation (Fig. 16-14); they are often quite sensitive to moving stimuli and may respond better when the properly oriented stimulus is moved in one direction rather than in the opposite direction (Fig. 16-14, *D*). Complex cells often receive input from both eyes and are thus called binocular. Remarkable as it may seem, the receptive fields of a given binocular complex cell are on corresponding parts of the two retinas and have identical receptive field properties. Complex cells are found in the cortical layers above and below layer IV of area 17, and only rarely in layer IV itself.[49,54]

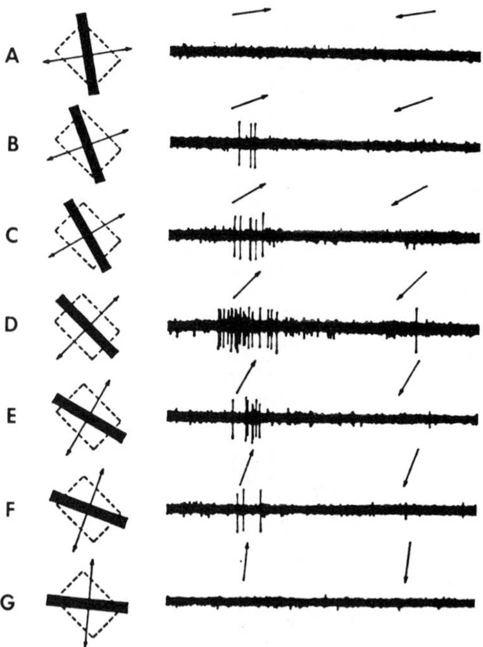

Fig. 16-14. Receptive field of complex cell from monkey striate cortex. Outline of receptive field is given by dashed line. Orientations of black bar stimulus are shown on left, and responses to that particular orientation are shown on right. Arrows indicate direction of bar movement (From Hubel, D. H., and Wiesel, T. N.: J. Physiol. [Lond] **195:**215, 1968.)

Hypercomplex cells. Hypercomplex cells, as originally defined, retain all of the properties of complex cells but also have the added feature of requiring the line stimulus to be of a specific length. If the properly oriented stimulus fits within the activating region of the receptive field (Fig. 16-15, *A*), the response is vigorous. If the stimulus extends outside the activating region on either side (Fig. 16-15, *B*), the response is greatly diminished or eliminated. Thus there are antagonistic regions evident on two sides of the activating region of the receptive field. The line stimulus within these antagonistic regions must be of the same orientation and direction of movement as required for the activating region. The antagonistic regions are only evident in reducing the response from the activating region; they do not produce a response when stimulated alone.[51]

Following the original description of hypercomplex cells as complex cells requiring a stimulus of specific length, a number of simple cells that also had length specificity were described. Subsequently, the rationale for defin-

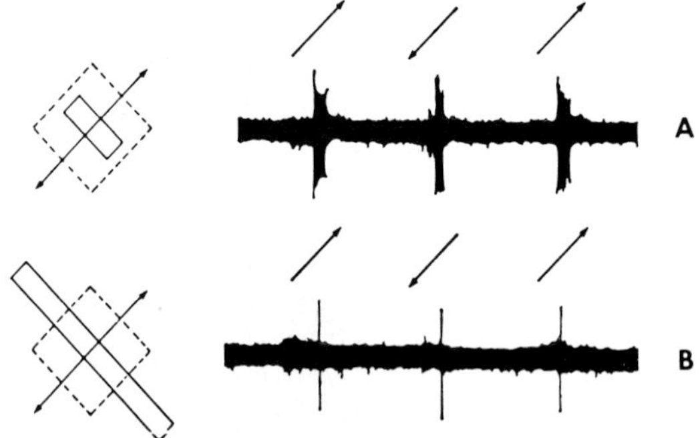

Fig. 16-15. Hypercomplex cell recorded in area 17 of monkey. Activating region of receptive field is outlined by broken line on left. Stimulus, in this instance a moving slit of white light, is represented by solid rectangle, and responses to movements in directions indicated by arrows are shown on right. In **A,** stimulus is contained within activating region; in **B** it extends into antagonistic regions on either side. (From Hubel, D. H., and Wiesel, T. N.: J. Physiol. [Lond] **195:**215, 1968.)

ing hypercomplex cells as a separate and distinct class has been challenged further by the observation that the extent of the decrease in response with increasing stimulus length varies along a continuum. Although there are some simple and complex cells that have a high degree of length specificity and others that have very little, there are apparently all gradations in between, making it difficult to form a separate class based on length specificity alone.[33,81,96]

Color-coded cells. Thus far we have concentrated on the spatial organization of the various receptive field types in striate cortex. Some of the cells in monkey striate cortex, especially in the part of the cortex representing the fovea, are also highly specific for color (Chapter 19). These cells, called color-coded cells, have the same types of spatial organization as non-color–coded concentric, simple, complex, and hypercomplex cells, but also respond preferentially to particular colors and do not respond to white light. A color-coded simple cell, for example, has parallel bands of excitation and inhibition that respond to particular colors. The central band might be excited by red and the flanking bands by green; thus the best response would be to a properly

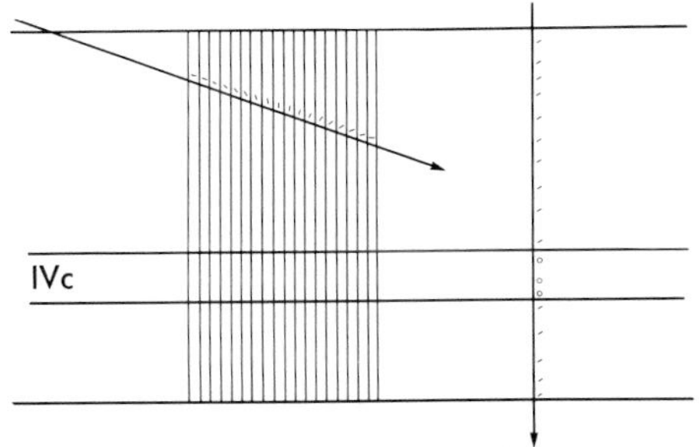

Fig. 16-16. Diagrammatic representation of orientation columns in monkey striate cortex. Vertical electrode penetration (arrow at right) encounters cells in layers above and below layer *IV C* that all have the same orientation specificity, shown by short lines. Cells of layer *IV C* with concentric, nonoriented receptive fields are shown by small circles. An oblique electrode penetration (arrow at left) encounters slight variations in orientation specificity as it passes from one column to the next. Clockwise or counterclockwise systematic variations in orientation take place in small steps of 10 degrees or less, occurring about every 50 μm. (From Hubel, D. H., and Wiesel, T. N.: Proc. R. Soc. Lond. Biol. **198:**1, 1977.)

oriented red bar that fills the center band while the adjacent bands are illuminated by green. Color-coded concentric and simple cells are primarily found in layer IV, while color-coded complex and hypercomplex cells are prevalent in layers II, III, V, and VI.*

Columnar organization

Orientation columns. It should be clear from the material presented thus far that most cortical cells are specialized to respond to lines, edges, or bars, and that the orientation of the line stimulus is critical for a particular cell. Repeated electrode penetrations of adjacent regions of the striate cortex have demonstrated that cells with receptive fields of a particular orientation are grouped together; an electrode penetration perpendicular to the cortical surface will encounter cells with receptive fields of only one orientation as it travels through the layers. This vertical grouping of cells with identical orientation specificity has been called an orientation column. Adjacent orientation columns differ slightly, and in a systematic way, in their orientation specificity, so that an electrode traveling nearly parallel to the cortical laminae will encounter cells with receptive fields of first one orientation and then cells with orientation specificity that differs by about 10 degrees in a clockwise or counterclockwise direction[58,60] (Fig. 16-16). The cortex is thus divided into tiny columns of cells with similar orientation specificity, although the columns are not actually "columns" in the strict architectural sense, but rather slabs or sheets with irregular, undulating configurations.[63]

Ocular dominance columns. A different, and apparently independent system of columns exists with respect to the binocular input to cortical cells. As pointed out earlier, individual concentric and simple cells in layer IV of monkey striate cortex receive input from only one eye, whereas most complex and hypercomplex cells in layers above and below layer IV receive input from both eyes. The receptive fields in the two eyes are identical for a given binocular cell in terms of size, orientation, motion sensitivity, and directionality. They also are in pre-

*See references 19, 24, 25, 54, and 76-79.

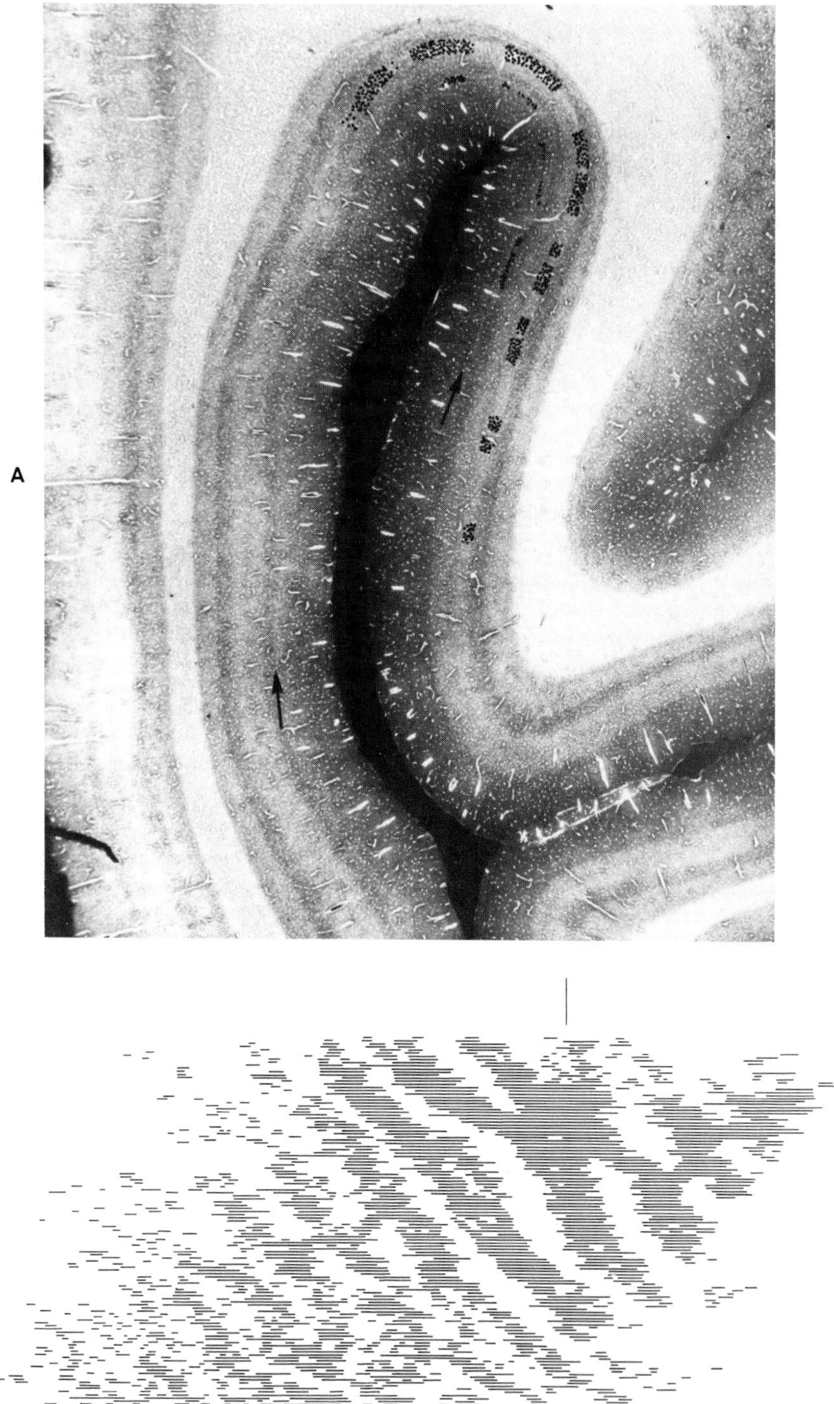

Fig. 16-17. A, Areas of degeneration in monkey striate cortex resulting from small lesion in one dorsal layer of lateral geniculate. Areas showing degenerating terminals of geniculate axons have been dotted in. Heavy degeneration is present in layer *IVC* and lighter degeneration in layer *IVA*. Note that degenerating terminals occur in patches, with clear areas between. **B,** Reconstruction of region showing degenerating lateral geniculate terminals resulting from small lesion in dorsal layer of lateral geniculate. Reconstruction is made from complete series of slides, of which **A** is a sample. Straight black lines represent areas of degeneration in layer *IVC* as seen from cortical surface. (From Hubel, D. H., and Wiesel, T. N.: J. Comp. Neurol. **146:**421, 1972.)

cisely corresponding parts of the contralateral visual field. However, the inputs from the two eyes are often not identical in their relative influence on the cortical cell. Input from one eye or the other is usually dominant; the dominant input produces a greater response to a given stimulus than the input from the other eye does. A group of binocular complex and hypercomplex cells in layers II, III, V, and VI that receive a stronger input from one of the two eyes, along with the cells in layer IV that receive input from the same eye, is called an ocular dominance column.[54]

One can get an idea of the shape of cortical ocular dominance columns by making small lesions in individual layers of the lateral geniculate in the monkey. Such a lesion will destroy a population of cells receiving direct input from only one eye. The degenerating terminals of these destroyed geniculate cells can be stained with special techniques to demonstrate the projection of a small bit of a monocular geniculate layer to the striate cortex. Just as the physiologic recordings have demonstrated regions in layer IV of striate cortex that receive input from only one eye, the experimental anatomy demonstrates that the individual monocular geniculate layers project to small regions of layer IV and not to immediately adjacent regions (Fig. 16-17, *A*). A two-dimensional reconstruction of the region of cortex receiving degenerating terminals following a small lesion in one geniculate lamina can be made from serial sections through the cortical region. Such a reconstruction (Fig. 16-17, *B*) gives a view of layer IV as seen from the surface of the cortex. It is apparent from this reconstruction that a given geniculate layer projects to the cortex in a series of connecting stripes; cells in the geniculate layer immediately beneath the lesioned layer receive input from the other eye and project to the region of cortex between the stripes shown in this reconstruction.[57]

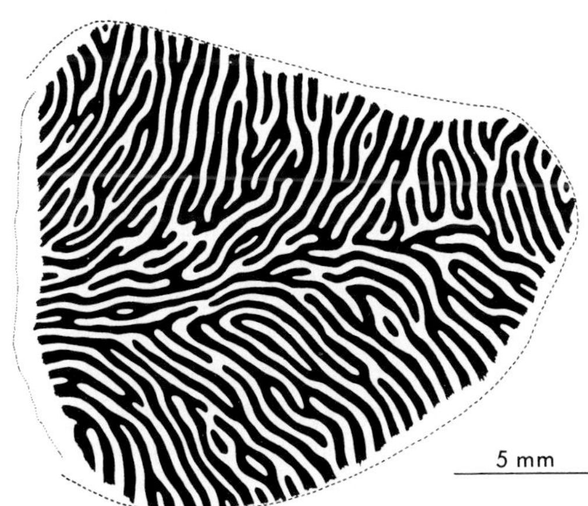

Fig. 16-18. Representation of ocular dominance columns in a relatively large segment of monkey striate cortex of the right occipital lobe. View is of layer *IVC* seen from above; ocular dominance columns for one eye are in black and those for the other eye in white. The foveal representation is to the right. (From Hubel, D. H., and Wiesel, T. N.: Proc. R. Soc. Lond. Biol. **198:**1, 1977.)

Three other anatomic methods have been used to demonstrate ocular dominance columns in monkeys, and one has been used to demonstrate them in human visual cortex.[41] All the methods have provided the same picture; ocular dominance columns are stripes or bands, appearing much like the stripes of a zebra[60,68] (Fig. 16-18). Although the general arrangement of the stripes is similar from one monkey to the next, there is a considerable amount of variation in the specific configuration; thus no two patterns are exactly alike. There does not seem to be any clear-cut relationship between ocular dominance columns and orientation columns, in that the two systems of sheets or bands intersect at many different angles in a seemingly unpredictable fashion.[63] Both ocular dominance and orientation columns would better be termed ocular dominance and orientation stripes or bands, but the word column was used first and will probably persist.

Uniformity of striate cortex. Despite the absence of a fixed angle of intersection between orientation columns and ocular dominance columns, there is a rather striking homogeneity of the striate cortex in all parts of the visual field. As discussed earlier, the representation of central vision in the striate cortex is greatly expanded. Thus the demands for processing the fine detail of central vision are handled by devoting a larger area of cortex that is uniform in thickness to each degree of central visual field than is devoted to each degree of peripheral field. This solution is quite different than the situation in the retina. Since the area of retina per degree of visual field must, for optical reasons, be the same in all parts of the retina, the retina subserving foveal vision is thicker and contains increased numbers of neuronal elements.

As a consequence of the increased area of cortical surface devoted to central vision, an electrode traveling parallel to the cortical layers must move farther to traverse 1 degree of the central visual field than to traverse 1 degree of the peripheral field. An electrode penetration that is perpendicular to the cortical surface encounters receptive fields that are all near a particular point in the visual field, but there is a certain amount of scatter in receptive field location around this point. Thus the representation of the visual field in the striate cortex is not strictly point to point. Any one point in the cortex has cells with receptive fields that cluster about a point in the visual field, but with some scatter. The total area of visual field represented at a given cortical point by the overlapping receptive fields has been called the aggregate field.[58,59] The amount of scatter within an aggregate field bears a close, apparently fixed relationship to the size of the receptive fields encountered at a particular cortical point. Receptive fields are quite small in the cortical representation of the fovea; the size of the aggregate field is a few times the size of an average receptive field. The same relationship holds in the cortical representation of the peripheral visual field; receptive fields are larger and the aggregate field is larger, but it is still a few times the size of an average receptive field. Placement for two-electrode penetrations perpendicular to the cortical layers must be 1 to 2 mm apart to observe a slight separation of their aggregate fields. The distance across the cortical surface that is required to separate the aggregate fields of two-electrode penetrations is the same in all parts of striate cortex, even though the movement within the visual field will be many more degrees in the peripheral field representation than it will be in the central field representation. Thus the volume of cortex contained beneath an area on the cortical surface 1 to 2 mm square contains all the neuronal circuitry required to analyze a particular part of the visual field, whether that part be a fraction of a degree of the central field or many degrees of the peripheral field.

The organization of columns is apparently also constant across striate cortex, with the obvious exception of the small part of cortex that receives input only from the extreme nasal aspect (temporal crescent of visual field, Fig. 16-2) of the contralateral eye and thus does not have ocular dominance columns. Throughout the remainder of striate cortex, ocular dominance columns are about 0.4 mm in width. A shift in receptive field orientation of about 10 degrees occurs approximately every 50 μm in an electrode traverse that is parallel to the cortical layers; thus an electrode travel of about 1 mm would be required to cover 180 degrees of receptive field orientations. Thus within the area of cortex subserving an aggregate field (1 to 2 mm^2) there is ample room for a left and right eye ocular dominance column (0.8 to 1 mm) and all receptive field orientations (1 mm). Hubel and Wiesel,[58-60] whose anatomic and physiologic studies provided these observations, have proposed the term hypercolumn for such areas of cortex. Hypercolumns are the same size throughout striate cortex, but many more are present per degree of visual field in the representation of the central visual field than are present in the representation of the peripheral field to provide for the much greater detail of analysis.

Receptive field hierarchy

The evidence from receptive field analysis indicates that cortical cells are much more exacting in their demands for specification of the visual stimulus than are the cells of the retina and geniculate. Receptive fields may be divided into classes on the basis of these properties, and the classes may be ordered in increasing degrees of complexity or in increasing demands for stimulus specificity. There can be no doubt that receptive field properties of a given cell are determined by the properties of the cells that connect to it as well as by the location, density, and transmitter action of the connections. The problem then becomes one of using the receptive field properties and the available anatomic evidence in an attempt to formulate a model for how visual information is processed. The model that Hubel and Wiesel have proposed suggests that geniculate cells project to simple cortical cells. In the simplest statement of the model, the elongated excitatory region of a simple cell like that in Fig. 16-12, *C,* would be made up of the on centers of several geniculate cell receptive fields in a row; the antagonistic off flanks of the simple cell would then be composed of the overlapping off regions of these same geniculate cells. The supporting evidence for this formulation comes from the observation that simple cells are mainly found in layer IV of the cat and monkey striate regions, where the geniculate axons terminate.[49,54] Furthermore, most cells in layer IV are stellate; recent elegant studies employing iontophoretic labeling of individual cells whose receptive fields have been studied demonstrate that most simple cells are stellate, and that these are most prevalent in layer IV.[106]

In the Hubel-Wiesel model a complex cell is thought of as receiving input from several simple cells of the same orientation whose receptive fields are overlapping to produce the complex cell receptive field.[49] This suggestion is strengthened by the observation that cells of the same orientation are located in the same cortical column[49,54] and that a major feature of cortical anatomy is the wealth of vertical interconnections between cells. In the monkey complex cells are binocular and simple cells are mainly monocular, adding support to the idea that complex cells are at a more advanced stage of processing.

Direct and complete proof of the hierarchical model of cell interconnections in the visual cortex will require highly painstaking experiments that may not be feasible at present. For example, to prove that a given complex cell gets input from simple cells would require a microelectrode study of the complex cell with subsequent search with other microelectrodes for all its inputs, followed ideally by labeling of all the studied cells for subsequent anatomic identification. Evidence that is contrary to the model in its simplest statement is contained in the finding that some complex cells receive at least a portion of their input directly from lateral geniculate cells,[42,100] but the relative strength of the geniculate input, in contrast to input from other cortical cells, remains unsettled. The hierarchical model is derived from a division of cells into types based on certain receptive field properties, especially the organization of excitatory and inhibitory regions. It is conceivable that other models could be derived if the cell classifications were on the basis of receptive field properties like motion sensitivity, speed of movement, or directionality. No doubt the Hubel-Wiesel model will require modification as the study of cortical physiology and anatomy continues. Whatever the ultimate validity of the model proves to be, it and the observations that led up to it will undoubtedly be an important milestone in the understanding of the function of the visual cortex.

EXTRASTRIATE VISUAL REGIONS

For many years, the traditional view of the organization of the visual system held that visual information reached the cortex solely by means of the geniculostriate pathway, and that striate cortex projected in turn to two adjacent cortical zones in the occipital lobe, areas 18 and 19, for subsequent processing. Recent advances in the techniques of neuroanatomy and neurophysiology have produced a number of alterations and revisions of this concept.*

Multiple visual field representations

The cortex of the occipital lobe immediately anterior to the striate cortex contains not two but many representations of the contralateral hemifield. These have been most extensively demonstrated in the owl monkey (Fig. 16-19), which has a cerebral hemisphere that is virtually free of the variable and complex convolutions that complicate the analysis of the rhesus monkey visual regions. Striate cortex (also called visual area 1, or V1) contains a complete representation of the contralateral visual hemifield, as described earlier. It is bordered anteriorly by a second representation of the visual field, V2. The representation of the vertical line passing through the center of the fovea, called the vertical meridian, is at the border between these two regions. The field representation in V2 is

*See references, 1, 17, 105, 119-121, and 126.

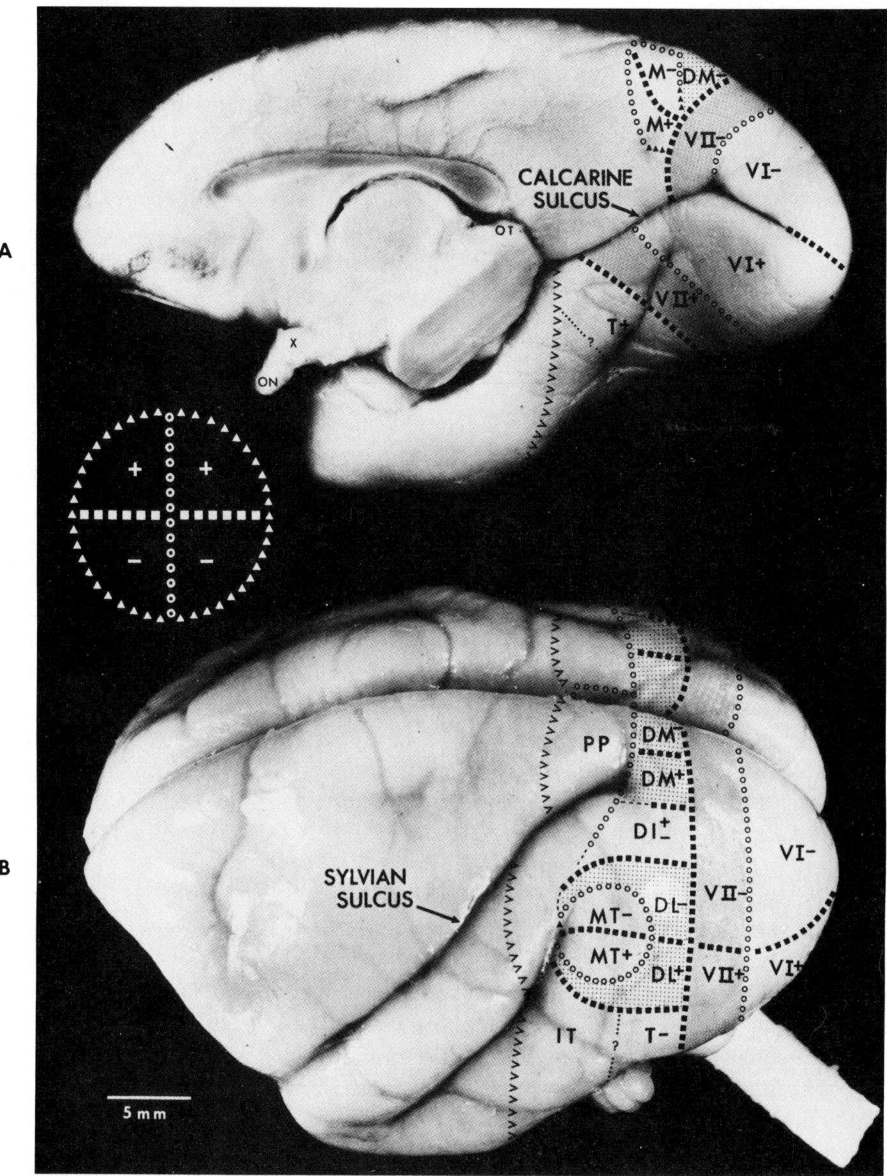

Fig. 16-19. Striate and extrastriate visual regions in the cerebral cortex of the owl monkey. **A,** Medial aspect of the right hemisphere; **B,** dorsolateral aspect of the brain seen from the left side. The perimeter chart between the photographs shows the symbols used to indicate the visual field representations in the cortex; open circles denote the vertical meridian, squares the horizontal meridian, and triangles the extreme periphery. Plus signs indicate the upper quadrants and minus signs the lower quadrants. The row of Vs indicates the limits of visually responsive cortex. *VI* (or V1), Striate cortex; *VII* (or V2) second visual area; *DI,* dorsointermediate area; *DL* dorsolateral crescent; *IT,* inferotemporal cortex; *M,* medial area; *MT,* middle temporal area; *ON,* optic nerve; *OT,* optic tectum; *PP,* posterior parietal cortex; *T,* tentorial area; *X,* optic chiasm. (From Allman, J.: Prog. Physiol. Psychol. **7:**1, 1977.)

split along the representation of the horizontal line passing through the fovea, called the horizontal meridian, which forms the border between V2 and the regions anterior to it (Fig. 16-19). In the cortex anterior to V2 there are numerous representations of the contralateral hemifield, some of them complete and some of them partial, which have been grouped together as the "third tier" of visual regions. They extend beyond the occipital lobe, into the posterior parietal, middle, and inferior temporal cortex.[1] The relationship between V1 and V2 is similar in the rhesus monkey and in many other mammals.[1,105,110] Multiple representations of the visual field, called V3, V3A, the V4 complex, and MT are present in the cortex anterior to V2 in the rhesus monkey. Although several of these areas were discovered before extrastriate visual regions were described in the owl monkey, the complex geometry of the convoluted rhesus monkey cortex has made it more difficult to define their precise retinotopic organizations. It is quite likely that eventually additional cortical areas will be discovered in as yet unexplored regions of rhesus monkey occipital cortex.[105,107,126]

Extrastriate visual regions receive visual input from two principal sources: thalamic connections from the pulvinar and corticocortical connections from striate cortex and from other extrastriate visual regions. The pulvinar of the rhesus monkey, a complex thalamic nucleus with several subdivisions, receives a small projection directly from the retina[13] and more extensive visual input from the superior colliculus. The pulvinar projects extensively to extrastriate visual cortical regions and also has a sparse projection to striate cortex[5,85,86,103]; many of these cortical regions project in reciprocal fashion back to the pulvinar. The striate cortex, the only cortical area in primates that receives direct input from the lateral geniculate, projects to many of the extrastriate visual regions, which in turn are extensively interconnected with striate cortex and with each other.[17,119,120,121]

Functional specialization

Delineation of the multiple representations of the visual field in the extrastriate cortical regions has thus far outpaced progress in understanding the functional significance of these regions. It would appear from present evidence that at last some of the regions are specialized for processing particular aspects of visual information. This specialization is evident in the variations from one region to the next in the relative prevalence of cells with particular receptive field properties. For example, opponent color cells are grouped together in clusters and are quite prevalent in the rhesus monkey V4 complex[122,125]; they are much less prevalent in the other extrastriate regions that have been studied. A few patients have been reported with cortical lesions that produced deficient color vision; other aspects of visual function such as acuity and stereopsis were preserved, suggesting that a region or regions specialized for processing color information may have been selectively damaged. Most of these patients have had lesions involving extrastriate visual cortex in the anteromedial aspect of the occipital lobe, at the occipitotemporal junction.[75,89]

The extrastriate region called MT in both rhesus monkey and owl monkey may be specialized in some way for the analysis of motion within the visual scene, since a striking proportion of cells in this region show a strong preference for stimuli moving in a particular direction. Many of these cells respond best to stimuli moving across the animal's field of view in one direction but not in other directions; other cells respond preferentially to stimuli moving toward or away from the animal.[123,124]

Another example of apparent regional specialization in extrastriate visual cortex is the prevalence of cells in V2 and probably also V3 that may be processing information for stereoscopic depth perception.[55,91] These cells, called binocular disparity cells, have receptive fields that are displaced a small amount horizontally from each other in the retinas of the two eyes. They respond poorly or not at all when the receptive field in one eye is stimulated alone, but respond quite well when both receptive fields are stimulated simultaneously. For a given disparity cell, the receptive field from the right eye may be displaced to the right of the receptive field from the left eye; such a cell is said to have uncrossed disparity. A cell with the

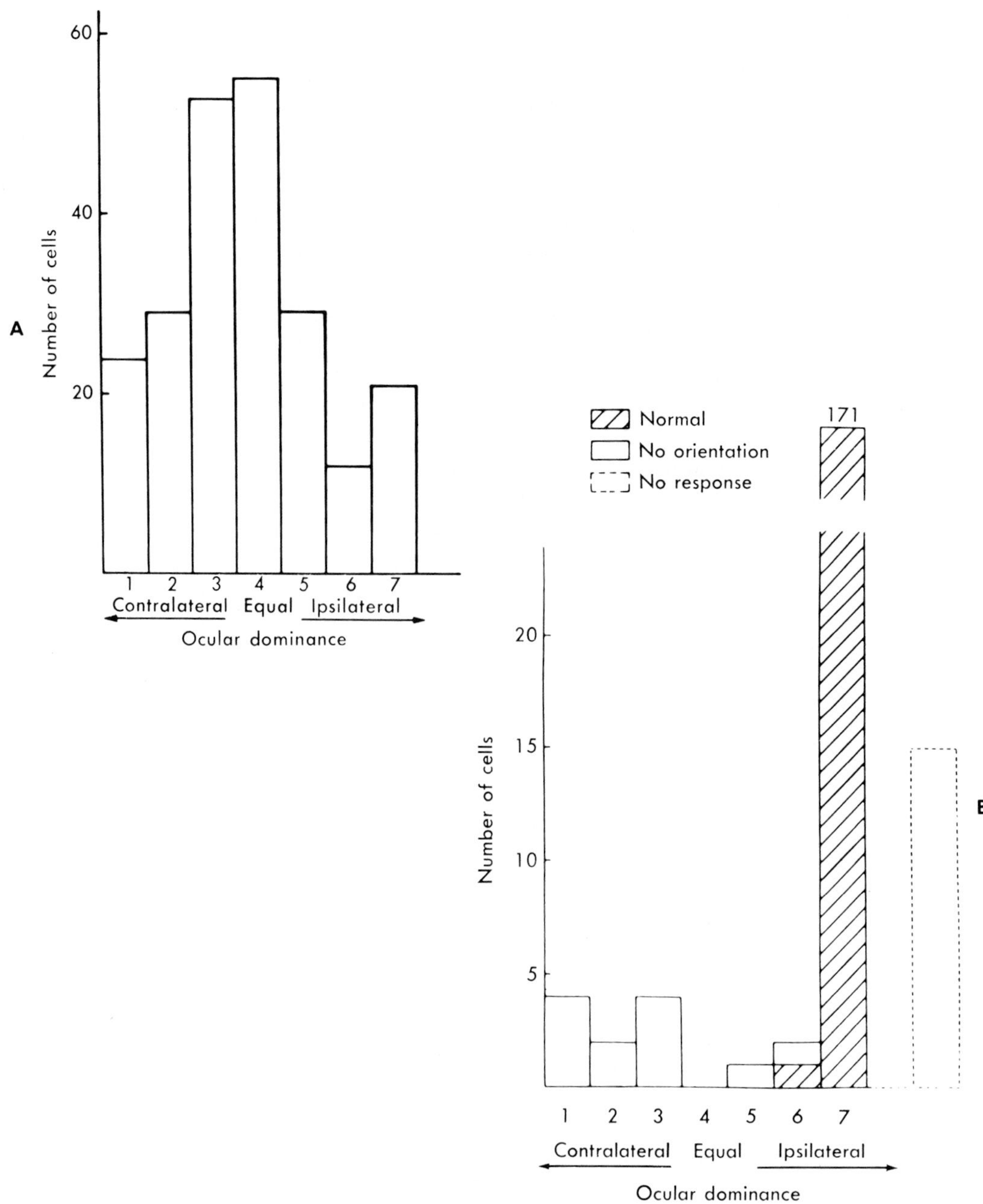

Fig. 16-20. A, Ocular dominance distribution of 223 cells recorded in striate cortex of normal adult cats. Each cell studied was assigned to an ocular dominance group, number *1* through *7*, depending on relative input from the two eyes. Those in group *1* receive input only from contralateral eye; those in group *4* receive approximately equal input from the two eyes; and those in group *7* receive input only from ipislateral eye. **B,** Ocular dominance distribution of 199 cells recorded in left visual cortex of five monocularly deprived kittens that had right eye closed from birth. Recordings were made when kittens were 8 to 14 weeks old. Shaded bars represent cells that had normal, highly specific receptive field properties. Nonshaded bars represent cells that lacked normal orientation specificity. Bar to far right indicates number of cells encountered that did not respond at all to visual stimuli. (**A** from Wiesel, T. N., and Hubel, D. H.: J. Neurophysiol. **26:**1003, 1963; **B** from Wiesel, T. N., and Hubel, D. H.: J. Neurophysiol. **28:**1029, 1965.)

right eye receptive field to the left of the left eye field has crossed disparity (Chapter 21). The amount of retinal disparity also varies from cell to cell. Cells with nearly the same horizontal disparity in receptive fields tend to be grouped together in the cortex in columns. Since the receptive fields in the two eyes are displaced slightly from one another in the visual field when the foveas of the two eyes are fixed on a given point, a disparity cell responds best when the stimulus is either closer or farther away than the plane of fixation, depending on the receptive field displacement for the particular cell. Such units with binocular disparity in the receptive fields could thus be functioning in stereoscopic depth perception.[56] It is also possible that they have a role in convergent and divergent eye movements.

VISUAL DEPRIVATION

Although we are far from a complete understanding of the interconnections between cells in the lateral geniculate and visual cortex, it is clear from the analysis of receptive fields that these interconnections are highly organized and precise. Many aspects of the complex organization of the geniculate and striate cortex are complete before birth and must therefore be determined largely by genetic factors and environmental influences that do not include visual sensory experience. On the other hand, during the early postnatal maturation of the cat and monkey a number of features of geniculate and cortical physiology can be altered by producing major alterations in the animal's visual experience. Several of the alterations in sensory experience that have been produced experimentally stem directly from various pathologic conditions that occur in humans. These experiments were initially designed in an attempt to understand the alterations in visual function that occur as a consequence of strabismus, congenital cataract, corneal scarring, ptosis, astigmatism, and other conditions that disturb visual sensory experience early in life.

The exact degree to which the complicated organization of receptive fields in the visual system is complete at the time of birth is not yet fully known, in part because of the difficulty in obtaining and working with newborn animals. The eyes of kittens open about 8 days after birth; at that time the ocular media are not clear, and it is rather difficult to anesthetize and maintain the animal for lengthy experiments. Nevertheless, a number of studies have been carried out in newborn kittens in an attempt to define the interaction between visual experience and what might be called "preprogramed" or genetically determined maturational events.[30,50,90]

For about the first 3 weeks of life, the receptive fields of cells in striate cortex develop in the same fashion whether the kitten is reared in complete darkness or in a laboratory environment with normal day-night cycles. Cells with orientation specificity are evident very soon after the eyes open, but a large proportion of cells have nonoriented receptive fields or else do not respond to visual stimuli at all. About half the cells have receptive fields that are binocular; the remainder are monocular, and most are responsive to stimulation of the contralateral eye.

Distinct differences between dark-reared and normally reared kittens become evident at about 4 weeks of age. Cells with finely tuned orientation specificity become more and more prevalent in normal kittens, whereas dark-reared kittens not only fail to develop more orientation specific cortical neurons, they also apparently lose those that were present earlier in life. On the other hand, the binocular input to most cortical cells that is characteristic of adult animals develops in both dark-reared and normal animals.[30] Thus it would appear that visual experience plays little or no role in the development of striate cortex for the first 3 weeks after the kitten is born, but that after that time visual experience is necessary for the continued development of oriented receptive fields and for the maintainance of those already present. The connections responsible for binocular input to cortical cells continue to develop in the absence of visual experience as long as both eyes are deprived; monocular deprivation, on the other hand, produces very marked alterations in both the lateral geniculate and the striate cortex.

Monocular deprivation

Suturing one eyelid closed in a newborn kitten or monkey deprives the animal of form vision in that eye and also appreciably reduces the amount of light reaching the retina.

After several weeks of deprivation the cell bodies in the lateral geniculate laminae that receive input from the deprived eye are smaller than those in the laminae receiving normal input,[65,109,113] but the cells in the deprived laminae continue to demonstrate the normal center-surround receptive field organization.[113] If one eye is occluded by a large translucent contact lens rather than by lid suture, the eye is deprived of form vision but receives nearly normal amounts of light. Cells in the deprived geniculate laminae are again smaller than normal, but the difference is less pronounced.[113] Of considerable interest is the observation that one class of cat geniculate cells, the Y-cells described earlier, are found much less frequently in layers receiving input from the deprived eye than in layers receiving normal input, while the X cells are normal in all layers.[97]

Cells of the striate cortex are more drastically affected by monocular deprivation. A cat raised for the first 2 to 3 months of life with one eye occluded will lose most of the excitatory input from that eye onto cells of the striate cortex.[114] This finding is best demonstrated by comparing the relative influence of the two eyes on cortical cells in the normal and monocularly deprived cat. As pointed out earlier, most cells in the striate cortex of the normal cat receive input from both eyes, although the excitatory input from one eye is frequently stronger than the input from the other eye. The eye providing the stronger input is called the dominant eye; a histogram showing the number of cells receiving the various relative inputs from the two eyes is called an ocular dominance histogram. The histogram for a series of cells from normal adult cat striate cortex is shown in Fig. 16-20, *A*, where it can be seen that the majority of cells are binocular, receiving input that is either approximately equal from the two eyes (group 4) or else dominated by the contralateral or ipsilateral eye (groups 2, 3, 5, and 6). A smaller number of cortical cells in the cat are monocular, receiving input from only the contralateral or ipsilateral eye (groups 1 and 7).

The ocular dominance histogram derived from tabulating cells in the striate cortex of monocularly deprived cats is quite abnormal (Fig. 16-20, *B*); most cells that respond receive input only from the nondeprived eye, and only a small number have binocular input. A few cells are identified that do not respond to stimuli in either eye. Both cats and monkeys with monocular visual deprivation for the first 2 to 3 months of life have severe defects in visual acuity when using the deprived eye, even when tested after the deprived eye has been reopened for several months.[23,109] Monocular deprivation early in life thus causes the normal binocular input to cortical cells that is present at birth to be lost; the loss of input from the deprived eye is manifest in the poor visual acuity evident on behavioral testing.

Very striking changes in the morphology of the ocular dominance columns of striate cortex also occur with monocular deprivation. The terminals of lateral geniculate afferents receiving input from the nondeprived eye cover much broader bands than normal in layer IVC of rhesus monkey striate cortex, (Fig. 16-21, *A* and *B*) whereas lateral geniculate terminals from the deprived eye are restricted to bands that are narrower than normal[61,62] (Fig. 16-21, *C*). Although not directly proved, the differences in cell body size between deprived and nondeprived lateral geniculate laminae are probably a reflection of this change in cortical ocular dominance columns; cells with extensive axon terminals presumably have larger cell bodies than cells with more restricted terminals.

Competition

A considerable body of evidence has been developed indicating that the loss of input from the deprived eye to cells in the striate cortex is the result of some type of competition between the inputs from the deprived and normal eyes. An early indication that such competition is playing a role came from the observations of kittens with both eyes sutured closed from birth. On the basis of the results from

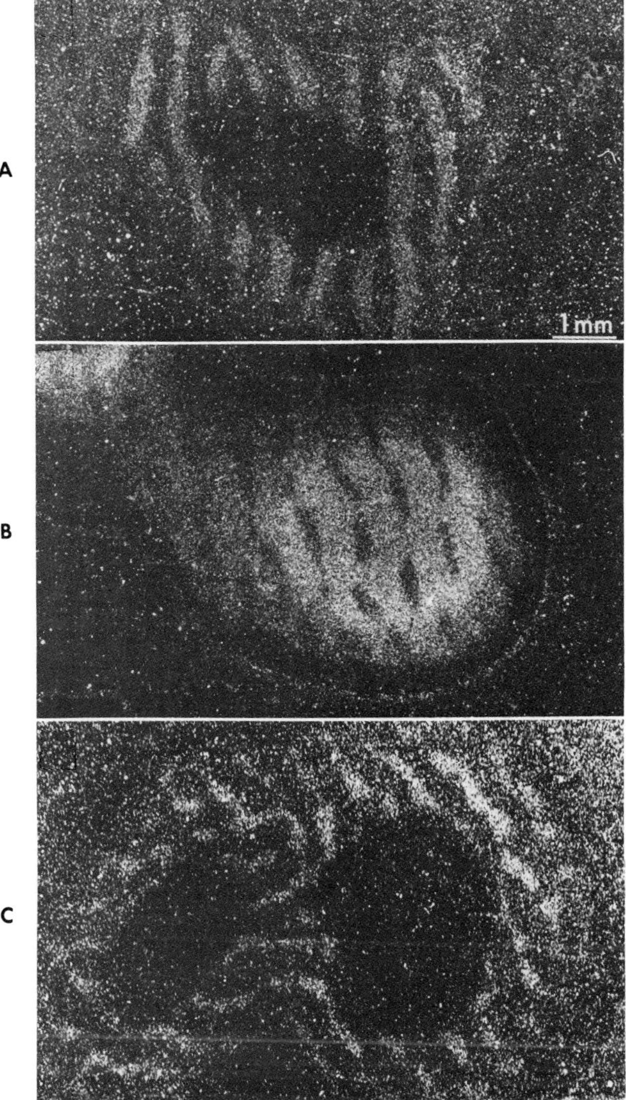

Fig. 16-21. A, Ocular dominance columns in a layer *IVC* of a normal rhesus monkey. One eye was injected with a large amount of radioactive proline and fucose, which were transported up the optic nerve to the lateral geniculate. Some of the label moved from the optic nerve terminals to the lateral geniculate cells innervated by the injected eye and was transported to the cortical terminals of these lateral geniculate cells. The plane of section is tangential through layer *IVC;* the white stripes are the silver grains that result from exposure to the radioactive label seen in dark-field microscopy. Bands of labeled terminals (white stripes) are approximately the same width as the bands of unlabeled terminals (dark stripes). **B,** Similar tangential section through layer *IVC* of a monkey whose right eye was closed at 2 weeks; the nondeprived eye was injected as in **A** several months later. Terminals of axons from the nondeprived geniculate cells occupy much wider bands than normal. **C,** Similar section through layer *IVC* of another monkey that had the right eye closed at 3 weeks of age, and the same eye injected several months later. Terminals of axons from the deprived geniculate cells occupy bands that are much narrower than normal. (From Hubel, D. H., Wiesel, T. N., and LeVay, S.: Cold Spring Harbor Symp. Quant. Biol. **40:**581, 1976.)

monocularly deprived kittens, it might be predicted that binocularly deprived animals would have large regions of striate cortex containing cells that do not respond at all to visual stimuli. Such is not the case; cells responsive to visual stimuli are present throughout the striate cortex of binocularly deprived kittens. The cells have a nearly normal ocular dominance histogram (Fig. 16-22), and many have normal receptive field properties. There is also a group of cells that do not have normal orientation specificity, and another group that is unresponsive to visual stimuli (Fig. 16-22). Thus the effect of eye closure on the visual cortex depends on the status of the other eye. It is as if the normal eye has a competitive advantage in the monocular deprivation situation, while neither eye has the advantage when both are closed.

Strabismus

Support for the idea that competition between inputs plays a role in the maintenance of cortical inputs comes from two types of experiment that disrupt normal binocular input, but do not produce severe visual deprivation. An artificial exotropia produced by cutting the medial rectus of a newborn kitten results in the two eyes looking in different directions, and therefore the afferent information arriving at a given cortical cell from the two eyes is not the same. In such

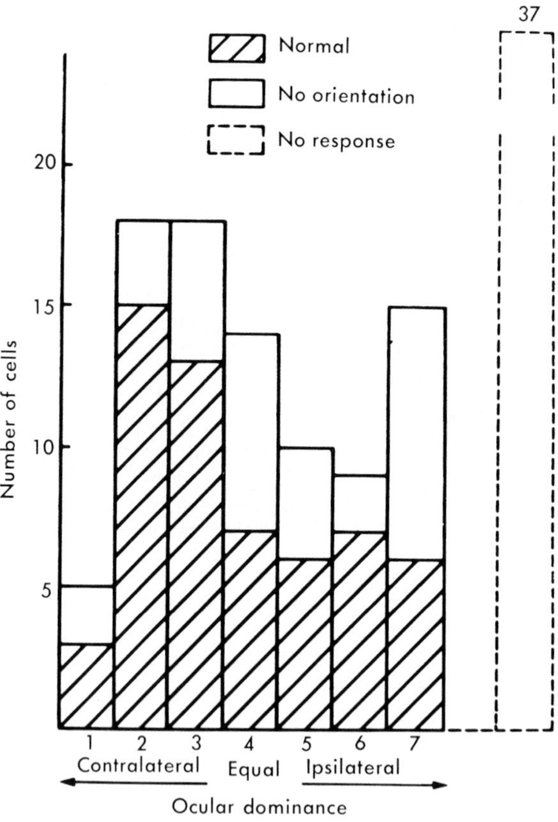

Fig. 16-22. Ocular dominance distribution for 126 cells in striate cortex of four cats with binocular lid suture from time of normal eye opening to 2½ to 4 months of age. Shaded bars indicate cells with normal receptive fields; open bars indicate cells lacking normal orientation specificity. Bar on far right gives number of cells that did not respond at all to visual stimuli. Ocular dominance groups are same as in Fig. 16-20. (From Wiesel, T. N., and Hubel, D. H.: J. Neurophysiol. **28:**1029, 1965.)

a situation cortical cells lose their normal binocular input and come to be influenced almost exclusively by one eye or the other, but not by both (Fig. 16-23).[115]

Cats raised with the artificial divergent strabismus produced by cutting one medial rectus muscle appear to have normal visual acuity in either eye.[115] They might be expected to have severely impaired binocular depth perception, but this has not been tested. Monkeys have been raised from birth with an artificial esotropia produced by cutting the lateral rectus and shortening the medial rectus of one eye. The visual acuity in the deviated eye is greatly reduced,[108] and the number of cells in striate cortex responding to stimulation of the deviating eye is also reduced.[3]

An ocular dominance histogram similar to that of the cat with artificial exotropia may also be produced by covering first one eye then the other with a translucent contact lens on alternate days. This procedure again disrupts the normal correspondence of the input onto cortical cells from the two eyes without producing severe deprivation in either eye.[52] Cortical cells respond to one eye or the other, but not both, and the cats have very poor stereoscopic depth perception.[87] Thus to maintain the normal binocular input to cortical cells that is present soon after birth, the input from the two eyes must

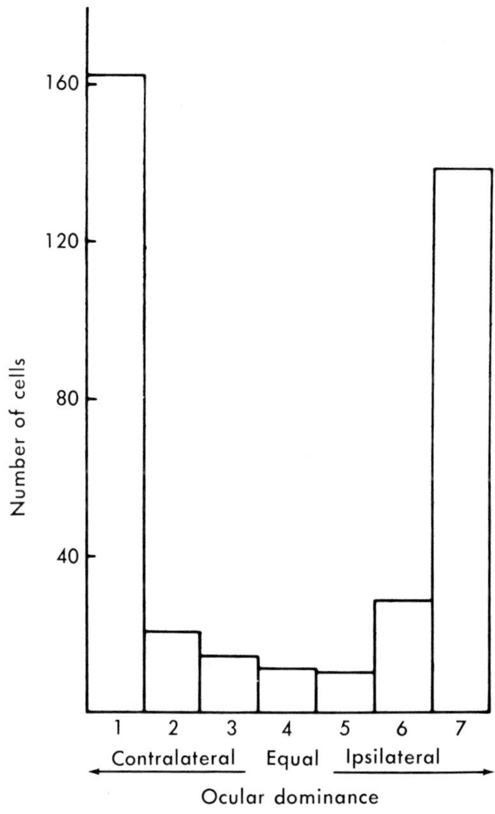

Fig. 16-23. Ocular dominance distribution for 384 cells recorded in striate cortex of four cats with artificial strabismus produced by cutting medial rectus muscle at time of normal eye opening. Recordings were made at 3 months of age in three cats and at 1 year in the fourth. All cells have normal receptive field characteristics, but most receive input from only one eye. Ocular dominance groups as in Fig. 16-20. (From Hubel, D. H., and Wiesel, T. N.: J. Neurophysiol. **28:**1041, 1965.)

be similar; if it is not, one eye or the other, presumably the one that was normally dominant in the binocular situation, comes to dominate completely, and the input from the other eye is lost.[52]

Finally, the competition hypothesis is strengthened by detailed histologic observations of the lateral geniculate in cats with monocular deprivation. As mentioned earlier, the cell bodies in the geniculate layers receiving input from the deprived eye are smaller than normal. The smallness of cell bodies is thought to be at least in part the result of the deprived geniculate cells losing out in the competition for synaptic sites on cortical cells, and therefore having fewer synaptic endings and axonal branches to support. If that were the case, then geniculate cells that do not have to compete with other geniculate cells for cortical synaptic sites should be of normal size. This prediction has been shown to be correct for the cells in the monocular region of the cat geniculate layer A that receive input from the most nasal region of the contralateral retina. Since this region is monocular, there are no cells in layer A1 that receive input from corresponding parts of the visual field, and thus no cells to compete for synaptic sites in the visual cortex. After monocular deprivation for the first 3 months of life the cells in the medial aspect of layer A of the contralateral lateral geniculate are small, but the cells in the lateral monocular segment are of normal size.[38] Another way to eliminate competition between the two eyes is to produce a small lesion in the temporal retina of the normal eye in a kitten with the lid of the other eye sutured. In this situation a small group of cells in geniculate lamina A1 on the same side as the retinal lesion undergoes severe atrophy because its retinal input has been destroyed by the lesion. Cells of lamina A are, of course, small because of the contralateral lid suture, but the cells directly above the group of severely atrophic cells of lamina A1 are of nearly normal size.[37]

Orientation deprivation

The visual cortex is also subject to alterations of receptive field properties other than ocular dominance during the early months of life. It is possible to restrict the orientation of visual stimuli a kitten is allowed to experience by putting goggles with vertical or horizontal stripes over the kitten's eyes[99] or by limiting the kitten's visual experience to several hours a day in a large tube with vertical or horizontal stripes covering its inside surface.[7,8] The orientation specificity of neurons in the visual cortex is drastically altered by such restrictions; after 5 months of viewing only horizontal stripes most of the cells in the striate cortex have the preferred orientation of their receptive field near the horizontal axis (Fig. 16-24), and no receptive fields have vertical orientation. The opposite is true for a kitten whose only visual experience has been vertical stripes (Fig. 16-24).[7] It would appear that the orientation of receptive fields in the visual cortex has been changed toward the orientation of the stripes viewed. The other alternative, that cells with receptive field orientations perpendicular to the stripes viewed are simply lost because of underactivity, does not seem to hold, since there are no areas of cortex that contain cells unresponsive to visual stimuli.[7]

An analogous, although less severe, form of orientation deprivation occurs in humans who have a high degree of uncorrected astigmatism during childhood. In astigmatism the refractive power of the eye differs in various meridians; the axes of maximum and minimum power, called the principal meridians, are perpendicular to each other. Contours parallel to one of the principal meridians are therefore continually blurred. When the astigmatism is corrected, these individuals continue to have reduced acuity for contours in the orientation that was most severely defocused before correction.[29,80]

Critical period

There is a clearly defined period during the early postnatal life of kittens and monkeys during which the visual system is susceptible to alteration by the types of visual deprivation just described. The susceptibility to the effects of monocular eye closure begins in the cat at the beginning of the fourth postnatal week. The kitten remains highly susceptible until sometime between the sixth and eighth weeks; the

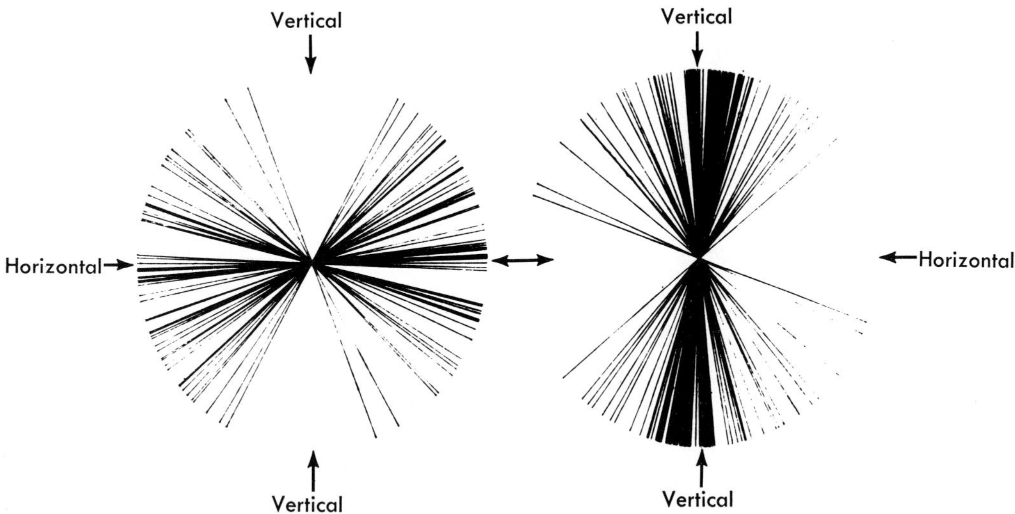

Fig. 16-24. On left are optimal receptive field orientations for 52 neurons in striate cortex of kitten that viewed only horizontal stripes during first 5 months of life. On right are receptive field orientations for 72 cortical neurons from cat that viewed only vertical stripes. Each line represents optimum orientation of single receptive field. (From Blakemore, C., and Cooper, G. F.: Nature **228:**477, 1970.)

susceptibility then declines gradually over the next 4 weeks. During the period of high susceptibility (fourth and fifth weeks) monocular eye closure for as little as 1 day produces a striking reduction in the number of cortical cells that respond to the deprived eye.[55,82] During the period when susceptibility is declining (6 to 12 weeks), longer periods of deprivation are required to produce similar results. Monocular deprivation for over 1 year in an adult cat produces no effect.[55] A similar period of susceptibility, or critical period, exists in the rhesus monkey, lasting from birth to about 12 weeks. A severe loss in visual acuity may result from eye closures as short as 2 weeks during this period.[108]

The critical period in humans has not been as clearly defined as in these animals, but it can be estimated from the study of carefully selected clinical cases. Significant loss of visual acuity can be demonstrated in children who have had monocular deprivation in early childhood as a result of lid closure, patching, or traumatic cataract. The critical period for susceptibility to such deprivation appears to last to at least age 4, and perhaps somewhat longer.[108] Monocular deprivation periods as short as 1 to 4 weeks in a child under 1 year of age have been shown to produce significant loss in visual acuity.[2,108] The critical period for the development of binocularity in humans has been estimated by studying infants who developed convergent strabismus at various ages; sensitivity to this form of deprivation begins in humans several months after birth, peaks between 1 and 3 years of age, and probably continues for several more years.[4]

Clinical implications

The experimental studies of visual deprivation have several reasonably clear implications for the treatment of children with ocular abnormalities. Even though the critical period for susceptibility to visual deprivation is not precisely known in humans, both the experimental studies and the studies of human case material indicate that pathologic conditions such as ptosis and cataract that cause monocular visual deprivation ought to be corrected as soon as possible. The visual impairment that results from the loss of cortical input from the deprived eye may not be reversible if the deprivation occurs briefly during the period of high susceptibility or if it is present for a long time during the period of reduced susceptibility. A further clear implication of both the experimental and clinical studies is that extreme caution is necessary when patching or lid suture is considered as a means of treatment in infants and young children.

The studies of artificially produced strabismus in animals may not be quite so directly applicable to strabismus in children, since it is not clear that producing strabismus in an animal by cutting an eye muscle is analogous to naturally occurring strabismus. Nevertheless, certain parallels are evident. Since in both artificial and naturally occurring strabismus the two eyes are not fixating on the same object, individual cortical cells will not receive corresponding signals from the two eyes. In the cat with exotropia this leads to the loss of input from one eye or the other, with the result that most cortical cells become monocular rather than binocular. If the result in human visual cortex is similar, one might expect difficulty in binocular fusion or the absence of stereoscopic depth perception when the strabismus is finally corrected. The infant monkey with an artificial esotropia develops a loss of visual acuity, or amblyopia, that may be analogous to the amblyopia that frequently develops in children with strabismus. The mechanism may well be the partial loss of cortical connections from the in-turning eye. Early surgical correction of strabismus in children is therefore indicated; once amblyopia develops, the best hope for recovery is early treatment at a stage when the cortex is still susceptible to alteration.

REFERENCES

1. Allman, J.: Evolution of the visual system in the early primates, Prog. Physiol. Psychol. **7:**1, 1977.
2. Awaya, S., et al.: Amblyopia in man, suggestive of stimulus deprivation amblyopia, Jpn. J. Ophthalmol. **17:**69, 1973.
3. Baker, F. H., Grigg, P., and von Noorden, G. K.: Effects of visual deprivation and strabismus on the response of neurons in the visual cortex of the monkey, including studies on the striate and prestriate cortex in the normal animal, Brain Res. **66:**185, 1974.
4. Banks, M. S., Aslin, R. N., and Letson, R. D.: Sensitive period for the development of human binocular vision, Science **190:**675, 1975.
5. Benevento, L. A., and Rezak, M.: The cortical projections of the inferior pulvinar and adjacent lateral pulvinar in the rhesus monkey *(Macaca mulatta):* an autoradiographic study, Brain Res. **108:** 1, 1976.
6. Bishop, P. O., et al.: The determination of the projection of the visual field on the lateral geniculate nucleus in the cat, J. Physiol. (Lond) **163:**503, 1962.
7. Blakemore, C., and Cooper, G. F.: Development of the brain depends on the visual environment, Nature **228:**477, 1970.
8. Blakemore, C., and Mitchell, D. E.: Environmental modification of the visual cortex and the neural basis of learning and memory, Nature **241:**467, 1973.
9. Brodmann, K.: Beitrage zur histologischen Localisation der Grosshirnrinde dritte Mitteilung: die Rinderfelder der niederen Affen, J. Psychol. Neurol. Leipzig **4:**177, 1905.
10. Brodmann, K.: Vergleichende Lokalisation lehre der Grosshirnrinde, Leipzig, East Germany, 1909, Johann Ambrosius Barth Verlagsbuchhandlung.
11. Brouwer, B., and Zeeman, W. P. C.: The projection of the retina in the primary optic neurone in monkeys, Brain **49:**1, 1926.
12. Burke, W., and Cole, A. M.: Extraretinal influences on the lateral geniculate nucleus, Rev. Physiol. Biochem. Pharmacol. **80:**106, 1978.
13. Campos-Ortega, J. A., Hayhow, W. R., and Cluver, R. F.: A note on the problem of retinal projections to the inferior pulvinar of pirmates, Brain Res. **22:** 126, 1972.
14. Clark, W. E. L., and Sunderland, S.: Structural changes in the isolated visual cortex, J. Anat. **73:**563, 1939.
15. Cleland, B. G., Dubin, M. W., and Levick, W. R.: Sustained and transient neurones in the cat's retina and lateral geniculate nucleus, J. Physiol. (Lond) **217:**473, 1971.
16. Cleland, B. G., Dubin, M. W., and Levick, W. R.: Simultaneous recording of input and output of lateral geniculate neurones, Nature [New Biol.] **231:**191, 1971.
17. Cragg, B. G.: The topography of the afferent projections in the circumstriate visual cortex of the monkey studied by the Nauta method, Vision Res. **9:**733, 1969.
18. Daniel, P. M., and Whitteridge, D.: The representation of the visual field on the cerebral cortex in monkeys, J. Physiol. (Lond) **156:**203, 1961.
19. Daw, N. W.: Neurophysiology of color vision, Physiol. Rev. **53:**571, 1973.
20. Daw, N. W., and Pearlman, A. L.: Cat colour vision: one cone process or several? J. Physiol. (Lond) **201:**745, 1969.
21. Daw, N. W., and Pearlman, A. L.: Cat color vision: evidence for more than one cone process, J. Physiol. (Lond) **211:**125, 1970.
22. De Valois, R. L., Jacobs, G. H., and Jones, A. E.: Response of single cells in primate red-green color vision system, Optik **20:**87, 1963.
23. Dews, P. B., and Wiesel, T. N.: Consequences of monocular deprivation on visual behavior in kittens, J. Physiol. (Lond) **206:**437, 1970.
24. Dow, B. M.: Functional classes of cells and their laminar distribution in monkey visual cortex, J. Neurophysiol. **37:**927, 1974.

25. Dow, B. M., and Gouras, P.: Color and spatial specificity of single units in rhesus monkey foveal striate cortex, J. Neurophysiol. **36:**79, 1973.
26. Dreher, B., Fukada, Y., and Rodieck, R. W.: Identification, classification and anatomical segregation of cells with X-like and Y-like properties in the lateral geniculate nucleus of old world primates, J. Physiol. **258:**433, 1976.
27. Enroth-Cugell, C., and Robson, J. G.: The contrast sensitivity of retinal ganglion cells of the cat, J. Physiol. (Lond) **187:**517, 1966.
28. Famiglietti, E. V., Jr., and Peters, A.: The synaptic glomerulus and the intrinsic neuron in the dorsal lateral geniculate nucleus of the cat, J. Comp. Neurol. **144:**285, 1972.
29. Freeman, R. D., Mitchell, D. E., and Millodot, M.: A neural effect of partial visual deprivation in humans, Science **175:**1384, 1972.
30. Fregnac, Y., and Imbert, M.: Early development of visual cortical cells in normal and dark reared kittens: relationship between orientation selectivity and ocular dominance, J. Physiol. **278:**27, 1978.
31. Fukada, Y.: Receptive field organization of cat optic nerve fibers with special reference to conduction velocity, Vision Res. **11:**209, 1971.
32. Garey, L. J., and Powell, T. P. S.: An experimental study of the termination of the lateral geniculo-cortical pathway in the cat and monkey, Proc. R. Soc. Lond. (Biol.) **179:**41, 1971.
33. Gilbert, C. D.: Laminar differences in receptive field properties of cells in cat primary visual cortex, J. Physiol. **268:**391, 1977.
34. Gilbert, C. D., and Kelly, J. P.: The projections of cells in different layers of the cat's visual cortex, J. Comp. Neurol. **163:**81, 1975.
35. Glaser, J. S.: Neuro-ophthalmology, New York, 1977, Harper & Row, Publishers, Inc.
36. Guillery, R. W.: Patterns of synaptic interconnections in the dorsal lateral geniculate nucleus of cat and monkey: a brief review, Vision Res. (suppl. 3), p. 211, 1971.
37. Guillery, R. W.: Binocular competition in the control of geniculate cell growth, J. Comp. Neurol. **144:**117, 1972.
38. Guillery, R. W., and Stelzner, D. J.: The differential effects of unilateral lid closure upon monocular and binocular segments of the dorsal lateral geniculate nucleus in the cat, J. Comp. Neurol. **139:**413, 1970.
39. Harrington, D. O.: The visual fields: a textbook and atlas of clinical perimetry, ed. 4, St. Louis, 1976, The C. V. Mosby Co.
40. Hickey, T. H., and Guillery, R. W.: Variability of laminar patterns in the human lateral geniculate nucleus, J. Comp. Neurol. **183:**221, 1979.
41. Hitchock, P. F., and Hickey, T. L.: Ocular dominance columns: evidence for their presence in humans Brain Res. **182:**176, 1980.
42. Hoffmann, K., and Stone, J.: Conduction velocity of afferents to cat visual cortex: a correlation with cortical receptive field properties, Brain Res. **32:**460, 1971.
43. Hoffmann, K., Stone, J., and Sherman, S. M.: Relay of receptive-field properties in dorsal lateral geniculate nucleus of the cat, J. Neurophysiol. **35:**518, 1972.
44. Hoyt, W. F.: Correlative functional anatomy of the optic chiasm, Clin. Neurosurg. **17:**189, 1969.
45. Hoyt, W. F., and Luis, O.: Visual fiber anatomy in the infra-geniculate pathway of the primate: uncrossed and crossed retinal quadrant fiber projections studied with Nauta silver stain, Arch. Ophthalmol. **68:**94, 1962.
46. Hoyt, W. F., and Luis, O.: The primate chiasm: details of visual fiber organization studied by silver impregnation techniques, Arch. Ophthalmol. **70:**69, 1963.
47. Hubel, D. H., and Wiesel, T. N.: Receptive fields of single neurones in the cat's striate cortex, J. Physiol. (Lond) **148:**574, 1959.
48. Hubel, D. H., and Wiesel, T. N.: Integrative action in the cat's lateral geniculate body, J. Physiol. (Lond) **155:**385, 1961.
49. Hubel, D. H., and Wiesel, T. N.: Receptive fields, binocular interaction and functional architecture in the cat's visual cortex, J. Physiol. (Lond) **160:**106, 1962.
50. Hubel, D. H., and Wiesel, T. N.: Receptive fields of cells in striate cortex of very young, visually inexperienced kittens, J. Neurophysiol. **26:**994, 1963.
51. Hubel, D. H., and Wiesel, T. N.: Receptive fields and functional architecture of two non-striate visual areas (18 and 19) of the cat, J. Neurophysiol. **28:**229, 1965.
52. Hubel, D. H., and Wiesel, T. N.: Binocular interaction in striate cortex of kittens reared with artificial squint, J. Neurophysiol. **28:**1041, 1965.
53. Hubel, D. H., and Wiesel, T. N.: Cortical and callosal connections concerned with the vertical meridian of visual fields in the cat, J. Neurophysiol. **30:**1561, 1967.
54. Hubel, D. H., and Wiesel, T. N.: Receptive fields and functional architecture of monkey striate cortex, J. Physiol. (Lond) **195:**215, 1968.
55. Hubel, D. H., and Wiesel, T. N.: The period of susceptibility to the physiological effects of unilateral eye closure in kittens, J. Physiol. (Lond) **206:**419, 1970.
56. Hubel, D. H., and Wiesel, T. N.: Cells sensitive to binocular depth in area 18 of the macaque monkey cortex, Nature **225:**41, 1970.
57. Hubel, D. H., and Wiesel, T. N.: Laminar and columnar distribution of geniculocortical fibers in the macaque monkey, J. Comp. Neurol. **146:**421, 1972.
58. Hubel, D. H., and Wiesel, T. N.: Sequence regularity and geometry of orientation columns in the monkey striate cortex. J. Comp. Neurol. **158:**267, 1974.
59. Hubel, D. H., and Wiesel, T. N.: Uniformity of monkey striate cortex: a parallel relationship between field size, scatter, and magnification factor. J. Comp. Neurol. **158:**295, 1974.

60. Hubel, D. H., and Wiesel, T. N.: Functional architecture of macaque monkey visual cortex, Proc. R. Soc. Lond. (Biol.) **198:**1, 1977.
61. Hubel, D. H., Wiesel, T. N., and LeVay, S.: Functional architecture of area 17 in normal and monocularly deprived macaque monkeys. Cold Spring Harbor Symp. Quant. Biol. **40:**581, 1975.
62. Hubel, D. H., Wiesel, T. N., and LeVay, S.: Plasticity of ocular dominance columns in monkey striate cortex, Philos. Trans. R. Soc. Lond. (Biol.) **278:**377, 1977.
63. Hubel, D. H., Wiesel, T. N., and Stryker, M. P.: Anatomical demonstration of orientation columns in macaque monkey, J. Comp. Neurol. **177:**361, 1978.
64. Kass, J. H., Guillery, R. W., and Allman, J. M.: Some principles of organization in the dorsal lateral geniculate nucleus, Brain Behav. Evol. **6:**253, 1972.
65. Kupfer, C., and Palmer, P.: Lateral geniculate nucleus: histological and cytochemical changes following afferent denervation and visual deprivation, Exp. Neurol. **9:**400, 1964.
66. LeVay, S.: On the neurons and synapses of the lateral geniculate nucleus of the monkey, and the effects of eye enucleation, Z. Zellforsch. Mikrosk. Anat. **113:**396, 1971.
67. LeVay, S.: Synaptic patterns in the visual cortex of the cat and monkey. Electron microscopy of Golgi preparations, J. Comp. Neurol. **150:**53, 1973.
68. LeVay, S., Hubel, D. H., and Wiesel, T. N.: The pattern of ocular dominance columns in macaque visual cortex revealed by a reduced silver stain, J. Comp. Neurol. **159:**559, 1975.
69. Levick, W. R., Cleland, B. G., and Dubin, M. W.: Lateral geniculate neurons of cat: retinal inputs and physiology, Invest. Ophthalmol. Vis. Sci. **11:**302, 1972.
70. Lund, J. S.: Organization of neurons in the visual cortex, area 17, of the monkey *(Macaca mulatta),* J. Comp. Neurol. **147:**455, 1973.
71. Lund, J. S., and Boothe, R. G.: Interlaminar connections and pyramidal neuron organization in the visual cortex, area 17, of the macaque monkey, J. Comp. Neurol. **159:**305, 1975.
72. Lund, J. S., et al.: The origin of efferent pathways from the primary visual cortex, area 17, of the macaque monkey as shown by retrograde transport of horseradish peroxidase, J. Comp. Neurol. **164:**287, 1975.
73. Malpeli, J. C., and Baker, F. H.: The representation of the visual field in the lateral geniculate nucleus of the Macaca mulatta, J. Comp. Neurol. **161:**569, 1975.
74. Marrocco, R. T.: Sustained and transient cells in monkey lateral geniculate nucleus: conduction velocities and response properties, J. Neurophysiol. **39:** 340, 1976.
75. Meadows, J. C.: Disturbed perception of colours associated with localized cerebral lesions, Brain **97:** 615, 1974.
76. Michael, C. R.: Color vision mechanism in monkey striate cortex: dual opponent cells with concentric receptive fields, J. Neurophysiol. **41:**572, 1978.
77. Michael, C. R.: Color vision mechanisms in monkey striate cortex: simple cells with dual opponent color receptive fields, J. Neurophysiol. **41:**1233, 1978.
78. Michael, C. R.: Color sensitive complex cells in monkey striate cortex, J. Neurophysiol. **41:**1250, 1978.
79. Michael, C. R.: Color sensitive hypercomplex cells in monkey striate cortex, J. Neurophysiol. **42:**726, 1979.
80. Mitchell, D. E., and Wilkinson, F.: The effect of early astigmatism on the visual resolution of gratings, J. Physiol. **243:**739, 1974.
81. Movshon, J. A.: Hypercomplexities in the visual cortex, Nature **272:**305, 1978.
82. Movshon, J. A., and Dursteler, M. R.: Effects of brief periods of unilateral eye closure on the kitten's visual system, J. Neurophysiol. **40:**1255, 1977.
83. O'Leary, J. L.: A structural analysis of the lateral geniculate nucleus of the cat, J. Comp. Neurol. **73:** 405, 1940.
84. O'Leary, J. L., et al.: Synapses in the lateral geniculate nucleus of the primate, Neurology **15:**548, 1965.
85. Ogren, M. P., and Hendrickson, A. E.: Pathways between striate cortex and subcortical regions in *Macaca mulatta* and *Saimiri sciureus:* evidence for a reciprocal pulvinar connection, Exp. Neurol. **53:**780, 1976.
86. Ogren, M. P., and Hendrickson, A. E.: The distribution of pulvinar terminals in visual areas 17 and 18 of the monkey, Brain Res. **137:**343, 1977.
87. Packwood, J., and Gordon, B.: Stereopsis in normal domestic cat, Siamese cat, and cat raised with alternating monocular occlusion, J. Neurophysiol. **38:** 1485, 1975.
88. Pearlman, A. L., and Daw, N. W.: Opponent color cells in the cat lateral geniculate nucleus, Science **167:**84, 1970.
89. Pearlman, A. L., Birch, J., and Meadows, J. C.: Cerebral color blindness: an acquired defect in hue discrimination, Ann. Neurol. **5:**253, 1979.
90. Pettigrew, J. D.: The importance of early visual experience for neurons of the developing geniculostriate system, Invest. Ophthalmol. **11:**386, 1972.
91. Poggio, G. F., and Fischer, B.: Binocular interaction and depth sensitivity in striate and prestriate cortex of behaving rhesus monkey, J. Neurophysiol. **40:**1392, 1977.
92. Polyak, S.: The vertebrate visual system, Chicago, 1957, University of Chicago Press.
93. Rodieck, R. W.: Visual pathways, Ann. Rev. Neurosci. **2:**193, 1979.
94. Sanderson, K. J., Bishop, P. O., and Darian-Smith, I.: The properties of the binocular receptive fields of lateral geniculate neurons, Exp. Brain Res. **13:**178, 1971.
95. Schiller, P. H., and Malpeli, J. G.: Functional specificity of lateral geniculate nucleus laminae of the rhesus monkey, J. Neurophysiol. **41:**788, 1978.
96. Schiller, P. H., Finlay, B. L., and Volman, S. F.:

Quantitative studies of single-cell properties in monkey striate cortex. I, Spatiotemporal organization of receptive fields, J. Neurophysiol. **39:**1288, 1976.

97. Sherman, S. M., Hoffmann, K., and Stone, J.: Loss of a specific cell type from dorsal lateral geniculate nucleus in visually deprived cats, J. Neurophysiol. **35:**532, 1972.
98. Singer, W.: Inhibitory binocular interaction in the lateral geniculate body of the cat, Brain Res. **18:** 165, 1970.
99. Spinelli, D. N., et al.: Visual experience as a determinant of the response characteristics of cortical receptive fields in cats, Exp. Brain Res. **15:**289, 1972.
100. Stone, J.: Morphology and physiology of the geniculocortical synapse in the cat: the question of parallel input to the striate cortex, Invest. Ophthalmol. **11:** 338, 1972.
101. Szentagothai, J.: The structure of the synapse in the lateral geniculate body, Acta Anat. **55:**166, 1963.
102. Tombol, T.: Two types of short axon interneurons in the specific thalamic nuclei, Acta Morphol. Acad. Sci. Hung. **17:**285, 1969.
103. Trojanowski, J. Q., and Jacobson, S.: Areal and laminar distribution of some pulvinar cortical efferents in rhesus monkey, J. Comp, Neurol. **169:**371, 1976.
104. Updyke, B. V.: The patterns of projection of cortical areas 17, 18 and 19 onto the laminae of the dorsal lateral geniculate nucleus in the cat, J. Comp. Neurol. **163:**377, 1975.
105. Van Essen, D. C.: Visual areas of the mammalian cerebral cortex, Ann. Rev. Neurosci. **2:**227, 1979.
106. Van Essen, D. C., and Kelly, J.: Correlation of cell shape and function in the visual cortex of the cat, Nature **241:**403, 1973.
107. Van Essen, D. C., and Zeki, S. M.: The topographic organization of rhesus monkey prestriate cortex, J. J. Physiol. **277:**193, 1978.
108. von Noorden, G. K.: Experimental amblyopia in monkeys. Further behavioral observations and clinical correlations, Invest. Ophthalmol. **12:**721, 1973.
109. von Noorden, G. K.: Histological studies of the visual system in monkeys with experimental amblyopia, Invest. Ophthalmol. **12:**727, 1973.
110. Wagor, E., Mangini, N. J., and Pearlman, A. L.: Retinotopic organization of striate and extrastriate visual cortex in the mouse, *Mus musculus* J. Comp. Neurol. (in press) 1980.
111. Walls, G. L.: The lateral geniculate nucleus and visual histophysiology, U.C.L.A. Publ. Physiol. vol. 9, 1953.
112. Walsh, F. B., and Hoyt, W. F.: Clinical neuro-ophthalmology, ed. 3, Baltimore, 1969, The Williams & Wilkins Co.
113. Wiesel, T. N., and Hubel, D. H.: Effects of visual deprivation on morphology of cells in the cat's lateral geniculate body, J. Neurophysiol. **26:**978, 1963.
114. Wiesel, T. N., and Hubel, D. H.: Single-cell responses in striate cortex of kittens deprived of vision in one eye, J. Neurophysiol. **26:**1003, 1963.
115. Wiesel, T. N., and Hubel, D. H.: Comparison of the effects of unilateral and bilateral eye closure on cortical unit response in kittens, J. Neurophysiol. **28:** 1029, 1965.
116. Wiesel, T. N., and Hubel, D. H.: Spatial and chromatic interactions in the lateral geniculate body of the rhesus monkey, J. Neurophysiol. **29:**1115, 1966.
117. Wilson, M. E., and Cragg, B. G.: Projections from the lateral geniculate nucleus in the cat and monkey, J. Anat. **101:**677, 1967.
118. Wilson, P. D., Rowe, M. H., and Stone, J.: Properties of relay cells in cat's lateral geniculate nucleus: a comparison of W-cells and X- and Y-cells, J. Neurophysiol. **39:**1193, 1976.
119. Zeki, S. M.: Representation of central visual fields in prestriate cortex of monkey, Brian Res. **14:**271, 1969.
120. Zeki, S. M.: Convergent input from the striate cortex (area 17) to the cortex of the superior temporal sulcus in the rhesus monkey, Brain Res. **28:**338, 1971.
121. Zeki, S. M.: Cortical projections from two prestriate areas in the monkey, Brain Res. **34:**19, 1971.
122. Zeki, S. M.: Colour coding in rhesus monkey prestriate cortex, Brain Res. **53:**422, 1973.
123. Zeki, S. M.: Functional organization of a visual area in the posterior bank of the superior temporal sulcus in the rhesus monkey, J. Physiol. **236:**549, 1974.
124. Zeki, S. M.: Cells responding to changing image size and disparity in the cortex of the rhesus monkey, J. Physiol. **242:**827, 1974.
125. Zeki, S. M.: Colour coding in the superior temporal sulcus of rhesus monkey visual cortex, Proc. R. Soc. Lond. Biol. **197:**195, 1977.
126. Zeki, S. M.: Uniformity and diversity of structure and function in rhesus monkey prestriate visual cortex, J. Physiol. (Lond) **277:**273, 1978.

Chapter 17

Electrical phenomena in the retina

ELIOT L. BERSON, M.D.

The rapid growth of knowledge about electrical phenomena in the retina both in normal subjects and in patients with retinal disease precludes a comprehensive review in a single chapter. Nevertheless, ophthalmologists have expressed a need for a summary of this subject to have a better understanding of retinal degenerations in humans and to have a basis for doing further research. This chapter has been written in an effort to fulfill this need; the text has been expanded from that in the previous edition to present some recent advances.

THE ELECTRORETINOGRAM

Origin of components

In 1865 Holmgren[159] showed that an alteration in electrical potential occurred when light fell on the retina. Dewar[85] recorded this light-evoked electrical response or electroretinogram (ERG) from humans for the first time in 1877. The response of the dark-adapted eye to white light was separated by Einthoven and Jolly[103] into an early cornea-negative a-wave, then a cornea-positive b-wave, a slower, usually cornea-positive c-wave, and in some mammals a small d-wave or off effect coincident with cessation of illumination (Fig. 17-1). Granit[145,146] identified three components or processes (P-I, P-II, and P-III) that disappeared successively from the ERG of a cat during deepening anesthesia and considered the ERG as a summation of these processes; his analysis (Fig. 17-2) has provided a frame of reference for more recent studies.

Tomita[274,275] showed that ERG components could be localized by depth recording with microelectrodes within the frog retina. Components of the ERG, recorded from within the retina, could also be recorded at the cornea with similar waveform. From a series of depth recordings in the cat retina Brown and Wiesel[59] found a correlation between amplitude maxima and retinal layers respectively for the a-, b-, and c-wave; they clearly demonstrated that at least part of the a-wave is generated in a layer distal or external to that responsible for the b-wave. Brown and Watanabe[58] found that a large a-wave could be recorded in the local foveal

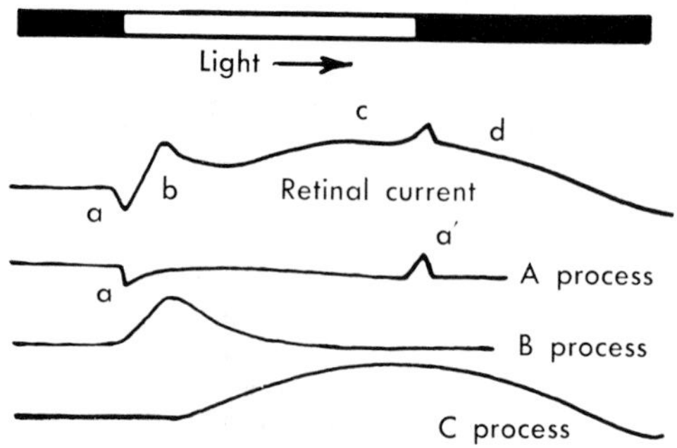

Fig. 17-1. Diagram of Einthoven and Jolly's analysis of retinal response (top tracing) into three components or processes, *A*, *B*, and *C*. (From Adrian, E. D., and Matthews, R.: J. Physiol. [Lond] **63:**378, 1927.)

□ This work was supported in part by the National Retinitis Pigmentosa Foundation, Baltimore, and the George Gund Foundation, Cleveland.

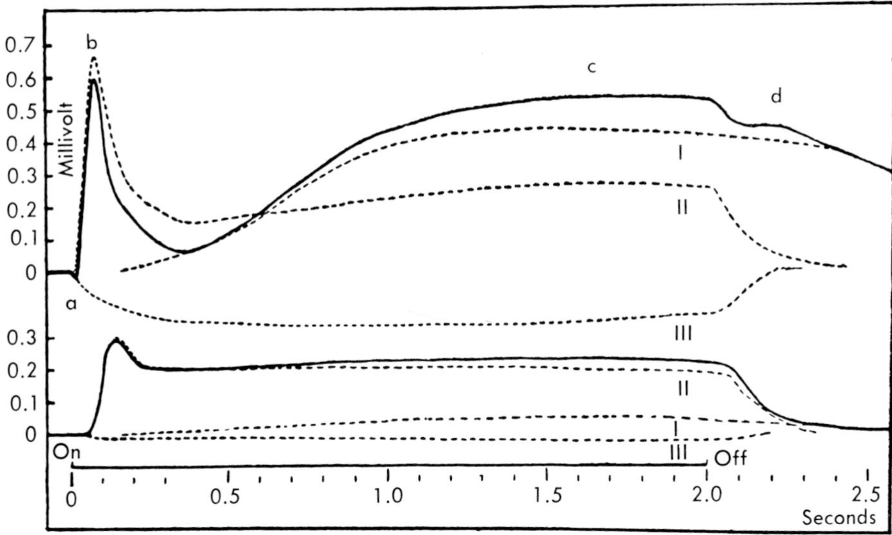

Fig. 17-2. Analysis of ERG at two stimulus intensities. Upper tracing, 14 mL. Lower tracing, 0.14 mL. The a-wave has been broadened slightly out of proportion to demonstrate its derivation more clearly. Duration of stimulus was 2 seconds. (From Granit, R.: Sensory mechanisms of the retina, London, 1947, Oxford University Press.)

ERG of the monkey where photoreceptors are in abundance and the remaining neuropile is scant, while the a-wave was much smaller and the b-wave relatively larger in the peripheral retina where the inner nuclear layer is more prominent. In retinas of cold-blooded vertebrates Murakami and Kaneko[208] divided P-III into a distal and proximal component; distal P-III was recorded in the receptor layer and proximal P-III in the inner nuclear layer. Distal P-III has been correlated with the onset of the a-wave of the ERG in the monkey.[60] Proximal P-III has not been unequivocally demonstrated in all mammalian species.[52] Witkovsky, Dudek, and Ripps[294] subdivided distal P-III in the carp into a faster component (fast P-III), which is generated by photoreceptors, and a slower component (slow P-III), which appears to be generated by the distal portions of the Müller cells.

Armington, Johnson, and Riggs[12] (Fig. 17-3) observed that the a-wave recorded at the cornea from the human eye under dark-adapted or scotopic conditions consists of two components; the earlier cornea-negative component (a_p) has the spectral sensitivity of the cone system and is little affected by light adaptation,

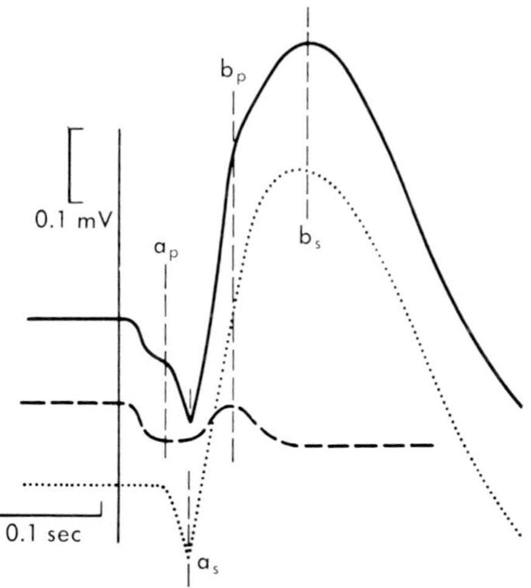

Fig. 17-3. Analysis of ERG in dark-adapted human eye as the resultant of photopic (dashed line) and scotopic (dotted line) components. The a-wave is composed of photopic (a_p) and scotopic (a_s) components, and the b-wave is similarly composed of photopic (b_p) and scotopic (b_s) components. (From Armington, J. C., Johnson, E. P., and Riggs, L. A.: J. Physiol. [Lond] **118:**289, 1952.)

while the later cornea-negative component (a_S) is much depressed by light adaptation and has the spectral sensitivity of the rod system.[12]

The cellular origin of the b-wave (P-II) of the ERG has been less clearly resolved. Following central retinal artery occlusion in monkeys[139] and humans,[155] cells in the inner nuclear layer are destroyed while photoreceptors appear intact, and the b-wave is eliminated while the a-wave is preserved. Large responses similar in waveform to the b-wave have been observed when the tips of microelectrodes are on opposite sides of the inner nuclear layer.[152] Microelectrode studies in the mud puppy *(Necturus)* by Miller and Dowling[204] (Fig. 17-4) provide evidence for the Müller cell as the site of gen-

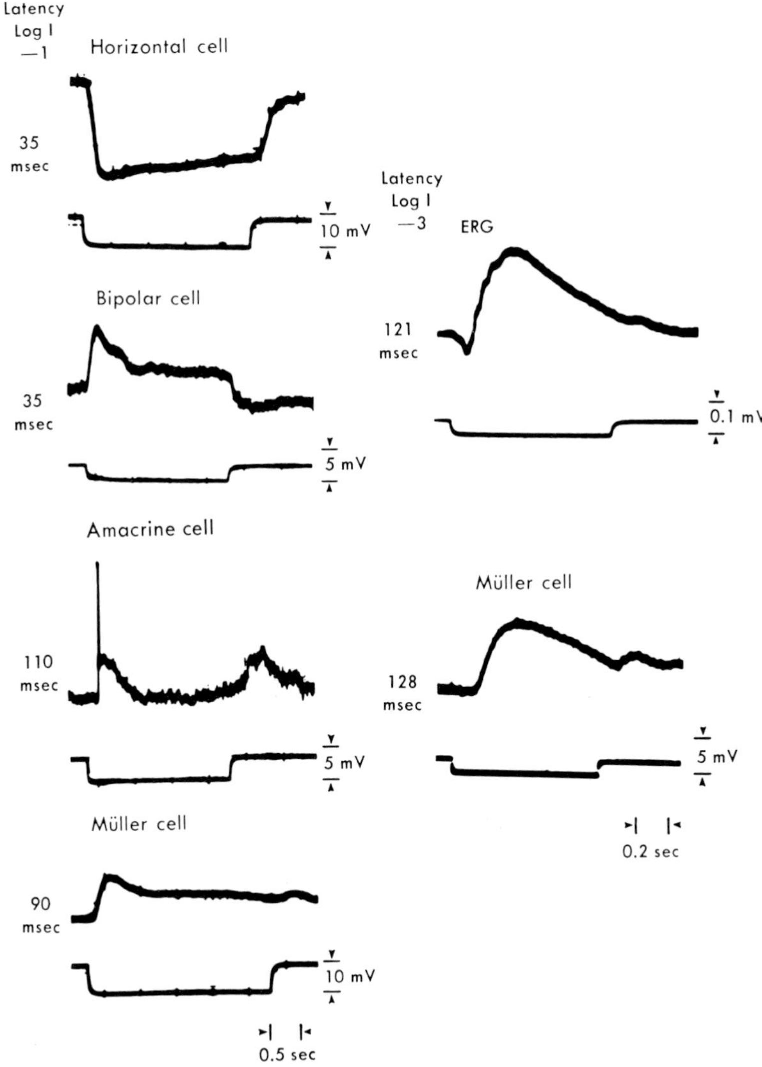

Fig. 17-4. Comparison of ERG with intracellular responses recorded from cells of inner nuclear layer of mud puppy. All responses were evoked by light flashes that illuminated entire retina evenly. Müller (glial) cell response closely matches b-wave in latency and waveform, especially at relatively low intensities (right). Latencies and waveforms of neuronal and Müller cell responses are compared on left. (From Miller, R. F., and Dowling, J. E.: J. Neurophysiol. **33:**323, 1970.)

eration of the b-wave. The waveform of the intracellularly recorded Müller cell potential and the extracellularly recorded b-wave of the ERG are similar, particularly at low stimulus intensities. The Müller cell does not generate an a-wave. Furthermore, the latency of the Müller cell response and the b-wave correspond closely over a wide range of stimulus intensities, and stimulus intensity–response amplitude relations agree over a stimulus range of 5 to 6 log units. Intracellular responses from other cells in the inner nuclear layer of this amphibian do not match the ERG b-wave (Fig. 17-4).[92,204]

Although the b-wave is generated proximal to the photoreceptors, rhodopsin content of the retina after exposure to bright light is linearly related to the log threshold of the b-wave in the normal rat. Furthermore, rats fed a vitamin A–free diet show a decline in rhodopsin levels in the retina and a parallel rise of the log threshold of the ERG b-wave.[91]

Adrian[1,2] recognized the separate contributions of the rod and cone systems in the generation of ERG responses. Stimulus conditions that favored cones (photopic) resulted in small b-wave responses of short latency, whereas stimulation of the rod receptor system under scotopic conditions produced relatively large cornea-positive b-wave responses of longer latency. Under conditions of complete dark adaptation the peak sensitivity of the major cornea-positive ERG b-wave in response to single flashes of light (presented at 2 second intervals) was near 504 nm, and the ERG spectral sensitivity curve approximated the absorption spectrum of rhodopsin. Under conditions of steady white light adaptation sufficient to eliminate the rod contribution to the ERG, the peak sensitivity of the ERG b-wave in response to single flashes shifted from 504 nm to about 555 nm (the so-called Purkinje shift).[167]

Psychophysical studies on humans by Stiles[265] and Wald[290] have shown that at least three cone mechanisms, defined by their spectral sensitivity curves, can be revealed by measurement of thresholds to narrow band light stimuli presented in the presence of intense chromatic backgrounds. Three cone mechanisms (Fig. 17-5), defined by their spectral

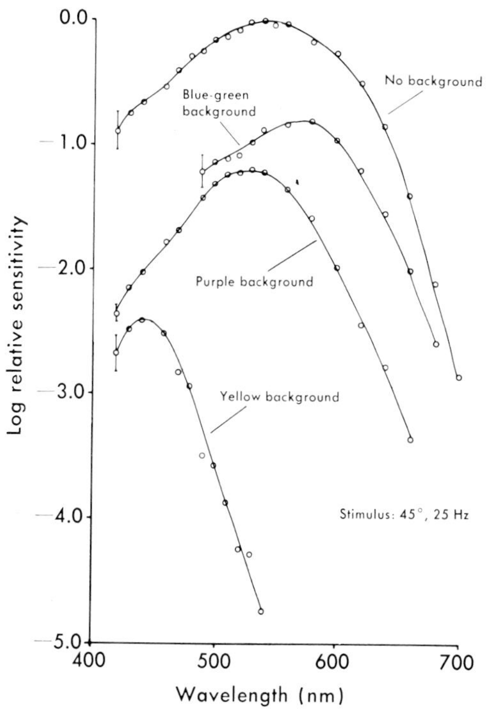

Fig. 17-5. Spectral sensitivity curves describing response of monkey cone mechanisms to 25 Hz stimulus under conditions of dark adaptation or in presence of intense chromatic backgrounds. Sector disc (50% duty cycle) was used to present flickering stimuli (25 Hz). Stimulus subtended visual angle of 45 degrees and was centrally superimposed on 68-degree background. Spectral sensitivity data were based on log relative quantum flux at retina necessary to elicit criterion amplitude in ERG. Red cone mechanism showed best separation on blue-green (Wratten 47) adapting field of 6.2 log troland intensity, green cone mechanism on purple (Wratten 35) adapting field of 5.8 log troland intensity, and blue cone mechanism on yellow (Corning 3482) adapting field of 7.2 log troland intensity. Data points (○) are an average from three animals. Vertical lines equal ±1 SD (average). (From Mehaffey, L., III, and Berson, E. L.: Invest. Ophthalmol. **13**:266, 1974.)

sensitivity curves, can also be separated in the ERG b-wave recorded at the cornea of the cynomolgus monkey by a similar approach.[203,230,282] ERG b-waves recorded in response to narrow band stimuli superimposed on an intense yellow background reveal the short wavelength (blue) mechanism with its maximum sensitivity near 440 nm. Narrow band stimuli presented on a purple background reveal a middle wavelength (green) mechanism, with maximum sensitivity near 540 nm. When the background is blue-green, narrow band stimuli elicit a long wavelength (red) mechanism, with maximum sensitivity near 580 nm. In each case the chromatic background eliminated the rod contribution and most, if not all, of the contributions from two cone mechanisms. The spectral sensitivity curves of these three mechanisms, derived from measurement of ERG b-waves, approximate cone pigment absorption data from individual cones obtained with microspectrophotometric measurements by Marks, Dobelle, and MacNichol[198] as well as Brown and Wald.[57] Therefore it is important to recognize that ERG b-waves can provide a measure of cone and rod system activity in monkeys and in humans, even though evidence from the mud puppy indicates that the b-wave reflects activity of the glial cells (Müller cells) and is not generated directly by retinal neurons.

Microelectrode studies, complemented by observations on the effects of various retinotoxic agents, have helped to define the cellular origin of the c-wave (P-I). Steinberg, Schmidt, and Brown[262] recorded a c-wave intracellularly from the pigment epithelium of the cat. Noell[219] showed that intravenously administered sodium iodate, which severely injured the pigment epithelium, selectively reduced the c-wave in the rabbit. Dowling and Ripps[97] found that application of sodium aspartate to the all-rod retina of the skate (Fig. 17-6, middle tracings) suppresses the responses of proximal elements but leaves relatively unaffected the electrical activity of the photoreceptors (a-wave) and pigment epithelium (c-wave). Removal of the

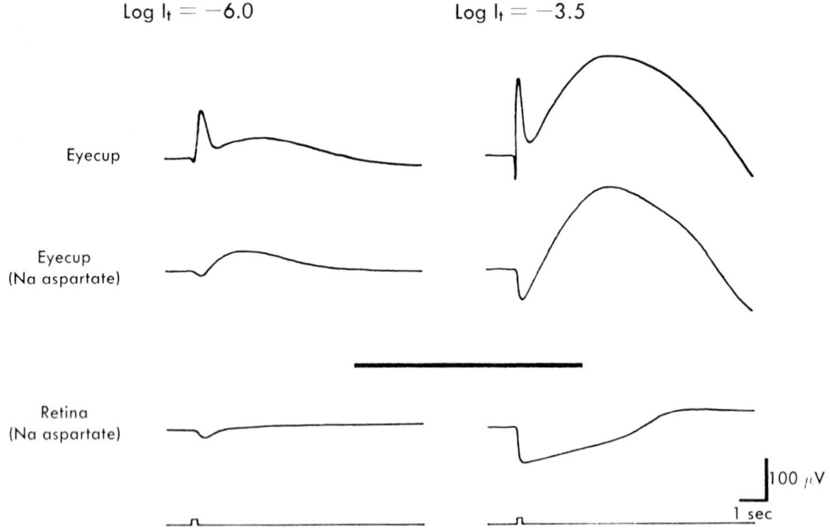

Fig. 17-6. Effects of Na (sodium) aspartate on electroretinographic responses from eyecup and isolated retina of skate. Responses were elicited by 0.2-second stimuli at two intensities (log I_t = −6.0 and −3.5). Upper traces were recorded from untreated eyecup preparation and show a-, b-, and c-waves of normal ERG. After immersion for 3 minutes in Ringer's solution containing 50 mmol L-sodium aspartate (middle tracings), b-wave was suppressed, but a- and c-waves were essentially unaltered. Removing aspartate-treated retina from eyecup eliminated c-wave. Log I_t values give filter density attenuating test beam (log $I_t \sim -D$), where log I_t =0 corresponds to 1.1 mW cm^{-2}. (From Dowling, J. E., and Ripps, H.: J. Gen. Physiol. **60:**698, 1972, copyright © 1972, Rockefeller University Press.)

aspartate-treated retina from the eyecup, which contains the pigment epithelium, eliminated the c-wave (Fig. 17-6, bottom tracings). Although the pigment epithelium must be present to generate a c-wave, the rods certainly contribute to or lead to the response because the spectral sensitivity of the dark-adapted c-wave response corresponds with the absorption spectrum of rhodopsin in the rods and not that of melanin in the pigment epithelium.[88,147]

Clinical usefulness of ERG testing in establishing the site of visual loss derives in part from the fact that the a-, b-, and c-waves are generated by cells distal to the ganglion cell. Electroretinographic testing can help to distinguish abnormalities in the pigment epithelium and outer and inner nuclear layers from abnormalities in the ganglion cell layer or optic nerve.[128,129,165] For example, the ERG is very small or nondetectable in patients with advanced retinitis pigmentosa[171] and congenital amaurosis of Leber[156] (diseases that involve the photoreceptors or pigment epithelium or both) but is normal in patients with Tay-Sachs' disease,[78] glaucoma, or neuromyelitis optica (diseases that involve ganglion cells or the optic nerve).

Physiologic basis

In 1896 von Helmholtz[285] recognized that current flowing from vitreous to sclera, which should hyperpolarize the synaptic terminals of rods and cones, caused light to appear brighter, whereas current flowing from sclera to vitreous, which should depolarize the terminals, caused light to look dimmer. Intracellular recordings from the inner segments of the carp and *Necturus* receptors revealed a resting potential of −30 to −40 mV and a hyperpolarizing response of the photoreceptors to light. Toyoda and co-workers[276] made the interesting observation in *Necturus* cone inner segments and *Gekko* rod outer segments that during the response to light the resistance of the cell membrane rose substantially, whereas passive hyperpolarization of the cell by passing current into it does not routinely affect membrane resistance. The dark current (steady flow of radial current that causes a standing voltage gradient in the interstices between rods of the rat retina) is reduced by the action of light, thus increasing the cell membrane potential.[148] Toyoda and associates[276] suggested that the action of light on receptors was to make their membranes less permeable to ions such as sodium whose passive entry tends to depolarize the membranes. Externally applied calcium mimics the effects of light and reduces the dark current of lizard cones.[301] Excessive cyclic guanosine monophosphate (c-GMP) injected intracellularly into rod photoreceptors of the isolated, perfused toad retina has resulted in depolarization of rods within milliseconds[205]; the mechanism by which this effect occurs remains to be defined. Replacement of sodium by an agent with limited or decreased membrane permeance, such as choline, abolishes the ERG of the isolated toad retina.[118]

Faber first suggested that the b-wave of the ERG reflected the activity of the Müller cells, the glial cells of the retina.[107] Recent evidence tends to support this idea, since it appears that the b-wave depends on current flow along the Müller cell in response to increases in extracellular potassium ion concentrations $[K^+]_o$. Kuffler and Nicholls[185] noted that glial cells in the optic nerve showed slow potential changes as potassium accumulated about them during optic nerve discharge and that the glial cell behaved like a potassium electrode. More recently, Dick and Miller[87] in the mud puppy and Kline, Ripps, and Dowling[174] in the skate, using potassium-sensitive electrodes, observed two sites of potassium ion efflux in response to light, one in the distal retina in the region of the outer plexiform layer and the second more proximally in the vicinity of the amacrine cells. The $[K^+]_o$ accumulation in the distal retina was faster and more transient in nature than that in the proximal retina.[172,174]

Kline, Ripps, and Dowling[174] proposed a model that attempts to explain how current flow in the extracellular space along the Müller cell generates the b-wave voltage (Fig. 17-7). Sites at which potassium is released by active neurons are at the interface of regions I and II and II and III respectively; at these sites, called current sinks, current flows into the Müller

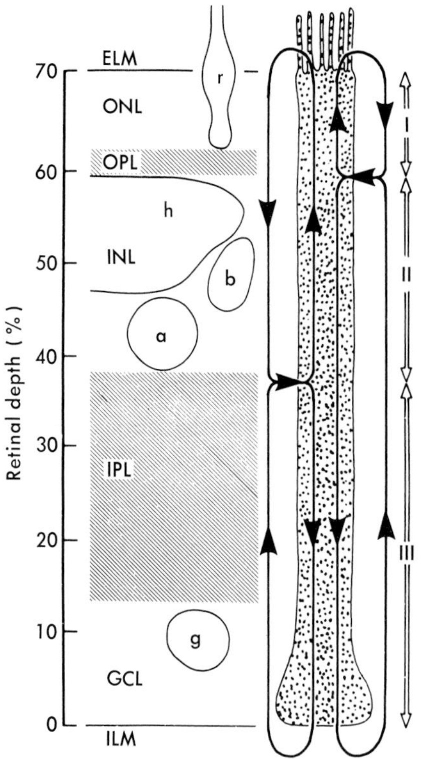

Fig. 17-7. Drawing of the skate retina. Relative retinal thickness is based on the dimensions observed in O_sO_4-fixed sections (70% retinal depth corresponds to about 105 μm). The Müller cell (stippled) extends from the internal limiting membrane *(ILM)* at the vitreal surface of the retina to the external limiting membrane *(ELM)* in the outer nuclear layer *(ONL)*. Proposed current flow set up around the Müller cell is indicated by arrows. There is a distal current sink at the interface of regions *I* and *II* and a proximal current sink at the interface of regions *II* and *III*. Cell types: *g,* ganglion; *a,* amacrine; *b,* bipolar; *h,* horizontal; *r,* receptor; *GCL,* ganglion cell layer; *IPL,* inner plexiform layer; *INL,* inner nuclear layer; *OPL,* outer plexiform layer. (From Kline, R., Ripps, H., and Dowling, J. E.: Proc. Natl. Acad. Sci. U.S.A. **75:**5727, 1978.)

cell, producing the various current paths shown in Fig. 17-7. Considering the b-wave as it is clinically recorded at the cornea as a transretinal potential, the contribution to the b-wave voltage from the proximal sink is relatively negligible (i.e., because of the location of the proximal sink midway along the Müller cell, there are two opposite currents of approximately the same magnitude). On the other hand, the distal current sink is asymmetrically placed along the Müller cell and therefore gives rise to the large cornea-positive potential recorded as the b-wave. The distal $[K^+]_o$ source is critical for the generation of the b-wave, and Dick and Miller[87] and Kline and co-workers[174] have suggested that the $[K^+]_o$ in the distal retina derives from the activity of the depolarizing bipolar cells. This model is under study using current source density analysis[216,217,284a] and using selective glial-toxic agents[270a] to determine whether other retinal cells are also contributing to the b-wave response.

The physiologic basis of the c-wave, a very slow response relative to the a- and b-waves, depends on depletion of potassium ions in the extracellular space between the photoreceptors and pigment epithelium distal to the external limiting membrane; following a flash of light, potassium ions presumably move into photoreceptor cells and the distal Müller cell processes thereby creating an ionic imbalance across the apical surface of the pigment epithelium. This change results in the large, slow cornea-positive potential, which is referred to as the c-wave of the ERG.[222,223]

Separation of rod and cone components

Review of the spectral sensitivity functions of the rod and cone systems reveals (Fig. 17-8) that the rods (solid curve) are more sensitive than the cones (dashed curve) across almost the entire visible spectrum; this difference diminishes in the long wavelength portion of the spectrum, and with very long wavelength (deep red) stimuli ($\lambda > 680$ nm), cone function may

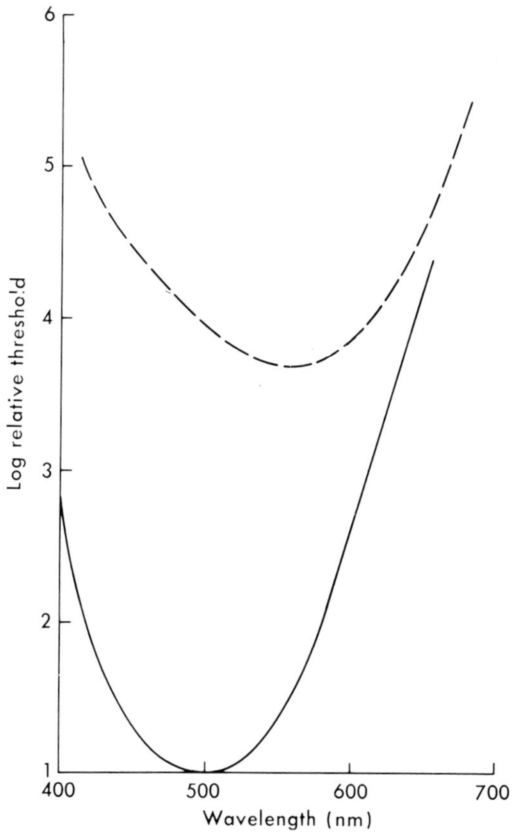

Fig. 17-8. Continuous line is CIE (Commission Internationale de l'Eclairage) scotopic luminosity curve (rod spectral sensitivity function) derived from psychophysical measurements and placed at level for normal human subjects; dashed line is Wald's photopic luminosity curve (spectral sensitivity function for the cone mechanisms under photopic conditions) derived from psychophysical measurements of peripheral retinal function. ERG spectral sensitivity curves for normal rod and cone systems also respectively approximate solid line and dashed line curves.

be isolated. The rod contribution to the ERG can be separated by recording from the dark-adapted subject and stimulating with a relatively *dim* short wavelength (blue) or *dim* long wavelength (orange-red) light stimulus. One would anticipate that the cones could be isolated simply by presenting single flashes of deep red light, but in fact, with the Grass photostimulator as the light source, single flashes of this deep red light are too dim to elicit an easily detectable cone ERG at the cornea from a normal subject, much less an ERG from a patient with retinal degeneration. In clinical testing, stimuli bright enough to elicit a cone response not only from normal subjects but also from patients with retinal disease can be achieved, without signal averaging, only with broad-band filters (for example, Wratten 26, $\lambda > 600$ nm). A broad band red (Wratten 26) stimulus, presented well above threshold, elicits not only an early cornea-positive b-wave (photopic x-wave)[206] from the cones but also a later cornea-positive b-wave from the rods (p. 474). If the stimulus is well above rod threshold, as is the usual white light stimulus used for testing patients with suspected retinal disease, the ERG a- and b-waves are generated by both the rod and cone systems.

The cone response can be seen in relative isolation by stimulating the eye in the presence of a background light sufficient to eliminate the rod contribution to the ERG.[131] White light flickering stimuli presented at 25 to 30 flashes/sec (25 to 30 Hz) separate the red and green cone responses from the rod response; rod responses fuse to a repetitive white light stimulus above 20 Hz, and this results in a nondetectable rod ERG.[89,90,127,153] Therefore the peak sensitivity of the dark-adapted ERG to a 25 Hz stimulus (a response from the cone system) is near 555 nm (Fig. 17-5). Flicker fusion frequency for red and green cones is 50 to 60 Hz.

The cone and rod contributions to the human ERG are respectively represented in recordings from a patient with dominant stationary night blindness (Fig. 17-9, top) and a patient with congenital rod monochromatism (Fig. 17-9, bottom). Responses (columns 1 and 2) were obtained to scotopically balanced light stimuli (long wavelength and short wavelength lights matched in brightness under conditions of complete dark adaptation to elicit equal-amplitude rod ERG b-waves near threshold from a normal subject).[35,36] These scotopically balanced stimuli, when presented well above threshold, elicit equal ERG responses from the rod monochromat (bottom, columns 1 and 2) but elicit unequal responses both from the patient with only cone function (top, columns 1 and 2) and the normal subject with cone and rod function (middle, columns 1 and 2). Fig. 17-9 (extreme right column) illustrates that the rod monochromat (bottom) has no detectable response to white flickering (30 Hz) stimuli in contrast to the response of the normal subject (middle) and the patient with dominant stationary night blindness (top). Long wavelength and short wavelength light stimuli can also be photopically balanced, that is, matched in brightness to elicit equal-amplitude 30 Hz responses from a normal subject near threshold. Single flashes of these photopically balanced lights, presented in the presence of a full-field background light sufficient to eliminate the rod contribution to the ERG (Fig. 17-9, columns 3 and 4, bottom row), will elicit matched responses from the normal subject (middle row) and the patient with night blindness and only cone function (top row). Lights matched in brightness for subjects with normal cone pigments will elicit mismatched responses from a protanope or deuteranope.[135]

ERGs are often described as photopic or scotopic responses; this usually refers to the fact that the responses were recorded under photopic (light-adapted) or scotopic (dark-adapted) conditions. This description does not necessarily define which receptor system(s) has generated the response, particularly in patients with retinal disease. Scotopic ERGs to relatively intense white light stimuli usually represent contributions from both the rod and cone systems; scotopic ERGs to relatively dim white light can be generated by the rods alone. Photopic ERGs may represent either responses from the light-adapted or dark-adapted cone[137] system, and the temporal aspects of the light-adapted and dark-adapted cone ERG responses

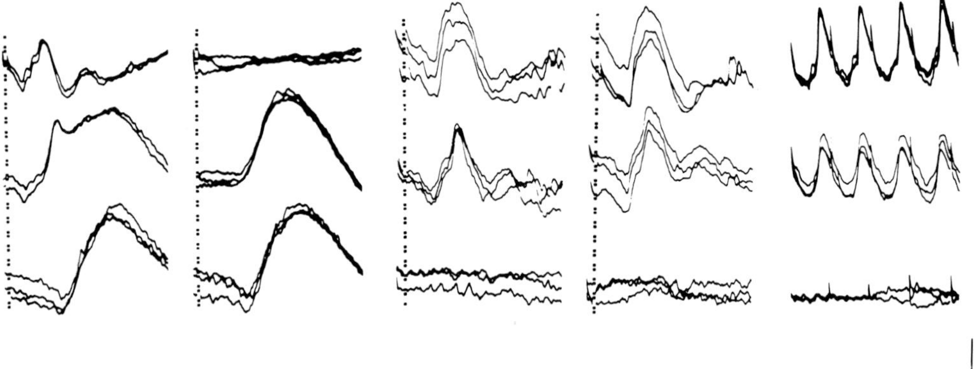

Fig. 17-9. ERG responses to scotopically balanced red ($\lambda > 600$ nm, column 1) and blue ($\lambda < 470$ nm, column 2) light stimuli, to photopically balanced orange ($\lambda > 550$ nm, column 3) and blue-green ($\lambda < 550$ nm, column 4) stimuli in presence of 5 to 10 ft-L background light, and to flickering (30 Hz) white stimuli (column 5) are shown successively from top to bottom for patient with night blindness (Nougaret type), normal subject, and congenital rod monochromat. Two or three responses to same stimulus are superimposed; calibration symbol signifies 60 milliseconds horizontally and 50 μV vertically for columns 1 and 2, 30 milliseconds horizontally and 50 μV vertically for columns 3 and 4, and 60 milliseconds horizontally and 100 μV vertically for column 5; corneal positivity is an upward deflection; stimulus onset, vertical hatched line for columns 1 to 4, and shock artifacts for column 5. (From Berson, E. L., Gouras, P., and Hoff, M.: Arch. Ophthalmol. **81:**207, 1969.)

are different (p. 478). In patients with retinal disease, dark-adapted cone thresholds can approximate dark-adapted rod thresholds,[197] and then the minimal white light stimulus required to elicit a near threshold ERG b-wave response will be almost the same for both receptor systems. Therefore stimulus wavelength, stimulus brightness, state of retinal adaptation, stimulus frequency, and application of the concept of balanced or matched light stimuli are all important in separating the rod and cone system contributions to abnormal ERG responses.[36,135]

Summation of rod and cone electroretinogram components

The human ERG recorded at the cornea in response to a full-field *(ganzfeld)* stimulus is a mass response generated by cells across the entire retina; loss of half the photoreceptors across the retina is associated with approximately a 50% reduction in ERG amplitude.[13] An ERG obtained in response to a constant test stimulus is proportional in size to the area of the image of the test field on the retina, and there appears to be no interaction between one area of the retina and another in generating response amplitudes even when regions illuminated are small and near to each other.[49,51,52,54]

Gouras recorded ERGs intraretinally (note that the polarity is reversed) in the perifovea of the rhesus monkey (Fig. 17-10) and found that responses (left column) to a dim, short wavelength light (i.e., responses only from the rod system) and responses (middle column) to a narrow band, deep, long wavelength light (i.e., responses in this experiment only from the cone system) algebraically summate when both light stimuli (right column) are presented together at varying time intervals. Because the responses were generated independently, they appear in any position relative to each other, depending on the time interval of the stimulus flashes, and algebraically summate under all conditions.[130]

ERGs recorded at the cornea in humans in response to stimuli bright enough to stimulate the cone and rod systems also represent an algebraic summation of cone and rod contributions. For example, Fig. 17-9 (left column) illustrates responses to long wavelength light ($\lambda > 600$ nm); the response from the normal dark-adapted subject (middle) represents the algebraic summation of the response from a dominant stationary night-blind patient (top) with only cone function and that from a rod monochromat (bottom) with only rod function.[36]

The peak-to-peak ERG amplitude recorded from a patient with dominant stationary night blindness (normal or near-normal cone function and no detectable rod function) and a rod monochromat or patient with advanced cone

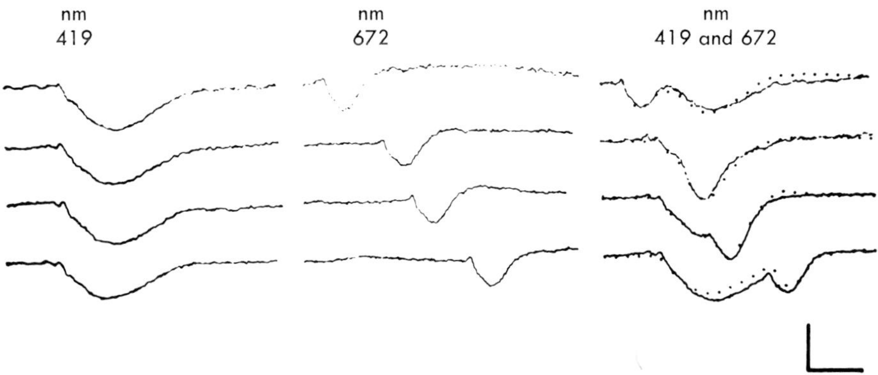

Fig. 17-10. ERG recorded from within perifoveal retina to 10-millisecond pulses of monochromatic light (419 and 672 nm) covering 24.6 mm² of retina. Each trace is sum of 10 responses. Traces on left are responses to 419 nm, alone, and begin with stimulus. Those in middle are responses to 672 nm, alone, and only upper trace begins with stimulus, the stimulus being delayed progressively in lower traces. Traces on right are responses to both stimuli together. Dotted lines show algebraic sum of separate responses to 672 and 419 nm. Calibration signifies 0.1 mV vertically and 0.1 second horizontally. (From Gouras, P.: J. Physiol. [Lond] **187:**455, 1966.)

degeneration (normal rod function and no detectable cone function) are respectively about 100 μV and 300 μV in response to white light stimuli, whereas the ERG to the same white light stimuli, recorded from a normal dark-adapted young adult, would be approximately 400 μV in amplitude. The ERG of a young adult with ocular albinism (reduced pigmentation with increased internal reflection of the stimuli within the eye) is usually well above 475 μV.[179] The amplitude of a child's ERG gradually increases during the first year of life and approaches that of a normal adult around age 1; rod and cone components were not separated in this study.[302] The ERG amplitude of a normal young child sedated with a barbiturate such as thiopental (Pentothal) can be reduced as much as 50% below that of a normal adult.

Although the rods outnumber the cones 13 to 1 in the normal human retina,[228,296] the cones account for 20% to 25% of the ERG response amplitude. The blue cone mechanism contributes maximally 2 to 3 μV to the total cone ERG response (to white light) of 75 to 100 μV in the normal subject, whereas the red and green cone mechanisms contribute the rest in approximately equal amounts.[203] Although the red and green cone contributions, as well as the rod contribution, can be detected in response to single flashes, present recording techniques require signal averaging to identify the small contribution from the blue cone mechanism. Therefore the ERG responses to single flashes of white light in the dark-adapted state from the rod monochromat (normal rod function, no detectable cone function) and the blue (π_1) cone monochromat (apparently normal blue cone and normal rod function, no red or green cone function) are approximately the same. Differences in ERG spectral sensitivity of the dark-adapted monkey eye obtained respectively in responses to 25 and 50 Hz flickering stimuli (Fig. 17-11) can be explained by a summation of the blue, green, and red cone mechanisms at the lower frequency and by the fact that only the green and red cone mecha-

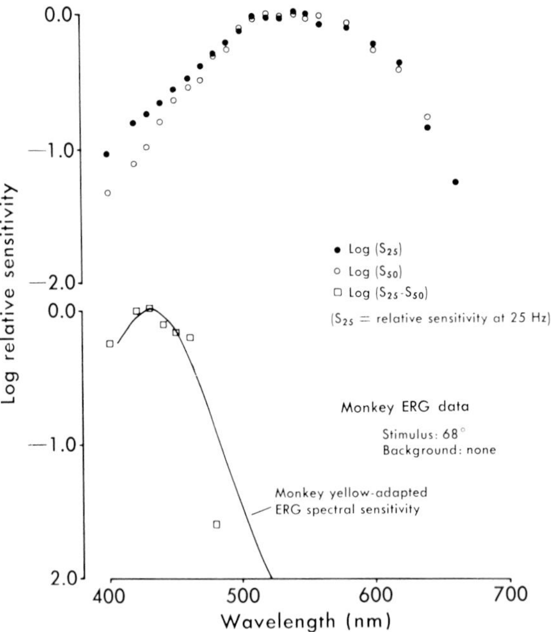

Fig. 17-11. ERG spectral sensitivity data describing monkey dark-adapted cone responses to 25 Hz (•) and 50 Hz (○) narrow-band stimuli subtending visual angle of 68 degrees. Subtraction, on linear basis, of the two sets of data yields difference points (□), shown here normalized on log plot. Curve drawn through points is spectral sensitivity curve of blue cone mechanism isolated with flickering (25 Hz) stimuli in presence of intense yellow adapting field (see Fig. 17-5). (From Mehaffey, L., III, and Berson, E. L.: Invest. Ophthalmol. **13:** 266, 1974.)

nisms contribute to responses at the higher frequency.[203]

Temporal aspects of electroretinogram

The temporal aspects of the cone and rod ERG responses depend on stimulus intensity and state of retinal adaptation. Fig. 17-12 illustrates that relatively dim short wavelength light elicits slow, small responses from the rod system, and more intense light stimuli result in faster and larger responses.[31,46] Fig. 17-13 (columns 1 and 2) shows cone ERG responses from a normal subject and a patient with dominant (Nougaret-type) stationary night blindness (no detectable rod function) to demonstrate that the cornea-positive b-waves are slower in re-

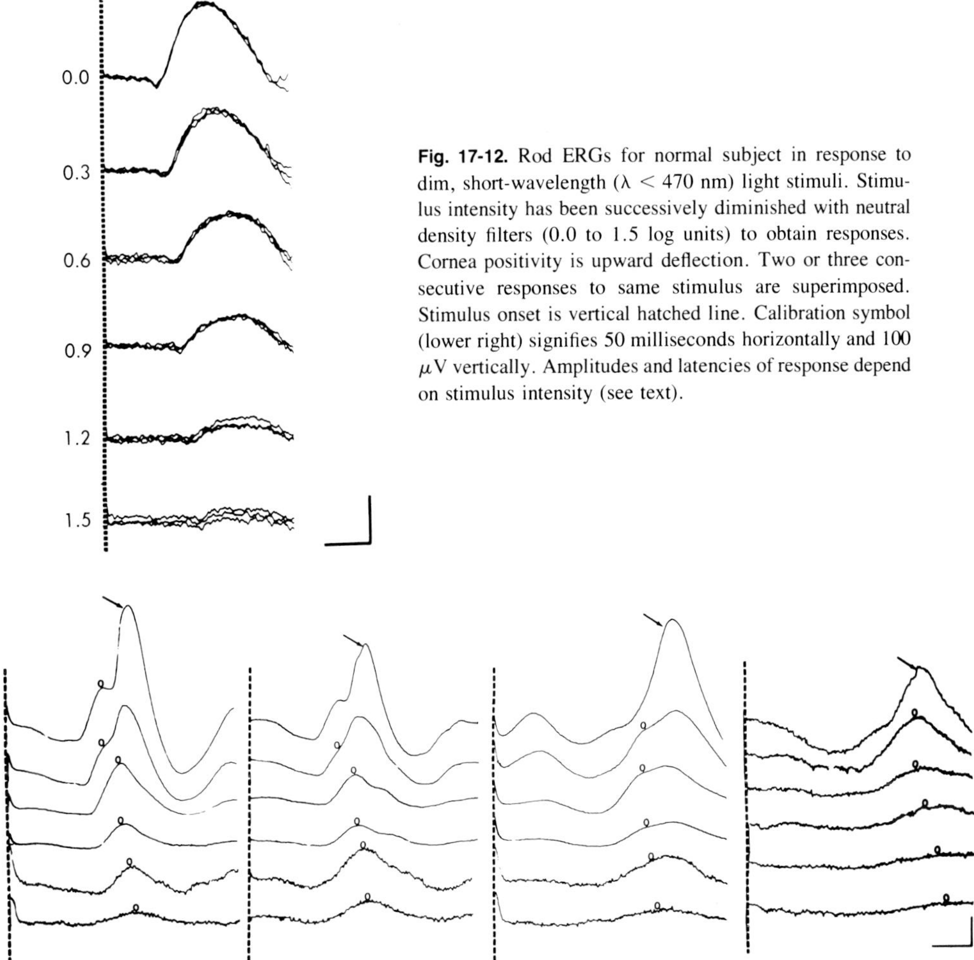

Fig. 17-12. Rod ERGs for normal subject in response to dim, short-wavelength ($\lambda < 470$ nm) light stimuli. Stimulus intensity has been successively diminished with neutral density filters (0.0 to 1.5 log units) to obtain responses. Cornea positivity is upward deflection. Two or three consecutive responses to same stimulus are superimposed. Stimulus onset is vertical hatched line. Calibration symbol (lower right) signifies 50 milliseconds horizontally and 100 μV vertically. Amplitudes and latencies of response depend on stimulus intensity (see text).

Fig. 17-13. ERG responses successively from left to right for normal subject, patient with Nougaret's nyctalopia, young patient with dominant retinitis pigmentosa with reduced penetrance, and older patient with same condition. In each column stimulus intensity has been successively diminished from top to bottom with neutral density filters (0.0, 0.3, 0.6, 1.0, 1.3, and 1.6 log units). Each trace is computer summation of 150 responses to flashes of yellow light (Wratten 15) in presence of background light of 5 to 10 ft-L. Stimulus onset, vertical hatched line; corneal positivity is upward deflection, and calibration symbol, lower right corner, is 10 milliseconds horizontally and 20 μV vertically for top four responses in columns 1 to 3, and 4 μV vertically for lower two responses in columns 1 to 3 and all responses in column 4. Stimulus onset to peak of responses (arrow) is cone implicit time. The peak may be h-wave off effect of Nagata.[211] Circles (o) indicate comparable peaks in responses for each patient and illustrate gradually prolonged timing of response with decreasing stimulus intensity. (From Berson, E. L., Gouras, P., and Hoff, M.: Arch. Ophthalmol. **81**:207, 1969.)

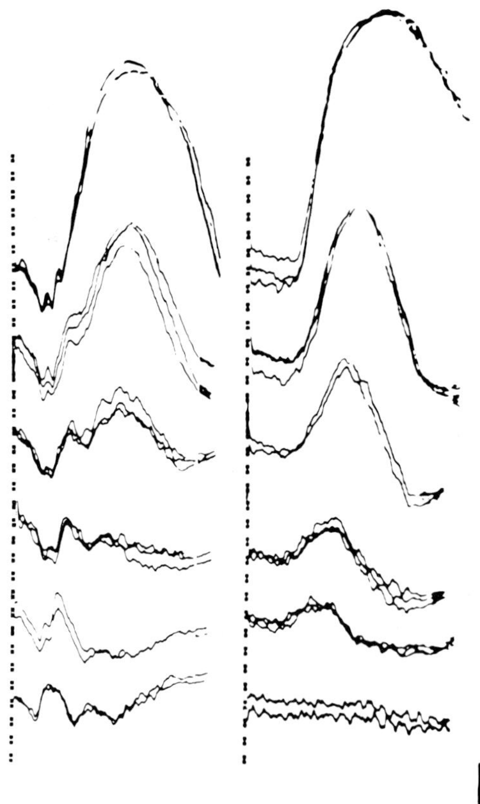

Fig. 17-14. ERG responses from normal subject elicited by scotopically matched long-wavelength (left column, $\lambda > 550$ nm) and short-wavelength (right column, $\lambda < 550$ nm) light stimuli. Top row of traces recorded in dark-adapted state. Responses to same two stimuli (i.e., stimulus intensity held constant) in presence of increasing white full-field steady background are shown successively from top to bottom—dark, 0.06, 0.16, 0.5, 0.8, and 5.0 ft-L. With increasing intensity of adapting lights, rod responses to scotopically balanced stimuli become diminished in amplitude, and implicit times become shorter. Early cone component is clearly visible only in response to long-wavelength stimulus in presence of background light. Stimulus onset is noted by vertical hatched line. Calibration symbol, lower right, signifies 60 milliseconds horizontally and 50 μV vertically. Corneal positivity is an upward deflection. (From Berson, E. L., Gouras, P., and Gunkel, R. D.: Arch. Ophthalmol. **80:**58, 1968.)

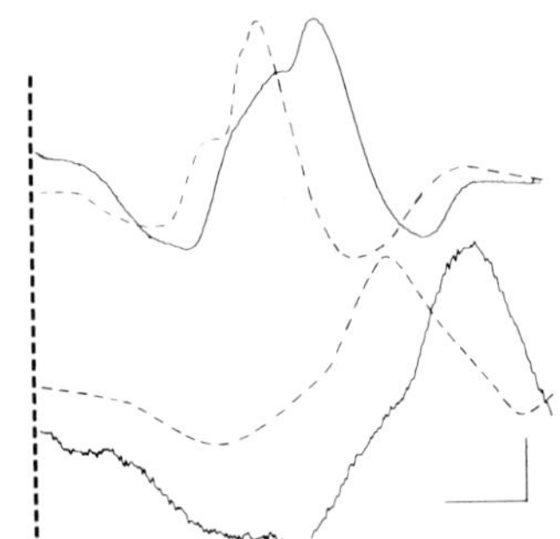

Fig. 17-15. ERG cone responses summed by computer for Nougaret's nyctalope (top) and patient with dominant retinitis pigmentosa with reduced penetrance (bottom). Dashed line for each patient is response to maximum yellow-light stimulation (noted in Fig. 17-13) in presence of 10 ft-L white background light; solid line represents cone response to this same yellow light under dark-adapted conditions; stimulus onset, vertical hatched line; corneal positivity is upward deflection, and calibration symbol (lower right corner) is 10 milliseconds horizontally for both patients, 20 μV vertically for Nougaret's nyctalope, and 4 μV vertically for patient with retinitis pigmentosa. (From Berson, E. L., Gouras, P., and Hoff, M.: Arch. Ophthalmol. **81:**207, 1969.)

sponse to dim light stimuli than comparable oscillations (designated 0) recorded from the same subjects in response to more intense light stimuli. Marked delays in cone responses are seen in two children with early retinitis pigmentosa (columns 3 and 4). With regard to the state of retinal adaptation, the light-adapted rod ERG b-waves in response to long (left column) and short (right column) wavelength stimuli (Fig. 17-14) are faster than the dark-adapted responses with stimulus intensity held constant. Similarly, for the cone system, ERG b-waves from the dark-adapted cone system are slower than the ERGs from the light-adapted cone system, again with stimulus intensity held constant (Fig. 17-15).[36]

Boynton and Riggs,[48] Asher,[14] and Fry and Bartley,[116] have emphasized the importance of stray light in interpreting ERG responses. Light is scattered by small heterogeneities of the transparent media and by the choroid; for example, it is estimated that the minimum scatter outside a focal image on the retina is about 10% of the incident light flux at the plane of the

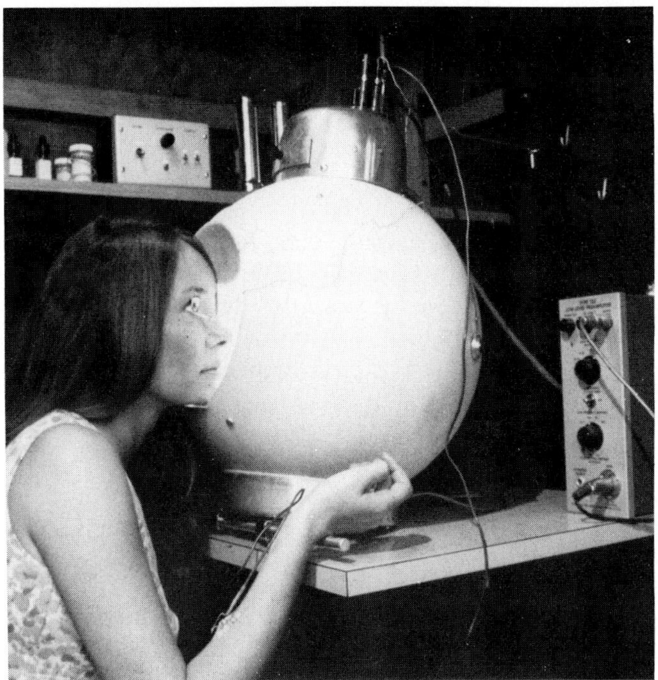

Fig. 17-16. Ganzfeld system. Stroboscope light (Grass PS 2) enclosed in case and attached to top of diffusing sphere illuminates inner white surface of this dome (40 cm in diameter), providing a full-field stimulus. Lights are recessed in top of dome so that patient can be tested in presence of steady full-field background light. During the test the patient, with contact lens electrode in place, sits with her head supported by chin rest. This dome can be mounted on a wall with the opening toward the floor so that an adult or young child can be tested in a reclining position. Wavelength of test flash can be modified by filters interposed between light source and dome. The original prototype[33,109] of this dome was constructed by Gunkel. Electrode attached to patient's forearm is grounded to amplifier common ground terminal through junction box located in case on top of dome. Responses are amplified (×1000) by AC coupled (0.8 to 250 Hz bandpass) battery-powered amplifier (right in photograph). Amplifier has high impedance differential input first stage, which provides 80 to 100 dB of common mode rejection. Responses are displayed on oscilloscope modified with low-persistence phosphor so that consecutive sweeps can be photographed in rapid succession (every 2 seconds) with Polaroid camera mounted to oscilloscope screen. Under usual recording conditions ERGs as low as 10 μV in amplitude (peak to peak) can be resolved from background noise. For analysis of responses less than 10 μV in amplitude, a computer of average transients is required. (From Rabin, A. R., and Berson, E. L.: Arch. Ophthalmol. **92**:59, 1974.)

pupil.[191] The importance of stray light was demonstrated in experiments in which a stimulus flash subtending a small angle at the eye produced a slightly larger ERG when it fell on the disc than when it fell on the retina.[14,48] With a focal light source the ERG recorded at the cornea would be expected to be a summation of responses of different amplitudes and latencies; the part of the retina stimulated with direct (more intense) light would generate faster, larger responses, and the part of the retina stimulated with stray light (less intense) would generate slower, smaller responses. Similarly, ERGs recorded in the presence of an overhead room light or focal background illumination could be expected to result in a similar phenomenon; that is, part of the retina would be exposed to bright focal background illumination and part of the retina would be exposed to relatively dim stray background light. Under these conditions some of the photoreceptors would respond more quickly in the presence of the more intense background, and some would respond more slowly in the presence of a less intense background.

With the above considerations in mind, a full-field (ganzfeld) system has been used to record the human ERG[33,135,236]; this system incorporates a full-field stimulus[221] and a full-field background (Fig. 17-16). All the receptors are stimulated in a relatively homogeneous manner. Under conditions of dark adaptation or in the presence of a full-field background, recordings become remarkably reproducible with respect to amplitudes and latencies, even in children with variable fixation.[33,62,236] Possible differences in retinal illumination due to the size of the pupil are minimized by dilating all patients' pupils with 10% phenylephrine hydrochloride and 1% cyclopentolate hydrochloride prior to testing.

Technique for recording the human electroretinogram

Electroretinographic testing became feasible clinically in 1941 when Riggs[241] discovered that a stable electrical connection with the cornea could be achieved for more than an hour with a silver disc electrode mounted in a scleral contact lens. When the lens was placed on the eye, the silver made contact with a physiologic saline solution between it and the cornea. A reference or neutral electrode was placed on the skin over the cheekbone. Some electrical noise could be eliminated by placing the recording electrode and reference electrode close together, that is, a double-electrode contact lens with the recording electrode as part of a corneal contact lens and the reference electrode as part of the lid speculum (Fig. 17-17).[190]

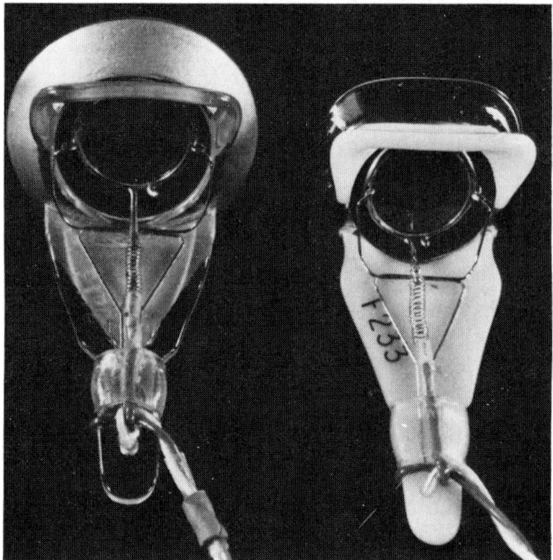

Fig. 17-17. Double-electrode (Burian-Allen) contact lenses used to obtain ERG responses. Although adult-sized lens (left) has been used successfully in most patients over age 6, small speculum (right) should be available to evaluate young children with small interpalpebral fissures. The two leads from lens are connected to differential inputs of an amplifier via junction box located in case on top of dome (Fig. 17-16) and are current limited by 2 mA fuses. (From Rabin, A. R., and Berson, E. L.: Arch. Ophthalmol. **92:** 59, 1974.)

A patient with the contact-lens electrode on the topically anesthetized cornea sits in front of the ganzfeld dome during the test (Fig. 17-16). Sedation or anesthesia is usually required to perform this test in children under age 6. Ambient electrical noise is minimized by surrounding the patient and the dome with a grounded copper enclosure. Recording equipment is placed outside the enclosure. With appropriate shielding of the patient and the dome, and with preamplifiers powered by batteries, signal averaging is not necessary for routine testing.

The rate of presentation and duration of the stimuli used to elicit the ERG can affect the amplitudes and latencies of responses to successive flashes of light. This problem can be obviated by using a brief flash (10 microseconds in duration) presented at 2-second intervals to evaluate rod or cone function. Furthermore, stimulation of the dark-adapted patient with suprathreshold white flicker will light adapt the patient, so the rods cannot be tested immediately thereafter. Sequential testing of first the rods and then the cones with reliable measurements of amplitudes and latencies can be accomplished by dark adapting the patient for 30 to 45 minutes and then presenting first single flashes of dim short and long wavelength stimuli, then single flashes of dim white light, then white flicker (30 Hz), and finally single flashes of short wavelength, long wavelength, and white light stimuli in the presence of a steady white background light sufficient to eliminate the rod contribution to the ERG.

Once the patient is dark adapted, the lens is placed on the topically anesthetized eye; the actual testing time with this protocol for a cooperative patient can be as little as 10 to 15 minutes per eye. Although published values can be used as a guideline,[32,38] it is recommended that normal ranges for amplitude and b-wave implicit times (time interval between stimulus onset and the peak of the major cornea positive component of the rod or cone b-wave response) be obtained for each ganzfeld system by testing normal subjects of different ages prior to evaluating patients with retinal disease or suspected retinal disease.

Full-field electroretinogram responses in retinal degenerations

The conventional ophthalmoscopic evaluation of the patient with retinal degeneration to determine how much of the macula and how much of the periphery appear visibly involved can be complemented, extended, and often redefined by an ERG evaluation of how much the cones are involved (both macular and peripheral cones) and how much the rods are involved (both macular and peripheral rods). Cone and rod system responses algebraically summate in ERGs from normal subjects and appear to algebraically summate in ERGs from patients with retinal disease. Separation of the rod and cone components in ERGs from normal subjects has provided a basis for separation of these components in the ERGs from patients with retinal disease.

When ERGs are obtained in response to a full-field test stimulus, generalized involvement of a photoreceptor system has been associated not only with a reduction in ERG amplitude but also with a delay in ERG b-wave implicit time. These reductions in amplitude and delays in implicit time have been demonstrated for the rod system in ERG b-wave responses from rats depleted of vitamin A[220] and for the cone system in ERG b-wave responses recorded from cats fed a taurine-free diet, with casein as the source of protein.[237,252,253] These reductions in amplitude and delays in b-wave implicit time occur in the early stages of these degenerations at a time when nearly all the rod (rat) and cone (cat) photoreceptors are partially affected. In contrast to the widespread but apparently partial involvement of the phororeceptor systems in the early stages of diet-induced retinal degenerations,[91,220,252] focal but complete destruction of photoreceptors (following photocoagulation) either in the periphery or in the macula of humans has been associated with reductions in ERG b-wave amplitudes without delays in ERG b-wave implicit times.[114]

Studies of the temporal aspects of the human ERG have provided criteria that allow separation of localized retinal diseases (localized chorioretinal scars or localized macular degenerations) from the early stages of widespread pigmentary retinal degenerations (degenerations that involve all or nearly all the photoreceptors across the entire retina). For example, patients with delimited chorioretinal scars secondary to old toxoplasmosis that are either localized in the macula or in the peripheral retina show reduced amplitudes of the rod (Fig. 17-18) and cone (Fig. 17-19) components in the ERG but have normal ERG b-wave implicit times.[36] In contrast, patients with widespread pigmentary retinal degenerations[33,36,42,43] have delayed rod (Fig. 17-18) or cone system (Fig. 17-19) b-wave implicit times or both.

Patients with dominantly inherited sector retinitis pigmentosa have bone spicule pigmentation and retinal arteriolar narrowing in one or more quadrants of the fundus and have normal

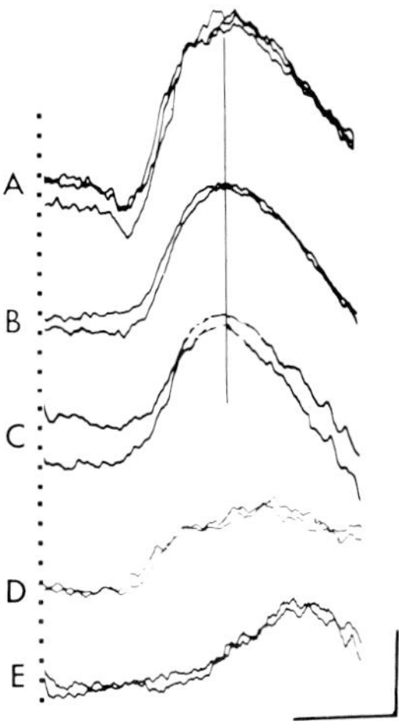

Fig. 17-18. Rod ERGs to blue light successively from top to bottom for normal subject, *A;* patient with central chorioretinal scar, *B;* patient with peripheral chorioretinal scar, *C;* patient with early stages of sex-linked retinitis pigmentosa, *D;* patient with early stage of dominant retinitis pigmentosa, *E.* Two or three consecutive responses to same stimulus are superimposed. Calibration symbol (lower right corner) signifies 60 milliseconds horizontally and 50 μV vertically; cornea positivity is upward deflection; stimulus onset is vertical hatched line. Vertical line has been extended from peak of normal rod b-wave response through b-waves of patients with local chorioretinal scars, *B* and *C*, to show that these rod implicit times are within normal range in contrast to delayed rod implicit times in patients with widespread hereditary retinitis pigmentosa, *D* and *E.* (From Berson, E. L., Gouras, P., and Hoff, M.: Arch. Ophthalmol. **81:**207, 1969.)

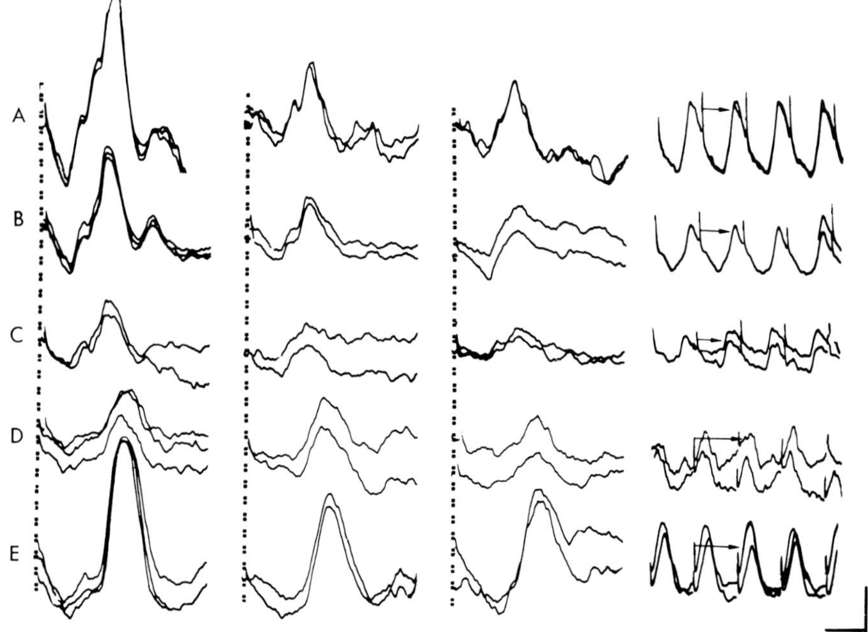

Fig. 17-19. Cone ERGs from top to bottom for normal subject, *A;* patient wtih central chorioretinal scar, *B;* patient with large peripheral chorioretinal scar, *C;* patient with moderately advanced sex-linked retinitis pigmentosa, *D;* patient with early stage of dominant retinitis pigmentosa with reduced penetrance, *E*. Left column of responses was obtained with single flashes of white light stimuli (32×10^3 ft-L) in presence of adapting field of 10 ft-L; second and third columns represent responses of long-wavelength ($\lambda > 550$ nm) and short-wavelength ($\lambda < 550$ nm) photopically matched lights in presence of same background light; fourth column illustrates responses to flickering white light stimuli (30 Hz) without background light. Two or three responses to same stimulus are usually superimposed; calibration symbol (lower right corner) signifies 30 milliseconds horizontally and 50 μV vertically for columns 1 to 3 and 60 milliseconds horizontally and 100 μV vertically for column 4. Cornea positivity is upward deflection. Stimulus onset is vertical hatched lines for columns 1 to 3 and vertical shock artifacts in column 4. Cone implicit times are shown with arrows in column 4. (From Berson, E. L., Gouras, P., and Hoff, M.: Arch. Ophthalmol. **81:**207, 1969.)

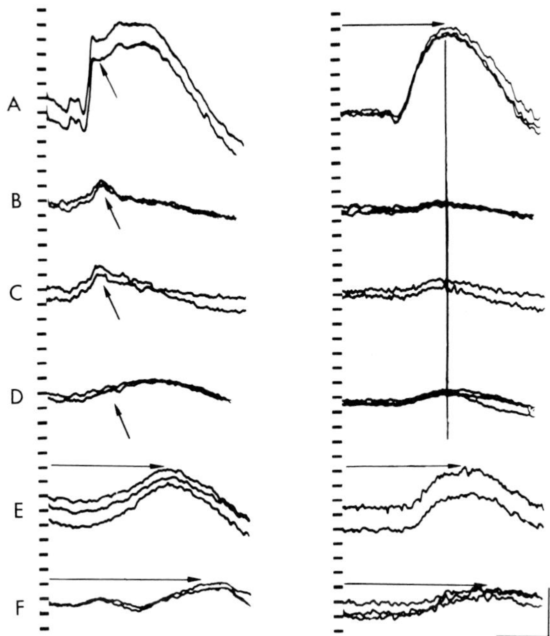

Fig. 17-20. ERGs to scotopically balanced light stimuli successively from top to bottom for normal subject, *A;* three patients with sector retinitis pigmentosa, *B, C,* and *D;* two children with recessively inherited retinitis pigmentosa, *E* and *F*. Responses on left are obtained with red light stimuli (Wratten 26, $\lambda > 600$ nm) and on right with scotopically matched blue light stimuli (Wratten 47, Wratten 47A, and Wratten 47B, with 0.6 neutral density filter, $\lambda < 470$ nm). Two to three responses to same stimulus are superimposed; calibration symbol (lower right corner) signifies 50 milliseconds horizontally and 100 μV vertically; cornea positivity is upward deflection; stimulus onset is vertical hatched line. Solid vertical line is extended from positive peak of normal rod ERG b-waves through b-waves of patients with localized retinitis pigmentosa; horizontal arrows designate rod b-wave implicit times. Oblique arrows (left column) designate splitting of early cornea-positive oscillation from cone system and later cornea-positive component from rod system. (From Berson, E. L., and Howard, J.: Arch. Ophthalmol. **86:**653, 1971.)

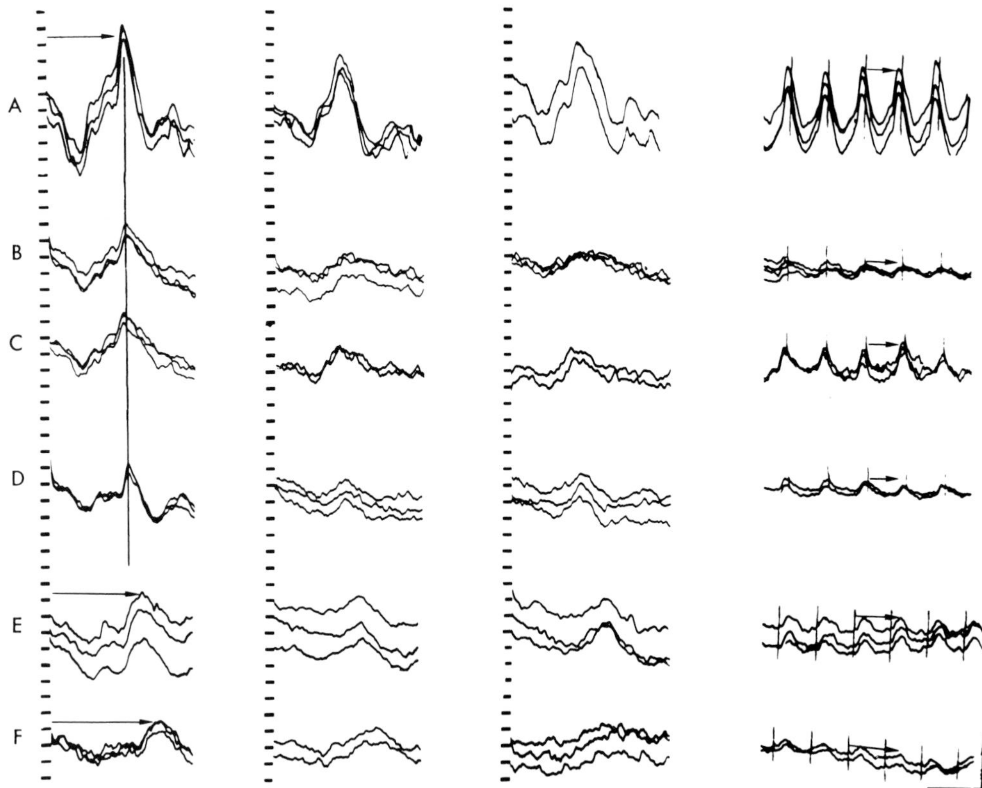

Fig. 17-21. Cone ERGs successively from top to bottom for normal subject, *A;* three patients with sector retinitis pigmentosa, *B, C,* and *D;* two children with widespread recessively inherited retinitis pigmentosa, *E* and *F*. Left column was obtained with single flashes of white light (32×10^3 ft-L) in presence of steady background light of 10 ft-L; second and third columns represent respectively responses to photopically matched long-wavelength (Cinnemoid 5, $\lambda > 550$ nm) and short-wavelength (Cinnemoid 16, $\lambda < 550$ nm) lights in presence of same adapting light; fourth column illustrates responses to flickering white light stimuli (30 Hz) without background light. Two or three responses to same stimulus are superimposed. Calibration symbol (lower right corner) signifies horizontally 25 milliseconds for columns 1 to 3 and 50 milliseconds for column 4; calibration signifies vertically 40 μV for columns 1 to 3 and 100 μV for column 4. Cornea positivity is upward deflection; stimulus onset is vertical hatched lines for columns 1 to 3 and vertical shock artifacts in column 4. Cone implicit times are shown with arrows in columns 1 and 4 for some patients. Solid line has been extended vertically from major positive peak of normal cone b-wave through b-waves of three patients with sector retinitis pigmentosa. (From Berson, E. L., and Howard, J.: Arch. Ophthalmol. **86:**653, 1971.)

psychophysical dark-adaptation rod thresholds in part of the retina and abnormal thresholds in other parts of the retina.[30,113,181] Studies of the ERG in dominant sector retinitis pigmentosa, which is minimally, if at all, progressive, have shown that this type of retinitis pigmentosa can be distinguished from widespread autosomal recessive retinitis pigmentosa, which is clearly progressive, on the basis of the temporal aspects of the ERG.[30,31] In sector retinitis pigmentosa, rod (Fig. 17-20) and cone (Fig. 17-21) ERG b-wave implicit times are normal (patients B, C, and D), whereas in widespread recessive retinitis pigmentosa the rod (Fig. 17-20) and cone (Fig. 17-21) ERG implicit times are markedly delayed (patients E and F), even at a stage when the amplitudes of the ERG in the sector and widespread types are comparably reduced.

Some observations on ERG amplitudes and implicit times in patients with retinal disease are summarized in Tables 17-1 and 17-2. Conclusions[22] based on these data are as follows:

1. Cone and rod ERG b-wave implicit times are normal in the ganzfeld ERG in sector retinitis pigmentosa or localized retinal disease.
2. Cone or rod ERG implicit times or both are markedly delayed in the early stages of four genetic types of widespread retinitis pigmentosa.
3. Substantially delayed cone ERG b-wave implicit times can be used to separate practically all patients with night blindness (i.e., those patients with elevated rod psychophysical thresholds to an 11-degree white test light in the Goldmann-Weekers dark adaptometer in all areas of the remaining visual field) and progressive forms of widespread retinal degenera-

Table 17-1. ERG* in sector or stationary retinal disease

		Cone ERG (b wave)		Rod ERG (b wave)	
Type	Cases	Amplitude	Implicit time	Amplitude	Implicit time
Sector retinitis pigmentosa	11	Reduced	Normal	Reduced	Normal
Stationary night blindness	20	Normal or reduced	Normal	Absent or normal for one flash	Normal if present
Macular degenerations	40	Reduced	Normal	Reduced	Normal
Chorioretinal scars	50	Reduced	Normal	Reduced	Normal

*Ganzfeld, clear media, dilated pupil.

Table 17-2. ERG* in widespread, progressive, pigmentary retinal degenerations

		Cone ERG (b wave)		Rod ERG (b wave)	
Type	Cases	Amplitude	Implicit time	Amplitude	Implicit time
Dominant retinitis pigmentosa with reduced penetrance	8	Normal or reduced	Delayed	Reduced	Delayed
Dominant retinitis pigmentosa with complete penetrance	11	Normal or reduced	Normal	Reduced	Delayed
Autosomal recessive retinitis pigmentosa	75	Reduced	Delayed	Reduced	Delayed or normal
Sex-linked recessive retinitis pigmentosa	7	Reduced	Delayed	Reduced	Delayed

*Ganzfeld, clear media, dilated pupil; large enough responses to separate into rod and cone components.

tions (Fig. 17-22, *B*) from the patients with stationary night blindness (p. 489) and localized retinal degenerations (Fig. 17-22, *A*).

4. A normal cone implicit time with abnormal rod function can be seen in stationary night blindness but does not exclude progressive retinal degeneration of the dominant type (Fig. 17-23).
5. The full-field ERG can be used to establish the diagnosis of widespread retinitis pigmentosa, often years to a decade before changes are clearly visible with the ophthalmoscope.

These separations among genetic types have been made in patients under age 15, for the most part. Individual patients have been followed over a decade from the stage of normal-appearing fundi and abnormal full-field ERGs to the stage of visible fundus abnormalities of retinitis pigmentosa and further reduction in ERG amplitudes.[32] In every family with retinitis pigmentosa in which an older affected relative has become nearly blind, the younger affected relatives have shown not only reduced ERGs but also delays in ERG b-wave implicit times.[22,23] Delayed cone b-wave implicit times have been observed not only in retinitis pigmentosa but also in other progressive widespread retinal degenerations with night blindness, including choroideremia,[22] generalized choroidal sclerosis,[22] and progressive albipunctate dystrophy. The delays in cone ERG b-wave implicit times seen in widespread forms of retinitis pigmentosa have not been seen in night-blind patients with chronic alcoholism and vitamin A deficiency, supporting the idea that the defect in retinitis pigmentosa does not involve a local deficiency of vitamin A in the retina.[250]

It is important to note that these delays in cone b-wave implicit times seen in the early stages of most types of retinitis pigmentosa are observed under conditions of light adaptation (i.e., in the presence of steady state white 30 Hz flickering stimuli or single flashes of white light in the presence of a steady white background sufficient to eliminate the rod contribution to the ERG). However, these same patients under dark-adapted conditions may have normal cone b-wave implicit times (Fig. 17-24). Under dark-adapted conditions, in response to blue and red test stimuli, matched in brightness for the rods, patients can have no clearly detectable rod function to blue or red light (Fig. 17-24, columns 1 and 2) but can have an easily detectable cone response to a red light stimulus that is usually normal in b-wave implicit time (Fig. 17-24, column 2). With light adaptation afforded by steady state white 30 Hz stimulus conditions, cone b-wave implicit times are substantially shorter for normal subjects and minimally shorter for patients; representative responses are illustrated in Fig. 17-24, column 3. Similar differences between normal sub-

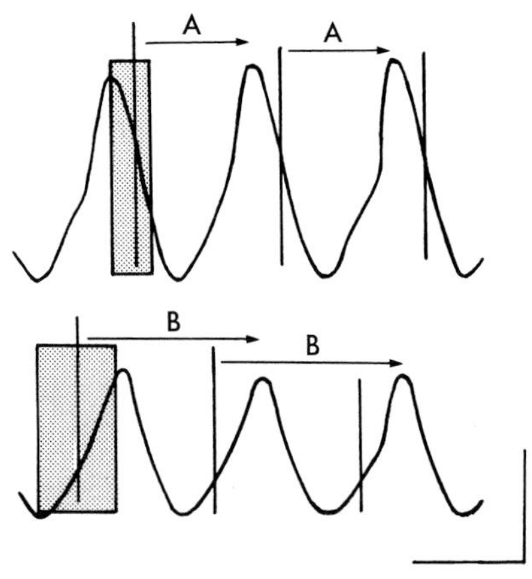

Fig. 17-22. Diagrammatic tracing to illustrate temporal aspects of cone ERG in response to 30 Hz white stimulus in normal subjects and patients with retinal disease. Stippled bar represents normal range (mean ±2 SD) for time of onset of stimulus flash in top tracing and abnormal range (mean ±2 SD) for time of onset in lower tracing. Vertical lines are representative for time of onset of each stimulus flash. Upper horizontal arrows, *A*, represent cone implicit time for patient with sector retinitis pigmentosa or localized retinal disease; and lower horizontal arrows, *B*, represent cone implicit time for patient with widespread recessively inherited retinitis pigmentosa. Note phase shift in relationship of stimulus artifact and response peak for *B* compared with *A*. For upper tracing, calibration symbol (lower right corner) signifies horizontally 25 milliseconds for all patients; calibration symbol signifies vertically 50 μV for normal patients and less than 50 μV vertically for patients with sector retinitis pigmentosa, chorioretinal scars, or retinal degenerations confined to macula. For lower tracing, calibration symbol (lower right corner) signifies horizontally 25 milliseconds and vertically 50 μV or less for patients with widespread recessively inherited retinitis pigmentosa.

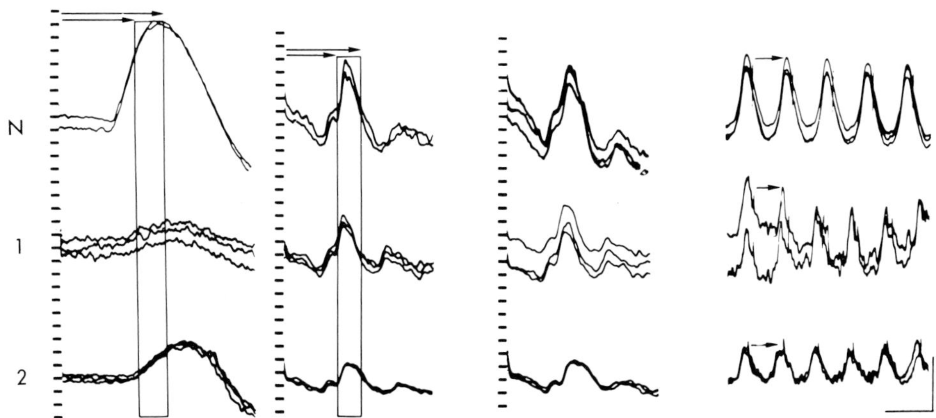

Fig. 17-23. ERGs for normal subject, *N*, and two children, *1* and *2*, with early stages of dominant retinitis pigmentosa with complete penetrance. Left column illustrates rod responses to dim, short-wavelength ($\lambda < 470$ nm) light. Second and third columns are cone responses to photopically matched long-wavelength ($\lambda > 550$ nm) and short-wavelength ($\lambda < 550$ nm) light flashes respectively in presence of 10 ft-L white background light. Fourth column illustrates cone responses to 30 Hz white flickering light. Horizontal arrows designate range for normal rod b-wave implicit times (left column) and cone b-wave implicit times (second column). Vertical bar defining this range (mean ±2 SD) in normal response has been extended through responses of these patients. Patients *1* and *2* have reduced rod and cone amplitudes, but implicit times of rod b-waves are delayed and cone b-waves are normal. Two or three consecutive responses to same stimulus are superimposed. Calibration symbol (lower right corner) signifies horizontally 50 milliseconds for columns 1 and 4 and 25 milliseconds for columns 2 and 3. Calibration symbol signifies vertically 100 μV for columns 1 and 4 and 40 μV for columns 2 and 3. Stimulus onset is vertical hatched line for columns 1 to 3 and vertical shock artifact in column 4.

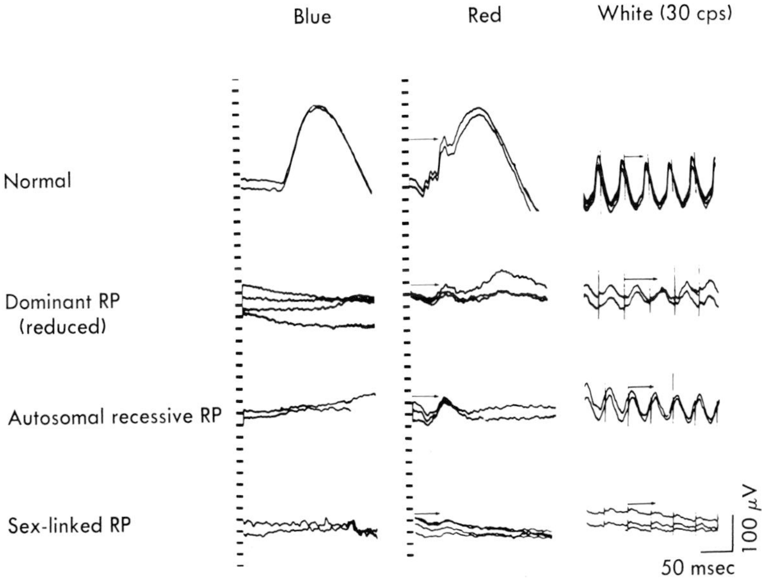

Fig. 17-24. ERGs in early retinitis pigmentosa (RP). Full-field ERG responses to blue ($\lambda < 470$ nm) and red ($\lambda > 600$ nm) light stimuli matched in brightness for the rods and white (30 Hz) light for a normal subject and representative young patients with dominant with reduced, autosomal recessive, and sex-linked retinitis pigmentosa. Responses in columns 1 and 2 were obtained after 45 minutes of dark adaptation. Cone b-wave implicit times (column 2) in normal subjects were confirmed by computer subtraction of the rod isolated response to blue light from the combined cone and rod response to red light with the residual component to red light being due to cone function. Stimulus onset is designated by vertical hatched lines in columns 1 and 2 and vertical shock artifacts in column 3. Arrows designate cone b-wave implicit times under dark-adapted (column 2) and light-adapted (column 3) conditions. Responses to 30 Hz flicker were considered delayed if greater than 32 milliseconds. Calibration symbol, lower right, designates 50 milliseconds horizontally and 100 μV vertically. (From Berson, E. L.: Light deprivation and retinitis pigmentosa, Vis. Res. 1980 [in press].)

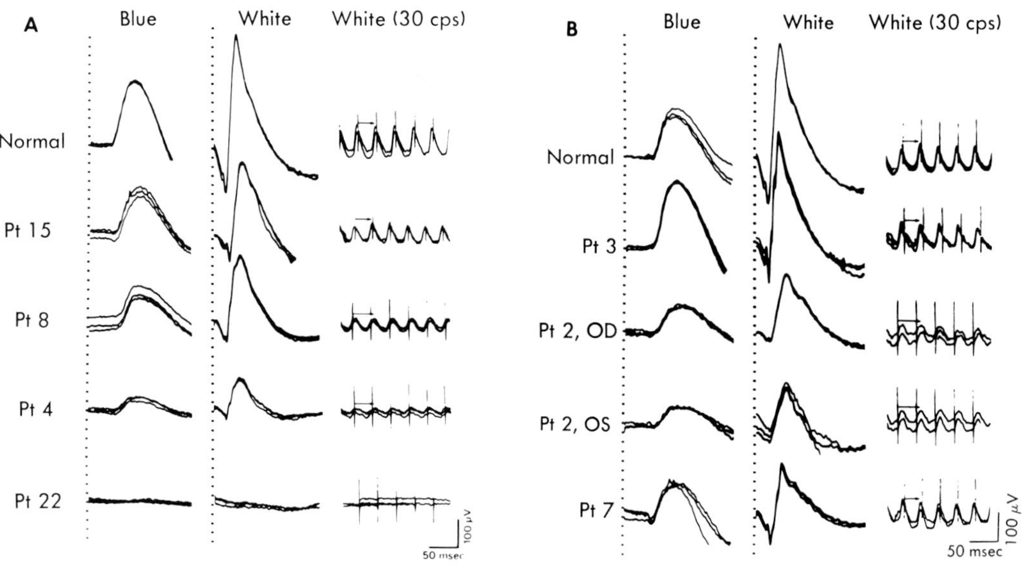

Fig. 17-25. A, ERG responses from a normal subject and four obligate female carriers of sex-linked retinitis pigmentosa. *Pt* (patients) *15,* age 39; *Pt 8,* age 41; *Pt 4,* age 51; *Pt 22,* age 70. **B,** ERG responses for a normal subject and three daughters of obligate carriers of sex-linked retinitis pigmentosa. *Pt 3,* age 23; *Pt 2,* age 21; *Pt 7,* age 20. *Pt 2* had bone spicule pigmentation only in OS. *Pt 3* and *Pt 7* had normal fundus examination results OU. For both **A** and **B** stimulus onset is designated by the vertical hatched lines for columns 1 and 2, and vertical shock artifacts for column 3. Cornea positivity is an upward deflection. Arrows in column 3 designate cone b-wave implicit times. (From Berson, E. L., Rosen, J. B., and Simonoff, E. A.: Am. J. Ophthalmol. **87:**460, 1979.)

jects and patients have been observed in cone b-wave implicit times recorded to single flashes of the same red light stimulus under dark-adapted and light-adapted conditions. These full-field ERGs would suggest that the patients have a defect in light adaptation of their cone system or, stated in another way, the cone system of these patients behaves as if relatively dark-adapted even in the light.[25,247]

Whereas dominant forms of retinitis pigmentosa (which are usually slowly progressive) have been separated in early life from autosomal recessive and sex-linked forms (which are characteristically more rapidly progressive), separation of autosomal recessive from sex-linked disease has not been possible on the basis of ERG waveforms alone. In the case of affected males with no affected female relatives, the question often arises as to whether they have sex-linked or autosomal recessive retinitis pigmentosa. In most cases, males with sex-linked disease are virtually blind by age 30 to 40, while males and females with autosomal recessive disease usually retain vision until age 45 to 60. A study of obligate female carriers of sex-linked retinitis pigmentosa, who were in most cases asymptomatic, has revealed that 22 out of 23 tested, or 96%, could be detected on the basis of abnormal full-field ERGs.[38] Obligate carriers (age 9 to 55 with 6 or less diopters of myopia) had ERGs that were either reduced in amplitude to single flashes of white light under dark-adapted conditions (i.e., $< 350\ \mu V$), or delayed in cone b-wave implicit time (i.e., >32 milliseconds), or both (Fig. 17-25, *A*) in one or both eyes. Rod-isolated responses to blue light proved less sensitive in detecting carriers than combined cone and rod responses to single flashes of white light under dark-adapted conditions. Some obligate carriers were detected on the basis of delays in cone ERG b-wave implicit times to 30 Hz flicker as the only abnormality. A few older obligate carriers with symptoms had very small or even nondetectable full-field ERGs.[38]

Among obligate carriers of childbearing age,

Table 17-3. Retinitis pigmentosa, the electroretinogram, and Mendel's laws*

Type	Families with at least one affected patient†	Siblings‡ with subnormal ERGs	Siblings§ with normal ERGs	Siblings affected with subnormal ERGs		Prediction of affected siblings‖
A (Dominant retinitis pigmentosa with complete penetrance)	6	5	6	$\frac{5}{11} = 45.5\%$	⟷	50%
B (Autosomal recessive retinitis pigmentosa)	18	10	29	$\frac{10}{39} = 25.6\%$	⟷	25%

*Age range of affected patients and their siblings was 6 to 20 years.
†Fundus examination showed visible changes of retinitis pigmentosa.
‡Fundus examination showed no visible abnormalities or minimal pigmentary changes.
§Fundus examination showed no visible abnormalities.
‖Percentages were predicted from laws of mendelian inheritance.

a patch of bone spicule pigmentation or an abnormal tapetal-like reflex in the macula was observed in the fundus with the ophthalmoscope in less than half of those who could be detected with the ERG. Daughters of obligate carriers had either normal ERGs (Fig. 17-25, *B,* patient 3) or abnormal ERGs (Fig. 17-25, *B,* patients 2 and 7) similar to those recorded from obligate carriers. The abnormal ERGs of carriers of sex-linked retinitis pigmentosa contrasted with the normal full-field ERGs and normal fundi observed in 20 obligate female carriers of autosomal recessive disease.[38]

Once carrier females of sex-linked disease are identified, affected male relatives would know they have sex-linked disease with a relatively poor visual prognosis and that all their daughters would be carriers and all their sons would be normal. Female relatives identified as carriers of sex-linked retinitis pigmentosa would know they have a 50% chance of having an affected son and a 50% chance of having a carrier daughter with each childbirth.

In families with autosomal dominant and autosomal recessive retinitis pigmentosa, younger siblings of clearly affected individuals have been evaluated with the full-field ERG at a stage when these siblings (age range 6 to 20 years) had minimal or no changes visible with the ophthalmoscope. The numbers of siblings with abnormal ERGs (reduced and delayed responses) expressed as percentages of the total number of siblings tested in families with the dominant and recessive types of retinitis pigmentosa agreed respectively with the percentages predicted from the mendelian laws that define dominant and recessive patterns of inheritance (Table 17-3). The mendelian patterns described previously for families in which some members had advanced stages of retinitis pigmentosa[17] have been confirmed in a study of families in which some members have the early stages.[20] Available data indicate that patients age 6 or over with normal full-field cone and rod amplitudes and normal cone and rod b-wave implicit times will not develop hereditary retinitis pigmentosa at a later time. Therefore the ERG, separated into rod and cone components, can be used in families with retinitis pigmentosa to establish the diagnosis of normality or abnormality relatively early in life and thus can help to determine visual prognosis in younger siblings of clearly affected individuals.[20,22,23,24]

At present no treatments are known for practically all types of retinitis pigmentosa (exceptions are the early stage of hereditary abetalipoproteinemia and possibly Refsum's disease). Nevertheless, early diagnosis has helped in genetic counseling, vocational orientation, and selection of candidates for possible therapeutic trials such as light deprivation of one eye with an opaque scleral contact lens.[19,25]

Electroretinograms in stationary forms of night blindness

Although the majority of patients with severe night blindness in all quadrants of the fundus have progressive retinal degenerations, some patients with night blindness due to retinal malfunction have stationary diseases. It is the normality or near normality of the cone system in full-field ERGs that allows separation of stationary forms of night blindness from practically all forms of night blindness associated with early stages of retinitis pigmentosa.[157]

One type of stationary night blindness is dominately inherited nyctalopia (Nougaret type). These patients (Fig. 17-9, top row) have a normal fundus appearance and have normal or nearly normal cone ERGs, but show no evidence of a rod ERG a-wave or b-wave. Fundus reflectometry studies have indicated that even though these patients are night blind, rhodopsin kinetics are normal[67] and cone pigment kinetics are also normal or nearly normal.[7] The absence of the rod a-wave as well as the absence of the electro-oculogram (EOG) light rise (p. 509) has suggested that this condition affects the proximal part of the rod photoreceptor or possibly the synaptic site between the rod photoreceptor and more proximal retinal cells.[64,67,133]

Another type of stationary night blindness is congenital nyctalopia with myopia (−3.5 to −14.5 D). This condition is inherited by either a sex-linked or autosomal recessive mode. These patients cannot attain normal dark-adapted rod thresholds and have fundus findings of myopia. They show no rod ERG b-wave in response to dim blue-light stimuli and have a characteristic cornea-negative response to white light in the dark-adapted state and a smaller cornea-negative response and a cornea-positive component in the light-adapted state (Fig. 17-26).[157,254] The waveforms in congenital nyctalopia with myopia can be explained by retention of the a-wave and b-wave from the cone system and the a-wave from the rod system with an absence of the b-wave from the rod sys-

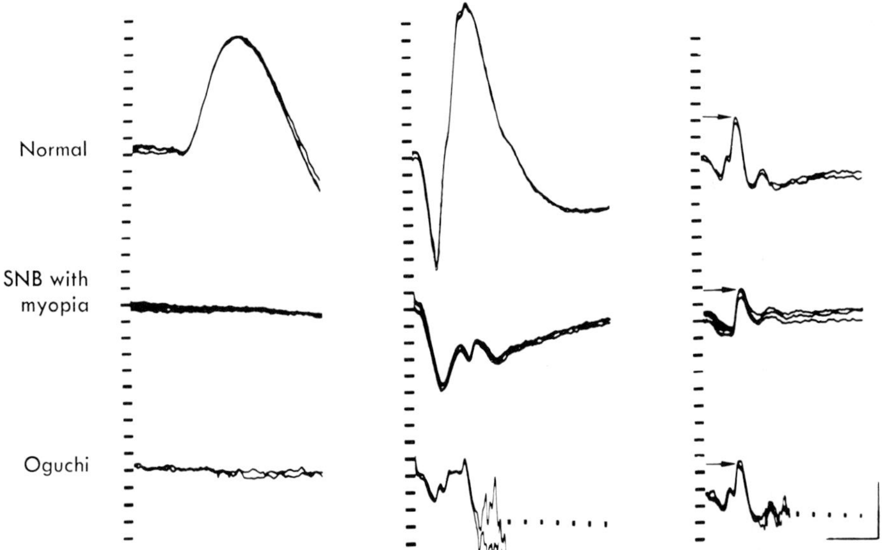

Fig. 17-26. ERGs from normal emmetropic subject (top row), patient with congenital stationary night blindness (SNB) with moderate myopia (middle row), and emmetropic patient with Oguchi's disease (bottom row). Responses were obtained to dim blue ($\lambda < 470$ nm) light (left) and white (16×10^3 ft-L) flashes (middle) after 1 hour of dark adaptation. Responses to white (16×10^3 ft-L) flashes in presence of steady 10 ft-L white background sufficient to eliminate rod contribution are illustrated in right column. Stimulus onset is vertical hatched line. Cornea positivity is upward deflection. Two or three successive responses are illustrated. Horizontal arrows (right column) designate cone b-wave implicit times. Responses from this patient with Oguchi's disease are often interrupted by reflex blinking, so latter part of some responses cannot be illustrated. Calibration symbol (lower right corner) designates 50 milliseconds horizontally and 100 μV vertically for all tracings. Stimulus flash duration was 10 microseconds.

tem; the ERG is a summation of all three components in the dark-adapted state and only the cone a- and cone b-wave in the light adapted state.[64,67] The defect appears to be some abnormality in intraretinal transmission of the response from the rod photoreceptors to the proximal retinal cells. Application of a high concentration of magnesium (which tends to inhibit the release of transmitter at nerve terminals) to the skate retina results in a selective loss of the b-wave and a waveform similar to that observed in patients with congenital nyctalopia with myopia.[98,244] Rhodopsin kinetics measured by fundus reflectometry are normal, and the light rise of the EOG (p. 509) is preserved, so this condition is thought to involve a defect that is more proximal in the retina than is the defect in the Nougaret nyctalope. ERG amplitudes are reduced (Fig. 17-26) in recessive nyctalopia with myopia compared with those from normal emmetropes; this probably occurs because of the known reduction of ERG amplitudes seen in patients with moderate axial myopia as the only finding.[45,86,179,233]

Recessively inherited stationary night blindness can occur without myopia. In one type the patient retains only a cone ERG, has a normal appearance to the fundus, and after prolonged dark adaptation attains only normal dark-adapted cone threshold on psychophysical testing. In another type, called Oguchi's disease, the patients usually require 2 to 12 hours to attain normal dark-adapted rod thresholds and show a characteristic change from a golden brown fundus in the light-adapted state to a normal color fundus in the dark-adapted state (the Mizuo phenomenon). Following 1 hour of dark adaptation (Fig. 17-26), patients with Oguchi's disease have no rod b-wave in response to dim blue light, a cornea-negative response to white light, and a normal cone response (Fig. 17-26, right column) to white light in the presence of a background light sufficient to eliminate the rod contribution to the ERG. Following complete dark adaptation (after 12 hours) some patients have a normal rod b-wave amplitude and normal rod b-wave implicit time, but only in response to one or two flashes of light.[135] The test flash used to elicit the ERG, although relatively dim, can be intense enough to light adapt the rod system. Rhodopsin kinetics are normal, and the light rise of the EOG is preserved; therefore the defect in Oguchi's disease, as well as in congenital nyctalopia with myopia, appears to be located more proximal in the retina than is the defect in the Nougaret nyctalope.

Another type of autosomal recessively inherited stationary night blindness is fundus albipunctatus. Patients have many yellow-white deposits in the deep retina, with greatest density in the posterior pole outside the macula.[66,199] Results of fundus reflectometry in one adult patient showed that both rod visual pigment and foveal cone pigment regenerate slowly. This patient had half-times of recovery for rod and cone pigments of about 60 and 20 minutes respectively, values approximately 20 and 16 times greater than those found in normal subjects. These changes in visual pigment levels paralleled the prolonged cone and rod limbs of his dark-adaptation curve. This patient's ERG and light rise of the EOG, abnormal after short-term dark adaptation, reached normal values within 3 hours of dark adaptation.[66] Other patients studied with the full-field ERG showed normal cone and rod amplitudes and normal b-wave implicit times after full dark adaptation.[199] The findings in fundus albipunctatus point to an abnormality in visual pigment regeneration and suggest an abnormality in the relationship between the photoreceptors and pigment epithelium.[66] The composition of the yellow-white lesions and their relationship to the abnormality of visual pigment regeneration are not known.

The defect in adaptation in Oguchi's disease differs from that of fundus albipunctatus in several respects. First, in Oguchi's disease the rod mechanism is selectively affected while the cone limb of the dark-adaptation curve as measured by psychophysical testing proceeds at a normal rate. Second, the rate of rhodopsin regeneration in Oguchi's disease as measured by fundus reflectometry appears normal because the defect presumably occurs proximal to the photoreceptors. Third, relatively weak illumination will delay the appearance of the rod branch of the dark-adaptation curve as measured by psychophysical testing by more than 30 minutes in Oguchi's disease, whereas exposure to a similar light produces a far less effect in fundus albipunctatus.[244]

For all types of stationary night blindness, correlation of the electrophysiologic findings with electron-microscopic studies remains to be done. It is interesting that these diseases, involving either the rod photoreceptor or cells proximal to the rod photoreceptor, are not associated with pigment migration or attenuation of the retinal vessels, whereas these latter findings are characteristically seen in patients with rod photoreceptor cell degeneration and retinitis pigmentosa.

Advances in electro-optical technology have resulted in night-vision devices that allow patients with stationary night blindness as well as patients with moderately advanced retinitis pigmentosa to use their cones to function under scotopic (starlight or moonlight) conditions or, if necessary, under dim photopic conditions (dusk). These light-amplifying devices have been incorporated into a monocular pocketscope or binocular goggle, which can be used as an aid to alleviate the symptom of night blindness.[21,37]

Hereditary diseases affecting the retina—some therapeutic considerations

Electroretinographic testing has allowed early detection of abnormal retinal function in some hereditary diseases affecting the retina and thereby has provided an opportunity for assessing the efficacy of therapeutic trials. The importance of early diagnosis was emphasized when the abnormal ERG (abnormal cone and rod function) in a patient with hereditary abetalipoproteinemia (Bassen-Kornzweig's disease) and night blindness was reversed to normal within hours with large doses of orally administered vitamin A. An older affected patient with a more advanced stage of this disease did not respond to this treatment.[143]

Another potentially treatable hereditary disease associated with night blindness and retinal degeneration is gyrate atrophy of the choroid and retina. Patients with this autosomal recessive disorder have myopia, constricted visual fields, elevated dark adaptation thresholds, small or nondetectable full-field ERGs, and chorioretinal atrophy distributed around the peripheral fundus and sometimes near the optic disc. The chorioretinal atrophy differs in extent among young patients of comparable age.[40] Patients develop cataracts and usually become virtually blind between ages 40 to 50 due to extensive chorioretinal atrophy. Biochemical abnormalities include 10- to 20-fold elevations of plasma ornithine,[39,40,201,257,272] hypolysinemia,[39,201] hyperornithinuria,[40,272] and virtual absence of ornithine ketoacid transaminase (OKT) in extracts of cultured skin fibroblasts[40,173,224,255,277] and in cultured lymphocytes.[169,279] Genetic heterogeneity appears to exist, since some patients have shown a 30% to 50% fall in plasma ornithine levels within a week when given orally administered vitamin B_6 (300 to 500 mg/day), while other patients have not responded.[40,292] All patients so far studied have shown a decline in plasma ornithine levels of 50% or more when placed on a low-protein (15 g/day), low-arginine diet.[41,280] It remains to be established whether any degree of biochemical responsiveness to diet, vitamin B_6, or both in patients with gyrate atrophy of the choroid and retina will alter ERG function or the course of this chorioretinal degeneration. About 2 years will be required to show whether or not stabilization has occurred in young patients, if one takes into account the natural history of the ocular disease.[40]

The retinal degeneration observed in Refsum's disease (heredopathia atactica polyneuritiformis) may also be amenable to treatment. Patients with this autosomal recessively inherited disease characteristically have a peripheral neuropathy, ataxia, and an elevated cerebrospinal fluid protein level with a normal cell count.[238] Ocular findings include an atypical retinitis pigmentosa–like appearance in the peripheral fundus and reduced or nondetectable ERGs.[239] Some have anosmia, deafness, pupillary abnormalities, lens opacities, electrocardiographic malfunction, skeletal abnormalities, and skin changes resembling ichthyosis. A diagnosis depends ultimately on the demonstration of an elevated serum phytanic acid.[238] Patients have shown a defect in alpha-hydroxylation of phytanic acid, and this enzyme defect can be detected in extracts of cultured skin fibroblasts.[263] Pathogenesis of this disease appears to involve a replacement of the long-chain fatty acids in phospholipids and triglycerides with phytanic acid derived from dietary sources. Accumulation of phytanic acid in many tissues then leads to consequent malfunction. Treatment with a low-phytol, low–phytanic acid diet (i.e., excluding green leafy vegetables, animal fats, and milk products) has led to lowered serum phytanic acid levels, improved nerve conduction times, and lowered cerebrospinal fluid protein levels.[105,238,239,264] At this time it is not known whether dietary treatment will result in improved or stabilized retinal function over the long term.

At the time of detection most patients so far studied with gyrate atrophy or Refsum's disease have shown full-field ERGs that have been very small or nondetectable. Recently a patient with gyrate atrophy and one with Refsum's disease have been detected in the early stages; these patients had large enough responses to separate into rod and cone components (Fig. 17-27). Interestingly, the full-field rod and cone responses from these patients, although reduced in amplitude, were within our normal range with respect to b-wave implicit times. Similarly, a patient with the early stages of hereditary abetalipoproteinemia was not reported to show marked delays in cone or rod b-wave implicit times.[143] The normal temporal aspects of these ERGs contrast with the substantially delayed cone or rod b-wave implicit times, or both, recorded from patients with early stages of the more common forms of wide-spread hereditary retinitis pigmentosa in whom no metabolic abnormalities have yet been found.

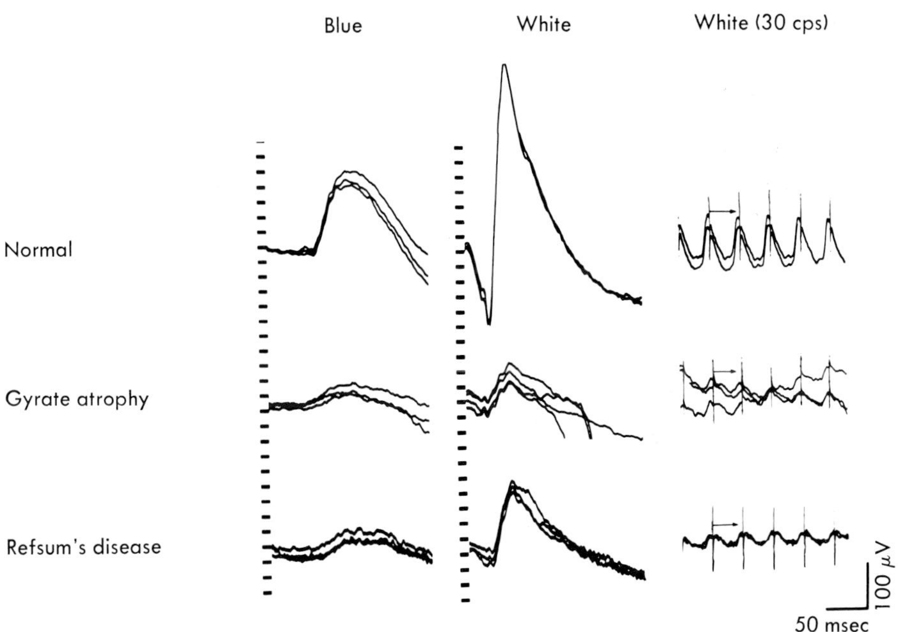

Fig. 17-27. Full-field ERGs to single flashes of dim blue ($\lambda < 470$ nm) light, single flashes of white light under dark-adapted conditions, and white 30 Hz flickering light for a normal subject, a 12-year-old female with gyrate atrophy of the choroid and retina, and a 30-year-old male with Refsum's disease. Stimulus onset is designated by the vertical hatched lines in columns 1 and 2 and the vertical shock artifacts in column 3. Arrows (column 3) designate cone b-wave implicit times. The responses of these patients, although reduced in amplitude, fall within the normal range with respect to b-wave implicit time for the rod isolated response to blue light (normal range, 71 to 108 milliseconds in this test system) and for the cone isolated response to 30 Hz flicker (normal range, 25 to 32 milliseconds).

Focal electroretinograms

The total cone population in the human retina is approximately 6.8×10^6; the number of cones in the macula (central 10 degrees) is maximally 4.4×10^5; therefore the macula contains maximally 7% of the total retinal cone population.[228,296] The foveola (about 1 degree) contains about 30,000 cones compared with 130 million receptors across the entire retina. This could explain the inability of the full-field ERG system to detect abnormalities confined to the foveola, to the fovea (5 degrees), or to the macula; a patient can lose foveal photoreceptor function and have 20/200 vision with a normal full-field ERG. Conversely, a patient with advanced retinitis pigmentosa and less than 10-degree central fields can still retain 20/20 vision and have a nondetectable full-field ERG.

One difficulty in recording the small foveal or macular cone ERG at the cornea is that scattered light from a focal test stimulus can evoke a response from many receptors outside the macula, and therefore the foveal or macular ERG cannot be seen in isolation. Furthermore, the macula contains as many rods as cones, and the macular rod ERG can thereby obscure the contribution from the macular or foveal cones. Biersdorf and Diller[44] used computer averaging with a small test stimulus (flickering red stimulus, $<$ 4-degree visual angle) superimposed on a large (50-degree) background sufficient to eliminate the rod contribution and obtained a local ERG at the cornea from the macular area. They compared the ratio of the ERG response produced by stimulation of the macula to that obtained by light flashed on the optic disc to try to account for the problem of stray light; they reported that this technique could be used to detect macular degenerations in patients when their visual acuity decreased to the 20/40 to 20/80 range.

Riggs, Johnson, and Schick[168,243] have isolated photopic (cone) responses by stimulating the eye with a barred stimulus pattern; each time the bars shift position the stripes that are dark become light and vice versa. With phase alternation of this fine barred stimulus pattern, the overall amount of light on the retina is always the same and the amount of stray light outside the stimulus area proper also remains relatively constant. The response from the cone system to a stimulus (19-degree visual angle) presented at 8 to 10 alternations/sec is a sinusoidal waveform with amplitudes ranging up to 7 to 8 μV. When the visual angle of the stimulus was $<$ 2 degrees and centrally placed on the fovea, Armington[11] showed that this response was barely detectable even with signal averaging.

Foveal and parafoveal cone ERGs have also

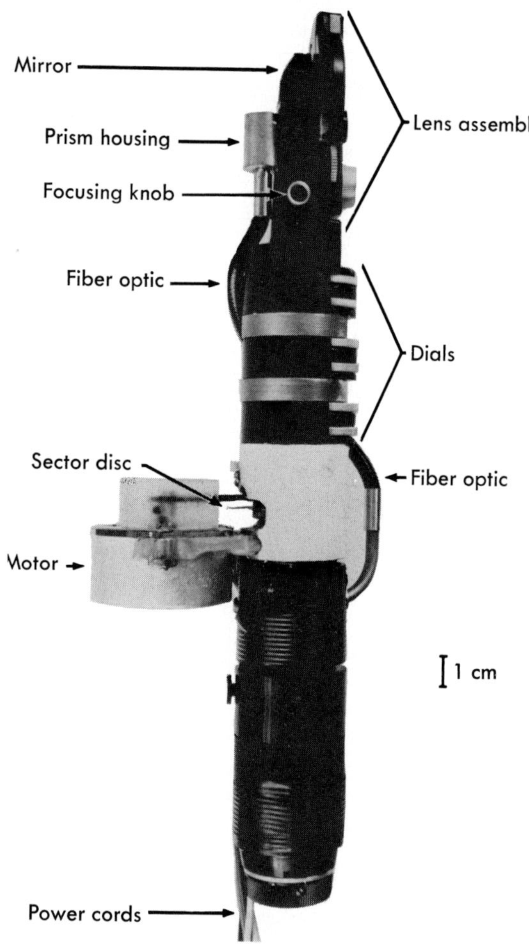

Fig. 17-28. Two-channel stimulator ophthalmoscope. Stimulus channel internal to instrument; surround (or background) channel conducted in fiber optic bundle as shown. A motor-driven sector disc is used to regulate frequency of stimulus. Dials can be used to modify size, wavelength, and intensity of stimulus. Instrument weighs little more than a standard direct ophthalmoscope. (From Sandberg, M. A., Berson, E. L., and Ariel, M.: Arch. Ophthalmol. **95:**1805, 1977.)

been elicited with a hand-held, two-channel stimulator ophthalmoscope (Fig. 17-28). This instrument has one light source for both stimulus and surround in the base of the instrument.[246] The stimulus light channel, contained within the body of the instrument, can be focused on the fundus by a focusing knob. The surround light channel is conducted from the base to the head of the instrument by a fiber optic and prism housing. The optics in the head of the instrument are arranged so that lights from the stimulus and surround channels are reflected by a mirror to enter the patient's pupil coaxially and in maxwellian view. The stimulus centered within the surround is viewed on the fundus by the examiner through the lens assembly in the head of the instrument. With this instrument the levels of stimulus retinal illuminance and surround retinal illuminance can be precisely adjusted. The surround light is used not only to minimize the effect of stray light from the stimulus but also to provide sufficient retinal illumination to visualize retinal landmarks.

The eye to be tested is dilated with a mydriatic to facilitate visualization of the fundus. A double-electrode contact lens is placed on the topically anesthetized cornea and a ground electrode on the forehead; the stimulus on the fundus can be visualized through the contact lens without difficulty. Responses are differentially amplified, tuned by a bandpass filter, and computer averaged. The computer is triggered by a photocell in the stimulator ophthalmoscope. The signal-averaging computer contains a bipolar artifact reject buffer so that voltages > 5 μV, presumably due to eye movements, are eliminated from the averaged response. The examiner can compensate for small eye movements by adjusting this hand-held instrument. If eye movements occur that are so large that the examiner cannot keep the stimulus on the area of interest, the examiner interrupts the computer averaging by releasing a foot pedal switch. For a given retinal area at least three consecutive computer summations are performed; this typically requires about 3 minutes of recording time.[249]

Representative focal cone ERGs to a 4-degree white (4.8 log troland) flickering stimulus centered within a 10-degree white (5.5 log troland) steady annular surround are illustrated (Fig. 17-29) for a normal subject (top row), a patient with a one-disc diameter central macular scar (second row), and a patient with strabismic amblyopia (bottom row) with the stimulus centered on the fovea or centered 5 degrees nasal to the foveola. Recordings done at two stimulus frequencies (42 Hz, illustrated, as well as 28 Hz, not illustrated) established the relationship of stimulus onset to the corresponding (i.e., next but one) response peak. Normal subjects showed smaller responses in the parafovea than the fovea consistent with the known fall in cone density with increasing eccentricity from the foveola. The patient with a macular scar showed responses indistinguishable from noise when the stimulus was centered within the scar but normal responses when the stimulus was centered outside of the scar in a parafoveal area that appeared normal on ophthalmoscopic examination. For the patient with strabismic amblyopia the stimulus could be maintained on the fovea or parafovea despite variations in eye position, and a normal ERG was recorded in both areas. A congenital rod monochromat showed responses (not illustrated) that were indistinguishable from noise. These findings helped to establish that these ERGs were focal responses generated by the cone system under these test conditions.[249]

Foveal cone ERGs are illustrated (Fig. 17-30) for a normal subject and three patients with juvenile hereditary macular degeneration with visual acuities ranging from 20/60 to 20/200. These patients had normal full-field flicker cone ERGs but foveal responses were reduced in amplitude (i.e., < 0.18 μV) without (patient 1) or with (patient 2) delays (i.e., > 38 milliseconds) in b-wave implicit time or indistinguishable from noise (patient 3). Focal cone ERGs elicited with the stimulator ophthalmoscope have proved useful for detecting and quantitating macular malfunction in all patients tested with early stages of juvenile hereditary macular degeneration with visual acuity reduced to 20/50 or below. Patients with visual acuity less than 20/100 had smaller and slower foveal cone ERGs than those with better visual acuity. Similar testing has been performed on

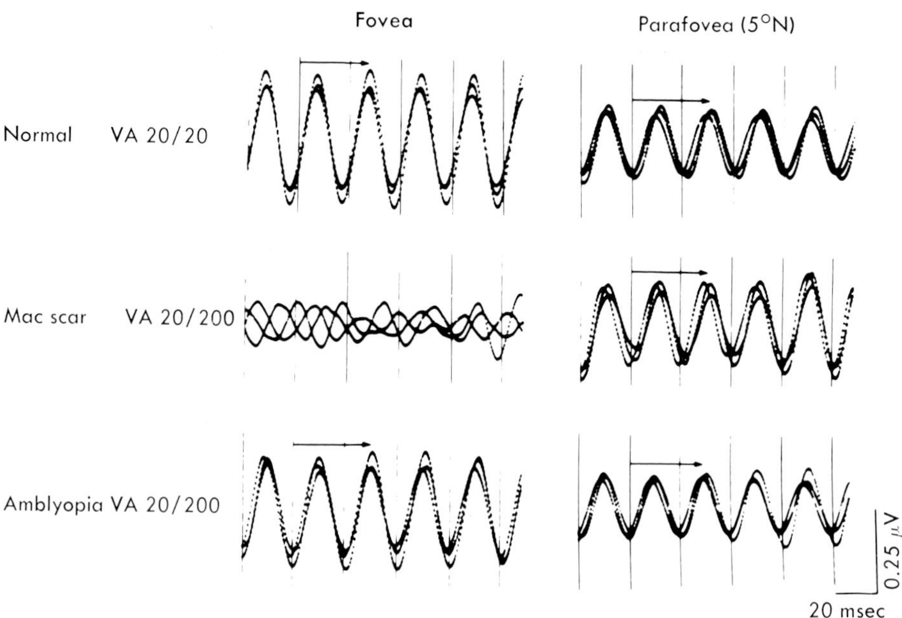

Fig. 17-29. Focal cone ERGs from a normal subject, a patient with a 1-disc diameter central macular scar, and a patient with strabismic amblyopia elicited with a 4-degree, 42 Hz white flickering stimulus centered on the fovea (left column) or centered 5 degrees nasal to the foveola (right column). Responses represent computer summation of 128 sweeps; three consecutive runs are illustrated. Arrows designate b-wave implicit times for detectable responses. (From Sandberg, M. A., Jacobson, S. G., and Berson, E. L.: Am. J. Ophthalmol. **88:**702, 1979.)

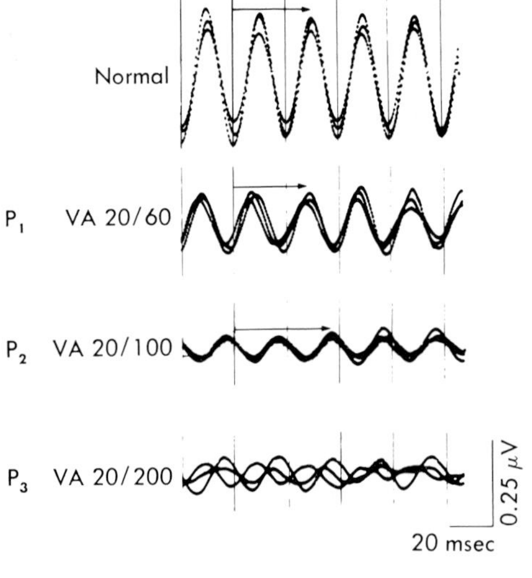

Fig. 17-30. Foveal cone ERGs for a normal subject and three patients (P_1, P_2, and P_3), ages 11, 13, and 23 respectively, with juvenile hereditary macular degeneration (Stargardt's disease). Responses represent computer summation of 128 sweeps; three consecutive runs are illustrated. Arrows designate b-wave implicit times for detectable responses. (From Sandberg, M. A., Jacobson, S. G., and Berson, E. L.: Am. J. Ophthalmol. **88:**702, 1979.)

patients with hereditary retinitis pigmentosa with reduced and delayed full-field cone ERGs, and all of those tested with 20/40 or below have shown foveal cone ERGs that are reduced in amplitude with normal b-wave implicit times[249] (Fig. 17-31).

Foveal cone ERGs elicited with a spot stimulus have been compared to midperipheral cone ERGs elicited by an annular stimulus concentric with the fovea in three young patients with dominant retinitis pigmentosa with reduced penetrance positioned on a bite bar before a maxwellian view optical system. Focal cone ERGs were recorded from the green and red cone systems in combination by a method of rod silent substitution. With this method two alternating lights of different wavelength, matched in brightness for the rods, were presented to the same retinal area; the brighter light for the green and red cones was designated as the stimulus increment (I_s) and the dimmer light as the background (I_b). This method was chosen so that cone ERGs unmodified by any rod contribution could be measured at low levels of cone adaptation; testing could be done at a low frequency to separate the b-wave of one response from the a-wave of the next response. In these three patients with full-field cone ERGs that were normal or nearly normal in amplitude but substantially delayed in b-wave implicit time, foveal cone ERGs elicited with rod silent substitution were normal in amplitude and b-wave implicit time while midperipheral cone ERGs were minimally reduced but substantially delayed (Fig. 17-32). These findings support the idea

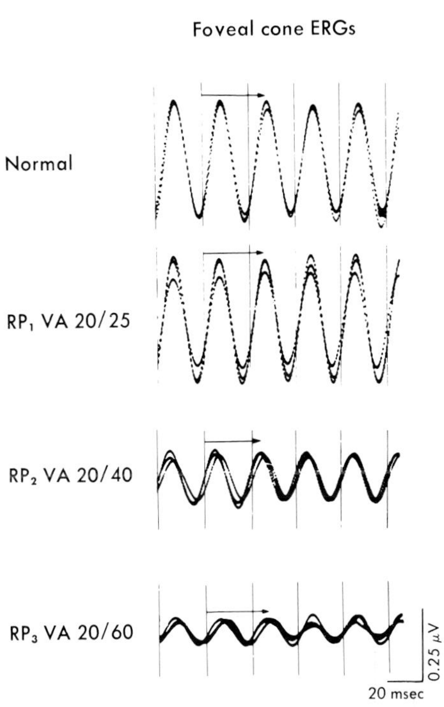

Fig. 17-31. Foveal cone ERGs from a normal subject, a 17-year-old patient with dominant retinitis pigmentosa with reduced penetrance *(RP_1)*, a 32-year-old patient with autosomal recessive retinitis pigmentosa *(RP_2)*, and a 24-year-old patient with sex-linked retinitis pigmentosa *(RP_3)*. Three consecutive computer summations ($n=256$) are shown. Arrows designate b-wave implicit times. (From Sandberg, M. A., Jacobson, S. G., and Berson, E. L.: Am. J. Ophthalmol. **88:**702, 1979.)

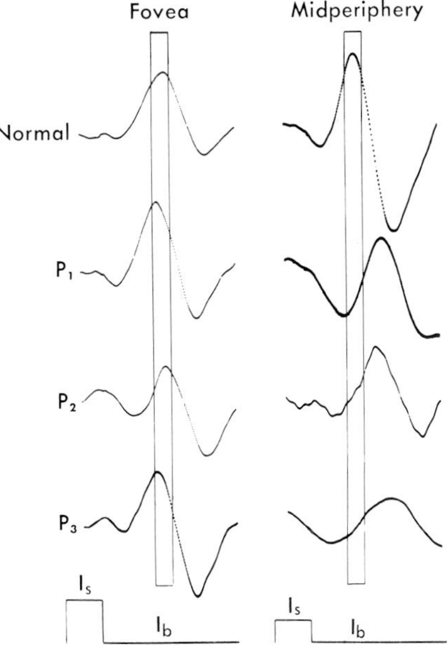

Fig. 17-32. Focal ERGs from the green and red cone systems in combination for a normal subject and three young patients (P_1, P_2, and P_3) with dominant retinitis pigmentosa with reduced penetrance in response to a 3.75-degree diameter spot centered on the foveola (left column) and a 38 to 44 degree diameter annulus concentric with the foveola (right column). I_s at 556 nm (1700 photopic trolands) is alternated at 5 Hz with I_b at 500 nm (120 photopic trolands). Vertical bars designate normal limits for b-wave implicit time. Calibration markers signify onset and duration of 20 milliseconds I_s horizontally and 0.25 μV vertically. (From Sandberg, M. A., and Effron, M. H., and Berson, E. L.: Invest. Ophthalmol, Vis. Sci. **17:**1096, 1978.)

that the abnormal full-field cone ERGs seen in these patients are due to an abnormal extrafoveal cone contribution to their full-field responses.[248]

Efforts have been made to simulate in normal subjects the abnormal midperipheral cone ERGs seen in these patients. Minimal reductions in amplitude with substantial delays in b-wave implicit time could be recorded from normal subjects in the midperiphery when the brightness of I_b was reduced more than the brightness of I_s while still maintaining rod silent substitution. This was not observed in normal subjects when the brightnesses of I_b and I_s were proportionately diminished (i.e., neutral density effect or simulation of decreased visual pigment in the photoreceptors) or when the area of I_b and I_s was decreased (i.e., reduced width of annulus or simulation of decreased numbers of photoreceptors). These findings support the idea that the abnormal midperipheral cone ERGs observed in young patients with dominant retinitis pigmentosa with reduced penetrance are due in part to a lower than normal state of light adaptation of the cone system and that reduced amount of visual pigment in the photoreceptors or reduced numbers of photoreceptors cannot be solely responsible for their abnormal cone ERGs.[248] These findings, in addition to those obtained with full-field ERG testing (Fig. 17-24), suggest that many patients with retinitis pigmentosa have a defect in the mechanism by which their cone system adapts to light.

Foveal function in the electroretinogram and the visually evoked cortical potential

Whereas the fovea contributes a small percentage (estimated $< 2\%$) of the response amplitude of the full-field ERG from all the cones across the entire retina, the visually evoked cortical potential (VECP) recorded at the occiput is dominated by the foveal cones.[234,242] This is undoubtedly due in part to the fact that the ratio of foveal to extrafoveal representation is increased by a factor of about 1000 in the area striata[242] compared with that in the retina. In addition, the foveal projection in the area striata is located superficially at the occipital pole, whereas the peripheral retinal projection is located deeper (more anteriorly) in the occipital lobe; therefore the foveal projection is relatively close to the scalp electrodes placed over the occipital pole to record the VECP.

The waveform of the VECP (also called the VER, or visually evoked response) is variable and complex; a typical waveform (Fig. 17-33) to a slowly repetitive stimulus contains the primary response (waves 1, 2, and 3), the secondary response (4 and 5), and the after discharge (rhythmic waves beginning with wave 6).[18,121] Rietveld and co-workers[240] found that a unit retinal area, 25 minutes of arc from the foveal center, contributed to the VECP only 0.1 as much as the fovea itself. It has been estimated that the central 2-degree area generates 65% of the response. The implicit times are related within limits to stimulus intensity, but are not much shortened by stimuli with field size greater than 1 degree.[283]

The effects of image blur of the test stimulus, attention of the subject, binocular rivalry, and other variables on the VECP are well described in other texts.[179,234,235] The problem of stray light has been minimized by use of pattern-reversal stimulation to elicit the VECP, and with this technique, abnormalities have been detected in patients with strabismic amblyopia.[259] The VECP is also abnormal in optic atrophy,[164] although no data exist to quantitate the degree of damage necessary to produce a significant alteration of the VECP. It has had limited, if any, value in assessing macular function behind a lens opacity because a smaller than normal response can result from the reduction of stimulus intensity and image blur on the retina produced by the opacity.[179]

Variable fixation of the eyes presents a technical problem in placing the stimulus on the fovea, and this problem had limited application of VECP testing to evaluation of central foveal (< 2 degrees) function in patients with suspected disease. This problem of variable fixation has been minimized by viewing the stimulus on the central fovea during testing.[104,158,163,166,246,256] As an example, the two-channel stimulator ophthalmoscope (Fig. 17-28) has been used to elicit focal cone visually evoked cortical responses.[246] For this testing a 5 Hz stimulus,

1.5 degrees or smaller, is centrally superimposed on a steady 10-degree background. Bipolar responses are recorded between the inion and vertex with a ground electrode attached to the patient's ear lobe. The sensitivity of this technique is illustrated in the detection of abnormal foveolar function in a patient with a foveolar cyst that subtended 30 minutes of arc and visual acuity reduced to only 20/25 (Fig. 17-34). When a stimulus larger than the cyst was centered on the cyst (area A), the patient could see the flickering stimulus, and focal cone VECPs were within the normal range. When a stimulus that subtended 10 minutes of arc was centered on the cyst, the patient could not see the flickering stimulus, and the focal cone VECP was indistinguishable from noise. When the 10-minute stimulus was positioned just nasal to the cyst in a portion of the foveola that appeared normal on ophthalmoscopic examination (area B), the patient could see the flickering stimulus and the VECP was clearly detectable. With a weak stimulus increment and a 1.5-degree flickering stimulus centrally superimposed on a steady 10-degree background, all patients who have been tested with visual acuity of 20/25 or below and juvenile hereditary macular degeneration have shown abnormal focal cone VECPs. Abnormal VECPs have been either shifted in phase with or without reductions in amplitude or so reduced in ampli-

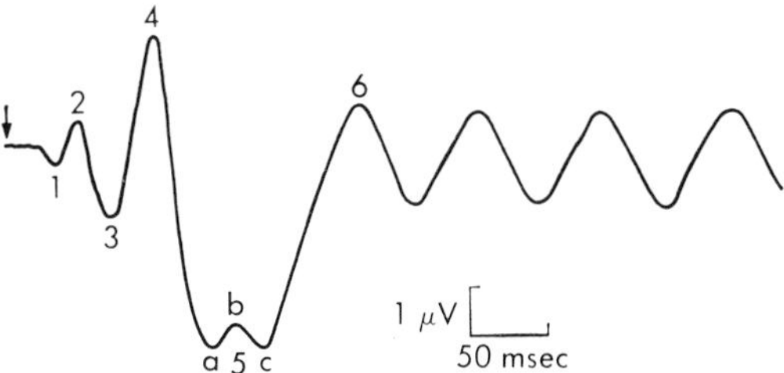

Fig. 17-33. Representative VECP recorded with monopolar electrode (inion-ear) in response to white light stimulus. Cumulation of 50 sweeps of computer. (From Bergamini, L., and Bergamasco, B.: Cortical evoked potentials in man, Springfield, Ill., 1967, Charles C Thomas, Publisher.)

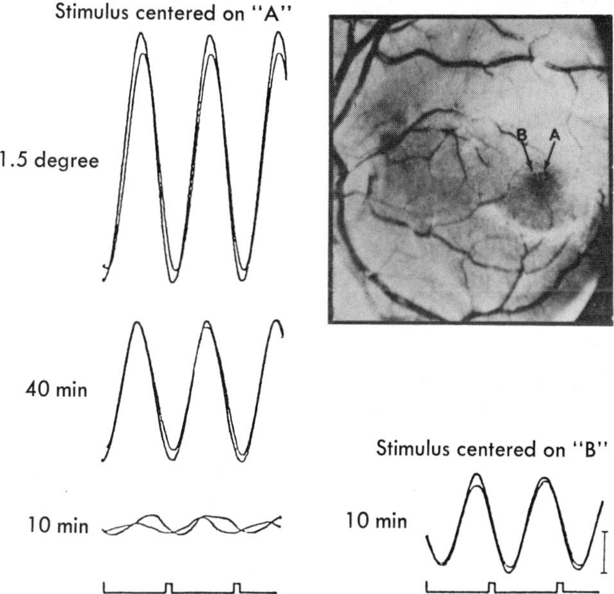

Fig. 17-34. Focal cone VECPs from left eye of patient with bilateral foveolar cysts; responses were obtained to 7.0 log troland white stimuli of varying diameter flickering at 5 Hz. Stimuli were centrally superimposed on steady 5.0 log troland white background. Fundus photograph of left eye from this patient illustrates foveolar cyst. The stimulus was centered on cyst (area *A*) or just nasal to cyst (area *B*). Two runs to same stimulus condition are superimposed; calibration (lower right) is 1 μV and vertical deflections (lower row) indicate onset and duration of each stimulus. (From Sandberg, M. A., Berson, E. L., and Ariel, M.: Arch. Ophthalmol. **95:**1805, 1977.)

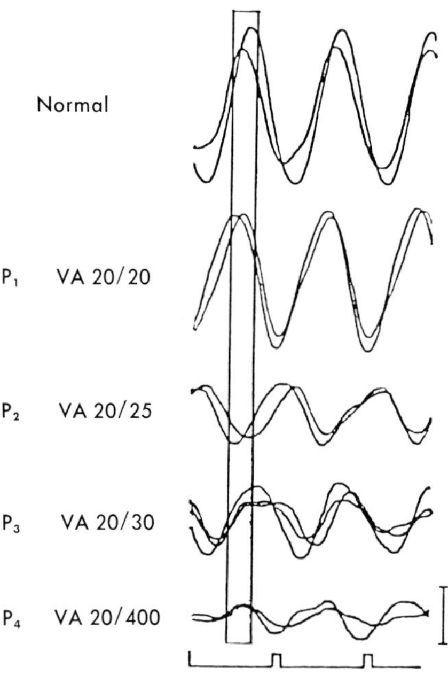

Fig. 17-35. Focal cone VECPs from a normal subject and members of a family with juvenile hereditary macular degeneration (i.e., Stargardt's disease); responses were elicited to a 1.5-degree, 5 Hz, 5.6 log troland white stimulus centered on 10-degree, 5.0 log troland steady white background. Stimulus was centered on foveola during testing. P_1 through P_4 were 14, 11, 13, and 17 years of age, respectively P_1 has normal fundus and normal VECP; P_2, P_3, P_4 had fundus abnormalities and abnormal VECPs. Vertical bar defines normal range of peak latencies. At least two runs to same stimulus condition are superimposed; calibration (lower right) is 1 μV and vertical deflections (lower row) indicate onset and duration of stimulus. (From Sandberg, M. A., Berson, E. L., and Ariel, M.: Arch. Ophthalmol. **95:**1805, 1977.)

tude as to be indistinguishable from noise (Fig. 17-35).[246]

Although the VECP can be a sensitive index of central visual malfunction, it has not been helpful in separating visual loss on a retinal basis from visual loss due to disease in the optic nerve or cerebral cortex. The VECP has been abnormal in macular degeneration, optic atrophy, and strabismic amblyopia. In contrast, the foveal ERG, if abnormal, helps to localize the site of visual loss to the preganglion cell retina; the foveal cone ERG has been abnormal in patients with hereditary macular degeneration[165,249] (visual acuity of 20/50 or less), but has been normal in patients with optic atrophy[165] or strabismic amblyopia.[165,249]

EARLY RECEPTOR POTENTIAL

Origin of components

The early receptor potential (ERP) is a rapid response (Fig. 17-36) that can be detected when the retina is stimulated with an intense flash of light.[56,74] Stimulus intensity required to generate the ERP is approximately 10^6 times brighter than that required to elicit the ERG (Fig. 17-37).[75] The human ERP is completed within 1.5 milliseconds and is followed by the leading edge of the a-wave of the ERG.[26,65,120,298] The human ERP recorded at the cornea has a waveform similar to that recorded at the cornea or intraretinally from other animals. The initial cornea-positive phase, R1, has been associated with the conversion of lumirhodopsin to metarhodopsin I,[231] and the later cornea-negative phase, R2, with the conversion of metarhodopsin I to metarhodopsin II in the rat retina.[76] In the rat the action spectrum of the ERP corresponds with the absorption spectrum of rhodopsin.[75] ERP amplitude is linearly proportional to the amount of unbleached rhodopsin in the rat retina. Although rhodopsin is the predominant visual pigment in frog, monkey, and human retinas, the ERP in these species is generated primarily by cone visual pigments.[123-125] In humans, Goldstein and Berson[126] found that 60% to 80% of the total amplitude of the ERP is generated by the cones and 20% to 40% by the rods, and Carr and Siegel[65] confirmed the cone dominance of the human ERP by showing that the ERP action

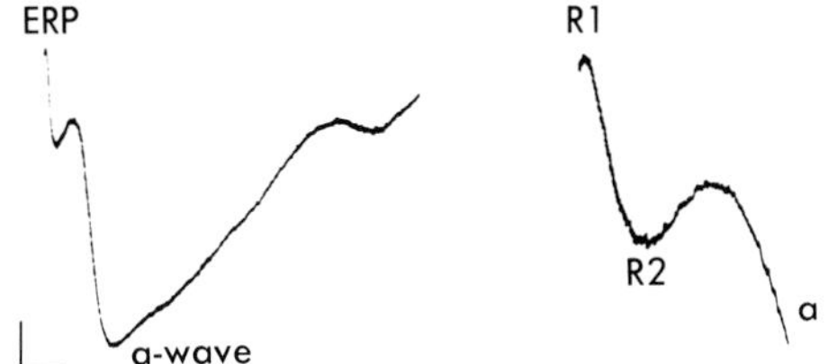

Fig. 17-36. Normal human ERP followed by a-wave of ERG (left tracing); cornea-positive peak, *R1*, and later cornea-negative, *R2*, of ERP are designated (right tracing). Stimulus onset is at beginning of each trace, and upward deflection from beginning represents cornea positivity. Calibration symbol signifies 2 milliseconds horizontally and 100 μV vertically for left tracing, and 0.5 milliseconds horizontally and 50 μV vertically for right tracing. (From Berson, E. L., and Goldstein, E. B.: Arch. Ophthalmol. **83:** 412, 1970.)

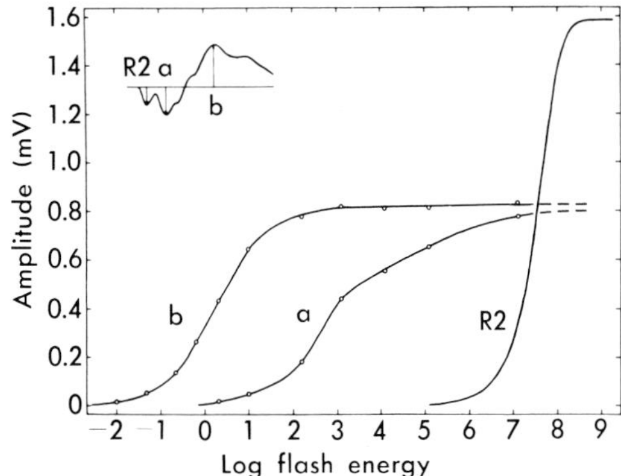

Fig. 17-37. Amplitudes of *R2*, a-wave, and b-wave in ERG of dark-adapted albino rat as functions of log of stimulus flash energy. On this semilog plot a response such as *R2*, which is proportional to flash energy, becomes a rapidly upward curving line. Data for a- and b-wave amplitudes were obtained from single rat, leaving sufficient time between flashes to allow eye to dark adapt. Data similar to these have been obtained from more than 10 animals. Uniform illumination of entire retina was obtained by placing small, uniformly illuminated section of Ping-Pong ball over eye. Pentobarbital, 50 mg/kg, was used for anesthesia; homatropine (1%), to dilate pupil. Flash duration 0.7 millisecond; white light. Log (flash energy) = 0 corresponds to 1 quantum/rod. Amplitudes were measured from baseline as shown in diagram. (From Cone, R. A.: Cold Spring Harbor Symp. Quant. Biol. **30:**483, 1965.)

spectrum of the dark-adapted eye matched the human photopic luminosity curve (with peak sensitivity near 555 nm).

The ERP response consists of a photostable component (probably about 5% of the response) and a photolabile component (about 95% of the response).[55] The photostable component can be recorded from the eyecup (without retina) preparation in the frog and is presumably generated by light stimulation of melanin granules in the pigment epithelium. The photolabile component is generated in the receptor outer segments, and regeneration of this component can be correlated with the regeneration rate of photoreceptor visual pigments.

Physiologic basis

The lack of a measurable latency, as well as the correlation of components of the ERP with the accumulation of specific photoproducts of bleaching, supports the idea that this response is generated in the receptor outer segments. ERP amplitude depends not only on the concentration of unbleached visual pigment in the receptor but also on the orientation of the visual pigments within the outer segments,[53,77] the orientation of the outer segment membrane,[9] and the chemical environment of the receptors.[53,232] When the retina is heated, the ERP disappears at just the temperature at which the visual pigment molecules lose their regular orientation.[77] The ERP is thought to depend on charge displacement in the outer segments during photochemical reactions. One possible reason for the cone dominance of the human ERP, even though rhodopsin is the predominant visual pigment, is that many cone disc lumina are continuous with the extracellular space (the vitreous and therefore the cornea), whereas the majority of rod outer segment discs are isolated from and surrounded by a plasma membrane and therefore are "shielded" from the vitreous and cornea. Charge displacement in the "unshielded" cone discs could explain the cone dominance of the human and monkey ERP recorded at the cornea.

Technique for recording the human early receptor potential

In contrast to the double-electrode contact lens used for recording the ERG, a special scleral lens with the electrode in a side arm

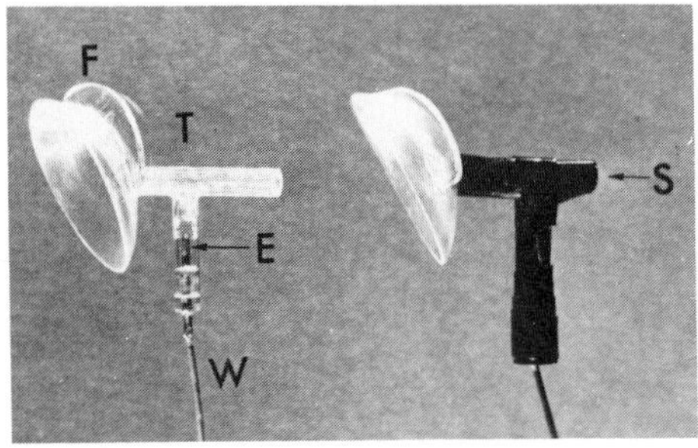

Fig. 17-38. Lenses for recording human ERP. Black tape has been removed from T-tube of left lens to show electrode, *E*. Letters designate silver recording electrode, *E*, mounted in plastic T-tube, *T*. T-tube is attached to inferior corneal curvature of lens. Electrode is shielded from light with black tape (right lens) to prevent photovoltaic artifacts. Contact between electrode and cornea is through saline bridge. Saline solution is introduced at *S* through polyethylene tube attached to T-tube. Movement of upper and lower lids was limited by lid flange, *F*, and T-tube. Wire lead, *W*, from electrode is plugged into junction box, which is in turn connected to preamplifier. Electrical resistance between front of cornea and electrode was less than 32,000 Ω. (From Berson, E. L., and Goldstein, E. B.: Arch. Ophthalmol. **83:**412, 1970.)

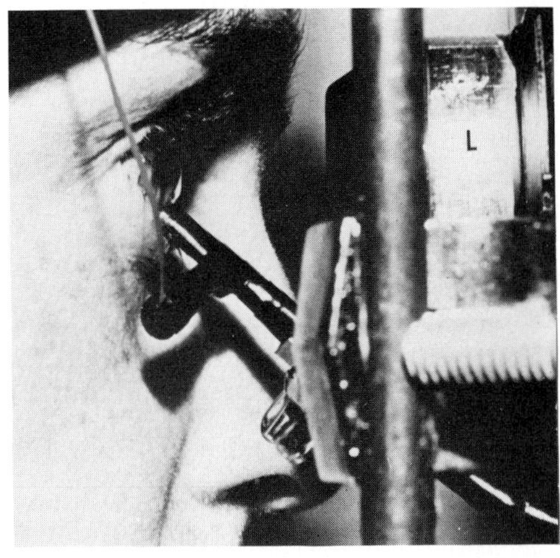

Fig. 17-39. Side view of subject with scleral contact lens electrode for recording ERP on topically anesthetized right eye. Stimulus from electronic flash gun is focused into eye through condensing lens, *L*. T-tube attached to lens, covered with black tape during testing, shields silver recording electrode from stimulus. (From Berson, E. L., and Goldstein, E. B.: Arch. Ophthalmol. **83:**412, 1970.)

(Fig. 17-38) is required to record the ERP.[26] This lens was designed to eliminate the photovoltaic artifact that occurs within the first millisecond after light strikes a metallic electrode. The lens and side arm are filled with saline solution to ensure a saline bridge between the cornea and the recording electrode, and the side arm is covered with black tape to protect the recording electrode from exposure to the stimulus flash as well as reflected light from the eye. The response is conducted to the electrode in the side arm. The reference electrode is taped to the forehead, and the ground is attached behind the ear. The routine ganzfeld, used to record the ERG, cannot be used to record the ERP because the light stimulus is too dim. The patient therefore sits in front of a flash gun (shielded), and the stimulus flash is presented in maxwellian view to maximize stimulus intensity (Fig. 17-39).

Early receptor potential responses in retinal disease

The ERP obtained under conditions of dark adaptation has been found to be subnormal in amplitude in young patients with dominant forms of retinitis pigmentosa (Fig. 17-40)[26] as well as sex-linked[27] and autosomal recessive retinitis pigmentosa.[273] The temporal aspects of R2 have not been delayed in patients with delays in ERG implicit times[26] (Fig. 17-41). The abnormal ERP amplitudes localized a defect in the receptor outer segments even at a time when affected children had 20/20 visual acuity, full fields with conventional perimetric testing, and minimal or absent changes on ophthalmoscopic examination. In patients with retinitis pigmentosa it has not been determined whether the photoreceptors are altered primarily or become altered secondary to some change in the pigment epithelium. Possible explanations[26] for the decrease in ERP amplitude could be a decrease in the concentration of unbleached visual pigment in the diseased receptors, a change in the orientation of the outer segment membrane, a change in the orientation of visual pigment molecules within the outer segments, a change in the chemical environment of the receptors, or possibly a decrease in the number of photoreceptors, or any combination of these.

In addition to the reduction in ERP ampli-

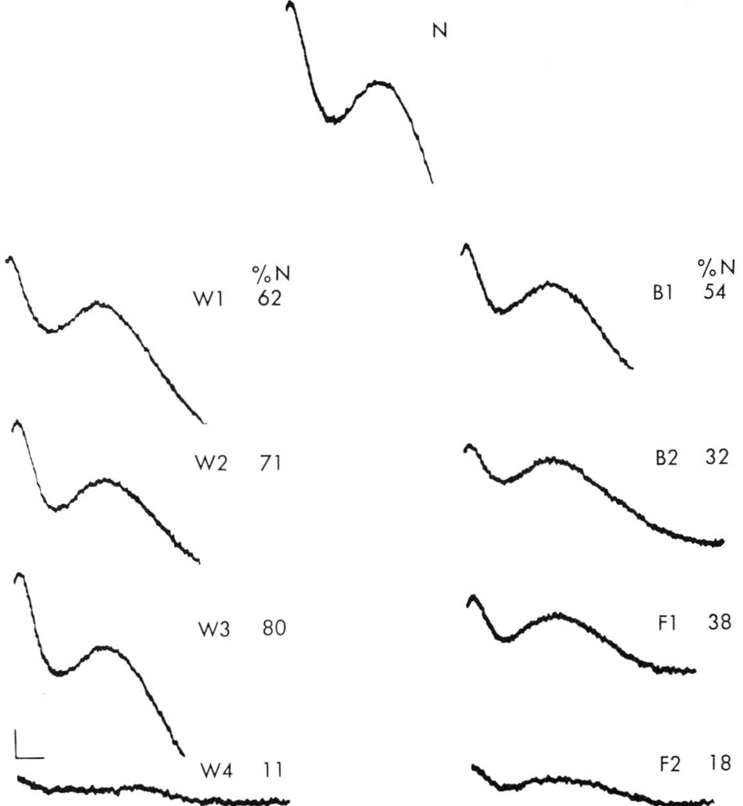

Fig. 17-40. Representative ERPs for normal subject, *N*, and eight patients with dominantly inherited retinitis pigmentosa, families *W*, *B*, and *F*. Responses to single flashes of white light recorded after patient had been dark adapted for a minimum of 1 hour. Amplitudes (from cornea-positive peak, *R1*, to cornea-negative peak, *R2*) for each patient are expressed as percentage of average normal response of 191 μV. Averages are based on 10 to 20 responses for each patient, except *F1* and *F2*, which are based on two or three responses. Stimulus onset is at beginning of trace. Calibration symbol (lower left corner) is 50 μV vertically and 0.5 millisecond horizontally. (From Berson, E. L., and Goldstein, E. B.: Arch. Ophthalmol. **83:**412, 1970.)

tude, patients with early retinitis pigmentosa have shown faster than normal ERP recovery rates during dark adaptation after a bleaching flash. These abnormal ERP recovery rates could be described by exponential functions, and the half-times of regeneration in children with dominant disease were about twice as fast as the half-times of regeneration recorded from normal subjects tested under the same bleaching conditions.[28] The half-times of regeneration have appeared to be characteristic for a given family. Although loss of rod function may contribute to the accleration of ERP recovery rates in patients with retinitis pigmentosa, it is difficult to explain the faster than normal recovery rates of patients with retinitis pigmentosa on the basis of loss of rod function alone.[29] Since the human ERP has been shown to be generated primarily by the cones and since ERP recovery rates have been correlated with regeneration rates of visual pigments, the faster than normal ERP recovery rate in patients with early retinitis pigmentosa suggested as one possibility that some abnormality in the cone pigment regeneration process had occurred in these diseased retinas. Faster than normal recovery rates were measured in both dominant and sex-linked retinitis pigmentosa.[28,29]

Although no complications have been reported in recording the human ERP, the response is technically more difficult to obtain than the ERG because it involves careful posi-

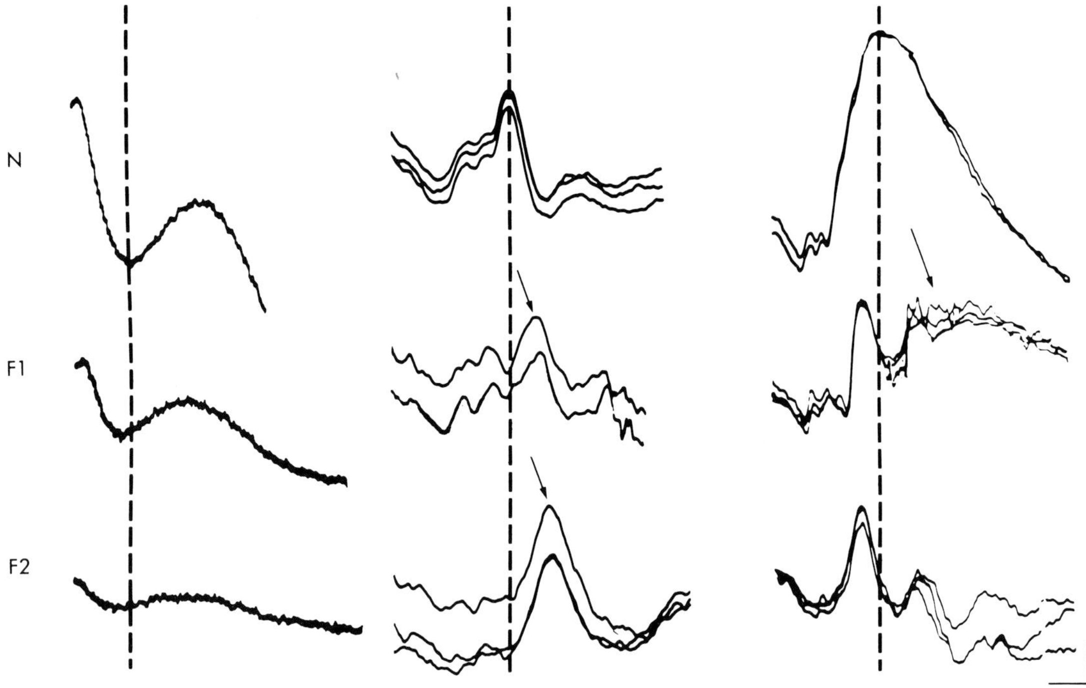

Fig. 17-41. ERPs (column 1) and ERGs (columns 2 and 3) for normal subject, *N*, and two children with dominant retinitis pigmentosa with reduced penetrance. Responses in column 2 are cone responses to 32 × 10^3 ft-L white light flash in presence of steady white background light of 10 ft-L. Responses in column 3 are dark-adapted ERGs to single flashes of long-wavelength light (λ > 550 nm) of sufficient intensity to elicit both cone and rod components in response. Arrows point to delayed cone (column 2) and rod (column 3) b-waves. Rod component in column 3 is so delayed in *F1* compared with normal that clear splitting of rod and cone components can be seen. Stimulus onset is at beginning of each trace. Two or three consecutive responses are superimposed for responses in columns 2 and 3. Dashed line has been vertically extended from peak of *R2* of normal subject through responses of *F1* and *F2* (see text). Calibration symbol (lower right corner) is 50 μV vertically and 0.5 millisecond horizontally for column 1, 12.5 milliseconds horizontally and 50 μV vertically for column 2, and 25 milliseconds horizontally and 30 μV vertically for column 3. (From Berson, E. L., and Goldstein, E. B.: Arch. Ophthalmol. **83**:412, 1970.)

tioning of the patient in maxwellian view and isolation of the recording electrode from the stimulus flash. Although the ERP has provided a valuable new dimension for clinical research, the ERG remains at this time the preferred electrophysiologic test for routine diagnostic evaluation of retinal function.

Clinical pathologic correlations

Postmortem examination of a donor eye from a 24-year-old male with sex-linked retinitis pigmentosa provided an unusual opportunity to correlate the ultrastructural abnormalities observed in remaining cones and rods with findings obtained 3 weeks prior to death with psychophysical and electroretinographic testing.[271] This patient noted a slight decrease in visual acuity and night blindness but did not report a deficiency in his visual field. Because of his poor general health, only limited testing could be done prior to death. Bone spicule pigmentation was distributed 45 to 60 degrees anterior to the fovea in all quadrants.

This patient had a 50% reduction in cone density in the foveola and a gradual diminution in cone density out to about 60 degrees, where the cone density increased; cone density appeared normal in the far periphery. Cone outer segments were considerably shortened in length in the foveola, and the inner segments appeared slightly increased in diameter (Fig. 17-42). In the parafovea cone inner segments

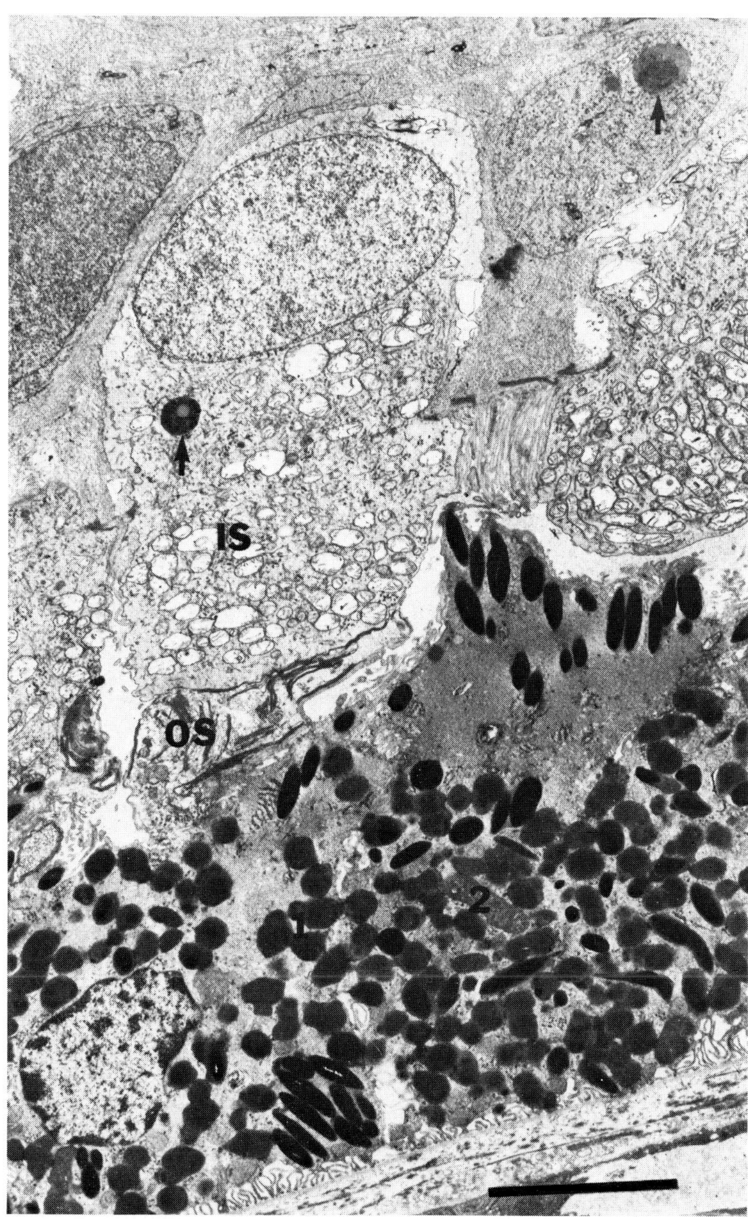

Fig. 17-42. Cones in the central fovea from donor eye with sex-linked retinitis pigmentosa have enlarged inner segments *(IS)* and distorted remnants of outer segments *(OS)*. Autophagic vacuoles (arrows) are seen in the perinuclear cytoplasm. Pigment epithelial cells contain large numbers of melanolysosomes *(1)*, lysosomes *(2)*, and few free melanin granules. Apical protrusions of these cells extend between cone inner segments. Horizontal bar (lower right) is 5μm. (From Szamier, R. B., et al.: Invest. Ophthalmol. Vis. Sci. **18:**145, 1979.)

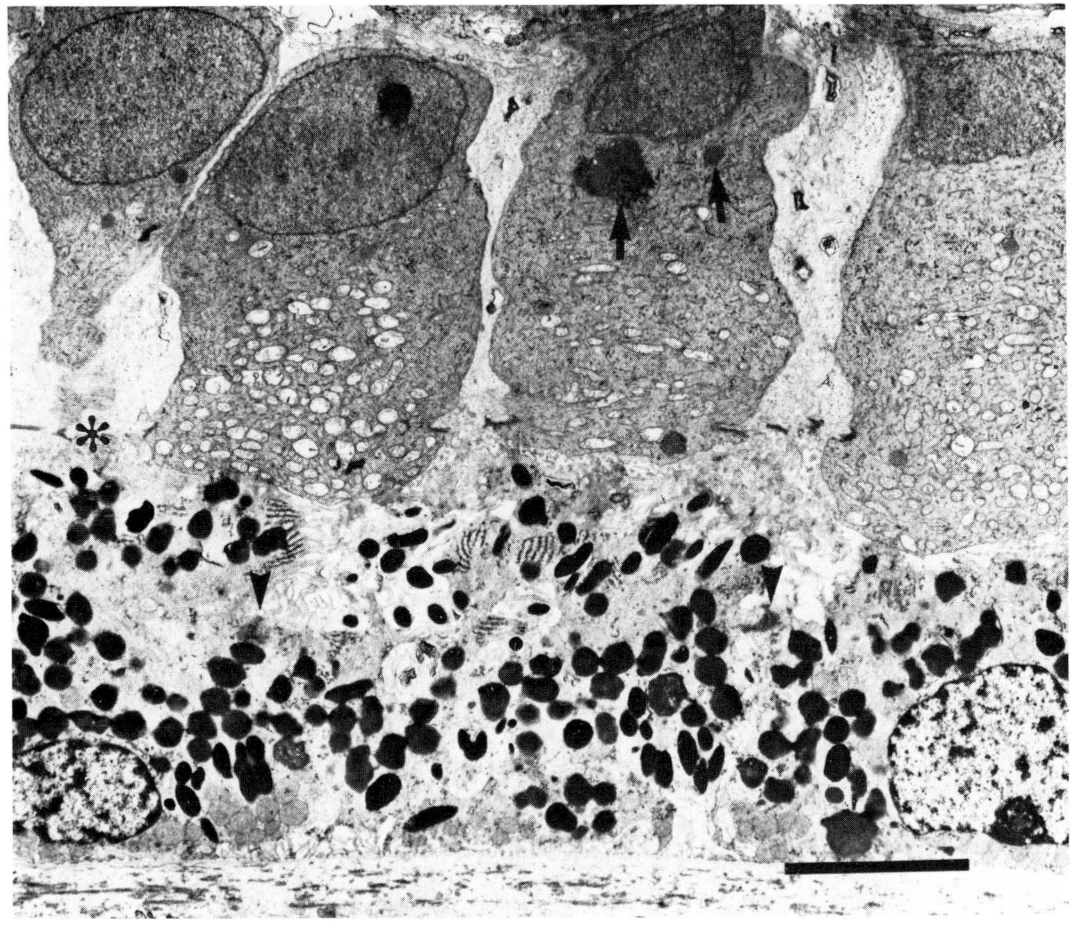

Fig. 17-43. Parafoveal cones have no organized outer segments and small portions of their inner segments extend beyond the external limiting membrane (asterisk). Autophagic vacuoles (arrows) are present in cone cell bodies. Protrusions of the pigment epithelial cells extend proximal to their apical tight junctions (arrowheads). Horizontal bar, lower right, represents 5 μm. (From Szamier, R. B., et al.: Invest. Ophthalmol. Vis. Sci. **18:**145, 1979.)

were directly apposed to the pigment epithelium, and no organized cone outer segments could be seen (Fig. 17-43). In the far periphery cones appeared slightly shortened with respect to outer segment length. Rods were first seen 60 degrees anterior to the fovea at the anterior edge of the bone spicule pigmentation and were slightly reduced in length in the far periphery. Although shortened about 25% in length, remaining peripheral rods had well-ordered discs (Fig. 17-44) and intact cell bodies.

Visual acuity reduction to 20/80 was consistent with the reduced cone density and reduced outer segment length in the central fovea. Psychophysical dark-adaptation threshold in the fovea was elevated 4 log units above normal rod threshold and 1.4 log units above normal cone threshold, and this was consistent with absence of rods in this region and reduced length of remaining cone outer segments. In the periphery dark-adaptation thresholds were minimally elevated and the rods were only slightly shortened in length. The patient had constricted visual fields with I-4 and III-4 white test lights and retained full kinetic visual fields only with a V-4 white test light in the Goldmann perimeter (even though no detectable organized cone outer segments were seen in the midperiphery

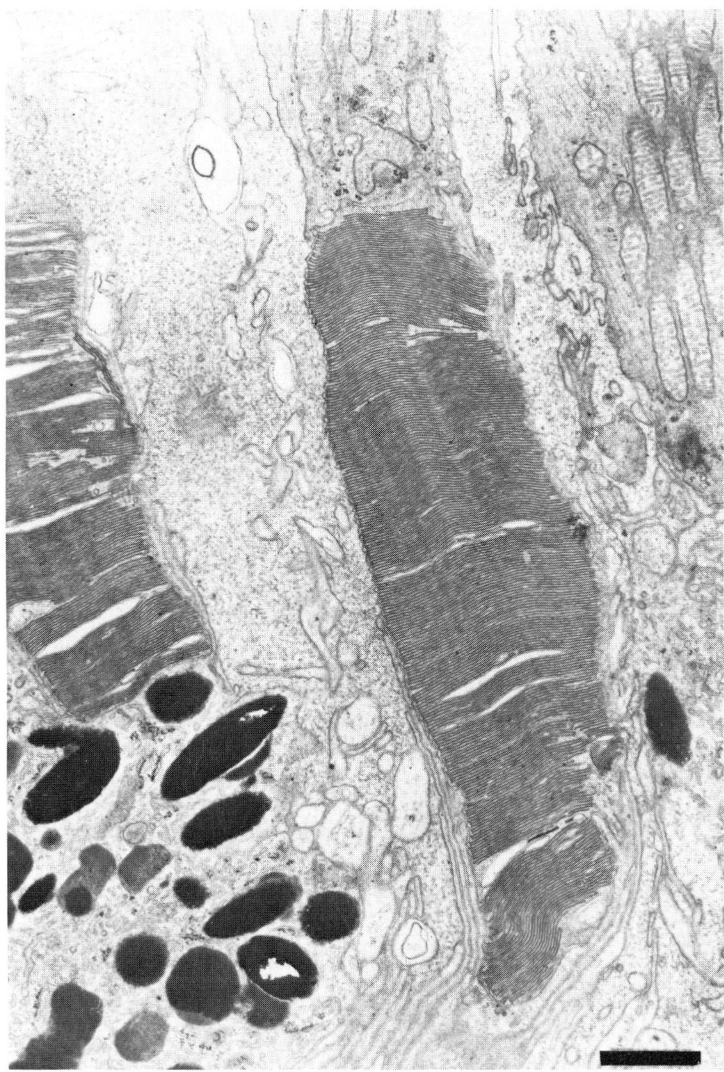

Fig. 17-44. Representative rods in the far periphery of donor eye with sex-linked retinitis pigmentosa have outer segments shortened in length with well-ordered discs. Outer segments are surrounded by microvillous processes of the pigment epithelium. Some processes extend up to the inner segments and are distended. Free melanin granules are prominent in the apical portion of the pigment epithelium. Horizontal bar (lower right) is 1 μm. (From Szamier, R. B., et al.: Invest. Ophthalmol. Vis. Sci. **18:**145, 1979.)

at the time of death); the gradual constriction of his visual field was consistent with the decreased cone density from the fovea through the midperiphery. ERG amplitudes were reduced more than 90% below normal, and this was compatible with widespread photoreceptor loss and structural abnormalities in the remaining photoreceptors. This correlation of premortem findings obtained with psychophysical and electroretinographic testing with postmortem ultrastructural observations supports the idea that these tests provide an index of the extent and type of photoreceptor involvement in a patient with hereditary retinitis pigmentosa.[24,271]

THE ELECTRO-OCULOGRAM

As early as 1848 Du Bois-Reymond reported that a difference in electrical potential of about 6 mV existed between the cornea and the back of the eye. The human eye behaves like a dipole, oriented along its anteroposterior axis with the cornea positive to the posterior pole. Movement of the eye will cause changes of the potential in one electrode placed near the inner canthus relative to another placed near the outer canthus of the eye. The record of eye movement obtained by this means is called the electro-oculogram, or the EOG. For eye excursions up to 30 degrees, a linear relationship with an accuracy of about 1.3 degrees exists between the EOG potential and the excursion of the eye. EOGs should not be confused with electromyograms, which are recordings of changes in potentials of individual contracting muscle fibers.

The EOG consists of at least two separate potentials, one that is insensitive and the other that is sensitive to light. Under usual recording conditions the subject with electrodes near the inner and outer canthus alternately observes fixation lights spaced 30 degrees apart, and the potential is measured under conditions of dark adaptation after prior exposure to room illumination.[10] The light-insensitive potential (or standing potential) per 30 degrees of eye excursion decreases slightly over a period of 8 or 9 minutes, at which time the so-called dark trough or lowest potential recorded in the dark-adapted state can be measured (Fig. 17-45). After exposure to light, the potential per 30 degrees of eye excursion gradually increases and reaches a peak in 10 to 15 minutes (light rise of the EOG), at which time the response under conditions of light adaptation may be more than twice that of the potential recorded under conditions of dark adaptation. The standing potential recorded from electrodes placed near the canthi can vary from 20 μV/degree to 5 μV/degree in normal subjects,[8] due in part to the placement of the electrodes. Therefore the EOG has been most reliably interpreted as a ratio of peak voltage obtained in the light-rise potential over the minimum voltage obtained in the dark-trough or standing potential without changing the placement of the electrodes during the test.[10] This ratio is usually greater than 1.9 in normal human subjects under age 50. After the light-rise peak, the cornea-positive oscillations vary in voltage in a damped sinusoidal manner.

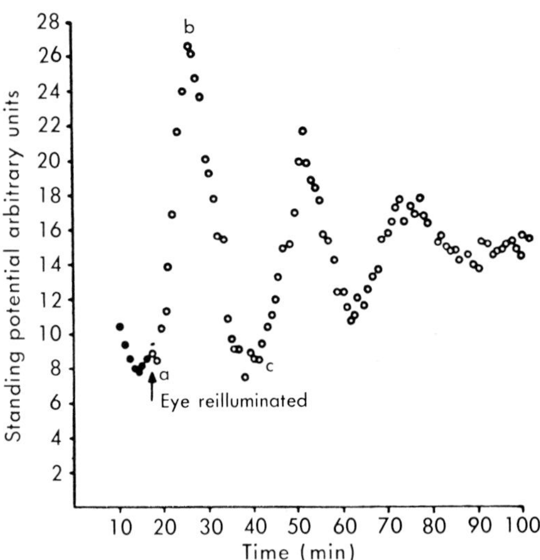

Fig. 17-45. Normal human EOG. Open circles are data describing oscillations of potential after reillumination of dark-adapted eye. Solid circles imply reading taken in darkness. Note initial transient fall in potential, *a,* first light peak, *b,* and first light trough, *c.* (From Arden, G. B., Barrada, A., and Kelsey, J. H.: Br. J. Ophthalmol. **46:**449, 1962.)

Origin of the components of the electro-oculogram

Microelectrode studies have located a large direct current (DC) component just distal to the outer margin of the outer nuclear layer of the retina,[59] and clinical studies suggest that the pigment epithelium is responsible for this standing potential. Lasansky and DeFisch[187] measured a transmembrane potential of 20 to 30 mV in the isolated pigment epithelium and choroid of the toad with the pigment epithelium surface positive with respect to the choroidal surface. The light-sensitive, large, slow cornea-positive component must depend on the activity of the photoreceptors as well as cell(s) in the inner nuclear layers of the retina because it is induced by light and also because it is eliminated by central retinal artery occlusion.[133,139] Following central retinal artery occlusion in the monkey, a small, light-sensitive cornea-negative potential is maintained for the duration of light stimulation and becomes apparent only when the large cornea-positive (light-rise) reponse is abolished. This response appears to be part of the a-wave of the ERG and the P-III component of Granit. The source of this response may be the receptor cells, although the horizontal cells have not been excluded as a site for this cornea-negative response in the EOG.

Initially the light rise of the EOG was considered to represent only rod activity, but it has since become clear that the cones also contribute to the response.[140] This latter conclusion is based not only on studies of the action spectrum of this response in the dark-adapted and light-adapted state, but also is consistent with the observation that human subjects without rod function can have a large light-rise response. More work is needed to understand not only the cellular origin(s) but also the physiologic basis of the EOG.

Electro-oculogram responses in retinal disease

The light-insensitive potential of the electro-oculogram provides a means of measuring the function of the pigment epithelium without having to stimulate the photoreceptors.[178] It is also a method for quantitating ocular movements. The test must be interpreted with caution if the patient has macular degeneration and poor fixation. In the early stages of retinitis pigmentosa the light rise of the EOG is reduced, but the standing potential is normal even at a time when the ERG is markedly abnormal or even nondetectable. In advanced stages the standing potential of the EOG becomes reduced. In advanced choroideremia, when only sclera is visible in large areas, both the light rise and the standing potential are absent. In patients with progressive cone-rod degeneration,[34] in whom both photoreceptor systems are abnormal, the light rise is affected early in the condition, but the standing potential is relatively preserved (Fig. 17-46). In contrast, in progressive cone degeneration[35] the light-rise to dark-trough ratio appears to be relatively preserved even though the patient has a widespread abnormality in the cone system. Patients with dominant, Nougaret-type, stationary night blindness with normal cone function and no detectable rod function have a markedly diminished light-rise response, whereas in recessive congenital stationary night blindness (both Ogu-

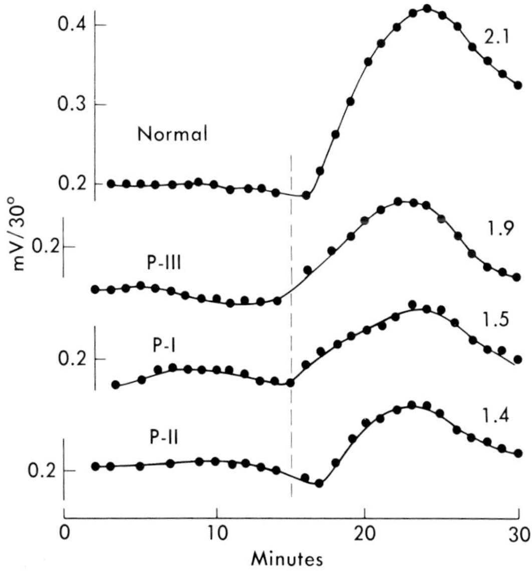

Fig. 17-46. EOG responses for normal subject and three patients, *P-I, P-II,* and *P-III,* with progressive cone-rod degeneration. Vertical hatched line indicates time of onset of 5 ft-L background light during test; ratio of light peak to dark trough shown on right above each response; scale (left) indicates amplitude of EOG response in mV/30 degrees of visual angle. (From Berson, E. L., Gouras, P., and Gunkel, R. D.: Arch. Ophthalmol. **80:** 68, 1968.)

chi's disease and recessive nyctalopia with myopia) the cornea-positive response of the EOG is normal. Therefore the EOG (as well as ERGs) can help to distinguish the dominant type of stationary night blindness from the recessive types. In chloroquine retinopathy[227] the EOG is abnormal in the advanced stages when large areas of the retinal pigment epithelium are abnormal, but in the very early stage, the EOG is normal.[141]

The ERG is usually abnormal in conditions in which the EOG is abnormal, but four exceptions have been observed clinically where the patient has a normal or nearly normal ERG with an abnormal EOG light-rise to dark-trough ratio. These conditions are butterfly-shaped pigment dystrophy of the fovea,[84] fundus flavimaculatus,[84] advanced drusen,[180] and vitelliform dystrophy or Best's disease.[83,115] More recent studies of patients with fundus flavimaculatus have shown that most patients have normal EOGs[110,218] particularly when the light rise is evaluated with a ganzfeld dome rather than with a small diameter x-ray viewing box.[112] Patients with dominant drusen have also been reported to have normal EOGs.[111] Vitelliform macular dystrophy deserves special attention because the EOG is abnormal even in asymptomatic "carriers" of this condition who have no abnormalities visible with the ophthalmoscope.[79,83] Therefore the EOG can serve as a genetic marker to identify this autosomal dominant disease in families in which a "skipped" generation is suggested by history or routine ocular examination. The EOG has also been considered as an aid in distinguishing true dysgenesis neuroepithelialis of Waardenburg from congenital amaurosis of Leber; in both types of retinal degenerations the ERG is absent, but the former has a normal or nearly normal EOG, and the latter has a nondetectable EOG.[156]

When the EOG is expressed as a ratio, it should be remembered that the ratio depends not only on the function of the pigment epithelium but also on the light-sensitive component generated in the outer retina. Therefore when EOG ratios are found to be abnormal, it does not necessarily follow that the pigment epithelial cells are abnormal; the contribution of the pigment epithelium is separated in the light-insensitive component or dark trough of the EOG but not in the ratio of the light rise to the dark trough.

MICROELECTRODE STUDIES OF THE VERTEBRATE RETINA

Electrophysiologic measurements have provided a direct approach to monitor and try to understand retinal function in normal human subjects and patients with retinal disease. Interpretations derived from these data are limited by the fact that large numbers of cells generate responses recorded at the cornea and by the fact that the physiologic bases for many abnormalities in these responses are not known. The pathogenesis of substantial delays in cone ERG b-wave implicit times in patients with retinitis pigmentosa is an electrical phenomenon that requires further study. More detailed understanding of the organization of the retina at the cellular level is needed in normal vertebrates. Further studies of animals with hereditary retinal degenerations (for example, *rd* mice,[68,188,189,194,281] Royal College of Surgeons[99,207] and Wag Rij rats,[186] Irish setters,[4,5,193] miniature French poodles,[4] Norwegian elkhounds,[4] baboons,[278] and Alaskan malamutes[245]) and further research on animals with nutritionally induced retinal degenerations (for example, vitamin A-deficient rats[91,220] and taurine-deficient cats[251-253]) should also help to provide a basis for further understanding of human retinal degenerations. This section will focus on microelectrode studies of the vertebrate retina.

Recordings from mud puppy retina

The ability to record from single cells with microelectrodes and identify with stains the cell from which the recording is taken represents a significant advance in the past decade that undoubtedly will help in understanding the organization of the vertebrate retina. The mud puppy has been studied in great detail by Werblin and Dowling[293] because this amphibian has cells with large perikarya that facilitate intracellular recording. Responses (Fig. 17-47) have been obtained to a spot of light about 100 μm in diameter focused on the electrode or centered on annuli 250 or 500 μm in diameter.

The photoreceptors (Fig. 17-47) increase

their membrane potential; they hyperpolarize for the duration of illumination. The photoreceptors respond poorly to both small and large annuli. Spot and annulus presented together show no differences compared with a spot stimulus alone. The photoreceptor cells appear to have a small field and respond relatively autonomously. In contrast, horizontal cells respond with large hyperpolarizing responses (L-type of S-potential, see below) over large retinal areas. When spot and annulus are presented simultaneously, summation is observed, and this is consistent with anatomic observations that horizontal cells have input from receptors over a wide field. Two types of bipolar cells have been demonstrated in the mud puppy retina. In one type, spot illumination elicits a sustained hyperpolarizing potential (Fig. 17-48, left). For the other type, spot illumination results in a sustained depolarizing poten-

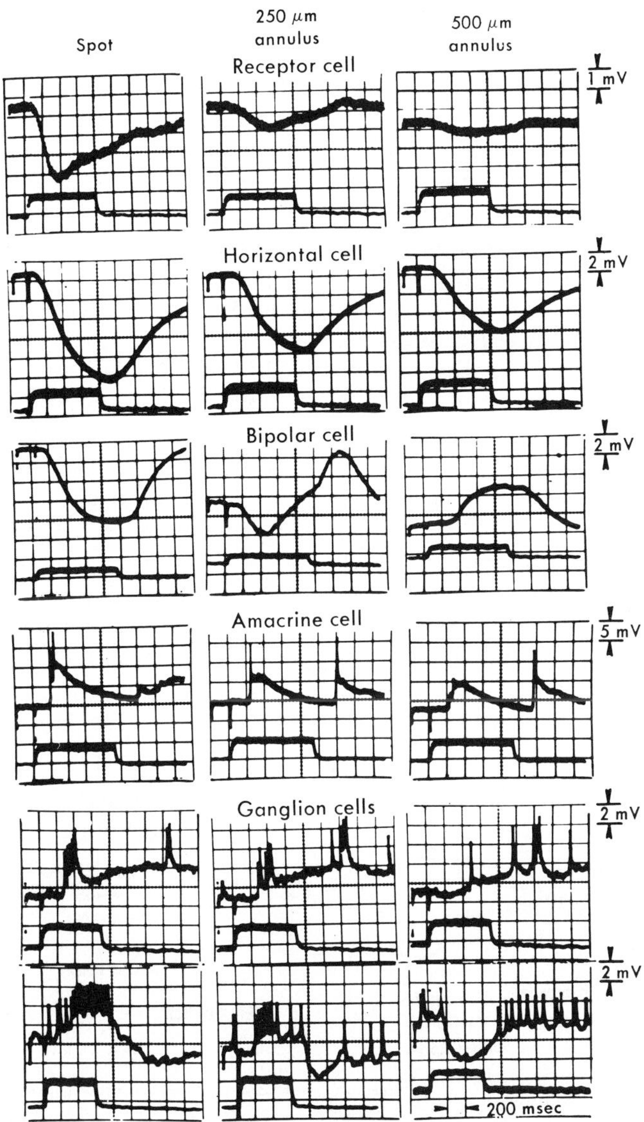

Fig. 17-47. Intracellular recordings from neurons of mud puppy retina. Responses were elicited with spot of light focused on electrode (left column) and with small and large annulus (center and right columns). Distal retinal neurons respond with slow, graded, mostly hyperpolarizing potentials: proximal neurons respond with depolarizing, mostly transient potentials. (From Werblin, F. S., and Dowling, J. E.: J. Neurophysiol. **32:**339, 1969.)

tial. For either bipolar cell type, when the central region is stimulated, addition of annular illumination antagonizes or reduces the sustained potential produced by the central spot (Fig. 17-48, right). Therefore, at the bipolar layer, an antagonistic center-surround organization can be observed with appropriate stimulus conditions.[92]

In contrast to the slow graded potentials seen in the distal retina (receptors, horizontal cells, and bipolar cells), the neurons in the proximal retina of the mud puppy respond mostly with depolarizing and transient potentials. Amacrine cells respond transiently to static retinal illumination with on and off responses to illumination anywhere in their receptive field (Fig. 17-47). Only a few spikes are seen riding on the transient depolarizations; thus it is not clear whether the slow potentials or the spikes are the important component of signal transmission of the amacrine cell. Reciprocal synapses of amacrine cell processes back onto bipolar terminals, just adjacent to bipolar ribbon synapses, raise the possibility of some feedback mechanism; the transient depolarizing response of the amacrine cell therefore could represent some negative feedback mechanism of amacrine cells on bipolar cells. Amacrine cells respond similarly to a bright spot on a dark background or a dark spot on a light background, a feature of many motion-sensitive and directionally sensitive cells. Amacrine cell responses may account for the directionally sensitive responses of on- and off-type ganglion cells (Fig. 17-47).

On the basis of microelectrode recordings from single cells, the outer plexiform layer of the mud puppy appears concerned with static or spatial aspects of illumination, whereas the inner plexiform layer is concerned more with dynamic or temporal aspects of illumination. Two physiologic types of ganglion cells may be related to the primary input into each type. One type receives direct input from bipolar cells and has a receptive field organization similar to bipolar cells; central illumination depolarizes the cell, while surround illumination inhibits this type in a sustained fashion (Fig. 17-47, lowermost responses). This type predominates in mammalian retinas. The second type of ganglion cell receives its major input from the amacrine cell and responds transiently to retinal illumination similar to the responses of amacrine cells (Fig. 17-47, fifth row). Therefore each type of ganglion cell may be carrying to the brain information processed in the two plexiform layers.[92]

S-potentials

When microelectrodes are inserted into certain regions of the inner nuclear layer of the fish retina, a response (Fig. 17-49) of large amplitude (10 to 50 mV) can be recorded superimposed on a negative resting potential (10 to 50 mV). This response, usually negative, is maintained for the duration of light stimulation. This response was named S-potential by Motokawa to acknowledge the discoverer, Svaetichin[270]; the cellular origin for

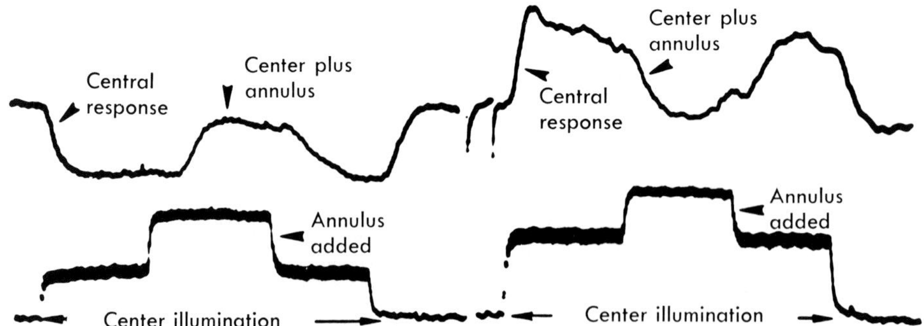

Fig. 17-48. Antagonistic effect of annular illumination on response of hyperpolarizing (left) and depolarizing (right) types of bipolar cells. In each case, central illumination was maintained steadily while annulus was flashed. (From Werblin, F. S., and Dowling, J. E.: J. Neurophysiol. **32:**339, 1969.)

this potential was not clear in 1959. Using intracellular marking techniques of Stretton and Kravitz,[268] Kaneko[170] showed in 1970 that these S-potentials are produced by horizontal cells. S-potentials have subsequently been recorded from many vertebrate horizontal cells, including the cat and monkey. One unusual characteristic of S-potentials is their large receptive field. Watanabe and Tosaka[291] showed that stimuli several millimeters away from the recording microelectrode influenced the size of S-potentials in the fish. The S-potential could be reduced by cutting the retina between the light stimulus and the recording electrode; this supported the idea that current and not scattered light was spreading from one region of the retina to the other. The receptive fields appeared larger than single horizontal cells, and this has raised the possibility that horizontal cells are electrically coupled to one another[213] or that perhaps horizontal cells synapsing on cones could in turn relay signals to other horizontal cells.

S-potentials can be separated into two distinct classes: one class produces a negative or hyperpolarizing response to all wavelengths (so-called luminosity or L-units)[212]; the other (Fig. 17-50) produces a hyperpolarizing response to long wavelength stimuli and a depolarizing response to short wavelength stimuli (so-called chromaticity or C-units).[195] The L-units have been thought to play a role in brightness perception, and the C-units have been thought to play a role in color vision. A red-adapting light not only depressed the contribution of a red-sensitive mechanism but also enhanced the contribution of a green-sensitive mechanism to the same L-unit[196] (Fig. 17-51). Specific cone mechanisms in C-units have been studied by determining action spectra in the presence of selective chromatic adaptation in fish; a green-sensitive and possibly a red-sensitive mechanism have been isolated in this way.[212] Rods also contribute to S-potentials in the mammalian retina, but it appears that rod and cone signals add their effects independently to this response.[260,261]

Microelectrode studies in the turtle by Baylor, Fuortes, and O'Bryan[16] showed that a depolarizing potential in a cone could be detected whenever a neighboring horizontal cell was hyperpolarized by current passed through a second microelectrode. This finding suggested that horizontal cell terminals continuously release a hyperpolarizing transmitter on cones that can be reduced by hyperpolarizing the horizontal cell.

Recordings from horizontal cells in the perfused retina of the turtle have also shown that low levels of calcium, high levels of magnesium, and cobalt hyperpolarize the horizontal cell membrane and suppress the response to light but only partially affect the response of receptor cells. These results are interpreted as consistent with the idea that a depolarizing transmitter is released by photoreceptors in darkness. The hyperpolarizing response to light of the horizontal cells would then result from a

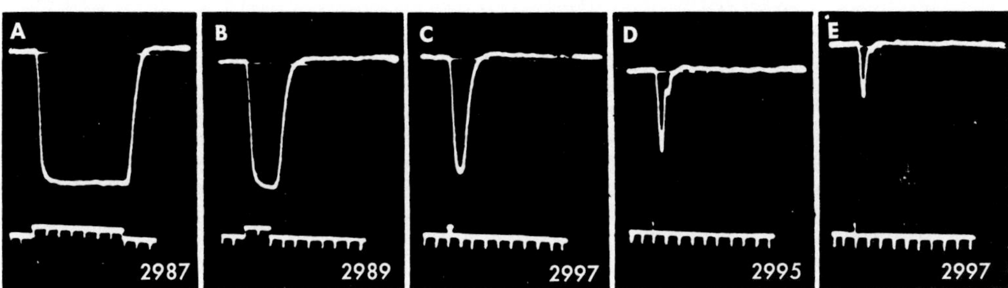

Fig. 17-49. Recordings of S-potentials from fish elicited at different millisecond durations (*A,* 500; *B,* 150; *C,* 20; *D,* 5; *E,* 0.5) of light stimuli (constant intensity and amplification). Time marks, 70 milliseconds. Light stimulus marked on time scale. (From Svaetichin, G.: Acta Physiol. Scand. **29** [suppl. 106]:565, 1953.)

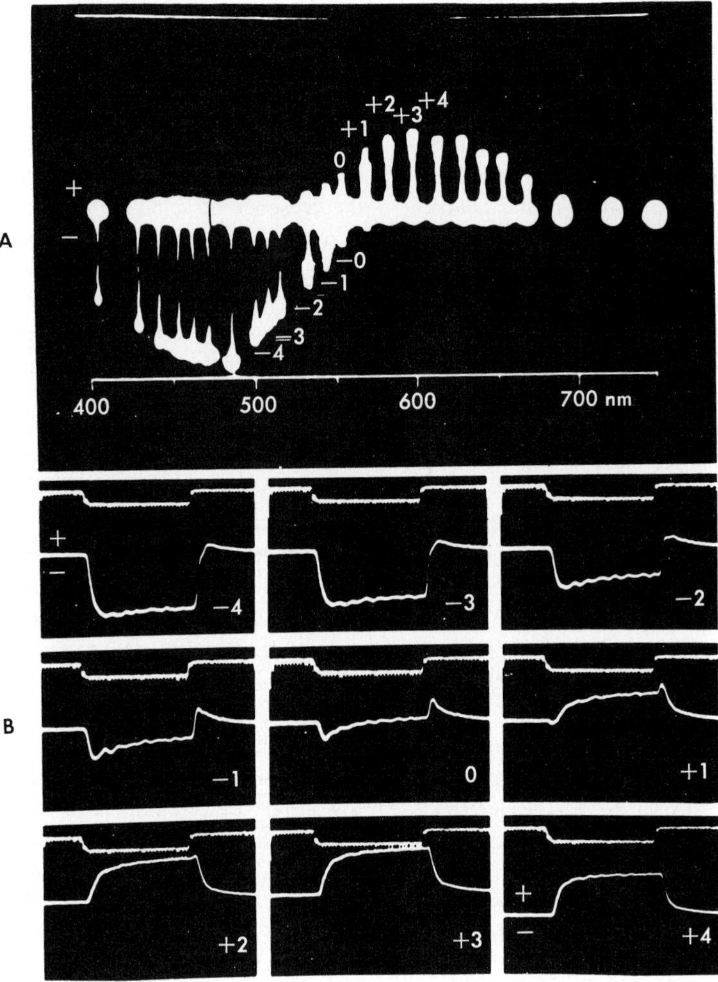

Fig. 17-50. Changes in time course of responses from C-unit of Mugil's retina as function of wavelength. **A,** Recording of response amplitude as function of wavelength. **B,** Responses recorded as function of time. Each numbered record was taken simultaneously with response peak bearing same number as in **A.** In **B,** top tracing in each record indicates time in tenths and hundredths of a second. Deflection of top tracing is due to output of photocell circuit used to monitor light flash (duration 0.3 second). Records taken at approximately 3-second intervals. (From MacNichol, E. F., Jr., and Svaetichin, G.: Am. J. Ophthalmol. **46:**26, 1958.)

reduction in the amount of transmitter released.[71]

The following arrangement is possible on the basis of available data. In the dark, cones may release a depolarizing transmitter on horizontal cells,[63] and horizontal cells may release a hyperpolarizing transmitter on cones.[16] One class of horizontal cells receives a direct input from one class of cones (C-unit), and another class receives an input from more than one class of cones and possibly rods (L-units). In this circuit the horizontal cells can provide a negative feedback mechanism. When a photoreceptor is stimulated by light, the photoreceptor membrane hyperpolarizes and the concentration of the depolarizing transmitter released by the photoreceptors becomes reduced. This leads to hyperpolarization of the horizontal cells and a resultant decrease in the release of hyperpolarizing transmitter back onto the photoreceptor cell. Therefore horizontal cells could increase the dynamic range over which cones function, could increase spatial contrast, and could even mediate color opponent properties to individual cones.[138]

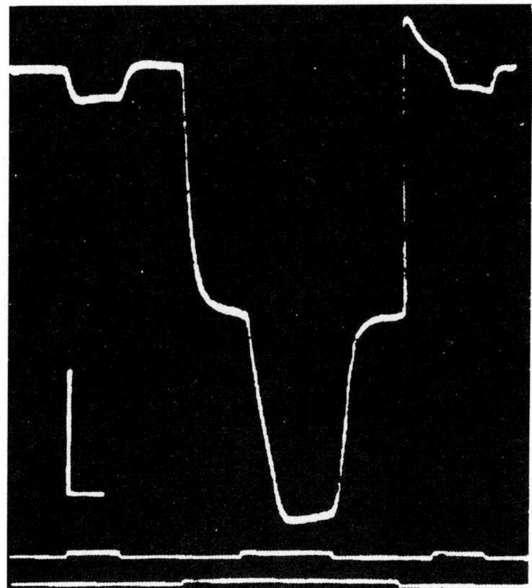

Fig. 17-51. Amplification of response to short-wavelength stimulation by red background light in L-unit of pike. Middle record shows duration of short-wavelength stimulus; lower one shows duration of red-adapting light. (From Maksimova, E. M., Maksimov, V. V., and Orlov, O. Y.: Biofizika **11**:472, 1966.)

Retinal neurotransmitters

Available evidence supports the idea that in darkness vertebrate photoreceptors are partially depolarized and release neurotransmitter onto second-order neurons (the horizontal and bipolar cells). Light hyperpolarizes the photoreceptors and depresses the release of transmitter; the light response of second-order neurons is therefore thought to occur secondary to a decrease of neurotransmitter secretion from the receptor terminal.

Several substances have been considered as possible candidates for the photoreceptor neurotransmitter. Atropine, an antagonist of muscarinic acetylcholine receptors, blocks synaptic transmission between photoreceptors and horizontal cells in the turtle retina.[122] Gamma-aminobutyric acid (GABA) has been reported to depolarize horizontal cells in the skate[69] and weakly hyperpolarize horizontal cells of the carp.[269] Aspartate and glutamate have effects on second-order cells that closely simulate effects of a receptor transmitter as horizontal[70,72,269] and hyperpolarizing bipolar cells are depolarized by these agents while depolarizing bipolar cells are hyperpolarized.[93,209,210] These acidic amino acids (particularly aspartate)[72,295] or very similar molecules may be the receptor transmitter, since no other substances so far tested have these effects. However, it remains to be established that these amino acids are released by photoreceptors in the dark or that light interrupts this release.[93]

In the inner plexiform layer five neurotransmitters (GABA, glycine, dopamine, acetylcholine, and an indoleamine, perhaps serotonin) have been identified by histochemical and other methods and associated with amacrine cells.[144,200,214,284] Certain amacrine cells appear to accumulate one or another of these substances, and it appears likely that different pharmacologic types of amacrine cells make different connections and perform different functions. For example, dopaminergic amacrine cells of the rabbit retina make synaptic contacts only with other amacrine cells, suggesting that they may play a modulating role in the retina.[95]

Interplexiform cell

In teleost fish and new-world monkeys fluorescence microscopy has revealed dopamine-containing retinal neurons whose perikarya are found in the inner portion of the inner nuclear layer among amacrine cells and whose processes extend not only in the inner plexiform layer but also to the outer plexiform layer.[102] These neurons, called interplexiform cells, have been found in a variety of species,[47,119] but only in the teleost fish and new-world monkeys have these cells shown dopamine fluorescence. The input to the interplexiform cells is exclusively from the amacrine cells, whereas the output is in both plexiform layers and is particularly noticeable in the outer plexiform layer with respect to synapses onto horizontal cells and some bipolar cell dendrites.[94,96,100] These cells do not contact ganglion cells or photoreceptor terminals. These interplexiform cells are of particular interest as they appear to provide a pathway for information flow from the inner to the outer retina.

The effects of dopamine on cells in the outer plexiform layer have been studied to consider the possible role of the interplexiform cell.[154] Dopamine depolarizes horizontal cells and depresses their responsiveness to light. Dopamine slightly hyperpolarizes a depolarizing bipolar cell, increases the amplitudes of responses to central spot illumination, and decreases the amplitudes of responses to annular illumination. These observations have led to the proposal that one role of the interplexiform cells is to regulate center-surround antagonism in the outer plexiform layer; activation of interplexiform cells could enhance bipolar cell center responsiveness while depressing lateral inhibitory effects initiated by horizontal cells.[93]

Oscillatory potentials

The vertebrate ERG elicited with high-intensity light stimuli shows not only a large a-wave but also a series of rhythmic oscillations superimposed on the b-wave (i.e., the b-wave is here defined as beginning at the trough of the cornea-negative a-wave). These rhythmic oscillations are called oscillatory potentials and representative responses from a normal human subject recorded at the cornea are shown in Fig. 17-52.[299] The oscillatory potentials are most easily detected in the mesopic state, and recordings from humans have revealed that the threshold and spectral sensitivity as well as temporal and adaptational characteristics differ from those of the b-wave.[286-289] Depth profile studies in the frog by Brindley[50] and others[297] have shown that the maximum amplitude of oscillatory potentials was in the inner nuclear layer. Ogden[225] and Ogden and Wylie[226] found that the maximum amplitude of the first three oscillatory potentials of the pigeon, chicken, and monkey ERG were at the level of the inner plexiform layer and suggested that axon terminals of bipolar cells, amacrine cell processes, and dendrites of ganglion cells could be involved in generation of these responses. Korol and co-workers[177] showed that intravitreal injection of glycine led to damage to amacrine cells and disappearance of oscillatory potentials in the ERG of the rabbit. In the mud puppy, Wachtmeister and Dowling[289] found that the oscillatory potentials reversed in polarity as a function of depth of their electrodes, thereby reflecting a radial flow of current in the retina; they also found that these potentials were selectively depressed

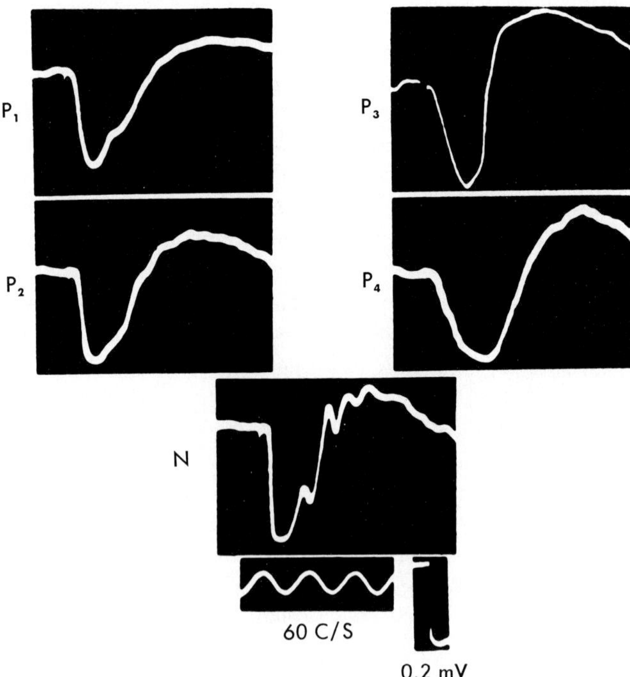

Fig. 17-52. Dark-adapted ERGs in four patients with early diabetic retinopathy visible with the ophthalmoscope. These four patients lacked the oscillatory potentials that are seen in the normal *(N)* dark-adapted ERG. Flash intensity was 37.5 J. (From Yonemura, D., Aoki, T., and Tsuzuki, K.: Arch. Ophthalmol. **68:**49, 1962.)

by GABA, glycine, glutamate, and dopamine but not by acetylcholine. They suggested that the oscillatory potentials are generated by inhibitory feedback synaptic circuits within the retina.

Oscillatory potentials have been considered of clinical importance as they have been shown to disappear in patients with diabetic retinopathy[6,61,175,299] (Fig. 17-52) as well as other diseases with known inner retinal ischemia. In prediabetic eyes with no visible abnormalities on ophthalmoscopic examination, the oscillatory potentials have been reported to be normal in amplitude,[299,300] subnormal in amplitude,[299,300] supernormal,[258] or delayed in peak latency with or without reductions in amplitude.[299] Comparisons of the relative sensitivity of the oscillatory potential, fluorescein angiography, and vitreous fluorophotometry in the same prediabetic eyes, to determine the earliest changes in diabetic retinopathy, remains to be done. Yonemura[299] has claimed, based on a retrospective study, that a rapid decline in the amplitude of the oscillatory potential would predict an impending progression of retinopathy into the proliferative stage. Further studies are needed to assess the predictive power of this response.

Recordings from ganglion cells and optic nerve fibers

In 1938 Hartline[149] observed three distinct types of responses from three separate fibers in records of action potentials recorded with microelectrodes from the optic nerve fibers of the frog eye (Fig. 17-53). In A the fiber responds with a rapid burst of impulses when the light is turned on. Although the light is kept constant, this soon dies down to a steady slower discharge. No response to cessation of illumination can be recorded from this fiber. In B a different fiber responds as the previous one, but the impulses stop completely, although the light is kept on. In C, another fiber gives no response when the light is turned on or throughout the entire duration of illumination, but when the light is turned off, there is a rapid burst of impulses. These fibers have been respectively termed "on," "on-off," and "off" units. In the frog each of these responses is peculiar to a particular fiber, which never gives any other type of response regardless of the conditions of stimulation or adaptation of the retina. An increase in illumination produces an increase in the frequency of discharge, and the sensitivity of the retina is diminished by light adaptation and increased by dark adaptation. The B fibers are extremely sensitive to any movement of the retinal image, whether it be a spot of light or a small shadow on the uniformly illuminated retina. The response in the C fibers to cessation of light stimulation usually subsides in a few seconds, and the discharge in these fibers can be abruptly suppressed at any time merely by reillumination of the retina.

Hartline[150] explored the retinal surface with

Fig. 17-53. Oscillographic records of action potentials in three single intraocular optic nerve fibers of frog's eye, showing three characteristic response types. *A,* Response to illumination of retina consisting of initial burst of impulses followed by maintained discharge lasting throughout illumination. No response to cessation of illumination in this fiber (off response in this record is partly due to retinal potential and partly to another fiber, which discharged several small impulses). *B,* Response only to onset and cessation of light. *C,* Response only to cessation of illumination. In this record, time is marked in fifths of seconds, and signal marking period of illumination fills white line immediately above time marker. (From Hartline, H.: Am. J. Physiol. **121:**400, 1938.)

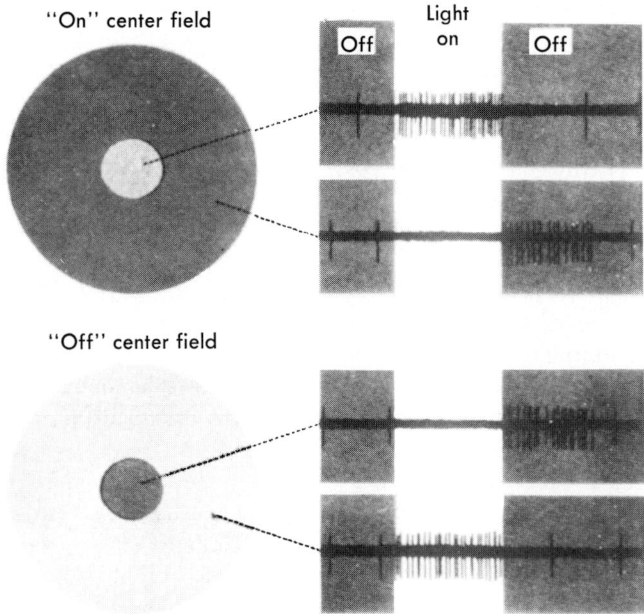

Fig. 17-54. Concentric fields are characteristic of retinal ganglion cells. *Top,* oscilloscope recording shows strong firing by on-center type of cell when spot of light strikes field center. If spot hits an off area, firing is suppressed until light goes off. *Bottom,* responses of another cell of off-center type. (From The visual cortex of the brain by D. Hubel. Copyright © 1963 by Scientific American, Inc. All rights reserved.)

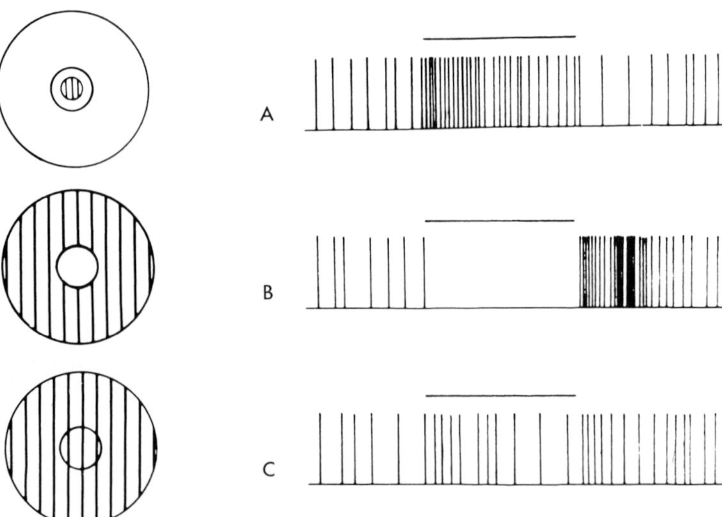

Fig. 17-55. On-center receptive field in the cat. **A,** Spot of light directed into center causes vigorous discharges. **B,** Surround illuminated by ring of light inhibits background discharge and causes responses when light is turned off. **C,** Light covering entire receptive field has relatively small effect when turned on and off because in this example antagonistic actions in center and surround almost cancel each other. Illumination, indicated by dark line, lasts 0.5 second. (From Kuffler, S. W.: Invest. Ophthalmol. **12:**794, 1973; drawn from unpublished records, Kuffler, 1953.)

a small spot of light and could map an area from which responses could be recorded in a single fiber when the spot was turned on or off. He termed this region the receptive field of the fiber. In the frog many fields were of the order of 1 mm in diameter—much larger than the diameter of rods (6 μm), although it roughly agreed with the spread of the dendritic tree of some ganglion cells. Hartline observed that the response type (on, on-off, or off) remained unchanged in all points of the receptive field of a particular fiber. The sensitivity of the receptive field was greatest in the central region and diminished in the peripheral region of the field. Over the central portion he was able to show that threshold intensity was inversely proportional to area of illumination. Furthermore, summation could be observed in one receptive field; a ganglion cell could show activity in response to two subthreshold stimuli presented in separate areas of its receptive field where either stimulus presented alone elicited no activity from the ganglion cell.[151]

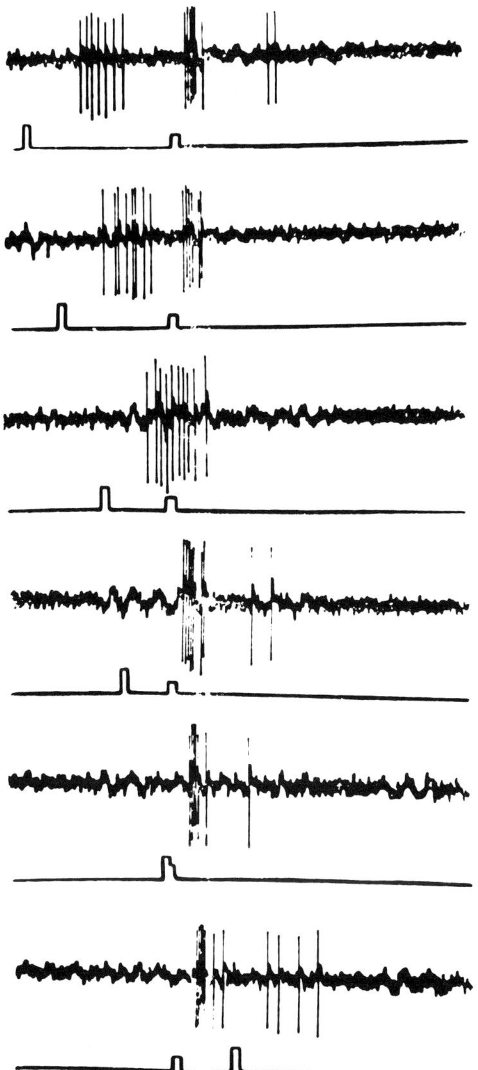

Fig. 17-56. Responses of dark-adapted perifoveal ganglion cell stimulated with two light pulses, one that affects rods (460 nm, larger photocell response) and one that affects cones (672 nm, smaller photocell response). Rod stimulus is delivered at different times relative to cone stimulus. Stimuli cover 24.6 mm^2 of retina, including center of cell's receptive field. Oscillograph of photocell's response is below each of cell's responses. Duration of each oscillograph is 0.5 second. Positivity is upward. (From Gouras, P., and Link, K.: J. Physiol. [Lond] **184:**499, 1966.)

The concept of a receptive field as described by Hartline has been modified on the basis of observations in the mammalian retina. Recordings from ganglion cells in the cat showed that different response-types could be obtained in the same ganglion cell by moving the same test spot to different positions. The receptive field could be subdivided into concentric zones (Fig. 17-54); a circular central zone of high sensitivity yielded either on or off responses, outside this central zone was a concentric annular zone of lower sensitivity that yielded the opponent-type of response (off responses if the center had on responses, on responses if the center had off responses).[161,183] Kuffler[184] showed interactions if the central zone and surround were stimulated simultaneously; the center and surround were functionally antagonistic (Fig. 17-55).

These antagonistic surrounds described by Kuffler are to be distinguished from silent surrounds described by Barlow[15] in the frog. Silent inhibitory surrounds do not themselves generate an impulse from a given ganglion cell, although light-on in a silent surround can reduce the response to light-on in the center of the receptive field of that ganglion cell. The concept of a receptive field is further complicated by such observations as the fact that some ganglion cells in the cat can be excited by a stimulus as distant as 45 degrees from the center of a conventional receptive field.[202] Therefore the definition of a receptive field of a ganglion cell must include some statement of the specificity of effects on that ganglion cell.[192]

The light-induced responses of the monkey retina show separate and independent contributions from both the rod and cone receptor systems and the ERG amplitude (Fig. 17-10) represents a summation of these contributions.[130] In contrast, the output of the ganglion cell represents activity usually from one or the other photoreceptor, and the ganglion cell becomes refractory to input from the other (Fig. 17-56). Interestingly, cone bipolar cells appear to synapse directly onto ganglion cells in the cat and probably in the monkey, whereas rod bipolar cells appear to synapse on an amacrine cell called the AII amacrine cell; this amacrine cell, which receives almost exclusive input from the rod system, in turn synapses on ganglion cells to form a major pathway for rod signals leaving the retina.[108,176,215] The A-II, or rod, amacrine cell, between rod bipolar and ganglion cells, could possibly account for the observation[142] that threshold signals from the rods reach ganglion cells later than those from the cones in the monkey perifovea.

Hubel and Wiesel found the receptive fields of the spider monkey had sharply demarcated on centers with antagonistic off surrounds or the reverse.[162] Those centers nearest the fovea had a diameter as small as 4 minutes of arc (20 μm) on the retina, and some in the periphery had a diameter of about 2 degrees. Responses were rarely color specific. In contrast, in the rhesus monkey color opponency could be easily demonstrated especially in the fovea and foveola.[136] Gouras showed that many on-center ganglion cells received excitatory signals from only one cone mechanism and contributions from another cone mechanism in the surround. One type of cell discharged transiently to maintained stimuli of any wavelength (so-called phasic); this type received signals from both green- and red-sensitive cone mechanisms, both of which excite in the center and inhibit in the periphery of the cell's receptive field. The second type discharged continuously to maintained stimuli of appropriate wavelength (so-called tonic); this type received excitatory signals from only one cone mechanism, either blue, green, or red sensitive in the center, and inhibition from another cone mechanism in the periphery of its receptive field.

Whereas most tonic cells showed color opponency, phasic cells did not show this phenomenon. Gouras[134] found that the tonic cells had slower conduction velocities than the phasic cells. Tonic cells tended to show linear summation and phasic cells nonlinear summation over their receptive fields.[82] Tonic cells were more common near the fovea where small ganglion cells predominate and phasic ganglion cells were more common toward the retinal periphery where ganglion cells are generally larger; however, both cell types were found adjacent to one another across the retina.[132] A small fraction of ganglion cells in the monkey retina did not have properties of tonic or phasic cells; these cells lacked center-surround organization.[80,81]

Monkey tonic and phasic retinal ganglion cells have several similarities respectively with cat X (sustained) and Y (transient) ganglion cells described by Cleland, Dubin, and Levick,[73] Enroth-Cugell and Robson,[106] and Fukada.[117] Both tonic and X cells predominate in the central retina and have slower conducting axons than phasic and Y cells. An important difference is that monkey tonic ganglion cells show color opponency, while cat X cells do not. Cats also have another ganglion cell system (called the W system) with conduction latencies longer than the X and Y systems;[266,267] no exact counterpart in the monkey retina has yet been described.

The major subdivision of ganglion cells into tonic or X-like and phasic or Y-like cells is particularly important, since Dreher, Fukada, and Rodieck[101] and Kruger[182] have found that lateral geniculate cells with X-like properties (strong, sustained responses to stationary stimuli, unresponsiveness to fast-moving stimuli, and relatively long response latency) receive excitatory input from retinal cells with relatively slow-conducting axons, while lateral geniculate nucleus cells with Y-like properties (transient, weak responses to stationary stimuli, responsiveness to fast-moving stimuli, relatively short response latency) receive excitatory input from retinal cells with relatively fast-conducting axons. Moreover, these cells appear anatomically segregated within the monkey lateral geniculate nucleus with X-like cells in the parvocellular layers and Y-like cells in the magnocellular layers.[101,182]

The X-Y (tonic-phasic) separation appears to be an advance in understanding two parallel systems that leave the retina and carry different aspects of the physical image of light on the photoreceptors to the central nervous system. It will be of interest to see if the X and Y systems are differentially affected in various diseases such as amblyopia, glaucoma, and toxic optic neuropathies.

Single cell approach

Although investigations of the retina are concerned with over 100 million cells, relatively few neuronal types are present, and repeating units are evident. Information derived from microelectrode recordings from single cells have begun to reveal arrangements within the retina and its projection through the optic nerve to the lateral geniculate nucleus and visual cortex.[184] Receptive fields become more complex as one ascends through successive cell layers. For example, within the monkey cortex some cells respond only to light oriented in a particular way on the retina, others respond to particular configurations, dimensions, or directions of movement, and others respond only to certain wavelength bands. One problem now is to understand further not only the patterns of connections within the retina and between the retina and cortex, but also the integrative mechanisms of these connections at synaptic sites. Another clinically relevant problem is the further definition of the relationship of the cone and rod photoreceptors to the pigment epithelium. Electrophysiologic and psychophysical studies combined with collaborative ultrastructural, histochemical, and biochemical investigations would seem to have the best chance for clarifying these problems. Such research could provide some of the new dimensions needed to help patients with diseases involving the neuronal units responsible for vision.

REFERENCES

1. Adrian, E. D.: Electric responses of the human eye, J. Physiol. (Lond) **104:**84, 1945.
2. Adrian, E. D.: Rod and cone components in the electric response of the eye, J. Physiol. (Lond) **105:**24, 1946.
3. Adrian, E. D., and Matthews, R.: Action of light on the eye; discharge of impulses in the optic nerve and its relation to electrical changes in the retina, J. Physiol. (Lond) **63:**378, 1927.
4. Aguirre, G. D.: Inherited retinal degenerations in the dog, Trans. Am. Acad. Ophthalmol. Otolaryngol. **81:**667, 1976.
5. Aguirre, G. D., et al.: Rod-cone dysplasia in Irish setters. A defect in cyclic GMP metabolism in visual cells, Science **201:**1133, 1978.
6. Algvere, P., and Gjotterberg, M.: The diagnostic value of the oscillatory potentials of the ERG and fluorescein angiography in diabetic prolif. retinopathy, Ophthalmologica **168:**97, 1974.
7. Alpern, M., Holland, M. G., and Ohba, N.: Rhodopsin bleaching signals in essential night blindness, J. Physiol. (Lond) **225:**457, 1972.
8. Arden, G. B., and Barrada, A.: Analysis of the electrooculograms of a series of normal subjects, Br. J. Ophthalmol. **46:**468, 1962.
9. Arden, G. B., and Ikeda, H.: Effects of hereditary degeneration of the retina on the early receptor potential and the corneofundal potential of the rat eye, Vision Res. **6:**171, 1966.
10. Arden, G. B., Barrada, A., and Kelsey, J. H.: New clinical test of retinal function based upon the standing potential of the eye, Br. J. Ophthalmol. **46:**449, 1962.
11. Armington, J. C.: The electroretinogram, the visual evoked potential, and the area-luminance relation, Vision Res. **8:**263, 1968.
12. Armington, J. C., Johnson, E. P., and Riggs, L. A.: The scotopic a-wave in the electrical response of the human retina, J. Physiol. (Lond) **118:**289, 1952.
13. Armington, J. C., et al.: Summation of retinal potentials, J. Opt. Soc. Am. **51:**877, 1961.
14. Asher, H.: The electroretinogram of the blind spot, J. Physiol. (Lond) **112:**40P, 1951.
15. Barlow, H. B.: Summation and inhibition in the frogs retina, J. Physiol. (Lond) **119:**69, 1953.
16. Baylor, D. A., Fuortes, M. G. F., and O'Bryan, P.: Receptive fields of cones in the retina of the turtle, J. Physiol. (Lond) **214:**265, 1971.
17. Bell, J.: Retinitis pigmentosa and allied diseases of the eye. In Pearson, K., editor: Treasury of human inheritance, London, 1922, Cambridge University Press, vol. 2.
18. Bergamini, L., and Bergamasco, B.: Cortical evoked potentials in man, Springfield, Ill., 1967, Charles C Thomas, Publisher.
19. Berson, E. L.: Experimental and therapeutic aspects of photic damage to the retina, Invest. Ophthalmol. **12:**35, 1973.
20. Berson, E. L.: Retinitis pigmentosa, the electroretinogram, and Mendel's laws, Trans. Pa. Acad. Ophthalmol. Otolaryngol. **26:**109, 1973.
21. Berson, E. L.: Night blindness: some aspects of man-

agement. In Faye, E., editor: Clinical low vision, Boston, 1976, Little, Brown & Co., pp. 301-306.
22. Berson, E. L.: Retinitis pigmentosa and allied retinal diseases: electrophysiologic findings, Trans. Am. Acad. Ophthalmol. Otolaryngol. **81**:659, 1976.
23. Berson, E. L.: Hereditary retinal diseases: classification with the full-field electroretinogram. In Lawwill, T., editor: Documenta Ophthalmologica proceedings Series, no. 13. Fourteenth ISCERG Symposium, Louisville, Ky., May 10 to 14, 1976, The Hague, 1977, Dr. W. Junk BV, pp. 149-171.
24. Berson, E. L.: Hereditary retinal diseases: applications of electrophysiological and psychophysical testing. In Symposium: Applications of psychophysics to clinical problems, San Francisco, 1979, Committee on Vision, National Academy of Sciences, National Research Council, Cambridge, England, 1980, Cambridge University Press (in press).
25. Berson, E. L.: Light deprivation and retinitis pigmentosa. In Symposium: Intense light hazards in ophthalmic diagnosis and treatment, 1979, Houston, Vision Res. 1980 (in press).
26. Berson, E. L., and Goldstein, E. B.: The early receptor potential in dominantly inherited retinitis pigmentosa, Arch. Ophthalmol. **83**:412, 1970.
27. Berson, E. L., and Goldstein, E. B.: The early receptor potential in sex-linked retinitis pigmentosa, Invest. Ophthalmol. **9**:58, 1970.
28. Berson, E. L., and Goldstein, E. B.: Recovery of the human early receptor potential during dark adaptation in hereditary retinal disease, Vision Res. **10**:219, 1970.
29. Berson, E. L., and Goldstein, E. B.: Cone pigment regeneration, retinitis pigmentosa, and light deprivation, Vision Res. **12**:749, 1972.
30. Berson, E. L., and Howard, J.: Temporal aspects of the electroretinogram in sector retinitis pigmentosa, Arch. Ophthalmol. **86**:653, 1971.
31. Berson, E. L., and Kanters, L.: Cone and rod responses in a family with recessively inherited retinitis pigmentosa, Arch. Ophthalmol. **84**:288, 1970.
32. Berson, E. L., and Simonoff, E. A.: Dominant retinitis pigmentosa with reduced penetrance; further studies of the electroretinogram, Arch. Ophthalmol. **97**:1286, 1979.
33. Berson, E. L., Gouras, P., and Gunkel, R. D.: Rod responses in retinitis pigmentosa, dominantly inherited, Arch. Ophthalmol. **80**:58, 1968.
34. Berson, E. L., Gouras, P., and Gunkel, R. D.: Progressive cone rod degeneration, Arch. Ophthalmol. **80**:68, 1968.
35. Berson, E. L., Gouras, P., and Gunkel, R. D.: Progressive cone degeneration, dominantly inherited, Arch. Ophthalmol. **80**:77, 1968.
36. Berson, E. L., Gouras, P., and Hoff, M.: Temporal aspects of the electroretinogram, Arch. Ophthalmol. **81**:207, 1969.
37. Berson, E. L., Rabin, A. R., and Mehaffey, L.: Advances in night vision technology: a pocketscope for patients with retinitis pigmentosa, Arch. Ophthalmol. **90**:427, 1973.
38. Berson, E. L., Rosen, J. B., and Simonoff, E. A.: Electroretinographic testing as an aid in detection of carriers of x-chromosome–linked retinitis pigmentosa, Am. J. Ophthalmol. **87**:460, 1979.
39. Berson, E. L., Schmidt, S. Y., and Rabin, A. R.: Plasma amino acids in hereditary retinal disease: Ornithine, lysine and taurine, Br. J. Ophthalmol. **60**:142, 1976.
40. Berson, E. L., Schmidt, S. Y., and Shih, V. E.: Ocular and biochemical abnormalities in gyrate atrophy of the choroid and retina, Ophthalmology **85**: 1018, 1978.
41. Berson, E. L., Shih, V. E., and Gargiulo, M.: Unpublished observations.
42. Berson, E. L., et al.: Rod and cone responses in sex-linked retinitis pigmentosa, Arch. Ophthalmol. **81**: 215, 1969.
43. Berson, E. L., et al.: Dominant retinitis pigmentosa with reduced penetrance, Arch. Ophthalmol. **81**:226, 1969.
44. Biersdorf, W. R., and Diller, D. A.: Local electroretinogram in macular degeneration, Am. J. Ophthalmol. **68**:296, 1969.
45. Black, R. K., Jay, B., and Kolb, H.: Electrical activity in the eye in high myopic, Br. J. Ophthalmol. **50**:629, 1966.
46. Bornschein, H., Goodman, G., and Gunkel, R. D.: Temporal aspects of the human electroretinogram, Arch. Ophthalmol. **57**:386, 1957.
47. Boycott, B. B., et al.: Interplexiform cells of the mammalian retina and their comparison with catecholamine-containing retinal cells, Proc. R. Soc. Lond. (Biol.) **191**:353, 1975.
48. Boynton, R. M., and Riggs, L. A.: The effect of stimulus area and intensity upon the human retinal response, J. Exp. Psychol. **42**:217, 1951.
49. Brindley, G. S.: The effect on the frog's electroretinogram of varying the amount of retina illuminated, J. Physiol. (Lond) **134**:353, 1956.
50. Brindley, G. S.: Responses to illumination recorded by microelectrodes from the frog's retina, J. Physiol. (Lond) **134**:360, 1956.
51. Brindley, G. S.: Additivity in the electroretinogram, J. Physiol. (Lond) **137**:51, 1957.
52. Brindley, G. S.: Physiology of the retina and visual pathway, ed. 2, Baltimore, 1970, The Williams & Wilkins Co.
53. Brindley, G. S., and Gardner-Medwin, A. R.: The origin of the early receptor potential of the retina, J. Physiol. (Lond) **182**:185, 1966.
54. Brindley, G. S., and Westheimer, G.: The spatial properties of the human electroretinogram, J. Physiol. (Lond) **179**:518, 1965.
55. Brown, K. T.: The electroretinogram: its components and their origin, Vision Res. **8**:633, 1968.
56. Brown, K. T., and Murakami, M.: A new receptor potential of the monkey retina with no detectable latency, Nature **201**:626, 1964.
57. Brown, P. K., and Wald, G.: Visual pigments in single rods and cones of the human retina, Science **144**:45, 1964.

58. Brown, K. T., and Watanabe, K.: Isolation and identification of a receptor potential from the pure cone fovea of the monkey retina, Nature **193:**958, 1962.

59. Brown, K. T., and Wiesel, T. N.: Localization of origins of electroretinogram components by intraretinal recording in the intact cat eye, J. Physiol. (Lond) **158:**257, 1961.

60. Brown, K. T., Watanabe, K., and Murakami, M.: The early and late receptor potentials of monkey cones and rods, Cold Spring Harbor Symp. Quant. Biol. **30:**457, 1965.

61. Brunette, J. R.: Oscillatory potentials. A clinical study in diabetics, Can. J. Ophthalmol. **5:**373, 1970.

62. Brunette, J. R.: A standardizable method for separating rod and cone responses in clinical electroretinography, Am. J. Ophthalmol. **75:**833, 1973.

63. Byzov, A. L., and Trifonov, Yu. A.: The response to electric stimulation of horizontal cells in the carp retina, Vision Res. **8:**817, 1968.

64. Carr, R. E., and Siegel, I. M.: Electrophysiologic aspects of several retinal diseases, Am. J. Ophthalmol. **58:**95, 1964.

65. Carr, R. E., and Siegel, I. M.: Action spectrum of the early human receptor potential, Nature **225:**88, 1970.

66. Carr, R. E., Ripps, H., and Siegel, I. M.: Visual pigment kinetics and adaptation in fundus albipunctatus. In Documenta Ophthalmologica proceedings series. Eleventh ISCERG Symposium, Bad Nauheim, West Germany, The Hague, Dr. W. Junk BV, 1974, pp. 193-204.

67. Carr, R. E., et al.: Rhodopsin and the electrical activity of the retina in congenital night blindness, Invest. Ophthalmol. **5:**497, 1966.

68. Carter-Dawson, L. D., LaVail, M. M., and Sidman, R. L.: Differential effect of the rd mutation on rods and cones in the mouse retina, Invest. Ophthalmol. Vis. Sci. **17:**489, 1978.

69. Cervetto, L., and MacNichol, Jr., E. F.: Pharmacology of horizontal cells in the isolated perfused skate retina, Biol. Bull. **141**(2):381, 1971.

70. Cervetto, L., and MacNichol, Jr., E. F.: Inactivation of horizontal cells in turtle retina by glutamate and aspartate, Science **178:**767, 1972.

71. Cervetto, L., and Piccolino, M.: Synaptic transmission between photoreceptors and horizontal cells in the turtle retina, Science **183:**417, 1974.

72. Chappell, R. A., et al.: Neurotransmission between rods and horizontal cells in the skate retina, Biol. Bull. **155:**421, 1978.

73. Cleland, B. G., Dubin, M. W., and Levick, W. R.: Sustained and transient neurones in the cat's retina and lateral geniculate nucleus. J. Physiol. **217:**473, 1971.

74. Cone, R. A.: Early receptor potential of the vertebrate retina, Nature **204:**736, 1964.

75. Cone, R. A.: The early receptor potential of the vertebrate eye, Cold Spring Harbor Symp. Quant. Biol. **30:**483, 1965.

76. Cone, R. A.: Early receptor potential: photoreversible charge displacement in rhodopsin, Science **155:**1128, 1967.

77. Cone, R. A., and Brown, P. K.: Dependence of the early receptor potential on the orientation of rhodopsin, Science **156:**536, 1967.

78. Copenhaver, R. M., and Goodman, G.: The electroretinogram in infantile, late infantile, and juvenile amaurotic family idiocy, Arch. Ophthalmol. **63:**559, 1960.

79. Cross, H. E., and Barr, L.: Electro-oculography in Best's macular dystrophy, Am. J. Ophthalmol. **77:** 46, 1974.

80. DeMonasterio, F. M.: Properties of ganglion cells with atypical receptive-field organization in retina of macaques. J. Neurophysiol. **41:**1435, 1978.

81. DeMonasterio, F. M., and Gouras, P.: Functional properties of ganglion cells of the rhesus monkey retina, J. Physiol. **251:**167, 1975.

82. DeMonasterio, F. M., Gouras, P., and Tolhurst, D. J.: Concealed colour opponency in ganglion cells of the rhesus monkey retina, J. Physiol. **251:**217, 1975.

83. Deutman, A. F.: Electro-oculography in families with dystrophy of the fovea, Arch. Ophthalmol. **81:**305, 1969.

84. Deutman, A. F., et al.: Butterfly-shaped pigment dystrophy of the fovea, Arch. Ophthalmol. **83:**558, 1970.

85. Dewar, J.: The physiological action of light, Nature **15:**433, 1877.

86. Dhanda, R. P.: ERG in myopic retinal degenerations. In Nakajima, A., editor: Jpn. J. Ophthalmol. **10** (suppl.):325, 1966.

87. Dick, E., and Miller, R. F.: Light-evoked potassium activity in mudpuppy retina: its relationship to the b-wave of the electroretinogram, Brain Res. **154:**388, 1978.

88. Dodt, E.: Ein Doppelinterferenzfilter-Monochromator besonders hoher Leuchtdichte, Bibl. Ophthalmol. **48:**32, 1957.

89. Dodt, E.: Cone electroretinogram by flicker, Nature **168:**783, 1957.

90. Dodt, E., and Wadenstein, L.: The use of flicker electroretinography in the human eye, Acta Ophthalmol. **32:**165, 1954.

91. Dowling, J. E.: Nutritional and inherited blindness in the rat, Exp. Res. **3:**348, 1964.

92. Dowling, J. E.: Organization of vertebrate retinas, Invest. Ophthalmol. **9:**655, 1970.

93. Dowling, J. E.: Information processing by local circuits: the vertebrate retina as a model system. In Schmitt, F. O., and Worden, F. G., editors: The neurosciences: fourth study program, Cambridge, Mass., 1979, The M.I.T. Press, pp. 163-181.

94. Dowling, J. E., and Ehinger, B.: Synaptic organization of the interplexiform cells of the goldfish retina, Science **188:**270, 1975.

95. Dowling, J. E., and Ehinger, B.: Synaptic organization of the dopaminergic neurons in the rabbit retina, J. Comp. Neur. **180:**203, 1978.

96. Dowling, J. E., and Ehinger, B.: The interplexiform

cell system. I, Synapses of the dopaminergic neurons of the goldfish retina, Proc. R. Soc. Lond. (Biol.) **201**:7, 1978.
97. Dowling, J. E., and Ripps, H.: Adaptation in skate photoreceptors, J. Gen. Physiol. **60**:698, 1972.
98. Dowling, J. E., and Ripps, H.: Neurotransmission in the distal retina: the effect of magnesium on horizontal cell activity, Nature **242**:101, 1973.
99. Dowling, J. E., and Sidman, R. L.: Inherited retinal dystrophy in rats, J. Cell Biol. **14**:73, 1962.
100. Dowling, J. E., Ehinger, B., and Heddon, W. L.: The interplexiform cell: a new type of retinal neuron, Invest. Ophthalmol. Vis. Sci. **15**:916, 1976.
101. Dreher, B., Fukada, Y., and Rodieck, R. W.: Identification, classification and anatomical segregation of cells with X-like and Y-like properties in the lateral geniculate nucleus of old-world primates. J. Physiol. **258**:433, 1976.
102. Ehinger, B., Falck, B., and Laties, A. M.: Adrenergic neurons in teleost retina, Z. Zellforsch. **97**:285, 1969.
103. Einthoven W., and Jolly, W. A.: The form and magnitude of the electrical response of the eye to stimulation by light at various intensities, Q. J. Exp. Physiol. **1**:373, 1908.
104. Eisenberg, M. F., and Copenhaver, R. M.: A fiber optic ophthalmoscope for focal retinal stimulation. Med. Res. Eng. **6**:23, 1967.
105. Eldjarn, L., et al.: Dietary effects on serum phytanic-acid levels and on clinical manifestation in heredopathia atactica polyneuritiformis, Lancet **1**:691, 1966.
106. Enroth-Cugell, C., and Robson, J. G.: The contrast sensitivity of retinal ganglion cells of the cat. J. Physiol. **187**:517, 1966.
107. Faber, D. L.: Dissertation, Buffalo, N.Y., 1969, State University of New York.
108. Famiglietti, E. V., and Kolb, H.: A bistratified amacrine cell and synaptic circuitry in the inner plexiform layer of the retina, Brain Res. **84**:293, 1975.
109. Finkelstein, D., Gouras, P., and Hoff, M.: Human electroretinogram near the absolute threshold of vision, Invest. Ophthalmol. **7**:214, 1968.
110. Fishman, G. A.: Fundus flavimaculatus: a clinical classification, Arch. Ophthalmol. **94**:2061, 1978.
111. Fishman, G. A., Carrasco, C., and Fishman, M.: The EOG in familial drusen, Arch. Ophthalmol. **94**:231, 1976.
112. Fishman, G. A., et al.: Electrooculogram testing in fundus flavimaculatus, Arch. Ophthalmol. **97**:1896, 1979.
113. Franceschetti, A., François, J., and Babel, J.: Les héredodégenérescences chorio-rétiniennes, Paris, 1963, Masson & Cie, Editeurs, vol. 1, pp. 325-338, 351-369.
114. François, J., and deRouck, A.: Behavior of ERG and EOG in localized retinal destruction by photocoagulation. In Burian, H. M., and Jacobson, J. H., editors: Clinical electroretinography. Proceedings of the Third International Symposium, Highland Park, Ill., 1964, Oxford, 1966, Pergamon Press, Inc., pp. 191-202.
115. François, J., de Rouck, A., and Férnandez-Sasso, D.: Electrooculography in vitelliform degeneration of the macula, Arch. Ophthalmol. **77**:726, 1967.
116. Fry, G. A., and Bartley, S. H.: The relation of stray light in the eye to the retinal action potential, Am. J. Physiol. **111**:335, 1935.
117. Fukada, Y.: Receptive field organization of cat optic nerve fibers with reference to conduction velocity. Vision Res. **11**:209, 1971.
118. Furukawa, T., and Hanawa, I.: Effects of some common cations on electroretinogram of the toad, Jpn. J. Physiol. **5**:280, 1955.
119. Gallego, A.: Horizontal and amacrine cells in the mammal's retina, Vision Res. **3**(suppl.):33, 1971.
120. Galloway, N. R.. Early receptor potential in the human eye, Br. J. Ophthalmol. **51**:261, 1967.
121. Gastant, H., and Régis, H.: Visually evoked potentials recorded transcranially in man. In Symposium: The analysis of central nervous system and cardiovascular data using computer methods, SP72, Washington, D.C., 1964, NASA.
122. Gerschenfeld, H. M., and Piccolino, M.: Muscarinic antagonists block cone to horizontal cell transmission in turtle retina, Nature **268**:257, 1977.
123. Goldstein, E. B.: Visual pigments and the early receptor potential of the isolated frog retina, Vision Res. **8**:953, 1968.
124. Goldstein, E. B.: Contributions of cones to the early receptor potential in the rhesus monkey, Nature **222**:1273, 1969.
125. Goldstein, E. B., and Berson, E. L.: Cone dominance of the human early receptor potential, Nature **222**:1272, 1969.
126. Goldstein, E. B., and Berson, E. L.: Rod and cone contributions to the human early receptor potential, Vision Res. **10**:207, 1970.
127. Goodman, G., and Bornschein, H.: Comparative electroretinographic studies in congenital night blindness and total color blindness, Arch. Ophthalmol. **58**:174, 1957.
128. Goodman, G., and Ripps, H.: Electroretinography in the differential diagnosis of visual loss in children, Arch. Ophthalmol. **64**:221, 1960.
129. Goodman, G., Ripps, H., and Siegel, I. M.: Cone dysfunction syndromes, Arch. Ophthalmol. **70**:214, 1963.
130. Gouras, P.: Rod and cone independence in the electroretinogram of the dark-adapted monkey's perifovea, J. Physiol. (Lond) **187**:455, 1966.
131. Gouras, P.: Visual adaptation: its mechanism, Science **157**:583, 1967.
132. Gouras, P.: Identification of cone mechanisms in monkey ganglion cells, J. Physiol. (Lond) **199**:533, 1968.
133. Gouras, P.: Relationships of the electro-oculogram to the electro-retinogram. In The clinical value of electroretinography. ISCERG Symposium, Ghent, Belgium, 1966, Basel, Switzerland, 1968, S. Karger AG, pp. 66-73.
134. Gouras, P.: Antidromic responses of orthodromically

identified ganglion cells in monkey retina, J. Physiol. **204:**407, 1969.

135. Gouras, P.: Electroretinography: some basic principles, Invest. Ophthalmol. **9:**557, 1970.
136. Gouras, P.: Color opponency from fovea to striate cortex, Invest. Ophthalmol. **11:**427, 1972.
137. Gouras, P.: Light and dark adaptation. In Fuortes, M. G. F., editor: Handbook of sensory physiology, New York, 1972, Springer-Verlag New York Inc., vol. 7/2, pp. 609-634.
138. Gouras, P.: S-potentials. In Fuortes, M. G. F., editor: Handbook of sensory physiology. New York, 1972, Springer-Verlag New York, Inc., vol. 7/2, pp. 513-529.
139. Gouras, P., and Carr, R. E.: Light-induced DC responses of monkey retina before and after central retinal artery interruption. Invest. Ophthalmol. **4:**310, 1965.
140. Gouras, P., and Carr, R. E.: Cone activity in the light-induced DC response of monkey retina, Invest. Ophthalmol. **4:**318, 1965.
141. Gouras, P., and Gunkel, R. D.: The EOG in chloroquine and other retinopathies, Arch. Ophthalmol. **70:** 629, 1973.
142. Gouras, P., and Link, K.: Rod and cone interaction in dark adapted monkey ganglion cells, J. Physiol. (Lond) **184:**499, 1966.
143. Gouras, P., Carr, R. E., and Gunkel, R. D.: Retinitis pigmentosa in abetalipoproteinemia: effects of vitamin A, Invest. Ophthalmol. **10:**784, 1971.
144. Graham, Jr., L. T.: Comparative aspects of neurotransmitters in the retina. In Davson, H., and Graham, L. T., Jr., editors: The eye, New York, 1974, Academic Press, Inc., vol. 6, pp. 283-342.
145. Granit, R.: The components of the retinal action potential in mammals and their relation to the discharge in the optic nerve, J. Physiol. (Lond) **77:**207, 1933.
146. Granit, R.: Sensory mechanisms of the retina, London, 1947, Oxford University Press.
147. Granit, R., and Munsterhjelm, A.: The electrical response of dark-adapted frog's eyes to monochromatic stimuli, J. Physiol. (Lond) **88:**436, 1937.
148. Hagins, W. A., Penn, R. D., and Yoshikami, S.: Dark current and photocurrent in retinal rods, Biophys. J. **10:**380, 1970.
149. Hartline, H.: The response of single optic nerve fibers of the vertebrate eye to illumination of the retina, Am. J. Physiol. **121:**400, 1938.
150. Hartline, H.: The receptive fields of optic nerve fibers, Am. J. Physiol. **130:**690, 1940.
151. Hartline, H.: The effects of spatial summation in the retina on the excitation of the fibers of the optic nerve, Am. J. Physiol. **130:**700, 1940.
152. Hashimoto, Y., Murakami, M., and Tomita, T.: Localization of the ERG by aid of histological method, Jpn. J. Physiol. **11:**62, 1961.
153. Hecht, S.: Rods, cones and the chemical basis of vision, Physiol. Rev. **17:**239, 1937.
154. Hedden, Jr., W. L., and Dowling, J. E.: The interplexiform cell system. II, Effects of dopamine on goldfish retinal neurones, Proc. R. Soc. Lond. (Biol.) **201:**27, 1978.
155. Henkes, H. E.: Electroretinography in circulatory disturbances of the retina, Arch. Ophthalmol. **51:**42, 1954.
156. Henkes, H. E., and Verduin, P. C.: Dysgenesis or abiotrophy? A differentiation with the help of the electro-retinogram (ERG) and electro-oculogram (EOG) in Leber's congenital amaurosis, Ophthalmologica **145:**144, 1963.
157. Hill, D. A., Arbel, K., and Berson, E. L.: Cone electroretinograms in congenital nyctalopia with myopia, Am. J. Ophthalmol. **78:**127, 1974.
158. Hirose, T., Miyake, Y., and Hara, A.: Simultaneous recording of electroretinogram and visual evoked response, Arch. Ophthalmol. **95:**1205, 1977.
159. Holmgren, F.: Method att objectivera effecten av ljusintryck på retina, Ups. Läkaref. Förh. **1:**177, 1865-1866.
160. Honda, Y.: Quantitative analysis of some effects of ouabain upon the electrical activity of mammalian retinas, Invest. Ophthalmol. **11:**699, 1972.
161. Hubel, D.: The visual cortex of the brain, Sci. Am. **209**(5):57, 1963.
162. Hubel, D., and Wiesel, T.: Receptive fields of optic nerve fibers in the spider monkey, J. Physiol. (Lond) **154:**572, 1960.
163. Inoue, J., Takeo, K., and Akiba, T.: The visual evoked potentials to focal illumination of the retina by direct view ophthalmoscopy, Acta Soc. Jpn. Ophthalmol. **77:**1149, 1973.
164. Jacobson, J. H., Hirose, T., and Suguki, T. A.: Simultaneous ERG and VER in lesions of the optic pathway, Invest. Ophthalmol. **6:**279, 1968.
165. Jacobson, S. R., et al.: Foveal cone electroretinograms in strabismic amblyopia; comparison with macular scars, juvenile macular degeneration, and optic atrophy, Trans. Ophthalmol. Soc. U.K. 1980 (in press).
166. Jacobson, S. G., and Sandberg, M. A.: Nasal-temporal asymmetry of visual thresholds from known retinal areas in strabismic amblyopia, Invest. Ophthalmol. Vis. Sci. **19**(suppl.):271, 1980.
167. Johnson, E. P., and Cornsweet, T. N.: Electroretinal photopic sensitivity curves, Nature **174:**614, 1954.
168. Johnson, E. P., Riggs, L. A., and Schick, A. M. L.: Photopic retinal potentials evoked by phase alternation of a barred pattern. In Burian, H. M., and Jacobson, J. H., editors: Clinical electroretinography. Proceedings of the Third International Symposium, Highland Park, Ill., 1964, Oxford, 1966, Pergamon Press, Ltd., pp. 75-91.
169. Kaiser-Kupfer, M. I., Valle, D., and Del Valle, L. A.: A specific enzyme defect in gyrate atrophy, Am. J. Ophthalmol. **85:**200, 1978.
170. Kaneko, A.: Physiological and morphological identification of horizontal, bipolar and amacrine cells in goldfish retina, J. Physiol. (Lond) **207:**623, 1970.
171. Karpe, G.: Basis of clinical electroretinography, Acta Ophthalmol. **24**(suppl.):84, 1945.

172. Karwoski, C. J., and Proenza, L. M.: Light evoked changes in extracellular potassium concentrations in mudpuppy retina, Brain Res. **145:**515, 1978.
173. Kennaway, N. G., Weleber, R. G. and Buist, N. R. M.: Gyrate atrophy of choroid and retina: deficient activity of ornithine ketoacid aminotransferase in cultured skin fibroblasts, N. Engl. J. Med. **297:**1180, 1977.
174. Kline, R. P., Ripps, H., and Dowling, J. E.: Generation of b-wave currents in the skate retina, Proc. Natl. Acad. Sci. U.S.A. **75:**5727, 1978.
175. Kojima, K., et al.: ERGs in diabetes. In Nakajima, A., editor: Retinal degenerations, ERGs and optic pathways. Fourth ISCERG Symposium, Hakone, Japan, 1965, Tokyo, 1966, Maruzen Co. Ltd., pp. 120-125; Jpn. J. Ophthalmol. vol. 10 (suppl.), 1966.
176. Kolb, H., and Famiglietti, E. V.: Rod and cone pathways in the inner plexiform layer of cat retina, Science **186:**47, 1974.
177. Korol, S., et al.: In vivo effects of glycine on retinal ultrastructure and averaged electroretinograms, Brain Res. **97:**235, 1978.
178. Krill, A. E.: The electroretinogram and electrooculogram: clinical application, Invest. Ophthalmol. **9:**600, 1970.
179. Krill, A. E.: Hereditary retinal and choroidal diseases, New York, 1972, Harper & Row, Publishers, pp. 248-249.
180. Krill, A. E., and Klien, B. A.: Flecked retina syndrome, Arch. Ophthalmol. **74:**496, 1965.
181. Krill, A. E., Archer, D., and Martin, D.: Sector retinitis pigmentosa, Am. J. Ophthalmol. **69:**977, 1970.
182. Kruger, J.: Stimulus dependent color specificity of monkey lateral geniculate neurons, Exp. Brain Res. **30:**297, 1977.
183. Kuffler, S.: Discharge patterns and functional organization of mammalian retina, J. Neurophysiol. **16:**62, 1953.
184. Kuffler, S.: The single-cell approach in the visual system and the study of receptive fields, Invest. Ophthalmol. **12:**794, 1973.
185. Kuffler, S., and Nicholls, J. G.: The physiology of neuroglial cells, Ergeb. Physiol. **57:**1, 1966.
186. Lai, Y. L., and Jonas, A. M.: Rat model for hereditary retinal degeneration. In Landers, M. B., III, et al., editors: Retinitis pigmentosa, New York, 1976, Plenum Publishing Corp., pp. 115-136.
187. Lasansky, A., and DeFisch, F. W.: Potential, current, and ionic fluxes across the isolated retinal pigment epithelium and choroid, J. Gen. Physiol. **49:** 913, 1966.
188. LaVail, M. M., and Mullen R. J.: Role of the pigment epithelium in inherited retinal degeneration analyzed with experimental mouse chimeras, Exp. Eye Res. **23:**227, 1976.
189. LaVail, M. M., and Sidman, R. L.: C57BL6J mice with inherited retinal degeneration, Arch. Ophthalmol. **91:**394, 1974.
190. Lawwill, T., and Burian, H. M.: A modification of the Burian-Allen contact lens electrode for human electroretinography, Am. J. Ophthalmol. **61:**1506, 1966.
191. Le Grand, Y.: Form and space vision, Bloomington, Ind., 1967, Indiana University Press.
192. Levick, W. R.: Receptive fields of retinal ganglion cells. In Fuortes, M. G. F., editor: Handbook of sensory physiology, New York, 1974, Springer-Verlag New York Inc., vol. 7/2, pp. 531-566.
193. Liu, Y. P., et al.: Involvement of cyclic GMP phosphodiesterase activator in an hereditary retinal degeneration, Nature **280:**62, 1979.
194. Lolley, R. N., and Farber, D. B.: Abnormal guanosine 3′, 5′-monophosphate during photoreceptor degeneration in the inherited retinal disorder of C3H/HeJ mice, Ann. Ophthalmol. **8:**496, 1976.
195. MacNichol, Jr., E. F., and Svaetichin, G.: Electric responses from the isolated retinas of fishes, Am. J. Ophthalmol. **46:**26, 1958.
196. Maksimova, E. M., Maksimov, V. V., and Orlov, O. Y.: Intensified interaction between signals of receptors in cells that are sources of S-potentials, Biofizika **11:**472, 1966.
197. Mandelbaum, J.: Dark adaptation. Some physiological and clinical considerations, Arch. Ophthalmol. **26:**203, 1941.
198. Marks, W. B., Dobelle, W. H., and MacNichol, Jr., E. F.: Visual pigments of single primate cones, Science **143:**1181, 1964.
199. Marmor, M. F.: Defining fundus albipunctatus. In Lawwill, T., editor: Documenta Ophthalmologia proceeding series, no. 13. Fourteenth ISCERG Symposium, Louisville, Ky., May 10-14, 1976, The Hague 1977, Dr. W. Junk BV, pp. 227-234.
200. Masland, R. H., and Livingstone, C. J.: Effect of stimulation with light on synthesis and release of acetylcholine by an isolated mammalian retina, J. Neurophysiol. **39:**1210, 1976.
201. McCulloch, J. C., et al.: Hyperornithinemia and gyrate atrophy of the choroid and retina, Ophthalmol. **85:**918, 1978.
202. McIlwain, J. T.: Some evidence concerning the physiological basis of the periphery effect in the cat's retina, Exp. Brain Res. **1:**267, 1966.
203. Mehaffey, III, L., and Berson, E. L.: Cone mechanisms in the electroretinogram of the cynomolgus monkey, Invest. Ophthalmol. **13:**266, 1974.
204. Miller, R. F., and Dowling, J. E.: Intracellular responses of the Müller (glial) cells of the mudpuppy retina: their relation to b-wave of the electroretinogram, J. Neurophysiol. **33:**323, 1970.
205. Miller, W. H., and Nicol, G. D.: Evidence that cyclic GMP regulates membrane potential in rod photoreceptors, Nature **280:**64, 1979.
206. Motokawa, K., and Mita, T.: Uber eine einfachere Untersuchungs Methode und Eigenschaften der Aktionsstrome der Netzhaut des Menschen, Tohoku J. Exp. Med. **42:**114, 1942.
207. Mullen, R. J., LaVail, M. M.: Inherited retinal dystrophy: primary defect in pigment epithelium de-

termined with experimental rat chimeras, Science **192:**799, 1976.
208. Murakami, M., and Kaneko, A.: Subcomponents of PIII in cold-blooded vertebrate retinae, Nature **210:** 103, 1966.
209. Murakami, M., Ohtsu, J., and Ohtsuka, T.: Effects of chemicals on receptors and horizontal cells in the retina, J. Physiol. (Lond) **227:**899, 1972.
210. Murakami, M., Ohtsuka, T., and Shimazaki, H.: Effects of aspartate and glutamate on the bipolar cells in the carp retina, Vision Res. **15:**456, 1975.
211. Nagata, M.: Photopic flicker ERG in cases of congenital night blindness and total color blindness. In Henkes, H. E., and van der Tweel, L. H., editors: Flicker. Proceedings of the symposium on physiology of flicker and flicker electroretinography, Amsterdam, Sept., 1963, The Hague, 1964, Dr. W. Junk BV, pp. 352-366.
212. Naka, K., and Rushton, W. A. H.: S-potentials from luminosity units in the retina of fish *(Cyprinidae),* J. Physiol. (Lond) **185:**587, 1966.
213. Naka, K., and Rushton, W. A. H.: The generation and spread of S-potentials in fish *(Cyprinidae),* J. Physiol. (Lond) **192:**437, 1967.
214. Neal, M. J.: Acetylcholine as a retina transmitter substance. In Bonting, S. L., editor: Transmitters in the visual process, Elmsford, N.Y., 1976, Pergamon Press, Inc., pp. 127-143.
215. Nelson, R., et al.: Neural responses in the rod and cone systems of the cat retina: intracellular records and Procion stains, Invest. Ophthalmol. Vis. Sci. **15:**946, 1976.
216. Newman, E. A.: Current source density analysis of the b-wave of the frog retina, Am. J. Neurophysiol. **43:**1335, 1980.
217. Newman, E. A.: B-wave currents in the frog retina, Vision Res. **19:**227, 1979.
218. Noble, K. G., and Carr, R. E.: Stargardt's disease and fundus flavimaculatus, Arch. Ophthalmol. **97:** 1281, 1979.
219. Noell, W. K.: The origin of the electroretinogram, Am. J. Ophthalmol. **38:**78, 1954.
220. Noell, W. K., Delmelle, M. C., and Albrecht, R.: Vitamin A deficiency effect on the retina: dependency on light, Science **172:**72, 1971.
221. Norden, L. C., and Leach, M. E.: Calibration of the ERG stimulus. In Lawwill, T., editor: Documenta Ophthalmologia Proceeding series, no. 13. Fourteenth ISCERG Symposium, Louisville, Ky., May 10-14, 1976, The Hague, 1977, Dr. W. Junk BV, pp. 393-403.
222. Oakley, B., II,: Potassium and the photoreceptor-dependent pigment epithelial hyperpolarization, J. Gen. Physiol. **70:**405, 1977.
223. Oakley, B., II, and Green, D. G.: Correlation of light-induced changes in retinal extracellular potassium concentration with c-wave of the electroretinogram, J. Neurophys. **39:**1117, 1976.
224. O'Donnell, J. J., Sandman, R. P., and Martin, S. R.: Gyrate atrophy of the retina: inborn error of L-ornithine; 2-ornithine; 2-oxoacid aminotransferase, Science **200:**200, 1978.
225. Ogden, T.: The oscillatory waves of the primate electroretinogram, Vision Res. **13:**1059, 1973.
226. Ogden, T., and Wylie, R.: Avian retina. I, Microelectrode depth and marking studies of local ERG, J. Neurophysiol. **34:**357, 1971.
227. Okun, E., et al.: Chloroquine retinopathy. A report of eight cases with ERG and dark-adaptation findings, Arch. Ophthalmol. **69:**59, 1963.
228. Østerberg, G.: Topography of the layer of rods and cones in the human retina, Acta Ophthalmol. **6** (suppl.):1, 1935.
229. Ostrovskii, M. A., and Dettmar, P.: The influence of ouabain on the electroretinogram of the isolated superfused retina of the frog, Biofizika **11:**724, 1968.
230. Padmos, P., and van Norren, D.: Cone spectral sensitivity and chromatic adaptation as revealed by human flicker electroretinography, Vision Res. **11:**27, 1971.
231. Pak, W. L.: Some properties of the early electrical response in the vertebrate retina, Cold Spring Harbor Symp. Quant. Biol. **30:**493, 1965.
232. Pak, W. L., Rozzi, V. P., and Ebrey, T. G.: Effect of changes in the chemical environment of the retina on the two components of the early receptor potential, Nature **214:**109, 1967.
233. Pallin, O.: The influence of the axial length of the eye on the size of the recorded b-potential in the clinical single-flash electroretinogram, Acta Ophthalmol. (suppl.) vol. 101, 1969.
234. Perry, N. W., and Childers, D. G.: The human visual evoked response, Springfield, Ill., 1969, Charles C Thomas, Publisher.
235. Potts, A. M., and Nagaya, T.: Studies on the visual evoked response. II, The effect of special cortical activity, Invest. Ophthalmol. **6:**657, 1967.
236. Rabin, A. R., and Berson, E. L.: A full-field system for clinical electroretinography, Arch. Ophthalmol. **92:**59, 1974.
237. Rabin, A. R., Hayes, K. C., and Berson, E. L.: Cone and rod responses in nutritionally induced retinal degeneration in the cat, Invest. Ophthalmol. **12:**694, 1973.
238. Refsum, S.: Heredopathia atactica polyneuritiformis: phytanic acid storage disease (Refsum's disease) with particular reference to therapeutic and pathogenetic aspects. In Tower, D. B., editor: The nervous system, New York, 1975, Raven Press, vol. 2, pp. 229-234.
239. Refsum, S.: Heredopathia atactica polyneuritiformis: phytanic storage disease (Refsum's disease) with particular reference to ophthalmological disturbances, Met. Ophthal. **1:**73, 1977.
240. Rietveld, W. J., Tordoir, W. E. H., and Duyff, J. W.: Contribution of the fovea and parafovea to the visual evoked response, Acta Physiol. Pharmacol. Neerl. **13:**300, 1965.
241. Riggs, L. A.: Continuous and reproducible records of the electrical activity of the human retina, Proc. Soc. Exp. Biol. Med. **48:**204, 1941.

242. Riggs, L. A., and Wooten, B. R.: Electrical measures and psychophysical data on human vision. In Jameson, D., and Hurvich, L. M., editors: Handbook of sensory physiology, New York, 1972, Springer-Verlag New York, Inc., vol. 7/4, pp. 690-731.
243. Riggs, L. A., Johnson, E. P., and Schick, A. M. L.: Electrical responses of the human eye to moving stimulus patterns, Science **144:**567, 1964.
244. Ripps, H.: Night blindness and the retinal mechanisms of visual adaptation, Ann. R. Coll. Surg. Engl. **58:**2, 1976.
245. Rubin, L. F.: Hemeralopia in dogs, Trans. Am. Acad. Ophthalmol. Otolaryngol. **81:**667, 1976.
246. Sandberg, M. A., Berson, E. L., and Ariel, M.: Visually evoked response testing with a stimulator-ophthalmoscope: macular scars, hereditary macular degeneration, and retinitis pigmentosa, Arch. Ophthalmol. **95:**1805, 1977.
247. Sandberg, M. A., Berson, E. L., and Effron, M. H.: A defect in light adaptation in retinitis pigmentosa, Invest. Ophthalmol. Vis. Sci. **19**(suppl.):259, 1980.
248. Sandberg, M. A., Effron, M. H., and Berson, E. L.: Focal cone electroretinograms in dominant retinitis pigmentosa with reduced penetrance, Invest. Ophthalmol. Vis. Sci. **17:**1096, 1978.
249. Sandberg, M. A., Jacobson, S. G., and Berson, E. L.: Focal cone electroretinograms in retinitis pigmentosa and juvenile macular degenerations, Am. J. Ophthalmol. **88:**702, 1979.
250. Sandberg, M. A., Rosen, J. B., and Berson, E. L.: Cone and rod function in vitamin A deficiency with chronic alcoholism and in retinitis pigmentosa, Am. J. Ophthalmol. **84:**658, 1977.
251. Schmidt, S. Y.: Biochemical findings in animals with retinal degenerations, Met. Ophthalmol. **2:**247, 1978.
252. Schmidt, S. Y., Berson, E. L., and Hayes, K. C.: Retinal degeneration in cats fed casein. I, Taurine deficiency, Invest. Ophthalmol. **15:**47, 1976.
253. Schmidt, S. Y., et al.: Retinal degeneration in cats fed casein. III, Taurine deficiency and ERG amplitudes, Invest. Ophthalmol. Vis. Sci. **16:**673, 1977.
254. Schubert, G., and Bornschein, H.: Beitrag zur Analyse des Menschlichen Elektroretinogramms, Ophthalmologica **123:**396, 1952.
255. Shih, V. E., et al.: Ornithine ketoacid transaminase deficiency in gyrate atrophy of the choroid and retina, Am. J. Hum. Genet. **30:**174, 1978.
256. Shipley, T.: The visually evoked occipitogram in strabismic amblyopia under direct-view ophthalmoscopy, J. Pediatr. Ophthalmol. **6:**97, 1969.
257. Simell, O., and Takki, K.: Raised plasma ornithine and gyrate atrophy of the choroid and retina, Lancet **1:**1031, 1973.
258. Simonsen, S.: ERG in diabetics. Sixth ISCERG Symposium, Ghent, Belgium, 1966, Basel, Switzerland, 1968, S. Karger AG, p. 403.
259. Sokol, S.: Visual evoked potentials to checkerboard pattern stimuli in strabismic amblyopia. In Desmedt, J. E. editor: Visual evoked potentials in man: new developments, Oxford, 1977, Clarendon Press, pp. 410-417.
260. Steinberg, R. H.: Rod and cone contributions to S-potentials from the cat retina, Vision Res. **9:**1319, 1969.
261. Steinberg, R. H.: Rod-cone interaction in S-potentials from the cat retina, Vision Res. **9:**1331, 1969.
262. Steinberg, R. H., Schmidt, R., and Brown, K. T.: Intracellular responses to light from cat pigment epithelium: origin of the electroretinogram c-wave, Nature **227:**728, 1970.
263. Steinberg, D., et al.: Studies on the metabolic error in Refsum's disease, J. Clin. Invest. **46:**313, 1967.
264. Steinberg, D., et al.: Phytanic acid in patients with Refsum's syndrome and responses to dietary treatment, Arch. Intern. Med. **125:**75, 1970.
265. Stiles, W. S.: Colour vision: the approach through the increment threshold sensitivity, Proc. Natl. Acad. Sci. U.S.A. **45:**100, 1959.
266. Stone, J., and Fukuda, Y.: Properties of cat retinal ganglion cells: a comparison of W-cells with X- and Y-cells, J. Neurophysiol. **37:**722, 1974.
267. Stone, J., and Hoffman, K.-P.: Very slow-conducting ganglion cells in the cat's retina: a major, new functional type? Brain Res. **43:**610, 1972.
268. Stretton, A. O. W., and Kravitz, E. A.: Neuronal geometry: determination with a technique of intracellular dye injection, Science **162:**132, 1968.
269. Sugawara, K., and Negishi, K.: Effects of some amino acids on the horizontal cell membrane potential in the isolated carp retina, Vision Res. **13:**977, 1973.
270. Svaetichin, G.: The cone action potential, Acta Physiol. Scand. **29**(suppl. 106):565, 1953.
270a. Szamier, R. B., Ripps, H., and Chappell, R. L.: On the glial-cell origin of the ERG b-wave, Invest. Ophthalmol. Vis. Sci. **19**(suppl.):39, 1980.
271. Szamier, R. B., et al.: Sex-linked retinitis pigmentosa: ultrastructure of the photoreceptors and pigment epithelium, Invest. Ophthalmol. Vis. Sci. **18:**145, 1979.
272. Takki, K., and Simell, O.: Gyrate atrophy of the choroid and retina with hyperornithinemia (HOGA), Birth Defects **XII:**373, 1976.
273. Tamai, A.: Studies on the early receptor potential in the human eye. III, ERP in primary retinitis pigmentosa, Yonago Acta Med. **18:**18, 1974.
274. Tomita, T.: Studies on the intraretinal action potential. I, Relation between the localization of micropipette in the retina and the shape of the intraretinal action potential, Jpn. J. Physiol. **1:**110, 1950.
275. Tomita, T.: Electrophysiological study of the mechanisms subserving color coding in the fish retina, Cold Spring Harbor Symp. Quant. Biol. **30:**559, 1965.
276. Toyoda, J., Nosaki, H., and Tomita, T.: Light-induced resistance changes in single photoreceptors of *Necturus* and *Gekko,* Vision Res. **9:**453, 1969.
277. Trijbels, J. M. F., et al.: L-ornithine-ketoacid-transaminase deficiency in cultured fibroblasts of a patient with hyperornithinemia and gyrate atrophy of the choroid and retina, Clin. Chim. Acta **79:**371, 1977.

278. Vainisi, S. J., Beck, B. B., and Apple, D. J.: Retinal degeneration in a baboon, Am. J. Ophthal. **78:**279, 1974.

279. Valle, D., Kaiser-Kupfer, M. I., and Del Valle, L. A.: Gyrate atrophy of the choroid and retina: deficiency of ornithine aminotransferase in transformed lymphocytes. Proc. Natl. Acad. Sci. U.S.A. **74:** 5159, 1977.

280. Valle, D., et al.:Gyrate atrophy: amino acids and correction of hyperornithinemia with low arginine containing diet, J. Clin. Invest. **65:**371, 1980.

281. Van Nie, R., Ivanyi, D., and Demant, P.: A new H-2-linked mutation, rds, causing retinal degenertion in the mouse, Tissue Antigens **12:**106, 1978.

282. Van Norren, D., and Padmos, P.: Spectral sensitivity of macaque cones determined with an ERG method, report no. IZF 1971-10, Amsterdam, 1971, Institute for Perception RVO-TNO.

283. Vaughn, H. G.: The perceptual and physiologic significance of visual evoked responses recorded from the scalp in man. In Burian, H. M., and Jacobson, J. H., editors: Clinical electroretinography. Proceedings of the Third International Symposium, Highland Park, Ill., 1964, New York, 1966, Pergamon Press, Inc., pp. 203-223.

284. Voaden, M. J.: Gamma-aminobutyric acid and glycine as retinal neurotransmitters. In Bonting, S. L., Transmitters in the visual process, Elmsford, N.Y., 1976, Pergamon Press, Inc., pp. 107-125.

284a. Vogel, D. A., and Green, D. G.: Potassium release and b-wave generation: a test of the Müller cell hypothesis, Invest. Ophthalmol. Vis. Sci. **19**(suppl.): 39, 1980.

285. von Helmholtz, H.: Handbuch der physiologischen Optik, ed. 2, Leipzig, East Germany, 1896, Voss.

286. Wachtmeister, L.: Luminosity functions of the oscillatory potentials of the human electroretinogram, Acta Ophthalmol. **52:**353, 1971.

287. Wachtmeister, L.: Stimulus duration and the oscillatory potentials of the human electroretinogram, Acta Ophthalmol. **52:**729, 1971.

288. Wachtmeister, L.: On the oscillatory potentials of the human electroretinogram in light and dark adaptation, Acta Ophthalmol. **116**(suppl.):1, 1972.

289. Wachmeister, L., and Dowling, J. E.: The oscillatory potentials of the mudpuppy retina, Invest. Ophthalmol. **17:**1176, 1978.

290. Wald, G.: The receptors of human color vision, Science **145:**1007, 1964.

291. Watanabe, K., and Tosaka, T.: Functional organization of the Cyprinid fish retina as revealed by discriminative responses to spectral illumination, Jpn. J. Physiol. **9:**84, 1959.

292. Weleber, R. G., Kennaway, N. G., and Buist, N. R. R.: Vitamin B_6 in management of gyrate atrophy of choroid and retina, Lancet **2**(8101):1213, 1978.

293. Werblin, F. S., and Dowling, J. E.: Organization of the retina of the mudpuppy. *Necturus maculosus*. II, Intra-cellular recording, J. Neurophysiol. **32:**339, 1969.

294. Witkovsky, P., Dudek, F. E., and Ripps, H.: Slow PIII component of the carp electroretinogram, J. Gen. Physiol. **65:**119, 1975.

295. Wu, S. M., and Dowling, J. E.: L-Aspartate: evidence for a role in cone photoreceptor synaptic transmission in the carp retina, Proc. Natl. Acad. Sci. U.S.A. **75:**5205, 1978.

296. Wyszechi, G., and Stiles, W. S.: Color science, New York, 1967, John Wiley & Sons, Inc., p. 206.

297. Yonemura, D., and Hatta, M.: Studies of the minor components of the frog's electroretinogram, Jpn. J. Physiol. **16:**11, 1966.

298. Yonemura, D., and Kawasaki, K.: The early receptor potential in the human electroretinogram, Jpn. J. Physiol. **17:**235, 1967.

299. Yonemura, D., and Kawasaki, K.: Electrophysiological study on activities of neuronal and non-neuronal retinal elements in man with reference to its clinical application, Jpn. J. Ophthalmol. **22:**195, 1978.

300. Yonemura, D., Aoki, T., and Tsuzuki, K.: Electroretinogram in diabetic retinopathy, Arch. Ophthalmol. **68:**49, 1962.

301. Yoshikami, S., and Hagins, W. A.: Light, calcium and the photocurrent of rods and cones, Biophys. Soc. Ann. Meeting Abstr. **11:**47a, 1971.

302. Zetterstrom, B.: Clinical electroretinogram in children during first year of life, Acta Ophthalmol. **29:** 295, 1951.

Chapter 18

Visual acuity

GERALD WESTHEIMER, Ph.D.

Visual acuity refers to the spatial limit of visual discrimination. It is surely the single most significant measure of functional integrity of the biologic apparatus to which the eye professions are dedicated. If restricted to expressing a patient's visual status in only one number, most practitioners would opt for one of the form 20/20.

Technically speaking, a visual acuity measurement involves the determination of a threshold, and our discussion will therefore have to deal with the problems typically associated with sensory thresholds: specification of the physical stimulus, transduction in the sense organ, anatomic and physiologic substrates, criteria and scales of measurements, techniques of obtaining threshold values, and influence of interacting variables.

Fechner,[8] who claimed to have invented his psychophysical law on October 22, 1850 (in the morning, in bed!), made the distinction between outer and inner psychophysics. By outer psychophysics he meant the relationship between the physical stimulus outside the body and the associated mental state; the term inner psychophysics he applied to the relationship between the physiologic state at the level of the sense cells and the mental state. Leaving aside the emphasis Fechner placed on the mental, or psychologic, aspect, the distinction between outer and inner psychophysics focuses attention on the role of transduction of the physical stimulus in the sense organ. Nowhere does this distinction assume greater significance than in spatial vision. Later generations of students of sensation have used the terms *distal* and *proximal* stimulus, the former denoting in our case the light distribution in object space, as a physicist might describe it, and the latter the pattern of quantal absorption in retinal receptors. The intervening processes of optical image formation, including refraction, diffraction, absorption, and scattering, are at least as important in understanding visual acuity and what can interfere with it as any of the succeeding stages: photochemical transduction in the retinal receptors, sorting and transmission of neural signals in the retina and visual pathways, and higher cortical processing. There is no need at present to define the end stage of visual acuity. Depending on the situation, we may seek a patient's verbal response, a "voluntary" motor response such as a gesture or a button press, or an "involuntary" response such as pupil contraction or optokinetic nystagmus or we may be satisfied with a set of electric signals from some part of the central nervous system.

SPECIFICATION OF THE STIMULUS (PHYSICAL BASIS)

The retinal receptor cells are not exposed directly to the light from the objects, but rather to the energy distribution in the image formed by the eye's optical system. The first step in dealing with visual acuity then is the specification of the relationship between objects and their retinal images.

The schema for this relationship has two aspects: the relative spacing of the images of sequential object points and the light spread in the image of each point. The discussion is facilitated by choosing the chief ray as the reference.

Of the light emanating from any object, only

the bundle that is admitted by the pupil matters as far as the retinal image is concerned. To save the effort of tracing all the rays through the cornea and checking which are intercepted by the iris, we employ the artifice of finding the image of the iris formed by the cornea: the entrance pupil. Any ray from any object that passes through the entrance pupil will also end up in the retinal image. The ray from the object to the center of the entrance pupil is called the chief ray because it will identify the center of the light bundle from that particular object. To examine the situation in any region of the retina in which reasonable homogeneity of imagery applies, it suffices to study the locations of the intercept of the chief rays with the retina because the light spread associated with the image of each point is centered around the chief ray.

The entrance pupil is situated about 3 mm behind the corneal vertex. When one looks at an eye and measures the pupil size, one is in reality measuring the size of the entrance pupil, which is about 10% larger. Each chief ray uniquely defines a retinal position. It is therefore a satisfactory procedure to transfer specification of retinal distances (in linear measure along the retina) into the measure of angular separation of object-sided chief rays. In a typical emmetropic eye, a retinal distance of 1 mm corresponds to angular separation of chief rays of about 3.5 degrees. The convention of expressing retinal distance in angular measure of chief rays in object space is almost universal.

A significant feature of the chief-ray method of specifying retinal image position is that, because the bundle of rays converging onto the image plane is centered on the chief ray, the method retains its validity even in the presence of focus changes due to lenses or accommodation (so long as the pupil center stays put). The image patch may change size with focus changes, but the location of its center (the chief ray intercept) does not move. If the eye were an ideal optical instrument without diffraction or aberrations, no more need be said about imagery than what angle the chief rays of the objects make with each other and where the objects are situated with respect to the plane conjugate to the retina. In actual practice, additional considerations center on the extent of departure of the image quality from the ideal of a point image for a point object.

In optical instrument design it is customary to partition the deviations from point imagery into a variety of classes. The significant datum, however, is the total effect of all of them, that is, the actual light spread in the image of a point object or the point-spread function. Once this basic information is available, it is possible to describe the light distribution in any object by merely superposing the spread functions centered on all the elements making up the object.

It is difficult to ascertain the value for the light spread in the retinal image of a given eye, but indirect measurements have shown it to have the general shape shown in Fig. 18-1. Several factors contribute to the spread[14]:

1. Diffraction. According to the wave theory of light, limitation of the aperture causes a spread of light even in a fully focused system. The Fraunhofer diffraction image of a point object by a circular aperture has the familiar bell shape with oscillating fringes shown in textbooks of physics (Fig. 18-2). It comes to its first zero at a radial distance of 1.22 λ/a radians (angle in object space), where λ is the wavelength of light and a is the diameter of the entrance pupil. The central patch of the pattern is called Airy's disk and contains most of the light energy; the height of the first ring is only 1.75% of the central peak. Whenever the eye's pupil is less than 2 mm in diameter, the actual image spread is substantially equal to the diffraction image, and the other factors (below) can usually be ignored.

2. Aberrations. Because of a variety of factors that may vary from one eye to the next, rays entering through the periphery of the pupil may not converge on the geometric image point, contributing, therefore, to the spread of light in the image beyond that caused by diffraction. These effects become more prominent as the pupil widens, more or less offsetting the resulting reduction of the diffraction effects. As a consequence the image quality usually does not improve much beyond that of a diffraction-limited instrument with a 2.5 mm

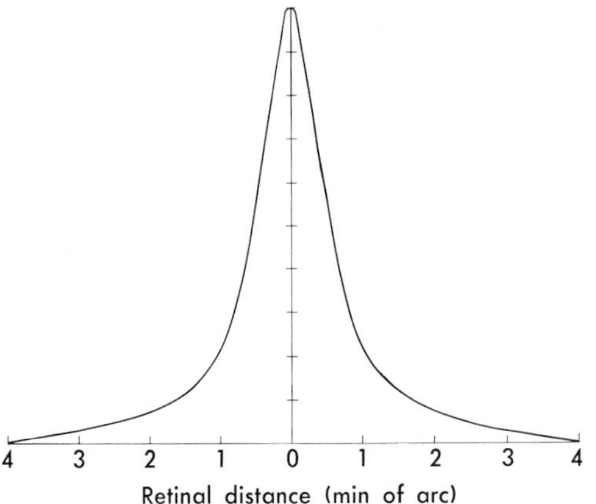

Fig. 18-1. Line-spread function, that is, light distribution in the image of a very thin line object, for a human eye in best focus and moderate pupil diameter.

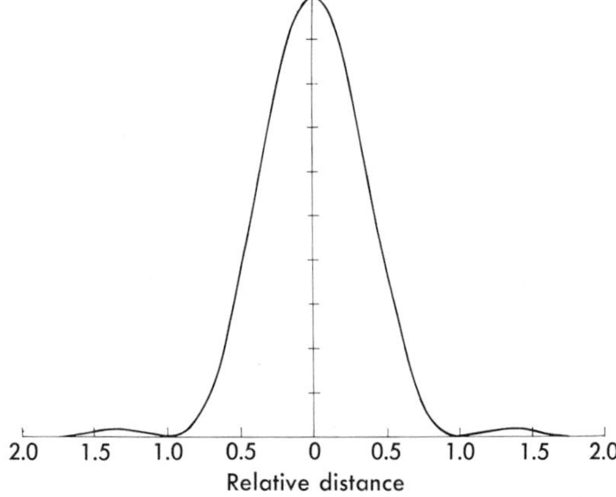

Fig. 18-2. Fraunhofer's diffraction pattern for a point object. Abscissae are in normalized coordinates, where 1.00 is equal to an angle of 1.22 λ/a radians in object space. Pupil diameter = a, wavelength of light = λ.

pupil diameter. For pupil diameters larger than 5 mm there is usually an increase of the spread because the peripheral regions of the cornea and lens are often afflicted with optical aberrations while contributing heavily to the total light entering the eye. For example, enlarging of the pupil from 6 to 7 mm contributes 6.5 times more additional light than enlarging it from 2 to 3 mm. For cone vision, the Stiles-Crawford effect would reduce this to a factor of about 2.

3. Scatter. Because the ocular media have some microscopic and ultramicroscopic structure, light is scattered in its passage from the cornea to the retina. Backward scatter is utilized by an examiner when examining an eye with the slit lamp, but forward scatter can be more serious. Its effect can be quite extensive, spreading light from even a narrow incident beam over a considerable portion of the retina. Complaints of glare can have their origin in scattered light, which increases in prominence with age. Shielding the eyes from the direct rays of intense light sources (wearing the green eyeshades of older days!) is good advice, though not easily put into effect for automobile headlights at night.

4. Absorption. The media are not uniformly transparent to incoming light. In general, the shorter the wavelength of the entering light, the smaller the proportion that reaches the retinal receptors.

5. Focus factors. The effect of defocus on visual acuity will be dealt with separately and has been adequately characterized. It must be remembered, however, that when a person has active accommodation, it cannot be taken for granted that the accommodative stance will always be appropriate to the stimulus distance. This is especially a problem when no sharply delineated targets are available to anchor accommodation. Night and instrument myopia are well-known instances of the phenomenon, but it can also occur when "fogging" during a refractive examination.

Altogether, then, it is only a conjecture to assume that a given eye under a given set of circumstances will display the optimal point-spread function.

The procedures needed to deduce the light spread in images other than points have been outlined elsewhere.[29] They have to be followed for each particular target when the relative effect of the various factors on visual acuity is being determined. Light-spread, the first in a sequence of transformations, needs special attention, lest phenomena be assigned to complex neural interaction when for example, they may have a simpler explanation in light scatter or accommodation instability.

RETINAL ANATOMY

One unavoidable bottleneck in the processing of spatial information in the retina is the finite size of the retinal receptors. In the fovea, the cones are packed approximately two to the linear minute of arc,[17] and each cone's local sign is indivisible. In principle it is not possible, therefore, to resolve patterns whose spacing demands separate sampling of intensity at intervals smaller than half a minute of arc.

Other limits are set by the neural connectivity of retinal cells at their various layers. This applies particularly in the retinal periphery, where many rod connections converge on a ganglion cell. Although the rods themselves are small, the fact that summation of their signals takes place over areas up to several degrees in diameter sets upper bounds to the partitioning of spatial information.

While there is no question of the role played by the elements of the retinal mosaic in limiting resolution, certain spatial distinctions can be made that appear to have a finer grain—the hyperacuities. Since the diameter of the retinal receptors constitutes an unavoidable limitation to partitioning spatial information except at defined intervals, just as does the extended dimension of the image of a point object, there must be sophisticated neural processing that interrelates signals from adjoining cells to provide the information as to location.

PHYSIOLOGIC FACTORS

Because visual acuity is conceptually and operationally well anchored in psychophysics, the detailed dissection of its physiologic substrates is not an issue central to its discussion. In fact, current electrophysiology has not yet demonstrated neural processing mechanisms that approach the best human thresholds.

Nevertheless, spatial differentiation is inevitably coupled to light difference detection. If for any reason this is deficient, spatial resolution suffers. Thus one finds that visual acuity follows pari passu intensity discrimination sensitivity for the small stimulus areas that are involved.[21] For example, the detection of a double star as composed of two separate stimuli requires that the trough between the two light peaks be deep enough that an intensity discrimination ($\Delta I/I$) for such a small stimulus area can be carried out. Because $\Delta I/I$ has to be larger when the luminance gets lower, resolution deteriorates, that is, the peaks have to be further separated to create a trough that delivers a $\Delta I/I$ value sufficiently large to be detected for that area at the prevailing luminance.

Many other variables that influence visual acuity, for example, adaptation and exposure duration, exert their effect predominantly on the light discrimination sense and only through it on resolution. In many cases the effects can be traced for some distances through the optical, anatomic, and physiologic stages and the threshold identified by a prevailing limitation along the way.

Before returning to a brief survey of some of these interacting variables, we need to outline the procedures involved in actual visual acuity measurements.

ACUITY CRITERIA

Within the general definition of visual acuity, that is, thresholds in which spatial dimension is the variable, some obvious subdivisions can be recognized, and they are best outlined by the different criteria set for the response of the observer (Table 18-1):

1. The criterion of the presence of a single feature *(minimum visible)*
2. The criterion of the presence, or internal arrangement, of identifying features in a visible target (ordinary visual acuity or minimum resolvable)
3. The criterion of the relative location of visible features (the spatial minimum discriminable)

Minimum visible

What is involved in minimum visibility is the detection of the presence of a visual stimulus, but the stimulus variation is carried out by manipulating the contrast of the target through the medium of varying its size. The most typical example of this remains the experiments in which one measures the minimum width of a telegraph wire that can be seen against a uniform sky. The threshold value is of the order of 1 *second* of arc, that is, a very small fraction of the diameter of a retinal receptor. The situation is, however, not as startling as it sounds, because it has a simple basis in the variation of the physical stimulus. The retinal light spread for a single thin dark line seen against a uniform bright background is a dimple with the

Table 18-1. Classification of visual acuity according to criteria

Criterion	Minimum visible	Minimum resolvable (ordinary visual acuity)	Minimum discriminable (hyperacuity)
Task	Determine presence or absence of a target	Determine presence of, or distinguish between more than one, identifying feature in a visible target	Determine relative location of two or more visible features with respect to each other
Typical forced choice psychophysical question	Is there a line in this field? If there was a line in the field, was it horizontal or vertical?	Is this a "C" or an "O"? Is the gap in the C up, down, right, or left?	Is the upper line to the right or the left of the lower line?
Physiologic basis	Local brightness difference threshold (ΔI)	Detection of brightness differences between several adjoining small areas	Assignment of relative local signs to two or more suprathreshold visual features
Method of measurement	Vary object size	Vary object size or spacing between object components	Vary relative location of features
Magnitude of best threshold	~1 second of arc	~30 seconds of arc	~3 seconds of arc
Effect of image degradation	Moderate	Serious	Slight (except in stereoacuity)

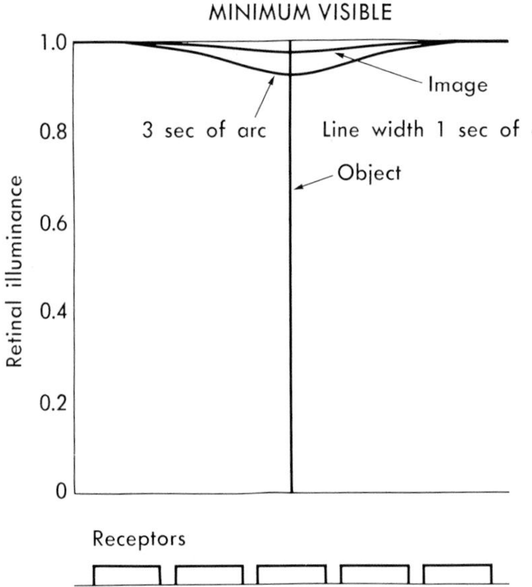

Fig. 18-3. Schematic diagram to illustrate that the *minimum visible* is a brightness rather than a spatial visual threshold. A single dark line seen against a uniform background is widened until it can be detected. As the line is widened, the retinal image pattern, which has the outline of the complement of the eye's line-spread function, increases in contrast, but remains invariant in shape. Detection occurs when the ΔI threshold is reached for the prevailing adaptation level. Although the object is changed in a spatial dimension, detection is purely that of a brightness change. For threshold line width (~1 second of arc) the retinal image contrast is just a few percentage points. (From Westheimer, G.: Invest Ophthalmol. **18:**893, 1979.)

cross-sectional outline of the line-spread function. In the human eye, this has a width at half height of at least 1 minute of arc. For all targets that have a width of about this value or less, the shape of the light distribution remains about the same; variations of target width will manifest themselves purely as variations of dimple depth. Threshold measurements for the minimum visible (where a dark line, for example, is progressively widened until its presence is detected) are in reality merely ΔI thresholds for a more or less fixed retinal light distribution whose contrast is varied by varying the target width (Fig. 18-3). The situation is analogous to one in the time domain, when one can let a light pulse reach threshold by increasing either the intensity of a stimulus or its duration, so long as the pulse length remains within the limits of the critical duration of the Bunsen-Roscoe-Bloch law.

In a definitive study, Hecht and Mintz[12] showed that variations in contrast sensitivity as a function of luminance, for example, can fully account for variations in the minimum visible spatial threshold.

Although the stimulus change is effected in the dimension of space, the minimum visible threshold does not really approach the essence of visual acuity because the subject's judgment does not demand the making of any spatial differentiations, except in the trivial sense of whether a field is uniform or not. The other two subdivisions of visual acuity, however, demand just that differentiation.

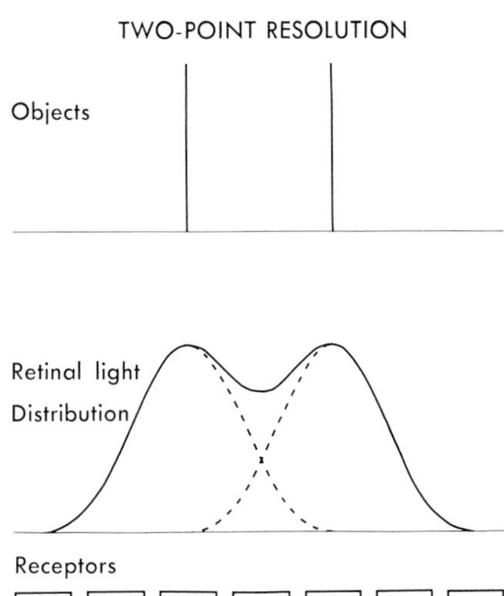

Fig. 18-4. Example of ordinary visual acuity is the detection of doubleness for a stimulus consisting of two points, each of which is imaged on the retina as a point-spread function. The amount of overlap depends on point separation. The essential elements for detecting doubleness, that is, for resolution, are (1) an underlying retinal image pattern with two peaks separated by a trough, (2) a retinal illuminance difference between the peaks and the trough that is within the ΔI capability of the visual system in the prevailing state of adaptation, and (3) separate localization of the differentially stimulated regions.

Minimum resolvable or ordinary visual acuity

Most commonly associated with visual acuity is the concept of Snellen's letters or Landolt's C's. A high-contrast, clearly visible target is shown and the subject has to make a spatial judgment best exemplified by the distinction between a P or an F, a B or an R, and a C or an O. That is, either the presence of a gap or the relative arrangement of components of a letter has to be detected. Since we are dealing with a resolution task, there are more pointed tests of this capacity than Snellen's letters: for example, a double star that is being separated till seen double or a grating whose spatial frequency is being reduced until its structure becomes apparent. However, Snellen's letters have the virtue of not requiring a binary decision that is subject to guessing; they belong to a moderately sized ensemble of well-known patterns and optimize information transfer between patient and examiner.

Conceptually, the simplest situation is one in which members of a point or line pair are moved apart until the observer can judge them to be separate. Each of the two bright points or lines would be imaged on the retina with the light distribution of the point- or line-spread function. Initially, the two spread functions will overlap thoroughly, but as the target pair is separated, overlap will be only partial. A pattern emerges, characterized by two humps with an intervening trough (Fig. 18-4). Resolution, that is, correct judgment of doubleness, can be achieved when the peak-trough ratio of retinal illuminance is accommodated by the $\Delta I/I$ ratio of the visual system for the stimulus area involved at the prevailing level of adaptation. In addition, the effective grain of the visual system has to be small enough for the peaks and trough to fall within separate detecting units, regardless of whether they are defined by the limitations of retinal anatomic structure or synaptic organization.

In a normal observer in best focus the resolution limit or, as it is usually called, the minimum angle of resolution (MAR) is between 30 seconds of arc and 1 minute of arc.

There is remarkable concordance between the observed minimum angle of resolution, the expected resolving capacity of the eye's optics, and the predicted performance of a system that has a set of receptive elements with the dimension of the retinal mosaic in the fovea. Most students of the subject do not regard this as coincidence but as evidence for evolutionary convergence. Animal studies tend to confirm this view; birds with acuity better than humans' have appropriate receptor and pupil size.

Visual acuity decreases whenever the conditions are no longer optimal. For a patient to manifest visual acuity of 20/20 or better, the requirements are good optical imagery, foveal fixation, intact receptor structure and function, photopic luminance levels, and, of course, full integrity of the involved neural pathways.

Minimum discriminable or hyperacuity

Certain spatial distinctions can be made by a normal observer when the threshold is much lower than ordinary acuity, and these must therefore have a fundamentally different basis. The best known of these is alignment or vernier acuity.[37] The tasks share with ordinary acuity the presence of a clearly delineated target and should therefore never be confused with the threshold for the minimum visible, where merely the presence or absence of a target is being judged. However, in a hyperacuity test the subject is asked to make a judgment as to location of an element, usually relative to another element of the same target. Typical examples of hyperacuity configurations are shown in Fig. 18-5. They all share the property of allowing spatial judgments whose thresholds in normal observers are 2 to 10 seconds of arc.

The mechanism subserving hyperacuity is still being explored,[31] but so much is clear; no contradiction is involved with the optical and receptor mosaic factors that limit ordinary visual acuity. Localization of a feature can be achieved with arbitrary precision so long as enough light quanta are available. The neural processing that is required for these judgments must, however, be quite sophisticated.

Stereoscopic acuity also has a threshold of a few seconds of arc and may therefore be included under this heading, but its processing probably differs somewhat from that in ordinary hyperacuity.

MEASUREMENT OF ORDINARY VISUAL ACUITY (MINIMUM ANGLE OF RESOLUTION)

A variety of patterns have been employed to ascertain a patient's minimum angle of resolution. The major feature of all of them is that a given pattern is bodily enlarged or reduced to find the threshold size at which the judgment can be correctly made.

The most familiar of these tests is the Snellen chart. In a normal observer in best focus the resolution limit is between 30 seconds of arc and 1 minute of arc. The standard testing procedure is to enlarge the pattern until resolution can be achieved. Commonly the overall size of the letter is five times the width of each limb (Fig. 18-6). The easiest description is afforded by the Landolt C. The reference letter is a ring

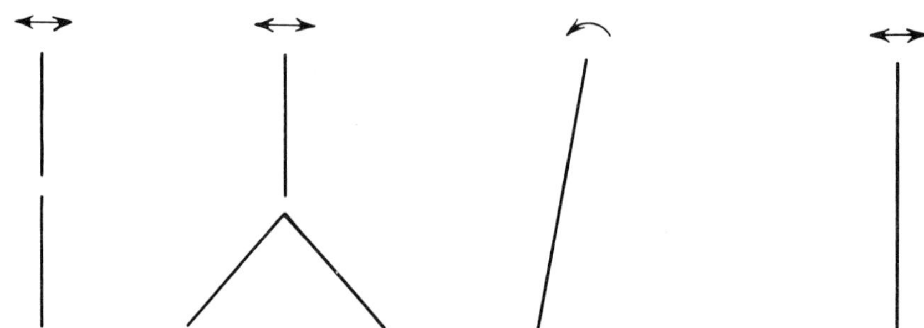

Fig. 18-5. Typical target configurations, which demonstrate the hyperacuity capability of the visual system, that is, the detection of very small differences in relative localization of features. Arrows indicate the direction of displacement, which in each case can be judged to a few seconds of arc under optimal conditions. From left to right, vernier offset detection of two vertical lines; offset detection of the tip of a chevron from a line; detection of orientation change of a line; detection of lateral displacement of a target.

with an outer diameter subtending 5 minutes of arc at the observer's eye and an inner diameter subtending 3 minutes of arc. A gap 1 minute of arc wide is made in the ring, and the ring is presented with its opening in one of, say, four possible positions, up, down, right, or left. The subject has to indicate in which direction the C is pointing. At an observation distance of 6 m (20 feet) the overall size of the letter is 8.73 mm and the gap is 1.75 mm. If this is the subject's threshold, that is, if his minimum angle of resolution is 1 minute of arc, the visual acuity is identified as 6/6 or 20/20. Suppose, however, he has a minimum angle of resolution of 0.75 minute of arc, that is, at 6 m he can resolve a letter with a feature that subtends 1.3 mm and whose overall size is 6.5 mm; such a letter has a gap that subtends 1 minute of arc at 4.5 m or 15 feet and would be resolvable by an observer with 1 minute of arc resolution at such a distance. The subject is then said to have visual acuity of 6/4.5 or 20/15.

Illiterate E and Landolt's C tests are based on the same principle as Snellen's charts. One may distinguish from these certain repetitive patterns such as gratings or checkerboards. In these the corresponding feature size must always be clearly understood; in a grating it is half the length of a period; in a checkerboard it is the side length of a square.

On the whole, single features such as individual-style letter charts are to be preferred because under certain conditions of defocus, repetitive patterns may at times be spuriously resolved at a size for which a patient cannot consistently make correct judgments on letters.

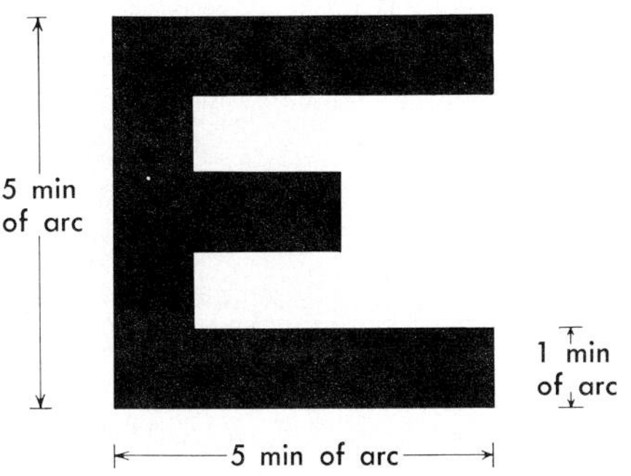

Fig. 18-6. 20/20 Snellen's letter.

All letters in the alphabet are not equally legible[24]; most charts do not use the whole alphabet. Test charts have been created with letters in other scripts.

Instruments have occasionally been designed that permit zooming of letters until they can be resolved, but the multiple choice psychophysical technique of letter charts with several letters to the row has never been bettered in the practical situation.

The procedure of requiring a patient to read a letter chart rests on facilities for verbal communication that may not always be satisfied. A patient may not be able to understand the request or may not be able to indicate to the examiner what his performance actually is. For this reason, several so-called "objective" techniques of measuring visual acuity have been devised. They utilize a nonverbal response mode, but it cannot always be guaranteed that the results will be equivalent because they may be channeled through different neural circuits. This has been made particularly clear by the case of one patient[2] who was manifestly blind by all observable criteria, but whose grating "resolution" measured by evoked potentials was normal.

The two most prominent objective measuring techniques for the minimum angle of resolution involve neural electrical potentials and oculomotor responses.

The evoked potential technique[18] is based on the presentation of a differentiated spatial pattern to the subject and measuring the changes in the electroencephalogram (EEG) that accompany the presentation. In the most sophisticated modern versions of this test, the stimulus consists of a change merely in the internal light distribution in the pattern without any overall changes in total light reaching the eye. Examples are the instantaneous exchange of light and dark squares in a checkerboard or the replacement of a uniform field with a grating of the same average luminance. Whenever the pattern dimension is too small to be resolved, no changes are expected in the evoked potential on stimulus presentation. Utilization of temporal repetition and a signal averager can allow good measurements of resolution.

Pursuit and optokinetic eye movements are

released by target movements, usually with little or no "voluntary" components. Here again, targets with spatial differentiations too small to be resolved would not be expected to lead to associated smooth eye movements, and we clearly have a good principle for acuity measurement.[10] A variant of the eye movement method, observation of the relative frequency of voluntary fixations, has recently been the primary vehicle for accumulating information about the development of visual acuity in infants.[7] In one of two equivalent but separated patches of the infant's visual field a grating is shown, and an observer notes the relative frequency of voluntary fixations on the two; the expectation is that a field with a differentiated pattern will be favored for fixation. One needs to remember, however, that this technique presupposes access to and integrity of the oculomotor pathways.

When there is a lack of concordance of results with the various measuring techniques, the origin has obviously to be sought in difference in the pathways utilized in them.[5] Evoked potential recordings with electrode placements favoring the visual cortex would probe the most elementary pathways, but letter recognition and verbalization involve many other cortical areas. Whatever technique is applied, attention has to be given to the method of identifying a threshold. Even when there is no resolution, an EEG blip may occasionally occur synchronously with a stimulus, a randomly occuring saccade (although a smooth movement is unlikely) land on the target, or a subject may correctly guess the letter presented or the target's direction if it is an illiterate E or a grating. Special procedures have been devised to ensure that objective techniques separate "signal" from "noise." Threshold determination is, of course, one of the time-honored problems of psychophysics. The method of constant stimuli with forced choice has been effectively applied, as also the method of staircases. In either case, the usual value of thresholds is the target size at which the subject responds correctly on 75% of occasions. In clinical practice it works well to indicate the number of letters a patient misses in a line of Snellen's letters or can read beyond it (e.g., 20/15 − 1 or 20/20 + 2).

Thresholds obtained with a rigorous psychophysical method not only give a mean value of, say, the minimum angle of resolution, but also a standard error of this mean. Such a number has the virtue of permitting conclusions about whether the threshold is significantly different statistically in one situation as compared with another. It has been found that the standard error of the minimum angle of resolution, in common with that of other sensory limits, remains an approximately constant proportion of the mean (Weber's law).[30]

Table 18-2. Visual acuity equivalents in different notations

MAR (minutes of arc)	Snellen's visual acuity		Snell-Sterling's visual efficiency (%)	Snellen's fraction	Log visual acuity relative to 20/20
	Feet	Meters			
0.5	20/10	6/3	109	2.0	0.3
0.75	20/15	6/4.5	104	1.33	0.1
1.00	20/20	6/6	100	1.0	0
1.25	20/25	6/7.5	96	0.8	−0.1
1.5	20/30	6/9	91	0.67	−0.18
2.0	20/40	6/12	84	0.5	−0.3
2.5	20/50	6/15	76	0.4	−0.4
3.0	20/60	6/18	70	0.33	−0.5
4.0	20/80	6/24	58	0.25	−0.6
5.0	20/100	6/30	49	0.2	−0.7
6.0	20/120	6/36	41	0.17	−0.78
7.5	20/150	6/45	31	0.133	−0.88
10.0	20/200	6/60	20	0.10	−1.0
20.0	20/400	6/120	3	0.05	−1.3

This finding suggests that a logarithmic scale be applied to the minimum angle of resolution, for example, the size of the letters in a visual acuity chart increase in geometric proportion as follows: 20/16, 20/20, 20/31, 20/39, 20/48, 20/61, 20/76, 20/95, 20/120, 20/149, 20/186, 20/232. In practice, variations of such a scheme have been followed for nearly 100 years.[22] There is medicolegal significance in the identification of the percentage loss of visual acuity as a consequence of disease or injury. An earlier attempt by Snell and Sterling[25] led to a system of measurement in which any increase by 1 minute of arc in the minimum angle of resolution was regarded as reducing the visual efficiency to 86% of its previous value. Table 18-2 indicates several ways for specifying visual acuity levels: minimum angle of resolution; Snellen's acuity; Snell-Sterling's efficiency rating; Snellen's fraction (that is, the reciprocal of the minimum angle of resolution); the logarithm of Snellen's fraction.

Identical numbers can give considerably different impressions, depending on the mode of presentation. As mentioned before, because the ratio

$$\frac{\Delta \text{ MAR}}{\text{mean MAR}}$$

is approximately constant, the logarithmic scale is the most appropriate. In it all reductions of acuity by a given factor constitute equivalent decrements. For example, the step from 20/10 to 20/20 is equivalent to the steps from 20/20 to 20/40, or from 20/200 to 20/400.

FACTORS INFLUENCING VISUAL ACUITY

Because all optical, anatomic, and physiologic elements are at or near their peak performance when a subject exhibits what we call normal visual acuity, a diminution of function in any of these constituent elements will manifest itself in a reduction in visual acuity. The list of factors influencing visual acuity is legion,[28] and no attempt will be made to give an exhaustive account of them here. There are some stimulus variables, however, whose effects are of more universal interest and have been well documented.

Refractive error. As soon as the optics of the eye are defocused, the point-spread function widens, and two stars, to be identified as separate, need to be farther apart than in the fully focused state. The width of the defocused point-spread function depends directly on the amount of defocus and inversely on the pupil size. For normal observation of a Snellen chart, the data in Fig. 18-7 are typical.[15] There are, however, some complications. Due to optical peculiarities, some repetitive patterns, for example, gratings or checkerboards, can occasionally be resolved with a defocused optical system when letters of a similar size cannot. Another important variable is pupil size. Depth of focus increases with reduction in pupil size; thus the curve in Fig. 18-7 will be less steep in a patient with a 2 mm pupil than shown. That a young hyperope can compensate for his refractive error by accommodation and hence may have no deficit in his unaided acuity needs no special mention. But it must also be remembered that retention of the state of zero accommodation is an active process. Thus blurring a young emmetropic eye with a +2 D lens may not make it exactly 2 D myopic—with good anchor points missing, accommodation may be active and the defocus could be more than 2 D and, more importantly, quite unstable. In astigmatic imagery, the point-spread function will have different dimensions in the various direc-

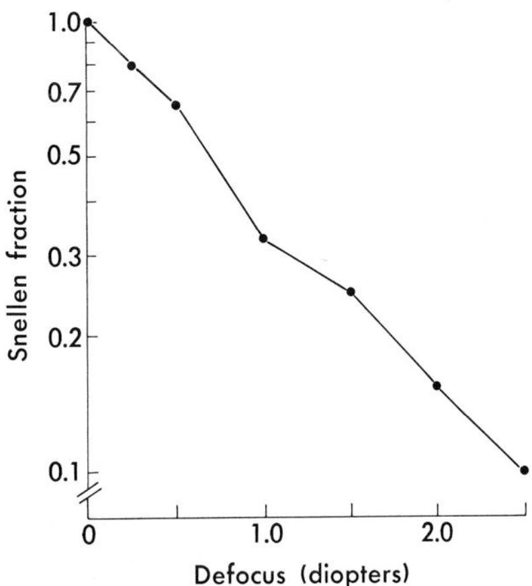

Fig. 18-7. Visual acuity (Snellen's fraction) as a function of defocus for a typical eye. The ordinates are plotted on logarithmic coordinates. (Data from Laurance.[15])

tions. There will be meridional variation in acuity leading to a choppy performance on letter charts.

Retinal eccentricity. Only in the center of the fovea are the conditions appropriate for maximum acuity. Even 1 degree away from it there is a reduction to about 60% of maximum[34]; the function of visual acuity with eccentricity is shown in Fig. 18-8.[26] There is consensus that this is due to neural connectivity rather than optical blur or receptor spacing. It is true that cones are farther apart as distance from fovea increases, but rods remain closely packed and could, in theory, have been connected to carry good spatial information. The reduction in image quality in the retinal periphery is certainly nowhere near as severe as the reduction in acuity. The immediate cause is an increase in retinal summation areas, that is, the area from which excitatory signals converge on a ganglion cell. Since the visual cortex appears to be organized in processing modules,[13] each presumably with a constant number of input lines, one seeks concordance with the cortical magnification ratio (i.e., the number of degrees of visual field represented in a millimeter of visual cortex).[19]

Luminance. Data from Shlaer's[21] definitive study of the effect of luminance on visual acuity are shown in Fig. 18-9. The evidence points to a separate rod and cone branch of the curve. Rods asymptote at a value of about 8 minutes of arc, and this is the acuity usually found in absence or complete dysfunction of cones.[23] Visual acuity remains constant over a wide range of photopic luminances, extending from the level of full moonlight to that of a bright sky on a sunny day. Very high luminances cause an unexplained reduction in acuity,[36] even in adjoining zones of moderate luminance.

Pupil size. With pupil size below about 2.5 mm, the eye's point-spread function in good focus becomes progressively wider, and a reciprocal relationship is expected between minimum angle of resolution and pupil size. Depending on the quality of the optics of the peripheral zones of the pupil, visual acuity remains approximately constant[16] in the range of 2.5 to 6 mm, beyond which the aberrations begin to widen the point-spread function again.

Exposure duration. Although there is progressive reduction of visual acuity with decreasing exposure duration in the millisecond range, this can be offset by increasing illumination[11] to ensure constancy of the number of absorbed quanta. Visual acuity in most observers is, however, not as good for target

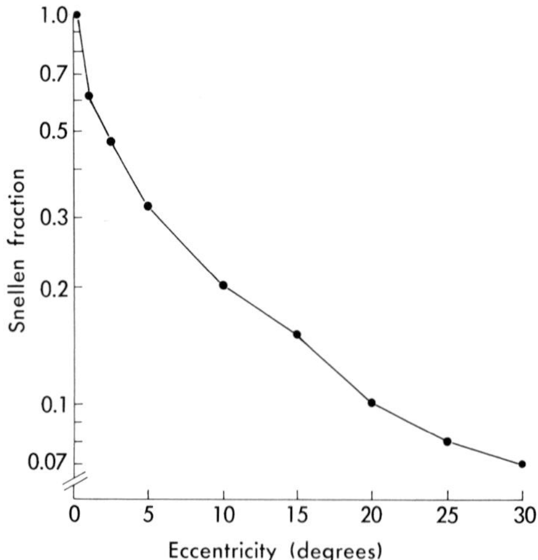

Fig. 18-8. Visual acuity (Snellen's fraction) as a function of retinal eccentricity. (Data from Wertheim.[26])

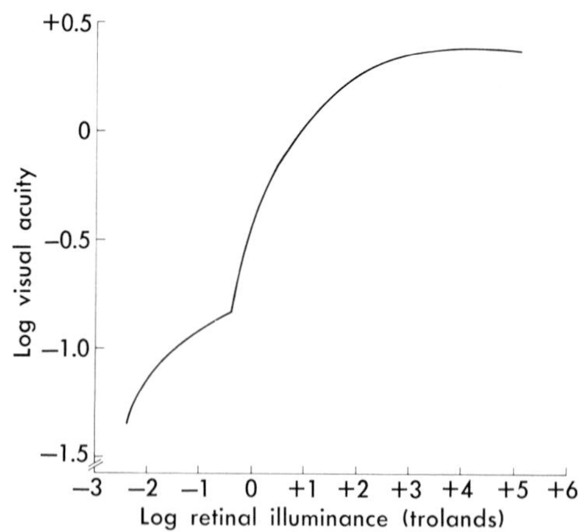

Fig. 18-9. Visual acuity (Snellen's fraction) as a function of luminance. (Data from Shlaer.[21])

exposures in the range of 100 to 500 milliseconds as for longer durations, even though retinal summation is no longer a factor here.[1]

Target and eye movements. There is a decrement in certain visual functions during saccades and also in the general case in which there is significant movement of the retinal image. A target moving at moderate velocities, a few tens of degrees per second for at least half a second, can induce good enough pursuit eye movement to ensure acuity not far from normal. Small movements of the image do not, however, detract from acuity[33]—strict stability of the retinal image is not a requirement for optimum resolution.

Meridional variations in acuity. Differences in acuity in the various retinal meridians have been widely reported. They require grating or similar targets that permit the selective evaluation of the function one meridian at a time. The usual finding is that horizontal and vertical meridians are favored, although this is not universally so. The differences rarely exceed 15%, and there are claims for an etiologic influence connected with uncorrected astigmatism.[6]

Interaction effects. Visual acuity suffers when targets are too close together. Thresholds rise and occasionally may even double when a competing target is presented within a few minutes of arc. That this is not purely optical in origin is demonstrated by the nonmonotic relationship between threshold and intertarget distance; the diminution of performance is maximal at a distance of 2 to 5 minutes of arc and disappears for larger and smaller distances. The effect can be observed with ordinary visual acuity[9] and also vernier[32] and stereoscopic acuity.[3]

Developmental aspects. The difficulties of assessing the resolving capacity of the infant had for a long time left the time course of visual acuity development unclear. By the time ordinary methods—illiterate E's, for example—can be utilized, visual acuity is normal. It takes several months after birth for the full development of pursuit eye movement, and it is therefore not easy to differentiate between inability to resolve a pattern and inability to execute the movement that would betray resolution. Preferential looking methods, utilizing the frequency of intersaccadic fixation in regions containing a pattern as compared with those not containing a pattern, have yielded fairly consistent results in the hands of dedicated experimentalists. Reasonable indication of the state of development of visual acuity with infant age is given in Fig. 18-10. The subject has recently been thoroughly reviewed by Dobson and Teller.[5] When evoked potential is used as the measure, the data routinely come up with thresholds that are lower. This may be because evoked potentials tap off at an earlier state of elaboration of the visual acuity signal than eye movements or verbal responses; it must not be forgotten, however, that evoked responses are based on an averaging procedure effectively summing often many hundreds of stimuli. Active research is in progress on the possible interference with normal development (for example, by patching one eye or failing to correct the optical defects engendered by aphakia or aniridia). The interest was sparked by Wiesel and Hubel's[35] observation in 1963 that cats

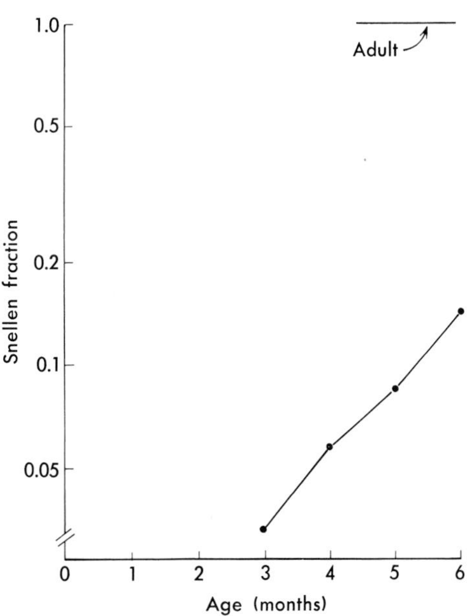

Fig. 18-10. Visual acuity as a function of age in the infant. (Data from Dobson and Teller.[5])

whose eyes were kept artificially closed in their first few weeks of life had visual behavior as well as single-cell responses in the visual cortex that differed from the norm. There appears to be a clear need for the visual system to be exposed to appropriate optical stimuli during a critical period of development, but it is as yet too early to identify the exact period and the nature of the needed stimuli in the human infant.

SINUSOIDAL GRATING TARGETS

We have seen earlier in this chapter that the optical spread in the eye constitutes one possible limit to resolution. It would be advantageous to have a test that probes the retinal and neural stages of the visual acuity processing alone, bypassing, so to speak, the eye's optics. As it happens, there is one mode of illuminating the retina that creates a pattern largely independent of the optical quality of the eye: Young's interference fringes (Fig. 18-11). When two monochromatic coherent point sources are imaged in the pupil, light diverging from them into the vitreous will form a system of interference fringes whose angular spacing in object space is given by the formula λ/a, where λ is the wavelength and a is the separation of the sources in the plane of the pupil. When laser light of wavelength 623 nm is used, fringe spacing of 1 minute of arc occurs with a source separation of 2.14 mm in the pupil, fringe spacing of 2 minutes of arc (minimum angle of resolution = 1 minute of arc) with a source separation of 1.07 mm, and so on. The retinal image is a sinusoidal grating, that is, a grating whose intensity profile is sinusoidal. Its major virtue is that it retains very high contrast and constant fringe spacing regardless of the state of refraction or aberrations. The procedure has been utilized to find the resolution limit of the visual system when the optical factors have been eliminated,[4,27] disclosing that retinal resolution matches that of good optics with 2 to 2.5 mm pupil. Gratings with finer fringes, created by separating the source images beyond 2.5 mm, are not resolved.

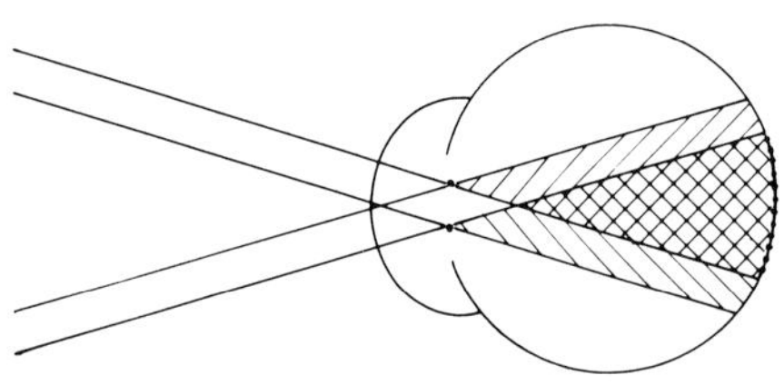

Fig. 18-11. Basis for the interference fringe method of measuring resolution of the retinal and central stages of the visual system, bypassing the refractive stage. A pair of coherent point sources is imaged in the plane of the pupil. Wherever the two bundles overlap in the image space, a system of interference fringes is formed. Spacing of the fringes in angular measure in the eye's object space is given by λ/a radians, where λ = wavelength of light and a = separation of the two point sources. When λ = 623 mm and a = 1.0 mm, fringe peaks are 2.14 minutes of arc apart.

The clinical utility of the test is, however, still an open matter. Careful ophthalmoscopic and retinoscopic examination will give an excellent indication of the optics of an eye, making it unnecessary to bypass the eye's optics just to ascertain whether they caused an acuity deficit. A much more important question concerns the capacity for resolution of the retinal and central stages when they cannot be reached via the standard optical imaging route because of opacities in the media. If it were possible to smuggle a system of interference fringes into the vitreous, one could in principle reach a conclusion as to the integrity of the neural stages prior to instituting surgical relief for opacities of the media. The difficulty in the way of this approach is apparent from Fig. 18-11. Fringes of a given spacing arise from the interference of coherent beams from two points in the pupil a fixed distance apart. For maximum contrast, the intensity of the two beams must be equal. When the effective intensity of one beam is reduced to k times the other ($1 > k > 0$), the interference fringe contrast is reduced to $\frac{2\sqrt{k}}{(1 + k)}$. The essential precondition for the application of a sinusoidal grating interference fringe test that bypasses the optics of the eye is that pairs of locations can be found in the media that allow the admission of twin coherent beams whose image-sided intensities are not excessively different. The psychophysical safeguards when using a grating ("Do you now see a set of stripes?") are somewhat more cumbersome than when using letters

("Please read the letters on this line of the chart"), but do not constitute a crippling disadvantage for such a test.

There has been a great deal of interest in sinusoidal grating targets used via the ordinary optics of the eye. In 1956 Schade[20] found that sinosoidal gratings with fringe spacing of 5 to 8 minutes of arc can be seen when they have a remarkably low contrast, often much less than 1%, while fringes with finer spacing (i.e., higher spatial frequency) and also with coarser spacing (i.e., lower spatial frequency) have to have higher contrast to be visible. The curve giving the relationship between sensitivity and spatial frequency is called the modulation sensitivity curve of the eye, and it exemplifies the band-pass characteristic of the visual system, which is also apparent in the temporal domain. The origin of the effect is largely in the center-surround organization of neural elements in the retina and the visual projection. An excitatory center flanked by inhibitory surrounds is a characteristic of retinal ganglion cells and also of cortical neurons, and these are well matched by light patterns such as gratings that can deliver the appropriate stimulus to the various components of the receptive field simultaneously. Unless only a few cycles are shown at a time, however, gratings are not sufficiently localized to probe, say, the fovea, which has a diameter of 30 minutes of arc. Tests of the modulation sensitivity may turn out to give useful pointers to the performance of certain neural stages of vision. Because such tests involve stimulus manipulation in the domain of contrast rather than of space, they are not, strictly speaking, visual acuity tests.

REFERENCES

1. Baron, W. S., and Westheimer, G.: Visual acuity as a function of exposure duration, J. Opt. Soc. Am. **63:**212, 1973.
2. Bodis-Wollner, I., et al.: Visual association cortex and vision in man: pattern evoked occipital potentials in a blind boy, Science **198:**629, 1977.
3. Butler, T., and Westheimer, G.: Interference with stereoscopic acuity: spatial, temporal and disparity tuning, Vision Res. **18:**1387, 1978.
4. Campbell, F. W., and Green, D. G.: Optical and retinal factors affecting visual resolution, J. Physiol. (Lond) **181:**576, 1965.
5. Dobson, V., and Teller, D.: Visual acuity in human infants: a review and comparison of behavioral and electrophysiological studies, Vision Res. **18:**1469, 1978.
6. Emsley, H. H.: Irregular astigmatism of the eye. Effect of correcting lenses, Trans. Opt. Soc. Lond. **27:**28, 1925.
7. Fantz, R. L.: Pattern vision in young infants, Psychol. Record **8:**43, 1958.
8. Fechner, G. T.: Elemente der Psychophysik, Leipzig, East Germany, 1860, Breitkopf & Härtel VEB.
9. Flom, M. C., Weymouth, F. W., and Kahneman, D.: Visual resolution and contour interaction, J. Opt. Soc. Am. **53:**1026, 1963.
10. Goldmann, H.: Objektive Sehschärfenbestimmung, Ophthalmologica **105:**240, 1942.
11. Graham, C. H., and Cook, C.: Visual acuity as a function of intensity and exposure time, Am. J. Psychol. **49:**654, 1937.
12. Hecht, S., and Mintz, E. U.: The visibility of single lines at various illuminations and the retinal basis of visual resolution, J. Gen. Physiol. **22:**593, 1939.
13. Hubel, H., and Wiesel, T. N.: Functional architecture of macaque monkey visual cortex, Proc. R. Soc. Lond. (Biol.) **198:**1, 1977.
14. Keesey, U. T.: Effects of involuntary eye movements on visual acuity, J. Opt. Soc. Am. **50:**769, 1960.
15. Laurance, L.: Visual optics and sight testing, ed. 3, London, 1926, School of Optics.
16. Leibowitz, H.: The effect of pupil size on visual acuity for photometrically equated test fields at various levels of luminance, J. Opt. Soc. Am. **42:**416, 1952.
17. Polyak, S.: The retina, Chicago, 1941, University of Chicago Press.
18. Regan, D.: Evoked potentials in psychology, sensory physiology and clinical medicine, London, 1972, Chapman & Hall Ltd.
19. Rolls, E. T., and Cowey, A.: Topography of the retina and striate cortex and its relationship to visual acuity in rhesus and squirrel monkeys, Exp. Brain Res. **10:**298, 1970.
20. Schade, O. H., Sr.: Optical and photoelectric analog of the eye, J. Opt. Soc. Am. **46:**721, 1956.
21. Shlaer, S.: The relation between visual acuity and illumination, J. Gen. Physiol. **21:**165, 1937.
22. Sloan, L. L.: Measurement of visual acuity, Arch. Ophthalmol. **45:**704, 1951.
23. Sloan, L. L.: Congenital achromatopsia: a report of 19 cases, J. Opt. Soc. Am. **44:**117, 1954.
24. Sloan, L. L., Rowland, W. M., and Altman, A.: Comparison of three types of test target for the measurement of visual acuity, Q. Rev. Ophthalmol. **8:**4, 1952.
25. Snell, A. C., and Sterling, S.: Percentage evaluation of macular vision, Arch. Ophthalmol. **54:**443, 1925.
26. Wertheim, T.: Uber die indirekte Sehschärfe, Z. Psychol. **7:**172, 1894.
27. Westheimer, G.: Modulation thresholds for sinusoidal light distributions on the retina, J. Physiol. **152:**67, 1960.
28. Westheimer, G.: Visual acuity, Ann. Rev. Psychol. **16:**359, 1965.

29. Westheimer, G.: Optical properties of vertebrate eyes. In Fuortes, M., editor: Handbook of sensory physiology. Physiology of photoreceptor organs, Berlin, 1972, Springer-Verlag, vol. 7/2, pp. 449-482.
30. Westheimer, G.: The scaling of visual acuity measurements, Arch. Ophthalmol. **97:**327, 1979.
31. Westheimer, G.: The spatial sense of the eye, Proctor lecture, Invest. Ophthalmol. **18:**893, 1979.
32. Westheimer, G., and Hauske, G.: Temporal and spatial interference with vernier acuity, Vision Res. **15:** 1137, 1975.
33. Westheimer, G., and McKee, S. P.: Visual acuity in the presence of retinal-image motion, J. Opt. Soc. Am. **65:**847, 1975.
34. Weymouth, F. W., et al.: Visual acuity within the area centralis and its relation to eye movements and fixation, Am. J. Ophthalmol. **11:**947, 1928.
35. Wiesel, T. N., and Hubel, D. H.: Single-cell responses in striate cortex of kittens deprived of vision in one eye, J. Neurophysiol. **26:**1003, 1963.
36. Wilcox, W. W.: The basis of the dependence of visual acuity on illumination, Proc. Natl. Acad. Sci. U.S.A. **18:**47, 1932.
37. Wülfing, E. A.: Über den Kleinsten Gesichtswinkel, Z. Biol. **29:**199, 1892.

Chapter 19

Color vision

NIGEL W. DAW, Ph.D.

The sensation of color is subjective. Individuals are taught names for their color sensations and subsequently use these names whenever the same sensation is obtained. The color that various observers will report for an object does not vary much from one observer to another, with the exception of color-deficient observers.* The color depends on a number of parameters, including the wavelength or band of wavelengths coming from the object, the wavelengths coming from other objects in the field of view, and the wavelengths that the observer was looking at before he looked at the object. Unfortunately the precise relationship between color and these parameters has not been established, despite centuries of work. However, some general points have been established. One is object color constancy; the fact that the color of an object does not change much when the illumination is altered over a surprisingly wide range. The usual explanation for object color constancy is that color depends primarily on the relative brightness of an object at various wavelengths, compared to other objects in the field of view, and that these relative values do not change when the illumination is altered, even though the absolute values do.[13]

*A number of psychologists are now recommending that the term "color deficient" be used, rather than the term "color blind." The majority of people who have been called color blind in the past can distinguish blue from red or green, although they cannot distinguish red from green. Therefore the term color blind overstates their deficiency. Also the term color blind has been found to raise unwarranted fears among less well-educated segments of the population who sometimes think that it is a form of blindness in the more usual sense of that word.

Although there is no one-to-one relationship between color and wavelength (λ), wavelengths of light are often described by the colors that they have when seen in isolation, or in the spectrum. The spectrum may be displayed by refracting sunlight into its components with a prism. If, like Newton, we make a narrow slit in an opaque window blind, sunlight falling on the slit will illuminate a narrow patch of floor or wall the shape and size of the slit. In a sense the illuminated patch is an image of the slit. A prism in the path of the light from the slit will bend, or refract, the light so that the image of the slit is no longer in the straight line joining sun and slit. The angle that each wavelength of radiation is displaced from its path by the prism depends on the type of glass and the angle between its surfaces, but in general the bending is inversely proportional to the wavelength. Since short wavelengths are bent the most by the glass, the image of the slit formed on the wall or screen by these wavelengths is farthest from the line joining sun and screen. Successive images of the slit are formed by radiation of increasing wavelength closer to the direct image of the slit (Fig. 19-1). The amount by which the radiation is bent or refracted is related to the nature of the glass or other refracting material and the angle of incidence by the index of refraction, n_m.

$$n_m = \frac{\text{sine } i}{\text{sine } r} \quad \text{(Snell's law)}$$

In this expression, i is the angle between the incident ray and the normal (perpendicular) to the glass surface, and r is the angle between the refracted ray and the normal. However, we

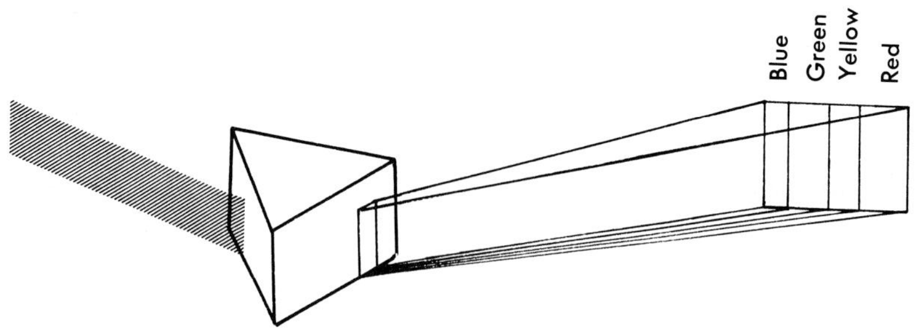

Fig. 19-1. Sunlight is bent from its straight path by prism. Short wavelength light is bent through greater angle than is long wavelength light.

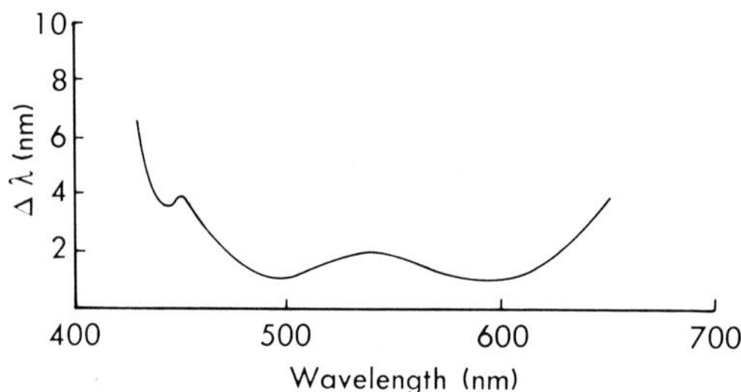

Fig. 19-2. Discrimination of wavelength by human eye. Change in wavelength, which can be just detected, $\Delta\lambda$, is plotted as a function of wavelength. (From Wright, W. D., and Pitt, F. H. G.: Proc. Physiol. Soc. **46:**459, 1934.)

have seen that different wavelengths are bent different amounts by the medium, so the index of refraction differs for different wavelengths, and the index for a given wavelength must be specified as $n_{m\lambda}$. Spreading the white light into its component wavelengths by the different refractive indices of the prism is said to be dispersion by the prism. The linear array of wavelengths is the spectrum. The color sensations these wavelengths produce in the visual system are the spectral colors.

Very approximately the spectral colors are as follows: wavelengths in the region around 430 nm give the sensation of violet; 460 nm, blue; 520 nm, green; 575 nm, yellow; 600 nm, orange; 650 nm, red. Intermediate wavelengths produce sensations that are often described as if they were mixtures—blue-green or yellow-green—even though the stimulus light is monochromatic.

Wavelength discrimination

The normal observer is able to detect a difference between two spectral lights that differ by as little as 1 nm in wavelength in the regions of 490 nm (blue-green) and 585 nm (yellow). In other regions it is necessary to have a greater difference in wavelength, but only in the violet and red is a difference greater than 4 nm necessary (Fig. 19-2).

OPTICAL FACTORS AFFECTING COLOR

A number of physical factors affect color. These have to do with the structure of the eye. The passage of light between the cornea and the molecules of visual pigment is affected by both absorption and refraction of different wavelengths by different amounts.

Lens

The lens of the eye absorbs shorter wavelengths. For a young person this absorption is only significant for wavelengths less than 450 nm, and the lens will appear to be quite transparent. In older people the absorption can be significant up to 550 or 600 nm (Fig. 19-3). Since the aqueous and vitreous also absorb light, these measurements underestimate the total effect. The absorption can also vary quite a lot from one individual to another of the same age. In some older people the lens will appear yellow; the wavelengths absorbed are primarily the blue ones, so the wavelengths passed are the red and green ones, which combine to give yellow.

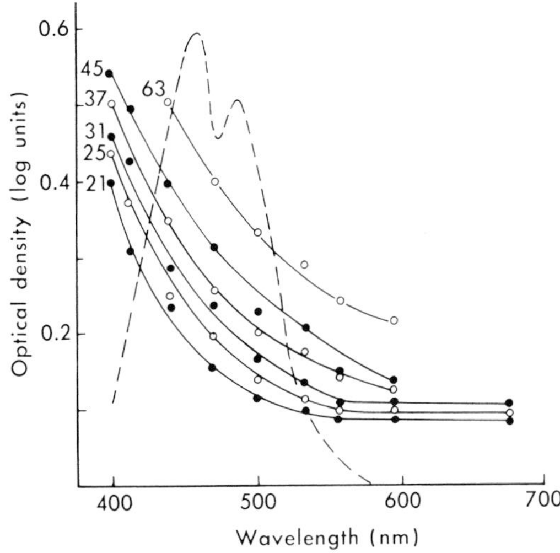

Fig. 19-3. Optical density of living human crystalline lenses plotted as function of wavelength (solid lines). Numbers refer to ages of observers. Broken line gives optical density of macular pigment as function of wavelength. (Modified from Said, F. S., and Weale, R. A.: Gerontologia [Basel] **3:**213, 1959, and Ruddock, K. H.: Vision Res. **3:**417, 1963.)

Macular pigment

The center of the retina also has a yellow spot, the *macula lutea*. This is easily seen in the dissected eye or, with more difficulty in life, by photographing or observing the retina with an ophthalmoscope in red-free light. The spot is due to a carotenoid pigment, xanthophyll, in an area up to 5 degrees from the fovea. The optical density of this pigment as a function of wavelength is also shown in Fig. 19-3. The molecules of pigment are arranged so that they absorb blue light polarized in the radial direction more than blue light polarized in the tangential direction. If one looks at a white card through a linear polarizer, one will see two blue sectors separated by two yellow sectors, the figure called Haidinger's brushes. The phenomenon disappears quickly because the image is stabilized on the retina. Fading can be prevented by rotating the polarizer. The figure is often used as a test of macular function or to locate the straight-ahead direction of an amblyopic eye subjectively in orthoptics. The macular pigment may also be seen as a transient inhomogeneity in a uniform field of blue or white light (Maxwell's spot).

Chromatic aberration

The lens of the eye is not corrected for chromatic aberration. Blue wavelengths are focused closer to the lens than green wavelengths are and green wavelengths closer than red wavelengths because the shorter wavelengths are bent more by the lens. Most people, when looking at a scene illuminated by monochromatic light, will alter their focus to be correct for that wavelength. When looking at a scene illuminated by white light, the eye is normally focused for a wavelength of about 580 nm. Red and green wavelengths are not badly out of focus, but blue wavelengths may be, particularly for myopic individuals looking at objects far away. A point source of blue light of 440 nm will give a blur circle on the retina of about 12 minutes of arc. This is particularly noticeable when looking at automobile taillights that have magenta plastic covers. From a distance these taillights appear to be red spots inside a blue circle. Recently, op artists have made use of the phenomenon in a number of paintings with red-blue boundaries that appear to shimmer.

Stiles-Crawford effect

Light entering near the edge of the pupil is less effective than light entering at the center of the pupil because of the shape of the receptors and the fact that they are embedded in a medium of a different refractive index (Chapter 11). This is known as the Stiles-Crawford effect, and it is wavelength dependent.[8] Another factor is that the receptors are long and thin and therefore denser for light hitting them axially than for light hitting them obliquely. Wavelengths between 500 and 550 nm entering near the edge of the pupil appear to be shorter than they really are, and other wavelengths appear to be longer. This effect is not significant for vision under normal circumstances, but it does have to be taken into account when careful psychophysical measurements are made. The effect can be observed by dilating the pupil of one eye and comparing the colors seen in one eye with the colors seen in the other.

TRICHROMACY OF COLOR VISION

Young is credited with being the first to suggest that human color vision is trichromatic, although others made similar suggestions 40 years before Young. In Young's words,

> Now, as it is almost impossible to conceive each sensitive point of the retina to contain an infinite number of particles, each capable of vibrating in perfect unison with every possible undulation, it becomes necessary to suppose the number limited, for instance, to the three principal colors, red, yellow and blue.[30]

The proof of Young's hypothesis had to wait 150 years for the development of reflection densitometers and microspectrophotometers. We know now that there are three classes of cones in the human retina with different but overlapping spectral sensitivities (Fig. 19-4). One class has a spectral sensitivity that peaks at 440 to 450 nm (blue receptors), one class peaks at 535 to 555 nm (green receptors), and one class peaks at 570 to 590 nm (red or yellow receptors). The overlap occurs because the sensitivity of all visual pigments falls off sharply on the long wavelength side of the peak, but much less sharply on the short wavelength side of the peak, due to the presence of a secondary band of absorption (the beta band or *cis*-peak) at wavelength two thirds of the primary band. The sensitivity on the short wavelength side of the peak never falls below 10% of the peak in the visible region. Therefore it is possible to stimulate only one class of cones (the 570 to 590 nm cones) with long wavelengths, but short wavelengths will always stimulate all three classes of cones.

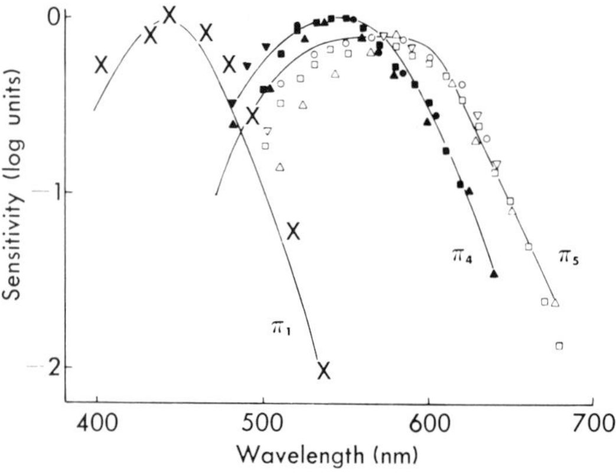

Fig. 19-4. Measurements of cone pigments and processes. Open symbols for red cones, closed symbols for green cones, crosses for blue cones. Measurements for red cones displaced downward 0.1 log unit for clarity. Curves from increment threshold measurements from Stiles.[25] ● and ○, Selective bleaching measurements by reflection densitometry, from Rushton[23] and Baker and Rushton.[2] ▼ and ▽, Selective bleaching measurements by transmission through isolated retina, from Brown and Wald.[5] ■ and □, Increment threshold measurements against bright backgrounds (artificial monochromacy), from Brindley.[3] ▲, △, and X, Also increment thresholds against bright backgrounds, from Wald.[26]

The most direct evidence for the three classes of cones comes from microspectrophotometry. An individual cone of a dissected retina is aligned under a small spot of light, and its absorption is measured at various wavelengths. This is done before and after bleaching the visual pigment to give a "difference spectrum." Since the photoproducts of bleaching a visual pigment absorb light, the difference spectrum does not accurately represent the spectral sensitivity of the receptor, but it does give some idea of the wavelength of peak sensitivity. So far only seven human cones have been measured, and a comparable number from the macaque monkey, which has color vision like the human.[6,14] The results confirm three groupings, with peak sensitivities at 437 to 458 nm, 520 to 543 nm, and 562 to 583 nm.*

The difference spectrum for large numbers of receptors may be measured by selective bleaching. If this is done in the fovea, where the rods are absent, the blue cone receptors are relatively scarce, and the macular pigment is involved, only red and green cone receptors are found. First the red cone receptors are bleached with a long wavelength that is effective for them only, to yield a difference spectrum for them. Then the green cone receptors are bleached with a shorter wavelength to yield another difference spectrum for them. The technique may be

*Some authors have attempted to describe the spectral sensitivity of cone pigments by taking the peak wavelength and drawing the Dartnall nomogram for that wavelength. The Dartnall nomogram is a curve that describes vitamin A_1 rod pigments, which all have the same shape of spectral sensitivity when plotted on a wave number axis. There is considerable evidence that it does not apply to cone pigments.

applied to the living human fovea, in which case light passes through the receptors twice (reflection densitometry[2,23,28]), or to the isolated retina, in which case light passes through the receptors once, and absorption by the optic media of the eye is not involved.[5] Bleaching techniques can also be used to determine an action spectrum, rather than a difference spectrum, by determining how much light at each wavelength is required to yield a fixed amount of bleaching.[2,23] The various bleaching techniques agree with each other tolerably well in measuring the spectra for the red cone receptors and green cone receptors (Fig. 19-4).

Some of the cone mechanisms may also be isolated by the psychophysical techniques developed by Stiles[25] (increment threshold measurements). The essence of the experiment is to use a background of a wavelength that is particularly effective for two of the three cone mechanisms and that will consequently adapt them more than the third cone mechanism. The threshold measured will then be the threshold for the third cone mechanism.

When the background is extremely bright, the results agree well with the spectra obtained from selective bleaching of the cone pigments[3,26] (Fig. 19-4). When the background is less bright, three basic mechanisms are isolated, which Stiles called π_1, π_4, and π_5. Microspectrophotometric measurements from the blue cone receptors agree well with π_1. Measurements of the pigment in the green cone receptors agree well with π_4. However, π_5 is rather broader than measurements of the pigment in the red cone receptors. Some authors have suggested that π_5 represents a neural process onto which red and green cone receptors converge.[26]

This is a powerful accumulation of evidence for the trichromacy of color vision. We can summarize it by saying that there are three types of cone pigment in the human. The three types of cone pigment are segregated into separate cones; therefore one can talk about three types of cone receptor.

Color matches

One of the consequences of trichromacy is that almost any wavelength or band of wavelengths can be matched by a mixture of three well-chosen monochromatic lights (usually 400 to 450 nm, 510 to 520 nm, and 630 to 700 nm). One can say any band of wavelengths rather than almost any if one allows negative intensities of the matching lights, that is, if one sometimes allows one of the matching lights to be mixed with the light to be matched rather than the matching lights. A match can be represented as

$$a_4\lambda_4 \equiv a_1\lambda_1 + a_2\lambda_2 + a_3\lambda_3$$

or

$$a_3\lambda_3 + a_4\lambda_4 \equiv a_1\lambda_1 + a_2\lambda_2$$

if a "negative intensity" has to be used

where

a_i = Intensity of light of wavelength λ_i
λ_4 = Wavelength to be matched
λ_1, λ_2, and λ_3 = Matching wavelengths

The trichromacy of color matches was first established by Helmholtz and Maxwell and was a compelling argument for the three-receptor theories of color vision.

The congruent sign is used in these equations, rather than the equal sign, because a color match holds in almost all circumstances. A color match is an equivalence for each of the three types of cone receptor. Therefore it is an equivalence for all other cells in the nervous system that receive their inputs from cone receptors. Color matches hold under various states of adaptation. They can also be added to each other to yield other color matches, for example, if

$$a_4\lambda_4 \equiv a_1\lambda_1 + a_2\lambda_2 + a_3\lambda_3$$

and

$$a_7\lambda_7 \equiv a_5\lambda_5 + a_6\lambda_6$$

then

$$a_4\lambda_4 + a_7\lambda_7 \equiv a_1\lambda_1 + a_2\lambda_2 + a_3\lambda_3 + a_5\lambda_5 + a_6\lambda_6$$

There are some circumstances in which color matches break down, for example, at very high intensities. Because of self-screening, the spectral sensitivity of a receptor is slightly different from the spectral sensitivity of the individual molecules of pigment within the receptor. Self-screening occurs when the density of pigment in a receptor is sufficiently large that the molecules of pigment at the near end of the receptor absorb an appreciable fraction of the light and act as a filter for the molecules of pigment farther down the receptor. The effect of self-screening is to make the spectral sensitivity of the receptor broader than the spectral sensitivity of the individual molecules in the receptor. (If the receptor were so long that all the incident light was absorbed by it, the spectral sensitivity would be flat.) There is not much self-screening in human cone receptors. There is enough, however, so that when some of the pigment is bleached away at high intensities, and the pigment in the receptor becomes more dilute, the spectral sensitivity of the receptor changes a little. This contributes to the breakdown of color matches established at lower intensities. Absorption of light by the photoproducts of the pigment may also be a factor (Chapter 15).

Color matches established in the fovea also break down in other areas of the retina. This is partly due to the macular pigment and partly to contributions from rods absent in the fovea but present elsewhere in the retina. Although there are four classes of receptor away from the fovea, three types of cones and the rods, vision is still trichromatic. This is usually explained by suggesting that the rods converge onto the same cells as the cones; thus there are only three channels for color at higher levels of the system.

Color triangle

A color triangle can be drawn to describe the trichromacy of color mixtures and is useful for deciding which bands of wavelengths are indistinguishable from each other. Three reference wavelengths are chosen, say 450 nm, 520 nm, and 650 nm, and placed at the vertices X, Y, and Z of a triangle. The position of other wavelengths is then determined as follows. Suppose we wish to position 500 nm. First a match is made

$$a_1\lambda_{450} + a_2\lambda_{520} \equiv a_3\lambda_{500} + a_4\lambda_{650}$$

with the intensities a_1, a_2, a_3, and a_4 usually expressed in units of energy. Next we draw the point W between X and Y (Fig. 19-5) so that

$$\frac{WX}{WY} = \frac{a_2}{a_1}$$

Finally, we place 500 nm at the point V on the line WZ so that

$$\frac{WV}{WZ} = \frac{a_3}{a_4}$$

The essential rule is a center of gravity rule. The point W is the center of gravity of weights (intensities) a_1 at X and a_2 at Y. It is also the center of gravity of weights a_3 at V and a_4 at Z.

The various wavelengths fall in the color triangle along a curve through the three reference wavelengths. The ends of the curve are usually joined by a straight line that describes the various blue-red mixtures of 400 nm and 700 nm. Any other band of wavelengths can be described by a point in the triangle. If the band of wavelengths is matched by energy a_1 of 450 nm, a_2 of 520 nm, and a_3 of 650 nm, it is represented by the center of gravity of weights a_1 at X, a_2 at Y, and a_3 at Z.

A color triangle does not describe the color of a band of wavelengths unless other circumstances are defined. A color triangle just describes what mixtures of wavelength can be substituted for each other without changing the color. For example, 590 nm is matched by equal parts of 570 nm and 626 nm. In circum-

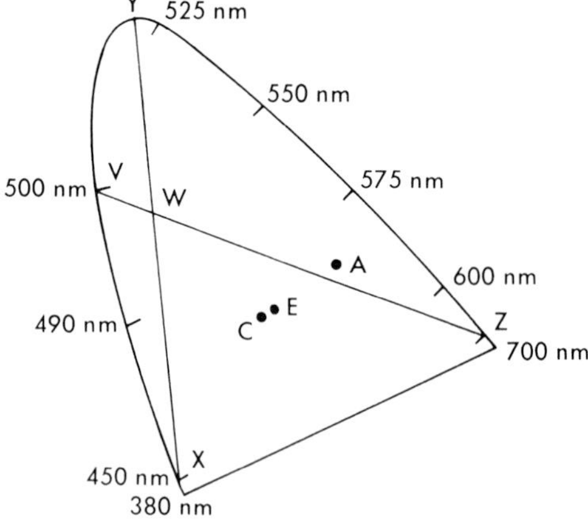

Fig. 19-5. CIE chromaticity diagram. Point E represents source radiating equal amounts of energy in equal intervals of wavelength throughout spectrum. Standard C is substitute for daylight. Standard A is tungsten lamp. (From The science of color, Washington, D.C., 1963, The Optical Society of America.)

stances where 590 nm appears yellow, the mixture will also appear yellow; in circumstances where 590 nm appears green, the mixture will be green; and where 590 nm is brown, the mixture will be brown, and so on.

The color triangle shown in Fig. 19-5 is the CIE* chromaticity diagram. Also shown are various broad bands of wavelength that have been defined as standards. Standard source *A* is a defined tungsten lamp run at a defined current and voltage. Standard source *C* is a substitute for daylight and consists of a tungsten lamp with a defined blue filter in front of it. Point *E* represents a source radiating equal amounts of energy in equal intervals of wavelengths throughout the spectrum.

Complementary wavelengths

Complementary wavelengths are those which, when mixed in appropriate proportions, make white. White is not precisely defined, but the point *E* on the CIE chromaticity diagram is one of several that can appear white. Thus a straight line through point *E* intersects the wavelength curve at two points that are complementary wavelengths. A wavelength of 455 nm is complementary to 575 nm, and 650 nm to 495 nm. The complementaries of wavelengths fcm 497 to 573 nm are mixtures of 400 and 700 nm rather than a single wavelength.

NEURAL FACTORS AFFECTING COLOR

We pointed out at the beginning of this chapter that color is a sensation. The sensation is determined by the connections of the receptors in the retina to other cells in the nervous system, as well as by the properties of the receptors themselves. To be absolutely correct, we should have talked entirely about wavelength up to this point and should only now be talking about color. While most of the phenomena in color vision have been known for some time, the neural connections have only recently been investigated, and many facts are still obscure. The psychologic phenomena will be discussed first, followed by what is known about the neural connections responsible for them.

*Commission Internationale de L'Éclairage.

Hue, saturation, lightness

A color is said to have hue, saturation, and lightness. Hue is the extent to which it is red, green, blue, or yellow, and so on. Saturation is the extent to which a color is strong or weak. Spectral wavelengths become desaturated when white light or their complementaries are added to them. Lightness to a certain extent is a self-explanatory attribute. Yellow is a light color, whereas brown is a dark one. Dark colors cannot be obtained unless there are lighter areas around.

Simultaneous color contrast

Simultaneous color contrast is usually demonstrated by observing the color of a spot in a surround. A gray spot in a red surround appears greenish. A gray spot in a green surround appears reddish, and so on. The general rule is that the color of the spot tends toward the complementary of the color of the surround. Simultaneous color contrast is an important phenomenon in color vision because it is related to the phenomenon of object color constancy referred to earlier.

Just to emphasize the lack of correlation between color and wavelength, we can list some of the colors that 585 nm can take on; 585 nm appears green in a surround of 650 nm, red in a surround of 540 nm, gray in a surround of 585 nm 0.7 log unit brighter, black in a surround of 585 nm 2.0 log units brighter, brown in a surround of 570 nm 1.0 log unit brighter, and so on.

Successive color contrast

Successive color contrast is more commonly described as colored afterimages. When one stares at a red spot for several seconds and then looks at a gray card, one sees a green spot on the card. As with simultaneous color contrast, the general rule is that the afterimage tends toward the complementary of the primary image. Successive color contrast is probably not an important phenomenon in color vision in normal circumstances. One reason is that the eye is continually moving around; thus one rarely stares at an object long enough to form an afterimage. Another reason is that if one does form an afterimage and then looks at a scene with contours in it, the afterimage is usually a low-contrast out-of-focus image superimposed on a higher contrast in-focus image, and the visual system does not notice it. However, it is possible that the continual small tremor of the eye moving boundaries backward and forward across the receptors may lead to successive color contrast effects that are important.

Color-contingent aftereffects

The first color-contingent aftereffect to be described was a color aftereffect contingent on orientation.[16] One forms the aftereffect by looking at a horizontal red-black grating, alternating with a vertical green-black grating every few seconds, for a period of several minutes. Subsequently, vertical white-black gratings appear reddish, and horizontal white-black gratings appear greenish. The colors are not nearly so striking as the colors seen in colored afterimages, but the effect can last much longer. Other aftereffects have also been described, including an orientation aftereffect contingent on color, a motion aftereffect contingent on color, a color aftereffect contingent on motion, and a color aftereffect contingent on spatial frequency, that is, the spacing between bars of a grating.

Color-coded cells

All cells in the visual system have a receptive field (Chapters 14 and 16). In anatomic terms, the receptive field comprises all the receptors connected, directly or indirectly, to the cell. In physiologic terms, the receptive field includes all the parts of the retina that can be stimulated to influence the activity of the cell. A color-coded cell is any cell with a response that seems to be specific for some aspect of color. This may mean that some wavelengths produce a response and other wavelengths do not, or it may mean that the type of response produced by some wavelengths is different from the type of response produced by other wavelengths. Two types of color-coded cells are found at more peripheral levels of the visual system, and they have been named opponent color cells and double opponent cells. More complex types of receptive field are found at more central levels, where cells are specific for both the color and orientation of the stimulus.

Opponent color cells

The first color-coded cells found in the nervous system after the receptors are opponent color cells. An opponent color cell is a cell that gives one polarity of response for some wavelengths and the opposite polarity of response for other wavelengths. For example, the activity of the cell may be increased by yellow light and reduced by blue light (Fig. 19-6). Diffuse white light has very little effect on such a cell, since the white light will be absorbed by some receptors tending to increase the activity of the cell and at the same time by other receptors tending to reduce its activity.

In a sense, opponent color cells are concerned with successive color contrast. When green is followed by white, the white appears reddish, because of successive color contrast. Consider an opponent color cell that gives an on response to red light, an off response to green light, and no response to white light. If it is stimulated by nothing followed by red, it will respond with an increase of activity just after the transition. The same is true if it is stimulated by green followed by white. As far as this cell is concerned, therefore, these two sequences of stimuli give similar responses just after the transition, and the sensations obtained at that time are also similar.

Double opponent cells

Double opponent cells are cells opponent for both color and space. An example is a cell that has both a center and a surround to its receptive field, and both are color coded. The response may be on to red light, off to green light in the center of the receptive field and off to red light, on to green light in the periphery of the receptive field (Fig. 19-7). Uniform illumination tends to give very little response, no matter what the wavelength. In most cases white light will also not elicit a response in either the center or surround of the receptive field.

In a sense, double opponent cells are concerned with simultaneous color contrast. A white spot in a green surround appears reddish

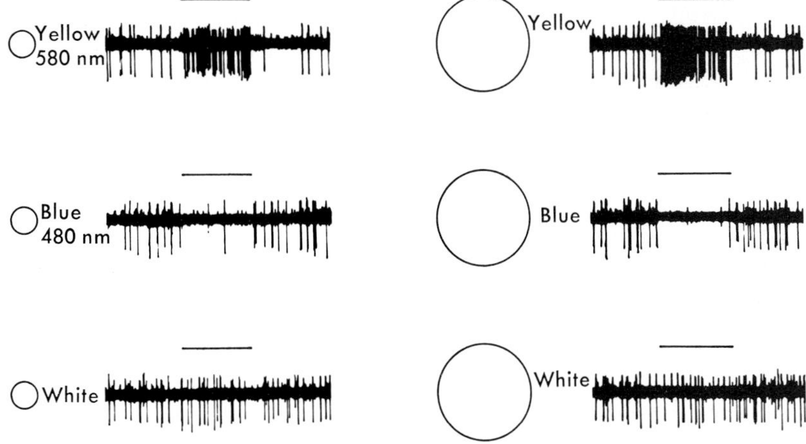

Fig. 19-6. Opponent color cell from the lateral geniculate nucleus of a rhesus monkey. Yellow light excites the cell, blue light inhibits it, and white light has little effect. Horizontal lines above the records indicate the duration of illumination. (From Wiesel, T. N., and Hubel, D. H.: J. Neurophysiol. **29:**1115, 1966.)

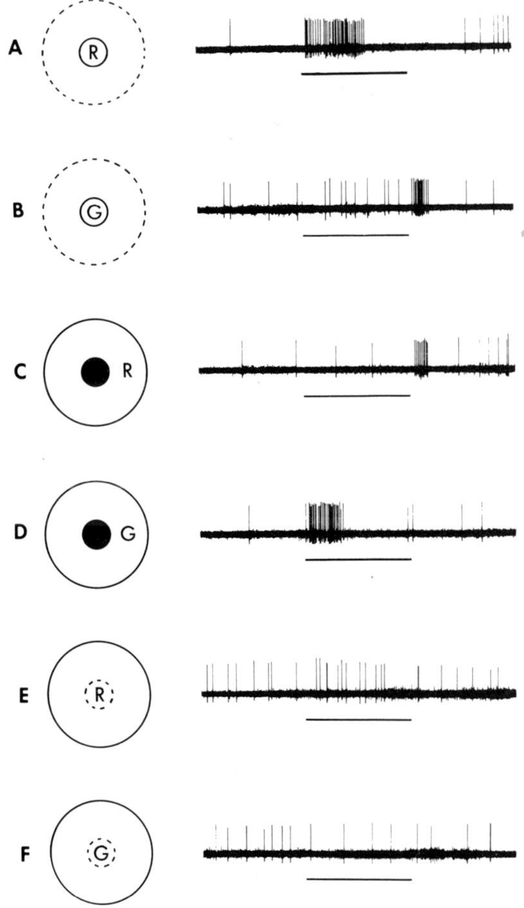

Fig. 19-7. Double opponent cell from striate cortex of rhesus monkey. Responses to **A,** red spot shone on the center of the receptive field; **B,** green spot; **C,** red anulus shone on the surround of the receptive field; **D,** green anulus; **E,** uniform red illumination of whole receptive field; **F,** uniform green illumination. (From Michael, C. R.: J. Neurophysiol. **41:**572, 1978.)

because of simultaneous color contrast. The cell described earlier will give an on response for a red spot in a white or black surround and also for a white or black spot in a green surround. The responses to these two stimuli are similar for this cell, just as the sensations for the central spot are similar for the whole animal. These analogies should not be pressed too far, but they do help to emphasize that neurons beyond the receptors are important in color perception.

Cells specific for color and orientation

In the rhesus monkey striate cortex there are a variety of cells that are specific for both color and orientation. They have been categorized as color sensitive simple,[19] complex,[20] and hypercomplex[21] cells because they have characteristics analogous to the simple, complex, and hypercomplex properties described for cells that are not color coded (Chapter 16). The simple cells have a bar-flank double opponent arrangement to their receptive fields.[19] They may, for example, give an on response to a bar of red light at the appropriate orientation, an off response to a bar of red light on either side of it, an off response to a bar of green light properly centered, and an on response to a bar of green light illuminating one of the flanks. Complex color-coded cells respond to color boundaries of the appropriate orientation, and the response is independent of the part of the receptive field being stimulated. For hypercomplex color-coded cells, the edge must be short.

Fig. 19-8 shows the response of a hypercomplex color-coded cell. It responded to an edge with red above and green below, moving upward (Fig. 19-8, *B*). Black-green and red-black edges produced much less of a response (Fig. 19-8, *A* and *C*). The response was abolished if the edge was lengthened, the orientation changed, or if it was moved downward instead of upward. Such cells provide an extremely detailed analysis of the various attributes of an object.

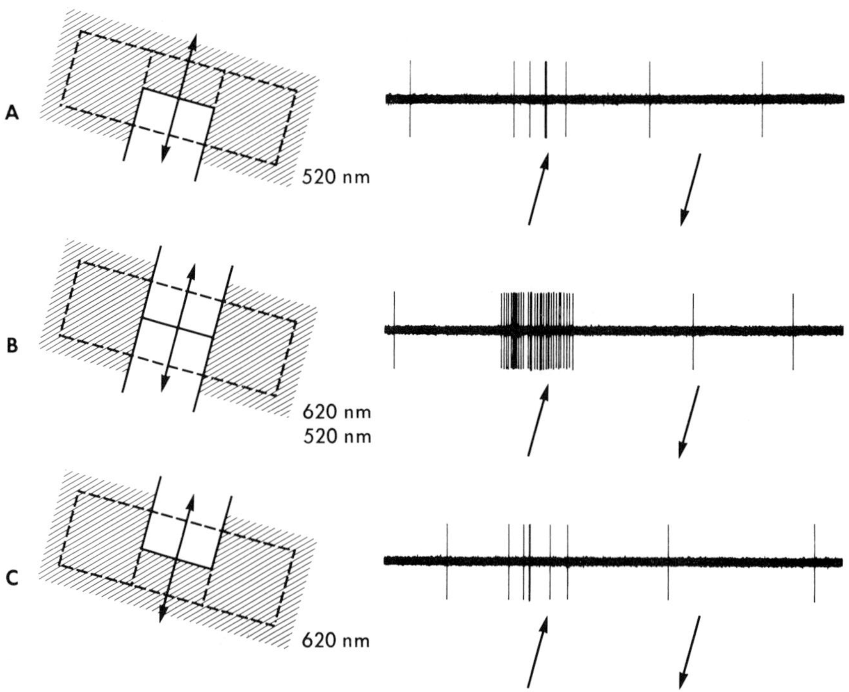

Fig. 19-8. Response of a hypercomplex color coded cell from rhesus monkey striate cortex. (From Michael, C. R.: J. Neurophysiol. **42:**726, 1979.)

Area in cortex specific for color

The simple, complex, and hypercomplex color-coded cells mentioned earlier are found in the striate cortex, interspersed with other cells that are not specific for color. Cells of the striate cortex project to several regions of extrastriate cortex, including an area where all the cells are color coded, which is on the posterior bank of the superior temporal sulcus in the rhesus monkey[31] (Chapter 16). In this region cells seem to be arranged in columns according to color and to have spectral sensitivities that are narrower than those of more peripheral cells.

Hierarchy of color-coded cells

The color-coded cells that have been described are arranged in a hierarchy. Opponent color cells are found among ganglion cells of the retina[10] and in the lateral geniculate nucleus.[7,29] Double opponent cells, with either center surround or bar-flank receptive fields, are found in the input layer, layer IV, of the striate cortex.[18,19] Complex and hypercomplex color-coded cells are also found in the striate cortex, in layers II, III, V, and VI.[20,21] The anatomic localization of these cells as well as physiologic recordings in which a cell body and one of its input fibres are recorded at the same time suggest that cells at one level of the hierarchy converge to form the receptive fields of the cells at the next higher level.

Basis of perception of color

Some aspects of the perception of color can now be understood on the basis of the physiology. In principle, both simultaneous color contrast and successive color contrast can be understood, as described earlier. In practice, more details are required. For example, the precise colors seen in simultaneous color contrast should depend on the relative contributions of the red, green, and blue receptors to the double opponent cells, and presumably several classes of double opponent cell are required. Getting this information requires recordings from several hundred, perhaps several thousand, double opponent cells, and this laborious task has not yet been tackled.

At higher levels of the visual system, it is clear that cells such as color-coded hypercomplex cells analyze the scene in incredible detail, each cell being responsible for some extremely specific aspect of the stimulus. Moreover, different aspects of the stimulus, such as color, are analyzed in different areas of the cortex. We do not have the vaguest idea at the present time as to how all this analysis in detail is put back together into the percept of an object of a particular color.

Rods and color vision

It is commonly supposed that rods play no part in color vision. This is certainly true in the fovea, where rods are absent, and is probably true outside the fovea above the level of rod saturation (Chapter 22), which occurs at 120 to 200 cd/m^2 seen through the normal pupil, or 2000 to 3000 scotopic trolands. However, color-coded cells in the visual system receive input from both rods and cones.[29] Also, color contrast is seen in images that stimulate only rods and the red cone receptors.[15] Undoubtedly contributions from the rods affect color vision at mesopic levels outside the fovea.

Peculiarities in blue cone system

Both acuity and sensitivity are poorer in blue light than in red or green light.[4,11] This shows up in a condition called small-field tritanopia, in which the visual system is essentially dichromatic and blue-blind for small objects. The usual explanation is that there is a relative scarcity of blue cone receptors, and also a greater convergence of blue cone receptors onto cells higher in the nervous system. The latter point is supported by some neurophysiologic evidence.[10] Of course the visual system does not need to have high acuity for the blue cone system, since the chromatic aberration of the lens ensures that the blue image is going to be out of focus anyway.

Retinal distribution of color vision

Trichromatic vision extends 20 to 30 degrees from the point of fixation. Peripheral to this, red and green become indistinguishable, and in the far periphery all color sense is lost, although cones are still found in this region of the retina (Fig. 19-9). As one brings a colored test object (red) into the visual field, the sub-

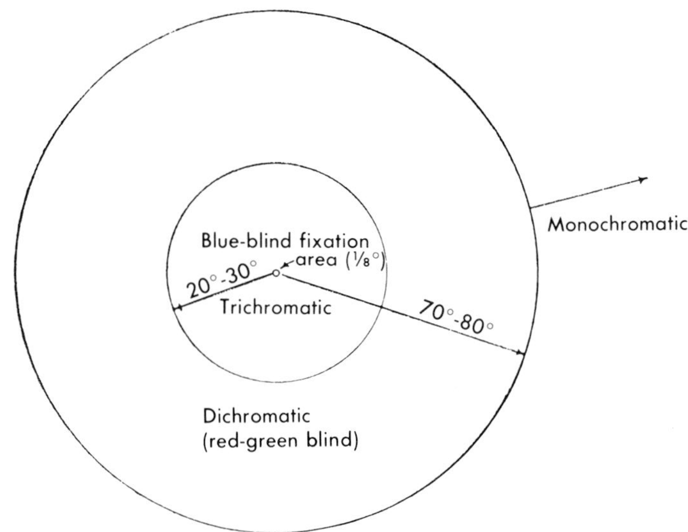

Fig. 19-9. Map of distribution of color vision in retina. (From Wald, G.: Science **162**:230, 1968.)

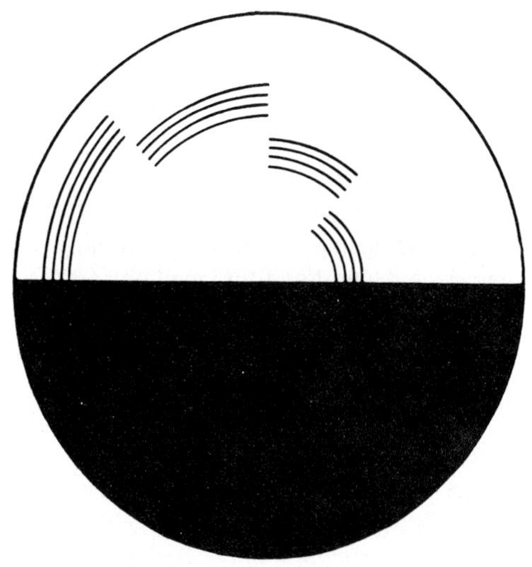

Fig. 19-10. Benham's top.

ject is first aware of a colorless object in the periphery. Then as the object is advanced, he sees the object successively as salmon pink or yellow and eventually as red.

The very center of the fovea (⅛ degree) is blue blind. This is a particular example of small-field tritanopia, and there is some suggestion that blue receptors may be completely absent in this area.

Color from sequences of black and white

Some temporal sequences of black and white give color (Benham's top, Fig. 19-10). If Benham's top is rotated clockwise at about 10 rps, the outer band appears to be bluish, the next green, and the innermost reddish. Reversal of the direction of rotation reverses the order of the colors. Although the colors are not very saturated, the phenomenon has been used in television displays. The explanation is not clear, but undoubtedly the sequence of black and white leads to a response in some color-coded cells similar to the response produced by unsaturated colors.

DEFECTIVE COLOR VISION

There are two types of defective color vision: congenital and acquired. Congenital color deficits are well understood. They are nearly all due to absence or alteration of one of the pigments in the photoreceptors. Acquired color deficits are much less well understood. People with congenital color deficits may be divided into classes according to whether they are red deficient (protans), green deficient (deutans), or blue deficient (tritans) (Table 19-1). They may also be divided into anomalous trichromats,

Table 19-1. Classification of color-deficient observers

	Red deficient	Green deficient	Blue deficient
Anomalous trichromats	Protanomal	Deuteranomal	Tritanomal
Dichromats	Protanope	Deuteranope	Tritanope
Monochromats	Blue cone monochromats		
	Rod monochromats		

dichromats, anomalous dichromats, or monochromats. Anomalous trichromats are people who generally require three wavelengths to match another wavelength but do not accept the color matches made by normal people. Dichromats require only two wavelengths to match another wavelength and will accept color matches made by normal people. They can match one wavelength in the spectrum to white light, and this is called the neutral wavelength (near 495 nm for protanopes and deuteranopes, near 570 nm for tritanopes). Anomalous dichromats are rare; they do not accept color matches made by normal people. The term monochromat has been used in a confusing way and will be discussed below.

Since dichromats accept color matches made by normal people, one would expect them to have two classes of cone receptor with normal spectral sensitivity, the third class being absent. Measurements of their pigments by reflection densitometry[2,23] and of their cone processes isolated by colored backgrounds[27] show that this is true. Protanopes have normal green and blue cones; deuteranopes, normal red and blue cones; tritanopes, normal red and green cones.

Anomalous trichromats have three classes of cone receptor, but one is abnormal. The latest evidence is that protanomalous people lack the red receptors, and instead have two pigments both peaking in the range of the normal green receptors.[1] Similarly, deuteranomalous people lack green receptors, and have two pigments peaking close to each other in the range of normal red receptors.

Protans really are red blind in the sense that they are comparatively insensitive to wavelengths beyond 650 nm. This is true for both protanopes, and protanomals, since both lack red cone receptors. The peak of the luminosity curve for protans is shifted toward shorter wavelengths, compared to the normal. Protans are therefore the easiest color-deficient observers to test and classify.

Deutans are not green blind, and tritans are not blue blind, in the same sense. This is because the red cone pigment is quite sensitive to green wavelengths, and both red and green cone pigments are quite sensitive to blue wavelengths, according to the general rule that the sensitivity of visual pigments does not fall off sharply on the short wavelength side of the peak. Very careful measurements are required to distinguish the luminosity curve of either deutans or tritans from normal. In fact, the variations in color of the lens and macular pigment from one observer to another make this task almost impossible. In general, color-deficient people confuse colors, rather than being blind to them. Protans and deutans confuse various shades of red with various shades of green. Tritans confuse shades of blue with longer wavelengths.

The term monochromat has been used to describe some people who are genuine monochromats, and some who are not. The genuine rod monochromat has only 500 nm pigment in his retina (some of it may be in cones, with fast kinetics like normal cone pigments). There are also blue cone monochromats who have 500 nm pigment and normal blue cone pigment but no red or green cone pigment (again some of the 500 nm pigment may be in cones). Both rod monochromats and blue cone monochromats have poor acuity and nystagmus, emphasizing that acuity for the blue cone system is not much better than acuity for the rod system. Blue cone monochromats are not really monochromats; they can distinguish blue from yellow and have a neutral point about 465 nm. Occasionally genuine cone monochromats are found, lacking rods and all color vision.

Genetics of color deficiencies

The genes for protans and deutans are recessive and sex linked. In the male population about 1% are protanopes, 1% protanomals, 1% deuteranopes, and 5% deuteranomals. The frequency in females is as expected from the genetics, about 0.4% in total for red-green defects. The gene for tritans is autosomal and incompletely dominant. The gene for some forms of monochromatism is autosomal and recessive, but for some blue cone monochromats it is sex linked and recessive, as would be expected if the trait were a combination of the protan and deutan defects.

Testing for congenital color deficits

Color-deficient individuals characteristically confuse colors that normal people can distinguish. Pseudoisochromatic color plates are patterns of colored and gray dots that reveal one pattern to the normal, another to the color deficient.[12] The Ishihara plates and Hardy-Rand-Rittler plates are examples. The illumination must be carefully controlled. Red-green defects (protans and deutans) can be detected quite easily, but detailed classification may be more difficult.

The Farnsworth 100-hue test is a simple and very useful test consisting of pastel chips arranged in a color sequence that is quite obvious to the normal.[12] Color-deficient persons make characteristic errors in arranging the chips. The results of the test are recorded on a circular graph. The greater the error in ordering the chips, the farther the score is plotted from the center of the circle (Fig. 19-11). This test is probably the best for detection of tritan defects.

Another test for red-green defects is the

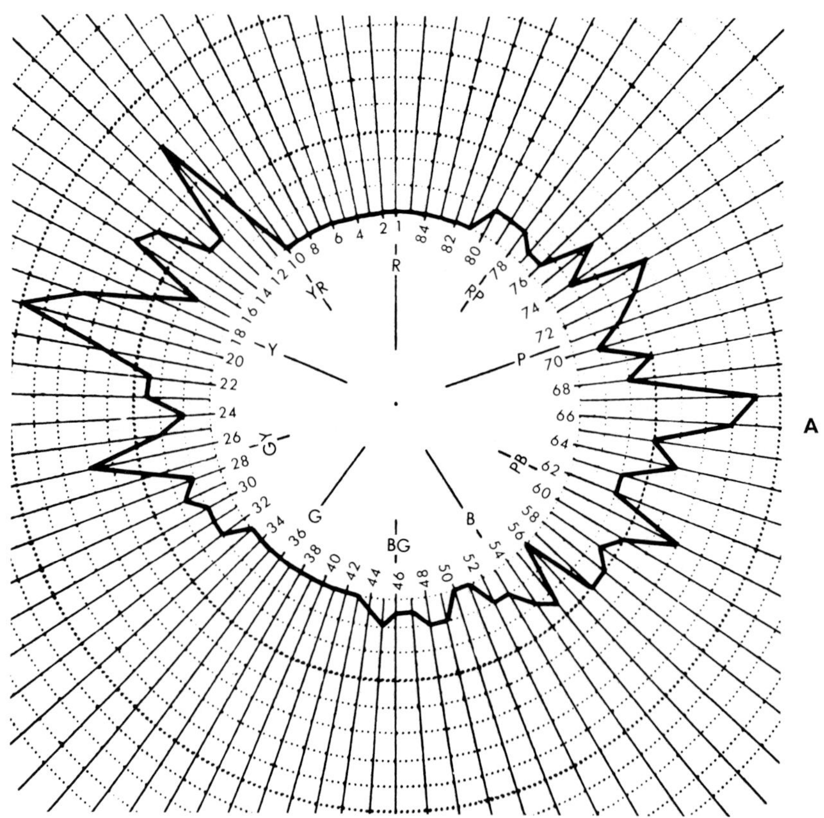

Fig. 19-11. Color vision scores on Farnsworth-Munsell 100-hue test. Magnitude of error in chip arrangement indicated by distance from center. **A,** Protanope. **B,** Deuteranope. (From Linksz, A.: In Whitcomb, M. A., and Benson, W., editors: Armed Forces Committee on Vision, Washington, D.C., 1968, National Academy of Sciences, National Research Council.)

anomaloscope. The observer is asked to match a mixture of red and green wavelengths to a yellow. One indication of a defect is the relative amounts of red and green required. Protanomals usually require too much red, and deuteranomals may require too much green. Another indication of a defect is the range of settings found acceptable. The range can be very wide for protanopes and deuteranopes. No anomaloscope has been devised for tritan defects because the variation of macular pigment and lens color in normal people is wide, and the normal distribution of acceptable settings is correspondingly wide when a blue wavelength is involved.

Proper diagnosis of color vision deficiencies is not easy because the tests that are cheap and comparatively easy to administer can give misleading results, while the tests that give accurate results require quite a lot of training for the tester. It is particularly important not to overdiagnose the deficiency and consequently exclude an individual from an occupation that he could well manage.[12] Various devices exist whereby color-deficient people can learn to distinguish objects that would otherwise appear the same to them, but the devices require considerable training and some of them are awkward to use.[24]

Acquired color vision deficits

Color vision deficits can be caused by various diseases and conditions, including toxins, inflammation or detachment of the retina, retinopathy, macular degeneration, and some optic nerve diseases. Pseudoisochromatic tests and anomaloscopes, which were devised specifically to detect the absence of one of the three classes of cone pigment, are not of much use

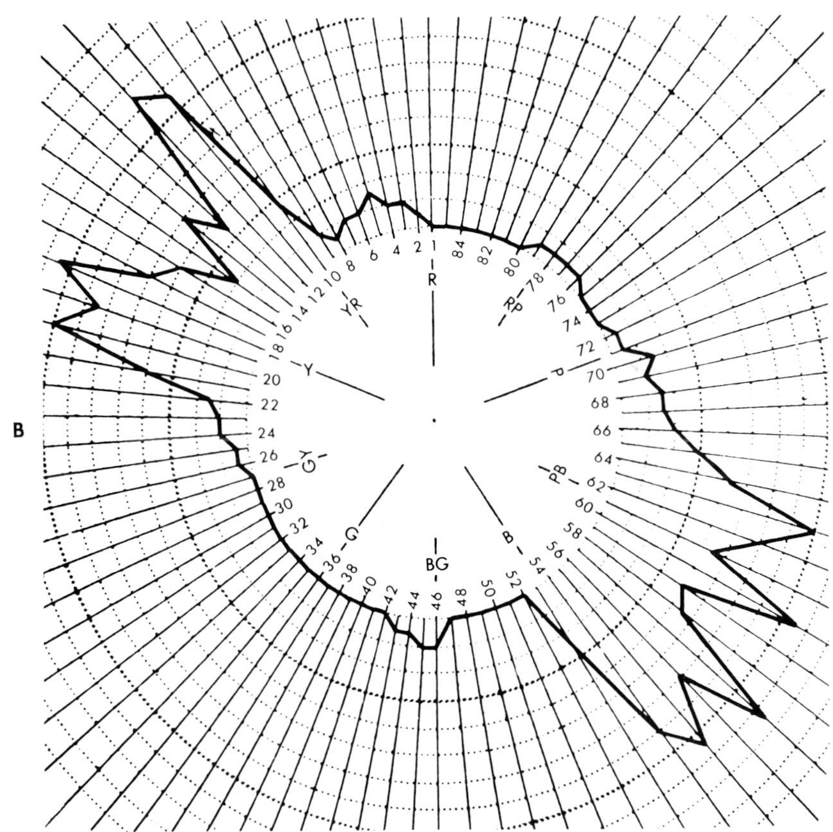

Fig. 19-11, cont'd. For legend see opposite page.

in detecting these acquired deficits in color vision. This is to be expected, because most diseases that affect the cones affect all cones, irrespective of the pigment inside, and diseases that affect more central cells affect other aspects of the physiology of the color vision pathways. The Farnsworth tests are the most useful generally available tests for acquired deficits. More specific tests have been used in particular laboratories for particular diseases.

There is a guideline, originally formulated by Koellner,[9] that lesions in the outer layers of the retina give rise to a blue-yellow defect, while lesions in the inner layers of the retina and the optic nerve give rise to a red-green defect.[9] Blue-yellow deficits have been found in retinitis pigmentosa, senile macular degeneration, siderosis retinae, myopia gravis, and choroidoretinitis, and red-green deficits have been found in multiple sclerosis and optic atrophy. However, the correlation is not always very good, and the defect is not always clearly just red-green or just yellow-blue. Moreover, dominant hereditary optic neuropathy, which affects the optic nerve, causes a blue-yellow defect. There are no clear physiologic reasons why such a correlation should exist, although further research may reveal some.

Color vision deficits that are found only in one eye, that change with time, that vary from one part of the retina to another, or that are associated with a change in visual acuity, dark adaptation, or flicker discrimination are nearly always acquired rather than congenital.

Some people with lesions in the cerebral cortex have color vision deficits.[17,22] These may involve naming of colors or, alternatively, may involve perception of colors. The latter is often associated with a scotoma of the upper visual field, and very likely represents a lesion in an area specific for color, like that found in the rhesus monkey (see p. 453).

REFERENCES

1. Alpern, M., and Moeller, J.: The red and green cone visual pigments of deuteranomalous trichromacy, J. Physiol. (Lond) **266:**647, 1977.
2. Baker, H. D., and Rushton, W. A. H.: The red-sensitive pigment in normal cones, J. Physiol. (Lond) **176:**56, 1965.
3. Brindley, G. S.: The effects on colour vision of adaptation to very bright lights, J. Physiol. (Lond) **122:** 332, 1953.
4. Brindley, G. S.: The summation areas of human colour-receptive mechanisms at increment threshold, J. Physiol. (Lond) **124:**400, 1954.
5. Brown, P. K., and Wald, G.: Visual pigments in human and monkey retinas, Nature **200:**37, 1963.
6. Brown, P. K., and Wald, G.: Visual pigments in single rods and cones of the human retina, Science **144:**45, 1964.
7. De Valois, R. L., Abramov, I., and Jacobs, G. H.: Analysis of response patterns of LGN cells, J. Opt. Soc. Am. **56:**966, 1966.
8. Enoch, J. M., and Stiles, W. S.: The colour change of monochromatic light with the retinal angle of incidence, Opt. Acta **8:**329, 1961.
9. François, J., and Verriest, G.: On acquired deficiency of color vision, Vision Res. **1:**201, 1961.
10. Gouras, P.: Identification of cone mechanisms in monkey ganglion cells, J. Physiol. (Lond) **199:**533, 1968.
11. Green, D. G.: The contrast sensitivity of the colour mechanisms of the human eye, J. Physiol. (Lond) **196:** 415, 1968.
12. Lakowski, R.: Theory and practice of colour vision testing: a review, II, Br. J. Ind. Med. **26:**265, 1969.
13. Land, E. H.: The retinex, Am. Sci. **52:**247, 1964.
14. Marks, W. B., Dobelle, W. H., and MacNichol, E. F., Jr.: Visual pigments of single primate cones, Science **143:**1181, 1964.
15. McCann, J. J., and Benton, J. L.: Interaction of the long-wave cones and the rods to produce color sensations, J. Opt. Soc. Am. **59:**103, 1969.
16. McCullough, C.: Color adaptation of edge detectors in the human visual system, Science **149:**1115, 1965.
17. Meadows, J. C.: Disturbed perception of colours associated with localized cerebral lesions, Brain **97:**615, 1974.
18. Michael, C. R.: Color vision mechanisms in monkey striate cortex: dual-opponent cells with concentric receptive fields, J. Neurophysiol. **41:**572, 1978.
19. Michael, C. R.: Color vision mechanisms in monkey striate cortex: simple cells with dual opponent-color receptive fields, J. Neurophysiol. **41:**1233, 1978.
20. Michael, C. R.: Color sensitive complex cells in monkey striate cortex, J. Neurophysiol. **41:**1250, 1978.
21. Michael, C. R.: Color sensitive hypercomplex cells in monkey striate cortex, J. Neurophysiol. **42:**726, 1979.
22. Pearlman, A. L., Birch, J., and Meadows, J. C.: Cerebral color blindness: an acquired defect in hue discrimination, Ann. Neurol. **5:**253, 1979.
23. Rushton, W. A. H.: A cone pigment in the protanope, J. Physiol. (Lond) **168:**345, 1963.
24. Schmidt, I.: Visual aids for correction of red-green colour deficiencies, Can. J. Optom. **38:**38, 1976.
25. Stiles, W. S.: Color vision: the approach through increment-threshold sensitivity, Proc. Natl. Acad. Sci. U.S.A. **45:**100, 1959.
26. Wald, G.: The receptors of human color vision, Science **145:**1007, 1964.
27. Wald, G.: Defective color vision and its inheritance, Proc. Natl. Acad. Sci. U.S.A. **55:**1347, 1966.
28. Weale, R. A.: Photosensitive reactions in fovea of nor-

mal and cone-monochromatic observers, Opt. Acta **6:** 158, 1959.

29. Wiesel, T. N., and Hubel, D. H.: Spatial and chromatic interactions in the lateral geniculate body of the rhesus monkey, J. Neurophysiol. **29:**1115, 1966.
30. Young, T.: On the theory of light and colours, Philos. Trans. R. Soc. Lond. (Biol.) **92:**12, 1802.
31. Zeki, S. M.: Colour coding in the superior temporal sulcus of rhesus monkey visual cortex, Proc. R. Soc. Lond. (Biol.) **197:**195, 1977.

GENERAL REFERENCES

Dartnell, H. J. A., editor: Photochemistry of vision. Handbook of sensory physiology, New York, 1972, Springer-Verlag New York, Inc., vol. 7/1.

Daw, N. W.: Neurophysiology of color vision, Physiol. Rev. **53:**571, 1972.

Jameson, D., and Hurvich, L. M., editors: Visual psychophysics. Handbook of sensory physiology, New York, 1972, Springer-Verlag New York, Inc., vol. 7/4.

Wyszecki, G., and Stiles, W. S.: Color science, New York, 1967, John Wiley & Sons, Inc.

Chapter 20

Entoptic and allied phenomena

ROBERT A. MOSES, M.D.

In the description of entoptic phenomena we are concerned with the visualization of certain structures within the eye through the proper arrangement of incident light. These structures may be normal to the eye, or they may be imperfections such as opacities in the vitreous or lens. Sensations arising from structures normally present in the eye, such as shadows from retinal blood vessels or the movement of blood cells in the capillaries, are usually not visualized by a person, either because his attention has never been called to them or because it takes some special conditions of illumination to make them apparent. Under this heading must be included other phenomena that result from unusual stimulation of the retina by inadequate stimuli, such as phosphenes from digital pressure on the globe and sparks and flashes of light from mechanical stimulation of the retina.

OPACITIES IN THE OCULAR MEDIA

When the ocular fundus is illuminated by a retinoscope or by an ophthalmoscope held a foot or so from the eye, the pupil is seen to be filled with reddish light. Any imperfection of the ocular media that prevents light from returning from the fundus to the examiner's eye is seen as a dark area in the patient's pupil and is commonly referred to as an opacity of the media. Thus a streak of mucus on the cornea is seen by the examiner as a dark line because it refracts light reflected from the fundus away from the observer's view. Dark areas seen against the fundus reflection may be divided into two types. The first type is made up of conditions that produce local changes of refraction. The mucous thread mentioned above is an example. Others are irregularity of the surface of the cornea or breaks in Descemet's membrane secondary to disease or injury. The effect of these irregularities of the media on the retinal image is to spread light and reduce contrast. Likewise, corneal scars and changes in the lens fibers (cataract) scatter light because of irregular refraction and degrade the retinal image contrast. The second type of dark areas seen in the red fundus reflection by the examiner is caused by truly opaque bodies, which absorb light rather than scatter it. An example is a clump of pigmented cells on the anterior lens capsule, pulled loose from the iris at a site of adhesion (posterior synechia). True opacities merely reduce the amount of light reaching the retina and scatter light only by diffraction at the edge of the opacity.

The effects of these two classes of optical defects on the patient's vision are very different. The translucent opacities, which spread or scatter light, are often damaging to vision, whereas the true opacities often are not appreciated by the patient. A treatment for light-scattering scars of the cornea, widely used before corneal transplantation was developed to its present excellence, was to convert the corneal translucency to a true opacity by tattooing the area, often with considerable improvement in visual acuity.

Both classes of opacity can be visualized subjectively with appropriate illumination. A small pinhole (0.1 mm) in a card held before

□ This work was supported in part by Research Grant EY 00256 from the National Eye Institute, National Institutes of Health.

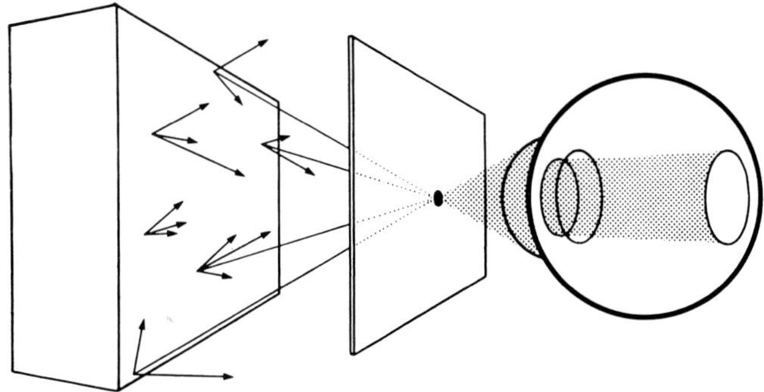

Fig. 20-1. Illuminated pinhole at anterior focus of eye acts as point source of light. Light entering eye is collimated by ocular refracting elements and casts sharp shadows on retina. (From Rubin, M. L., and Walls, G. L.: Studies in physiological optics, Springfield, Ill., 1965, Charles C Thomas, Publisher.)

a bright background such as the sky or a bright wall is adequate. As the pinhole approaches the eye, it is seen as a bright spot that grows in size. When the pinhole, a small source of light, is at the anterior focus of the eye (about 17 mm in front of the cornea), the refractive media of the eye bend the diverging light rays to parallel (Fig. 20-1). The subject then sees a patch of light. The border of the patch is the shadow of the pupillary margin of the iris; the patch varies in size with changes in pupil diameter. Optical discontinuities are seen as shadows or bright areas within the patch according to how the light is blocked out or refracted.

ENTOPTIC PHENOMENA CONNECTED WITH THE TEAR FILM AND CORNEA

The pinhole method of viewing allows a number of interesting entoptic phenomena from the cornea and the tear film on its surface to be subjectively visualized. Superficial horizontal bands, caused apparently by folds in the corneal epithelium, may be seen in the entoptic field. They run in an unbroken line across the entire width of the pupil and change their positions as the lids are slowly approximated or separated, advancing as the lids advance and receding slightly as the lids recede. They are quite distinct from the entoptic image of the tear film. The lacrimal fluid that adheres to the upper lid margin produces a longitudinal stripe that can be visualized, especially if the palpebral fissure is narrowed.

Droplets of tear fluid and mucus on the cornea can likewise be seen as bright spots surrounded by a dark ring. These move up and down as the palpebral fissure is widened or narrowed. Various folds and channels may be visualized. These consist of vertical lines that are not exactly straight, but that are gentle curvatures in some spots; in other regions the bending is more sharply angular. The individual channels are soon recognized and remembered from the peculiarities of their shape. It will be noted that the same configurations are recognized in successive attempts to visualize the phenomenon on different days, which speaks for their definite anatomic structure. Their anatomic origin is not certain, but they may represent vitreous membranes.

ENTOPTIC PHENOMENA CONNECTED WITH THE LENS

The fact that a star is not visualized as a point of light but as a star figure results from the structure of the lens, which breaks up the rays of light. Probably this phenomenon is related intimately to the suture lines of the normal lens.

Opacities in the lens are observed very easily by the pinhole method (Fig. 20-2).

Under ordinary circumstances one is not aware of the imperfections in the ocular media. Most persons have small opacities in the lens, but these do not interfere in any way with vision. None of the ocular media, with the exception of the aqueous humor, is perfectly

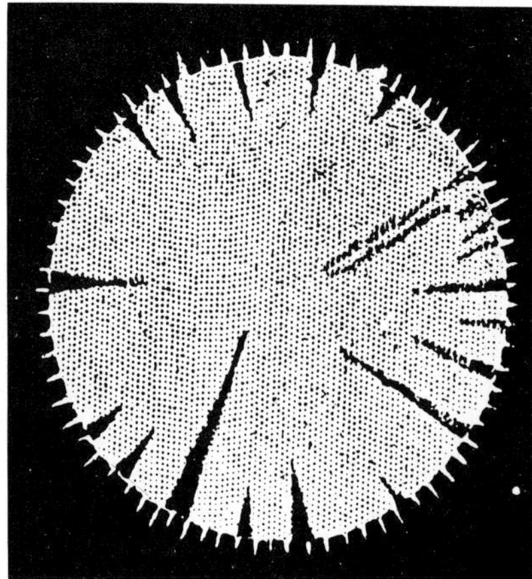

Fig. 20-2. Early cortical cataract as it appears entoptically in collimated light. (After Darier.)

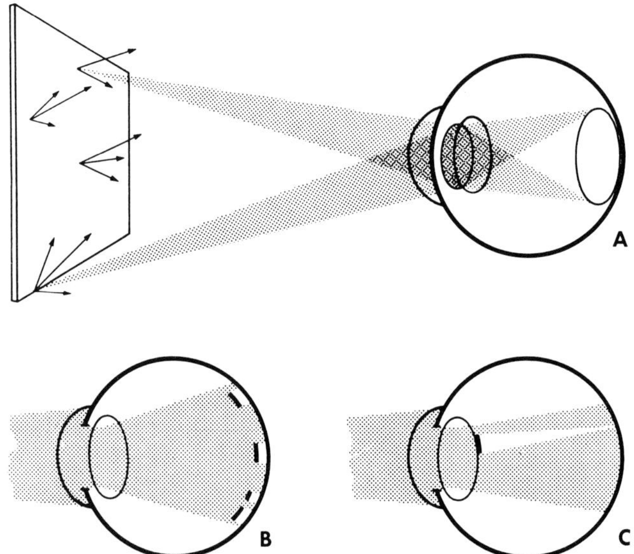

Fig. 20-3. A, Since light from any point in extended source can reach pupil, brightly illuminated field gives large brightly illuminated image, most of eyeball being filled with light. **B,** If opacity is near retina, shadows are sharply seen. **C,** Small opacity directly behind lens, for example, will not cast sharp shadow (umbra) on retina, since some light from extended source can still reach spot in retina where shadow should be, thus blotting it out. This explains severe annoyance caused by even very small opacities lying near retina (shadows cast being sharply defined), and "inconspicuousness" of much larger opacities situated more anteriorly in vitreous or lens, such as early cataractous changes—opacities within lens (shadow in latter case being only blurrily outlined, if at all visible.) (From Rubin, M. L., and Walls, G. L.: Studies in physiological optics, Springfield, Ill., 1965, Charles C Thomas, Publisher.)

transparent, since they are composed of fibers and cells with nuclei; yet, under ordinary circumstances one is quite unaware of these imperfections. As a matter of fact, even large opacities in the cornea and lens may not give rise to any imperfections in the retinal image if they are dense and block out the light rather than scatter it. This results from the fact that the opacity lies so far in front of the retina that its shadow does not interfere with the formation of the retinal image (Fig. 20-3, *C*).

Fig. 20-3 shows that the effect of blocking

out part of the lens is reduction of the light reaching the retina.

The closer an opacity is to the retina, the more likely is its shadow to interfere with the retinal image, as shown in Fig. 20-3, *B*. Its umbral cone will be larger in direct proportion to its backward displacement. At a given location, the larger the opacity, the longer and broader will be its umbra. Thus a small opacity close to the retina may cause a larger section of its umbral cone to impinge on the retina than that of an extensive opacity that happens to lie farther forward. An opacity that casts a sharp umbra on the retina may be outlined as a positive scotoma against an evenly white background. Uniformity of the background provides no distracting details that might interfere with the recognition of the shadow.

If the opacity lies well forward in the vitreous, it will have to be relatively large to disturb the retina by its umbra; a small opacity at the same level will cast its umbral cone far in front of the retina and may cause little or no annoyance. The closer the opacity lies to the plane of the retina, the less evident is the penumbra, and the larger and more sharply defined is the umbra. One may readily illustrate this fact by holding a pencil between a light source and a white paper; the closer the pencil is held to the paper, the larger and more clearly defined is the umbra, and the less evident is the penumbra. There is much greater contrast between the dark shadow and the rest of the receiving surface when the umbra is at its maximum and the penumbra at its minimum.

This statement also applies to the retina, and it explains the extreme annoyance caused by even a very small opacity near the retina and the indifference to a much larger opacity situated anteriorly. Patients frequently complain of small specks that can be seen against a bright background such as a page of a book. These move about as the eye is moved. The specks may be single, round globules or sometimes chains of little bright spots resembling somewhat a string of pearls. Fine dark lines and amorphous masses are also seen occasionally and may become annoying when they cross the line of sight. They usually disappear after a few moments, but sometimes they last for days and become a source of discomfort and alarm to the patient. They are called muscae volitantes because they seem to flit about like flies as the line of sight is directed in various directions.

Most of the vitreous opacities seen entoptically by patients are entirely harmless, and the patient needs nothing more than reassurance about them. However, vitreous opacities may be of serious import. When a patient complains of vitreous opacities, especially if they are accompanied by flashes of light, they are always suspect of heralding a separation of the retina. One form of vitreous opacity that occurs frequently and that patients notice entoptically consists of a large, formed particle which, with the ophthalmoscope, can be seen floating in the posterior part of the vitreous and is large enough to produce a definite shadow on the underlying retina. It may be caused by detachment of the posterior part of the vitreous and possibly represents a torn piece of the posterior part of the hyaloid canal.

RETINAL BLOOD VESSELS

The blood vessels of the retina lie in front of the rods and cones; hence they should cast a shadow on these elements and be perceived. That they are not noticed under ordinary conditions of illumination is due to the fact that the visual elements underlying the vessels become adapted to this pattern of illumination. Only when the light is thrown from one side so that the shadow falls on elements unaccustomed to it do they become visible. The vessels are seen easily when the beam of a slit lamp is focused on the sclera as far back from the limbus as possible. Patients will often remark on the picture they see as the beam sweeps over an eye in bringing it to focus on the anterior segment. The interlacing branches of the retinal vessels are seen as black lacework against a red background.

Visualization of the vascular tree of the retina by transillumination is a commonly used clinical test of retinal function. A small flashlight or transilluminator lightly placed against the lid skin near the outer canthus or directly on the conjunctiva is oscillated in position to prevent local adaptation to the shadows of the retinal vessels. The vascular tree is readily visualized by most subjects, and many can describe the branches arching over the straight-ahead direction to the empty-appearing blind spot in the temporal field. Although the patient's visualization of the vascular net is evidence of retinal function and is somewhat reassuring to the examiner who cannot see the ocular fundus because of corneal opacity or cataract, the test has two severe limitations. The first of these is that some persons with normal retinal functions cannot be made to understand the test, so a negative response is not conclusive evidence of lack of function. The second limitation is that the avascular fovea is not tested.

Another method of observing the blood vessels is to look at a bright background through a pinhole disc kept in motion in front of the eye. The blood vessels then appear as dark branching lines on a bright background. If the pinhole is not kept in constant motion, the image disappears; the motion of the disc keeps the shadows falling on different sensitive elements and thus avoids adaptation. When the image falls on one set of sensitive elements for even a short period of time, they rapidly adapt to such an image.

If pressure on the eye is made, the vessels can be seen to pulsate. The pulsations of the retinal vessels can sometimes be seen entoptically, especially after physical exercise, according to Friedman.[7] The first change observed is a sharp, rapid expansion of that part of the arterial tree involved, which is synchronous with the cardiac systole. This is followed immediately by a second phase, which is a slower contractile movement along the same path. If an individual has not exercised preceding examination, the excursions of the vessels are usually too small to allow entopic visualization.

The explanation of this phenomenon must be in the retinal circulation and not in the choroidal circulation because the macula is free from the pulsating figures. This fact is in keeping with our knowledge of the avascular structure of the macula. If the choroid were implicated, the pattern would be seen in the macula equally well because the choroidal circulation is essentially the same behind the macula as in the regions adjacent to it.

CAPILLARY CIRCULATION

If one looks at a brightly illuminated surface such as a white sheet of paper or at the cloudless sky, small, dancing spots can be seen. These usually appear as bright circles against a somewhat darker background. Occasionally their shape is more oval, and one side is brighter than the other. These spots are highly motile and pursue more or less definite paths, which are usually short and somewhat curved. The moving spots are seen best if one looks at a background of monochromatic light in the region from 350 to 450 nm. This is the region of the spectrum that is absorbed by hemoglobin. Because of this it has been suggested that the moving particles are red blood cells passing through the retinal capillaries. During their passage the hemoglobin absorbs the blue end of the spectrum; hence a slight shadow is cast on the underlying rods and cones. Naturally this shadow will be more dense in light that hemoglobin absorbs.

The movement of the corpuscles cannot be seen exactly at the point on which the subject fixates, which is to be expected if the particles are blood corpuscles in the retinal capillaries, since the fovea is a blood-free area as far as retinal circulation is concerned. It has been suggested that the shadows of the blood corpuscles are cast on the underlying rods and cones only when the corpuscles pass down into the deep capillary network. They vanish as they ascend into the inner layers of the retina.

Some authors consider the corpuscles to be white blood cells and not erythrocytes. The chief argument for this point of view is the fact that the spots seen in any one field selected for observation are not sufficiently numerous to be red corpuscles. To account for the small number of spots, it has been suggested that the red corpuscles overlap one another in a capillary and therefore cast a continuous shadow on the retina. But when a white cell is present among the red cells, it does not absorb the blue light, since it has no hemoglobin, and therefore produces a rift in the shadow figure. This rift corresponds to what is seen as a fine, bright moving point. Thus one sees the white corpuscles by contrast with the red corpuscles and sees them best in light that casts the best shadow (makes the greatest contrast). Since red light passes almost equally well through both erythrocytes and leukocytes, a contrast does not appear.

At the present time it cannot be said which of these theories is correct. There can be scarcely any doubt that the phenomenon is connected intimately with the circulation of the blood. The particles move in more or less definite paths, having the arrangement of capillary loops, and the movement is pulsatile and shows marked acceleration when the heart is speeded up. Finally, the region of the fovea, known to contain no retinal blood vessels, fails to show any of these particles. The relatively few particles seen in the field would incline one to accept the explanation that the spots are white corpuscles, especially since they appear light against a darker background.

Granted that these particles are in the bloodstream, to what exact situation of the retinal circulation should they be assigned? There are two possible locations where the phenomenon may take place. Either the corpuscles are in the precapillary arterioles, or they are in the capillary loops in the deep retinal layers. The patterns made by the particles correspond with the finest branchings of the retinal vessels, which can be seen entoptically. These are probably the precapillary arterioles. The phenomenon has served a useful purpose in measuring the blood-free area around the fovea in the living subject.

The measurements of the area devoid of these entoptically seen corpuscles, according to several different authors, are given in Table 20-1 (compare the values with the area of the functional fovea, p. 578).

Although relatively wide variations between subjects exist, the two eyes of the same person are usually identical. The method, therefore, may be applied clinically to study the effects of drugs or other changes on the fovea. It has been found that physostigmine and pilocarpine change the entoptic appearance of the fovea and that many of the circumfoveal capillaries disappear under the influence of these drugs. Pressure on the eye slows up the movements of the corpuscles until they finally stop. When the pressure is removed, the circulation becomes quicker than before. By the use of the ophthalmodynamometer it has been determined that the intraocular pressure at which cessation of movement of the particles occurs is about 50 mm Hg.

Marshall[9] has demonstrated that the corpuscles are not in the retinal circulation of the nerve fiber layer. Their size and velocity are too great for this position, and their course is not that of the main retinal vessels. He determined this by observing the luminous bodies against the reflected mercury light with one eye and produced a map of the main vessels entoptically by eccentric rotation of a pinhole in a piece of black cardboard placed in front of the other eye. It could be seen that the two tracings were on different planes, the luminous points appearing farther away and giving the impression of being more highly magnified. Although the pathways in which the luminous points moved could not be made visible by this method, their jerky movements in what appeared to be a network of channels suggested that the cells were circulating in the capillaries of the inner nuclear layer.

CHORIOCAPILLARY CIRCULATION

Marshall[9] has been able to visualize the choriocapillary circulation in his own eye, but Friedman[7] was unable to confirm his observations. It is difficult on theoretical grounds to see how one could visualize entoptically the choroidal circulation. The spongy character of the choroidal network with its superimposed layers of blood vessels would mask any possible shadows of individual vessels. Furthermore, the light would have to be incident to the posterior

Table 20-1. Blood-free area of retina determined entoptically

Author	Measurement of blood-free area	
Abelsdorff and Nagel	1° 30′	0.410 mm (author's eye)
Gescher	1° 29′	0.420 mm (author's eye)
Sperling, Miller, and Adler		0.40-0.50 mm (38 normal eyes)

part of the sclera and not come through the pupil.

BLUE ARCS OF RETINA

If the observer in a dark room fixates a point slightly to the temporal side of a small source of light with one eye, he will see two small bands, or arcs, of light, bright blue in color, radiating from the stimulating light toward the blind spot. These arcs are always in the horizontal plane and appear almost as soon as the stimulating light is turned on. Various authors have observed and described them, but their actual cause is still unknown. The stimulating light may be any color or even white, but red is the best color to use, probably because there is less scattering of the rays. The light source may be any shape—a rectangular source is usually employed. When the temporal side of the source is fixated, both arcs converge toward the blind spot. If the upper edge only is fixated, only the lower arc appears, and similarly when one fixates the lower edge of the source, only the upper arc appears. Fixation on the nasal side of the source produces a blue haze between the previous regions where the arcs appeared that has been referred to as the blue spike.

The arcs (including the spike) have been the subject of much study, and various theories of their origin have been proposed.* At the present time it appears that bioluminescence as a cause can be ruled out and that secondary electrical excitation, of neighboring retinal nerve fibers or neurons, by active nerve fibers is the probable origin of the phenomenon.

SELF-ILLUMINATION OF RETINA; PHOSPHENES

When the eyes are kept in the dark and dark adaptation has become complete, a person does not have a sensation of black, but of a definite grayness that is much lighter than the sensation produced by looking at a black velvet surface in the light. This sensation of grayness or of light probably arises from both the retina and the cortex. It is called self-illumination of the retina.

Pressure on the eye produces the impression of a dark circular spot in the field of vision directly opposite the point of pressure on the globe. These visual sensations, produced by inadequate retinal stimuli, are called phosphenes. Prolonged digital pressure over the eye produces a pressure ring in the form of a broad circular band of blue.[7] The pressure has to be strong enough to cause slight discomfort and, according to some authors, must be maintained for 3 minutes. Two spots of blue, one nasally and one temporally, appear, and these slowly expand and assume the form of a broad arc. The figure continues to grow until the arcs coalesce into a blue circle, the center of which is oval and devoid of color. It is sharply demarcated and corresponds approximately to the macular limits. The cause of the phenomenon is not known, but the situation of the pressure oval corresponds roughly to the zone of maximum rod population described by Osterberg.[13]

If gentle pressure is applied to the temporal side of a completely dark-adapted eye, an immediate bright ring of blue appears, seemingly on the nasal side. This is almost instantaneous and disappears even though the pressure is continued. This phenomenon is noticed best after a night's sleep if it is elicited while the room is still completely dark.

PHYSIOLOGIC AND PATHOLOGIC HALOS

Under certain physiologic conditions, colored halos can be seen around small white lights viewed from a distance. These colored rings are caused by the breaking up of white light by the various layers of cells of the media through which the light must pass on its way to the retina. The most common site of the halos under normal conditions is the lens. The lens fibers act as a radially arranged diffraction grating. The halos are not very bright as compared to those that can be produced experimentally or are seen in pathologic conditions. They are thought by some to have a smaller angular diameter than pathologic halos. The colors of both physiologic and pathologic halos are the same; the short waved lights, blue and violet, are next to the stimulating light, with the long waves on the outside, and red the outermost ring. The angular diameter of physiologic halos

*References 1, 2, 4, 6, 8, 11, and 15.

ranges from 7 to 8 degrees, measured to the outer ring. Colored halos are easily produced by steaming a piece of glass or dusting lycopodium powder over a piece of glass and then observing a candle flame through it.

Pathologic halos occur from several conditions. Chronic conjunctivitis with mucous secretion is very likely to produce them, particularly early in the morning after mucus has collected during the night. Patients arising early in the morning and seeing halos around lights may become alarmed. Since this is the time of day when the pressure is elevated in most patients with glaucoma, the condition may alarm the physician also. Halos are complained of frequently by persons who have suffered from too intense exposure to light, such as in snow blindness, and presumably result from the conjunctivitis caused by the exposure to light.

The chief interest in halos is centered in their occurrence in glaucoma. There can be no doubt that they are due directly to increased pressure in the eye because they come and go as the pressure increases and decreases. They probably are caused by the breaking up of light as it passes through the edematous corneal epithelium. When the intraocular pressure is increased, small spaces filled with fluid appear in the cornea just anterior to Bowman's membrane. The light is broken up into the spectral colors by droplets of fluid in the epithelium in the same way that light passing through droplets of rain is broken up to form a rainbow.

MOORE'S LIGHTNING STREAKS

In 1935 and again in 1940, Moore[10] described an entoptic phenomenon consisting of flashes of light as follows:

> [they are] most frequently likened to lightning, seen to the temporal side of the eye, never to the nasal side, and vertical in direction. These flashes are accompanied by the simultaneous development of a crop of opacities in the vitreous. They seldom occur before middle age and are more frequent in the female sex. They do not imply any serious disease of the eye, either at the time or subsequently.[10]

In 1954, Berens and co-workers[3] laid stress on the fact that this syndrome was of more serious import than Moore believed. The question is of some importance to ophthalmologists, since a number of people report these streaks and are quite bothered about them. The phenomenon is probably the result of vitreous detachment, which occurs in many persons of middle age. As the vitreous shrinks forward toward its base, traction on the retina can produce phosphenes. Traction on the temporal retina near the ora serrata may not be noted subjectively since the periphery of the temporal retina is blind, whereas similar traction on the seeing peripheral nasal retina may be observed as a phosphene referred to the temporal visual field. The patient reporting a ''lightning streak'' should not be dismissed lightly; a good look at the peripheral retina and ora serrata is in order. Fortunately most persons who report such phosphenes do not get retinal detachment, but some do.

PHOSPHENE OF QUICK EYE MOTION

A phenomenon not recorded previously is described by Nebel,[12] who believes that it is related mechanically to, but is definitely distinct from, Moore's lightning flashes. It is described by the term flick phosphene. It is observed best by the dark-adapted eye and is seen by most people on awakening from sleep just before dawn. Then, if one moves the eyes rapidly from one side to the other, one observes in each monocular field a bright pattern having the general shape shown in Fig. 20-4. Each eye produces its own phosphene, which is short lived and rapidly dies out; thus it cannot be repeated as the retina fatigues. The phosphenes are seen in the two eyes simultaneously and are two separate images (Fig. 20-5). Each has the shape of a sheaflike pattern, and in most persons the apex is not pointed but truncated. When it is first elicited, the pattern is bright yellow or orange, and details of the sheaves are sharp. With repeated eye movements, however, the details become indistinct and blurred. It is ascribed to instantaneous transient deformation of the posterior surface of the vitreous close to the optic disc. When the eye is flicked suddenly, the inertial lag of the vitreous causes the deformation, which is transmitted directly to the retina and causes the fibers in this region

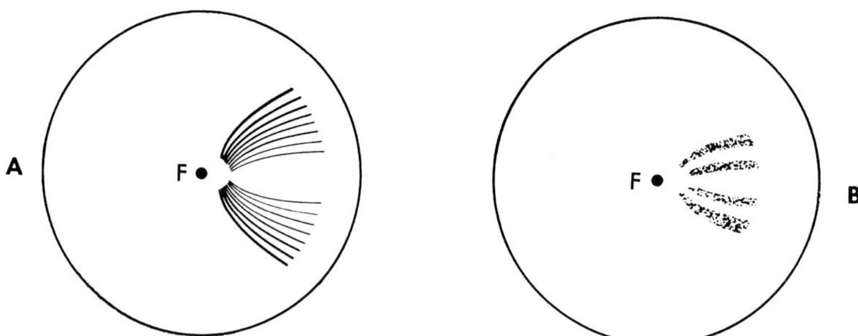

Fig. 20-4. Phosphene of right eye in right-to-left flick. **A,** Phosphene pattern in rested eye. **B,** Pattern after fatigue. *F,* Fovea, as identified by afterimage of fixated inducing light. (From Nebel, B.: Arch. Ophthalmol. **58:**236, 1957.)

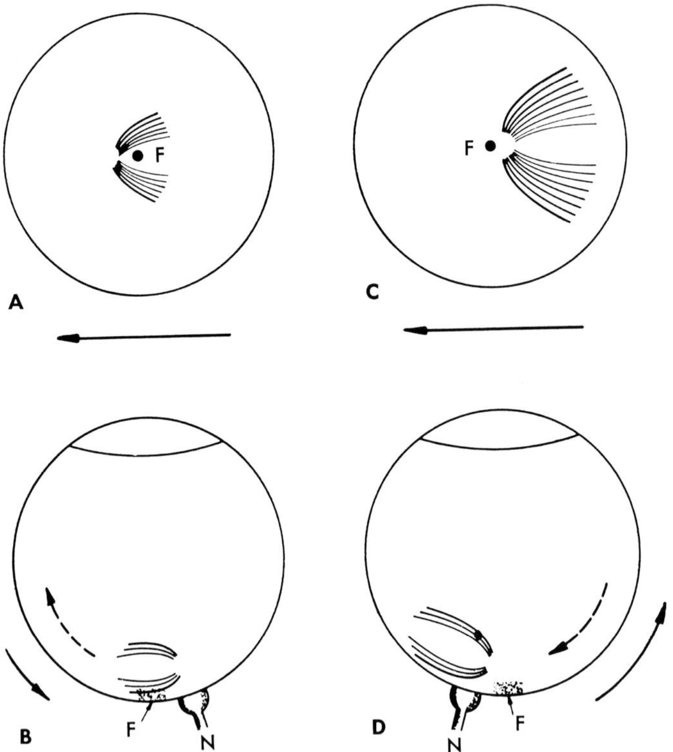

Fig. 20-5. Phosphenes of right-and-left flick and interpretation relating to topography of their origin. **A** and **C,** Phosphenes as seen by person. **B** and **D,** Nebel's projection of phosphenes back to retina. Opposing arrows, shearing forces set up by acceleration of wall of globe and inertial retardation of vitreous. *N,* Nerve; *F,* fovea. Projected phosphene is tilted and displaced slightly to make it visible. (From Nebel, B.: Arch. Ophthalmol. **58:**237, 1957.)

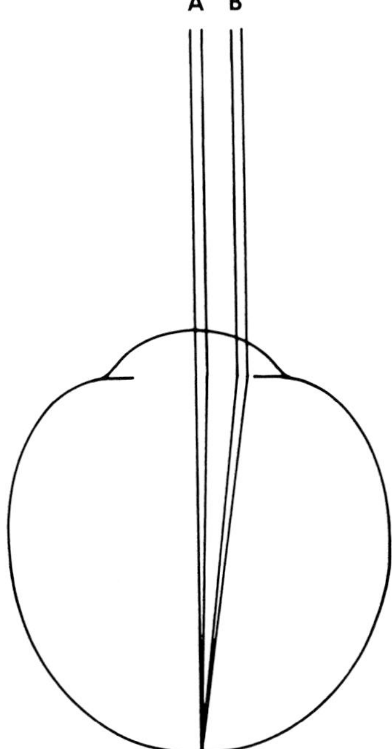

Fig. 20-6. Normal retinal directional sensitivity effect. Light in bundle entering through center of the pupil, *A*, is more effective in stimulating retinal cones than in bundle coming into eye near edge of dilated pupil and hence reaching retinal cones obliquely, *B*. (From Westheimer, G.: Arch. Ophthalmol. **79:**584, 1968.)

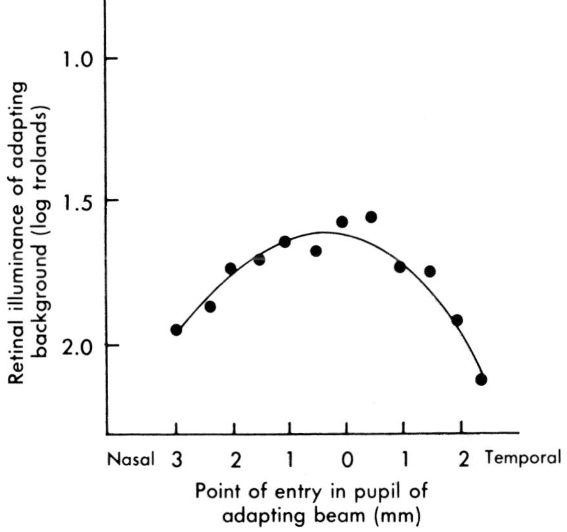

Fig. 20-7. Normal retinal directional sensitivity pattern. Abscissae: point of entry along horizontal meridian of pupil, of bundle of light providing background; ordinates: relative sensitivity of retina, measured by inverse of amount of light necessary in background field to bring constant ΔI stimulus to threshold. Background (7.5 degrees) was exposed continuously and ordinates give its retinal illuminance in trolands of green light (tungsten source filtered by gelatin filter and neutral wedge). Incremental stimulus was circular field, 12 minutes of arc in diameter, 0.05 second in duration, 30 trolands of red light, and was placed at threshold by adjustment of wedge in background beam. (From Westheimer, G.: Arch. Ophthalmol. **79:**584, 1968.)

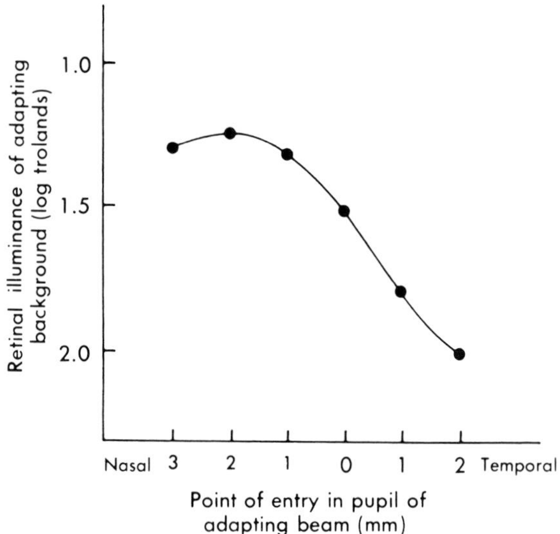

Fig. 20-8. Asymmetrical retinal directional sensitivity pattern in author's right eye, foveal vision. Abscissae: point of entry of bundle of light providing background; ordinates: relative sensitivity of retina, measured by inverse of amount of light necessary in background field to bring constant ΔI stimulus to threshold. (From Westheimer, G.: Arch. Ophthalmol. **79:**584, 1968.)

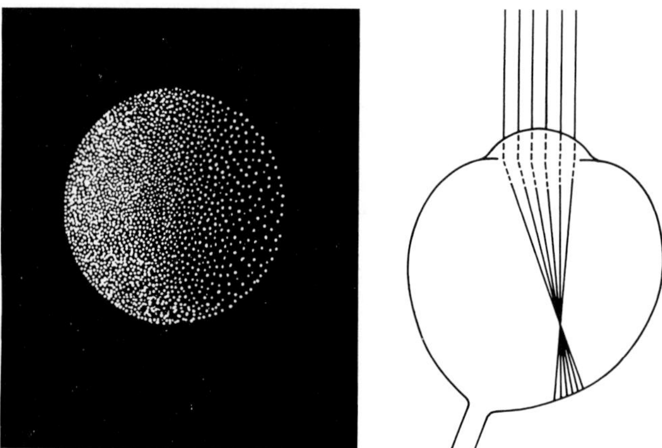

Fig. 20-9. Left, appearance of foveal blur patch of bright star against dark sky seen with uncorrected myopic right eye with large pupil. Stiles-Crawford pattern of this eye, illustrated in Fig. 20-8, is asymmetrical. Right, schematic diagram illustrating path of rays making up blur patch. Fact that seen pattern is brightest near its left border implies that receptors in region of fovea point in direction of nasal edge of dilated pupil. (From Westheimer, G.: Arch. Ophthalmol. **79:**584, 1968.)

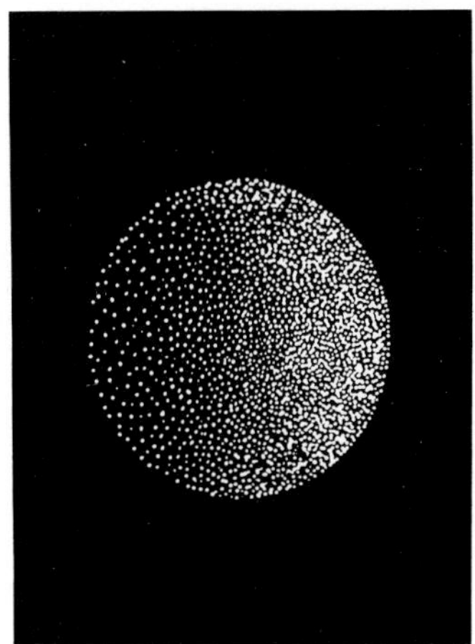

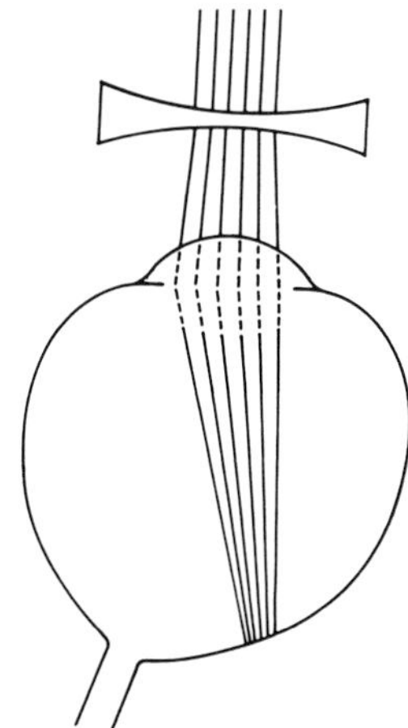

Fig. 20-10. Left, appearance of foveal blur patch of bright star against dark sky seen under artificially hyperopic condition with large pupil in right eye with asymmetrical Stiles-Crawford effect (Figs. 20-8 and 20-9). Right, schematic diagram illustrating path of rays making up blur patch. Right edge of seen pattern corresponds to nasal edge of retinal blur patch. Fact that it is brightest implies that receptors in foveal region are pointing to nasal edge of dilated pupil. (From Westheimer, G.: Arch. Ophthalmol. **79:**584, 1968.)

to fire off. It may be an early senescent sign of normal slight shrinkage of the vitreous.

HAIDINGER'S BRUSHES

If the normal eye observes a surface illuminated by plane-polarized white light, yellow or blue brushes or sheaves radiating from the fixation point are seen. De Vries and co-workers[5] and later Stanworth and Naylor[16] showed that the brushes are caused by variations in absorption by oriented macular pigment in the foveal region.

Since the effect is caused by an oriented pigment in front of the layer of photoreceptors, any process that upsets this orientation, even though it does not disturb the photoreceptors themselves, may lead to disappearance of the brushes. Therefore the test for Haidinger's brushes may show early changes in the macula such as edema before they reach the point of interfering with visual acuity or of being visible ophthalmoscopically. The test forms a simple, rapid means of demonstrating macular edema in the early stages of macular disease.

Maxwell's spot is seen as a circle surrounded by a blue halo in intermittent diffused blue light. The spot appears at the point of fixation. The phenomenon may be related to the yellow macular pigment, but this is not certain. It also may be related to the fact that the central fovea is devoid of blue color receptors.

STILES-CRAWFORD EFFECT

Entoptic visualization of the Stiles-Crawford effect of the first type provides interesting evidence of retinal contour in myopia.[18] The usual Stiles-Crawford demonstration (Figs. 20-6 to 20-8) may be simulated by deliberately throwing the image of a point source of light (a bright star) out of focus (Figs. 20-9 and 20-10). The entoptic appearance in some myopes is what is expected from foveal cone tilt. Reference has been made to x ray and other phosphenes on pp. 357 to 358.

REFERENCES

1. Alpern, M., and Dudley, D.: The blue arcs of the retina, J. Gen. Physiol. **49:**405, 1966.
2. Amberson, W.: Secondary excitation in the retina, Am. J. Physiol. **69:**354, 1924.
3. Berens, C., et al.: Moore's lightning streaks, Trans. Am. Ophthalmol. Soc. **52:**35, 1954.
4. Boehm, G.: The entoptic phenomenon of the "blue arcs" (observations on subjects affected with night blindness and total color blindness), Ophthalmologica **4-5:**276, 1949.
5. De Vries, H., Spoor, A., and Jielff, R.: Properties

of the eye with respect to polarized light, Physica **19:**419, 1953.
6. Friedman, B.: The blue arcs of the retina, Arch. Ophthalmol. **6:**663, 1931.
7. Friedman, B.: Observations on entoptic phenomena, Arch. Ophthalmol. **28:**285, 1942.
8. Ladd-Franklin, C.: Alternative theories to account for the reddish blue arcs and the reddish blue glow of the retina, J. Opt. Soc. Am. **16:**333, 1928.
9. Marshall, C.: Entoptic phenomena associated with the retina, Br. J. Ophthalmol. **19:**177, 1935.
10. Moore, R. F.: Subjective lightning streaks, Br. J. Ophthalmol. **19:**545, 1935.
11. Moreland, J. D.: Threshold measurements of the blue arcs phenomenon, Vision Res. **8:**1093-1106, 1968.
12. Nebel, B.: The phosphene of quick eye motion, Arch. Ophthalmol. **58:**235, 1957.
13. Obsterberg, G.: Topography of the layer of rods and cones in the human retina, Acta Ophthalmol. **6:**1, 1935.
14. Pischel, D.: Detachment of the vitreous as seen with slitlamp examination, Trans. Am. Ophthalmol. Soc. **50:**329, 1952.
15. Snell, P.: The entoptic phenomenon of the blue arc, Arch. Ophthalmol. **1:**475, 1929.
16. Stanworth, A., and Naylor, E.: The measurement and clinical significance of the Haidinger effect, Trans. Ophthalmol. Soc. U.K. **75:**67, 1955.
17. Verhoeff, F.: Are Moore's lightning streaks of serious portent? Am. J. Ophthalmol. **41:**837, 1956.
18. Westheimer, G.: Entoptic visualization of Stiles-Crawford effect, Arch. Ophthalmol. **79:**584-588, 1968.

GENERAL REFERENCE

Rubin, M. L., and Walls, G. L.: Studies in physiological optics, Springfield, Ill., 1965, Charles C Thomas, Publisher.

Chapter 21

Binocular vision

PETER O. BISHOP, M.B., B.S., D.Sc. (Sydney) F.A.A., F.R.S.

Binocular vision is that state of simultaneous vision with two seeing eyes, neither of which need necessarily be normal, that occurs when an individual fixes his visual attention on an object of regard.[164] Normal binocular vision implies binocular single vision (fusion) and a high level of stereoacuity. The emphasis in this chapter will be directed toward stereopsis, since this is, for all practical purposes, the only advantage of binocular vision. Binocular visual acuity (as opposed to stereoacuity) is only very marginally better than monocular acuity. Stereopsis is the binocular perception of depth made possible by the fact that the two eyes view the external world from different vantage points. Because of the horizontal separation of the eyes, the two retinal images are slightly different (binocular parallax), and it is the horizontal retinal image disparities that provide the essential data for the binocular depth estimates (Fig. 21-1). Stereopsis provides a vivid and accurate relative depth discrimination, and the depth experience has the same immediate perceptual quality as does color. It is a unique phenomenon, no hint of which exists in monocular vision, and one which cannot adequately be described to a person unable to experience it. In addition to binocular disparity cues to depth there are many powerful monocular cues, such as object overlay, linear and aerial perspective, light areas and shadows, texture density gradients, and motion parallax. Depth estimates based on these so-called secondary cues do not, however, have the same perceptual quality and accuracy as do those due to stereopsis. Nevertheless, secondary cues do make possible a fairly high level of depth discrimination. Following the sudden loss of vision in one eye and hence loss of stereopsis, a person may experience difficulty in judging depth. In time, however, the patient learns greater reliance on secondary cues and may become almost as skillful as he was before he lost his eye. These monocular cues to depth will not be considered further here.[27,60,61]

In clinical practice the goal of therapy at the present time is not so much directed toward achieving the advantages of normal binocular depth perception (stereopsis) as it is toward removing or alleviating the disadvantages that result when binocular vision becomes disordered, namely the double vision and the loss of visual acuity in the deviating eye (suppression and amblyopia). It is indeed surprising that the advantages of binocular vision should weigh so little and the disadvantages press so heavily. In the clinic, therefore, relatively little attention has been given to stereopsis, and the emphasis has been directed toward binocular single vision, or fusion. The term fusion is used in two ways: motor fusion and sensory fusion. The term *motor fusion* refers to the reflex vergence movements needed to bring about and maintain the correct binocular alignment, whereas *sensory fusion* is the process by which binocular single vision is achieved from two similar but nevertheless disparate retinal images.

Evolution of binocular vision

Stereoscopic depth perception requires the correlation of information from the two eyes, and certain features of the organization of the visual system are usually regarded as essential

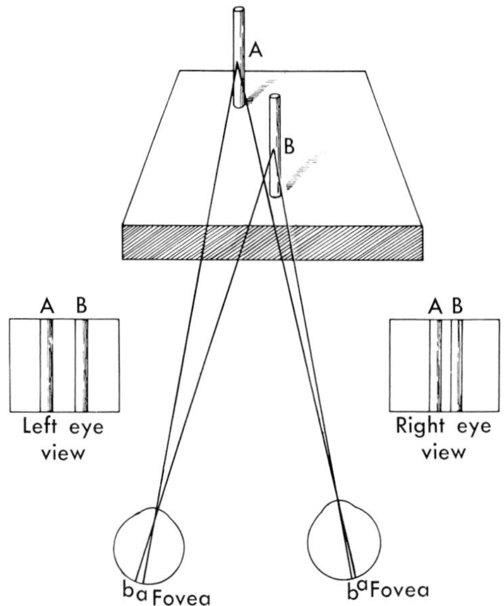

Fig. 21-1. Retinal image disparity based on horizontal separation of the two eyes.

for this correlation. These include frontally placed eyes to allow a considerable overlap of the visual fields of the two eyes, the binocular coordination of eye movements so that objects stimulate corresponding retinal points, and a semidecussation of the optic tract to permit the interaction of inputs from corresponding parts of each retina. Since only the higher mammals, such as the carnivores and primates, have developed all these features to any significant degree, the widely accepted view has been that only these animals possess stereopsis. The belief that vertebrates below the mammals were without stereopsis was based largely on the fact that these animals all lack even a partial decussation at the optic chiasm; instead, all the fibers from one eye cross to the opposite side of the brain. These conclusions regarding stereopsis now have to be drastically revised.

Although naturalistic observations have long suggested that many vertebrates below the mammals have excellent depth perception, it is only recently that convincing evidence has become available for a stereoscopic depth mechanism in some of these animals, namely the toad[31] and some birds. Anatomic studies,[101] physiologic recording experiments,[152,153] and behavioral studies[57] all indicate that owls and kestrels possess stereopsis. In birds there are two large parallel pathways transmitting information from the retina to the external striatum in the telencephalon. One pathway proceeds via the optic tectum and thalamus (nucleus rotundus) to the noncortical ectostriatum and the other passes directly through thalamic nuclei (opticus principalis thalami, OPT) to a region (the Wulst) that forms a transition zone between the hyperstriatum and the pallial (cortical) mantle. Each Wulst receives both ipsilateral and contralateral retinal projections, the ipsilateral projection being formed by OPT axons that cross back to the cerebral hemisphere of the same side via the supraoptic decussation. Thus the supraoptic decussation is the avian analog of the mammalian optic chiasm, the partial decussation occurring at least one synapse further along the optic pathway. It seems that the thalamo-Wulst pathway in birds is analogous to the mammalian geniculostriate system, whereas the avian tectothalamoectostriatal pathway resembles the mammalian conduction route that extends from the superior colliculus via the thalamus (nucleus lateralis posterior and/or pulvinar) onto the extrastriate and inferotemporal cortex. Nauta and Karten[125] have suggested that in the course of vertebrate evolution the cells in the external striatum of reptiles and birds have migrated into the pallial mantle to form part of the mammalian neocortex, the hyperstriatum being analogous to the mammalian striate cortex and the ectostriatum being comparable to the extrastriate cortex.

Each eye of bifoveate lateral-looking birds, such as the kestrel and vulture, has two visual axes, one through a central fovea close to the optical axis for lateral monocular viewing and the other through a temporal fovea, well away from the central fovea, for use in binocular vision. Pettigrew[151] has suggested that, whereas the main retinal projection to the visual Wulst in these birds is from the temporal fovea, the projection to the optic tectum comes mainly from the central fovea. This proposed division of function correlates with the two patterns of eye movements. The independent chameleon-like eye movements of monocular vision use the central fovea for tracking a target, while the conjugate movements of binocular vision simultaneously track the target onto the two temporal foveae.[151] Thus the tectoectostriatal system provides the main neural mechanism for monocular pattern vision in birds, whereas the thalamo-Wulst system seems to be primarily concerned with binocular visual processing. Physiologic recording has shown that the Wulst has a precise retinotopic organization, a high degree of binocular interaction, and neurons that appear to be selectively tuned to stereoscopic depth cues.

In the course of mammalian evolution, particularly in the carnivores and primates, there has been a gradual recession of the snout and movement of the eyes from a lateral sideways looking position to a frontal binocular location. In the process these animals have lost the ability to control the movement of each eye independently of the other. The two eyes now become carefully yoked together, moving at all times in fairly precise register, one with the other. Along with the change in eye position and the pattern of eye movement there has been a progressive development of a partial decussation at the chiasma whereby the fibers from the temporal retina of the ipsilateral eye and the nasal fibers from the contralateral eye come together in an optic tract, thereby enabling the retinocerebral pathways from corresponding retinal areas in the two eyes to project to a common terminus in the striate cortex in a fairly precise topographic manner. These changes in the primary visual pathway brought with them a massive reorganization of the visual centers in the brain. All mammals still have the two main visual pathways from retina to cortex, but the geniculostriate system now becomes the main pathway for pattern vision as well as stereopsis. Even so, the striate cortex requires interaction with the superior colliculus to be effective.[40] Although the relative size of the superior colliculus becomes considerably less than it is in nonmammals, the colliculo-pulvinar-cortex system remains and continues to play an important part in visual processing. Visual movement and contour detection persist in monkeys after removal of the striate cortex,[39] but even this remaining visual capacity is abolished by lesions involving the cortical projection of the inferior pulvinar.

The evolutionary changes in the brain were correlated with another important advance, namely the development of eye-hand coordination. The adoption of the upright posture freed the hands from their role in postural support and enabled the development of manipulative skills under visual guidance. This guidance is best served by frontal vision, since the finest manipulate skills require the coordinated use of the two hands. Frontal vision is also essential for stereopsis, and the evolution of binocular vision was undoubtedly a major factor in bringing manual skills to their highest level of performance in the human. Under normal viewing conditions the quality of stereoscopic vision improves as the objects regarded are brought closer to the eyes, up to a point where the ability to focus begins to fall off. There is maximal benefit, therefore, for close work involving delicate eye-hand coordination.

For the modern human, however, stereoscopic perception no longer seems to have the important place it is widely held to have had in the evolution of our species. For us, stereopsis can to a very large extent be replaced by higher order secondary cues to depth that are operative in monocular as well as binocular vision. Nevertheless, in the evolutionary process one important advantage of stereopsis could have been that it enabled predators to ''see through'' the camouflage by which hunted animals sought to blend in with their surroundings. Objects effectively camouflaged to the monocular view may nevertheless stand out quite clearly when viewed stereoscopically. Furthermore, as we shall see, the stereoscopic processes that made this discimination possible would not have required any prior familiarity with, or recognition of, the objects concerned. Hence the essential steps in the evolution of stereopsis could well have taken place either before or at the same time as the development of the complex neural mechanisms needed to make use of secondary cues to depth. Quite apart from the matter of camouflage, however, there is the possibility that stereopsis also played an important part in the development of the ability to use those higher order secondary cues. Once developed, these secondary mechanisms could then have operated independently of and largely substituted for the stereoscopic processes that were partly responsible for their development in the first place.

Some definitions

The strabismic literature suffers a major handicap through the use of undefined or ill-defined anatomic terms, such as fovea and macula, to express physiologic and pathophysiologic concepts. While there is every reason for the continued use of these anatomic terms, much of the vagueness and confusion that exists would be eliminated if the regions referred to had assigned to them agreed on locations and dimensions in terms of degrees of visual angle relative to the center of the foveola. Thus if the angle of the strabismus is not to be given in terms of degrees of visual angle, it should be given in terms of retinal (anatomic) landmarks previously defined in terms of visual angles with respect to the visual axis. The visual axis is the line from the center of the foveola through the nodal point of the eye.

Ophthalmic prisms are calibrated in terms of prism diopters, and in clinical practice ocular deviations are usually measured in terms of equivalent prism diopters. One prism diopter (1^{Δ}) causes the lateral displacement of a light beam by 1 cm on a screen at 1 m distance. However, in this account the degree of visual angle will be used, since it is a more generally useful unit for expressing not only ocular deviations but also spatial relations within the visual field ($1^{\Delta} = 0.57$ degrees). Ophthalmic prisms could, of course, be equally well calibrated in degrees of deviation.

Macular region

As we remarked earlier, any discussion of binocular vision inevitably involves the use of anatomic terms relating to structures in the macular region. It is *not* the aim of this section to provide an anatomic description of this region, but rather to define the various zones concentric with the center of the foveola in terms of visual angles from the center. To do this each zone is given a brief anatomic description and then defined in terms of visual angles.

Fig. 21-2 is a scale diagram giving the location and dimensions of the principal regions in the central retina. The values given represent a consensus of those available in the literature.[47,73,156] It is to be hoped that these values, or others close to them, will eventually become an acceptable basis for description and discussion. In the remainder of this chapter a consistent attempt will be made to quantify the various phenomena and concepts and also to relate them to the various retinal regions wherever this is possible.

In line with the preceding discussions the various retinal regions will now be described and defined. The term *macular region* refers to that part of the central retina having a diameter of 5.5 mm and limited by the outer boundary of the perifovea where the ganglion cells are reduced to a single layer. It is here defined as the region of the retina within a radius of 9.2 degrees from the center of the foveola. The term macular region is to be carefully distinguished from macula lutea. The latter is a rather ill-defined region surrounding the fovea but somewhat smaller than the macular region just defined. It is characterized by the presence of a yellow pigment in both the ganglion cells and bipolar cells in this region. The *foveola* (0.35 mm across) is the small central region in which the thickness of the retina is reduced so as to contain only photoreceptors, glial cells, and Müller's cells. It is here defined as having a radius of 0.6 degrees. The *fovea* (fovea centralis, central fovea; 1.5 mm across) is a small depression where the retina is reduced to about half its normal thickness. By definition here, its outer boundary has a radius of 2.5 degrees from the center of the foveola. Moving toward the center of the retina, the inner nuclear layer is reduced to a double row of cells at the edge of the fovea, and within the fovea itself the bipolar cells practically disappear. The internal plexiform, ganglion cell, and nerve fiber layers are also absent from the fovea. The *parafovea* is characterized by the densest accumulation of nerve cells in the entire retina, especially those of the ganglion cell and inner nuclear layers, and by the thick outer plexiform layer. The foveal region and parafovea together have a diameter of 2.5 mm. The outer boundary (radius 4.2 degrees) is the point where the gan-

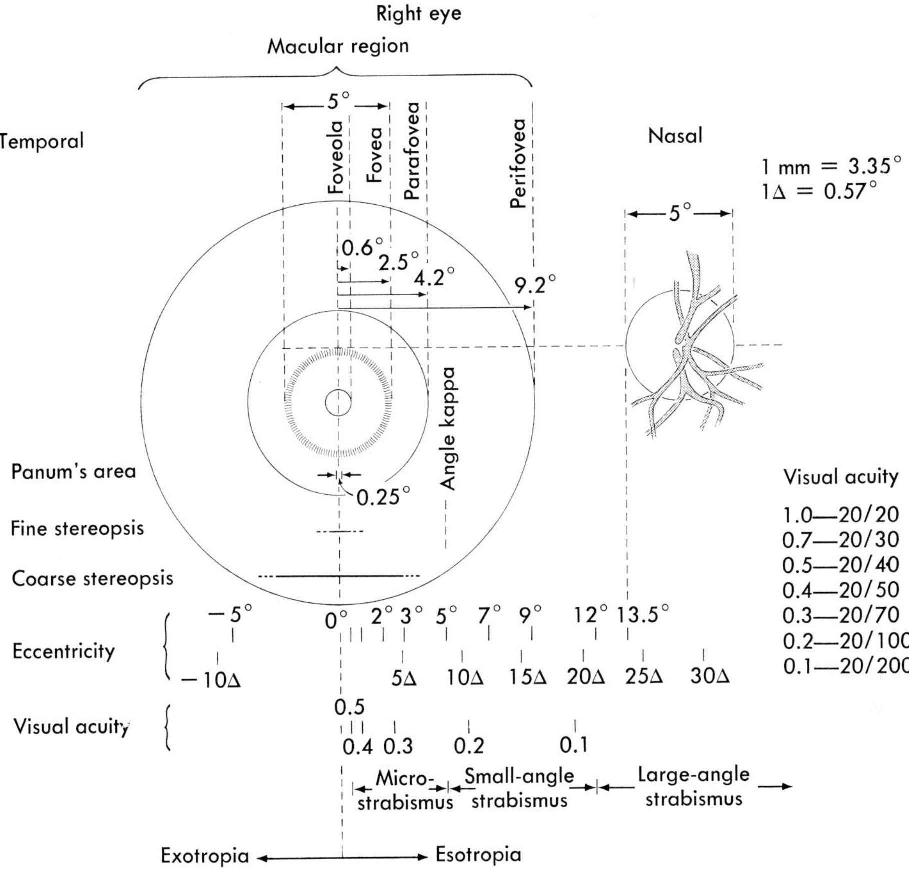

Fig. 21-2. Scale diagram of fundus of right eye showing location and dimensions of various structural and functional areas in macular region.

glion cell layer has four rows of nuclei. The *perifovea* (1.5 mm in width) ends where the ganglion cells are reduced to a single layer. Its outer boundary has a radius of 9.2 degrees from the center of the foveola.

In addition to the definitions just given for the various retinal regions, terms such as bifoveolar fixation may still need further clarification. Thus bifoveolar fixation could refer to a number of different states: the visual axes of the two eyes intersecting at the fixation point, the visual axis of one eye falling somewhere within the projection of the foveola of the other eye, or, finally, only some measure of overlap of the projections of the two foveolas.

Estimate of angle of strabismus

To relate strabismus to binocular physiology, it is important to have an appreciation of the range of the deviation that may be observed in the squinting eye. For distant vision the visual axes should normally be very close to parallel, and therefore any deviation from parallelism represents an angle of squint (positive for convergence, esotropia; negative for divergence, exotropia). For purposes of clinical description it has been found convenient to make a strabismus classification having three categories: microstrabismus (< 5 degrees), small-angle strabismus (5 to 12 degrees), and large-angle strabismus (> 12 degrees) (Fig. 21-2).

The diagnosis and clinical definition of strabismus is based primarily on the cover test, using the corneal reflections to detect a movement of the eye. Although 6 to 9 degrees of esotropia may not be recognized by the casual untrained observer, particularly in the presence of a positive angle kappa, it is readily distinguished by the trained observer using the cover test. The angle kappa is the angle between the central pupillary line and the visual axis. The smallest shift or deviation that can be detected by the cover test is about 1 degree.[163] While the ultimate aim of strabismic therapy is to achieve fine stereopsis and fusion, the common end result of surgery is a microstrabismus of up to 5 degrees. The angle kappa is an important factor influencing the cosmetic appearance of any ocular deviation. The average value of the angle kappa is generally given as +5 degrees. A positive angle kappa diminishes the cosmetic effect of an esotropia, whereas it increases that of an exotropia, so that, according to Hugonnier and Clayette-Hugonnier,[89] the cosmetic defect is slight or absent within the arbitrary limits of −4 to +14 degrees. Most authors, however, have rather smaller limits for a satisfactory cosmetic result, namely a horizontal deviation less than ±6 degrees and a vertical deviation less than ±3 degrees.[49,158,176] A rough estimate of the angle of deviation may be obtained by inspection of the corneal reflections as the patient looks directly at a light held by an observer immediately in front of him (Hirschberg's method). The locations of the pinpoint images of light as reflected by the two corneas are compared in relation to the pupil and to the cornea as a whole (Fig. 21-3). Because the angle kappa is normally positive, the reflections do not coincide with the centers of the pupils, but rather are slightly to the nasal side in each case. A 1 mm displacement of the corneal reflection roughly corresponds to an angle of 7 degrees. In esotropia inward movement of the nonfixing eye gives the appearance of the reflection moving outward toward the temporal edge of the pupil and vice versa for exotropia. In Fig. 21-3 the relative positions of the reflections in the two situations have been drawn to scale for a pupil diameter of 4 mm.

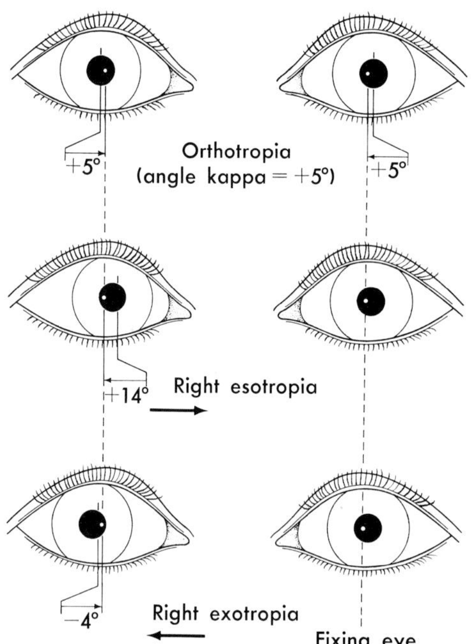

Fig. 21-3. Scale diagram of corneal reflections for 4 mm pupil in normal subject (orthotropia) and their use to determine angle of deviation in strabismus (Hirschberg's method).

PSYCHOPHYSICS

Analysis of visual space*

Oculocentric visual direction (monocular vision)

We owe to Hering (1861)[68,69] the first detailed analysis of visual space using the concept of *visual lines*. When we view a particular object with steady gaze, the image of the object of regard falls on the foveola. Consider first monocular vision. The visual direction of the object of regard can be represented by a line (visual line) joining the object of regard to the center of the foveola. This line is the *principal visual line* or *visual axis*. We fix the position of all the other objects in the monocular visual field by their oculocentric visual directions with respect to the visual axis. In other words, the frame of reference is oculocentric. Thus each point on the retina can be thought of as having its own particular visual direction or visual line passing out through the nodal point of the eye. A visual line is, therefore, the locus of all points fixed relative to the eye whose images stimulate a given point on the retina. From this it follows that for a given position of the eye, objects having superimposed retinal images will be seen as being in alignment in the visual field (law of oculocentric visual direction), al-

*See references 24, 77, 111, 136, and 190.

though at different distances from the eye along the same visual line.

Egocentric visual direction (binocular vision)

The idea of monocular visual lines leads to the binocular concept of corresponding retinal points. The latter concept will be introduced at this stage, but dealt with in greater detail subsequently. Retinal points in the two eyes are said to be corresponding if, *when stimulated separately,* they appear to have the same visual direction. Therefore corresponding points have, by definition, the same monocular visual direction. Because each eye sees the world from a different viewpoint, the oculocentric frame of reference is necessarily different for the two eyes. In binocular vision a single system of visual directions is needed, whose frame of reference is related to the head (egocentric) rather than to the eyes. When we use two eyes, we seem to see visual space as though from an imaginary single eye (cyclopean eye; after Helmholtz) situated in the head midway between the two eyes. The concept of a cyclopean eye is a useful although limited one for introducing the ideas of binocular vision.

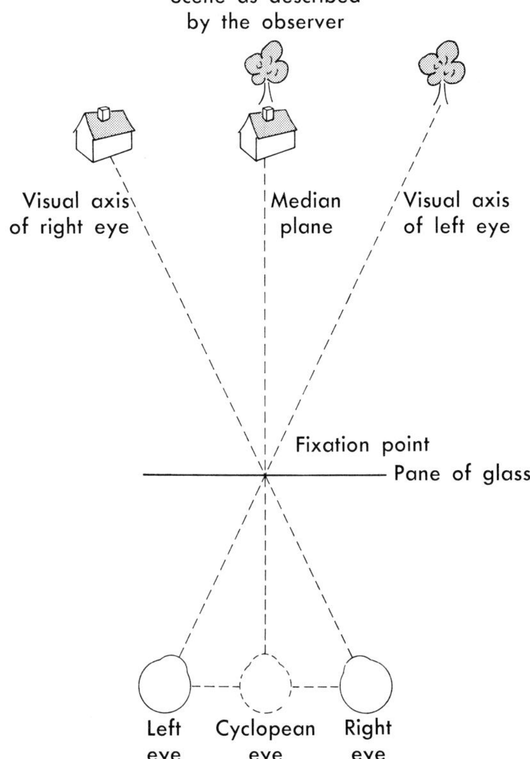

Fig. 21-4. Diagrammatic representation of Hering's law of identical binocular directions. (From Howard, I. P., and Templeton, W. B.: Human spatial orientation, London, 1966, John Wiley & Sons Ltd.)

Fig. 21-4 illustrates Hering's classic experiment in which he demonstrated that in binocular vision the two visual axes have the same visual direction (law of identical visual direction). He described the experiment as follows:

> Let the observer stand about ½ meter from a window which affords a view of outdoors, hold his head very steady, close the right eye, and direct the left to an object located somewhat to the right. Let us suppose it is a tree which is well set off from its surroundings. While fixing the tree with the left eye a black mark is made on the window pane at a spot in line with the tree. Now the left eye is closed and the right opened and directed at the spot on the window, and beyond that to some object in line with it, e.g., a chimney. Then with both eyes open and directed at the spot, this latter will appear to cover parts of the tree and the chimney. Both will be seen simultaneously, now the tree more distinctly, now the chimney, and sometimes both equally well according to which eye's image is victor in the conflict. One sees therefore, a spot on the pane, the tree and the chimney in the same direction.[69]

Hering generalized the law of identical visual directions to apply to all corresponding retinal points. For any given pair of corresponding visual lines there is a single visual line in visual space on which appears every object that actually lies on each of the two monocular lines. Since all the visual lines of both eyes are judged to point to one and the same projection center (egocenter), they can all be regarded as passing through the nodal point of the imaginary cyclopean eye.

Fig. 21-5 provides a dramatic illustration of the drastic restructuring of the frames of reference that occurs in the changeover from monocular to binocular vision. A card, marked on each side as shown, is placed in line with the bridge of the nose, and each eye fixates the superimposed point F. To monocular view each line (dashed and full) appears to have its true location in the median plane of the head. Lines

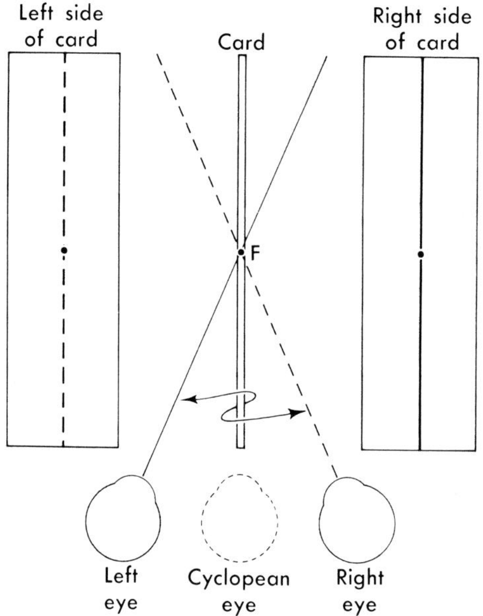

Fig. 21-5. Diagram to illustrate nature of restructuring of visual space that occurs in changeover from monocular to binocular vision.

that actually lie in the median plane of the head (as in Fig. 21-5) will, however, at every point except the fixation point, lie across noncorresponding visual lines of the two eyes. Our analysis indicates that they should therefore be seen as a cross whose arms intersect at the fixation point, and this is clearly seen to be the case. It is as though the oculocentric frame of reference of each eye had pivoted scissor-fashion about the fixation point, bringing the visual axis of each eye to the median plane of the head and thereby constituting a single visual axis for the cyclopean eye. Objects seen monocularly to be in the median plane will move over in binocular vision to appear to lie along the visual axis of the opposite eye.

As we shall see, the perceptual restructuring of visual space that occurs in the changeover from monocular to binocular vision is much more complex and subtle than is implied by this analysis. The analysis does, however, emphasize that an adequate neural model must make allowance for a considerable functional plasticity both in respect to the scope of the changes and the speed with which they occur.

Singleness of binocular vision

Horopter concept[132,136,171,190]

The singleness of binocular vision had been appreciated as a problem since very early times, but there could hardly have been any understanding of its basis before Kepler's first correct description of the retinal image in 1604.[104] Very soon afterward, in 1613, Aguilonius[1] introduced both the concept and the term "horopter" as the locus of points seen as single with the two eyes. Although the concept of the horopter (Greek *horos,* boundary) implied the notion of corresponding points on the two retinas, the idea of correspondence as an anatomic concept was first proposed by Newton in 1704.[129] His suggestion that a partial decussation at the optic chiasm would explain the singleness of binocular vision, however, remained without experimental foundation (clinical or anatomic) until the nineteenth century. The partial decussation was not finally established anatomically until the work of von Gudden in 1874.[182] It was these early notions and the later anatomic observations extending into the twentieth century that established the concept of corresponding points largely in anatomic terms and based it on the idea that each retinal point, together with its projection to the cerebral cortex, has one, and only one, visual direction, which is fixed and cannot be changed. As we shall see, however, these ideas are no longer acceptable.[127]

Vieth in 1818[180] first clearly specified the geometry of corresponding points, putting forward the idea that points seen as single in the horizontal plane of the fixation point (fixation plane) should lie on a circle that passes through the fixation point and the optical centers of the two eyes. In 1826, Müller[124] made Vieth's ideas widely known and the geometric construction, originally due to Vieth, subsequently became known as the Vieth-Müller circle (Fig. 21-6).

Thus the horopter, defined as the locus of points seen as single with the two eyes, led to the concept of corresponding points as those points in the two retinas that have the same *monocular* visual directions. If two retinal points have the same monocular visual direc-

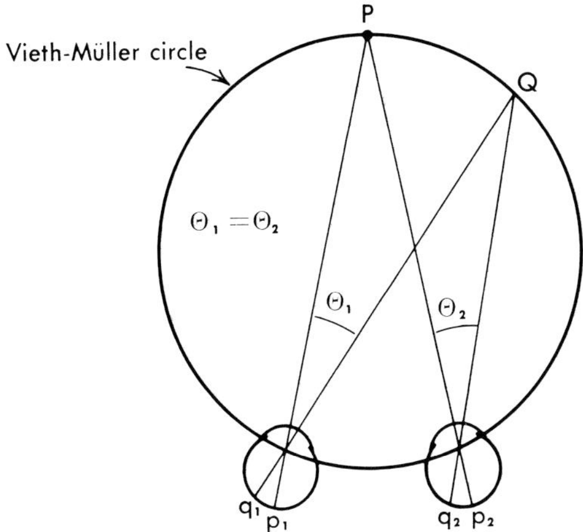

Fig. 21-6. Vieth-Müller's horopter circle. Fixation point, *P*, and point *Q* fall on geometrically corresponding points in the two retinas.

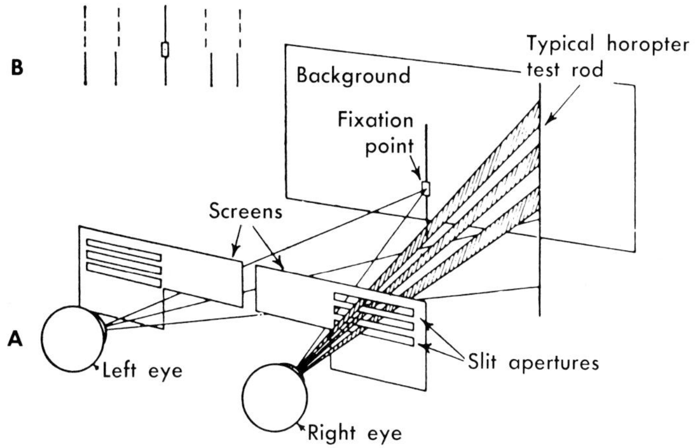

Fig. 21-7. A, Nonius method for determining longitudinal horopter by equating monocular visual directions. **B,** Observer's view of fixation point and test rods. (From Ogle, K. N.: Researches in binocular vision, Philadelphia, 1950, W. B. Saunders Co.)

tions, it seemed entirely reasonable to assume that in the binocular view a given object will be seen as single if the images fall on these two retinal points. The horopter was conceived as a surface in space, but as generally used both then and now the term refers to a line in the fixation plane. In a recent historical review, Shipley and Rawlings[171] have emphasized that, *based on the above definitions,* the nonius horopter must be regarded as the only valid horopter. The nonius (or vernier) horopter is so called after Nuñez, a sixteenth century Portuguese mathematician who invented a contrivance for the graduation of mathematic instruments. The Frenchman Vernier subsequently introduced an improved version of the Nuñez instrument.

The nonius method for determining the horopter requires the subject to fixate a target and observe a vertical rod to one side of the fixation target. The rod is masked by screens so designed that only the upper half of the rod is seen by one eye and the lower half by the other (Fig. 21-7, *A*). The screen has slit apertures (grid); thus the part of the rod seen through them appears as an interrupted line

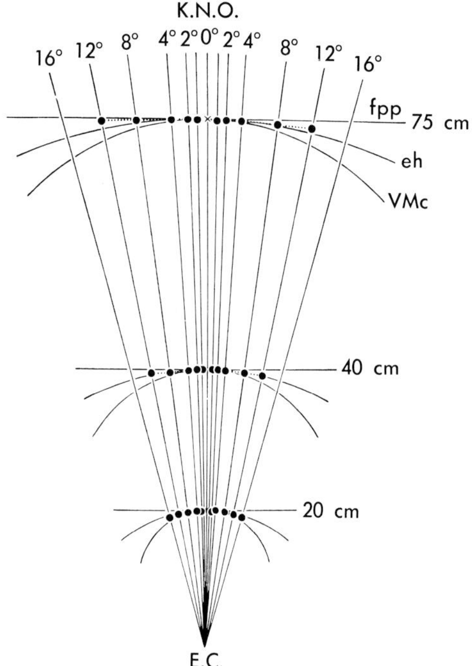

Fig. 21-8. Nonius horopter (filled circles and dotted lines) plotted at three different viewing distances, together with three geometrically constructed reference curves. Vieth-Müller's circle, *VMc;* equidistant horizon, *eh;* and true frontoparallel plane, *fpp*. (Redrawn by Shipley, T., and Rawlings, S. C.: Vision Res. **10:**1225, 1970; from Ames, A., Jr., Ogle, K. N., and Glidden, G. H.: J. Opt. Soc. Am. **22:**599, 1932.)

(Fig. 21-7, *B*) (grid nonius method). Both eyes see the fixation point, and a position of the rod is found for which the upper and lower halves appear aligned. For that position the rod is seen by both eyes in the same monocular visual directions. The horopter is plotted by moving the rod to a series of different positions, although the loss of visual acuity in the periphery limits the plot to eccentricities of about 12 degrees. Only horizontal retinal image disparities are used for stereoscopic depth discrimination; vertical disparities as such do not give rise to any perception of depth. This being the case, experimenters determining the horopter have usually used plumb lines or thin vertical rods as test objects, giving rise to the name "longitudinal horopter." Such a horopter plot therefore represents mean values over vertical sections of the retinas rather than retinal points.

Fig. 21-8 shows a nonius horopter plotted at three different viewing distances and, for comparison, three geometrically constructed reference curves: the Vieth-Müller circle (VMc), the equidistant horizon (eh), and the true frontoparallel plane (fpp). The horopter is concave toward the observer and lies between the Vieth-Müller circle and the frontoparallel plane. The lack of coincidence between the horopter and the Vieth-Müller circle is called the Hering-Hillebrand horopter deviation (Fig. 21-8).

There are, however, problems with the nonius criterion in its relation to binocular vision. Although the horopter concept was originally introduced as a basis for an understanding of the singleness of binocular vision, it largely lost this relevance with the recognition by Panum[144] in 1858 that singleness was not confined to a single surface but extended over a volume in space (p. 587). Since that time, the application of the horopter concept has shifted away from a special concern for binocular single vision toward more general problems of stereopsis. Nevertheless, the construction of the horopter has still continued to be based on the idea of corresponding points, and the nonius criterion has retained a prominent place in the literature. In line with the changing nature of the concept, however, the following new horopter criteria were introduced:

1. The apparent frontoparallel plane
2. The center of the region of binocular single vision
3. The locus of points about which stereoacuity is maximal
4. The region least excitatory to fusional movements

The frontoparallel plane criterion was introduced by Hering in 1861,[68] and since it is the easiest to use and gives the most consistent results, most horopter studies have been based on it. While fixating a central rod, the observer adjusts other eccentrically located rods until they appear to lie in the same frontal plane as the fixation rod. Among the difficulties associated with this criterion is the ambiguous relationship between the apparent frontal plane and the equidistant surface (Fig. 21-8). A detailed description and analysis of the various criteria

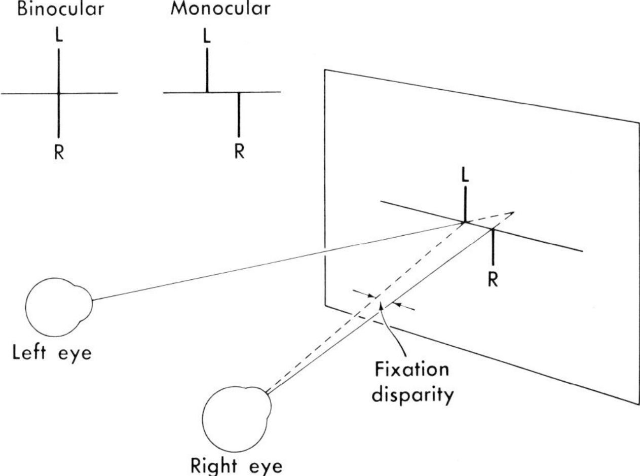

Fig. 21-9. Perspective drawing of use of nonius test lines to determine fixation disparity.

is, however, beyond the scope of this chapter.*

The variety of criteria that have been proposed as bases for the horopter testify to the fact that the horopter concept has yet to find a clearly defined place in the theory of binocular vision. Nevertheless, it remains an important concept. Both from a theoretical and a physiologic point of view the horopter is probably best defined as the locus of maximal stereoacuity. The *stereoacuity horopter* should be determined on the basis of relative depth judgments, that is, peripheral-peripheral comparisons of object points (p. 596). There may, however, be practical difficulties with such a definition, and no systematic observations are available in this regard.[19] In line with this definition the horopter can also be conceived of as the zero-disparity reference plane (core plane, nuclear plane) that contains the fixation point and relative to which stereoscopic depth estimates are made.

As far as the other criteria are concerned, the frontoparallel plane has neither a sound theoretical basis nor any clear relationship to the other proposals. The nonius criterion, although doubtless in accord with the original definition of the horopter, suffers the grave disability of being based on corresponding points that have fixed monocular visual directions. We have already observed that binocularly seen directional values differ from those seen monocularly, and therefore it is not surprising that the nonius horopter should differ from those based on essentially binocular criteria. Because it is devoid of normal binocular cooperation, the nonius criterion is open to the same kind of criticism that can be leveled at nearly all the clinical tests used for determining retinal correspondence. Clinical tests that disturb normal binocular vision are said to be dissociating. These matters will be discussed again later on.

*See particularly references 127, 132, 136, 171, and 190.

Fixation disparity

The phenomenon of fixation disparity[132,139] further illustrates the fact that binocularly seen directional values may differ from those seen monocularly. The perspective drawing in Fig. 21-9 illustrates the phenomenon and shows schematically how the disparity is estimated. Two nonius test lines, one underneath the other, can be viewed in one of two ways; either both lines may be seen by both eyes or each line may still be seen simultaneously, but only monocularly, the upper by the left eye (L) and the lower by the right eye (R). First the two lines are observed binocularly and adjusted to be in vertical alignment. Then the two test lines are observed monocularly, and if they no longer appear to be in vertical alignment, a fixation disparity is said to exist. The amount of the disparity is expressed as the angular separation of the two lines required to restore the apparent vertical alignment (Fig. 21-9). This is achieved by keeping one test line constant and moving the other (here the right).

In this example the two test lines were located at the fixation point, and to monocular viewing the effective fixation point was displaced distal to the binocularly seen fixation point (negative or exofixation disparity). Since a binocularly seen fixation point is used in the determination of the nonius horopter, the presence of a fixation disparity will necessarily displace the nonius horopter as a whole proximal or distal to the fixation point.

The relative importance of motor and sensory factors in the generation of fixation disparity is far from clear. In the binocular and strabismic literature the tendency has always been to stress the motor factors. The objective measurements by Hebbard[67] have shown that in the change from binocular to monocular viewing there may be a change in convergence equal to the fixation disparity. Ogle[132] and co-workers[139] regard fixation disparity as primarily an oculomotor muscle imbalance leading to a small error in convergence. They studied the change in fixation disparity with forced vergences using prisms. It seems more likely, however, that the change in convergence is an adaptation to a different (binocular) set of visual directions rather than primarily an oculomotor imbalance leading to a retinal disparity that the stereopsis system must absorb. Fixation disparity is a variant of normal binocular vision. The disparities are, in fact, very small, being usually in the range of −5 to +3 minutes of arc.[132] In any case, they cannot be greater than Panum's fusional area if diplopia is not to result. A fixation disparity is to be found in the majority of subjects who are heterophoric, the direction of the disparity usually being the same as that of the phoria. The magnitude of the disparity cannot be predicted, however, from the magnitude of the phoria, and the direction of the disparity may sometimes differ from that of the phoria. The term heterophoria (or phoria) refers to the tendency of the eyes under dissociating (monocular) conditions to deviate from the binocular position maintained by the fusion reflex. Thus a phoria is regarded as a latent deviation of the eyes that is compensated for by the fusion reflex.

Panum's fusional area

By his invention of the stereoscope in 1838, Wheatstone[193] was the first to recognize that

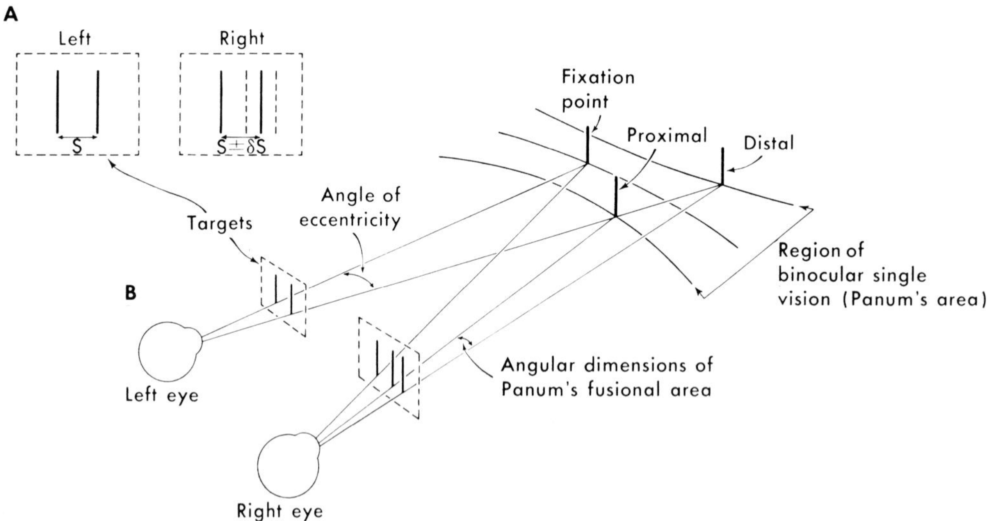

Fig. 21-10. Perspective drawing, **B**, showing use of vertical rods to determine proximal and distal limits of Panum's fusional area and equivalent thin line stereotargets, **A.** Fixation is maintained on left line of fused images. Angular separation ($\pm\delta S$) of vertical dashed lines in **A** corresponds to horizontal extent of Panum's area.

disparate retinal images provide the essential cue for binocular depth perception. He also recognized that his observations, by their implication that stimulation of disparate or noncorresponding retinal points could still give rise to single vision, contradicted the theory of corresponding points. Before Wheatstone's observations it had been widely held that only objects whose images fall on corresponding points would be seen as single and that objects proximal and distal to the horopter would therefore be seen double. In 1858, Panum[144] resolved this dilemma by using stereotargets similar to those in Fig. 21-10, *A*. He found that two pairs of vertical lines could be fused into a single pair provided that the difference in the separation of the lines in each target did not exceed a certain value ($\pm\delta S$). If the difference was greater than δS, fusion no longer occurred and three lines were seen. He proposed that for any point on one retina there is a small *circle* or *area* of points on the other retina, stimulation of which will lead to fusion of the two monocular inputs. Thus binocular single vision is not limited to the immediate vicinity of the horopter, but rather extends for a small distance proximal and distal to it, the region for binocular single vision having since come to be known as Panum's fusional area. Beyond the limits of this fusional area objects appear double (physiologic diplopia). Although we are not ordinarily aware of the double vision, it is easy to demonstrate its existence. Hold a finger up before the eyes but fixate a point or object some little distance farther away, and the finger will be seen as double. To get this effect it is essential to give the major share of one's visual attention to the point some distance beyond the finger and only a minor share to the finger. By closing first one eye and then the other, it is easy in this example to show that the left finger of the pair is being seen by the right eye and the right finger by the left eye (crossed disparity).

Measurements of the size of Panum's fusional area have usually been made either by the use of stereotargets as mentioned previously or by the use of plumb lines or thin vertical rods in real space. The relationship between these two methods is shown schematically in Fig. 21-10, *B*. While fixating one rod, the observer moves a second rod proximal and distal along a given visual line for the left eye. This second rod can be moved only over a short distance in either direction before it will be seen double. The transverse (horizontal) extent of Panum's fusional area for that particular angle of eccen-

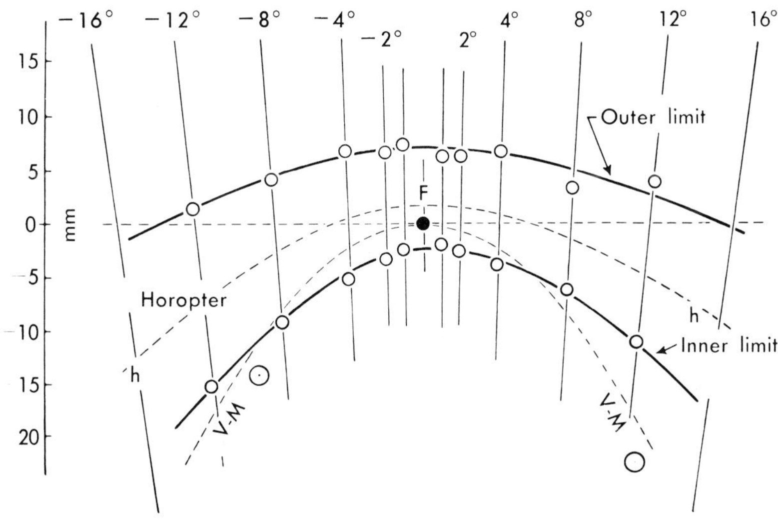

Fig. 21-11. Region of binocular single vision as determined experimentally for one subject. Vertical scale indicates horizontal distances with respect to fixation point, *F* (filled circle), negative values being toward observer and positive values away from observer. Scaling in degrees refers to angles that various projection lines make with respect to straight-ahead line through fixation point. (From Ogle, K. N.: Researches in binocular vision, Philadelphia, 1950, W. B. Saunders Co.)

tricity is expressed as the visual angle subtended at the right eye by the limiting positions at which the rod appears double. The full extent of the fusional area is determined by repeating the procedure at different eccentricities, although the quality of peripheral vision limits estimates to eccentricities within about 10 degrees.

Fig. 21-11 shows the region of binocular single vision as determined experimentally for one subject at an observation distance of 40 cm; the longitudinal horopter curve (h . . . h) near the center of the region of binocular single vision has been drawn as the locus of the intersection of any given visual line for one eye and the line bisecting the angle subtended at the other eye by the inner and outer limits of Panum's area taken along that visual line. For this criterion the horopter does not pass through the fixation point, F. The backward displacement of the whole of Panum's area in relation to the fixation point is to be attributed to an exofixation disparity (p. 586) in this subject.

Limits of Panum's fusional area

Most of the estimates of the limits of Panum's area[117] have been complicated by the fact that increasing retinal disparities stimulate vergence movements in the interests of maintaining fusion. Recent studies by Mitchell[118] and Palmer[143] have avoided the influence of these fusional movements by using flash presentations of the test target having durations of exposure (120 milliseconds) less than the latency for vergence. These authors report that at the fixation point the extent of Panum's area is about 15 minutes of arc between the inner and outer limits (about 7 minutes of arc in front of and behind the fixation point). This is well within the bounds of the foveola (Fig. 21-2). For 11 subjects Mitchell found a range of 13.5 to 23 minutes of arc. These results are similar to those that can be inferred from studies of fixation disparity. The maximum amount of fixation disparity that can be measured before diplopia occurs in response to forced vergence movements of the eyes using prisms gives a measure of the size of disparity threshold for diplopia and therefore the size of Panum's area. In contrast to most previous investigations these authors also found relatively little interindividual or intraindividual variation in Panum's areas and little evidence of modification by training. Furthermore, they found that the extent of the fusional area was the same for vertical as for horizontal disparities.

The extent of Panum's fusional area increases with increasing eccentricity away from the fixation point. Most authors, however, report a much more rapid increase in the size of the area than that given by Ogle[133] in Fig. 21-16. Thus Mitchell[118] found that the size of the area had increased to 1 degree at an eccentricity of 5 degrees. Further work is needed to establish the normal limits of binocular single vision in the periphery, particularly in view of the report by Bagolini[5] and Bagolini and Ravalico[6] of a considerable increase in this area in small-angle strabismus (pp. 601 to 602).

There are other recent observations indicating that, even in the center, Panum's fusional area may have dimensions larger than those given by earlier accounts and to have a degree of plasticity not hitherto accepted. Thus Fender and Julesz[55] have again drawn attention to the fact that in binocular fixation the motions of the two visual axes are not perfectly correlated; thus they wander with respect to one another over a range of disparities the same as, or even larger than, the extent of Panum's fusional area as classically described (p. 607). These errors of binocular fixation are, however, not associated with diplopia even during prolonged viewing. It is also important to appreciate that when flash presentations of the test target are used to determine the limits of Panum's area, these errors of binocular fixation are still effective even if the influence of fusional vergences is avoided. Thus Panum's area must be large enough to absorb these errors of fixation.

Studies on the extent of Panum's area have invariably been carried out with fusional stimuli (thin rods, threads) that are probably only minimally effective. However, using binocularly stabilized retinal images, Fender and Julesz[55] have recently shown that the dimensions of Panum's area depend greatly on the class of stimulus, increasing as the quantity of disparity cues provided by the stimulus increases. Binocularly stabilized random-dot stereoscopic images (p. 604), which provide a wealth of disparity cues, can be pulled apart symmetrically on the two retinas by up to about 2 degrees in the horizontal direction without loss of stereopsis or fusion. Thus fusion is still possible under these conditions with the equivalent of a divergent image disparity of 2 degrees, the midpoint of each image being located 1 degree from the center of its respective foveola. In addition, Fender and Julesz observed that the disparity limit at the breakpoint for diplopia was always greater than that for the recovery of fusion, a hysteresis effect that also increased with the quantity of disparity cues in the stereotargets. Confirmation and further development of these studies are needed. It is not clear at this stage what changes in the classic accounts are needed to incorporate this new knowledge.

Binocular visual direction in Panum's area

We have already described the global restructuring of visual space that occurs in the changeover from monocular to binocular vision, oculocentric visual directions giving place to egocentric directions. In addition, there are rather more subtle local changes in visual direction within Panum's fusional area. Provided the horizontal disparities between the images on the two retinas are not too great (i.e., within Panum's area), instead of two images, a single image is seen either in front of or behind the reference plane. The single binocular image seems to have a visual direction which is not that of either monocular image but appears, as a compromise, in the mean visual direction between them. von Tschermak-Seysenegg[190] has applied the term *allelotropia* to this fusional displacement of the two monocular visual directions. It is a phenomenon that applies not only in the vicinity of the fixation point but also throughout the whole of Panum's area both with respect to distance in depth and eccentricity from the center.

Fig. 21-12 shows a pair of simple line stereograms, in each of which a small square is asymmetrically located within a larger one. When viewed stereoscopically, the small squares fuse and go into depth. It now appears as a single square located centrally within a larger square and deep to it in space. Thus the apparent positions (visual directions) of the two small monocularly seen squares change when they fuse and go into depth. It is worth noting that the small inner square has a *relative* depth value with respect to the reference plane provided by the larger outer square and an *absolute* depth value with respect to the head.

Just as two different monocular visual directions can fuse binocularly into one, so too, one can imagine, it might be possible for two contours with the same monocular directions to be made to appear to have different binocular visual directions. Linschoten[112] has, in fact, presented a number of stereograms that demonstrate such a shift in binocular visual direction from identical monocular directions in the two eyes.

Binocular ("retinal") rivalry

Binocular rivalry[69,70,109] (sometimes incorrectly called retinal rivalry) occurs when dissimilar pattern elements are presented to the two eyes at the same time. There may be suppression of one element by the other, so only

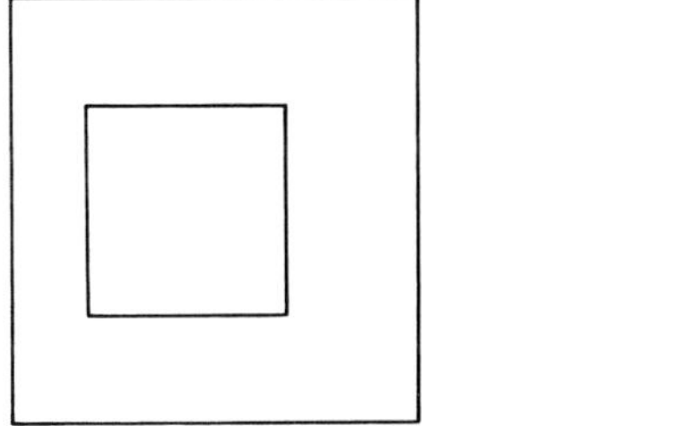

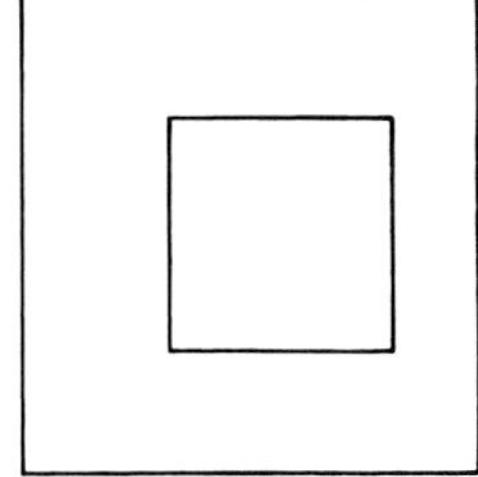

Fig. 21-12. Stereogram to illustrate displacement of monocular visual directions under binocular fusion. After fusion, two small inner squares become centrally located in depth within larger outer square.

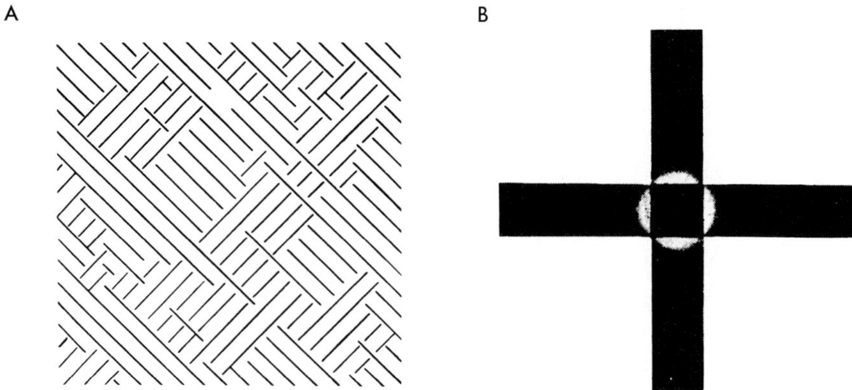

Fig. 21-13. Binocular appearance of rivalrous stereotargets. **A,** Rivalry and suppression associated with thin lines. **B,** Suppressive effect of contours. (After Panum.)

one is seen, or else an irregularly alternating suppression of first one and then the other. Binocular rivalry is not evident under normal viewing conditions because of the similarity of the two retinal images. The properties of the suppression that occurs in binocular rivalry, however, strongly suggests that it is of importance in normal fusion and stereopsis, but a clear understanding of its role in this respect has yet to emerge. Relatively little is known about what determines whether only local features in the retinal image will be suppressed, whether these local features are suppressed at random, or whether a large part of the input to an eye will be suppressed as a whole. There are, however, many striking parallels between the psychophysics of binocular rivalry and recent neurophysiologic observations concerning possible stereoscopic mechanisms.

Two classic stereograms, originally due to Panum, will be briefly described. The first (Fig. 21-13, *A*) illustrates the powerful effects of rivalry and suppression that are associated with thin lines; two targets of equal size consist of parallel diagonal lines so arranged as to be in opposite directions for the two eyes. The binocular impression is that of a continually changing mosaic, with some areas having lines running in one direction and other areas where the direction is opposite. The areas seem to vary randomly, with now one diagonal, now the other, in any given area. At no time do the diagonals seem to cross one another. The second example (Fig. 21-13, *B*) illustrates the effect of contours. A vertical black bar is presented to one eye, and a horizontal black bar to the other. The binocular appearance is of a small, central black square surrounded by a white halo spreading out from the overlapping edges of the bars.

Suppression tends to occur whenever the lines presented to the two eyes differ sufficiently in orientation, length, or thickness, and dots and thin lines are more effective than thicker lines.[72] Moving contours are particularly effective.[62] Provided their orientations are different, a sharp contour in one eye will suppress weaker contours in the other, and the suppression will tend to remain constant throughout the observation period. In contrast to the behavior of thin lines and strong contours, when two uniform areas of different brightness are viewed binocularly, they tend toward an average brightness, provided the areas are in register (binocular brightness averaging[109]). Thus, whereas lines and edges rival, areas of uniform brightness tend to average.

Rivalry in line patterns occurs at the intersection of the lines, and suppression spreads out to include line segments to either side of the dominant line. Thus each line or contour viewed with one eye has associated with it a band of suppression in respect to patterns in the other eye. Kaufman[102] has measured the spread of this suppression by presenting a horizontal line to one eye and two vertical lines to the other. The suppression of the horizontal line segment between the two vertical lines in the binocular view was studied as a function of the separation of the vertical lines. The spread of the suppression associated with *each* vertical line is then half the distance between the two verticals. The suppression remained at a maximal level up to about 7 minutes of arc from the point of overlap in the horizontal direction and thereafter decreased rapidly to reach a minimum value at about 1 degree from the intersection. The full extent of the region of suppression to either side of the intersection is given by doubling these values. Suppression of a similar magnitude also occurs in the vertical direction. It is interesting that although the suppression associated with a line spreads well beyond the limits of Panum's fusional area, it is nevertheless at a maximal value over a region comparable in size to Panum's area.

Theories of fusion

Apart from the recent concepts that have come from neurophysiologic observations (p. 615), the various theories that have been proposed to account for the singleness of binocular vision all have a fairly long history. At one time the idea was widely held that each eye is alternately suppressed so that we see with only one eye at a time. More recently the idea of global suppression of an eye has been replaced by the concept of alternating local suppressions of the retinal images so that only portions of these images are perceived alternately by one eye and then the other.[3,179] Hochberg[72] also assumed a cyclopean field constructed piecemeal out of the contents of one eye or the other. Le Grand,[105] on the other hand, has argued that singleness is due to a failure in the detection of the diplopia because it lies beyond the limits of binocular visual acuity. Again, theories have been advanced that give eye movements a fundamental role in stereopsis. Thus it has been suggested that a depth plane may be defined by the set of vergence changes needed to bring a given set of image disparities to zero. There are, however, a variety of observations that invalidate eye movement theories of binocular vision. Fusion and stereopsis occur despite, and not because of, eye movements. Thus normal binocular single vision can be obtained with binocular stabilization of the retinal images.[55] None of the preceding theories of fusion are at all satisfactory.

Corresponding points and fusion: a recapitulation

We have seen that the singleness of binocular vision turned out to be a much more complex problem than was envisaged by both the original concept of the horopter and the associated idea of anatomically fixed corresponding points on the two retinas. The horopter can, however, still retain its usefulness as a concept if it is considered not so much as a basis for an understanding of the singleness of binocular vision, but rather as a reference plane for binocular depth discriminations (stereopsis). Various phenomena such as Panum's fusional area, fixation disparity, allelotropia (fusional displacement), and the constantly varying binocular disparities due to inequalities in the eye motions (drifts and saccades, p. 606) clearly invalidate the concept of fixed corresponding points, at least as far as binocular visual direction and depth values are concerned. If we are to retain the definition of corresponding points as being those retinal points which have the same monocular visual direction, a new definition or term is needed for the analogous points in binocular vision, a definition or term that allows sufficient plasticity to take account of the fact that points or contours which are corresponding in one circumstance may not be so in another.

Stereoscopic depth perception

Stereopsis and fusion: some general considerations

Horizontal retinal image disparities give rise to stereopsis; vertical disparities do not. The features in the two retinal images essential for stereopsis are local contours in which the direction of the brightness change across each matching contour pair must be the same in both eyes. Provided the contours are visible, stereoscopic threshold depends only on the level of the retinal adaptation to light and is virtually independent of the ratio of the contrast across the contour. Although only horizontal disparities are used for stereopsis, vertical disparities are fused, as well as those in the horizontal direction. However, stereoscopic depth can still be experienced over a relatively limited range even if the retinal images are not fused, and this is the case whether the doubling is vertical or horizontal.

By observing stereoscopic depth during the flash of an electric spark, Dove, in 1841,[45] demonstrated that eye movements are not essential for stereopsis, and virtually the same experiment was performed by Wheatstone in 1838,[193] when he reported stereopsis with afterimages. The latter are, of course, fixed retinal images. Recently, Fender and Julesz[55] have confirmed Wheatstone's observation by using experimentally stabilized retinal images. Nevertheless, vergence eye movements are needed to bring the stereo images within the critical limits of disparity, although it is doubtful whether convergence per se makes any significant contribution to depth perception. Retinal image disparity is the essential stimulus for fusional movements, and neither diplopia nor any stereoscopic effect resulting from disparity is necessary. Thus fusional movements occur without a break in single vision when a very weak prism is placed before one eye while an individual binocularly fixates a given object.

Stereoscopic acuity

Under any given conditions, stereoscopic acuity is the inverse of the depth threshold, the latter being measured as the least difference in sagittal distance that can be discriminated between two objects in space. The lower the threshold, the higher the acuity. Consider two objects, points A and B (Fig. 21-14), situated close to the median plane and located at slightly different distances in space. The line LR joining the nodal points of the two eyes subtends angle α at point A and angle β at point B. The *binocular disparity* or *stereoscopic parallax* of the two points is measured by the difference between the angles α and β and hence $\Theta_1 - \Theta_2$. The stereoscopic acuity corresponds to the threshold of detection of the binocular disparity, $\Theta_1 - \Theta_2$, and is usually expressed in seconds of arc. Like other psychophysical thresholds, there is, however, a range of uncertainty over which there is only a probability of detection; therefore the threshold must be assessed on a statistical basis. Two different psychophysical methods[52] are commonly used for the experimental determination of the threshold. With the *method of constant stimuli* the frequency of correct responses to a series of equally spaced disparities between the images in the two eyes is used to construct a frequency of detection curve. The threshold is then expressed either as the standard deviation of the normal distribution curve fitted to the data or, more commonly, as the value corresponding to the criterion of 75% correct answers. Alternatively, by the *method of average error or adjustment* the subject can adjust the test object so as to appear the same distance in depth as a standard or reference object. The stereoscopic threshold is then expressed as the mean deviation (or, alternatively, the standard deviation) from the mean value of the settings. In comparing the reports of different investigators, allowance must therefore be made not only for differing experimental methods but also for the varying ways of expressing the results.

From Howard[76] (1919) onward, investigators

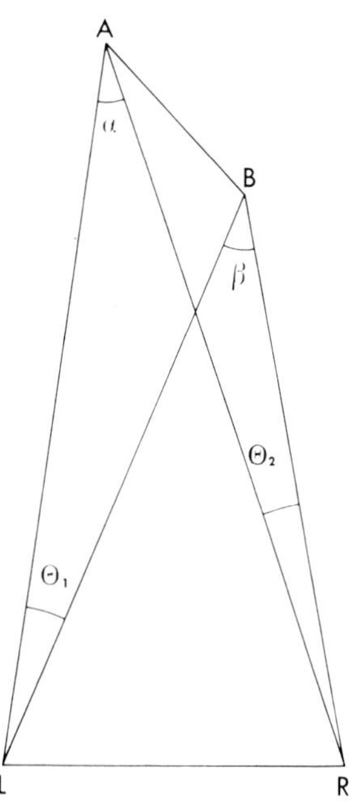

Fig. 21-14. Definition of stereoacuity. Line *LR* joining nodal points of two eyes subtends angles α and β at object points *A* and *B*. Stereoacuity corresponds to threshold detection of binocular disparity $\beta - \alpha$ or $\theta_1 - \theta_2$.

have reported stereoscopic thresholds down to about 2 seconds of arc, although, depending on the type of test targets used and the experimental conditions, values up to 40 seconds of arc may be considered as being in the normal range. Vertical rods make possible a much lower stereoscopic threshold than do object points. Not surprisingly, there is a marked correlation between stereoscopic acuity and monocular visual acuity,[11] although some subjects with good monocular vision may have very poor binocular vision. Stereoscopic acuity and vernier acuity have about the same threshold values, and the various factors such as luminance level and contrast affect the two thresholds in very much the same way. Thus the curve relating stereoscopic threshold to retinal illumination has the classic break between photopic and scotopic vision. In addition, vernier acuity and stereoscopic acuity decrease in a similar manner with retinal eccentricity (p. 596). Hence it is very likely that the two visual abilities involve the same mechanisms. Berry[11] compared the threshold acuities for vernier and stereoscopic tasks under similar conditions. Using two vertical rods, one above the other in the median plane, Berry showed that for small rod separations, vernier acuity is better. However, as the vertical separation of the rods is increased beyond 134 seconds of arc, stereoscopic acuity surpasses vernier acuity.

At any given level of background (adapting) luminance both stereoscopic acuity and the depth assessment remain fairly constant, provided the object details can still be seen by both eyes. The quality of stereopsis is proof against considerable impairment in the vision of one eye. Darkening one eye with a filter is without much effect until the acuity of the covered eye has been reduced to 0.3, and depth perception only ceases when the acuity in that eye falls to 0.1.[116] Even if one retinal image is considerably blurred, either artificially or as a result of poor accommodation, the binocular percept is not only still seen in depth but also retains all the detail of the sharper image.

Tests for stereopsis

Basically there are two kinds of tests used for the assessment of stereoacuity: real depth instruments using three-dimensional test objects and haploscopic devices by which separate two-dimensional targets can be presented to each of the two eyes. The two most common real depth instruments, the Verhoeff stereopter and the Howard-Dolman apparatus, although rarely used in clinical practice, are occasionally employed for screening tests for specific occupations such as aviators in the armed services. In these two tests the subject is required to judge the relative depths of vertical wires or rods.

Of the many haploscopic devices, the most widely used tests employ Polaroid vectograph cards or projector slides.[146,147] The vectographs produce the image disparity, the cards or slides being so imprinted that each target is polarized at 90 degrees with respect to the other. Properly oriented Polaroid spectacles are then worn that present each target separately to each eye. The distant vectograph slides (A-O Project-O-Chart System), projected onto a screen at 6 m, only provide for gradations of stereothreshold over the range of 240 to 30 seconds of arc. The vectograph cards for near tests (Titmus; Wirt) are viewed at 40 cm. The Titmus stereotest has a range of thresholds from 1000 to 40 seconds of arc, whereas the Wirt stereotest provides a rather wider range of 3000 to 14 seconds of arc. Subjects with normal bifoveolar binocular vision have stereoacuity thresholds in the range of 14 to 40 seconds of arc, with an average of 24 seconds of arc.[146,164] According to Parks,[146] stereoacuity tests provide a clear separation between subjects with normal binocular vision and those with microstrabismus. The latter have stereothresholds greater than 70 seconds of arc. Because it has a threshold range down to 14 seconds of arc, the Wirt near test is the most reliable of the various stereoacuity tests in use at present. It provides the most distinct separation of the microstrabismic patient from the normal. As Burian[27] has pointed out, stereothresholds should always be expressed in seconds of arc of retinal disparity, and the use of artificial assessment scales (e.g., the Sheppard scale) and percentage estimates of normal should be discontinued.

Although no detailed systematic knowledge is available concerning the incidence of strabismus in the community, a figure of about 3% is commonly quoted. Stereoblindness would doubtless have an incidence at least comparable to that of strabismus and that of stereopsis deficiency may be as high as 15%.[99] In clinical practice relatively little attention has so far been given to the assessment of stereoacuity. There is a clear need for the development of more satisfactory quantitative tests both for routine clinical use and for the determination of the distribution within the community of the various levels of stereoacuity. It is not known, for example, how common it is for normal visual acuity to be associated with stereopsis deficiency or stereoblindness. However, our expanding knowledge of strabismus, particularly of very small–angle strabismus, has led to a more general appreciation that the measurement of stereoacuity provides a sensitive test for grading the quality of binocular cooperation and that it is one of the most effective methods of screening for microstrabismus.* Julesz[99] has recently developed a series of random-dot stereograms (p. 604) as an objective test for determining stereopsis deficiency; stereograms of this kind will doubtless find an important place, not only as a routine test in clinical practice, but also as a tool in strabismic research.

Spatial range of stereopsis

When considering the spatial range over which stereopsis operates for a given fixation point, it is important to make a clear distinction between the actual depth discrimination of one object point being farther or nearer than another and the spatial location or eccentricity at which the discrimination is made, either to left or right, above or below, or closer or farther than the fixation point (p. 596). Fig. 21-15 shows in diagrammatic form that, like Panum's fusional area, the range over which stereopsis operates forms a curved layer or zone in visual space that is roughly parallel to a frontal plane.

It has been known since the middle of the last century[69,183] that the stereoscopic perception of depth is still possible even though the retinal images are sufficiently disparate to be seen as double (Fig. 21-15). Ogle[133] was, however, the first to make a systematic investigation of the limiting disparities within which such stereopsis exists.

Moving one rod in depth away from the fixation point, he described two levels of stereoscopic perception (Fig. 21-16). The first level, referred to as *patent stereopsis,* operates over a relatively small range of disparities. The depth experience is evident and compelling, the increase in depth being clearly proportional to the increase in disparity. Since the patent stereopsis of Ogle's description extends for about 10 minutes of arc disparity at the fixation point, it is operative for a small disparity range after the appearance of diplopia. With further increase in disparity the strong sense of depth fades and the second level called *qualitative stereopsis* is reached. At this level the target can, however, still be reliably located as farther or nearer than the fixation point, but the disparity-depth relationship is lost. Beyond the range of qualitative stereopsis the sense of depth finally fades altogether and the double images are only indefinitely localized. For patent stereopsis Ogle reported disparities that ranged from about 10 minutes of arc at the fixation point to 70 minutes of arc at a peripheral angle of 6 degrees. The qualitative sense of depth of the double images ceases at disparities of about 15 minutes of arc at the fixation point but up to nearly 2 degrees at a peripheral angle of 6 degrees (Fig. 21-16).

As a result of recent work mainly concerned with the midline of the visual field, Ogle's account has been refined and his concepts of patent and qualitative stereopsis replaced by the allied notions of fine and coarse stereopsis, respectively. For the moment we are concerned only with the range over which stereopsis operates; the question of stereoacuity within that range will be considered later. Ogle's[133] technique of continuous exposure of the test object may have been subject to eye-movement artifact and possibly adaptive processes. Later workers have avoided these problems by using short exposures of a target presented in random sequence. Using a vertical line stereoscopically viewed by each eye, Richards[162] studied perceived depth as a function of binocular disparity. The initial region in Fig. 21-17, where perceived depth increases monotonically

*References 27, 53, 114, 146, 163, and 176.

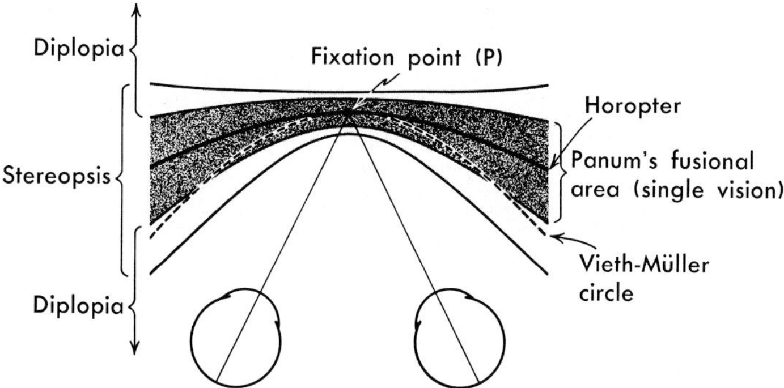

Fig. 21-15. Diagram showing relative positions of empirical horopter and Vieth-Müller's circle, overlapping areas for stereopsis and for binocular single vision and regions where diplopia begins.

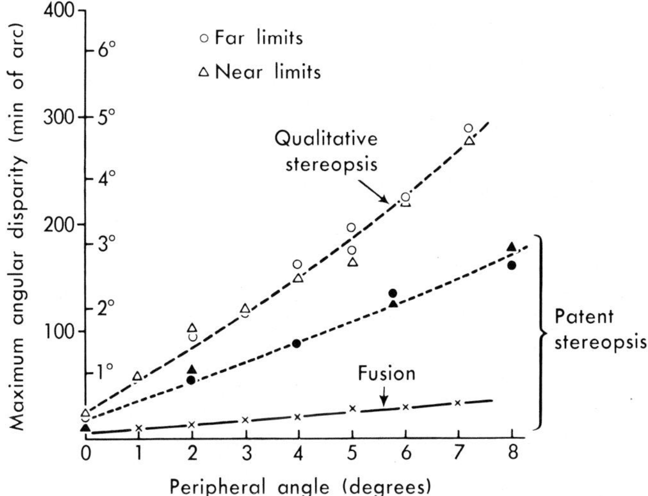

Fig. 21-16. Disparity limits for fusion and for patent and qualitative stereopsis plotted against retinal eccentricity. (From Ogle, K. N.: In Davson, H., editor: The eye, New York, 1962, Academic Press, Inc., vol. 4, pp. 271-324.)

with increasing disparity, is presumably equivalent to Ogle's patent stereopsis, and the later region where perceived depth decreases until it disappears is to be equated with qualitative stereopsis. Within the initial region at smaller disparities the filled circles in Fig. 21-17 indicate disparity values that yield a single fused percept of a vertical line in depth. This region of filled circles, equivalent to Panum's fusional area, will be called the region of fine stereopsis. Outside this region the open circles indicate disparities at which the line appears double. This is called the region of coarse stereopsis. Although they appear to have a somewhat similar connotation, fine and coarse stereopsis must not be confused with the quite distinct concepts of local and global stereopsis[100] (p. 605).

In the midline of the visual field, Westheimer and Tanzman[192] showed that the upper limit for reliable qualitative localization of a target (coarse stereopsis) as nearer or farther than the fixation point is about 6 degrees of convergent disparity and about 10 degrees of divergent disparity, and even larger disparities can be recognized in the periphery of the visual field. Furthermore, Mitchell[119,120] has shown that similarity of the two retinal images is not a neces-

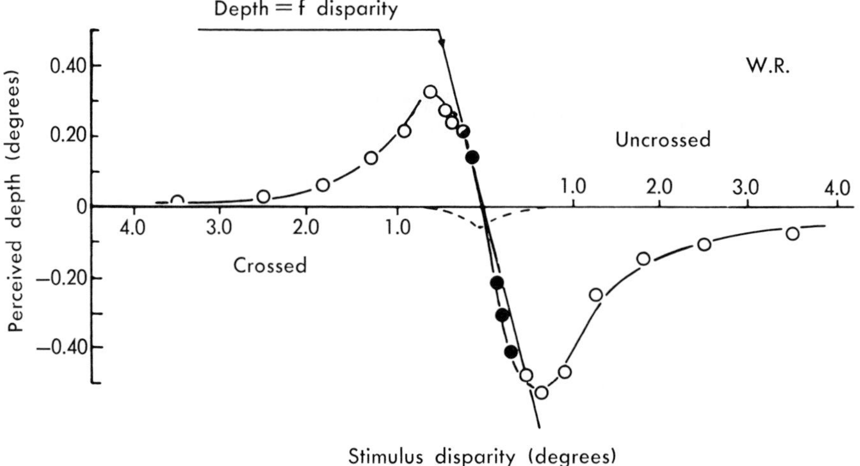

Fig. 21-17. Perceived depth elicited by a bar flashed with different crossed and uncrossed disparities relative to the fixation point. Filled circles indicate disparity values that yield a single fused percept in depth. Open circles show which stimuli appeared double. The dashed line shows the perceived depth of the monocular control stimuli. (From Richards, W.: J. Opt. Soc. Am. **61:**410, 1971.)

sary requirement for qualitative localization in depth, at least for large disparities, and for targets presented briefly in relatively empty fields. Thus coarse stereoscopic depth perception is possible with diplopic images of quite dissimilar shape.

Variation of stereoacuity with spatial location

Both stereoacuity and visual acuity decrease in a similar manner with increasing retinal eccentricity. In Fig. 21-18, Le Grand[105] has gathered together the data from ten different investigations to show the variation in relative visual acuity (Vr) with eccentricity in photopic vision. Using logarithmic coordinates for both variables, he obtained two straight lines intersecting at an eccentricity of about 5 degrees. Ogle[132] obtained a rather similar result with stereoacuity. He examined the threshold disparity necessary to give rise to a sense of depth between two test point sources of light located at equal distances to either side of a fixation light (Fig. 21-19). The lateral separation of the two outer targets could be changed as desired, but they were always kept equally separated from the center light. Two important points emerge from Fig. 21-19. First, up to about 5 degrees from the fixation point there is only a fairly gradual increase in stereothreshold as the two targets are each moved laterally away from the fixation point into the peripheral visual field. In keeping with visual acuity (Fig. 21-18), however, there is, beyond the 5-degree mark, a much more rapid increase in stereothreshold. The second point is the remarkable fact that a relatively low threshold obtains even though depth comparisons are being made between targets whose images are separated by as much as 10 degrees on the retina. Thus accurate depth judgments can be made over relatively large spatial separations.

Both Ogle[134] and Rawlings and Shipley[160] studied relative disparity judgments as a function of the absolute disparity of the targets for various peripheral angles. In both cases stereoacuity was examined in terms of peripheral-peripheral comparisons at different eccentricities. However, in considering the variation of stereoacuity with spatial position, Blakemore[19] seems to have been the first to recognize the importance of differentiating between absolute and relative image disparities. Practical considerations doubtless led Blakemore to use the Vieth-Müller circle as a reference, but the most appropriate reference surface is clearly the horopter of maximal stereoacuity. Object points, including the fixation point, that lie in

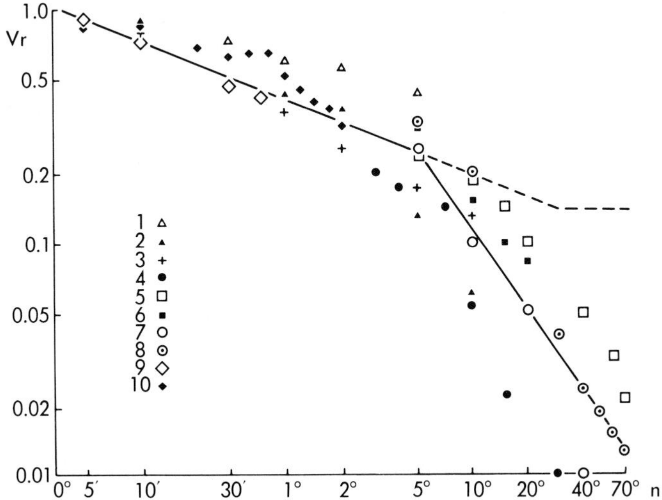

Fig. 21-18. Variation of relative visual acuity Vr with eccentricity in photopic vision. The two scales are logarithmic, but by convention, abscissa begins at $n = 0$. Replotted from Le Grand, who gathered the data from ten different investigations. (Reprinted from Form and space vision by Yves Le Grand, translated by Michel Millodot and Gordon G. Heath. Copyright © 1967 by Indiana University Press, Bloomington, Ind. Reprinted by permission of the publisher.)

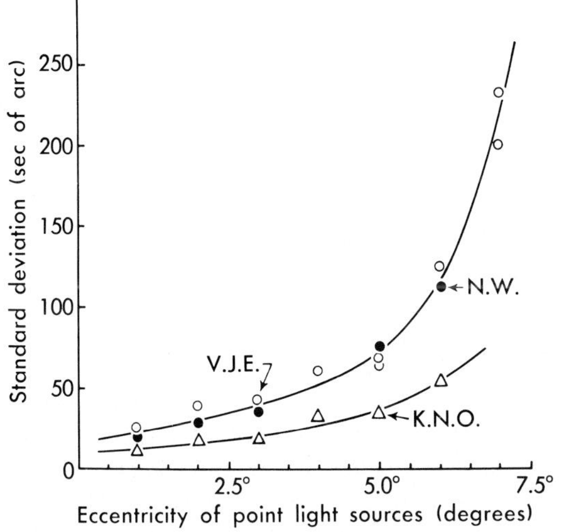

Fig. 21-19. Stereoscopic thresholds for two horizontally separated point light sources plotted against retinal eccentricity. The lights were always located at equal distances on either side of fixation light. (Replotted and abscissa rescaled from Ogle, K. N.: Researches in binocular vision, Philadelphia, 1950, W. B. Saunders Co.)

that horopter surface can be said to have zero *absolute disparity*. Objects closer than the horopter have convergent absolute disparity and those farther away, divergent absolute disparity. The difference in absolute disparity of two objects is called the *relative disparity* between them; one object can be convergent or divergent relative to another. At any eccentricity from a given fixation point and at any absolute disparity at that eccentricity the stereoacuity can be measured as a relative depth discrimination between targets placed at the location in question. In keeping with earlier studies, Blakemore[19] showed (Fig. 21-20) that

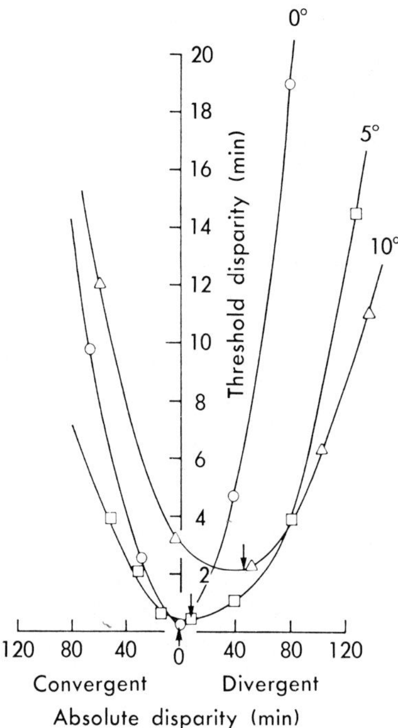

Fig. 21-20. Mean convergent-divergent stereothreshold plotted against absolute disparity for one subject. Circles, squares, and triangles represent data for peripheral angles of 0, 5, and 10 degrees, respectively. Zero on abscissa lies on Vieth-Müller's circle. Small arrows indicate disparity of stereoacuity horopter relative to Vieth-Müller's circle. (From Blakemore, C.: J. Physiol. [Lond] **211:**599, 1970.)

there is only a gradual increase in stereothreshold, at least up to an eccentricity of 5 degrees, as the targets move out along the horopter away from the fixation point. In addition, Fig. 21-20 shows that at any eccentricity the threshold also rises, but now approximately exponentially, as the targets are moved nearer and farther in depth or absolute disparity away from the horopter. Furthermore, Fig. 21-20 shows that relative depth discriminations can also be made over a fairly wide band in visual space up to about 1.5 degrees of absolute disparity to either side of the horopter, even in the center of the visual field, and over a rather larger range in the periphery.

In view of the range over which stereopsis operates, it is perhaps not so surprising that Fender and Julesz[55] (p. 589) were able to preserve stereopsis when binocularly stabilized stereotargets had a divergent absolute disparity of 2 degrees. At that absolute disparity, subjects were still able to make a relative disparity discrimination of 8 minutes of arc. What is surprising is not so much the ability to make a depth discrimination but that fusion was still present with divergent images as much as 1 degree from the center of the foveola.

Stereoacuity and viewing distance

The limiting range of stereoscopic perception is the greatest distance at which an object can just be detected as nearer than an object at infinity. It is likely that when secondary cues to depth are eliminated, stereoacuity is independent of viewing distance. In practice, however, secondary cues are difficult to avoid, and as the limiting distance is approached, the relative importance of these secondary cues increases. Largely on theoretical grounds, the limiting viewing distance is generally taken to be about 500 m. Beyond this distance, depth estimates are based entirely on secondary cues.

Time relations: simultaneous and successive presentations

Stereoacuity continues to improve with increasing duration of exposure of the test objects. Ogle and Weil[138] reported a threshold of about 40 seconds of arc for an exposure time of 200 microseconds, much the same as it was at 6 milliseconds. Thereafter a fourfold decrease in stereothreshold occurred with increasing exposure time from 6 milliseconds to 1 second. The suggestion has been made that the improvement is due to eye movements. Thus Rønne[165] has proposed that the subject chooses by trial and error the optimal horopter or plane of reference that gives the best approximation to the average minimal disparities analogous to curve fittings to a number of points. Visual acuity also improves with time, but experiments with stabilized vision have shown that eye movements are probably not responsible for the improvement either in visual acuity[103] or in stereoacuity.[172]

For stereoscopic perception to occur it is not necessary for the targets to be presented to the two eyes at the same time.[51,137] Provided the exposure time is brief, stereoscopic depth is still experienced up to interstimulus intervals of about 100 milliseconds, although with a steadily increasing rise in stereothreshold over that period.

Size, shape, and color differences in the retinal images

For normal binocular vision the two retinal images should be equal in size and shape or very nearly so. With good normal vision, image sizes are significantly different only when objects are viewed at near distance in asymmetric convergence. Otherwise, overall size differences in the two images are invariably the result of dioptric abnormality. The term *aniseikonia* is used for the condition in which the difference in the size or shape is abnormally large. For those with good binocular vision this difference is less than 0.5%, although in an unselected population differences up to 3% or more may be found. With appropriate apparatus a size difference of 0.25% can be appreciated by subjects with good visual acuity. However, the difference that can give rise to symptoms is rather large; it may be as little as 1%, but a difference of up to 5% may be tolerated by the visual system, although only at the expense of a reduction in the quality of stereopsis. The size difference that can be tolerated depends on the nature of the binocular stimulus. Thus random-dot stereograms provide a much more effective stimulus than do simple line drawings, and stereopsis can still be obtained when one of the random-dot targets is expanded by as much as 15%.[98]

A difference in retinal image size occurs in *anisometropia,* a congenital abnormality in which the two eyes have unequal total refractions. Impairment of binocular vision results both from the size difference and the fact that the uncorrected image is necessarily blurred. Anisometropia is one of the commonest causes of microstrabismus. In the condition of *aphakia,* in which the lens is absent (nearly always as a result of surgical removal of cataract), the eye is greatly hypermetropic. Refractive correction by a strong convex lens placed at the usual spectacle distance (12 mm in front of the cornea) leads to an increase in retinal image size of about 25%. This degree of aniseikonia makes fusion impossible, and the resulting diplopia is very disturbing. With contact lens correction the image magnification varies between 4% and 10%, and for some patients this may afford comfortable binocular vision.

When the retinal image size difference is limited to either the horizontal or vertical dimension, opposing effects occur. Although vertical disparities cannot in themselves give rise to stereoscopic depth perception, they can, nevertheless, significantly modify the depth experience. When the image in one eye is magnified in the horizontal dimension, as by a meridional lens with the axis vertical, there is a characteristic distortion of the stereoscopic perception of visual space.[132] If, however, the image is magnified in the vertical dimension by placing the axis of the lens horizontal, the effect is rather surprising. The distortion that results in this case is as though the lens had been placed before the other eye at axis vertical so as to magnify the image in that eye in the horizontal dimension. This so-called *induced effect*[131] is in keeping with the fact that an overall magnification lens before one eye causes little or no distortion. The changes caused by the horizontal and vertical magnifications compensate for each other. The induced effect also finds expression in the observation that we are unaware of the vertical size differences that occur when objects are viewed at near distance in asymmetrical convergence. Depending on the eccentricity of the convergence, the retinal images may differ as much as 10% or more in their vertical dimensions under these circumstances.

The induced effect occurs within definite limits of image size difference. The threshold for the appearance of the effect is of the same order as that for stereoscopic depth. The induced effect is maximal when the vertical size difference reaches about 7% and thereafter declines slowly, although it is still appreciable at magnifications of 20% or more. The disparities at which the maximal effect occurs are larger than those at the limit of Panum's fusional area in the vertical direction; thus the subject may observe doubling of the vertically separated contours. Complex effects occur when the asymmetries involve an oblique axis or when there is a difference in retinal image shape.[132]

Binocular vision research has paid relatively

little attention to the place of color in stereopsis. Stereograms in which colored areas are used to generate monocularly distinguishable contours readily produce the percept of depth even when the colors are matched for equal luminance. Lu and Fender[113] have, however, shown that depth perception is possible with colored random-dot stereograms only when the individual elements are delineated monocularly by a luminance contrast and not when the monocular contours are created by color differences alone. They used colored random-dot stereograms in which randomly selected red and green squares could be matched for equal luminance. Thus stereopsis with Julesz-type patterns appears to be color blind, but the significance of this observation is by no means clear. The important variable for the generation of depth perception may be contour length. Disparity detection based solely on color differences may require longer contours than are to be found in the usual random-dot stereograms.

Influence of vertical disparity

Ogle[135] studied the combined effect of horizontal and vertical disparities so as to determine the limiting effects of vertical disparities on stereoscopic depth perception and fusion. The horizontal disparity threshold for a point source of light to be seen as nearer or farther than the fixation point was determined in the presence of increasing vertical disparities between the two retinal images. Near the fixation point vertical doubling of the images took place when the vertical disparity exceeded about 7 minutes of arc, which corresponds to the vertical extent of Panum's fusional area. Despite the onset of doubling, however, stereoacuity was relatively unaffected, the double images being almost as accurately localized in depth as before. Thereafter stereoacuity decreased only slowly with increase of vertical disparities up to as much as 25 to 30 minutes of arc.

Orientation disparity

The term orientation disparity refers to the angular difference between line images in the two eyes. There has been controversy about the place of orientation disparities in binocular vision ever since Wheatstone[193] observed a single line tilted in depth in the median plane when he stereoscopically viewed targets each consisting of a line rotated from the vertical in opposite senses for the two eyes. The matter of the controversy was whether the fusion into a single image was achieved by a central neural mechanism or by cyclorotation of the eyes. There seemed little doubt that central fusion does occur, since there are many examples in the literature in which fusion has been shown to take place, even though it would have required impossible eye movements.[14] Marlowe[115] showed that stereograms composed of identical line segments in random orientations may still give rise to fusion and depth perception even though the line segments may be rotated from alignment by at least ±15 degrees. Despite the random orientation disparities, the individual line segments fuse and appear to be tilted in depth. The cyclodisparity range for fusion is, however, greater for the vertical than for the horizontal line segments. At disparities of ±15 degrees, the horizontal line segments were often seen as double.

The role of eye movements in cyclofusion has largely been resolved by recent objective observations demonstrating the importance of line length.[35,175] It has been shown that cyclofusion has both a central neural component as well as a motor response in the form of compensatory eye movements. For line lengths subtending less than 10 degrees the central component is the major factor, and fusion occurs with relatively little eye rotation. For line lengths above 10 degrees there is a relative increase in the motor component, so that at 50 degrees it may become the major factor. These results demonstrate the importance of peripheral retinal stimulation in evoking a substantial motor component as part of the cyclofusional response. The role of orientation disparity as a cue for stereoscopic depth will be discussed on p. 617.

Limiting cases for fusion and stereopsis

The stereogram known as Panum's limiting case (Fig. 21-21, *A*) is generally regarded as presenting the minimum requirements for the perception of stereoscopic depth. One eye sees

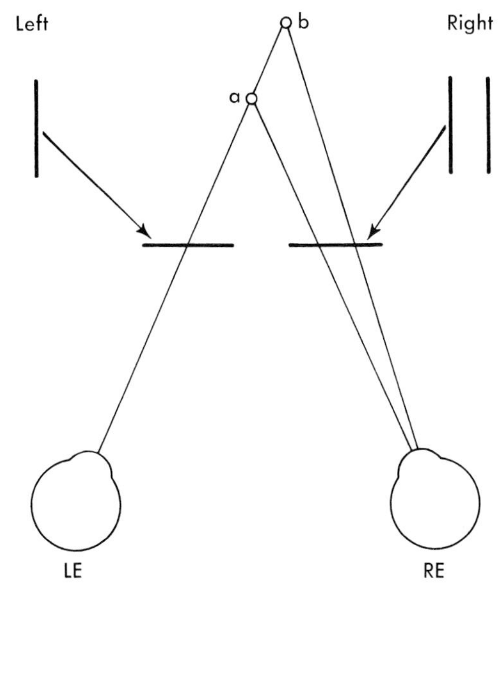

B

Fig. 21-21. **A,** Panum's limiting requirements for perception of stereoscopic depth. **B,** Dot stereotargets showing that image components on same horizontal level are not needed for perception of stereoscopic depth; only horizontal vector is needed.

a single vertical line, and the other eye two slightly separated vertical lines. With binocular viewing only two lines are seen, one of the two appearing nearer than the other. In many ways a more informative limiting case is that shown in Fig. 21-21, *B*. Stereopsis depends on horizontal retinal image disparities, but Shipley and Rawlings[171] have recently drawn attention to the fact that it is not essential for the image disparity of spots to be in a horizontal line; only a horizontal vector between the spots is needed. Thus any two pairs of spots such as those in Fig. 21-21, *B*, will fuse and go into depth, provided there is a disparity between the horizontal vectors between the spots in each target. Both of these limiting cases have a ready explanation in terms of the neural theory of fusion and stereopsis.

Fine and coarse stereopsis and fusion*

Bishop and Henry[15] have reviewed the evidence for considering stereopsis as a dual process, consisting of fine and coarse mechanisms for disparity detection, the two mechanisms acting, to some extent at least, independently of one another. Each type of stereopsis seems to have associated with it a fusional vergence reflex mechanism and the equivalent terms central (fine) and peripheral (coarse) stereopsis and fusion are in use in clinical practice.

Fine stereopsis is a highly specific pattern-matching process involving very local features of the two retinal images and operating over a relatively narrow range of spatial disparities—probably no more than about 0.5 degrees. Dissimilar images cannot be simultaneously perceived. If binocular single vision is to occur, the two retinal images must be closely similar, otherwise there is retinal rivalry and suppression of one or other of the antagonistic image features. By contrast, *coarse stereopsis* is a much less specific process that can operate on visual images quite dissimilar in form, luminance, and contrast and separated by several degrees in spatial position—up to as much as 7 to 10 degrees of retinal image disparity.[119,120,192] Since coarse stereopsis acts beyond the range of retinal rivalry, dissimilar images can be simultaneously perceived. For coarse fusion (single vision), however, the images must again be similar, but the disparity limits over which fusion can take place (Panum's area of up to about 7 degrees) are much larger than they are for fine stereopsis. Coarse single vision may occur in the absence of fine fusion, but fine single vision is always accompanied by coarse fusion.

Microstrabismus[48,148,186]

The idea that there are two distinct stereoscopic and fusional vergence mechanisms has arisen out of a study of microstrabismus (monofixation syndrome).[146,147,164,176,177] Microstrabismus is a very small–angle squint of about

*References 15, 95, 100, 146, 147, 155, 162, and 164.

5 degrees or less that can come about as the result of a number of quite separate etiologic factors. Many patients with microstrabismus are symptomless and quite unaware of their visual deficiency until it is revealed by routine clinical investigation. An important group of patients with microstrabismus is, however, made up of those who have been successfully treated for congenital strabismus, since the end result of successful treatment is almost invariably a very small-angle squint of 5 degrees or less. To avoid seeing double in the central part of their binocular visual field, patients with microstrabismus develop a facultative central suppression scotoma in the nonfixating eye. This scotoma, 1 to 3 degrees across, is facultative because it is present only under conditions of binocular viewing and disappears on monocular gaze. Because of the suppression scotoma and the disruption of normal binocular vision, these patients have poor stereoacuity. Nevertheless, despite the poor stereoacuity, a strong fusional vergence reflex persists.

In normal subjects it has been shown that the limiting disparities that can still initiate vergence eye movements lie between 5 and 10 degrees.[191] Furthermore, vergence movements indistinguishable from normal can still be initiated in the appropriate direction by targets so dissimilar as to be incapable of sensory fusion, although the vergence produced by these dissimilar targets does not proceed to completion, presumably because the targets cannot be fused.[120] Hence the degree of similarity of the disparate images that permits the initiation of vergence is obviously different from that which permits its completion. These observations suggest that different central mechanisms are involved in the control of these two phases of eye movement. Furthermore, the disparity limits that still elicit vergence eye movements are the same as those that also permit the images to be subjectively localized in depth. Hence the two mechanisms, for coarse stereopsis and for vergence eye movements respectively, probably share a common afferent limb.

It is the coarse (peripheral) fusional vergence reflex that holds the eyes in binocular alignment after successful surgery for strabismus. The aim of therapy is to bring the eye alignment within the 7-degree fusional vergence range so that binocular single vision and coarse stereopsis can be established and maintained. From a clinical point of view, however, the quality of the stereopsis that results is of lesser importance than the cosmetic appearance of good eye alignment and the establishment of binocular single vision. Nevertheless, the measurement of stereoacuity does provide a sensitive assessment of the level of binocular cooperation. The suggestion is that fine stereopsis and central fusional vergence are largely or wholly suppressed in microstrabismus, apparently leaving the coarse mechanisms to operate independently. On this basis coarse stereopsis gives patients with microstrabismus a level of stereoacuity no better than about 70 seconds of arc and usually rather worse than this, whereas subjects with normal stereopsis have a stereoacuity better than about 40 seconds of arc.

Anomalies of stereopsis and vergence

The distinction between fine and coarse stereopsis has received additional support from observations on the anomalies both of stereopsis and of fusional vergence that are now known to be of common occurrence in the apparently normal general population. Richards[161,162] investigated individual differences in stereopsis for the range of disparities from 0.5 to 4 degrees (coarse stereopsis). By requiring his subjects to report targets flashed on a screen as "nearer than," "farther than," or "the same depth as" the screen, he assessed their ability to discriminate relatively large crossed, uncrossed, or zero retinal image disparities. He found that individuals may have discrete anomalies of stereopsis limited to one or more of these discrimination tasks. For example, a person may be blind for "nearer than" or crossed disparities while still maintaining a normal ability to perceive "farther than" or uncrossed disparities. Richards found all combinations of stereoanomaly, the frequency of occurrence of one or more disability being about 30% of the population. Jones[95] has recently replicated Richards' findings concerning the occurrence of specific types of stereoanomaly, but added the important observation that all his (Jones') subjects had otherwise normal binocular vision as determined by standard clinical tests with a stereoacuity better than 20 seconds of arc in each case. This indicates that these forms of stereoanomaly involve only the coarse stereoscopic mechanisms and not those for fine-disparity, high-resolution stereopsis. Jones' experimental procedure and range of target disparities were similar to those used by Richards, but, in addition, the subject's horizontal eye movements were monitored photoelectrically so that vergence responses could be correlated with the subject's depth experience. Jones found an incidence of vergence anomalies in his subjects as high as 20%, but the anomaly was apparently not detrimental under normal seeing conditions, since it was present in persons with adequate binocular vision and normal fine stereopsis. He found two forms of vergence anomaly in response to briefly exposed stimuli; either convergence responses to near or

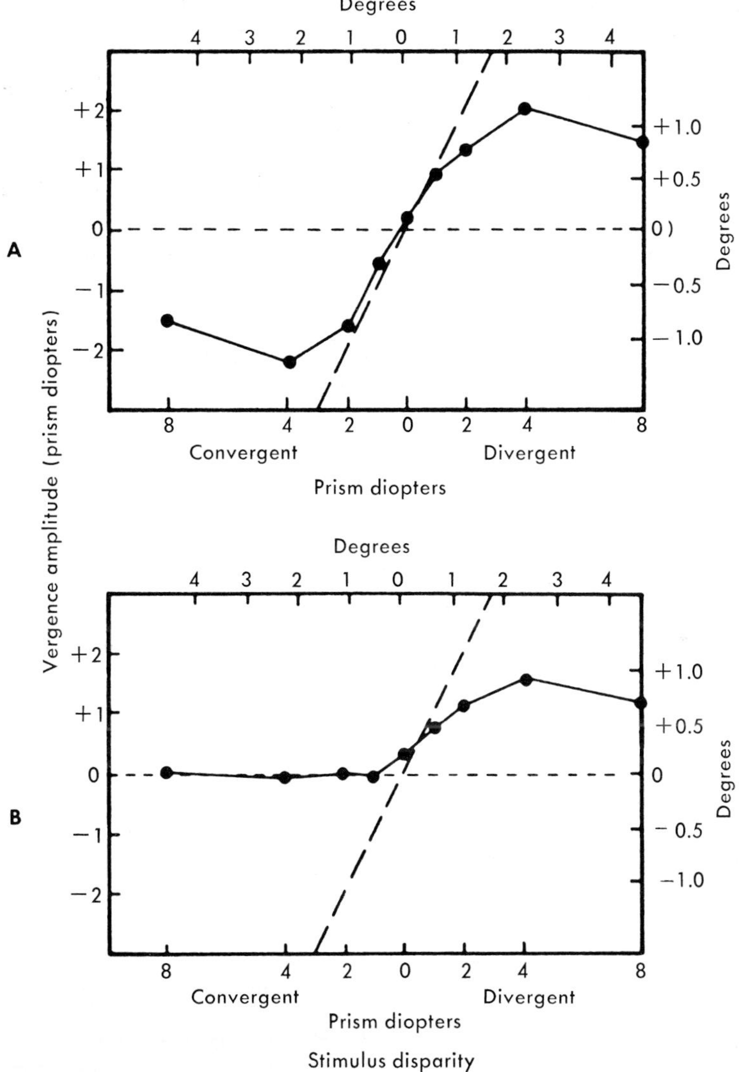

Fig. 21-22. Vergence-disparity profiles for a normal, **A,** and a convergence-anomalous subject, **B,** obtained by determination of the average peak-vergence amplitude for each disparity value. The dashed diagonal line indicates the amplitude of the vergence required for complete responses. In **A** the responses are incomplete except for 1 prism diopter of uncrossed disparity. In **B** there is a complete absence of convergence, but the divergent responses are of normal amplitude. (From Jones, R.: J. Physiol. [Lond] **264:**621, 1977.)

divergent response to far could be absent (or extremely attenuated) while the retained vergence appeared normal in all respects. The vergence-disparity profile in Fig. 21-22, *A*, is from a normal subject; it shows the average peak vergence amplitude for various stimulus disparity values. Since the stimulus duration was only 200 milliseconds, the vergence responses are incomplete. Complete eye movements would be described by the dashed diagonal line through zero. Fig. 21-22, *B*, shows the corresponding profile from a convergence anomalous subject. There is a complete absence of convergence, yet the divergent responses are of normal amplitude. This suggests that convergence and divergence are normally initiated independently and that the vergence anomaly is the result of a functional deficit in one component. The presence of a vergence anomaly was always associated with the occurrence of a coarse stereoanomaly of the kind originally described by Richards. The converse was, however, not always true. The occurrence of a stereoanomaly did not necessarily mean that a vergence deficiency would also be present.

The stimuli used in Jones' experiments were symmetric about the fixation point, so that the subject's task involved the detection of disparities to either side of the vertical midline in the visual field. However, the stereoanomalies and vergence anomalies are not specially related to this region and may occur in relation to parts of the visual field wholly within either hemifield.

The psychophysical[95,162] and oculomotor[95] responses just described suggest the existence in the human visual system of three distinct populations of disparity-specific neurons subserving, respectively, convergent and divergent disparities for coarse stereopsis and disparities in the immediate vicinity of the horopter for fine stereopsis. It may be that, in a proportion of the normal population, one or more of these classes of neurons may be absent or nonfunctional. Strong support for these ideas have come from recent observations of Poggio and Fischer.[155] They have described three classes of depth-sensitive cells in the striate and prestriate cortex of the monkey: one class (tuned excitatory and tuned inhibitory neurons) specialized for making fine depth discriminations and two other classes (near and far cells) for detecting relatively coarse disparities in front of the fixation point (near cells) and behind it (far cells). For further details see p. 617.

Random-dot stereograms[97,99]

Julesz[97] was the first to consider using random-dot stereograms as a way of studying stereoscopic perception without the complication due to monocular (secondary) cues to depth. When viewed monocularly, each of the two stereotargets in Fig. 21-23, *A*, appears as a random array of black and white squares having a roughly uniform texture without any form or depth to be seen. The two random-dot patterns are identical except that a central square region in one field has been displaced laterally with respect to the same region in the other field. This displacement cannot be seen monocularly, and it is only appreciated by viewing the targets stereoscopically. On binocular fusion there is a vivid depth impression, with the central square (figure) floating out in front of the unshifted surround. Each stereotarget by itself contains no information whatever about the central square. This information is contained only in the *relation* between the two arrays, and it is only because it is perceived in depth that the square is seen at all.

Observations using random-dot patterns have many important consequences for an understanding of binocular vision. It is still widely held that in binocular vision each uniocular mechanism independently develops a complete perceptual image of its own and that the brain somehow compares these fully elaborated uniocular percepts before using the difference information to add the sense of depth. This idea is clearly false. Stereoscopic depth can be perceived in the absence of any monocular cues to depth or monocular form perception and also without any binocular cues except retinal image disparity. Thus the neural mechanisms for stereoscopic depth must come into play before those responsible for form perception.

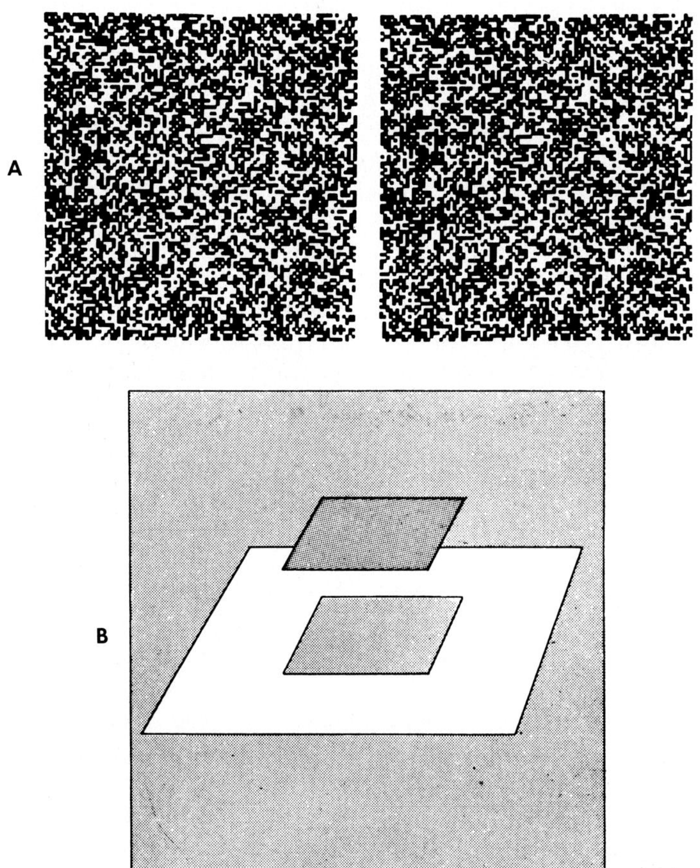

Fig. 21-23. Random-dot stereogram. When **A** is stereoscopically fused, a square appears in vivid depth above randomly textured surround, as indicated by perspective diagram, **B.** (From Julesz, B.: Foundations of cyclopean perception, Chicago, 1971, University of Chicago Press.)

There is a sequential analysis of the stereospatial data whereby depth information is extracted first and then used for the recognition of form.

With random-dot targets, such as Fig. 21-23, *A,* there is no monocular cue for the perception of the test shape (i.e., raised surface) so that double images of the test shape cannot exist. It is only after fusion that binocular shape is perceived. For random-dot stereograms at least, binocular fusion is also achieved on a mosaic dot-by-dot basis and not by bringing together elaborated forms. Fusion and stereopsis are part and parcel of the same process. Furthermore, random-dot stereopsis suggests that the mechanisms in the brain responsible for the depth discrimination need not have available to them information concerning the specific contribution each eye makes. In other words, eye of origin discrimination is probably not essential for stereopsis.

Local and global stereopsis

Presumably, with Julesz-type stereograms, the depth information is first extracted on the basis of a dot-by-dot or square-by-square correlation, each dot pair being assigned a depth value. Julesz[99,100] has applied the term *local stereopsis* to this local correlation process. Because there are a large number of possible ways in which the black and white squares can be paired, such a local process operating on the basis of a square-by-square comparison cannot by itself provide a satisfactory basis for stereoscopic depth perception. When there is ambiguity as to which elements in the two retinal images correspond to each other, a global process is needed that evaluates different possible sets of corresponding pairs and selects one set of matched pairs that, by their depth values, can provide the data for the recognition of a three-dimensional form such as the center square and its surround in Fig. 21-23. Julesz refers to this additional mechanism as *global stereopsis,* since the problem of ambiguities can only be resolved on a global basis. Nevertheless, the local stereoscopic mechanism can operate on element pairs that are not local in the sense of one pair being closely adjacent to another. Classical stereograms (e.g., Fig. 21-21) may consist of only a few widely scattered dots or line segments, in which no ambiguities arise as to which of the dots or line segments correspond to one another. Such a process of unambiguous depth localization may require no more than a local stereoscopic mechanism regardless of the amount of the binocular disparity of the corresponding elements. A satisfactory theory for binocular depth perception must obviously take account of both local and global stereopsis.

Binocular thresholds for luminance and acuity

Under normal testing conditions binocular visual acuity shows no improvement over the better of the two eyes tested monocularly. Under laboratory conditions and at the limits of the subject's resolving powers, binocular vision is, however, significantly better both for absolute luminance threshold and for visual acuity. In experiments designed to test absolute luminance thresholds, Pirenne[154] found that binocular sensitivity was greater than monocular sensitivity by a factor of 1.25. This improvement can be explained by probability summation between independent channels as though the two eyes belonged to two different persons. Quite apart from this probability summation, there is clear physiologic evidence that summation of signals from the two eyes does occur on single cells in the visual cortex. We may expect, therefore, that in addition to and apart from the probability summation at absolute luminance thresholds there would also be psychophysical evidence of the cortical summation of monocular signals at higher levels of ambient illumination. Campbell and Green[28] have indeed found a binocular improvement factor of $\sqrt{2}$ for the contrast sensitivity for sinusoidal gratings at a high light level (80 cd/m^2). Their finding may be explained if it is assumed that when identical images are fed to two matched eyes, their outputs are summed and that these outputs are noisy. Because the standard error of the sum of n independent measurements of a random noise process decreases as $\sqrt{n}$, an observer using two eyes can obtain two measurements that thus permit a $\sqrt{2}$ lower contrast to be detected. Recently, however, Home[75] found for 20 subjects at high-adaptation luminances a binocular/monocular contrast sensitivity ratio of about 1.5 when the preferred eye was used for monocular viewing and a ratio greater than 1.6 for the nonpreferred eye. The reduced effectiveness of binocular summation at the lowest light levels (binocular/monocular ratio about 1.3) is probably due to the weakness of the vergence cues that would make it difficult to ensure that corresponding retinal areas are being stimulated. Under these conditions the binocular improvement approaches a probability summation between independent channels, as just noted. Even at the higher luminance levels the binocular superiority is reduced the more the visual task involves the resolution of spatial structure in a given stimulus. For high contrast two-dimensional stimuli under photopic conditions and using the preferred eye, the binocular/monocular acuity ratio is only about 1.05.[75]

Binocular disparities due to eye movements

Although the small involuntary movements of the eyes in steady fixation have been known since the work of Dodge,[44] it is only relatively recently that binocular movements have begun to be studied.[42] In normal vision the small involuntary motions (flicks and drifts) of the eyes are not perfectly correlated; thus there is a constantly varying amount of binocular disparity. Any theory of binocular vision must therefore take into account not only the reflex vergence movements that establish fusion in the first place, but also the errors both of vergence and vertical alignment during fixation.

Using a "suck-on" contact lens arrangement, Fender and St-Cyr[56] studied the interplay of flicks and drifts in binocular fixation. Using their data, Fender (personal communication) subsequently examined how one visual axis wanders with respect to the other during steady binocular fixation as a result of the uncorrelated components in the flicks and drifts of each eye. A binocular disparity existed whenever the divergence of each eye from its respective mean position differed in the two eyes. By regarding one eye as fixed at its mean position and transferring all these differences to the other eye, it was possible to express the way in which one visual axis wandered with respect to the other. The binocular disparities at the various sampling times over the 2-minute viewing period could be considered as a bivariate normal distribution centered on the fixed eye. The standard deviations of the distribution varied with the nature of the fixation target and were consistently smaller for vertical line targets than for horizontal lines. Average values for the standard deviations for a variety of targets in two subjects were horizontal 4.8 minutes of arc and vertical 8.9 minutes of arc. Further studies along these lines are urgently needed, and there has been as yet no systematic examination of the small involuntary eye movements in strabismus.

NEUROPHYSIOLOGY*

Neural mechanisms for binocular depth discrimination in mammals have been studied most extensively in the cat† but also in the monkey[85,155,198] and in the sheep.[30]

It is important for these neurophysiologic studies that both cat and monkey have been shown by behavioral methods to have stereoscopic vision. Using a three-rod binocular depth discrimination task, Blake and Hirsch[17] have found the threshold disparity angle for the cat to be less than 4 minutes of arc,[142] which was about 10 times better than the smallest depth detectable when the animal was forced to use monocular vision. This level of stereoacuity is to be expected in view of the cat's relatively poor visual acuity, about one sixth that of the human. Further confirmation that the cat has stereopsis is the demonstration that animals which lack binocularly activated cortical cells display a marked deficit in binocular depth discrimination despite normal visual acuity in both eyes.[17] The disruption of the normal binocular innervation of cortical cells in these animals was brought about by subjecting them as kittens to a relatively prolonged regimen of alternating monocular deprivation, which included the whole of the critical period. Thereafter the cortical cells virtually ceased to be binocularly activated and could be discharged only from one eye or the other (p. 629). Although the cat's stereoacuity is poorer than the human's, it can, however, tolerate greater disparities than can the human before failing to fuse the images in the two eyes. Using the same fusion task in each case, Packwood and Gordon[142] found that the largest horizontal disparity fused by the cat (about 50 minutes of arc) was about five times greater than that by the human. By contrast, the human can tolerate much larger disparities beyond the range of fusion than can the cat before losing the ability to make stereoscopic depth discriminations, although the cat can also make depth judgments based on disparities too large for fusion (up to about 2 degrees).

Bough[25] provided qualitative evidence of stereopsis in the macaque using a successive discrimination task based on random-dot stereograms with disparities of about 40 minutes of arc.[32] More recently the stereoacuity of the macaque monkey has been evaluated quantitatively using the Howard-Dolman test and shown to be comparable to stereopsis in the human, that is, a best threshold disparity angle of about 2 seconds of arc[167] when the measured values were corrected for differences in the interpupillary distances in the two species.

The following outline of the neural theory for binocular single vision and stereopsis is based very largely on data from the cat and it should be emphasized that it provides a reasonably

---

*See references 9, 12, 13, 16, 128, and 155.
†See references 9, 16, 18, 96, 127, and 130.

satisfactory account only for those early stages in the binocular process broadly referred to by Julesz as *local stereopsis* (p. 605). The local stereoscopic process is conceived of as acting first to extract from the visual scene small elements of pattern or very localized features of objects, to which it then assigns relative depth values indicating one object feature as being nearer or farther away than another. There must then be further stages of neural processing, first recognizing that certain pattern elements or features taken together form a set with a common spatial regularity, and thereafter combining the various sets to represent a global pattern. Only local neural circuits need be involved in local stereopsis, but the further stages of neural processing would require larger scale cooperative interactions between these local neural elements.[100,126] There is, however, as yet little understanding of the neural mechanisms underlying the process of *global stereopsis* by which we achieve the perception of whole objects in stereoscopic depth. Pattern recognition is largely a learned process involving memory. Nevertheless, binocular vision is an attractive field for neurophysiologic research because many of the mechanisms involved in global stereopsis must operate without recourse to memory and be less complex than those involved in monocular form perception.

Receptive fields of cortical cells

Our present understanding of the neurophysiology of binocular vision has largely come from microelectrode recordings of single cells in the visual cortex. The basic features of the visual pathways in the cat relevant to the present account are shown in Fig. 21-24. The essential anatomic arrangements by which the brain discriminates retinal image disparities is provided by the partial decussation at the chiasma. This enables the fiber pathways from corresponding, or nearly corresponding, points in the two retinas to come together in the visual cortex. The pathways from the two eyes remain relatively independent of one another through the lateral geniculate nucleus and effectively come together for the first time by synapsing on cells in the striate cortex.

Most of the cells in the striate cortex can be independently driven from both eyes, and Hubel and Wiesel[78,79] were the first to observe that each binocularly activated neuron has two receptive fields, one for each eye. The idea of a *receptive field* is the basic concept in visual neurophysiology at the present time. In experimental terms the receptive field is the small area on the tangent screen in front of the cat that is specific for each cell, although there is considerable overlapping of the areas of the various cells. For a striate cell to fire, a specific set of stimulus features must be present in the cell's receptive field, such as a light-dark border or contour line having a particular orientation and perhaps length, which must be moved in

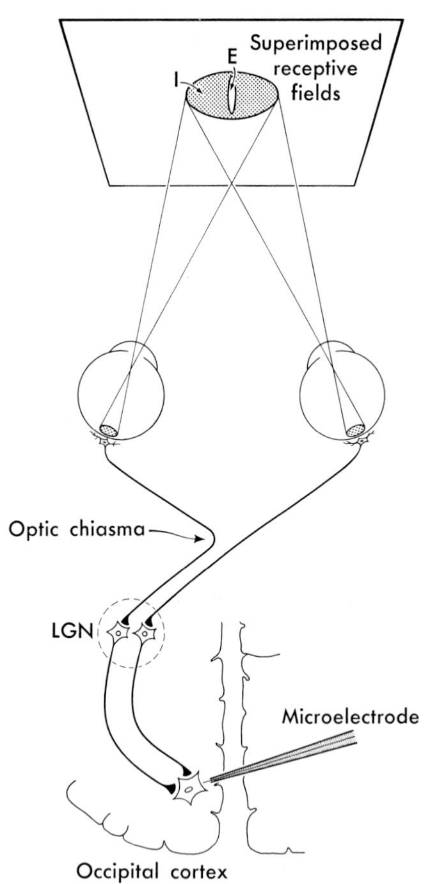

Fig. 21-24. General experimental arrangement for recording from single cells in cat striate cortex showing two receptive fields of binocularly activated neuron in register in plane of tangent screen. *E,* Discharge center; *I,* inhibitory sidebands; *LGN,* lateral geniculate nucleus. (From Bishop, P. O.: Aust. J. Sci. **32:**383, 1970.)

the right direction and at the correct speed, otherwise the cell will remain silent or its steady discharge be unaffected. The receptive fields of a cortical cell are conventionally represented as rectangles, and for many purposes this is adequate. Most receptive fields are, however, roughly circular or elliptical and have a fairly complex spatial organization, which varies from one cell type to another. Aspects of this spatial organization are shown in Fig. 21-24, where the two receptive fields of the cortical cell are in precise register on the tangent screen. Each cell therefore monitors specific features of objects, but only if they lie within the cell's receptive field.

The two receptive fields of a cortical cell are in approximately corresponding positions in the contralateral hemifield. The two fields have about the same size and spatial arrangement; thus whatever stimulus is most effective in one eye — in form, orientation, direction, and rate of movement—is generally the most effective in the other eye.[79,81] For the most part the only major difference between the two fields relates to the phenomenon of eye dominance, one eye frequently driving the cell more effectively than the other eye, although the dominant eye varies from cell to cell. Occasionally the two fields can differ in other respects such as direction selectivity (p. 615).

Local stereopsis: essential steps

Small horizontal displacements between the images in the two eyes provide the cues for binocular depth estimates. The evidence of psychophysics suggests that in the initial stages these cues are used on a mosaic, feature by feature basis. It is the small size of the receptive fields, especially in the center of gaze, that enables these small local features to be discriminated in the first place, and accurate feature by feature pairing takes place because the two receptive fields of the one cell have the same highly specific stimulus requirements. Each cortical cell responds to those particular parts of the images in the two eyes that belong to the one local feature of the external object. This is brought about not only because the monocular stimulus requirements for each receptive field are the same, but also because the cell will fire only when this monocularly optimal stimulus is applied at that distance from the animal where the two receptive fields are in precise register (p. 611). In the natural environment, depending on the size of the receptive fields and the precision required of the registration, it is most unlikely that in such a small region in space there will be two separate but identical object features. The dot-by-dot pairing ambiguities that occur with random-dot stereograms do not ordinarily arise in the natural environment, since with normal bifixation every position of convergence will always provide the cortical cells with a large number of appropriate local matches in the images of the two eyes. No doubt many cells respond to inappropriate image matches, but these responses will be eliminated in the further stages of neural processing that select the particular global sets of local matches out of which the perception of form or pattern can emerge. With random-dot stereograms there is usually only one position of convergence that provides the particular sets of local matches out of which the global stereoscopic process can extract significant form. This probably explains why in the natural environment the perception of stereoscopic depth seems to be immediate, whereas with random-dot stereograms it may take several seconds to develop.

By the act of discharging binocularly, a cortical cell not only discriminates a local object feature but also assigns to it a particular depth value. The depth value assigned depends on the phenomenon of *receptive field disparity*.

Receptive field disparity

Only some binocularly driven visual cortex neurons have their receptive fields for the two eyes in exactly corresponding positions in the visual field; the majority of cells show varying degrees of disparity in the locations of their fields. Under natural viewing conditions and with normal bifixation, the movements of the two eyes are highly coordinated (Hering's law) so that the two receptive fields for each cell are nearly always in register either in the depth plane of the fixation point or close to it. To study the location and properties of the receptive fields of visual neurons it is, however, essential that the eyes remain as stationary as possible, and with present techniques this is achieved by fully paralyzing the experimental animal. When the extraocular muscles are paralyzed, the eyes diverge slightly and then take up a fixed position. Because the eyes are fixed in slight divergence, the two receptive fields of a cell also have fixed positions horizontally separated from one another on the tangent screen. By recording from a series of single cells from one small region of the visual cortex it is possible to plot a number of pairs of receptive fields, one pair for each of the cells that the microelectrode encounters on its path through the thickness of the gray matter. Since the eyes have diverged, two groups of receptive fields will be plotted, and because of the orderly topographic projection of the visual field onto the cerebral cortex, the members of each group will all lie fairly close to one another on the tangent screen.

If all the receptive field pairs were exactly corresponding, under natural viewing conditions any position of the eyes that brought one receptive field pair into register on the tangent screen would necessarily bring all the remaining pairs into register. This possibility can be tested in the paralyzed preparation by moving the centers of the receptive fields in the group for one eye to a single point and observing whether the same translations applied to the group for the other eye will also bring them to a single point. It is invariably found, however, that the fields in the second group do not translate to a single point but form a two-dimensional Guassian distribution. Thus the greatest density of receptive fields is to be found at the center of the scattergram, the density rapidly falling off in all directions from the center. By definition the receptive fields that lie at the center of the distribution are said to be exactly corresponding; the remainder show *receptive field disparity*.

These observations are diagrammatically represented in the upper part of Fig. 21-25. Only three pairs of receptive fields are considered, each conventionally represented in the form of a rectangle, although of distinctive shape so that the pairs can be readily recognized as pairs. The group for the right eye have been moved to a single point, thereby giving them a common visual direction. Appropriate shifts to the left eye fields still leave them out of alignment. By way of simplification, only horizontal receptive field disparities are illustrated, but the fields would also be scattered in the vertical direction. The histogram under the left eye (Fig. 21-25, *A*) shows the experimentally determined horizontal receptive field disparities for a large number of receptive field pairs. It can be seen that the disparities have a fairly normal distribution about the mean position. When the vertical direction is taken into account, the scatter is seen to be two-dimensional (Fig. 21-25, *B*), as referred to earlier. In the cat, for receptive fields near the

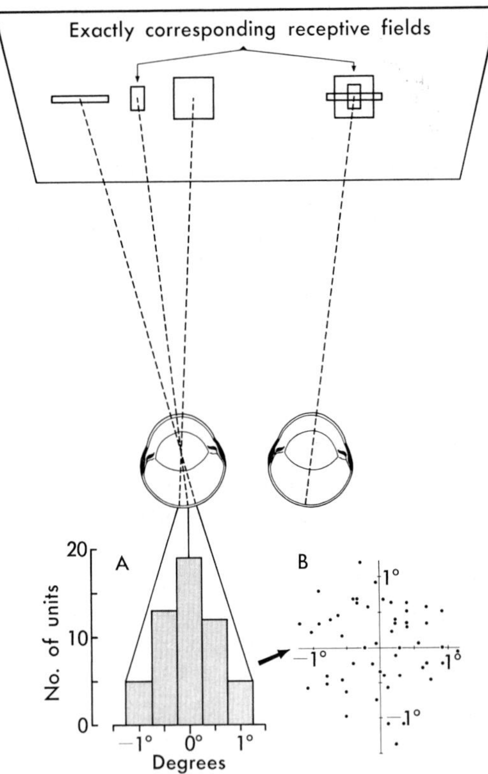

Fig. 21-25. Diagram of method used to obtain quantitative measure of receptive field disparities. **A,** Histogram showing actual distribution of horizontal receptive field disparities of 54 binocular receptive field pairs within about 5 degrees of visual axis.[130] **B,** Scattergram of horizontal and vertical receptive field disparities of same units as in **A.**[130] (From Bishop, P. O.: In Jung, R., editor: Handbook of of sensory physiology, Berlin, 1973, Springer-Verlag, vol. 7/3A, p. 268.)

center of gaze, the scatter of disparities is radially symmetric, having a standard deviation of 0.5 degree.[96,130] There is some question in the literature concerning the magnitude of this scatter, but it is generally recognized that the disagreement stems from technical problems involved in monitoring residual eye movements in the paralyzed preparation. Recently, however, von der Heydt and co-workers[181] have also found a disparity scatter with a standard deviation of 0.5 degree.

Neurophysiologic definition of horopter

In monocular vision the eye is directed toward an object point when the maximum number of receptive fields are located at that point. This would be the case when the object point is imaged on the center of the foveola (or area centralis in the cat). In binocular vision it may be assumed that the eyes are directed toward a fixation point when the maximum number of binocular receptive field pairs are in register at that point. It is possible that the phenomenon of fixation disparity has its basis in the different positions of the eyes that are required to have either the monocular or the binocular set of receptive fields in register at the fixation point. There is also a surface in space for which the maximal number of binocular receptive field pairs are in accurate register. This surface, which by definition includes the fixation point, also provides for the horopter a neurophysiologic definition analogous to the stereoacuity horopter of binocular psychophysics. Receptive fields are corresponding when they are in precise register in this surface, and fields that are in register in planes either closer to the animal or farther away show receptive field disparity.

As a simplification, the horopter can be considered as a curved line in the fixation plane. It was pointed out earlier that paralysis of the animal leads to divergence of the eyes. In such a preparation, recording from a population of single cells in a localized region of the cortex produces two closely grouped clusters of receptive fields, one for each eye, horizontally spaced apart on the tangent screen. If similarly localized recordings are made in cortical regions farther away from the central representation, the receptive fields are now more eccentrically located, and it is found that the mean separation of the two clusters is less than it is for the cells with centrally located fields. There is, in fact, a progressive decrease in the mean receptive field separation with increasing retinal eccentricity (Fig. 21-30, *A; F* $> F'$).

The preceding observation has an important bearing on the shape of the horopter. Under natural conditions the receptive fields are brought into register by movements of both eyes. As a simplification, in Fig. 21-30, *A*, the two receptive fields near the fixation point can be brought into register by moving only the left eye, and hence the left eye receptive field, by a distance *F* as indicated by the arrow. This movement necessarily translates the left peripheral receptive field as well by approximately the same distance *F* in the frontoparallel plane of the fixation point, so that it comes to lie on the other side of its companion field ($F > F'$). Thus the peripheral fields will then be in register in a plane closer to the animal than the tangent screen. This causes the horopter to be concave toward the animal.

Neural mechanism for binocular depth discrimination

As noted earlier, each receptive field of a cortical neuron has the same highly specific stimulus requirements, but if the cell is to fire under binocular conditions a further stimulus specificity must be met; the monocularly optimal stimulus must now be applied to the two receptive fields at the same time. Under natural viewing conditions, simultaneous application will occur only when the stimulus is applied at the crossing point of the lines of sight through the centers of the two receptive fields. At the crossing point the two receptive fields are in accurate register (Fig. 21-26), and when this is so, the firing of the cell is greatly facilitated (Fig. 21-26, *A*, 2). At frontoparallel planes closer to (*A*, 3) or farther away from (*A*, 1) the animal than the crossing point, when the two receptive fields become offset with respect to one another, mutual inhibition occurs and the firing may cease altogether. The two receptive fields have only to be out of reigster by

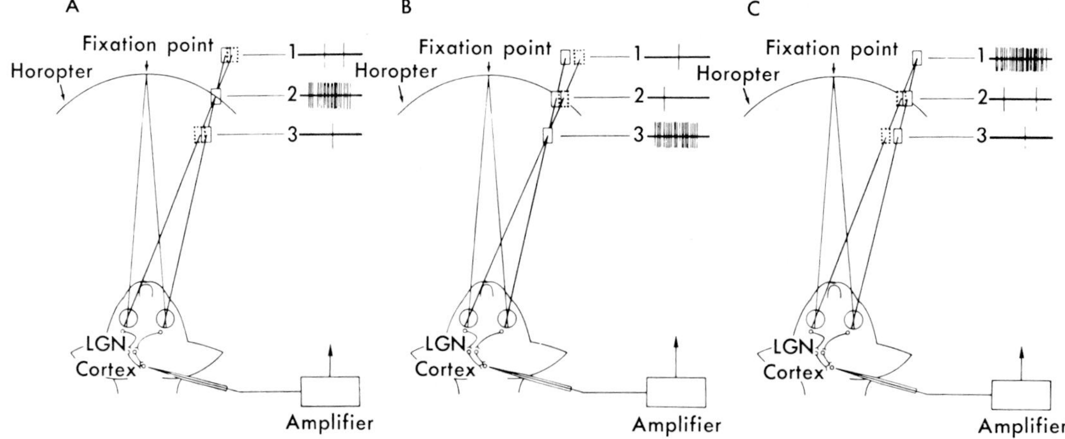

Fig. 21-26. Scheme showing how binocular depth information can be coded by neurons in cat visual cortex. **A,** Maximal firing of binocularly activated neuron occurs when the two monocular receptive fields are in register and can be stimulated simultaneously, tracing *2*. Being in register on horopter, this receptive field pair has zero disparity; that is, they are corresponding. **B,** Receptive field pair having convergent disparity; maximal firing, tracing *3*. **C,** Receptive field pair having divergent disparity; maximal firing, tracing *1*. (Modified from Bishop, P. O.: In Jung, R., editor: Handbook of sensory physiology, Berlin, 1973, Springer-Verlag, vol. 7/3A, p. 281.)

a small amount compared to the width of the receptive field for the mutual inhibition to be complete. Thus each cortical cell will respond only when the object feature specific for it is located at that depth in space where its two receptive fields are in reigster. For a given position of the eyes, the distribution of receptive field disparities means that different pairs of fields will be in register at different depths in space, some closer than the horopter (Fig. 21-26, *B*) and some farther away (Fig. 21-26, *C*). Because they respond to stimuli at different depths, the cortical cells can assign relative depth values to particular object features in the field of view. The values are referenced to the fixation point and horopter and depend on the location of the particular disparity within the receptive field disparity distribution. The small size of the receptive fields, particularly in respect to central vision, means that depth values are assigned to localized features of the visual scene, thereby providing the kind of mosaic element-by-element analysis that has already been presumed to occur on the basis of the results obtained with random-dot stereograms. Furthermore, the distribution of receptive field disparities ensures that over a limited spatial range determined by the spread of the disparities, there are enough receptive field pairs available to respond to all the significant features in the visual scene. The spread of receptive field disparities is a major factor in determining the limits of Panum's fusional area (p. 615).

Binocular interaction on cortical cells

Of the several cell types to be found in the striate cortex, the simple cell is probably the one most directly concerned in the early stages of binocular depth discrimination (local stereopsis). The other types of cell will not be considered here. A brief account of the receptive field organization of a simple neuron is needed to appreciate the nature of the binocular interaction that occurs on the cell when its two receptive fields are stimulated. The main features of this spatial organization are shown in the scale diagram in Fig. 21-27. The receptive field is roughly circular, with a centrally located long narrow excitatory region (discharge center) and much larger inhibitory sidebands to either side of the central region. These cells respond best to the movement of elongated light or dark bars or edges, provided they are optimally oriented. For the cell in Fig. 21-27, the optimal stimulus orientation is vertical, as shown by the long

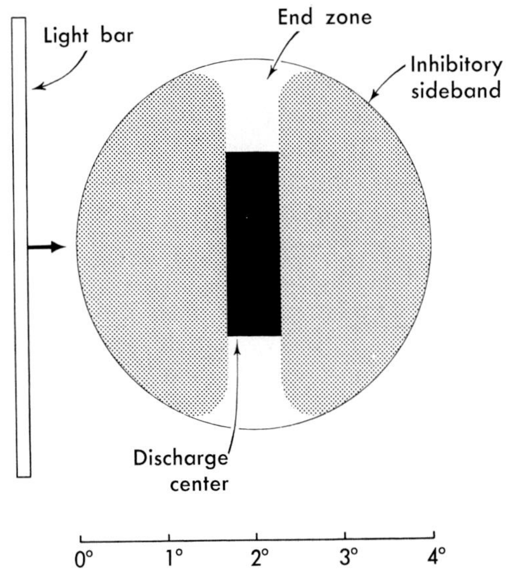

Fig. 21-27. Scale diagram of receptive field of simple striate neuron showing how long, narrow bar of light is used to stimulate cell.

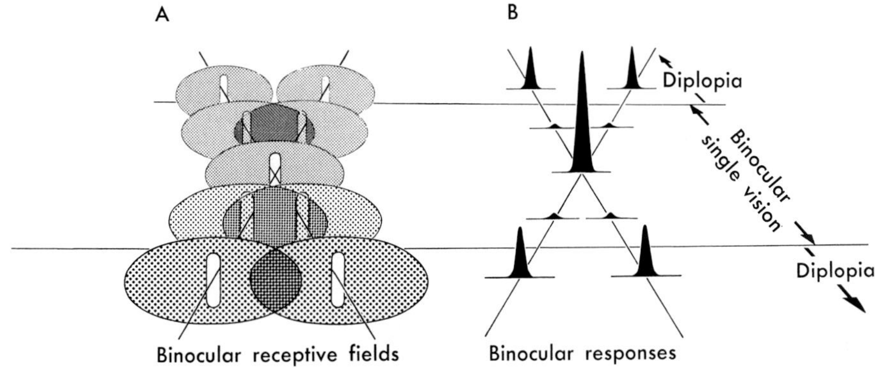

Fig. 21-28. Schematic diagram of essential neural mechanisms underlying binocular single vision and stereopsis. **A,** Superimposition of monocular receptive fields of single binocularly activated striate neuron at various frontoparallel planes in vicinity of intersection of respective lines of sight through fields. **B,** Schematic representation of average response histograms showing facilitation and depression of binocular responses to stimuli situated at same frontoparallel planes as receptive fields in **A.** (From Bishop, P. O.: In Schmitt, F. O., editor: The neurosciences: second study program, New York, 1970, Rockefeller University Press, p. 479.)

narrow bar to the left of the receptive field. The optimal stimulus orientation is always parallel to the long axis of the discharge center as it lies between the two inhibitory sidebands. When the bar or edge is angled away from the optimal orientation, the discharge center is suppressed and the whole of the receptive field becomes purely inhibitory. Cells differ in respect to their optimal orientation. All optimal orientations are to be found, with possibly a slight preference for the vertical and horizontal. In this account only vertical stimulus orientations will be considered, since these make the major contribution to binocular depth discrimination.

Fig. 21-28 illustrates in a purely diagrammatic way the binocular interaction of the two receptive fields of a simple neuron both when they are in register and when they are progressively farther and farther out of alignment. In

Fig. 21-28, *A*, the excitatory regions of the receptive field are represented as clear vertical strips, and the inhibitory sidebands are dotted. The two receptive fields have been plotted as a pair at each of five different frontoparallel planes along their respective lines of sight. They are in accurate register where the two lines of sight cross, and immediately in front of and behind this crossing point the two receptive fields are offset with respect to one another so that now each excitatory region falls within an inhibitory sideband of its fellow field. Still farther away, the receptive fields separate from each other, so in these frontoparallel planes each receptive field may now be stimulated independently of the other.

Fig. 21-28, *B*, is a highly simplified diagram showing the kind of responses that are to be obtained from the cell as a result of the interaction of their two receptive fields. The various peaks in Fig. 21-28, *B*, represent the responses to a stimulus, such as a bar of light, as it passes over the two receptive fields when they are located at the different frontoparallel planes shown in Fig. 21-28, *A*. When the two excitatory regions are in accurate register, the binocular response is markedly facilitated. Just in front of and behind this position the receptive fields

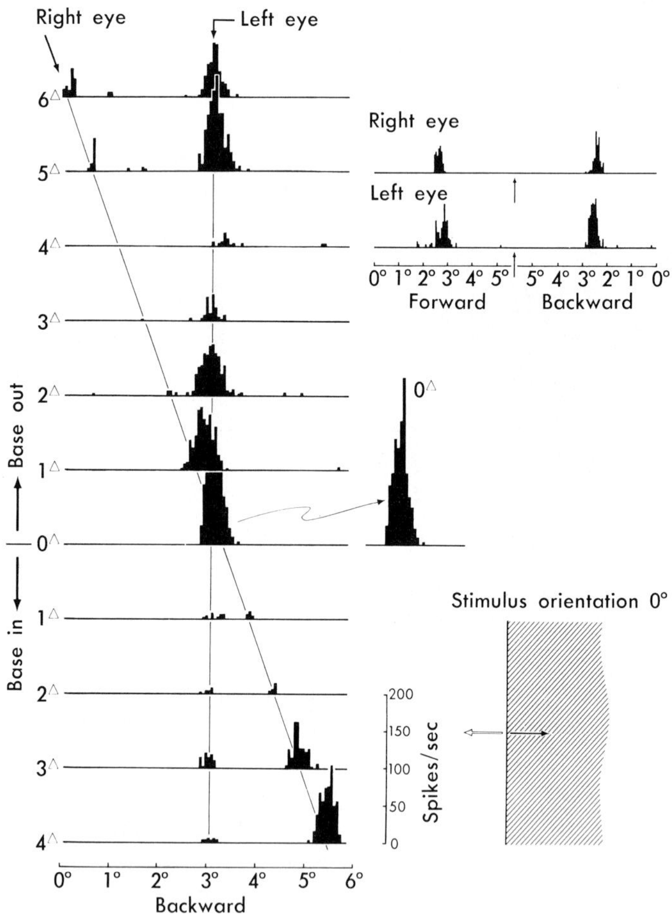

Fig. 21-29. Upper right, monocular responses from simple striate neuron to horizontal movements of vertically oriented light-dark border passing across each receptive field from left to right and back again. Left column, binocular responses to vertical border from same unit over 10Δ range of receptive field alignments, keeping left eye receptive field constant, and moving right eye field with prisms. Full extent of facilitated response when receptive fields were in register (0Δ) is shown at right of column. (Modified from Bishop, P. O., Henry, G. H., and Smith, C. J.: J. Physiol. [Lond] **216:**39, 1971.)

are just sufficiently out of register for mutual inhibition to occur, the excitatory region of one field falling within an inhibitory sideband of the other. Here the response is either greatly depressed or abolished. When the receptive fields become separated, each produces a response independently of the other.

The experimental observations in Fig. 21-29 were obtained by interacting the two receptive fields of a simple striate neuron in a manner closely analogous to that shown diagrammatically in Fig. 21-28. The stimulus was a vertical light-dark border moved broadside from left to right across the receptive fields and back again. The separate monocular responses to the border are shown in the upper right. The monocular responses are about equal for the two eyes, and for the two directions of stimulus movement in the one eye. In the left hand column in Fig. 21-29 the two receptive fields have been interacted, although only the responses to the backward movement of the light-dark border are illustrated. The receptive fields were first moved into accurate register (0^{Δ}) by means of a prism in front of the left eye, the prism power for this eye being thereafter held constant. With the receptive fields in register, the binocular response was greatly facilitated, being 84% greater than the sum of the monocular responses to the same stimulus. Then, with a variable prism in front of the right eye, the receptive field for the right eye was moved horizontally out of register with the left eye field, first to the left and then to the right. The receptive fields were misaligned in prism steps of 1^{Δ}, and binocular responses were obtained at each prism setting. For positions on either side of exact register the responses from the two eyes were virtually abolished. Farther again to one or the other side the responses recovered, but although the prism settings covered a total range of 4 degrees, this was still not sufficient to allow the responses to recover their normal level. Despite this, the general pattern of binocular interaction in Fig. 21-29 closely resembles the schematic representation in Fig. 21-28.

In the type of binocular interaction just described (Figs. 21-28 and 21-29) each of the two receptive fields had similar properties including particularly the same direction selectivity. Very occasionally, however, cells are encountered in both cat[16,36,149] and monkey[198] visual cortex, which have the opposite direction selectivity in the two eyes. A vigorous binocular response can be obtained from these cells only when the retinal images in the two eyes move in opposite directions. Since this occurs only when objects move in depth, either toward or away from the animal, these cells appear to be specialized to signal object motion in depth. When objects move in a frontoparallel plane at right angles to the line of sight, the images move in the same direction in the two eyes, and for these cells there is then mutual inhibition from the eyes in both directions of stimulus motion.[16]

Binocular interaction on cortical cells can also be analyzed in terms of the differential latencies of the impulses from the two eyes.[13,37,96] This form of analysis is relevant to the Pulfrich phenomenon[157] in human psychophysics. If a plumb bob oscillating in a frontal plane is viewed with a neutral density filter before one eye, the bob appears to move in an elliptical path that seems to be nearer the subject for movement in one direction and farther away for movement in the opposite direction. The generally accepted explanation for the stereo effect is that the dark glass increases the latency of the impulses from the affected eye.

Binocular single vision: Panum's fusional area

With normal binocular vision, even the optimal stimulus will excite a cortical cell only if it is presented to its two receptive fields at the same time. This will occur only when the stimulus is applied at that depth in space where these two receptive fields are in accurate register. The phenomenon of binocular single vision, or fusion, is a necessary consequence of the fact that both receptive fields are reporting the *same* stimulus feature at the *one* locus. Thus when stimulation occurs with the receptive fields in register, the question of double vision does not arise. If the visual feature is situated either closer to or farther away from the animal than the depth at which the receptive field pairs concerned are in register, it must stimulate these fields at slightly different times. Double vision is prevented under these circumstances by the marked mutual inhibition that occurs between the two members of a receptive field pair. The mutual inhibition, however, only occurs over a relatively restricted region in space determined by the dimensions of the inhibitory sideband components of the receptive fields. This inhibitory action provides an explanation for the absence of diplopia within Panum's fusional area. Outside this local area the possibility of double vision arises because the receptive field pairs become sufficiently misaligned to act independently of one another. Even so, in the natural environment mutual inhibitory influences will still occur because the receptive field pairs will now be continually presented with dissimilar, and therefore inappropriate, stimuli. An inappropriate stimulus leads to inhibition of the cell even under monocular conditions. This presumably accounts for the fact that we are usually unaware of double vision outside Panum's fusional area.

The phenomenon of fusional displacement (allelotropia, p. 589) can also be accommodated by the theory just outlined. Even though two receptive fields are involved, only the one object feature is detected by each cortical cell and assigned by it to a single locus in space, which is where the two fields are in precise register. This single object feature must necessarily have a visual direction different from either or both of the visual directions of the two receptive fields.

The spread of receptive field disparities forms the basis of Panum's fusional area, since it allows object features to be assigned to single loci (i.e., seen as one) over a limited range of depths in space. Although it is easy to conceive the limits of binocular single vision as being a function of the spread of receptive field disparities, the limited knowledge presently available makes it difficult to put forward proposals as to the nature of this function. Fig. 21-30 shows in diagrammatic form the construction of a possible region of binocular single vision in the cat based on the arbitrary assumption that for any given visual direction the limits of the region are set at the one standard deviation on either side of the mean of the receptive field disparity distribution. The construction is completed by repeating the procedure over a range of different visual directions (i.e., at different retinal eccentricities). The horopter was defined earlier as the locus of the

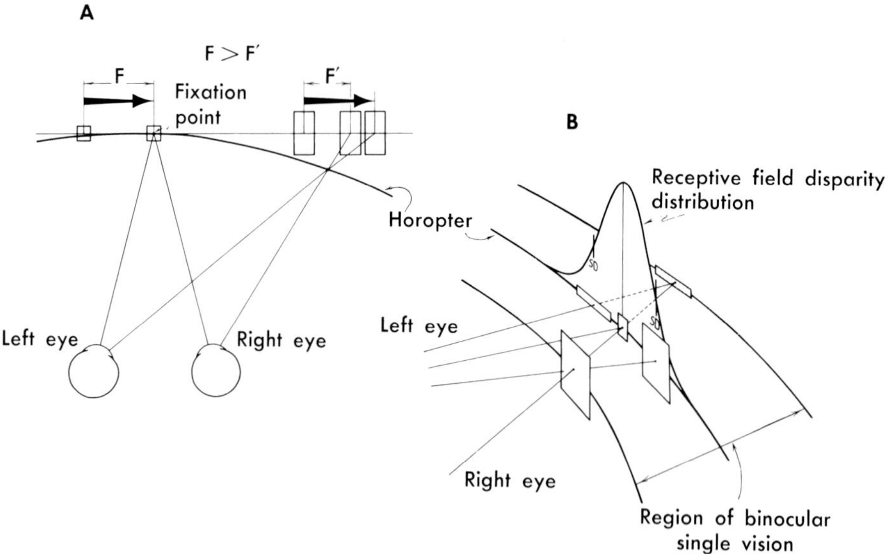

Fig. 21-30. A, Construction for horopter in cat. Decrease in mean receptive field separation with increasing retinal eccentricity, $F > F'$, causes horopter to be concave toward animal. **B,** Construction of region of binocular single vision in cat analogous to Panum's fusional area in the human. (From Joshua, D. E., and Bishop, P. O.: Exp. Brain Res. **10:**389, 1970.)

mean of the various receptive field disparity distributions.

Preferred stimulus orientation disparities

Angular differences between the two retinal images (orientation disparities) occur whenever tilted lines and edges are viewed binocularly, and it has been proposed that the detection of these orientation disparities may act as a second neural mechanism for binocular depth discrimination.[22] One observation in support of this suggestion is that a striate cell does not necessarily have the same optimal stimulus orientation for the two eyes. Both in cat[22,128] and monkey[23] the cells exhibit a range of different preferred orientations in the two eyes of about ±20 degrees. However, the relatively poor orientation disparity discrimination possessed by striate cells, at least in the cat,[128] makes it unlikely that these cells do operate as orientation disparity detectors. Nevertheless, a striking feature of the range of optimal stimulus orientation disparities found among cat striate neurons is that it is approximately the same as the range of retinal image orientation disparities the animal is likely to experience in normal viewing. Thus for any tilted line that the cat observes there stands available a striate neuron whose preferred stimulus orientation in the two eyes matches the orientation disparity of the line concerned. While tilt detection appears to be based on position disparities in the two retinal images rather than any difference in their orientations,[128] striate cells make position disparity discriminations most effectively (i.e., fire maximally) when the preferred stimulus orientation disparity of the striate cell matches the orientation disparity of the retinal images to which it is responding. Furthermore, a range of preferred stimulus orientation disparities is doubtless essential for the central cyclofusion of orientationally disparate retinal images in the interests of binocular single vision.

Local stereoscopic mechanisms in monkey

This account of the neural mechanisms for binocular depth discrimination, based on data obtained from the cat, can now be greatly augmented by important new discoveries in the monkey. Hubel and Wiesel[85] had observed cells sensitive to binocular depth in area 18 of the macaque monkey cortex but, failing to find any such cells in area 17, concluded that mechanisms subserving stereoscopic depth must lie outside the striate area. The elegant new experiments of Poggio and Fischer[155] have, however, again focused attention on area 17 as the primary cortical area for stereopsis in the monkey. They recorded from single cells in the foveal striate and prestriate cortex of the awake performing macaque monkey trained to maintain fixation on a small luminous spot (Fig. 21-31, fixation spot F) for several seconds at a time. During the periods of fixation, single moving light or dark bars were presented on a cathode-ray tube (CRT) screen at a series of different depths relative to the fixation plane. The front oscilloscope screen (CRT) and the rear screen (S) in Fig. 21-31 were superimposed both for the monkey's eyes and the television camera (CCTV) by a beam splitter (BS). During the periods of fixation the position stability of the eyes was monitored by an infrared corneal reflection method. The stability was sufficiently good to enable Poggio and Fischer to test single cells for their ability to discriminate the location of a stimulus in depth by moving the CRT screen along the center line of the system, as indicated by the double-headed arrow in Fig. 21-31.

Nearly all neurons in the foveal striate and prestriate cortex of the alert and behaving macaque receive a binocular input, and the majority of these (84%) are sensitive to the location of stimuli in depth. In both cortical areas Poggio and Fischer[155] recognized four types of depth-sensitive cells and, although the properties of these cell types were much alike in the two cortical areas, the following account refers specifically to the striate cortex.

Tuned excitatory neurons. Tuned excitatory neurons were the most common type, comprising 48% of the depth-sensitive cells in the striate cortex. Although each gave an excitatory response to monocular stimulation of either eye, they responded rather differently to binocular stimulation. When stimulated binocularly they gave an enhanced excitatory response, but only at a particular depth in space at or not

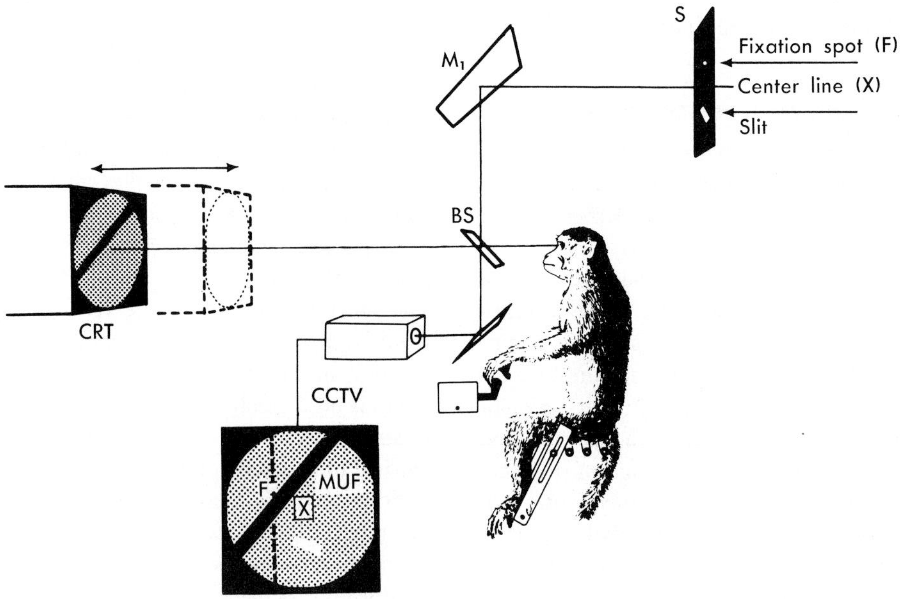

Fig. 21-31. Experimental setup for studying the depth sensitivity of cortical neurons showing the position of the monkey relative to the front oscilloscope screen *(CRT)* and rear projection *(S)* superimposed for the monkey's eye and the television camera *(CCTV)* by a beam splitter *(BS)*. Stimulus depth was introduced by changing the position of the front screen along the center line of the system. (From Poggio, G. F., and Fischer, B.: J. Neurophysiol. **40:**1392, 1977.)

far from the plane of the fixation point. Nearer or farther there was a more or less pronounced inhibition over sidebands that could extend to more than 1 degree of disparity in both directions. The precise depth to which the cell was sensitive varied from cell to cell over a narrow range.

Tuned inhibitory neurons. Tuned inhibitory neurons were much less common (18%), but they were in many ways similar to the tuned excitatory cells except that their responses were suppressed at or close to the plane of the fixation point.

Near and far cells. The two remaining types made up 34% of the depth sensitive cells. Both had asymmetric response profiles to either side of the plane of the fixation point. The *near cells* were binocularly excited by stimuli in front of the fixation point and suppressed by stimuli behind. The *far cells* had the opposite kind of response profile: excited behind the fixation point and suppressed in front. Unlike the tuned excitatory and tuned inhibitory neurons, the binocular excitatory and inhibitory interactions on near and far cells takes place over a wide range of depths in space, and for many of the cells the full extent of the range could not be estimated because of limitations in the apparatus. For each cell, however, the transition from excitation to suppression always took place within about 0.2 degree disparity from the plane of the fixation point.

Fine stereopsis. The properties of the tuned excitatory and tuned inhibitory neurons clearly fit them to operate as part of a fine stereopsis mechanism. Each cell is capable of a fine depth discrimination at its preferred retinal disparity, and a high stereoacuity is achieved by the sharpness of the position disparity tuning. In some cells the tuning was less than 3 minutes of arc, which was the limit of the accuracy of Poggio and Fischer's method. Furthermore, the population of tuned cells show a range of different crossed and uncrossed position disparity preferences, so a range of different stimulus disparities can be detected. Even so the range is quite small, ±0.2 degree at the center of gaze and expanding out somewhat away

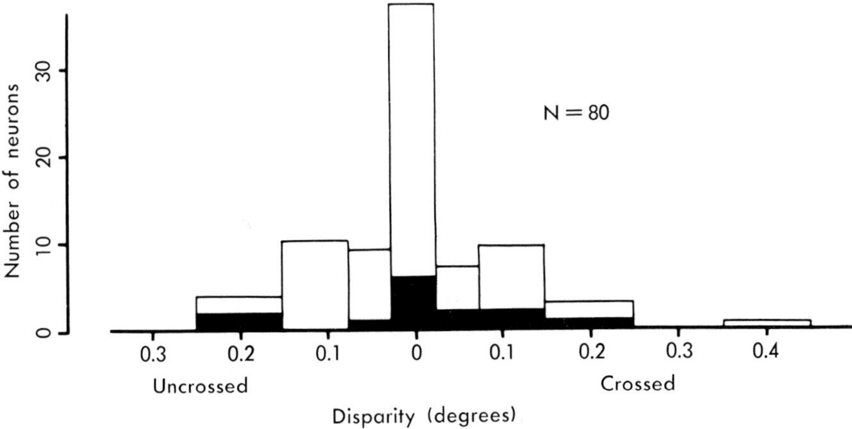

Fig. 21-32. Distribution of preferred disparities of tuned excitatory (open areas) and tuned inhibitory (filled areas) neurons recorded from monkey cortical areas 17 and 18. (From Poggio, G. F., and Fischer, B.: J. Neurophysiol. **40:**1392, 1977.)

from the center. The distribution of the preferred stimulus disparities for the two types of tuned neurons in Fig. 21-32 has a standard deviation of ±0.1 degree. The properties of the tuned cells in the monkey are obviously closely similar to those that have been studied in the cat. The distribution of the preferred disparities in the monkey are analogous to the distribution of the receptive field disparities in the cat, and the response profiles for the tuned cells resemble the position disparity tuning curves that have been obtained from striate cells in the cat. The distribution of receptive field disparities in the cat has, however, a range of a little more than ±1 degree, whereas the distribution of preferred disparities in the monkey has a range of only ±0.2 degree. The spread of the receptive field disparities determines the extent of Panum's fusional area. On the assumption that Panum's fusional area is the same in the macaque monkey as in the human, the distributions in cat and monkey just described are in keeping with the observation that the cat has a fusional range about five times greater than the human (p. 607).

Coarse stereopsis. The far and near cells could operate as part of a coarse stereopsis mechanism. The existence of two types of neuron differentially encoding convergent and divergent disparities has been suggested by Richards[161] as a result of his psychophysical observations on partially stereoblind subjects (p. 602). In addition, the far and near cells could be used for the binocular servocontrol of vergence eye movements. Poggio and Fischer suggest that the far neurons might activate the convergence system and the near neurons the divergence system in a reciprocal way; at the fixation distance the cortical cells for the two systems are balanced and eye vergence remains stationary. The long range of binocular interaction displayed by the far and near cells could explain why depth judgments can be made and vergence eye movements initiated by stimuli with 5 to 10 degrees disparity, grossly exceeding the width of Panum's fusional area. It also explains how the different kinds of stereoanomalies and disorders of vergence eye movements might come about.

A striking feature of layer IVC in monkey cortex (and layer IV in the cat) are the alternating regions dominated by the left eye or the right eye, known as ocular dominance "columns" (monkey[83,107]; cat[108,169]). The bulk of the geniculocortical afferents segregate themselves into roughly parallel alternating stripes, one set from the left eye, the other from the right. Microelectrode recordings show that the cells in each region are discharged only monocularly, the segregation of the afferents determining the eye preference. It was therefore an unexpected observation by Poggio and Fischer[155] when they failed to demonstrate a binocular input in only 4 out of 199 cells in areas 17 and 18, particularly as the regions sampled included layer IVC. They did, however, find cells that gave an excitatory response to stimulation of one of the two eyes and not of the other. About 13% of the cells in the two striate areas were of this type with the majority (20 out of 25) located in area 17 rather than 18 and principally in layer IVC. These cells would have been classified as monocular had they not been tested with stimuli presented in depth. The main cell of this type was the tuned inhibitory neuron. Presumably the segregated geniculate terminals in layer IVC, coming from one eye or the other, are all excitatory and, being direct geniculate afferents, would be selectively revealed by autoradiographic and reduced silver methods. If the input from the nondominant eye is inhibitory, it would presumably operate through an intracortical interneuron and therefore would not be made evident by current histologic methods. These observations are important because they mean that there is a total binocular mixing of information from the two eyes as soon as the impulses reach the striate cortex. In binocular vision there is no separation elaboration of monocular form perception.

Binocular fixation: involuntary eye movements

Stereopsis and fusion are not affected by the constantly varying binocular disparities due to the residual inequalities in the saccades and drifts in the two eyes during binocular fixation, despite the fact that these disparities have a spatial distribution equal to or greater than Panum's fusional area (p. 607). The horizontal and vertical spread of receptive field disparities provides a mechanism for offsetting this residual incoordination. During binocular fixation the two visual axes wander both about the fixation point and with respect to one another. Thus the locus of maximal receptive field alignment will also wander with respect to the fixation point. Nevertheless, within the limits of the spread of receptive field disparities there will always be other receptive field pairs to take over the signaling of information about object features at the fixation point when random eye motions have displaced the receptive field pairs that were formerly in register at that site. Furthermore, the stereoscopic mechanism is relatively insensitive to the retinal image disparities caused by these small nonconjugate eye movements because it is concerned only with relative depth estimates: stimulus A in front of stimulus B, which in turn is in front of stimulus C and so on. These movements shift the retinal images bodily from one part of the receptive field disparity distribution to another without in any way altering the depth sequences of the various image features. The effect of relatively large nonconjugate eye movements is to shift the discrimination of the depth sequence to the tails of the receptive field disparity distribution. There will be a momentary reduction in the level of the stereoacuity before the fixation reflex returns the object of regard to the center of the receptive field disparity distribution.

Stereopsis about the vertical midline

While the nature of the neural organization in the region of the vertical midline of the retina poses a number of important problems, it also provides valuable insights into possible mechanisms for fine and coarse stereopsis. The classic description of the nasotemporal partition of the retina is that it takes place along a sharp vertical line passing through the center of the foveola. According to this idea the visual hemifields, in their projection to the cerebral hemispheres, are strictly limited by the vertical through the visual axis of each eye, information from the two right hemifields going to the left cerebral hemisphere and vice versa for the two left hemifields. It can be seen from the upper diagram in Fig. 21-33 that objects lying between the visual axes and in front of the fixation point are imaged only on the two temporal retinae, while objects between the visual axes and behind the fixation point are imaged only on the two nasal retinae. Were a strict nasotemporal partition to hold as described earlier, it is also clear from Fig. 21-33 that information about objects between the two visual axes would be available to each cerebral hemisphere from only one eye, closer than the fixation point only via the ipsilateral eye and farther than the fixation point only via the contralateral eye. Thus stereopsis

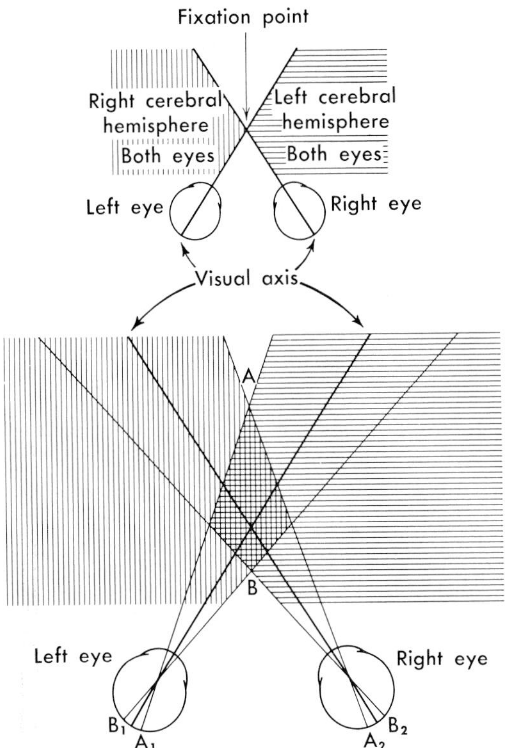

Fig. 21-33. Diagrams to show that neural mechanism for binocular depth discrimination requires nasotemporal overlap.

would not be possible in just those regions in the visual field where stereoacuity is known to be maximal. Stereoacuity, like visual acuity, is maximal at the fixation point; that is, stereothresholds are lowest for objects just nearer to or farther than the fixation point. Furthermore, neither in ordinary experience nor in psychophysical observation is there any evidence to suggest that there is a qualitative difference between the nature of stereoscopic perception when it involves the vertical midline of the two retinae as opposed to corresponding regions away from the midline. The same neural mechanisms appear to operate in both cases. The phenomenon of receptive field disparity as applied to the retina in general is based on a partial decussation at the optic chiasm whereby ganglion cells located in the area of retina nasal to the foveola send their axons to the contralateral tract, and cells located temporal to the foveola send their axons to the ipsilateral tract. This enables the retinocerebral pathways from corresponding retinal area in the two eyes to come together in an optic tract, thereafter traveling together, undergoing the same, although largely independent, processing in the lateral geniculate nucleus, and finally synapsing on common cells in the striate cortex in a fairly precise topographic manner. The essential feature of this organization is that the same neural transformations can be readily applied to the two visual pathways from corresponding retinal areas on their way to the cerebral cortex. On the basis of a strict nasotemporal partition of the retina such a common neural processing could not be applied to the afferents of a binocular striate neuron were its two receptive fields to be located on noncorresponding sides of the vertical midline.

There is still another problem posed by the concept of a sharp nasotemporal partition when the binocular disparities due to involuntary eye movements are taken into consideration (p. 606). For a given binocular fixation point these movements are continually displacing the visual axis of one eye into the "wrong" hemifield of the other eye, and during these intervals of retinal misalignment stereopsis would again be impossible because information in the region of overlap would be available from only one eye.

Two neuronal mechanisms exist that, when taken together, provide a solution for these problems. One mechanism is a nasotemporal overlap in the retina whereby ganglion cells that project to ipsilateral and contralateral optic tracts intermingle in a vertically oriented median strip of retina centered on the foveola. From the lower diagram in Fig. 21-33 it can be seen that this enables each visual cortex to receive information via both eyes from a median region (AB) in the visual field centered on the fixation point. Such an overlap also provides the necessary basis for the receptive field disparity mechanism to operate in the vicinity of the midline. Thus if one receptive field of a binocularly activated striate neuron is located on the vertical midline, it is just as likely to have its companion field for the other eye located to the left as to the right of the midline. This overlap also makes the stereoscopic mechanism relatively insensitive to the binocular

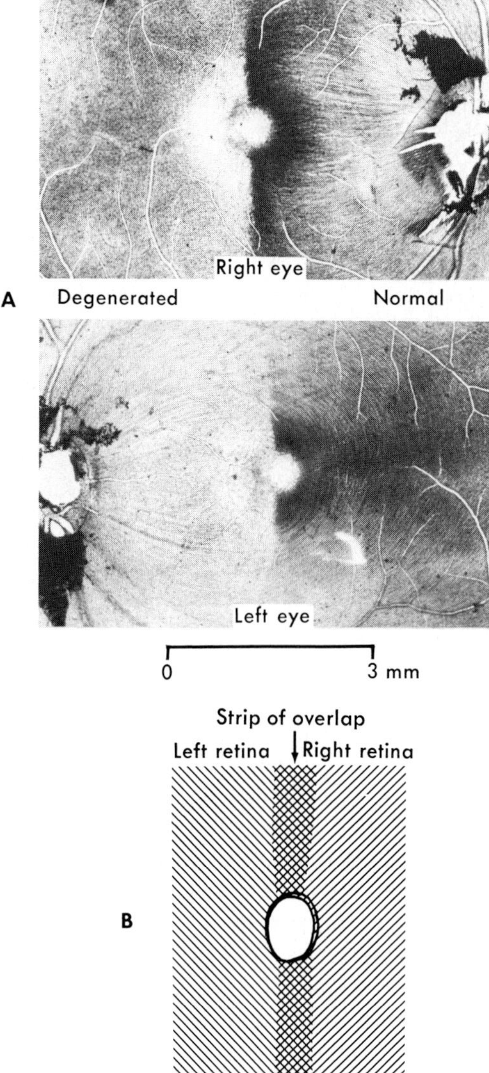

Fig. 21-34. A, Two retinas from one monkey seen in methylene-blue whole mounts. Right optic tract had been sectioned 6 months previously, causing degeneration of ganglion cells in temporal part of right retina and nasal part of left retina. Nasotemporal overlap can be appreciated by comparing in the two retinas, vertical line of demarcation between normal and degenerated areas as shown in **B. B,** Scale drawing showing region of nasotemporal overlap produced by superimposing the two retinas after left-to-right inversion of left retina. (From Stone, J., Leicester, J., and Sherman, S. M.: J. Comp. Neurol. **150:**333, 1973.)

disparities resulting from the involuntary eye movements of bifixation. The narrow vertical strip of retina projecting to both cerebral hemispheres is probably mainly concerned with fine stereopsis. The other mechanism mentioned earlier, involving fibers crossing in the corpus callosum, may provide a mechanism for coarse stereopsis and vergence eye movements.[10]

By sectioning one optic tract in the cat and observing the pattern of degeneration of retinal ganglion cells, Stone[173] was the first to demonstrate that both optic tracts were supplied by ganglion cells in both eyes from a median strip of nasotemporal overlap. Stone and his collaborators[174] have subsequently shown that this is also the case in the monkey. Fig. 21-34, *A*, shows methylene-blue whole mounts of the two retinas from one monkey whose right optic tract had been sectioned 6 months previously. Degeneration of ganglion cells is clearly evident in the temporal part of the right retina and the nasal part of the left retina. The nasotemporal overlap can be appreciated by comparing, in relation to the fovea in each retina, the vertical line of demarcation between normal and degeneration areas in the two eyes. The overlap is shown more clearly in the scale drawing in Fig. 21-34, *B*, which has been produced by superimposing outlines of the two retinas after a left-to-right inversion of the left retina. Considering only the cell bodies, the median strip of overlap, centered on the fovea, is about 1 degree wide. When the cell dendrites and corresponding receptive fields are taken into consideration, the overlap would be somewhat wider than 1 degree. This overlap has also been demonstrated neurophysiologically in the cat at the level of the lateral geniculate nucleus,[166] and in the vicinity of the 17-18 border in the visual cortex.[18,106] Furthermore, the overlap at the cortex survives section of the corpus callosum.[106]

The lesion-degeneration technique used by Stone and co-workers[174] is beset by the difficulty of distinguishing surviving ganglion cells from neuroglia. This problem is avoided by the use of the newer horseradish peroxidase tracing technique. Bunt, Minckler, and Johanson[26] made unilateral injections of horseradish peroxidase into the dorsal lateral geniculate nucleus and optic tract of monkeys and subsequently examined the pattern of retrograde labeling of retinal ganglion cells. The glial cells were distinguished by the absence of peroxidase-positive granules. Like Stone and his colleagues, they found that a 1-degree wide vertical strip is present in the monkey retina in which ganglion cells projecting ipsilaterally as well as contralaterally are intermixed. In addition, they found a mixing of labeled and unlabeled ganglion cells in a band approximately 0.5-degree wide along the nasal and temporal rims of the foveal pit which in the macaque monkey is 2 degrees in diameter. Thus the 1-degree vertical strip of overlap is widened to a total of 3 degrees around the fovea. Apart from an occasional extrafoveal labeled ganglion cell as far as 2 degrees from the vertical midline in the otherwise unlabeled hemiretinae, there was none of the very broad overlap of different classes of ganglion cells that is to be found in the cat retina.[173] Although the overlap in the monkey is restricted to a very narrow band that straddles the vertical midline, it is, nevertheless, more than ample to account for the distribution of preferred retinal image disparities that Poggio and Fischer[155] found for tuned excitatory and tuned inhibitory neurons in the monkey cortical areas 17 and 18. Thus the overlap can provide for fine stereopsis in the vicinity of the fixation point the same kind of receptive field disparity mechanism that is to be found for regions of the visual field away from the vertical midline. In addition, it is enough to make the fine stereopsis mechanism relatively insensitive to the retinal image disparities that result from asymmetries in the involuntary eye movements of steady bifixation.

Evidence for a nasotemporal overlap in the human has been provided by Huber,[88] who has shown that following removal of one occipital lobe there is a resulting homonymous hemianopia with a separation line between the normal and blind halves of the field that lies about 1 degree from the fixation line toward the blind half. Occipital lobectomy removes the contribution that fibers in the corpus callosum make to the bilateral representation of the vertical midline region. Sparing of vision in the otherwise blind hemifield as a result of unilateral damage to the occipital cortex is, however, more commonly described as being restricted to the foveal region where it is known as "macular" or more accurately as "foveal sparing." It seems likely that the more restricted kind of sparing and that described by Huber[88] are different manifestations of the same nasotemporal overlap mechanism, the difference between the two sets of observations being possibly due to Huber's more adequate control of the patient's fixation.

The vertical strip of overlap of ipsilaterally and contralaterally projecting ganglion cells in the monkey is, however, quite inadequate to provide a basis either for a coarse stereopsis mechanism or for fusional vergence reflexes. These processes require receptive field disparities up to 7 degrees or more. Hence, in the monkey at least, and doubtless also in the human, the only likely basis for coarse stereopsis would be via corpus callosum connections, since this seems the only way in which coarse disparities of 5 to 10 degrees could influence cells in the visual cortex. Thus Blakemore[20] found that coarse stereopsis was present in a patient who had sustained a traumatic lesion of the optic chiasm that was diagnosed to have completely interrupted the crossed fibers. Stereopsis in this case would have completely depended on fibers crossing in the corpus callosum. Although coarse stereopsis has been demonstrated in a patient with a split chiasm, it has yet to be shown that fine stereopsis can occur in the absence of callosal connections. Tests specifically designed to resolve this possibility have yet to be applied. There is the further possibility that fine stereopsis requires both mechanisms if only to ensure adequate binocular fixation.

There is much anatomic and neurophysiologic evidence (review[10]) that cells in a strip of cortex in the vicinity of the projection of the vertical midline of the retina along the 17-18 border in one occipital lobe send their axons in a topographically organized manner to the corresponding region in the occipital lobe of the other side. This arrangement also provides each visual cortex with information via both eyes from both sides of the vertical midline. Present evidence is, however, inadequate to say whether even these connections are sufficiently broad to provide the large receptive field disparities that are needed for coarse stereopsis and fusional vergence reflexes.

DEVELOPMENT OF BINOCULAR VISION[2]

Normal postnatal development

In the past, the assessment of visual acuity in preverbal infants has been neglected because of the difficulty in eliciting a response properly indicative of the level of the acuity. Dobson and Teller[43] have recently reviewed the data that are now becoming available concerning the development of visual acuity in the first 6 postnatal months. Three different objective techniques have been developed—optokinetic nystagmus (OKN), preferential looking (PL), and the visually evoked potential (VEP). All three techniques show that acuity improves rapidly between birth and 6 months of age, although the acuity values with VEP are typically higher than those measured with PL and OKN. When comparable scoring criteria are used to estimate acuity, the combined results of the three procedures indicate that acuity increases from about 20/300 in the newborn period to close to 20/20 by 6 months. Little systematic attention has, however, been paid so far to the question of establishing statistically valid age norms.

Because the macaque monkey provides a good animal model for human vision, the relative time scales for the development of visual acuity in the two species is a matter of considerable interest. Using the OKN[140] and the PL[178] techniques, the grating acuity in the newborn macaque monkey has been found to be about the same as the human infant at birth (20/300), but subsequently the monkey's acuity develops much more rapidly. The improvement takes place over approximately the same range in the two species, but it does so in the first 6 weeks in the monkey rather than the 6 months needed by the human infant. Thus at least some macaque and human infants achieve adult acuity as early as 6 weeks and 6 months respectively.

Since stereoacuity depends on monocular visual acuity, stereopsis must develop either at the same time as or subsequently to appropriate levels of visual acuity, with possibly coarse stereopsis and fusional vergence reflexes developing first and then fine stereopsis. The level of visual acuity attained during the first few months of infancy is probably adequate for the development of coarse stereopsis, but such a development can hardly take place in the absence of fusional vergence reflexes. It has been widely held that during the early weeks of infancy, eye movements are "independent, irregular and unconjugated . . . [and that] conjugate fixation does not become accurate until the age of 6 months by which time convergence is firmly established."[46] These views are now undergoing drastic revision. Recent evidence emphasizes the substantial sensory competence of the newborn, actively seeking visual information by controlled eye scanning rather than behaving as a passive reflex organism.[66] Our knowledge of the development of binocular reflexes is, however, still only rudimentary. Dayton and his colleagues[38] have shown, both by electro-oculography as well as by observation, that nearly a third of the newborn infants they examined could locate and follow a moving target in the visual field with close conjugation of the two eyes. Although the saccadic movements were of larger amplitude than in the adult, the time and direction of each movement were similar, again demonstrating close conjugation of the eyes.

Only fragmentary observations are available concerning the subsequent development of binocular fixation, although direct observation suggests that fusional vergence reflexes are well developed by the age of 6 months. The stages in the development of stereoscopic depth perception are even more obscure. It has become clear that the earlier views on the relative immaturity of vision in the newborn were the result not only of inadequate test procedures but also of a failure to distinguish the postnatal development of the basic visual mechanisms and functions from the relatively slower development of the visual learning involved in object recognition and in visually guided behavior.[141] One important outcome of the study by Ordy and his colleagues[141] is that in the postnatal development of vision in the macaque monkey optokinetic nystagmus acuity develops much earlier than behavioral discrimination acuity, and this same pattern is to be observed in the human infant. In the light of these reports on the development of visual acuity it is likely that the human infant also achieves adult levels of stereoacuity much earlier than the age suggested by Amigo[2] as a result of his observations using the Titmus stereotest. He applied the test to infants between the ages of 3 and 5½ years and concluded that adult stereoacuity was not reached until after 5 years of age. Fox and co-workers[58] have now described an objective stereotest that minimizes the need for cooperation by the subject and hence can be applied to preverbal and preliterate infants. These observers induced optokinetic nystagmus in adult human subjects by means of a vertical array of stereoscopic contours formed from dynamic random-element stereograms and appearing to move continuously in a horizontal direction. Since stereopsis is a prerequisite for the perception of stereoscopic contours, their ability to induce optokinetic nystagmus provides the basis for an objective test for stereopsis and stereoacuity. Such a test, specific for stereopsis, is also important when the development of binocular vision is considered because it provides a way of distinguishing between depth perception based on secondary monocular cues and true stereopsis.

Plasticity during maturation: human observations

It has long been a fact of clinical observation that visual deprivation has a much more marked effect in the very young than it does in later childhood and adult life. Corneal clouding or cataract may prevent form vision and allow only diffuse light to enter the eye. In the very young and up to the age of about 5 years this visual deprivation usually leads to severe and irreversible amblyopia *(amblyopia ex anopsia),* which will persist despite surgical correction of the underlying condition and prolonged enforced usage of the amblyopic eye. There is also a direct relationship between the onset of esotropia and the occurrence of amblyopia *(strabismic amblyopia).* The earlier the onset the more rapidly is the amblyopia acquired and the less susceptible it is to modification in later childhood and adult life. In older children and adults amblyopia does not develop.

While these abnormalities acquired by the very young may persist throughout life, clinical observation also indicates that considerable plasticity exists during the maturation of the visual processes. It is only within the maturation period for stereopsis that the young strabismic patient retains sufficient plasticity to develop the sensory adaptation of suppression and abnormal retinal correspondence that permit single binocular vision. After the age of about 9 years, children are incapable of acquiring these adaptations and may suffer persistent double vision. Thus abnormalities having their onset in the first few months or years of life may still be reversed, provided treatment is not delayed. The almost universally adopted treatment for strabismic amblyopia is constant and complete occlusion of the preferred eye, the aim being to restore maximal visual acuity to the amblyopic eye. Occlusion of the fixating eye forces the use of the amblyopic eye and prevents the inhibition that the fixating eye ordinarily exerts on the deviating eye when both eyes are open. Maximal visual acuity in the previously amblyopic eye is achieved when children are occluded before the age of 4 years, and the treatment becomes increasingly less effective between 4 and 9 years of age and is almost without effect after the age of 9 years. Occlusion well before the age of 4 years may not improve the final visual acuity, but the younger the patient, the more rapidly the formerly amblyopic eye responds to the occlusion. Occlusion therapy can be started as early as 6 months. However, one of the hazards of occlusion is the development of *occlusion amblyopia* in the preferred eye. In very young children, occlusion amblyopia can develop within a few weeks, the formerly amblyopic eye gaining full vision as quickly as the originally fixating eye acquires a deep amblyopia. Such a development is prevented and equality of vision is maintained in the two eyes by alternating the occlusion from one eye to the other. Animal experiments (to be described on p. 629) suggest, however, that while occlusion can assist the development of good monocular visual acuity, it may at the same time hinder the development of normal binocular vision and stereopsis. Other forms of treatment are needed both to institute and maintain stereopsis and to avoid the adverse effects of abnormal binocular vision. In the surgical treatment of congenital esotropia, Taylor[176] found that surgery is effective in producing binocular single vision with a reasonably high level of stereopsis, ranging from 40 to 400 seconds of arc, only when it is accomplished before the age of 2 years. After the age of 2 years a good cosmetic result was obtained, but a residual small manifest strabismus always remained.

Plasticity during maturation: animal experiments

Although indicating that there is considerable plasticity both in the immediate postnatal period and in early infancy, observations on the human infant have not provided much insight into the nature of the mechanisms responsible for the plasticity. Beginning with the pioneering studies of Wiesel and Hubel[194,195] however, there have been striking developments in our understanding of these mechanisms mainly through the application of behavioral techniques combined with single-unit electrophysiology and experimental neurohistology to the cat and the monkey. The large body of literature that has now accumulated in this field, however, has been primarily concerned with the general problem of plasticity and only somewhat secondarily with plasticity as it affects binocular vision. Attention is here concentrated on binocular vision.

Cat experiments

Ocular dominance of cortical cells. Hubel and Wiesel[79] had earlier shown that about four fifths of cells in the striate cortex of the adult cat can be influenced from both eyes. They used the term *ocular dominance* to refer to the fact that the relative influence of the two eyes on cortical cells varies from one cell to the next. Fig. 21-35, *A,* shows the ocular dominance distribution of 223 cells that they recorded from the striate cortex in a series of normal adult cats.[79] The cells in the seven dominance groups ranged from those influenced only by the contralateral eye (group 1), through those having mixed influence (groups 2 to 6), to those exclusively ipsilateral (group 7). For the cells in group 4 there was no obvious difference between the two eyes. Like retinotopic order, this pattern of ocular dominance is innately determined. Thus a dominance histogram similar to the adult pattern is also to be found in normal kittens close to the time of birth before they have had patterned visual experience.[80] During the first months after birth these binocular connections are, however, very vulnerable and can be disrupted by various procedures of visual deprivation such as closing the lids of one or both eyes, producing a strabismus surgically, or occluding each eye alternately day by day.[82,84,195]

Monocular and binocular deprivation. Eye closure by lid suture prevents patterned vision but leads to only minor light deprivation. Nevertheless, monocular eyelid suture in a young kitten causes dramatic alteration in the physiologic organization of the visual cortex. The cortical cells suffer an almost total loss of binocularity. Functional connection with the deprived eye is lost, and virtually every cell comes to be driven only through the normal nondeprived eye. Except in respect to binocularity, however, nearly all the cells are still found to respond normally. Fig. 21-35, *B,* shows the dominance histogram of 199 cells recorded in the left visual cortex of monocularly deprived kittens. At the time of recording the animals were 8 to 14 weeks old, and all had had the right eye closed by lid suture from the time of normal eye opening at 8 to 10 days postnatal. Dominance by the ipsilateral nondeprived eye is almost complete. As might be expected from the physiologic observations, animals such as these appear behaviorally blind when forced to use the deprived eye.[41,194,196]

Contrary to what might be expected, binocular deprivation caused much less disruption of cortical connections than did monocular deprivation, and a large measure of binocularity still persisted. Fig. 21-35, *C,* shows the dominance histogram of 126 cortical cells recorded from kittens, all of whom had both eyes closed from soon after birth till the time of recording

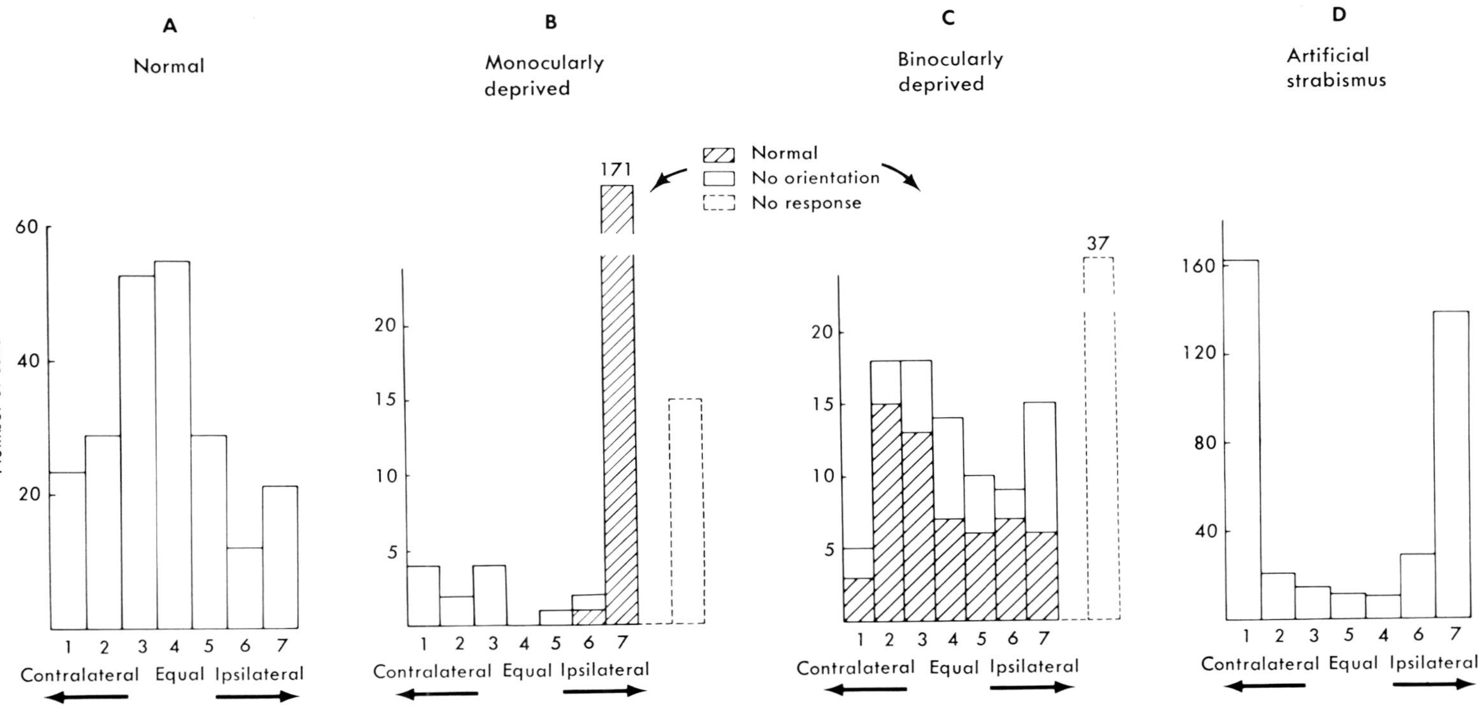

Fig. 21-35. Ocular dominance distributions of single cells recorded from striate cortex of cats under various conditions. **A,** Normal adult cats. **B,** Kittens 8 to 14 weeks old, all of which had one eye closed by lid suture from 8 to 10 days after birth. **C,** Kittens 2½ to 4½ months old that had bilateral eye closure from 6 to 18 days after birth. **D,** Kittens 3 months to 1 year old that had divergent squint produced by section of medial rectus of one eye at 8 to 10 days after birth. (**A** from Hubel, D. H., and Wiesel, T. N.: J. Physiol. [Lond] **160:**106, 1962; **B** and **C** from Wiesel, T. N., and Hubel, D. H.: J. Neurophysiol. **28:**1029, 1965; **D** from Hubel, D. H., and Wiesel, T. N.: J. Neurophysiol. **28:**1041, 1965.)

3 to 4 months later. Although most of the cells were still binocularly influenced, the cortex, however, was not normal, and while many cells did respond normally, there were many that responded abnormally and many that were completely unresponsive. The physiologic abnormality in the cortex is reflected in the fact that binocularly deprived animals appear behaviorally blind.[41,196]

It is clear that the loss of cortical input from the deprived eye depends on the absence of patterned images in that eye, not just the reduction in illumination.

Experimental strabismus, alternating occlusion, and anisometropia

Divergent strabismus. By sectioning the medial rectus muscle in one eye 8 to 10 days after birth, Hubel and Wiesel[82] produced a marked divergent squint in kittens. Except for this procedure, the animals were allowed to develop normally until the time of cortical recording at either 3 months or 1 year later. The recording revealed that the squint had produced a severe decline in the number of cells in area 17 that could be driven from both eyes. The ocular dominance histogram of 384 cells that was recorded from these animals is shown in Fig. 21-35, *D*. Except for the loss of binocularity, the cells behaved normally in every other respect. Hubel and Wiesel concluded that the loss of binocularity was due, not to a loss of cells, but rather to changes in the overall ocular dominance distribution. It was apparently the lack of synergy between the afferents from the two eyes that caused the ocular dominance of the cells to change with an overall increase in the numbers in groups 1 and 7 at the expense of the others. The maintenance of normal synaptic connections seems therefore to depend not only on the amount of incoming impulse activity, but also on a normal interrelationship between activity in the different afferents. Despite the marked loss of binocular innervation by the neurons in area 17, the cats with divergent squint showed no obvious behavioral abnormalities suggestive of defective vision, presumably because the squint was divergent and the cats developed the ability to fixate with either eye alternately.

Convergent strabismus. The known greater incidence of amblyopia in human subjects with esotropia as opposed to exotropia subsequently led Ikeda and her colleagues[91,94] to study the effects of a persistent monocular convergent squint produced by removing the lateral rectus and superior oblique muscles and the nictitating membrane in kittens at different postnatal ages from 3 to 16 weeks. For purposes of comparison, they also prepared kittens with a divergent squint. The visual performance of these animals was then examined at 4 to 10 months of age both by neurophysiologic[91] and behavioral[94] methods. Recordings were made both from the striate cortex, to give an estimate of the proportion of binocularly innervated cortical cells, and from the lateral geniculate nucleus so as to compare the spatial resolving power (visual acuity) of geniculate neurons receiving their input from the normal retina with that of cells with their input from the equivalent area of the retina in the squint eye. Regardless of whether the squint was convergent or divergent, they found that the proportion of binocularly driven striate cells was greatly reduced. Recording from the lateral geniculate nucleus, however, revealed important differences between the two kinds of squint. The visual acuity of geniculate cells was estimated in terms of the highest spatial frequency of a sinusoidal grating stimulus to which the cells responded with modulated firing. In cats with a divergent squint, the visual acuity of the geniculate cells was normal and this was the case also in cats with a convergent squint, but only for those cells that had their input from the normal eye. By contrast, the geniculate cells connected to the central retina of the esotropic eye had a significantly poorer visual acuity than those with their afferents from the central retina of the normal eye. The major defect was suffered by the sustained (X) cell population and the loss of visual acuity was restricted to central vision, since the spatial resolution of geniculate cells connected to the peripheral retina of the esotropic eye was normal. Analysis of the sensitivity loss at different spatial frequencies showed that the higher spatial frequencies suffered the greater loss, a result similar to that found for strabismic amblyopia in the human. As shown in Fig. 21-36, the loss of visual

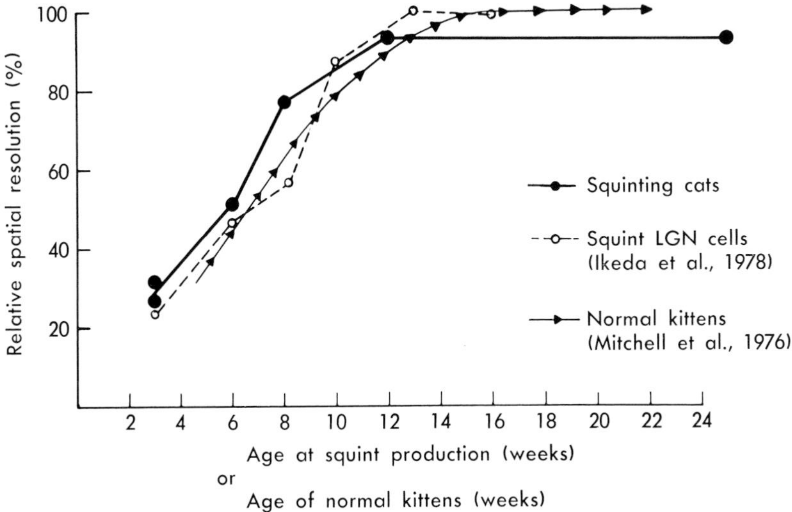

Fig. 21-36. Visual acuity of adult cats with squint produced at different ages[94] compared to that of normal kittens of different ages.[122] The behaviorally determined visual acuity of the squinting eye (relative to the nonsquinting fellow eye) is plotted as a function of age of squint production. Open circles, neurophysiologically determined visual acuity of sustained lateral geniculate neurons driven from the area centralis of squinting eyes of adult cats reared under similar conditions.[93] (From Jacobson, S. G., and Ikeda, H.: Exp. Brain Res. **34:**11, 1979.)

acuity in the esotropic eye also depended on the age at which the squint was produced. Both behavioral observations on the cats (filled circles) and electrophysiologic tests on their geniculate cells (open circles) indicate that the spatial resolution was lowest for animals reared with a squint from 3 weeks of age. Thereafter, the later the onset of the squint, the closer the final spatial resolution of the squinting eye in the adult cat approximates that of the normal eye. When a squint was produced in a cat at or beyond 12 weeks of age, there was no significant difference between the acuities of the squinting and the normal fellow eye. The period of susceptibility to the effects of the convergent squint therefore extends from about 3 to 12 weeks of age. This sensitive period will be discussed in more detail on p. 630. Also shown in Fig. 21-36 is the behaviorally determined[122] curve for the development of visual acuity in the normal kitten, the postnatal age of the kitten being plotted against spatial resolution relative to the usual adult threshold of 5 cycles/degree. The similarity between the curve for normal development and those from cats with convergent squint suggests that spatial resolution in the esotropic eye is arrested at the age of squint onset. Possible mechanisms responsible for the loss of visual acuity in the squinting eye will be discussed on p. 638.

Alternating occlusion. In kittens, Hubel and Wiesel[82] placed an opaque contact occluder over one eye one day and the other eye the next, alternating each day from shortly after birth to an age of 10 weeks. This kept the eyes from working together without introducing the possibility of antagonistic interaction between them. No defects of visual behavior were noted, but cortical recordings gave results similar to those obtained in squint animals, the shift in ocular dominance being even more extreme. Thus alternating eye occlusion leads to a failure of development of the normal binocular connections, further demonstrating that these require synchronous input from the two eyes during the sensitive period (p. 628).

Experimental anisometropia. Ikeda and Tremain[90] have simulated the effects of anisometropia in the developing kitten by daily application of atropine eyedrops to one eye onward from day 21 to 23 after birth. Atropine not only dilates the pupil, but also paralyzes accommodation so that the atropinized eye receives blurred images. Neurophysiologic recordings were then made both from the striate cortex and from the lateral geniculate nucleus when the kittens were 5 to 8 months of age. Once again there was a loss of cortical binocularity and a shift in ocular dominance away from the atropinized eye and toward the normal eye, although the effect was much weaker than that due to eyelid suture or strabismus described earlier. Atropine applied to both eyes, however, did not disturb cortical binocularity. In contrast to the cortex, recordings from sustained (X) cells in the lateral geniculate nucleus with their input from the area centralis region of an atropinized eye showed a significant loss of visual acuity as compared to cells driven from a normal eye. This loss of visual acuity by geniculate cells driven from an atropinized eye occurred regardless of whether one eye or both eyes received blurred images during the sensitive period. Thus the alteration in the ocular dominance pattern in the cortex that is found when only one eye is atropinized seems not to be correlated with the loss of visual acuity by geniculate cells, since the latter occurs with both monocular and binocular application of atropine. This observation leads to the conclusion that amblyopia and the loss of binocularity have separate causes (p. 641).

From all these observations it is clear that any form of visual deprivation that prevents a normal input to one eye in a very young kitten will produce a loss of visual acuity by the cells in the lateral geniculate nucleus that have their input from the deprived eye, and this loss will be accompanied by a shift in the normal balance in the striate cortex between the two eyes in favor of the nondeprived eye.

Sensitive period and recovery from deprivation. For the deprivation effects described earlier to occur, the deprivation must take place during a fairly limited period in the first few months of a kitten's life known as the *sensitive period*. In a series of experiments, Hubel and Wiesel[84] have defined in some detail the sensitive period for unilateral eye closure. Their technique was to close one eye for various periods of time at different ages, reopen the eye, and then record from the cortex at various times later. They found that susceptibility to the effects of eye closure begins near the start of the fourth week, remains high until some time between the sixth and eighth weeks, and then declines, disappearing finally around the end of the third month. After 4 months cats seem to be insensitive even to long periods of monocular deprivation as tested by behavioral, physiologic, and morphologic criteria. The time course of the sensitive period as defined by behavioral methods paralleled that given by neurophysiologic procedures.[41]

Normal binocular viewing is only one of a number of experimental regimens that may be instituted following relief from a period of monocular deprivation produced by eyelid suture. During the period of deprivation, the nondeprived eye appears to gain an advantage that it still preserves, to some extent, after the deprivation has been relieved, thereby tending to suppress the recovery of the reopened eye. Closure of the normal eye when the deprived eye is opened (*reverse closure* or *reverse suture*) not only prevents any such suppressive effect, but also forces maximal use of the recovering eye. In this way it may be possible to discover the limits to which recovery of normal monocular vision can take place, but only, as we shall see, at the expense of the development of binocular vision. Other experimental procedures are possible, for example, alternating periods of reverse occlusion could be interspersed with periods of binocular viewing, but these have yet to be explored experimentally. There are obvious parallels in the occlusion therapy used in clinical practice.

Monocular deprivation followed by binocular viewing. Unilateral eye closure in kittens for the first 3 to 4 months of life leads to severe and permanent changes; Wiesel and Hubel[197] found that few, if any, cortical cells recovered responses to the formerly deprived eye even when, following opening of the eye, extended periods of binocular viewing were allowed. Behavioral recovery from apparent blindness was also very limited.[41] The severity of the permanent defect is, however, graded, depending on the duration of the period of deprivation and the time within the sensitive period at which the deprivation occurs.

Since the effects of monocular deprivation may be seen following only a few days of eye closure,[84] it is possible to examine the extent to which behavioral and physiologic recovery may be seen *within* the sensitive period. Contrary to the limited recovery reported in an earlier study,[84] Mitchell and co-workers[121] found that considerable behavioral and physiologic recovery can take place in kittens when a period of monocular eye closure, extending from the age of 8 days to 45 or 60 days, is followed by a further period of simple binocular viewing. Behavioral recovery was assessed by measuring visual acuity through the formerly deprived eye, and physiologic recovery was estimated from the eye dominance distributions and response properties of cortical cells. When first opened, the deprived eye appeared blind in all respects, but within a few days, depending on the animal, there was a dramatic recovery in the animal's behavior when using the deprived eye. After 8 weeks of binocular viewing, kittens deprived to day 45 attained nearly normal visual acuity in their deprived eyes. Had the kittens been examined at the time their sutured eyes were opened, their cortical cells would presumably have shown a marked shift in eye dominance towards the nondeprived eye. As Wiesel and Hubel[195,196] have shown, monocularly deprived kittens given no recovery period have fewer than 5% of cortical cells dominated by the deprived eye. By contrast, cortical recordings made by Mitchell and associates[121] after 8 weeks of binocular viewing revealed that 31% of the cells were now dominated by the initially deprived eye.

Unlike an earlier report,[84] Mitchell and colleagues[121] also found that following the period of binocular viewing the majority of cortical cells, irrespective of their eye dominance, possessed qualitatively normal orientation selective receptive fields. The striking feature about the cells was, however, that very few of them could be binocularly activated, even in kittens that showed substantial shifts in cortical eye dominance. The switch in eye dominance that occurs during the recovery period apparently takes place without an intermediate stage of binocular innervation. The initial period of monocular deprivation produces a squint that persists during the period of recovery, thereby preventing or greatly reducing the cortical binocular interactions needed for the development of binocular vision.

Monocular deprivation followed by reverse closure. As mentioned earlier, closure of the normal eye when the deprived eye is opened not only prevents any suppressive effect by the nondeprived eye, but also forces maximal use of the recovering eye. Blakemore and Van Sluyters[21] have, in fact, found that complete neurophysiologic recovery of the formerly deprived eye does occur if reverse suture takes place early enough in the sensitive period. Fig. 21-37 shows the results Blakemore and Van Sluyters[21] obtained from six kittens, all of whom had unilateral lid suture 8 to 10 days after birth as indicated by the vertical dashed line. After varying periods of deprivation, five of the kittens had reverse closure at the times, indicated by the filled circles. Thereafter, in each case a period

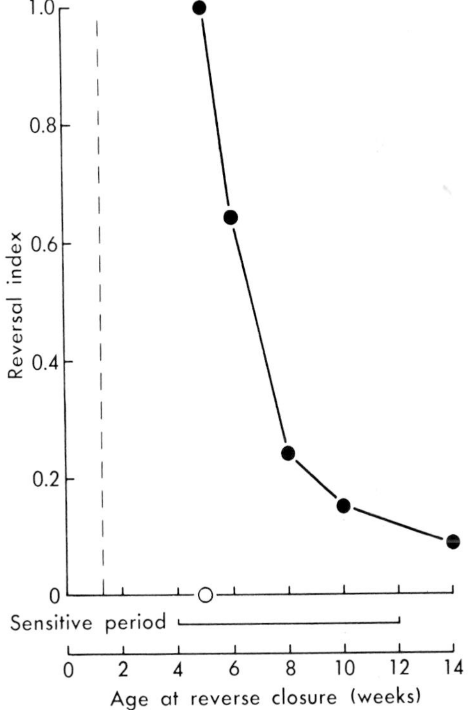

Fig. 21-37. Effect of reverse closure of normal eye on recovery of deprived eye in five kittens (filled circles), all of which had unilateral lid suture from about 9 days (vertical dashed line) for varying periods (5, 6, 8, 10, and 14 weeks) up to time of reverse closure. Ordinate scale indicates degree of recovery by initially deprived eye as tested 9 weeks after reverse closure. Reversal index of 1.0 indicates complete switch in ocular dominance from normal to initially deprived eye. Control animal (open circle) with unilateral eye closure from 9 days to 5 weeks had complete eye dominance by normal eye when tested at 5 weeks. (Modified from Blakemore, C., and Van Sluyters, R. C.: J. Physiol. [Lond] **237:**195, 1974.)

of 9 weeks was allowed before the reopened eye was tested for recovery. The remaining kitten (open circle), without any reverse closure, acted as a control. When this animal was tested at the end of 5 weeks of monocular deprivation, dominance by the normal eye was already complete and no cortical cells could be driven from the deprived eye. By contrast, the animal that had had reverse closure at the end of the same period of deprivation (5 weeks) showed complete recovery of the deprived eye when tested 9 weeks later. There was now complete dominance by the formerly deprived eye and no cortical cell could be driven by the initially normal eye. Presumably the deprived eye in this animal lost all functional connections with cortical cells during the period of deprivation but regained them during the recovery period, completely taking over from the initially normal eye (reversal index of 1.0). The reversal index in Fig. 21-37 is an indication of the degree of takeover by the originally deprived eye.

After reverse closure, the deprived eye can recapture cortical neurons almost as fast as it lost control during the period of deprivation,[21] although the time taken in each case depends very much on when within the sensitive period the respective intervals occurred. The longer the period of deprivation and the later the time at which reverse closure is carried out, the less the degree of recovery. In Fig. 21-37 the curve traced out by reverse closure is presumably an indication of the degree of residual plasticity of binocular properties in the cortex. It agrees well with the form and time course of the sensitive period (Fig. 21-37, horizontal bar) as described by Hubel and Wiesel.[84] The roughly exponential decline of the sensitive period suggests that some plasticity remains in the adult cat's cortex. Certainly extremely prolonged periods of reverse closure can lead to some recapture of the cortex by a deprived eye, even in the adult.[29]

Mitchell and co-workers[121] monocularly deprived kittens by lid suture from day 8 to 45 or 60. They then compared the behavioral and physiologic recovery of the monocularly deprived eye when, on eye opening and during the recovery period, either binocular viewing was allowed or the initially nondeprived eye was reverse sutured. After a recovery period of 8 weeks, the visual acuity in the reverse sutured kittens was either the same as or close to that of normal animals of comparable age and only marginally higher than that of kittens with binocular viewing. On cortical recording, the majority of the cells, whether or not they changed their ocular dominance, had receptive fields with qualitatively normal response properties. As with kittens given binocular viewing, cortical recording rarely encounters binocular cells in these kittens. According to earlier reports,[21,123] these binocular cells sometimes had grossly dissimilar receptive field properties in the two eyes. However, when kittens are allowed a period for normal postnatal development of binocular vision before initial monocular deprivation and then reverse sutured, the receptive fields that reappear in the originally deprived eye are well matched to those in the other.[23]

Hence, in the cat at least, it seems that reverse suturing produces a recovery in the initially deprived eye that is only marginally faster and better than when binocular viewing is allowed, and further work is needed to discern the nature of the benefit of reverse suturing. But again, a marked squint is produced that is obvious to casual observation.[123] This squint is more commonly divergent than convergent and, when divergent, is greater in degree than when convergent. These observations parallel those made by Awaya and associates[4] in human infants (p. 642).

Normal anatomy and effects of visual deprivation in cat. The well-marked physiologic effects of monocular eyelid suture that appear in the cortex are not accompanied by obvious histologic changes, at least to routine staining. By contrast, at the geniculate level there are easily recognized histologic changes that are, however, not associated with any significant change in the responses of the cells to the visual stimuli,[54] except for some impairment in spatial resolution. A much more severe reduction in the spatial resolving power of geniculate neurons has, however, been reported as a result of a surgically produced esotropia[91] (p. 628).

The lateral geniculate nucleus in the cat has a laminated binocularly innervated medial portion and a much smaller monocularly innervated lateral portion, the border between the two portions being indicated by the thick arrows in Fig. 21-38. The two frontal sections in Fig. 21-38, one through the right lateral geniculate nucleus, A, and the other through the left nucleus, B, are from a kitten whose right eye had been sutured for 89 days. In a normal animal it is the laminated medial portion that drives the binocular cells in the cortex. Each geniculate lamina has an input from only one eye, the laminae for one eye alternating with those for the other eye. The binocular cortical cells, however, receive their afferents from two or more layers and from cells closely adjacent to one another in adjoining laminae.

In the normal postnatal development of the lateral geniculate nucleus in kittens there is a pronounced rapid phase of cell growth, especially marked between the second and fifth weeks. After 2 months there is relatively little further growth of cells in the monocular segment, but those in the binocular segment continue to grow until about 4 months when they reach adult size, some 20% larger than the monocular.[59] Cragg[33] has shown that synaptic development proceeds at about the same rate in the visual cortex as in the lateral geniculate nucleus, with synapses in the geniculate developing approximately 2 days earlier than those in the cortex. In both areas there is a rapid increase in the number of synapses between 8 and 37 days after birth. Thus the onset of the sensitive period in the kitten occurs toward the end

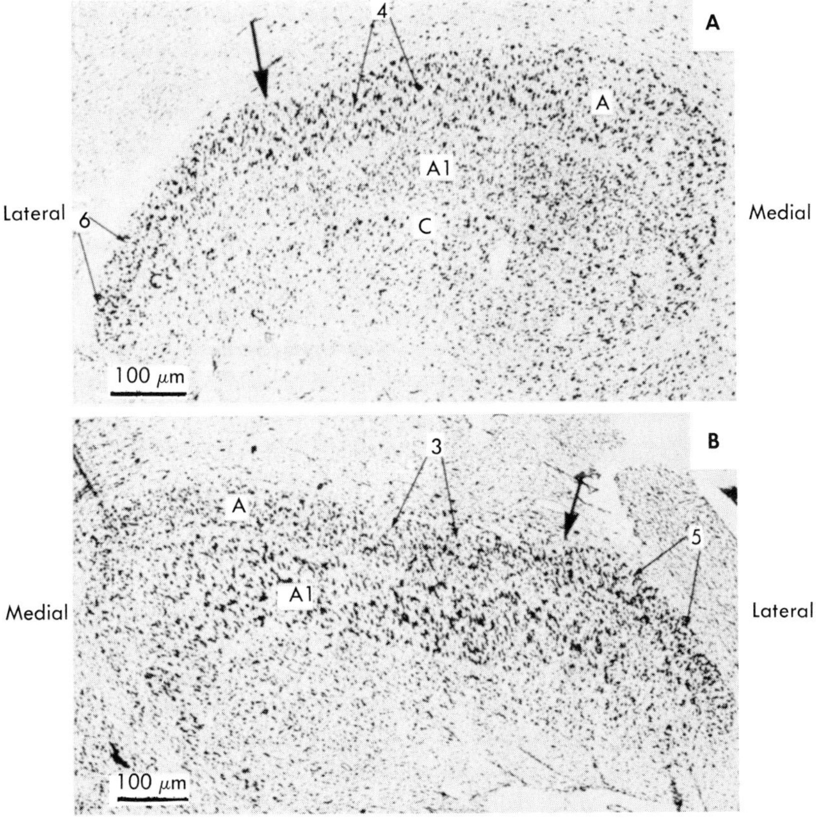

Fig. 21-38. Frontal sections through right, **A**, and left, **B**, lateral geniculate nuclei of kitten whose right eye had been sutured postnatally for 89 days. In each case thick arrow indicates border between binocularly innervated laminated medial portion and monocularly innervated lateral portion of nucleus. Due to visual deprivation, cells in lamina *A1* of right nucleus and lamina *A* of left nucleus are abnormally small. (From Guillery, R. W., and Stelzner, D. J.: J. Comp. Neurol. **139:**413, 1970.)

of the phase of rapid geniculate cell growth and the period of most synapse formation in both the geniculate and visual cortex. Although not so well defined, the sensitive period ends at about the same time as the geniculate cells in the binocular segment of the nucleus attain their adult size.

The effects of unilateral lid suture on the cells in the lateral geniculate nucleus of a young kitten are most readily seen in the two principal laminae labeled A and A1 in Fig. 21-38. Layer A is innervated by the contralateral eye and A1 by the ipsilateral eye. There is reduced cell growth in the visually deprived laminae[194] the reduction in mean cell size varying from 25% to 40% according to different observers. However, this effect occurs only in the binocular segment of the nucleus where the deprived laminae lie opposite a normally innervated lamina.[65] The cells in the monocular segment are not affected.

In normal cats laminae A and A1 are comprised chiefly of large Y cells and medium-sized X cells. Although there is general agreement that monocular eyelid suture leads to a reduction in cell size in the lateral geniculate nucleus, the anatomic evidence that the development of Y cells is more seriously affected than the development of X cells[110] is not in satisfactory accord with the physiologic evidence based on the relative frequency of recording from X and Y cells in normal and pattern-deprived animals.[54,170]

In contrast to eyelid suture, rather less is known about the effects on cells in the lateral geniculate nucleus of the kitten of a squint produced during the sensitive period. Unlike the monkey, only the effects of a convergent squint have so far been examined. Ikeda and co-workers[92] have reported that in the geniculate layers fed by the squinting eye the mean cell size is smaller and the cell density higher than in the layers fed by the normal eye. Their analysis of the cell size distribution also led them to conclude that there was a reduction in size of all classes of cell and not necessarily a differential effect on the large cells, as suggested earlier.

It seems unlikely that sensory deprivation by itself is the factor responsible for the change in geniculate cell growth following monocular deprivation, since retinal ganglion cells appear normal in animals that show marked geniculate changes.[194,196] Furthermore, comparison of the monocular segments of the two lateral geniculate nuclei in the one animal following unilateral lid suture shows no significant cell differences. The monocular segment of a nucleus receives its input from the monocular crescent of the visual field via the contralateral eye. Nevertheless, recent work by Ikeda and her colleagues suggests that, although the deprived retinal ganglion cells may appear morphologically normal, their weak responses possibly do not provide the necessary stimulus for the normal development of the geniculate cells during the sensitive period. The changes that occur in the visual cortex are then assumed to be a consequence of the changes that have taken place in the lateral geniculate nucleus.

The cat striate cortex is divided into somewhat ill-defined regions dominated by one eye or the other.[169] Monocular deprivation appears to cause the same characteristic changes in these columns in the cat[123] as are to be found in the monkey,[23,87] but since the ocular dominance columns are much more clearly defined in the monkey striate cortex an account of the changes that occur in them is best left until later.

Monkey experiments

Species differences preclude any simple application to humans of the data obtained on kitten. For this reason work on subhuman primates is of special importance particularly, since the macaque monkey appears to have a visual system that is virtually identical to that of the human.

Ocular dominance of cortical cells. The ocular dominance distribution of cells in the monkey striate cortex differs from that in the cat. In particular, the cells in layer IVC in the monkey have an eye dominance distribution that is strikingly different from that in the other layers.[23,86] These cells are unusual in that they lack any orientation preference and, being segregated into eye dominance bands, have an ocular dominance distribution that is highly monocular. Overall, they are approximately equally divided between the two eyes, but the mixed eye dominance categories (2 to 6) are virtually absent. The ocular dominance distribution of the cells in the remaining laminae are, like those in the cat striate cortex (Fig. 21-35, *A*), much more evenly distributed between the eye dominance categories.[23,83,168]

Visual deprivation

Monocular deprivation: physiologic observations. Monocular lid suture at about 4 weeks of age for periods of 2 to 4 weeks causes a marked shift in ocular dominance toward the nondeprived eye for cells in all layers of the striate cortex whether the cell recordings are carried out at the time the deprived eye is opened[23] or later, after allowing binocular viewing for periods extending up to 4 years.[34] Strongly binocular cells are preferentially affected, but small regularly periodic regions dominated by the deprived eye and presumably reflecting vestiges of the normal ocular dominance columns tend to persist. Each of these regions extends through the thickness of the cortex, and the cells in them have apparently normal receptive field properties. Extension of the period of deprivation accentuates the shift in eye dominance and, with prolonged monocular deprivation for the first 2 years of life, dominance by the normal eye is virtually complete. When the monocular deprivation is delayed until 11 to 16 months of age, there is no detectable influence on layer IVC, but outside IVC there still appears to be a small shift in ocular dominance. Prolonged deprivation in an adult is without effect.

Blakemore and co-workeers[23] studied the effects of reverse suture carried out after varying periods of monocular deprivation, all of which began soon after birth but extended until times ranging from 5½ to 38½ weeks of age. To allow the consequences of reverse suture to be complete, the monkeys were allowed to use the initially deprived eye for a period of 4 months before cortical recording. Early reverse suture (at 5½ weeks) led to a recapture of cortical cells by the newly experienced eye that was just as complete as the original degree of takeover by the initially experienced eye. Furthermore, the cells regain in the newly experienced eye approximately the same normal receptive field properties that they are in the process of losing in the initially experienced eye. Once again, small regularly recurring patches dominated by the newly deprived eye still persist, and the cells within them have normal receptive field properties. Binocular neurons are, however, extremely rare. Late reverse suture (at 38½ weeks) does not seem to change cortical ocular dominance, and reversal at the end of the second month has an intermediate effect.

The predominant effect of monocular deprivation in early life is the disruption it causes in the binocular innervation of cortical neurons, a condition that persists after reverse suture. Very few binocularly innervated neurons remain; thus binocular depth perception should be profoundly affected. There are, however, some surprising features revealed by the effects of reverse suture. The changeover in eye dominance seems to take place without any intermediate stage of binocular innervation. Furthermore, the changes produced both by the initial period of monocular deprivation and by reverse suture *seem* to be accomplished, at least in the short term, by some kind of switching that leaves the fundamental organization of the cortex relatively unaffected, with parts of the neural circuitry persisting in the background in an inactive state.[23]

Monocular deprivation: behavioral observations. Unlike the physiologic observations described earlier, no assessment has yet been made in the monkey of the extent to which reverse suturing can bring about behavioral recovery in an eye monocularly deprived for short periods *within* the sensitive period when, on eye opening, the normal eye is closed for varying periods. So far in behavioral studies, when monocular lid suture has been carried out for short periods within the sensitive period,[184] opening the deprived eye has always been followed by relatively prolonged periods of binocular viewing. Recovery in the deprived eye may therefore have been suppressed by action of the normal eye. In the studies made by von Noorden,[184] the short periods of monocular deprivation were produced by lid suture commencing between the fourth and the eighth week of age and extending over further periods of 2 to 4 weeks. As mentioned earlier, eye opening was then followed by extended periods of binocular viewing. Monocular lid closure commencing at the fourth and the fifth weeks of age caused severe and permanent amblyopia with visual acuity always worse than 20/290. Beyond the period of binocular viewing, even enforced usage of the previously deprived eye still produced no evidence of any recovery of visual acuity. Four-week monocular lid closure from 8 to 12 weeks of age had no residual effect on visual acuity at least after the period of binocular viewing, although visual acuity testing was only to a 20/39 level.

von Noorden and his colleagues[184,188,189] have studied the behavioral effects of relatively prolonged periods of monocular deprivation. Commencing between the first and the twelfth week of life, monkeys had monocular lid suture for periods ranging from 30 weeks to 2 years. During the period of lid suture the animals were trained with the open eye to respond to the procedures involved in visual acuity testing. Then, at the end of the period of monocular deprivation and with the visual test criterion achieved, reverse suture was performed, the deprived eye being opened and the nondeprived eye closed. With reverse suture now forcing the use of the initially deprived eye, visual acuity testing was continued for periods of 6 to 24 months. Irreversible amblyopia occurred in all animals whose lids were sutured between birth and 9 weeks of age. Not a single animal recovered visual acuity. Although visual acuity testing was only to a 20/39 level, lid closure at the age of 12 weeks did not produce any permanent deficit; visual function in the formerly deprived eye returned rapidly and completely after closure of the normal eye. Thus, in so far as comparable studies have been made, the behavioral effects of monocular deprivation are in keeping with the physiologic observations. Unfortunately, none of the behavioral studies of monocular deprivation have commented on the effects it has on binocular eye alignment and the presence of strabismus.

Sensitive period. Rather surprisingly, the monocular deprivation and reverse suture studies indicate that the time of onset and duration of the sensitive period in the monkey are very similar to those of the cat, that is, from about the fourth to the twelfth weeks of age, although some residual effects of deprivation may still be detectable as late as 1 year. There is, however, a rapid decline in sensitivity at the end of the second month. Nevertheless, despite the similarities, there are also significant differences between the effects of monocular deprivation in the cat and monkey, particularly when binocular viewing is allowed after a short period of deprivation early in the sensitive period. Although differing observations have been made,[84,121,197] it seems that in the cat binocular viewing still allows considerable physiologic and behavioral recovery in the formerly deprived eye. By contrast, the deprived eye in the monkey is much more vulnerable to suppression by the normal eye, and no physiologic or behavioral recovery appears to take place.

Binocular deprivation. Crawford and associates[34] recorded from neurons in the striate cortex of an adult monkey that had had bilateral lid suture from 2 to 10 weeks of age. The most striking features were the large number of unresponsive neurons, detected only because of their spontaneous activity, and the severe reduction in the proportion of units falling into the strongly binocular categories. Many of the remaining binocular cells had abnormal properties with vague receptive field boundaries and lacking any degree of stimulus specificity. In these respects the visual cortex of the monkey seems even more vulnerable to binocular deprivation than that of the cat. Both eyes of the binocularly deprived monkey were amblyopic.[184]

Experimental strabismus. Experimental esotropia produces physiologic and behavioral effects quite different than those of exotropia. Using surgery on the extraocular muscles, von Noorden and Dowling[187] produced esotropia in three monkeys at the postnatal ages of 1 day, 7 days, and 17 months, respectively, and exotropia in a further two monkeys at 6 days and 6 months, respectively. Then, during a period when they were allowed to use both eyes, the monkeys were trained to respond to visual acuity testing. During this training period, the two esotropic animals operated on at the beginning and end of the first week appeared to fixate exclusively with the undeviated eye, whereas the exotropic animals could hold fixation with either eye. After the training period, the lids of the eye not operated on were sutured together and visual acuity testing was resumed with the strabismic eye alone. The time for closure of the normal eye was always after the end of the sensitive period as determined by monocular deprivation experiments.

Strabismic amblyopia appeared only in the two monkeys made esotropic in very early life. Immediately after closure of the normal eye these esotropic animals appeared essentially blind. Amblyopia was absent in the third esotropic animal whose onset of surgical strabismus was at 15 months, and it was absent in the two exotropic monkeys. Over an extended 2-year period, with the sound eye still firmly sutured, one esotropic animal achieved a 20/360 level of visual acuity but the other failed to improve to better than gross light perception.

Baker and co-workers[7] subsequently made cortical recordings on the two esotropic animals with amblyopia and the animal made exotropic at 6 days of age. In both types of strabismic animal very few neurons were binocularly activated and the number of unresponsive neurons was higher than normal, but those that could be driven all had normal receptive field properties. In the esotropic animal, with only light perception in the deviated eye, there was a marked shift in ocular dominance toward the undeviated eye. The result was similar, although less extreme, in the esotropic eye with the 20/360 level of visual acuity. In the exotropic animal the numbers of cells driven by one eye or the other were much more nearly equal.

The lack of synergy in the binocular stimulation of the cortical cells in these animals apparently prevented the development of a population of binocularly innervated neurons and, when the squint was convergent, suppression from the nondeviating eye seems to have been responsible for the amblyopia in the deviating eye. As with monocular lid suture, amblyopia developed only when esotropia was induced within the first 3 months of life, suggesting that the period of sensitivity of the visual system to abnormal stimulation is similar for both forms of amblyopia.

Normal anatomy and effects of visual deprivation in monkey. The more highly organized laminar structure of the primate cortex and the regularity with which the geniculocortical axons terminate within it make it possible to follow the structural changes caused by visual deprivation much more readily in the monkey than in the cat. The bulk of the geniculocortical afferents terminate in the deep part of layer IV (layer IVC), where they segregate themselves into a series of roughtly parallel alternating stripes, one set connected to the left eye, the other to the right.[107] This organization is, however, not fully achieved until some time after birth. The fetal eye-injection experiments of Rakic[159] have shown that initially the optic tract fibers from both eyes and the corresponding geniculocortical axons overlap completely in their respective projections to the lateral geniculate nucleus and layer IV in the visual cortex. The segregation of the inputs to the lateral geniculate nucleus occurs about midway through gestation, but the segregation of the geniculocortical inputs is still not complete at birth. Anatomic evidence indicates that in the first week or so of life layer IVC probably still contains a mixture of inputs from the two eyes, perhaps at most points along its length, and physiologic recording shows that many nonoriented fourth layer neurons can be driven by either eye.[87] Hubel and co-workers[87] propose that in the few weeks before and after term segregation of the inputs takes place by a process of competitive interaction in which underactive and relatively sparse terminals at any point tend to regress. In normal animals this retraction of the two sets of terminals would, it is assumed, lead to the formation of ocular dominance stripes of equal size for the two eyes.

Prolonged visual deprivation in monkeys by monocular lid suture carried out in the first few weeks of life leads to cell shrinkage in the deprived layers of the lateral geniculate nucleus.[7,87,185] In contrast to the cell shrinkage in the lateral geniculate nucleus, the cortex appears normal to Nissl's stain. Other histologic methods, however, show a marked change in the relative sizes of the ocular dominance stripes in layer IVC, with a shrinkage of stripes receiving input from the deprived eye and a corresponding expansion of those with input from the normal eye, thus preserving the normal repeat distance (left-eye column plus right-eye column). The changes seem more severe when the closure is done at 2 weeks rather than later, but cell body size and packing density appear normal in the shrunken stripes. On the basis of these changes, the marked reduction in geniculate cell body sizes in the laminae receiving input from the deprived eye would simply reflect the reduced metabolic requirements of their smaller terminal arborizations.

The idea that the axonal segregation in layer IV is normally perfected during the first few weeks of life by competitive rejection of the weaker input to any region provides an explanation for the presence and duration of the sensitive period, this being the time during which the sorting out process is taking place. It remains to be seen whether the process of segregation remains incomplete sufficiently long to account for the deprivation effects that may still occur as late as 3 months or more after birth. Some form of axonal growth or regeneration may be required as well (p. 640).

Sensitive period: general observations

Genetic control and visual experience

At the time of their early experiments on visual development in kittens, Hubel and Wiesel[80] reported that they had found cortical neurons with normal adult-type specificity of responsiveness in animals that had not opened their eyes. They therefore concluded that much of the complex organization of the visual cortex responsible for receptive field properties, binocular interaction, and functional architecture is developed solely under genetic control and is present in very young kittens without visual experience. They further concluded that the effects of visual deprivation are not simply due to a failure in forming synaptic connections but rather to a disruption of connections that were there from the start.

These early views have had to be modified in the light of more recent work by Pettigrew[150] and others who have been unable to confirm that a normal adult-type organization is present in the kitten visual cortex soon after birth and in the absence of visual experience. Innate factors do ensure that a high proportion of cortical neurons are binocularly innervated at or soon after birth, but their properties are immature. They have selectivity for direction of stimulus motion and some weak orientation specificity, but the two receptive fields of a cortical neuron are only roughly matched in orientation preferences, and they lack sensitivity to binocular disparity. Visual experience is apparently essential not only to match the properties of the two receptive fields but also to refine their disparity selectivity. A close similarity in the feature selectivity of the two eyes is an essential requirement if cortical cells are to achieve the precise binocular interactions that are needed for depth discrimination. This matching process may be the major development during the sensitive period as far as the visual system is concerned, especially when the size, optics, and position of the eyes appear to be changing independently during this time.

Possible factors operating during the sensitive period

Many factors that may act during the sensitive period have been proposed to account for the effects of visual deprivation, but only those factors receiving particular attention in current research can be mentioned here. The various proposals are, however, by no means mutually exclusive.

Binocular competition. Many of the diverse observations described earlier can be linked together by the idea that during the sensitive period the afferent paths from the two eyes compete for control over the cortical cell.[63-65,196] It is suggested that geniculocortical axons arising from adjacent geniculate laminae compete with each other for synpaptic surfaces on binocular cortical neurons. Unilateral lid suture upsets the balance of this competition, and the reduced cell growth in the lateral geniculate nucleus is a consequence of the unbalanced axonal development and the smaller number of active synapses made by each axon. It has already been noted (p. 634) that retarded cell growth occurs only in the binocular segment of the nucleus. Binocular competition can occur in this segment because it is here that adjacent laminae receive their input from homonymous retinal areas and send their axons to a single cortical locus. Competition does not occur in the monocular segment, since it receives from only one retina.

Two further observations strongly support the idea of binocular competition.

Binocular lid suture. The competition hypothesis predicts that bilateral lid suture should be less effective than monocular lid suture, since in the former case all the geniculate laminae would be equally deprived. Despite earlier observations to the contrary, it now appears[64] that binocular lid suture has relatively little effect on geniculate cell growth either in the binocular or monocular segments of the nucleus.

Cell growth after relief from binocular competition. Guillery[63] has studied the effects of binocular competition on geniculate cell growth in a particularly elegant way. Competition in part of the binocular segment can be abolished by making a small lesion in the retina of one eye. This produces a zone of transneuronal degeneration in a geniculate lamina corresponding to the damaged eye. If at the same time the other, normal, eyelid is sutured, those cells in the lamina corresponding to the sutured eye that lie opposite the zone of transneuronal degeneration will suffer the effects of visual deprivation without competition from cells in the adjacent lamina. This experimental design has the added advantage that within the same nucleus, parts of a deprived lamina that do suffer binocular competition can be compared with that part in the same lamina that is relieved of competition because it is opposite a zone of degeneration. Guillery found that the part of the deprived lamina lying opposite the zone of degeneration contained cells that were larger than would have been expected had there been a competitive effect from the other lamina. The rest of the deprived lamina showed the reduced cell growth typical of unilateral lid suture.

Hubel and associates[87] have applied the concept of binocular competition to the visual deprivation effects that occur in layer IV of the monkey cortex. Their work suggests that the sensitive period in the development of the monkey visual system corresponds to the time during which the geniculocortical axon terminals are becoming segregated into left eye and right eye stripes in layer IV. As has been mentioned earlier, they propose that in the normal development of the visual system competition between the geniculate terminals leads to their segregation from one another, the process occurring by a retraction of the two sets of terminals so that each stripe comes to be dominated by one eye or the other. They further propose that the same competitive mechanism may also be responsible for the deprivation effects that occur in layer IV. Lack of use of a set of terminals puts them at a competitive disadvantage, so that at any point along IVC the terminals from the opposing normal eye set take over, provided they have not already retracted. Where no competition is possible because the other set has already disappeared, the deprived set survives and is apparently intact. The end result of the deprivation thus depends on the amount of overlap of the terminals that still existed at the time of eye closure. This explains the greater severity of the changes in the early closures and suggests that once the process of segregation is complete eye closure will be without effect. Further evidence in favor of the idea of competition is the anatomic observation that the temporal-crescent input to the monkey cortex is largely spared on the side opposite the closed eye, compared to the severe effects on cortical stripes in the binocular part of area 17.

Once the ocular dominance stripes within IVC are shrunken through deprivation, a crucial test for the binocular competition hypotheses of Hubel and co-workers[87] is whether the stripes are capable of reexpansion following early reverse suture. The hypothesis would be contradicted if reexpansion of the stripes is possible. Blackmore and associates[23] have provided somewhat inconclusive physiologic evidence that following early reverse closure the originally deprived eye stripes reexpand and those of the newly deprived eye shrink until they are again roughly equal in width.

A further difficulty for the binocular competition hypothesis concerns the effect of reverse suture on the sizes of cell bodies in the lateral geniculate nucleus. Just as in the cat,[50] early reverse suture can correct or even invert the size differences between deprived and nondeprived laminae, the initially small deprived geniculate cells becoming relatively enlarged. These observations again suggest that either some regeneration of axon terminals or a recovery of function has taken place in the formerly shrunken ocular dominance stripes.

Regeneration of axon terminals and the restoration of functional connections. Considerable recovery of function can take place without the need for reversal of lid suture when the eye deprived by lid closure is opened. This observation is incompatible with the simple notion of binocular competition, since under these circumstances the formerly deprived eye does not gain a competitive advantage.[121] Genetic factors are presumably largely responsible for the basic pattern of the geniculate terminals in the ocular dominance columns, since very young monkeys and kittens seem to possess columns that are qualitatively similar to those in adult animals although perhaps not fully formed.[87,159] Mitchell and co-workers[121] have suggested that in the process of regeneration following visual deprivation there is an innate tendency for the growing fibers to return to the innervation sites they once occupied or to which they were originally predisposed.

An alternative suggestion has been put forward by Blackemore and colleagues.[23] Rather than have the geniculate terminals physically displaced during a period of visual deprivation and then undergo regeneration when the deprivation is relieved, they have proposed that the axons of the deprived afferent pathways or associated cortical interneurons might retain their cortical distributions but simply have their postsynaptic influence repressed. Thus in areas invaded by the normal eye terminals, the terminals from the deprived eye remain but are nonfunctional. A process of derepression of existing but ineffective synapses then accounts for the observation that the input restored after reverse suture leads to receptive field properties very similar to those that could reasonably be assumed to have been there originally. This seems a more attractive possibility than the total regeneration of the same complex pattern of interconnections as were displaced during the period of deprivation.

There is, as yet, however, no evidence that axonal sprouting is responsible for the restoration of functional connections in the cortex, and the application of a variety of techniques[87] has so far failed to demonstrate the continued presence of nonfunctional fibers and synapses. Cell proliferation cannot be a factor, since it ceases altogether in the cortex 2 months before birth.[159]

Habitual blurring of the retinal image. Ikeda and her colleagues[90] propose that habitual exposure of the fovea or area centralis to defocused images during the sensitive period is the common factor responsible for the effects of visual deprivation and strabismus. They argue as follows: The defocused image provides an inappropriate stimulus for the sustained ganglion cells in the central retina. The weak (or absent) firing of these ganglion cells then fails to provide the input necessary for the normal development of the cells in the lateral geniculate nucleus, resulting in their retarded or arrested growth. The geniculate cells therefore become amblyopic, losing contrast sensitivity and visual acuity. In turn, the inadequate growth and development of the geniculate cells leads to a deficiency in the normal development of their axon terminals in the visual cortex, and this allows the terminals from the normal eye to dominate the responses of the cortical cells.

Kittens reared with esotropia in one eye develop amblyopia, but those reared with exotropia in one eye do not, although cortical binocularity is reduced in both conditions. Exotropic kittens develop alternating fixation, but not those with esotropia. Apparently alternating fixation is possible when the visual axes are divergent but difficult to achieve when the two visual axes cross over in front of the animal. Thus the area centralis of an exotropic eye receives clearly focused images during the sensitive period, whereas in esotropia the squinting eye receives only blurred images of objects at different distances from the fixation point, since accommodation is determined by the normal nonsquinting eye.

There is no correlation between the development of amblyopia and the changes in ocular dominance in the visual cortex. Regardless of whether the squint is convergent or divergent, binocular function must be affected since a misalignment of the two eyes is present in both cases. It is the synergy between the inputs from the two eyes that is required for the development of normal binocularity.

These arguments, based on experiments in kittens, can obviously be applied to human clinical conditions. In anisometropia, for example, it is the hypermetropic eye with the blurred images that becomes amblyopic.

Sensitive period: clinical application

Clinical experience interpreted in the light of the animal experiments just described leaves little doubt that during postnatal development the human visual system also passes through a sensitive period in which it is highly susceptible to visual deprivation. Bishop, in the last edition of this book, made an estimate of the age at onset and time course of this sensitive period in humans on the basis of various clinical observations. These observations, already referred to earlier may be briefly summarized. Thus stimulus deprivation amblyopia due to a corneal opacity acquired as late as 2 years of age may lead to a permanent reduction in visual acuity to as low as 20/800, and an irreversible amblyopia may still occur even when the obstruction to vision takes place as late as 5 years of age.[188] In the treatment of esotropia occlusion of the preferred eye remains effective in restoring the visual acuity of the amblyopic eye up to the age of 4 years and thereafter becomes increasingly less effective until, at about the age of 9 years, it is without effect.[145] A similar time course is to be observed in the ability of the young strabismic patient to develop the sensory adaptations of suppression and abnormal retinal correspondence that permit single binocular vision. Finally, a functional cure of congenital esotropia can be obtained if the deviation is eliminated completely by surgery performed during the first 2 years of life.[176] These observations suggest that the sensitive period has a later onset and a more prolonged time course in humans than it has in the cat or the monkey. It was proposed that the time course of the susceptibility to deprivation begins during the first 6 months of life, when binocular fixation if established, has passed its peak by the age of 2 years or possibly earlier, is well down at 4 years, and thereafter undergoes a slow decline to cease about the age of 9 years.

These deductions from clinical observations received strong independent support from studies using the interocular transfer of the tilt aftereffect as a measure of cortical binocularity. Binocular innervation of cortical neurons is an essential prerequisite for the development of normal binocular single vision and stereopsis after the surgical correction of a squint. The existence and time course of the sensitive period for the development of binocularity in humans has been defined by means of the tilt aftereffect test. A high-contrast grating tilted slightly clockwise appears less tilted after prolonged viewing and a vertical grating may then appear to be tilted slightly counterclockwise. Both of these aftereffects transfer from an adapted eye to the unadapted eye to a substantial degree, presumably via binocularly innervated cortical neurons. The tilt aftereffect test of binocularity is highly correlated with stereoacuity and persons with a history of esotropia exhibit poor stereoacuity and a low interocular transfer of the effect. Banks and co-workers[8] used the tilt aftereffect test to assess the binocularity of subjects who had had a convergent strabismus (> 10 degrees) during different periods of their lives followed by surgical correction of the deviation to better than 5 degrees. They found that strabismus begins to affect binocularity when the onset of the squint is at about 4 months of age. The effect peaks between 1 and 3 years and thereafter decays exponentially until about the age of 6 years. Hohmann and Creutzfeldt[74] used a similar technique. They concluded that the sensitive period for the development of binocular neurons extends at least to the third year.

It is also possible to define a sensitive period in the development of the human visual system by examining the period during which growth occurs, since in organisms generally sensitivity to outside influences is maximal at times of rapid growth. Hickey[71] has described two partially overlapping periods of postnatal cell growth in the human lateral geniculate nucleus. For cells in the parvocellular layers there is a period of rapid growth that ends about 6 months after birth, although the cells do not reach their adult size until near the end of the first year. Cells in the magnocellular layers continue to grow rapidly for about a year after birth, but do not complete their growth for a further 12 months. This time course of development is in keeping with other estimates of the sensitive period in humans. On the basis of the time taken for the postnatal completion of general bodily growth and maturation, it is also to be expected that the sensitive period in the human would be considerably longer than it is in the cat and monkey.

Awaya and co-workers[4] have made observations on the effects of visual deprivation in human infants that show a remarkable parallel to those produced by lid suture and reverse closure in animal experiments (p. 635). They collected a series of 15 infants with unilateral reduction of visual acuity below 20/100, all of whom had undergone occlusion for approximately 1 week after surgery for palpebral entropion. Typically, one eye was operated on first and remained patched with daily changes for about 1 week. Soon after the patched eye was uncovered the other eye was then operated on, and it also remained occluded in the same manner for about a week. The occlusions occurred in most cases within the first year of life, but in 5 cases the age was from 13 to 36 months. When examined in later infancy they all had, in addition to the unilateral loss of visual acuity, a marked squint (usually > 14 degrees) as well as eccentric fixation. The amblyopia was most severe in those occluded early in infancy during the first year, and occlusion of only 1 week was needed to produce marked effects. Exotropia was usually observed; 12 cases showed exotropia and esotropia was found in only three cases. For 12 of the 15 cases there was a complete record as regards the eye laterality and the order of surgery. Of these, four cases had unilateral surgery and all showed amblyopia on the operated side. In the remaining 12 cases bilateral surgery was performed, and all had amblyopia on the side operated on last. The fact that amblyopia was found on the side operated on second suggests that the eye operated on first was able to recover during the second postoperative occlusion, the whole sequence being a close parallel to the reverse closure carried out in animal experiments.

These observations indicate that when one eye is occluded in early infancy there may be damage to binocular cooperation together with the development of a strong dominancy by the nonoccluded eye and a reduction in visual acuity in the occluded eye. The effects are graded, depending on the age of onset and the duration and severity of the visual deprivation. Thus the severe and prolonged deprivation due to a corneal opacity may lead to a permanent amblyopia, but the amblyopia due to a relatively brief unilateral occlusion may be reversed. The human observations are in keeping with animal experiments that lead to the conclusion that amblyopia and the loss of cortical binocularity have separate causes (p. 630, 640). In human deprivation amblyopia the strabismus is a consequence of the amblyopia, whereas in strabismic amblyopia the amblyopia is the result of the strabismus.

Although the plasticity of the visual system during the sensitive period offers many therapeutic possibilities, the treatment of visual disabilities in early life is complicated by the many interlocking factors that can come into play. Thus the visual deprivation caused by congenital cataract should be relieved by surgery at the earliest possible age and followed by immediate aphakic correction to avoid the gross blurring of the retinal image that would act as a further continuation of the visual deprivation and hence perpetuate the amblyopia. This deprivation amblyopia, coupled with the anisometropia, would in turn lead to secondary strabismus, and then strabismic amblyopia may become superimposed on the existing amblyopia.

Because the proper alignment of the two eyes during the sensitive period is essential for the normal development of binocular vision, there seems little doubt that where corrective strabismic surgery is indicated, it should again be undertaken as early as possible. It is, however, a much more difficult task to define the place of occlusion in treatment, particularly in relation to surgery, and its use presents a therapeutic dilemma. Even after surgical intervention has achieved good alignment of the eyes, the presence of amblyopia in the formerly squinting eye may prevent the development of a level of binocular sensory cooperation adequate to maintain that alignment. Furthermore, recovery of acuity in the amblyopic eye may not take place unless the possible inhibitory effect of the preferred eye is prevented by its occlusion. Nevertheless, occlusion of the preferred eye clearly prevents that synergistic action of the two eyes essential for the development of normal binocular vision. This would particularly be the case where, over an extended period, alternating occlusion is used to maintain equality of vision in the two eyes. It seems that although equal monocular vision of good acuity can be achieved in this way, it is, nevertheless, achieved at the expense of the development of binocular vision. Although occlusion amblyopia is usually reversible if it develops in the formerly dominant eye of older children up to the age of 5 years during the treatment of strabismic amblyopia, this may not be the case with younger infants. For this reason, the dominant eye of children under 2 years should not be occluded for more than 1 week without reexamination of the patient.[147,184] If longer periods of occlusion are required, the patch should be switched to the amblyopic eye for 2 days each week to enforce the use of the occluded eye.

von Noorden[185a] has emphasized the need for caution in interpreting the results of animal experiments with respect to the management of congenital esotropia. As he points out, there is at present no useful primate model available for the study of congenital esotropia. All information regarding the consequences of strabismus on the structure and function of the visual system in primate animal models has been based on artificial and often incomitant strabismus in animals with presumably normal innate binocular connections. The possibility cannot be excluded that so-called congenital esotropia is indeed the result of a congenital defect in respect to innate binocular connections.

To evaluate the various effects described earlier there is an urgent need for much more detailed *quantitative* information about the final level of stereoacuity and binocular cooperation (including particularly the angle of any residual squint) that is to be found in those patients for whom a satisfactory and stable alignment has been achieved by surgery and occlusion. It seems that persistent occlusion in early life has a differential effect on fine and coarse stereopsis, disrupting the fine but having a much less powerful effect on the coarse mechanism. Certainly every effort should be made to avoid interference with normal visual experience in early life such as would occur if an eye is covered for any significant period.

ACKNOWLEDGMENT

I am indebted to Miss J. Livingstone for considerable secretarial assistance and to Mr. K. Collins for help in the preparation of the figures for publication.

Peter O. Bishop

REFERENCES

1. Aguilonius, F.: Opticorum libri sex. Antwerp, Belgium, 1613. (Discussed in von Rohr, M.: Auswahl aus der Behandlung des Horopters bei Fr. Aguilonius um 1613, Z. Ophthalmol. Opt. **11**:41, 1923.)
2. Amigo, G.: Visuo-sensory development of the child, Am. J. Optom. Physiol. Opt. **49**:991, 1972.
3. Asher, H.: Suppression theory of binocular vision, Br. J. Ophthalmol. **37**:37, 1953.
4. Awaya, S., et al.: Amblyopia in man, suggestive of stimulus deprivation amblyopia, Jpn. J. Ophthalmol. **17**:69, 1973.
5. Bagolini, B.: Anomalous correspondence: definition and diagnostic methods, Doc. Ophthalmol. **23**:346, 1967.
6. Bagolini, B., and Ravalico, G.: The binocular visual field in small-angle strabismus. In Mein, J., Bierlaagh, J. J., and Brummel-kamp-Dons, F. E. A., editors: Orthoptics. Proceedings of Second International Orthoptic Congress, Amsterdam, 1972, Excerpta Medica BV, pp. 43-49.
7. Baker, F. H., Grigg, P., and von Noorden, G. K.: Effects of visual deprivation and strabismus on the responses of neurons in the visual cortex of the monkey, including studies of the striate and prestriate cortex in the normal animal, Brain Res. **66**:185, 1974.
8. Banks, M. S., Aslin, R. N., and Letson, R. D.: Sensitive period for the development of human binocular vision, Science **190**:675, 1975.
9. Barlow, H. B., Blakemore, C., and Pettigrew, J. D.: The neural mechanisms of binocular depth discrimination, J. Physiol. (Lond) **193**:327, 1967.
10. Berlucchi, G.: Anatomical and physiological aspects of visual functions of corpus callosum, Brain Res. **37**:371, 1972.
11. Berry, R. N.: Quantitative relations among vernier, real depth and stereoscopic depth acuities, J. Exp. Psychol. **38**:708, 1948.
12. Bishop, P. O.: Beginning of form vision and binocu-

lar depth discrimination in cortex. In Schmitt, F. O., editor: The neurosciences: second study program, New York, 1970, Rockefeller University Press, pp. 471-485.
13. Bishop, P. O.: Neurophysiology of binocular single vision and stereopsis. In Jung, R., editor: Handbook of sensory physiology, Berlin, 1973, Springer-Verlag, vol. 7/3A, pp. 255-305.
14. Bishop, P. O.: Orientation and position disparities in stereopsis. In Cool, S. J., and Smith, E. L., III, editors: Frontiers in visual science, New York, 1978, Springer-Verlag New York, Inc., pp. 336-350.
15. Bishop, P. O., and Henry, G. H.: Spatial vision, Ann. Rev. Psychol. **22:**119, 1971.
16. Bishop, P. O., Henry, G. H., and Smith, C. J.: Binocular interaction fields of single units in the cat striate cortex, J. Physiol. (Lond) **216:**39, 1971.
17. Blake, R., and Hirsch, H. V. B.: Deficits in binocular depth perception in cats after alternating monocular deprivation, Science **190:**1114, 1975.
18. Blakemore, C.: Binocular depth discrimination and the nasotemporal division, J. Physiol. (Lond) **205:** 471, 1969.
19. Blakemore, C.: The range and scope of binocular depth discrimination in man, J. Physiol. (Lond) **211:** 599, 1970.
20. Blakemore, C.: Binocular depth perception and the optic chiasm, Vision Res. **10:**43, 1970.
21. Blakemore, C., and Van Sluyters, C.: Reversal of the physiological effects of monocular deprivation in kittens: further evidence for a sensitive period, J. Physiol. (Lond) **237:**195, 1974.
22. Blakemore, C., Fiorentini, A., and Maffei, L.: A second neural mechanism of binocular depth discrimination, J. Physiol. (Lond) **226:**725, 1972.
23. Blakemore, C., Garey, L. J., and Vital-Durand, F.: The physiological effects of monocular deprivation and their reversal in the monkey's visual cortex, J. Physiol. (Lond) **283:**223, 1978.
24. Boring, E. G.: Sensation and perception in the history of experimental psychology, New York, 1942, Appleton-Century-Crofts.
25. Bough, E. W.: Stereoscopic vision in the macaque monkey: a behavioural demonstration, Nature **225:** 42, 1970.
26. Bunt, A. H., Minckler, D. S., and Johanson, G. W.: Demonstration of bilateral projection of the central retina of the monkey with horseradish peroxidase neuronography, J. Comp. Neurol. **171:**619, 1977.
27. Burian, H. M.: Practical aspects of depth perception. In Potts, A. M., editor: The assessment of visual function, St. Louis, 1972, The C. V. Mosby Co., pp. 161-186.
28. Campbell, F. W., and Green, D. G.: Monocular versus binocular visual acuity, Nature **208:**191, 1965.
29. Chow, K. L., and Stewart, D. L.: Reversal of structural and functional effects of long-term visual deprivation in cats, Exp. Neurol. **34:**409, 1972.
30. Clark, P. G. H., Donaldson, I. M. L., and Whitteridge, D.: Binocular visual mechanisms in cortical areas I and II of the sheep, J. Physiol. (Lond) **256:** 509, 1976.
31. Collet, T.: Stereopsis in toads, Nature **267:**349, 1977.
32. Cowey, A., Parkinson, A. M., and Warwick, L.: Global stereopsis in rhesus monkeys, Q. J. Exp. Psychol. **27:**93, 1975.
33. Cragg, B. G.: The development of synapses in the visual system of the cat, J. Comp. Neurol. **160:**147, 1975.
34. Crawford, M. L. J., et al.: Physiological consequences of unilateral and bilateral eye closure in macaque monkeys: some further observations, Brain Res. **84:**150, 1975.
35. Crone, R. A., and Everhard-Halm, Y.: Optically induced eye torsion, Albrecht v. Graefes Arch. Klin. Exp. Ophthalmol. **195:**231, 1975.
36. Cynader, M., and Regan, D.: Neurones in cat parastriate cortex sensitive to the direction of motion in three-dimensional space, J. Physiol. (Lond) **274:**549, 1978.
37. Cynader, M., Gardner, J., and Douglas, R.: Neural mechanisms underlying stereoscopic depth perception in cat visual cortex. In Cool, S. J., and Smith, E. L., III, editors: Frontiers in visual science, New York, 1978, Springer-Verlag New York, Inc., pp. 351-372.
38. Dayton, G. O., et al.: Developmental study of coordinated eye movements in the human infant. II, An electro-oculographic study of the fixation reflex in the newborn, Arch. Ophthalmol. **71:**871, 1964.
39. Denny-Brown, D., and Chambers, R. A.: Physiological aspects of visual perception. I, Functional aspects of visual cortex, Arch. Neurol. **33:**219, 1976.
40. Denny-Brown, D., and Fischer, E. G.: Physiological aspects of visual perception. II, The subcortical vision direction of behavior, Arch. Neurol. **33:**228, 1976.
41. Dews, P. B., and Wiesel, T. N.: Consequences of monocular deprivation on visual behaviour in kittens, J. Physiol. (Lond) **206:**437, 1970.
42. Ditchburn, R. W., and Ginsborg, B. L.: Involuntary eye movements during fixation, J. Physiol. (Lond) **119:**1, 1953.
43. Dobson, V., and Teller, D. Y.: Visual acuity in human infants: a review and comparison of behavioral and electrophysiological studies, Vision Res. **18:** 1469, 1978.
44. Dodge, R.: An experimental study of visual fixation, Psychol. Rev. **8:**1, 1907.
45. Dove, H. A.: Die Combination der Eindrück beider Ohren und beider Augen zu einem Eindrück, Ber. Preuss. Akad. Wiss., 1841. (Cited in Boring, E. G.: Sensation and perception in the history of experimental psychology, New York, 1942, Appleton-Century-Crofts.)
46. Duke-Elder, S., and Cook, C.: Embryology. In Duke-Elder, S., editor: System of ophthalmology, St. Louis, 1963, The C. V. Mosby Co., vol. 3, part 1.
47. Duke-Elder, S., and Wybar, K. C.: The anatomy of the visual system. In Duke-Elder, S., editor: System

of ophthalmology, St. Louis, 1961, The C. V. Mosby Co., vol. 2.

48. Duke-Elder, S., and Wybar, K.: Ocular motility and strabismus. In Duke-Elder, S., editor: System of ophthalmology, St. Louis, 1973, The C. V. Mosby Co., vol. 6.
49. Dunlap, E. A.: Overcorrection in horizontal strabismus surgery. In Transaction of New Orleans Academy of Ophthalmology: Symposium on strabismus, St. Louis, 1971, The C. V. Mosby Co., pp. 255-267.
50. Dürsteler, M. R., Garey, L. J., and Movshon, J. A.: Reversal of morphological effects of monocular deprivation in kitten's lateral geniculate nucleus, J. Physiol. (Lond) **261:**189, 1976.
51. Efron, R.: Stereoscopic vision. I, Effect of binocular temporal summation, Br. J. Ophthalmol. **41:**709, 1957.
52. Engen, T.: Psychophysics. I, Discrimination and detection. In Kling, J. W., and Riggs, L. A., editors: Woodworth and Schlosberg's experimental psychology, ed. 3, New York, 1971, Holt, Rinehart & Winston, Inc., pp. 11-46.
53. Epstein, D. L., and Tredici, T. J.: Use of the four-diopter base-in prism test in micro-exotropia, Am. J. Ophthalmol. **74:**340, 1972.
54. Eysel, U. Th., Grüsser, O.-J., and Hoffmann, K.-P.: Monocular deprivation and signal transmission by X- and Y-neurons of cat lateral geniculate nucleus, Exp. Brain Res. **34:**521, 1979.
55. Fender, D., and Julesz, B.: Extension of Panum's fusional area in binocularly stabilized vision, J. Opt. Soc. Am. **57:**819, 1967.
56. Fender, D. H., and St-Cyr, G. J.: The interplay of drifts and flicks in binocular fixation, Vision Res. **9:**245, 1969.
57. Fox, R., Lehmkuhle, S., and Bush, R. C.: Stereopsis in the falcon, Science **497:**79, 1977.
58. Fox, R., Lehmkuhle, S., and Leguire, L. E.: Stereopscopic contours induce optokinetic nystagmus, Vision Res. **18:**1189, 1978.
59. Garey, L. J., Fisken, R. A., and Powell, T. P. S.: Observations on the growth of cells in the lateral genicular nucleus of the cat, Brain Res. **52:**359, 1973.
60. Gibson, J. J.: The perception of the visual world, Boston, 1950, Houghton Mifflin Co.
61. Graham, C. H.: Visual space perception. In Graham, C. H., editor: Vision and visual perception, New York, 1965, John Wiley & Sons, Inc., pp. 504-547.
62. Grindley, G. C., and Townsend, V.: Binocular masking induced by a moving object, Q. J. Exp. Psychol. **17:**97, 1965.
63. Guillery, R. W.: Binocular competition in the control of geniculate cell growth, J. Comp. Neurol. **144:**117, 1972.
64. Guillery, R. W.: The effect of lid suture upon the growth of cells in the dorsal lateral geniculate nucleus of kittens, J. Comp. Neurol. **148:**417, 1973.
65. Guillery, R. W., and Stelzner, D. J.: The differential effects of unilateral lid closure upon the monocular and binocular segments of the dorsal lateral geniculate nucleus in the cat, J. Comp. Neurol. **139:**413, 1970.
66. Haith, M. M.: Visual competence in early infancy. In Held, R., et al., editors: Handbook of sensory physiology, Berlin, 1978, Springer-Verlag, vol. 8, pp. 311-356.
67. Hebbard, F. W.: Comparison of subjective and objective measurements of fixation disparity, J. Opt. Soc. Am. **52:**706, 1962.
68. Hering, E.: Vom Ortsinne der Netzhaut, Beiträge zur Physiologie, Leipzig, East Germany, 1861, Wilhelm Engelmann, pp. 9-80.
69. Hering, E.: Der Raumsinn und die Bewegungen der Augen. In Hermann, L., editor: Handbuch der Physiologie, Leipzig, 1879, Vogel, vol. 3/1, pp. 343-601. (Radde, C. A., translator: Spatial sense and movements of the eye, Baltimore, 1942, American Academy of Optometry.)
70. Hering, E.: Grundzüge der Lehre vom Lichtsinn. In Graefe, A., and Saemisch, T., editors: Handbuch der gessammten Augenheilkunde, ed. 2, Leipzig, East Germany, 1925, Wilhelm Engelmann, vol. 3. (Hurvich, L. M., and Jameson, D., translators: Outlines of a theory of the light sense, Cambridge, Mass., 1964, Harvard University Press.)
71. Hickey, T. L.: Postnatal development of the human lateral geniculate nucleus: relationship to a critical period for the visual system, Science **198:**836, 1977.
72. Hochberg, J.: Depth perception loss with local monocular suppression: a problem in the explanation of stereopsis, Science **145:**1334, 1964.
73. Hogan, M. J., Alvarado, J. A., and Weddell, J. E.: Histology of the human eye, Philadelphia, 1971, W. B. Saunders Co.
74. Hohmann, A., and Creutzfeldt, O. D.: Squint and the development of binocularity in humans, Nature **254:** 613, 1975.
75. Home, R.: Binocular summation: a study of contrast sensitivity, visual acuity and recognition, Vision Res. **18:**579, 1978.
76. Howard, H. J.: A test for the judgment of distance, Am. J. Ophthalmol. **2:**656, 1919.
77. Howard, I. P., and Templeton, W. B.: Human spatial orientation, New York, 1966, John Wiley & Sons, Inc.
78. Hubel, D. H., and Wiesel, T. N.: Receptive fields of single nuerones in the cat's striate crotex, J. Physiol. (Lond) **148:**574, 1959.
79. Hubel, D. H., and Wiesel, T. N.: Receptive fields, binocular interaction and functional architecture in the cat's visual cortex, J. Physiol. (Lond) **160:**106, 1962.
80. Hubel, D. H., and Wiesel, T. N.: Receptive fields of cells in striate cortex of very young visually inexperienced kittens, J. Neurophysiol. **26:**994, 1963.
81. Hubel, D. H., and Wiesel, T. N.: Receptive fields and functional architecture in two nonstriate visual areas (18 and 19) of the cat, J. Neurophysiol. **28:**1029, 1965.
82. Hubel, D. H., and Wiesel, T. N.: Binocular interaction in striate cortex of kittens reared with artificial squint, J. Neurophysiol. **28:**1041, 1965.

83. Hubel, D. H., and Wiesel, T. N.: Receptive fields and functional architecture of monkey striate cortex, J. Physiol. (Lond) **195:**215, 1968.
84. Hubel, D. H., and Wiesel, T. N.: The period of susceptibility to the physiological effects of unilateral eye closure in kittens, J. Physiol. (Lond) **206:**419, 1970.
85. Hubel, D. H., and Wiesel, T. N.: Cells sensitive to binocular depth in area 18 of the macaque monkey cortex, Nature **225:**41, 1970.
86. Hubel, D. H., and Wiesel, T. N.: Functional architecture of macaque monkey visual cortex, Proc. R. Soc. Lond. (Biol.) **198:**1, 1977.
87. Hubel, D. H., Wiesel, T. N., and LeVay, S.: Plasticity of ocular dominance columns in monkey striate cortex, Philos. Trans. R. Soc. Lond. (Biol.) **278:** 377, 1977.
88. Huber, A.: Homonymous hemianopsia after removal of one occipital lobe, Excerpta Medica **222:**1333, 1970.
89. Hugonnier, R., and Clayette-Hugonnier, S.: Strabismus heterophoria, ocular motor paralysis: clinical ocular muscle imbalance, Veronneau-Troutman, S., translator and editor, St. Louis, 1969, The C. V. Mosby Co.
90. Ikeda, H., and Tremain, K. E.: Amblyopia resulting from penalisation: neurophysiological studies of kittens reared with atropinisation of one or both eyes, Br. J. Ophthalmol. **62:**21, 1978.
91. Ikeda, H., and Wright, M. J.: Properties of LGN cells in kittens reared with convergent squint: a neurophysiological demonstration of amblyopia, Exp. Brain Res. **25:**63, 1976.
92. Ikeda, H., Plant, G. T., and Tremain, K. E.: Nasal field loss in kittens reared with convergent squint: neurophysiological and morphological studies of the lateral geniculate nucleus, J. Physiol. (Lond) **270:** 345, 1977.
93. Ikeda, H., Tremain, K. E., and Einon, G.: Loss of spatial resolution of lateral geniculate nucleus neurones in kittens raised with a convergent squint produced at different ages of development, Exp. Brain Res. **31:**207, 1978.
94. Jacobson, S. G., and Ikeda, H.: Behavioural studies of spatial vision in cats reared with convergent squint: is amblyopia due to arrest of development? Exp. Brain Res. **34:**11, 1979.
95. Jones, R.: Anomalies of disparity detection in the human visual system, J. Physiol. (Lond) **264:**621, 1977.
96. Joshua, D. E., and Bishop, P. O.: Binocular single vision and depth discrimination. Receptive field disparities for central and peripheral vision and binocular interaction on peripheral single units in cat striate cortex, Exp. Brain Res. **10:**389, 1970.
97. Julesz, B.: Binocular depth perception of computer-generated patterns, Bell System Tech. J. **39:**1125, 1960.
98. Julesz, B.: Stereopsis and binocular rivalry of contours, J. Opt. Soc. Am. **53:**994, 1963.
99. Julesz, B.: Foundations of cyclopean perception, Chicago, 1971, University of Chicago Press.
100. Julesz, B.: Global stereopsis: cooperative phenomena in stereopscopic depth perception. In Held, R. et al., editors: Handbook of sensory physiology, Berlin, 1978, Springer-Verlag, vol. 7, pp. 215-256.
101. Karten, H. J., et al.: Neural connections of the "visual Wulst" of the avian telencephalon. Experimental studies in the pigeon *(Columba livia)* and owl *(Speotyto cunicularia)*, J. Comp. Neurol. **150:**253, 1973.
102. Kaufman, L.: On the spread of suppression and binocular rivalry, Vision Res. **3:**401, 1963.
103. Keesey, U. T.: Effects of involuntary eye movements on visual acuity, J. Opt. Soc. Am. **50:**769, 1960.
104. Kepler, J.: Ad vitellionem paralipomena, quibus astronomiae pars optica traditur, Francofurti, 1604.
105. Le Grand, Y.: L'Espace visuel (Form and space vision), Millodot, M., and Heath, G. G., translators, Bloomington, Ind., 1967, Indiana University Press.
106. Leicester, J.: Projection of the visual vertical meridian to cerebral cortex of the cat, J. Neurophysiol. **31:**371, 1968.
107. Le Vay, S., Hubel, D. H., and Wiesel, T. N.: The pattern of ocular dominance columns in macaque visual cortex revealed by a reduced silver stain, J. Comp. Neurol. **159:**559, 1975.
108. Le Vay, S., Stryker, M. P., and Shatz, C. J.: Ocular dominance columns and their development in layer IV of the cat's visual cortex: a quantitative study, J. Comp. Neurol. **179:**223, 1978.
109. Levelt, W. J. M.: On binocular rivalry, Soesterberg, The Netherlands, 1965, Institute for Perception RVO-TNO.
110. Lin, C. S., and Sherman, S. M.: Effects of early monocular eyelid suture upon development of relay cell classes in the cat's lateral geniculate nucleus, J. Comp. Neurol. **181:**809, 1978.
111. Linksz, A.: Physiology of the eye. Vision, New York, 1952, Grune & Stratton, Inc., vol. 2.
112. Linschoten, J.: Strukturanalyse der binokularen Tiefenwahrnehmung, Groningen, The Netherlands, 1956, Wolters-Noordhoff BV.
113. Lu, C., and Fender, D. H.: The interaction of color and luminance in stereoscopic vision, Invest. Ophthalmol. **11:**482, 1972.
114. Manley, D. R.: Classification of esodeviations. In Manley, D. R., editor: Symposium on horizontal ocular deviations, St. Louis, 1971, The C. V. Mosby Co., pp. 3-48.
115. Marlowe, L. H.: Orientation of contours and binocular depth perception, Providence, R.I., 1969, Brown University (doctoral thesis).
116. Matsubayashi, A. Forschung über die Tiefenwahrnehmung, IX, Acta Soc. Ophthalmol. Jpn. **42:**1920, 1938. (Cited in Graham, C. H.: Visual space perception. In Graham, C. H., editor: Vision and visual perception, New York, 1965, John Wiley & Sons, Inc., pp. 504-547.)
117. Mitchell, D. E.: A review of the concept of "Panum's fusional areas," Am. J. Optom. Physiol. Opt. **43:** 387, 1966.

118. Mitchell, D. E.: Retinal disparity and diplopia, Vision Res. **6:**441, 1966.
119. Mitchell, D. E.: Qualitative depth localization with diplopic images of dissimilar shape, Vision Res. **9:**991, 1969.
120. Mitchell, D. E.: Properties of stimuli eliciting vergence eye movements and stereopsis, Vision Res. **10:**145, 1970.
121. Mitchell, D. E., Cynader, M., and Movshon, J. A.: Recovery from the effects of monocular deprivation in kittens, J. Comp. Neurol. **176:**53, 1977.
122. Mitchell, D. E., et al.: Visual resolution in young kittens, Vision Res. **16:**363, 1976.
123. Movshon, J. A.: Reversal of the physiological effects of monocular deprivation in the kitten's visual cortex, J. Physiol. (Lond) **261:**125, 1976.
124. Müller, J.: Zur vergleichenden Physiologie des Gesichtsinnes des Menschen und der Thiere, Leipzig, 1826, Cnobloch. (Cited in Boring, E. G.: Sensation and perception in the history of experimental psychology, New York, 1942, Appleton-Century-Crofts.)
125. Nauta, W. J. H., and Karten, H. J.: A general profile of the vertebrate brain, with sidelights on the ancestry of cerebral cortex. In Schmitt, F. O. et al., editors: The neurosciences. Second study program, New York, 1970, Rockefeller University Press, pp. 7-26.
126. Nelson, J. I.: Globality and stereoscopic fusion in binocular vision, J. Theor. Biol. **49:**1, 1975.
127. Nelson, J. I.: The plasticity of correspondence: aftereffects, illusions and horopter shifts in depth perception, J. Theor Biol. **66:**203, 1977.
128. Nelson, J. I., Kato, H., and Bishop, P. O.: Discrimination of orientation and position disparities by binocularly activated neurons in cat striate cortex, J. Neurophysiol. **40:**260, 1977.
129. Newton, I.: Opticks, query 15, book 3, part 1, 1704; third edition, 1730. (Reprinted New York, 1952, Dover Publications, Inc.)
130. Nikara, T., Bishop, P. O., and Pettigrew, J. D.: Analysis of retinal correspondence by studying receptive fields of binocular single units in cat striate cortex, Exp. Brain Res. **6:**353, 1968.
131. Ogle, K. N.: Induced size effect. I, A new phenomenon in binocular space perception associated with the relative sizes of the images of the two eyes, Arch. Ophthalmol. **20:**604, 1938.
132. Ogle, K. N.: Researches in binocular vision, Philadelphia, 1950, W. B. Saunders Co.
133. Ogle, K. N.: Disparity limits of stereopsis, Arch. Ophthalmol. **48:**50, 1952.
134. Ogle, K. N.: Precision and validity of stereoscopic depth perception from double images, J. Opt. Soc. Am. **43:**906, 1953.
135. Ogle, K. N.: Stereopsis and vertical disparity, Arch. Ophthalmol. **53:**495, 1955.
136. Ogle, K. N.: The optical space sense. In Davson, H., editor: The eye, New York, 1962, Academic Press, Inc., vol. 4/2, pp. 211-417.
137. Ogle, K. N.: Stereoscopic depth perception and exposure delays between images in the two eyes, J. Opt. Soc. Am. **53:**1296, 1963.
138. Ogle, K. N., and Weil, M. P.: Stereoscopic vision and the duration of the stimulus, Arch. Ophthalmol. **59:**4, 1958.
139. Ogle, K. N., Martens, T. G., and Dyer, J. A.: Oculomotor imbalance in binocular vision and fixation disparity, London, 1967, Henry Kimpton, Publishers.
140. Ordy, J. M., et al.: Visual acuity in newborn primate infants, Proc. Soc. Exp. Biol. Med. **115:**677, 1964.
141. Ordy, J. M., et al.: Postnatal development of vision in a subhuman primate *(Macaca mulatta),* Arch. Ophthalmol. **73:**674, 1965.
142. Packwood, J., and Gordon, B.: Stereopsis in normal domestic cat, Siamese cat, and cat raised with alternating monocular occlusions, J. Neurophysiol. **38:**1485, 1975.
143. Palmer, D. A.: Measurement of the horizontal extent of Panum's area by a method of constant stimuli, Opt. Acta **8:**151, 1961.
144. Panum, P. L.: Physiologische Untersuchungen über das Sehen mit zweiaugen, Kiel, 1858, Schwering.
145. Parks, M. M.: Management of eccentric fixation and ARC in esotropia. In Manley, D. R., editor: Symposium on horizontal ocular deviations, St. Louis, 1971, The C. V. Mosby Co., pp. 81-87.
146. Parks, M. M.: Sensory adaptations in strabismus. In Transactions of New Orleans Academy of Ophthalmology: Symposium on strabismus, St. Louis, 1971, The C. V. Mosby Co., pp. 93-120.
147. Parks, M. M.: The monofixation syndrome. In Transactions of New Orleans Academy of Ophthalmology: Symposium on strabismus, St. Louis, 1971, The C. V. Mosby Co., pp. 121-153.
148. Parks, M. M.: Ocular motility and strabismus, New York, 1975, Harper & Row Publishers, Inc.
149. Pettigrew, J. D.: Binocular neurones which signal change of disparity in area 18 of cat visual cortex, Nature **241:**123, 1973.
150. Pettigrew, J. D.: The effect of visual experience on the development of stimulus specificity by kitten cortical neurones, J. Physiol. (Lond) **237:**49, 1974.
151. Pettigrew, J. D.: Comparison of the retinotopic organization of the visual Wulst in nocturnal and diurnal raptors, with a note on the evolution of frontal vision. In Cool, S. J., and Smith, E. L., III, editors: Frontiers in visual science, New York, 1978, Springer-Verlag New York, Inc., pp. 328-335.
152. Pettigrew, J. D., and Konishi, M.: Effect of monocular deprivation on binocular neurones in the owl's visual Wulst, Nature **264:**753, 1976.
153. Pettigrew, J. D., and Konishi, M.: Neurones selective for orientation and binocular disparity in the visual Wulst of the barn owl *(Tyto alba),* Science **193:**675, 1976.
154. Pirenne, M. H.: Binocular and uniocular threshold of vision, Nature **152:**698, 1943.
155. Poggio, G. F., and Fischer, B.: Binocular interaction and depth sensitivity in striate and prestriate cortex of behaving rhesus monkeys, J. Neurophysiol. **40:**1392, 1977.

156. Polyak, S.: The vertebrate visual system, Chicago, 1957, University of Chicago Press.
157. Pulfrich, C.: Die Stereoskopie im Dienste der isochromen und heterochromen Photometrie, Naturwissenschaften **10:**553, 1922.
158. Raab, E. L., and Parks, M. M.: Immediate postoperative overcorrection after recession of the lateral recti: effect on the final result. In Manley, D. R., editor: Symposium on horizontal ocular deviations, St. Louis, 1971, The C. V. Mosby Co., pp. 174-182.
159. Rakic, P.: Prenatal development of the visual system in the rhesus monkey, Philos. Trans. R. Soc. Lond. (Biol.) **278:**245, 1977.
160. Rawlings, S. C., and Shipley, T.: Stereoscopic acuity and horizontal angular distance from fixation, J. Opt. Soc. Am. **59:**991, 1969.
161. Richards, W.: Stereopsis and stereoblindness, Exp. Brain Res. **10:**380, 1970.
162. Richards, W.: Anomalous stereoscopic depth perception, J. Opt. Soc. Am. **61:**410, 1971.
163. Romano, P. E.: The quality of binocular co-operation and the 4-dioptre prism test. In Fells, P., editor: The First Congress of the International Strabismological Association, St. Louis, 1971, The C. V. Mosby Co., pp. 289-300.
164. Romano, P. E., and Romano, J. A.: Fusion: a new classification and methods for determining the level of sensory binocular cooperation, Survey Ophthalmol. **17:**458, 1973.
165. Rønne, G.: The physiological basis of sensory fusion, Acta Ophthalmol. **34:**1, 1956.
166. Sanderson, K. J., and Sherman, S. M.: Nasotemporal overlap in visual field projected to lateral geniculate nucleus in the cat, J. Neurophysiol. **34:**453, 1971.
167. Sarmiento, R. F.: The stereoacuity of macaque monkey, Vision Res. **15:**493, 1975.
168. Schiller, P. H., Finlay, B. L., and Volman, S. F.: Quantitative studies of single-cell properties in monkey striate cortex. II, Orientation specificity and ocular dominance, J. Neurophysiol. **39:**1320, 1976.
169. Shatz, C. J., Lindstrom, S., and Wiesel, T. N.: The distribution of afferents representing the right and left eyes in the cat's visual cortex, Brain Res. **131:**103, 1977.
170. Sherman, S. M., Hoffmann, K.-P., and Stone, J.: Loss of a specific cell type from dorsal lateral geniculate nucleus in visually deprived cats, J. Neurophysiol. **35:**532, 1972.
171. Shipley, T., and Rawlings, S. C.: The Nonius horopter. I, History and theory, Vision Res. **10:**1225, 1970.
172. Shortess, G. K., and Krauskopf, J.: Role of involuntary eye movements in stereoscopic acuity, J. Opt. Soc. Am. **51:**555, 1960.
173. Stone, J.: The naso-temporal division of the cat's retina, J. Comp. Neurol. **126:**585, 1966.
174. Stone, J., Leicester, J., and Sherman, S. M.: The naso-temporal division of the monkey's retina, J. Comp. Neurol. **150:**333, 1973.
175. Sullivan, M. J., and Kertesz, A. E.: Binocular coordination of torsional eye movements in cyclofusional response, Vision Res. **18:**943, 1978.
176. Taylor, D. M.: Is congenital esotropia functionally curable? Trans. Am. Ophthalmol. Soc. **70:**529, 1972.
177. Taylor, D. M.: Congenital esotropia: management and prognosis, New York, 1973, Intercontinental Medical Book Corp.
178. Teller, D. W., et al.: Development of visual acuity in infant monkeys *(Macaca nemestrina)* during the early postnatal weeks, Vision Res. **18:**561, 1978.
179. Verhoeff, F. H.: A new theory of binocular vision, Arch. Ophthalmol. **13:**151, 1935.
180. Vieth, G. U. A.: Ueber die Richtung der Augen, Ann. Physiol. **58:**233, 1818. (Cited in Boring, E. G.: Sensation and perception in the history of experimental psychology, New York, 1942, Appleton-Century-Crofts.)
181. von der Heydt, R., et al.: Disparity sensitivity and receptive field incongruity of units in the cat striate cortex, Exp. Brain Res. **31:**523, 1978.
182. von Gudden, B.: Ueber die Kreuzung der Fasern im Chiasma Nervorum opticorum, Albrecht v. Graefes Arch. Klin. Exp. Ophthalmol. **20:**249, 1874.
183. von Helmholtz, H.: Handbuch der physiologischen Optik, Leipzig, East Germany, 1867, Voss; third German edition, 1909-1911. (Southall, J. P. C., translator: Treatise on physiological optics, New York, 1925, Columbia University Press; reprinted New York, 1962, Dover Publications, Inc.)
184. von Noorden, G. K.: Experimental amblyopia in monkeys. Further behavioral observations and clinical correlations, Invest. Ophthalmol. **12:**721, 1973.
185. von Noorden, G. K.: Histological studies of the visual system in monkeys with experimental amblyopia, Invest. Ophthalmol. **12:**727, 1973.
185a. von Noorden, G. K.: A primate model for amblyopia. In Reinecke, R. D., editor: Strabismus, New York, 1978, Grune & Stratton, Inc., pp. 23-29.
186. von Noorden, G. K.: Burian-von Noorden's binocular vision and ocular motility: theory and management of strabismus, St. Louis, 1979, The C. V. Mosby Co.
187. von Noorden, G. K., and Dowling, J. E.: Experimental amblyopia in monkeys. II, Behavioral studies in strabismic amblyopia, Arch. Ophthalmol. **84:**215, 1970.
188. von Noorden, G. K., and Maumenee, A. E.: Clinical observations on stimulus-deprivation amblyopia (amblyopia ex anopsia), Am. J. Ophthalmol. **65:**220, 1968.
189. von Noorden, G. K., Dowling, J. E., and Ferguson, D. C.: Experimental amblyopia in monkeys. I, Behavioral studies of stimulus deprivation amblyopia, Arch. Ophthalmol. **84:**206, 1970.
190. von Tschermak-Seysenegg, A.: Einführung in die physiologische optick, Munich, West Germany, 1942, Bergmann. (Boeder, P., translator: Introduction to physiological optics, Springfield, Ill., 1952, Charles C Thomas, Publisher.)

191. Westheimer, G., and Mitchell, D. E.: The sensory stimulus for disjunctive eye movements, Vision Res. **9:**749, 1969.
192. Westheimer, G., and Tanzman, I. J.: Qualitative depth localization with diplopic images, J. Opt. Soc. Am. **46:**116, 1956.
193. Wheatstone, C.: Contributions to the physiology of vision. I, On some remarkable, and hitherto unobserved, phenomena of binocular vision, Philos. Trans. R. Soc. Lond. **128:**371, 1838.
194. Wiesel, T. N., and Hubel, D. H.: Effects of visual deprivation on morphology and physiology of cells in the cat's lateral geniculate body, J. Neurophysiol. **26:**978, 1963.
195. Wiesel, T. N., and Hubel, D. H.: Single-cell responses in striate cortex of kittens deprived of vision in one eye, J. Neurophysiol. **26:**1003, 1963.
196. Wiesel, T. N., and Hubel, D. H.: Comparison of the effects of unilateral and bilateral eye closure on cortical unit responses in kittens, J. Neurophysiol. **28:** 1029, 1965.
197. Wiesel, T. N., and Hubel, D. H.: Extent of recovery from the effects of visual deprivation in kittens, J. Neurophysiol. **28:**1060, 1965.
198. Zeki, S. M.: Cells responding to changing image size and disparity in the cortex of the rhesus monkey, J. Physiol. (Lond) **242:**827, 1974.

Chapter 22

Visual adaptation

W. A. H. RUSHTON, Sc.D., F.R.S.

As every photographer knows, if you are to obtain a good contrasty picture of any scene, you must get the light exposure right. Too little light (because the exposure is too brief or the iris stop too small) will result in a very dark picture with only the highlights showing. Too strong a light, on the other hand, causes overexposure, as though the whole picture were caught in a luminous fog. Automatic cameras ensure the correct exposure for any scene with a photocell that measures the average brightness and then makes that "brightness signal" from the photocell change the iris stop so that the average light falling on the film lies near the midpoint of that film's sensitivity range. The human eye in its ordinary use throughout the day adapts to the continued changes in illumination very much in this way, though pupil size is not the chief regulator.

Although the problems of eye and camera are much the same, the eye is much more urgent. We have *got* to see. A photographer often can afford to wait for good conditions, and very bad photographs are not worth taking. But most animals depend on their eyes to find their food and to escape their enemies. Without sight they starve or are eaten. And to see even a vague shadow moving in the gloom may mean an escape from death.

An animal's eye must detect over the most enormous range of illumination—something like 10 billion to 1. And at each level it must do the best it can to discriminate in space (acuity), time (flicker), and wavelength (color).

No photographer would think of using only one kind of film for all conditions in which we see. At the lowest brightness levels he must use a sensitive film with a coarse grain. This could not be a color film, which of necessity needs much more light than a black-white film to register the extra colored information. At the brightest levels he could use a fine-grain color film and with a brief snap could register colored detail and rapid movement without smudging.

Our eyes likewise have two different films, the sensitive colorless vision of the rods and the rapidly responding chromatic vision of the cones. We do not actually change film; rods and cones remain side by side on the retina, and both send what information they have to the brain. But in dim light most of the information comes from rods, and in bright light most from cones.

We have spoken of the eye as a photographic camera, but it resembles more closely a television camera. The optical image is not fixed on the retina; it is instantly encoded into a complex of nerve signals that are transmitted to the brain. There they are not displayed on any kind of screen to regenerate the original picture. They are fed into the brain's complex of computers that process the visual signals to give the proper sensorimotor output.

With a photographic camera the only way to adjust the response to light is to adjust the actual light that falls on the film. But the electric output from a television camera allows much more versatile adjustments by manipulation of the electric signals. It is principally by processing the retinal nerve signals that human light and dark adaptation occurs.

It is true that when the change from light to darkness is enormous, for instance, on going from sunlight into a dark lecture theatre, the change is mainly from the cone film to the rod film, and chemical complications occur. These are important, but quite exceptional features of our daily visual life. Nearly all ordinary adaptation results simply from controlled attenuation of the nerve signals on their way from the rods and cones to the optic nerve.

QUANTUM CATCH AND N SIGNALS

We must remember that light, like matter, has an atomic structure, the atoms of light being called "quanta." So in the photochemical action of light on a photographic film, single quanta of light are caught by single molecules of silver bromide and produce each a single molecular change.

It is the same in the eye. Single quanta are caught by single molecules of rhodopsin in the rods, and each catch produces a change that can excite the whole rod. The photosensitive pigment of rod vision is a chromoprotein called rhodopsin. It may easily be seen in the rods of a frog's retina quickly dissected out under a photographic safelight. If the retina is spread on paper and brought into the light of the room, it is seen to be pink. But the pinkness rapidly fades away when the retina is exposed to a bright light. Rhodopsin has been *bleached*. The bleaching of a rhodopsin molecule is associated with the excitation of the rod that contains it. But although a single quantum caught is sufficient to excite the rod that catches it, it needs about six quanta to be caught in a rod cluster for the flash to be reliably detected. If in complete darkness we examine what we see (with eyes either open or shut), we see a sort of scintillation—the intrinsic light of the retina, or "retinal noise." The activity of *one* rod is not detectable amidst this noise, it needs about six rods acting in unison to be seen. Thus the threshold for detection is about six quanta close together in space and time.

If a microelectrode is inserted into a rod, for example, in the excised retina of an axolotl (Fig. 22-1), an electric signal can be recorded in response to a flash of light.[10] The white circles in Fig. 22-1 show the size of signal V for flashes of various energies I plotted on a logarithmic scale to cover a millionfold range.

Before we consider the special features of this curve, it is of interest to ask whether the curve of Fig. 22-1, which relates the quantum catch I with the size of nerve signal V, applies not only to axolotl rods but also to human rods. Actually that same curve was established in the human about 2 years *before* the axolotl results. One cannot record from single rods in the living human, but by a special psychophysical technique in which excitatory and inhibitory flashes were balanced against each other,[1] the size of nerve signal N could be plotted against I, the number of quanta in the flash that caused it. The relation found was

$$N/N_{max} = \frac{I}{I + \sigma} \quad (1)$$

where σ is a large constant of about 1000 quanta absorbed per rod per flash.

It was highly satisfactory when the size of nerve signal that had been inferred from psychophysical analysis, N/N_{max}, turned out to be the same as the voltage ratio actually recorded from the axolotl rods, V/V_{max}, in another laboratory 2 years later, and the curve through the white circles in Fig. 22-1 is identically our equation 1.

We may therefore take the white circles of Fig. 22-1 to represent the signals generated in

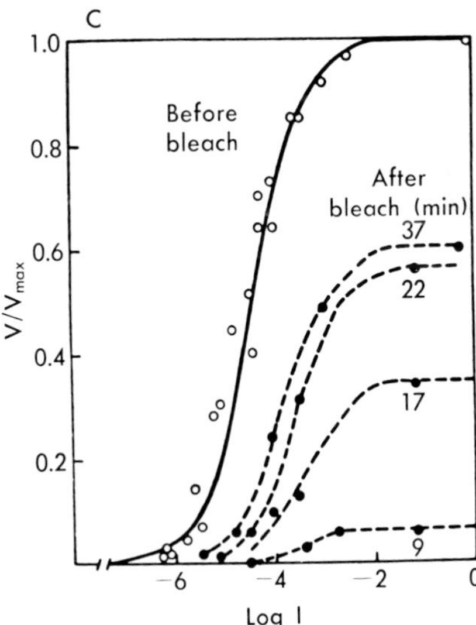

Fig. 22-1. Electric response from single rod in excised eye of an axolotl resulting from flash of intensity, I. Intracellular recording. (From Grabowski, S. R., Pinto, L. H., and Pak, W. L.: Science **176:**1240, 1972.)

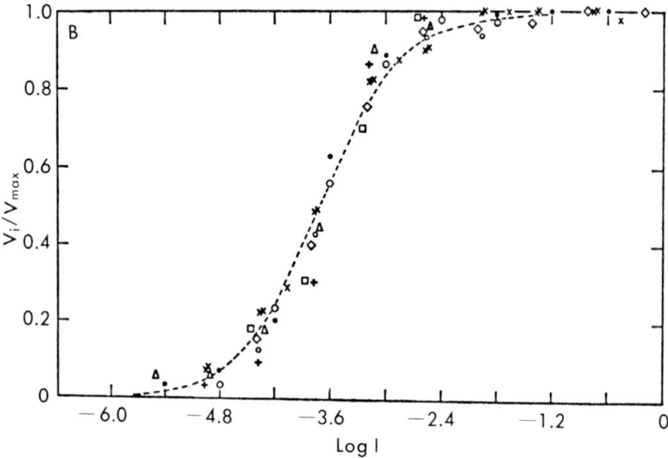

Fig. 22-2. Electric response from single cone in excised eye of a turtle, resulting from flash of intensity, I. Intracellular recording. (From Baylor, D. A., and Fuortes, M. G. F.: J. Physiol. [Lond] **207**:77, 1970.)

human rods by flashes of various intensities. No light ($\log I = -\infty$) gives no signal. As $\log I$ increases (moving to the right), V increases at first slowly, later steeply, and finally reaches a summit, V_{max}, where rods are "saturated" and cannot generate a larger signal however large the light. At this level a subject possessing only rod vision (the so-called rod monochromat or typical achromatopsia) will not be able to discriminate at all, since every rod is excited to its full saturated amount. These subjects therefore say that at this level of illumination (about 1000 troland*), they are bathed in a uniform bright luminous mist through which they can see no features of the scene. They always carry with them dark glasses, so that $\log I$ may be reduced and their V signals slide down the curve of white circles (Fig. 22-1) to the region where V rises sharply with $\log I$. At this dimmer level of illumination the bright mist vanishes and they can see with good rod vision.

Though it is convenient to plot on a scale of $\log I$, this distorts the I-V relation. As may be seen from equation 1, when I is small compared to σ, N is proportional to I, and in fact, in the human the rod signals N increase linearly with the quantum catch from threshold to a flash about 1000 times as stong. The relation

*If a white screen is illuminated by a standard candle 1 m distant in front, an observer viewing this screen through a pupil of radius 1 mm will receive retinal illumination of 1 troland (td) (p. 369).

$$V/V_{max} = \frac{I}{I + \sigma} \tag{1a}$$

which fits our psychophysical results (equation 1) and also the curve of Fig. 22-1, applies equally to the results from cones. Baylor and Fuortes[7] (Fig. 22-2) have found it with intracellular recording from turtle cones, and our psychophysical analysis gives the same thing. But with cones, thresholds and σ are about 100 times as strong as with rods.

Since with any small area, A of test flash $A \times I$ is constant for threshold (Ricco's law), the signals from neighboring receptors are pooled, and N is preportional to the total quantum catch in the area independent of its distribution in the cluster.

BACKGROUND ADAPTATION

It is common experience that if a screen upon which lantern slides are projected is also illuminated by a shaft of sunlight from a badly curtained window, the impression of the slides becomes very faint and is diminished to invisibility if the light shaft is strong enough. This is an example of the well-known Weber-Fechner relation, which may be stated as follows: If a steady light I_1 shines upon the screen and an added light patch must have intensity ΔI_1 just to be seen, it is found that when the steady light is changed to I_2, the added patch ΔI_2 will just be seen when

$$\Delta I_1/I_1 = \Delta I_2/I_2 = k \tag{2}$$

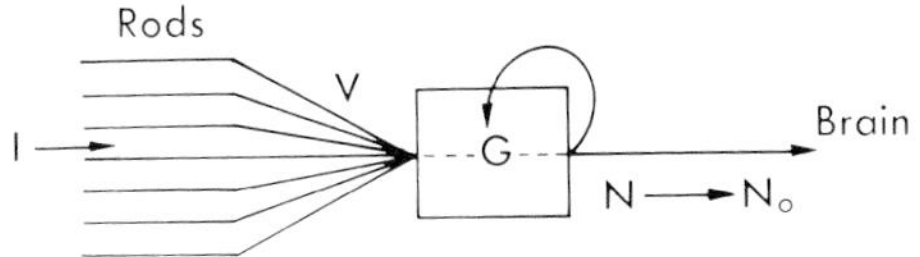

Fig. 22-3. Schematic automatic gain control for rod signals.

What, we may ask, is going on with N, the nerve signals, to account for this familiar Weber-Fechner's effect? What is going on is exactly what would be expected of an automatic camera that aimed to keep the contrast optimal. As we have seen, it adjusts the mean light input so that the intensity on the film is in the midsensitivity range. But a television camera may achieve the same more simply by an automatic gain control on the output signal (Fig. 22-3).

Fig. 22-3 shows a schematic model of the way the eye probably adjusts the size of its signal N to the brain. I is the light intensity falling on the rods whose output is V, which (for lights that are not saturatingly strong) is proportional to I. V enters G, the automatic gain box, whose output feeds back to the box and soon attenuates the output signal N to a fixed value N_0 near the middle of the brain's operating range. If V remains steady at any value, N becomes N_0; if V changes, G soon changes so as to compensate, and N becomes N_0 again. Thus as V alters from moment to moment throughout the day, N alters correspondingly but always drifting back toward N_0, the position of optimal signal contrast. It is easy to see how this mechanism will result in the Weber-Fechner relation (equation 2).

A steady screen luminance, I, will produce a steady rod output, V, proportional to it ($V = aI$), and the resulting G box output is GaI, where G is the signal gain on passing through the G box. As a result of the feed-back, G soon changes to G_1 where $G_1aI = N_0$.

Now if to this equilibrium set up by the background I we add a flash ΔI, the increase of G box input will be $a\Delta I$ and of output $G_1a\Delta I$. However, we have just seen that $G_1a = N_0/I$, hence the increase of output is $N_0\Delta I/I$.

If the condition for detection of a flash ΔI is that the resulting increase of output reaches some criterion level, K, then $N_0\Delta I/I = K$. Since N_0 is also a constant, we conclude that:

$$\Delta I/I = \text{Constant} \tag{3}$$

This is the Weber-Fechner law (equation 2). It results simply from the automatic G mechanism that keeps the visual signals properly scaled for optimal contrast without overloading the brain.

Another consequence, and a very surprising one, follows if we can *stabilize* the image on the retina, that is, fix it there so that it cannot move relative to the pattern of rods and cones. A sharp afterimage is one example. Then N, the signal to the brain, will soon be N_0 in every nerve fiber and the subject will experience the sensation of being in a dense luminous mist with no details visible of any kind—but a mist that is instantly dispelled by a tiny relative movement, for that will bring back all features with twice their contrast. All this is found to be the case!

A well-known instance is the Purkinje figures, which are the shadows of the retinal vessels. Since these blood vessels lie between the receptors and the incoming light, their shadows must fall on the rods and cones all the time. But when we look at the clear sky or a uniform illuminated screen, we cannot see any trace of them. For these shadows are *stabilized* on the retina and move exactly with it so that the G mechanism soon brings the signal from each nerve fiber to the fixed size N_0 and the field appears uniform.

If, on the other hand, the retina is illuminated by light concentrated near the limbus and moved about, then the shadows cast by this light also move about over the rods and cones beneath the layer of blood vessels. So the shadows now fall on receptors that had just before been adapted to light, and the vessels are strikingly displayed (Purkinje figures) by the N signals that never have time to reach their N_0 equilibrium.

Equation 1 gave the size of N signals for various energies of light flash in the absence of any backgrounds. If we suppose that the flash falls upon a steady background of intensity Θ that operates according to the mechanism of Fig. 22-3, the formula becomes

$$N/N_{\max} = \frac{I}{I + \sigma} \times \frac{\Theta_D}{\Theta_D + \Theta} \tag{4}$$

where Θ_D is the intrinsic light of the retina or receptor noise. It is the feature mentioned earlier that requires not one but some six quanta to be caught in order to be detectable against it. The relation is shown graphically in Fig. 22-4 from Alpern, Rushton, and Torii.[2]

The curve marked H_1 plots for human rods N, the nerve signal, against log I, the light flash. This is the same relation plotted in Fig. 22-1 from axolotl rods, and the curves are identical. Curve A is the same as H_1, but log N is ordinate (scale to left) instead of N(scale to right). Curve A running straight at 45 degrees shows that N is proportional to I throughout a 1000-fold range, as mentioned earlier.

The other curves show the N values when I falls on a steady background of 0.3, 0.9, and 2.1 log td, respectively. It is seen in the upper set of curves that change of backgrounds simply displaces the curve vertically by an amount equal to log $(\Theta + \Theta_D)$. This accords with equation 4, which may be written

$$\log N = \log N_{\max} + \log\left(\frac{\Theta_D \times I}{I + \sigma}\right) - \log(\Theta + \Theta_D) \quad \textbf{(4a)}$$

Consequently a change in log $(\Theta + \Theta_D)$ only changes log N by a fixed amount for every point on the curve, that is, displaces it vertically. On the linear N scales below, the same is displayed by a change in the vertical *scaling* factor.

The good fit between points (not shown here for A) and the theroretical curves means that equation 4 holds accurately over a range of many log units of intensity of N, I, and Θ.

Cone adaptation to backgrounds

Nearly all that has been said concerning rod adaptation to backgrounds applies also to cones, but whereas in the human there is only one class of rods, namely, rhodopsin-filled rods, there are three classes of cones filled with erythrolabe (red catching), chlorolabe (green catching), and cyanolabe (blue catching) cone pigments. Each class behaves fairly independently of the others. For instance, the red-sensitive cones satisfy the general equation 4, but the constants σ and Θ_D are about 100 times as great as with rods. To apply the formula, all lights must be expressed in units of quantum catch by the cone in question. Erythrolabe absorbs strongly in the red, but chlorolabe absorbs weakly, and quanta that fall upon a cone but are not absorbed contribute nothing toward its ex-

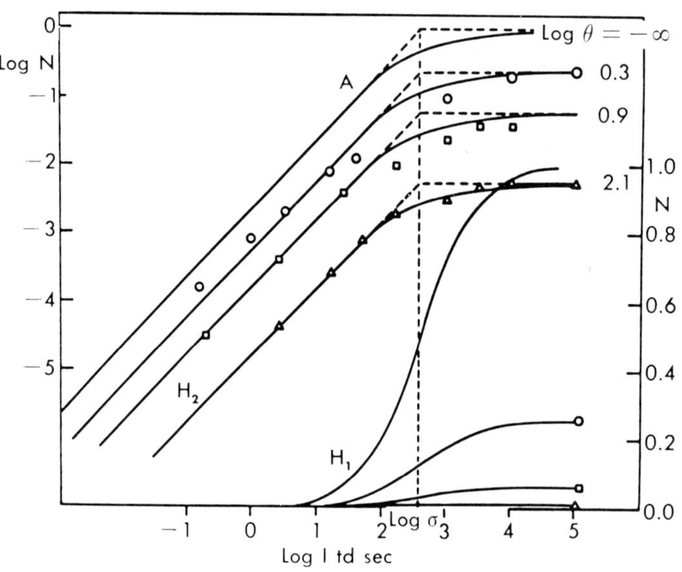

Fig. 22-4. Lower set of curves, H_1, plot size of rod signals (scale right) resulting from flashes of intensity, I, projected on steady background θ indicated in log trolands by numbers on right, namely log $\theta = -\infty$, 0.3, 0.9, and 2.1 for the four curves in descending order. Upper set of curves, H_2, are same results but with rod signals, N, plotted on log scale (left). (From Alpern, M., Rushton, W. A. H., and Torii, S.: J. Physiol. [Lond] **206:**209, 1970.)

citation. Thus the energy of each light must be multiplied by the absorption coefficient of the cone pigment studied, at that wavelength, to get the figure to be used in equation 4.

The two-color Weber-Fechner relations (which have been very extensively studied by Stiles[16,17]) show in a striking way the *independence* of the various cone types. The threshold of each is raised by the background light absorbed by neighboring cones *of the same class* and is absolutely unaffected by the absorption in rods or any other class of cones. This cone independence is in striking contrast to the way that cone signals add in their contribution to brightness and divide one by the other in their contribution to color. This probably signifies that each class of cone has its own private *G* box for Fechner's scaling, but farther down the line there are then quite different processings for brightness and color.

Where is the G box situated?

The main signal attenuation that underlies the Weber-Fechner law is likely to take place in the eye rather than in the brain. The eye of a frog remains excitable many hours after excision, and microelectrode records can be taken from single ganglion cells in such an isolated eye. When a steady light I falls on such an eye, and a flash ΔI is added just sufficient to give an added response, it is found the $\Delta I/I$ = constant (Weber's law). The same thing is found from ganglion cell records in the eye of a decerebrate cat, and in the human it may be seen in the electroretinogram.

Where in the retina the scaling occurs is not so easily answered, but it is likely to be performed by the horizontal cells. There are large cells ramifying across the retina through the nerve tangle of rod and cone pedicles and their connections with bipolar dendrites.

It is known that horizontal cells are hyperpolarized (or depolarized) by rod and cone activity and that they in turn modify the activity of the receptors when artifically hyperpolarized (or depolarized). This situation appears to contain the feedback element postulated in Fig. 22-3. The large horizontal cells, like the photocell of the automatic camera, send a signal proportional to the *average* local illumination, and this signal reduces all the receptor-to-bipolar outputs to the proper intensity level, and this without any loss of contrast.

LATERAL INHIBITION

For simplicity I have discussed so far the signals from receptors to brain as though the message from each part of the retina was independent of activity in neighboring channels. The phenomenon known as contrast shows that the sensation from any part of the retina may depend greatly on the illumination of the surround.

In Fig. 22-5, for instance, the gray squares are identical in tone (reflectivity), but that surrounded by black looks lighter than that surrounded by white. There are many well-known "optical illusions" that depend on this *lateral inhibition* between each channel and its neighbors.

Fig. 22-5. A gray square looks lighter on a black background than it does on a white background.

Fig. 22-6 shows diagrammatically a more surprising example. We have a strip that is black at one end, white at the other, and in between changes gradually from black to white as shown by the dotted line.[14] We might expect that the appearance of brightness would follow this dotted line and give a true impression of the "amount of black" at each point. But the actual appearance follows the heavy line, which at the dark end goes darker than the black end of the strip and at the bright end goes brighter than the white end of the strip. These two bands, the dark and the bright, are called Mach's bands from the famous scientist who first noted them.

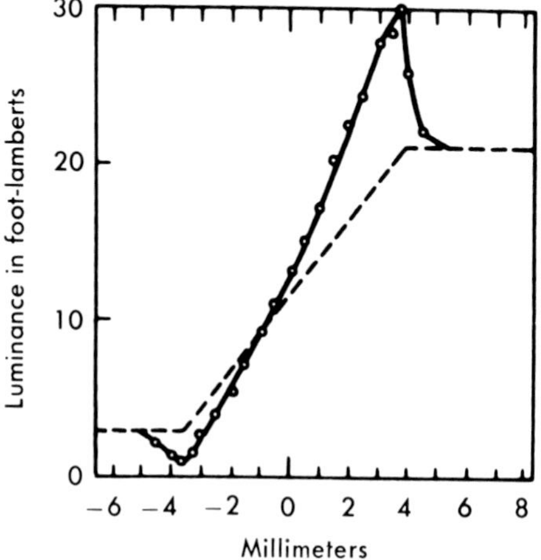

Fig. 22-6. Mach bands at junctions of brightness gradient with uniform fields. Broken line, luminance distribution; solid line, subjective distribution. (From Lowry, E. M., and DePalma, J. J.: J. Opt. Soc. Am. **51**:740, 1961.)

A still more striking illusion discovered by Kenneth Craik is seen in Fig. 22-8. The inner half of the circular disc appears much lighter than the outer half, but this is an illusion; the reflectivity of each half is the same. There is a simple way by which any doubter can convince himself of this (and I hope you are a doubter). Take a card and cut out a circular ring that will just cover the transition circle between the inner and outer halves of the disc. With the ring in place the transition circle is covered, and only the outer and inner halves are exposed to view. They are now seen to be identical in lightness, as indeed they are physically. The disc in Fig. 22-7 shows how disc in Fig. 22-8 was formed. Disc in Fig. 22-7 was spun, and Fig 22-8 is a photograph of this spinning disc. The transition circle is where there is too much white on one side and too much black on the other, and this produces a lateral spread of information (or rather misinformation) throughout the whole disc. Cover this transition circle and the misinformation is cut out, and the two halves seem to be truly equal.

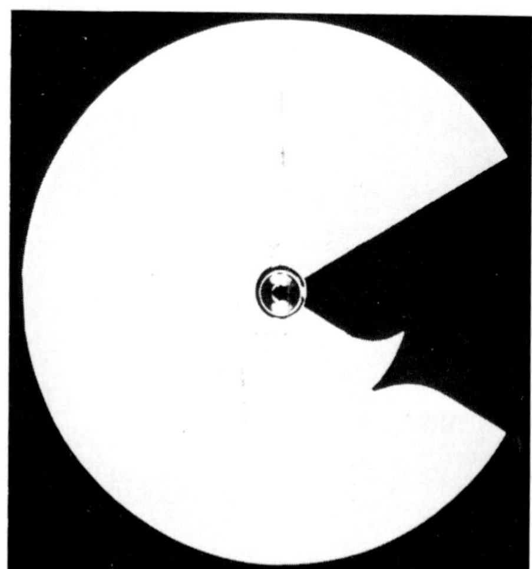

Fig. 22-7. White circular disc with 60-degree sector cut out of form shown. (Courtesy Kenneth Craik.)

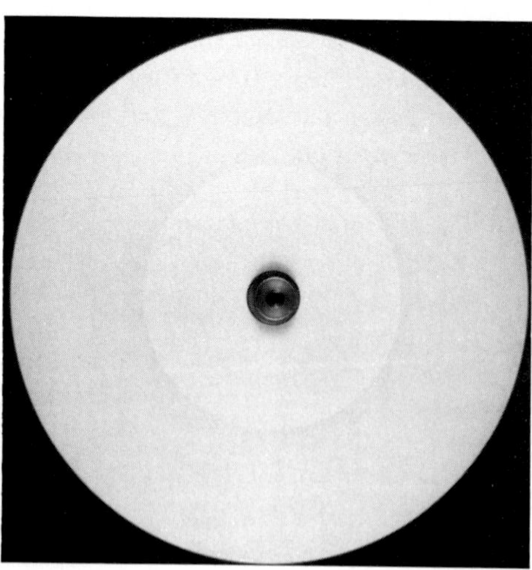

Fig. 22-8. Disc shown in Fig. 22-7 spun fast and photographed. Except for discontinuity near midradius, whole disc is 5/6 white and 1/6 black, but outer half looks darker. (Courtesy Kenneth Craik.)

These contrast phenomena clearly result from a lateral interaction between nerve channels of an inhibitory nature. A strong signal in one nerve fiber will cause depression of the signals in neighboring fibers, but weak signals will cause little lateral inhibition. Thus in Fig. 22-5 the bright surround inhibits strongly the brightness of the central gray and makes it blacker, but the black surround inhibits the central gray less than if the whole large square was uniform gray, and in that sense may be said to make the center whiter.

There is much electrophysiologic evidence for lateral inhibition in the retina. If records are taken from single ganglion cells from the retina, for example, of a frog,[11] it is found that a small spot of light will elicit a discharge if it falls upon a limited area of retina, the receptive field of that ganglion cell. Barlow[5,6] and Kuffler[13] discovered that outside the area where light provoked a ganglionic discharge there lay an inhibitory area where nothing happened when light fell if the ganglion cell was quiescent, but if the cell was already responding to light from the central area, this peripheral light stopped the discharge.

It appears that all retinal ganglion cells have a receptive field composed of a central excitatory area with an inhibitory surround, and in general if exposed to a uniform field that covers the whole area, inhibition just balances excitation so that no cell activity results.

It is easy to imagine how a bright edge advancing across a dark field invades first the inhibitory surround and makes the darkness blacker, then falls on the excitatory center and signals brightness, and finally covers the whole receptive field uniformly and leaves signaling to the new position of the moving edge. Mach's bands and other contrast phenomena seem to be the result of this kind of lateral inhibition.

What is the function of lateral inhibition in the eye? The optical image is far from perfect. Owing to light scatter and aberrations of various kinds the image is smudged. Lateral inhibition sharpens contours and goes some way toward undoing the smudging.

Perhaps we may go further and claim that an enhanced contour picture is "better" than a perfect representation. Receptors are newspaper reporters collecting what they see and reporting back to the brain in code. "No change" whether in time or over space is no news, and they do not report it. It is the change movement of contour that is worth noting, and lateral inhibition enhances it.

ROD ADAPTATION TO BLEACHING

When a subject who has adapted to bright sunlight enters a rather dark room, the utter extinction of rod sensitivity and the slowness of its return is so striking that for long it has exercised a powerful influence on visual theory. The so-called "photochemical theory" of Hecht and Wald attributed much of visual performance to the bleach-regeneration kinetics of rhodopsin. They never measured the kinetics but invented arbitrary constants to suit the theory and, unfortunately, guessed the values about a million times wrong. We now know, largely through work started by Hecht himself, that visual performance is controlled by single quanta and single molecules not by the mass equilibrium of thousands.

But Hecht was correct in his surmise that the desensitization that occurs after exposure to strong light was the result of the bleaching of rhodopsin, and recovery was due to rhodopsin regeneration. Hecht never knew what in fact the relation was between the amount bleached and the log threshold rise, but it certainly was more complicated than Wald's simple claim that when half the rhodopsin was removed by bleaching it would need twice the amount of incident light for the same quantum catch. On that view 50% bleaching should double the threshold, but in fact the threshold is raised some 8 log units. So Wald's claim was in error by a factor of 100 million. The actual relation between bleaching and the rise of log threshold is given by

$$\log(\Delta I/\Delta I_0) = 10HB \quad (5)$$

where

ΔI = Threshold
ΔI_0 = Fully dark threshold
B = Fraction of rhodopsin still in bleached state
H = Constant of about 2

H is the rise in log threshold due to 10% bleaching. This was first determined in the human by reflection densitometry[8] and confirmed by ERG thresholds on rats.[9] The recovery of sensitivity in the dark is easily measured on patients and can give us valuable information about the performance of rods and cones, so I shall now describe how this is done.

DARK-ADAPTATION CURVES

The subject's head is fixed by chin and brow rests, one eye is covered, and the other fixates on a small red lamp to align the eye. While gazing there he sees a flashing light either at the fixation point or at some other place in the visual field. The intensity of the flashing light can be altered by shifting a gray photometric wedge placed in the beam, attenuating the light more, the thicker the portion interposed. The wedge is shifted until the flash is reduced just to threshold for the subject, and the time and wedge setting noted.

Fig. 22-9 shows a set of dark-adaptation curves obtained in this way.[12] Consider the white squares where the flashes fell 5 degrees from the fovea (fixation point). First the retina was exposed to a bright light that bleached away some of the visual pigments. Almost immediately (time, 0 minute) the subject started to make threshold measurements in the apparatus. At first the threshold is very high, but it rapidly falls and begins to level out. At 9 minutes (white squares) a new feature appears, and the threshold falls more quickly again. It takes over half an hour to reach the full dark-adapted value, and that is 3 log units (1000 times) more sensitive than the first point.

There are very strong grounds for believing that the upper branch of the dark-adaptation curve measures the thresholds of cones, and the lower part, rods. As seen in Fig. 22-9, when the flash falls on the fovea (0 degree curve), which is devoid of rods, the lower curve is missing, whereas it becomes more prominent at 10 degrees, where the retina contains mostly rods. On the lower branch the flashes look colorless whatever the test wavelength, confirming the achromatic nature of rod vision, but on the upper branch the color of the test flash is usually recognized. Finally, the spectral sensitivity of the two branches can be measured by using test flashes of various wavelengths in succession and finding for each the threshold energy. The lower branches have the spectral sensitivity of

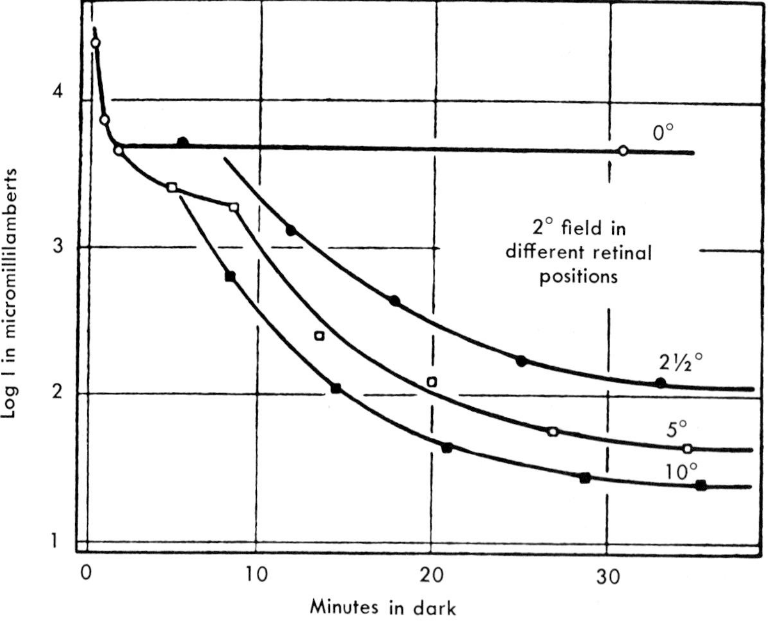

Fig. 22-9. Dark adaptation as measured with a 2-degree field placed at different distances from center. (From Hecht, S., Haig, C., and Wald, G.: J. Gen. Physiol. **19:**321, 1935.)

rods, which corresponds to the absorption spectrum of rhodopsin; the upper branch corresponds to the spectral sensitivity of cones.

The simple versatility of the dark-adaptation apparatus makes it a valuable clinical instrument. By moving the position of the fixation point, we can investigate the sensitivity of rods and cones in any part of the visual field (as Fig. 22-9 shows), and the test flash can also be altered in area, duration, and color.

Hecht suggested that the slow recovery of rod sensitivity in dark adaptation was due to the regeneration of rhodopsin that had been bleached; this was shown to be correct in the human when the course of bleaching and regeneration of rhodopsin could be measured.

Campbell and Rushton[8] were able to do this by using an application of the ophthalmoscope in which the light from the subject's illuminated retina, instead of being returned to the eye of the investigator, was deflected onto a photomultiplier cell, which therefore measured the amount of light reflected from behind the retina. Since this light has been twice through the retina, it must have suffered absorption by the visual pigments there, and the more these regenerate, the more they absorb the light and the smaller the photocell signal.

Fig. 22-10 from Rushton and Powell[15] shows in the human the relation between dark-adaptation curves and the regeneration of rhodopsin. A bleaching exposure was given that bleached 99%, 42%, 24%, or 13% as labeled, and the subject immediately (time zero) started to dark adapt, making measurements of two kinds: (1) log threshold measurements that gave ordinary dark-adaptation curves shown by circles and triangles and (2) retinal reflection measurements in the densitometer that gave the percent pigment still in the bleached state. These are plotted as the continuous lines with scale in the middle of the illustration.

It is seen that after each bleach, rhodopsin regenerates along the same exponential curve and that at every stage, points coincide with curves. This means that log threshold is always raised a fixed multiple of the fraction of rhodopsin in the bleached state; which is what is expressed in equation 5.[3]

Since normally we cannot see the rod branch of the dark-adaptation curve earlier than the moment when it appears beneath the cone branch, we cannot correlate the amount of regeneration with log threshold rise earlier than the 7% still bleached in Fig. 22-10. However, by the special psychophysical technique that enabled us to measure the *N* signals we were able to measure rod thresholds much earlier, and Fig. 22-11 shows (circles) and rod dark-adaptation curve in a normal man extending over a millionfold range of thresholds with no cones involved. The rhodopsin regeneration (tri-

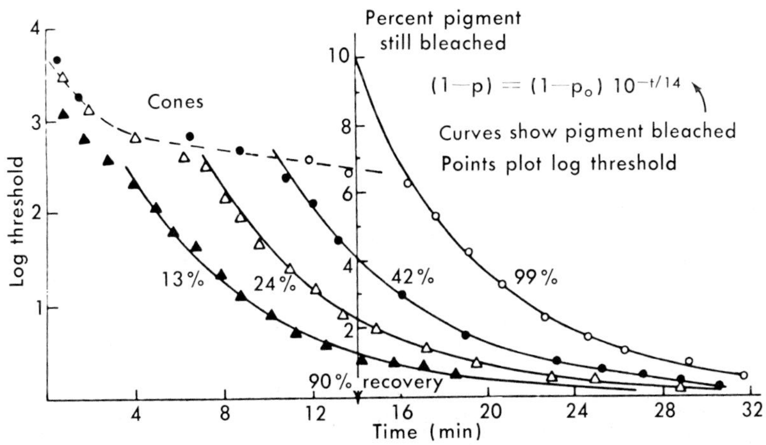

Fig. 22-10. Circles and triangles show dark adaptation curves. Log threshold plotted against time in the dark following a light that bleached away 99%, 42%, 24%, or 13% of rhodopsin, as labeled. Continuous lines show percent of rhodopsin still in bleached state (central scale) as measured by retinal densitometry. (From Rushton, W. A. H., and Powell, D. S.: Vision Res. **12:**1073, 1972.)

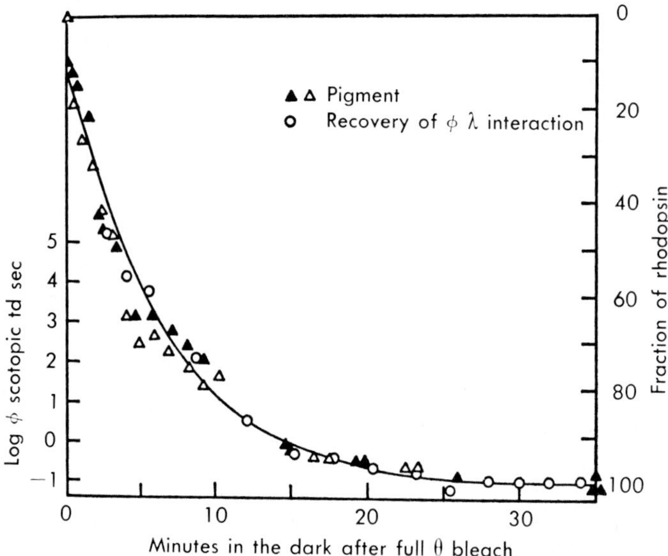

Fig. 22-11. Rod dark adaptation and pigment regeneration in human. Circles show rod dark adaptation curve (log threshold curve on left). Triangles show fraction of rhodopsin present (two runs) at that point in recovery. Threshold was determined by special technique that excluded intrusion of cones, so rod thresholds are measured uncontaminated over millionfold range. (From Alpern, M., Rushton, W. A. H., and Torii, S.: J. Physiol. [Lond] **207:**449, 1970.)

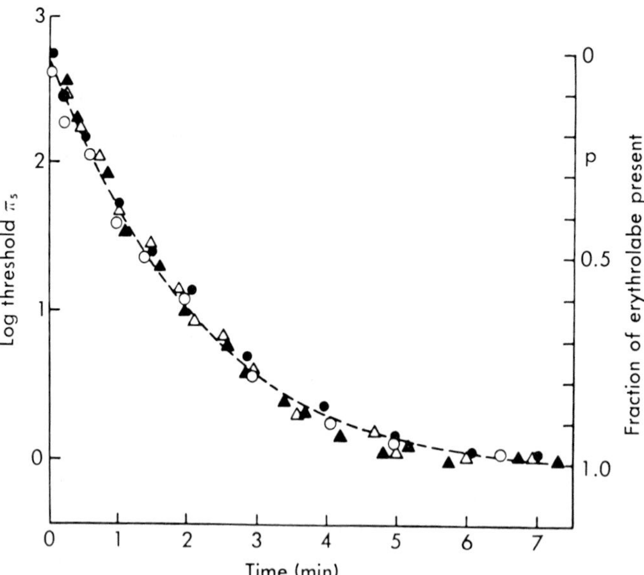

Fig. 22-12. Cone dark adaptation and pigment regeneration. Black and white triangles show log threshold (two runs) using red test flash following strong adaptation, which at 0 minutes had bleached away 95% of erythrolabe, the red pigment. Black and white circles show fraction of erythrolabe present (two runs) at that stage of recovery in the dark measured by retinal densitometry. (From Baker, H. D., and Rushton, W. A. H.: J. Physiol. [Lond] **176:**56, 1965.)

angles) measured by reflection densitometry in the same experiment shows over the whole range that log threshold rise is proportional to fraction of rhodopsin bleached as expected from equation 5.[3]

This result also applies to cone thresholds and cone pigment regeneration, as may be seen from Fig. 22-12 from Baker and Rushton.[4] After a strong light had bleached all the cone pigments away, the dark-adaptation curve was obtained using a red test flash. It therefore plots the recovery of red-sensitive cones. Black and white triangles give the results of two runs. Circles show two runs where, by retinal densitometry, the fraction of erythrolabe, the red-sensitive cone pigment, was measured at each moment during recovery (scale on right). It is seen that the rise of log threshold is proportional to the amount of pigment in the bleached state at each moment, as expected of equation 5.

The constant H in equation 5 seems to vary a good deal from animal to animal, and we do not know on what it depends. It is much smaller with human cones than with human rods. For instance, 50% bleaching raises the human rod threshold about 8 log units (6 in Fig. 22-11) but cones by only some 1.5 log units (Fig. 22-12), and this explains why we are hardly ever so light adapted but that we can see well with our cones.

If the light is too strong, there is great advantage in operating on the nerve signals (Fig. 22-3) rather than reducing the light itself. Light *excluded* from entering the eye cannot give information, but *processing* of nerve signals may use most of the information of the full light. Consider the analogy of a sensitive galvanometer. If the current to be measured is too strong, it may be shunted; that is, only 10% of the current passes through the galvanometer, and 90% passes through the shunt and is there degraded into heat. But in principle a better arrangement is to make the galvanometer suspension ten times as stiff. The steady galvanometer readings will be the same as with the shunt, but the stiff suspension will move the pointer much faster so that the time resolution of the galvanometer will be far superior, and quick changes of current will be faithfully recorded. In this case, sensitivity has been exchanged for time resolution.

In analogous fashion in our eyes, sensitivity is *exchanged* for improved resolution in time (flicker) and space (acuity), as well as reliability (signal to noise ratio). If we look at a brightly illuminated transparent photograph of some scene, at first we are dazzled. But in a few seconds the brightness fades and the detail appears. We can watch the "stiffening of our suspension."

Thus an important part of adaptation is the way that more light is organized to give improved discrimination in many directions, but I shall say no more of that here.

It is of interest to compare adaptation to backgrounds and adaptation to bleaching. Background adaptation is precisely what would be expected of an automatic television camera designed to adjust the signals so they give optimal contrast without overloading. Bleaching adaptation, per contra, seems a pointless nuisance, and no one yet has made sense of it. To put the rods out of action as soon as 7% of their rhodopsin is bleached looks like the design of a madman, but our eyes are superbly designed. If I cannot tell you the point and purpose of bleaching adaptation, it is because I have failed to understand something fundamental in the mechanism of vision.

REFERENCES

1. Alpern, M., Rushton, W. A. H., and Torii, S.: The size of rod signals, J. Physiol. (Lond) **206:**193, 1970.
2. Alpern, M., Rushton, W. A. H., and Torii, S.: The attentuation of rod signals by backgrounds, J. Physiol. (Lond) **206:**209, 1970.
3. Alpern, M., Rushton, W. A. H., and Torii, S.: The attentuation of rod signals by bleaching, J. Physiol. (Lond) **207:**449, 1970.
4. Baker, H. D., and Rushton, W. A. H.: The red-sensitive pigment in normal cones, J. Physiol. (Lond) **176:** 56, 1965.
5. Barlow, H. H.: Action potentials from the frog's retina, J. Physiol. (Lond) **119:**58, 1953.
6. Barlow, H. B.: Summation and inhibition in the frog's retina, J. Physiol. (Lond) **119:**68, 1953.
7. Baylor, D. A., and Fuortes, M. G. F.: Electrical responses of single cones in the retina of the turtle, J. Physiol. (Lond) **207:**77, 1970.
8. Campbell, F. W., and Rushton, W. A. H.: Measure-

ment of the scotopic pigment in the living human eye, J. Physiol. (Lond) **130:**131, 1955.
9. Dowling, J. E.: The chemistry of visual adaptation in the rat, Nature **188:**114,1960.
10. Grabowski, S. R., Pinto, L. H., and Pak, W. L.: Adaptation in retinal rods of axolotl: intracellular recordings, Science **176:**1240, 1972.
11. Hartline, H. K.: The receptive fields of optic nerve fibers, Am. J. Physiol. **130:**690, 1940.
12. Hecht, S., Haig, C., and Wald, G.: Dark adaptation of retinal fields of different size and location, J. Gen. Physiol. **19:**321, 1935.
13. Kuffler, S. W.: Discharge patterns and functional organization of mammalian retina, J. Neurophysiol. **16:**37,1953.
14. Lowry, E. M., and DePalma, J. J.: Sine-wave response of the visual system. I, The Mach phenomenon, J. Opt. Soc. Am. **51:**740, 1961.
15. Rushton, W. A. H., and Powell, D. S.: The rhodopsin content and the visual threshold of human rods, Vision Res. **12:**1073, 1972.
16. Stiles, W. S.: Increment thresholds and the mechanisms of colour vision, Doc. Opthalmol. **3:**138, 1949.
17. Stiles, W. S.: Colour vision: the approach through increment threshold sensitivity, Proc. Natl. Acad. Sci. U.S.A. **45:**100, 1959.

GENERAL REFERENCES

Cornsweet, T. N.: Visual perception, New York, 1970, Academic Press, Inc.

Ratliff, F.: Mach bands: quantitative studies on neural networks in the retina, San Francisco, 1965, Holden-Day, Inc.

Chapter 23

Time-dependent functions in vision

GERARD M. SHICKMAN, Ph.D.

The visual system tries to make sense of an incredibly complex physical world with which it has almost no direct contact. To do this arduous job the system samples, stores, integrates, differentiates, erases, and performs other operations on whatever information about the world reaches it as a pair of retinal images. Since these images, imperfect though they are, usually contain much more information than the system can assimilate or the observer can use, vision must be selective; and because the information differs from moment to moment, operations on it must take place in what the users of computers call "real time." Of course, this term does not imply instant response. The system requires a certain amount of time to collect and to process information received, but this time must be limited to keep pace with changes in the observer's environment and to give him a coherent picture of its currently important features. In turn, the observer selects and emphasizes the components of this picture in ways that depend on his developmental and adaptational history.

This chapter examines several ways in which time and time-related variables influence vision, and some temporal characteristics of the normal visual system.

EDITORIAL FUNCTION OF VISUAL SYSTEM

As a rule, visual experience conforms fairly well to the spatiotemporal order of the physical world, but it is never a perfect representation because the visual system itself organizes the information it receives. By various methods the system enhances relevant, discards irrelevant, and condenses redundant information. The substance of information is "something different"; thus boundaries and other changing elements of the light distribution, both spatial and temporal, convey most of the information in the retinal image. Accordingly, the visual system responds with special emphasis to borders and spatial gradients as well as to temporal fluctuations of retinal illuminance and spectral composition. On the other hand, it is also alert to opportunities for grouping and simplification based on uniformity. In short, the system is a differentiator, but it is also an integrator, often a comparator, sometimes an extrapolator. Its continued function depends on a changing pattern of stimulation but consists largely of a search for invariances and orderly relations within and between two retinal images.

EXAMPLES OF SHORT-TERM AND LONG-TERM VISUAL TIME DEPENDENCE

Time, a parameter of all visual processes, influences sight in diverse ways. Different temporal orders of magnitude are important for different phases of vision. In general the processes that initiate vision are quite rapid, as can be inferred from the kinetics of pigment bleaching and regeneration[6] and from the electrical potentials recorded from receptors exposed to flashes of light.[42,54,172,191] Absorption of a single quantum of light by a molecule of photoreceptor pigment, the first step in visual excitation, is almost instantaneous and leads within microseconds to changes in the shape of the molecule, followed by complete dissociation of the molecule into all-*trans* retinal and opsin. Since the dissociative process can occur even in darkness and is relatively slow, it is evident that excitation of the visual receptor takes place before the molecular changes have run their course.[265]

Although, ideally, absorption of a single quantum can trigger a neural event,[108] several quanta must ordinarily be absorbed by even the most completely dark-adapted eye to elicit a sensation of light.[115] These absorptions must occur within a restricted retinal area during a limited period to constitute an adequate stimulus. During this interval of hundredths or possibly tenths of a second the retina summates the effects of individual quantal absorptions at both photochemical and neural levels; this so-called energy summation, or temporal summation, is an example of the integration performed by the visual system.

In brighter light the duration of a fixatory pause during which the eye takes in information is task related, being about 200 to 300 milliseconds in reading.[252] Saccadic movements from one fixated point to another require approximately 10 to 80 milliseconds, depending on the angle between initial and final lines of sight, and can be repeated as often as three to five times/sec as the viewer scans an object or a painting.[282] Of particular interest in this context is that this piecemeal input, obtained in brief successive glimpses, gives rise to perceived continuity and stability, again a result of the visual capacity for storage and integration.

Other examples of this property include the growth of brightness as a function of the length of time a light is presented[38,137] and the increase, with the duration of exposure, of the number of dots that can be correctly reported after a single glance (the so-called span of apprehension).[11,128]

The ability to store information rapidly is important for efficient reading. It is limited by the capacity of the short-term memory; old information must be displaced to allow the entry of new information, a process Averbach and Coriell[12] call erasure. In general, the upper limiting time for visual storage and other forms of temporal integration varies from about 0.01 to 1.0 second, depending on the nature of the observer's task, state of his ocular adaptation, and kind of retinal receptors most active under the prevailing conditions of stimulation.

Photochemical light and dark adaptation occupy a different time scale. The concentration of visual pigment continues to change throughout several minutes or, in dark adaptation, several hours. Interestingly enough, these slow changes produce little visual sensation. Instead, they control the neural organization and activity level within the retina and so establish visual sensitivity, determine the delay, magnitude, and time course of electrophysiologic as well as psychophysical responses to light, and mediate the interactions among lights presented at different times or in different parts of the visual field.

An interplay of neural elements is also apparent in several other types of adaptive processes, some faster and some slower than light and dark adaptation. For example, a light presented to some fixed part of the peripheral retina tends to fade from view after several seconds but can be restored by slight eye movements. This phenomenon demonstrates the reliance of the visual system on change. If the two retinal images differ appreciably in structure, spectral composition, or contour brightness, the visual system tends to deal with them one at a time, alternating one with the other. This solution to the incompatibility of retinal images is called binocular, or retinal, rivalry. The alternation frequency is of the order of 5 to 10 times/min, but depends on the amount and kind of disparity,[156,274] a point of special significance for the ophthalmologist whose patient has aniseikonia. With sufficiently reduced disparity of the images, information from the two eyes is fused rather than alternated, and the residual difference is used as the basis of stereoptic depth perception.

Time can affect vision in yet another way if it determines the importance of stimulation and consequently the attention or vigilance the observer devotes to it. Hernández-Peón has shown that both neural and overt behavioral responses to repeated clicks or flashes decline with the number of repetitions if the individual events have no specific importance.[118] This process, requiring a few minutes, is called habituation. If the event acquires meaning (for example, if it warns of punishment or presages reward), the observer again becomes responsive at both behavioral and neural levels. We may interpret habituation as a response to redundancy, since the first flash, representing a departure from what has gone before, may carry more information than an established pattern of repetitive flashes. If this information proves to be significant, sensitivity remains high and response persists; if not, sensitivity and the likelihood of response diminish.

To round out this section let us take one more complex, but important, example of time dependence in the visual system. A new pair of glasses not only improves a patient's visual acuity but also, because of its curvatures, acts as a pair of prisms whose strength varies according to the direction of the patient's line of sight through them. This variable prismatic action accounts for the "swimming" of the field reported by some patients with new glasses and is related to the "jump" often seen through bifocals as the patient shifts from far to near vision. Shifts of fixation from the center to the periphery of the field enclosed by the glasses cause deflection of the line of sight through an angle that changes with eccentricity; likewise, rays of light from different parts of the field are bent through different angles by the lenses so that either eye movement or head movement alters the apparent magnification of and spatial relations among seen objects. After a few days, however, these distortions and the accompanying tendency to reach in the wrong direction for objects disappear as the patient reorganizes his visual and visuomotor response patterns. Once the adjustment is complete, the patient can remove and replace his glasses and encounter no visual difficulty, as the appropriate set of responses automatically comes into play. This capacity of the system to compensate for time-varying distortions of the visual field is one of the most remarkable and useful of its functions.[110,194]

TEMPORAL ORGANIZATION OF RETINOCORTICAL PATHWAYS AND OTHER VISUAL AND VISUOMOTOR CENTERS

Through the analysis of retinal cytoarchitecture begun by Sjöstrand,[231] Dowling,[29,68-71] Werblin,[267] and others, it has become clear that retinal synaptic structure and function are orderly and that excitation and inhibition within the cellular matrix are precisely organized temporally and spatially. Functional organization is similarly precise within the visual cortex,[125,126] as well as within the lateral geniculate[88] and the superior colliculus.[238] Although anatomists have been largely concerned with spatial organization, recent studies have shown that the visual pathways also have well-organized temporal properties.

Rods and cones

A difference of time scales has particular significance for the functional separation of rod and cone mechanisms. It has long been known that rods adapt more slowly than cones to light and to darkness; in addition, the rod response has, on the average, greater latency and duration than that of the cones. Gouras has shown that "with all patterns of retinal stimulation and states of adaptation, threshold signals of the rods reach a ganglion cell later and those of the cones earlier than approximately 50 msec after a light stimulus."[97] Since the same ganglion cell can transmit either rod or cone signals in the dark-adapted retina, the fact that only the cone signal is transmitted in the light-adapted retina appears the result of this temporal separation of the two signals, whereby cone signals preempt the response of the ganglion cell and render it refractory to rod signals. It seems likely that the functional independence of rod and cone systems in duplex retinas is due at least in part to the incompatible timing of responses within these systems.

Late receptor potentials

Analysis of the cone and rod late receptor potentials (RP) of cynomolgus monkey supports this position (Fig. 23-1).[271] Within the scotopic range, rod late RP cannot be made to resemble cone late RP by any variation of stimulus intensity. Under mesopic conditions the two RPs can be reliably distinguished by their decay rates, the rod RP being prolonged. Above low photopic levels the rod component disappears, suggesting that rod response is lost, rather than saturated, as cones become increasingly active.[272]

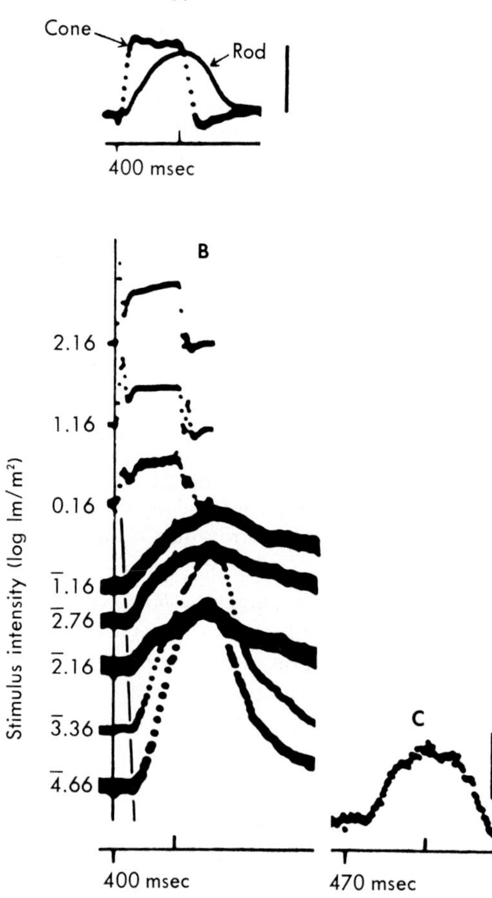

Fig. 23-1. Cone and rod late receptor potentials compared. **A,** Potentials elicited by stimuli equalized for number of quanta per unit retinal area, wavelength 560 nm (cone) and 508 nm (rod). **B,** Relative latencies of cone late receptor potentials (top three records) and rod late receptor potentials (lower five records) as functions of stimulus intensity. Latency of rod late receptor potential decreases approximately linearly with increase of log stimulus intensity. **C,** Pure rod late receptor potential, peripheral retina, low stimulus intensity. (Data of Whitten and Brown.[271])

Retinal ganglion cells: transient and sustained responses

The response of a given retinal ganglion cell depends on the spatial and temporal distribution of light within the concentric excitatory and inhibitory regions of its receptive field, as well as on the intrinsic properties of the neural network that supplies its input. Werblin[267] has shown that in the amphibian *Necturus,* horizontal cells respond to ongoing illumination, but amacrine cells respond only to changes of illumination. When the stimulating light goes on, a given ganglion cell (mammalian retina) may respond with sustained or with transient excitation and is correspondingly classified as a "sustained" or a "transient" cell.[50] Different authors designate these cells or functionally similar ones as tonic and phasic,[98,169] type X and type Y,[77,220] type II and type I,[90] or group II and group I,[120] insofar as several criteria for identification are met other than the time course of the discharge pattern. Cleland and Levick[48,49] distinguish "brisk-sustained" and "brisk-transient" cells, respectively identifiable with X and Y cells, from "sluggish-sustained" and "sluggish-transient" cells whose long latency and slow conduction are consistent with the behavior of the so-called W cells. Rowe and Stone compare X, Y, and W cells, arguing persuasively for this noncommittal designation rather than singling out the temporal response pattern as an identifier. Their information is reproduced here as Table 23-1. Similar comparisons, mainly between transient and sustained cells, are presented by Ikeda and Wright[131] and by Cleland, Levick, and Sanderson.[51] Part of the summary of Ikeda and Wright is included in Table 23-2.

The cat has been predominant as the experimental animal, but neural organization in the visual system of other species is generally comparable, however, not entirely so. For example, Caldwell and Daw[43] show that in rabbits, unlike cats, "one cannot make the generalization that cells with more complex receptive fields have a slow conduction velocity [or] that cells which respond only to slow movements and have low spontaneous activity have a slow conduction velocity."

From poststimulus histograms of the activity of cat retinal ganglion cells, whose receptive fields received intermittent exposure to spots of light with variable intensity and area, Ikeda and Wright[131] concluded that the excitation of sustained cells consists of both transient and sustained components. Transient cells have no sustained components, but at the onset of the stimulating light emit a burst of excitation whose magnitude and shape change with stimulus intensity and with the part of the receptive

Table 23-1. Some properties of cat retinal ganglion cells*

	Y cells	X cells	W cells
Receptive field center size	Large, 0.5 to 2.5 degrees	Small, 10 min of arc to 1 degree	Large, 0.4 to 2.5 degrees
Linearity of center-surround summation	Nonlinear	Linear	Not tested
Periphery effect	Present	Usually absent	Absent
Axonal velocity	Fast, 30 to 40 m/sec	Slow, 15 to 23 m/sec	Very slow, 2 to 18 m/sec
Soma size, peripheral retina	Large, >22 μm	Medium, 14 to 22 μm	Small, <15 μm
Proportion of population	<10%	Approximately 40%	Approximately 50% to 55%
Retinal distribution	Concentrate near area centralis, more numerous relatively in peripheral retina	Concentrate at area centralis	Concentrate at area centralis and in streak
Central projections	To laminae A, A_1, and C of LGN, to MIN, and, via branching axon, to SC from the A laminae of LGN to cortical areas 17 and 18, also by branching axon; from MIN to areas 17, 18, and 19	To laminae A, A_1, and C of LGN; thence to area 17; to midbrain (a minority), but probably not to SC	To SC, to C laminae of LGN; thence to visual cortex area 17 and/or 18, and 19
Nasotemporal division	Nasal cells project contralaterally; most temporal cells ipsilaterally; strip of intermingling centered slightly temporal to area centralis	Nasal cells project contralaterally; temporal cells project ipsilaterally; narrow strip of intermingling centered on area centralis	Nasal cells project contralaterally; most temporal cells also project contralaterally; about 40% of temporal cells project ipsilaterally
Responses to standing contrast	Phasic or transient in most cells; some are tonic or sustained especially near area centralis; all are tonic when dark adapted	Most give tonic responses in mesopic conditions, many are transient when light adapted	Either tonic or phasic
Receptive field "layout"	On-center/off-surround or off-center/on-surround	Same as for Y cells	Some have same layout as Y and X cells, others have on-off centers, some have purely inhibitory centers, some are directionally selective or colour coded
Morphologic correlates	α Cells	β Cells	γ Cells

*From Rowe, M. H., and Stone, J.: Brain. Behav. Evol. **14:**185,1977.

field stimulated. Varying the position of a small stimulating spot within the receptive field shows that the two types of cells have different sensitivity gradients. For the on-center sustained cell an inhibitory zone surrounds a narrow, steep-sided excitatory zone, whereas the transient cell has a broader, less abrupt sensitivity gradient without obvious surround inhibition.

The cells also differ as to spatial response characteristics. For sustained cells, threshold luminance first decreases, then increases, then levels off as the area of the stimulating spot increases, whereas transient cells do not exhibit the secondary rise in threshold.[51] With small stimulating spots, excitation in the receptive field center of the transient as well as the sustained cell builds up according to Ricco's law; a constant product of stimulus area and stimulus luminance generates a standard level of excitation. Ikeda and Wright[131] suggest that the "center" of the receptive field be defined as the

Table 23-2. Other properties of cat retina ganglion cells*

	Transient cells	Sustained cells
Probable inputs	Predominantly amacrine cells	Predominantly bipolar cells
Response to small spot at receptive field center, 0.05 Hz	Transient excitation with recovery to the maintained rate	Sustained excitation
Maintained activity	Low	High
Effect of defocusing	Does not require a sharp focus for response	State of focus critical for response
Response to spot size	Large spot optimal	Small spot optimal
Spatial frequency sensitivity	Nonlinear, high frequency not resolved	Linear, high frequencies resolved
Temporal resolution	Good	Poor
Latency	Long latency to small spot, but decreases with increasing flux; decreases with supraoptimal stimuli	Short latency to small spot; no decrease with supraoptimal stimuli
Functional role	Analysis of temporal aspects of visual stimuli, movement detection, orientation responses to visual stimuli	Analysis of spatial contrast, form recognition

*From Ikeda, H., and Wright, M. J.: J. Physiol. (Lond) **227**:769, 1972.

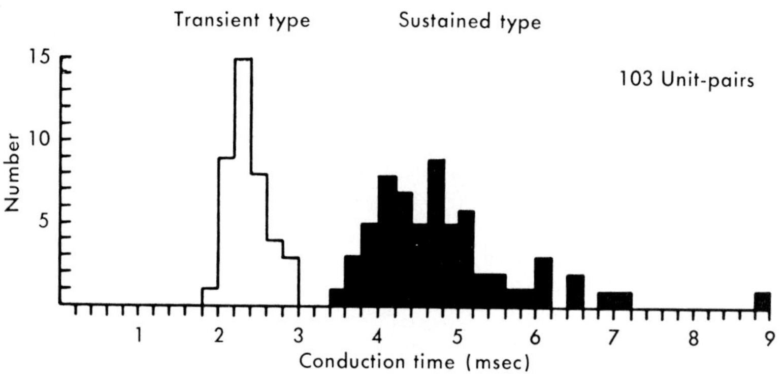

Fig. 23-2. Frequency histogram of retinogeniculate conduction times. Note that all transient retinal ganglion cells (white bars) fell into group with short conduction times; all sustained cells (black bars) made up group with longer conduction times. (Data of Cleland, Dubin, and Levick.[50])

area over which Ricco's law of spatial summation holds, since with large stimulating spots the conventionally defined centers and surrounds of transient and sustained fields interact dissimilarly.

High levels of stimulation saturate the response of both types of cells, with latency decreasing as stimulus intensity increases for transient cells but decreasing until saturation and then tending to increase for sustained cells. Both stimulus and background intensity determine the response latency,[47] which is generally shorter for sustained than for transient cells when the stimulating spot is small. However, the sustained cells respond optimally to small spots, and the transient cells to larger spots, in keeping with the greater inhibitory contribution from the receptive-field periphery of sustained cells.[131]

The behavior and sensitivity profile of the sustained cells suggest that the surround mechanism of their receptive fields does not fully overlap the field center, whereas those properties of the transient cells are consistent with a complete overlap of the two components.[120] Enoch reviews psychophysical techniques he

Table 23-3. Calculation of retinogeniculate and retinocortical conduction times*

	Transient	Sustained
Latency of ganglion cell response to photic stimulation (present results, standard spot)	38 to 56 milliseconds	19 to 24 milliseconds
Intraretinal axon lengths for these cells (present results)	5.6 to 8.0 mm	3.5 to 4.0 mm
Intraretinal conduction velocity[248]	3.5 to 4.9 m/sec	1.5 to 1.8 m/sec
Estimated intraretinal conduction time for these cells (present results)	1.1 to 2.3 milliseconds	1.7 to 2.6 milliseconds
Extraretinal axon lengths for these cells (present results)	34.0 to 35.5 mm	33.0 to 34.5 mm
Extraretinal conduction velocity[248]	29 to 39 m/sec	9 to 14 m/sec
Estimated extraretinal conduction time for these cells (present results)	0.9 to 1.2 milliseconds	2.3 to 3.8 milliseconds
Total estimated retinogeniculate conduction time (present results)	2.0 to 3.5 milliseconds	4.0 to 6.4 milliseconds
Measured retinogeniculate conduction time[50]	1.8 to 3.0 milliseconds	3.4 to 8.6 milliseconds
Impulse initiation in LGN[122]	0.3 milliseconds	0.3 milliseconds
Time from LGN (optic radiation) to cortex[122]	1.0 to 1.8 milliseconds (complex cell)	1.6 to 3.6 (simple and hypercomplex cells)
Total estimated retinocortical conduction time for these cells (present results)	3.0 to 5.3 milliseconds	5.6 to 10.0 milliseconds
Total estimated latency to photic stimulation at cortex (assuming a cortical response) i.e., ganglion cell latency + retinocortical conduction time (present results)	41.0 to 61.3 milliseconds	24.6 to 34.0 milliseconds
Measured latency of the cortical cells to photic stimulation	Not available (complex cell)	Not available (simple and hypercomplex cells)

*From Ikeda, H., and Wright, M. J.: J. Physiol. (Lond) **227**:769, 1972.

and his co-workers have used clinically to differentiate syndromes associated with sustained-like and transient-like responses in human vision.[22,73]

Fast and slow optic-nerve and optic-tract fibers

Corresponding to the two "brisk" classes of retinal ganglion cells are two populations of optic nerve fibers distinguished by conduction velocity—fast for transient cells, slower for sustained cells (Fig. 23-2).[50,90] These respective axon groups with occasional exceptions couple transient retinal ganglion cells to transient receptive fields in the lateral geniculate nucleus, and sustained retinal cells to sustained geniculate fields.[50] Both geniculate types have concentric organization, with a large "suppressive" field that envelops antagonistic central and surround components and inhibits the responses of both. Responses of the geniculate fields are tightly synchronized with discharges of the retinal fields that drive them.[158] Surrounds as well as centers of the geniculate fields have the response characteristics of, and presumably receive input from, retinal field centers.[170] If several on-center retinal fields activate the center of a given geniculate field, several off-center retinal fields activate its surround, and vice versa. This "neural funneling" arrangement[27] allows a single relay cell of the lateral geniculate to collect, process, and transmit modal information about a spatial and temporal pattern of retinal activity. The separation of transient from sustained information is maintained as transient geniculate cells project to the striate cortex via rapidly conducting pathways and sustained cells via slower pathways.[249]

Conduction velocity differences among monkey geniculostriate neurons appear to be important for color coding.[173] Such differences are not confined to afferent pathways, but are also observed in both excitatory and inhibitory corticofugal paths returning to the lateral geniculate nucleus.[256]

Signal transmission times for successive stages of the retinocortical pathways are related to conduction velocities and path lengths in Table 23-3.[131] The responses of sustained cells have shorter latencies than those of transient cells, but take longer to reach the visual cortex; hence the cortical arrival times for the two types of signals differ by somewhat less than their retinal latencies. When both cones and rods are active, cone signals tend to dominate in the transient component and rod signals in the sustained component of the ganglion cell response.[109]

Cortical information processing

The coaction of fast-fiber and slow-fiber groups also modifies the timing of events within the hierarchy of cortical receptive fields. Hubel and Wiesel[125,126] showed that cortical cells with simple receptive fields can combine their outputs to generate the complex receptive fields of later cells, groups of which in turn drive cells with hypercomplex receptive fields. This anatomic sequence[62] suggests serial information processing, but the dual innervation of the cortical receptive fields by fast and slow fibers renders parallel information processing possible. In the cat, for example, fast fibers from the lateral geniculate provide input to complex cortical receptive fields, while slow fibers innervate simple and hypercomplex receptive fields.[122] Intracortical connections provide the basis for both serial and parallel information processing.[164,247]

Studies of spatial and temporal contrast sensitivity in cat cortex[183] demonstrate parallel processing of pattern and movement information in visual areas 17 and 18, which contain, respectively, a predominance of X (sustained-type, rapidly conducting axons) and Y (transient-type, more slowly conducting axons) cells. The relatively high spatial frequencies to which X cells respond, coupled with a response mainly to low and moderate temporal frequencies, give them an advantage for pattern discrimination, whereas the Y cells, with a preference for relatively low spatial frequencies and moderate to high temporal frequencies, seem best suited for reaction to movement. Unlike area 17, which receives axons with a broad spectrum of conduction velocities from W, X, and Y cells, area 18 lacks input from X cells. The fast Y-cell input to area 18 permits an early response to movement information entering this area, more or less contemporaneous with pattern-information processing in area 17.

Superior colliculus

Dual innervation also extends to the superior colliculus via direct retinocollicular and recursive corticocollicular pathways, labeling the colliculus as a possible comparator of information at early and late stages of processing.[176] With certain exceptions,[278] the axons of retinal W cells bypass the lateral geniculate and project directly to the superior colliculus.[250] Conduction velocity along these axons is very slow, from 5.5 to 14 m/sec. In addition, the fast (35 to 45 m/sec) retinogeniculate fibers branch to the colliculus, and fast fibers complete a retinogeniculocorticocollicular loop that originates in transient retinal cells. There is no evidence for either direct or indirect projections from retinal sustained cells to the colliculus. In the context of temporal factors the superior colliculus merits detailed description, being a major center for visual control of the eye, head, and body movements that orient an animal to new and moving objects in its visual field.

Like the lateral geniculate,[174] the superior colliculus has both visual and nonvisual connections. Superficial layers of the colliculus receive inputs from the contralateral and ipsilateral retinas (excluding, in monkeys, the central 7-degree area) and from the visual areas of the ipsilateral cortex (including the macular representation). Their outputs feed thalamic nuclei, nucleus lateralis posterior, ventral lateral geniculate, and deep collicular layers. In addition to receiving input from the superficial layers, the deep layers also receive fibers from the somatosensory and auditory systems, from nonvisual areas of the cortex, and from the contralateral superior colliculus. Outputs of the deep layers extend to the posterior thalamus, to the pulvinar, and to the brain stem, especially to those nuclei involved in coordination of eye and head movements.[102,165] The extent and variety of these anatomic linkages attest to the functional importance of collicular integrity.

In this respect, important differences among species exist. Major collicular lesions produce deficits of fixation, tracking, and pursuit eye movements, which undergo partial recovery in rats, cats, and monkeys, but not in hamsters or tree shrews. Furthermore such lesions impair space perception and reduce discrimination of depth, movement, pattern, brightness, and color, again with interspecific variations.[237]

In cats and rabbits neurons of the superior colliculus and pretectum can differentiate the visual results of head or eye movement from stimulus movement. These neurons discharge in conjunction with eye movements, apparently controlling the direction and velocity of saccades and tracking movements that culminate in foveal fixation. They may also be involved in generation of the so-called corollary discharge, which gives feedback about the expected result of oculomotor signals.[168,237]

Receptive fields of the superior colliculus resemble, in degree of specialization, those of the visual cortex. Again, interspecies differences are important. In the cat, which has been most extensively studied, four movement-sensitive types of collicular receptive fields have been described[240]:

1. With clearly defined preferred and null directions of movement
2. With two well-defined, contiguous preferred directions opposed diametrically by two nearly null directions
3. With axis selectivity
4. With no apparent directional selectivity

A fifth type is not movement sensitive. Collicular fields driven by fast fibers respond to stimulus velocities greater than 200 degrees/sec in the visual field and are generally nonselective as to stimulus direction, whereas those fields driven by slowly conducting fibers from retinal W cells respond to velocities under 100 degrees/sec and show directional selectivity. Cortical cells that project to the superior colliculus have complex receptive fields, respond, in general, to high stimulus velocities, and are mostly direction selective.[121] Many are driven binocularly, as are sizable numbers of collicular cells.[276] Cortical ablation greatly reduces the number of binocularly driven cells in the colliculus and also abolishes directional selectivity, indicating that the properties of the cat's collicular units are largely imposed by their cortical connections. In rabbits, however, the receptive-field characteristics of collicular units are not affected by cortical ablation[243] but are indigenous to the colliculus or possibly derived from the specialized retinal receptive fields in this animal. Eight different collicular receptive-field structures and trigger features have been described for the rabbit[243]; these include uniform, concentric, movement-sensitive, directionally selective, and hypercomplex types.[234] Monkeys are reported to have no directionally selective collicular units,[224] but ground squirrels[179,180] and cats[175,240] have many such units.

The directionally nonselective receptive fields of the superior colliculus in monkeys respond best to smooth stimulus movement (superficial cells) or jerky movement (deeper cells), with bursts of discharges at the beginning and the end of the movement. The deep cells also play a role in saccadic eye movements and fire ahead of such movements. A given cell discharges only before saccades having a particular extent and direction. Electrical stimulation of such cells in alert cats or monkeys induces conjugate saccades, again of specific amplitude and direction unaffected by the duration or frequency of the electrical pulse train. Prolonged stimulation brings about repeated saccades, similar in size, and eventually head and body movements. Schiller[224] interprets the role of the superior colliculus in generating these movements to be one of orientation and target acquisition. However, it has been argued that the colliculus may mediate a shift of attention toward a target prior to a saccade, rather than control the actual movement.[281] Supporting evidence includes the following:

1. Removal of superior colliculi in monkeys does not produce permanent gross defects of visually guided eye movement
2. Collicular receptive fields are too large to be consistent with accurate control of the eye movement

3. Cell discharge in deep collicular layers precedes any eye movements

It is reemphasized, however, that the effects of collicular lesions or removal vary rather widely from one species to another and depend on the task (discrimination versus orientation) imposed on the animal in behavioral tests.[237] Identification of, or discrimination among, visual stimuli depends on the integrity of visual cortex or forebrain structures and is little affected by collicular lesions, whereas visuomotor localization requires intact collicular and other tectal structures and is relatively unimpaired by cortical lesions.[226]

Many of the directionally selective collicular receptive fields of cats are bimodally excitable. They respond to auditory or somatosensory as well as to visual signals, provided the target or sound source moves laterally away from the animal.[275] These fields may serve to coordinate orienting responses to targets heard, seen, or both, although the possibility has been raised that they respond merely to an attempt to follow such a target.

Bimodal receptive fields also are found in the cat's visual cortex. Usually with binocular inputs and precise axis orientation, they may be classified as simple, complex, or hypercomplex.[84] A colliculooculomotor relay from such bimodal cortical fields could mediate eye movements toward a sound source. It is not known whether the audiovisual response of receptive fields in the superior colliculus is indigenous or is imposed by the characteristics of these cortical units.

Pathways to pontine nuclei

Also participating in the visual control of movement are pathways extending to pontine nuclei from the superior colliculus (cat,[8] monkey[213]), and from area 18 of the visual cortex.[96] Lateral geniculate projections carry parallel representations of the visual field to ipsilateral areas 17 and 18; in area 18 are neurons especially sensitive to target movement, whose axons activate receptive fields in the pons. Although a flash of light excites these fields (latency about 25 milliseconds), their best response is to lines or edges of a particular orientation, moving in a preferred direction. Information about the direction and velocity of moving objects is relayed to the cerebellum, where it is coordinated with proprioceptive input.

Visual influences on biologic rhythms

In addition to the spatiotemporal patterns of light that activate the neurologic mechanisms just described, the cycle of daylight itself exerts an important influence on the physiology and behavior of many animal species. It does so in part through the agency of a retinohypothalamic pathway.[153] Only the suprachiasmatic nuclei of the hypothalamus receive retinal input, but they do so in every species studied from rat to monkey. This input in the female rat appears to control the neuroendocrine cycle of vaginal cornification. In rats and other animals the retinohypothalamic pathway provides a means of synchronizing circadian rhythms of drinking, activity, and adrenal cortisone production, as well as the cycle of pineal enzyme levels, with the diurnal fluctuation of light.

FUNCTIONAL RELATIONS OF VISUAL PROCESSES TO SHORT-TERM TEMPORAL VARIABLES OF STIMULATION

Possibly the simplest situation that can be dealt with by the eye is a single test light on an otherwise dark background. Even this configuration is not entirely simple, because many characteristics of the lighted area can affect its detectability or visibility and must therefore be specified. Among them are the luminance, spatial extent, and spectral composition of the test stimulus, its location in the visual field, its shape, the luminance gradient at its border, and its internal uniformity, as well as its duration, speed of onset and extinction, and temporal relation to other lights. Despite such complications, fairly simple general rules describe the visual response to isolated lights.

Temporal summation

How long does a light have to shine to be just seen against a dark background? For a given size and retinal locus of the test field, the answer depends mainly on luminance. An intense light need not remain on as long as a weak one to reach threshold visibility. To determine the threshold, let us start with an intense light, having a luminance of 10,000 mL, and adjust the duration of its exposure on repeated trials until, on the average, an observer detects only 50% of its presentations. Under these conditions only a short flash will be needed. If we now reduce the test luminance from 10,000 to 5000 mL, the observer will detect fewer than 50% of the flashes presented. To restore his performance level, we must double the duration of the flash. Threefold reduction of luminance requires three times the duration, and conversely, threefold increase of luminance can be compensated by a threefold reduction of duration. In general, such reciprocity holds for flashes shorter than a certain limiting duration; over a substantial range a constant product of luminance *(B)* and duration *(t)* achieves (statistically speaking) a constant effect on the visual system. Since this product *(Bt)* is proportional to the total luminous energy of the test flash, we say that flashes equal in energy produce equal effects. The relation $Bt = k$ is Bloch's law for vision. More precisely, a given amount of luminous energy has the same effect, regardless of its distribution throughout a limited interval of time. The retina accumulates the results of sufficiently rapid energy input until the visual system meets a criterion of performance such as detection. This process, called temporal summation, has both photochemical and neural components.

Photochemical summation continues throughout much longer intervals than does neural summation.[37] Within the limit imposed by neural action, temporal summation at the retinal level is nearly complete. Beyond the limit, called the critical duration, neural summation ceases, and the visual effect of the test light becomes dependent on luminance alone rather than on the luminance-duration product. Thus it is necessary to specify, if a visual threshold is to be measured, whether this threshold shall be minimum energy or minimum luminance. Measurement of threshold energy requires flashes shorter than the critical duration,[242] whereas measurement of threshold luminance, probably more useful for the assessment of everyday visual function, requires relatively long test flashes.

A critical duration is also measurable with suprathreshold stimuli. On a physiologic level, Levick and Zacks[157] propose (for cat ganglion cell response) that it be defined as that duration within which an arbitrary temporal distribution of light energy produces a response indistinguishable from the response to a brief flash of equal energy. This definition is more rigorous than the usual psychophysical one, since an observer can usually distinguish a long flash from a short one of equal energy at the threshold of detection.

Visual capacity for temporal summation is not constant but depends on stimulus parameters such as size and background luminance, on the state of ocular adaptation,[242] and quite crucially on the visual task required of the observer. Fig. 23-3, a plot of energy versus duration for various test stimuli, illustrates some of these determinants. Logarithmic scales allow a wide range of variation to be presented in a limited space and also facilitate comparison of data obtained under different conditions. Each curve in Fig. 23-3 represents a set of conditions that are equivalent in that they produce a particular level of visual performance, for example, detection of a test flash against a background (*A* and *B*), matching of test brightness to a criterion (*C*, open circles), or 50% correct identifications of binary digits presented in triads (*C*, filled circles). Note that for short presentations of the test stimuli the horizontal segments of the curves show that a constant energy or luminance-duration product is needed to produce a given effect. A break in each curve defines the critical duration, which is seen to be longer for small than for large test areas,[219] for dark than for lighter backgrounds,[218] and for complex perceptual tasks than for detection or brightness matching.[136] A slope of unity above the break would indicate the production of equivalent effects by luminance-duration products that increase as rapidly as the duration of the test flash itself: in other words, the luminance or retinal illuminance needed for a particular effect (threshold, a brightness match, or some other measure) becomes independent of test durations that exceed the critical value.

Kahneman's demonstration that temporal summation interacts with visual acuity illustrates the task dependence of temporal summative power in a way that may have particular meaning for the ophthalmologist.[133] Test stim-

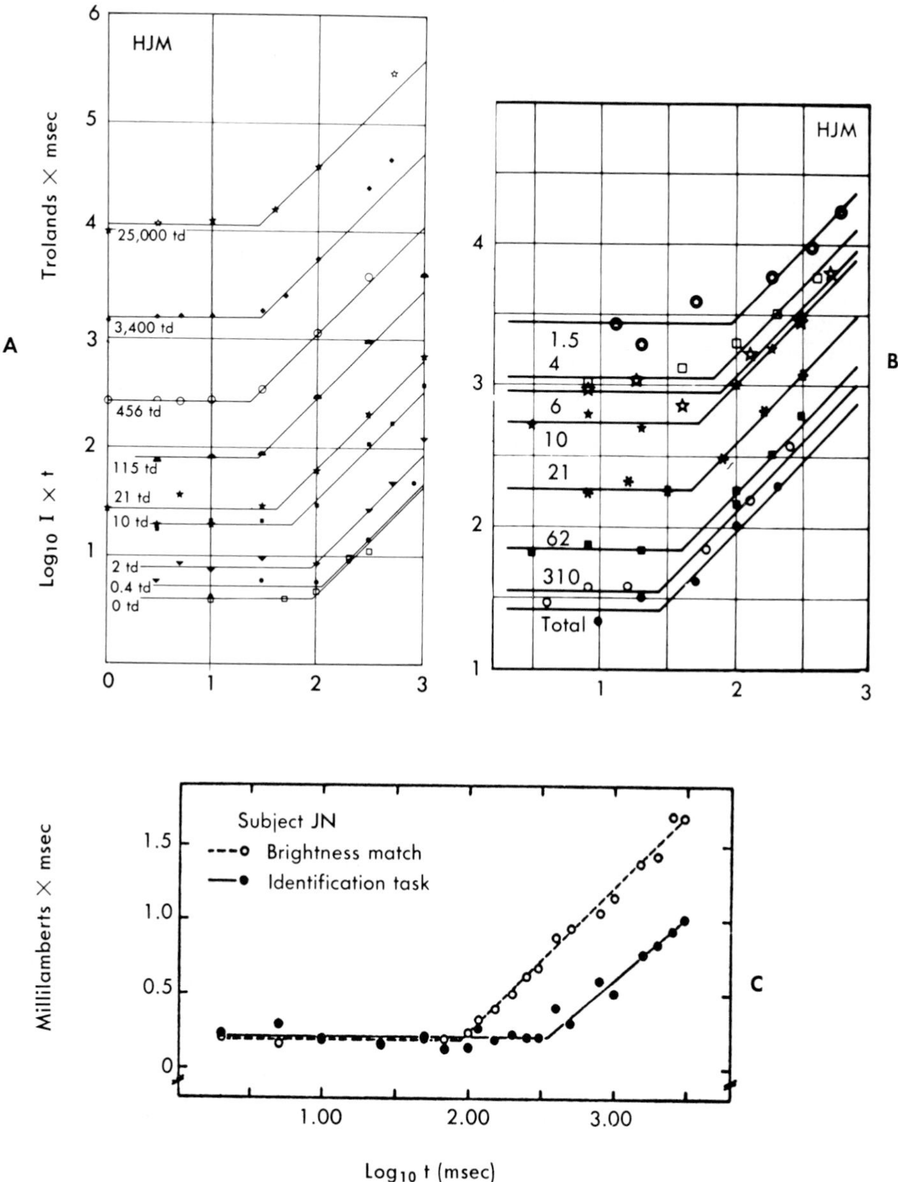

Fig. 23-3. Luminance-duration reciprocity and its breakdown above stimulus-dependent and task-dependent critical duration. **A,** Variation of threshold energy (retinal illuminance × duration) with duration of 1-degree foveal test field on different backgrounds (Data of Roufs.[218]) **B,** Variation of threshold energy with duration for foveal test fields having different diameters in minutes of arc, and for whole-eye stimulation (total). (Data of Roufs and Meulenbrugge.[219]) **C,** Open circles indicate test field energy (luminance × duration) required to match brightness of 10-degree test flashes of different durations to that of standard 10-millisecond flash. Filled circles indicate background energy required for 50% correct reading of binary digits (black on white) as function of background duration. (Data of Kahneman and Norman.[136])

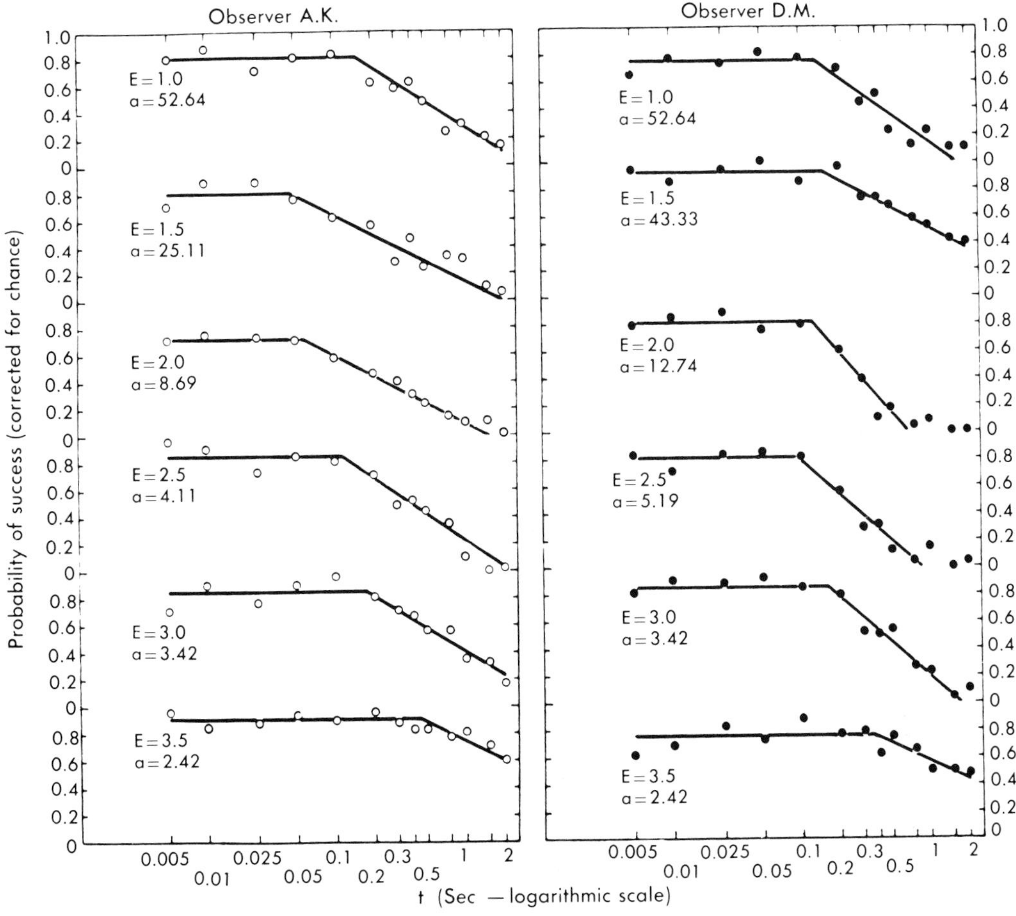

Fig. 23-4. Luminance-duration reciprocity. Probability that observer correctly reported orientation of gap in black Landolt C, as related to duration of its white 8-degree background. For each curve, product (background luminance × duration) was held constant at value *E*; *a* is width of gap in minutes of arc for C presented on each background. (Data of Kahneman.[133])

uli were Landolt C targets of various sizes, whose background luminances were adjusted until observers could detect the critical detail (the opening of the C) with a probability of about 80% when the exposure duration was 10 milliseconds. Thereafter the luminances and exposure times were reciprocally varied to maintain a constant product. Six target sizes, allowing six different energy levels, were presented, and the probability of detection plotted as a function of duration for each; Fig. 23-4 shows the results. As in Fig. 23-3, the data fit curves having a horizontal segment and a sloping segment; the break between them defines the critical duration. For short exposures, constant energy preserves a constant level of detectability, but detectability then drops as exposures longer than the critical duration fail increasingly to compensate for decrements of luminance. The critical duration diminishes with increasing background luminance at first, in keeping with its behavior as shown in Fig. 23-3, but it increases at still higher background levels to approximately 1 second, a value in striking contrast to the 0.1 second that represents the upper limit on summation of brightness or threshold energy.

Subsidiary experiments ruled out the possibility that this increase is related to target size or the retinal location of the critical detail and

Table 23-4. Critical durations of adapted color (π) mechanisms[257]

Color	Mechanism	Background wavelength (nm)	Test wavelength (nm)	Critical duration range (milliseconds)
Red	π_5	638	561	60 to 130
Green	π_4	503	561	63 to 166
Blue	π_1	442	503	189 to ?
Blue	π_2	442	503	Not reported
Blue	π_1	442	602	>200

led Kahneman[133] to conclude that the required level of detection determines the critical duration.

> Similar values of t_c will be found for equally visible targets presented at a single energy level, even when the targets are physically quite different. . . . On the other hand, different values of t_c are found for a given target when presented at energy levels which produce different levels of performance. . . . Equal-energy stimuli which vary greatly in duration may be encoded differently at the periphery and decoded as equivalent messages at some central locus. This suggestion is in good accord with what appears to be the main conclusion of the present work: that characteristics of the visual response are to be considered together with characteristics of the visual stimulus as determinants of temporal summation.[133]

Table 23-5. Critical duration for wavelength pulses and luminance pulses[205]

Retinal illuminance (td)	Wave-length (nm)	Critical duration (milliseconds)	
		for $\Delta\lambda$	for ΔL
0.1	527	—	96
	600	—	90
10.0	600	110	64
	580	160	62
	527	250	68
	480	250	56
110.0	527	94	40

Exposure duration as such, not total energy, is reported to be the relevant variable for detection of the gap in a Landolt ring, which improves with time up to 400 milliseconds at photopic luminances.[18] Results are similar for single exposures up to 400 milliseconds and for monocular and dichoptic double exposures of 100 milliseconds separated by a dark interval. These findings, like Kahneman's,[133,136] suggest the participation of a central nervous mechanism.

Color mechanisms: temporal summative characteristics

The degree to which mechanisms responsible for color vision differ as to temporal characteristics has been a matter of controversy.[258]

Individual color mechanisms (which Stiles noncommittally has labeled π mechanisms) may be partially isolated for study by selective retinal adaptation to background light of one wavelength and probed by test flashes of a different wavelength.[244-246] Increasingly intense backgrounds progressively desensitize the π mechanisms, in order of decreasing sensitivity to the background wavelength. Test threshold is determined by whichever mechanism remains most sensitive to the test wavelength. A plot of threshold versus background radiance (TVR curve) consists of two or more upcurving segments, positioned vertically as a function of test wavelength and horizontally as a function of background wavelength. Families of such curves generated with different combinations of test and background wavelengths may be transformed to delineate the spectral sensitivity of the individual π mechanisms: π_1 with peak sensitivity at 440 nm ("blue"), π_2 440 to 480 nm ("blue"), π_3 440 nm ("blue"), π_4 540 nm ("green"), π_4' 540 nm ("green," high luminance), π_5 575 nm ("red"), π_5' 587 nm ("red," high luminance). For details of the analysis see Stiles[244-246] and Enoch.[72]

Recent measurements of the critical duration

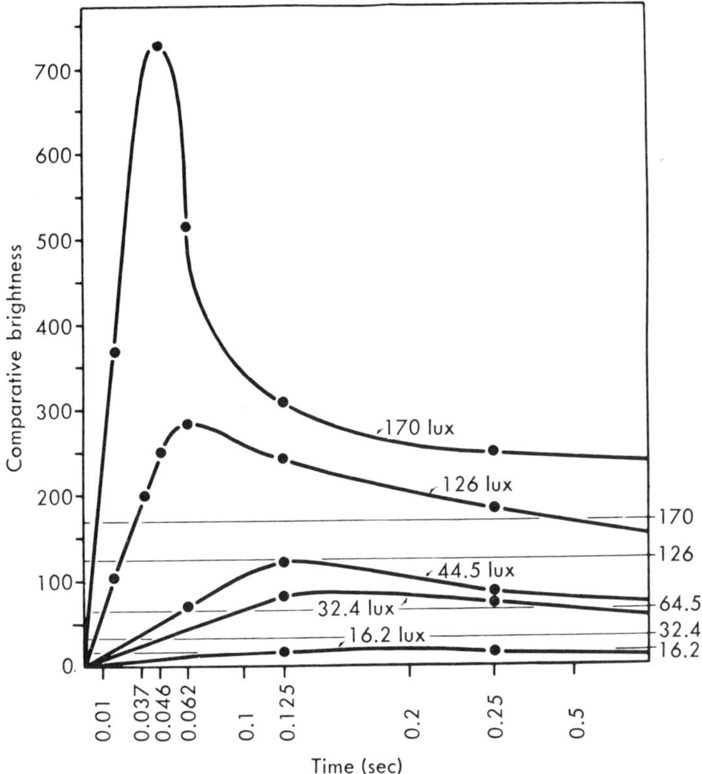

Fig. 23-5. Brightness of flashes having various luminances, as functions of flash duration. Such curves have been thought to parallel variation of brightness during first 500 milliseconds of prolonged light stimulus. (Data of Broca and Sulzer.[38])

for different π mechanisms have shown that short-wavelength mechanisms integrate energy over longer intervals than long-wavelength mechanisms. The separability of the color channels in this respect indicates that temporal summative properties are at least partially determined at an early stage of the visual system.[148] Adaptation decreases the critical duration, within a range specific to each mechanism as shown in Table 23-4 for 6-degree backgrounds, 30-minute test fields, located 45 minutes from the fovea in the temporal visual field.[262]

Critical durations for threshold monochromatic increments of luminance and threshold changes of wavelength under otherwise identical conditions are compared in Table 23-5.[205] Threshold detection of wavelength pulses (2-degree field, central fixation) requires a constant product of wavelength change times duration up to a critical duration, then depends on wavelength change alone. This rule also applies to increments of luminance; however, for each spectral region examined the critical duration for wavelength pulses is markedly longer than that for luminance pulses and depends on the part of the spectrum involved as well as the stimulus level. The data fit the hypothesis for which physiologic supporting evidence has already been presented, that separate channels convey chromatic and luminance information to the visual cortex.

Brightness as a function of duration

When a light is turned on, it is not seen instantly but requires a certain amount of time to reach threshold, and still longer to attain its full brightness as its excitatory effects undergo temporal summation in the visual system. Since temporal summation is limited, we expect brightness to level off as flash duration increases. In fact, we observe in Fig. 23-5 a transient brightening followed by a decline of brightness, the so-called Broca-Sulzer effect.[38]

A brief flash can appear brighter than a longer flash having the same luminance, although it will of course seem dimmer if sufficiently short. The luminance of a short flash adjusted to match the brightness of a continuous standard light decreases to a minimum as flash duration approaches 50 to 100 milliseconds and increases again for longer flashes. The luminance of a long flash may also be varied to match the brightness of short flashes having fixed luminance, and if plotted as a function of the duration of the short flash is seen to peak when the short flashes are 50 to 100 milliseconds long. These complementary experiments show that a duration of 50 to 100 milliseconds is optimal for flash brightness. Since the Broca-Sulzer effect occurs "under light and dark adaptation, in fovea and periphery, with white and with colored light, with stimuli of different size, uniocularly and binocularly, and whether presented simultaneously or asynchronously,"[137] we must look for an underlying mechanism with a broad range of operation. It seems likely that the specific sensitivity of the visual system to change and the Broca-Sulzer effect share a common mechanism, seen in most sensory end-organs as a pattern of neural discharge that is most rapid at the onset of stimulation and tapers off as the stimulation continues. Such a pattern characterizes the responses of many retinal ganglion cells, but is complicated by border effects[10] and by the asynchronous development of antagonistic excitatory and inhibitory action within the neural network that constitutes the receptive field of each ganglion cell.[150,215] Inhibition probably contributes to the fact, pointed out by Boynton,[30] that as the duration of a short flash increases, slightly more luminance is required to keep its brightness constant than would suffice to maintain a constant luminance-duration product. This evidence of imperfect temporal summation shows that the visual system is most efficient when excited by very short flashes.

As flash duration increases, visual efficiency declines. Troxler's phenomenon, also called local adaptation, illustrates this principle dramatically. If an observer fixates carefully on a small stationary object but pays attention to a steady light presented to the peripheral retina, he will notice that the peripheral stimulus fades away within several seconds while the fixated object remains visible. The dimmer the target, the sooner it disappears. Pirenne and Marriott[196] have shown that a dim target fades away before light from it has bleached an appreciable fraction of the retinal photopigment. Both the prolongation of visibility despite increased rates of bleaching and the continued availability of unbleached pigment at the time of disappearance show that Troxler's phenomenon depends mainly on neural rather than photochemical action.[45] Persistence time apparently increases with the number of retinal receptive fields excited by the stimulus, since the disappearance takes longer with increasing target size at a given eccentricity and, for a given target size, with proximity to the fovea, which is correlated with a gradient of receptive field size from large to small.[124] As the target disappears, its background seems to fill in the area it occupied, as if the visual system extrapolated the conditions beyond the target boundary to create a perceptual continuity. Such extrapolation also seems to explain why an observer cannot ordinarily detect his own blind spot, or certain scotomas.[92,266]

Once the target has disappeared, a slight shift of the observer's line of sight restores its visibility. For complete reappearance the shift need not be as large as the angle subtended by the target but only large enough to locate one of its borders on a relatively unexposed adjacent part of the retina. It is evident, therefore, that some motion of the eye or the retinal image is needed to maintain the visibility of a peripheral target and that borders or discontinuities of illumination between adjacent retinal areas have special significance for vision. The high sensitivity of the eye to movement in the peripheral field, which has great biologic value for many animals, is related to these properties of the system.

Eye motion during fixation

The foregoing discussion emphasized a visual requirement for changing stimulation. Fortunately, the normal oculomotor system precludes perfect fixation, which amounts to perfect stabilization of the image of the fixated object on the retina.[1] The innervation of the extraocular muscles pits one against another in tonic balance, but never so perfectly that residual drifts, microsaccades, and tremors of the eyeball are not present.[221,261] Cornsweet[55] has found the drifts to be mainly due to instability of the oculomotor mechanism and to be uncorrelated for the two eyes, whereas the microsaccades or "flicks" are binocularly correlated and corrective, returning the retinal image to an optimal fixation point after a drift. Although the various movements that can be lumped under the name physiologic nystagmus—drifts up to 6 minutes of arc,* microsaccades of 1 to 20 minutes of arc within approximately 0.025 seconds, and tremors of 10 to 30 seconds of arc, with frequencies of 30 to 80 cps[65,66]—are tiny, they result in an oscillatory displacement of any detail of the retinal image over several cone diameters as fixation continues for several seconds.[211] Movement, relative to any receptor, of a contour element or other nonuniformly illuminated small region of the image results in a continuous fluctuation of light on that receptor, therefore an unsteady excitatory state. Information about the location of contours and other details of the retinal image must therefore enter the visual system as differences of time-average excitation patterns in neighboring populations of receptors. It turns out, however, that steady state differences equal to these time-averages are not adequate to stimulate the system for more than a few seconds.

The functional importance of physiologic nystagmus becomes evident when it is counteracted by optical means that stabilize the image of the fixated object on the retina. Under stabilized image conditions, centrally viewed targets fade from view just as do peripheral targets under the influence of Troxler's phenomenon. Destabilization restores their visibility, as well as the visibility of entoptic structures that are normally fixed with respect to the retina. The reader may demonstrate this point by looking through a pinhole in a card held close to the eye while shaking the card to move the pinhole across the pupil. As the pinhole moves, light through it casts shadows of the retinal blood vessels on normally unshadowed retinal areas and makes the entire retinal vessel tree visible, only to disappear again if the pinhole is removed or held stationary. Sophisticated optical systems[64,212] permit the stabilization and controlled movement of extraocular objects through the use of mirrors mounted on tight-fitting contact lenses, or alternatively, objects may be fixed directly to the contact lens by means of a stalk.[199,282] Eye-controlled movement of a spot displayed on an oscilloscope has also proved to be a versatile technique.[233]

Strict immobilization of any target, black, colored, or white, including one as bright as the glowing filament of an incandescent lamp, by the image-stabilizing apparatus results in a so-called empty field appearance,[282] which observers often describe as black. In fact, a suddenly moved black object in such a field becomes visible and looks blacker than its background. With complete stabilization the contours or objects within the field do not reappear, although movement of the contact lens or other components of the optical system can cause them to do so. Disappearance of a stabilized image supports the principle that change is needed to sustain vision. Introduction of change by flicker or gradual increment or decrement of background luminance slows fading or restores visibility; for example, Yarbus[282] has demonstrated that a narrow line (subtending 3 minutes of arc) that fades to give the empty field appearance if background luminance remains steady can be held at the threshold of visibility by continuous variation of the background at a rate of about 30%/sec. This percentage remains constant throughout a range of luminances of about 1000 to 1. Slow flicker also maintains target visibility, but as flicker frequency increases, small details fade, followed by larger ones.[55,282] An interesting point in this context is that once the empty field appearance has been achieved with a stabilized lamp filament, a momentary dazzling impression of the filament recurs when the lamp is turned off.[282] The release of inhibition accompanying the retinal off effect may be responsible.

There is much evidence that stabilized image effects have central as well as retinal determinants.[93,94] The rate of fading depends on

*For comparison, the head of a thumbtack at 10 feet subtends about 10 minutes of arc, the full moon 30 minutes of arc. At the fovea, 1 minute of arc subtends a distance of about 5 μm, and centers of adjacent cones are about 0.6 minute of arc apart.

whether stimulation is monocular or binocular[149] and on the structure of the (unstabilized) stimulus presented to the contralateral eye,[53] on the attention an observer pays to a particular detail, and on the complexity and meaningfulness of the figure viewed. Whereas individual lines disappear as units, not piecemeal, different lines of complex figures fade independently or in organized patterns if the stimulus object has meaningful parts.[111,200] Pritchard and coworkers describe the appearance of the word BEER as changing to BE, BEE, PEEP, BEEP, PEER, or other organized constellations during the course of fading.[200] In the same way, B may disappear as a 1 and a 3 in independent phases, and a line seen as a face in profile takes longer to disappear than a wavy line without specific connotations. Observations such as these imply that more than fatigue or some such mechanism is involved in the fading of stabilized retinal images. The neural elements responsible for fading do not act independently; rather, they act as groups, possibly having the form of "property filters" sensitive to edges, motion, or other specific attributes of the stimulus. A comparison with the receptive fields of the visual system, whose increasing specialization from retina to cortex is well known,[124,127,277] seems inevitable. Additional indirect argument for the specificity of fading mechanisms may be based on results of Riggs and Whittle,[210] who demonstrated no difference between averaged occipital potentials (or retinal potentials) recorded with a rapidly flickering, spatially stabilized grid target, whether this target was visible or had faded from view.

Aside from clarifying the role of physiologic nystagmus in visual function, stabilized image–like effects introduce certain problems for tonography and set limits on exposure durations in perimetry. Ideally the patient's eye should act as a stable platform for the tonographic pressure transducer. For this reason the patient lies supine and is asked to fixate his uncovered eye on a target mounted in the ceiling. Despite this maneuver the covered eye does not remain stable but tends to drift through increasingly large angles with time[55] because it is essentially in the dark and has no fixation point. Meanwhile the fixated target becomes blurred and loses contrast as tears evaporate from the uncovered eye, which is anesthetized to reduce the blink rate. Eventually the target fades and the uncovered eye drifts until it again becomes visible, whereupon the patient, whose line of sight has deviated from proper fixation, makes a corrective saccadic movement. Because the saccades are binocularly coordinated, this movement is reflected in the covered eye and introduces an artifact into the tonographic record. It is difficult, if not impossible, for the patient to keep his eyes still throughout the 4-minute recording period, although he is usually not aware of their small movements during normal fixation.

Saccades, tracking, and optokinetic nystagmus

The probability that a reflexive saccadic movement will be made increases with the eccentricity of the target relative to the fovea,[55] whether the target is moving or stationary. If the target moves, the observer instructed to track it makes an initial saccade to locate its image on the fovea and thereafter can perform smooth pursuit movements that are effectively locked to the target if its motion is not too rapid. Slow, cyclic target movement along a fixed path generates tracking movements that have negligible lag or even anticipate those of the target, evidence that a predictive component exists in the oculomotor control system.[239] When target motion is irregular, tracking movements lag as a rule, but include corrective saccades that return the line of sight at least momentarily to the target. Tracking error becomes large and saccadic size and frequency increase with increasing target velocity, until a limit is reached at which tracking breaks down entirely.

Although tracking of isolated moving objects is largely voluntary, it is almost unavoidable if a large part of the visual field moves as a unit. Two recently described populations of cells in the (monkey's) visual cortex may account for this difference. The excitability of one population by relative movement of elements within the retinal image and of the second population by movement of the image as a whole across the retina seems to explain the ability of the visual system to distinguish external object movement from eye movement.[36] If involuntary tracking, which tends to compensate for overall image movement, represents the oculomotor response to events that activate cells of the second population, a possible role for these cells in maintaining the position of the image upon the retina is also indicated.

Reflexive tracking of extended movement is the basis of several useful clinical tests that can be conducted with a grid of parallel vertical stripes that move across the observer's visual field. A convenient form of such a grid is the optokinetic drum, a rotating cylinder that may be a few inches in diameter or large enough to surround the head of the observer at a viewing distance of several feet. As the stripes move past the observer, his eyes exhibit a series of reflex oscillations called optokinetic nystagmus (OKN), in which tracking alternates with rapid retrace or flyback movements. This reflex has great clinical utility: I shall mention only one example, an objective test of visual acuity applicable to infants and other linguistically restricted patients as well as to animals.[14,161,162,207,232] Its use with malingerers can be valuable, since the optokinetic movements cannot be completely inhibited in the absence of a fixation point. Two forms of this test are often used; one requires a stationary object adjusted in width until fixation on it overrides the optokinetic response; the other consists of measuring the smallest line width or checkerboard pattern that elicits the response.

Use of OKN to obtain scotopic luminosity functions in close agreement with psychophysical data demonstrates its value as a sensitive objective indication of visual threshold.[178]

How are the two phases of OKN related to fixation? Under bright illumination the rapid phase of eye movement begins approximately at the midline and is directed toward the origin of drum motion, which suggests that an object moving toward the fovea attracts fixation movements more strongly than one moving away. However, when foveal function is eliminated by the use of very low-level illumination or when it has been lost, as in cases of central scotoma, the rapid phase, although still directed in opposition to the drum movement, ends rather than begins in the vicinity of the midline.[123] In other words, the eyes deviate from the midline in the direction of the rapid component when foveal function is present and in the direction of the slow component when it is absent. This difference has led Hood to postulate separate foveal and peripheral mechanisms for OKN.[123] As for the slow phase, although it appears to be an attempt to maintain fixation on the drum and so minimize the error signal attendant to displacement of a fixated target from the fovea, it fails in that the angular velocity of the eye movement is usually less than that of the drum. For a normal eye the peak velocity of tracking recorded by Hood was about 40 degrees/sec reached when drum velocity was 80 degrees/sec. The eye/drum ratio of velocities diminished as drum speed increased. For the other eye of the same observer, which had a central scotoma, tracking velocity peaked at a higher value (90 degrees/sec), corresponding to a drum speed of 120 degrees/sec. It appears from these data that the tendency of the eye to maintain a target image as nearly fixed on the retina as possible is an extrafoveal function limited by foveal participation in the optokinetic response. Similarly, the accuracy of foveal tracking of small targets deteriorates at velocities above 30 degrees/sec.[14] Nevertheless, the combined function of the fovea and periphery allows the utilization of positional difference signals to initiate saccades that displace image detail onto the fovea.

Saccadic suppression

As saccades move the line of fixation from one detail of the environment to another, there is little impression of blur or even movement of the external scene,[168,169] as if somehow vision were being suppressed during the saccade. In fact, several experiments have shown that saccades are accompanied by reduced visual efficiency. For example, test flashes presented during saccades are less likely to be seen, regardless of their position in the visual field, than similar flashes in similar positions presented while the eye is stationary. One hypothesis advanced to explain the reduced visibility is that the displacement of the test area on the retina during the course of the saccade, or "retinal smear," effectively spreads the test flash over a larger retinal area and so diminishes the retinal illuminance at any point in that area. Two observations militate against this hypothesis. First, visibility is reduced even when the test flashes are too brief to allow appreciable retinal movement to occur during their presentation[262]; second, suppression begins before the eye movement.[154,283] This second observation seems to rule out as explanations (1) the curvature of saccadic movements,[255] which might displace the image of the test flash away from the expected position, or (2) possible reductions of efficiency, sensitivity, or acuity of the peripheral retina as contrasted with the fovea. The precedence of suppression has suggested that it may act to initiate the saccade rather than result from it; hence the statement of Latour that "it is not the proprioceptors of the eye muscles which cause the visual system to be 'blind' during eye movements, but rather that the 'blind mind' allows the eye to start its movement."[154]

Most authors have agreed that saccadic suppression probably has a central nervous origin, but Richards[209] has argued in favor of a retinal origin; if a saccade occurs during the retinal processing time between the arrival of the test flash and the release of impulses in the optic nerve and if it interferes with excitation during this delay time, for example, by exerting shear forces on the retina, the flash may precede the eye movement and may still be suppressed. Richards augments this argument with data showing that a passive movement induced by a finger tap on the eye has the same suppressive effect as a saccade of similar size and speed.

Image movement, rather than eye movement as such, can bring about suppressive effects.[16,132,167,181] The more complex the image, the greater the suppression (after correction for retinal smear).[181] Whether the mechanism of suppression is retinal or central has not been unequivocally established. The possibility that inhibition of sustained by transient cells in the lateral geniculate nucleus may be involved in saccadic suppression has been mentioned.[35,230]

There is evidence that cells with complex receptive fields in area 18, which are subserved by a rapidly conducting pathway and respond to abrupt motion, may be the only neurons of the visual cortex that remain excitable during saccades.[189]

Saccadic suppression of about 1 log unit has been shown to accompany (and antedate) voluntary saccades and the saccadic phase of vestibular nystagmus,[283] which is generated during and after rotation of the observer and results from displacement of the cupulae of his semicircular canals. Latour[155] has demonstrated suppression during the fast phase of optokinetic nystagmus. Although Richards' argument for a retinal origin may hold true for suppression that precedes large saccades such as these, it seems less likely to be valid for suppression that has been shown to precede involuntary microsaccades.[26,283] Beeler,[26] with a stabilized retinal image confined to the fovea, showed that brightness thresholds started to change 60 milliseconds before an involuntary microsaccade, rose as much as 0.5 log unit (a factor of 3), and remained high until 75 milliseconds after the movement. When target motions of 15 minutes of arc were timed to coincide with microsaccades during normal fixation, observers failed to detect them, although more than 90% of such displacements could be seen at other times. Motion detection began to deteriorate about 40 milliseconds before the microsaccade and continued to be impaired until about 50 milliseconds after the eye movement.

Evidence is ample that saccadic suppression affects foveal as well as peripheral vision, but under ordinary viewing conditions the time course of foveal suppression is hard to measure because voluntary saccades larger than a degree or so displace the target image onto parafoveal regions. One way to overcome this difficulty is to stabilize the retinal image. Another, developed by Volkmann, Schick, and Riggs,[262] makes ingenious use of the relative sensitivities of the fovea and parafovea. The observers looked on command from one fixation point to another 6 degrees away, both located on a white background against which a matrix of small dots could be flashed. The 0.5-degree spacing of the dots ensured that some of them were always presented to the fovea whether the eye was stationary or moving. The exposure time was 6 microseconds to avoid retinal smear. Dot luminance was such that observers could easily detect these exposures with the fovea but not 2 degrees peripherally, and with the stationary eye but not during a saccade. Eye movements were photographed, and the relative timing of saccades and test flashes was determined from the records. The results showed that probability of detection is reduced to 50% for flashes presented 20 milliseconds before the saccade, becomes essentially zero during the saccade, and then improves to 50% by 75 milliseconds after the saccade.

This experiment, and also Beeler's, indicates that minimal detection of foveal flashes occurs at the same time as the saccade. Such a coincidence is not found with peripheral test flashes,[154,283] which are maximally suppressed some 40 milliseconds before the eye movement begins. The difference is apparently related to the retinocortical latency, which is longer for peripheral stimulation under scotopic conditions than for foveal stimulation under photopic conditions. To account for suppression, Volkmann and co-workers[263] also assume that the oculomotor command signal induces a central inhibitory state that interferes with excitation due to the test flash. Although some evidence of such inhibition exists, its role as a mechanism for saccadic suppression has not been proved.

Visibility is not all that is affected by saccadic suppression. The ability of short flashes to evoke a pupillary response is also reduced greatly if the flashes precede or accompany saccadic eye movements.[284] Fig. 23-6 shows the time course of pupillary suppression in conjunction with that of flash detectability occasioned by saccades in the same experiment. Remember that although the curves are superimposed, the response of the pupil comes after the saccadic movement has been completed. The pupil suppression curve is based on the timing of the test flashes in relation to the saccade, but shows the extent of the pupil response measured later. It seems that suppression of the pupil response begins well before flash detectability becomes impaired, but the time difference may indicate only that preparation for the saccade tends to suppress the brightness of the test flash by amounts too small, at the outset, to reduce its detectability below 100%. A second remarkable feature of the data is that flashes observers fail to report having seen give rise to

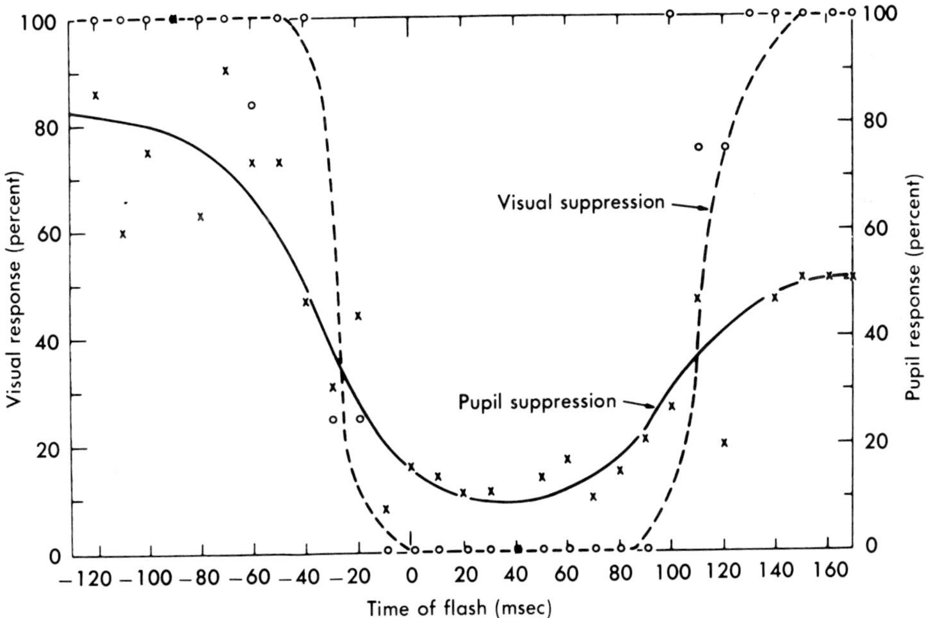

Fig. 23-6. Saccadic suppression. Time course of visual and pupillary responsiveness to test flashes presented before, during, or after 8-degree saccadic eye movements that start at time 0. (Data of Zuber, Stark, and Lorber.[284])

pupillary responses. Possibly the central mechanism for flash detection becomes less sensitive than the oculomotor part of the system during saccadic suppression. This observation would be inconsistent with a purely retinal locus for suppression.

Masking

As an observer scans his visual environment, he makes saccadic movements that quickly substitute, at the fovea, one detail for another within the retinal image. Saccadic suppression implies that mechanisms exist which tend to cancel the stimulation from the old detail and prepare the retina to receive the new. A similar, although probably not identical, effect frequently occurs when lights or patterns overlap or coincide in the visual field simultaneously or within restricted intervals of time, roughly comparable to the duration (10 to 80 milliseconds) of a saccade.[282] Such a coincidence often results in an interaction of the stimuli that may alter the visibility, detectability, or threshold energy of at least one of them. Interactions of this kind, which are described by the generic term visual masking, take place when two uniformly illuminated spots of light impinge simultaneously or successively on the same foveal or extrafoveal retinal area, as well as under more complex conditions. One of the lights, variously called the inducing, conditioning, or masking stimulus, brings about a change of sensitivity that is measured with the other, or test, stimulus. This measurement involves a specialized application of increment-threshold technique, which has been widely used to define such characteristics of vision as rod-sensitivity functions,[3,106] the wavelength sensitivity of individual color mechanisms,[72,244-246] and the shape, size, and functional organization of human normal[39,177,268-270] and abnormal[74-76] receptive fields. Evidence exists that some receptive field properties so defined may be those of lateral geniculate neurons,[185] although some have been localized at the retina.[74-76]

The usual test procedure in masking experiments is to present the paired stimuli repeatedly at intervals sufficient to minimize interaction between members of successive pairs and to adjust the luminance of the test light after each presentation so as to approach a level at which it is just visible (or satisfies some other criterion, such as a brightness match to a standard light). This procedure is essentially the same as that used for tests of dark adaptation following preadaptation to a bright stimulus, except that repetition allows measurement, by sampling,[236] of very rapid sensitivity changes that take place in the immediate temporal vicinity of a masking flash. Under the assumption that each presentation of the masking stimulus alters sensitivity in the same way, thresholds can be determined for brief test flashes paired with masking stimuli at each of several fixed time relations.[56] Threshold plotted as a function of the time interval between the onsets of the test flash and the masking stimulus in each pair is called a masking curve and is presumed to depict the variation of sensitivity brought about by any one of the masking stimuli.

The luminance, duration, spectral composition, retinal location, size, repetition rate of the masking stimulus, and similar parameters of the test flash exert important influence on the amount and time course of this variation. Masking also depends on the relative sizes of the test and masking stimuli and on the proximity of contours, if any, within the two flashes, or of the borders of the two. In addition, delivery of one of the flashes to each eye (dichoptic presentation) produces results that differ in certain important ways from those of monocular presentation.[25,81,82] Dichoptic masking of a steadily presented test object in one eye's visual field by a moving contour presented to the other eye[104] and a comparable monocular effect[166] have been described. Sensitization to the test flash during certain parts of the masking cycle has also been reported.[253,254]

In general four significant categories of masking can be distinguished:

1. Masking of light by light
2. Masking of pattern by structurally related pattern
3. Masking of pattern by structurally unrelated pattern
4. Masking of pattern (or darkness) by light

A further distinction is made between masking, which describes the interaction of spatially overlapping stimuli, and metacontrast, an inter-

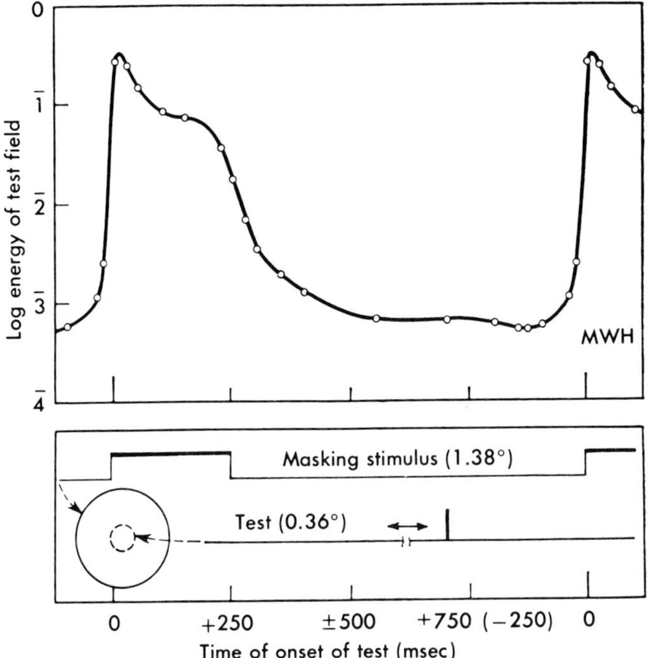

Fig. 23-7. Masking. Time course of threshold variation due to 50 ft-L, 250 millisecond masking flash, measured with "impulse" test flash. Lower panel shows flash diameters and field configuration. Last seven data points are same as first seven. (Data of Sperling.[236])

action between nonoverlapping stimuli.* Each type of masking follows different and rather complex rules.[34,145] Kahneman[135] has provided an excellent review of recent experiments and theory in masking. A possible physiologic basis involving interactions among transient and sustained neural elements is presented by Breitmeyer and Ganz.[34] The following description will be mainly concerned with masking of light by light but will also refer to contour and border interactions.

Fig. 23-7 illustrates a typical masking curve in relation to the 250-millisecond flash, repeated once per second, that produced it. Although the logarithmic calibration of the ordinate is advantageous for the display of data that vary over a wide range, it has the disadvantage of exaggerating small values at the lower end of the scale while compressing large values at the upper end. The masking curve in Fig. 23-7 encompasses a range of test-flash energies of approximately 500 to 1, justifying the logarithmic ordinate, but a linearization of the data may be an aid to interpretation. Reading from left to right, the first 15 points on the curve represent approximate relative energies of 2.0, 4.1, 9.2, 1000 (maximum), 890, 540, 310, 260 (just before the downturn), 135, 65, 25, 12.5, 7, 4.5, and 2.5 (where the curve becomes flat). This set of data has several quite noteworthy features. First, threshold begins to change almost 100 milliseconds before the masking flash is presented and has increased almost fivefold at a point in time 20 milliseconds prior to its onset. Second, threshold is maximal for test flashes that appear simultaneously with the onset of the masking flash. Third, it drops within 150 milliseconds to a fairly steady level, which would diminish only slightly if the masking flash were prolonged. Fourth, the ratio of the peak threshold to this steady value is substantial, about 4 to 1 for this particular set of data. Fifth, threshold begins to drop again just before the masking flash is turned off; it reaches its dark value after a few hundred milliseconds. (The course of threshold near the end of the flash depends on

*References 4, 5, 12, 105, and 225.

stimulus luminance. With intense masking flashes the threshold rises, rather than drops, at this point just before the recovery phase.[13,56]

Physiologic considerations

It is clear that the photochemical aspects of retinal function cannot account for all the complexities of this masking curve, particularly the anticipatory changes of sensitivity near the onset and offset of the masking flash and the threshold drop during the course of the flash. Nevertheless it is likely that receptor properties underlie some of the phenomena of masking. Brown and co-workers[42] have pointed out the striking resemblance between the time course of the late receptor potential elicited by a small foveal stimulus and that of foveal masking, particularly with respect to the initial peak of the threshold. These authors also relate the shape of the recovery curve after the masking flash, which includes an early rapid phase and a later slow phase that represents the beginning of conventionally measured dark adaptation,[13] to the respective decay rates of cone and rod late receptor potentials. Furthermore, the observation of Brown and Murakami[41] that the steady phase of cone late receptor potentials evoked by intense 20-microsecond flashes is extended in proportion to the logarithm of flash energy, is suggestive in view of Sperling's[236] description of a qualitatively comparable prolongation (Fig. 23-8) of masking by extremely brief flashes of increasing energy.

Since features of this kind are seen at the receptor level, the question may be raised whether receptors or some higher level of the retina or visual pathway should be considered the primary site of masking. Actually, something other than receptors must be involved, if only because masking occurs even at scotopic levels—luminances so low as to render negligible the probability that any one receptor will absorb quanta derived from both the masking and the test flashes, as would be necessary for interaction of the two flashes at the receptor level. Receptor properties alone cannot be responsible for contour interactions or for the threshold changes that take place in apparent anticipation of the onset or extinction of the masking flash. Both physiologic and functional mechanisms have been proposed to account for this so-called backward masking, or Crawford effect, as well as other phenomena of masking. (See Breitmeyer and Ganz[34] for a review.)

A physiologic mechanism that can account for backward masking, a combination of differential latency and lateral inhibitory interaction, is clearly illustrated by Ratliff and co-workers,[202] although Crawford[56] first suggested it in principle. If response latency varies reciprocally with the intensity of stimulation, a weak stimulus must precede a strong one to bring about a temporal coincidence or overlap of the two responses, which then interact. The strong response may simply preempt or saturate neural channels that would otherwise be open to the weak response,[31,223] or else it may actively inhibit the weak response itself. A lateral inhibitory effect seems implicit in the larger amounts of masking that occur as the area of the masking flash decreases,[70,89] but the inhibition is evidently one of contour on contour and not one of area on area.

Where is the interaction? Taken as a whole, available data appear to support the theoretical position that interaction in masking takes place partly at the cortical level, but involves neural signals that have been preconditioned by properties of the receptors and by interactions within the retinal neural network. Since lateral inhibition occurs at the retina, it seems reasonable to postulate a retinal locus for the interaction of responses to monocular masking and test stimuli. This view finds support in the work of Frumkes and Sturr,[89] who tested at the fovea and at eccentricities of 7 and 15 degrees in the nasal field. They noted that the highest thresholds at each position were obtained with a masking flash whose size increased with decreasing luminance and with increasing retinal eccentricity, much as do the diameters of retinal receptive-field centers. Nevertheless, Battersby and Wagman,[25] as well as Frumkes and Sturr, conclude that much of the interaction in masking must occur at a retrochiasmal locus, since the time course of backward masking effects and the variation of peak threshold with mask-

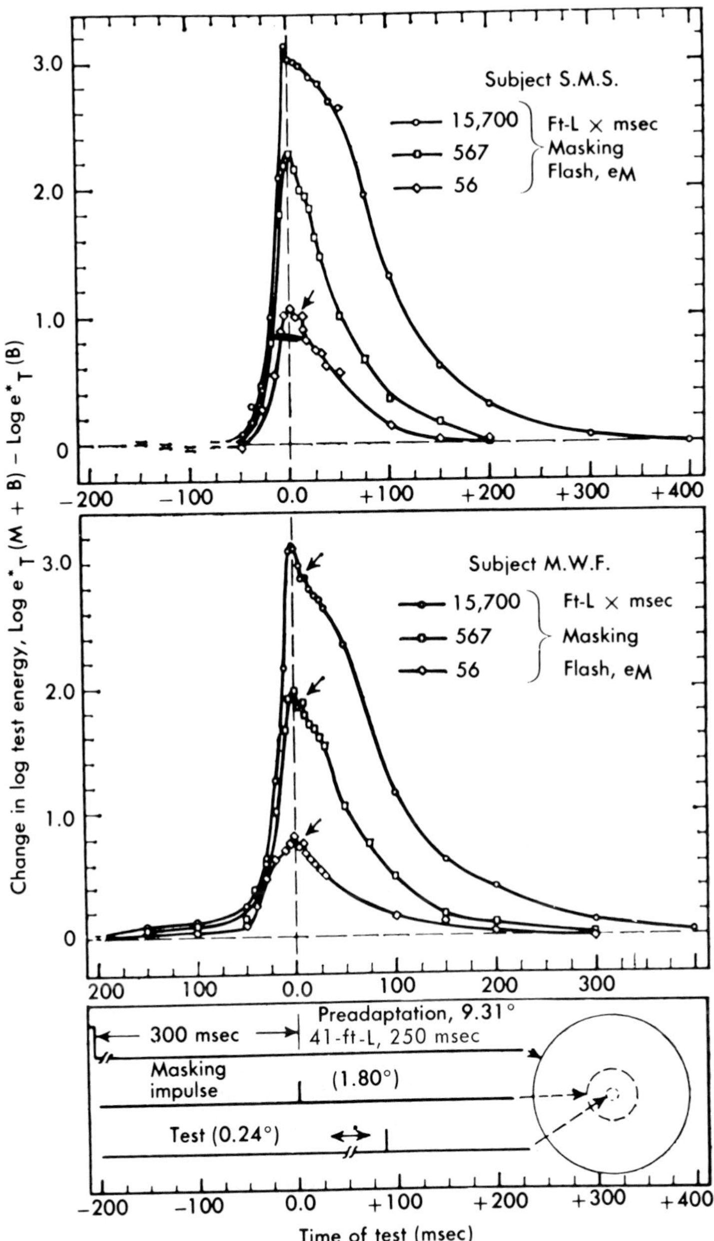

Fig. 23-8. Masking. Time course of threshold variation due to "impulse" masking flashes of different energies, measured with "impulse" test flash. Two observers. Lowest panel defines stimulating conditions. (Data of Sperling.[236])

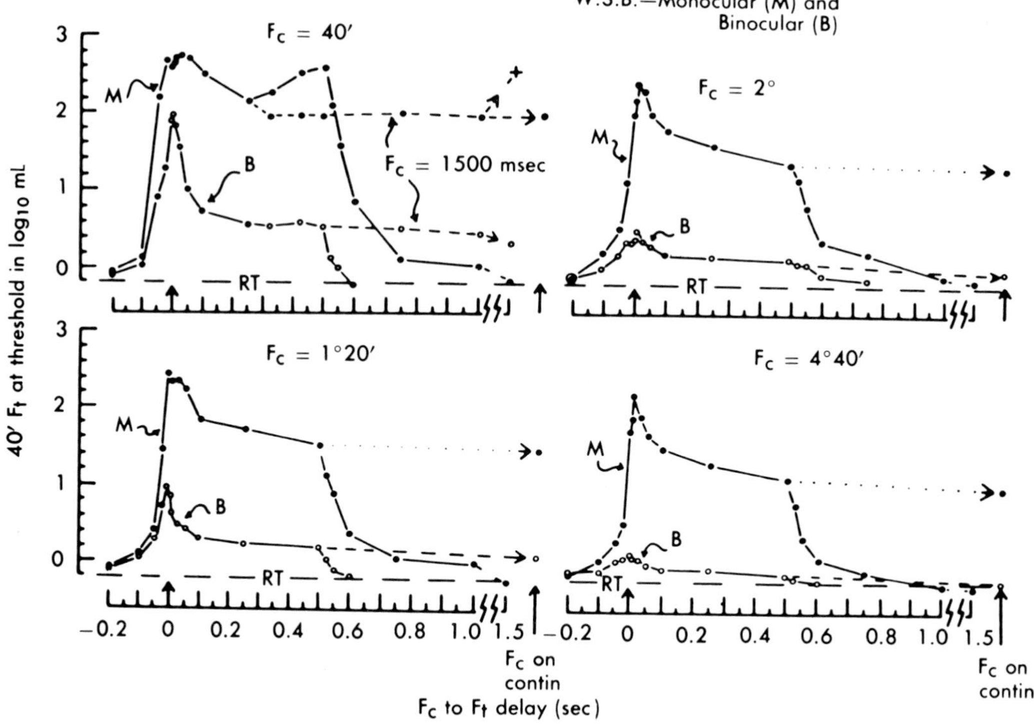

Fig. 23-9. Masking. Time course of parafoveal threshold variations caused by 100 mL, 500-millisecond masking flashes of four sizes, measured with 5-millisecond test flash equal in diameter to smallest masking flash. Filled circles indicate monocular presentation; open circles indicate dichoptic presentation; *RT*, resting threshold in absence of masking flash. Monocular and dichoptic test-flash thresholds measured with prolonged masking flashes also shown. (Data of Battersby and Wagman.[25])

ing-field size are in certain respects similar for both dichoptic and monocular masking.[25] Masking curves for the two conditions of stimulation and four field sizes are compared in Fig. 23-9.

Interaction between the two eyes that raises the dichoptic masked threshold would have to occur at a locus such as the visual cortex or possibly the lateral geniculate nucleus,[222] where excitation from both eyes is combined. Hubel and Wiesel[127] have demonstrated the existence of cortical cells that can be stimulated by light presented to either eye and therefore seem to be appropriate as sites for masking interactions. If such cells are involved in masking, it should be possible to test the integrity of the entire visual pathway by proper use of masking as a clinical technique.

An examination of Fig. 23-9 reveals evidence of interactions at retinal as well as cortical levels of the visual system. Each curve shows the typical initial peak of threshold in approximate coincidence with the onset of the masking flash and its subsequent decline to a relatively stable end-point value, just before the recovery to resting threshold. Except in the special case of the 40-minute of arc masking flash, which is the same size as the test flash, this end-point threshold is quite similar to what it would be if measured on a steady luminance equal to that of the masking flash. The similarities found between monocular and dichoptic masking curves in Fig. 23-9 strongly suggest that both share a common mechanism, but certain important differences show up when the relative amounts of masking within and between the various curves are calculated.

The ratio of peak threshold to end-point

threshold increases with increasing masking flash size in monocular masking but decreases in dichoptic masking. This difference can probably be attributed to the direct action of the retinal on effect in monocular masking. If a test flash is superimposed on the masking flash in the same eye, retinal responses to the two flashes interact. In dichoptic masking, retinal interaction would seem to be ruled out, but the shape of the dichoptic masking curve indicates that the form of the on response at the retina stimulated by the masking flash is preserved to some extent, though much attenuated, at higher levels.

The ratio of dichoptic to monocular peak thresholds varies differently from the corresponding ratio of end-point thresholds as masking field size changes. Whereas increasing the field size from 40 minutes to 4 degrees, 40 minutes greatly diminishes the dichoptic peak threshold luminance relative to the monocular peak (it is reduced from 17% to 1% of the monocular peak in Fig. 23-9), the change of field size from 1 degree, 20 minutes to 4 degrees, 40 minutes has almost no effect on the ratio of end-point thresholds (dichoptic end-point threshold is about 5% of the monocular end-point threshold, essentially the ratio observed with steady backgrounds equal to the masking flash luminance). Hence it appears that cortical responses related to onsets of masking flashes have variable proportions with respect to such responses at the retina, depending on field size, whereas the comparative retinal and cortical activity levels resulting from later parts of the masking flash are independent of field size for a given luminance. This conclusion, of course, requires the assumption that test thresholds are directly proportional to activity levels.

Enlargement of the masking flash separates the cortical representation of its border from that of the test flash. The accompanying falloff of dichoptic peak threshold suggests that near the onset of the masking flash, borders must interact at the cortex (or lateral geniculate) if they do not do so at the retina to produce significant masking. From the small effect of field size on the ratio of dichoptic to monocular thresholds measured later in the masking flash, we may infer that border interaction becomes less important for masking as the transient response to the onset of the masking flash dies down. These observations emphasize once again the contrasting properties of static and dynamic visual responses. Fiorentini and her associates[83] present related data that imply a subcortical (lateral geniculate) site for static interactions, but a central site for (dichoptic) dynamic interactions.

Background flicker enhances the dynamic aspects of the response, with concomitant effects on masking. Thresholds for test flashes superimposed on a rapidly flickering background change according to the relative timing of the test and background flashes and exceed thresholds measured against a steady background having the same time-average luminous energy.[32] The importance of rate of change for masking has been demonstrated with low-frequency sinusoidally fluctuating backgrounds, which raise thresholds most for test flashes superimposed during rapid luminance increase, not during the peak of luminance.[229] Steady light added to the sinusoidal variation introduces secondary threshold maxima during the downswing of luminance, possibly by accentuating off effects within the visual pathway.

A theoretical analysis by Sperling[236] also distinguishes onset effects from later effects of the masking flash. Sperling examined the interaction of very brief photopic "impulse" masking and test flashes (Fig. 23-7), (Hallett[107] has performed similar experiments using rod vision.) To a first approximation, masking flashes of equal energy produce equal changes of threshold energy, regardless of the backgrounds on which they are superimposed. This rule can be generalized to the peaks of masking produced by prolonged masking flashes; the peak threshold, if measured with an impulse test flash, depends almost entirely on the luminous energy of the first 60 milliseconds of the masking flash and is the same as if this energy were concentrated in an impulse of light. This finding has a twofold implication: first, perfect temporal energy summation takes place within the first 60 milliseconds of the masking flash; second, the rest of the flash can be thought of as a background light having no influence on the initial threshold peak. Both these statements, of course, are approximations that hold only within limited conditions.

The duration of the test flash is important for the validity of these generalizations. This is one implication of an attempt by Boynton and Kandel[31] to isolate the threshold changes caused by masking at the onset of a flash (neural factors) from the effect of a preadapting light used to desensitize the eye, supposedly by photochemical depletion. These authors measured the time course of response to a masking flash with a 40 millisecond test flash (1000 times as long as that of Sperling) and also assessed the loss of sensitivity caused by preadaptation by presenting the test flash without a masking flash. The preadapting light was turned off briefly during each test, but continuous backgrounds gave compatible results. As increasing levels of preadaptation reduced sensitivity, the peak response to the masking flashes dropped, as predicted, below the value obtained with masking flashes on a dark background and remained lower until the adapting light became intense enough to raise the threshold independently. This change is inconsistent with Sperling's expectation for equal flashes. Comparison of masked thresholds after light adaptation with absolute thresholds raised by adaptation alone allowed a calculation of what Boynton and Kandel took to represent the physiologic on response to the masking flashes under the different adapting conditions. Battersby and Wagman[24] used a similar scheme to compare the effects of neural and photochemical mechanisms of masking at a peripheral retinal location. Their 5-millisecond test flash gave results that Sperling[236] showed to be in fair agreement with his prediction.

Functional considerations

Different authors have answered the question, "What happens to the test stimulus in backward masking?" in more or less elaborate ways, depending on their theoretical biases and the complexity of their data. Differential latencies and transmission times, temporal overlap of excitation, preemption, and inhibition are sensory processes that have been proposed to account for backward masking. These processes may be conceived of as precursors or components of higher nervous functions, perceptual processes, by which a state of excitation is interpreted as a seen object. Because some time is always needed for a perceptual process to develop, it has been argued that a masking flash can disrupt it in a way that prevents its completion and keeps the test flash from being perceived.[160,264] Alternatively, the test flash can be registered in a previsual short-term memory, but the process of readout that would make it visible is interrupted by the masking flash.[12]

One of the simplest functional mechanisms of masking that has been proposed is one of apparent contrast reduction; energies of two flashes separated by a limited interval are wholly or partially additive, making the larger of the two act as a background for the other, regardless of their order of appearance.[78] The result is a reduction of effective contrast, or apparent relative brightness, between the two, thus the smaller flash, normally the test flash, need not be dimmed as much to attain its threshold of effective contrast as would be necessary to bring it to absolute threshold in the absence of a masking flash. Contrast reaches a minimum, which necessitates the least reduction of test energy to bring it to threshold, when the onsets of the test and masking flashes coincide. In its simplest form the contrast-reduction hypothesis ignores physiologic on effects and latency differences prior to interaction. It can account in part for both "backward" and "forward" masking of either light or dark test areas but does not always predict the detailed shape of the masking curve. (For example, the curves in Fig. 23-8 would be symmetric according to this hypothesis.)

The contrast-reduction hypothesis sometimes falls short because contour formation as well as energy summation influences apparent contrast.[134] Subjects asked to match the contrast of a standard target, a square on a white background, by adjusting the duration of a briefly exposed similar target between white preexposure and postexposure fields required longer exposures to match specified low contrast levels than were predicted by a simple contrast-reduction hypothesis based on energy summation. The discrepancy arose when exposure time was insufficient to allow the target to develop a perceptually distinct boundary, as was demonstrated by adding a thin (5 minutes of arc) black border to the white preexposure and postexposure fields in spatial coincidence with the edge of the target square. When this contour information was available, the predicted durations of exposure produced a contrast match. If contour formation must precede or accompany contrast perception, its impairment by masking or short exposure should increase the energy difference between masking and test flashes that corresponds to threshold contrast.

The formation time of a contour also affects the threshold in its vicinity when the contour is a part of the masking stimulus rather than the test stimulus. A relevant experiment is that of Novak and Sperling,[187] who measured the threshold of small (13 minutes of arc), 10 millisecond test flashes viewed against a continuous white background, and located 7 degrees, 33 minutes horizontally in the nasal field of the right eye. A black bar 53 minutes across appeared at various distances from the test flash, either continuously or for 10, 50, or 500 milliseconds. When the edge of the bar was adjacent to the test spot, thresholds exceeded values measured on the white background alone if the bar was presented continuously or for 500 milliseconds, but not for shorter exposures. A significant aspect of the results is that these thresholds were raised whether the test flashes appeared on the white background or on the black bar itself, provided they were close to its border. Ikeda and Boynton[130] have observed an analogous temporal effect, attributed to visual off and on responses; test flash thresholds rose at the beginning and at the end of a flash of darkness that interrupted an otherwise continuous light. These findings are additional examples of visual change specificity; a spatial or temporal discontinuity of luminance determines threshold more powerfully than luminance as such.

In relation to this point, it is interesting that light and contours presented dichoptically interact very little, although light can mask a contour if both are within the same retinal area of one eye. The difference implies a functional separation of responses to light and contours in the visual pathways, which is again consistent with the dichotomy of continuity and discontinuity that seems typical of vision. On the other hand, spatially continuous contours within masking and test flashes can interact both monocularly and dichoptically, even if they do not overlap. Such interference, like that of adjacent, suitably timed, uniformly luminous fields, can be called metacontrast. An example of metacontrast involving contours is the erasure phenomenon of Averbach and Coriell,[12] which reduces legibility of a letter if an adjacent bar or surrounding circle appears shortly after the letter. This effect increases if the letter in question is not isolated but is included in a matrix or array of letters. A related interaction among parts of acuity targets such as Snellen charts is not unknown to ophthalmologists, some of whose patients, particularly amblyopic or retinally impaired ones, can read isolated letters of the chart even though these letters are illegible in the context of the whole chart. Similar effects have been reported for more or less crowded matrices of E's[184] and for Landolt C's; discrimination of the critical detail decreases to a minimum and subsequently increases as a square of enclosing bars is brought closer to the C. The critical distance of the bars from the C for minimum discriminability is the same whether the entire display is presented to one eye,[85] or the C to one eye and the square to a corresponding retinal location of the other.[86] Interactions of this sort for dichoptic viewing probably have a cortical locus, and the similarity of monocular to dichoptic results may imply a cortical basis for both.

Metacontrast interactions occur when two nonoverlapping contours or lights are presented successively within a limited range of spatial and temporal separations. Similar combinations or timing and spacing may also produce apparent movement, as if one contour or light crossed the space between the two and became the other. This kind of apparent movement, called phi movement, has been studied extensively, and its laws have been worked out by Korte.[146] If lights are used to produce the effect, the brighter they are, or the longer the interval between them, the more distant they must be from each other in order to give optimal-phi (also called beta) movement.

The knowledge that a continuous displacement of the retinal image is not required for sensations of movement has become the foundation of the television and motion picture industry. Successive displacements of stationary details of the motion picture, frame by frame, give rise to perceptual continuity, much as do the successive samples of the physical world that come to us during fixations. A significant aspect of the motion picture is that despite the apparent continuity of movement within the image structure the temporal discontinuity of the light that creates the structure on the screen is still evident as flicker. The visual system in a sense abstracts temporospatial discontinuity from purely temporal discontinuity and processes the two differently. Keesey[138] presents related data.

Flicker does not imply a lack of interaction between successive flashes, but the number of the flashes and the rate at which they arrive at the eye determine the extent to which they interact. As an example of the interaction among three flashes, Robinson[214] has shown that the masking of one flash by another can be counteracted by a supplementary flash presented thereafter, a process he calls disinhibition. Such effects have been attributed to interactions between retinal sustained and transient pathways.[33,34] Rather complex interactions occur among successive flashes whenever the interval between them is short.

Such interactions are often studied with flashes presented sequentially throughout an extended interval. When the flashes are about 1 second or more apart, their visibility, brightness, and other characteristics appear similar to those of isolated flashes having the same luminous energy if shorter than a critical duration, or the same luminance if longer. Increased frequency of recurrence brings about interactions of responses to the successive flashes that can be detected subjectively, as well as electrophysiologically at various levels of the retinocortical pathway. The intermittent light, which appears simply as a discrete succession of flashes at low repetition rates, begins to flicker more and more rapidly as frequency continues to rise, until eventually the flicker is too rapid to be seen and the light seems to be steady. Subjective color effects[186,188] and spatial patterns[208] are often seen in conjunction with these changes. More complex stimulus sequences have suggested the possibility of a temporal "code" for subjective colors.[80] Supporting this view are data concerning the color-specific temporal patterns of cat optic tract responses to colored flashes.[147] There is also an increase of brightness as the repetition rate of flashes approaches a value near 10/sec, and a decline of brightness at higher frequencies. Since maximal brightness enhancement occurs at flash frequencies similar to that of the electroencephalographic alpha waves, Bartley[19] suggested that synchronization of responses to the flashes with the central nervous activity recorded as alpha waves might underlie the increase of brightness. Despite the possibility of other interpretations,[20,144,201] such synchronization is known to occur, as flashes in the frequency range around 10 flashes/sec tend to entrain the alpha rhythm with their own rate of presentation.[2] This so-called photic driving can induce seizures in some patients and has been used to detect latent epilepsy,[52] which should be remembered by those who intend to study flicker responses clinically or experimentally.

The transition, at some frequency beyond the point of maximal brightness enhancement, from a flickering appearance to one of steadiness defines the flicker-fusion threshold. Here and at higher frequencies, brightness conforms to Talbot's law[151]; if subjectively fused intermittent light and objectively steady light of the same color are equally bright, they have exactly the same average luminous energy per unit time. This relation is one of the most precisely valid laws of visual science and demonstrates the complete integration by the visual system of the stimulus energy in each cycle of the fused intermittent light. The so-called Talbot brightness of the flash train equals that of a steady light created by spreading the luminous energy of each flash uniformly throughout the interval from its onset to the onset of the next flash. The resulting luminance would equal the flash luminance times the fraction of each light-dark cycle occupied by light. This fraction is variously called the on-time fraction, the pulse-to-cycle fraction, the light-time fraction, or the duty cycle. We shall adopt the term duty cycle in subsequent discussion.

Critical flicker frequency

The changeover from the appearance of flicker to fusion is not abrupt but takes place over a transitional range of frequencies. Within this range a frequency may be found by experimental methods that may be taken as the boundary between the two modes of appearance. This frequency is the critical flicker fusion (CFF) frequency. On a simple level we may interpret the CFF as a measure of the temporal resolving power of the visual system under the particular conditions of stimulation. Alternatively, since each flash gives rise to a response that must decay detectably before the next response is separately discernible, we may think of flicker fusion as a measure of differential brightness sensitivity. In a later section of this chapter we shall see that these alternative interpretations are closely related.

Determinants of CFF

Many variables of the stimulus and of the observer interact to determine the value of the CFF[152,195] Some of them are listed in the following sections. Brown[40] presents a more comprehensive treatment.

Test field

Luminance. As a rule, CFF increases with the luminance of the intermittent stimulus. Ferry[79] first attempted to determine the relation of these variables by means of a black-and-white sector disc, which rotated at various speeds under a quite limited range of light intensities. Porter,[198] unaware of Ferry's work, attacked a similar problem with better equipment and observed that CFF rose in proportion to the logarithm of flash luminance within a fairly wide range of luminances. The relation, applicable to flicker rates above roughly 10 flashes/sec, has come to be known as the Ferry-Porter law. Fig. 23-10 displays data that verify this law for a foveal test field over a 10,000 to 1 range of retinal illuminance and show its inapplicability at very high levels.[116] The turnover at high luminances may be due to prolonged cone late receptor potentials, as suggested by Whitten and Brown.[273] A practical consequence of the rise of CFF with luminance is the possibility of relief for patients who complain that

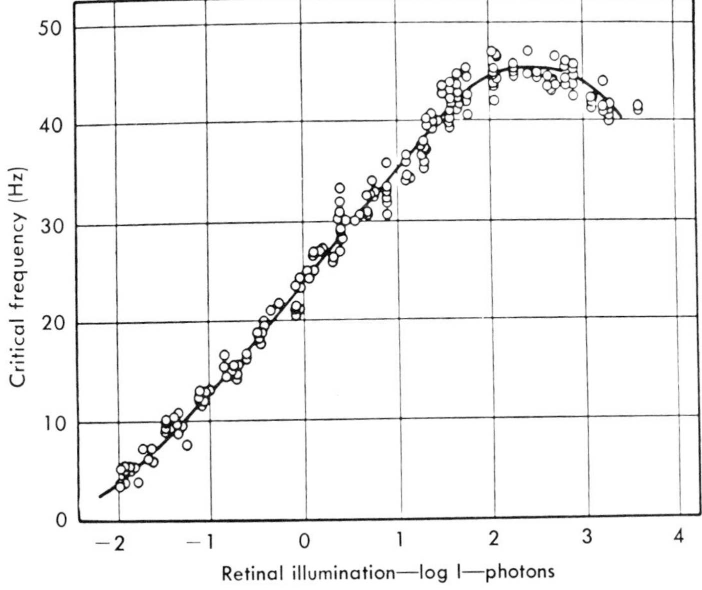

Fig. 23-10. Critical flicker frequency at rod-free fovea as influenced by retinal illuminance (photons = trolands) of test field, showing conformity to the Ferry-Porter law over range of four logarithmic units. (From Hecht, S., and Verrijp, C. D.: J. Gen. Physiol. **17:**251, 1933.)

lights around them flicker. A pair of sunglasses may be all that is needed.

Spectral composition. Different wavelengths of light do not stimulate the rods and cones of the retina with equal efficiency if their radiant energies are equated. For this reason, equal-energy spectral lights have different luminances, which according to the Ferry-Porter law should result in different CFF values. If the energy of the different wavelengths is adjusted to match them for brightness under photopic conditions, their respective CFF's should be alike for any particular brightness and should follow the Ferry-Porter logarithmic function as brightness increases. Hecht and Shlaer[112] tested this prediction with seven wavelengths and found it generally correct for their 19-degree flickering field, as may be seen in Fig. 23-11. Except for small differences of maximal CFF at high brightness, the upper (cone) branch of the flicker function is the same for all seven wavelengths. (Similar experiments by Pokorny and Smith[197] have shown that the slope of the photopic CFF–log *I* curve increases in deuteranopes and decreases in protanopes relative to the normal function.) The lower (rod) branch is variable with wavelength because photopically equal brightnesses are not equal scotopically. Wavelengths for which rod sensitivity greatly exceeds cone sensitivity determine CFF over longer ranges of retinal illuminance at the low end of the scale than wavelengths for which rod and cone sensitivities are more nearly the same. This explains the broomlike shape of the graph in Fig. 23-11. Most bipartite visual functions of this sort (such as dark-adaptation curves) have as their basis the duplex composition of the retina.

Retinal position. Since the CFF function for rods differs from that for cones, the CFF for a test stimulus confined to a limited area of the retina depends on the relative numbers of rods and cones stimulated in the area. Population densities of rods and cones vary from one part of the retina to another,[190] and the extent to which each population is activated by light is a function of retinal illuminance. If the test light is confined to the fovea, it activates only cones, and the CFF function rises according to the Ferry-Porter law as luminance changes throughout a wide range. If the same test light is presented to an extrafoveal area, it stimulates rods

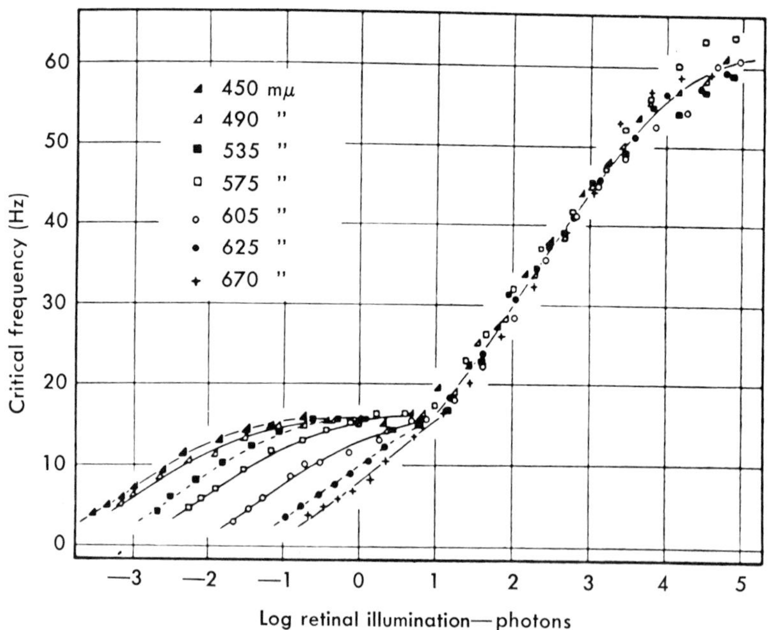

Fig. 23-11. Critical flicker frequency of 19-degree test field as related to retinal illuminance (photons = trolands), for monochromatic lights of different wavelengths having equal photopic brightness. (From Hecht, S., and Shlaer, S.: J. Gen. Physiol. **19**:965, 1936.)

at low luminances and cones at higher luminances, giving a bipartite CFF function. Such functions, obtained by Hecht and Verrijp,[114] are plotted in Fig. 23-12. The lower maximum CFF values found with increasing retinal eccentricity result from the depletion of cone populations, leaving mainly rods to determine CFF. Below the retinal illuminance denoted by log $I = -1$, rods alone limit the CFF for the 2-degree test field (within a 10-degree surround) used. The range of applicability of the Ferry-Porter law also shrinks with reduction of the available cone population and is greatest in the fovea.

Size. Granit and Harper[101] using test flashes that subtended 0.92 degree to 5 degrees, found that CFF increased linearly with the logarithm of stimulus area. This relation, the Granit-Harper law, held for stimuli within a 1000 to 1 range of luminances, and retinal eccentricities as much as 10 degrees. Later authors who tested the law under wider ranges of test field size, luminance, and retinal location generally confirmed its validity. The law holds within the fovea,[87] and although it is applicable to field sizes up to nearly 50 degrees,[216] the total area of the field does not determine CFF. Roehrig[217] measured the same value of CFF for a complete 49.6-degree field and for an annular field of the same diameter with 66% of its central area blanked out. For flicker, as for masking, the borders of the field have special importance.

CFF depends on the proportions of rods and cones within the stimulated area at photopic luminances, as Hecht and Smith[113] showed with field sizes from 0.3 to 19 degrees. When rods and cones are stimulated, the CFF-versus-area curve has two branches, of which only the cone branch conforms to the Granit-Harper law.

As Fig. 23-12 shows, the CFF for small targets is greatest in the fovea at any given photopic level of retinal illuminance. However, large targets in the periphery may have higher values of CFF than those closer to or surrounding the fovea. This difference may explain why a television set seems to flicker less when viewed directly than it does when the viewer looks off to the side. Different relative sensitivities and summative ability of the rod and cone subsystems, and also receptive field density, may be responsible for this interaction of field size and retinal eccentricity. The dominance of transient receptive fields in the peripheral retina may be a factor.[91] In addition, a flash of a given size occupies a large number of small receptive fields in the central visual field but a smaller number of larger receptive fields in the periphery. A train of such flashes, therefore, may have a greater chance to evoke synchronous

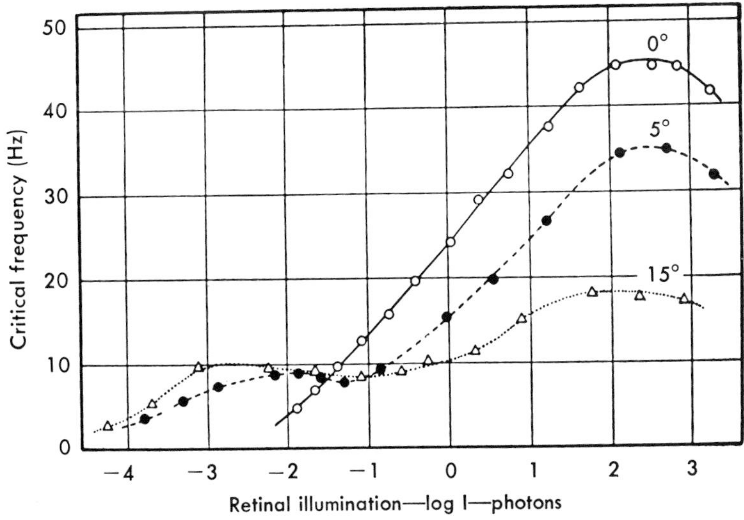

Fig. 23-12. Critical flicker frequency of a 2-degree white test field as related to retinal illuminance (photons = trolands), measured at fovea, 5-degrees above fovea, and 15-degrees above fovea. (From Hecht, S., and Verrijp, C. D.: J. Gen. Physiol. **17**:251, 1933.)

activity in the limited number of peripheral receptive fields than in the large number of central ones, most of which have "sustained" characteristics.

Temporal waveform. As a rule, flicker studies have been performed with lights that were turned on and off abruptly at regular intervals. Other kinds of luminance variation in time were tried, but the results could not be simply interpreted until DeLange systematized matters.[59-61] DeLange's work, which revolutionized the study of flicker, is described later in this chapter.

Duty cycle. The proportion of light to darkness during a flicker cycle has a complex influence on CFF that varies with stimulus luminance and area. According to Piéron,[195] CFF varies in proportion to the logarithm of the *dark* fraction of the cycle, provided that flash trains of different duty cycles are held at the same time-average luminance (Talbot level). This restriction makes it necessary to change flash luminance to compensate for changes of duty cycle. Constant flash luminance yields variations with duty cycle that are difficult to interpret without a unifying principle. Bartley and Nelson[22,23] have discussed the dependence of CFF on duty cycle at length, and Kelly[141] has provided an interpretation based on DeLange's analysis.

Duration of flashes. If flash duration is held constant, duty cycle increases with flicker frequency. As a consequence the average retinal illuminance from flashes of constant luminance rises, and with it, the ocular adaptation level. Relation of CFF to adaptation of the eye is discussed below.

Number of flashes in the train. An initially dark-adapted eye becomes partially light adapted during the first several flashes of a flickering stimulus. In addition, the response of the eye becomes increasingly well synchronized with the flashes as more of them arrive at the retina.[19,21] These changes bring about an increase of CFF with the overall duration of the flash train[9,100] and with the total number of flashes presented, up to approximately 10.[171] The cone late receptor potential may be involved in the increase of CFF under these conditions, since it is prolonged by high-level stimulation only until a state of light adaptation is achieved.[273]

Monocular versus binocular presentation. We have seen that masking occurs if test and masking flashes are presented separately to the two eyes, with the implication that interactions take place beyond the chiasm. The neural pathways have been thought not to mix at or ahead of the lateral geniculate body,[63] but weak interactions at that level have recently been described.[222]

Similarly, two flash trains interact if the flashes are dichoptically presented. Sherrington[228] was the first to observe that synchronized (in phase) flash trains at the two eyes fused at a higher frequency than did alternating (out of phase) flash trains, which stimulated each eye during the dark intervals at the other. The effect was quite small under these conditions; however, Sherrington found significant binocular interactions. Perrin[193] and Lipkin[163,164] demonstrated that steady light presented to one eye reduces CFF measured for the other. Furthermore, Perrin showed that intermittent light having a frequency above CFF at one eye reduced CFF for the other eye more than did a steady light of equal brightness. This effect implies a cortical, rather than retinal, limit on CFF, since some fluctuating excitation must get to the cortex from the "fused" eye to account for the difference.

Figuration (shape, striation, pattern). The shape of a test field determines the ratio of its circumference to its area, so that manipulation of shape allows investigation of border effects than can be varied while area remains constant. Some workers, particularly Crozier and Wolf,[57] have added stripes or other contours within the area of the test field and have found that CFF for a high-luminance striated field with relatively large duty cycle (75% or higher) increased more rapidly with luminance than that of an unpatterned field. This result was not strictly due to border interactions, as Crozier and Wolf[58] proved; their method of interrupting the light in the earlier experiment caused movement of shadows over the retina, but border interactions were not ruled out entirely. Other experiments[99] gave evidence for such interactions; CFF for a striated test field was generally lower than that of a uniform test field of equal overall size; the CFF decreased to a minimum and then increased again as stripe width was varied; the minimal CFF was below that of an unpatterned field equated in area to the lighted part of the striped field. These effects may reasonably be attributed to border interactions.

Background

Luminance. If the test field is surrounded by darkness, the CFF depends not only on the luminance and other characteristics of that field but also on stray light, scattered within the ocular media or reflected from the illuminated area onto other parts of the retina. An observer may respond to flicker in the large halo of stray light even when the test field appears steady. In addition, dissimilar states of retinal adaptation within and beyond the test area can affect the CFF. To control the situation, the test field is often presented within a lighted surround, which can be used to manipulate the adaptation level, to mask stray light, and to vary the contrast at the border of the test field. The effects of changing surround luminance are somewhat complex and depend on test field size as well as on surround size.[28] In general, the highest CFF is obtained with a surround matched to the Talbot brightness of the test area, a condition that minimizes border contrast.

Size. CFF varies with the area of the surround; for example, Foley[87] showed that for a small foveal test field, CFF increased in proportion to the logarithm of this area for surround diameters up to 4 degrees. One effect of a large surround is to light-adapt a large part of the retina, thus reducing possible stray light effects. It also tends to make the test situation less stressful for the observer than it is when only a small test field is visible.[112,114]

Observer: visual properties

Adaptive state. Adaptation determines CFF largely because it governs the relative sensitivity of rod and cone mechanisms. Also, light adaptation reduces the size and changes the organization of retinal receptive fields in a way that enhances the likelihood of inhibitory interactions.[15] In general, the higher the level of light adaptation, the higher the CFF for a given test stimulus. If a train of flashes is presented to an initially dark-adapted eye, light adaptation develops gradually. As the results of masking experiments show, even one flash radically changes visual sensitivity. It can therefore be expected that the CFF will change rapidly during the early part of the flash train, and more slowly thereafter. This effect has been investigated with abbreviated trains of flashes.

Adaptation can depend on the method by which test flashes are generated. Two types of apparatus have come into general use for production of repeated flash trains; these are the episcotister, or rotating sector disc, and the gas discharge tube, or stroboscope. The electrically driven Sylvania R1131C or R1130B glow-modulator tube may be used to produce complex flicker sequences conveniently. Television-type (cathode-ray) displays allow manipulation of both temporal and spatial patterning of the test field. At one time, flicker was studied by means of discs on which alternate black and white sectors were printed, but the modern version of the rotating sector disc has open and closed sectors, usually of constant proportions, that interrupt light at a focal point within the optical system used to stimulate the eye. The relative angular sizes of the sectors control the duty cycle of the flash train, which is therefore independent of the speed of the disc. The result of a constant duty cycle is a stable Talbot brightness above CFF and therefore a stable ocular adaptation level proportional to the time-average luminance (or luminous energy) of the flash train. A stroboscope, on the other hand, produces very short flashes whose durations do not change appreciably as their repetition rates increase, so that higher flash rates raise the time-average energy at the retina. Hence, whereas tests with an episcotister are made under essentially constant adaptation, tests with a stroboscope involve an adaptation level proportional to the rate of flash presentation and can therefore influence the recorded level of CFF. It is necessary to know which technique is used to interpret clinical literature on flicker testing.

It is of some interest, also, that preexposure to a flickering light can change the CFF by so-called local adaptation, not to be confused with Troxler's phenomenon (p. 678).[7,192,260] This process can introduce variability into measured CFF values. High preexposure frequencies tend to increase CFF, and low ones decrease CFF.

Pupil size. Light evokes a pupillary constriction, which lowers the retinal illuminance and may reduce CFF. To prevent interaction between pupil size and CFF an artificial pupil is often placed before the observer's eye, or a mydriatic or cycloplegic can be administered.

Pathology. Many attempts have been made to use flicker and the CFF for clinical purposes. Flicker electroretinograms can give information about rod and cone function, and flicker-evoked cortical potentials testify to the integrity of the retinocortical pathway. However, the neurophysiology of flicker as inferred from evoked responses recorded from scalp electrodes is complicated, and the relation of these responses to psychophysical aspects of flicker is not clear.[259]

A variety of perimetry depending on detection of flicker at various parts of the visual field has been developed, which locates contours of equal CFF analogous to the isopters of sensitivity determined with steady perimetric targets.[129] It turns out that CFF is depressed near the blind spot, but is enhanced in the temporal retina as if to compensate for the deficiency in the nasal retina of the other eye.[279] According to Wolf and Vincent,[280] a target of 1 to 2 degrees yields optimal results in flicker perimetry.

Pathologic conditions of all kinds have been examined by the use of flicker. The literature is too extensive to review here. For a sampling of recent work in the field related to visual function, see Henkes and Van der Tweel.[117]

Observer: nonvisual properties

Age. The consensus of many investigations is that CFF declines with age. Changes of the ocular media and macular pigment, as well as retinal and central nervous mechanisms, may be responsible for the decline. Domey[67] has published a statistical analysis which shows the average change of CFF to be from 44.8 Hz at age 13 to 19, to 33.1 Hz at age 80 to 89, when the duty cycle is 2% and the surround, 17 degrees in diameter, has a slightly higher luminance (23.8 mL) than the foveal test target. Duty cycle affects the correlation found between age and CFF quite significantly.

Fatigue, hunger, drugs, body state, general health. Reduced body efficiency and depletion of vitality tend, in general, to lower CFF, as do certain drugs such as alcohol. Physostigmine applied topically diminishes CFF, although other autonomic drugs (pilocarpine, homatropine, epinephrine) do not. Hypoxia and inhalation of carbon dioxide lower CFF, whereas hyperventilation increases CFF.

Personality. Many workers have described correlations between observers' attitudes or personality traits such as anxiety, rigidity, neurosis, or perseverative tendency and the measured value of CFF. Do inherent personality-related variations of sensory capacity exist that can be revealed by CFF, or do individuals differ simply as to how much near-threshold change they must see to respond positively, for example, to the question, "Do you see flicker"? Clark has demonstrated that, in fact, response bias can affect the value of CFF.[44] On the other hand, an indirect effect of personality on sensory function as such is also possible, since autonomic activity, which can be related to personality and attitude, controls pupil size and responsiveness and therefore may have a bearing on CFF.[119] More direct relations of CFF to sensory variables determined purely by personality have not been conclusively demonstrated.

Brain damage. CFF (measured in cats; behavioral discrimination) depends on interactions within extensive portions of the brain and can be altered by lesions within numerous structures, some of which are not included in the visual pathway as such.[227] These structures appear to be organized as two systems, concerned respectively with high-frequency and low-frequency flicker discrimination. Lesions of the high-frequency system lower CFF, whereas lesions within the low-frequency system raise CFF. On the basis of these findings, Schwartz and Chaney[227] suggest that the low-frequency system, which is located in the tectal region, interacts with the high-frequency system to determine CFF and also may inhibit that system to limit the discrimination of high flicker rates. The high-frequency system includes the visual cortex and the lateral geniculate body as well as nearby thalamic nuclei. Because the CFF depends on the integrity of both systems, it seems possible, in principle, to use CFF as an aid to detection of brain lesions or damage, but this technique obviously cannot pinpoint the location or nature of the defect. From the standpoint of diagnostic utility, determining the best combination of the many stimulus variables that control CFF may not be an easy task.

Clinicians may find it beneficial to follow a suggestion by DeLange[60] that flicker sensitivity of patients be tested with sinusoidally modulated light (p. 701) at about 10 Hz. DeLange reported relatively large variability of results at this frequency, which is close to the alpha rhythm and often facilitates brightness enhancement, and concluded that neural function or integrity may be assessed with greater sensitivity than with the usual high-frequency tests of CFF.

Method and instructions. Conventional measurement of CFF requires that the frequency of stimulation be varied, either upward through a transition from obvious flicker to fusion, or downward through a complementary transition from a steady appearance to just detectable flicker. These transitions rarely take place at the same frequency, and it is customary to use the mean of the two limiting values as the CFF. Unfortunately, these values are not absolute; what the observer is told to do, and how he is told to do it, can change them. He may himself control the flicker source. If instructed to increase the frequency until the flicker seems "just about to disappear" or "just at the threshold of disappearance," he is likely to stop at a lower value than he would with instructions that used the words, "just disappears." Depending on his tolerance for residual flicker, the observer may interpret such instructions liberally or conservatively; even when the instructions say "just disappears," he may turn the control knob until he is certain that no flicker remains, or else until he is sure that if he turns it any more no flicker will remain. The same interpretative problems can arise, whether the equipment, the experimenter, or the observer controls the flicker frequency. Hence it is necessary to instruct observers in a way that defines the permissible tolerance. As Clark[44] points out, instructions that increase the observer's willingness to tolerate residual effects can remove the differences of CFF that have been thought to distinguish the sensory capacities of different personality types and of mentally disturbed and normal subjects.

Modern interpretations of flicker fusion

Although there have been literally thousands of studies[95,151] related to flicker and fusion dating from 1740, when Segner first attempted to measure the "persistence of vision" following a flash of light,[151] only in recent years have orderly rules been devised that predict the value of CFF from properties of the flickering stimulus such as flash luminance and the way this luminance varies as a function of time. The work of DeLange,[59,60] based on research of Ives done 30 years before, placed the problem of CFF in the context of mathematical waveform analysis and linear filter theory.[159,235] A short explanation is in order.

The way any quantity varies as a function of time can be called its temporal waveform. It is possible to express every recurrent temporal waveform as a sum of elementary sinusoidal oscillations (sine waves, cosine waves, or both). One of these components, called the fundamental, is equal in frequency to the repetition rate of the original waveform. All the other components, called harmonics, are integral multiples of the fundamental frequency. Any periodic nonsinusoidal waveform, being reducible to a particular mixture of these elementary sinusoids, is a complex waveform. The property of a sinusoid that expresses how much of it is used is its amplitude, which is the maximum amount by which it deviates above or below its average value. Since mathematically speaking the fundamental and all harmonics have average values of zero, their sum, the complex waveform, will also have this average value unless a steady, nonzero component, called the DC component, is included in the mixture. The DC component determines the average value of the complex waveform. This provision is particularly relevant for light flashes, since light cannot have negative values, and hence a train of flashes or other recurrent waveforms of luminance must have an average value greater than zero (which accounts for its Talbot brightness). The general method by which an arbitrary recurrent waveform can be analyzed into its sinusoidal and DC components is Fourier analysis, and the components are often called Fourier components.[251]

Summation is a linear process, which is to say it can be performed by a linear system. What is a linear system? For our purposes we may think of a system as a signal-processing device, which admits an input signal at one end and delivers an output signal at the other. The hallmark of linearity is operation according to the principle of superposition, which states that for any combination of inputs the output is the same as it would be if the inputs were applied individually and the resulting outputs summed. When the input is a sinusoid, the output of a linear system is a sinusoid of the same frequency. When the input is a complex waveform, equivalent to a sum of sinusoids having particular frequencies, the output is likewise a sum of sinusoids having those frequencies.

Two characteristics of the system may cause the output waveform to differ from a complex input. First, system gain (defined as the ratio of output amplitude to input amplitude) may not be the same for sinusoids at all frequencies. Second, the system may introduce a time difference between the input and the output that may also be frequency dependent. Time shift expressed in units equal to the duration or period of one cycle and converted to angular measure (one period equals 2π radians or 360 degrees) is referred to as phase shift. A linear system with frequency-dependent gain and phase shift can be called a linear filter and can be described completely by plots of its gain and phase shift as measured with sinusoidal inputs throughout a wide range of frequencies. Gain plotted against frequency, both in logarithmic coordinates, defines what is called the attenuation characteristic of the filter. Determination of an attenuation characteristic for the visual system was the first part of DeLange's attack on the problem of CFF. However, DeLange faced a peculiar circumstance, which was the inaccessibility of a measurable output signal, and, in addition, he knew that the visual system does not necessarily behave linearly.

In general, the output waveform of a linear filter need not resemble the input, except that it must when the input is a sinusoid. A system is nonlinear if, given a sinusoidal input, it produces either a nonsinusoidal output or a sinusoid of different frequency. However, the system may act linearly enough with low-amplitude inputs to allow its analysis to proceed as if it were actually linear. The assumption of small-signal linearity is an important tool for the analyst and is essential to DeLange's approach to CFF.

DeLange's argument runs as follows. We know that the visual system is not linear, but it may be possible to treat it as a combination of linear and nonlinear parts. The validity of Talbot's law suggests that near the threshold of flicker fusion the system behaves linearly, at least up to the point at which brightness is determined. If the signal delivered to the brightness evaluator were proportional to some nonlinear transformation of the input—say, its logarithm—a steady light would produce one brightness, and a fluctuating light with equal time-average luminance, another brightness. DeLange supposed a logarithmic transformation, but the principle can be more easily illustrated by assuming a cube-root transformation, that is, that brightness is proportional to the cube root of luminance[241] and that the transformation takes place before the brightness of flashes whose rate exceeds CFF is averaged. If the luminance of each flash were 8 units and the duty cycle were $1/8$ (one eighth of each cycle occupied by light), then its physical average luminance would be $8/8 = 1$ unit. But the cube-root transformation reduces flash brightness to $8^{1/3} = 2$ units, and the brightness averaged over the cycle is $2/8$ or $1/4$ units. On the other hand, a steady light of 1 unit of luminance, the same physical average value as the flashing light, has a brightness of $1^{1/3}$, or 1 unit. With the values of this example, then, flashes having the same time-average luminance as a steady light would produce a Talbot brightness only $1/4$ that of a steady light. Since the Talbot brightness and the steady light brightness are, in fact, found experimentally to be equal, our assumption of a nonlinear transformation prior to the averaging process is invalid. Hence averaging takes place before any nonlinear transformation, and before brightness is evaluated.

Linearity may be assumed for the visual system only when inputs are small, as they can be said to be when they produce threshold outputs at all frequencies. However, a threshold implies a nonlinear detector, since its output is one of only two states, "yes" or "no," regardless of its input. On this basis DeLange reasoned that the visual system, operating near the threshold of flicker-fusion, could be divided conceptually into a small-signal–linear filter followed by a nonlinear detector. If the threshold sensitivity of the detector did not vary with frequency, all inputs to the visual system having the proper amplitude would be reduced to the same threshold value by the filter. Therefore measurement of the amplitudes of sinusoidal luminance variation that just reached the flicker threshold at different frequencies would allow the gain of the filter to be estimated (on a relative scale), and a plot of gain versus frequency in logarithmic coordinates would give the attenuation characteristic sought by DeLange.

Gain, in this context, is a measure of sensitivity to the sinusoidal variation of luminance. The more sensitive the visual system, the smaller the luminance variation needed to arouse a threshold sensation of flicker. Since gain, or sensitivity, and threshold amplitude are reciprocally related, the attenuation characteristic is plotted with threshold amplitude increasing downward, and therefore flicker sensitivity upward, along the logarithmic ordinate. Fig. 23-13 shows several attenuation characteristics obtained under different conditions.[139] To understand them we must distinguish two ways to express threshold amplitude.

Variation of light about its average or DC luminance level is called modulation, and thus DeLange's attenuation characteristics were obtained with sinusoidally modulated light. Although the amplitude of sinusoidally modulated light—its maximum luminance variation either above or below the DC level— cannot be larger than the DC value without necessitating physically unrealizable negative values of luminance, it may take on any smaller value, which may be related to the DC level by a factor called the modulation ratio. The value m of this ratio in percentage is defined as 100 times the ratio, amplitude of sinusoidal change/DC or average value. Since both parts of this fraction are in the same units, which may be luminance, retinal illuminance, or other photometric quantities as conditions require, the modulation ratio is a dimensionless, pure number. It expresses threshold amplitude as a percentage deviation from the average stimulus luminance. The alternative of course is to express threshold amplitude directly in absolute luminance or retinal-illuminance units.

We may now turn to the curves in Fig. 23-13, which show how each of these two types of threshold specification is useful for the interpretation of sinusoidal flicker. The left half of the figure shows how the threshold modulation ratio varies as a function of frequency. Each curve was obtained by adjusting the amplitude of sinusoidal modulation to threshold at various frequencies, with the average retinal illuminance held constant, then dividing each threshold amplitude by the average value. Since the average retinal illuminance controls the adaptation level of the eye, each of the curves represents the visual attenuation characteristic under a constant condition of adaptation, and constitutes the flicker-fusion boundary for that condition. Appropriately adapted, an observer sees any combination of frequency and modulation percentage below a given curve as flicker, and any combination lying above the curve as steady light. Increasing the adaptation level, then, enlarges the "flicker" area, which is consistent with the higher values of CFF predicted for higher luminances by the Ferry-Porter law. However, the attenuation characteristic far transcends the Ferry-Porter law in generality.[139] Kelly[142] has shown, in fact, that the Ferry-Porter law describes physically unrealizable performance.

It may be seen that different attenuation characteristics in the left half of Fig. 23-13 are well separated at high frequencies but come together at low frequencies. This difference indicates that low-frequency sinusoidal flicker reaches its threshold at a nearly constant value of m regardless of adaptation level in the photopic range, whereas high-frequency flicker involves threshold values of m that change both with frequency and with adaptation level. The right half of Fig. 23-13 shows that, in fact, m is the wrong measure for high frequencies; flicker at every frequency in the high range

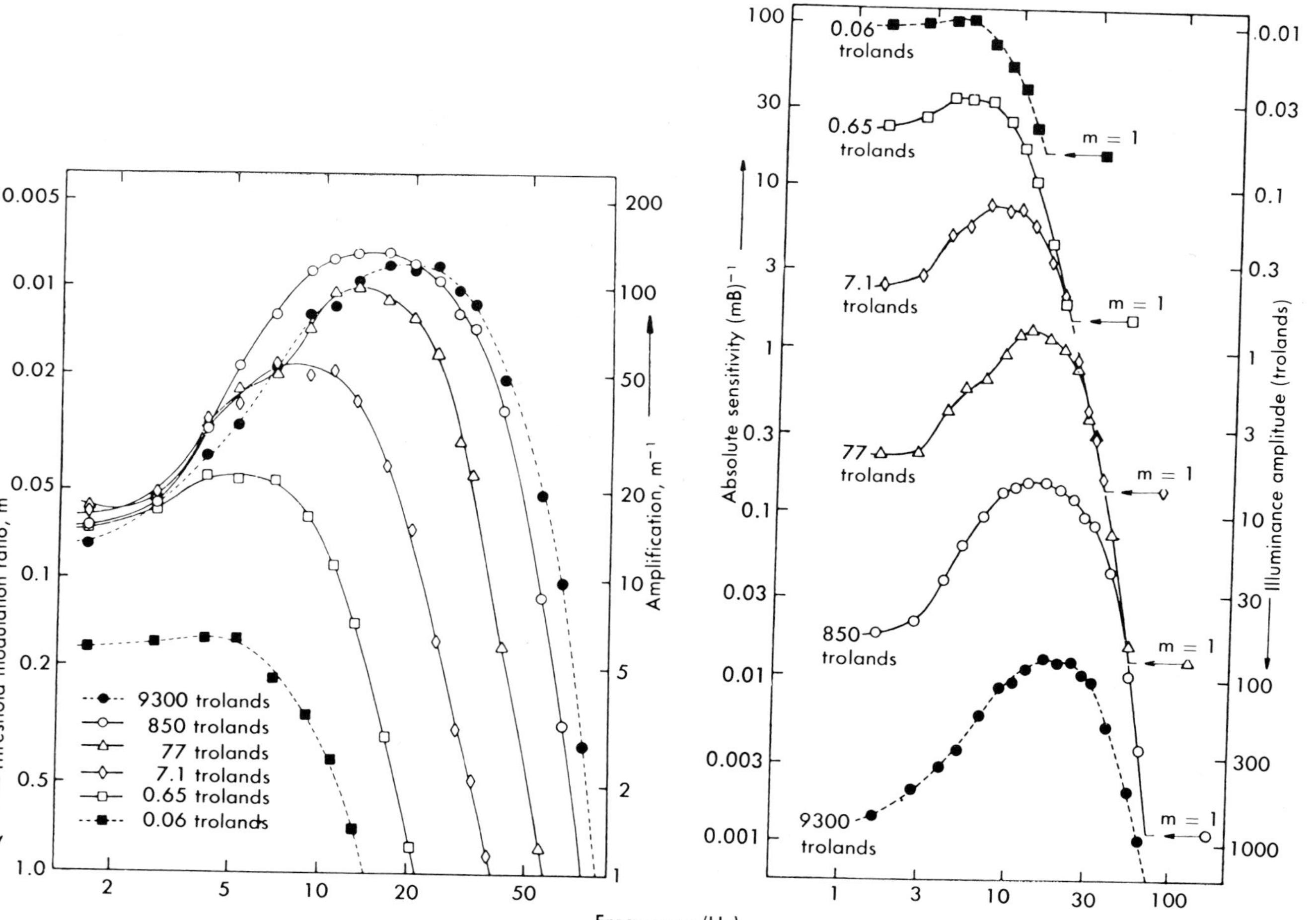

Fig. 23-13. Flicker sensitivity. Visual attenuation characteristics measured with sinusoidal variation about different average values of retinal illuminance. Left panel shows modulation ratios, *m*, increasing downward, that correspond to just detectable flicker at each of frequencies tested. Right panel shows absolute amplitudes, increasing downward, of retinal illuminance variation required for just noticeable flicker at various frequencies. Absolute sensitivity, increasing upward, is reciprocal of threshold absolute amplitude at each frequency. (Data of Kelly.[139])

has a unique threshold amplitude. The continuity of the frequency-amplitude relation (Kelly[141] calls it a "master curve") throughout this range reveals its essential independence of adaptation level, which is a necessary condition for linear response.

In the low-frequency range, since any of several input amplitudes (m times adaptation level) may be just detectable as flicker at a given frequency, the response is not linear. The DeLange filter-detector model assumes that all these amplitudes are reduced to a constant value of excitation at the detector. Filter gain (AC response) therefore depends on adaptation level (DC bias) at low frequencies, and this condition is inconsistent with linearity.

A simple two-stage theoretical model that fits the experimental data quite precisely has been devised by Kelly.[143] The first stage, accounting for linear high-frequency behavior of photopic flicker thresholds, involves diffusion or a similar process at or near the receptor level. The second stage, accounting for nonlinear low-frequency behavior, involves lateral inhibition in the neural network of the retina. A fixed threshold amplitude detector assesses the output of these stages, as in DeLange's model.

How does the attenuation characteristic predict the CFF for nonsinusoidal waveforms? In the high-frequency range where the system acts linearly, we can say that each of the sinusoidal components of such a waveform is attenuated independently of the others. As frequency increases, the gain of the filter in this range falls very rapidly (Fig. 23-13), so that the harmonic components of the complex input, which usually have smaller amplitudes than the fundamental, anyway, suffer greater attenuation than the fundamental and become insignificant as the input variation is reduced to its threshold amplitude. For this reason the flicker threshold for the complex waveform becomes the threshold of the fundamental component alone. Fourier analysis includes a calculation of the amplitude of this component, which, if located along the "master curve" in Fig. 23-13, allows the CFF to be read directly from the frequency scale. Alternatively, the retinal illuminance waveform may be integrated over one cycle to find its average value, the calculated amplitude of the fundamental divided by this value to get m, the value of m located on the appropriate curve in the left half of Fig. 23-13 as determined by the average retinal illuminance, and the CFF read from the frequency scale directly below.

The second method, used by DeLange, relies on the equivalence (with respect to CFF) of all complex waveforms that have equal fundamental amplitudes and equal average values. The first method is an improvement by Kelly,[141] who showed that, in fact, equal average values are not required in the frequency region encompassed by the "master curve." Within this region any waveform is equivalent to any other with the same fundamental amplitude.

Flicker sensitivity in relation to color, field size, and other parameters

Throughout the foregoing discussion we have assumed a constant size and spectral composition of a test field centered on the fovea. However, the shape of the attenuation characteristic varies systematically with color[60,61,140] and with chromatic adaptation. Green[103] has examined the attenuation characteristics of color mechanisms as isolated by the technique of Stiles. His data show that the blue (π_1) system has much lower modulation sensitivity than green (π_4) or red (π_5) systems, in keeping with its low CFF, low acuity, and other anomalous response characteristics. Low-frequency behavior is highly dependent on field size. Kelly's[139] attenuation characteristics (Fig. 23-13) were all obtained with a white 68-degree test field that was uniformly illuminated at any moment throughout a central area of 50 degrees, beyond which the luminance dropped smoothly to zero. In the frequency range under 10 to 20 Hz, they lie well below DeLange's which were obtained with a white 2-degree test field at the center of a 60-degree surround matched to its average luminance. A comparison of DeLange's high-frequency data with Kelly's, for similar levels of ocular adaptation, confirms the expectation from the Granit-Harper law that increased field size raises the CFF for otherwise equivalent test fields. This is the same as increasing flicker sensitivity at a given frequency. However, the attenuation characteristics show the opposite effect at the low-frequency end. Kelly has suggested by way of explanation that large test areas enhance inhibitory interactions among different retinal regions. It is conceivable that the lack of a centrally viewed border bears on this question, since uniform fluctuation over the entire extent of a large field minimizes differences of excitation that might otherwise exist in adjacent parts of the retina. If this is so, the flicker sensitivity becomes a relatively pure function of time (or frequency) without the possibility of differentiation based on spatial gradients or contours. On the other hand, configuration of a large field does not change the situation much; Thomas showed that low-frequency sensitivity measured in a room in which all lights were sinusoidally modulated was quite comparable with the value obtained with Kelly's unfigured large field, even though the observer was reading a book in presumably well-structured surroundings typical of a normal visual environment.[141] The visual response to wide-area flicker seems to be more or less independent of steady environmental information on which the flicker may be superimposed. Such independence would be consistent with a separation of static from dynamic response functions in the visual system, as it also implied by the binocular interactions of masking discussed earlier.

Flicker sensitivity in relation to properties of retina and visual pathways

As frequency increases, attenuation characteristics obtained with white light rise to a so-called pseudoresonance peak,[60] whose height and center frequency increase with the adaptation level. The peak is located within the 5 to 20 Hz range, in which brightness enhancement occurs, suggesting that the two effects may have related mechanisms.

When wavelength, rather than luminance, is the modulated variable, the pseudoresonance peak is absent. This finding[206] is consistent with other evidence for separate color and luminosity channels from retina to cortex. As an example, the visually evoked cortical potential contains contour-related, color-related, and brightness-related components that arrive over parallel pathways.[204] When a retinal image is stabilized, these three aspects of the target fade out separately in the order named.[94]

There is evidence that the mechanism of pseudoresonance is retinal at least in part. Fig. 23-14 displays data Cleland and Enroth-Cugell[46] obtained by recording from single ganglion cells of a cat's retina whose receptive fields were stimulated by small spots of sinusoidally modulated light. In *A* the maximum impulse frequency of the ganglion cell discharge was plotted (on a relative scale) as a function of stimulating frequency for a constant modulation ratio of 50% (sinusoidal fluctuation between 0.5 and 1.5 times the average retinal illuminance of 5×10^4 td). It is seen that a peak appears near 10 Hz for both on-center and off-center ganglion cells. *B* shows that the mean number of impulses per second recorded from the same two ganglion cells also attains a maximum when stimulus frequency is about 10 Hz, although the discharge of the on-center cell is more rapid than that of the off-center cell under the stimulating conditions used. *C* contains an attenuation characteristic for the on-center cell of *A* and *B*, with a peak at about 6 Hz. Possibly, the position of this peak depends as much on the experimenter's ability to hear fluctuations of impulse frequency synchronous with different frequencies of stimulus modulation as it does on the responsiveness of the cell itself, since the discharge of the cell was monitored by ear to determine its threshold. Nevertheless it is clear that the response of cat retinal ganglion cells parallels certain features of the human attenuation characteristic. Additional evidence that suggests a relation of flicker sensitivity to retinal properties is that of Ratliff and associates,[203] who recorded the frequency of impulses from the compound eye of the horseshoe crab *(Limulus)* and found, with average level and modulation amplitude held constant, a variation with stimulus frequency very similar to that depicted in Fig. 23-14, *A*. These authors examined the effect of changing field size. Increase of the diameter of the stimulating spot from 0.25 mm (just larger than the facet of a single ommatidium of the compound eye) to 1.5 mm (spanning 20 ommatidia) increased the maximum discharge frequency recorded and also raised from 2.5 to 3.5 Hz the frequency of stimulation that elicited this maximum response. Concomitantly, the increase of field size reduced the number of im-

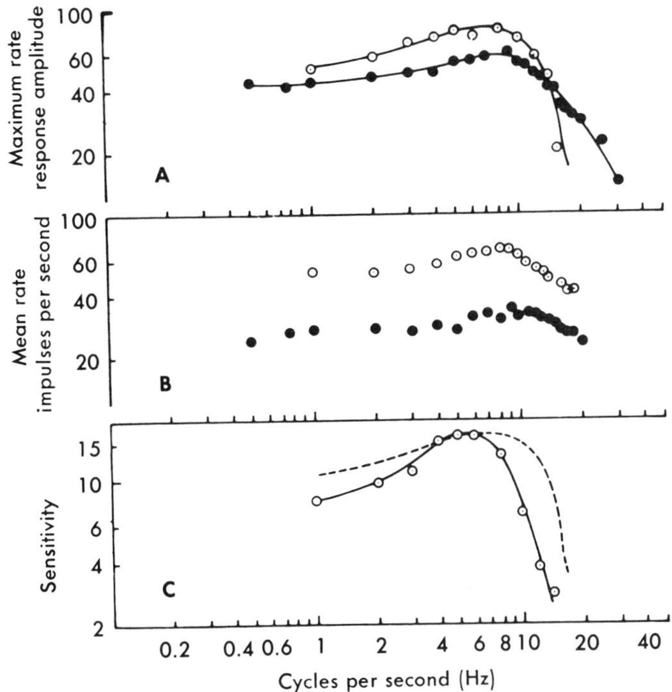

Fig. 23-14. Flicker responses of retinal ganglion cells of cat. Open circles indicate on-center cell; filled circles indicate off-center cell. **A,** Maximum rate of cell discharge for different frequencies of sinusoidally modulated light having constant modulation ratio of 0.5. **B,** Mean rate of cell discharge (conditions same as in **A**). **C,** Sensitivity function (attenuation characteristic). Sensitivity is reciprocal of modulation ratio, *m,* needed to produce just noticeable fluctuation of ganglion-cell discharge rate at each frequency of sinusoidal stimulation tested. Dashed line is maximum-response function of on-center cell in **A.** (Data of Cleland and Enroth-Cugell.[46])

pulses per second elicited by lower stimulus frequencies, much as flicker sensitivity is reduced in the low-frequency range of the attenuation characteristic. Both these effects in the eye of *Limulus* must result from lateral inhibition, which is the only significant form of interaction among ommatidia in this eye. Although interactions may be more complex in human vision, it is likely that a similar inhibitory mechanism underlies the changes of human flicker sensitivity related to field size.

Using cynomolgus monkeys, Baron and Boynton[17] tested this hypothesis directly. They eliminated postreceptor potentials by infusing the eye with sodium aspartate, then recorded the local electroretinogram from foveal cones during stimulation by sinusoidally modulated light, and found it to be flat up to 20 Hz at high adaptation levels. Hence the depression of low-frequency flicker sensitivity observed either psychophysically or at the ganglion cell level appears to result from postreceptor inhibitory interactions and is not a property of the cones as such.

REFERENCES

1. Adler, F. H., and Fliegelman, M.: Influence of fixation on the visual acuity, Arch. Ophthalmol. **12:**475, 1934.
2. Adrian, E. D., and Matthews, B. H. C.: The Berger rhythm: potential changes from the occipital lobes in man, Brain **57:**355, 1934.
3. Aguilar, M., and Stiles, W. S.: Saturation of the rod mechanism at high levels of stimulation, Opt. Acta **1:**59, 1954.
4. Alpern, M.: Metacontrast, J. Opt. Soc. Am. **43:**648, 1953.
5. Alpern, M., and Rushton, W. A. H.: The nature of rise in threshold produced by contrast flashes, J. Physiol. (Lond) **189:**519, 1967.
6. Alpern, M., and Sugiyama, S.: Photic driving of the critical flicker frequency, J. Opt. Soc. Am. **51:**1379, 1961.
7. Alpern, M., Maaseidvaag, F., and Ohba, N.: The kinetics of cone visual pigments in man, Vision Res. **11:**539, 1971.

8. Altman, J., and Carpenter, M. B.: Fiber projections of the superior colliculus in the cat, J. Comp. Neurol. **116:**157, 1961.
9. Anderson, D. A., Huntington, J., and Simonson, E.: Critical fusion frequency as a function of exposure time, J. Opt. Soc. Am. **56:**1607, 1966.
10. Arend, L. A., Jr.: Spatial factors in the Broca-Sulzer phenomenon, J. Opt. Soc. Am. **63:**879, 1973.
11. Averbach, E.: The span of apprehension as a function of exposure duration, J. Verb. Learn. Behav. **2:**60, 1963.
12. Averbach, E., and Coriell, A. S.: Short-term memory in vision, Bell System Tech. J. **40:**309, 1961.
13. Baker, H. D.: Initial stages of dark and light adaptation, J. Opt. Soc. Am. **53:**98, 1963.
14. Bárány, R.: The clinical aspects and theory of train nystagmus, Arch. Augenheilkd. **88:**139, 1921.
15. Barlow, H. B., Fitzhugh, R., and Kuffler, S. W.: Change of organization in the receptive fields of the cat's retina during dark adaptation, J. Physiol. (Lond) **137:**338, 1957.
16. Barlow, H. B., et al.: The effects of remote retinal stimulation on the responses of cat retinal ganglion cells, J. Physiol. (Lond) **269:**177, 1977.
17. Baron, W. S., and Boynton, R. M.: Temporal-response characteristics of primate cones, abstract, J. Opt. Soc. Am. **63:**1311, 1973.
18. Baron, W. S., and Westheimer, G.: Visual acuity as a function of exposure duration, J. Opt. Soc. Am. **63:** 212, 1973.
19. Bartley, S. H.: Visual sensation and its dependence upon the neurophysiology of the optic pathway, Biol. Symp. **8:**87, 1942.
20. Bartley, S. H.: Intermittent photic stimulation at marginal intensity levels, J. Psychol. **32:**217, 1951.
21. Bartley, S. H.: Central mechanisms of vision. In Magoun, H. W., editor: Handbook of physiology, Neurophysiology, Washington, D.C., 1959, American Physiological Society, vol. 1.
22. Bartley, S. H., and Nelson, T. M.: Some relations between pulse-to-cycle fraction and critical flicker frequency, Percept. Mot. Skills **10:**3, 1960.
23. Bartley, S. H., and Nelson, T. M.: A further study of pulse-to-cycle fraction and critical flicker frequency: a decisive theoretical test, J. Opt. Soc. Am. **51:** 41, 1961.
24. Battersby, W. S., and Wagman, I. H.: Neural limitations of visual excitability. I, The time course of monocular light adaptation, J. Opt. Soc. Am. **49:**752, 1959.
25. Battersby, W. S., and Wagman, I. H.: Neural limitations of visual excitability. IV, Spatial determinants of retrochiasmal interaction, Am. J. Physiol. **203:**359, 1962.
26. Beeler, G. W., Jr.: Visual threshold changes resulting from spontaneous saccadic eye movements, Vision Res. **7:**769, 1967.
27. Békésy, G. von: Funneling in the nervous system and its role in loudness and sensation intensity on the skin, J. Acoust. Soc. Am. **30:**399, 1958.
28. Berger, C.: Illumination of surrounding field and flicker fusion frequency with foveal images of different sizes, Acta Physiol. Scand. **30:**161, 1954.
29. Boycott, B. B., and Dowling, J. E.: The primate retina: light microscopy, Philos. Trans. R. Soc. (Biol.) **255:**109, 1969.
30. Boynton, R. M.: Some temporal factors in vision. In Rosenblith, W. A., editor: Sensory communication, Cambridge, Mass., 1961, The M.I.T. Press.
31. Boynton, R. M., and Kandel, G.: On responses in the human visual system as a function of adaptation level, J. Opt. Soc. Am. **47:**275, 1957.
32. Boynton, R. M., Sturr, J. F., and Ikeda, M.: Study of flicker by increment threshold technique, J. Opt. Soc. Am. **51:**196, 1961.
33. Breitmeyer, B. G.: Disinhibition in metacontrast masking of vernier acuity targets: sustained channels inhibit transient channels. Vision Res. **18:**1401, 1978.
34. Breitmeyer, B. G.: Metacontrast with black and white stimuli: evidence for inhibition of on- and off-sustained activity by either on- or off-transient activity. Vision Res. **18:**1443, 1978.
35. Breitmeyer, B. G., and Ganz, L.: Implications of sustained and transient channels for theories of visual pattern masking, saccadic suppression, and information processing. Psychol. Rev. **83:**1, 1976.
36. Bridgeman, B.: Visual receptive fields sensitive to absolute and relative motion during tracking, Science **178:**1106, 1972.
37. Brindley, G. S.: The discrimination of after-images, J. Physiol. (Lond) **147:**194, 1959.
38. Broca, A., and Sulzer, D.: La sensation lumineuse en fonction du temps, J. Physiol. Path. Gen. **4:**632, 1902.
39. Brown, D. R., et al.: Human receptive field characteristics: probe analysis of stabilized images, Vision Res. **13:**231, 1973.
40. Brown, J. L.: Flicker and intermittent stimulation. In Graham, C. H., editor: Vision and visual perception, New York, 1965, John Wiley & Sons, Inc.
41. Brown, K. T., and Murakami, M.: Delayed decay of the late receptor potential of monkey cones as a function of stimulus intensity, Vision Res. **7:**179, 1967.
42. Brown, K. T., Watanabe, K., and Murakami, M.: The early and late receptor potentials of monkey cones and rods, Symp. Quant. Biol. **30:**457, 1965.
43. Caldwell, J. H., and Daw, N. W.: New properties of rabbit retinal ganglion cells, J. Physiol. (Lond) **276:** 257, 1978.
44. Clark, W. C.: The psyche in psychophysics: a sensory decision theory analysis of the effect of instructions on flicker sensitivity and response bias, Psychol. Bull. **65:**358, 1966.
45. Clarke, F. J. J., and Belcher, S. J.: On the localization of Troxler's effect in the visual pathway, Vision Res. **2:**53, 1962.
46. Cleland, B. G., and Enroth-Cugell, C.: Cat retinal ganglion cell responses to changing light intensities: sinusoidal modulation in the time domain, Acta Physiol. Scand. **68:**365, 1966.

47. Cleland, B. G., and Enroth-Cugell, C.: Quantitative aspects of gain and latency in the cat retina, J. Physiol. (Lond) **206:**73, 1970.
48. Cleland, B. G., and Levick, W. R.: Brisk and sluggish concentrically organized ganglion cells in the cat's retina, J. Physiol. (Lond) **240:**421, 1974.
49. Cleland, B. G., and Levick, W. R.: Properties of rarely encountered types of ganglion cells in the cat's retina and an overall classification, J. Physiol. (Lond) **240:**457, 1974.
50. Cleland, B. G., Dubin, M. W., and Levick, W. R.: Sustained and transient neurones in the cat's retina and lateral geniculate nucleus, J. Physiol. (Lond) **217:** 473, 1971.
51. Cleland, B. G., Levick, W. R., and Sanderson, K. J.: Properties of sustained and transient ganglion cells in the cat retina, J. Physiol. (Lond) **228:**649, 1973.
52. Cobb, S.: Photic driving as a cause of clinical seizures in epileptic patients, Arch. Neurol. Psychiatry **58:**70, 1947.
53. Cohen, H. B.: The effect of contralateral visual stimulation on visibility with stabilized retinal images, Can. J. Psychol. **15:**212, 1961.
54. Cone, R. A.: The early receptor potential of the vertebrate eye, Symp. Quant. Biol. **30:**483, 1965.
55. Cornsweet, T. N.: Determination of the stimuli for involuntary drifts and saccadic eye movements, J. Opt. Soc. Am. **46:**987, 1956.
56. Crawford, B. H.: Visual adaptation in relation to brief conditioning stimuli, Proc. R. Soc. Lond. (Biol.) **134:**283, 1947.
57. Crozier, W. J., and Wolf, E.: Theory and measurement of visual mechanisms. X, Modifications of the flicker response contour, and the significance of the avian pecten, J. Gen. Physiol. **27:**287, 1944.
58. Crozier, W. J., and Wolf, E.: Theory and measurement of visual mechanisms. XI, On flicker with subdivided fields, J. Gen. Physiol. **27:**401, 1944.
59. DeLange, H.: Relationship between critical flicker frequency and a set of low-frequency characteristics of the eye, J. Opt. Soc. Am. **44:**380, 1954.
60. DeLange, H.: Research into the dynamic nature of the human fovea-cortex systems with intermittent and modulated light. I, Attenuation characteristics with white and colored light, J. Opt. Soc. Am. **48:**777, 1958.
61. DeLange, H.: Research into the dynamic nature of the human fovea-cortex systems with intermittent and modulated light. II, Phase shift in brightness and delay in color perception, J. Opt. Soc. Am. **48:**784, 1958.
62. Denney, D., Baumgartner, G., and Adorjani, C.: Responses of cortical neurones to stimulation of the visual afferent radiations, Exp. Brain Res. **6:**265, 1968.
63. De Valois, R. L.: Color vision mechanisms in the monkey, J. Gen. Physiol. **43:**115, 1960.
64. Ditchburn, R. W.: A new apparatus for producing a stabilized retinal image, Opt. Acta **10:**325, 1963.
65. Ditchburn, R. W., and Foley-Fisher, J. A.: Assembled data in eye movements, Opt. Acta **14:**113, 1967.
66. Ditchburn, R. W., and Ginsborg, B. L.: Involuntary eye movements during fixation, J. Physiol. (Lond) **119:**1, 1953.
67. Domey, R. G.: Statistical properties of foveal CFF as function of age, light/dark ratio, and surround, J. Opt. Soc. Am. **54:**394, 1964.
68. Dowling, J. E.: Synaptic organization of the frog retina: an electron microscopic analysis comparing the retinas of frogs and primates, Proc. R. Soc. Lond (Biol.) **170:**205, 1968.
69. Dowling, J. E., and Boycott, B. B.: Neural connections of the retina: fine structure of the inner plexiform layer, Symp. Quant. Biol. **30:**393, 1965.
70. Dowling, J. E., and Boycott, B. B.: Organization of the primate retina: electron microscopy, Proc. R. Soc. Lond. (Biol.) **166:**80, 1966.
71. Dowling, J. E., and Werblin, F.: Organization of retina of the mudpuppy, *Necturus maculosus*. I, Synaptic structure, J. Neurophysiol. **32:**315, 1969.
72. Enoch, J. M.: The two-color threshold technique of Stiles and derived component color mechanisms. In Jameson, D., and Hurvich, L. M., editors: Handbook of sensory physiology. Visual psychophysics, Berlin, 1972, Springer-Verlag, vol. 7/4, 537-567.
73. Enoch, J. M.: Quantitative layer-by-layer perimetry, Invest. Ophthalmol. **17:**208, 1978.
74. Enoch, J. M., Berger, R., and Birns, R.: A static perimetric technique believed to test receptive field properties: extension and verification of the analysis, Doc. Ophthalmol. **29:**127,1970.
75. Enoch, J. M., Berger, R., and Birns, R.: A static perimetric technique believed to test receptive field properties: responses near visual field lesions with sharp borders, Doc. Ophthalmol. **29:**154, 1970.
76. Enoch, J. M., and Sunga, R. N.: Development of quantitative perimetric tests, Doc. Ophthalmol. **26:** 215, 1969.
77. Enroth-Cugell, C., and Robson, J. G.: The contrast sensitivity of retinal ganglion cells of the cat, J. Physiol. (Lond) **187:**517, 1966.
78. Eriksen, C. W., and Hoffman, M.: Form recognition at brief durations as a function of adapting field and interval between stimulations, J. Exp. Psychol. **66:**485, 1963.
79. Ferry, E. S.: Persistence of vision, Am. J. Sci. **44:** 192, 1892.
80. Festinger, L., Allyn, M. R., and White, C. W.: The perception of color with achromatic stimulation, Vision Res. **11:**591, 1971.
81. Fiorentini, A., and Bittini, M.: Binocular interaction effect. II, Investigation on the time behavior of monocular sensitivity during contralateral stimulation, Atti Fond. G. Ronchi **17:**286,1962.
82. Fiorentini, A., and Bittini, M.: Time course of monocular sensitivity during contralateral stimulation, Opt. Acta **10:**55, 1963.
83. Fiorentini, A., Bayly, E. J., and Maffei, L.: Peripheral and central contributions to psychophysical spatial interactions, Vision Res. **12:**253, 1972.
84. Fishman, M. C., and Michael, C. R.: Integration of

auditory information in the cat's visual cortex, Vision Res. **13:**1415, 1973.

85. Flom, M. C., Weymouth, F. W., and Kahneman, D.: Visual resolution and contour interaction, J. Opt. Soc. Am. **53:**1026, 1963.
86. Flom, M. C., Heath, G. C., and Takahashi, E.: Contour interaction and visual interaction: contralateral effects. Science **142:**979,1963.
87. Foley, P. J.: Interrelationships of background area, target area and target luminance in their effect on the critical flicker frequency of the human fovea, J. Opt. Soc. Am. **51:**737, 1961.
88. Freund, H.-J.: Neuronal mechanisms of the lateral geniculate body. In Jung, R., editor: Handbook of sensory physiology, Visual centers in the brain, Berlin, 1973, Springer-Verlag, vol. 7/3B, pp. 177-246.
89. Frumkes, T. E., and Sturr, J. F.: Spatial and luminance factors determining visual excitability, J. Opt. Soc. Am. **58:**1657, 1968.
90. Fukada, Y: Receptive field organization of cat optic nerve fibers with special reference to conduction velocity, Vision Res. **11:**209, 1971.
91. Fukada, Y., et al.: Functional significance of conduction velocity in the transfer of flicker information in the optic nerve of the cat, J. Neurophysiol. **29:**698, 1966.
92. Gassel, M. M., and Williams, D.: Visual function in patients with homonymous hemianopia. III, The completion phenomenon, Brain **86:**229, 1963.
93. Gerrits, H. J. M., and Vendrik, A. J. H.: Simultaneous contrast, filling-in process and information processing in man's visual system, Exp. Brain Res. **11:**411, 1970.
94. Gerrits, H. J. M., de Haan, B., and Vendrik, A. J. H.: Experiments with retinal stabilized images; relations between the observations and neural data, Vision Res. **6:**427, 1966.
95. Ginsburg, N.: Flicker fusion bibliography, 1953-1968, Percept. Mot. Skills **30:**427, 1970.
96. Glickstein, M., Stein, J., and King, R. A.: Visual input to the pontine nuclei, Science **178:**1110, 1972.
97. Gouras, P.: The effects of light-adaptation on rod and cone receptive field organization of monkey ganglion cells, J. Physiol. (Lond) **192:**747, 1967.
98. Gouras, P.: Identification of cone mechanisms in monkey ganglion cells, J. Physiol. (Lond) **199:**533, 1968.
99. Graham, C. H., and Landis, C.: Effect of striated fields on critical flicker frequency, J. Opt. Soc. Am. **49:**580, 1959.
100. Granit, R., and Hammond, E. L.: Comparative studies on the peripheral and central retina. V, The sensation-time curve and time course of the fusion frequency on intermittent stimulation, Am. J. Physiol. **98:**654, 1931.
101. Granit, R., and Harper, P.: Comparative studies on the peripheral and central retina. II, Synaptic reactions in the eye, Am. J. Physiol. **95:**211, 1930.
102. Graybiel, A. M.: Some extrageniculate visual pathways in the cat, Invest. Ophthalmol. **11:**322, 1972.
103. Green, D. G.: Sinusoidal flicker characteristics of the color sensitive mechanisms of the eye, Vision Res. **9:** 591, 1969.
104. Grindley, G. C., and Townsend, V.: Binocular masking induced by a moving object, Q. J. Exp. Psychol. **17:**97, 1965.
105. Growney, R. and Weisstein, N.: Spatial characteristics of metacontrast, J. Opt. Soc. Am. **62:**690, 1972.
106. Hallett, P. E.: Rod increment thresholds on steady and flashed backgrounds, J. Physiol. (Lond) **202:** 355, 1969.
107. Hallett, P. E.: Impulse functions for human rod vision, J. Physiol. (Lond) **202:**379, 1969.
108. Hallett, P. E.: Quantum efficiency and false positive rate, J. Physiol. (Lond) **202:**421, 1969.
109. Hammond, P.: Chromatic sensitivity and spatial organization of LGN neurone receptive fields in cat: cone-rod interaction, J. Physiol. (Lond) **225:**391, 1972.
110. Harris, C. S.: Perceptual adaptation to inverted, reversed and displaced vision, Psychol. Rev. **72:**419, 1965.
111. Hebb, D. O.: The semiautonomous process: its nature and nurture, Am. Psychol. **18:**16, 1963.
112. Hecht, S., and Shlaer, S.: Intermittent stimulation by light. V, The relation between intensity and critical frequency for different parts of the spectrum, J. Gen. Physiol. **19:**965, 1936.
113. Hecht, S., and Smith, E. L.: Intermittent stimulation by light. VI, Area and the relation between critical frequency and intensity, J. Gen. Physiol. **19:**979, 1936.
114. Hecht, S., and Verrijp, C. D.: Intermittent stimulation by light. III, The relation between intensity and critical fusion frequency for different retinal locations, J. Gen. Physiol. **17:**251, 1933.
115. Hecht, S., Shlaer, S., and Pirenne, M. H.: Energy, quanta, and vision, J. Gen. Physiol. **25:**819, 1942.
116. Hecht, S., Shlaer, S., and Verrijp, C. D.: Intermittent stimulation by light. II, The measurement of critical fusion frequency for the human eye, J. Gen. Physiol. **17:**237, 1933,
117. Henkes, H. E., and Van der Tweel, L. H.: Flicker, Doc. Ophthalmol. **18:**83, 1964.
118. Hernández-Peón, R.: Reticular mechanisms of sensory control. In Rosenblith, W. A., editor: Sensory communication, Cambridge, Mass., 1961, The M.I.T. Press.
119. Hess, E. H.: Attitude and pupil size, Sci. Am. **212**(4):46, 1965.
120. Hickey, T. L., Winters, R. W., and Pollack, J. G.: Center-surround interactions in two types of on-center retinal ganglion cells in the cat, Vision Res. **13:** 1511, 1973.
121. Hoffmann, K.-P.: Conduction velocity in pathways from retina to superior colliculus in the cat: a correlation with receptive-field properties, J. Neurophysiol. **36:**409, 1973.
122. Hoffmann, K.-P., and Stone, J.: Conduction velocity of afferents to cat visual cortex: a correlation with

cortical receptive field properties, Brain Res. **32:**460, 1971.

123. Hood, J. D.: Observations upon the neurological mechanism of optokinetic nystagmus with especial reference to the contribution of peripheral vision, Acta Otolaryngol. **63:**208, 1967.

124. Hubel, D. H., and Wiesel, T. N.: Receptive fields of optic nerve fibers in the spider monkey, J. Physiol. (Lond) **154:**572, 1960.

125. Hubel, D. H., and Wiesel, T. N.: Receptive fields, binocular interaction and functional architecture in the cat's visual cortex, J. Physiol. (Lond) **160:**106, 1962.

126. Hubel, D. H., and Wiesel, T. N.: Receptive fields and functional architecture in two nonstriate visual areas (18 and 19) of the cat, J. Neurophysiol. **28:**229, 1965.

127. Hubel, D. H., and Wiesel, T. N.: Receptive fields and functional architecture of monkey striate cortex, J. Physiol. (Lond) **195:**215, 1968.

128. Hunter, W. S., and Sigler, M.: The span of visual discrimination as a function of time and intensity of stimulation, J. Exp. Psychol. **26:**160, 1940.

129. Hylkema, B. S.: Examination of the visual field by determining the fusion frequency, Acta Ophthalmol. **20:**181, 1942.

130. Ikeda, M., and Boynton, R. M.: Negative flashes, positive flashes and flicker examined by increment threshold technique, J. Opt. Soc. Am. **55:**560, 1965.

131. Ikeda, H., and Wright, M. J.: Receptive field organization of 'sustained' and 'transient' retinal ganglion cells which subserve different functional roles, J. Physiol. (Lond) **227:**769, 1972.

132. Jeannerod, M., and Chouvet, G.: Saccadic displacements of the retinal image: effects on the visual system in the cat, Vision Res. **13:**161, 1972.

133. Kahneman, D.: Temporal summation in an acuity task at different energy levels—a study of the determinants of summation, Vision Res. **4:**557, 1964.

134. Kahneman, D.: Exposure duration and effective figure-ground contrast, Q. J. Exp. Psychol. **17:**308, 1965.

135. Kahneman, D.: Method, findings and theory in studies of visual masking, Psychol. Bull. **70:**404, 1968.

136. Kahneman, D., and Norman, J.: The time-intensity relation in visual perception as a function of the observer's task, J. Exp. Psychol. **68:**215, 1964.

137. Katz, M. S.: Brief flash brightness, Vision Res. **4:**361, 1964.

138. Keesey, U. T.: Flicker and pattern detection: a comparison of thresholds, J. Opt. Soc. Am. **62:**446, 1972.

139. Kelly, D. H.: Visual responses to time-dependent stimuli. I. Amplitude sensitivity measurements, J. Opt. Soc. Am. **51:**422, 1961.

140. Kelly, D. H.: Visual responses to time-dependent stimuli. IV, Effects of chromatic adaptation, J. Opt. Soc. Am. **52:**940, 1962.

141. Kelly, D. H.: Sine waves and flicker fusion, Doc. Ophthalmol. **18:**16, 1964.

142. Kelly, D. H.: Diffusion model of linear flicker responses, J. Opt. Soc. Am. **59:**1665, 1969.

143. Kelly, D. H.: Theory of flicker and transient responses. I, Uniform fields, J. Opt. Soc. Am. **61:**537, 1971.

144. Kohn, H., and Salisbury, I.: Encephalographic indications and brightness enhancement, Vision Res. **7:**461, 1967.

145. Kolers, P. A.: Intensity and contour effects in visual masking, Vision Res. **2:**277, 1962.

146. Korte, A.: Kinematoskopische Untersuchungen, Z. Psychol. **72:**193, 1915.

147. Kozak, W. M., and Reitboeck, H. J.: Color-specific temporal spike sequences in single optic-tract fibers of the cat, abstract, J. Opt. Soc. Am. **63:**1278, 1973.

148. Krauskopf, J., and Mollon, J. D.: The independence of the temporal integration properties of individual chromatic mechanisms in the human eye, J. Physiol. (Lond) **219:**611, 1971.

149. Krauskopf, J., and Riggs, L. A.: Interocular transfer in the disappearance of stabilized images, Am. J. Psychol. **72:**248, 1959.

150. Kuffler, S. W.: Discharge patterns and functional organization of mammalian retina, J. Neurophysiol. **16:**37, 1953.

151. Landis, C.: An annotated bibliography of flicker fusion phenomena covering the period 1740-1952, Washington, D.C., 1953, Armed Forces, National Research Council.

152. Landis, C.: Determinants of the critical flicker-fusion threshold, Physiol. Rev. **34:**259, 1954.

153. Laties, A. M.: The visual connection: enter nucleus suprachiasmaticus, Invest. Ophthalmol. **12:**237, 1973.

154. Latour, P. L.: Visual thresholds during eye movements, Vision Res. **2:**261, 1962.

155. Latour, P. L.: Cortical control of eye movements, thesis, Soesterberg, The Netherlands, 1966, Institute for Perception RVO-TNO.

156. Levelt, W. J. M.: On binocular rivalry, thesis, Soesterberg, The Netherlands, 1965, Institute for Perception RVO-TNO.

157. Levick, W. R., and Zacks, J. L.: Responses of cat retinal ganglion cells to brief flashes of light, J. Physiol. (Lond) **206:**677, 1970.

158. Levick, W. R., Cleland, B. G., and Dubin, M. W.: Lateral geniculate neurons of cat: retinal inputs and physiology, Invest. Ophthalmol. **11:**302, 1972.

159. Levinson, J. Z.: Flicker fusion phenomena, Science **160:**21, 1968.

160. Lindsley, D. B.: Electrophysiology of the visual system and its relation to perceptual phenomena. In Brazier, M. A. B., editor: Brian and behavior, Washington, D.C., 1961, American Institute of Biological Sciences, vol. 1.

161. Ling, W., and Gay, A. J.: Optokinetic nystagmus: a proposed pathway and its clinical application. In Smith, J. L., editor: Neuro-ophthalmology, vol. 4, St. Louis, 1968, The C. V. Mosby Co.

162. Linksz, A.: Visual acuity in the newborn with notes

on some objective methods to determine visual acuity, Doc. Ophthalmol. **34:**259, 1973.

163. Lipkin, B. S.: Monocular flicker discrimination as a function of the luminance and area of contralateral steady light. I, Luminance, J. Opt. Soc. Am. **52:** 1287, 1962.
164. Lipkin, B. S.: Monocular flicker discrimination as a function of the luminance and area of contralateral steady light. II, Area, J. Opt. Soc. Am. **52:**1296, 1962.
165. Lund, R. D.: Anatomic studies on the superior colliculus, Invest. Ophthalmol. **11:**434, 1972.
166. Luria, S. M.: Effects of continuously and discontinuously moving stimuli on the luminance threshold of a stationary stimulus, J. Opt. Soc. Am. **55:**418, 1965.
167. MacKay, D. M.: Elevation of visual threshold by displacement of retinal image, Nature **225:**90, 1970.
168. MacKay, D. M.: Visual stability, Invest. Ophthalmol. **11:**518, 1972.
169. MacKay, D. M.: Visual stability and voluntary eye movements. In Jung, R., editor: Handbook of sensory physiology. Central processing of visual information: integrative functions and comparative data, Berlin, 1973, Springer-Verlag, vol. 7/3A, pp. 307-331.
170. Maffei, L., and Fiorentini, A.: Retinogeniculate convergence and the analysis of contrast, abstract, Brain Res. **31:**371, 1971.
171. Mahneke, A.: Fusion thresholds of the human eye as measured with two or several light flashes, Acta Ophthalmol. **36:**12, 1958.
172. Mainster, M. A., White, T. J., and Stevens, C. C.: Mathematical analysis of rhodopsin kinetics, Vision Res. **11:**435, 1971.
173. Marrocco, R. T., and Brown, J. B.: Correlation of receptive field properties of monkey LGN cells with the conduction velocity of retinal afferent input, Brain Res. **92:**137, 1975.
174. McIlwain, J. T.: Nonretinal influences on the lateral geniculate nucleus, Invest. Ophthalmol. **11:**311, 1972.
175. McIlwain, J. T., and Buser, P.: Receptive fields of single cells in the cat's superior colliculus, Exp. Brain Res. **5:**314, 1968.
176. McIlwain, J. T., and Fields, H. L.: Interactions of cortical and retinal projections on single neurons of the cat's superior colliculus. J. Neurophysiol. **34:** 763, 1971.
177. McKee, S. P., and Westheimer, G.: Specificity of cone mechanisms in lateral interaction, J. Physiol. (Lond) **206:**117, 1970.
178. Metz, J. W., and Balliet, R. F.: Visual threshold: human scotopic luminosity functions determined with optokinetic nystagmus, Vision Res. **13:**1001, 1973.
179. Michael, C. R.: Integration of retinal and cortical information in the superior colliculus of the ground squirrel, Brain Behav. Evol. **3:**205, 1970.
180. Michael, C. R.: Visual response properties and functional organization of cells in the superior colliculus of the ground squirrel, Vision Res. **3:**(suppl.) 299, 1971.
181. Mitrani, L., Mateeff, S., and Yakimoff, N.: Is saccadic suppression really saccadic? Vision Res. **11:** 1157, 1971.
182. Movshon, J. A.: The velocity tuning of single units in cat striate cortex, J. Physiol. (Lond) **249:**445, 1975.
183. Movshon, J. A., Thompson, I. D., and Tolhurst, D. J.: Spatial and temporal contrast sensitivity of neurones in areas 17 and 18 of the cat's visual cortex, J. Physiol. (Lond) **283:**101, 1978.
184. Müller, P.: Über das Sehen der Amblyopen, Ophthalmologica **121:**143, 1951.
185. Nakagama, K.: Local adaptation in cat LGN cells: evidence for a surround antagonism, Vision Res. **11:** 501, 1971.
186. Nelson, T. M.: Stimulus intermittency and color studies in review, J. Gen. Psychol. **84:**17, 1971.
187. Novak, S., and Sperling, G.: Visual thresholds near a continuously visible or a briefly presented light-dark boundary, Opt. Acta **10:**187, 1963.
188. Nyström, S. H. M.: Effects of photic stimulation on neuronal activity and subjective experience in man, Acta Neurol. Scand. **42:**505, 1966.
189. Orban, G. A., and Callens, M.: Influence of movement parameters on area 18 neurones in the cat, Exp. Brain Res. **30:**125, 1977.
190. Østerberg, G. A.: Topography of the layer of rods and cones in the human retina, Acta Ophthalmol. suppl. VI, 1935.
191. Pak, W. L.: Rapid photoresponses in the retina and their relevance to vision research, Photochem. Photobiol. **8:**495, 1968.
192. Pantle, A.: Flicker adaptation. I, Effect on visual sensitivity to temporal fluctuations of light intensity, Vision Res. **11:**943, 1971.
193. Perrin, F. H.: A study in binocular flicker, J. Opt. Soc. Am. **44:**60, 1954.
194. Pick, H. L., Jr., and Hay, J. C.: Gaze-contingent adaptation to prismatic spectacles, Am. J. Psychol. **79:**443, 1966.
195. Piéron, H.: Vision in intermittent light, Hurvich, L. M., and Jameson, D., translators. In Neff, W. D., editor: Contributions to sensory physiology, New York, 1965, Academic Press, Inc., vol. 1.
196. Pirenne, M. H., and Marriott, F. H. C.: Visual functions in man. In Davson, H., editor: The eye, New York, 1962, Academic Press, Inc., vol. 2.
197. Pokorny, J., and Smith, V. C.: Luminosity and CFF in deuteranopes and protanopes, J. Opt. Soc. Am. **62:** 111, 1972.
198. Porter, T. C.: Contributions to the study of flicker, Proc. R. Soc. A. **70:**313, 1902.
199. Pritchard, R. M.: A collimator stabilizing system, Q. J. Exp. Psychol. **13:**181, 1961.
200. Pritchard, R. M., Heron, W., and Hebb, D. O.: Visual perception approached by the method of stabilized images, Can. J. Psychol. **14:**67, 1960.
201. Rabelo, C., and Grüsser, O.-J.: Die Abhängigkeit der subjektiven Helligkeit intermittierenden Lichtreize von der Flimmerfrequenz. (Brücke-Effekt, "brightness enhancement"): Untersuchungen bei verschie-

denen Leuchtdichte and Feldgrösse, Psychol. Forsch. **26:**299, 1961.
202. Ratliff, F., Hartline, H. K., and Miller, W. H.: Spatial and temporal aspects of retinal inhibitory interaction, J. Opt. Soc. Am. **53:**110, 1963.
203. Ratliff, F., et al.: Enhancement of flicker by lateral inhibition, Science **158:**392, 1967.
204. Regan, D.: An evoked potential correlate of colour: evoked potential findings and single-cell speculations, Vision Res. **13:**1933, 1973.
205. Regan, D., and Tyler, C. W.: Temporal summation and its limit for wavelength changes: an analog of Bloch's law for color vision, J. Opt. Soc. Am. **61:** 1414, 1971.
206. Regan, D., and Tyler, C. W.: Some dynamic features of color vision, Vision Res. **11:**1307, 1971.
207. Reinecke, R. D.: Review of optokinetic nystagmus from 1954-1960, Arch. Ophthalmol. **65:**609, 1961.
208. Remole, A.: Subjective patterns in a flickering field: binocular vs. monocular observation, J. Opt. Soc. Am. **63:**745, 1973.
209. Richards, W.: Visual suppression during passive eye movement, J. Opt. Soc. Am. **58:**1159, 1968.
210. Riggs, L. A., and Whittle, P.: Human occipital and retinal potentials evoked by subjectively faded visual stimuli, Vision Res. **7:**441, 1967.
211. Riggs, L. A., Armington, J. C., and Ratliff, F.: Motions of the retinal image during fixation, J. Opt. Soc. Am. **44:**315, 1954.
212. Riggs, L. A., et al.: The disappearance of steadily fixated visual test objects, J. Opt. Soc. Am. **43:** 495, 1953.
213. Robinson, D. A.: Eye movements evoked by collicular stimulation in the alert monkey, Vision Res. **12:** 1795, 1972.
214. Robinson, D. N.: Disinhibition of visually masked stimuli, Science **154:**157, 1966.
215. Rodieck, R. W., and Stone, J.: Analysis of receptive fields of cat retinal ganglion cells, J. Neurophysiol. **28:**833, 1965.
216. Roehrig, W. C.: The influence of area on the critical flicker-fusion threshold, J. Psychol. **47:**317, 1959.
217. Roehrig, W. C.: The influence of the portion of the retina stimulated on the critical flicker fusion threshold, J. Psychol. **48:**57, 1959.
218. Roufs, J. A. J.: On the relation between the threshold of short flashes, the flicker-fusion frequency and the visual latency, Institute for Perception Research, Eindhoven, Netherlands. Annual Progress Report No. 1, pp. 69-77, 1966.
219. Roufs, J. A. J., and Meulenbrugge, H. J.: The quantitative relation between flash threshold and the flicker fusion boundary for centrally fixated fields, Institute for Perception Research, Eindhoven, Netherlands, Annual Progress Report No. 2, pp. 133-139, 1967.
220. Rowe, M. H., and Stone, J.: Naming of neurones: classification and naming of cat retinal ganglion cells, Brain Behav. Evol. **14:**185, 1977.
221. St-Cyr, G. J.: Signal and noise in the human oculomotor system, Vision Res. **13:**1979, 1973.
222. Sanderson, K. J., Darian-Smith, I., and Bishop, P. O.: Binocular corresponding receptive fields of single units in the cat dorsal lateral geniculate nucleus, Vision Res. **9:**1297, 1969.
223. Schiller, P. H.: Single unit analysis of backward visual masking and metacontrast in the cat lateral geniculate nucleus, Vision Res. **8:**855, 1968.
224. Schiller, P. H.: The role of the monkey superior colliculus in eye movement and vision, Invest. Ophthalmol. **11:**451, 1972.
225. Schiller, P. H., and Smith, M. C.: Detection in metacontrast, J. Exp. Psychol. **71:**32, 1966.
226. Schneider, G. E.: Two visual systems. Science **163:** 895, 1969.
227. Schwartz, A. S., and Chaney, C.: Neural mechanisms involved in the critical flicker frequency of the cat, Brain Res. **1:**369, 1966.
228. Sherrington, C. S.: The integrative action of the nervous system, New Haven, Conn., 1906, Yale University Press.
229. Shickman, G. M.: Visual masking by low frequency sinusoidally modulated light, J. Opt. Soc. Am. **60:** 107, 1970.
230. Singer, W., and Bedworth, N.: Inhibitory interaction between X and Y units in the cat lateral geniculate nucleus, Brain Res. **49:**291, 1973.
231. Sjöstrand, F. S.: Ultrastructure of retinal rod synapses of the guinea pig eye as revealed by three dimensional reconstructions from serial sections, J. Ultrastruct. Res. **2:**122, 1958.
232. Smith, J. L.: Optokinetic nystagmus, Springfield, Ill., 1963, Charles C Thomas, Publisher.
233. Smith, K. U., Putz, V., and Molitor, K.: Eye movement-retina delayed feedback, Science **166:**1542, 1969.
234. Spear, P. D., et al.: Ontogenesis of receptive field characteristics of superior colliculus neurons in the rabbit, Brain Res. **45:**67, 1972.
235. Sperling, G.: Linear theory and the psychophysics of flicker, Doc. Ophthalmol. **18:**3, 1964.
236. Sperling, G.: Temporal and spatial visual masking. I, Masking by impulse flashes, J. Opt. Soc. Am. **55:** 541, 1965.
237. Sprague, J. M.: The superior colliculus and pretectum in visual behavior, Invest. Ophthalmol. **11:**473, 1972.
238. Sprague, J. M., Berlucchi, G., and Rizzolatti, G.: The role of the superior colliculus and pretectum in vision and visually guided behavior. In Jung, R., editor: Handbook of sensory physiology. Visual centers in the brain, Berlin, 1973, Springer-Verlag, vol. 7/3B, pp. 27-101.
239. Stark, L.: Neurological control systems: studies in bioengineering, New York, 1968, Plenum Publishing Corp.
240. Stein, B. E., and Arigbede, M. O.: A parametric study of movement detection properties of neurons in the cat's superior colliculus, Brain Res. **45:**437, 1972.
241. Stevens, S. S.: Duration, luminance and the brightness exponent, Percept. Psychophys. **1:**96, 1966.

242. Stewart, B. R.: Temporal summation during dark adaptation, J. Opt. Soc. Am. **62:**449, 1972.
243. Stewart, D. L., Birt, D., and Towns, L. C.: Visual receptive-field characteristics of superior colliculus neurons after cortical lesions in the rabbit, Vision Res. **13:**1965, 1973.
244. Stiles, W. S.: The directional sensitivity of the retina and the spectral sensitivities of the rods and cones, Proc. R. Soc. Lond. (Biol.) **127:**64, 1939.
245. Stiles, W. S.: Incremental thresholds and the mechanisms of colour vision, Doc. Ophthalmol. **3:**138, 1949.
246. Stiles, W. S.: Color vision: the approach through incremental threshold sensitivity, Proc. Natl. Acad. Sci. U.S.A. **45:**100, 1959.
247. Stone, J.: Morphology and physiology of the geniculocortical synapse in the cat: the question of parallel input to the striate cortex, Invest. Ophthalmol. **11:** 338, 1972.
248. Stone, J., and Freeman, R. B., Jr.: Conduction velocity groups in the cat's optic nerve classified according to their retinal origin, Exp. Brain Res. **13:** 489, 1971.
249. Stone, J., and Hoffmann, K.-P.: Conduction velocity as a parameter in the organization of the afferent relay in the cat's lateral geniculate nucleus, Brain Res. **32:**454, 1971.
250. Stone, J., and Hoffmann, K.-P.: Very slow-conducting ganglion cells in the cat's retina: a major, new functional type? Brain Res. **43:**610, 1972.
251. Stuart, R. D.: An introduction to Fourier analysis, New York, 1961, John Wiley & Sons, Inc.
252. Taylor, E. A.: The spans: perception, apprehension and recognition as related to reading and speed reading, Am. J. Ophthalmol. **44:**501, 1957.
253. Teller, D. Y.: Sensitization by annular surrounds: temporal masking properties, Vision Res. **11:**1325, 1971.
254. Teller, D. Y., et al.: Sensitization by annular surrounds: sensitization and masking, Vision Res. **11:** 1445, 1971.
255. Thomas, E. L., and O'Bierne, H.: Curvature in the saccadic movement, Arch. Ophthalmol **77:**105, 1967.
256. Tsumoto, T., Creutzfeldt, O. D., and Legend, C. R.: Functional organization of the corticofugal system from visual cortex to lateral geniculate nucleus in the cat, Exp. Brain Res. **32:**345, 1978.
257. Uetsuki, T., and Ikeda, M.: Adaptation and critical duration for Stiles π mechanisms, J. Opt. Soc. Am. **61:**821, 1971.
258. Uttal, W. R.: Chromatic and intensive effects in dot-pattern masking: evidence for different time constants in color vision, J. Opt. Soc. Am. **63:**1490, 1973.
259. Van der Tweel, L. H.: Relation between psychophysics and electrophysiology of flicker, Doc. Ophthalmol. **18:**287, 1964.
260. Vega, A., Costiloe, J. P., and Parsons, O. A.: Flicker threshold shifts as a function of frequency of interposed stimulation, the local adaptation phenomenon, J. Exp. Psychol. **77:**609, 1968.
261. Verheijen, F. J.: A simple after-image method demonstrating the involuntary multidirectional eye movements during fixation, Opt. Acta **8:**309, 1961.
262. Volkmann, F. C.: Vision during voluntary saccadic eye movements, J. Opt. Soc. Am. **52:**571, 1962.
263. Volkmann, F. C., Schick, A. M. L., and Riggs, L. A.: Time course of visual inhibition during voluntary saccades, J. Opt. Soc. Am. **58:**562, 1968.
264. von Noorden, G. K., and Burian, H. M.: Perceptual blanking in normal and amblyopic eyes, Arch. Ophthalmol. **64:**817, 1960.
265. Wald, G., and Brown, P. K.: The molar extinction of rhodopsin, J. Gen. Physiol. **37:**189, 1953.
266. Walls, G. L.: The filling-in process, Am. J. Optom. **31:**329, 1954.
267. Werblin, F. S.: Functional organization of a vertebrate retina: sharpening up in space and intensity, Ann. N.Y. Acad. Sci. **193:**75, 1972.
268. Westheimer, G.: Spatial interaction in the human retina during scotopic vision, J. Physiol. (Lond) **181:**881, 1965.
269. Westheimer, G.: Spatial interaction in human cone vision, J. Physiol. (Lond) **190:**139, 1967.
270. Westheimer, G.: Rod-cone independence for sensitizing interaction in the human retina, J. Physiol. (Lond) **206:**109, 1970.
271. Whitten, D. N., and Brown, K. T.: The time courses of late receptor potentials from monkey cones and rods, Vision Res. **13:**107, 1973.
272. Whitten, D. N., and Brown, K. T.: Photopic suppression of monkey's rod receptor potential, apparently by a cone-initiated lateral inhibition, Vision Res. **13:**1629, 1973.
273. Whitten, D. N., and Brown, K. T.: Slowed decay of the monkey's cone receptor potential by intense stimuli, and protection from this effect by light adaptation, Vision Res. **13:**1659, 1973.
274. Whittle, P.: Binocular rivalry and the contrast at contours, Q. J. Exp. Psychol. **17:**217, 1965.
275. Wickelgren, B. G.: Superior colliculus: some receptive field properties of bimodally responsive cells, Science **173:**69, 1971.
276. Wickelgren-Gordon, B.: Some effects of visual deprivation on the cat superior colliculus, Invest. Ophthalmol. **11:**460, 1972.
277. Wiesel, T. N., and Hubel, D. H.: Spatial and chromatic interactions in the lateral geniculate body of the rhesus monkey, J. Neurophysiol. **29:**1115, 1966.
278. Wilson, P. D., and Stone, J.: Evidence of W-cell input to the cat's visual cortex via the C laminae of the lateral geniculate nucleus, Brain Res. **92:**472, 1975.
279. Wolf, E., and Gardiner, J. S.: Sensitivity of the retinal area in one eye corresponding to the blind spot in the other eye, J. Opt. Soc. Am. **53:**1437, 1963.
280. Wolf, E., and Vincent, R. J.: Effect of target size on critical flicker frequency in flicker perimetry, Vision Res. **3:**523, 1963.
281. Wurtz, R. H., and Goldberg, M. E.: The primate superior colliculus and the shift of visual attention, Invest. Ophthalmol. **11:**441, 1972.

282. Yarbus, A. L.: Eye movements and vision, Haigh, B., translator, New York, 1967, Plenum Publishing Corp.

283. Zuber, B. L., and Stark, L.: Saccadic suppression: elevation of visual threshold associated with saccadic eye movements, Exp. Neurol. **16:**65, 1966.

284. Zuber, B. L., Stark, L., and Lorber, M.: Saccadic suppression of the pupillary light reflex, Exp. Neurol. **14:**351, 1966.

GENERAL REFERENCES

Bartley, S. H.: Temporal features of input as crucial factors in vision. In Neff, W. D., editor: Contributions to sensory physiology, New York, 1968, Academic Press, Inc., vol. 3, pp. 82-135.

Brooks, B., and Jung, R.: Neuronal physiology of the visual cortex. In Jung, R., editor: Handbook of sensory physiology, Visual centers in the brain, Berlin, 1973, Springer-Verlag, vol. 7/3, pp. 325-440.

Brown, J. H. U., and Gann, D. S.: Engineering principles in physiology, New York, 1973, Academic Press, Inc., vol. 1, chapters 2, 4, 6.

Gersuni, G. V., editor: Sensory processes at the neuronal and behavioral levels, Rose, J., translator, New York, 1971, Academic Press, Inc., chapters 1-5.

Greve, E. L.: Single and multiple stimulus static perimetry in glaucoma; the two phases of perimetry, Doc. Ophthalmol. **36:**1, 1973.

Grüsser, O.-J., and Grüsser-Cornehls, U.: Neuronal mechanisms of visual movement perception and some psychophysical and behavioral correlations. In Jung, R., editor: Handbook of sensory physiology. Central processing of visual information: integrative functions and comparative data, Berlin, 1973, Springer-Verlag, vol. 7/3A, pp. 333-429.

Hyvärinen, J.: CNS: afferent mechanisms with emphasis on physiological and behavioral correlations, Annu. Rev. Physiol. **35:**243, 1973.

Ingle, D., and Schneider, G. E., editors: Subcortical visual systems. Proceedings of conference held at Massachusetts Institute of Technology, Cambridge, Mass., June 26-28, 1969, Brain Behav. Evol. **3:**1, 1970.

Jameson, D., and Hurvich, L. M., editors: Handbook of sensory physiology. Visual psychophysics, Berlin, 1972, Springer-Verlag, vol. 7/4.

Kelly, D. H.: Flicker. In Jameson, D., and Hurvich, L. M., editors: Handbook of sensory physiology, Berlin, 1972, Springer-Verlag, vol. 7/4, pp. 273-302.

Levick, W. R.: Receptive fields of retinal ganglion cells. In Fuortes, M. G. F., editor: Handbook of sensory physiology. Physiology of photoreceptor organs, Berlin, 1972, Springer-Verlag, vol. 7/2, pp. 531-566.

MacLeod, D. I. A.: Visual sensitivity, Annu. Rev. Psychol. **29:**613, 1978.

McIlwain, J. T.: Central vision: visual cortex and superior colliculus, Annu. Rev. Physiol. **34:**291, 1972.

Regan, D.: Evoked potentials in psychology, sensory physiology and clinical medicine, New York, 1972, John Wiley & Sons, Inc.

Ronchi, L.: An annotated bibliography of variability and periodicities of visual responsiveness, Florence, Italy, 1972, Fond. G. Ronchi.

Van de Grind, W. A., Grüsser, O.-J., and Lunkenheimer, H.-U.: Temporal transfer properties of the afferent visual system: psychophysical, neurophysiological and theoretical investigations. In Jung, R., editor: Handbook of sensory physiology. Central processing of visual information: integrative functions and comparative data, Berlin, 1973, Springer-Verlag, vol. 7/3A, pp. 431-573.

Van Hof, M. W., and Collewijn, H., editors: Vision in the rabbit. Proceedings of Symposium, Medical Faculty of Rotterdam, June 18 and 19, 1970, Doc. Ophthalmol. **30:**1, 1971.

Witkovsky, P.: Peripheral mechanisms of vision, Annu. Rev. Physiol. **33:**257, 1971.

Index

D

E

F

G

H

M

N

P

Q

R

S

U

W

X

Y

Z

Biological index